4th International Conference on Energy Materials and Environment Engineering (ICEMEE 2018)

E3S Web of Conferences Volume 38 (2018)

Kuala Lumpur, Malaysia
13 - 15 April 2018

Editor:

Mohamed M.H. Mostafa

ISBN: 978-1-5108-6393-4

Printed from e-media with permission by:

Curran Associates, Inc.
57 Morehouse Lane
Red Hook, NY 12571

Some format issues inherent in the e-media version may also appear in this print version.

Printed by Curran Associates, Inc. (2018)

For additional information, please contact EDP Sciences – Web of Conferences at the address below.

EDP Sciences – Web of Conferences
17, Avenue du Hoggar
Parc d'Activité de Courtabœuf
BP 112
F-91944 Les Ulis Cedex A
France

Phone: +33 (0) 1 69 18 75 75
Fax: +33 (0) 1 69 28 84 91

contact-edps@webofconferences.org

Additional copies of this publication are available from:

Curran Associates, Inc.
57 Morehouse Lane
Red Hook, NY 12571 USA
Phone: 845-758-0400
Fax: 845-758-2633
Email: curran@proceedings.com
Web: www.proceedings.com

TABLE OF CONTENTS

Foreword

On behalf of the organizing committee of the 2018 4th International Conference on Energy Materials and Environment Engineering (ICEMEE 2018), I would like to express a warm 'Thank You" to all the participants, for their important contribution, they brought to the Conference! I strongly appreciate the contribution of the authors, who submitted valuable papers and agreed to do successive revisions of their papers, following the recommendations received from the reviewers.

The aim of the ICEMEE 2018 Conference was to bring together experts in energy materials and environment engineering, in order to exchange ideas and make new connections to extend and develop partnerships between the researchers from all the world.

We received more than 350 articles, which were peer reviewed by the International Scientific Committee or reviewers and they were also cross checked for similarities, for good scientific practice in publishing. Only 169 articles are accepted for publication in the ICEMEE 2018 Proceedings.

This conference provides an excellent opportunity for everyone involved to learn and discuss about sustainable development and green technology to the following areas:

Environmental Science and Environmental Engineering

Material Science and Technology

Water Conservancy and Civil Engineering

Energy Engineering

Many thanks to the members of the Scientific Committee and reviewers, who accepted to do the reviewing free of charge, which spent their time and effort to bring a substantial contribution to the ICEMEE 2018 Conference, by providing professional evaluation of the papers and suitable recommendations.

My partners have done their best and without their excellent work it would have been impossible to organize this event. Thank you very much to the Organizing Committee!

We have received excellent feedback up to now, from all the participants, saying they will be having fruitful scientific activities and they will spend enjoyable days, during the ICEMEE 2018 Conference!

Prof. Mohamed M H Mostafa
Chairman of ICEMEE 2018 conference

Committee

Conference Chair:
Prof.Sivakumar Manickam, University of Nottingham, Malaysia
Prof.Mohamed M H Mostafa,Central University of Technology,South Africa

Organizing Committee:
Prof. Leony Ortiz, University Politécnica Salesiana ,Ecuador
Prof. Alexander Aguila Téllez, University Politécnica Salesiana,Ecuador
Prof. Cris Edward F.Monjardin, Mapua University, Philippines
Prof. Samo Kralj, University of Maribor, Slovenia
Prof. Vizureanu, Technical University Iasi, Romania
Prof. Dongfang Yang, Guizhou Minzu University,China
Prof. Min Peng, Northeast Petroleum University,China
Prof. Lars P. Mikkelsen, Technical University of Denmark, DTU Wind Energy,Denmark
Prof. Jinlong Zhang, East China University of Science and Technology, China
Prof. Marcello De Falco, University "Campus Bio-Medico" of Rome, Italy
Prof. Chris Bowen, University of Bath, UK
Prof. Laurent Ruhlmann, University of Strasbourg, France
Prof. Alan J Simson, Leeds Beckett University, UK
Prof. Jarot S. Suroso, Bina Nusantara University, Indonesia
Prof. Bingqing Wei, University of Delaware, USA
Prof. Li Liao, Huazhong University of Science & Technology (HUST), China
Prof. Yuesong Shen, from Nanjing Tech University, China
Prof. Qiang Chen, Yunnan Academy of Forestry China
Prof. Xinhua Qi, Chinese Academy of Agricultural Sciences, China
Prof. Himanshu Narayan, from National University of Lesotho, Lesotho
Prof. Dezhi Ning, Dalian University of Technology, China
Prof. Xiusheng Tang, Nanjing Hydraulic Research Institute, China
Prof. Sharen Wang, Guizhou University, China
A.Prof. Fredy Kurniawan, Institut Teknologi Sepuluh Nopember, Indonesia
A.Prof. Hakan Caliskan, Usak University, Turkey
A.Prof. Abdul Lateef Olanrewaju, Universiti Tunku Abdul Rahman, Malaysia
A.Prof. Yi Zhang, Chinese Academy of Sciences, China
Dr. Shuqin Wang, Department of Central Asia & Russia E&P,China
Dr. Ali Hassan, Beijing University of Technology, China
Dr. Fengji Luo, University of Sydney, Australia
Dr. Gang Liu, Hohai University, China
Dr. Feng Kong, China Meteorological Administration Development Research Center, China
Dr. Jaroslaw Jakubowicz, Poznan University of Technology, Poland
Dr. Goh Wan Inn, Universiti Tun Hussein, Malaysia
Dr. Kai Sun, Anhui Agricultural University, China
Dr. Mohamad Anuar Kamaruddin, Univesity Sains Malaysia, Malaysia
Dr. Leon Mishnaevsky Jr., Technical University of Denmark, Denmark
Dr. Paolo Beccarelli, University of Nottingham, UK
Dr. Azharul karim, Queensland University of Technology, Australia
Dr. Lluc Canals Casals, Universitat Politecnica de Catalunya, Spain
Dr.Masa Noguchi, Faculty of Architecture, Building and Planning, Australia
Dr. Xingxing Zhang, University of Nottingham Ningbo, China
Dr. Uttam Roy, National Institute of Technology Agartala, India
Dr. Tze Ching Ong, Politeknik Kuching Sarawak, Malaysia

Dr. George Filis, Bournemouth University, UK
Dr.Ali Mansour LAGOUN, Scientific and Technical Research Center on Physical And Chemical Analyses, CRAPC, Bousmail, Algeria
Dr.Nathan Vinod Kumar, National Institute of Oceanography, India
Dr.Diana Iruretagoyena Ferrer, Imperial College London, United Kingdom
Dr.Narottam Das, University of Southern Queensland, Australia
Dr.Babak Salamatinia, Monash University School of Engineering, Malaysia
Dr.Salam Alrekabi, Al-Mustaqbal University College,Iraq
Dr. Cuiju Zhang,Chongqing Technology and Business University,China
Dr.Ya Tao, Chinese Academy for Environmental Planning, China
Dr.Iulian Agape, echnical University, Romania
Dr.Yanru Fang, China Agricultural University,China
Mr.Chenhao Shi, CNNC Beijing Research Institute of Uranium Geology, China
Mr.Alper Kerem, Osmaniye Korkut Ata University, Turkey
Mr.Tochukwu A.S. Ugwoke, Michael Okpara University of Agriculture, Nigeria

edp sciences

Web of Conferences

Statement of Peer review

In submitting conference proceedings to *Web of Conferences*, the editors of the proceedings certify to the Publisher that

1. They adhere to its **Policy on Publishing Integrity** in order to safeguard good scientific practice in publishing.
2. All articles have been subjected to peer review administered by the proceedings editors.
3. Reviews have been conducted by expert referees, who have been requested to provide unbiased and constructive comments aimed, whenever possible, at improving the work.
4. Proceedings editors have taken all reasonable steps to ensure the quality of the materials they publish and their decision to accept or reject a paper for publication has been based only on the merits of the work and the relevance to the journal.

Title, date and place of the conference

2018 4th International Conference on Energy Materials and Environment Engineering
April 13-15, 2018
Kuala Lumpur, Malaysia

Proceedings editor(s):

Prof. Mohamed M H Mostafa, Central University of Technology, South Africa
Prof.Sivakumar Manickam, University of Nottingham, Malaysia

Date and editor's signature

Mohamed Mostafa

17, avenue du Hoggar - PA de Courtabœuf - BP 112 - 91944 Les Ulis Cedex A (France)
Tél. : 33 (0)1 69 18 75 75 - Fax : 33(0)1 69 07 45 17 – www.edpsciences.org

Method of Harmonic Source Identification Based on Harmonic Apparent Power

Lin XU[1], Rui Xing LIN[1], Shuang Ting XU [2,*]

[1]State Grid Sichuan Electric Power Research Institute, Chengdu, Sichuan, 610072, China
[2]College of Electrical Engineering and Information Technology, Sichuan University, Chengdu, Sichuan, 610065, China

Abstract: Based on the harmonic apparent power defined by the IEEE 1459-2010 power standard, the paper proposed a new method of identifying the main harmonic source at the point of common coupling (PCC). Using the reference impedance method, the utility and customer harmonic voltage and current contributions are calculated. And then combined with the IEEE 1459-2010 standard, defined the harmonic apparent power caused by the independent action of the equivalent harmonic source on the system side and the user side, by directly comparing the size of the two to determine the location of the main harmonic source. The method can consider the comprehensive influence of harmonic voltage and harmonic current, then identify the main harmonic source at PCC point. Finally, the rationality of the method is verified by the simulation of the IEEE 13 node system.

1 Introduction

With the rapid development of modern industry and the gradual maturation of power electronics technology, a large number of nonlinear loads are connected to the power grid, injecting massive harmonic and affecting the power quality.

Identification the harmonic source correctly is the basis of harmonic control, and it have been a lot of research. Traditional harmonic active power method in [1], it may have a large error in that the result is influenced by the phase difference of the harmonic source. The method that identified harmonic source by the size of Budeanu distortion power [2], which did not set the threshold to distinguish between the small harmonic source and the non-pure resistance linear load. Based on load parameter segmentation, identifying whether the load is harmonic source by the degree of non-linearity in [3]. However, it is assuming that linear load as RL model caused the linear capacitive load be considered as harmonic source. Different non-active power component comparison method in [4], it compared the size of the fundamental reactive power Q_1, Fryze reactive power Q_F and Sharon reactive power apparent S_Q to determine the main harmonic source. Then the team did further improving in [5], and introduced the fictitious reactive power Q_x instead of S_Q. However, the accuracy of this method is not high when the system nominal power is small and the value of the fictitious reactive power Q_x is between Q_1 and N.

2 Methodology for Harmonic Source Identification

2.1. The Power Definition

Under steady conditions, non-sinusoidal voltage and current can be decomposed as follows:

$$\begin{aligned} v &= v_1 + v_h \\ i &= i_1 + i_h \end{aligned} \tag{1}$$

Where v_1 and i_1 are fundamental components, v_h and i_h are harmonic components, with the fundamental voltage phase as a reference, then

$$\begin{aligned} v_1 &= \sqrt{2}V_1 \sin wt \\ v_h &= \sqrt{2}\sum_{h\neq 1} V_h \sin(hwt + \alpha_h) \end{aligned} \tag{2}$$

$$\begin{aligned} i_1 &= \sqrt{2}I_1 \sin(wt - \theta_1) \\ i_h &= \sqrt{2}\sum_{h\neq 1} I_h \sin(hwt + \beta_h) \end{aligned} \tag{3}$$

Where h is the number of harmonic, V_1 and V_h are the RMS value of fundamental voltage and the hth harmonic voltage, I_1 and I_h are the RMS value of fundamental current and the hth harmonic current, α_h and β_h are the phase of the hth harmonic voltage and current.

Apparent power can be decomposed as follows:

$$\begin{aligned} S^2 &= V^2 I^2 = (V_1 I_1)^2 + (V_1 I_H)^2 + (V_H I_1)^2 + (V_H I_H)^2 \\ &= S_1^2 + D_I^2 + D_V^2 + S_H^2 \end{aligned} \tag{4}$$

Where D_I, D_V, S_H are current distortion power, voltage distortion power and harmonic distortion power, respectively. This decomposition method was proposed by Professor Emanuel of Worcester Polytechnic Institute in [6], and adopted by the IEEE 1459-2010 standard [7].

Combined with the definition of THD, the relationship between harmonic apparent power S_H and THD is can be shown as:

$$\begin{aligned} S_H &= V_H I_H = V_1 I_1 \frac{V_H I_H}{V_1 I_1} \\ &= S_1 \mathrm{THD_V THD_I} \end{aligned} \tag{5}$$

Where THD_V and THD_I are the rate of voltage and current distortion, respectively. From the equation (5), harmonic apparent power S_H, as a part of apparent power, only contains the interaction of the non-fundamental voltage and current, therefore S_H is not affected by fundamental components, which is the same with the harmonic distortion power. In addition, the harmonic apparent power is the result of each order of harmonic voltage and current. If the power component is taken as the key index for identifying the main harmonic source, the effects of each harmonic voltage and current can be comprehensively considered.

2.2. The Reference Impedance

According to reference impedance method proposed in [8], the harmonic contribution of utility and customer at PCC point can be obtained. The method can take into account the impact of impedance variation on the decomposition of contribution. The principle can be described as, setting a utility reference impedance and customer reference impedance, making the difference between the true impedance and the reference impedance equivalent as an additional harmonic source, and the original harmonic source plus additional harmonic source, is the new equivalent harmonic source. The specific process shown in Figure 1. Where Z_u and Z_c are harmonic impedance of utility and customer, Z_{ur} and Z_{cr} are the reference impedance of utility and customer, V_{pcc} and I_{pcc} are the voltage and current of PCC point.

According to figure 1, the new equivalent harmonic source of utility and customer after the conversion can be expressed as:

$$\begin{aligned} V_{u-new} &= V_{pcc} - Z_{ur} I_{pcc} \\ I_{c-new} &= V_{pcc} / Z_{cr} + I_{pcc} \end{aligned} \tag{6}$$

The harmonic reference impedance of utility Z_{ur} can be equivalent with the system short-circuit impedance, and the harmonic reference impedance of customer Z_{cr} can be equivalent with load resistance.

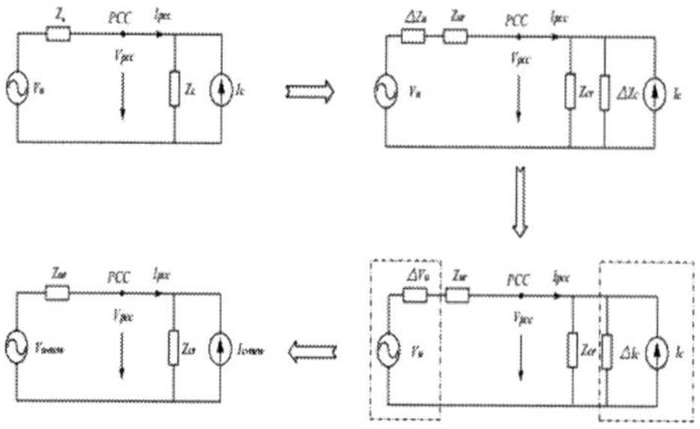

Fig. 1. The Diagram of Reference Impedance Method.

By using the superposition theorem, the harmonic voltage at PCC point caused by the independent action of the equivalent harmonic source on utility and customer is calculated with (7)

$$\begin{aligned} V_{pcc-u} &= \frac{Z_{cr}}{Z_{ur} + Z_{cr}} V_{u-new} \\ V_{pcc-c} &= \frac{Z_{ur} Z_{cr}}{Z_{ur} + Z_{cr}} I_{c-new} \end{aligned} \tag{7}$$

Similarly, the harmonic current at PCC point caused by the independent action of the equivalent harmonic source on utility and customer is calculated with (8)

$$\begin{aligned} I_{pcc-u} &= \frac{1}{Z_{ur} + Z_{cr}} V_{u-new} \\ I_{pcc-c} &= -\frac{Z_{cr}}{Z_{ur} + Z_{cr}} I_{c-new} \end{aligned} \tag{8}$$

2.3. Harmonic Source Identification

The harmonic voltage and current at PCC point caused by the independent action of the equivalent harmonic source on utility are defined as THD_{Vu} and THD_{Iu}, respectively.

$$\mathrm{THD}_{Vu} = \frac{\sqrt{\sum_{h=2} V^2_{pcch-u}}}{V_{pcc1}}; \mathrm{THD}_{Iu} = \frac{\sqrt{\sum_{h=2} I^2_{pcch-u}}}{I_{pcc1}} \tag{9}$$

Where V_{pcc1} and I_{pcc1} are the RMS value of the fundamental voltage and current, respectively. V_{pcch-s} and I_{pcch-s} are the RMS value of the hth harmonic voltage and current at PCC point caused by the independent action of the utility harmonic source.

Similarly, the harmonic voltage and current at PCC point caused by the independent action of the customer are defined as THD_{Vc} and THD_{Ic}, respectively.

$$THD_{Vc} = \frac{\sqrt{\sum_{h=2} V^2_{pcch-c}}}{V_{pcc1}} \tag{10}$$

$$THD_{Ic} = \frac{\sqrt{\sum_{h=2} I^2_{pcch-c}}}{I_{pcc1}} \tag{11}$$

Where V_{pcc-c} and I_{pcc-c} are the RMS value of the hth harmonic voltage and current at PCC caused by the independent action of customer. Harmonic apparent powers S_{Hu} and S_{Hc} at PCC point caused by the the independent action of the equivalent harmonic source on the system side and the user side can be calculated with equation (12) and (13):

$$S_{Hu} = S_{pcc1}\mathrm{THD}_{Vu}\mathrm{THD}_{Iu} \quad (12)$$

$$S_{Hc} = S_{pcc1}\mathrm{THD}_{Vc}\mathrm{THD}_{Ic} \quad (13)$$

Where S_{pcc1} is the fundamental apparent powers at PCC point.

In summary, the main harmonic source at PCC point can be identified by comparing the size of S_{Hs} and S_{Hc}, the concrete process of the method are as follows:

(1) Measure voltage and current at PCC point, and analyze harmonic component

(2) Based on the reference impedance method, the reference impedance of utility and customer are equivalent to the system short circuit impedance and the load resistance, respectively. Then estimate the harmonic contribution of utility and customer that $V_{pcc\text{-}u}$, $V_{pcc\text{-}c}$, $I_{pcc\text{-}u}$, $I_{pcc\text{-}c}$, according to (6)-(8)

(3) According to (9)-(10), calculate the harmonic voltage and current at PCC point caused by the independent action of the utility and customer that THD_{Vu}, THD_{Iu}, THD_{Vc}, THD_{Ic}.

(4) Calculate the harmonic apparent powers caused by the utility and customer harmonic source that S_{Hu}, S_{Hc}, according to (11)-(12).

(5) Compare the size of S_{Hu} and S_{Hc}, if the former is larger, the main harmonic source is the upstream user side at the PCC that the utility. If the latter is larger, the main harmonic source is considered as the downstream user side at the PCC that the customer. If $S_{Hu}=S_{Hc}$, the utility and the customer contributions on the overall harmonic pollution at the measurement point are comparable.

3 Case Studies

In order to validate the performance of the proposed method, simulation tests are carried on the 13-bus IEEE test system [9] through PSCAD/EMTDC, as shown in figure 2, and simulation data are analyzed through MATLAB. In this test system, five measurement points are selected as the PCC of interest: 32, 33, 45, 92 and 84.

The test system is powered by the node 50, and it includes voltage regulators, single-phase and three-phase loads. The harmonic source loads include the fluorescent lamp group, the frequency converter (ASD) and other household loads. The system reference impedance can be determined by the system short-circuit impedance.

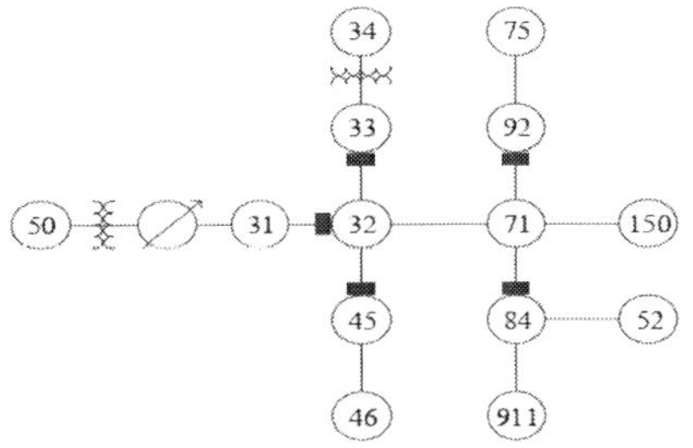

Fig. 2. The 13-bus IEEE test system.

A. Case 1

The harmonic source is connected to node 75 and node 92, the proposed method and the non-active power comparison method [8] are applied at each measurement point, and the results are shown in Table 1 and Table 2. At this case, the actual main harmonic source of node 32 and node 92 is downstream of the measurement point (the customer side), and the rest of the measurement points are the upstream of the measurement point (the utility side).

Tab. 1. The Results of Case 1 (Non-active Power Comparison Method)

	Q_1/kvar	Q_x/kvar	N/kvar	Result
32	1027.45	1034.66	1040.75	Customer
33	178.43	178.46	178.69	Utility
45	229.38	229.41	229.43	Utility
84	10.47	11.99	18.46	Utility
92	381.6	405.83	426.7	Customer

Tab. 2. The Results of Case 1 (Proposed Method)

	S_{Hs}/kVA	S_{Hc}/kVA	Result
32	0	2.88	Customer
33	0.14	0	Utility
45	0.05	0	Utility
84	0.26	0	Utility
92	0	5.33	Customer

According to Table 1 and Table 2, it shows that in the proposed method, the value of S_{Hc} in node 33, 45 and 84 are approximate to 0, far less than the value of SHu, so identified the upstream of measurement point (the utility side) as the main harmonic source identification for measurement points (the system side), and the utility side should responsible for harmonic pollution of the three measuring points. The value of S_{Hc} in node 32 and 92 are greater than the value of SHu, and SHu value is approximately equal to 0, therefore identified downstream of the measuring point (the customer side) is the main harmonic source, and the customer side should take full responsibility for the harmonic pollution of the two measurement points. The proposed method in all measurement points are given the correct result.

In the non-active power comparison method, the fictitious reactive power Qx in node 32 and node 92 are closer to non-active power N, therefore can identified downstream of the measuring point (the customer side) as the main harmonic source. For the remaining measurement points, since the fictitious reactive power Qx is closer to the fundamental reactive power Q_1, the upstream of the measuring point (the utility side) is identified as the main harmonic source. This method also gives correct results at all measurement points.

B. Case 2

The harmonic source is connected to node 75, 92, 45, 46, and the simulation results are listed in Table 3 and Table 4. At this case, the actual main harmonic source of node 32, 45 and 92 is downstream of the measurement point (the customer side), and the rest of the measurement points are the upstream of the measurement point (the utility side).

Tab. 3. The Results of Case 2
(Non-active Power Comparison Method)

	Q_1/kvar	Q_x/kvar	N/kvar	Result
32	1072.66	1080.86	1086.01	Customer
33	178.30	178.34	178.57	Utility
45	266.07	226.09	278.16	Utility
84	12.63	14.80	17.69	Utility
92	376.98	401.20	422.24	Customer

Tab. 4. The Results of Case 2
(Proposed Method)

	S_{Hs}/kVA	S_{Hc}/kVA	Result
32	0	2.74	Customer
33	0.14	0	Utility
45	0.25	0.59	Customer
84	0.16	0	Utility
92	2.08	3.13	Customer

In the non-active power comparison method, since the values of the three powers are very close, it can be considered that there is no non-linear load at the downstream of the measurement point 33, which is consistent with the fact. The fictitious reactive power Qx is closer to the fundamental reactive power Q1 than to the non-active power N in node 45, therefore given a false result that does not conform to the facts.

For the proposed method, the value of S_{Hs} at measurement point 32 is about equal zero, so the downstream of the measurement point (the customer side) is fully responsible for harmonic pollution. Similarly, at measurement points 33 and 84, the value of SHu is approximate to zero, so the upstream of the measurement point are main harmonic source. For the measurement points 45 and 92, since the value of S_{Hc} is larger than SHu, the main harmonic source is the downstream of this measurement point (the customer side), which is consistent with the fact that actual harmonic sources are injected into the system.

5 Conclusion

In this paper, a method of main harmonic source identification is proposed based on harmonic apparent power. According to the reference impedance method, the harmonic voltage and current at the PCC point when the utility and the customer function separately are calculated, and it can avoid the influence of the impedance variation on the contribution decomposition. Based the harmonic apparent power component in IEEE 1459-2010, a new definition of harmonic apparent power S_{Hu} and S_{Hc} which generated when the utility and customer individual action is proposed, and identified the main harmonic source by comparing the size of the two. Simulation analysis verifies the effectiveness of the proposed method.

The proposed method starts from the view of the apparent power decomposition, and consider not only the harmonic voltage or current of a concern order, but also the comprehensive influence of various harmonic order voltage and current, which provides a new way to distinguish the comprehensive harmonic pollution responsibility between the the utility side and the customer side. The method proposed in this paper is simple in operation and can be easily applied to the metering device, which is of engineering significance.

Acknowledgement

The work was supported by project: Research on Harmonic Quantization Method for Nonlinear Loads(NO: 52199716001P), State Grid Sichuan Electric Power Research Institute.

References

1. Yang H, Porotte P, Robert A. *Proceedings of the 3rd International Conference on Power Quality.* **2**, 160 (1994)
2. D Stevanović, P Petković. Electrical Engineering. **97**, 165 (2015)
3. Antonio Dell'Aquila, Maria Marinelli. IEEE Transactions on Power Delivery. **19**, 1284 (2004)
4. Cataliotti A, Cosentino V, Nuccio S. IEEE Transactions on Instrumentation and Measurement. **57**, 1554 (2008)
5. Cataliotti A, Cosentino V, Nuccio S. IEEE Transactions on Instrumentation and Measurement. **57**, 85 (2008)
6. Emanuel A E. IEEE Transactions on Instrumentation and Measurement. **3**, 1457 (2005)
7. IEEE Standard definitions for the measurement of electric power quantities under sinusoidal , nonsinusoidal , balanced , or unbalanced conditions. IEEE Std 1459, 2010.
8. Tomaž Pfajfar, Boštjan Blažiþ, Igor Papiþ. *9th International Conference on Electrical Power Quality and Utilization*(2007).
9. IEEE Working Group on Nonsinusoidal Situations. IEEE Transactions on Power Delivery. **11**, 79 (1996)

PFC Performance Improvement of Ultra-supercritical Secondary Reheat Unit

Jun LI[1,*], *WeiWei* LI[2], *XingWang* Hu[3], *WenChuan* Sun[3] and *Gang* Kong[1]

[1]State Grid Shandong Electric Power Research Institute, Ji'nan, Shandong 250002, China
[2]Network Center, Shandong University of Arts, Ji'nan, Shandong 250014, China
[3]State Grid Shandong Electric Power Company, Ji'nan Shandong 250001, China

Abstract. Ultra-supercritical secondary reheat unit has been widely used in the world because of its advantages of large capacity, low consumption and high efficiency etc., but rapid load change ability of the turbines to be weakened which caused by its system organization, cannot meet the requirements of power grid frequency modulation. Based on the analysis of the control characteristics of ultra-supercritical once-through reheat unit, the primary frequency control based on feed-water flow overshoot compensation is proposed. The main steam pressure generated by the feed-water is changed to improve the primary frequency control capability. The relevant control strategy has been applied to the 1000MW secondary reheating unit. The results show that the technology is feasible and has high economical efficiency.

1 Introduction

To develop clean energy and improve the conversion efficiency of fossil energy is the current trend. Facing this trend of development, there are large changes in China's power system power structure. First, the large amount of new energy generators have worked. Second, small capacity thermal power unit due to the high power generation coal consumption, low energy efficiency gradually quit running, large capacity and high parameters of supercritical and ultra-supercritical units is promoting operation. Supercritical units are more efficient. The supercritical unit is 2% higher than the subcritical unit, and the supercritical unit is 3% higher than the supercritical unit, which has great advantages in energy saving and emission reduction [1,2].

The state planning commission requires that the new 600MW and above units should adopt supercritical technology. The cost of participating in primary frequency compensation(PFC) is huge. The economy and safety of supercritical and ultra-supercritical thermal power units are often adversely affected by frequency modulation. For large-capacity and high-parameter thermal power units, the problem is more serious unit safety. Under the background that the existing bearing capacity of generator set has not been greatly improved, the main steam pressure and temperature of the unit are greatly raised, so, the thermal stress caused by the boiler steam pressure fluctuation becomes more prominent. In recent years, many power plants in China have repeatedly suffered serious accidents such as the explosion of the boiler tube [3-5]. This not only brings huge economic losses to the power plant, but also causes large disturbance to the power system due to the sudden shutdown of large-capacity generating units. Some units in order to run safely, the main steam pressure and temperature have to be lowered to a sub-critical level to prevent the boiler accident, but after the parameter is reduced, the meaning of supercritical and ultra-supercritical units have been lost. Therefore, how to make the thermal power unit fully participate in frequency modulation and reduce the cost of frequency modulation by means of technical means is a difficult problem.

2 Features of Secondary Reheat Ultra-supercritical Unit

Supercritical and ultra-supercritical boilers are composed of heated pipes without steam drum. Main steam valve control the steam turbine power, at the same time, in turn affect the once-through boiler outlet resistance characteristics[6]. Because of the lack of the steam drum cushion, dynamic characteristics which influenced by main steam valve opening is far greater than the drum boiler, therefore, the characteristics of once-through boiler and steam drum boiler is completely different. For example, when the load of the turbine increases, the turbine power regulation system drives the main steam valve to open, the steam flowing into the turbine increasing, the large lag and inertia of boiler cannot quickly generate enough steam to match the turbine, so the main steam pressure drop, weaken the regulator flow increase. Because of the power balance, the turbine power regulation system will continue to function, further opening the valve to get more steam, which will

* Corresponding author: lijunsepco@163.com

make the main steam pressure continue to deteriorate, such a vicious circle.

Drum boilers have steam drum, even if the turbine flow greater changes, the changes of main steam pressure is not too large. Due to the relatively small regenerative heat of the boiler, the once-through boiler is more sensitive to the disturbance, which makes the coupling between the turbine and the boiler serious. As a result, the pressure of the main steam changes greatly, which not only reduces the control quality but also increases the risk of the once-through boiler operation. Therefore, under the traditional strategy of PFC, fluctuation of main steam pressure and temperature caused by random load changes will have a very adverse effect on the operation of supercritical and ultra-supercritical units.

To meet the requirements of energy saving and environmental protection, ultra-supercritical secondary reheat boiler has a secondary reheat tower structure, which is reheated once more than the conventional unit. With the heating surface increased, energy storage is greatly increased. Due to the secondary reheat unit has one extra high pressure cylinder than the conventional unit, the work ratio of the turbine is 19.8% of the ultrahigh pressure cylinder, 21% of the high pressure cylinder and 59.2% of the intermediate and low pressure cylinder. The conventional primary reheat turbine basic work is 33.3% of high pressure cylinder , 33.3% of intermediate pressure cylinder and 33.3% of low pressure cylinder. Compared with primary reheat unit, the load by the high pressure cylinder of secondary reheat unit accounts for nearly half of it, which reduces the throttling loss of the turbine to a certain extent and improve the turbine operating efficiency, but also cause the rapid load change ability of turbine to poor.

3 Control Characteristics of Secondary Reheat Ultra-supercritical Unit

The emphasis of traditional machine-network coordination is the interaction between the grid and the generator, but often neglect the safety and dynamic characteristics of the boiler. Under the background of serious coupling between the once-through boiler and turbine and the low heat storage of the once-through boiler, the safety and life issues need to be refocused on. The traditional control of machine-network coordination means that when the external load changes, the power change command is sent to the boiler and turbine control system at the same time, and the valve opening and the boiler are adjusted synchronously. The core idea of the control is to improve the response speed of the unit to the load change of the power grid, and not fully consider the pressure fluctuation of the boiler.

In other words, the traditional machine-network coordination and machine-boiler coordination control focus on the turbine and power grid, PFC loop only reflects the negative feedback control strategy for power grid frequency, does not consider the impact on the boiler, thus there is a big security risk. Considering the characteristics of supercritical and ultra-supercritical units, the problem of PFC will develop into a coordinated control problem of boiler, machine and network. The control method that takes into account the economy, safety and operational safety of the unit is considered. So that the supercritical units not only fully participate in the power system frequency control, but also greatly reduces the frequency modulation parameters caused by the huge volatility, achieve the goal of network and source win-win.

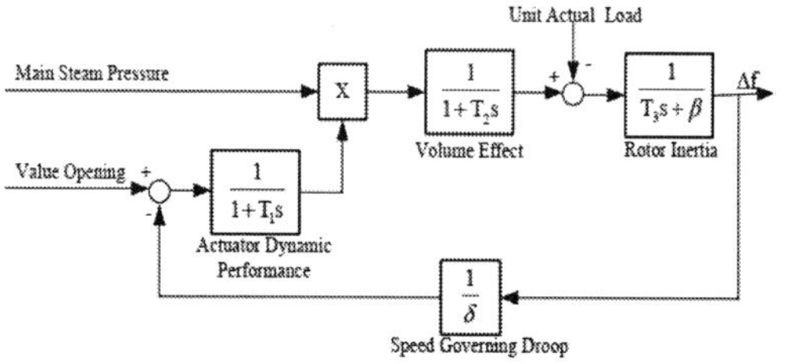

Fig. 1. PFC Mathematical Model of the Turbine

PFC is the unit near the working point of active power of small-scale adjustment, so for the primary frequency control analysis of turbine model is linear model, so the dynamic properties can be expressed by the transfer function. The coupling of the turbine and the once-through boiler is main steam flow regulating valve, which is the main steam valve opening and main steam pressure. Therefore, the turbine model should include the main steam flow and the main steam pressure. The dynamic characteristics of PFC is need to consider the actuator dynamics, high-pressure cylinder steam volume and rotor inertia. According to the input and output relationship of each component, the above links is the first-order inertia link which is shown in Fig.1.

According to the principle of gas flowing through the main steam valve, the main steam flow is related to the main steam pressure and the opening degree of main steam valve, which is proportional to the product.

$$G_s = kP_s S_v$$

In the formula, G_s is the main steam flow, k is constant coefficient, P_s is the main steam pressure, S_v is the opening degree of main steam valve.

In order to make the turbine actual power tracking the given power, the power feedback is need to be sent to the control loop. Regardless of the main steam pressure fluctuations, the schematic diagram shown in Fig.2. On the basis of the pure speed regulation, the electric power output from the generator is measured, and then fed back to the input, which is the valve control signal of the turbine after the sum of the given power and the primary frequency.

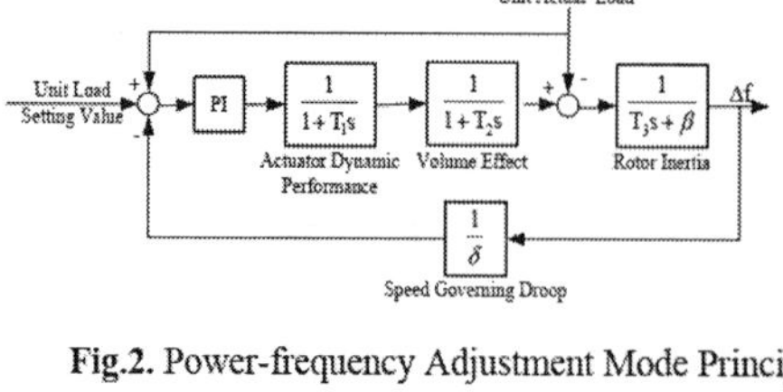

Fig.2. Power-frequency Adjustment Mode Principle

 https://doi.org/10.1051/e3sconf/20183801002

4 Optimization Design and Result

The grid-connected power plant must be equipped with one frequency modulation function, and the primary frequency modulation investment/withdrawal signals should be connected to the power dispatching agencies. The artificial dead zone, the rate of change of the speed regulation system and the maximum adjustment load limit of the primary frequency regulation, and the speed control system of delay rate, response speed and so on should meet the requirements of the primary frequency regulation technology management of the regional grid[7,8]..Units running in the grid must be put into primary frequency modulation function, when the grid frequency fluctuations should automatically participate in frequency modulation, and the grid power plant must not withdraw from the unit's frequency modulation function.

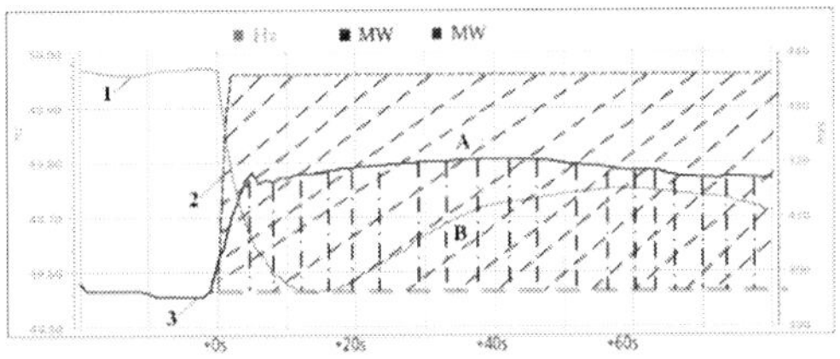

Fig.3. Schematic of PFC Assessment Criteria

The unit's load response when grid frequency in the fall is shown in Fig.3. Among them, 1 is the grid frequency, 2 is the standard load output, and 3 is the actual load output. The shadow area A is the integral area that the standard load output minus the initial load output on the assessment time. The shadow area B is the integral area that the actual load output minus the initial load output on the assessment time.

P_0 is the unit actual load at t_0 moment. Pr(t) is the unit actual load at different time. Ps(t) is the unit standard load at different time. According to the Chinese primary frequency modulation standards. The value of η needs to exceed 80% to meet the standard requirements. Because of the different heat storage capacity, the action time of PFC to different types of thermal power unit is 0.5 to 2 minutes inequality. PFC is adjusted by the control methods according to differ characteristics. So, it is possible to meet the standard requirements by increasing the amount of load in the later period.

4.1. Method Design

According to 'Guide of primary frequency control test and performance acceptance for thermal power generation units' and other technical standards, unit's speed governor droop is no more than 5%, rated grid frequency is 50Hz, dead zone of PFC is ±0.033Hz.

When the grid frequency deviation value does not exceeded the setting range or the unit regulating command is not fully opened, the unit uses the regenerative heat and the regulating valve to adjust the unit load. When the grid frequency deviation exceeds the setting range and the regulating command is fully open, the feed-water flow overshoot compensation can be used. The secondary reheat ultra-supercritical cannot satisfy the requirement of primary frequency load modulation when the frequency difference of the power grid is large In actual operation, if the unit's water flow changes within a certain range in a certain period of time, it will not affect the steam temperature of the unit, but will cause the main steam to change. However, the change of the main steam pressure will have obvious effect on unit power, that is, the main steam pressure can changed by adjusting the water flow.

As a feasible implementation, when the power grid frequency deviation exceeds the set range, firstly, if the unit regulating command is more than 99%, namely the

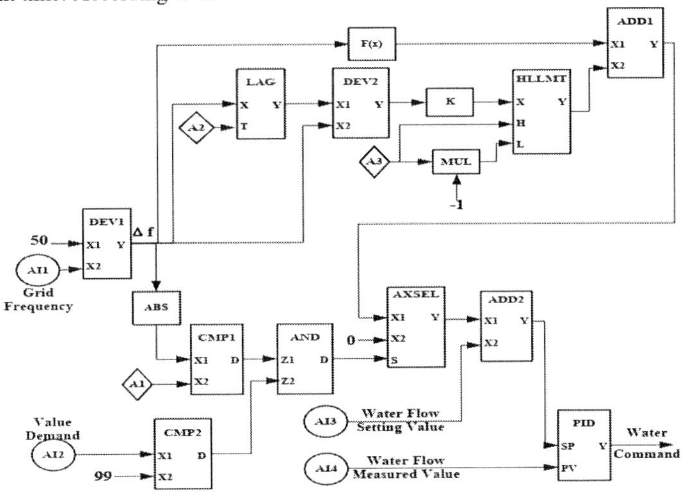

Fig.4. Schematic Diagram of PFC Optimal Control System Based on Feed-water Flow Overshoot Compensation

unit's value is fully open, the system will be according to the power grid frequency deviation value calculated by the superposition of feed-water flow overshoot offsets to the original, generate a new feed-water flow value. Secondly, if the unit regulating command is not greater than 99%, the unit shall be deemed to be tone still has to adjust allowance, use tone to adjust the opening value, namely the normal regulation of feed-water flow set point to keep the original.

As shown in Fig.4, DEV is the subtracter module, F(x) is a dead zone nonlinear function, LAG is a delay module, HLLMT is the high and low limit limiting module, AXSEL is the analog selector, A is the analog generator, CMP is the comparator module. Through the value of frequency modulation measurement, it is effective to use the heat storage and feed-water supply to ensure the frequency modulation capability of the unit.

4.2 Application Effect

Take one 1000MW ultra-supercritical secondary reheat unit as an example, host parameters were 31MPa/600℃ /620 ℃ /620 ℃ , the parameter is in the international leading level. Considering the heat storage and pressure changes, set the grid frequency deviation value 0.1 in A1, and the lag time set in A2 is 10s, the high and low limit limiting set in A3 is 50.According to the relationship between the change of feed-water flow and power change,

$$\Delta t / H = 3 \times \Delta MW + 50$$

that is, multiply the primary frequency variation by 3 times and add 50 to get the value of the change of feed-water flow, that is, the total ADD1 output is this value. The assessment indicators of before and after the technical transformation is shown in Table1.

Table 1. Monthly Statistics of Grid Frequency Assessment

Date	Electricity Assessment (MWh)	Date	Electricity Assessment (MWh)
2016-1	500.30	2017-1	16.02
2016-2	155.16	2017-2	137.25
2016-3	529.30	2017-3	18.28
2016-4	669.30	2017-4	0.34
2016-5	878.00	2017-5	31.72
2016-6	675.40	2017-6	67.81
2016-7	684.09	2017-7	273.02
2016-8	512.20	2017-8	255.83
2016-9	189.75	2017-9	57.56
Total	4793.50		857.83

Comparing between January to September of 2016 and 2017, the grid assessment of PFC are 4793.50MWh and67.81MWh respectively, which improve the performance by more than 80%. According to the electricity price 0.3924 yuan/kWh calculation, 1.54 million yuan can be obtained. The economic benefits were significant.

5 Conclusion

By analyzing the control characteristics of the thermal power units, especially the ultra-supercritical secondary reheat units, combined with the frequency control requirements of the power grid with large-scale power gaps, the performance improvement technique of PFC was proposed. Under the premise of ensure the stable operation, by changing the feed-water flow to further improve the quality of PFC, the method can improve unit's dynamic adjustment ability of power system frequency, and ensure its stable operation at the same time. In the smart grid environment, it is of great significance to improve the coordination ability of all kinds of units, to improve the stable operation of power grids.

References

1. G. Nowak , A. Rusin. Shape and operation optimization of a supercritical steam turbine rotor[J]. Energy Conversion and Management, 2013,74:417-425.
2. Dai Yiping, Zhao Ting, Tian Yunfeng. Research on the primary frequency control characteristics of generators in power system[J]. Second IEEE Conference on Industrial Electronics and Applications, 2007, 569-574.
3. Mo Yaowei. Research of pipe explosion of 1000MW ultra-supercritical unit boilers[J]. East China Electric Power, 2008,36(2):16-22.
4. AntonioVisioli. A new design for a PID plus feedforward controller[J]. Journal of Process Control, 2004,14:457-463.
5. F.Shinskey. Process Control Systems Application Design Adjustment[J]. McGraw-Hill, New York, 1996.
6. Huang, H.P. & Jeng, J.C. Monitoring and assessment of control performance for single loop systems[J]. Ind. Eng. Chem. Res., 2002, 41:1297-1309.
7. Yu Daren, Guo Yufeng, Xu Jiyu. The primary frequency regulation stability of parallel turbo-generators[J]. Proceedings of the CSEE, 2000, 20(9):59-63.
8. Li Duanchao, Chen Shi, Chen Zhongyuan. Real-time measurement and reward method of the efficiency of generator unit primary frequency regulation[J]. Automation of Electric Power Systems, 2004, 28(2): 70-72.

* Corresponding author: lijunsepco@163.com

Study on the fiber fouling in drying exhaust heat utilization of wood industry

Yadi Pan *, *Shuya* Shan, *Yao* Wei, *Feng* Ji, *Jinping* Weng, *Yulan* Tian and *Jin* Qian

Nanjing Forestry University;

Abstract: Based on the viscoelastic and creep deformation properties, a new deposition mechanism model for slender wood fiber particles in wast heat utilization is proposed in this paper. And the equivalent sphere method is used to describe the particle feature size in the model. With the proposed deposition model of flexible slender particles, the critical criteria are obtained. The influence of particle size, aspect ratio and damping factor on particle deposition has been investigated. The results indicate that particle deposition increases with the particle size decrease, aspect ratio and damping factor increase. According to the present deposition model, a coupling simulation with FLUENT and EDEM method was carried out for the flow field of fiber drying tail gas in heat piping exchanger, which indicated that particle deposition mainly occurred at the central windward area of fin due to the direction changer and the magnitude decrease of collision velocity between fiber particles and wall. Experiment of heat recovery of drying tail gas revealed that using the H fin tubes instead of rectangular fin tubes can greatly relieve the deposition of wood fiber particles, which provided a useful way to save energy in wood industries.

1. Introduction

In the wood fabrication industry, there are large amount of fiber drying exhaust, where the temperature is usually about 80 °C. An annual output of exhaust gas could be produced up to about 250,000 m^3/h in a 100,000 m^3 middle density fiberboard production line, which can get about 180,000 kW waste heat if they are cooling to 40 °C. However, the deposition of fiber particles in the heat exchanger has increased the cost of equipment operation, and decreased the efficiency of heat recovery. Therefore, the mechanism of fiber deposition is one of hot topics in the research of energy saving and emission reduction in the current wood industry.

There have been relevant research of the collision deposition mechanism for elastic particles on the rigid wall and such mechanism was used as guideline in practical applications. However, as a kind of polymer, the creep deformation of viscoelastic wood fiber would occur under the external force load. Furthermore, fiber particle is slender, whose aspect ratio is greater than 1. So far, the deposition mechanism of such flexible slender particles has rarely been reported. Feng et al. [1] studied the deposition of elastoplastic particles on the wall, and analyzed the influence of adhesion between the particles and the wall on the deposition of elastic plastic particles under different velocity. It concluded that the influence of the interface adhesion is more important than that of the plastic deformation at the low speed. Marshall et al. [2] analyzed the behavior of the particles after its collision with the wall in suspension flow, such as rebound, deposition and tumbling. And it pointed out that the particle will roll instead of sliding after its collision with the wall because of the impact of air flow. The recovery coefficient of particle-wall collision was studied by Troiano and co-workers [3], which including coke sawdust and wood chips. In literature [4], the fluidization and agglomeration of the slender particles that was divided into a plurality of rigid segments by hinges were researched. Lin et al. [5] put forward the control equations of particle sedimentation in static water based on the cylindrical particle oscillation frequency ω, which considered the influences of the oscillation, the torque and inter force of particles. Based on the lattice Boltzmann method, Lin et al. [6-8] studied the influences of the aspect ratio, wall constraint and particles interaction on the settlement of slender particles, which reveals the orientation and motion characteristic of particles.

Aewald algorithm, based on slender body theory, was used in the research of Saintillan et al.[9] to simulate the settlement of non-Brown rigid slender particles under the action of gravity and a Reynolds number of 0. The numerical results showed that the fibers with different sizes were aggregating continuously during the settlement process. And the particles would separate from the cluster when the cluster reached a certain size, and then aggregated with other particles to another clusters. Based on JKR theory and adhesive elastic contacts, Marshallin et

*Correspondence: 59647664@qq.com; Tel.: +86-138-158-74276-; +86-138-090-32833

al. [10] studied three-dimensional simulations of micro-particle deposition and aggregation on an individual fiber of an array using the discrete element method (DEM). The simulation results showed that the fluid viscosity and velocity would affect the particle capture, and the particle deposition occurred in a narrow area near the front fiber line due to the effect of high shear force and particle shadow on the surface of fiber. Helzel et al. [11] proposed columnar Brown particles settlement kinetic model of low Reynolds number flow in dilute suspension, using the multi-scale dynamics model to describe the suspended particles, which could be used to describe the cluster formation for high concentration of particle flow. All of these studies provide valuable references for the analysis on particle deposition mechanism of wood fiber, but there are still a lot of work needs to be further researched on the deposition mechanism of slender particles considering its viscoelastic properties due to the unique stress relaxation and energy consumption characteristics.

Due to the variety of the components, size and shape of the wood fiber particles emitted by wood enterprise, with a large number of water vapor. The process of fiber deposition is affected by many factors and deposition mechanism is really complex. To simplify the analysis, this paper would do some exploratory research on the deposition mechanism of wood particles.

2. Deposition model

Some assumptions were made to simplify the analysis model: (1) the deformation of the metal wall was ignored due to the deformation of viscoelastic creep of wood fibers is much bigger than that of the metal wall; (2) The mass and radius of wall was considered infinite with the comparison of the fine dust particles; (3) The collision of particle clusters with the wall surface was not considered as the particles are not easily collided with each other in the low dust concentration. We only analyzed the collision the single particle and wall.

2.1. The force analysis of particles in the flow field

The dust concentration distributes in the range of 120~180mg/m^3 in the fiber dry exhaust heat utilization, in which the particle concentration is relatively low. Stokes model was taken in the study the collision between particles and walls. Particles of 80~160 mesh size were analyzed in this paper. The screening statistical analysis shows that the aspect ratio is less than 4. Select volume equivalent sphere model was used for deposition mechanism of particle analysis, and we used the volume shape factor to modify this model. Volume equivalent sphere diameter fiber particle d_V is shown by the equation (1).

$$d_V = \sqrt[3]{6V/\pi}, \quad (1)$$

Where V in (1) is the actual volume of a particle. And the particle volume shape coefficient φ_V is shown as following:

$$\varphi_V = V/d^3 = \pi d_V^3/(6d^3) \quad (2)$$

Where d in (2) is the actual transverse particle size of particle.

When fiber dust moved and collides with the wall in a flow field, it suffers multitudinous forces including the collision normal impact force F_{in}, gravity F_g, buoyancy F_b, lift F_{lv}, adhesion F_{adh}, surface normal force F_{tn}, tangential impact force F_{it}, drag force F_d, friction force F_f, which are shown in Figure 1.

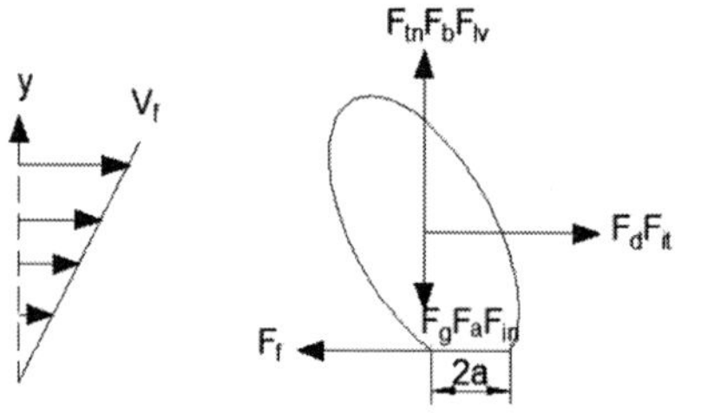

Fig. 1. The force analysis of the target particles.

The particle size of 95~200μm (mesh size: 80-160) are selected in this study, whose equivalent spherical diameter ranges from 140 to 360μm by formula (1). These particles belong to micron-sized powder particles, their magnitude of the contact surface adhesion force is 1000 times of gravity [12-13], so the influence of gravity of particles F_g in deposition is very small, which can be neglected. The air density is much smaller than the particles', so the buoyancy is also negligible. Lift Flv can be expressed as: $F_{lv} = 1.625\rho_f d_p^2\sqrt{v_f}u_r\sqrt{\partial u/\partial y}$, it is related to the fluid viscosity v_f, the velocity gradient along the direction of the fluid particles, and the relative speed u_r, eta [14-15], and V_f is the air flow rate. The lift of particles is usually ignored due to small air velocity gradient u along the y direction for such size particles. By analysis of force on the particle, the force balances the equation of the particles in the vertical direction can be expressed as equation (3):

$$F_{tn} - F_{in} - F_{adh} = 0, \quad (3)$$

In the same way, the forces balance, i.e. the drag force F_d, tangent impact force F_{it} and friction force F_f, acting on the particles in the horizontal direction, are shown by equation (4).

$$F_d + F_{it} = F_f, \quad (4)$$

Where flag fore is: $F_d = C_D \pi\rho_f u_r d_p^2/8$, CD is the drag coefficient, and friction force is$F_f = f^*F_n$, in which $F_n = -F_t$, and f^* is effective friction coefficient.

2.2. Normal deposition model

High polymer has typical characteristics of the viscoelastic. We proposed a viscoelastic damping contact mechanics model for wood fiber based on the lag contact mechanics model called Hertz-Flores [16]. And the normal contact force of the particle when it collides with wall can be expressed as:

$$F_{in} = K_{in}\delta + \chi\delta V_{in}, \quad (5)$$

The first item in the right side of the equation is elastic force, K_{in} is the stiffness coefficient, $K_{in} = \frac{4}{3}E^*(R^*)^{\frac{1}{2}}$, where R*is the effective radius in collision system, $1/R^* = 1/r_1 + 1/r_2$, Where r_1 and r_2 represents particle and peace wall radius respectively, and E^* is effective elastic modulus, $1/E^* = (1 - v_1^2)/E_1 + (1 - v_2^2)/E_2$, where v_1 and v_2 are particles and wall Poisson ratio respectively, E_2 is wall elastic modulus, E_1 is relaxation modulus for viscoelastic fiber particles, which could be describe by complex form: $E_1 = E' + iE''$, where E' is storage modulus, E'' is loss modulus. The second item is the viscous dissipation force, where χ is hysteresis damping factor. It is related to the structure of the material, viscosity, water content, temperature and so on, "δ" is the normal contact displacement and V_{in} is the normal relative velocity.

According to JKR theory and Hertz-Flores [16] theory, when wood fiber particles collide with the wall, the energy conservation equation is shown as equation (6) when the normal compression distance of particle collision reaches the maximum value.

$$Q_{in} + Q_{adh} = Q_e + Q_{nh}, \quad (6)$$

Where Qin is the impact kinetic energy before the collision of particles, Q_{adh} is the adhesion energy between the particles and the wall during the approaching process, $Q_{adh} = 2\pi a_0^2\omega_1$, in which a0 is the initial radius of the contact surface between the contact surface only by the adhesion force F_{adh}, $F_{adh} = 3\pi R^*\omega_1$, where ω_1 is the surface energy of the two collision during the approaching process, Q_e is elastic storage of wood fiber particles in the energy dissipation after the collision, $Q_e = \int_0^{\delta_{max}} F_n d\delta = \frac{2}{5}K\delta_{max}^{5/2}$; and Q_{nh} is the dissipated energy which means irreversible work of wood fiber particles to overcome the internal friction and the molecular chain segment movement, $Q_{nh} = \chi E^*(R^*)^{0.5}\delta_{max}^{2.5}/2$, including high elastic deformation and viscous friction flow friction.

After collision, energy dissipation will become heat existing in the collision system, while the elastic energy would be stored in two ways, one of which is used to overcome the adhesion between the collision bodies and restore the deformation, the other part is transformed into the particle kinetic energy which makes the particle rebound off the wall in the velocity of V_{rn}. The process meets the energy conservation following equation (7).

$$\frac{1}{2}m^*V_{rn}^2 = Q_e - Q'_{adh}, \quad (7)$$

Where m^* is the effective collision mass: $1/m^* = 1/m_1 + 1/m_2$; Q'_{adh} is the adhesion energy that particles must be overcome during the separating process, $Q'_{adh} = 2\pi a^2\omega_2$, where ω_2 is the particle interface energy in the separating process. Literature [1] studied that ω_2 in the separating process is usually much bigger than ω_1 in the approaching process, especially for viscoelastic particles, the number between the them is relatively large; and a is equivalent radius of contact surface area after the collision.

For meet the requirement of equation (7), that is, the elastic energy storage of particles is equal to the adhesion energy that particles need to overcome when they leave the deposition surface, the normal velocity components of particles impinging on the wall are defined as the critical velocity V_{in_cri}, described by equation (8) as following:

$$V_{in\_cri} = f(\varphi_V, R^*, m^*, E^*, \chi, \omega_1, \omega_2), \quad (8)$$

2.3. Tangential deposition model

When the particle tangential impulse is larger than the rolling friction resistance, which satisfies equation (9), the particles will roll down and get out of the wall. Otherwise, the particles will deposit.

$$F_t = F_d + F_{it} \geq f^*(F_{in} + F_{adh}), \quad (9)$$

Where F_{it} is the tangential component of particle contact force. According to the literature [17]: $F_{it} = \min(f^*F_{in}, K_tV_{it}\Delta t)$, where Δt is the contact time. And according to the article [18], the recovery of tangential motion after the particles collision on the premise that action time of the tangential impact force is at least equal to the impact time of the normal's. In this paper, the correction coefficient a is used to multiply the normal contact time: $\Delta t = 1.47a(5m^*/4n)^{0.4}V_{in}^{-0.2}$, $n = 4E^*\sqrt{R^*}/3$, and the incident angle θ is defined: $\tan\theta = V_{it}/V_{in}$. According to the definition of drag force, the air drag force on particle is very small, which can be ignored. While due to such small size of fiber particles, the effect of adhesion force can also be ignored comparing to the inertial impact force in the collision process.

As a result, the equation (9) can be transformed into a function of the impact angle. The minimum impact angle of the collision particles rolling away from the wall is defined as the critical impact angle: θ_{it_cri}, and equation (10) could be gotten:

$$\theta_{it\_cri} = f(f^*, \chi, K_{it}, K_{in}), \quad (10)$$

In summary, the criteria for collision deposition of wood fiber particles and walls could be described as follows:

(1) when $\theta > \theta_{it\_cri}$, the particles will continue to move along with the fluids and will not be deposited on the wall;

(2) when $\theta < \theta_{it\_cri}$ and $V_{in} \geq V_{in\_cri}$, the particle will rebound away from the wall;

(3) when $\theta < \theta_{it\_cri}$ and $V_{in} < V_{in\_cri}$, the particles will deposit on the wall.

3. Analysis of influencing factors

3.1. predictive parameters of model

For analysis the influence of some factors on slender viscoelastic particle deposition, such as particles size, aspect ratio, damping factor and so on, the critical criteria of a case collision between wood fiber and wall is predicted based on the proposed model, with mesh size between 80 to 160. And the calculation parameters of the wood fiber and wall are shown in Table 1.

Table 1. Calculation parameters of collision between particles and wall.

element		value
wall	elasticity model E2/(Mpa)	230
	Poisson ratio v_2	0.3
	Effective friction f*	0.1
Wood fiber particles	Relaxation model E1/(Mpa)	37.7
	Poisson ratio v_1	0.4
	Density P/(kg·m-3)	500

3.2. influence factors

Based on the model, fiber particles whose damping factor is 0.1, the particle sizes of 100, 150 and 200 μm are selected to analyze the influence of the aspect ratio on the critical speed. The result is shown in Figure 2.

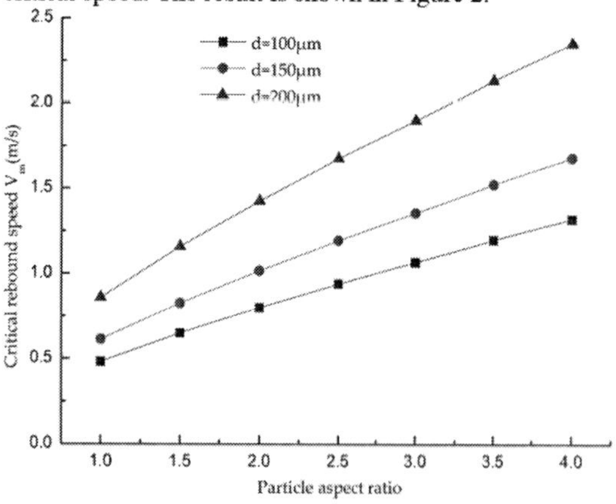

Fig. 2. Influence of aspect ratio on critical rebound velocity.

Theoretically, the large aspect ratio means particles are slenderer, and the viscosity would increase. With the same collision force, the dissipated energy is bigger, the adhesion energy needed to overcome during the separating progress from the wall surface is greater, which means the initial impact kinetic energy needed for critical rebound would be greater. The prediction results by the model shown in Figure 2 also showed that the critical rebound speed increase with the particle aspect ratio increases, which is consistent with the theoretical analysis.

The damping factor of wood fiber is different with the materials, structures and moisture content. According to the model, the influence of damping factor on critical speed was analyzed at a ratio of 2 with different particles diameter, and the result is shown in Figure 3.

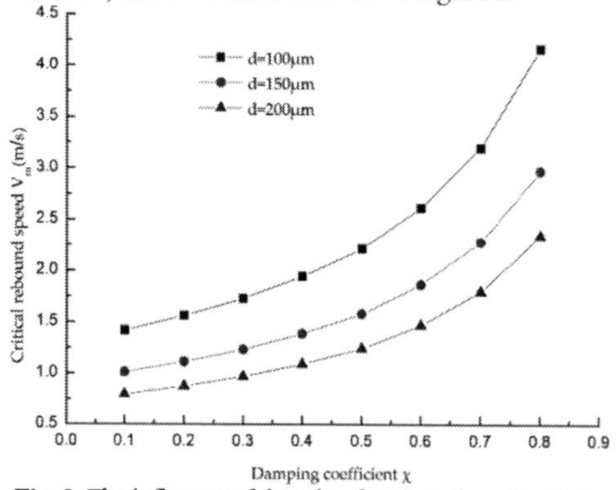

Fig. 3. The influence of damping factor on the critical rebound velocity.

Based on the viscoelastic theory proposed by Flores [16], when the damping coefficient of particles increases, the energy consumption in the process of collision is bigger, the elastic energy stored is smaller, and the required critical velocity of particle collision rebound would be bigger. the critical velocity in Figure 3 increases with damping coefficient, under the same particle size, which conforms to the theoretical analysis.

What's more, under the same aspect ratio and damping coefficient, the larger particle size, the greater particle mass and impact energy with the same impact velocity, which lead to the bigger the elastic energy and be easier for particles to rebound, namely, critical rebound speed is smaller. The model shows that the critical velocity is inversely proportional to the 1.2th power of particle size, and the critical velocity decrease with the increase of particle size as shown in Figure 2 and 3, which consistent with the theoretical analysis. As for the particles with aspect ratio bigger 4, we need further study for those more flexible and curl morphology.

4. Experiment and simulation analysis

4.1. simulation analysis

A coupling simulation method with the FLUENT and EDEM software was used to simulate the movement of dry exhaust gas in the field of heat piping exchanger. The saw dust particle with the aspect ratio of 2,3,4, and size 0.15 mm are used in the simulation. Through the axial

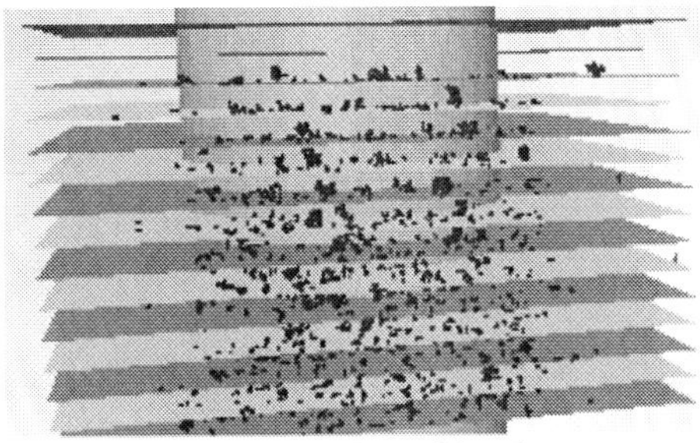

(a) view in the front of the tube

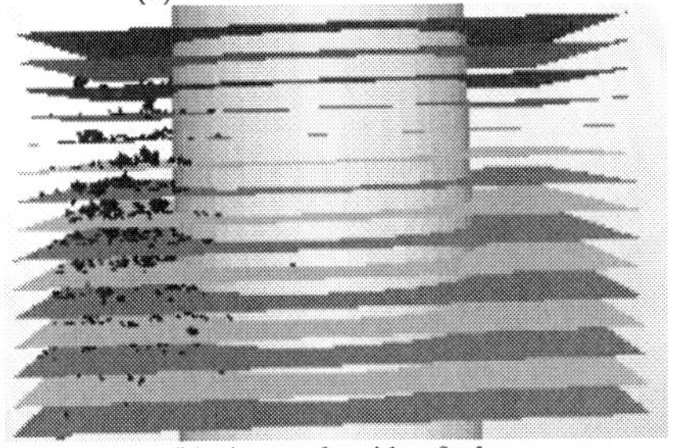

(b) view at the side of tube

Fig. 5. Simulation results of fibrous particle deposition.

superposition of the based particle, the slender geometry of the particle can be created with EDEM software, which surface smoothness can be adjusted by the superposition displacement and particle size. Figure 4 shows the simulation particle with size 0.15 mm and aspect 4 in EDEM.

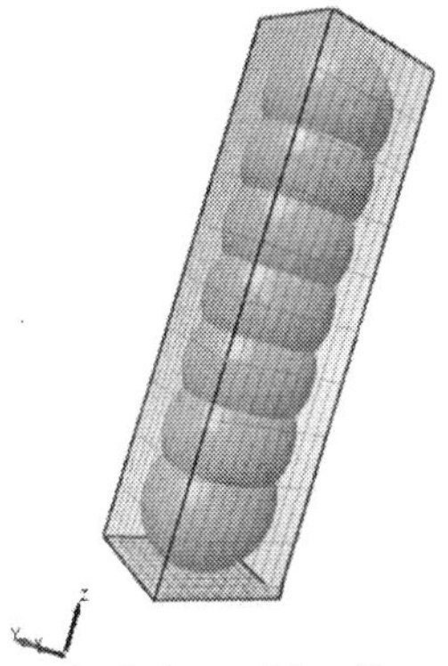

Fig. 4. simulation particle with aspect ratio of 4

According the slender saw particles, hysteretic damping factor of 0.1J/m2 is set up for measure the viscoelastic energy consumption in simulation. And the standard k-Ɛ model was used to simulate the flow of fluid. Based on the present slender particulate deposition model, a User-defined function is embed in FLUENT software, which can obtain the impact velocity V and impact angle θ of particles near wall, as the flowing:

```
DEFINE_DPM_EROSION(dpm_accr,p,pt,f,normal,alpha,Vmag,Mdot)\
{......
Vn=Vmag*sin(alpha);   /*obtain the normal velocity of impact particle*/
If (alpha>cita_cir)   /*judging the impact angle of the particle*/
return;
if (Vn>Vin_cir)   /*judging the critical velocity of impact particle*/
  return;
else
  F_UDMI(f,t,0)+=P_MASS(P);   /*storage the mass of deposition particles*/
F_UDMI(f,t,1)=num_in_data+1;   /*storage the number of deposition particles*/
......
}
```

When the dry tail gas flow from to the heat piping, its flow characteristics changer greatly. When the particle flow to the central windward area, the velocity reduced and the direction changed for the formation of boundary layer along the cylinder, while these changes mach smaller in windward side region for the boundary layer along the plate. And the impact angle and velocity of the collision between particles and wall would decrease for the change of flow direction in the field. If the impact angle is less than the critical the impact angle, and the velocity of particle is less than the critical rebound also, which means $\theta < \theta_{it\_cri}$ and $V_{in} < V_{in\_cri}$, the particles will be deposited on the wall. Fiber particles around the the central windward area of tubes would easily match deposition critical than those around the side region of tubes. Therefor, the amount of fiber particle deposition on the central windward area is greater than on the windward edge area, as shown in figure 5 from simulation.

4.2. experience and analysis

At present, heat piping exchanger are usually applied in the waste heat utilization device of fiber drying desiccation in Chinese wood industry. In this research, the gravity heat pipe is used for experiment. The experimental device is shown in Figure 6.

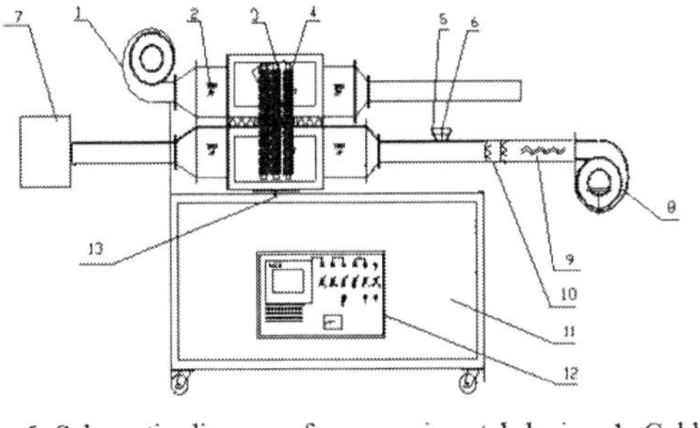

Fig. 6. Schematic diagram of an experimental device. 1. Cold fan; 2. Temperature, humidity, pressure test holes; 3. heat pipe experiment section; 4. The rotary wheel; 5. Flashboard; 6. Ash adding port; 7. Dust elimination device; 8. hot end fan; 9. Electric heating device; 10. Humidifier; 11. Working table; 12. Control cabinet; 13. condensate discharge port

As the above simulation results shown, wood fiber particles mostly deposited at the central windward area of fin, both the rectangular fin tub and the H fin tub with a gap at the center were used to experiment. The size of wood fiber particles in the experiment is less than 0.2mm. When the inlet velocity of the dry tail gas t is 3m/s, the ash distribution on the surface of the fin surface of the heat pipe is shown in Figure 8(a) and (b) separately after 5 hours of operation.

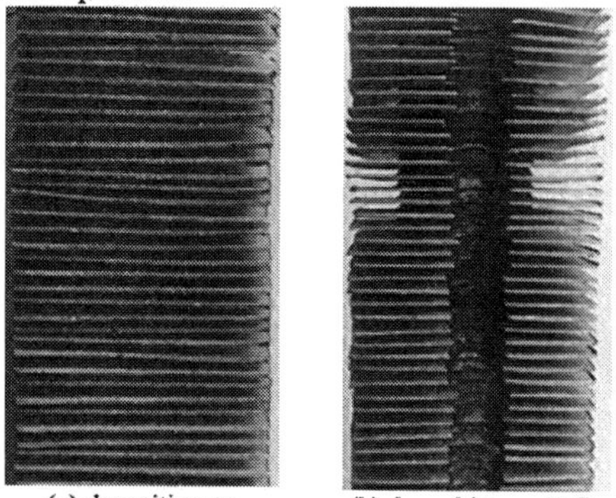

(a) deposition on rectangular fin (b) deposition on H fin

Fig. 7. Experimental results.

As shown in Figure 7(a), at the central windward area of rectangular fin, the particulate fouling is serious than the side of fin. While in Figure 7(b), as the central area of the fin is removed, the particle fouling on H fin is seldom, which can be guide for the waste heat recovery of middle density fiberboard production enterprises.

5. Conclusions

In this paper, the characteristics of wood fiber particles with size less than 0.2 mm under the external force load are studied. Combined with viscoelastic and creep deformation properties of wood fiber, a deposition mechanism model for slender fiber particles is proposed for the first time, and the equivalent sphere method is used to solve the model of blanket deposition mechanism of wood particles with aspect ratio less than 4.

The predicted model indicates that the critical velocity of the wood fiber particles increases with the decrease of particle size, the increases of the aspect ratio and the damping factor, which are consistent with the theoretical analysis.

The coupling simulation with the FLUENT and EDEM method, using the present deposition mode, was carried out. The simulation results showed that the fouling deposition mainly occurred at the center fin of the front tube. And experiments on waste heat utilization of fiber drying tail gas were performed as well. Using the H fin tubes instead of rectangular fin tubes in heat exchanger can greatly relieve fouling deposition of wood fiber particles, which provides a useful way to solve the problem of the ash deposition. Although we do not have sufficient data to show the real profits we can get by our research in the real production lines in wood industries, we hope to contribute to successful development of the utilization of low temperature waste heat and efficiency improvement of heat exchanger for the energy-saving target.

Acknowledgments

This work was supported by the National Natural Science Foundation of China (Grant Number 51406088).

References

1. Xiqiao Feng, Huan Li, Hongping Zhao, Sholi-Wen Yu. Numerical simulations of the normal impact of adhesive microparticles with a rigid substance. POWDER TECHNOL, **189**, 34(2009).
2. Marshall, J.S., Particles aggregation capture by walls in a particulate aerosol channel flow. J Aerosol Sci, **38**, 333(2007).
3. Maurizio Troiano, Fabio Montagnaro, Piero Salatino, Roberto Solimene. Experimental characterization of particle-wall interaction relevant to entrained-flow gasification of biomass. FUEL, **19**, 674(**2017**).
4. Fan Geng, Yingchao Wang, Yimin Li, Longji Yuan, Xinyong Wang, Min Liu, Zhulin Yuan. Numerical simulation on mixing dynamics of flexible filamentous particles in the transverse section of a rotary drum. PARTICUOLOGY, **11**, 594(**2013).**
5. Jianzhong Lin, Weixiong Wang. Numercial simulation of the sedimentation of cylindrical particles. J Eng Thermophysrus Therm, **23**, 631(2002).
6. Jianzhong Lin, Shi Xing, Zhenjiang You. Effects of the aspect ratio on the sedimentation of a fiber in Newtonian fluids. J Aerosol Sci, **34**, 909(2003).
7. Jianzhong Lin, Changbin Wang, Xing Shi. Study on the effect of wall constraint on the settlement of columnar particles in Newton fluid. Chinese Journal of Applied Mechanics, **20**, 2(2003).
8. Jianzhong Lin, Xueming Shao, Xing Shi, Zhaosheng Yu. Study on the interaction of sedimenting cylindrical particles in still fluid. Acta Mech Sinica, **20**, 32(2004).
9. David Saintillan; Eric Darve; Eric S. G. Shaqfeh. A smooth particle-mesh Ewald algorithm for Stokes suspension simulations: The sedimentation of fibers. Phys Fluids, **17**, 033301(2005).
10. Li S.-Q., Marshall J.S. Discrete element simulation of micro-particle deposition on a cylindrical fiber in an array. J Aerosol Sci, **38**, 1031(2007).
11. Christiane Helzel, Athanasios E. Tzavaras. A kinetic model for the sedimentation of rod-like particles. Multiscale Model Sim, **15**, 500(2017).
12. Abd-Elhady M.S., Rindt C.C.M, van Steenhoven A.A. Contact time of an incident particle hitting a 2d bed of particles. Powder Technol, **191**, 315(2009).
13. Anandarajah A. Multi-mechanism anisotropic model for granular materials. Int J Plasticity, **24**, 804(2008).
14. Evgeny Rabinovich, Haim Kalman. Incipient motion of individual particles in horizontal particle-fluid systems: A. Experimental analysis. Powder Technol, **192**, 318(2009).
15. Akhatov I. S., Hoey J. M., Swenson O.F., Schulz D.L. Aerosol focusing in micro-capillaries: Theory and experiment. J Aerosol Sci, **39**, 691(2008).
16. Paulo Flores, Margarida Machado, Miguel T. Silva, Jorge M. Martins. On the continuous contact force models for soft

https://doi.org/10.1051/e3sconf/20183801003

materials in multibody dynamics. Multibody syst dyn, **25,** 357(2011).

17. Schwager T., Becker V., Pöschel T. Coefficient of tangential restitution for viscoelastic spheres. *The European Physical Journal E, 27*, 107(2008).
18. Guangcheng Zhang, Huiming Tang, Xin Xiang. Characteristic parameters theoretical analysis of rockfall impact on ground. *Chinese Journal of Rock Mechanics and Engineering*, **31**, 2839(2012).

 https://doi.org/10.1051/e3sconf/20183801004

Pollution data analysis and characteristics of volatile organic compounds in the environment

Qi WANG[1], *Chēn* WANG[1], *Lujian* HOU[2], *Bo* LV[2], *Chén* WANG[1,*]

[1]School of Environmental Science and Engineering, Qilu university of technology (Shandong Academy of Sciences), Jinan, Shandong 250353, China
[2]Jinan Environmental Monitoring Center, Jinan, Shandong 250000, China

Abstract. Volatile organic compounds (VOCs) have a wide range of sources and have a significant impact on the ecological environment and human health, which have attracted wide attention of many researchers. In this paper, the pollution characteristics of VOCs, the role of VOCs in atmospheric chemistry including OH reaction reactivity (L_{OH}), Ozone Formation Potential (OFP) and SOA generation potential (SOAP), VOCs source apportionment were discussed and reviewed.

1 Introduction

As an important precursor of ozone and two organic particulate matter, Volatile Organic Compounds (VOCs) play an important role in the atmospheric chemical reaction process, and have attracted more and more attention. VOCs generally refer to volatile organic compounds with higher saturated vapor pressure (greater than or equal to 0.01 kPa at 20 ℃), low boiling point, small molecular weight and at room temperature[1]. Usually, VOCs are divided into non methane hydrocarbons including alkanes, alkenes, alkynes and aromatic hydrocarbons (NMHCs), oxygen containing organic compounds including aldehydes, ketones, alcohols, ethers and halogenated hydrocarbons (OVOCs), nitrogenous compounds, sulfur compounds and other categories[2].

As a kind of important gaseous pollutants in the atmosphere, VOCs not only have a direct impact on human health and ecological environment, but also can participate in photochemical reactions generated two pollutants such as ozone (O_3), peroxyacetyl nitrate and organic aerosol, so VOCs are one of the important precursors of O_3 and $PM_{2.5}$[3]. Hydrocarbons in volatile organic compounds react with nitrogen oxides to produce ozone, which can cause atmospheric photochemical smog and cause harm to human health and plant growth. Ozone is a typical pollutant of photochemical smog. VOCs are one of the most important reasons for the increase of atmospheric ozone concentration and the formation of regional photochemical smog, acid rain and haze.

At present, the natural and anthropogenic sources of VOCs are widely studied at home and abroad. As far as the global scale is concerned, the contribution of natural sources to the VOCs is more than the human source. Transportation is the world's largest VOCs anthropogenic sources, using solvent is the second largest emitter.

2 Study on the pollution characteristics of VOCs

It is very important to study the distribution of ambient VOCs to take effective air pollution control measures.

2.1 Ambient concentration

From the level of pollution, the VOCs ambient concentration in different areas has a certain difference. Wang[4] measured the concentration of VOCs in Beijing road site at a high level, and road air pollution is more serious than the air pollution in the region. Pankow[5] measured the concentration of benzenes at 13 sampling points in the United States (including the suburbs and cities) and find that lower than that in other countries. Thus, the emission of motor vehicles has become a major factor in the quality of the road air and the main source of volatile organic compounds in urban areas is motor vehicle exhaust. Table 1 presents a comparison between VOCs compounds in different cities. It can be seen from the table that in the seven cities, the concentrations of alkanes and aromatics in Beijing and Guangzhou are higher in domestic cities, which are related to the large emission of motor vehicles and more manufacturing industry in Beijing and Guangzhou areas. Compared with foreign cities, the concentration of isoprene and toluene in Singapore were higher than those in foreign cities. North and American were the next, Charleston and Los Angeles were the minimum while the mixing ratio of propane were higher than those in foreign cities. The VOCs emission characteristics between different cities were significantly different. This difference is related to

*Corresponding author: shanqing123@126.com [1]Email: wxqccfen1221@163.com

the change of the source of each city, the observation period and the method of observation and experiment.

Table 1 Comparison of the concentration of VOCs in different cities (ppbv)

Region	Propane	n-Butane	i-Butane	Ethene	Isoprene	Benzene	Toluene
Beijing 2014[6]	10.8	3.27	2.32	7.18	0.11	1.98	3.31
Guangzhou 2011-2012[7]	3.66	3.07	2.67	2.99	1.14	0.62	4.59
Hongkong 2010.09-10[8]	2.34	4.37	2.60	1.85	0.12	0.73	2.55
North American 2016.5[9]	-	0.86	-	-	-	2.2	1.1
Singapore 2011[10]	-	-	-	-	3.56±1.42	1.06±0.18	6.39±2.11
Charleston 2008[11]	3.54±1.02	1.01±0.35	0.44±0.17	0.3±0.15	1.82±0.85	0.084±0.033	0.12±0.055
Los Angeles 2007[12]	6.05±3.58	2.34±1.31	1.24±0.6	2.4±1.36	0.27±0.13	0.48±0.24	1.38±0.72

2.2 Time distribution characteristics

In terms of time distribution, the peak value varies from daily and season. The time distribution of VOCs in the atmosphere changed significantly, with obvious seasonal and diurnal variations.The strong emission source and meteorological conditions are the most important factors that affect the time distribution of VOCs in the atmosphere. Guo[13] investigated spatial variations of volatile organic compounds (VOCs) in the atmosphere and found that the concentration of VOCs in the atmosphere of Hongkong has significant seasonal difference and composition, which control the seasonal variation of VOCs concentration in urban atmosphere by a source of emissions, dichloromethane, xylene and trimethyl benzene concentration in summer was slightly higher than in winter, and methyl chloride, benzene and four vinyl chloride concentration peak appeared in the winter atmosphere. Shao[14] studied the daily variation of total volatile organic compounds (TVOC) in summer in Guangzhou. The results showed that the early and late peak concentrations of alkanes and olefin species appeared at 8:00 to 10:00 and 18:00 to 22:00 respectively, which were in accordance with the peak traffic time. The estimate is mainly related to the emission of motor vehicles. Moreover, the peak level of TVOC in the evening is higher than that in the morning because of the decrease in the height of the boundary layer at night.

2.3 Spatial distribution characteristics

In terms of spatial distribution, the concentration of VOCs varies with the change of height and distance. The total energy consumption in urban areas is higher, fuel volatilization, incomplete combustion and industrial activities will discharge a large amount of VOCs, which is closely related to human activities. For example, the high concentration of VOCs in large urban areas is alkane with a carbon number less than 4 ppbv, followed by alkenes and aromatics. The composition and concentration of VOCs in the atmosphere of remote areas are mainly influenced by the emission of natural sources and the long distance transmission of the atmosphere. Wei[15] has calculated unsaturated hydrocarbons and benzene in the atmosphere of Guizhou, Guangxi, Sichuan and other places in China and the results showed that the proportion is higher than that of the eastern developed cities (Beijing, Tianjin, Shanghai, Zhejiang, etc), which is related to the distribution difference between the fixed burning source and the road vehicle emission source in the provinces.

3 The role of VOCs in atmospheric chemistry

3.1 Reactive activity of VOCs

Volatile organic compounds (VOCs) are important precursors of ozone products in the atmosphere, and their concentrations in the atmosphere usually directly affect the level of ozone pollution. Owing to the great variety of VOCs and the great difference in the chemical structure of the different VOCs, the contribution of different VOCs in ambient air to atmospheric oxidation capacity is very different[16]. At present, there are two ways to calculate the chemical activity of VOCs in ambient air. One is the reactivity of VOCs to OH radical (L_{OH}), the other is the increment reaction activity of VOCs (MIR) to calculate Ozone Formation Potential (OFP).

3.1.1 OH reaction reactivity

The method of measuring the reactive activity of VOCs in the photochemical reaction process of the atmosphere mainly includes the activity of OH radical reaction. The OH reaction activity analysis method considered the reaction rate constant of the concentration of each VOCs species and the OH radical, which is usually a OH consumption rate method. The reaction rate of the VOCs species and the OH free radical reaction L_{OH} reflects the speed of the chemical reaction of the VOCs species in the atmosphere. The concentration of the species and the rate constant of its reaction with the OH radical determine L_{OH} and the calculation formula of L_{OH} is as follows:

$$L_{OH} = [VOC]_i \times K_{iOH} \quad (1)$$

In the formula, $[VOC]_i$ is the concentration of i in the species in VOCs, K_{iOH} is a constant of the rate of the reaction between the species i and the OH radical, L_{OH} is the rate of OH consumption of the species i. Shao[17] used the reaction rate of VOCs species in the atmosphere and OH free radical to analyze the chemical activity of all kinds of VOCs species in Beijing. The results show that the atmospheric VOCs mixture ratio of 15% olefin provides about 75% of the atmospheric chemical activity in the atmospheric VOCs. The contribution of the olefin components of C_4 and C_5 is the greatest

3.1.2 Ozone Formation Potential (OFP)

L_{OH} neglects the production of O_3, which is consumed by the oxidation of other pollutants. In order to be able to synthetically measure the effects of the chemical reaction activity of the VOCs species on the formation of ozone. Carter[18] proposed VOCs incremental reactivity (IR), that is, a change in the O_3 concentration produced by by adding or removing a unit specific VOCs in a mixture of a given gas mass VOCs. In general, the greater the IR is, the faster the VOCs species reacts. The maximum value of IR is achieved by changing the ratio of VOCs / NOx, which is called the maximum incremental reactivity (MIR)[18]. In the study, the ozone generation potential (OFP) is usually used to indicate the maximum contribution of the VOCs species to the ozone generation under the optimal conditions.Its calculation formula is as follows:

$$\mathrm{OFP} = [VOC]_i \times MIR \qquad (2)$$

Among them, $[VOC]_i$ is the concentration of i in the VOCs species, and OFP is the ozone generation potential of i of the VOCs species. Czader[19] uses the maximum increment reactivity (MIR) and the concentration of VOCs to study the ozone formation potential (OFP) in the Houston region. The results show that propylene, ethylene and formaldehyde are the larger species of OFP.

3.2 SOA generation potential (SOAP)

Many studies have shown that VOCs is an important precursor for the formation of Secondary Organic Aerosol (SOA)[20, 21]. The SOAP can be used to estimate the concentration of SOA generated by VOCs in the atmosphere. There are two common ways to estimate SOA generation at present: One is to calculate the rate of consumption in the atmosphere according to the measured VOCs concentration, and then to estimate the generation rate of SOA (formula 3). In the formula, t ($\mu g{\cdot}m^{-3}$) and t ($molecule{\cdot}cm^{-3}$) are the concentrations of oxidants ($\cdot OH$, O_3 or NO_3) in t at the t time, respectively. The reaction rate constant ($cm^3{\cdot}molecule^{-1}{\cdot}s^{-1}$); The other way is to deduce the initial emission concentration according to the measured VOCs concentration in the environment, assuming that all the VOCs emissions of these emissions are all consumed, and the potential of SOA is calculated by yield (formula 4).

$$\frac{\Delta SOA}{\Delta t} = \frac{\Delta VOC}{\Delta t} \times Y = k_{OX}[VOC]_t[OX]_t Y \qquad (3)$$

$$SOA_{potential} = [VOC]_{initial} \times Y \qquad (4)$$

In the global scale, the natural source VOCs is the main contributor to the SOA generated by the VOCs transformation. Lane[22] uses a model based on the VOC source list in the United States to estimate SOA. The results showed that the contribution of human source VOCs to SOA was higher in urban areas than in the suburbs, but it is still lower than the contribution of natural sources. In human precursors, the contribution of aromatic hydrocarbons, alkenes and alkanes to SOA is 36%, 32%, 32%, respectively.

4 VOCs source apportionment

Because of the variety of VOCs sources and the complexity of their emissions, it is of great importance to understand the proportion of main VOCs emissions and the proportion of each source of pollutants in the ambient air.

4.1 Benzene series ratio method

The ratio of benzene to toluene (B/T) is often used as a basis for judging the main sources of VOCs in the atmosphere[23]. When the B/T value is close to 0.5, the VOCs in the atmosphere mainly comes from the emission of vehicle exhaust[24, 25]. When the B/T value is more than 0.5, the VOCs in the atmosphere may come from biofuel or charcoal[26, 27]. It may also come from the use of petrochemical or paint.When the B/T value is more than 1, the VOCs in the atmosphere mainly come from coal combustion[28]. When the B/T value is less than 0.5, there may be other sources of VOCs in the atmosphere in addition to the motor vehicle[29]. Barletta[30] studied 27 road air samples in 25 cities in China and concluded that B/T was (0.58 + 0.21). Therefore, when the B/T value is close to 0.6, the vehicle exhaust emission is the main source of VOCs pollution in the atmosphere, which is more in line with the actual situation in China. The results of Song's[31] study also demonstrate this view.

4.2 Source appointment

Currently, the commonly used receptor models include CMB, PCA/APCS and PMF models[32]. CMB and PMF models are the source apportionment technology recommended by the United States Environmental Protection Agency. Common sources of pollution have characteristics of tracer, such as ethylene and acetylene as combustion source tracer; gasoline volatile source often tracer compounds to isopentane, n-butane and isobutene; high content of industrial emission sources in halogenated toluene, ethylbenzene and xylene; aromatic hydrocarbons are often from solvent use.

The analysis of VOCs sources in foreign countries began in 1970s. At that time, the American EPA used a model of the CMB1.0 version. By using the CMB receptor model, Vaga[33] found that the air VOCs from Mexico accounted for 58.7% of vehicle exhaust and

24.2% of LPG, which accounted for 81.9% of the total VOCs. Mahmoud[34] found that the sources of VOCs were mainly from the mobile source, the smelting of lead and the use of LPG.

Cai[35] used PCA/APCS (principal component analysis/absolute principal component scores) receptor model to analyze the source of air VOC in summer in Shanghai. The results show that there are five main sources of VOCs in the atmosphere of the central city. They are mainly vehicle emissions, fuel volatilization (liquefied petroleum gas / natural gas leakage and gasoline evaporation), solvent use, industrial production and biomass / biofuel combustion and marine sources. Their contribution rates are 34%, 24%, 16%, 14%, 12%, respectively.

In 1994, Paatero and Tapper[36] proposed the application of PMF model to solve source resolution problems for the first time. The PMF model is a matrix decomposition method that combines the error estimation in the data to solve a restricted weighted least square linear model. Maryam Sarkhosh[37] use PMF model to analyze the source of urban atmosphere in Tehran, Iran. The result showed that there were five sources of VOCs in Tehran, which accounted for 61% of vehicle emissions, 12% of fuel evaporation, 17% of urban natural gas, 8% of bio resources and 2% of industrial solvent sources.

5 Conclusion

In this paper, the research status of VOCs pollution at home and abroad is reviewed, and the following conclusions can be drawn: the main components of VOCs in the atmosphere are aromatic hydrocarbons, alkanes and alkenes; the concentration of volatile organic compounds in summer and winter in urban atmosphere is higher than that in spring and autumn; the diurnal variation of VOCs in urban air was in bimodal characteristic, which appeared at the peak of traffic in the morning and evening; natural sources are mainly from plant emissions and the main sources of human sources are automobile exhaust, fuel volatilization, and coating use.

It is of great significance to accelerate the study of the pollution characteristics of VOCs, and VOCs reactive activity which is to identify VOCs optimal control species and VOCs source apportionment. It is also of great significance to the protection of atmospheric environment and the prevention and control of VOCs pollution.

Acknowledgments

This work was supported by the Shandong Provincial Natural Science Foundation of China [grant numbers ZR2017LD010].

References

1. A. Laaksonen, M. Kulmala, C D. O'Dowd, et al. The role of VOC oxidation products in continental new particle formation[J]. Atmospheric Chemistry & Physics, 2008, 8(10):7819-7841
2. A. Kansal. Sources and reactivity of NMHCs and VOCs in the atmosphere: a review.[J]. Journal of Hazardous Materials, 2009, 166(1):17-26
3. K J. Kim, J C. Kim, J. Kim , et al. Development of hybrid technology using E-beam and catalyst for aromatic VOCs control[J]. Radiation Physics & Chemistry, 2005, 73(2):85-90
4. Y. Wang, Q I. Fei, X X. Lun. Temporal and Spatial Distribution Rule of Volatile Organic Compounds in Ambient Air around Urban Traffic Roads in Beijing[J]. Research of Environmental Sciences, 2010, 23(5):596-600
5. J F. Pankow, W.Luo, D A. Bender, et al. Concentrations and co-occurrence correlations of 88 volatile organic compounds (VOCs) in the ambient air of 13 semi-rural to urban locations in the United States[J]. Atmospheric Environment, 2003, 37(36):5023-5046
6. J. Li, S D. Xie, L M. Zeng, et al. Characterization of ambient volatile organic compounds and their sources in Beijing, before, during, and after Asia-Pacific Economic Cooperation China 2014[J]. Atmospheric Chemistry & Physics, 2015, 15(8):12453-12490
7. Y. Zou, X J. Deng, D. Zh, et al. Characteristics of 1 year of observational data of VOCs, NOx and O_3 at a suburban site in Guangzhou, China[J]. Atmospheric Chemistry & Physics, 2015, 15(12):6625-6636
8. S H M. Lam, S M. Saunders, H. Guo, et al. Modelling VOC source impacts on high ozone episode days observed at a mountain summit in Hong Kong under the influence of mountain-valley breezes[J]. Atmospheric Environment, 2013, 81(2):166-176
9. Wentworth G R, Aklilu Y A, Landis M S, et al. Impacts of a large boreal wildfire on ground level atmospheric concentrations of PAHs, VOCs and ozone[J]. Atmospheric Environment, 2018
10. R. Betha, V. Selvam, D R. Blake, et al. Emission characteristics of ultrafine particles and volatile organic compounds in a commercial printing center[J]. Journal of the Air & Waste Management Association, 2011, 61(11):1093-101
11. A K. Baker, A J. Beyersdorf, L A. Doezema, et al. Measurements of nonmethane hydrocarbons in 28 United States cities[J]. Atmospheric Environment, 2008, 42(1):170-182
12. S G. Brown, A. Frankel, H R. Hafner. Source apportionment of VOCs in the Los Angeles area using positive matrix factorization[J]. Atmospheric Environment, 2007, 41(2):223-237
13. H. Guo, S C. Lee, P K K. Louie, et al. Characterization of hydrocarbons, halocarbons and carbonyls in the atmosphere of Hong Kong[J]. Chemosphere, 2004, 57(10):1363-1372

14. X. Xin, S. Min, L. Ying. The diurnal variation of ambient VOCs and their role in ozone formation:case study in summer in Guangzhou[J]. Acta Scientiae Circumstantiae, 2009, 29(1):54-62
15. W. Wei, S. Wang, J. Hao, et al. Trends of chemical speciation profiles of anthropogenic volatile organic compounds emissions in China, 2005–2020[J]. Frontiers of Environmental Science & Engineering, 2014, 8(1):27-41
16. S. Sillman. The relation between ozone, NO x, and hydrocarbons in urban and polluted rural environments[J]. Atmospheric Environment, 1999, 33(12):1821-1845
17. J. Zhang J, M. Shao, University P, et al. Study on Composition of Ambient Volatile Organic Compounds (VOCs) in Beijing City[J]. Research of Environmental Sciences, 2004, 17(5):1-5
18. W P L. Carter. Development of Ozone Reactivity Scales for Volatile Organic Compounds[J]. Journal of the Air & Waste Management Association, 1994, 44(7):881-899
19. B H. Czader, B S T. Kim, W P L. Carter. A study of VOC reactivity in the Houston-Galveston air mixture utilizing an extended version of SAPRC-99 chemical mechanism[J]. Atmospheric Environment, 2008, 42(23):5733-5742
20. S. Xie, X. Tian. Formation Mechanism of Secondary Organic Aerosols from the Reaction of Volatile and Semi-Volatile Compounds[J]. Progress in Chemistry, 2010, 22(4):727-733
21. S J. Sjostedt, J G. Slowik, J R. Brook, et al. Diurnally resolved particulate and VOC measurements at a rural site: indication of significant biogenic secondary organic aerosol formation[J]. Atmospheric Chemistry & Physics, 2011, 11(12):5745-5760
22. T E. Lane, N M. Donahue, S N Pandis. Simulating secondary organic aerosol formation using the volatility basis-set approach in a chemical transport model[J]. Atmospheric Environment, 2008, 42(32):7439-7451
23. Chan, Lo-Yin, Chu, Kam-Wah, Zou, Shi-Chun, et al. Characteristics of nonmethane hydrocarbons (NMHCs) in industrial, industrial-urban, and industrial-suburban atmospheres of the Pearl River Delta (PRD) region of south China[J]. Journal of Geophysical Research Atmospheres, 2006, 111(D11):-
24. R. Perry, I L. Gee. Vehicle emissions in relation to fuel composition[J]. Science of the Total Environment, 1995, 169(1–3):149-156
25. P A. SCHEFF, R A. SCHEFF. Receptor modeling of volatile organic compounds. I: emission inventory and validation[J]. Environmental Science & Technology, 1993, 27(4):617-625
26. M O. Andreae. Emission of trace gases and aerosols from biomass burning[J]. Global Biogeochemical Cycles, 2001, 15(4):955-966
27. D. Brocco, R. Fratarcangeli, L. Lepore, et al. Determination of aromatic hydrocarbons in urban air of Rome[J]. Atmospheric Environment, 1997, 31(4):557-566
28. C Y M D. Santos. Atmospheric distribution of organic compounds from urban areas near a coal-fired power station[J]. Atmospheric Environment, 2004, 38(9):1247-1257
29. Y M. Zhou, Z P. Hao, H L. Wang. Pollution and Source of Atmospheric Volatile Organic Compounds in Urban-rural Juncture Belt Area in Beijing[J]. Environmental Science, 2011, 32(12):3560-3565
30. B. Barletta, S. Meinardi, F S. Rowland, et al. Volatile organic compounds in 43 Chinese cities[J]. Atmospheric Environment, 2005, 39(32):5979-5990
31. Y. Song, M. Shao, Y. Liu, et al. Source apportionment of ambient volatile organic compounds in Beijing[J]. Environmental Science & Technology, 2007, 41(12):4348-4853
32. X. Wang. Analysis of Ambient VOCs Levels and Potential Sources in Windsor[J]. University of Windsor, 2014
33. E. Vega, V. Mugica, R. Carmona, et al. Hydrocarbon source apportionment in Mexico City using the chemical mass balance receptor model[J]. Atmospheric Environment, 2000, 34(24):4121-4129
34. M. Abu-Allaban. A preliminary apportionment of the sources of ambient PM_(10), PM_(2.5), and VOCs in Cairo[J]. Atmospheric Environment, 2002, 36(35):5549-5557
35. C. Cai, F. Geng, Q. Yu, et al. Source apportionment of VOCs at city centre of Shanghai in summer[J]. Acta Scientiae Circumstantiae, 2010, 30(5):926-934
36. P. Paatero U. Tapper. Positive matrix factorization: A non-negative factor model with optimal utilization of error estimates of data values[J]. Environmetrics, 1994, 5(2):111-126
37. M. Sarkhosh, A H. Mahvi, M. Yunesian, et al. Source apportionment of volatile organic compounds in Tehran, Iran.[J]. Bulletin of Environmental Contamination & Toxicology, 2013, 90(4):440-445

 https://doi.org/10.1051/e3sconf/20183801005

Road Maintenance in Africa: Approaches and Perspectives

Mostafa Hassan M

Sustainable Roads, Urban and Transportation (SURT) research group, Department of Civil Engineering, Central University of Technology, South Africa

Abstract: In Africa, roads are the dominant mode of passenger and freight transport, for which the need is growing rapidly. It is noticeable that most of the African countries do not do enough to ensure the sustainability of road infrastructure as it has been widely reported that roads are affected, to varying degrees, by premature deterioration. Most of the African countries have adopted institutional reforms, notably entailing the creation of road funds and road agencies, and made significant progress on road maintenance. However, many challenges remain to be addressed in all of them to ensure appropriate maintenance. Although spending on road maintenance has increased over time in all African countries it remains insufficient to cover the needs. Poorly maintained roads constrain mobility, significantly raise vehicle operating costs, increase accident rates and their associated human and property costs, and aggravate isolation, poverty, poor health, and illiteracy in rural communities. This paper focuses, in particular, on road maintenance in some African countries considering types of road maintenance and the different approaches aiming at a comparison to reflect on similarities and differences.

1 Introduction

After independence, some African countries inherited road networks which were developed for the sake of colonial manipulation and military ascendancy [1]. These networks were developed and maintained to a certain extent because of the limited availability of funds. Road density, although low in general, started to increase and this increase was vividly noticed in some countries e.g. Madagascar, Malawi, Mozambique, and Niger [2]. Generally, road conditions have improved in most African countries in recent years as governments have strived to increase the density of their road networks. Remarkable progress has been made in founding organizations to manage and maintain African roads. Nonetheless, only one in three rural Africans have access to an all-season road and in cities, road construction has not kept pace with urbanization. In many countries, road maintenance remains inadequate. This is, mainly, because of lack of proper funding. Most African countries have followed a steady path in the roads sector, with the focus on improving the availability of funds for road maintenance and the capability to execute public works. One of such initiatives is the Sub-Saharan African Transport Policy Program. The main focus has been on the creation or enhancement of road funds which provide ring-fenced revenues for road maintenance, based on a user charge concept expressed through fuel levies. Another area of interest has been in the formation of independent Road Agencies in most of the African countries. This is all to achieve the main target of executing and maintaining good roads. One example is The South African National Road Agency Ltd (SANRAL) in South Africa. This paper looks at the different maintenance schemes and trends in some of the African countries. For the sake of length of the paper, only one country from each region in Africa is presented. The aim is to investigate the similarities and differences in the African maintenance perspectives.

2 East African Countries: Ethiopia Approach

Ethiopia is classified as am underdeveloped country and over the years has increased the size of the road network due to the approach to road maintenance and rehabilitation. Ethiopian road network has increased approximately from 6,400 km in 1951, 46,812 km in 2010 and in 2015 to 85,966 km [3]. Earlier, a lack of a comprehensive approach to maintenance in Ethiopia has led to an accumulation of backlog of road maintenance. However, with the reformation of the building blocks of road management initiative; which are: management, ownership, financing and responsibility; Ethiopia is gradually coming out of the shadow of poor road network [4].

The increase in road networks in Ethiopia resulted from the controls and regulations put in place by the RTA and the attention given by the Ethiopian government, as this was an endeavor taken by the government to help improve the country's economy. RTA was established in the 1967 to oversee matters related to roads and vehicles using roads in Ethiopia [5]. RTA's vision statement is; "*...ensure the provision of a modern, integrated and safe*

* Corresponding author email: mmostafa@cut.ac.za

road transport services to meet the needs of all the communities for strong and unitary economic and political system in Ethiopia". Therefore, RTA sees maintenance as the core in achieving this vision.

It is known that the road management aspect is always hampered by funding, but in the case of Ethiopia, this was resolved by the introduction of a road fund. The road fund in Ethiopia was established in 1997, after great consideration to resolve the maintenance issues. The road fund in Ethiopia was developed on the fee-for-service principle. However, to a great extent this reformation has been encouraging and thus, Ethiopia was classified as a good performer in terms of road maintenance. Overall, the maintenance culture in Ethiopia has greatly increased over the years, as a result of proper funding through the road fund model and the reformation made in the road management sector.

3 West African Countries: Nigeria Approach

Nigeria is one of the fastest growing economies in Africa and has the largest road network in West Africa and the second largest in Sub-Saharan Africa. The road network in Nigeria is estimated to be almost 200, 000km in length, however, the performance of the Nigerian road sector has not been satisfactory. Thus, this has contributed negatively and retards the economic growth in the country [6]. Although, a huge sum of money has been directed towards road development in Nigeria, due to increase in traffic volumes; roads are still not in good condition.

Federal Road Maintenance Agency (FERMA) in Nigeria is saddled with the maintenance issues at the federal level while the state's ministries handle the state and further helps with the local government roads [7]. Nevertheless, the financial and technical requirements for effective maintenance, rehabilitation and reconstruction are so staggering that the rate of maintenance could not match the deterioration [2]. Thus, there is a need to look into the maintenance schemes adopted by Nigerian's road agency which brought about the ineffectiveness. Unlike Ethiopia, despite some maintenance activities performed in Nigeria, routine and special maintenance are still behind schedule; this is as a result of roads not begin properly constructed and inadequate funding for maintenance activities. Overall, maintenance problems in Nigeria can be attributed to the lack of proper or consistent funding. Adding up to that is the issue of the method of contracting out road maintenance work, which FERMA and State ministries still rely on traditional methods; e.g. in-house teams; thus results in failure. However, privatization of maintenance work with performance-based contracts is considered as a cost-effective alternative.

4 Central African Countries: Angola Approach

Angola's road network consists of 7,777km paved, 28,018km gravel, and 36,528km earth roads, totaling 72,323km. Although, Government currently spends around 4.3 billion US dollar per year to rebuild the infrastructure, which is the equivalent to 14% of its GDP. Yet, there is still a lot of work to do to pull Angola from the shadows of war. Instituto Nacional de Estradas de Angola (INEA) is responsible for planning and managing Angola's national highway and between 2008 and 2009, a total of 5,600 km of roads were reconstructed.

In addition, much of the road network has received little or no maintenance, as many roads are located in the former war zone and have little or no traffic. Thus, INEA has limited knowledge of the true condition of many roads and of the likely rehabilitation costs. Therefore, it is necessary for INEA to improve data collection in order to understand the present condition of roads and bridges and the nature and volume of road traffic to be able to better plan highway and roads development and supply information to the private sector. The Angolan government has given priority to the rehabilitation of the national road network. However, at present, financing road rehabilitation and maintenance in Angola is directly dependent on the state budget as the road fund is not yet in operation [8].

The concept of a Road Fund (RF) was proposed to have independent management with the participation of different stakeholders, constituting the main source of revenue to INEA. Nevertheless, until the road user charges (tolls levies) are reviewed in Angola, the RF will be unable to serve its intended purpose to fund road rehabilitation and maintenance. Thus, INEA is proposing a strong increase, particularly in fuel and circulation taxes, with all proceeds reverting directly to the RF [8]. Overall, progress to complete Angolan road infrastructure is likely to take decades, but substantial efforts are already being made in the right direction.

5 North African Countries: Egypt Approach

The National Roads project is the largest Egyptian infrastructure project, with an adequate investment (0.7% of the GDP). However, road maintenance is said to be neglected, as it constitutes only 0.15% of the Egypt's GDP; this is small in comparison with that of Ecuador (0.23%), Morocco (0.24%) and 0.45% for Ukraine [9]. This negligence in-turn results in road accidents and consequently accounted for 1.5% loss of the GDP or even more. Since 2001, the Egyptian government has however increased the funding allocated to road and railway to improve transportation sector. Further, in 2009 an economic stimulus package increased the spending, putting transport infrastructure catch-up high on the political agenda. Yet, 75% of the roads need maintenance; as nearly 700 bridges out of the 1,706 bridges in Egypt are on the verge of collapse due to insufficient maintenance and this has contributed adversely to the domestic transport system, as well as the international transport system.

In harnessing this situation, the Government of Egypt set up an organization for the development and maintenance of roads. General Authority of Roads, Bridges and Land Transport (GARBLT) is in-charge of not only maintenance of roads in Egypt, but also its development. The organisation's mission is focussed on safety, efficiency and a well-managed road network and has jurisdiction over 23,000 km of the road network. Recently GARBLT started a program for bridge maintenance/management and repair. However, the program has not been as successful, due to a lack of human resources in terms of technical know-how and equipment for appropriate inspection and evaluation of bridges. In general, the state of maintenance of the road network in Egypt is still a dilemma.

Overall, considering the state of the road network in Egypt at present, it needs rehabilitation rather than maintenance. However, one of the major stumbling blocks is funding. Through GARBLT's initiative, funds are generated through tolls, advertisement, fines on overloaded trucks etc. However, this structure changed with a new regime taking over in Egypt since 2013. This change resulted in major projects running fast to expand the motorways under the supervision of the military. No data is available on these new projects, the new structure, or the maintenance plans.

6 Southern African Countries: South Africa Approach

South Africa has the longest total road network in Africa and the 10th longest in the world, similarly the 18th longest paved road network in the world. The total length is approximately 746,978 km, with only 21% paved (153,719 km). However, despite the fact that South Africa is economically foremost on the continent, it has roads in a fair to poor road condition of about 38%. Some of these conditions are due to a lapse in maintenance actives, although it is better when compared with that of other African countries [10].

Road maintenance in South Africa is handled by SANRAL; the creation of SANRAL in 1998 was a result of the government commitment towards transforming public sector [11]. SANRAL has jurisdiction over 92% of the national road network, of which 81% are non-toll and the remaining tolled (over 19,000 km). Further, the remaining 8% of the total road network is also tolled and is developed and managed by private bodies. Management by these private bodies is still overseen by SANRAL based on concession contract agreements. The concession contract duration is thirty years, after which the road is handed over to SANRAL. The non-toll road in South Africa are funded directly by tax revenues generated by the national government, but as a result of the insufficient funding from the government, in 1995, the toll road was introduced to support the development and maintenance of roads in South Africa. These are maintained with toll revenues and capital market borrowing [11]. Overall, while toll road conceding is rare in Africa, South Africa in the Sub-Sahara region has the most, which is approximately 0.1%.

It can be said that SANRAL is very close to achieving its aim and was categorised as a top performer alongside with Namibia. Yet, South Africa has a road maintenance funding backlog of approximately 80-149 billion Rand [12]. Apart from funding, SANRAL is getting it right with the strategies of maintenance such as; routine, periodic and special (urgent) maintenance which are in place. Only the municipal roads are not well managed, this results from poor data collection (only 4%) of the present conditions available, and thus municipality coordination and responsibility are confused. Nonetheless, the manner of contracting maintenance work, which is performance-based, has greatly contributed to the success achieved in South Africa [13].

7 Discussion on African Road Maintenance

Basically, the types of maintenance activities carried out are more of the same in most African countries, however, the approach differs. Nevertheless, the importance of road maintenance cannot be overemphasized because of its enormous contribution to the nation's regional integration and economic growth, social development, effective public administration and security. Additionally, maintenance sustains the quality and safety of the road condition close to the original design, and thus, minimizes the road user costs, accident rate, poverty level and even its reconstruction cost. However, delay in maintenance is tantamount to its reconstruction as it results in high direct and indirect costs. Yet, Burningham and Stankevich [14] reported that many countries, not only in Africa, usually spend just about 20-50% of the actual amount for road maintenance. However, according to World Bank reports, neglecting maintenance at any stage may result in spending 3 times or more of maintenance costs.

The overall comparison between the covered countries is presented in table 1. This table shows clearly that only Ethiopia has a running road funding model independent of the governmental support. However, the case of South Africa shows good progress in this regard while all other countries are dependent on the governmental budget. This budget, in all cases, is constraint by the development needs of the African countries. Therefore, maintenance always comes last on the priority list of these governments. The way forward is to develop different road funding models as per the special needs and cultures of the African countries. Moreover, it is evident that maintenance culture is not widely propagated in Africa as most countries do not give it the proper attention especially in countries with a vast population.

Focusing on the problems related to maintenance, it is obvious that these problems are the same everywhere as lack of funding and deficiency of data and information are the main problems. It is of utmost importance to the African countries to have a database connected to all countries to help in minimizing the costs connected to road maintenance and also construction.

Table 1 Comparison Between The Maintenance Perspectives of the Studied African Countries

Country	Road fund	Maintenance culture	Main problems
Ethiopia	Road fund model	Good	Insufficient fund
Nigeria	Annual fund from the government	Very poor	Lack of funds Lack of proper administration
Angola	Annual fund from the government + Road fund model (Partial)	Poor	Lack of information Lack of fund Need to develop network
Egypt	Annual fund from the government	Very poor	Lack of fund Lack of information
South Africa	Annual fund from the government + Road user charge model	Fair	Poor data at municipalities

8 Conclusions

Considering the current situation in most African countries, to a great extent, the level of road maintenance's effort can be said to be comparatively high to the overall economic size, yet these efforts remain little when compared with those of other developing countries around the world. A considerable percentage of the African countries can be said to have gotten it right on the aspect of road ownership and responsibility and creation of road management agencies, yet, the financing aspects still remain lacking, as most still depends directly on the government funding which is absolutely insufficient. Generally, to a large extent the road agencies know what to (with regards to maintenances) but the funding is always handicapping them.

As mentioned in the case of Ethiopia, the giant step in the establishment of the road fund has been a great contribution to the success recorded. While, Egypt, Nigeria and South Africa have established road agencies, the implementation of a road fund is not yet in place. However, the situation of South Africa is far better than that of Nigeria, because of the implementation of toll roads, which is one of the largest in the continent. Toll roads can still be considered to be equivalent to road fund, as this effort in South Africa is coupled with the private partnership concept which thus place South Africa above in the continent.

In conclusion, there is a need for African countries to look into the following as a way-out of the current road maintenance predicament:

• Incorporation of maintenance into road project and sector strategies

• Consideration of establishment the road fund concept

• Expansion of toll roads network

• Investment in other forms of transportation such as rail transport and integrated public transport system to reduce congestion of the roads

• Consideration of performance-based contract for contracting out maintenance work

• Involvement of public-private partnership concession concept

References

1. Fage, J D A History of Africa 2002 *Taylor and Francis*, fourth edition.
2. Foster, V and Briceno-Garmendia, C 2010 Africa's Infrastructure: A Time for Transformation. The International Bank for Reconstruction and Development / the World Bank.
3. The World Bank Ethiopia Road Sector Support Project 2015 Report No PAD700.
4. Road Sector Development Program 2001 Road Fund in Ethiopia: From Inception to Realization Mid - Term Review Meeting of RSDP, 6-7 February, 2001.
5. RTA: Ethiopian road transport authority 2014 [Online] available at http://www.rta.gov.et/ [Accessed 24 April, 2015].
6. The United Nations 2014 Economic Development in Africa: catalysing Investment Growth in Africa eISBN 978-92-1-056677-3.
7. Federal Roads Maintenance Agency (FERMA) 2014 Federal Roads Maintenance Agency (FERMA). [Online] Available at: http://www.ferma.gov.ng [Accessed April 2014].
8. The World Bank 2005 Private Solutions for Infrastructure in Angola, the World Bank, Washington, DC.
9. MENA-OECD Initiative 2010 Infrastructure: Business Climate Development Strategy: Phase 1 Policy Assessment: Egypt.
10. Jones, D Paige-Green, P and Sadzik, E Development of Guidelines for Unsealed Road Assessment 2003 *Transportation Research Record*. 1819. DOI:

10.3141/1819a-42.

11. SANRAL (2014). 2014 Annual Report. [Online]. Available at: http://www.nra.co.za/content/SANRAL_Annual_Report_2014_PRESS_14_9_15.pdf [Accessed 25 April, 2015].
12. Krygsman, S. (2011). Road maintenance: Toll roads part of a package of solutions. [Online]. Available at: http://blogs.sun.ac.za/news/2011/12/23/ [Accessed July 2015].
13. Goitom, H. (2014). National Funding of Road Infrastructure: South Africa, [online]. Available at: http://www.loc.gov/law/help/infrastructure-funding/southafrica.php#skip_menu [Accessed June 2015].
14. Burningham, S and Stankevich, N 2005 Why road maintenance is important and how to get it done, Transport Notes, The World Bank, Washington, DC.

Study on test of coal co-firing for 600MW ultra supercritical boiler with four walls tangential burning

WU Ying [1*], *ZHONG Yong-lu* [1], and *YIN Guo-mingi*[1]

[1]Jiang Xi Electric Power Science & Research Institute,Nanchang,Jiangxi 330096,China

Abstract. On account of nine commonly used coals in a Jiangxi Power Plant,two kinds of coal were selected to be applied in coal co-firing test through industrial analysis,elementary analysis and thermogravimetric analysis of coal.During the coal co-firing test,two load points were selected,three coal mixtures were prepared.Moreover,under each coal blending scheme,the optimal oxygen content was obtained by oxygen varying test.At last,by measuring the boiler efficiency and coal consumption of power supply in different coal co-firing schemes,the recommended coal co-firing scheme was obtained.

1 Introduction

A power plant in Jiangxi is affected by the location, capacity and so on. The coal entering the furnace is mostly local automobile coal, and the coal quality is changeable. Mixing coal with coal becomes the inevitable choice of the plant. However, the previous attempts to mix coal in the power plant were carried out with a simple weighted average, and the limitations were large. The mixed coal is too simple and rough, resulting in a lot of cases, boiler load capacity fluctuations, but also easy to bring the boiler reheat steam temperature low, low combustion efficiency, affect the economic operation of units [1-2]. Therefore, this paper adopts the method of theoretical research and field test to study the coal blending and burning test of No. 1 boiler in a power plant in Jiangxi, which provides a scientific basis for the power plant coal, coal and boiler operation.

2 Equipment Profile

No. 1 boiler of a power plant in Jiangxi is designed and manufactured for Harbin boiler works. The model is HG-2035/26.15-YM3. The boiler shape layout, single furnace, improved low NOX PM MACT low NOx main burner and staged air combustion system, wall tangential firing mode. The boiler adopts one intermediate reheat, balanced ventilation, tight closed arrangement, solid slag discharge, all steel frame and full suspension structure. The main design parameters of boiler are shown in table 1.

Table 1. Main design parameters of boiler.

Parameter	Unit	BMCR	BRL
Evaporation capacity	t/h	2035	1938
Main steam temperature	℃	605	605
Main steam pressure	MPa	26.15	26.04
Feed water temperature	℃	296	293
Flue gas temperature (before correction)	℃	127	125
Flue gas temperature (after correction)	℃	122	120
Boiler calculation efficiency (low)	%	93.97	94.05
Evaporation capacity	t/h	2035	1938

3 Test characteristics of coal

The combustion characteristics of pulverized coal have an important influence on the operation of boiler [3]. According to a large number of coal blending tests show that by different kinds of coal mixed coal, although in the industry, the elemental analysis of coal quality index is not significantly different, but the coal ignition, combustion and burnout, stable combustion and combustion characteristics of different [4-5]. Therefore, industrial analysis, elemental analysis and thermogravimetric analysis of 9 commonly used coal-fired power plants are provided. The results are shown in table 2~4.

Table 2. Proximate analysis of test coal.

Coal sample	Mt /%	Mad /%	Vad /%	Aad /%	St,ad /%	Qb,ad MJ/kg	HGI
Xinji Coal	8.1	0.9	29.71	28.73	0.26	21.04	72
XingyeCoal	8.1	0.63	9.54	56.72	0.47	11.95	76
Xing Coal No.1	9.90	1.57	27.45	24.56	1.31	21.06	66
Lu'an Xiadian Coal	10.10	1.50	9.91	26.24	0.26	22.36	98
Yanan Zhongsheng Coal	15.70	8.02	27.75	15.68	1.91	22.18	88
Datong Coal	6.90	3.27	26.48	24.77	0.64	21.56	69
Pankuang Coal	6.2	1.95	26.24	29.82	0.23	21.38	59

* Corresponding author: wuying3582@163.com

 https://doi.org/10.1051/e3sconf/20183801006

Yangou Coal	10.3	0.79	8.44	48.52	0.67	14.42	98
Indonesian Coal	15.23	6	39.36	8.95	0.81	23.09	50

Table 3. Elemental analysis of test coal.

Coal sample	Mt /%	Mad/%	Aad /%	Cad /%	Nad/%	Had/%	Oad/%
Xinji Coal	8.1	0.9	30.79	53.69	0.98	3.74	8.24
XingyeCoal	8.1	0.63	54.08	36.76	0.78	1.7	3.99
Xing Coal No.1	9.90	1.57	24.56	59.09	1.04	3.74	8.69
Lu'an Xiadian Coal	10.10	1.50	26.24	64.17	1.14	2.84	3.85
Yanan Zhongsheng Coal	15.70	8.02	15.68	63.64	0.84	3.48	6.43
Datong Coal	6.90	3.27	24.77	57.74	0.97	3.47	9.14
Pankuang Coal	6.2	1.95	29.82	57.01	0.9	3.64	6.45
Yangou Coal	10.3	0.79	48.52	41.02	0.84	2.02	6.14
Indonesian Coal	15.23	6	8.95	61.84	1.38	4.82	16.2

Table 4. Thermogravimetric analysis of test coal.

Coal sample	Ignition temperature Ti /℃	Burnout temperature Th /℃	Remarks
Xinji Coal	457.3	635.0	Easy to catch fire, easy to burn out
XingyeCoal	521.6	728.8	Hard to catch fire, hard to burn out
Xing Coal No.1	449.7	659.3	Easy to catch fire, easy to burn out
Lu'an Xiadian Coal	510.0	680.1	Hard to catch fire, easy to burn out
Yanan Zhongsheng Coal	444.3	631.1	Easy to catch fire, easy to burn out
Datong Coal	437.5	650.2	Easy to catch fire, easy to burn out
Pankuang Coal	456.9	652.4	Easy to catch fire, easy to burn out
Yangou Coal	516.6	717.6	Hard to catch fire, easy to burn out
Indonesian Coal	375.9	590.5	Easy to catch fire, easy to burn out

There are a lot of coal combustion characteristics of evaluation indicators, such as the dynamics of three factor[6], stable combustion and burnout characteristics index R[7] index[8], Fu Zhang index Fz[9], the comprehensive combustion characteristic index[10], the influence of the combustion characteristics of the pulverized coal is decomposed into two indicators to discuss: one is on fire, burning out is two. The former index considers the safety and stability of pulverized coal combustion in boiler, and the latter index considers the economy of combustion of pulverized coal in boiler [11-13].

According to the relevant data and using thermogravimetric analyzer in the laboratory for multiple kinds of coal combustion characteristics test of the statistical data shows that the ignition of coal difficult situation is divided into three categories: the[14] temperature are less than 450℃ fire for fire easily; the temperature at 450℃ ~500℃ range of fire is easy to fire the temperature more than 500℃; the fire is not easy to fire. Similarly, the coal burning out difficult situation is divided into three categories: the burnout temperature are less than 650 ℃ for easy burnout; the burnout temperature at 650℃~700℃ range is easy to burn; the burnout temperature more than 700℃ is not flammable.

Thermogravimetric analysis of 9 coal samples was carried out by using a comprehensive thermal analyzer. Because of the limited space, here only to Xinji coal is analyzed, the thermal analysis curves as shown in figure 1.

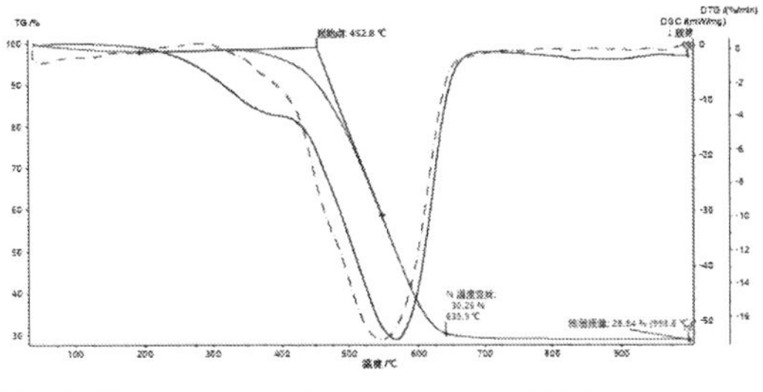

Fig. 1. Thermogravimetric analysis curve of Xinji coal.

Seen from Figure 1, Xinji coal ignition temperature is 452.8 ℃ , combustion temperature is 639.9 ℃ , is relatively easy to fire, easy to coal burning out.

4 Blending burning plan of coal blending

At present, blended coal mixing mainly has 2 ways: mixing before furnace, mixing in furnace (referred to as mixing before furnace) and sub grinding, mixing furnace (referred to as "grinding, mixing, burning")[15-18]. In this paper, the method of "mixing before furnace" and "mixing in furnace" is adopted.

Mixed coal is usually mixed with 2 or 3 single coals. If the mixture of coal is too much, it may lead to higher costs and more complicated combustion. Therefore, a combination of 9 coal industry analysis, elemental analysis and thermogravimetric analysis results and the existing coal yard power plant, the on-site mixed coal experiment selected 2 kinds of coal blending, which is outside the local coal quality ratio of lean coal.

Because the Xingye coal belongs to low calorific value, high ash content of inferior coal, in order to ensure the normal boiler load capacity, the power plant and consultation, and determine the Xinji Coal maximum mixing ratio does not exceed 30% (weighted value of 18314 kJ/kg). According to the results of coal quality analysis, a blending scheme for coal blending is worked out, as shown in table 5.

Table 5. Blending scheme.

Serial number	**Experiment content**	**Mixed coal scheme**	**Coal loading mode**
1	Oxygen change 1, Oxygen change 2, Oxygen change 3	Scheme 1 (Xinji coal: Xingye coal=8:2), four grinding operation; 600MW	Xinji coal and Xingye coal mixed in BCEF mill
2	Oxygen change 1, Oxygen change 2, Oxygen change 3	Scheme 2(Xinji coal: Xingye coal =7:3),five grinding operation; 600MW	Xinji coal and Xingye coal mixed in BCDEF mill
3	Oxygen change 1, Oxygen change 2, Oxygen change 3	Scheme 3(Xinji coal: Xingye coal =7:3),four grinding operation; 450MW	Xinji coal and Xingye coal mixed in BCEF mill
4	Oxygen change 1, Oxygen change 2, Oxygen change 3	Scheme 4(Xinji coal: Xingye coal =8:2),five grinding operation; 450MW	Xinji coal and Xingye coal mixed in BCDEF mill
5	Oxygen change 1, Oxygen change 2, Oxygen change 3	Scheme 5,four grinding operation; 450MW	Xinji coal mixed in BEF mill,Xingye coal in C mill

5 Test result analysis

The blending test of blending coal is carried out under the 600MW and 450MW loads according to table 5. The main test results of each working condition are shown in table 6.

Table 6. Test results of different conditions.

Mixed coal scheme	**O2 (%)**	**Boiler efficiency (%)**	**Auxiliary power consumption(%)**	**Power consumption of main auxiliary machine (%)**	**Coal consumption of power supply(g/kW.h)**
1	2.52	92.07	4.489	1.750	298.42
	2.95	92.88	4.500	1.774	294.63
	3.46	92.48	4.524	1.794	298.11
2	2.44	91.61	4.025	1.623	295.49
	3.05	92.80	4.031	1.646	292.94
	3.48	91.74	4.062	1.664	294.30
3	4.01	92.85	4.799	2.033	303.06
	4.42	92.43	4.802	2.091	304.44
	4.95	92.34	4.831	2.112	304.83
4	4.01	93.47	4.435	1.953	299.90
	4.47	92.89	4.471	1.997	301.88
	4.94	92.78	4.483	2.006	302.28
5	4.01	93.23	4.937	2.121	302.26
	4.48	93.08	4.942	2.146	302.76
	4.99	92.86	4.953	2.164	303.51

As shown in table 6:

1)Under the 600MW load, the minimum coal consumption is 292.94 g/kW.h, the maximum is 298.42

g/kW.h, the maximum and minimum difference is 5.48 g/kW.h. The highest coal coal consumption for Xinji Coal: Coal Industrial 7:3 ratio, outlet oxygen economizer is set to 2.5; the minimum coal consumption of coal blending for Xinji Coal: Coal Industrial 8:2 ratio, outlet oxygen economizer is set to 3.

2)Under the 450MW load, the minimum coal consumption is 301.88 g/kW.h, the maximum is 304.83g/kW.h, the maximum and minimum difference is 2.95g/kW.h. The highest coal coal consumption for Xinji Coal: Coal Industrial 7:3 ratio, outlet oxygen economizer is set to 5; the minimum coal consumption of coal blending for Xinji Coal: Coal Industrial 8:2 ratio, outlet oxygen economizer is set to 4.5.

3)In the same load, the Xinji coal coal blending scheme: Industrial 8:2, plant auxiliaries #1unit boiler side power rate and the electricity rate is lower than the Xinji Coal: Coal Industrial 7:3 mixed mode; in 450MW load, using C single industrial coal grinding. The furnace side power plant was the highest.

4)The boiler coal blending combustion method and oxygen change test have little influence on the power consumption of the plant, and the important factor affecting the coal consumption in the test is boiler efficiency.

5)Considering the economy of the unit, the proposed 450MW uses Xinji Coal: coal blending scheme of industrial =8:2. 600MW uses Xinji Coal: Coal Industrial 8:2 ratio, outlet oxygen economizer is set to 4.5.

6 Epilogue

In this paper, theoretical research and field experiments are combined to study the coal blending burning test of boiler No. 1 in a power plant in Jiangxi. The test results show that the boiler coal blending method and the oxygen change test have little influence on the power consumption of the plant, and the important factor affecting the coal consumption in the test is the boiler efficiency. 600MW load under different coal blending scheme under the boiler efficiency maximum difference of 1.14%, coal consumption is 5.48 g/kW.h maximum, 450MW load, different coal blending scheme under the boiler efficiency, the biggest difference is 1.13%, the maximum difference of 2.95 g/kW.h power supply coal consumption. Therefore, the reasonable coal blending scheme can significantly improve the boiler safety and economy.

References

1. WANG Wenhuan,PAN Bingchao,et al. Analysis of the Economy and Environmental Protection Features Based on Blending Lignite in Coal-fired Boiler,Boiler Technology, 2014,45(2):67-71.
2. CHEN Guodong, YUAN Yichao, WANG Bo. Application of Power Plant Boiler Blended Coal Combustion Technology,Boiler Technology, 2014, 45(5):53-57.
3. WANG Lingling, LU Zhimin, et al.Thermo-gravimetric study on dynamic characters in combustion reaction of Indonesia coal blended with bituminous coal.Thermal Power Generation, 2009, 38(3): 69-71.
4. LIU Jingyan,WANG Changan,et al. Investigation on Coal Properties and Combustion Characteristics of Blended Coals,Boiler Technology, 2012, 43(2):37-42.
5. WU Ying, CHEN Linguo, LI Haishan, et al. Study on test of coal co-firing for 700MW supercritical once-through. Power System Engineering, 2014, 30(3):24-26.
6. HU Rongzu, SHI Qizhen, et al. Thermal analysis kinetics[M]. The Science Publishing Company, 2001, 12-13.
7. ZENG Lingda, CHEN Qijuan, XIONG Jie, et al. Thermogravimetric test on combustion characteristics of blend coal and type selection suggestions of 680MW ultra-supercritical boiler[J]. Journal Of Power Engineering, 2008, 28(5):672-676.
8. LOU Bo, WANG Xiaocong. Thermogravimetric experimental study of mixed combustion of sewage sludge and coal [J]. Journal of Engineering for Thermal Energy and Power, 2011,26(1) :114-116.
9. HEN Bingyun, XUN Hua, HAN Jianchun. Analysis of combustion characteristics of the mixture of coal washing and coal gangue[J].Journal of Engineering for Thermal Energy and Power, 2011,26(5) :571-575.
10. XIAN Xiaohong, DU Yungui, ZHANG Guanghui. Combustion characteristics of coal blending by TG-DTG/DTA [J].Coal Conversion, 2011,34(3) :67-70.
11. XIANG Yi,ZHAO Weidong,et al.Thermogravimetric experiments and reaction kinetics analysis on combustion of indonesia lignite and shanxi bituminous coal. Thermal Power Generation, 2015, 44(11):14-19.
12. GONG Dehong,ZHANG Zhongxiao,et al. Effects of pulverizing method and oxygen content on combustion characteristics of mixed coal and the kinetic analysis.Thermal Power Generation, 2016, 45(10):40-49.
13. CHANG Jiaxing,YANG Zhongcan. Study on Zhundong Coal's Ash Fouling Indicator to Ensure Safe Operation of Boilers,Boiler Technology, 2013, 44(6):17-19.
14. MAO Xiaofei, WU Ying, ZENG Guofang, et al.Experimental research on blended coal combustion in a 700MW unit boiler. Thermal Power Generation, 2014, 43(5):69-74.
15. DUAN Xuenong, ZHU Guangming, JIAO Qingfeng, et al. Research and practice of blended coal combustion technology in power plant boiler[J]. Electric Power, 2008, 41(6):51-54.
16. ZHOU Zhen, LIU Liang, YAN Xiaozhong.Adjustment test and analysis of off-

design coal combustion of 1025t/h boiler,Power System Engineering, 2008,24(4):35-38.

17. WANG Qunying,SU Pan,et al. Influence of blending lignite with bituminous coal on coal feeding safety and operational costs for a 660 MW unit boiler. Thermal Power Generation, 2015(10):9-14.
18. HUA Xiaoyu,WANG Ruikun,et al.Anti-balancing efficiency and its uncertainty analysis of boiler burning mixed coal according to ASME and GB standard. Thermal Power Generation, 2016, 45(3):34-40.

The Detection Method of Fire Abnormal Based on Directional Drilling in Complex Conditions of Mine

Duan Huijun[1,*], Hao Shijun[1], Feng jie[2,3]

[1] Xi'an Research Institute of China Coal Technology &Engineering Group Corp, Xi'an,710077 China;
[2] Department of Geology and Environment Engineering, Xi'an University of Science and Technology, Xi'an, 710054, China;
[3] Shaanxi Coal and Chemical Technology Institute Co., Ltd, Xi'an, 710065 China)

Abstract: In the light of more and more urgent hidden fire abnormal detection problem in complex conditions of mine, a method which is used directional drilling technology is put forward. The method can avoid the obstacles in mine, and complete the fire abnormal detection. This paper based on analyzing the trajectory control of directional drilling, measurement while drilling and the characteristic of open branch process, the project of the directional drilling is formulated combination with a complex condition mine, and the detection of fire abnormal is implemented. This method can provide technical support for fire prevention, which also can provide a new way for fire anomaly detection in the similar mine.

1 Introduction

China is one of the countries with the most serious coal fire disasters. Spontaneous combustion of coals not only affects social economy, destroys coal resources, impacts eco-environment, but also causes series of problems to human survival and safety. As coal mining scale, strength and depth keep increasing, detection of concealed fire source in abnormal area for mine fire treatment has become the most difficult technical problem[1-3].

The main method for locating coal fire area is coal fire temperature measurement. Drilling hole for temperature measurement is generally adopted for detection, i.e., thermodetector and temperature sensor are arranged in the temperature measurement drill hole to measure the heat flow rate or the coal temperature. Currently, as mining conditions become more and more complex and diversified, fire areas primarily concentrate in abandoned roadways and gobs as well as in merged small pit mines, as a result, it is difficult to detect and measure the temperature of abnormal areas using conventional drilling technology. In this paper, directional drilling technology was adopted to bypass gob old pits to finally realize detection of abnormal fire source areas, and successfully implement temperature measurement and fire prevention & extinguishing using directional drilling[4-5].

2 Coal Mine Underground Directional Drilling Technology

Coal mine underground directional drilling technology is to drive downhole screw motor with sludge pump to drive the bit to revolve and cut, during drilling, the drill pipe does not revolve. The design trajectory and the actual drilling trajectory are displayed on the explosion-proof computer at the orifice, the tool face angle is changed by adjusting the PDM drill with elbow joint to control the bit dip angle and azimuth angle, hence controlling the trajectory[6-8]. Comparing to conventional revolving drilling, directional drilling not only can precisely control drill hole trajectory, but also can realize high angle bending of drill hole or in-hole branching through trajectory control, hence bypassing or avoiding obstacle[9]. Directional drilling technology has been widely used in coal mine underground gas extraction, geological structure detection and water control, whereas it's the first time for it to be used in the field of mine underground fire prevention & extinguishing, especially in abnormal fire source area detection and temperature measurement. Results show that directional drilling technology can perfectly complete fire source detection in abnormal area with good performance. Fig. 1 shows the directional drill rig for mine underground abnormal fire source area detection.

Fig 1 Directional drilling equipment of fire abnormal detection

2.1 Principle of mine underground directional

* Corresponding author: ypdhj@163.com

drilling technology

The coal mine underground directional drilling system comprises directional drill rig, cabled drill pipe, MWD instruments, sludge pump, and screw motor etc. The connection of the drilling system is as shown in Fig. 2.

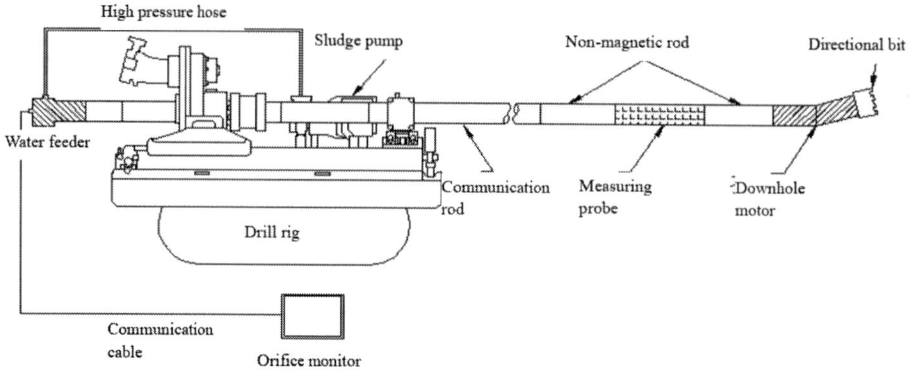

Fig 2 Connection diagram of MWD system in directional drilling

2.2 Characteristics of directional drilling technology

2.2.1 Measurement & calculation of directional drilling trajectory

The spatial pattern of the directional drilling axis can be manifested with some parameters of the axis. The spatial elements of the directional drilling hole axizzs include: the dip angle, azimuth angle and hole depth at different points of the drilling hole axis, vertical depth, horizontal displacement and horizontal deviation of the hole, and the curvature or bending strength of the curve section etc. Based on the basic elements of the drilling trajectory and using a certain calculation method, the spatial coordinate of each relevant point on the trajectory can be obtained[10-11]. Wherein the dip angle, azimuth angle and hole depth are the main basis for the design and control of directional drilling hole axis trajectory.

The MWD system of the directional drilling jig collects a group of drilling data every 3 meters. On this basis, the accurate coordinate of a certain measurement point on the drilling hole trajectory in the drilling hole design coordinate system can be obtained by calculation using formula (1), formula (2) and formula (3), complete drilling hole trajectory is obtained through replacing the arc of the measurement points with chord. Fig. 3 is the calculation principle.

$$X = \sum_{i=1}^{n} \Delta L_i \times \cos\left(\frac{\alpha_{i-1}+\alpha_i}{2}\right) \times \cos\left(\frac{\theta_{i-1}+\theta_i}{2} - \theta_0\right) \quad (1)$$

$$Y = \sum_{i=1}^{n} \Delta L_i \times \cos\left(\frac{\alpha_{i-1}+\alpha_i}{2}\right) \times \sin\left(\frac{\theta_{i-1}+\theta_i}{2} - \theta_0\right) \quad (2)$$

$$Z = \sum_{i=1}^{n} \Delta L_i \times \sin\left(\frac{\alpha_{i-1}+\alpha_i}{2}\right) \quad (3)$$

Where, ΔL is the distance between two measurement points during drilling; αi is the dip angle of point i of the drilling hole; θi is the azimuth angle of point i; θ0 is the main design azimuth angle of the drilling hole (i.e, azimuth angle of axis X); X is the horizontal displacement of point i; Y is the left-right displacement of point i; Z is up-down displacement of point i.

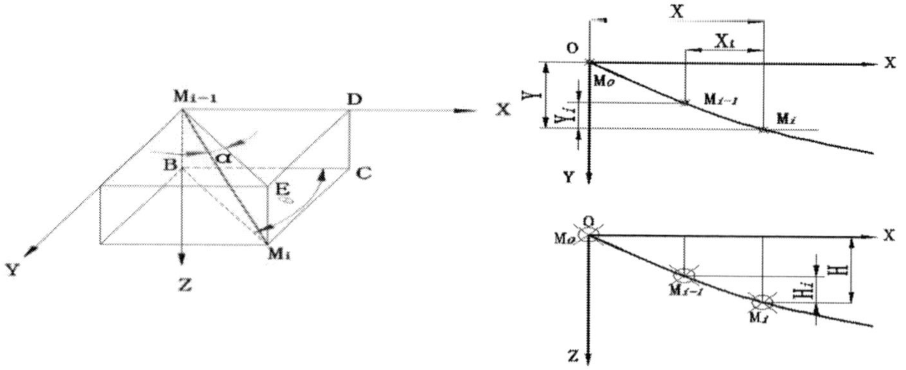

Fig 3 Calculation principle of surveying point coordinates for borehole track

2.2.2 Control of directional drilling trajectory

In the directional drilling system, screw motor is adopted to realize transmission of rotary speed, torque and trajectory control. It is a volumetric power conversion device and downhole motor that translates hydraulic pressure into mechanical energy. Through real-time monitoring of the hole-bottom trajectory by the MWD system while adjusting the tool face angle and the direction of the outer pipe elbow of the screw motor, the drilling hole trajectory control is realized.

2.2.3 Branching of directional drilling

In addition to controlling the trajectory, the PDM drill adopted by the directional drilling system can bypass or avoid encountered obstacle areas such as gob, old pits & small roadways and collapse columns through branching for safe drilling.

For the purpose of bypassing and avoiding obstacle, branching of directional drilling primarily adopts bypassing side drilling method, i.e., branch against the tool face angle at the hole section around the branching point, adopting depressurized (feed pressure) and timed drilling. Cut key grooves on the main hole wall and keep on making grooves, as the key groove gets deeper and deeper, new eyelets are gradually formed, the dip angle and azimuth angle are detected by the MWD instrument to compare with the original drilling hole trajectory, and when the feed pressure increases, and the sludge pump pressure increases slightly, and the returned rock (coal) powder at the orifice increases, it can be determined that the branching is successful, as shown in Fig. 4.

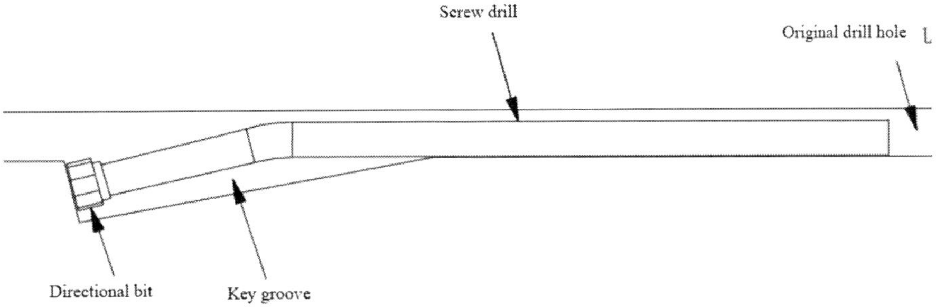

Fig 4 Open branch process of directional drilling

3 Mine Overview

The mine of Shanxi Huanjin Hanju Coal Industry Co., Ltd. is located at the south edge of Lvliang Mountain in Xipo Township, Xiangning County, Shanxi Province. The mine area is 26.5467km2, the geologic reserve is 307 million t, the recoverable reserve is 198 million t, the main mining beds are coal beds No.2-10, the design production capacity is 3 million t/a. Hanju Mine consists of multiple integrated old mines. There are many surrounding mines and small pits, there are exposed coal beds at the south east edge of the mine, there may be small pit mining area at the location where the coal bed is exposed, however, the location of the old pit entrance is unclear, the underground abandoned gob roadways proven already are complex, multiple times of spontaneous coal combustion and serious high CO concentration accidents occurred in the gob, resulting in multiple times of mine shutdown for rescuing, seriously affecting the safe production of the mine[12].

In order to detect the condition and temperature of the abnormal fire source area, conventional drilling method is generally adopted to drill to the suspicious area to arrange temperature sensor for temperature detection, so as to draw the temperature field map of the high temperature area, in a bid to locate the fire source location to guide subsequent fire prevention & extinguishing work. Hanju Mine has complex exploitation conditions, the abnormal areas and the abandoned roadway gobs cross with each other, the old pits are in chaos mining conditions, multiple times of conventional drilling have been attempted, encountering the upper layer of abandoned roadways and the lower layer of gobs etc., ending up in failure. To overcome this difficulty, in this paper, directional drilling technology and jig are utilized for temperature measurement and hazard elimination of the abnormal areas.

4 Design of Directional Drilling for Abnormal Fire Source Area Detection

4.1 Principle of design

(1) The directional drilling area for detection of abnormal fire source area is complex, therefore, potential hazards shall be fully taken into consideration in the design, for example, whether hole collapse may occur when crossing the coal-rock boundary, drill jamming when encountering unknown matters in abandoned roadway, so as to ensure safe drilling.

(2) Due to the importance and uncertainty of detection drilling , and abnormal gas may erupt when hitting the target fire source area, emergency response plan and measures must be provided in directional drilling design, e.g., arrange hazard gas interception device at the orifice of the directional drilling hole.

(3) Due to uncertainty and possible sudden occurrence situation in complex area detection and the need of temperature sensor arrangement, smooth design is adopted as the directional drilling trajectory design, this also facilitates emergency drilling accident handling. Specifically, limit the basic bending angles, horizontal

direction (azimuth angle): 0.75-1.0 degrees, vertical direction (dip angle): 0.75-1.0 degrees, common curvature radius: r=57.3/(1 /6)=344m.

4.2 Design of directional drilling

Mining engineering plan (consisting of coal bed roof & bottom contour line, roadway distribution and geological structure distribution etc.) accurately reflecting the coal bed space tendency of the hole arrangement area as well as the coal bed histogram are required for the directional drilling design, first, parameters of the control points of the drilling hole trajectory need to be calculated, then the directional drilling trajectory is obtained through interpolation calculation. As to the detection area of this paper, relevant geological data are barely available, only the coal bed dip angle and a few abandoned roadways are determined based on deduction. The opening location is determined at the central substation, the elevation of the roadway roof at the opening point and coal 2 roof is 667m, the design opening elevation is 665.2m, the drilling trajectory lowers first then rises in a concave shape (Fig. 5), the design hole depth is 230m, 2 directional drilling holes are subsequently designed.

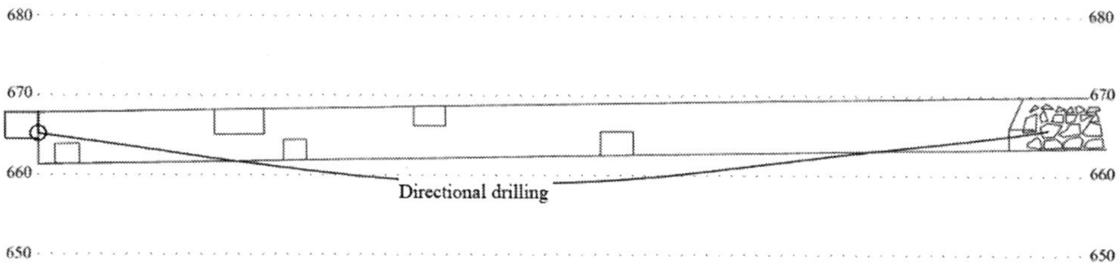

Fig 5 Design sketch of directional drilling trajectory

In addition, when the directional drilling reaches the abnormal fire source area, abnormal gas may erupt, therefore, at the opening section, straight drilling shall be ensured, 12m steel casing of φ127 shall be inserted, and fast sealing materials shall be injected to seal the hole and pressure test shall be satisfied, then check valve shall be arranged at the casing head to close the orifice to prevent hazard gas from erupting. For the hole structure and drilling parameters of the directional drilling, refer to Table 1.

Table 1 Borehole structure and drilling parameter of directional drilling

Drilling sequence	Hole diameter mm	Bit pressure KN	Rotary speed rpm	Pump displacement L/min	Remark
Surface section opening	Φ98	10~20	55~65	230	Opening
Surface section reaming	Φ153	10	25~30	230	Insert orifice casing
Second section orientation	Φ98	30~50		320	Directional drilling

For specific drilling design parameters, refer to Table 2.

Table 2 The design parameter of directional drilling

Hole No.	Design opening azimuth angle	Design opening dip angle	Design finished hole azimuth angle	Design finished hole dip angle	Design hole depth
1-1	171°	-7°	167°	20°	230m
1-2	182°	-7°	167°	20°	230m

5 Directional drilling for detection

5.1 Hole opening and sealing

Φ98mm straight drill bit + Φ73mm drill pipe assembly is adopted for hole opening to drill to the depth of 13m, then change to reaming drill rig assembly, i.e., Φ98mm/Φ153mm reamer bit + Φ73mm straight packed hole assembly to ream to the design depth, then insert Φ127mm orifice pipe×12m, inject special hole sealing material to seal the hole, after curing for 24 hours, conduct pressure test, and install orifice device and check valve.

5.2 Directional drilling

After drilling is started, in order to ensure the drilling reaches the target area safely, drilling must be strictly controlled in accordance with the design concave drilling trajectory, in addition, drilling shall also facilitate conditions for the final arrangement of temperature measurement lead and sensor, under the precondition of hitting the target by the trajectory, the drilled hole shall be as smooth as possible without high fluctuation and steps.

During work progress, due to lack of geological data, the abandoned roadways and gobs in the trajectory design are all obtained from deduction, this makes the directional drilling work very difficult.

(1) Directional drilling 1-1 encountered abandoned roadways respectively at 30m and 123m, causing drill jamming, almost resulting in near-miss drilling hole accident. Reopening had to be conducted after first time roadway penetration. In the second time of attempt, directional drilling branching is adopted to ensure safe drilling of directional drilling 1-1;

(2) The location of the suspicious abnormal area was unclear, no abnormality was observed at 230m depth of the hole of the original design. Through branching and expanding scope of detection, the target was finally detected during the third time of branching drilling.

(3) The detection under complex conditions determines the trajectory of the directional drilling is a concave one, assuming demanding test on smooth slagging and safe drilling. During the directional drilling, even bit pressure was maintained, pressure increase or decrease was conducted continuously and evenly, ream down frequently, advance moderately and impact frequently, drill rig operator shall closely observe drill torque, pump pressure and the returned water and slags, for any abnormality, stop drilling immediately, and promptly lift the drill rig to safe area for handling. The static duration of the drill rig in the mine was strictly controlled, except for measuring data, the static duration of the drill rig didn't exceed 10 minutes.

(4) Abnormal area was detected through directional drilling 1-1, directional drilling 1-2 conducted temperature measurement directional drilling by utilizing the geological conditions detected by hole 1-1, finally, cabled temperature measurement lead and sensor were arranged in the common drill pipe to complete temperature measurement of abnormal fire source area.

From design scheme preparation to safety measures implementation, to emergent equipment transport and to directional drilling 1-1 and 1-2 detection and completion of temperature measurement, it took 7 days and 20 shifts totally. For the actual drilling parameters, refer to Table 3.

Table 3 The actual parameter of directional drilling

Hole No.	Actual drilling opening azimuth angle	Actual drilling opening dip angle	Actual drilling finished hole azimuth angle	Actual drilling finished hole dip angle	Finished hole hole depth
1-1	168.2°	-7.5°	169.1°	8.4°	303m
1-2	171.5°	-9°	172.3°	11.2°	228m

The actual drilling horizontal trajectory and profile trajectory of directional drilling 1-1 and 1-2 are as shown in Fig. 6, Fig. 7 and Fig. 8.

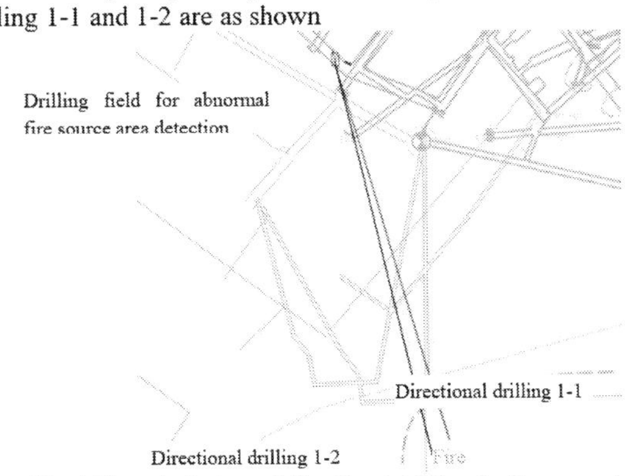

Fig 6 Planar trajectory of directional drilling for fire anomaly detection

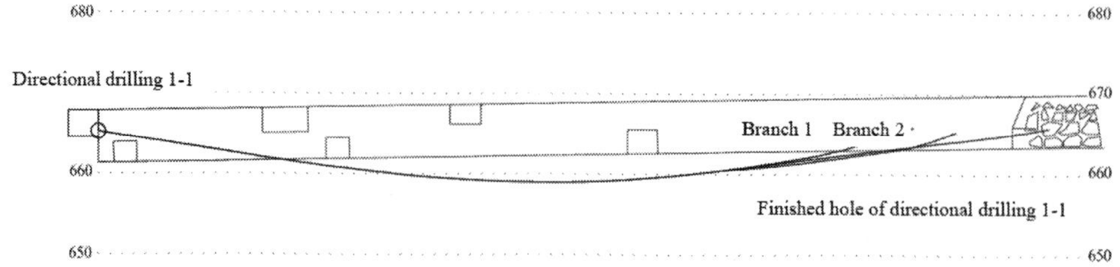

Fig 7 Profile trajectory of directional drilling 1-1

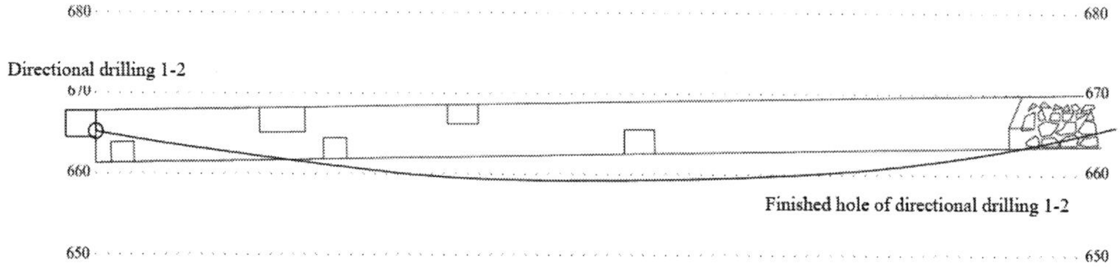

Fig 8 Profile trajectory of directional drilling 1-2

After the directional drilling was completed, temperature information of the abnormal fire source area was accurately obtained through arranged temperature measurement lead and sensor, then further fire prevention & extinguishing measures were taken, i.e., pour yellow mud, totally abut 50000m³ was poured, the effluent water temperature decreased from 110°C to 18°C gradually, and the CO eruption amount also decreased to normal value. Implementing abnormal fire source area detection and temperature measurement using directional drilling technology provides support for conducting targeted fire prevention & extinguishing work, ensuring safe production of mines with complex conditions.

6 Conclusion

Directional drilling technology was applied in the field of detection and temperature measurement of abnormal fire source area with complex conditions in coal mine underground for fire prevention and extinguishing for the first time. Through delicate organizing and bold innovation under unusual conditions, and artificially controlling the drilling trajectory by utilizing directional drilling technology to bypass and avoid abandoned roadways and gobs, through branching process, wide scope detection of the target abnormal area was completed. Finally the abnormal fire source area was detected, and the temperature was successfully measured through directional drilling, not only obtaining necessary condition information for subsequent fire prevention & extinguishing, but also providing a new concept for handling similar coal mine underground complex problems in the future.

References

1. Liang Yuntao. Gas Index Method to Predict Coal Spontaneous Combustion [J]. Coal science and technology, 2008, 36(6): 5-8
2. Wen Hu,Guo Jun,Jin Yongfei. Research Status and Trend of CO Generation Mechanism and Control Technology in Coal Mines[J]. Safety in coal mine, 2015, 46(6): 155-177.
3. Wang Liancong,Sun Yong,Feng Wenbin. Field Experiment Study on CO Generation and Sources Law in Coal Seam Mining[J]. Safety in coal mine, 2014, 45(12): 43-45.
4. Wang Xu. Study on Prevention and Control Technology of Spontaneous Combustion for Large Area Goaf of Nantun Coal Mine[D]. Xi'an University of Science and Technology, Xi'an, China, 2013.
5. Wang Liancong,Sun Yong,Feng Wenbin. Application of synthetically fire-extinguishing technology in Mine Spontaneous Fire [J]. Safety in coal mine, 2008, 39(4): 49-52.
6. Du Limeng,Shi Hao,Yao Panpan. Application of Horizontal Directional Drilling Technology in Geological Structure Exploration in Hudi Coal Mine [J]. Exploration Engineering (Rock & Soil Drilling and Tunneling), 2014, (6): 38-40.
7. Chi Guoming. Scouring zone crossing technology in high cutting fully mechanized face of Bulianta coal mine [J]. Coal Engineering, 2016, 48(8): 50-53.
8. Zhao Yongzhe. Research on Controlled Theory of Positive Displacement Motor Drilling in Coal Mine for nearly Horizontal Directional Borehole [D]. Mining Research Institute Xi'an Branch, Xi'an, China, 2008.
9. Yang Zhong,Duan Huijun,Bai Gang. Application of Directional Drilling Technology for Water Drainage in Goaf of Coal Mine [J]. Zhongzhou Coal, 2015, (8): 16-18.
10. Sun Rongjun. Research on Measurement-while-drilling Technology in Underground Mine and Data Processing Method of

https://doi.org/10.1051/e3sconf/20183801007

Borehole Trajectory [D]. Mining Research Institute Xi'an Branch, Xi'an, China, 2009.
11. Lu Jun,Xiong Kejian. Application of Horizontal 1000m Directional Drilling Rig to Mine Gas Drainage [J]. Coal Science and Technology, 2011, 39(12): 92-95.
12. Bai Gang,Duan Huijun,Wang Qing. Application of Directional Drilling Technology for Fire Prevention and Extinguishing in Coal Mine Goaf [J]. Safety in Coal Mines, 2016, 47(6): 132-135.

https://doi.org/10.1051/e3sconf/20183801008

Study on Spatio-Temporal Change of Ecological Land in Yellow River Delta Based on RS&GIS

GuoQiang An[1,2,*]

[1]Shandong Land Surveying and Planning Institute, 250014 Jinan, Shandong, China
[2]Shandong Scientific Research Base for Key Laboratory of Land Use Ministry of Land and Resources, Jinan, Shandong 250014, China

Abstract. The temporal and spatial variation of ecological land use and its current distribution were studied to provide reference for the protection of original ecological land and ecological environment in the Yellow River Delta. Using RS colour synthesis, supervised classification, unsupervised classification, vegetation index and other methods to monitor the impact of human activities on the original ecological land in the past 30 years; using GIS technology to analyse the statistical data and construct the model of original ecological land area index to study the ecological land distribution status. The results show that the boundary of original ecological land in the Yellow River Delta had been pushed toward the coastline at an average speed of 0.8km per year due to human activities. In the past 20 years, a large amount of original ecological land gradually transformed into artificial ecological land. In view of the evolution and status of ecological land in the Yellow River Delta, related local departments should adopt differentiated and focused protection measures to protect the ecological land of the Yellow River Delta.

1 Introduction

The construction of ecological civilization is an important part of socialism with Chinese characteristics in the new era. Studying the evolution and layout of ecological land is not only an important scientific issue, but also a social issue that concerns the ecological and environmental protection of China. Ecological land with water protection, soil conservation, biodiversity conservation and many other ecological functions, is the quality of regional ecological and environmental "barometer" [1]. Many Chinese scholars deeply studied the concept of ecological land use [2], classification [2,3,4] and the distribution pattern of ecological land in China [1,4], and laid the foundation for further studies as evolution of regional ecological land use [5], the driving forces [6] and etc. However, the related literature on the ecological land in the Yellow River Delta is still rare. This paper makes use of the data of remote sensing and land survey to study the evolution trend and distribution of ecological land in the Yellow River Delta in recent 30 years, which is of great significance for the protection of the ecological environment in the Yellow River Delta and the sustainable development of the region. The typical demonstration of related research on regional ecological evolution and layout of land use offered as well.

2 Study areas

The Yellow River Delta is located in the North-east of Shandong Province, on the south bank of the Bohai Gulf and the west coast of Laizhou Bay. It is one of the three major delta estuaries in China, and is the most concentrated area for ecological use in Shandong Province. The total land area is 16165.56Km2, including all of Dongying City and the most of Binzhou City, with a total population of 5.21 million. There are mainly three rivers, including the Yellow River, Xiaoqing River, Tuhai River, Magu River here. In recent years, human activities have a very strong impact on the Yellow River Delta region, and the area and catigories of ecological land have undergone fundamental changes. Among them, wetland, saline-alkali land and waters which have the leading function of ecology and environment have rapidly reduced, while artificial ecological land, non-ecological land continue to increase. To a certain extent, the tendency of ecological land change in the Yellow River Delta reflects the trend of the national ecological land use, especially in the east coast where the economy and society have developed rapidly in recent years.

3 Data and processing

3.1. Remote sense data

Landsat 4-5 TM data of six periods of 1985, 1990, 1995, 2000, 2005, 2010, and Landsat 8 OLI_TIRS data of 2015 were used to analyse the impact of human activities on the boundary of original ecological land in the Yellow River Delta for nearly 30 years. Sequence data of six periods were formed for the dynamic monitoring. The data set is provided by Geospatial Data Cloud site,

* Corresponding author: an_tiger_@163.com

Computer Network Information Centre, Chinese Academy of Sciences. (Http://www.gscloud.cn). In order to establish a uniform standard for interpretation, Landsat4-5 TM data was simulated by real colour synthesis using the red (band3), the green (band2) and the blue (band1).For the Landsat 8 OLI_TIRS data, the true colour synthesis is performed in Red(Band4), Green (Band3) and Blue (Band2).

3.2. Land surveying data

According to the survey data of land use change from 1996 to 2016, the change of ecological land use in recent 20 years was calculated; the evolution trend of ecological land in the past 7 years was carefully analysed according to the 2009 Land Use Status Survey Database and the Land Use Change Survey Database from 2010 to 2016 the study.

3.3. Basic unit

Based on the village-level administrative area as the basic unit, the database of land use survey in 2016 is used to study the ecological land use status quo in the Yellow River Delta.

3.4. Land classification

Based on research results of the classification of ecological land in China by the researcher Long Hualou [3] and Professor Wang Jing [4], this paper divides and merges the land use types of the Yellow River Delta, including three first-level ecological land use categories and 10 second-level Land use categories, as shown in Table 1.

Table 1. The Ecological Land Classification of Yellow River Delta Area.

Level 1 class code	Level 1 class name	Level 2 class code	Level 2 class name
1	Original ecological land use	11	Woodland
		12	Grassland
		13	Water land
		14	Mudflats
		15	Saline-alkali land
2	Artificial ecological land use	21	Arable land
		22	Orchard land
		23	Artificial woodland
		24	Artificial grassland
3	Non-ecological land use	31	Construction land

4 Method and model

4.1. NDVI

Normalized Difference Vegetation Index(NDVI), unsupervised classification and others were used as remote sensing image data enhance mental methods for the interpretation. NDVI was calculated by formula as:

$$NDVIi=(NIRi-Ri)/(NIRi+Ri)[7] \qquad (1)$$

$NDVIi$ is the i-th period Normalized Difference Vegetation Index. $NIRi$ and Ri correspond to the near infrared band and the red band of TM and OLI_TIRS images, respectively. After the enhancement of the image information, the boundary of human activities on the primary ecology is very clear, mainly including: building roads, building sea dams, reclamation of tidal flats, development of saline-alkali lands, breeding ponds, sun salt fields, reservoirs and residential area. The farthest boundary that these kinds of activities construct for the first time is the boundaries of original ecological land influenced by human activities.

4.2. Ecological land area index

Village-level administrative areas were used as the evaluation unit, and the ecological land area index (Ie) were used for the study of the distribution of ecological land in the Yellow River Delta, by the formula as:

$$Iei=Ei/Li \qquad (2)$$

Where Iei is the index of primary ecological land for the i-th unit; Ei is the primary ecological land area of the i-th village; Li is the sum of original ecological land, artificial ecological land and non-ecological land in the i-th village, which is the total area of the unit.

5 Findings

5.1 The Impact of Human Activities in the Past 30 Years

The distance between the boundary of original ecological land influenced by human activities and the coastline was calculated by the method according to the research results of Huang Bohua et al. [8]. Figure 1 show that in 1985, the maximum distance from the coastline affected by human activities was 50.41km, the minimum distance was 6.25km and the average distance was 28km. By 2015, the maximum distance is 26.23km, the minimum distance is 0Km, and the average distance is 7.86Km. In the past 30 years, the influence boundary of the original ecological land was push to coastline by an average speed 0.8 km per year.

Original ecological land such as mudflats, Saline-alkali land, Water land, Woodland and Grassland had been gradually replaced by artificial ecological land and Non-ecological land as arable land and construction land. Ecological balance had been destroyed year by year.

 https://doi.org/10.1051/e3sconf/20183801008

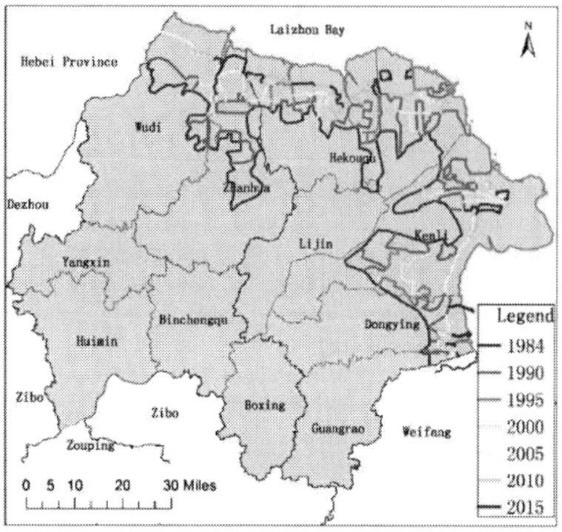

Fig. 1. 1984-2015 Changes in the boundaries of original ecological land in the Yellow River Delta

5.2 The Changes of Ecological Land in the Past 20 Years

Land use data from land survey statistics in 1996-2016 clearly shows the change of land use in the Yellow River Delta. It is clear that in the past 20 years, the total area of original ecological land in the study area has decreased by 177250.33hm^2, accounting for 30.89% of the total original ecological land use in 1985, with an average annual decrease of 8862.52hm^2, as shown in Chart 1.

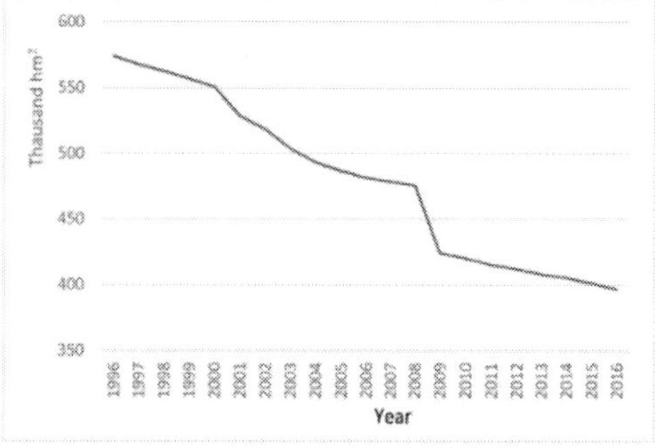

Chart 1. 1996-2016 Changes in the Area of Original Ecological Land in the Yellow River Delta

Among them, the forestland decreased by 3739.93hm^2, the mudflats decreased by 36285.33hm^2, and the saline-alkali land decreased by 31581.6hm^2, as shown in Table 2.

5.3 The Assessment on Status of Ecological Land and Zoning

According to the 2016 land use survey data, the primary ecological land area in the Yellow River Delta is 396594.81hm^2, the artificial ecological land area is 926810.52hm^2, and the non-ecological land area is 293143.23hm^2, the ratio of the three is 24.5: 57.3: 18.1. In the Yellow River Delta, there are 6,421 village-level units. According to the ecological land area index, with reference to the influence boundary of human activities on the original ecological land in recent thirty years and the change trend of ecological land in the past 20 years, the Yellow River Delta was divided into three zones. They are the general protection of ecological land zone, medium protection zone and key protection zone, as shown in Fig. 2.

Table 2. Comparison Table of Some Ecological Land Areas in the Yellow River Delta between 1996 and 2016

Year	Woodland	Grassland	Water land
1996	29761.28	143178.1	29056.1
2016	26021.35	27871.88	27848.78
Reduced area (hm^2)	3739.93	115306.3	1207.32
Reduced percent(%)	12.57	80.53342	4.16
Year	**Mudflats**	**Saline-alkali land**	
1996	150127.4	169385.04	
2016	113842.1	137803.44	
Reduced area (hm^2)	36285.33	31581.6	
Reduced percent(%)	24.17	32.79	

5.3.1 General Protection Zone.

There are 4605 units with the Ie values less than 0.10, accounting for 71.7% of the total. Such zones, with less original ecological land, are widely distributed in the Yellow River Delta inland areas. They are mainly distributed in areas far away from the coastline, well-developed agriculture and rapid expansion of construction land, such as southwest Guangrao, Boxing, Bincheng, southern Zhanhua, Yangxin, and southern Wudi. The general protection zone should pay attention to the ecological land around the towns and villages, according to the principle of ecological compensation, control construction occupancy, land consolidation and other human activities that may reduce the original ecological land. The region should increase the artificial ecological land such as forestland and garden land to make up for the shortage of original ecological land.

5.3.2 Medium Protected Zone.

There are 1718 units with the Ie values between 0.10-0.50, accounting for 0.26% of the total. Such zone have relatively more ecological land, and are mainly distributed along the rivers and boundaries of human activities such as Huimin, north-eastern Lijin, south western Kenli, southern estuary, central Zhanhua and central Wudi. The region should strengthen the protection of original ecological land use, with particular emphasis on the protection of original ecological land along the river, control of wastewater discharge, to prevent river water pollution. Strict measures should be taken for protection the original ecological land as to control the expansion of residential land, limit agricultural production which lead to the reduce of original ecological land.

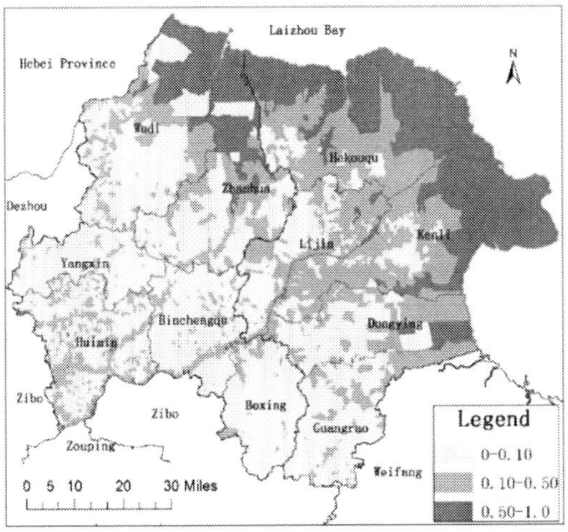

Fig.2. Zoning Map of Ecological Land Use in the Yellow River Delta

5.3.3 Key Protected Zone.

There are only 98 units with the Ie values above 0.50, accounting for 1.5% of the total. Such zones mainly located in the coastal areas, including the eastern part of Dongying District, Kenli East, northeast estuary, northern Zhanhua, Wudi North. The region is the most affected area for human activities and the most affected by human activities for the last 30 years.

Efforts should be made to increase the protection of ecological land such as river wetlands, mudflats and saline-alkali land, and strictly control the destruction of original ecological land by such activities as pit pond cultivation, beach reclamation, land consolidation and land reclamation from the sea. Under the conditions without affecting the functions and species diversity of ecological land, appropriate improvement or development of saline-alkali land into garden land or woodland in order to increase the green vegetation are permitted.

6 Conclusion and discussion

In this paper, based on information as Landsat TM image data, land use statistics and land use status survey database were used on the research of the impact of human activities on the Yellow River Delta, the ecological changes in the Yellow River Delta and Yellow River Delta ecological land use status. The results show that the boundary of original ecological land in the Yellow River Delta had been pushed toward the coastline at an average speed of 0.8km per year due to human activities. In the past 20 years, the area of original ecological land decreased by 8862.52hm^2 hectares, and a large amount of original ecological land gradually transformed into artificial ecological land. The current proportions of original ecological land, artificial ecological land and non-ecological land in the Yellow River Delta are 25:57:18.

In view of the evolution and status of ecological land in the Yellow River Delta, related local departments should adopt differentiated and focused protection measures to protect the ecological land of the Yellow River Delta. For General Protected Areas, artificial ecological land such as garden land and forestland should be appropriately increased, and the protection of the ecological land around the towns and villages around the village should be emphasized. For Medium-sized Protected Areas, land use as construction land shall be strictly controlled, pollutant discharge should be firmly forbidden, and the protection should focus on original ecological land along rivers; Key Protected Areas should strictly control the pit pond farming, beach reclamation, land reclamation and other activities on original ecological land. Protection of river wetlands, beaches, saline land and other ecological sites should be improved. Appropriate improvement or development of saline-alkali lands for the garden or woodland is worthy of consideration. Conclusions as below:

(1) The results show that the Yellow River delta area is strongly affected by human activities. The area of original ecological land decreases rapidly. Nowadays, 71.7% of the total area of original ecological land in village-level units accounts for less than 5% in their total area. The distribution of ecological land use is also unbalanced. Overall, there are many coastal distributions and the more inland distributions are less. Differentiated and focused protection measures should be taken to protect the ecological land in the Yellow River Delta.

(2) Using RS-GIS technology and original ecological land area index can reflect the trend and distribution of ecological land in the Yellow River Delta scientifically and accurately, which can provide scientific reference for the formulation of ecological land use protection and land use management policy.

(3) This paper observes and evaluates the ecological land of the Yellow River Delta from the macroscopic level. The change of land use types such as beach, wetland and saline-alkali land in ecological land can be regarded as the key issue in the future.

References

1. F. Yu, X.B. Li, L.J. Zhang, W.H. Xu, R. Fu, H. Wang. Acta Ecologica Sinica, **35(14)**, 4931 (2015).
2. H.B. Deng, C.D. Chen, X. Liu, G. Wu, Acta Ecologica Sinica, **29(3)**: 1519 (2009).
3. H.L. Long, Y.Q. Liu, T.T. Li, J. Wang, A.X. Liu, Chinese Journal of Eco-Environment, **1**, 1 (2015).
4. J. Wang, W. Wang, Y. Qi, Y. Qi, T. He, R.J. Wu, Y.Y. Chen, Geogr. Res. **36 (3)**, 453 (2017).
5. A.B. Wu, Y.X. Zhao, Tran. Chinese Soc. Agr. Eng., **33 (2)**,283-290 (2017).
6. A. Tang, X. S. Li, Z. F. Bian, Q.W. Yan, Eco. Eco., **32 (5)**, 136 (2016).

7. N.P. Robinson, B.W. Allred, M.O. Jones, A. Moreno, J. S. Kimball, D. E. Naugle, T.A. Erickson, A. D. Richardson, Rem. Sen., **9**,863 (2017).
8. B.H. Huang, F. Wu, J.K. Xu, Geo. And Inf. Sci. of Wu. Uni., **42 (3)**,395 (2017).

Study on the Enhanced Operation of Self-Ventilation-Based Coupling System for Domestic Wastewater Treatment

Lingwei Kong[1], *Lu* Wang[3], *Yi* Zhang[2], *Rongwu* Mei[1], and *Yu* Zhang[1]

[1]Environmental Science Research and Design Institute of Zhejiang Province, Hangzhou 310007;
[2]Institute of Hydrobiology, Chinese Academy of Sciences, Wuhan, 430072;
[3]Zhejiang Greentown LEAP Architecture Design Consulting Company, Hangzhou, 310005

Abstract. In this study, a new coupling system of biological filter bed and subsurface-flow constructed wetland based on the self-ventilation network was proposed, and the comparative pollutant removal efficiency at low and high influent concentration of the pilot coupling system with different substrates configurations were investigated. The study found that: The comparison system (b) had better removal rates than that of the original system (a), and the removal rate when treating low influent concentration was 74.10%, 94.14%, 73.57% and 69.53%, while in high influent concentration case was 81.30%, 90.28%, 88.57% and 75.36% for COD_{Cr}, NH_4^+-N, TN and TP, respectively. The removal of the above main water indexes of the comparison system (b) promoted by 11.00%, 11.55%, 2.69% and 8.09% respectively in low influent concentration case and 4.20%, 9.20%, 7.66% and 13.61% respectively in high influent concentration case when comparing to the original system (a), which showed that the optimized configuration of various kinds of substrates was significant and was more beneficial to the degradation and removal of pollutants. The adsorption and interception function of substrates in the constructed wetland was the main way of phosphorus removal. The function of self-ventilation ensured the amount of DO in the coupling system, making the phosphorus removal was less affected comparing to structure of traditional wetland.

1 Introduction

Different treatment process has different effects on wastewater treatment, which will have a directly impact on the environment [1,2]. The main method of rural sewage treatment in Japan is membrane process submerged in a small domestic sewage purification device, which has the merit of small aera demand, low cost, simple operation etc..; The Filter process used in Australia for rural domestic wastewater treatment is a combination wastewater reuse system of filtration, land treatment and culvert drainage [3]; In South Korea, soil filtration-plant system is used for rural sewage treatment, and after purification the effluent is used for farmland irrigation [4]. For rural domestic sewage treatment is mainly a single process, in recent years, some scholars have found their own ways and develop more practical combined treatment processes, such as the combination process of anaerobic biological filter and constructed wetland [5], bio-ecological combination system [6] and biological contact oxidation process-constructed wetland technology [7].

The rural domestic sewage has the characteristics of complex composition, high nitrogen and phosphorus, thus the single-stage constructed wetland cannot effectively degrade nitrogen and phosphorus in the actual treatment process. In this study, we used a new self-ventilation network biological filter bed and subsurface-flow constructed wetland coupling process, and selected different substrates and gradation to carry out the pilot construction to investigate the actual operation effect and influence factors, accumulating engineering experiences and design parameters for the further demonstration project.

2 Experimental materials and methods

2.1 Experimental materials and process options

The pilot project is to treat domestic sewage collected from Zhong village, which locates in Qiushi village, Jingshan town, Yuhang district, Hangzhou city, Zhejiang province. The new self-ventilation networking biological filter bed and subsurface-flow constructed wetland coupling process was made up of the upper layered biological filter bed and lower layered subsurface-flow constructed wetland. The flow chart and the scene of pilot-scale project is shown in Fig.2-1 and Fig.2-2, respectively:

* Corresponding author: 77378154@qq.com, 50991460@qq.com, zhangyi@ihb.ac.cn, mrw0518@sina.com, zhangyu0103@126.com,

Fig.2-1. The flow chart of the pilot project of the biological filter and subsurface-flow constructed wetland coupling process

Fig.2-2. The scene of pilot-scale project in Yuhang district of Hangzhou.

Under the same conditions of influent, ventilation and plant selection, the pilot project selected different types of substrates and gradation, designed two parallel coupling systems to investigate the effects of different substrates and configuration on the removal of pollutants, finally get a set of optimal operating parameters. As shown in Fig. 2-2, two sets of biological filter beds and subsurface-flow constructed wetland coupling system are separated by partitions, the right side is the original system (a) with blended filter stone, and on the other side is the comparison system (b) with different substrates. As can be seen from Fig. 2-2 that the upper layer is a biological filter bed, the lower layer is a subsurface-flow constructed wetland, and the left part of the subsurface-flow constructed wetland is covered by the biological filter bed unit of the same area in the upper layer, forming a local overlapping superposition.

The specific design parameters of pilot project are as follows: Grille well: 1.00 m×1.00 m×1.50 m, steel mixed (buried); Regulating pond: 2.50 m×4.80 m×3.00 m, steel mixed (buried), there format structure, similar to the role of septic tank hydrolysis acidification pool; Biological filter bed: 109.20 m^2, high:1.20 m , concrete cofferdam, geomembrane seepage control, placed above the subsurface-flow constructed wetlands; Biological filter substrates: 120 m^3, plantation of iris and ryegrass with a density of 30 plies/m^2 ; Subsurface-flow constructed wetland: 140.20 m^2, high: 1.50 m, concrete cofferdam, geomembrane seepage control; Wetland filters: 225 m^3, planting canna in subsurface-flow wetland without being covered by biological filter bed, the density of 9 plants/m^2; Out of the well: brick: 0.60 m × 0.60 m × 0.70 m.

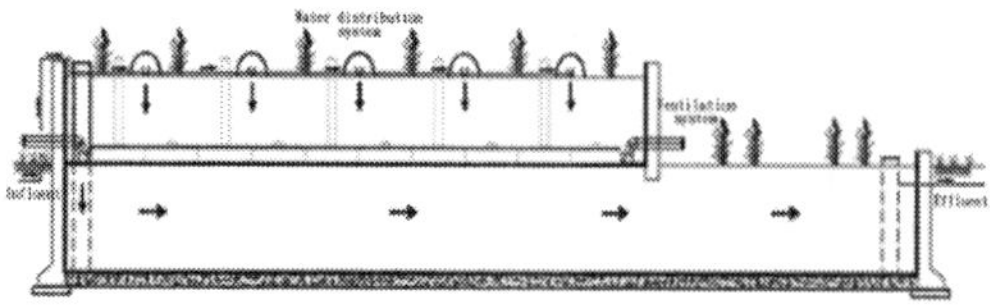

Fig.2-3. Structure and flow pattern of pilot project

Fig. 2-3 is a schematic diagram of the specific structure of the pilot project, coupled with the process of influent, effluent and the flow direction of sewage in the system. As shown in Fig. 2-3, the sewage is pretreated by the grating and the regulating pond, then lifted to the biological filter bed unit. The sewage is uniformly distributed through the water distribution pipe groove installed on the surface of the biological filter bed filter layer. Treated sewage at the bottom of the biological filter bed along the slope to the left of the water gathering well, and then through the water gathering well installed in the subsurface-flow constructed wetland water distribution system, making the sewage from left to right through the design of different filters area after filtration treatment effluent.

The original system (a) substrates composition: 1) The biological filter bed unit selected the filter stone with calcium (limestone, etc.). Vertical laying on three floors, Upper (H = 700 mm, ϕ = 2-5 mm), middle (H = 100 mm, ϕ = 13 mm), the lower layer (H = 400 mm, ϕ = 25 mm) ; 2) Subsurface-flow constructed wetland units selected the filter stone with calcium (limestone, etc.); From left to right laying length and size specifications were (L = 2500 mm, ϕ = 40 mm), (L = 1600 mm, ϕ = 25 mm), (L = 1600 mm, ϕ = 13 mm), (L=5600 mm, ϕ = 2-5 mm). In addition, two groups of horizontal and vertical blocks laid at the end of the subsurface-flow constructed wetlands are L = 1600 mm (H_{upper}= 200 mm, ϕ = 2-5 mm; $H_{lower\ layer}$ = 1300 mm, ϕ = 13 mm) and L = 2500 mm (H_{upper} = 200 mm, ϕ = 2-5 mm; H_{middle} = 100 mm, ϕ = 13 mm; $H_{lower\ layer}$ =1200 mm, ϕ= 25 mm).

The comparison system (b) substrates composition: 1) The biological filter bed unit selected three types of material of melon gravel, zeolite, ceramsite, vertical laying on three floors, Upper (melon broken gravel, H = 700 mm, ϕ= 3-5), middle (zeolite, H = 100 mm, ϕ = 10 -15 mm), the lower layer (ceramsite, H = 400 mm, ϕ = 20-40 mm) ; 2) Subsurface-flow constructed wetland units selected the filter stone with calcium; From left to right laying length and size specifications were (ceramsite, L = 2500 mm, ϕ = 20-40 mm), (zeolite, L = 1600 mm, ϕ = 10-15 mm), (Melon-gravel, L = 7200 mm, ϕ = 3-5 mm). In addition, two groups of horizontal and vertical blocks laid at the end of the subsurface-flow constructed wetlands are L = 1600 mm (Melon-gravel, H_{upper} = 200 mm, ϕ = 3-5 mm; zeolite, $H_{lower\ layer}$ = 1300 mm, ϕ = 10-15 mm) and L = 2500 mm (Melon-gravel, H_{upper} =200 mm, ϕ= 3-5 mm; zeolite, H_{middle} = 100 mm, ϕ = 10-15 mm; ceramsite, $H_{lower\ layer}$ = 1200 mm, ϕ = 20-40 mm).

2.2 Process parameters

The influent volume of pilot project design is 50 m^3/d, the two sets of coupling system for the operation of a comparative study from May of 2016 to December of 2016 for a period of low concentration and June of 2017 to September of 2017 for a series of high concentration influent period. In the stable operation stage, the influent parameters of pilot project of the new biological filter bed and subsurface-flow constructed wetland coupling system are shown in table 2-1.

Table 2-1 Influent quality indicators (Unit: mg/L)

Water quality indicators	Low concentration	High concentration
COD_{Cr}	7.87 ~ 45.60	24.00 ~ 442.00
NH_4^+-N	4.72 ~ 34.40	11.10 ~ 58.00
TN	7.59 ~ 35.50	16.70 ~ 73.60
TP	1.24 ~ 3.44	1.82 ~ 6.98

The new biological filter bed system and the subsurface-flow constructed wetland system are operated in intermittent ways, and the biological filter bed has a HRT of 18.87 h while for the subsurface-flow constructed wetland system is 57.54 h.

2.3 Analysis of projects, sampling and testing methods

The test indexs mainly including COD_{Cr}, TN, NH_4^+-N and TP. Measurement items and methods are shown in table 2-2.

Table 2-2 Test indicators and methods of the experiment

Index	Test Methods	Instrument
COD_{Cr}	Dichromate method HJ 828-2017	Taizhou Meixu HCA-100
TN	Alkaline potassium persulfate digestion UV spectrophotometric method HJ 636-2012	Beijing Puxi TH1810
HN_4^+-N	Nessler' reagent spectrophotometric HJ 535-2009	Beijing Puxi T6
TP	Ammonium molybdate spectrophotometric method GB/T 11893-1989	Beijing Puxi T6

3 Results and discussion

3.1 Study on the pollutant removal of pilot project coupling system with different substrates configurations at low influent concentration

Different natural conditions, water habits and the level of economic development can lead to the differences in the water quality of rural sewage. Some scholars [8] in their study of ecological filter bed technology of rural domestic sewage in the west mountainous area of Zhejiang found that the influent concentration of main water quality indicators like COD_{Cr}, NH_4^+-N, TN and TP was 32.10 mg/L, 1.13 mg/L, 1.33 mg/L and 0.61 mg/L respectively, and operation results showed that the above indexes of removal rate reached to 43.30%, 56.42%, 42.11% and 19.88% respectively, which even lower than the micro-polluted surface water [9].

Fig.3-1 shows the COD_{Cr}, NH_4^+-N, TN and TP removal performance of two coupled systems. The effluent of the system (a) and the system effluent (b) are the effluent from the original system (a) and the comparison system (b), respectively.

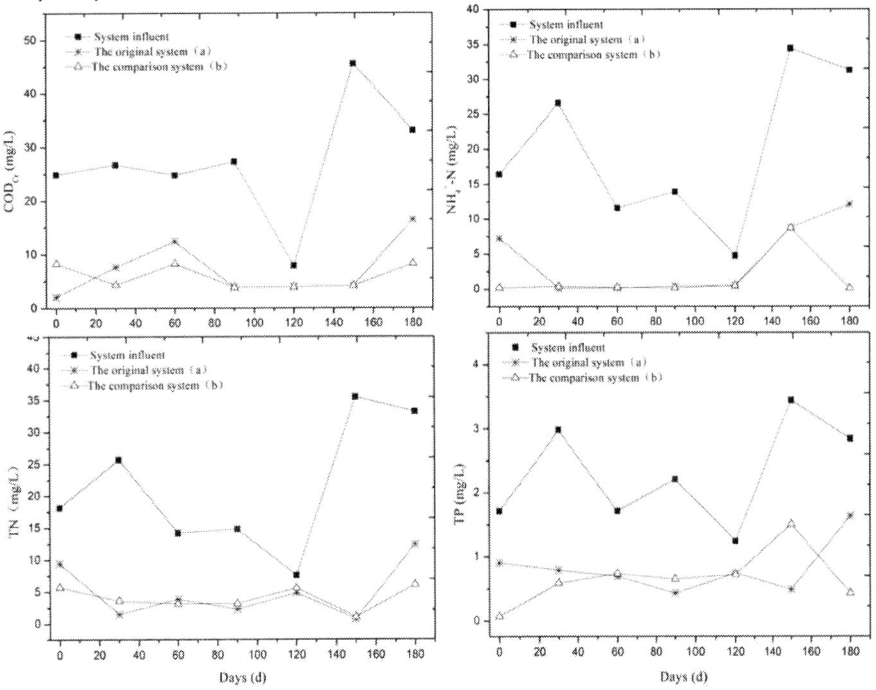

Fig.3-1 Removal of low influent concentration of COD_{Cr}、NH_4^+-N、TN and TP by coupling system

As can be seen from Fig.3-1, the concentration of effluent COD_{Cr} has been cut down to a certain degree after the pretreatment of grille and adjust pool, however, there has been a great change. It can also be seen from the figure that the influent of the coupling system is irregular, and the influent concentration comes to the lowest and the highest at 120th and 150th days respectively. The original system (a) and the comparison system (b) both maintain relatively satisfactory removal efficiency despite the relatively large fluctuations in water influent, of which the total average effluent COD_{Cr} concentration reaches to 7.23 mg/L and 5.88 mg/L while the removal rate is 63.10% and 74.10% respectively.

Although the NH_4^+-N concentration of the system fluctuates in a certain range, the original system (a) and the comparison system (b) reaches to 4.16 mg/L and 1.45 mg/L, respectively, and both maintain a good overall removal effect. The comparison system (b) has better effluent effect and the removal rate is 94.14%. The sampling of the effluent NH_4^+-N concentration of the original system (a) and the comparison system (b) are the same in the 30 d~90 d, and at the end of the sampling the NH_4^+-N concentration of the comparison system (b) is lower than that of original system (a).

The variation trend of TN concentration in the pilot coupling system is similar to that of NH_4^+-N, and both the effluent NH_4^+-N concentration of the original system (a) and the comparison system (b) basically keep decreasing from the initial to the 150th day, then both of the effluent concentrations begin to increase as the temperature decreases. The average effluent concentration of the original system (a) and the comparison system (b) reaches to 5.02 mg/L and 4.06 mg/L respectively, and in accordingly the TN removal rate reaches to 70.88% and 73.57% respectively, maintaining a good overall removal effect.

It is also known from Fig.3-1 that the fluctuation range of the influent TP concentration is larger (1.24 mg/L~3.44 mg/L), and there is no obvious change in the variation. The effluent TP concentration of the original system (a) has a slightly decrease trend in the beginning 150 d, and the effluent TP concentration increases to1.63 mg/L according to the decrease of temperature that after, which may be related to absorption ability of the calcium-contained substrate in the original system (a); The effluent TP concentration of the comparison system (b) shows a rising trend at the early stage and reaches to highest at 150 d, then decreases, and has an average removal rate of 69.53%.

3.2 Study on the pollutant removal of pilot project coupling system with different substrates configurations at high influent concentration

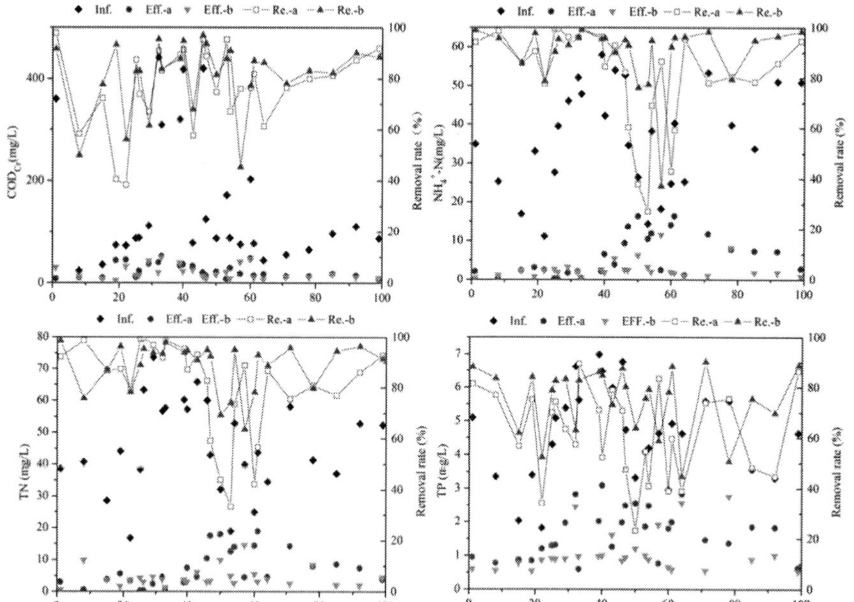

Fig.3-2. Removal of high influent concentration of COD_{Cr}、NH_4^+-N、TN and TP by coupling system

It can be seen from Fig.3-2 that the influent of the pilot project is irregular, and the influent concentration reaches to the lowest and the highest respectively on the 8th and the 32th days. In spite of large influent fluctuations, the original system (a) and the comparison system (b) coupled with the new biological filter bed and subsurface-flow constructed wetland technology all maintained an efficient removal effect, the average COD_{Cr} concentrations of the total effluent reached 23.63 mg/L and 20.22 mg/L, the removal rates were 77.10% and 81.30% respectively. COD_{Cr} removal rate of the comparison system (b) is more stable than the original system (a), and the average removal rate is better than the original system (a). Vymazal [10] studied the COD_{Cr}

removal efficiencies of horizontal-vertical wetland and multi-stage composite constructed wetlands, which were 84.00% and 83.80% respectively, similar to the study.

Fig.3-2 reflects the contribution of the coupling system to the removal of NH_4^+-N. As can be seen from the figure, although the influent concentration of NH_4^+-N fluctuated within a certain range, the average effluent concentrations of the original system (a) and the comparison system (b) reached 5.93 mg/L and 2.70 mg/L respectively, both of which remained better overall removal effect. Compared with the two, the comparison system (b) has better effluent effect and the removal rate is 90.28%.

The original system (a) had the lowest removal rate of NH_4^+-N on the 53rd day, only 27.08%, and its corresponding influent concentration was also lower at 14.40 mg/L. The comparatively high removal rate of the comparison system (b) shows that the removal efficiency of constructed wetlands with a variety of substrates configurations are less affected by the influent pollution load, own to the synergistic effect among various substrates [11]. Some research results [12-16] show that the removal rates of NH_4^+-N and TN are 57%~71% and 50%~75% in composite constructed wetland, respectively. Relative to the original system (a), the removal effect is relatively stable but the average removal rate is lower, the removal rate of the original system (a) is not stable may be related to the single configuration of the substrates.

The removal mechanism of N by constructed wetlands is very complicated [17], mainly including ammonia volatilization, matrix adsorption, plant absorption and microbial nitrification /denitrification. The main role is nitrification/denitrification [18,19], plant absorption from the long-term results to see only the second. As can be seen in Fig. 3-4, the average TN concentration of the effluent from the original system (a) and the comparison system (b) reached 7.27 mg/L and 4.40 mg/L, respectively, meeting A level in the standard of 《Discharge standard of pollutants for municipal wastewater treatment plant》 (GB 18918—2002), the original system (a) and the comparison system (b) on the TN removal rate reached 80.91% and 88.57%. Chen [20] selected anaerobic contact oxidation pond and vertical flow constructed wetland wastewater treatment system of rural sewage treatment showed that the TN removal efficiency was 75.60%, the removal rate slightly lower than the system. As can be seen from Fig. 3-3 and Fig.3-4, the trends of NH_4^+-N and TN during the whole monitoring period are very similar, this is consistent with the study by Liu [21], which is related to the main component of TN is NH_4^+-N.

Adsorption and retention of substrates in constructed wetland are the main ways to remove phosphorus [22], and different substrates show different adsorption characteristics and removal effects. As can be seen from Fig. 3-5, the influent TP concentration fluctuated in a wide range (1.82 mg/L~6.98 mg/L) with no obvious change rule. The influent concentrations at the 22nd and 39th days reached the lowest and the highest, respectively. The original system (a) and the comparison system (b) effluent TN average concentrations reached 1.66 mg/L and 1.09 mg/L, all maintained a good overall removal effect, the original system (a) and the comparison system (b) of TN removal efficiency reached 61.75% and 75.26%, respectively, which are better than that of Asuman [23] using slag and sand as substrates in wetland system. It can be seen that the optimal configuration of a variety of substrates is better for the removal of phosphorus.

3.3 Study on the improvement of removal efficiency by the optimized configuration of substrates at different influent concentrations

The original system (a) and the comparison system (b) of the new biological filter bed and subsurface-flow constructed wetland coupling process are both equipped with self-ventilated pipe network system, the only difference is the substrate selection and configuration. Therefore, it is necessary to study the effect of two sets of coupling systems on the removal of pollutants from the aspects of the selection and layout of substrates.

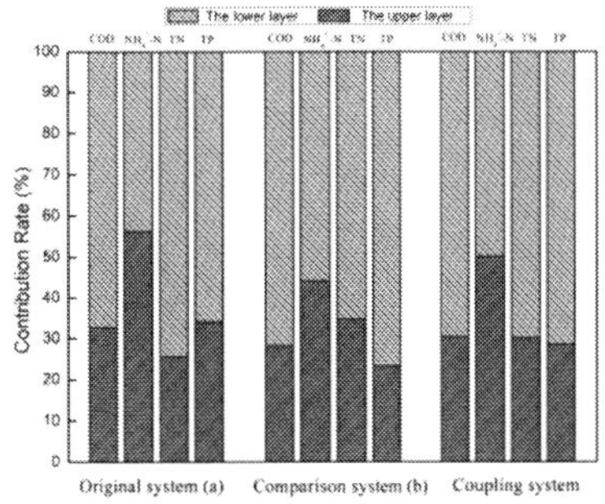

Fig.3-3. Contribution rates of removal of COD_{Cr}、NH_4^+-N、TN and TP by lower and upper layer of different coupling systems at low influent concentration.

As a whole that can be seen from Fig.3-3, the contribution rate of the biological filter bed and the subsurface-flow constructed wetland in low influent concentration is 30.40% and 69.60% respectively in the removal of COD_{Cr} by each unit of the pilot coupling system. In the removal of NH_4^+-N from each unit, the contribution rate of the biological filter bed and the subsurface-flow constructed wetland is 50.15% and 49.85%, respectively. The contribution rates of removing TN is 30.22% and 69.78%, respectively, while to TP the rate is 28.58% and 71.42%, respectively.

Table 3-1 Removal effects of the new biological filter bed and subsurface-flow constructed wetland coupling system when treating low influent concentration wastewater

Water quality index	The original system (a) removal rate (%)	The comparison system (b) removal rate (%)	Removal rate increase (%)
COD_{Cr}	63.10	74.10	11.00

NH_4^+-N	82.59	94.14	11.55
TN	70.88	73.57	2.69
TP	61.44	69.53	8.09

Table 3-1 lists the removal effects of the new biological filter bed and subsurface-flow constructed wetland coupling system when treating low influent concentration wastewater, and increase extent is also calculated. The comparison system (b) has significantly improved the main water quality index reduction than that of the original system (a), and the main water quality index is increased by 11.00%, 11.55%, 2.69% and 8.09% respectively when compared with the original system (a).

Table 3-2 Removal effects of the new biological filter bed and subsurface-flow constructed wetland coupling system when treating high influent concentration wastewater

Water quality index	The original system (a) removal rate (%)	The comparison system (b) removal rate (%)	Removal rate increase (%)
COD_{Cr}	77.10	81.30	4.20
NH_4^+-N	81.08	90.28	9.20
TN	80.91	88.57	7.66
TP	61.75	75.36	13.61

As can be seen from Table 3-2, the comparison system (b) has a significant increase in the concentration reduction of the main water quality index than the original system (a), The removal rates of COD, NH_4^+-N, TN and TP in the comparison system (b) were 4.20%, 9.20%, 7.66% and 13.61% higher than that of the original system (a) respectively. Thus, different substrates and configuration have a significant impact on the removal efficiency of the new biological filter bed and subsurface-flow constructed wetland coupling system, and the most obvious improvement for phosphorus removal efficiency. Different substrates have different adsorption characteristics and microbial adhesion properties, thus affecting the sewage treatment effect. Xu [24] selected four kinds of substrates to study their microbial activity changes and their effects on nitrogen removal, the results showed that the order of the microbial activity was sand/soil/peat mixture > soil > soil/sand mixture > sand. Gu [25] studied the characteristics of microbial attachment to different substrates, it was found that there were 709, 777, 583 and 568 species of bacterial colonies belonging to volcanic rocks, coke, zeolite and ceramsite, respectively, and the main microbial types and dominant species of different substrates were also different. Studies have shown that a variety of substrates combinations can improve the decontamination effect of wetlands. The types, physicochemical properties and disposition of substrates in constructed wetlands affect the growth of plants, and also affect the microbial activity in plant rhizosphere, ultimately affect the purification ability of constructed wetlands [11].

To sum up, it is easy to come to a conclusion that the comparison system (b) has higher removal rate in treating high influent concentration wastewater than treating low concentration case by the original system (a).

3.4 Effect of temperature on the coupling system treatment effect at low and high influent concentrations

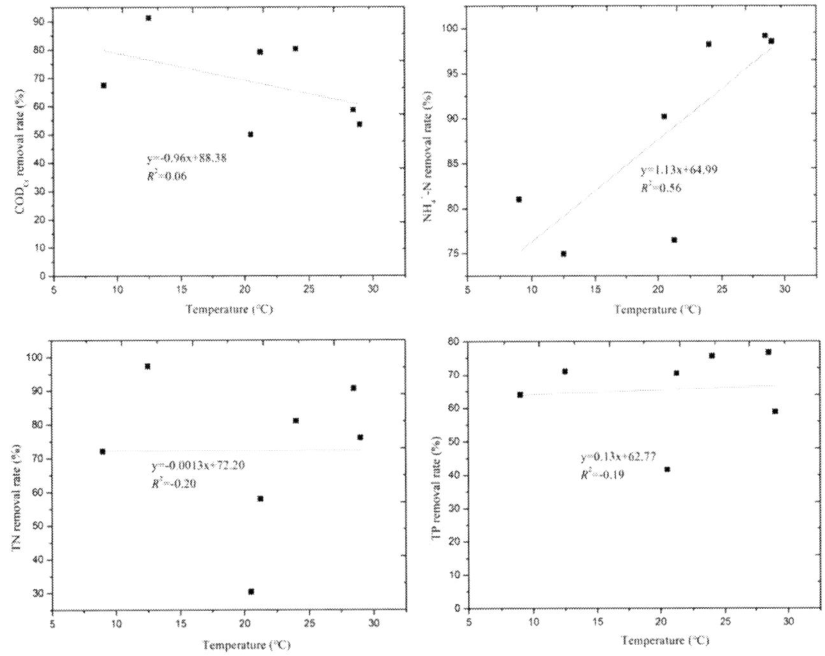

Fig.3-4 The impact on removal of main water index of coupling system by temperature at low influent concentration.

Fig.3-4 is a fitting chart of the effect of temperature on removal of main water quality index by a pilot coupling system treating low influent concentration wastewater. It can be seen that temperature has a negative correlation with the removal of COD_{Cr} from the coupling system. Meanwhile, the temperature has a significant positive correlation with the removal of NH_4^+-N by the coupling system, which is positively correlated with other researchers' conclusions [26]. In addition, it is also found from Fig.3-4 that the effect of temperature on the removal of TN and TP is not obvious.

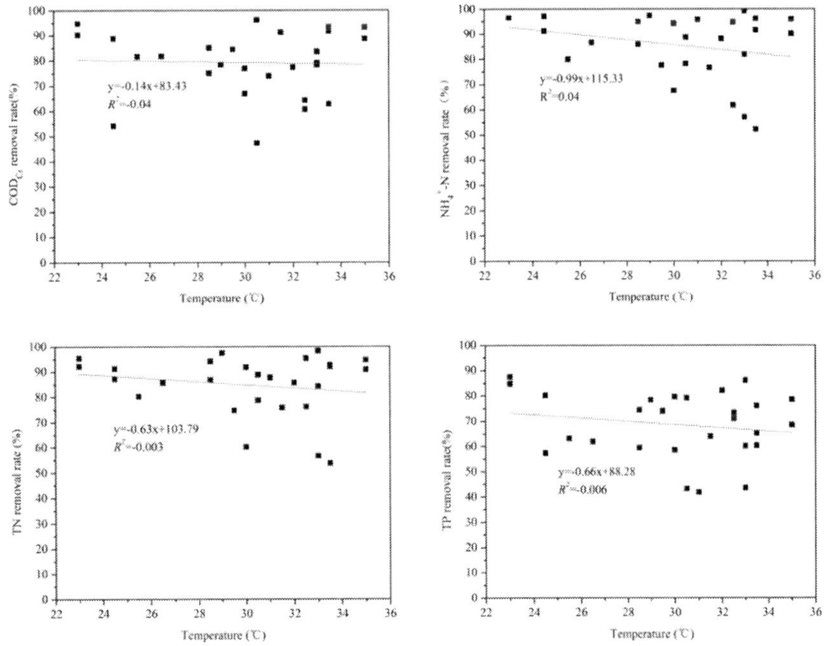

Fig.3-5. The impact on removal of main water index of coupling system by temperature at high influent concentration.

Fig.3-5 is a correlation fit plot of the effect of temperature on the operating performance of coupled system as a whole. It can be seen from Fig.3-5 that the temperature has a negative correlation with the removal of COD_{Cr} from the coupled system; Meanwhile, it is consistent with the conclusions of other scholars [27,28]. Some studies [29,30] show that temperature has little effect on COD_{Cr} removal rate in composite constructed wetlands, which is consistent with the results of this system (R^2=-0.04). As the sampling time in the period of June to September, the temperature fluctuations in the 23~35℃, it is a suitable temperature for nitrification and denitrification activities of micro-organisms [31], in addition, the influent concentration fluctuates greatly, and it may be difficult to show the actual fitting effect of NH_4^+-N, TN removal rate and temperature. When the temperature is decreased, the dissolved oxygen (DO) concentration in the wetland will generally decrease, affecting the redox reaction and TP removal rate, however, the coupling system has the function of self-ventilation to ensure the amount of DO in the system, so the fitting degree of the removal rate and temperature is low (R^2=-0.006), less affected by the outside temperature.

4 Summary

Based on the self-ventilated pipe network biological filter bed and subsurface-flow constructed wetland coupling system pilot project construction and its treatment of domestic sewage research draw the following conclusions:

1) The comparison system (b) had better removal rates than that of the original system (a), and the removal rate when treating low influent concentration was 74.10%, 94.14%, 73.57% and 69.53% for COD_{Cr}, NH_4^+-N, TN and TP, respectively. The removal of the above main water indexes of the comparison system (b) promoted by 11.00%, 11.55%, 2.69% and 8.09% respectively in low influent concentration.

2) The influent concentration fluctuates greatly in the coupling system. The concentrations of COD_{Cr}, NH_4^+-N, TN and TP in the comparison system (b) were 20.22 ± 13.37 mg/L, 2.70 ± 2.49 mg/L, 4.40 ± 3.05 mg/L and 1.09 ± 0.62 mg/ L. The removal of the above main water indexes of the comparison system (b) promoted by 4.20%, 9.20%, 7.66% and 13.61% respectively in high influent concentration. The optimal configuration of different substrates is more conducive to the degradation and removal of pollutants.

3) The optimized configuration of various kinds of substrates was significant and was more beneficial to the degradation and removal of pollutants. The adsorption and interception function of substrates in the constructed wetland was the main way of phosphorus removal. The function of self-ventilation ensured the amount of DO in the coupling system, making the phosphorus removal was less affected comparing to structure of traditional wetland.

Acknowledgements

This research was supported by Zhejiang Provincial Natural Science Foundation of China under Grant No.LY18E080005, the Major Science and Technology Program for Water Pollution Control and Treatment of China 12th Five-Year Plan under Grant No.2014ZX07101-012, and Scientific Research Institute of Zhejiang provincial science and technology department No.2018F10031.

References

1. Dixon A, Simon M, Burkitt T. Assessing the environmental impact of two options for small-scale wastewater treatment: comparing a reedbed and an aerated biological filter using a life cycle approach[J]. Ecological Engineering, **20**, 297-308 (2003).
2. Selma C. Ayaz, Özgür Aktaş, Nur Findik, et al. Phosphorus removal and effect of adsorbent type in a constructed wetland system[J]. Desalination & Water Treatment, **37**, 152-159 (2012).
3. Ceng Lingfang. A brief comment on new methods of rural domestic sewage treatment in foreign countries[J]. China Rural Nater and Hydropower, **9**, 30-31 (2001). (In Chinese)
4. Ham J H, Yoon C G, Jeon J H, et al. Feasibility of a constructed wetland and wastewater stabilisation pond system as a sewage reclamation system for agricultural reuse in a decentralised rural area[J]. Water Science & Technology A Journal of the International Association on Water Pollution Research, **55**, 503 (2007).
5. Ye Yaling, Lu Mengjiang. Downflow anaerobic biofilter combined with constructed wetland to treat domestic sewage in small towns[J]. Environmental Research and Monitoring, **1**, 43-45 (2004). (In Chinese)
6. Suo Yanli, Pan Jing, Chen Yongqiang. Research on decentralized domestic sewage treatment by combined biological and ecological system[J]. Journal of Shenyang Normal University (Natural Science Edition, **25**, 376-380 (2007). (In Chinese)
7. Wu Caibin, Xiang Sulin, Lu Xiuguo. Combined technology of biological contact oxidization and constructed wetland for processing rural domestic sewage[J]. Hubei Agricultural Sciences, **47**, 44-46 (2008). (In Chinese)
8. Qiu Zhi, Hu Zhifeng, Kong Lingwei, Zhang Chi. Research on engineering application of laminated ecological filter bed in the treatment of domestic sewage[J]. Environmental Pollution & Control, **36**, 43-45 (2014). (In Chinese)
9. Kong L.W., He F., Xia S.B., et al. Research on the application of the modified aquatic mat pond and integrated vertical-flow construction wetland coupling process based on luffa sponge for micro-polluted water treatment[J]. Desalination and Water Treatment, **76**, 155-165 (2017).

10. Vymazal J, Kröpfelová L. Multistage hybrid constructed wetland for enhanced removal of nitrogen[J]. Ecological Engineering, **84**, 202-208 (2015).
11. Wan Jiajing, Wan Zhan, Li Jun, et al. Function of the Substracts in the constructed wetlands[J]. Environmental Protection Science, **35**, 16-19 (2009). (In Chinese)
12. Travis M J, Weisbrod N, Gross A. Decentralized wetland-based treatment of oil-rich farm wastewater for reuse in an arid environment[J]. Ecological Engineering, **39**, 81-89 (2012).
13. S.Ç. Ayaz, Ö. Aktaş, N. Fındık, et al. Effect of recirculation on nitrogen removal in a hybrid constructed wetland system[J]. Water Science & Technology A Journal of the International Association on Water Pollution Research, **40**, 1-5 (2012).
14. Foladori P, Ortigara A R, Ruaben J, et al. Influence of high organic loads during the summer period on the performance of hybrid constructed wetlands (VSSF + HSSF) treating domestic wastewater in the Alps region[J]. Water Science & Technology A Journal of the International Association on Water Pollution Research, **65**, 890 (2012).
15. Zapater-Pereyra M, Ilyas H, Lavrnic, S, et al. Evaluation of the performance and space requirement by three different hybrid constructed wetlands in a stack arrangement[J]. Ecological Engineering, **82**, 290-300 (2015).
16. Zhang X, Inoue T, Kato K, et al. Performance of hybrid subsurface constructed wetland system for piggery wastewater treatment[J]. Water Science & Technology A Journal of the International Association on Water Pollution Research, **73**, 13 (2016).
17. Zhang D Q, Gersberg R M, Tansoon K. Constructed wetlands in China[J]. Ecological Engineering, **35**, 1367-1378 (2009).
18. Jamieson T S, Stratton G W. The use of aeration to enhance ammonia nitrogen removal in constructed wetlands[J]. Canadian Biosystems Engineering, **45**, 9-14 (2003).
19. Chung A K C, Wu Y, Tam N F Y. Nitrogen and phosphate mass balance in a sub-surface flow constructed wetland for treating municipal wastewater[J]. Ecological constructed wetland for treating municipal wastewater[J]. Ecological Engineering, **32**, 81-89 (2008).
20. Chen Heping, Zhang Chen, Zhu Heping, et al. Treatment of rural domestic wastewater by anaerobic-contact oxidation and vertical flow man-made wetland[J]. Journal of Ningbo University (Natural Science & Engineering Edition), **21**, 568-570 (2008). (In Chinese)
21. Liu Shentan, Wang Guofang, Xie Xiangfeng, et al. Effect of matrix on denitrification efficiency and distribution of nitrifying and denitrifying bacteria in constructed wetlands[J]. Journal of Southeast University (Natural Science Edition), **41**, 400-405 (2011). (In Chinese)
22. Vymazal J. Constructed wetlands for wastewater treatment in the Czech Republic[J]. Water Science & Technology A Journal of the International Association on Water Pollution Research, **7**, 1-14 (1996).
23. Ban M. Comparison of the treatment performances of blast furnace slag-based and gravel-based vertical flow wetlands operated identically for domestic wastewater treatment in Turkey[J]. Ecological Engineering, **24**, 185-198 (2005).
24. Xu Defu, Xu Jianming, Li Yingxue.The Microbial Activity of the Substrates in Constructed Wetland and Its Effect on the Removal of Nitrogen from Wastewater[J]. Journal of Agro-Environment Science, **27**, 753-757 (2008). (In Chinese)
25. Gu Yonggang, Jin Pengkang, Li Zhaoxin, et al. Effect of different substrates on the species formation in the water bodies supplemented by reclaimed water[J]. Environmental Pollution & Control, **39**, 1122-1126 (2017). (In Chinese)
26. Kong Lingwei, He Feng, Xia Shibin, et al. Studies on construction and performance of the Qiantang River water diversion de-nitrification demonstration project[J]. Environmental Pollution & Control, **36**, 60-66 (2014). (In Chinese)
27. Shen Z, Zhou Y, Hu J, et al. Denitrification performance and microbial diversity in a packed-bed bioreactor using biodegradable polymer as carbon source and biofilm support[J]. Journal of Hazardous Materials, **8**, 250-251 (2013).
28. Kong Lingwei, He Feng, Xia Shibin, et al. Studies on construction and performance of the Qiantang River water diversion de-nitrification demonstration project. Environmental Pollution & Control, **36**, 60-66 (2014). (In Chinese)
29. Mæhlum, T., Stålnacke, P. Removal efficiency of three old-climate constructed wetlands treating domestic wastewater: Effects of temperature, seasons, loading rates and input concentrations[J]. Water Science & Technology, **40**, 273-281 (1999).
30. Vanier S M, Dahab M F. Start-up performance of a subsurface-flow constructed wetland for domestic wastewater treatment[J]. Environmental Technology, **22**, 587-596 (2001).
31. Vymazal J. Horizontal sub-surface flow and hybrid constructed wetlands systems for wastewater treatment[J]. Ecological Engineering, **25**, 478-490 (2005).

Quantitative analysis of factors that affect oil pipeline network accident based on Bayesian networks: A case study in China

Chao Zhang[1,*], *Ting Xin* QIN[1], *Shuai* Huang[1], *Jian Song* Wu[2], and *Xin Yan* Meng[1]

[1]China National Institute of Standardization, 100191 Beijing, China
[2]School of Resources and Safety Engineering, China University of Mining and Technology, 100083 Beijing, China

Abstract. Some factors can affect the consequences of oil pipeline accident and their effects should be analyzed to improve emergency preparation and emergency response. Although there are some qualitative analysis models of risk factors' effects, the quantitative analysis model still should be researched. In this study, we introduce a Bayesian network (BN) model of risk factors' effects analysis in an oil pipeline accident case that happened in China. The incident evolution diagram is built to identify the risk factors. And the BN model is built based on the deployment rule for factor nodes in BN and the expert knowledge by Dempster-Shafer evidence theory. Then the probabilities of incident consequences and risk factors' effects can be calculated. The most likely consequences given by this model are consilient with the case. Meanwhile, the quantitative estimations of risk factors' effects may provide a theoretical basis to take optimal risk treatment measures for oil pipeline management, which can be used in emergency preparation and emergency response.

1 Introductions

Oil pipelines are used more and more widely and their accidents may result in serious consequences. Several types of factors, including hazard property, environment condition, hazard-affected carriers and emergency response, can affect oil pipelines accident consequences. The effect of these factors should be analyzed to identify their importance. And furthermore, the optimal risk treatment plan which can reduce the probability and consequences of an accident, can be developed based on the effect analysis.

Researchers have developed some models to analyze the effect of risk factors. Some diagram methods, including fault tree models, event tree models and bow-tie models, are effective to analyze these factors qualitatively. The risk factors can be identified by these diagram methods but it's difficult for quantitative analysis. Incident simulation models can analyze the factor effect quantitatively. But the simulation and calculation are complex and it's difficult to refer all the risk factors in a simulation model. Bayesian network (BN) is an effective method for probabilistic analysis and diagnose analysis and has been used to analyze oil accident issues.

In this paper, an integrated model is established to analyze the risk factors' effects of an oil pipeline accident. An oil pipeline accident case that happened in China has been taken as an example. Risk factors are identified qualitatively by the incident evolution diagram. Then they are deployed as the nodes deployment rule of BN model and initialed according to the conditional probabilities given by statistics and experts' estimation. Based on the BN model of this oil pipeline accident, the effects of risk factors can be analyzed quantitatively and the consequences can be predicted according to the most likely consequence situations.

2 Quantitative Analysis Model of an Accident Case

2.1 A Brief Description of Oil Pipeline Accident Case

An oil pipeline explosion accident case has been taken as the example because it refers to some important risk factors. The accident occurred at 10:25 p.m., Nov. 22, 2013, in an industrial district in Qingdao city, Shandong province in China. An oil pipeline broke, and the oil leaked into a municipal drainage culvert. Because the leakage to the culvert was not observed, the oil gas density increased continuously until it exploded; 62 people died, 136 were injured and more than 75 million RMB was lost in this explosion accident.

2.2 Identify Risk Factors Based on Incident Evolution Diagram

An incident evolution diagram is built to identify risk factors. This diagram comprises five kinds of nodes: Si (incident Situation), Ei (key Environment condition), Ti (emergency Target), Mi (response Mission) and Ci (consequence). The initial incident state S1 means the

* Corresponding author: zhangchao@cnis.gov.cn

incident occurs. Different environment condition Ei will lead to different Si. The emergency response target Ti is fixed to address the incident situation Si, and the mission that needed to be executed is Mi. If the target Ti is achieved, the final consequence Ci will occur in the condition that Ti is the final emergency target.

The incident evolution diagram for the oil pipeline accident case is shown in **Figure 1**. We can see that the environment condition (E_1-confined space & E_2-water area) will lead to serious consequences (C_2-ground and water pollution & C_4-explosion, casualties, economic loss). Meanwhile, emergency management (M_1, M_2, M_3 and M_4) can affect consequences obviously.

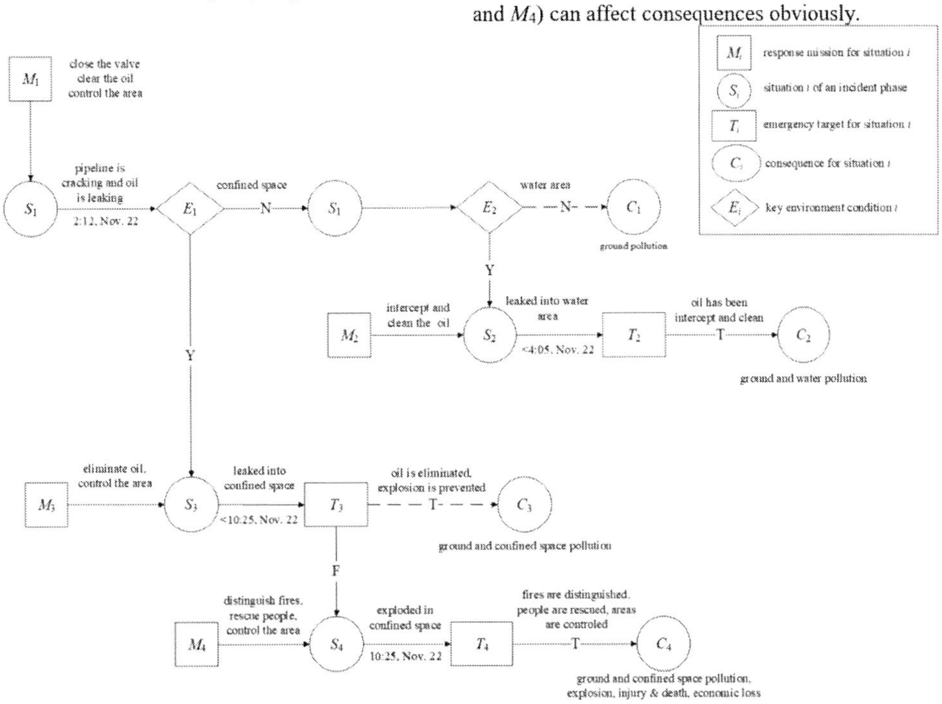

Fig. 1. Incident Evolution Diagram of the Oil Pipeline Accident Case

2.3 Factor Effect Analysis Based on Bayesian Network

2.3.1 Bayesian Network Method

The basic of BN is Bayesian condition probability theory. It contains conditional independence and joint probability distribution:

$$P(V_1, V_2, \ldots, V_k / v) = \prod_{1}^{k} P(V_i / v) \quad (i = 1, 2, \ldots, k) \qquad (1)$$

$$P(V_1, V_2, \ldots, V_k / v) = \prod_{1}^{k} P(V_i / Parent(V_i)) \quad (i = 1, 2, \ldots, k) \qquad (2)$$

Where V_1, V_2, …, V_k represent various variables, v is the normal node, which facilitates the expression of the conditional probability, and Parent(V_i) is the parent nodes of V_i.

2.3.2 Dempster-Shafer Evidence Theory

The Dempster-Shafer (DS) evidence theory is a method for quantitative analysis of system uncertainty. It can be used to calculate the conditional probabilities of BN factors **错误!未找到引用源。**. In this BN model, the conditional probabilities of risk factors' are calculated based on DS theory according to experts' estimation.

2.3.3 Bayesian Network Model for Oil Pipeline Accident

For an oil pipeline accident, the risk factors may refer to environment, hazard, incident, hazard-affected carries, emergency response, and consequences. As an integrated effect analysis model, these risk factors nodes should be considered and deployed according to the nodes deployment rule of BN model as **Figure 2**.

https://doi.org/10.1051/e3sconf/20183801010

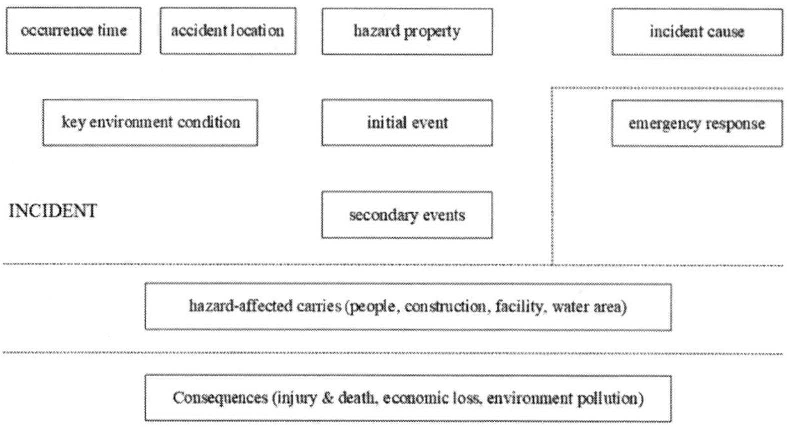

Fig. 2. Deployment Rule of the BN Nodes for an Oil Pipeline Accident

BN nodes include independent nodes and dependent nodes. The factors of occurrence time, accident location, hazard property (pipeline pressure, pipeline flux), incident cause and key environment condition (confined space nearby, water area nearby) are independent nodes. Their occurrence probabilities are based on experts' estimation and statistics. The factors of initial event, secondary events (flame and fire ball, water pollution), hazard-affected carries (threatened persons, buildings, infrastructure and lifeline), emergency response and consequences (casualties, economic loss, environment pollution) are dependent nodes. Their conditional probabilities are based on experts' estimation using DS methods. As the factor nodes have been identified and deployed, and the occurrence probability and conditional probability have been estimated, then the BN can be constructed as **Figure 3**. It is a snapshot by the BN software Netica. The effects of risk factors for consequences can be quantitative analyzed by this BN model.

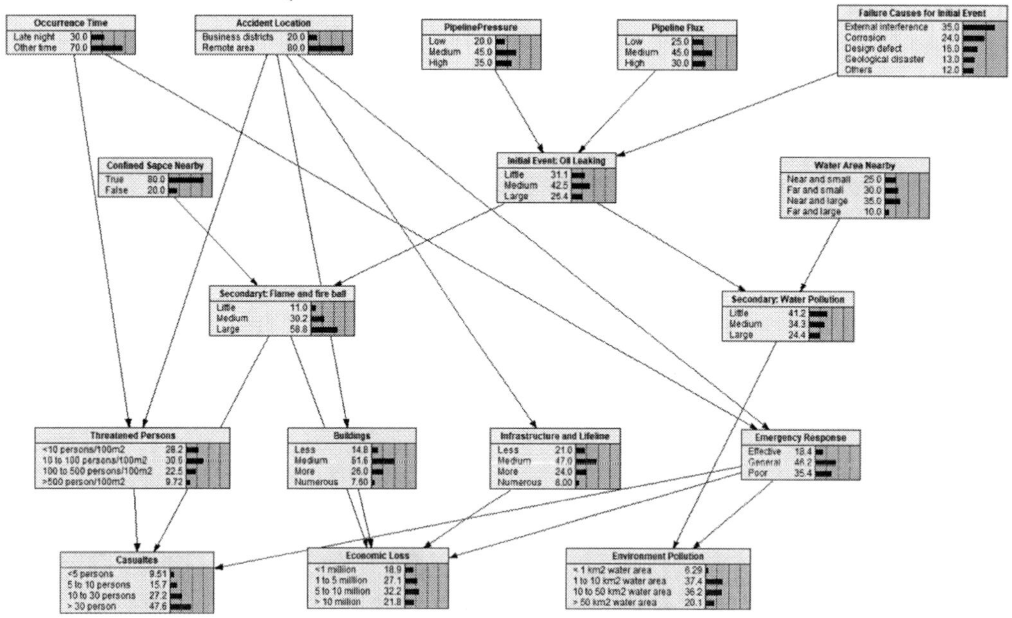

Fig. 3. The BN Model for the Oil Pipeline Accident Case

3 Results and Discussion

3.1 Scenario Setting and Consequences Estimation

As the basic of effect analysis, the situations of scenario nodes should be set according to the oil pipeline accident case. The situation setting of scenario nodes, including Time & Location nodes, Hazards Property nodes, Incident Case node, and Critical Environment Conditions nodes, are shown as Table 1, and the corresponding BN model is shown as Figure 4.

Table 1. Situation Setting of Scenario Nodes in the Oil Pipeline Accident.

Scenario Node	Situation Setting

* Corresponding author: zhangchao@cnis.gov.cn

Time & Location	① Occurrence Time	Late night
	② Accident Location	Business districts
Hazards Property	① Pipeline Pressure	High
	② Pipeline Flux	High

Incident Cause	① Failure Causes for Initial Event	External Interference
Critical Environment Conditions	① Confined Space Nearby	True
	② Water Area Nearby	Near and large

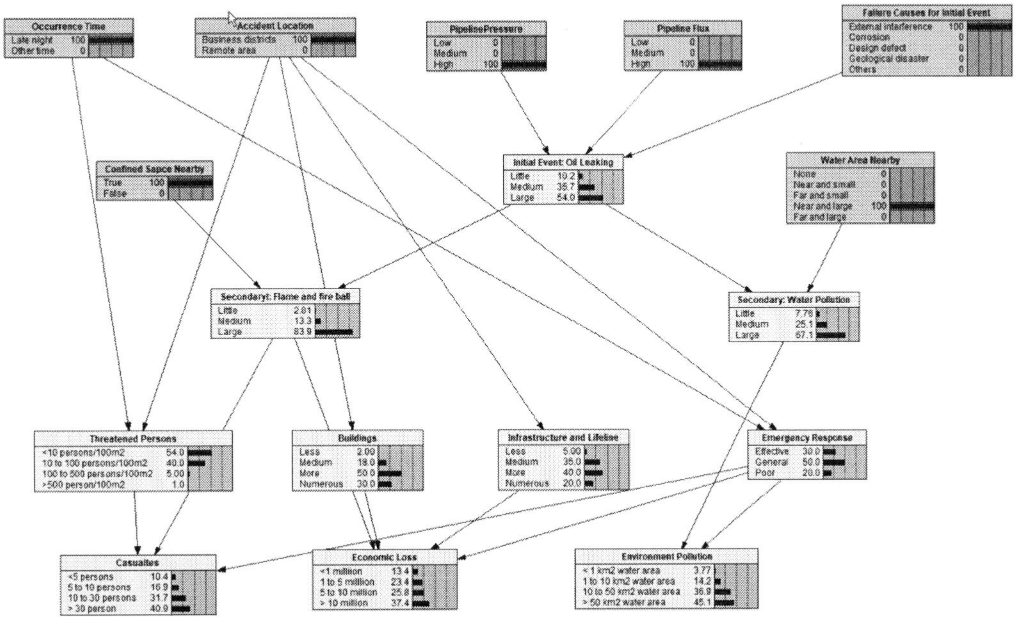

Fig. 4. The BN Model for the Oil Pipeline Accident Case Based on Scenario Setting

Figure 4 shows the most likely consequences are: more than 30 persons casualties, more than 10 million economic loss, and larger than 50 km^2 water area pollution. They are consilient with the case. According to this oil pipeline accident case scenario, the effects of risk factors can be quantitative analyzed.

3.2 Confined Space

One of the key environment conditions is Confined Space Nearby. As the oil leaked into a municipal drainage, the oil gas density has been increasing continuously until it exceeded the explosion limit and the exploded. The explosion was so rapid and violent that it resulted in serious casualties and economic loss. The effects of Confined Space Nearby factors to Casualties and Economic Loss are shown as **Figure 5**.

Figure 5 shows the critical effect of Confined Space Nearby condition. Both of the most serious of Casualties (> 30 persons) and Economic Loss (> 10 million) will be predominant in condition that there is a confined space nearby. And the corresponding probabilities increases obviously. For Casualties, it increases from 26% to 40.9%. And for Economic Loss, it increases from 20.8% to 37.4%.

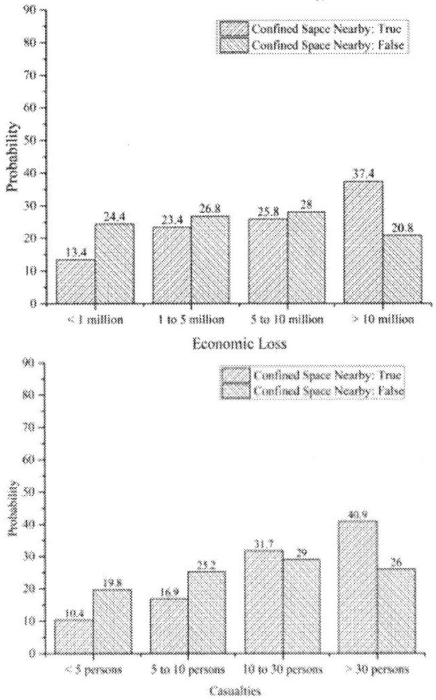

Fig. 5. Effects of Confined Space Nearby factors to Casualties and Economic Loss

3.3 Water Area

It is more difficult to deal with the pollution in water than on land. Therefore, Water Area Nearby is another key environment condition. The water pollution will be easier in condition that the water area is nearer, and will be more serious in condition that the water area is larger. The effect of Water Area Nearby factor to Environment Pollution is shown as **Figure 6**.

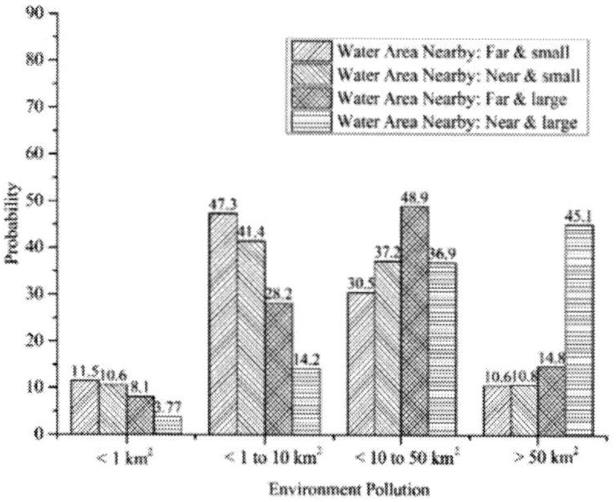

Fig. 6. Effect of Water Area Nearby factor to Environment Pollution

Figure 6 shows the critical effect of Water Area Nearby condition to Environment Pollution. Because the result of Initial Event Oil Leaking is Large (54.0%, as **Figure 4**), the probabilities of little environment pollution (< 1 km^2) is small (3.77% ~ 11.5%). And the probability of the consequence "1 to 10 km^2" is more than 40% both in "Far & small" condition and "Near & small" condition. It may result in serious water pollution if there is large water area near the incident site, especially for the consequence state ">50km^2". The probability of the most serious environment pollution (> 50km^2) increases significantly from the condition "Far & Large" (14.8%) to "Near & Large" (45.1%).

3.4 Emergency Response

Emergency response aims to protect people, reduce economic loss and avoid environment pollution. It plays an important role for consequences because it can control the incident, investigate the key environment conditions and prevent secondary event, and protect hazard-affected carriers. Take three kinds of consequences Casualties, Economic Loss and Environment Pollution as the targets to analyze the effect of "Emergency Response". The result is listed in **Table 2**. It is obvious that the more effective emergency response is, the less consequences will be caused.

Table 2. The Effect of Emergency Response to Consequences.

Probability (%)		**Emergency Response**		
		Effective	General	Poor
Casualties	① < 5 persons	14.9	12.0	9.21
	② 5 to 10 persons	22.0	16.9	17.4
	③ 10 to 30 persons	31.1	31.7	30.1
	④ > 30 persons	32.0	39.4	43.3
Economic Loss	① < 1 million	45.0	3.86	0.66
	② 1 to 5 million	43.2	19.4	7.25
	③ 5 to 10 million	11.3	38.8	17.4
	④ > 10 million	0.51	38.0	74.4
Environment Pollution	① < 1 km^2 water area	21.2	3.72	0
	② 1 to 10 km^2 water area	30.0	33.3	33.5
	③ 10 to 50 km^2 water area	34.5	38.1	34.4
	④ > 50 km^2 water area	14.2	24.9	32.2

4 Conclusion

This paper introduces a quantitative analysis model of the risk factors that affect oil pipeline network accident. It uses the incident evolution diagram to illustrate the process of incidents and to identify risk factors. Based on Bayesian model, the incident consequences can be estimated, and the risk factors' effects can be analyzed quantitatively.

In the oil pipeline accident case, the most likely consequences are consilient with the case. As the emergency resources are limit, the quantitative analysis results may provide a theoretical basis to take effective risk treatment measures for oil pipeline management, which can be used in emergency preparation and emergency response.

Acknowledgments

The authors deeply appreciate support for this paper by the National Natural Science Foundation of China (Grant No. 71673161), and the Fundamental Research Funds for the Central Business Unit (Grant No. 512015Y-4002).

References

1. W. Liang, J. Hu, C. Guo, et al, Eng. Appl. Artif. Intell **25**, 3 (2012)
2. S. Doug, V. Brian, K. Faisal, et al, J. Loss Prev. Process Ind 45 (2017)
3. M. Bertolini, M. Bevilacqua, J. Qual. Maint. Eng 12, 2 (2016)
4. D. Wang, E. Liu, L. Huang, Open Civ. Eng. J **11**, 1 (2017)
5. P. Jia, Y.U. Xiao, Q. Song, Ind. Saf. Environ. Prot 40, 2 (2014)
6. M.V.P. Muniz, G.B.A. Lima, Process Saf. Prog 7 (2017)
7. K. Peter, L. Buzna, D. Helbing, Int. J. Crit. Infrastruct, 4 (2008)
8. S.A. Joseph, B.J. Adams, B. McCabe, J. Infrastruct. Syst 16 (2010)
9. Z. Yuan, N. Khakzad, F. Khan, et al, Risk Anal **35** 2 (2015)
10. G. Kabir, R. Sadiq, S. Tesfamariam, Struct. Infrastruct. Eng 12 (2016)
11. X.H. Li, G.M. Chen, H.W. Zhu, Process Saf. Environ. Prot 103 (2016)
12. A.P. Dempster, *Classic Works of the Dempster-Shafer Theory of Belief Functions* (Springer, Berlin Heidelberg, 2008)
13. J. Wu, R. Zhou, S. Xu, et al, J. Loss Prev. Process Ind 46 (2017)

Analysis on the Channel for China's Textile and Garment Enterprises to Implement the Environmental Strategies

Wang Shi[1,*],

[1]School of Fashion, Dalian Polytechnic University, Dalian, Liaoning 116034, China

Abstract. The textile and garment industry is a traditional industry in China, and the environmental issues like its sewage, wastes, noise and other problems arising therefrom are becoming increasing prominent while bring economic benefits, which causes concern as well. This paper sums up the channels for China's textile and garment enterprises to fulfill the environmental strategies by analyzing the disclosure of the environmental information from 75 sampling companies of the textile and garment enterprises in China from 2012 to 2014 on the basis of Guidance on Environmental Information Disclosure of Chinese Listed Company(Draft for Public Review) and the Measures for the Publicity of Environmental Information (for Trial Implementation) issued by the Ministry of Environmental Protection of the People's Republic of China.

1 Introduction

Sanjay Sharma and Harrie Vredenburg proposed forward-looking environmental strategies in 1998. They divided the environmental strategies into two categories, namely the reactive environmental strategy and proactive environmental strategy respectively. [1] Different enterprises holds different attitudes towards the environmental issues, and of course, they would select different environmental strategies. With the strengthening of public awareness on environmental protection and the reinforcement on the supervision of the enterprises' environmental activities by the government, more and more enterprises are starting to focus on their environmental activities and their environmental consequences, and they are also starting and enlarge the efforts they make on the disclosure of the environmental information in order to prove its effective implementation of environmental responsibility, establish and maintain its fine social image.

As a whole, the textile and garment enterprises have a low quality in environmental information disclosure, and the majority of companies simply describe the performance of its environmental responsibility qualitatively, with little quantitative information, and such information is only available from the notes to the statement. Most enterprises do not integrate the environmental strategies into its overall business strategy, and gain more profits at the costs of environmental pollution and environmental resource consumption through the traditional approaches of production and operation.

2 Analysis on the Disclosures of Environmental Information by China's Textile and Garment Enterprises

The indicator system for environmental information disclosure is established on the basis of Guidance on Environmental Information Disclosure of Chinese Listed Company (Draft for Public Review) and the Measures for the Publicity of Environmental Information (for Trial Implementation) as well as the actual situations of the environmental information disclosed by the textile and garment enterprises in China.[2]The disclosed information contains 6 aspects: 1. Information on Environmental Strategy; 2. Information on Environmental Resource Consumption and Emission Pollution; 3. Information on the Implementation of Environmental Strategy; 4. Information on Environmental Responsibility; 5. Adequate Disclosure of Environmental Information and Clarity; 6. The Independent Report, a total of 24 disclosure items of environmental information. As shown in (Table 1) the following table, the indicator range of 19 items is (0-3), that of 4 items is (0-2), that of one indicator is (0-1), namely the range is (0-66). The level of the environmental information disclosed by the sampling enterprises can be obtained by adding the scores for such information disclosure made by the textile and garment enterprises.

* Corresponding author: wangs85@126.com

https://doi.org/10.1051/e3sconf/20183801011

Table 1 Rating Criteria for Environmental Information Disclosure

	Environmental Information Disclosure Items	**Rating Criteria**
	Environmental Strategy Information	
1	1.1 Enterprise-related environmental regulations and policies	Failure to disclose: 0 point, Qualitative disclosure: 1 point Quantitative disclosure: 2 points; Qualitative and quantitative disclosure: 3 points
2	1.2 Operator's environmental philosophy and values	
3	1.3 Environmental measures and policies that are planned to take	
4	1.4 Objectives and plans for environmental protection and pollution control.	
5	1.5 Global predication about the environmental conditions in the future	
	Information on Environmental Resource Consumption and Discharged Pollutants	
6	2.1 Categories and number of the consumed natural resources	Failure to disclose: 0 point, Qualitative disclosure: 1 point Quantitative disclosure: 2 points; Qualitative and quantitative disclosure: 3 points
7	2.2 Categories and number of the discharged pollutants	
8	2.3 Whereabouts of the discharged pollutants	Failure to disclose: 0 point; Simply disclosure: 1 point Detailed disclosure: 2 points
9	2.4 Whether the pollutant emission meets the standards	
	Information on the Implementation of the Environmental Protection	
10	3.1 Investment of the Enterprises in environmental governance (including greening)	Failure to disclose: 0 point, Qualitative disclosure: 1 point Quantitative disclosure: 2 points; Qualitative and quantitative disclosure: 3 points
11	3.2 Treatment and disposal of the wastes generated by the enterprise during the production	
12	3.3 Information on energy conservation and emission reduction as well as cleaner production; waste recovery and utilization and energy conservation	
13	3.4 Government subsidies, incentives, tax reduction and exemption	
14	3.5 Other environmental control measures and matters	
15	3.6 Disclosure of environmental technology and facilities	
16	3.7 Scientific research and development	
17	3.8 Environmental assessment on the project construction	
18	3.9 Information on whether passing the environmental quality certification	Yes: 2 points; No: 0 point
	Information on environmental responsibility	
19	4.1 Obligation, responsibility or litigation that may bear due to the environmental problem	Failure to disclose: 0 point, Qualitative disclosure: 1 point Quantitative disclosure: 2 points; Qualitative and quantitative disclosure: 3 points
20	4.2 Impact of the products made by or other productive activities conducted by the enterprises on the environment	
21	4.3 Costs and losses borne due to the environment pollution	
22	4.4 Impact of the environmental Issues on the financial status of the enterprise	
23	Centralize the Environmental Information Disclosure into Paragraphs	1 points if the environmental information is centralized into paragraphs for disclosure; otherwise, 0 points.
24	Independent Environmental Report, Social Responsibility Report, Sustainability Report or Environmental Report	The independent environmental report: 2 points; otherwise, 0 point

The score for the disclosure level of the environmental information is divided into 7 sections, recording the score distributions for the 225 samples from the 75 listed companies in the textile and garment industry in recent 3 years, as shown in Figure 1. There were 167 enterprises getting 0-10 points in 3 years from 2012 to 2014, 41 enterprises getting 10-20 points, 14 enterprises getting 20-30 points and 1 enterprise getting 30-40 points, 2 enterprises getting 50-60 points, and no enterprise getting 40-50 points as well as 60-66 points. Under the premise of 66 points in total, 74.22% of the enterprises got the scores of less than 10 points, and it can be seen from Figure 1 that, the vast majority of the enterprises did not attach importance to environmental issues, but fulfilled their environmental responsibilities passively, and these enterprises lacked the initiative for environmental management, and implemented the reactive environmental strategy.

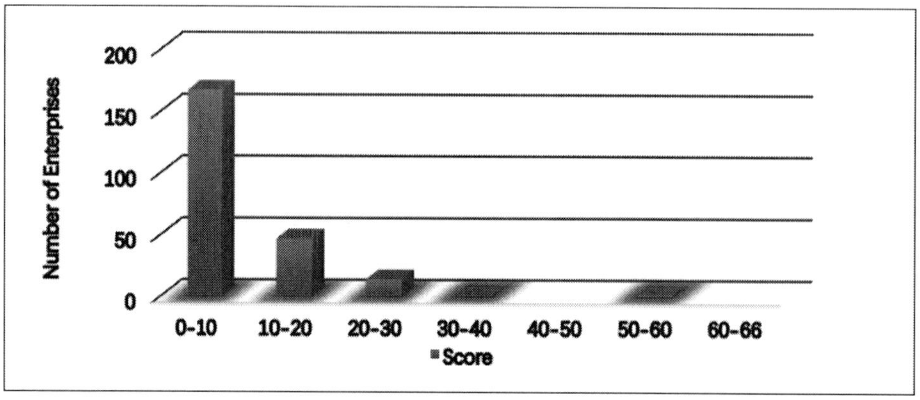

Fig. 1. Enterprise Score Distribution Diagram.

It could be seen from the average value for the environmental information disclosure, their score was no more than 10 points from 2012-2014, among which the score in 2014 was the lowest, and the average score obtained by the enterprises only accounted for about 10% of the total scores, indicating that the majority of the enterprises did not implement the proactive environmental strategies. See Table 2 for more information:

Table 2 Rating Criteria for Environmental Information Disclosure for the Desired Situation in 2012-2014

Year	2012	2013	2014
Average Score for Environmental Information Disclosure	7.87	7.96	6.97
Percentage Accounting for the Desired Situation(%)	11.92	12.06	10.56

The status quo for implementing the environmental strategies by the enterprises can be summed up roughly by analyzing the scores obtained by the textile and garment enterprises for the disclosure of the environmental information as well as the mode of disclosure. Based on the above criteria, the 75 textile and garment enterprises were scored according to the environmental information disclosed in the annual report of 2012-2014 and the independent report. Table 3 makes a summary, indicating the number of the enterprises implementing the corresponding environmental strategies, scores and percentage of the enterprises of the environmental responsibility accounting for the total enterprises. As shown in Table 3:

2.1. Disclosure Related to Environmental Strategy Information

The disclosed environmental strategy information contains the enterprise-related environmental regulations and policies, the operator's environmental philosophy and values, the environmental measures and policies that are planned to take, the objectives and plans for environmental protection and pollution control as well as the global predication about the environmental conditions in the future. The enterprises disclosing the environmental strategy information are less than half. Under the premise that the qualitative disclosure is awarded with 1 point, the quantitative disclosure is awarded with 2 points and the qualitative and quantitative disclosure will be awarded with 3 points as the rating criteria, the number of the enterprises disclosing the information is close to the score, indicating that the disclosing enterprises make qualitative description, and there are few textile and garment enterprises that own environmental philosophy and values and make plans for the environmental issues during the operations.

2.2 Disclosure Related to Information on Environmental Resource Consumption and Discharged Pollutants

There are only about 10 enterprises disclosing the "categories and number of the consumed natural resources as well as the types, quantity and content of the discharged pollutants", and there are only 2 or 3 enterprises disclosing the whereabouts of the pollutants. It is indicated that the enterprises do not attach importance to the consumption of the natural resources, and do not "conserve the energy"; they seldom disclose the whereabouts of the pollutants and the wastes, or do not attach importance to the impact of the pollutants on the environment, and do not "reduce the emissions".

2.3 Disclosure Related to Information on Implementation of Environmental Strategy

It can be seen from the disclosure that there are only 10 enterprises making investment in environmental governance and training the employees on environmental protection in recent 3 years, and only 35 enterprises conduct energy conservation and emission reduction as

Table 3 Number and Scores of the Enterprises for the Disclosure of the Environmental Information from 2012 to 2014

	Environmental Information Disclosure Items	2012			2013			2014		
		No*.	%*	Score *	No.	%	Score	No.	%	Score
	Environmental Strategy Information									
1	1.1	36	46.75	38	37	48.05	41	34	44.16	34
2	1.2	34	44.16	36	38	49.35	40	39	50.65	39
3	1.3	29	37.66	33	32	40.26	35	28	36.36	28
4	1.4	22	28.57	26	23	29.87	25	21	27.27	23
5	1.5	20	25.97	22	18	23.38	20	16	20.78	16
	Information on Environmental Resource Consumption and Discharged Pollutants									
6	2.1	9	11.69	15	12	15.58	16	8	10.39	10
7	2.2	10	12.99	18	8	10.39	14	7	9.09	13
8	2.3	3	3.90	4	2	2.60	3	2	2.60	3
9	2.4	21	27.27	25	22	28.57	27	18	23.38	20
	Information on the Implementation of the Environmental Protection									
10	3.1	11	14.29	19	12	15.58	23	9	11.69	14
11	3.2	17	22.08	24	13	16.88	19	13	16.88	19
12	3.3	33	42.86	55	36	46.75	58	34	44.16	48
13	3.4	32	41.56	65	33	42.86	65	42	54.55	64
14	3.5	11	14.29	13	13	16.88	15	9	11.69	9
15	3.6	24	31.17	34	29	37.66	41	30	38.96	46
16	3.7	18	23.38	30	17	22.8	30	18	23.38	27
17	3.8	8	10.39	12	7	9.09	11	4	5.19	6
18	3.9	28	36.36	56	28	36.36	54	27	35.06	54
	Information on environmental responsibility									
19	4.1	–	–	–	–	–	–	–	–	–
20	4.2	1	1.3	2	1	1.30	2	–	–	–
21	4.3	3	3.9	8	2	2.60	5	1	1.3	2
22	4.4	–	–	–	–	–	–	–	–	–
23	Centralize the Environmental Information Disclosure into Paragraphs	21	27.27	41	19	24.68	37	20	25.79	39
24	Independent Environmental Report, Social Responsibility Report, Sustainability Report or Environmental Report	17	22.08	34	17	22.08	34	14	18.18	28

Note：No* refers to Number of Enterprises
* refers to Percentage of Disclosing Enterprises
Score* refers to Project Score

well as cleaner production. There are only 18 technical research and development enterprises, and 25 enterprises owning environmental protection technology and facilities. The implementation measures for environment protection are the important elements for implementing the proactive environmental strategies, but there are only 30 out of 75 enterprises implementing the measures for environmental protection.

2.4 Disclosure Related to Information on Environmental Legal Liability

In general, the disclosed contents of the environmental liability assumed due to the violation of the environmental laws and regulations shall include the obligations, liabilities or litigations that shall be borne arising from the environmental issues, the impact of the products made by or other productive process conducted by the enterprises on the environment, the cost loss assumed due to the environmental pollution, like the sewage charges and governance fees, the environmental damage compensation paid to the state, and the impact of the environmental issues on the financial status of the enterprise. The environmental liability that shall be assumed due to the violation of environmental laws and regulations is often the consequences for non-implementation of the proactive environmental strategies, and there are only no more than 5 enterprises disclosing the environmental liability. The causes for such few enterprises disclosing the information are that the enterprises won't disclose the enterprise-related negative environmental information in order to maintain its fine social image.

3 Conclusion

Currently, the textile and garment industry in China still takes a fancy to short-term economic benefits, and gains economic growth at the costs of consuming the environmental resources and endangering the ecological environment by analyzing the status quo of the disclosure mode of the environmental information by and the scores obtained by the textile and garment enterprises, in combination of the proactive environmental strategies performed by the enterprises. Such situation is badly in need of changes, and the economic growth mode shall be transformed so as to realize the sustainable development of the economy and the environment.

References

1. Zhang Teng. Proactive Environmental Strategy - The Environmental Strategy that Shall be Taken by the Large-scale State-owned Enterprises[J]Market Modernization, 2012(32):91-92.
2. Fang Xiaojuan. Study on the Disclosure Level of Environmental Information, Environmental Performance and Financial Performance of the Listed Companies from Manufacturing Industry[D]. Anhui University, 2013

Evaluation of Water Resources Carrying Capacity in Shandong Province Based on Fuzzy Comprehensive Evaluation

Qiang Zhao, *Qian* Gao, *Mingyue* Zhu, and *Xiumei* Li*

School of Water Conservancy and Environment, University of Jinan, Jinan 250022, China

Abstract. Water resources carrying capacity is the maximum available water resources supporting by the social and economic development. Based on investigating and statisticing on the current situation of water resources in Shandong Province, this paper selects 13 factors including per capita water resources, water resources utilization, water supply modulus, rainfall, per capita GDP, population density, per capita water consumption, water consumption per million yuan, The water consumption of industrial output value, the agricultural output value of farmland, the irrigation rate of cultivated land, the water consumption rate of ecological environment and the forest coverage rate were used as the evaluation factors. Then,the fuzzy comprehensive evaluation model was used to analyze the water resources carrying capacity Force status evaluation. The results showed : The comprehensive evaluation results of water resources in Shandong Province were lower than 0.6 in 2001-2009 and higher than 0.6 in 2010-2015, which indicating that the water resources carrying capacity of Shandong Province has been improved.; In addition, most of the years a value of less than 0.6, individual years below 0.4, the interannual changes are relatively large, from that we can see the level of water resources is generally weak, the greater the interannual changes in Shandong Province.

1 Introduction

Water resources carrying capacity refers to the specific areas within the premise of a certain level of living production and the ecological function of water environment subsystem, through scientific and reasonable distribution measure, the maximum capacity to support the needs of the population, social, economic and ecological environment that the water supply can provide[1-5]. Shandong province is one of the northern water shortage regions. Based on Shandong province economic-socio-environment basic research date from 2001 to 2015, the application of fuzzy comprehensive evaluation model to evaluate the water resources carrying capacit, which aimed to provide a basis for the rational development and utilization of water resources in the leap-forward development of Shandong province.

2 Overview of Water Resources in the Study Area

Shandong province is located in the eastern seaboard of China, the northernmost end of east China area, there are three drainage areas of Yellow River, huai river and hai river. Precipitation is the main water resources in Shandong, and is greatly affected by rainfall. The Yellow River is the main external water source in Shandong, and the nansi lake is the largest lake in Shandong province. The average annual precipitation of the whole province is 679.5mm, and the yearly average total water resources is 30582 million m^3; The per capita water resource is $344m^3$, less than 1/6 of the national average, and 1/25 of the world average, the third lowest per capita water resources in all provinces (cities and autonomous regions). According to the international water shortage standard, per capita water resource occupancy less than $500m^3$ would be considered as extreme water shortage area. While Shandong is only 344 m^3 per capita, which shows that Shandong is in extreme water shortage.

3 The Construction of Fuzzy Comprehensive Evaluation Model

The fuzzy comprehensive evaluation method can provide an effective way for the research of regional water resources carrying capacity, the essence of this method is to affect the carrying capacity of water resources, on the basis of single factor evaluation, through comprehensive evaluation matrix expansion multi-factor comprehensive evaluation of water resources carrying capacity, thus it is concluded that the size of the water resources carrying capacity. In this paper, the fuzzy comprehensive evaluation model was applied to evaluate the overall water carrying capacity of Shandong province.

3.1. The Selection of Evaluation Factors, Classification and Grading

* Corresponding author: Xiumei Li@ stu_lixm@ujn.edu.cn

According to the characteristics of regional water resources, referring to the analysis of supply and demand of water resources in the index system, on this basis, fully considered the differences between the stock of natural water resources assigned and the different way of development, and used the water resources supply and demand analysis index system of the national and Shandong province as the basis, the evaluation index was divided into three levels (table 1). In this paper, the carrying capacity of water resources was divided into 3 levels, and the V_1 level is excellent, which indicates that the water resources in this region still have a large carrying capacity, and the water resources supply is adequate. The V_3 level is poor, which indicates that water resources carrying capacity is to allow the lowest, it is difficult to develop in depth, has restricted the social economy and stable development; the V_2 level status is somewhere in between, which indicates that this region basically can adapt to the economic development of the basin water resources, water resources development and utilization to the larger size, but still has certain potential of development and utilization.

Table.1 Rating Criteria for Each Evaluation Index

Index layer	Rule layer	Index layer	V_1	V_2	V_3
Water resources carrying capacity O	Water resources condition A_1	Per capita water resources u_1 (m³/per)	>600	300~600	<300
		utilization rate of water resources u_2 (%)	<30	30~80	>80
		Water supply modulus u_3 (ten thousand m³/km²)	>39	15~39	<15
		precipitation u_4 (mm)	>800	500~800	<500
	Socio-economic condition A_2	Per capita GDP u_5 (ten thousand yuan)	>4	2~4	<2
		Per capita density u_6 (a/km²)	<550	550~700	>700
		Water consumption per capita. u_7 (m³)	<30	30~100	>100
		Water consumption of ten thousand yuan. u_8 (m³/ten thousand yuan)	<70	70~200	>200
		Water consumption of industrial output of ten thousand yuan. u_9 (m³/ten thousand yuan)	<180	80~500	>500
		Water consumption of agricultural output of ten thousand yuan. u_{10} (m³/ten thousand yuan)	<400	400~1000	>1000
	Ecological environmental conditions A_3	Arable irrigation rate u_{11} (%)	>80	40~80	<40
		Water rate of ecological environment. u_{12} (%)	>5	1~5	<1
		Forest coverage rate u_{13} (%)	>35	10~35	<10

To better reflect the level of water resources carrying capacity, to take 1 point of numerical evaluation level: a_1 =0.05, a_2 =0.5, a_3 =0.95, after quantification can quantitatively reflect the level degree of the influence of the carrying capacity factors, the higher the value, the greater the water resources development capacity. According to the value of a_j and the value of membership b_j in B matrix, the comprehensive evaluation score of water resources carrying capacity in the comprehensive evaluation matrix B was calculated according to formula (1).

$$a = \frac{\sum_{j=1}^{3} b_j a_j}{\sum_{j=1}^{3} b_j} \quad (1)$$

In the formula, a is the comprehensive evaluation value of water resources carrying capacity based on the comprehensive evaluation result matrix B.

3.2 The Calculation of the Evaluation Matrix

According to the actual value of each evaluation factor, the value of the subordinate function γ_i of the judging matrix was evaluated by comparing the classification index. Fuzzy processing was used to construct membership function to smooth transition between all levels. For the evaluation factors γ_1, γ_2 and γ_3 ...... the evaluation grade membership function was calculated as follows:

$$\gamma_1=\begin{cases}0.5\left(1+\frac{k_1-u_1}{k_2-u_1}\right) & u_1>k_1\\ 0.5\left(1-\frac{u_1-k_2}{k_2-k_1}\right) & k_2<u_1\le k_1\\ 0 & u_1\le k_2\end{cases}\quad(2)$$

$$\gamma_2=\begin{cases}0.5\left(1-\frac{k_1-u_1}{k_2-u_1}\right) & u_1>k_1\\ 0.5\left(1+\frac{u_1-k_2}{k_2-k_1}\right) & k_2<u_1\le k_1\\ 0.5\left(1+\frac{k_3-u_1}{k_2-u_1}\right) & k_3<u_1\le k_2\\ 0.5\left(1-\frac{k_3-u_1}{k_3-k_2}\right) & u_1\le k_3\end{cases}\quad(3)$$

$$\gamma_3=\begin{cases}0.5\left(1+\frac{k_3-u_1}{k_2-u_1}\right) & u_1\le k_3\\ 0.5\left(1-\frac{u_1-k_3}{k_2-k_3}\right) & k_3<u_1\le k_2\\ 0 & u_1>k_2\end{cases}\quad(4)$$

Where: k_1 is the critical value of the evaluation level V_1 and V_2 ; k_3 is the critical value of rating V_2 and V_3 ; k_2 is the midpoint of the evaluation level V_2 . The critical value of each evaluation level corresponding to each indicator was shown in table 1.

3.3 Calculation of Index Weight

First of all, according to the index level of the partition, referring to China's water resource evaluation standard, the "1-9" scale method commonly used in AHP [6-8], was adopted to construct the judgment data matrix. Secondly, according to the judgement matrix, used the mathematical matrix to calculate the importance of the indicator layer (U) relative to the principle layer (A), the importance of the criterion layer relative to the target layer (O), and then end up with a judgment matrix. Note that the water resources condition is the foundation of water resources carrying capacity, so water condition here than the social and economic condition and ecological environment condition more important, got 3:1:1. Finally, the weight of the indicator layer to the principle layer multiplied by the weight of the principle layer to the target layer, so the weight of the indicator layer to the target layer was obtained (W), and the weight calculation results can be gotten.

3.4 Calculation of Comprehensive Evaluation of Water Resources Comprehensive Capacity

According to the weight vector W and the judgment matrix R calculated above, the calculation formula for comprehensive evaluation result of the water carrying capacity is as follows:

$$B=W\times R=\begin{bmatrix}W_1 & W_2 & \bullet & W_{13}\end{bmatrix}\times\begin{bmatrix}\gamma_{11} & \gamma_{12} & \gamma_{13}\\ \gamma_{21} & \gamma_{22} & \gamma_{23}\\ \bullet & \bullet & \bullet\\ \gamma_{13,1} & \gamma_{13,2} & \gamma_{13,3}\end{bmatrix}=\begin{bmatrix}b_1 & b_2 & b_3\end{bmatrix}\quad(5)$$

4 Integrate Assessment of Water Resource Carry Capacity in Shandong Province

According to the evaluation index of carrying capacity of table 1, the evaluation index of water resources carrying capacity in Shandong province was quantified. Through formula (2) (3) (4), we can find the values of each index belonging to V_1 , V_2 and V_3 : γi1, γi and γi3. The calculated results of each year were in the form of a matrix, which referred to the total number of fifteen matrices, namely [γi1、γi2 和 γi3], which was R. The results in 2002, 2005 and 2015 were shown in table 2.

Table.2 Each Indicator Corresponds to Each Level of Membership of Shandong Province in 2001、2005、2010、2015

Iindic ators	2001			2005			2010			2015		
	γ_{i1}	γ_{i2}	γ_{i3}	γ_{i1}	γ_{i2}	γ_{i3}	γ_{i1}	γ_{i2}	γ_{i3}	γ_{i1}	γ_{i2}	γ_{i3}
u_1	0	0.404	0.597	0	0.999	0.001	0	0.576	0.424	0	0.269	0.731

u_2	0	0.738	0.262	0.962	0.038	0	0.818	0.182	0	0.530	0.470	0
u_3	0	0.546	0.454	0.603	0.397	0	0	0.468	0.532	0	0.446	0.554
u_4	0	0.836	0.164	0.533	0.467	0	0.654	0.346	0	0	0.752	0.248
u_5	0	0.256	0.744	0.000	0.499	0.502	0.550	0.451	0	0.854	0.146	0
u_6	0.828	0.172	0	0.740	0.260	0	0.599	0.401	0	0.507	0.493	0
u_7	0.988	0.012	0	0.582	0.418	0	0.550	0.450	0	0.551	0.449	0
u_8	0	0.245	0.755	0.660	0.341	0	0.586	0.414	0	0.679	0.321	0
u_9	0	0.275	0.725	0.862	0.138	0	0.680	0.320	0	0.735	0.265	0
u_{10}	0	0.136	0.864	0	0.444	0.556	0.656	0.344	0	0.947	0.053	0
u_{11}	0.655	0.345	0	0.676	0.324	0	0.562	0.438	0	0.692	0.308	0
u_{12}	0	0.397	0.603	0	0.530	0.470	0	0.773	0.228	0.560	0.440	0
u_{13}	0	0.852	0.148	0.560	0.440	0	0.600	0.400	0	0	0.732	0.268

Table.3 Comprehensive Evaluation Results of Water Resources Carrying Capacity

B value	b_1	b_2	b_3	a
2001year	0.0266	0.5249	0.4485	0.3101
2002year	0.0652	0.3142	0.6205	0.2501
2003year	0.4031	0.4048	0.192	0.595
2004year	0.3395	0.4341	0.2263	0.5509
2005year	0.3752	0.5032	0.1216	0.6141
2006year	0.2114	0.461	0.3275	0.4477
2007year	0.3446	0.5116	0.1438	0.5904
2008year	0.3477	0.4784	0.1739	0.5782
2009year	0.3184	0.4704	0.2112	0.5482
2010year	0.3833	0.4744	0.1423	0.6085
2011year	0.4656	0.4132	0.1211	0.6550
2012year	0.453	0.3602	0.1867	0.6198
2013year	0.3848	0.4348	0.1803	0.5920
2014year	0.1485	0.442	0.4094	0.3825
2015year	0.3668	0.3921	0.241	0.5566

By the data in table 2 and formula (1) to (5) for carrying capacity of water resources in Shandong province comprehensive evaluation results were shown in table 3. For example, in 2014, the comprehensive evaluation result set B=[0.1485, 0.442, 0.4094], and the comprehensive ability a=0.3825, which shown that the capacity of water resources in Shandong province was not optimistic in 2014. It was not difficult to see the annual rainfall data of Shandong province in 2014, and the water resources capacity of Shandong province was very weak. But the value of B control was in good condition of $b_2 > b_3$, which indicated that a series of protective measures in recent years had effectively avoided the problem of insufficient carrying capacity of water resources brought by the climate problem, and improved the threat defense capability posed by natural disasters.

The change trend chart of water resources grading evaluation factor B and comprehensive evaluation factor a were shown in figure 1,2. Overall, the water resources carrying capacity gradually ascending steadily at the same time, through the growth of the 2001 to 2009, the a value gradually began to more than 0.6, the main above 0.6 in 2010-2015, which showed that Shandong province water resources carrying capacity has improved in general; The ranking coefficient b_2 remains the highest, indicated that the overall performance of water resources carrying capacity was in good condition, and the coefficient b_1 was gradually increasing at the same time, and the proportion b_3 was decreasing, it indicated that the water resource carrying capacity of Shandong province has a tendency to improve the membership of V_1, which was inseparable from the rational development of these years and scientific management. Secondly, the capacity of water resources was closely related to the amount of groundwater resources. In 2001, 2002, 2006 and 2014, the a value was significantly smaller than other years, again through the Shandong province water resources statistics showed that the groundwater resources in the past few years was lower than other years, and the carrying capacity of water resources in Shandong province were associated with groundwater reserves, namely once encountered drought years (dry year), groundwater can play the main water supply capacity.

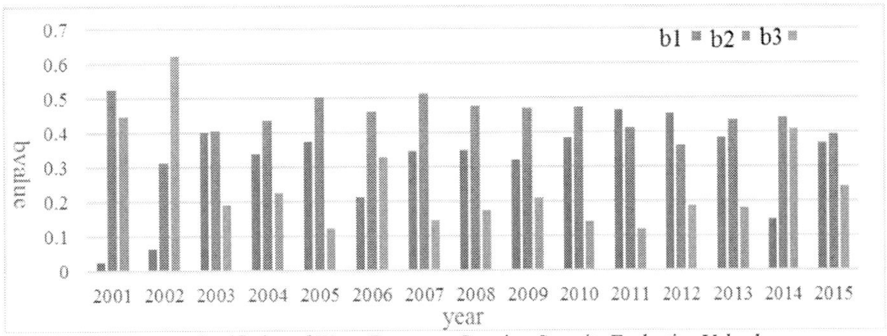

Fig. 1. Classifiction of Water Resources Carrying Capacity Evaluation Value b

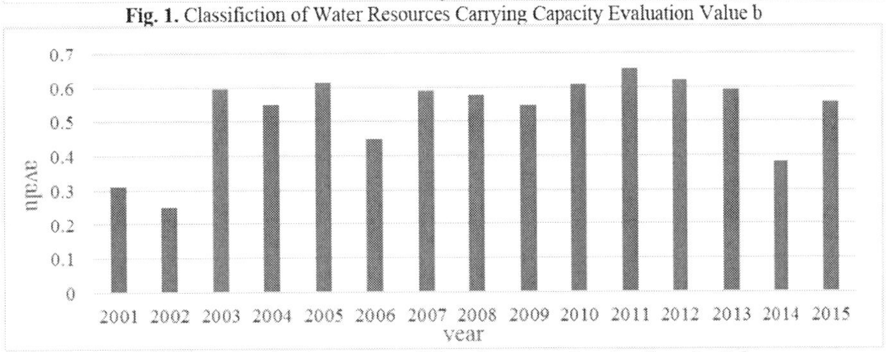

Fig. 2. Comprehensive Assessment of Water Resources Carrying Capacity Value a

5 Conclusion

This paper started from the characteristics of Shandong province economy-society-environmental composite system, comprehensive analyzed the main influencing factors of regional water resources carrying capacity as evaluation factors(to table 1), and the current status of water resources carrying capacity of Shandong province was evaluated by fuzzy comprehensive evaluation model. Results shown that the carrying capacity of water resources in Shandong province level gradually guaranteed, annual change gradually narrowed, but the contradiction between supply and demand of water resources is outstanding, the water shortage area is widely distributed, and in the later development should adhere to the open source throttling, pay equal attention to practical ability of to increase the supply capacity of water resources.

Acknowledgements

This study was financially supported by the Natural Science Foundation of China (NO.41471160)

References

1. Soft science research of water resources in Xinjiang province, Water Resources and Hydropower Engineering, 6:2-9 (1989) (in Chinese)
2. Y Feng, W Han, H Wang, J Lian, J Huang, Advances in Water Science, **14**, 1:109-113 (2003) (in Chinese)
3. Q Duan, C Liu, X Chen, W Liu, H Zheng, Acta Geographica Sinica, **65**, 1:82-90 (2010) (in Chinese)
4. Y Sun, Y Gao, Y Shi, F Xu, Water resources protection, **27**, 1:20-23.(2011) (in Chinese)
5. A Ding, D Chen, C Pan, X Xu, South-to north water diversion and water conservancy technology, **8**, 3:71-73 (2010) (in Chinese)
6. Water conservancy bureau of hydraulic power department. *Evaluation of water resources in China* (hydraulic power press, 1987) (in Chinese)
7. J Song, Y Gu, Natural Science Journal of Harbin Normal University, **31**, 1:60-62 (2015) (in Chinese)
8. J Dai, Y Pan, Q Liu, H Tang, Meteorological science, **34**, 4:428-434 (2014) (in Chinese)

Research of Variable Application Control System Based on Fuzzy Control

JinXia Huang[1], *Yu* Zhang[2], *XiaoGuang* Su[1], *SiQing* Tian[1,*]

[1]Institute of Information Electronic Technology, Jiamusi University, Heilongjiang Province154007, China.
[2]School of science, jiamusi university.

Abstract. The AT89C52 microcontroller and the fuzzy control algorithm is used to design the automatic control system of variable Spraying pesticides. The system adjusts automatically the electric regulating valve according to given Spraying pesticides per unit area so that the actual amount of Spraying pesticides equal to the set value. The simulation model of automatic control system of variable pesticide application was established, and the controller was simulated by MATLAB / SIMULINK software fuzzy control toolbox. The simulation results show that the structure and algorithm of the proposed automatic control system of pesticide application is feasible, The system improves greatly pesticide utilization, reduce pesticide residues and environmental pollution.

1 Introduction

The planting area of crops continues to increase with the development of agricultural machinery automation and intelligence, and people attach great importance to the problems of environmental pollution and pesticide residue. Conventional application of pesticides has could satisfy the needs of agricultural production. The droplet diameter can not be adjusted with operating environment changes. Variable spraying system implement variable fertilization depending on the number of pests and diseases. In this way, It not only improves the pesticide utilization, but also reduces the waste of pesticides and Adverse pesticides on the environment adverse effects. The system fully reflects its Significant economic, social and ecological benefits. Based on the principle of variable design a set of variable spraying automatic control system, and the controller was simulated by MATLAB software fuzzy control toolbox. The simulation results show that the system is stable and reliable, maneuverability is good, and can meet the requirements of agricultural production [1-2].

2 system description and function implementation

In order to realize automatic control, the design based on real-time sensing technology, and use the method of inputting the amount of spraying, that set a unit area spray volume.The current value of the flow rate is measured by the flow sensor, the deviation of the current flow value and the rate of change of the deviation are taken as the input of the control system, The input quantity is fed into the AT89C52 input after isolation amplification and analog-to-digital converter processing. It outputs the control quantity U (voltage) to control the actuator regulating valve after fuzzification, fuzzy reasoning and anti-fuzzification, so that it will chang to spray amount and the actual amount of spray equal to the set value.

3 System hardware design

The control system the main control unit uses 8-bit microcontroller AT89C52, Peripheral circuits include, sensor, isolation amplifier, MAX232 serial interface circuit, A / D, D / A conversion circuit, 4 × 4 keyboard and display circuit, alarm circuit etc [3], hardware structure shown in figure 1.

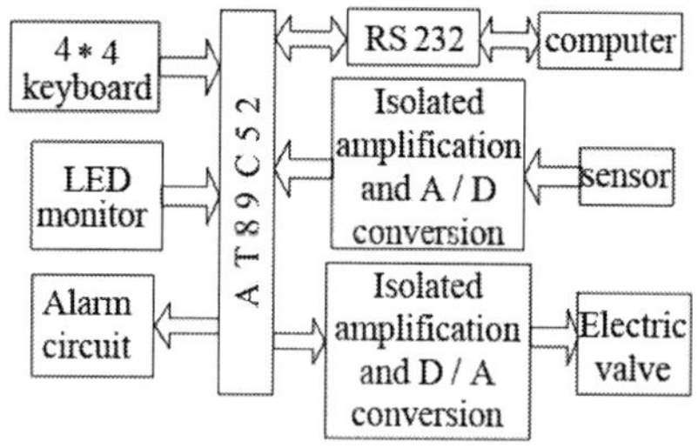

Fig. 1. variable application control system hardware structure

3.1 SCM choice

AT89C52 microcontroller is a low voltage, high performance CMOS 8-bit microcontroller by American company production. The chip contains 8K bytes of rewritable flash ROM and 256 bytes of random data memory, fully compatible with the MCS-51 product instruction, the device is manufactured using ATMEL's high-density, non-volatile memory technology [4].

* Corresponding author: SiQing Tian, hjxlcj2006@sina.com

3.2 Electric valve selection

Electric valve is used to control actuator of the system flow, by controlling the size of the opening angle of the electric valve to control the size of the system flow,The system uses the electric valve of DC0-10V standard voltage signal, The valve has a high control accuracy and fast response.The opening of the valve(0-100%) is proportional to the control signal. When the opening is 0, corresponding to 0V or 4mA ,when the opening is 100%, corresponding to 10V or 20mA. Travel time 60s, Max travel 25mm, Power voltage 24VAC [5].

3.3 Sensor selection

Taking into account the actual operating environment of the machine, the system were selected piezoresistive pressure sensors, turbine flow meters and Hall speed sensor[6]. The three types of sensors have wide speed range, strong anti-interference ability and wide temperature adaptation range, which can meet the requirements of agricultural working environment.

4 Structure Design and Simulation of Fuzzy Controller

4.1 Structure Design of Fuzzy Controller

The function of the fuzzy controller is to imitate the artificial control of thinking way of the human brain, to identify and decide the fuzzy phenomenon, to give precise control quantity and to control the controlled object. The design Use two-dimensional fuzzy controller according to the actual situation [7-8]. The working principle diagram was shown in Figure 2.

In the variable spraying fuzzy controller, The Spray pesticides amount per unit area is set to q_0, measured by the flow sensor recorded as $q(t)$. Then the deviation e and deviation changes ec as follows.

$$e(t) = q_0 - q(t) \quad (1)$$

$$ec(t) = e(t) - e(t-1) \quad (2)$$

According to the field practical experience. Determining the basic domain of Spraying pesticide deviation and deviation rate of change are [-5,5] L/hm^2, The output is the voltage applied to the motorized valve. The design set universe of deviation and deviation change amount for{-5, -4, -3, -2, -l, 0, l, 2,3,4,5} 11 levels. Fuzzy subset is divided into { negative large, negative middle, negative small, zero, positive small, positive middle, positive large } seven levels, The system uses trigonometric functions, half-trapezoidal function and half-trapezoidal function as the membership function of the input language value and uses the pulse function as the membership function of the output language value[9].

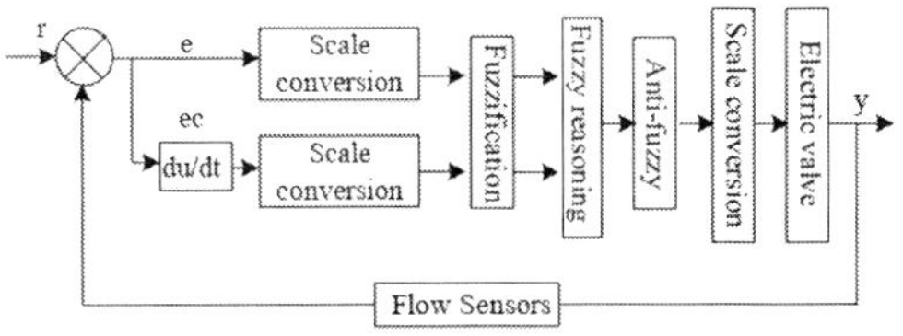

Fig. 2. Variable spraying pesticides fuzzy controller schematic

The fuzzy control rules are determined according to several fuzzy conditional statements drawn by experienced operators in practice many times.These fuzzy statement can be expressed by the fuzzy control rules table, As shown in Table 1.

Table 1. fuzzy control rules table

E \ EC	NB	NM	NS	ZO	PS	PM	PB
NB	PB	PB	PB	PB	PM	PS	ZO
NM	PB	PB	PB	PM	PS	ZO	ZO
NS	PB	PM	PS	PS	ZO	NS	NS
ZO	PM	PS	PS	ZO	NS	NM	NM
PS	PS	PS	ZO	NS	NS	NM	NB
PM	ZO	ZO	NS	NM	NB	NB	NB
PB	ZO	NS	NM	NB	NB	NB	NB

When the Spraying pesticide amount deviation E is larger and the deviation change EC is positive, Control amount should eliminate the deviation, U take Zhengda. When the deviation E is positive and the error change EC is negative. System should eliminate the deviation and the system does not produce overshoot, control amount U should take smaller [8-9].

4.2 Fuzzy Control Computer Simulation

In order to verify the effect of the fuzzy controller, this paper uses Matlab / Simulink software to simulate the variable pesticide control system based on fuzzy controller. Simulation module shown in Figure 3. Among them, Signal is the input signal Generator, Scope is the output oscilloscope, Adjust the simulation control parameters, run the Start control button, double-click the oscilloscope module, the system can be dynamic simulation[10]. Simulation results was shown in Figure 4.

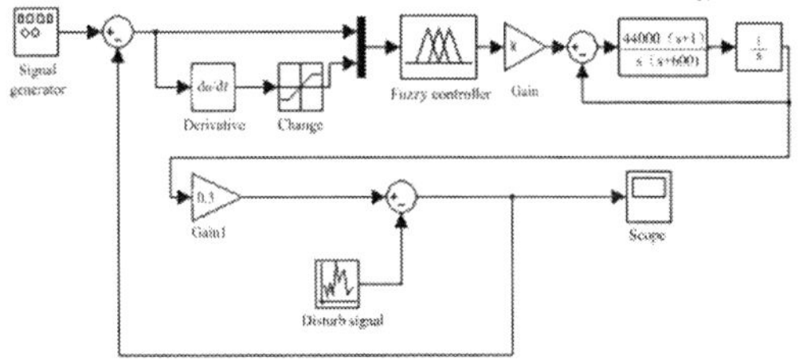

Fig. 3. SIMULINK Dynamic Simulation Model of fuzzy Control System

It can be seen from the figure that the system does not fluctuate much at the initial time, the operation is

https://doi.org/10.1051/e3sconf/20183801013

stable, the overshoot is small, the adjustment time is short, and the steady-state error is zero during the stationary period. The control system has achieved good results through field practical application to meet the control requirements of variable application. It improves the control accuracy and stability of the system.

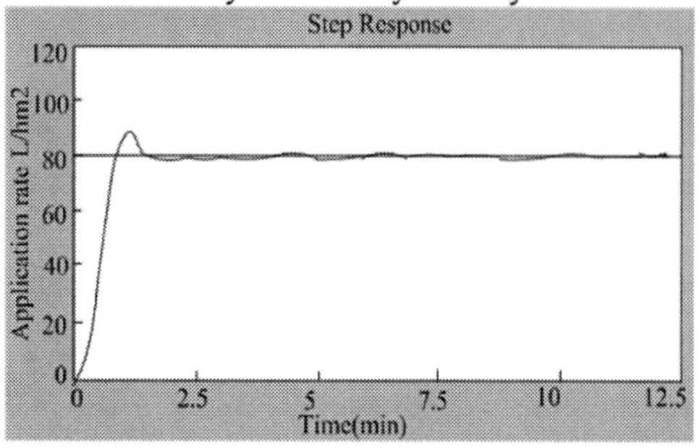

Fig. 4. Fuzzy control simulation curve

To contrast the superiority of the control algorithm design, Figure 5 is a simulation diagram using the traditional PID control. As can be seen from the figure, conventional PID control will have a larger overshoot, larger steady-state error and longer response time.

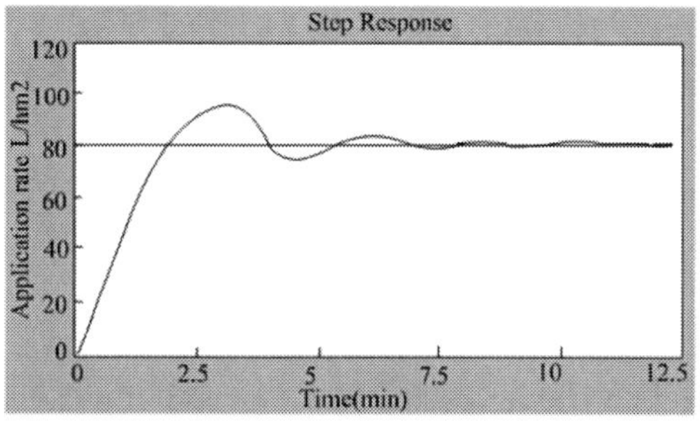

Fig. 5. PID control simulation curve

4 Conclusion

The microcomputer technology and fuzzy control are applied variable spray control system of vehicle sprayer. The simulation results of MATLAB / SIMULINK software show that the fuzzy controller has better control performance than the conventional PID controller, which can realize fast response, small overshoot, strong adaptability and antijamming ability to parameter variation, Fundamentally greatly improve the system's dynamic quality and steady-state accuracy, improve the overall performance of the system. The control system after the farm actually reached a very good control effect, easy to popularize, and has potential market application prospect.

Acknowledgements

Heilongjiang Provincial Department of Education Natural Science Funding Project (2016-KYYWF-0557); Jiamusi University application of key projects(12Z2201526).

References

1. Yang Li.Automatic variable spraying system design and algorithm research based on ARM7[D].Jilin University Master's degree thesis, 2008,06.
2. Kevin P.Gillis,D.Ken Giles,David C.Slaugh ter,et al.Injection and fluid handling syste for machine-visioncontrolledspraying[J].Sancramento,California, USA,ASAE Annual International Meeting , 2016:67-88.
3. Shi Xinmin, Hao Zhengqing. Fuzzy control and MATLAB simulation [M]. Beijing: Tsinghua University Press.2008,03.
4. Huang Jinxia. Fuzzy control in the Chinese herbal medicine harvester holding belt speed control [J]. Agricultural Mechanization Research, 2010.08
5. Wang Junhong. Variable Spray Controller Design Based on AT89C52 Microcontroller[J]. Chinese Journal of Microcomputer Information Volume 22, Issue 3, 2006.11.
6. Li LiJun. Design of Variable rate sprayer controller based on ARM[D], Jilin University,2016.03
7. Kevin P. Gillis, D. Ken Giles, David C. Slaugh ter,et al. Injection and fluid handling system for machine-visioncontrolledspraying[R]. Sancramento, California, USA: ASAE Annual Inte- rnational Meeting,2001.
8. Chen Yong, Zheng Jiaqiang.Study on Variable Spraying Control System with Accurate Application [J]. Journal of Agricultural Engineering, 2015, (5): 69-72.
9. Control of Variable Application Based on Fuzzy Control Journal of Agricultural Mechanics[J]. Agricultural Mechanization Research.2014, (6).
10. Zhou Hongming.Design of Agricultural Multiple Sprayer Design Parameters[J]. Agricultural Mechanizati-on Research.2014, (3):53-64.

Temporal and spatial evolution characteristics of water environment quality in Heze

Qian Gao, *Qiang* Zhao*, and *Xiumei* Li

School of Water Conservancy and Environment, University of Jinan, Jinan 250022, China

Abstract. The evolution of water environment is relatively complicated. The study of its characteristics is helpful to grasp the general direction of spatial and temporal evolution of water environment in Heze city, and to carry out water resources development and water environment protection more rationally. The comprehensive pollution index method for calculation, and the observed data are handled by Excel. In order to facilitate the analysis of the basin, Arcgis is utilized to map the watershed map. In addition, for the spatial evolution, surfer12 is used to analyze the spatial evolution characteristics the spatial evolution, and to draw the pictures of spatial evolution of chemical oxygen demand and water quality evolution. The study shows that: (1) In Heze, the quality of water environment has been improved year by year from 2006 to 2013. In 2014, the water environment has deteriorated. The content of volatile phenol has increased greatly, and the evolution trend of COD is close to the trend of the comprehensive pollution index. (2) In terms of Spatial state of water environment, the water quality of Zhuzhao New River and Wanfu River is poor, and Dongyu River water quality is better. Zhuzhao New River and Wanfu River water qualityis often worse than grade V or V, and Dongyu River water quality is mostly maintained in the grade Ⅳ. Through the analysis on the spatial revolution characteristics of water quality and chemical oxygen demand(COD),as a result, water quality is poor in the northern region,and the water quality in the southern region is better in Heze. Although the water quality has changed in recent years, the overall pattern is relatively stable.

1 Introduction

The quality of water environment refers to the appropriate degree of the water environment on the survival and reproduction of the population and socio-economic development, and usually refers to the degree of pollution in the water environment [1,2]. With the rapid development of the global economy and the growth of the population, the shortage of water resources has become a key problem that restricts the sustainable development of social and economic development in many countries and regions [3]. Over the past few decades, the change in globle climate and human activities have produced significant changes on the circulation of water resources. Therefore, it is important to research systemically the ecological and environmental protection of Heze city basin, and understand the sustainable development and utilization of water environment and the formation and evolution of water resources. It has important guidance on learning about the temporal and spatial characteristics of water cycle elements under the conditions of crowd activities and meteorological changes effect [4-6].

Through the study on the evolution of the water environmental quality in time and space, the root causes of the problem of water environment in Heze is explored. It can actively promot the construction of ecological civilization, and improve the water quality of the environment. Moreover, it is contirbute to find a suitable for the local economic and social development [7-10].

2 Materials and methods

2.1 Study area

Heze is situated between 34°33′33″-35°44′00″N and 114°47′30″-16°24′15″E along the southwest of Shandong Province, with the total area of 12239 km2[11]. It borders Henan Province to the south and the Jining City to the east. Heze is located in the West Lake system in the Yellow River and the Yellow River Old Course Triangle area due to the formation of the Yellow River burst alluvial plains. Most of the water near the east-west parallel migration and exit, remitted to Nansi Lake. In addition to the Yellow River, the inland rivers are mainly Dongyu River, Zhuzhao New River and Wanfu River (Figure 1).

2.2 Research methods

Chemical oxygen demand (COD), permanganate index, biochemical oxygen demand (BOD5) and ammonia nitrogen (NH3-N) were selected among the evaluation indexes. The research method adopts the integrated pollution index method, and the method used in this

* Corresponding author: Qiang Zhao@zhaoqiang8827@sina.com

evaluation is the relative quality of the parameters overlay type index method [12-14].

2.3 Date sources

From 2006 to 2014, Heze Municipal Environmental Monitoring Center Station routinely monitored 3 major rivers such as Zhao River, Dongyu River and Wanfu River. According to the condition of section monitoring, three sections are selected for each river respectively, and a total of nine monitoring sections are monitored (Figure 1).

Fig. 1. Geographical distribution of monitoring points

3 Results and discussion

3.1 Interannual Variability of Water Environmental Quality

3.1.1 Chemical oxygen demand (COD)

The chemical oxygen demand of Zhuzhao New River is decreasing from 2006 to 2014, and the maximum value is 45.6mg / L in 2007. From 2008 to 2012, the annuai content of COD decreased obviously. The chemical oxygen demand of Dongyu River reached the maximum in 2008 and changed steadily from 2009 to 2014. The chemical oxygen demand of Wanfu River increased from 2014 to 2006. By 2012, the chemical oxygen demand changes more smoothly and all the values are below 30 mg / L (Figure 2).

3.1.2 Biochemical oxygen demand (BOD_5)

As shown in Figure 2, the total oxygen demand of Zhuzhao New River decreased from 2006 to 2014 year by year, and rebounded in 2011 and 2014, but not higher than 11.52mg / L in 2006. The content of Biochemical Oxygen Demand in Dongyu River fluctuated greatly from 2006 to 2010, with a relatively stable change from 2010 to 2014. The change of Biochemical Oxygen Demand of Wanfu River was relatively stable from 2006 to 2012. A significant decrease in 2013 and a sharp increase in 2014 with a content of 8.9 mg / L.

3.1.3 Ammonia nitrogen (NH_3-N)

Analysis of Figure 3 shows: The content of ammonia nitrogen in Zhuzhao New River decreased obviously, and the change was more stable since 2009. The content of ammonia nitrogen in Dongyu River reaches its maximum in 2007, which is 1.41mg/L, and the content of ammonia nitrogen is relatively stable since 2008. The ammonia nitrogen content of Wanfu River showed a downward trend as a whole, reaching the lowest value of 0.78mg/L in 2009

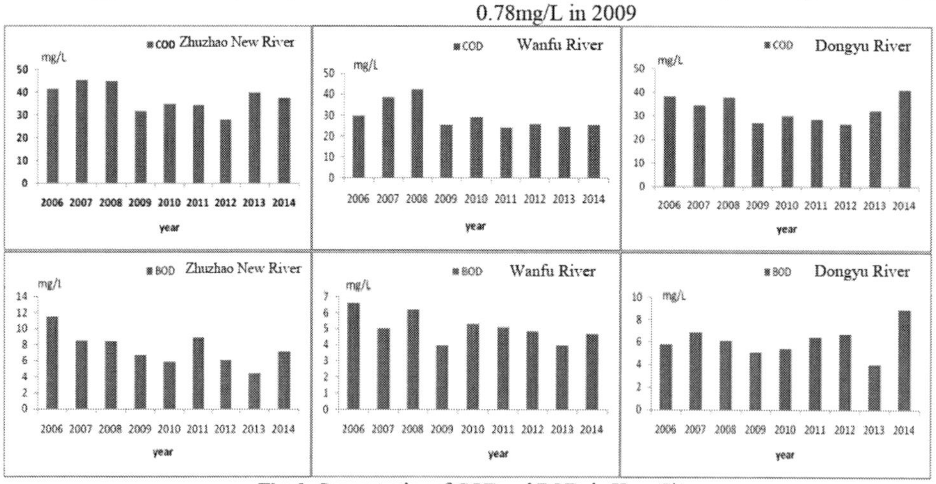

Fig. 2. Concentration of COD and BOD_5 in Heze City

 https://doi.org/10.1051/e3sconf/20183801014

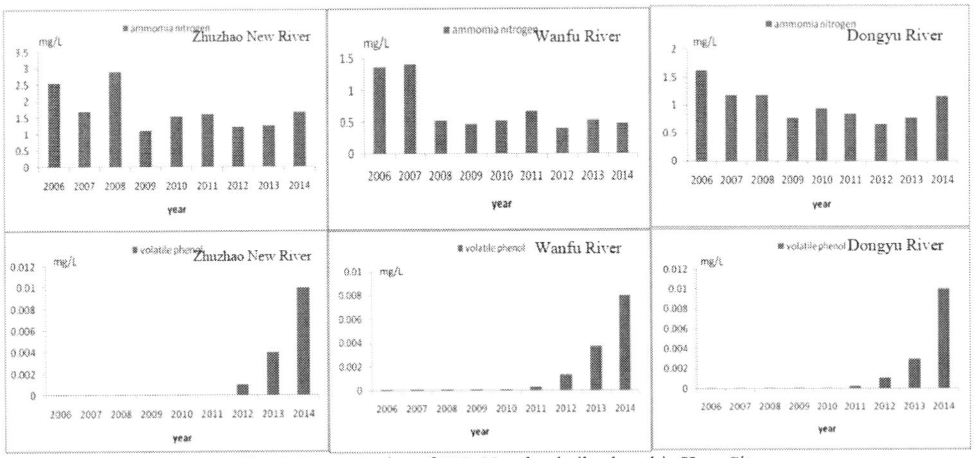

Fig. 3. Concentration of NH_3-N and volatile phenol in Heze City

3.1.4 Volatile phenol

The content of volatile phenol in Zhuzhao New River maintained at 0.001mg/L from 2006 to 2012, and an significant growth in 2012-2014; The volatile phenol content of Dongyu River has been changed steadily from 2006 to 2011, and it has been increased obviously since 2012. The volatile phenol content of Wanfu River was stable from 2006 to 2012. An obvious growth from 2012 to 2014, reaching the maximum value of grade IV water quality (Figure 3).

3.2 Interannual Variability of Water Environmental Quality

3.2.1 Chemical oxygen demand (COD)

As shown in Figure 4, the chemical oxygen demand in space can be roughly divided into three stages. From 2006 to 2011, the chemical oxygen demand of Zhuzhao New River and Wanfu River are larger, and they are the major pollution centers. In 2012, the water environment pollution center shifted downward, with Wanfu River and Dongyu River as the main pollution centers. By 2014, the chemical oxygen demand of three rivers in Heze has risen again. In this year, the comprehensive index of water environment pollution has also risen sharply. The pollution center has also been transferred to the upper reaches of Zhao Xin River and Wanfu River.

3.2.2 Biochemical oxygen demand (BOD_5)

In 2006-2008, Zhuzhao New River had a higher biochemical oxygen demand and was the main pollution center. In 2010, the biochemical oxygen demand of the three rivers gradually approached and the pollution center began to change. From 2011 to 2014, the contents of the Zhuzhao New River and Wanfu River catchments have been increasing year by year.

3.2.3 Ammonia nitrogen (NH_3-N)

Ammonia nitrogen is mainly concentrated in the Zhuzhao New River Basin, and Wanfu River in the Dongyu River Branch of the river due to the river branch of the pollutants. So its ammonia nitrogen content is relatively high. The Dongyu River fish ammonia content decreased, the overall pollution of the river also gradually weakened.

3.2.4 Volatile phenol

From 2006 to 2010, the volatile phenol contents of three rivers in Heze City were lower than 0.0001mg/L. Beginning in 2011, the volatile phenol content in rivers increased mainly in the Zhuzhao New River basin. In 2012 and 2013, the content of three rivers increased; in 2014, the content of three rivers increased greatly.

3.2.5 Water quality status

The water quality of three rivers in Heze improved from 2006 to 2014. The water quality of Dongyu River is in a good condition. Most of the water quality in the past nine years has remained at the level IV. Zhuzhao New River and Wanfu River water quality is poor, and most of the river pollution centers in Zhuzhao New River and Wanfu River nearby. The spatial evolution of water quality is consistent with the spatial evolution of the application of COD.

 https://doi.org/10.1051/e3sconf/20183801014

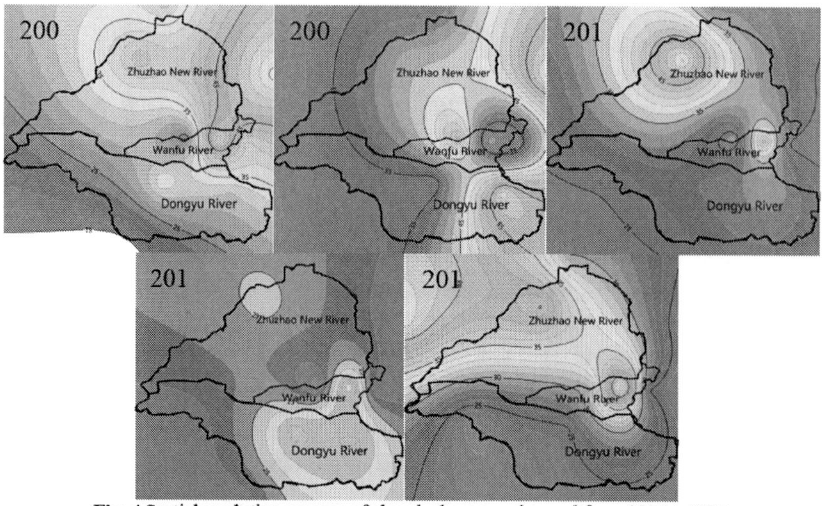

Fig. 4. Spatial evolution process of chemical oxygen demand from 2006 to2014

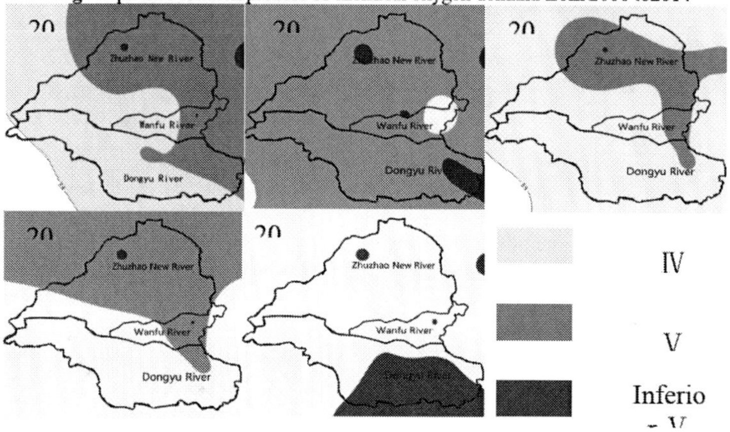

Fig. 5. Spatial evolution of river water quality during 2006 - 2014

3.3. Spatial and temporal change characteristics summary

By analyzing the temporal and spatial changes of the water environment of the three rivers in Heze City, such as the Suzhao River, the Dongyu River and the Wanfu River, the interannual variation characteristics of the water environment of the three rivers in Heze City are obtained. Zhuzhao New River comprehensive index of river pollution is the largest of the three rivers, the change is also greater. The pollution index of Dongyu River is relatively small, and the change range is relatively stable. From 2006 to 2013, the water quality of three rivers in Heze has gradually improved, and the water quality has been aggravated by 2014. Among the three rivers, the Dongyu River has better water quality. The pollution center is mainly in the lower reaches, and the water quality is dominated by grade IV and V. The water quality of Wabfu River was second. Due to the small drainage area in Heze City, the water quality is dominated by Grade V and Grade IV. Zhuzhao New River Pollution Center is mainly concentrated in the downstream often appear V category or Inferior V class water [15].

4 Conclusion

By monitoring water quality of monitoring sections of Zhuzhao New River, Wanfu River and Dongyu River in Heze City, The spatial and temporal evolution of water environment in Heze City over the past 9 years has been systematically studied and analyzed. The results show that from 2006 to 2014 overall water quality is gradually improving in Heze City. Among them, the quality of water environment has been improved year by year from 2006 to 2013, and indicators of water pollution in rivers decreased year by year. By 2014, the quality of water environment will decrease and the concentration of pollutant index substances in rivers will rise to a certain extent. Among the three rivers, the water quality of Zhuzhao New River in is worse, followed by Wanfu River, and the water quality of Dongyu River is the best. Based on the spatial evolution of water quality and the evolution of pollutants, as a result, the water quality in the northern part of Heze City is poor and the water quality in the southern part is better.

Acknowledgements

This study was financially supported by the Natural Science Foundation of China (NO.41471160)

References

1. J Cai, O Varis, H Yin, J. Clean. Prod. **142**, 4:2901-2910 (2016)
2. S Y Korkanç, S Kayıkçı, M Korkanç, J. Afr. Earth. Sci. **129**:481-491 (2017)
3. G Cao, D Han, J Moser, Environ. Manage. **51**, 6:1109-1125 (2013)
4. Q Zhou, W Wang, Y Pang, Z Zhou, H Luo, Water Science and Engineering, **8**, 4:263-272 (2015)
5. China Environmental Yearbook Environmental Monitoring Branch editorial board, *The China Environmental Yearbook 2005* (The China Environmental Yearbook, 2005) (in Chinese)
6. S K Barik, P R Muduli, B Mohanty, A T Behera, S Mallick, Cont. Shelf. Res. **136**:39-56 (2017)
7. W Luo, H Li, H Mu, *Water environment protection* (China Water Conservancy and Hydropower Press, 2009) (in Chinese)
8. S Su, D Li, Q Zhang, R Xiao, F Huang, Water. Res. **45**, 4:1781-1795 (2011)
9. W Sun, C Xia, M Xu, J Guo, G Sun, Ecol. Indic. **66**:306-312 (2016)
10. L Ren, E Cui, H Sun, Environ. Sci. Pollut, Res. Int. **21**, 23:13646-13655 (2014)
11. Z Zhang. *Shandong Geography* (Beijing Normal University Press, 2014) (in Chinese)
12. L Tong, Heilongjiang science and technology information, 15:113-113 (2016) (in Chinese)
13. G Xu, Z Jing, Journal of Tianjin University, **50**, 7:710-716 (2017)
14. T Sun, M Zhang, M Li, X Wang, Environmental Science and Technology, **v. 37**, 4:185-190 (2014) (in Chinese)
15. N C Ujjania, M Dubey, Curr. Science. **108**, 8:1430-1433 (2015)

https://doi.org/10.1051/e3sconf/20183801015

Intelligence Control System for Landfills Based on Wireless Sensor Network

Qian Zhang[1], *Chuan* Huang[1,*], and *Jian* Gong[1]

[1]State Key Laboratory of Coal Mine Disaster Dynamics and Control, Chongqing University, Chongqing 400044, China

Abstract. This paper put forward an intelligence system for controlling the landfill gas in landfills to make the landfill gas (LFG) exhaust controllably and actively. The system, which is assigned by the wireless sensor network, were developed and supervised by remote applications in workshop instead of manual work. An automatic valve control depending on the sensor units embedded is installed in tube, the air pressure and concentration of LFG are detected to decide the level of the valve switch. The paper also proposed a modified algorithm to solve transmission problem, so that the system can keep a high efficiency and long service life.

1 Introduction

Landfills is a mature disposal of municipal solid waste (MSW) nowadays because of its security and simplicity [1]. Due to the landfill gas (LFG) is the main product from relevant reactions, which consists of four primary components such as methane, carbon dioxide, oxygen and nitrogen. The landfills had a lasting effect that must be supervised over decades [2]. LFG is a promising renewable energy source provided additional incentive particular the methane in it, while the potential hazards associated is inevitable for its flammability [3-4]. Therefore, gas exhausting pipes arranged in landfills for collecting in general [5]. But the gas was pumped always with a uniform pressure, it leads difficultly to ensure the gas sucked evenly or entirely in time. On account of the problems mentioned above, we propose a new scheme aiming to achieve an efficient procedure to control the gas extraction in the landfill. Furthermore, the intelligence approach can be automatic instead of superfluous manual tasks.

Nowadays, with the development of the technology of the sensors and network technology, the devices with the wireless sensors embedded can achieved in engineering [8]. This paper proposed an air valve with a unit of wireless sensors set on the top of tubes, not only for receiving the data in tube, but also for controlling the switch of the suction depending on data analytics. The tools are available for supervision and alarm, which the thresholds can be set by the experts taking into account the landfill typology.

2 Intelligence Control System

2.1. Profile of the Model

The rationale of the control system proposed in this work is outlined in Figure 1. In general, the key of control part is air valves which will be set on the top of gas tubes with a series of modules. The first stage is data acquisition, data is collected by masses of wireless sensors. The unit consists of sensors can observe the percentage of CH_4, CO_2, O_2 and the air pressure in tube in a certain period. The captured data will be transmitted via wireless communication to the terminal. The computer terminal analyzes and generates a control signal on the basis of comparison between the captured data and predefined threshold set by staff. Staff can set and modify the predefined threshold by computer terminal. Then the control signal send back to each unit without delay and microprocessor implemented receive signals for managing the air valve devices. While after the data transfer procedure, data is uploaded and stored in dedicated database server, and is ready for the following processing such as download and analyze by the staff. So the entire process dominates by wireless sensors automatically.

2.2 Overview of the Algorithm

* Corresponding author: linette0325@163.com

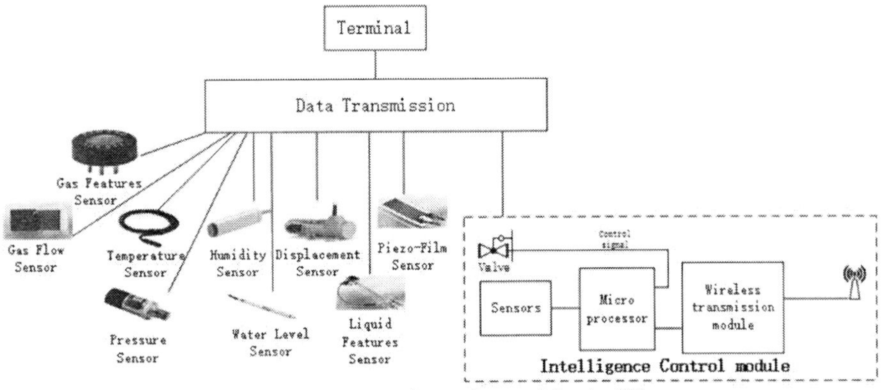

Fig. 1. Components of control network in a landfill.

Despite the wireless sensor units can collect data and receive control signal in a short period of time, the sensors will deal with a large quantity of information in our scenario of landfill in reality. However, each unit consists of several different kinds of sensors, the energy of one unit depending on the battery power and the transmission bandwidth is limited in network. In the previous study by author [11], we found compressive sensing theory demonstrates that it is possible to reduce the number of sampling and reconstruct the original signal accurately with high probability. So the method applied to the scenario to overcome associated problems firstly. However, the herein more detailed key elements of the method must be improved for our system based on wireless sensor networks.

In this paper, the compressive sensing procedure is introduced specific to the problems above. First of all, we use the sparse measurement matrix which the large and dense matrix multiplication operations in data acquisition are replaced by the summation of signal elements in sensor nodes to improve the algorithm. Thus, the storage of the measurement matrix and the computation complexity are decreased. Then, a hard thresholding algorithm with light computationally and high accuracy was proposed to reconstruct the original signal. The process of the proposed approach is provided in detail below.

Each node transmits data gathered to the terminal in a fixed time, while the node saves the original information as a matrix,

$$X=\begin{bmatrix} x_1 & x_2 & \text{L} & x_N \end{bmatrix}^T \tag{1}$$

where N is the number of sensors. According to the sparsity M of the original information, the corresponding number of the measurement obtained and matrixes Φ generated. The sparse $M\times N$-dimension measurement matrix Φ with entries is defined as follows

$$\theta_i=\sqrt{\frac{\alpha}{M}}\times\begin{cases} 1 & \text{with probability } \frac{1}{2\alpha} \\ 0 & \text{with probability } 1-\frac{1}{\alpha} \\ -1 & \text{with probability } \frac{1}{2\alpha} \end{cases} \tag{2}$$

with zero-mean value, $E\left[\theta_i^2\right]=\frac{1}{M}$, and $|\theta_i|\le\sqrt{\frac{\alpha}{M}}$.

It is obviously to see that with the value of α increased, the "0" elements increased, and the matrix Φ will be sparser. This matrix is very easy to compute with only a uniform random generator. For one $M\times N$ matrix, we only have to store a $\frac{M\times N}{2\alpha}$-dimensional matrix on the sensor.

Measure process in each node is expressed by

$$Y=\Phi X \tag{3}$$

as a result $Y=\begin{bmatrix} y_1 & y_2\text{L} & y_M \end{bmatrix}$. At last, nodes translates the measurement information Y.

After the transmitting via wireless local area network, Hard Threshold Iteration method is used in terminal to reconstruct each signal of the original X. The proposed algorithm is summarized as follows:

Initialize: $x_1=0$, residual **r**=**y**, iterations n=0;

x_n is the result of original signal which converted by CSSA with parameter m that refers to the maximum number; Supports $\Lambda=\emptyset$, $J=\emptyset$;

While halting criterion false do n=n+1;

$$x_{n+1}=x_n+\Phi^T r \tag{4}$$

$$J\leftarrow M(x_{n+1},2m) \tag{5}$$

$$T=\text{supp}(x_n)\cup J \tag{6}$$

$$b|_T=\Phi_T^{\dagger}y \tag{7}$$

$$x_{n+1}\leftarrow M(b,m) \tag{8}$$

$$\mathbf{r}_{new}=\mathbf{y}-\Phi x_{n+1} \tag{9}$$

return $X=X_n$.

3 Simulation of Algorithm

In this part, simulations by Matlab were employed to examine the effectiveness of the proposed scheme. We analyze the improved transmission algorithm in a laptop with an Intel i5-3210M CPU, 8GB RAM and Windows 10 operating system. First of all, our modified sparse measurement matrix was simulated by comparison with random Gaussian matrix and sparse binary matrix which used in compressed sensing algorithm usually nowadays. We test the simulation with the data length N=1024, measurements M=4K (K represents the sparse of signal), D=4, and α=M/4 for 1000 times. The results are shown in Table 1, where the space complexity refers to the storage space. The processing time is the interval from the generation.

Table 1. Comparison of complexity of the matrixes

	Space complexity (bytes)	Processing time (s)
Random Gaussian matrix	2990080	0.0103
Sparse binary matrix	2990080	0.0058
Modified sparse matrix	32768	0.0007

From Table 1, the storage of random Gaussian matrix and sparse binary matrix can easily overflow on some small memory node. The modified sparse matrix considerably reduces the need of memory size. Furthermore, the averages results of processing time over 1000 times can be seen that the capacity of our proposed matrix outstrips the others significantly.

To quantitatively evaluate the proposed reconstruction method, we simulate experiment for the performance of algorithm the algorithm in terms of root mean square error (RMSE). The RMSE denotes the root mean square error estimated between the original signal and the recovery result. Final result is the mean value of a large number of iterations. The analysis of performance comparison in terms of RMSE is shown. Computational complexity is calculated by these parameters: 1) the length of test data is 1600, 2) according to the randomness in the process, convergence time is an average value of 500 times. Figure 2 reflects the accuracy of the proposed method.

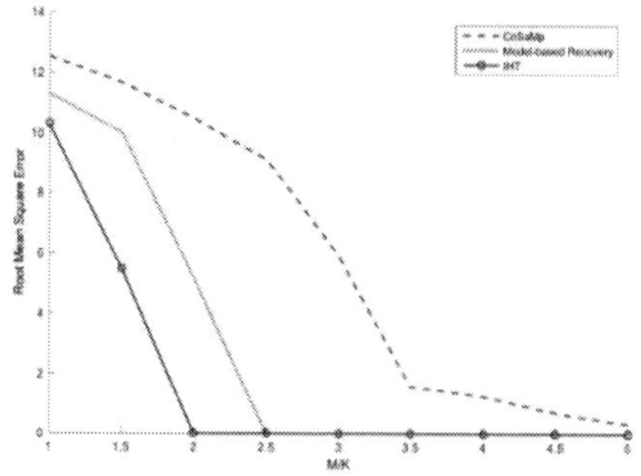

Fig. 2. Comparison of different algorithms. CoSaMp (dashed), Model-based Recovery (dash-dotted) and our method IHT (solid -ring).

Experiments results beyond these algorithms, it can be seen that proposed method can reconstruct the original signal more accurately compared than CoSaMp, Model-based Recovery. Our method can also guarantee reconstruct original signal in a higher accuracy with less measurements.

4 Conclusion

This paper presents an innovative intelligence system for controlling the air pump of the landfills, solved many types of engineering problems, such like LFG can be extracted timely, every air exhausting pipe has its adaptive pressure individually, and it is an automatic control system instead of manual work. The procedure includes two particular features. Firstly, control extraction automatically depending on a variety of wireless sensors units set on the valve, the air pressure and concentration of LFG detected which can serve the flow rate of valve. Staffs pay more attention to the result display on the screen and make decision to the valves in the landfills. Secondly, CS-based compression was found to WSN due to its lower complexity and best overall energy efficiency. This paper proposed a CS-based system within a sparse measurement matrix and a iterative hard thresholding recovery algorithm, which is expressed to improve the transmission effectively and precisely of mass data. The results show that the modified method effectively in the scenario. In our future work, we will make further improvement on the system to implement controlling and monitoring in reality.

References

1. Arigala S G, Tsotsis T T, Yortsos Y C, et al. Gas Generation, Transport, and Extraction in Landfills[J]. Journal of Environmental Engineering, 1995, 121(1):33-44.
2. Young A. Mathematical modeling of landfill gas extraction[J].Journal of Environmental Engineering,1989, 115(6): 1073.
3. Peter Kjeldsen, Morton A. Barlaz, Alix P. Rooker, et al. Present and Long-Term Composition of MSW Landfill Leachate: A Review[J]. Critical Reviews

in Environmental Science & Technology, 2002, 32(4):297-336.
4. Townsend T G, Wise W R, Jain P. One-Dimensional Gas Flow Model for Horizontal Gas Collection Systems at Municipal Solid Waste Landfills [J]. Journal of Environmental Engineering, 2005, 131(12): pages. 1716-1723.
5. Stark T D, Choi H. Methane gas migration through geomembranes[C]// 2005:120-125.
6. Tchobanoglous G, Theisen H, Vigil S. Integrated solid waste management: engineering principles and management issues [J]. Water Science & Technology Library, 1993, 8(1):63–90.
7. Yu L, Batlle F, Carrera J, et al. Gas flow to a vertical gas extraction well in deformable MSW landfills.[J]. Journal of Hazardous Materials, 2009, 168(2):1404-1416.
8. E. A. Garich, Wireless, Automated Monitoring for Potential Landslide Hazards, Master Thesis, Texas A&M University, May 2007.
9. D L Donoho. Compressed sensing [J].IEEE Trans. on Information Theory.2006, 52(4): 1289-1306.
10. Joel Tropp, Anna Gilbert, Signal Recovery from Random Measurements via Orthogonal Matching Pursuit, IEEE Transactions on Information Theory, 2008, 53(12): 4655-4666.
11. Liu G, Zhang Q, et al. Model-Based Adaptive Iterative Hard Thresholding Compressive Sensing in Sensor Network for Volcanic Earthquake Detection [J]. International Journal of Distributed Sensor Networks, 2015, 2015(3).

A Review on the Historical Development of Environmental Specimen Banks

Yu Xin LUO

Department of Environmental Science, College of Environmental Science and Engineering, Tongji University, 200092 Shanghai, China

Abstract. Environmental Specimen Banks (ESBs) are facilities that archive samples from the environment for future researching and monitoring purposes. In addition, the long-term preservation of representative specimens is an important complement to environmental studies and monitoring practice. This article presented detailed information about ESBs in the world, and reviewed the history and current status of ESBs in China. Besides, crucialshould leave 8 mm of space above the abstract and 10 mm after the abstract.

1 Introduction

Over the past few decades, as a result of rapid population growth, intensive agriculture activities, industrialization and urbanization, large amounts of persistent organic pollutants (POPs) have been released into the environment. These pollutants not only have adverse effect on the health of human beings, but also pose threats to the well-being of generations in the long run [1] . Thus, understanding the sources, fate and behavior of POPs and their impact on human beings requires long-term intensive monitoring efforts. Environmental monitoring studies have been indispensable in the exploring and understanding of the environmental fate and behavior of POPs [2] .

The idea of archiving biological and environmental samples for retrospective analysis has been recognized as a vital component for assessing the risks from chemical substance released into the environment. The chemical analytical approaches are expected to provide the database with deeper insights into chemical processes and hazardous situations. Such a detailed knowledge could further be used for environmental protections, regulations and pollution control measures, in which direction and to which extent technological developments or modifications are made. Therefore, the function and significance of ESBs have been acknowledged widely.

ESBs are facilities that archive various samples which have been treated with standardized procedures and passed the unity test. Environmental samples that are stored can be used to monitor the environmental state of pollutants, identify environmental problems, and study the correlations between cause and effect [3] . The sample materials can be used many years after the samples were collected. Therefore, when new or improved analytical techniques become available or a new interest arises for a certain contaminant that was not considered important in the past, samples from ESBs can be used for analyzing its temporal trends and predicting its exposure risks. The basic premise of a ESB is to collect and process environmental samples for archiving without changing their original chemical composition and preserve them in a stable environment for future analysis [4] . Thus, ESB can be considered as an important long-term observation system for taking measures to protect the global environment as well as for checking the efficacy of the actions taken by governments and international organizations to curtail or control the pollution by persistent contaminants [2] .

2 Historical development of ESBs in the world

The history of specimen banks dates back to the 1960s when the first bank was established in Sweden at Swedish Museum of Natural History, since then the importance of proper preservation of environmental specimen started to be realized, resulting in the establishment of several other ESBs worldwide. Currently, major ESBs in the world include German Environmental Specimen Bank at Federal Environment Agency, U.S. Environmental Specimen Bank Group at National Institute of Standards and Technology, Canada's National Aquatic Biological Specimen Bank and Database at the Canada Centre for Inland Waters/Environment Canada, Canadian Wildlife Service Specimen Bank at National Wildlife Research Centre, Japanese Environmental Specimen Bank for Global Monitoring (es-BANK) at Center for Marine Environmental Studies and Japanese Environmental Specimen Time Capsule Program at National Institute of Environmental Science (NIES) .

The ESB at the Swedish Museum of Natural History (SMNH) has been established for more than forty years, and acted as a powerful tool for national contaminant monitoring and eco-toxicological researches [5] . In cooperation with the Swedish Environmental Protection

Agency (SEPA), contaminant monitoring programs were set up in 1980 for studies of residue levels of persistent pollutants and their effects on biota in terrestrial, freshwater and marine environments [6] . The ESB also stores older samples of animal tissues and organs from different research projects on environmental effects of noxious substances. The homogeneous and continuous series of samples are also used for retrospective chemical analysis to analyze the levels of new or recently discovered contaminants. The ESB has tissue samples from more than 260,000 organisms, mostly from animals but also from plants (moss). The majority of the samples are stored at -30 °C and -80 °C, and certain types of samples are stored dry at room temperature.

German ESB started its development in the late 1970s, and stored environmental specimens from representative ecosystems in conditions which would prevent the chemical decomposition over a period of decades. The sampling started in 1985 and the types of sampling areas include ecosystems close to conurbation, riverine ecosystems, marine ecosystems, forestry ecosystems, nearly natural terrestrial ecosystems and agrarian ecosystems [7] . The ESB also archives various human samples, including blood, blood plasma, urine, saliva, and hair. Scientists have made extensive attempts to track the pathways of chemicals in the human bodies as well as their fate and transformation. The human samples aid in the analysis of which pollutants are likely to cast adverse effects on human health, and how these effects change over time. Since 2007, with the safety of commercial chemicals being more and more realized, the German ESB has become irreplaceable in evaluating the effectiveness of regulations related to hundreds of thousands of commercial chemicals, and further protecting the health of humans and the environment against adverse effects caused by these chemicals [8] .

U.S. Environmental Specimen Bank Group, based out of the Hollings Marine Laboratory in Charleston, South Carolina, has initialized a diverse set of programs and researches. The group includes NIST's Marine ESB and Reference Material Production Facility with long-term archival of well documented and preserved specimen for both retrospective and comparative environmental health analysis; a nuclear magnetic resonance (NMR) facility to support structural biology, metabolomics, and natural product researches; informatics; and NIST's Pacific Islands Project. The group has been involved in government, academic, and private partnerships resulting in programs geared towards promoting the development of cooperative and collaborative scientific researches for a better understanding of marine resources and environmental health. The Marine ESB include marine mammal tissues, mussels and oysters, fish tissues, seabird eggs, and peregrine falcon eggs and feathers. Many of these specimens have been analyzed retrospectively to determine time trends of various emerging contaminants of in the environment and as part of a multi-agency effort to determine health trends in marine animals.

The Canada's National Aquatic Biological Specimen Bank (NABSB) is located in a dedicated facility at the Canada Centre for Inland Waters in Burlington, Ontario. The NABSB holds more than 37,000 samples of fish and invertebrates collected over the last 30 years of environmental monitoring in Canada. Researches conducted using samples from the NABSB have produced more than 60 scientific publications, reports and book chapters. The National Wildlife Specimen Bank stores over 900,000 biological sub-samples taken from 200,000 tissue samples of over 800 species. Scientists use these samples to conduct researches on contaminant monitoring and effects of toxic chemicals on the environment. These researches help develop science-based regulations controlling the use, distribution, application, and disposal of chemicals. Established in 1963 as the Canadian Wildlife Service Specimen Bank, the National Wildlife Specimen Bank gives scientists the unique ability to use historical environmental specimens to: screen for new chemical contaminants; retrospectively analyze emerging contaminants of concern; conduct analysis in the future using more sophisticated analytical equipment.

The Japanese es-BANK is an orderly system of collection, archiving, cataloging and management of biological and environmental specimens from all over the world and provide them to a wide range of researchers within and outside the university for monitoring studies on environmental toxic substances. 1,300 species and 100,000 specimens were collected from Asia, Europe, Australia, North/South America and Antarctic, Arctic, North/South Pacific, North/South Atlantic, Indian and Antarctic Oceans during the last 50 years [9] . The pilot ESB was constructed at the National Institute for Environmental Studies(NIES) in 1979 and started a collection of samples under the program "Studies on the method for long-term environmental monitoring" in 1980 [10] . During the pilot phase, the specimens were mainly stored at -20 °C. In 2004, long-term storage using liquid nitrogen vapor (-150 °C) was completed in the Environmental Specimen Time Capsule Building. Also, there is a Time Capsule program for the systematic collection and banking of various kinds of environmental specimens representing the current environmental status in Japan.

The Norway ESB has been collecting environmental samples from across the nation, and acted as an important tool for the pollution status monitoring of varied organic pollutants. The Norway ESB is owned by the Ministry of Climate and Environment and managed by the Norwegian Environment Agency. The Norway ESB contains frozen samples of animals, plants, air and mud from across Norway and the Arctic. The samples become time capsules preserving the present environmental state, and they can provide valuable information in the future. The Italy Mediterranean Marine Mammal Tissue Bank collects and preserves biological samples collected from marinemammals along the Italian coasts of the Mediterranean Sea, in cooperation with the University of Padova, the Italian Ministry of the Environment, the Institutes for Animal Health, and several other non-profit Italian organizations dedicated to the marine mammal researches. Those ESBs all mentioned collected typical specimens are shown in Table1

Table 1. Comparison Between Various Country's ESBs

Name of ESB	Established time	Types of archived specimen
Swedish ESB	mid-1960s	otter, grey seal, harbor seal, ringed seal, osprey, white-tailed sea eagle, peregrine falcon
German ESB	1970s	algae, mussels, fishes, herring gulls, human specimens (blood and urine)
U.S. marine ESB	1989	marine mammal tissues, mussels and oysters, fish tissues, seabird eggs, and peregrine falcon eggs and feathers.
Canada's National Aquatic Biological Specimen Bank and Database	1977	trout, walleye, whitefish, alewife, rainbow smelt, slimy sculpin, plankton
Canadian Wildlife Service Specimen Bank	1963	birds, mammals, fishes, reptiles, amphibian, invertebrates
Japanese Environmental Specimen Time Capsule Program	1979	fishes, shellfishes, marine sediments, airborne particulate matter and human breast milk
Japanese Environmental Specimen Bank (es-BANK)	1960s	birds, fishes, reptile & amphibian, seal & sea otter, terrestrial mammals, cetacean & sea cows, invertebrate & others
Norwegian Environmental Specimen Bank	1993	animals, plants, air and mud from across Norway and the Arctic
Italy Mediterranean Marine Mammal Tissue Bank	2002	fin whale, sei whale, Risso's dolphin, Ziphiidae, arbour porpoise, striped dolphin, killer whale, humpback whale, common bottlenose dolphin

3 The development of ESB in China

In the past few decades, the economy of China is in the golden development period, and conflict between the environment and economy is also in the most pointed and sensitive stage. China's social and economic development has brought a lot of pollutants to the environment [11]. Furthermore, the volume of wastewater has increased dramatically with the booming of economy, and large amounts of wastewater, gas and solid have caused great burdens on China and the global environment. In recent years, China is facing serious environmental problems such as water and air pollution and biodiversity loss, which have seriously threatened human health, ecological security and the sustainable development of China [12]. Recognizing importance of the environmental pollution caused by extensive economic growth, Chinese government has taken a series of measures for environmental protection [13]. At the same time, some scholars are devoted to the study of environmental issues in China, analyzing the risk of various chemical pollutants, including heavy metals, persistent organic pollutants and endocrine disruptors, which have showed bioaccumulation, biomagnification, and long-range transport potential [14]. However, there are very few data available in China for temporal trends of these substances in the environment. However, due to the lack of archived samples, almost no retrospective studies on emerging substances have been conducted in environmental samples, which is unfavorable to find the sources of contamination and make effective countermeasures to protect the environment.

Since 1990s, six ESBs have been established in China for varied purposes, and most of them are located in Beijing and Shanghai. Soil and sediment are main specimen types, probably due to the convenience in sample collection and storage. In 1992, China's first ESB was established by China Environmental Monitoring Center in Beijing, and soil samples were collected from 29 provinces in China，along with sediment samples which were collected from the Yangtze River, Songhua River and the Pearl River [15]. In 1994, the second ESB named the Biological Environmental Specimen Bank (BESB) was built by the Laboratory of Nuclear Analytical Techniques, Institute of High Energy Physics, Chinese Academy of Sciences [16], and human tissue samples from Tianjin were archived and stored at -70 °C, including 112 liver samples, 20 renal samples, 20 cardiac samples and other tissue samples from accidental death. In 1998, the third ESB was built in Shanghai under the cooperation of the Shanghai Academy of Environmental Sciences and Shanghai Institute of Nuclear Research [17], and water, sediment and dust samples were the major specimen types.

In 2006, there were two ESBs established in Shanghai and Beijing separately. One is the Resource-Sharing Platform of Polar Samples called BIRDS, which was organized by the Polar Research Institute of China. The BIRDS consisted of Biology Depository, Ice-Snow Depository, Rock Depository, Deep Space Depository, and Sediment Depository. From 2007 to 2013, a total of 6,744 samples were collected continually every year from two polar areas, which can be used to conduct studies on polar environment. The other ESB was affiliated to the Research Center for Eco-Environmental Sciences, Chinese Academy of Sciences, which is the first 1,000-Level Purification ESB in China, and sample types include sediment, soil, water, dust, fish, and bird.

China Ocean Specimen Bank was built in 2001, and put into use in 2007. China Ocean Museum consists of sample warehouse, laboratory, database, exhibition hall and office and other parts, with a total area of more than 3000 square meters. The sample warehouse consists of three types of professional warehouses: normal temperature storehouse, 4 ℃ sample storehouse and -20 ℃ sample storehouse, which keeps the cobalt-rich crusts, polymetallic nodules, hydrothermal sulfides obtained from various oceanic survey voyages since 2001

The Yangtze ESB (YESB) program started in 2011 with the approval of the Ministry of Science & Technology and the Ministry of Finance. It is devoted to the long-term storage of representative environmental and human specimens from the Yangtze River Basin. The project started back in 2010 and became operational in 2014. Recently, the YESB has started the establishment of a long-term environmental monitoring program in the Yangtze River Delta, but not limited to this area. The YESB is expected to provide invaluable specimen resources for real-time and retrospective analyses of conventional and emerging pollutants as well as for ecotoxicological researches, and to provide scientific basis for environmental management and decision making. The YESB has an area over 2,000 square meters for specimen preparation, storage, pretreatment, and analysis, and has archived more than 3,000 specimens, including soil, lake and marine sediment, sewage sludge, human hair, and fish.

4 Perspective

ESB has been playing vital roles in the sound chemical management processes by supporting environmental monitoring and revealing long-term changes of pollutant levels in the environment. Now, it is expected to contribute to the international chemical management processes more efficiently by networking and collaborations among ESBs as well as supporting establishment of new ESBs particularly in developing countries. Environmental specimen banks have become an essential part of the fundamental research infrastructure for environmental sciences. Compared to the developed countries, China's Environmental specimen bank is clearly in its infancy. It is necessary to establish pilot studies to test the collection, storage and evaluation of representative environmental samples in regional ecosystems in China. Implementing research projects based on environmental specimen bank is the basis for updating sample collection specifications, getting more government support, understanding changes in environmental quality, and better setting environmental policies. Extensive cooperation with other international organizations in response to global environmental issues. Therefore, more systematic collection, scientific research, and international cooperation are fundamental ways to better develop ESB for illustrating the intensive environmental changes in China and the world.

References

1. A. Beyer, D. Mackay, M. Matthies, F. Wania and E. Webster, *Environmental Science & Technology*, **34**, 699-703., (2000) .
2. S. Tanabe and K. Ramu, *Marine Pollution Bulletin*, **64**, 1459-1474., (2012) .
3. K. S. Subramanian and G. V. Iyengar, *American Chemical Society*, 288., (1997) .
4. R. Day, P. Becker, O. Donard, P. RS and W. SA, *Environmental Science Processes & Impacts*, **16**, 10-27., (2013) .
5. L. Robert and K. Barbara, *Toxicological & Environmental Chemistry*, **27**, 251-266., (1990).
6. T. Odsjö, *Science of the Total Environment*, **B** 147-156., (1993) .
7. C. SchröterKermani, in *Presentation held at the 9th FECS Conference on Chemistry and the Environment.*, Bordeaux, France, (29.08.-01.09.2004) .
8. A. Bassanelli, in *Federal Ministry for the Environment, Nature Conservation and Nuclear Safety, Bonn (Germany)*, (2012) .
9. S. Tanabe and K. Ramu, *Marine Pollution Bulletin*, **64**, 1459-1474., (2012) .
10. Z. Karube, A. Tanaka, A. Takeuchi, Y. Takazawa, T. M, K. A, S. H and S. Y, *Environmental Science & Pollution Research*, **22**, 1587-1596., (2015) .
11. X. D. Li, W. X. Wang and Y. G. Zhu, *Science of the Total Environment*, **421-422**, 1-2., (2012) .
12. J. Liu and P. H. Raven, *Critical Reviews in Environmental Science & Technology*, **40**, 823-851., (2010) .
13. Y. Eddy and T. Sun, *Environmental Toxicology & Chemistry*, **27**, 1-3., (2010) .
14. K. Zhang, Y. L. Wei and E. Y. Zeng, *Science of the Total Environment*, **463-464**, 1093-1110., (2013) .
15. Yue wei , Zheng Daying. *China Enviromental Monitor*, 48-50., （1993）.
16. Zhang Peiqun, Chai Zhifang. *Nuclear Technology*, 24, 755-759., （2001）.
17. Y. X. Zhang, Y. S. Wang, Y. E. Qian, T. M. Guang, Z. Ming, L. I. De-Yi, F. Cheng and Z. Gui-Lin, *Journal of Environmental Sciences*, **12**, 398-401., (2000) .

The International Treaty on Global Warming: Is it Good or Bad for the Economy?

Xinyu Wei* and *Wenbin* Bao[1]

[1]Nanjing University of Science & Technology, Xiaolinwei 200, Nanjing, Jiangsu Province, China

Abstract. Global warming is one of the hottest topics all over the world. International authorities have worked together to negotiate the Paris Agreement on global warming. This Agreement has its supporters and critics. The key question is whether on balance is the Paris Assignment good or bad for the United States economy. This paper begins with some background information leading up to the passage of the treaty. Next, I outline what is in treaty. I then critically analyze the arguments in support of and against the Assignment. Finally, I explain the basis for my opinion that in the long run the treaty will benefit the United States economy.

1. Introduction

Scientific studies show that human beings cause global warming. Our voracity and endless craving eventually lead to a more vulnerable planet. Fortunately, we have just awoken from ignorance, searching for a way to better save this homeland. Consequently, "Green GDP", "Green Development" and other "green" initiatives have become popular. Because we share earth's environment, global environmental problems cannot be solved by one country. Instead, all the countries who are responsible for polluting the environment most act together to reduce overall pollution. Although signing an international treaty on global warming is a logical first step to combat global warming, the treaty could have some potential adverse consequences for the United States economy.

As a developed country, the U.S relies more on high-tech industries. Therefore, the issue arises: whether an international treaty on global warming is good or bad for the United States economy?

Opinions are divided on this issue. Some people believe that the treaty will overall be good for American business, and ultimately lead to a strong economy. Others argue that the treaty will hurt the United States economy because reducing greenhouse gases emission will increase the cost of production, thereby making American products more expensive and placing them at a competitive disadvantage in the global marketplace. Critics also argue that those countries who do not sign this burden on developed nations under the principle of "common but differentiated responsibilities" (UNFC,1). The Kyoto Protocol has three main mechanisms: International Emissions Trading, Clean Development Mechanism (CDM), and Joint implementation (JI). treaty would share the benefit of environment protection without paying for it, thus raising moral issues and damaging the relationships among nations.

2. Background

2.1 What is the treaty?

The United Nations Framework Convention on Climate Change (UNFCCC) is an international treaty that establishes a framework for how specific international treaties ("Kyoto Protocols" and "Paris Agreement") may be negotiated to set binding limits on greenhouse gas emissions.

2.2 Kyoto Protocol

Under the Kyoto Protocol, signatory nations agreed to binding emission reduction targets by controlling emissions of the main anthropogenic (i.e., human-emitted) greenhouse gases (GHGs) in ways that reflect underlying national differences in GHG emissions, wealth, and capacity to make the reductions (Grubb, M. ,2004).Recognizing that developed countries are principally responsible for the current high levels of GHG emissions in the atmosphere as a result of more than 150 years of industrial activity, the Protocol places a heavier

2.3 The Paris Agreement

The Paris Agreement sets out a global action plan to put the world on track to avoid dangerous climate change.

The Paris Agreement requires the signing nations to rapidly reduce their mission of greenhouse gases using the best available science to keep the increase in the average

* Corresponding author: 544914276@qq.com

global temperature to well below 2°C above pre-industrial levels (European Commission, 1).And the INDC's require each individual country to make individual contributions in order to achieve the worldwide goal. Each nation must report their individual contributions every five years and are to be registered by the UNFCCC Secretariat. Countries can cooperate and pool their nationally determined contributions.

3. The Opposing Views on the Paris Treaty

3.1The Arguments in favor of the Treaty

Advocates in favor of the treaty argue that the treaty will improve the United States economy because it requires participating nations to come up with ways to address global warming. The treaty uses a two-handed approach on the one hand, it requires the participants stop activities which would aggravate global warming. On the other hand, countries must also develop technology to better face global warming.

Pro-treaty advocates argue that limiting gas emissions will create a green economy. The rule in the treaty restricts maximum greenhouse gases emissions in every participating country, so the country has to limit its economic activities to meet this standard. This means more and more factories which aim at manufacturing will be closed, more coal-powered industries will be forbidden. The trend will be that the manufacturing-driven economy will change into a service industry-driven economy. On the other hand, the decrease in fossil fuel usage will certainly increase the potential of clean energy, including wind energy, water energy, tide energy and bioenergy. All these practices will not do harm to the nation's development, instead, it will promote the economy developing in a green manner.

According to pro-treaty advocates, this emission standard will propel new technology. A high gas emission industry can still exist as long as it complies with the rule. So if a heavy industry wants to survive, it must innovate. It can improve the performance of devices—increase the production efficiency—or come up with new high efficient and environment-friendly production modes. Both of these approaches will require new technology.

3.2The Argument against the Treaty

Just as there are many people who support the Paris treaty, there are many others who oppose it. Anti- treaty advocates believe that there is no global warming problem and therefore argue that the Paris treaty is unnecessary. Some of the more radical scientists who share this believe are referred to by the derogatory term "deniers." Other individual scientists, institutions, and organizations believe there was a break or "hiatus" in the global warming theory.

Anti-treaty advocates point to the most reliable temperature data, from orbiting weather satellites, which show that there has been no warming for nearly two decades. Despite the constant barrage of hyperventilating headlines of a melting planet and the unceasing clamor of climate catastrophists and computer modelers, global temperatures have not been rising as predicted — except in the always-wrong computer models.

If opponents take the position that there is no such thing as "global warming", they claim that signing a global treaty concerning global warming is vain. There is still another part of opponents, who show their disagreement by emphasizing the damage brought by such an international treaty.

Former President George W.Bush rejected for two reasons. First, he believed that the costs of reducing greenhouse gases would impose disproportionate disbenefit on the American economy in pursuit of a still-uncertain benefit. Second, he disliked the treaty not binding poor or developing countries to curb emissions (Lee Lane, 2006).Bush thought the treaty was a Zero-Sum Game, and in this game, China would be the winner, while the U.S would be the loser.

The U.S. government's own research has confirmed that domestic programs to reduce greenhouse gas emissions would wreak havoc on the economy, sending jobs overseas to countries such as Mexico and China. The Argonne National Laboratory in the Department of Energy studied the economic effects of proposed greenhouse gas emissions cuts on six domestic industries (wood and allied products, steel, petroleum refining, aluminum, chemical manufacturing, and cement production) and found that they would be devastating. Numerous studies have shown that meeting any new treaty commitments would result in a dramatic decline in U.S. gross domestic product (GDP) (Trisko, 1997).

What is more, in 1992 the U.S. Department of Commerce released a study by DRI, Inc., the study found that job losses resulting from the treaty would average between 520,000 and 1.1 million per year, depending on whether the CO2 emission goal was 1990 levels or 10% below 1990 levels. More than 5 million additional jobs would be at risk due to these policies, with Texas, California, Pennsylvania, Ohio, Illinois and Michigan facing the greatest job losses (Trisko, 1997).

4. My position

After much research and thought, I have concluded the advantages of the international treaty on global warming outweigh its disadvantages. The purpose of the international treaty should be to solve while at the same time advance the overall economy in the end. Signing the international treaty would ensure that many countries will come together to protect the planet. As Obama says, "Paris Agreement is the best chance we have to save the one planet we have" (Elizabeth, 2015). Most importantly, it is a good way to contribute to the sustainable development which is always the core issue among the nations.

The treaty stimulates the environment-friendly business, benefiting a sustainable economy. As the emerging of sustainable development, the "green" company is standing out among his competitors. According to a UCLA-led study, companies that voluntarily adopt international 'green' practices and standards have employees who are 16 percent more productive than the average(Jon Simmons, 2015) .High productive employees make the company more energetic and competitive, and this kind of company is more attractive to customers since they also want to make green choices. In the United Kingdom, 54% of consumers buy more environmentally friendly products compared to two years ago (Meglena, 2009). The international treaty is not merely a task among nations; the majority of the citizens want to play a part in reducing global warming.

Those who oppose the treaty claim that the global warming being on pause for two decades, and that the high cost of enforcement of the treaty making it not worth the candle are short-sighted. Although I admit that we should use the term "climate change" as opposed to "global warming" to describe the environmental devastation, we have to attach great importance on the environment degradation. It seems that we have spared our effort on something that not exactly exists, but the decreasing rainforests, rising sea level prove such efforts we make are necessity. We should put environment protection in the leading position when doing business for the good of the very people and future generations. As for the complaint that the treaty places uneven requirements on participating nations, I think it should no longer be a reason for the treaty's hurting American economy. Under the Paris Agreement, the INDCs mechanism will ensure that every country contributes its part in the context of its own national circumstances.

5. Conclusion

Global warming is a topic that we cannot ignore. As humans, we must maintain the delicate balance between the environment and the economy. The international treaty provides the tool we can use to achieve this goal. We need to protect the vulnerable environment, but also contribute to promote a strong economy.

However, people are divided on the effect on American economy that the international treaty has. Those who against the treaty claim that the high cost and unequal contribution brought by the treaty would harm the economy as a whole; some in favor of the treaty hold that the joint effort would totally create global cooperation.

I embrace the international treaty, because it will benefit the United States economy in the long run. To be honest, I was neutral when I first began to write this paper, believing that the treaty was a double-edge sword—it brought benefit, but also hurt the economy to some extent. However, after researching the treaty, I changed my mind because my research shows that the treaty will be beneficial. Even if global warming is on pause as proved in some scientists' research, we should never gain economic growth at the expense of the environment degradation. Our ambition to cope with the climate change will eventually turn into an economic opportunity, shared by all the participants who fulfill their responsibility stipulated in the treaty. In the meantime, I highly recommend that the treaty be modified to enable all nations to participant voluntarily and without a compulsory provision. This will refuse the argument that the practice is a zero-sum game.

On my honor, I have neither received nor given unauthorized assistance in any manner on this paper.

References

1. Gillis, Justin. NYT,(2016).
2. Rhein, M.; Rintoul, S. R. IPCC WGI AR5 (Report). p. 257. (2013).
3. Battisti, David; Naylor, Rosamund L. Science 323, 5911(2009).
4. US National Research Council. Washington, D.C., U.S.A.: National Academy Press. p.3. (2001)
5. Grubb, M. 2(2004). . International Review for Environmental Strategies 5 (1): 2 *(PDF version).*
6. http://unfccc.int/kyoto_protocol/items/2830.php
7. http://ec.europa.eu/clima/policies/international/negotiations/paris/index_en.htm
8. William F. Jasper New American (08856540) Retrived from University Libraries database (2016).
9. Trisko, Eugene M., Burnett, H. Sterling. Retrived from Points of View Reference Center database (1997)..
10. http://www.aei.org/wp-content/uploads/2014/07/-strategic-options-for-the-bush-administration_160635893399.pdf
11. http://www.nbcnews.com/news/us-news/obama-climate-deal-best-chance-we-have-save-one-planet-n479026
12. http://www.monster.com/career-start/a/most-unique-environmentally-friendly-benefits-at-green-companies
13. http://www.theguardian.com/commentisfree/cif-green/2009/nov/29/climate-change-european-commission

Analysis on the Trade Comparative Advantage of Manufacturing Industry in Guangzhou to the Belt and Road Area: Based on the RSCA Index

XiaoFei Luo[1], *YongHui* Han[2*], *DongMing* Wei[3]

[1]Faculty of Logistics, Guangdong Mechanical and Electrical Technical College, Guangzhou, Guangdong 510515, China
[2]Guangdong Institute for International Strategies, Guangdong University of Foreign Studies, Guangzhou, Guangdong 510420, China
[3]School of Business, Guangdong University of Foreign Studies, Guangzhou, Guangdong 510006, China

Abstract: After the economic development has entered the new normal stage, the manufacturing industry in Guangzhou faces new challenges and opportunities, and it urgently needs to change the way of its economy development. This paper uses the RSCA index to analyze Guangzhou's problems existing in the manufacturing industry, and to compare the situation to six different areas along the B&R. The results show that, Guangzhou has a long and steady comparative advantage in the garment and textile sector, metal products and leather products sector while it is comparative disadvantage in metal smelting and processing sector, chemical products sector and non-metallic mineral sector. From the perspective of regional comparison, Guangzhou has the strongest competitiveness with ASEAN and South Asia while it is complimentarily with Central Asia and West Asia.

1 Introduction

As one of the earliest coastal open cities, Guangzhou has gradually become an export-oriented city and integrated into the global value chain division system led by the developed countrieswith its competitive advantages such as geographical location, business culture, labor and natural resources. Since the 90s of last century, Guangzhou has set upthe automobile manufacturing, electronics manufacturing and petrochemical manufacturing as the three backbone industries of the city and this structure has not changed so far.

To date, the manufacturing industry in Guangzhou is still lagging behind in the R & D, design and marketing strategy. It is in the low-end position in the interest distribution map of the global value chain and gets a small share of interest. The slowdown of both the domestic and abroad economy has great impact on Guangzhou's manufacturing industry. The three traditional backbone industries cannot provide sufficient motivation to the manufacturing industry and its overall growth has slow down. In 2015, Guangzhou's total output value of manufacturing industry was 1.63 trillion Yuan, an increase of 2% over the same period of previous year, accounting for 13.8% of the total output value of the province's manufacturing industry. The added value of manufacturing industry was 0.4 trillion Yuan, an increase of 2.2% over last year.

Although the scale of manufacturing in Guangzhou is constantly expanding, the growth rate has gradually shifted into the steady growth of low speed (see Figure 1). However, the initiative put forward by the Central Government of China such as 'Belt and Road', 'Going Out Strategy' and 'Made in China 2025' has brought new opportunities to enhancing Guangzhou's manufacturing industry. Facing the opportunities, the first task is to clarify the comparative advantages of manufacturing industriesin Guangzhou to the Belt and Road area.

This paper uses the RSCA index toanalyze the manufacturing industry situation in Guangzhou to the B&R area.In order to enhance its innovation capacity and its position in the global value chain, Guangzhou's manufacturing industry shall follow the rules of comparative advantages of international production capacity and broaden its cooperation space.

* Corresponding author : hanyonghui2006@foxmail.com

 https://doi.org/10.1051/e3sconf/20183801018

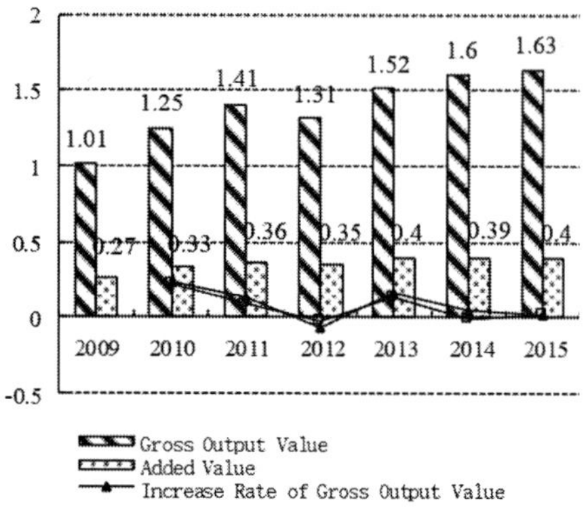

Fig. 1. Analysis on the development of Guangzhou's manufacturing industry[1] (Unit: Trillion Yuan)

2 Methods on comparative advantages analysis

2.1. RSCA Index Classified by Sectors

The RCA index (Revealed Comparative Advantage Index) is a commonly used measuring method in the literature of analysis on comparative advantage.It was first proposed by Balassa [1]. The formula is: $RCA_{it}=(X_{ij}/X_{tj})/(X_{iw}/X_{tw})$, where X_{ij} is the export of product i from area j, X_{tj} is the total export from area j, X_{iw} is the world's export of product i and X_{tw} is world's total export.

The RCA index can measure the comparative advantage level of different trade sectors in an area. When RCA index of a specific trade sector is above 1, it indicates that the sector in this area has comparative advantage, otherwise it has comparative disadvantage. However, Benedictis[2] found in his further research that, the RCA index has the defect of asymmetric on measuring comparative advantage. When the index value is between 0-1, it is considered as of comparative disadvantage, while the value range is [1,+∞], it is considered as of comparative advantage.The judgment value ofthe two rangesis not symmetrical, which makes the measured indicators distribution skew and directly affects the intuitive judgment on the comparative advantages of various industrial sectors.

In view of this, this paper uses the RSCA index to correct the RCA index deviation [3]. The formula is: $RSCA=(RCA-1)/(RCA+1)$ .By modifying the asymmetry bias of the RCA index, the RSCA index can be narrowed down the value range to [-1,1]. If the value of RSCA index is [0, 1], it indicates that the industrial sector has comparative advantage, while the value of [-1, 0] has comparative disadvantage.

[1] Data collected from Guangzhou statistical information network: http://www.gzstats.gov.cn/

The variables used in this paper are from 1995 to 2014. The data are from the World Trade Database, the UN Trade Database, the World Bank Database and the Guangzhou Statistical Yearbook.

According to China's industry classification for national economic activities(GB-T4753-2011), the related manufacturing industry is divided into 11 sectors: food processing, wood products, non-metallic mineral, metal smelting and processing, chemical products, leather products, paper and printing, garment and textile, metal products, sports and entertainment and equipment manufacturing.The calculation results of RSCA index of the above manufacturing industry sectors are shown in Table 1.

Table 1. Results of RSCA index of Guangzhou manufacturing industryclassified by sectors (1995-20174)

	Food processing	Wood products	Metal smelting and processing	Chemical products	Non-metallic mineral	Metal products
1995	0.373	-0.663	-0.983	-0.044	-0.599	0.630
1996	0.326	-0.689	-0.985	-0.602	-0.584	0.580
1997	0.280	-0.687	-0.988	-0.504	-0.568	0.679
1998	0.233	-0.548	-0.990	-0.406	-0.552	0.688
1999	0.186	-0.409	-0.993	-0.309	-0.536	0.737
2000	0.139	-0.269	-0.995	-0.211	-0.520	0.606
2001	0.038	-0.234	-0.993	-0.165	-0.346	0.815
2002	-0.102	-0.144	-0.994	-0.328	-0.218	0.574
2003	-0.008	-0.352	-0.995	-0.423	-0.386	0.229
2004	-0.077	-0.426	-0.995	-0.299	-0.496	0.506
2005	-0.206	-0.296	-0.998	-0.444	-0.607	0.497
2006	-0.237	-0.108	-0.997	-0.372	-0.654	0.651
2007	-0.370	-0.343	-0.997	-0.291	-0.631	0.268
2008	-0.456	-0.840	-0.990	-0.403	-0.702	0.227
2009	-0.626	-0.460	-0.987	-0.519	-0.666	0.863
2010	-0.726	-0.736	-0.983	-0.505	-0.811	0.738
2011	-0.761	-0.711	-0.990	-0.439	-0.818	0.651
2012	-0.781	-0.329	-0.988	-0.521	-0.746	0.280
2013	-0.818	-0.817	-0.986	-0.572	-0.619	0.825
2014	-0.816	-0.830	-0.988	-0.562	-0.202	0.792

	Equipment manufacturing	Leather products	Garment and textile	Paper and printing	Sports and entertainment
1995	-0.782	0.501	0.537	-0.844	0.501
1996	-0.556	0.419	0.234	-0.868	0.645
1997	-0.809	0.559	0.686	-0.878	0.479
1998	-0.681	0.700	0.371	-0.881	0.778
1999	-0.549	0.840	0.674	-0.862	0.671
2000	-0.417	0.981	0.976	-0.842	0.632
2001	-0.364	0.837	0.565	-0.813	0.440
2002	-0.157	0.677	0.742	-0.793	0.800
2003	0.290	-0.366	0.378	-0.724	0.719
2004	0.543	-0.176	0.594	-0.601	0.597
2005	0.710	-0.228	0.321	-0.545	0.507
2006	0.777	-0.083	0.603	-0.432	0.328
2007	0.864	-0.604	0.486	-0.333	0.662
2008	0.807	0.027	0.672	-0.133	0.140
2009	0.666	0.109	0.734	-0.239	-0.266
2010	0.644	-0.338	0.735	-0.556	-0.256
2011	0.551	-0.820	0.939	-0.547	-0.645
2012	0.717	-0.228	0.799	-0.591	-0.489

2013	0.875	0.607	0.354	-0.535	-0.186
2014	0.756	0.546	0.398	-0.529	-0.420

From Table 1 we see: First of all, fromthe changesof index of Guangzhou's manufacturing industries divided by industrial sectors, the comparative advantage indexarepositivein garment and textile sector and metal products sector, while they are negative on metal smelting processing sector, chemical processingand manufacturing sector, non-metallic mineral products sector and paper and printing sector.Thatmeans that Guangzhou has a long and steady comparative advantage in the labor-intensive industries such as garment and textile and metal products. However, Guangzhou shows a long-term disadvantage in manufacturing industries such as metal smelting processing, chemical processing and manufacturing,non-metallic mineral products.

Secondly, the average RSCA index value of Guangzhou's metal smelting sector is -0.9 (even in some years the index value is -0.998) and the average RSCA index value of the paper printing sector, wood processing sector and non-metallic mineral products sectorare around -0.7. These important manufacturing sectors are at a relatively significant disadvantage.

Thirdly, from the time trend of RSCA index, the RSCA index of Guangzhou's equipment manufacturing sectorraised from the -0.782 to 0.756, which shows that over the past 20 years, the equipment manufacturing sector with high technology contentin Guangzhouhas beengrowing strongly and theeffect of industrial transformation and upgrading is obvious.

2.2. Comparative Analysis to theB&R Area

In order to provide a clearer analysis of the comparative advantages of manufacturing industries in Guangzhou to the 65 countries along theBelt and Road area[2], we need to conduct a horizontal comparative analysis. We make X_{tw} and X_{iw} in the formula $RCA_{it}=(X_{ij}/X_{tj})/(X_{iw}/X_{tw})$ representing the B&R area and we will see the comparative advantage in Guangzhou to B&R area. The Table 2 shows the RSCA index of manufacturing industries in Guangzhou to B&R area from 1995 to 2014.

[2]According to the national data along the Belt and Road Network of China (https://www.yidaiyilu.gov.cn), 65 countries were selected as the research samples, including India, Ethiopia, Bosnia and Herzegovina, Montenegro, Turkmenistan, Lithuania, Latvia, Palestine, Albania Afghanistan, Estonia, Pakistan, Slovenia, Croatia, Lebanon, Oman, Bahrain, Yemen, Egypt, Jordan, Syria, Indonesia, Philippines, Singapore, Israel, Azerbaijan, Armenia, Czech Republic, Bangladesh, Belarus, Cambodia, Georgia Myanmar, Brunei, East Timor, Bhutan, United Arab Emirates, Thailand, Vietnam, Hungary, Iraq, Iran, Kyrgyzstan, Laos, Kazakhstan, Qatar, Kuwait, Moldova, Maldives, Malaysia, Macedonia, Mongolia, Nepal, Poland, Bulgaria, Romania, Serbia, Saudi Arabia, Slovakia, Tajikistan, Russia, South Africa, Sri Lanka, Turkey, Ukraine and Uzbekistan.

Table 2. RSCA index of manufacturing industries in Guangzhou to B&R area(1995-2014)

	Food processing	Wood products	Metal smelting and processing	Chemical products	Non-metallic mineral	Metal products
1995	-0.136	0.626	-0.989	-0.911	-0.890	-0.556
1996	-0.221	0.455	-0.990	-0.827	-0.879	-0.785
1997	-0.461	0.194	-0.991	-0.744	-0.869	-0.513
1998	-0.701	0.333	-0.991	-0.660	-0.859	-0.242
1999	-0.556	0.672	-0.992	-0.577	-0.848	-0.390
2000	-0.411	0.411	-0.993	-0.493	-0.838	-0.537
2001	-0.739	0.474	-0.990	-0.449	-0.747	0.157
2002	-0.632	0.276	-0.989	-0.542	-0.713	0.503
2003	0.095	0.401	-0.988	-0.583	-0.791	0.519
2004	0.379	-0.736	-0.990	-0.527	-0.827	0.523
2005	0.468	-0.886	-0.995	-0.599	-0.872	0.238
2006	0.309	-0.941	-0.991	-0.517	-0.866	0.147
2007	0.652	-0.160	-0.994	-0.424	-0.804	0.012
2008	-0.674	0.329	-0.965	-0.645	-0.893	0.104
2009	0.225	-0.326	-0.955	-0.612	-0.864	0.118
2010	0.239	-0.579	-0.957	-0.629	-0.873	0.256
2011	0.371	-0.624	-0.959	-0.653	-0.874	0.491
2012	0.318	-0.292	-0.950	-0.689	-0.873	0.388
2013	0.239	-0.691	-0.950	-0.700	-0.836	0.197
2014	0.432	-0.725	-0.969	-0.728	-0.714	0.298

	Equipment manufacturing	Leather products	Garment and textile	Paper and printing	Sports and entertainment
1995	-0.613	0.299	0.357	0.336	0.613
1996	-0.586	0.347	0.223	0.091	0.571
1997	-0.559	0.396	0.395	-0.155	0.813
1998	-0.653	0.445	0.286	-0.400	0.591
1999	-0.551	0.494	0.417	-0.646	0.101
2000	-0.448	0.543	0.549	-0.891	-0.389
2001	-0.496	0.551	0.642	-0.417	-0.196
2002	-0.432	0.349	0.690	-0.357	0.724
2003	-0.151	0.040	0.042	0.419	0.151
2004	-0.024	0.082	0.216	-0.336	-0.319
2005	0.046	0.044	0.439	0.459	0.024
2006	0.025	0.667	-0.010	-0.377	-0.720
2007	0.030	0.794	0.017	-0.299	0.676
2008	0.293	0.781	-0.014	-0.426	0.350
2009	0.291	0.657	0.586	-0.030	0.322
2010	0.423	0.626	0.541	-0.711	0.080
2011	0.418	0.679	0.577	-0.957	0.466
2012	0.519	0.771	0.473	-0.868	-0.099
2013	0.460	0.843	0.434	-0.057	0.013
2014	0.449	0.724	0.821	-0.222	-0.302

As shown in Table 2, from view of the change of manufacturing RSCA index of Guangzhou to the B&R area, the resultsare positive in garment and textile sector, metal products sector, leather productssector and sports and entertainment sector for most of the time, which reveals a long-standing comparative advantage, while they are negative in metal smelting and processing sector, chemical products sector and non-metal mineral products sector, which shows a long-standing comparative disadvantage. From the view of value of the RSCA index, the value of Guangzhou's metal smelting index wanders around -0.97 and the index value of non-metallic mineral products is around -0.84, which indicates that these important manufacturing sectors are of a significant comparative disadvantage. From the view of trend of the index, the index of equipment manufacturing sector rose from -0.613 of comparative disadvantage to 0.449 of comparative advantage. The index of food processing sector also fluctuated from a comparative disadvantage point to a comparative advantage point. This comprehensively explains that the transformation effect is significant in the equipment manufacturing sector and food processing sector in Guangzhou in recent years.

We group the 65 countries alongthe B&R area into six major regions as ASEAN, West Asia (WA), Central Asia (CA), South Asia (SA), the Commonwealth of Independent States (CIS), Central and Eastern Europe (CEE)(see Figure 2).

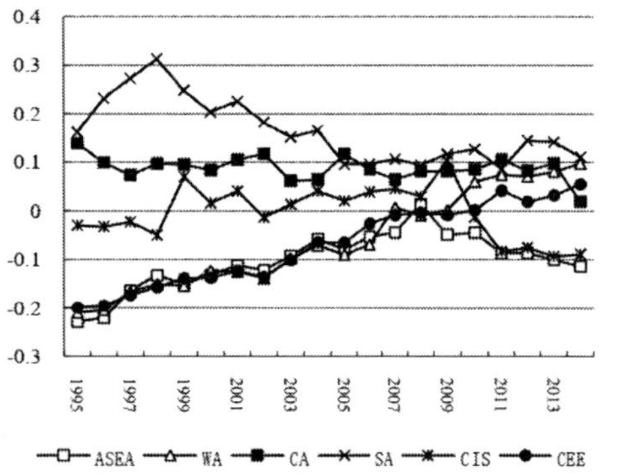

Figure 2 Comparative Advantages of Manufacturing Industries in Guangzhou to the Six Groups of B&R Area

The Figure 2 shows change trend of the comparative advantage of Manufacturing Industries in Guangzhou to the Six Groups ofB&R Area. From 1995 to 2014, Guangzhou's manufacturing industry has maintained its comparative advantage to South Asia (SA) and Central Asian (CA). However, there have been signs of declining in recent years, especially to South Asian. Since 1998, a significant downward trend has appeared. Although Guangzhou's manufacturing industry has more advantages in technology and high technological development than South Asia and Central Asia which makes Guangzhou maintaining a comparative advantage for a long time, as the cost of labor increases and the demographic dividends disappeared in Guangzhou, labor cost advantage in South Asia and Central Asia gradually appears [4], and the comparative advantage of Guangzhou's manufacturing industry is becoming weakened.

From 1995 to 2008, Guangzhou has been at a disadvantage position to ASEAN, West Asia (WA) and Central and Eastern Europe (CEE), but the disadvantage was constantly diminishing. The main reason may be that economic development of West Asia and Central and Eastern Europe and manufacturing exports were severely affected were the global financial crisis.

To ASEAN, Guangzhou's manufacturing industry turned from a comparative disadvantage to a comparative advantage in 2008, but turned into a comparative disadvantage in 2009 and continued to expand.After the global financial crisis in 2008, the domestic economic growth slowed down and the demand in the international market was sluggish. All these unfavorable factors weakened the international competitiveness of Guangzhou's manufacturing industry. Guangzhou and ASEAN countries have generally similar manufacturing structures, but ASEAN countries have lower labor costs [5], which enlarges the comparative disadvantage of manufacturing in Guangzhou to ASEAN.

In recent years, Guangzhou's manufacturing industry is under double pressure: the internal labor costs and operating costs continue to rise and theexternal economic conditions have continued to slump. In such situation, the trade condition is deteriorating. The low-end manufacturing industries have been hit by low-cost impacts from the Southeast Asia countries; while in the domestic market, the high-end manufacturing industries were suffered from the competitiveness from developed countries.

As a result, Guangzhou's manufacturing industry has begun to enter a new normal situation under the dual pressures from inside and outside.

3 Conclusion and suggestion

Guangzhou needs to develop the science and technology industries and promote the relevant policies on improving the development of modern manufacturing industry to promote the relevant policies as follows:

First of all, Guangzhou should make reasonable preparation of its advanced manufacturing industry development plan to promote technological innovation in industrial enterprises. Take advantages of the opportunities of Industry 4.0 era and 'Made in China 2025' strategy to deepen the innovation of industry and information transformation.

Secondly, Guangzhou shouldformulatethe high-end manufacturing development plans to strengthen the support of emerging industries. In order to support the breakthroughs of core manufacturing technology and its carrier construction and to cultivate the backbone enterprises, Guangzhou should increase the proportion of R&D expenditures in the city. Guangzhoucanmake full use of the capital resources like integrating the existing government funds, introducing the social capital, establishing the industry development funds or even establish the investment funds with Guangdong Province to attract more innovative investment enterprises.

Thirdly, a combination of policiesshould be introduced to promote manufacturing enterprises to reduce the cost of production factors and the business burden.

Finally, the government should give full play to its role as a platform for foreign exchange and cooperation such as the Guangzhou Overseas Exchange Association to guide the resources of overseas Chinese businessmen and overseas Chinese experts and scholars to flow into

Guangzhou and helpthe advanced manufacturing projects, personnel and funds to settle down in Guangzhou.

Acknowledgements

This work is partially supported by Key Grant Projects of Philosophy and Social Sciences Research, Ministry of Education of China (Grant No.16JZD018). Yonghui Han is the corresponding author, the e-mail is hanyonghui2006@foxmail.com.

References

1. B. Balassa, *Trade Liberalisation and "Revealed" Comparative Advantage*. Manchester School, **33**(2):99-123(1965).
2. L D.Benedictis, Three Decades of Italian Comparative Advantages. *World Economy*, **28**(11):1679-1709(2005).
3. T. Shang, Empirical Study on Comparative Advantage and Trade Pattern Change of Trade in Services of China: Analysis Based on RSCA and Lafay Indices. *Journal of International Trade*, **12**:70-77(2010).
4. Y. H. Han, X. F. Luo, Study on Regional Trade co-goverance between China and Central Asia: Feasibility of Establishing FTA Based on the "Belt and Road" Initiative. *International Economics and Trade Research*, **33**(02):72-84(2017).
5. S. H.Wang, X. J. Yuan, Analysis of the Locational Selection of theIndustiresDerectly Invested by Guangdong Enterprises in ASEAN. *International Economics and Trade Research*, **27**(10):46-52(2011).

Some Enlightenments of "Beautiful Rural Construction" on Rural Energy Policy in Beijing——Applying Informatization Means

Wang Zhi[1,*], *Wu* Kongan[2]

[1]Institute of Agricultural Information and Economics, Beijing Academy of Agricultural and Forestry Sciences, China
[2]Beijing's Tongzhou District Lucheng Towns Government, China

Abstract. "Beautiful rural construction" is a systematic project, rural energy is one of the important contents of its construction. In accordance with the concept of eco-friendly construction, Beijing carried out a thorough "structural adjustment of rural energy optimization," "Earthquake energy-saving projects of rural housing" and other measures. By conventional heating technology research in Beijing 13 counties and 142 villages, we predict the future of rural energy will further the implementation of solar heating, electric heating and other new green energy technologies. It is suggested to establish the "Beijing Rural Information Service Platform" and "Beautiful Rural Information Resource Bank" through the means of informatization, which will greatly strengthen the regulation and control of rural people-land relationship and realize the systematic optimization, making the cities and villages have. Space for human survival and sustainable development.

1. Introduction

Digital City is the use of spatial information to build virtual platforms, including urban natural resources, social resources, infrastructure, cultural, economic and other information about the city, in order to obtain and load up the digital form, so as to provide a wide range of aspects of government and society service [1-3]. Through advanced means of information, support Beijing new rural planning, construction, operation, management and emergency, can effectively improve the government management and service levels, improve the new rural management efficiency[4], conserve resources, promote the development of the beautiful countryside.

2. "Beautiful Rural Construction" is an important basis for the construction of "ecological civilization and beauty of China"

Chinese capital of Beijing, an area of 16,400 square kilometres in rural areas accounted for 1.5 million square kilometres, if rural development is not good, ecological Beijing will be no barrier, there will be no space environment, and sustainable development will be on the lack of foundation [5].

3. The Main Measures and Measures of Rural Energy Construction in "Beautiful Rural Construction"

We recognize the importance and urgency of the beautiful countryside construction, but also, perceive that the complexity and difficulty to build a beautiful countryside. In recent years, Beijing launched a series of livelihood projects, including farmhouse seismic energy-saving projects, risk village and insurance households from mountain relocation project, Clean Air Initiative in rural areas reducing coal, Agricultural non-point source pollution control projects, Rural Ecological Environment Construction Engineering, and so on, rural environment continues to improve, rural infrastructure and public service facilities gradually improved, the rural energy structure was further optimized, farmer housing conditions and income levels are rising[5,6]. New Rural Construction of Beijing has been built a beautiful countryside and laid a good foundation. At the same time we should also see that although the new rural construction of Beijing has made remarkable achievements, yet, with the urgent needs of the majority of farmers, and the requirements of building world-class city, there is still a big gap, but also need to be improved in the following areas.

3.1. Rural energy structure adjustment and optimizations

In recent years, the Beijing government has been to actively promote the use of new energy sources, new technologies in rural areas, and started to implement the " Reducing Coal and Replaced Coal, Clean air" Initiative. Preliminary accounting, in 2013 heating season, the Beijing rural areas completed 2954 tons of sulphur dioxide emissions, nitrogen oxides 3512 tons emissions,

* Corresponding author: angelwang1009@163.comg

and 7397 tons of carbon dioxide emissions, accounting for 22% of the city's rural areas, 4.3 million tons of total coal-fired. but there are still nearly 80% of farmers are still using low-grade coal[5], as well as a large number of public facilities in rural areas, enterprises and institutions were not clean energy transformation. Therefore, we should continue to promote the rural areas "Reducing Coal and Replaced Coal, Clean air" Initiative. Through the joint action of government, market and society, at a rate of about 1 million tons per year, with a further three years to complete rural clean alternative, while by replacing heating coal to electricity, sending natural gas to home, as well as using liquefied petroleum gas, appliances solar, biomass, air energy, geothermal energy utilization, etc., to realize gasification of rural areas for cooking, heating and clean [7, 8].

Through the survey of rural areas in Beijing found that the residents of the current application of the main heating technologies are electric heating, heat pump heating, gas heating, clean coal heating and solar heating in five ways[5]. According to the survey data, combined with the application of the Beijing area and product technology maturity, the existing mainstream rural areas in Beijing can be used for rural use of clean energy technologies in Table 1.

Table 1. Commonly used in rural areas heating technology

Serial number	Heating mode				
	category	**Heat source (priority)**	**Auxiliary heat source (alternate)**	**Terminal**	
1	Heat pump	Low temperature air source heat pump	—	Floor heating	Low temperature hot water terminal
2		Ground source heat pump	—	Floor heating	
3	Solar composite heating category	Solar hot water heating	Clean coal boilers	Floor heating	
4		Solar hot water heating	Gas boiler	Floor heating	
5		Solar hot water heating	Low temperature air source heat pump	Floor heating	
6		Solar hot water heating	Ground source heat pump	Floor heating	
7		Solar hot water heating	Electric heating	Floor heating	
8	Gas category	Gas boiler	—	Radiator / floor heating	High temperature / low temperature hot water terminal
9	Coal-fired class	Clean coal heating stove	—	Radiator / floor heating	
10	Electric heating	Electric hot water boiler	—	Radiator / floor heating	
11		Heat storage type electric boiler	—	Radiator / floor heating	
12		Carbon crystal plate class	—	—	—
13		Energy storage electric heating	—	—	

3.2. Rural Residential Seismic energy-saving projects

Beijing government started farmhouse seismic energy-saving projects since 2007, it has supported 380,000 rural residents in the implementation of energy-saving residential earthquake. Through the implementation of seismic energy-saving projects, farmers housing winter indoor temperature has increased by 4 to 6 degrees, saving more than 30% of coal-fired[9,10]. In summer, obviously feeling the cool interior, indoor temperature is relatively low outside temperatures 5 to 6 degrees [11]. According to statistics, only this thermal insulation, has an annual saving of coal-fired nearly 55 million tons, a significant reduction in carbon dioxide emissions, effectively improve air quality. According to surveys, there are still 530,000 rural residents in Beijing have seismic energy-saving requirements. Therefore, to further increase the Farmhouse seismic energy-saving efforts, using of 5 years, at a rate of about 100,000 a year, it will complete the city's Farmhouse seismic energy-saving task.

3.3. Sewage treatment works in rural areas

In recent years, new rural construction in Beijing, more than 500 villages have been built sewage treatment facilities in 1010, but due to complex process, facilities scattered layout, unfocused, factory network is not complete, the high cost of operation and maintenance, management operation is not in place and other reasons, rural sewage treatment facilities, 30% did not run or intermittent operation. In Beijing rural area, annual output of sewage is more than 100 million tons, but, treatment rate is only about 50%. Large amounts of sewage is discharged directly [12]. With the rapid development of urbanization and rural economic and social, supply and demand contradictions in rural city sewage treatment facilities have become increasingly prominent. Therefore, the reality requires us to rural sewage treatment as an important part of the construction of the beautiful countryside. For facilities that have been built to run on to assess the situation, according to local conditions to transform and strengthen the operation and maintenance, and ensure that the newly built sewage treatment facilities operating normally, the basic objectives of rural sewage discharge standards may successfully achieve in 2020.

4. Beijing's rural energy demand forecast

4.1. Analysis of energy - using factors

4.1.1 Farmers own needs

With the development of rural economy, farmers' concept of energy use is gradually changing. The higher income level of farmers, the more emphasis on the convenience and comfort with energy, and then health, the last is the economy. Therefore, in cooking, the farmers are more willing to use the cleanliness of liquefied petroleum gas or biogas, rather than coal. In the heating, according to affordability, farmers will focus on electricity, solar and other clean energy heating.

4.1.2 The Impact of Urbanization Policy

With the acceleration of urbanization in the rural areas and the implementation of ecological resettlement, the rural energy structure has also been greatly affected.

(1) Heating will be converted into centralized heating. The Beijing municipal government proposed that by 2020 the city will make about two-thirds of the farmers into urban residents, the living conditions of farmers will be dispersed from the living, into a centralized way to live, therefore, the corresponding heating will focus on the form heating mode change, which will help save energy consumption of heating [13-15].

(2) Farmers' choice of energy will become more commoditized and cleaner. Due to the concentration of living forms and the difficulty in obtaining biomass resources such as straw, farmers will tend to favour commercialized energy sources for energy selection [13-15]. With the popularization and cleaning of public facilities such as heat and gas, the proportion of the use of energy will be a substantial increase.

(3) Housing construction will become more energy-saving. With the process of urbanization, the new farmers will focus on energy-saving housing and comfort [16], thus directly reduce the heating load and reduce the use of heating energy.

(4) Energy use will become more low-carbon and comfortable [17]. Urbanization directly promote the increase of farmers' income, and with the construction of new houses, solar energy and geothermal energy and other low-carbon new energy and renewable energy in the proportion of domestic energy consumption will gradually increase.

4.2. Potential evaluation of coal adjustment in rural energy structure

In the heating, through the existing energy-saving transformation of rural households, the annual heating coal can be saved about 62.76 million tons of standard coal (in accordance with the energy-saving transformation to save 40% of the heating coal to calculate). On this basis, by 2020, 952,000 farmers will live in the cities and towns of residence, so that the original decentralized heating will transform into a central heating. This way can reduce heating coal-burning 94,100 tons of standard coal per year.

In addition, for the remaining decentralized heating farmers, the Government can actively promote solar heating, electric heating and other new green energy-saving technologies, which can reduce the heating coal-fired 25.10 million tons of standard coal.

From the adjustment potential, based on full use of existing resources and actively promote green energy-saving technology, respect for heating and cooking, the total coal pressure reduction will be more than 1.1451 million tons of standard coal per year.

5. Reflections on "Beautiful Rural Construction" by Using Information

In the capital, to promote the digital city construction process, the author has deeply thinking for rural energy policies and beautiful countryside construction as follows:

First, the establishment of Beijing rural information service platform and the beautiful countryside of information resource library. It will cover many aspects of rural infrastructure, rural economy, industry characteristics, population health, culture, education, grass-roots organizations and other resources for effective integration of information and data, and establish a cover Beijing, connecting Beijing, Tianjin city, county, township and village "digital village" information network platform.

The second is the establishment of "Beijing Digital Village" platform, its function should have a strong practical and advanced. It can form in a variety of vivid visual image, data, icons and text, etc., it can provide information on all aspects of rural work for the government at all levels in Beijing, so as to provide a reliable basis for the scientific and democratic decision-making of new rural construction. Beautiful countryside construction in every measure can be demonstrated and information management, the establishment of relevant information platform. Moreover, with the continuous improvement platform, which will provide reliable first-hand information for all agricultural sectors and investment units, it lay a solid foundation for the economic development of Beijing Rural Development and Investment, the same time, the Present Situation can be shown by the platform the beautiful countryside as a window, it can be to promote the rural areas this emerging e-commerce business model, so as to promote the exchange of agricultural products and accelerate rural economic development has laid a good foundation.

The third, pay attention to the establishment of several information platform, focusing on people's livelihood and ecological protection. The main measures "optimal adjustment of rural energy structure, agricultural nonpoint source pollution control, rural residential seismic energy-saving projects, sewage treatment works in rural areas" and other beautiful countryside construction are to be monitored and managed through the information platform to maximize for the construction of ecological civilization provide effective protection, as well as the national new rural construction to enhance digital play a leading and exemplary role.

References

1. Meng xiang-lin.. CITY, (2011) （4） .
2. HE Ling, WANG Jun,DONG qian.. JIANGSU AGRICULTURAL SCIENCES, 2011, 39(2).
3. Digital City Network. http://www.4d3s.com/
4. http://baike.baidu.com/view/2282762.htm.
5. WANG Zhi. Mechanical Engineering and Control Systems. 2016,PP:490-493
6. CHEN Zhang, YAN Ping, SHAO Wei. Beijing, Tianjin Economic Contacts[M].DATA, 2011, (3).
7. Yang X, Jiang Y, Yang M, et al. Energy and environment in Chinese rural housing: Currentstatus and future perspective. Front. Energy Power Eng. China 2010, 4(1): 35-46.
8. Ho-Lin H, Murray M, Cohen T, Colijn C, et al. modeling study. Lancet, 2008, 372:1473-1483.
9. Al-Homoud M S. Building and Environment, 2005, 40: 353–366.
10. Omakli K, Yuksel B. Applied Thermal Engineering,2003,23:473-479.
11. Erlandsson M. Building andEnvironment,1997,32(2):129-136.
12. WANG Zhi, ZHOU Lian-di, LIU Shi-rong, et al. JOURNAL OF ANHUI AGRICULTURAL SCIENCES, 2009, 37(10).
13. Ahmad.A, Saini JS, Varma HK. Energy Conversion and Management, 1995, 36(12): 1185-1195.
14. Hegazy AA. Energy Conversion and Management,2000, 41(4):401-417.
15. Varol Y, Oztop HF. Building and Environment, 2008, 43(9): 1535-1544.
16. Varol Y, Oztop HF. Building and Environment, 2007, 42(5): 2062-2071.
17. Wang Gehua, Chen Yanbini, Shi Deming. Transactions of the CSAE.1992(9):60-66.

Research characteristics and status on Environmental Psychology: a bibliometric and knowledge mapping analysis

Jun Hua XU[1], *Lei Ye* ZHU[1,*], and *Hai Bin* WANG[2,1,*]

[1]School of Educational Science, Huangshan University, Huangshan, 245041, China
[2]School of Business Administration, Zhejiang Gongshang University, Hangzhou 310018, China

Abstract. Environmental psychology is an interdisciplinary field that focuses on the interplay between individuals and their built and natural environments. Great progress has been made in the areas of environmental psychology by researchers form many countries. However, a thorough quantitative analysis to the emergent research trends and topics has not been found. To reveal the research characteristics and status on Environmental Psychology, 853 related papers from Web of Science core collection were analysed by CiteSpace II. The results show that: (1) the domain of Environmental Psychology was started in 1960s and showed a low growth over the past half century, which reaches a historical peak in 2017. Gifford, USA and the Journal of Environmental Psychology top the list of contributing authors, country and publication respectively. (2) "Environmental Psychology" is the most frequently keywords and has the longest span of the bursts. "ambient scent", "recycling", "children as outsiders" and "ambient temperature" are the top four largest clusters, which are the popular research topics in the domain of environmental psychology.

1 Introduction

Environmental psychology is an interdisciplinary field that focuses on the interplay between individuals and their built and natural environments [1]. The environments were broad, encompassing natural environments, social settings, built environments, learning environments, informational environments, and so on. Especially as people spend more time on network, the virtual environments are coming under increasing research attention [1]. The discipline is considered to be a key component of human, animal, and environmental welfare and is essential for policy-making [1]. It was in 1967 that "environmental psychology" was mentioned in the article, which was published by Rohles [2], but the birth of the discipline was marked by the establishment of "Journal of Environmental Psychology", which was created by Canter in 1979[3]. Subsequently, environmental psychology attracts many researchers' attention, and the domain includes theory research and practice aiming at improving human relations with the natural environment and making the built environment more humane [4]. In the theoretical analysis of the field, researchers have explored the core concepts [5], such as place identity, place attachment, environmental consciousness, behavior settings and so on. For instance, the ontology, epistemology and methodology proposed by Brown lay an important foundation for the research of pro-environmental psychology [6]. In the practical analysis of the field, researchers have explored the impact on the built environments to improve the environments, such as the design of nursing home [7], school environments [8], military camp [9], modern interior [10], architecture [11], and even the design of mobile application stores form the environmental psychology perspective [12]. As mention in the literature review, great progress has been made in the areas of environmental psychology from 1967 to 2017, however, very little was found in the literature to review the fruitful achievements in the field through a quantitative analysis by bibliometric and Knowledge Mapping. The purpose of the current study was to determine a quantitative review of Environmental Psychology researches based on the bibliometric analysis and knowledge mapping with the help of the CiteSpace II, which can explore the hot spot transformation in the research evolution of Environmental Psychology and then further promote the development the field of environmental psychology.

2 Methodology

2.1 data sources

The sources and cited papers for analysis were from the Web of Science core collection, which is the largest and most comprehensive academic information resources in the world covering over 8,700 core academic journals with the most influential research fields, for instance, science, engineering, biomedicine and other fields. The search conducted in the Web of Science core collection yielded 853 papers of all types under the WoS topic of

* Corresponding author: zhuleiye@hsu.edu.cn or wanghaibin@hsu.edu.cn

"environmental psychology", which contains 556 articles, 127 proceedings paper, 75 book reviews, 57 reviews, 24 meeting abstracts, and so on.

2.2 data processing

853 papers were saved in RIS format which includes the crucial information, encompassing title, authors, affiliations, publication name, abstract, keywords and so on. The bibliometric and knowledge mapping analyses were conducted with Java CiteSpace II software [13], which can used to bibliometric analysis and knowledge mapping analysis [14].

3 Results

3.1 Bibliometric analysis results

3.1.1 Chronology statistics

Based on the yearly analysis, the annual number of environmental psychology study published from 1967 to 2017 is shown in Fig. 1. The domain of Environmental Psychology was started in 1960s and showed a low growth over the past half century. Until 2002, the publishing trend of the field demonstrated a year-on-year growth, and reached a historical peak in 2017.

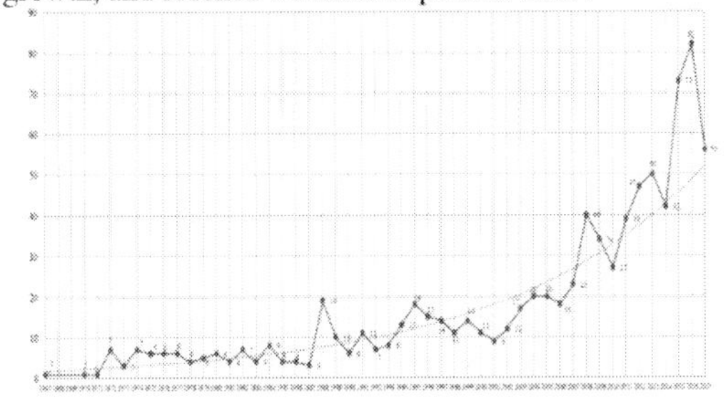

Fig.1. The annual number of the research (1967-2017)

3.1.2 Author statistics

Table 1 outlines the top 10 contributing authors and the numbers of their published papers. As shown, all of them have published more than five articles and Gifford tops the list with a total of ten articles, followed by Garling and Rioux, each of which has more than seven articles.

Table 1. Top 10 contributing authors

Rank	Author	FREQ
1	Gifford R	10
2	Garling T	8
3	Rioux L	8
4	Spencer C	7
5	Joye Y	6
6	Pol E	6
7	Stokols D	6
8	Wapner S	6
9	Anonymous	5
10	Bonaiuto M	5

3.1.3 Country statistics

Table 2 depicts the top 10 countries contributing to the studies on Environmental Psychology. As seen, USA is the best performer, followed by England, Canada and Australia, and each country has been contributed to more than fifty articles.

Table 2. Top 10 contributing countries

Rank	Country	FREQ
1	USA	240
2	England	89
3	Canada	52
4	Australia	50
5	Germany	47
6	Netherlands	37
7	Peoples R China	37
8	Sweden	34
9	France	33
10	Italy	30

3.1.4 Source publication statistics

Table 3 depicts the top 10 source publications contributing to the studies on Environmental Psychology. As shown, Journal of Environmental Psychology tops the list with a total of ninety articles, followed by Environment and Behavior, which has been published thirty-six articles.

Table 3. Top 10 contributing source publication

Rank	source publication	FREQ
1	Journal of Environmental	90
2	Environment and Behavior	36
3	Contemporary Psychology	16
4	International Journal of Psychology	16
5	American Psychologist	12
6	Journal of Business Research	12
7	Annual Review of Psychology	10
8	Revista Latinoamericana De	10
9	Frontiers in Psychology	9
10	Landscape and Urban Planning	9

3.2. Knowledge mapping results

3.2.1 High frequency keywords

Table 4 describes the top 10 high frequency keywords in the domain of the Environmental Psychology, as well as their frequencies. As shown, "environmental psychology" is the most frequently keywords. And "behavior", "environment", "perception" and "model" are positioned in the 2^{ed}, 3rd 4th and 5th places on the list. Fig 2 visualizes the same result (The higher the frequency, the larger the circle). It indicates that the interplay between individuals and environments attracts researchers' attention continuously in the field of environmental psychology.

 https://doi.org/10.1051/e3sconf/20183801020

Table 4. Top 10 keywords

Rank	Keywords	FREQ
1	environmental psychology	236
2	behavior	94
3	environment	68
4	perception	55
5	model	51
6	identity	44
7	health	44
8	psychology	37
9	experience	36
10	community	35

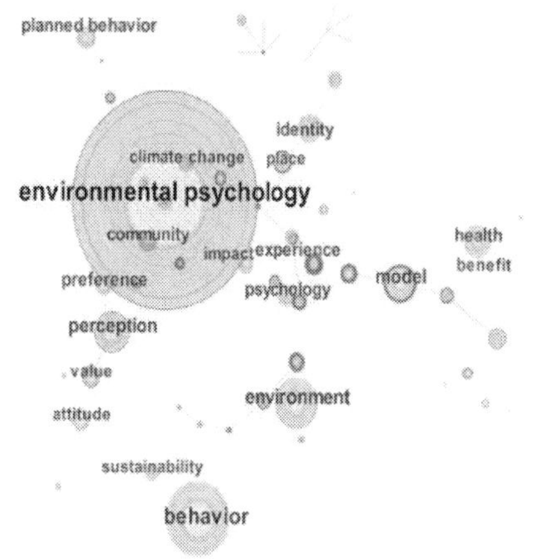

Fig.2. Top 10 keywords

3.2.2 Popular research trends

Fig 3 shows the top five keywords with the strongest citation bursts. As indicated, we discover that "environmental psychology" and "behavior" have the longest span, each of which has more than thirty-six years, which indicates that the relationship between environment and behavior is the eternal topic in the domain. In addition, the topic of "climate change" is one of the typical recent research frontiers, which started since 2014 and continued until 2017.

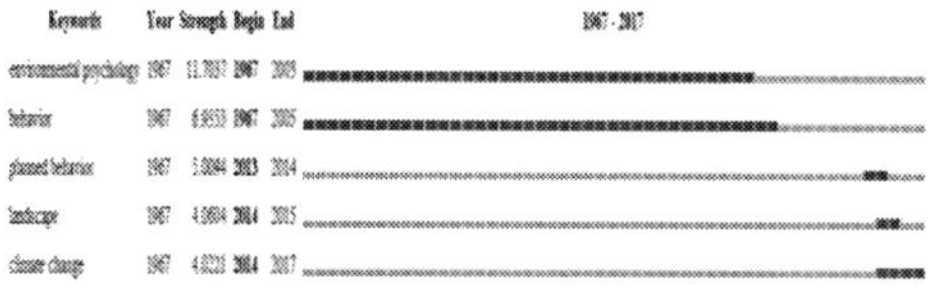

Fig.3. Top five keywords with bursts

3.2.3 Popular research topics

Fig 4 shows the top four largest clusters, which were "ambient scent", "recycling", "children as outsiders" and "ambient temperature". As shown, the largest cluster is "ambient scent" (#0), which contains 30 articles and its Silhouette value is 0.715. The high-frequency keywords of this cluster contain "environmental psychology", "model", "satisfaction" and "impact". The second largest cluster is "recycling" (#1), which contains 23 articles and its Silhouette value is 0.846. The high-frequency keywords of this cluster contain "sustainability". The third largest cluster is "children as outsiders" (#2), which contains 19 articles and its Silhouette value is 0.900. The high-frequency keywords of this cluster contain "impact", "attitude", "value" and "experience". The last largest cluster is "ambient temperature" (#3), which contains 18 articles and its Silhouette value is 0.995. The high-frequency keywords of this cluster contain "perception", "preference", "benefit" and "health".

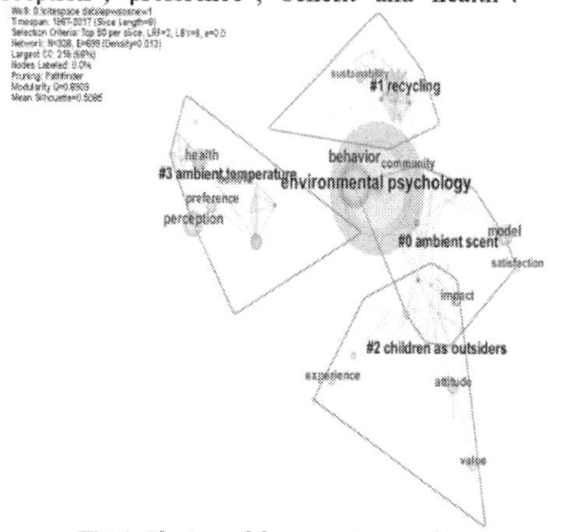

Fig.4. Clusters of the current researchers

4 Conclusions and further research

During the past half century, great progress has been made in the areas of environmental psychology by researchers form many countries, such as USA, England, Canada, Australia, Germany, Netherlands, China, and etcetera. However, a thorough quantitative analysis to the emergent research trends and topics has not been found. To fill this gap, the present study was to determine a quantitative review of Environmental Psychology researches based on the bibliometric analysis and knowledge mapping with the help of the CiteSpace II, which can explore the hot spot transformation in the research evolution of the field. In the bibliometric analysis section, the results show that the domain of Environmental Psychology was started in 1960s and showed a low growth over the past half century, 2017 reached a historical peak. Gifford tops the list with a total of ten articles, followed by Garling and Rioux. Similarly, USA is the best performer, followed by England, Canada and Australia. Moreover, the Journal of Environmental Psychology tops the list of the contributing source publications, echoing with the birth of environmental psychology.

In the section of Knowledge mapping analysis, the high frequency keywords are obtained by CiteSpace II. As shown in table 4 and fig 2, "environmental psychology" "behavior", "environment", "perception" and "model" are the most frequently keywords, indicating that the interplay between individuals and environments attracts researchers' attention continuously. Most importantly, the major popular research trends and topics in the field are obtained through keywords bursting and clustering by CiteSpace II. As shown, "environmental psychology" and "behavior" have the longest span, each of which has more than thirty-six years , indicating that the relationship between environment and behavior is the eternal topic in the domain. In addition, the topic of "climate change" is one of the typical recent research frontiers, which started since 2014 and continued until 2017. The result of clusters analysis shows that "ambient scent", "recycling", "children as outsiders" and "ambient temperature" are the top four largest clusters, which are the popular research topics in the domain of environmental psychology.

Certainly, there are still some limitations in our study. The database employed is only form the Web of Science core collection. In fact, there are many excellent articles about environmental psychology that are not included in this database. Not only need we to expand the scope of the in search, but also, we should further explore the status quo of the environmental psychology research.

5 Acknowledgement

This research was supported by the Key Project of Humanities and Social Sciences in Anhui Province of China (Grant NO. SK2015A519 and SK2014A390) and the Revitalization Plans of Anhui Province in China (Grant NO. Szzgih3-20 and Grant NO. Szzjih1-2-2018-19).

15.

References

1. R. Gifford, ANNU REV PSYCHOL **65**,541 (2014)
2. F. H. Rohles, PSYCHOL TDY **1**, 54 (1967)
3. A. Gill, B BR PSYCHOL SOC **35**, 74 (1982)
4. J. L. Nasar, BEHAV SCI **30**, 375 (2015)
5. C. S. Nanninga, L. Meijering, M. C. Schonherr, DISABIL REHABIL **37**, 1125 (2015)
6. P. Brown, ENERG RES SOC SCI **31**, 215 (2017)
7. Z. M. Moghadam, R. Mirzaei, INT J COMPUT SCI NET SCY **17**, 289 (2017)
8. G. Amicone, I. Petruccelli, M. Bonaiuto, PSICO SOC **12**, 131 (2017)
9. R. Cervinka, K. Roederer, I. Haemmerle, J ARCHIT PLAN RES **31**, 43 (2014)
10. K. Walker-Springett, R. Jefferson, K. Boeck, J ENVIRON MANAGE **166**, 525 (2016)
11. J. L. Nasar, J PLAN LIT **30**, 375 (2015)
12. J. Song, J. Kim, D. R. Jones, DECIS SUPPORT SYST **59**, 37 (2014)
13. C. Chen, J DATA INF SCI **2**, 1 (2017)
14. C. Chen, L. Leydesdorff, J ASSOC INF SCI TECH **65**, 334 (2014).

Spatial and Temporal Characteristics of Precipitation in Yungui Plateau during 1961-2015

Wu Jianfeng[1,2], Yu Luqin[1], Cao Guangjie[1], LI Wei[3]

[1]School of Geography and Tourism, Guizhou Education University, Guiyang 550018, China;
[2]Guizhou Provincial Key Laboratory of Geographic State Monitoring, Guiyang 550018, China;
[3]Institute of Mountain Resource, Guizhou Academy of Sciences, Guiyang 550001, China;

Abstract: In this paper, the daily precipitation data of 19 meteorological stations in Guizhou Province from 1961 to 2015 are used. Using GIS spatial analysis method and linear trend analysis method, the distribution characteristics of annual and seasonal precipitation in Guizhou province were analyzed from space and time. The results show that: (1) From the perspective of spatial distribution, annual precipitation is generally less in the south and north and less in the east and west. The precipitation in winter and spring is east-west distribution and decreases from east to west in the four seasons. The precipitation in summer is roughly north-south and south-north less. The precipitation in autumn is mainly concentrated in southwest Guizhou and Chishui valley. (2) From the time distribution point of view, the precipitation in Guizhou Province showed a general downward trend, of which the precipitation in spring, autumn and winter showed a decreasing trend, the downward trend in autumn was more obvious, and the precipitation in summer increased but not obviously.

1 Introduction

Precipitation is an important element that reflects the climatic characteristics of a region and is an important climatic factor affecting or reflecting the ecological environment[1]. With the development of science and technology, people's requirements for meteorological services are increasing year by year. Research on the spatial distribution of precipitation must be intensive and meticulous in order to meet the needs of modern society[2]. Therefore, it is more and more urgent to evaluate and predict long-term precipitation characteristics [3].

Most foreign experts and scholars believe that the average precipitation in the world and the northern hemisphere is a periodical change, but the trend of change in each stage is different. Hulme study shows that between 1950 and 1980, there was no significant decrease in the global average precipitation and a significant decrease in precipitation after the 1990s[4]. Vinnikov's study of precipitation in different regions of the Northern Hemisphere shows that precipitation increased by nearly 6% in the mid-latitudes of the Northern Hemisphere from the early 19th century to the 1990s[5].Kattsov proposed that the precipitation in the northern hemisphere mid-latitudes has been increasing strongly since the 1920s[6]. Diaz et al. Obtained the study on the precipitation in the southern hemisphere, and the trend of precipitation in the mid-high latitudes and low latitudes showed consistent[7]. Zuo Hongchao et al. Conducted a study on the precipitation in our country, showing that the regional average precipitation characteristics are obvious and the trend of change is not obvious[8]; Wang Yanjiao et al. The regional precipitation and decadal variation characteristics of precipitation in China during 1960-2010 show the precipitation There are obvious stages of change[9]; Li Yonghua et al. Analyzed the spatio-temporal features of the summer precipitation in the eastern part of Southwest China. The summer precipitation in the eastern part of Southwest China is mainly consistent with the whole region, and the frequency of occurrence is above 60%[10].

In recent years, although some scholars have done some research on precipitation in Guizhou Province, the characteristics analysis of precipitation in Guizhou Province in time and space is still relatively weak. Therefore, by applying the daily precipitation data of Guizhou Province in recent 55 years, the paper analyzes the precipitation characteristics of Guizhou Province in recent years by using ArcGIS technology and linear trend analysis method from different time scales and spatial analysis.

2 Study Area Overview

* Corresponding author: [a]wujianfeng74623@sina.com; [b]110695145@qq.com; [c]caoguangjie@gznc.edu.cn; [d]624360114@163.com;

Guizhou Province is located in the southwestern part of China and is located in the southeast of the Qinghai-Tibet Plateau. It is a part of the Yunnan-Guizhou Plateau with an average elevation of about 1,100 m. It belongs to a subtropical monsoon climate and is humid and humid throughout the year with an average annual temperature of about 15-18° C[11]. The terrain is high in the west and low in the east, with complex and diverse topography. 90% of the area is mountainous and hilly, with 73% of the area being karst.

3 Research Data and Methods

3.1 Data and Processing

In this paper, the daily precipitation data of 19 meteorological stations in Guizhou province for 1962-2015 years are selected as the research data, and the data are derived from the China Meteorological Data Network (http://data.cma.cn). The quarterly and annual precipitation data used in this study are calculated by accumulating daily precipitation.

3.2 Research Methods

3.2.1 Linear trend method

The linear trend analysis refers to the use of n to express the sample size, the time is expressed in t_i , the precipitation variable is expressed by x_i, and the linear regression equation between x_i and t_i is established.

$$\hat{x}_i = a + bt_i \qquad (i = 1,2,...n)$$

In the upper form, the precipitation factor is expressed by $\hat{x}$, the a is the regression constant, the b indicates the regression coefficient and the t represents the time [10].

3.2.2 Kriging

Kriging is one of the most commonly used spatial interpolation methods, the general formula is as follows:

$$\hat{Z}(x_0) = \sum_{i=1}^{n} \lambda_i z(x_i)$$

Among them: $Z = (x_i)$ the observations, they are located in the region x_i position; X_0 is the observed point to be estimated; λ_1 is the weighted coefficient; n is the total number of known observation points[12].

4 Results and Analysis

4.1 Precipitation spatial distribution

4.1.1 Spatial distribution of annual precipitation

Guizhou precipitation affected by topography, topography, altitude, slope and other factors. Precipitation performance of more than North South, East and West less. Among them, the precipitation in the eastern part of Tongren region, the southwestern part of Southeast Guizhou, the eastern part of Qiannan, most of Anshun and most of southwestern Guizhou are more than those of the other regions, and the average annual precipitation is basically above 1200mm; In Bijie region, southern Zunyi, the average annual rainfall is not higher than 1000mm (Figure 1).

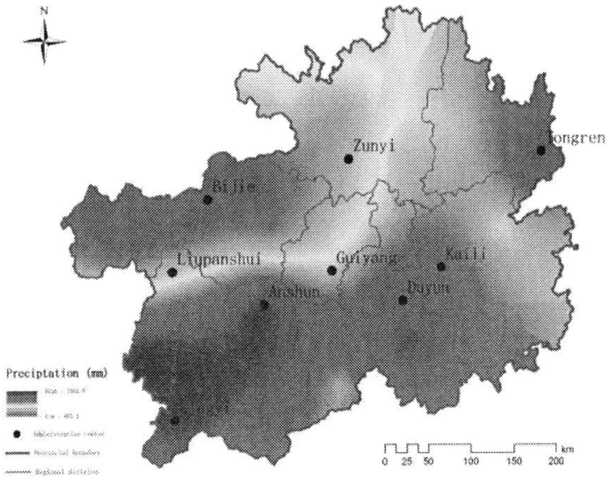

Figure 1 1961-2015 year average annual precipitation distribution map of Guizhou

4.1.2 Seasonal precipitation distribution

On the whole, the precipitation in spring decreased gradually from east to west, and the precipitation was more in the eastern part of Tongren, the eastern part of Guizhou Province and the southern part of Guizhou Province, with the precipitation reaching 390mm. The second part was the west of Tongren, the eastern part of Zunyi, Anshun And western Qiannan, the average precipitation is 360.3mm; the average precipitation in western Zunyi, Guiyang and western Guizhou is 311.0mm; in Bijie area, the average precipitation is only

173.1mm(Figure 2 (a)).

Summer is the year with the heaviest precipitation, which is affected by the summer monsoon from the Indian Ocean and the warm and humid currents over the Pacific Ocean, plus the uplift of the Yunnan-Guizhou Plateau. Precipitation changes roughly north-south, of which precipitation is mainly concentrated in Anshun, southern Liupanshui, most of Qianxinan, the average precipitation is 701.4mm. Followed by Qiannan Prefecture and Qiandongnan Prefecture. The average precipitation is 540.9mm. The less precipitation is the northern part of Bijie, the southwest part of Zunyi and the eastern part of Qiandongnan. The average precipitation is 456.1mm(Figure 2 (b)).

In the fall, precipitation rose from Siberia cold air and the Yunnan-Guizhou Plateau to the area north of Guiyang, except for the precipitation in the Chishui River with more precipitation (297.9mm), the precipitation decreased from north to south, southwestern Qianxinan Prefecture and southern Liupanshui The precipitation in the vast area is relatively concentrated. Precipitation is relatively small areas: most of Bijie area, southwest of Qiannan, and the eastern part of Qiandongnan(Figure 2 (c)).

East and West less precipitation in winter, something significant difference. Affected by the south cold air is relatively large, Tongren southeast of most parts of southeastern Guizhou, Qiannan Prefecture, more precipitation, the average rainfall of 104mm; precipitation in the central region of Guizhou in less than 65.86mm; Bijie more Less, the average precipitation is less than 50mm, at least Weining area, only 29.89mm(Figure 2 (d)).

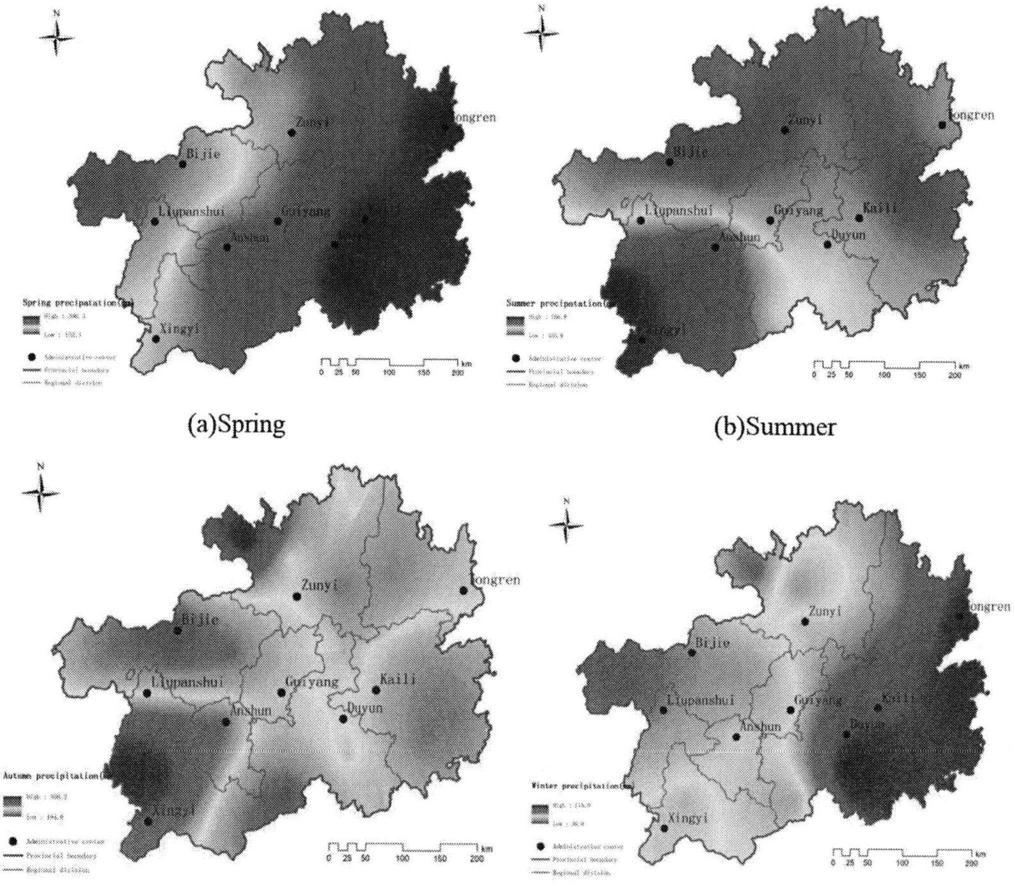

(a)Spring (b)Summer

(c)Autumn (d)Winter

Figure 2 The spatial distribution of seasonal precipitation in Guizhou in the past 1961-2015 years

4.2 Analysis of time trend characteristics of precipitation

4.2.1 Characteristics of annual precipitation time

From the linear trend, the trend of annual precipitation in Guizhou province is decreasing with the tendency of 14.18mm/10a. From the 5 year moving average precipitation can be seen in the past 55 years, precipitation of Guizhou province from 1961 to 1980, volatility is not, in 1980 to 1990 between the annual precipitation decreased significantly, from 1991 to 2000 completed a reduction by increasing from 2001 to 2015, precipitation showed a decreasing trend. From the numerical point of view, the annual average precipitation in Guizhou province is 1147.2mm in 1961~2015. The year with more rainfall is 1967, the annual precipitation is as high as 1357mm, followed by 2014, nearly 1340mm. The lowest annual precipitation appears in 2011, only 818.2mm, and the annual precipitation is the largest and the least year is 538.8mm (Figure 3).

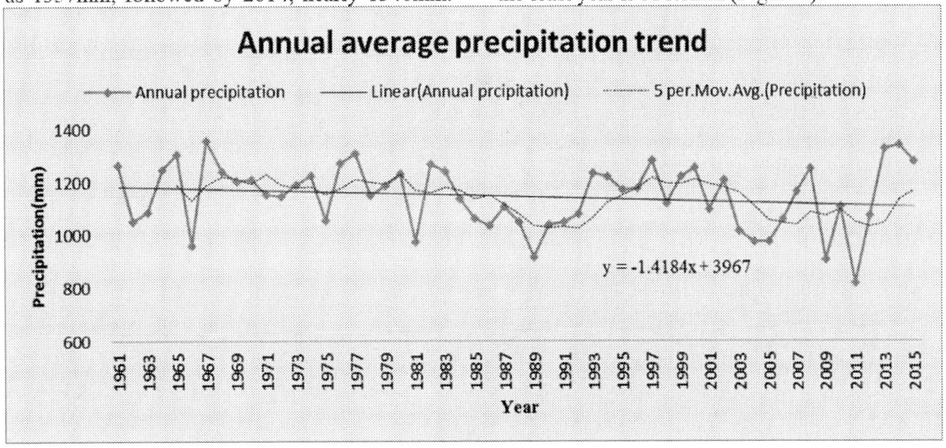

Figure 3 Change trend of annual precipitation in Guizhou province 1961~2015

4.2.2 Seasonal distribution of precipitation

The study area obtained by the seasonal variation trend of precipitation can be (Figure omitted), summer precipitation in Guizhou province has increased a weak trend, increasing the size of 4.0mm/10a; in the spring and autumn winter three season precipitation showed a downward trend, the average rainfall in the three quarter decreased in size respectively is 6.4mm/10a, 10.9mm/10a, 1.7mm/10a.In addition, according to the 5-year sliding average of the seasonal precipitation:

The average precipitation in spring shows a continuous increase trend from 1961 to 1975; there is a decreasing trend from 1976 to 1990; among them, the decrease extent is the most obvious since 1980. Statistics show that precipitation in 1980 than in 1991 decreased by 106.5mm.

Summer precipitation showed a decreasing trend from 1961 to 1967; Precipitation increased steadily from 1967 to 1971; Precipitation decreased obviously from 1971 to 1977; Precipitation changed little from 1978 to 1986; Precipitation decreased from 1992 to 2001. From 1992 to 2001, a small precipitation peak appeared. However, precipitation tends to decrease from 2006 to 2015. Overall, summer precipitation in Guizhou Province has been on the rise for the past 55 years.

Precipitation decreased from 1961 to 1971 in autumn; slightly increased from 1972 to 1983; declined again from 1982 to 1992; precipitation increased from 1992 to 2001; from 2001 to 2015 The annual precipitation slowly recovered from its lowest level in 55 years.

There are three small peaks and three small tumbles in winter, with the three small peaks appearing in 1971-1976, 1983-1985 and 1993-1998 respectively. Three minor ebb occurred in 1961-1971, 1976-1983 and 1985-1992, respectively. However, the precipitation has been decreasing from 1996 to 2012, with a slight increase but not significant from now to 2013.

5 Conclusion

Based on the daily precipitation data from 1961 to 2015 in Guizhou Province combined with ArcGIS spatial analysis method and linear trend analysis method, this study analyzed the precipitation characteristics of Guizhou Province in recent 55 years and reached the following conclusions:

(1) Spatial distribution: The average annual precipitation in Guizhou Province has been mostly distributed in the south and the north less and east and west less in 55 years, of which the annual average precipitation is more than 1100mm. The precipitation in the four seasons has obvious difference. The precipitation in winter and spring is less in the east and west, and the precipitation is decreasing from east to west, of which the precipitation in winter is the least, the summer is less than the north and the south is mainly concentrated in the southwest of Guizhou, In the most seasons, the precipitation in autumn is mainly concentrated in the southwest of Qianxinan Prefecture and the southern part of Liupanshui. The areas with less precipitation distribute in the Bijie area and the southwest of Qiannan.

(2)The time trend: Guizhou province 55 years precipitation overall downward trend. During the year, the seasonal variation of precipitation is large, and the precipitation in spring, autumn and winter has a significant downward trend, especially in autumn precipitation.

Acknowledgments

Guizhou provincial education department natural science foundation “Multi - source Remote Sensing Information Characteristics Analysis of Impervious Surfaces in Karst Watershed (QJHKY[2017]206) ” ; Guizhou province science and technology project “Poverty Evolution in Extremely Poverty Peaks and Poduks and Its

Man-Ground Relationship Response (QKHJ[2018]1123)"; National natural science foundation " The Coupling Mechanism of Population - settlement and Ecology in Typical Rural Areas of Karst Mountains and Its Optimization And Control Mode (41461041)" ; Guizhou province brainstorm project of social development project "Mechanism and Risk Assessment and Early Warning Study of Guizhou Flood and Drought Disaster in Changing Environment (QKH [2016] 2845)" .

References

1. Jin L Y, Fu J L, Chen F H. Over the past 44 years, Chinese northwest Precipitation Regional Difference and the impacts of global warming on the [J]. Geography Science, 2005:567-572.
2. Sun Q, Zhou S Q, Kang N, et al. GIS based precipitation analysis of the upper and middle upper reaches of the Yangtze River [J]. Journal of Nanjing Meteorological Institute, 2007,30 (2): 201-209.
3. Guizhou Meteorological Bureau. Guizhou short term weather forecast Handbook [M], Guizhou: Meteorology Press, 1987:369- 4761.
4. Hulme.M,R.Marsh,P.D.Jones.Global changes in a humidity index between 1931-1960 and 1961-1990[J].Climate Res.1992(2):1-22.
5. Vinnikov.K.Y,P.Y.Groisman,K.M.Lugina.Empirical data on contemporary climate chang-es(temperature and precipitation)[J].J.Climate.1990(3):662-677.
6. Kattsov.V,J.E.Walsh.Twentieth-century trends of Arctic precipitation from observationaldata and a climate model simulation[J].J.Climate.2000,13(4):1362-1370.
7. Diaz,H.F,R.s.Bradley,J.K.Eischeid.Precipitation fluction over global land areas since the late 1800's[J].J.Geophys.Res.1989(94):1195-1210.
8. Zuo H C, Lv S H, Hu Y Q. The change trend of temperature and precipitation in recent 50 years Chinese analysis [J]. Plateau Meteorology, 2004,23 (2): 238-244.
9. Wang Y J, Yan F.1960-2010 Precipitation Regional Differentiation and interdecadal changes in China, [J]. geo scientific progress, 2014,33 (10): 1354-1363.
10. Li Y H, Xu H M, Bai Yingying and so on. The spatial and temporal characteristics of summer precipitation in the eastern part of the southwest China are [J]. Plateau Meteorology, 2010, 29 (2): 523-530.
11. Yang R F. The existing problems and Development Countermeasures for the utilization of water resources in Guizhou province [J]. Henan water conservancy and south to North Water Transfer, 2013,57 (16): 19-20.
12. Duan J F, Wang Q C, Wang G L,et al. Analysis of spatial interpolation method of precipitation in Yongding River Basin of Shanxi based on GIS; [J]. Shanxi agricultural science, 2010,38 (3): 44-47.

Cyclic Flexural Behavior of Reinforced Concrete Beams

Prafulla B MALLA[1], Hong ZHOU[2], Yi CHE[1]

[1] State Key Laboratory of Costal and Offshore Engineering, Dalian University of Technology, Dalian, Liaoning 116024, China
[2] Department of Architectural Engineering, Shandong Water Conservancy Vocational College, Rizhao, Shandong Code, China

Abstract: The present study aims at investigating the cyclic flexural behavior of reinforced concrete beams with varying depths. Five reinforced concrete beams with beam depth ranging from 250 mm to 750 mm were tested under reversed cyclic loading and the influence of beam depth on the flexural strength and ductility of reinforced concrete beams was investigated. In addition, OpenSees was used to model the test specimens and the analytical results were compared with the experimental reuslts. It is shown that there is no apparent size effect on the normalized ultimate flexural strength of the tested beams, while for the displacement ductility factor, a significant size effect is observed. Load-deflection hysteric curves of test specimens obtained by the fiber-based element of OpenSees with Concrete03 and Hysteric models are in good agreement with those from experimental tests.

1 Introduction

The size effect has been experimentally evidenced for most brittle types of failures of plain and reinforced concrete members, for example, shear and torsional failure of beams, punching shear failure of slabs and pullout failure of bars etc. [1, 2]. In addition, the effect of size on flexural behavior of reinforced concrete beams was also investigated [3-9]. Early work on flexural behavior of reinforced concrete beams indicated that there was no apparent size effect on flexural strength and rotational capacity [3, 4]. Appa et al. [5] and Yi et al. [6] reported from their studies that size effect is significant for both flexural strength and ductility. Most of the research on under-reinforced concrete beams confirmed the existence of size effect on rotational capacity [7-9]. Some researchers also investigated behavior of over-reinforced and prestressed concrete beams failing due to concrete compressive failure and found that significant size effect on rotational capacity or ductility of beams exists [10-12].

However, most of the studies reported in the literature are concerned with the size effect of flexural behavior of reinforced concrete beams that contain only tensile reinforcement and are subjected to monotonic loading. In reinforced concrete moment-resistance frames, the cross section of the beam end to resist the positive and negative moment induced by lateral earthquake and vertical loads, locates in the potential plastic hinge region where sufficient ductility and rotational capacity is required. Therefore, a better understanding of size effect on flexural behavior of doubly reinforced concrete beams subjected to cyclic loading is necessary to ensure the safety of reinforced concrete structures.

In view of this, the influence of beam depth on the cyclic flexural behavior of RC beams was investigated experimentally in this study. Besides, nonlinear analysis of test specimens was carried out using OpenSees models.

2 Experimental Program

2.1 Test Specimens and Material Properties

In total, five specimens with beam depth ranging from 250 mm to 750 mm were tested under reversed cyclic loading. The specimens used in the experimental program consisted of two beam elements with identical section cast integrally with a stub as the column element at mid-span. The details of specimens are shown in figure 1 and table 1.

All specimens were cast from the same batch of concrete supplied by a local ready-mix plant. Concrete cubes of 150×150×150 mm^3 were casted along with the specimens, and compressive strength tests were performed at time of cyclic loading tests. Hot-rolled deformed bars with diameters of 12, 20, 22 and 25 mm were used as longitudinal reinforcement giving longitudinal reinforcement ratios of 1.08 % and 0.69%, respectively. Hot-rolled plain bars with 6, 8 and 10 mm diameters were used as vertical stirrups, with a constant transverse reinforcement ratio equal to 0.52%. Properties of steel bars are presented in table 2.

*Corresponding author: Email: cheyi@dlut.edu.cn

 https://doi.org/10.1051/e3sconf/20183801022

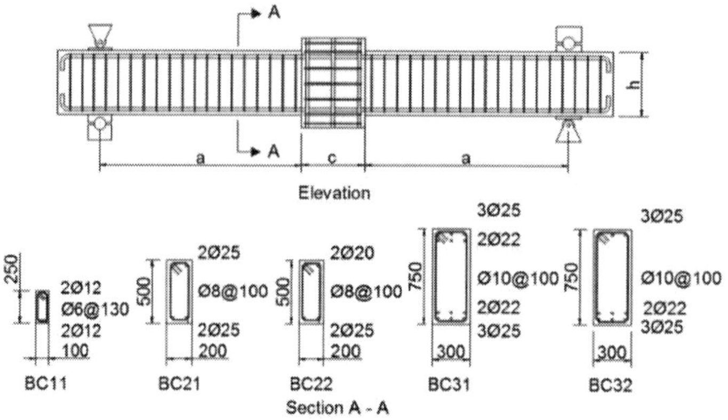

Figure 1 Details of test specimens (mm)

Table 1 Parameters of specimens

Specimen	Dimension (mm)				Concrete cubic strength f_{cu} (MPa)	Longitudinal rebar ratio (%)		Stirrups ratio ρ_{sv} (%)
	Width b	Depth d	Effective depth d_0	Span a		Top bar ρ_s	Bottom bar ρ_s'	
BC11	100	250	210	3	73.6	1.08	1.08	0.52
BC21	200	500	456	3	73.8	1.08	1.08	0.52
BC31	300	750	688	3	72.0	1.08	1.08	0.52
BC22	200	500	456	3	73.8	1.08	0.69	0.52
BC32	300	750	688	3	72.0	1.08	0.69	0.52

Table 2 Properties of steel bars

Diameter of steel bar (mm)	Yield strength (MPa)	Tensile strength (MPa)
6	407	578
8	467	507
10	384	535
12	420	618
22	423	598
25	470	660

2.2 Loading Protocol and Test Setup

Figure 2 shows the planned loading protocol for the reversed cyclic loading test. Each specimen was subjected to progressively increasing load cycles. After yielding of tensile reinforcement, progressively increasing displacement cycles were applied until the specimen failed completely. For each displacement step, three fully reversed cycles were applied. The test setup is shown in figure 3.

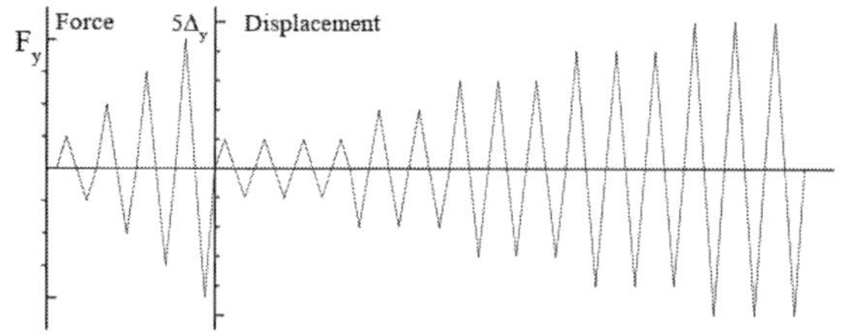

Figure 2 Loading protocol

Figure 3 Test setup

3 Test results and Size Effect

3.1 General Behavior

Similar failure patterns were also observed in other specimens regardless of specimen sizes. Flexural cracks were firstly observed at sections adjacent to the central stub. The failure of the specimen resulted from the yielding of tensile reinforcement followed by the crushing of concrete. Figure 4 shows the typical failure patterns of Specimens BC11, BC21 and BC31 at the end of the test.

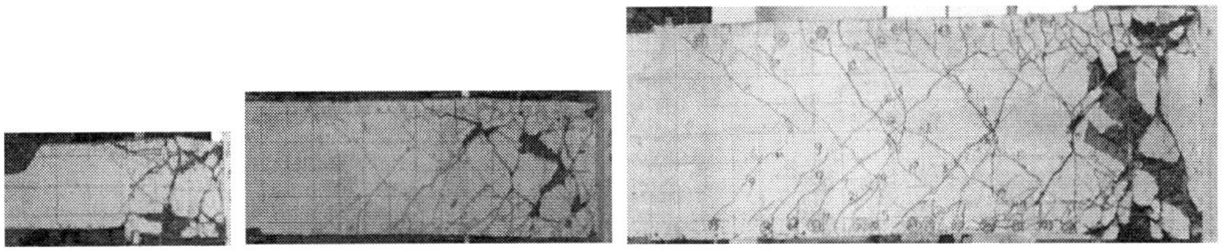

(a) Specimen BC11 **(b) Specimen BC21** **(c) Specimen BC31**
Figure 4 Failure patterns of specimens

3.2 Test Results

Test results are listed in table 3, in which the ductility factor for each specimen is calculated by

$$\mu = \Delta u / \Delta y \qquad (1)$$

Where μ is the displacement ductility factor, and Δy and Δu are displacements at yield load and ultimate load, respectively.

Table 3 Experimental results

Specimen	Loading direction	At yield		At ultimate		Ductility factor μ
		Moment My [kN.m]	Displacement Δy [mm]	Moment Mu [kN.m]	Displacement Δu [mm]	
BC11	+	17.46	3.32	20.12	29.23	8.80
	-	17.42	3.00	20.29	-	-
BC21	+	194.14	7.55	235.48	43.46	5.76
	-	192.30	7.62	223.43	-	-
BC31	+	656.43	13.01	754.49	65.11	5.00
	-	652.31	19.03	752.41	-	-
BC22	+	190.05	8.97	236.00	46.83	5.22
	-	114.70	4.66	137.68	-	-
BC32	+	662.22	12.24	779.62	59.16	4.83
	-	481.04	9.40	533.01	-	-

3.3 Size Effect

The normalized flexural ultimate strength, m, for each specimen is calculated by

$$m = \frac{M_u}{f_{cu} bd} \qquad (2)$$

Figure 5 shows the variation of normalized flexural strength with beam depth. As shown, there is no apparent size effect on the normalized ultimate flexural strength of the tested beams. Figure 6 plots the ductility factor varying with the depth of beams. For beams with symmetric longitudinal reinforcement, i.e. Specimens BC11, BC21 and BC31, the ductility factor decreases with increase of the depth of beams exhibiting a significant size effect. However, for beams with unsymmetric longitudinal reinforcement, i.e. Specimens BC22 and BC32, the size effect cannot be confirmed due to the insufficiency of test data.

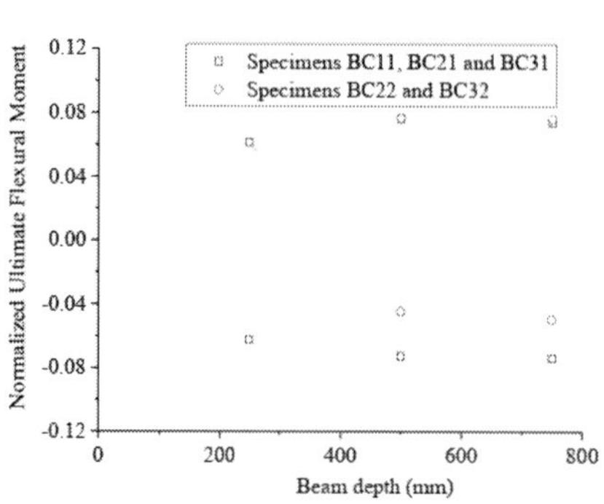

Figure 5 Normalized ultimate flexural moment

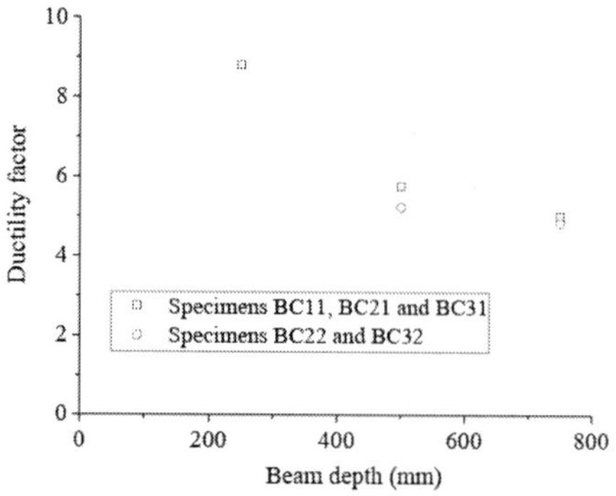

Figure 6 Ductility factor

4 OpenSees Modeling

4.1 The FE Model

Nonlinear analysis of the cyclic experiments was conducted by using OpenSees software which is based on fiber based finite element analysis. In the analysis, the influence of shear deformation was considered, whereas the slippage of reinforcement was ignored.

The uniaxial material concrete03 for concrete and the hysteretic model for reinforcement provided in OpenSees, as shown in figures 7 and 8, were used to simulate the nonlinear behavior of concrete and reinforcement, respectively. Detailed information of the FE models and the analysis of test specimens can be found elsewhere [13, 14].

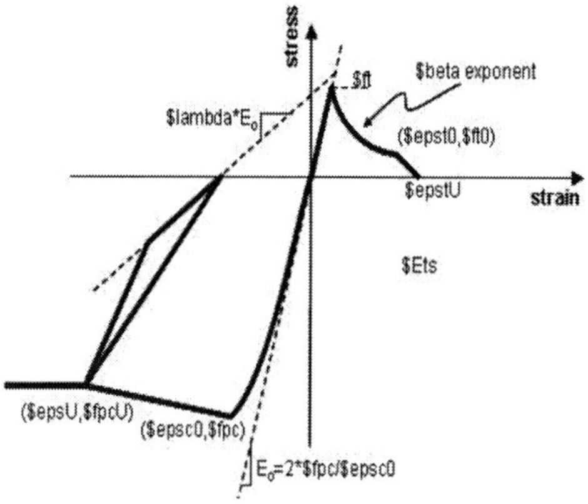

Figure 7 Concrete 02 model

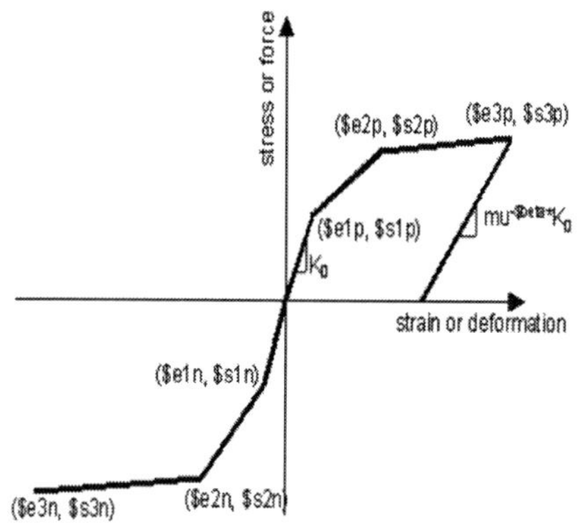

Figure 8 Hysteretic model

4.2 Analytical Results and Comparison

Figure 9 shows the load-deflection hysteric curves of Specimens BC11, BC21 and BC31obtained from the experiments and analysis by OpenSees. As can be seen, the results of the analysis are in good agreement with the experimental results, and the pinching effects of experimental curves are well captured by the FE analysis. The fiber-based element of OpenSees with Concrete03 model and Hysteric steel bar model is suitable to simulate cyclic behavior of RC beams and is able to consider the pinching effect.

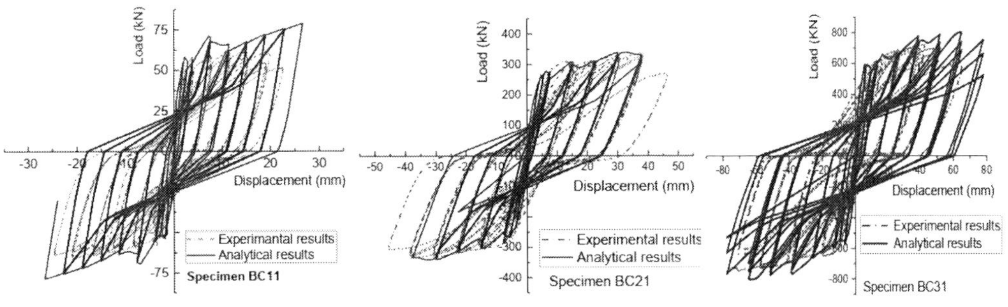

Figure 9 Comparison of load-deflection hysteric curves

5 Conclusion

Based on the present study, the following conclusions may be drawn:

(1) Test results show that there is no apparent size effect on the normalized ultimate flexural strength of the tested beams.

(2) For the displacement ductility factor, a significant size effect is observed experimentally in beams with symmetric reinforcement.

(3) The load-deflection hysteric curves obtained by the OpenSees analysis are in good agreement with those from experimental tests. It is found that the fiber-based element of OpenSees with Concrete03 and Hysteric models is suitable to simulate cyclic behavior of RC beams and is able to consider the pinching effect.

Acknowledgement

This research was financially supported by the Science Foundation of the State Key Laboratory of Costal and Offshore Engineering.

References

1. Bazant Z P and Planas J. Fracture and size effect in concrete and other quasibrittle materials [M]. 1st ed., CRC Press, London, 1998.
2. Kim J K, Yi S T and Kim J H J. Effect of specimen sizes on flexural compressive strength of concrete [J]. ACI Structural Journal, 2001, 98 (3): 416-424.
3. Corley W G. Rotational capacity of reinforced concrete beams, Journal of Structural Division [J]. 1966, 92 ST5: 121-146.
4. Alca N, Alexander S D B, and MacGregor J G. Effect of size on flexural behavior of high-strength concrete beams [J], ACI Structural Journal, 1997, 94 (1): 59-67.
5. Appa R G, Vijayanand I and Eligehausen R. Studies on ductility and evaluation of minimum flexural reinforcement in RC beams [J]. Materials and Structures, 2008, 41: 759-771.
6. Yi S T, Kim M S, Kim J K and Kim J H J. Effect of specimen size on flexural compressive strength of reinforced concrete members [J]. Cement and Concrete Composites, 2007, 29 (3): 230-40.
7. Bigaj A and Walraven J. Size effect in plastic hinges of reinforced concrete members [J]. Heron, 2002, 47 (1): 53-75.
8. Carpinteri A, Corrado M, Mancini G and Pagi M. Size-scale effect on plastic rotational capacity of reinforced concrete beams [J]. ACI Structural Journal, 2009, 106 (6): 887-895.
9. Khatieb M D. Experimental investigation of scale influence on plastic rotational capacity of reinforced concrete beams [J]. Engineering, 2016, 8:196-203.
10. Ozbolt J, Mestrovic D, Li Y and Eligehausen R. Compression failure of beams made of different concrete types and sizes [J]. Journal of Structural Engineering, 2000, 126 (2): 200-209.
11. Belgin C M and Şener S. Size effect on failure of overreinforced concrete beams [J]. Engineering Fracture Mechanics, 2008, 75(8): 2308-2319.
12. Vorel J, Gattu M and Bazant Z P. Size effect in flexure of prestressed concrete beams failing by compression softening [J]. Journal of Structural Engineering, 2014, 140 (10): 04014068.
13. S. Mazzoni, F. Mckenna, M. H. Scott, et al. OpenSees command language manual [R]. Pacific Earthquake Engineering Research (PEER) Center, 2007.
14. P. B. Malla P B. Nonlinear analysis of reinforced concrete based on Abaqus and OpenSees [D]. Dalian University of Technology, Dalian, China, 2016.

Corrosion effect of microorganisms and their metabolite on cement mortar lined pipelines in reclaimed water distribution systems

Fan YANG[1*], *Minning* CHEN[1]

[1]College of Engineering and Technology, Tianjin Agricultural University, 22, Jinjing Road, Tianjin300384, China

Abstract. The reclaimed water containing high salinity, great amounts of organic matters and high nutrients can easily lead to growth of biofilms in reclaimed water distribution systems (RWDSs). The microbes colonize the cement surface and microbial metabolites can cause cement biodeterioration. To understand the effect of microbial involvement in the degradation, this study investigated the transformation characteristics of cement-mortar lining and microbial biomass in the simulated RWDS for 1 year by X-ray diffractometer (XRD), X-Ray Fluorescenc (XRF), Heterophic bacteria count (HPC) and DAPI staining. Microbial metabolites were analyzed by GC/MS. The result shows that the carbonation reaction took place in the surface of the eroded cement-mortar lining where the content of $CaCO_3$ was continuously increasing while the content of hydrated compounds were decreasing. The depositing layer of $CaSO_4 \cdot 2H_2O$, $CaAl_2Si_2O_8 \cdot 4H_2O$ and $Mg_4Al_2(OH)_{14} \cdot 3H_2O$ on the lining surface were formed by minerals such as Ca, Si, Al and Mg lost from the degraded hydrated compounds. Microbial biomass in the RWDS has maintained an increasing trend during the study. The main microbial metabolites of the biofilm on the cement surface are fatty acids, amino acids, and carbohydrate.

1 Introduction

The utilization of reclaimed water is an effective ways to alleviate water resources shortage, reduce pollution load of urban water environment, improve the ecological environment. The ductile iron pipe with cement-mortar lining are widely used in the reclaimed water distribution systems (RWDSs) because of its cost-effective, stable mechanical property and durability. Although the reclaimed water meets the regulatory standards, a large amount of organic matters and inorganic ions in reclaimed water can cause microorganism regrowth, biofilm induced corrosion and water quality changes during transport through the distribution system to the use point [1-4].

Cement hydration products $Ca(OH)_2$, C_3S, C_2S, C_3A and calcium silicate hydrates (CSH) can be stable in the high alkaline environment (pH 12.5-13.5). When the cement-mortar is immersed in the reclaimed water, erosion media such as CO_2, sulfate and chloride ion may all reacted with cement hydration products by diffusing into the pores of cement-mortar lining. Carbonic aggressivity is one type of interaction with water concerns the calcium carbonate saturation state and carbonate speciation of the water, which can lead to the rapid degradation of the cement material(eq.1-2). High sulfate reacts with $Ca(OH)_2$ in the cement-mortar lining to form gypsum (eq.3) which can cause internal expansion and softening [5,6].The microbes colonized on the concrete surface and in its pores, capillaries and micro-cracks cause damage through biodeterioration, such as sulphur bacteria, nitrifying bacteria, acid producing fungus and their microbial metabolites. Sulfur-oxidizing bacteria produce biogenic sulfuric acid which can react with the $Ca(OH)_2$ in the cement-mortar lining to form gypsum (eq.4, 5). Microbial metabolites like organic acid, carbon dioxide, mineral acids like sulfuric acid and nitric acid all can cause biodeterioration by reacting with the components of cement (eq.6) [7].

$$Ca(OH)_2 + CO_2 \longrightarrow CaCO_3 + H_2O \quad (1)$$

$$CaCO_3 + CO_2 + H_2O \longrightarrow Ca(HCO_3)_2 \quad (2)$$

$$Ca(OH)_2 + Na_2SO_4 + 2H_2O \longrightarrow CaSO_4 \cdot 2H_2O + 2Na^+ + 2OH^- \quad (3)$$

$$H_2S + 2O_2 \xrightarrow{SOB} H_2SO_4 \quad (4)$$

$$Ca(OH)_2 + H_2SO_4 \longrightarrow CaSO_4 \cdot 2H_2O(gypsum) \quad (5)$$

$$Ca(OH)_2 + 2C_2H_4(OH)COOH + 3H_2O \longrightarrow Ca[C_2H_4(OH)COO]_2 \cdot 5H_2O \quad (6)$$

Yang et al. have found that reclaimed water had carbonic aggressivity to the cement-mortar liner, and the erosion products on the lining surface can not form a effective protective layer [8]. Jin et al. extracted extracellular polymeric substances (EPS) on biofilm in a reclaimed water distribution system from different culturing stages contained different proportions of

* Corresponding author:fanyang20062800@163.com

protein and polysaccharide with similar functional groups. Excessive amount of EPS promoted anodic dissolution through EPS-Fe binding to accelerate the pipe corrosion [9]. In this study, the erosion process of micro-organisms and their metabolites on cement-mortar liner in a simulated RWDS was investigated. The changes of physicochemical characteristics of eroded cement-mortar lining were studied by XRD and XRF and the microbial biomass in RWDS was regularly monitored. Combined with the above results, the microbial metabolites were analyzed to reveal the chemical and microbiological erosion of cement lining in RWDS.

2 Materials and methods

2.1. Pipe loop set-up and experimental operation

The simulated reclaimed water distribution system consists of three lm-length DN100 new cement mortar lined ductile iron pipe sections, the water tank, metering pump and flow meter connected by PVC plastic pipe (DN20) to form a recirculating pipe rig system (Figure 1). Five 5cm-length DN100 new cement-mortar lined ductile iron pipe sections (referred to as "tested coupon") (shown in Figure 1) are also placed in the pipe loop system to provide cement-mortar samples and biofilm samples for subsequent tests. Reclaimed water was circulated from the water tank to testing loops using a metering pump at flow rate of 10 $L \cdot h^{-1}$ with the hydraulic retention time 48h. There were no disinfectant added during the experimental period.

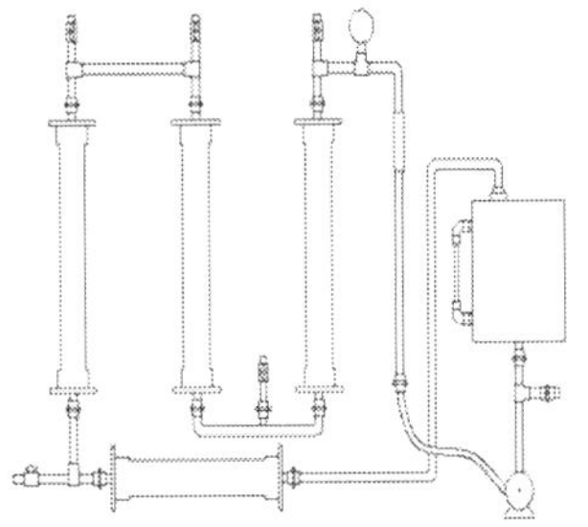

Fig.1. Simulated reclaimed water distribution system used for experiment

2.2 Analytical methods and Reagents

Water quality was measured according to Standard Methods [10]. The pH, DO and conductivity were measured using Portable Multiparameter meter (Sension156 HACH, USA) and the turbidity was measured using a 2100P Turbidimeter (HACH, USA). The BOD_5 was measured using a Biochemical Oxygen Demand Trak II Analyzer (HACH, USA). The CODcr, TN, Ammonia-N, TP and Residual chlorine were measured using using HACH DR2800 spectro-photometer. Alkalinity and total hardness were measured by chemical titration. The sulfate and chloride were analyzed by Ion Chromatography System-1000 (Dionex, USA). The total iron and SiO_2 were analyzed by the Inductively Coupled Plasma Optical Emission Spectro-meter (Perkin-Elmer Optima 2000, USA). As the reclaimed water in system was replaced every two weeks, sampling for water quality analysis was performed at one time per week with three parallel samples and the water parameter values were the average values of the three parallel samples.

Table 1. Water quality parameters of reclaimed water (units: conductivity in $\mu s \cdot cm^{-1}$, alkalinity in $mg \cdot L^{-1}$ as $CaCO_3$, others in $mg \cdot L^{-1}$)

pH	DO	Conductivity	Turbidity
6.95	7.04	891.50	0.34
BOD_5	COD_{Cr}	Total Nitrogen	Ammonia Nitrogen
2.15	14.95	14.60	2.38
Total Phosphoru s	Residual Chlorine	Alkalinity	Total Hardness
0.26	0.70	108.20	243.45
Sulfate	Chloride	Total Iron	SiO_2
211.55	239	<0.016	1.19

2.3 Experimental water

The reclaimed water used in the system was taken from a water reclamation plant in an northern city of China from December 2014 to December 2015. The average values of the monitored parameters biweekly were summarized in Table 1. Water quality parameters of reclaimed water which contained high levels of conductivity, hardness, inorganic ions and organic matter were inferior to tap water.

2.4 Physicochemical characterization of cement mortar lining composition analysis

Cement-mortar lining samples collected from the pipe coupons connected in pipe loop systems were pretreated for X-ray diffraction (XRD) and X-ray Fluorescence (XRF), as soon as possible. The pretreatment procedures were as follows: samples scraped from the pipe coupons were pulverized in an agate mortar and passed through a 150μm mesh sieve, then vacuum-freeze-dried. The surface layer sample (SL) of cement-mortar liner was the top layer with 2-3mm thickness scraped off by a spatula and the rest of cement-mortar liner was the inner layer s amples (IL). The XRD (D/max-rA, Rigaku, USA) operation parameters were: Cu Kα radiation at 40KV and 100mA, the 2o ranged from 3°C to 70°C with a 0.02°C step, and a 0.15s count time at each step. Crystalline phase was identified using the Jade XRD software, and crystalline phase composition of the cement-mortar liner was quantitatively determined by contrasted parameters of intensity method [11]. The elemental composition of cement-mortar liner samples was determined by XRF (Advant' XP, Thermo Electron, Switzerland). The XRF spectrometer with a Lawrencium

 https://doi.org/10.1051/e3sconf/20183801023

(Lr) excitation tube was employed with voltage of 0-70 KV, and with current of 0-120mA.

2.5 Microbial biomass analysis

Heterotrophic plate counts (HPC) and total DAPI counts were used to enumerate cells on the cement-mortar liner surface at different experimental period in the system. For enumeration by HPCs (CFU/L), bacteria were collected by filtration on 0.45-μm porosity, 45-mm diameter, acetated cellulose filters. The filters were placed on R_2A agar plates for cultivation and incubated for 7 days at 22 ± 2°C. Duplicate counts were performed through single-plate dilution for each measurement [12]. Biomass samples were washed twice and diluted (10^{-3} or 10^{-4}) in PBS medium, vortexed and homo-genized. Suspensions were then filtered using 0.2mm pore-size black polycarbonate membrane filters. The cells were then fixed in the dark by dipping the filter into a 4% formaldehyde solution for 2-3h at room temperature, briefly washed with sterile deionized water, and air-dried in the dark. Filter were then placed on a glass slide and stained with 60-ml of a DAPI solution (1mg·ml^{-1} in distilled water) for 5min at room temperature in the dark [13]. Washed and air-dried filters were mounted on a glass slide with an anti-fading mounting medium (Vectashield H-1000) and examined using an epifluorescence microscope (BX60; Olympus) equipped with a×100 objective lens and a set of DAPI-specific filters (ref. U-MWU, Olympus: Dichroic Mirror 400nm, BandPass330-385nm, and Barrier 420nm). The procedure was applied to duplicate samples.

2.6 Microbial metabolites extraction

Fresh biofilm sample scraped from cement-mortar liner under aseptic conditions was immersed in an appropriate amount of PBS solution, and well-dispersed by sonicator. The biofilm solution was centrifuged at 12000rpm for 10 minutes, 4°C. With supernatant liquid removed, the residue and the indicated amount of cation exchange resin (DOWEX 50X8, 20-50 mesh, Fluka 44445) were incubated in oscillator and stirred for 2h at 4°C. The cation exchange resin was pre-washed in extraction buffer (2mM Na_3PO_4, 4mM NaH_2PO_4, 9mM NaCl, 1mM KCl, pH 7.0; 1kg DOWEX in 2L extraction buffer, 1h with stirring). Then the mixture was centrifuged at 12000rpm for 10 minutes 4°C and the supernatants were metabolites for analysis [14].

2.7 Metabolomic GC/MS analysis

An Agilent7890A/5975C GC-MS chromatograph with a 30m HP-5MS capillary column, i.d. 250μm, film thickness (df) 0.25μm separation column was used for analysis of microbial metabolites. All injections were performed in split less mode with 1.0μl volume; the oven was held at an initial temperature of 80°C for 2.0min before increasing to 300°C at 10°C min^{-1}; the final temperature was held for 6.0min. The transfer line was held at 280 °C and the detector voltage at -1624 V. Mass spectra were acquired from 35 to 780 m/z, at an acquisition frequency of 4 spectra s^{-1}. Instrument control, data acquisition and processing were performed using MSD Chemstation (Rev E02.00) (Agilent Technologies, Mulgrave, Australia) [15].

3 Results and discussion

3.1 The condition of cement mortar lining deterioration

3.1.1 XRD analysis of compositions of cement mortar lining

X-ray diffraction analysis (XRD) analysis of cement-mortar liner samples (Table 4) showed that the content of $CaCO_3$ in all the samples obviously increased with the experimental time prolonging. However, the hydration products of cement: Portlandite ($Ca(OH)_2$), Calcium Silicate (Ca_3SiO_5, abbrev. C_3S), Larnite (Ca_2SiO_4, abbrev. C_2S), Brownmillerite ($Ca_2(Al, Fe^{3+})_2O_5$) and Tricalcium aluminate ($Ca_3Al_2O_6$, abbrev. C_3A) had gradually decreased contents. When the experiment was carried out to the 296 days and the 564 days, all the crystalline materials in the SL samples were $CaCO_3$ because of the carbonation reaction. Calcium which escaped from the hydration products of cement-mortar reacted with CO_3^{2-} or HCO^{3-} in water leading to deterioration of the cement material. A large amount of white materials deposited on the surface of the lining in the day 296 were crystalline composition: $CaSO_4 \cdot 2H_2O$, $CaAl_2Si_2O_8 \cdot 4H_2O$ and $Mg_4Al_2(OH)_{14} \cdot 3H_2O$. These materials generated by the dissolved Ca, Si, Al and Mg minerals in hydration products were unstable and they had low strength.

Table 2. Relative percentages of the crystal material of cement lining test slice based on X-ray diffraction measurement.

Sample Name	$CaCO_3$	$Ca(OH)_2$	C_3S	C_2S	$Ca_2(Al,Fe^{+3})_2O_5$	C_3A	$CaSO_4 \cdot 2H_2O$	$CaAl_2Si_2O_8 \cdot 4H_2O$	$Mg_4Al_2(OH)_{14} \cdot 3H_2O$
IL sample0 d	39.2%	13.7 %	15. 7%	7.9%	13.7 %	9.8%	0%	0%	0%
white insoluble remainder on the surface-296d	0%	0%	0%	0%	0%	0%	16%	49%	36%
SL sample-296d	100%	0%	0%	0%	0%	0%	0%	0%	0%
IL sample-296d	38%	12%	50 %	0%	0%	0%	0%	0%	0%
SL sample-564 d	100%	0%	0%	0%	0%	0%	0%	0%	0%
IL sample-564d	81%	9%	10 %	0%	0%	0%	0%	0%	0%

3.1.2 XRF analysis of compositions of cement mortar lining

Table 3 shows that in the day 296 the results of the XRF analysis of the skin samples with the thickness of 2-3mm on the cement mortar lining surface and the remaining internal samples. From this result, it can be found that the content of Ca in the lining skin sample is higher than that in the inner layer sample in the day 296, but the contents of major elements of the surface layer sample such as Si, Al, Mg and Fe are lower than those in the inner layer sample. Cement mortar specimens were eroded by chemical and microbial factors of water quality, so the main elements reduced and the elements of reduced contents lost into water.According to the results of XRD analysis, all the main crystalline materials in the inner surface samples are $CaCO_3$. But the XRF results show that there are amorphous minerals containing Si, Al, Mg, Fe and other elements in the surface layer. The previous cement hydration products changed into these minerals after the lining suffering erosion.

Table 3. XRF Semi-quantitative Analysis of the main elements in the Surface and Inner of cement lining samples.

Sample ID	Element	Relative percentage %
Surface -296d	Ca	72.94
	Si	9.23
	Al	3.85
	Mg	3.50
	Fe	3.35
	S	3.70
	Ti	0.64
	Zn	0.96
	Na	0.73
	K	0.83
	Mn	0.27
Inner layer -296d	Ca	65.76
	Si	16.66
	Al	5.75
	Mg	4.19
	Fe	2.85
	S	1.25
	Ti	1.52
	Zn	0.66
	Na	0.57
	K	0.41
	Mn	0.37

3.2 Microbial biomass of simulate reclaimed water network and changes in related indicators of water quality

The Total DAPI counts showed that the microbial biomass of the reclaimed water in system had been rising slowly during the 1 year-experiment due to no disinfectant added (Fig. 2A). Microbiological growth curve of the effluent water of simulate RWDS (Fig. 2B) shows that the microorganism undergone a acclimation phase, logarithmic growth phase, stationary phase and endogenous respiration phase within a month. High nutrients and organic matter levels in combination with low chlorine residuals in reclaimed water initiate bacterial growth in the RWDS. Both the SL and IL samples contained high percentage of sulfur elements indicated by XRF analysis, which confirmed the biological activity of the sulfur related bacteria (SOB/SRB).

3.3 GC-MS analysis of microbial metabolites

GC-MS showed that the majority of the metabolites identified were a combination of fatty acids, amino acids and carbohydrates. The organic acids mainly included undecanoic acid, hexadecanoic acid, octadecanoic acid, fatty acid 16 : 0, lactic acid, oxalic acid, octadecenoic acid, myristic acid, adipic acid, pelargonic acid, etc. Non-organic acid materials mainly included glucose, fructose, glycerin, urea, pyroglutamic acid, putrescine, etc. Organic acids in microbial metabolites can react with cement hydration products $Ca(OH)_2$ and CSH forming soluble substances or gypsum, which could result to the deterioration of cement materials.

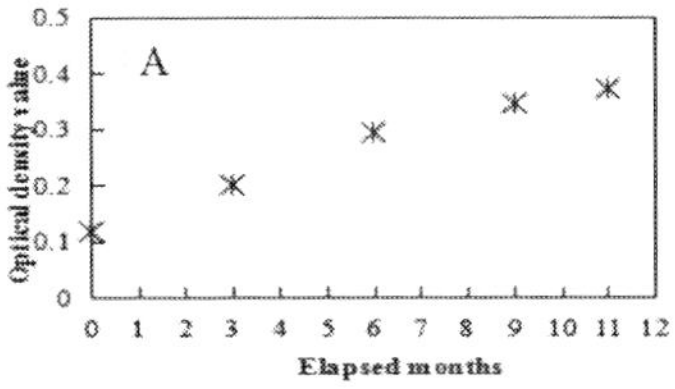

* Corresponding author:fanyang20062800@163.com

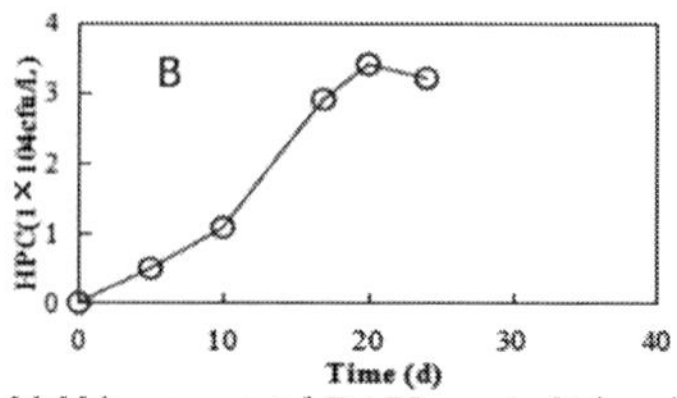

Fig.2. Variation curve of effluent microbial biomass : total DAPI counts (A) and Heterotrophic plate counts (HPC) (B)

4 Conclusion

XRD and XRF analysis results showed that reclaimed water had strong chemical and microbiological attack on cement-mortar lining. Reclaimed water had lower pH than of that of cement-mortar lining. A calcium carbonate deposit was formed on the liner surface because of the carbonation reaction of $Ca(OH)_2$ and CSH in cement-mortar, and Ca, Si, Al, Mg and other elements in cement hydration products also diffused into the transported water due to erosion. Nitrification of microorganisms in the lining biofilm reduced the pH of the interface and promoted the erosion reaction. The metabolites of microorganisms mainly included fatty acids, amino acids and carbohydrates. Organic acids can react with cement hydration products and result to the deterioration of cement material. Therefore, the use of effective disinfectants to inhibit microbial regrowth and improve the physicochemical property of cement-mortar lining for better corrosion resistance must be employed to maintain the operational safety in recycled water distribution system.

5 Funding information

This work was financially supported by National Natural Science Foundation of China (No. 51308392). Also, the project was funded by the Research Fund of Tianjin Key Laboratory of Aquatic Science and Technology. The funders had no role in study design, data collection and interpretation, or the decision to submit the work for publication.

References

1. F. Teng, Y.T. Guan, W.P. Zhu, Corros. Sci., **50**, 2816-2823 (2008)
2. H.B. Wang, H. Chun, X.X. Hu, M. Yang, J.H. Qu, Water Res., **46**, 1070-1078 (2012)
3. P. Sarin, V.L. Snoeyink, J. Bebee, K.K. Jim, M. A. Beckett, W.M. Kriven, J.A. Clement, Water Res., **38**, 1259-1269 (2004)
4. J.T. Jin, G.X. Wu, Y.T. Guan, Water Res., **71**, 207-218 (2015)
5. P. Leroy, M.R. Schock, I. Wagner, H. Holtschulte, *Cement-Based Materials. In Internal Corrosion of Water Distribution Systems*. Cooperative Research Report. AWWA Research Foundation, Denver, CO, 313-388 (1996)
6. J. Monteny, E. Vincke, A. Beeldens, N.D. Belie, L. Taerwe, D.V. Gemert, W. Verstraete, Cem. Concr. Res., **30**, 623-634 (2000)
7. R.P. George, V. Vishwakarma, S.S. Samal, U.K. Mudali, Concrete Research Letters, **3**, 491-506 (2012)
8. Y.Yang, Study on the effective of water quality and hydraulic characteristics on the erosiveness of pipe inner wall for the distribution systems of reclaimed water. Xi'an University of Architecture and Technology, Xi'an, China, (2012)
9. J.T. Jin , G.X. Wu, Z.H. Zhang, Y.T. Guan, Bioresour. Technol., **165**, 162-165 (2014)
10. L.S. Clesceri, A.E. Greenberg, A.D. Eaton, *Standard method for the examination of water and wastewater*. American Public Health Association, Washington, DC (1999)
11. F.H. Chung, J. Appl. Crystallogr., **8**, 17-19 (1975)
12. M. Batté, L. Mathieu, P. Laurent, M. Prévost, Can. J. Microbiol., **49**, 741-753 (2003)
13. H.S. Kim, P.R. Jaffé, water res., **41**, 2089-2100 (2007)
14. G. Guibaud, N. Tixier, A. Bouju, M. Baudu, Chemosphere. **52**, 1701-1710 (10)
15. D. J. Beale, M. S. Dunn, Corros. Sci., **55**, 272-279 (2012)

 https://doi.org/10.1051/e3sconf/20183801024

The calibration analysis of soil infiltration formula in farmland scale

Tao QIAN, Na Na HAN※, and Shuan Ling CHANG

Department of Hydraulic Engineering, Tianjin Agricultural University Tianjin, China

Abstract. Soil infiltration characteristic is an important basis of farmland scale parameter estimation. Based on 12 groups of double-loop infiltration tests conducted in the test field of tianjin agricultural university west campus. Based on the calibration theory and the combination of statistics, the calibration analysis of phillips formula was carried out and the spatial variation characteristics of the calibration factor were analyzed. Results show that in study area based on the soil stability infiltration rate A calculate calibration factor α_A calibration effect is best, that is suitable for the area formula of calibration infiltration and α_A variation coefficient is 0.3234, with A certain degree of spatial variability.

1 Introduction

Soil infiltration of rainfall, surface water, soil water and groundwater in the process of transformation is an important link, its infiltration characteristics is to design the ground irrigation and hydrological runoff calculation of an important base, in the migration of nutrients, many aspects, such as efficient utilization of water resources and hydrological conversion are of great significance. Many studies have shown that there is significant spatial variability in soil characteristics [1,2,3]. In terms of soil infiltration characteristics, domestic and foreign scholars have been using the traditional methods such as statistics, soil transfer function and calibration theory of spatial variation of soil infiltration characteristics and scale effect are analyzed, at the same time to study the different factors such as topography and land use ways for the influence of the spatial variation of soil infiltration characteristics, and some achievements have been made. Such as Bai meijian[4] studied the characteristics of spatiotemporal variability of soil infiltration during winter wheat growth under ground irrigation conditions. The spatial variation of soil stability infiltration rate in small watershed was studied by Yuan jianping[5].Jia hongwei[6] in Shiyang river basin, such as the establishment of a single parameter and soil infiltration model transfer function, for the soil infiltration characteristics on large scales of spatial variability study provides a method for reference;He Dan[7] systematically studied the spatial variation intensity, spatial variation structure and main variation sources of soil infiltration characteristics under different land use methods.However, the study on spatial variability of soil infiltration characteristics in field scale is relatively rare.Therefore, this paper studies the hydrological cycle experimental base of the west campus of Tianjin agricultural college. According to the small scale of farmland, soil infiltration test was carried out, and the soil infiltration law of small scale soil was analyzed, and the infiltration parameters were calibrated and the spatial variation characteristics were analyzed.

2 Study area overview and test method

The experiment was carried out in the experimental field at the west campus of tianjin agricultural college. The experimental field is located in the west suburb of tianjin, and the soil texture is light loamy soil, also known as tidal soil. Chao soil is a kind of zonal soil, by development of river alluvium materials under the action of groundwater formed half water into the soil, buried depth of groundwater in 1 to 1.5 meters, and with the rainy season and dry season lift, groundwater along the capillary can rise to the surface soil and soil have "night tide" phenomenon, so named chao soil. The soil in the western suburb of tianjin is a kind of moist soil, the soil color is mostly light brown, the PH value is 7.0-8.5, and the dry and wet alternation forms a lot of plastic film, embroidery pattern, rust stain, sand ginger, and a small amount of iron manganese nodules. The organic matter content of tidal soil is low, only about 1%. According to the actual situation of regional topography, In the field of 60m 80m mtitian, it is selected to carry out the two-ring infiltration method seepage test of the 12 series. And at each measuring point for collecting soil samples, determination of the initial soil water content (drying method), saturated water content (drying method), field capacity (two-ring infiltration method) and dry density (two-ring infiltration method). And record the temperature with a thermometer.

3 Philip's empirical formula

* Corresponding author: hnn23144@163.com.

Philip (1957) puts forward the empirical formula of soil infiltration. A large number of studies have shown that the Philip formula is simple in form, with few parameters, and has a clear physical foundation. It can well describe the one-dimensional infiltration process under homogeneous short duration conditions. It has important application value. The formula is as follows:

$$I(t) = St^{\frac{1}{2}} + At \quad (1)$$

The infiltration rate *i (t)* formula is as follows:

$$i(t) = \frac{1}{2} St^{-\frac{1}{2}} + A \quad (2)$$

In the formula: *I (t)* is the cumulative infiltration (cm) on the unit area; *t* is infiltration diachronic (min); *S* is soil infiltration rate (cm/min$^{0.5}$); A is a stable infiltration rate (cm/min).

In the initial stage of infiltration, parameter *S* plays a major role, which is equivalent to horizontal infiltration. With the increase of infiltration time, parameter *A* becomes the main factor affecting infiltration. When the Philip formula is applied to solve the problem, the infiltration parameters *S* and A are often obtained by the field infiltration test.

4 Results and Analysis

4.1 Calibration of Philip's empirical formula

The theory of calibration is developed on the basis of the law of similarity, that is, the soil in the fields of the field is similar to each other and neglects the influence of other factors. The calibration principle for the Philip infiltration formula can be found in the literature [1]. According to the calibration principle, the following formula is used to calibrate the Philip infiltration type. as follows:

$$\alpha_{si} = (S_i / \overline{S})^2; \ \alpha_{Ai} = (A_i / \overline{A})^{1/2} \quad (3)$$

$$I^{\#} = \alpha I; t^{\#} = \alpha^3 / t \quad (4)$$

In the formula, *i* is the number of measuring points. *Si* and *Ai* are the infiltration parameters of the corresponding measurement points. α_{Si} is based on the calibration factor value calculated on the basis of the imbibition rate S, and $\alpha_{Ai}$is the calibration factor value based on the calculation of the steady seepage rate of *A*. The calibration factor of each test point is shown in Table 1; α is the calibration factor, $I^{\#}$and $t^{\#}$ are the cumulative infiltration and infiltration diachronic time after calibration, and S and A are the mean values of S and *A*, respectively. Using the next type calculation:

$$\overline{S} = \frac{1}{n}\sum_{i=1}^{n} S_i; \overline{A} = \frac{1}{n}\sum_{i=1}^{n} A_i \quad (5)$$

Table 1 Infiltration parameters and scaling factors in survey points

Test point number	S (mm/min0.5)	A (mm/min)	α_S	α_A
1	4.7105	0.1208	0.65065	1.41753
2	2.5005	0.0046	0.18334	0.101393
3	8.6744	0.6477	2.20644	1.203136
4	10.892	0.7278	3.47879	1.275363
5	14.028	0.8317	5.77039	1.363362
6	2.8671	0.5096	0.24104	1.067192
7	14.028	0.8317	5.77039	1.363362
8	1.9611	0.3021	0.11264	0.821681
9	0.6591	0.0201	0.01273	0.211946
10	1.4098	0.0528	0.05828	0.343514
11	1.3947	0.0214	0.05703	0.218693
12	6.9528	1.2991	1.41753	1.703919
Average value	5.8397	0.4475		

Type (3) and (4) is the calibration relation used in infiltration analysis. By introducing it into 1 (Philip), combined with the calibration principle, we can get the calibrated formula of infiltration in the study area.

$$I^{\#} = 5.8397t^{\#0.5} + 0.4475t^{\#} \quad (6)$$

4.2 Optimal selection of calibration factor

According to the field measured data, the original cumulative infiltration curve of each measuring point is drawn, as shown in Figure 1. It can be seen from the diagram, Because of the spatial variation of soil soil infiltration characteristics, the seepage curves of each measuring point are very different, so calibration and analysis are needed.

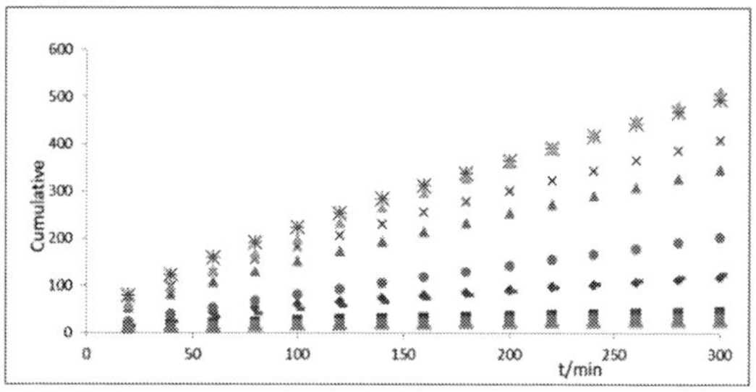

Fig. 1. Original cumulative infiltration curves of survey points

In order to obtain a better $I^{\#}$ - $t^{\#}$ relationship for each test point, the spatial variability of infiltration parameters is studied quantitatively. By using the calibration factor α_A and α_S based on the infiltration parameters A and S, the infiltration curves of each test point in the study area are calibrated. The results of the calibration are shown in Figure 2 and figure 3. It is visible from the diagram, There is a clear difference between the curve of the α_S calibration and the curve described in formula 6, the effect of calibration is poor. The curve of the α_A calibration is not exactly the same as the curve described in formula 6, but there is a high consistency, the effect of calibration is better. This is in line with the conclusions of Nie Weibo's [8] study.

https://doi.org/10.1051/e3sconf/20183801024

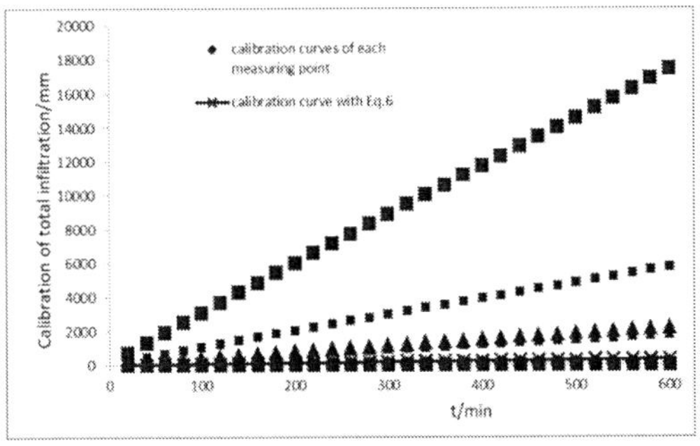

Fig. 2. $\alpha_s$calibration curve and formula (6) curve

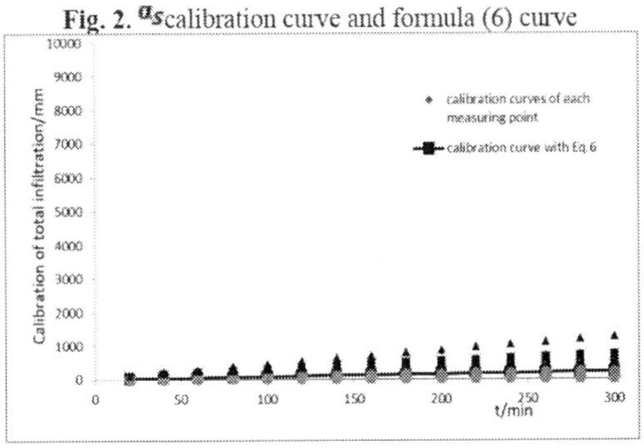

Fig.3. $\alpha_A$calibration curve and formula (6) curve

Therefore, in practical application, α_A can be selected to calibrate the infiltration curve of different measuring points. Based on the calibration principle, the selected calibration factor X and formula (6) are combined to obtain the relationship between the cumulative infiltration quantity and the infiltration time of any measurement points in the region:

$$I^{\#} = 5.8397\alpha_A^{0.5}t^{\#0.5} + 0.4475\alpha_A^{2}t^{\#} \qquad (7)$$

4.3 Analysis of variation characteristics of calibration factor α_A space.

Equation (7) contains only one unknown quantity, Therefore, it is necessary to analyze its spatial variation characteristics to reflect the spatial variation characteristics of soil infiltration. SPSS software was used for statistical analysis of 12 sets of α_A values obtained from measured data. The results are shown in table 2, According to the data in the table, the calibration factor α_A is the skewed distribution, with the variation coefficient of 0.3234, indicating that the calibration factor α_A in the selected region has certain spatial variation characteristics.

Table 2 Statistical results of scaling factor α_A

Measuring point number	12
Distribution type	Skewness distribution
The maximum value	1.7
The minimum value	0.1
The average	0.8494
standard deviation	0.5512
skewness	-0.043
kurtosis	-1.579
Coefficient of variation	0.3234

5 Conclusion

Philip's formula can better describe the water infiltration characteristics in the field. On the basis of calibration theory, Philip's formula is calibrated. It turns out that α_A is the best, And based on it, the formula of the optimal calibration calculation for the cumulative infiltration of soil in the experimental plot of the west campus of tianjin agricultural college is obtained. The analysis results show that the calibration factor α_A has a certain degree of spatial variation.

References

1. Lei ,Z.D.,Yang S.X,Xie S.C. One-step method of scaling the soil hydraulic properties in the filed [J]. Journal of Hydraulic Engineering,1986,17 (12) : 1 - 10.
2. Huang ,G.H. A review of experimental study on spatial variability of soil hydraulic properties [J]. Advance in WaterScience,1999,10(4) : 450- 457.
3. Li,X.Y,Lei ,T.W,Wang W. Spatial variability of field soil properties and Kriging method[J]. Agric.Borealioccidentalis,2000,28(6) : 30- 35.
4. Bai, M.J, Xu D, Li,Y.N, Li,J.S. Study on spatiotemporal variability of soil infiltration parameters in ground irrigation [J].Journal of soil and water conservation, 2005,19 (5):120-151
5. Yuan,J.P. Zhang,S.L. Zhang,C.Y. Jiang,D.S. Spatial variation of soil stability infiltration rate in small watershed in loess hilly region [J]. Journal of soil science, 2001,38 (4) 579-583
6. Jia,H.W. Kang,S.Z. Zhang,F.C. Tong,L Yao,L.M. Study on spatial variation of soil infiltration characteristics in plain area of Shiyang river basin [J]. Advance in Water Science, 2006,17(4) 471-476
7. He,D. Ma,D.H. Zhang,X.Z. Zhang,J.B. Zheng, Z.C. Spatial variation of soil infiltration characteristics and its source of variation [J]. Water science progress, 2013,24 (3) 340-348
8. Nie,W.B. FeiL.J. Ma， X.Y. Study on Spatial Variability of Infiltration Characteristics in Yangling District[J]. Soils, 2013, 45(1): 152–158.

 https://doi.org/10.1051/e3sconf/20183801025

Design and Implementation of Green Construction Scheme for a High-rise Residential Building Project

Yong Zhou[1,*], *You Zhen* Huang[1]

[1]School of Architecture, Huang Gang Normal University, 438000 Huang Gang, Hubei, China

Abstract. This paper mainly studies the green construction scheme of a high-rise residential building project. From "four sections one environmental protection"， saving material, water saving, energy saving, economical use of land and environmental protection conduct analysis and research. Adopting scientific, advanced, reasonable and economical construction technology measures, implementing green construction method. Promoting energy-saving technologies in buildings, ensuring the sustainable use of resources, Maximum savings of resources and energy, increase energy efficiency, to reduce pollution, reducing the adverse environmental impact of construction activities, ensure construction safety, build sustainable buildings.

1 Introduction

A high-rise residential building project is 36 stories, the floor area is 3686.68 square meters. It consists of two parts: the aboveground house and the underground garage. The underground garage (including the civil defense project) has a total height of 5.1m. The height of the main building is 109.9m, the total construction area is 14286 square meters. The basic structure of the project is a pile raft. The diameter of the pile is 600mm, the pile end holding layer is silty clay layer. The main structure form is reinforced concrete shear wall, the seismic fortification intensity of the project is 7 degrees. The seismic grade is level 2. The environment category of the concrete structure on the ground is class I, underground part of the concrete structure environment categories III class.

Considering that the construction period of the project is relatively tight, the construction period of the project is relatively long. There are many construction during the winter rain. The requirements for civilized construction, environmental protection and safety are very high. The aim of this project is to create "provincial quality project", "municipal civilized construction site" and "bureau observation site", to meet the requirements of "four sections, environmental protection and occupational health and safety" as stipulated by the relevant national standards, to meet the requirements of owners and social welfare to the maximum extent. According to GB/T50905-2014, Green construction code for construction projects, GB50325-2010（2013）, the standard of indoor environmental pollution control of civil construction engineering, (GB/T50378-2006), green building evaluation standard, GB/T50640-2010, evaluation criteria for green construction of construction projects, GB50411-2007, Specification for acceptance of construction quality of building energy conservation projects, GB12523-2011, emission standard of environment noise for boundary of construction site. Indoor air quality standards GB/T18883-2016, and the bidding documents for the general project contracting project, tenders, construction contracts and related annexes, tenders, construction contracts and related annexes, special plan for this green construction.

2 Text

Green construction is refers to engineering construction, under the premise of ensuring quality, safety and other basic requirements, through scientific management and technological progress, minimize saving resource, reducing construction activities that adversely affect the environment. Realize "four sections of environmental protection", saving material, water saving, energy saving, economical use of land and environmental protection and so on. According to the bidding documents, construction contract and design drawings of the project, combined with the design of the construction organization and the actual conditions of the site and the main features of the project, fully consider the engineering quality and construction safety guarantee system, organization and management of engineering projects, layout of construction site, total construction schedule control and so on, we will formulate a special plan for green construction. The control items and targets of the green construction scheme are shown in table 1.

* Yong Zhou: 183337681@qq.com

Table 1. Project green construction project control project and target.

Serial number	Control project	Control objectives
1	Saving material and material resources utilization	The actual loss rate of material is reduced by 30%
2	Water conservation and water resources utilization	The allocation ratio of water-saving appliances reached 80%, The actual water consumption is reduced by 30%
3	Energy conservation and energy utilization	Save 10% on actual electricity consumption
4	Economical use of land and construction land	The effective utilization ratio of temporary facilities is greater than 90%
5	Environmental protection	Dust control, Noise control, Light pollution Control, Sewage control, Construction waste control

2.1 Saving material and material resources utilization

Saving material and material resources utilization, it is controlled mainly from five aspects, including economy of structural materials, economy of palisade materials, the economy of decorative materials and savings of working materials. All kinds of material saving in this project are shown in table 2.

Table 2. Various materials saving control list of this project.

Serial number	Classification material saving control
1	Economy of structural materials
2	Economy of palisade materials
3	The economy of decorative materials
4	Savings of working materials

Economy of structural materials, control steel scrap rate is less than 1.5%, optimize the reinforcement plan, the rebar connection method adopts straight thread form, and use the reinforcement to process the excess tail material to make the lower horse stool, short support, etc. The template is made of aluminum alloy, which can save wood completely. Strengthen control over the quality of formwork. Avoid stitching between templates, the failure of the slurry and the support to strengthen the bulging mold, affecting the quality of concrete. Scientifically and reasonably arrange the concrete supply plan, implement active control, control and dynamic control, ensure the concrete saving rate of concrete is greater than 3%.

Economy of palisade materials, to strengthen the management of the transportation, stacking, transshipment and custody of the surrounding materials such as bricks and blocks, pay attention to the light, prevent damage. Before masonry walls, first, try to arrange the dry brick (block), reasonably determine the formation and thickness of dry brick (block). It is strictly forbidden to cut bricks at random during the masonry process. Avoid using whole bricks less than 1/3 of length to avoid wasting material. For aerated concrete blocks must be cut by cutting machine, to reduce the damage to the remaining bricks. Using pre-mixed mortar technology, reduce the waste of mortar.

The economy of decorative materials. Make overall arrangement before construction, for large decorative blocks, it can be pre-arranged, minimize the use of non-monolithic materials. The order of interior decoration construction, should be strictly according to the first canopy, then metope, finally ground, organize construction from top to bottom. Avoid improper contamination or damage to the veneer due to process arrangement. For paint, latex paint and other liquid decorative materials, should be based on the amount of each shift and the actual demand of the working face, using a separate barrel for use, ensure that the day or shift is used up, avoid material deterioration or cross contamination after opening. Various semi-finished products used in construction, such as stone, glass and wood decorative components, the construction project department shall be responsible for providing specific dimensions, the supplier shall process the products according to relevant standards and requirements.

Savings of working materials. The annual loss rate of the steel tubes used to strictly control scaffolds and scaffolds is within 1%. The annual loss rate of connection coupler is less than 2%. To protect the entrance, guard rail, enclosure, step Angle, etc. We should make full use of the remaining waste templates and wood beams, take recycling. For structures full of scaffolding, it is best to use a button-down scaffold with a combination of pipe fittings. The main keel is adopted to shape the back and shape lock system, the secondary keel adopts galvanized small square steel, reduce the construction turnover material to the wooden side a lot of use.

2.2 Water conservation and water resources utilization

Water conservation and water resource control include water management, recycling water, secondary use of water resources, water-saving systems and water-saving appliances, collection and utilization of unconventional water sources, etc. Rational determination of water conservation and water resources control objectives, ensure that the reuse of non-traditional water sources and recycled water is greater than 30%. Municipal water saving rate is greater than 10%. The water supply system of construction site is arranged according to living area and production area respectively. The living area water pipe network is the PPR pipe hot melt connection, the director diameter is 50mm, the diameter of the branch pipe is 25mm. half ball valve faucet is set at the end of each branch pipe. Setting up rainwater sedimentation tank outside the construction site, and set the water gun, the rainwater collection is used for secondary recycling. Construction site equipment, equipment, vehicle washing water set up circulating water device. Water-saving systems and water-saving appliances are adopted in the construction site and living areas. Improve the allocation ratio of water-saving appliances. Water saving products should be used for temporary projects. Install metering device, adopt specific water-saving measures. Using a fire pool or sedimentation tank, collecting rainwater and surface water, used for construction and production of water. Site equipment, equipment, vehicle washing, spray the road, greening and irrigation water, preferential use of unconventional water sources, try not to use municipal tap water. The water supply pipe network should be designed according to the water consumption, reasonable diameter, the pipeline is simple, take effective measures, reduce leakage of pipe network and water equipment.

2.3 Energy conservation and energy utilization

Energy conservation and energy utilization is mainly controlled from three sub-items, such as machinery and equipment and production, living and office temporary facilities and construction electricity and lighting.

Energy saving measures for machinery and equipment, to ensure the low consumption, efficiency and safety of mechanical equipment, the completion rate of mechanical equipment is 100%. The maintenance of construction machinery and equipment must be done in time. The power of the construction machinery is required to match the load. The electromechanical installation project should adopt the energy-saving mechanical equipment such as low energy consumption, high efficiency inverter type welding machine and hand-held power tool. For the site construction tower crane, construction elevators, searchlights, welding machines and other mechanical equipment of the electricity metering, it should be a machine that sets a single meter, the meter is measured separately.

Energy-saving control measures for production, living and office temporary facilities, living and office temporary facilities, site production, life temporary housing layout to the south and north, in a zigzag pattern, to obtain good lighting and ventilation effect. The wall, roof and other surrounding materials should adopt energy-saving and heat-insulating materials. The office layout is reasonable and orderly. Living areas are heated by bio-fuel boilers, the heating equipment is provided by the manufacturer, demand the economy beautiful and practical. In the field construction project department and construction team to carry out a wide range of energy saving assessment, strengthen the staff to save electricity consciousness. For the living area, the electricity shall be measured separately. Statistical power consumption, and comparative analysis. The canteen should also be measured separately, special statistics canteen USES electricity.

Energy saving measures for construction electricity and lighting, in addition to the lighting fixtures used during nighttime construction, the utilization rate of energy-saving lamps in the field is 100%. For offices, dormitories and other indoor lighting power density value control in the 5 w / ㎡, the basement construction area and other special lighting power density value control in the 4 w / ㎡. Temporary building lighting in office area and living area is illuminated by fluorescent lamps. Set up the "save electricity" sign in an eye-catching indoor location. Indoor lamps are controlled by each switch with no more than 2 lamps. The construction process can be arranged reasonably, avoid using high-power equipment at the same time, to reduce the peak effect of power load.

2.4 Economical use of land and construction land

According to the site conditions and the actual needs of the project, the temporary construction should be reasonably arranged. The floor area of all kinds of temporary buildings should be controlled according to the minimum area of land use index. The control index of construction land rate of unit construction area is 25%. Optimizing the construction plan of deep foundation pit, control the width of the soil in the foundation pit, minimize excavation and backfill of earthwork, protect the surrounding natural environment. Construction site warehouse, material storage yard, rebar processing room, work shed and other layout, should be near the temporary road, reduce transport distance. The temporary office and living room adopt the standardized prefabricated structure, such as double deck light steel activity board room. Site area as much as possible using ground greening, reduce site hardening area.

2.5 Environmental protection control measures

Environmental protection control measures include dust control, noise and vibration control, light pollution control, water pollution control and construction waste control.

Dust control measures. During the construction phase of earthwork, must control the construction site visual dust height less than 1.5m, and it cannot spread beyond the construction site. In the stage of construction, installation and decoration, it is necessary to control the site of the operation area to measure the dust height less than 0.5m. The difference between the average concentration of the total suspended particulates (TSP) and the urban background value of the total suspended particulates (TSP) measured in the surrounding area of the construction site must be less than 0.08mg/m3. There is no dust in the area, such as office area, living area, parking lot and activity area. The construction area is arranged into a circular road. The hardened pavement is not less than 6 m wide. The driving speed of the construction site should be controlled, the speed limit is

5 km/h. In addition to the hardening part of the construction site, the rest of the ground should be green, reduce dust pollution. Construction waste is sealed in bags, Effective prevention of dust pollution during transportation. For bagged powder materials such as cement, gypsum powder, putty powder, etc. sealed warehouse and sealed container shall be set on site. Handle with light in loading and unloading, dust can be avoided.

Noise and vibration control measures. Reasonable selection of construction machinery for noise and vibration, such as pile driver, forklift, excavator, dump truck, vibration machine, etc. Reasonable arrangement of mechanical time, ensure the smooth and efficient operation of mechanical construction. Vehicles are prohibited from honking in the field. Measure and control the noise level of the field every day. The wood processing shed around the site should be boarded up. When concrete is poured, there are no vibration bars, collision bars or templates. The concrete after pouring belt, construction joint, structural expansion die and so on, try to use artificial methods, minimize the use of mechanical equipment such as pneumatic picks. Reference to national standard GB12523-2011 environmental noise emission standards for construction field, the noise control limit for each construction stage of the project is shown in table 3.

Table 3 Noise control table for each construction stage of this project.

The construction phase	Daytime noise limit (unit: fb)	Night noise limit (unit: fb)
Earthwork construction	65	55
Pile foundation construction	80	Ban on the construction
Structure construction	65	55
Decorate construction	55	50

Light pollution control measures. For light intensity of night light lamps and lanterns should set hood. In the field of welding construction, special shade cloth is used to cover the area, set a fire fighting position at the lower part of the weld. In the flash butt welding machine field operation, except for the artificial side, the remaining three sides are closed with waste templates. For office area, living area night outdoor lighting all USES energy-saving lamps and lanterns.

Water pollution control measures. One septic tank is set up in the living area and office area respectively, set up a grease trap. The septic tank is cleaned by the professional cleaning department once a month. The grease trap is cleaned by the professional cleaning department once a month. To discharge water quality on site, each month, the relevant departments conduct a test. The site antifreeze and equipment lubricating oil are placed on the special shelf of the warehouse, to avoid spills and pollution.

Construction waste control measures. This project controls less than 400 tons of construction waste per 10,000 square meters. Solid waste includes concrete slags, block edges, brick edges and so on, and the reuse rate is not less than 30%. The recycling rate of ground ash is not less than 70%. The construction waste is classified and stacked, classification process. Close storage of toxic and hazardous waste. At the site construction waste recycling station, construction of a simple solid waste treatment plant, the part of the solid waste to take classification recycling, separate organic matter removal and fragmentation treatment. And then I'm going to put it in, it can be use again. To advance the design of the sub-item of the fine decoration division, as far as possible to reduce the masonry material, floor tile and ceiling board materials such as the use of non – block, control produces more waste and garbage.

3 Conclusion

Through the planning and implementation of the green construction scheme of this project, mainly includes "four sections one environmental protection", saving material, water saving, energy saving, economical use of land and environmental protection and so on. Adopting scientific, advanced, reasonable and economical construction technology measures, implementing green construction method, implementing green construction management. Promoting energy-saving technologies in buildings, ensuring the sustainable use of resources, build sustainable buildings. Maximum savings of resources and energy, increase energy efficiency, to reduce pollution, reducing the adverse environmental impact of construction activities, ensure construction safety, realize harmonious development with nature and society Achieve high quality, high efficiency, safety, civilization, low consumption and other green construction targets.

References

1. Wang Y. To explore the application of green construction technology in construction engineering [J]. Green Environmental Protection Building Materials, 222-225, 1 (2018)
2. Li Y S. Application of green construction technology in high-rise building construction [J]. Housing and Real Estate, 179-183, 12 (2017)
3. Xu B. Construction of green construction technology and its application research [J]. Chinese & Overseas Architecture, 157-160, 12 (2017)
4. Wang H B, Bao Z M, Wang W Z. The present situation of green construction technology of domestic construction engineering [J]. Soil Engineering and Foundation, 490-494, 4 (2017)
5. Zhu X X. Quality control study of housing construction [J]. China Building Materials Technology, 78-80, 2(2016)

6. Liu A F. To resolve the quality control of industrial and civil construction projects [J]. Low Carbon World, 166-167, 1 (2017)
7. Wang C X. Research on the quality control measures of construction projects [J]. Jiangxi Building Materials, 78, 4 (2016)
8. Zhang S F. Building project quality control in the whole process analysis [J]. Manager' Journal, 293, 11 (2013)
9. Live D H. Series the building project construction process quality control [J]. Manager' Journal, 93, 9 (2010)
10. Chen F. How to control the whole process of construction quality [J]. Friend of Science Amateurs, 76-77, 5 (2010)
11. Goo R. Building engineering construction whole process quality management research [J]. Industrial C, 7, 23 (2015)
12. Chen B. Building engineering construction quality management and control [J]. Guangdong Science and Technology, 44-45, 15 (2012)
13. Wei W D Shallow theory of construction engineering project construction phase quality control key [J]. The High-quality Goods, 45-46, 2 (2016)
14. Liao H. Introduction to the construction of engineering quality control [J]. The Green Building, 62-64, 3 (2014)
15. Yu L J. Introduction to the quality control of construction engineering construction stage [J]. Architecture and Budget, 28-30, 4 (2014)
16. [16. GB. 50325-2010（2013）, Standard of indoor environmental pollution control for civil construction projects [S]. Beijing: China Planning Press, (2010)
17. GB/T50905-2014, Green construction code for construction projects [S]. Beijing: China Construction Industry Press, (2014)
18. GB12523-2011, Emission standard of environment noise for boundary of construction site [S]. Beijing: China Construction Industry Press, (2011)

The consistency of Arsenic's high value regions in surface and bottom waters in Jiaozhou Bay

Dongfang Yang[1,2,3,*], *Jianxun* Chai[1,2], *Xiaoye* Gao[1,2], *Yunjie* Wu[1,2], and *Sixi* Zhu[1,2]

[1]Research Center for Karst Wetland Ecology, Guizhou Minzu University, Guizhou Guiyang, Guizhou Guiyang, China
[2]College of Chemistry and Environmental Science, Guizhou Minzu University, Shanghai, 550025, China
[3]North China Sea Environmental Monitoring Center, SOA, Qingdao 266033, China

Abstract. This paper analyzed the contents and distributions of As in Jiaozhou Bay in 1982. Results showed that As's contents in bottom waters in coastal waters in the southwest of the bay and the bay mouth in Jiaozhou Bay in July and October 1982 were 0.88-4.48 μg L^{-1} and 0.84-1.16 μg L^{-1}, respectively. As's contents were much lower than the guide line of Grade I, and the pollution level of As in bottom waters in Jiaozhou Bay in 1982 was still very slight. There was high value region in coastal waters in the southwest of the bay in July 1982 (4.48 μg L^{-1}), while in October 1982 high value region was in the bay mouth (1.16 μg L^{-1}). The source input of As from overland runoff in the southwest was strong in July, resulted in relative high value region in surface waters in coastal waters in the southwest of the bay, and therefore resulted in high value region in bottom waters in same region by means of rapid sedimentation of As. October was wet season and there was little As input to the bay, resulted in low As's contents in surface waters, as well as bottom waters. At any time, the terrigenous As was foremost transported to surface waters, and was subsided to sea bottom through water body by vertical water's effect. Hence, the high value regions of As's contents in surface and bottom waters had characteristic of consistency in Jiaozhou Bay.

1 Introduction

As is one of the widely used elements in industry and agriculture, yet certain form of As such as arsenic trioxide is high toxic. Nowadays, many marine bays have been polluted by As due to the rapid development of economic and population size ([1-2]). Hence, understanding the horizontal distributions and vertical transferring processes of As in marine bays is essential to pollution control ([3-4]). The changes of As's contents were mainly determined bay source input and vertical water's effect, and resulted in the varions of As's distributions in surface and bottom waters ([3-6]).

Jiaozhou Bay is a semi-closed bay located in Shandong Province China, and has been polluted by various pollutants including As since China's Reform and Opening-up ([1-2]). This paper the anzlyzed contents and distributions of As in Jiaozhou Bay in 1982. Results found that the terrigenous As was foremost transported to surface waters, and was subsided to sea bottom through water body by vertical water's effect. The high value regions of As's contents in surface and bottom waters had characteristic of consistency in Jiaozhou Bay. These results were important basis for scientific research and environment management decision-making.

2 Materials and method

2.1 Study area

Jiaozhou Bay is located in the south of Shandong Province, eastern China (35°55′-36°18′ N, 120°04′-120°23′ E). The total area, average water depth and the width of the bay mouth are 446 km2, 7 m and 3km, respectively. The bay mouth is connecting to the Yellow Sea in the south, and is surrounded by Qingdao, Jiaozhou and Jiaonan in the east, north and west, respectively (Fig. 1). There are a dozen of rivers, all of which are seasonal rivers ([7-8]).

2.2 Data collection

The investigation on As in surface waters in Jiaozhou Bay was carried on in July and October 1982 in four sampling sites (i.e., 083, 084, 122 and 123) (Fig. 1). As in waters was sampled and monitored follow by National Specification for Marine Monitoring ([9]).

* Corresponding author: dfyang_dfyang@126.com

 https://doi.org/10.1051/e3sconf/20183801026

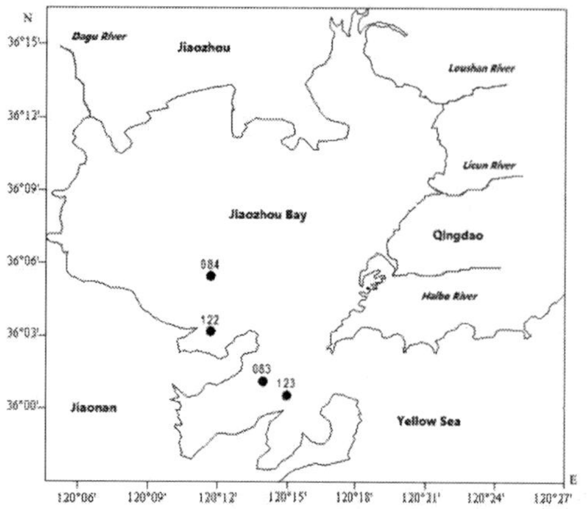

Fig. 1. Geographic location and monitoring sites in Jiaozhou Bay.

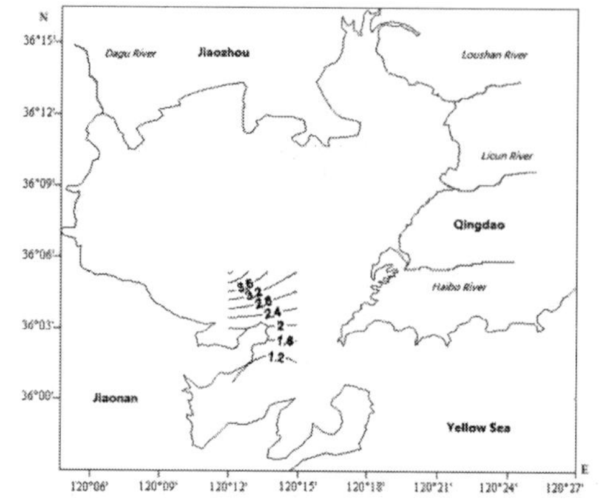

Fig. 2. Horizontal distribution of As in bottom water in Jiaozhou Bay in July 1982/μg L^{-1}

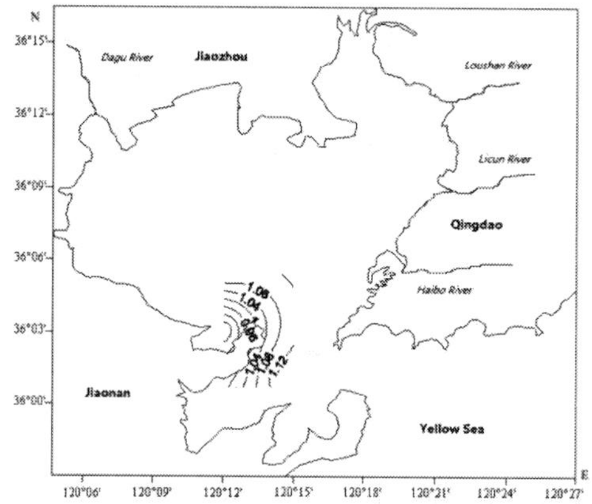

Fig. 3. Horizontal distribution of As in bottom water in Jiaozhou Bay in October 1982/μg L^{-1}

3 Results

3.1 Contents and pollution level of As in bottom waters

As's contents in bottom waters in coastal waters in the southwest of the bay and the bay mouth in Jiaozhou Bay in July and October 1982 were 0.88-4.48 μg L^{-1} and 0.84-1.16 μg L^{-1}, respectively. In according to Sea Water Quality Standard (GB 3097-1997) for As, As's contents were much lower than the guide line of Grade I for As (20.00 μg L^{-1}). In general, the pollution level of As in bottom waters in Jiaozhou Bay in 1982 was still very slight (Table 1).

Table 1. Contents and pollution level of As in bottom waters in Jiaozhou Bay 1982

	July	October
Content/μg L^{-1}	0.88-4.48	0.84-1.16
Grade	I	I

3.2 Horizontal distributions of As in bottom waters

In July 1982, high value of As's contents was in Site 084 in coastal waters in the southwest of the bay, and the countour lines of As's contents were forming a series of parallel lines that decreasing from the coastal waters in the southwest of the bay (4.48 μg L^{-1}) to the bay mouth (0.88 μg L^{-1}) (Fig. 2). In October 1982, high value of As's contents was in Site 123 in the bay mouth, and the countour lines of As's contents were forming a series of semi-circles that decreasing from the bay mouth (1.16 μg L^{-1}) to the coastal waters in the southwest of the bay (0.84 μg L^{-1}) (Fig. 3).

4 Discussion

4.1 High sedimentation regions of As

In July 1982, As contents were decreasing from the southwest of the bay to the bay mouth, indicated that high sedimentation process was occurring in coastal waters in the southwest of the bay (4.48 μg L^{-1}). In October 1982, As contents were decreasing from the bay mouth to the southwest of the bay, indicated that high sedimentation process was occurringin the bay mouth (1.16 μg L^{-1}). The sedimentation strength in coastal waters was much stronger than in the bay mouth. At any time, the terrigenous As was foremost transported to surface waters, and was subsided to sea bottom through water body by vertical water's effect ([4-5], ([10])). The horizontal distributions of As's contents were determined by source input and vertical water's effect.

4.2 Transferring processes of As

The horizontal distribution of As's contents in surface waters in July 1982 showed that As's contents were decreasing from the coastal waters in the southwest to the bay mouth, and indicated that overland runoff was the major source. Hence, there was high value region of As's contents in surface waters in coastal waters in southwest of the bay, and by means of vertical water's effect ([4-5]), As was transferred to and accumulated in bottom layer of waters, resulted in high value region in bottom waters in this area. The horizontal distribution of As's contents in surface waters in October 1982 showed that As's contents were showing no trends but disorder patches, and therefore indicated that there was little souce input. Hence, there was no high value region of As's contents in surface waters in the bay, nor high value region in bottom waters in this area. However, in case of relative strong source input, there was high value regions in surface waters, as well as in bottom waters, and the high value regions had characteristic of consistency by means of vertical water's effect.

5 Conclusions

As's contents in bottom waters in coastal waters in the southwest of the bay and the bay mouth in Jiaozhou Bay in July and October 1982 were 0.88-4.48 μg L^{-1} and 0.84-1.16 μg L^{-1}, respectively. As's contents were much lower than the guide line of Grade I, and the pollution level of As in bottom waters in Jiaozhou Bay in 1982 was still very slight. The horizontal distributions of As's contents were determined by source input and vertical water's effect. In case of relative strong source input, there was high value regions in surface waters, as well as in bottom waters, and the high value regions had characteristic of consistency by means of vertical water's effect.

Acknowledgments

This research was sponsored by Research Projects of Guizhou Nationalities University ([2014]02), Research Projects of Guizhou Province Ministry of Education (KY [2014] 266), Research Projects of Guizhou Province Ministry of Science and Technology (LH [2014] 7376).

References

1. D.F. Yang, W.P. Song, S.T. Chen, J.H. Guo, L. Zheng, Coastal Engineering, **31**, 4(2012)
2. D.F. Yang, Y.H. Zhao, Z.G. Piao, Ocean Development and Management, **31**, (2014)
3. D.F. Yang, S.X. Zhu, F.Y. Wang, S.T. Chen, Z.J. Xu, S.X. Zhu, Y.J. Wu, Meterological and Environmental Research, **5**, (2014)
4. D.F. Yang, F.Y. Wang, H.Z. He, Y.J. Wu, S.X. Zhu, Proceedings of the 2015 international symposium on computers and informatics, (2015)
5. D.F. Yang, F.Y. Wang, X.L. ZhaoY.J. Wu, S.X. Zhu, Sustainable Energy and Enviroment Protection, (2015)
6. D.F. Yang, F.Y. Wang, X.Q. Yang, Y.J. Wu, S.X. Zhu, Advances in Computer Science Research, **3252**, (2015)
7. D.F. Yang, Y. Chen, Z.H. Gao, J. Zhang, F. Wang, Chinese Journal of Oceanology and Limnology, **23**, 1(2005)
8. D.F. Yang, F.Y. Wang, Z.H. Gao, W.L. Cui, S.X. Huo, Marine Science, **28**, 6(2004)
9. China's State Oceanic Administration, *The specification for marine monitoring* (Ocean Press, Beijiang, 1991)
10. Yang DF, Miao ZQ, Xu HZ, Y. Chen, J.Y. Sun, Marine Environmental Science, **32**, 3(2013)

Restoration and Purification of Dissolved Organic Nitrogen by Bacteria and Phytoremediation in Shallow Eutrophic Lakes Sediments

Xin Li and *Yi Yue*

Laboratory of Waterway Environmental Protection Technology, Tianjin Water Transport Engineering Science Research Institute, Tianjin 300456, China

Abstract. Endogenous organic nitrogen loadings in lake sediments have increased with human activity in recent decades. A 6-month field study from two disparate shallow eutrophic lakes could partly reveal these issues by analysing seasonal variations of biodegradation and phytoremediation in the sediment. This paper describes the relationship between oxidation reduction potential, temperature, microbial activity and phytoremediation in nitrogen cycling by calculation degradative index of dissolved organic nitrogen and amino acid decomposition. The index was being positive in winter and negative in summer while closely positive correlated with biodegradation. Our analysis revealed that rather than anoxic condition, biomass is the primary factor to dissolved organic nitrogen distribution and decomposition. Some major amino acids statistics also confirm the above view. The comparisons of organic nitrogen and amino acid in abundance and seasons in situ provides that demonstrated plants cue important for nitrogen removal by their roots adsorption and immobilization. In conclusion, enhanced microbial activity and phytoremediation with the seasons will reduce the endogenous nitrogen loadings by the coupled mineralization and diagenetic process.

1 Introduction

In recent years, the global problems of lake-eutrophication, especially nitrogen distribution or migration, have been greatly concerned due to their wide spreading. This issue is critical in understanding the mechanism of nitrogen cycling and freshwater ecosystem responses to anthropogenic mediated change. Sediments are the main destinations as storing pool of nutrients by diagenetic transformation while some endogenous organic nitrogen release into the sediment leading to a eutrophic restoration under some conditions. In addition to mostly studied exogenous inputting, conversely, some cases had focused mainly in form, content, distribution and seasonal variation of dissolved organic nitrogen (DON) degradation and migration; and others in situ, including the profile of amino acids and its cycle by realizing the comparisons in mineralization, degradation and diagenesis in sediments[1-2].

In some specific cases, eutrophic lake is hot due to its widely encountered. In sediment-water interface, the occurrence and decomposition of DON is accurate understanding by geochemical dynamics as well as some related impact factors. The mineralization of degradation and phytoremediation was unclearly in calculation and interrelation. The redox potential (Eh) in sediment is closely related with hypoxia condition. Besides Eh, the seasonality of temperature and microbial activity are another factors. Likewise, phytoremediation is also a research point in sediments restoration. All these mutual relations have many uncertainties to further cognition yet.

To illustrate the issue, the mainly part of dissolved organic nitrogen, amino acids degradation could be explained by specific calculation [3]. The current laboratory incubation was to estimate this process under the assumption which the statics analyse in situ is less and the cycling mode proved to be studied by theoretical formula [4]. These incubations suggest that bacterial and photochemical degradation play a critical role in the composition of DON in some suitable conditions [5]. Because the total hydrolysable amino acid (THAA) in situ frequently comprises the largest part of DON in sediments and they are the nutrition sources to heterotrophs [6]. In this paper, our aim was to determine whether the remove or inhibit of endogenous DON cycling by factors of temperature, Eh, microbial activity (fluorescein-diacetate-ester, 'FDA' to represent) and phytoremediation to control eutrophic restoration. A further aim was to provide a powerful tool to study the sources and biogeochemical cycling of labile organic nutrient in aquatic system.

2 Material and Methods

2.1. Sample Collection

* Corresponding author Yi Yue: lixin1609@tju.edu.cn

Samples were obtained from Lake Qingnian (latitude 39° 06′ 38′ ′ N, longitude 117° 06′ 59′ ′ E) and Lagoon Qilihai (latitude 39° 17′ 21′ ′ N, longitude 117° 35′ 32′ ′ E) in Tianjin. Qingnian Lake is a typical shallow urban eutrophic lake that had no large stationary sources of pollution, whilst the lake is subject to discharges by sewage in previous years. Lagoon Qilihai was evolved by a process of marine regression at the late Holocene in North China Plain which is a marsh wetland with agriculture and fisheries features. In field study, we monitored depth, temperature, pH, turbidity, chlorophyll- α of overlying water and moisture content, Eh of sediment. The results in situ were showed in Table. 1. Samples of Lake Qingnian were obtained from two contrasting spots named 'QNP' and 'QNC' while Lagoon Qilihai had only one sampling spots named 'QLH'. 'QNP' is a zone which has luxuriant plant near the edge of bay and 'QNC' is a deep open zone at central east of the lake. Every sample was mixing five cores by standardized method collected from the surface sediments (10 cm depth) by a self-made cylindrical core sampler. Sediment and water samples were removed in polypropylene bags or acid-washed PE bottles back to the laboratory. Within 12 h after sample collection, the pore water was obtained by the centrifugal drainage technique [7]. The sediment samples were tiled to dry and sieved by 60 meshes. And then it stored in the refrigerator.

Table 1 Values of depth, temperature, pH, turbidity, chl- α , and moisture content in field study.

Core	Depth (m)	T (°C)	pH	Moisture content (%)	Turbidity (NTU)	Chlorophyll-α (μg/L)
QNP	1.7	23.8	8.4	55-64	61-73	63.9
QNC	3.9	23.8	8.4	61-72	55-70	23.1
QLH	4.2-4.8	23.4	8.2	50-59	42-53	15.3

2.2 Sample Extraction and Analysis

Dissolved inorganic nitrogen (ammonium, nitrate and nitrite) and urea were determined calorimetrically by UV-2800 ultraviolet spectrophotometer (Unico, China) [8]. Total dissolved nitrogen (TDN) in water was determined by the method of Persulfate Oxidation (PO) while its concentration in sediment was monitored by Kjedahl method. DON was calculated as a margin between the TDN and DIN (NH4+-N, NO3--N and NO2--N) readings. Chlorophyll- α was determined calorimetrically by Jin and Tu' method [9]. The Eh was monitored by the national standardized method (SL94-1996) by a platinum electrode. The THAA and hexamine (amino sugar) was determined by CEC chromatography and ninhydrin-postcolumn derivation [10]. And they were monitored by amino acid analyser (Hitachi L-8900, Japan) after sample filtration. The FDA value was monitored in this method: the surface sediment samples were cultured after pre-treatment; after adding FDA and centrifugation, cores were determined colorimetrically at 490 nm using the external standard method [11].

3 Results and Discussion

3.1 Nitrogen Components in Two Lakes

To investigate the mechanism behind, we first monitor each component of nitrogen in water and sediment. The average concentrations of five objects in 6-month experiments were observed in Tab. 2. The TDN of Lake Qingnian was characterized 5053 mg/Kg and 4483 mg/Kg in average in sediment. Meanwhile, the DON were 4641 mg/Kg and 4020 mg/Kg respectively while significantly higher than theirs in overlying water and in pore water. Comparing with inorganic nitrogen, DON represented the dominant nitrogen pool, being about 90% and 55% of TDN in sediment and overlying water respectively. Lagoon Qlihai had a less severe eutrophication condition and 'QNP' had a less eutrophication condition than 'QNC'. It once again proved that phytoremediation can be a method of lakes eutrophication. Meanwhile, the TDN/DON ratios in two lakes were similar to other cases. All above revealed that sediments are the main nitrogen pool due to the internal release and Lake Qingnian had a severe nitrogen loading than Lagoon Qilihai. In terms of DON, THAA-N (Nitrogen in THAA) was the main component in sediment and in pore water while urea and hexamine slightly accounting some of weights. These phenomena may be explained by the dynamics of nitrogen release: DON was ammoniated to NH4+ and some amino acids being decomposition in this process. An anaerobic condition that contributed to amino acids biodegradation by the production of its own and hexamine was derived from microbial decomposition. Besides, external input had lead urea concentration higher in overlying water than in pore water.

Table 2 Average contents of TDN, DON, THAA-N, Urea-N and Hexosamine-N in two lakes. (mg/Kg).

	Overlying Water		Pore Water			Sediments		
	QNP and QNC	QLH	QNP	QNC	QLH	QNP	QNC	QLH
TDN	6.10	8.87	14.42	17.81	10.42	4641	5053	1562
DON	3.85	4.60	8.38	8.92	5.43	4020	4483	1355
THAA-N	1.88	1.61	3.79	3.98	2.50	1569	1777	579

Urea-N	0.83	0.89	0.53	0.58	0.96	1.46	2.20	2.31
Hexosamine-N	--	--	0.55	0.61	0.46	145	163	77

The average contents of TDN, DON and THAA-N in 'QNP' were lower than 'QNC'. It is suggested that plant root and microbial activity promote organic nitrogen convert to inorganic via ammonification, especially in reed and ceratophyllum prosperity. Being more nitrogen in sediment, an enhanced internal loading leads to a higher DON concentration in Lake Qingnian than Lagoon Qilihai. Meanwhile, it could be explained a more activated microbial status and a principle which heterotrophs in sediments is prosperity by more ON sources.

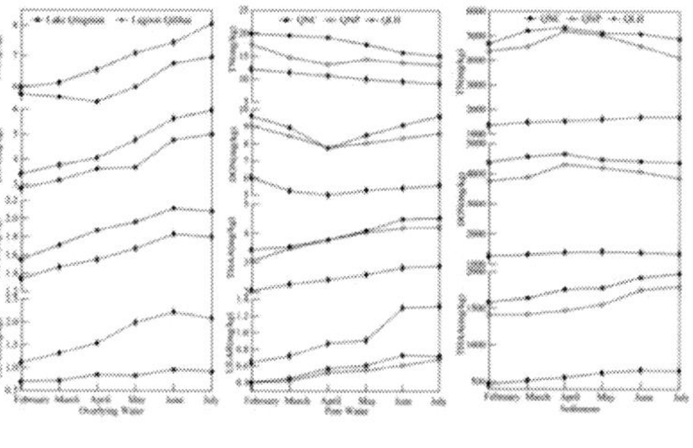

Fig 1 Seasonal dynamics of TN, DON, THAA and urea in two lakes in 6-month field study.

Seasonal dynamics for TDN, DON, THAA and hexamine measured were given in Fig. 1. It depicted that Lake Qingnian and Lagoon Qilihai have similar trends. In Lake Qingnian, 'QNP' had greater seasonal amplitude than 'QNC'. After April, TDN and DON decreased gradually in sediment while their contents had an opposite trend in overlying water. Comparing with DON concentration in pore water and overlying water, there were no regular gradient differences between them. The external input of DON may be the main reason to explain these phenomena. The THAA values in three phases had similar increasing trends. It showed that urea inputting make great affect to its content especially in 'QLH'. Referring DON and THAA values in Lagoon Qilihai were higher than Lake Qingnian may be due to exogenous inputs. The seasonal variability of urea was increased in pore water also confirmed the above process.

3.2 Degradative Index of Amino Acid in Sediment-Water Interface

The recognition of degradative index (DI) of amino acid in organic nitrogen cycling allows conclusions to be drawn to the mineralization and biodegradation in the sediment-water interface. The index is an indicator of organic matter diagenesis. DI was calculated with a formula based on principal components analysis. It is a diversity of samples ranging from fresh phytoplankton to highly degraded turbidity sediment samples:

$$DI = \sum_i [\frac{var_i - AVG\,var_i}{STD\,var_i}] \times fac.Coe.f_i$$

Where DI is the degradation index; vari is the individual amino acids (mole percentage); AVG vari and STD vari are the mean, standard deviation respectively. The fac.Coe.fi is the factor coefficient by Dauwe's experienced projection from Tab. 3 (2). The more negative for DI value is the more degraded condition for the sample. And a positive DI value indicatives fresh materials in sediment. Correlation analysis was performed using a Spearman Rank Correlation coefficient. Statistically significant differences between the two lakes were determined using the 'Execl Statistical Package'. Means are reported with a 95% confidence interval. DI values in two lakes measured are given in Fig. 2 respectively. Alanine, threonine, aspartate, glycine and glutamic acid take larger weight for DI. They are also closely related to the biomass and oxygen condition in sediment. Despite the large percentage of arginine and lysine, it is because they are more nitrogen molecules amino acids. DI ranged from 1.99 to -1.77 with an apparent downward trend. That means amino acid had an apparent accelerated degradation tendency in sediment. The downward rate of Lagoon Qilihai was more significant. It fully illustrated that, with less artificial interference, the lake has a more effective self-purification capacity by biodegradation and phytoremediation. Meanwhile, the 'QNP' had a higher degradability than the 'QNC'. A reasonable explanation to above phenomena: Expansion of plant root, microbial activity and the loose surface sediment with new coverage of organism leading to an increasing tendency of sediment porosity and further strengthens the fluxes values calculably. Furthermore, it suggested that the seasonable change of phytoremediation is another main related factor to amino acid accumulation or degradation in sediments.

Table 3 Mean deviation, standard deviation and factor coefficient for different amino acids in two lakes.

	Plant Area in Lake Qingnian		Centre Area in Lake Qingnia		Lagoon Qilihai		
	AVG var_i	STD var_i	AVG var_i	STD var_i	AVG var_i	STD var_i	Fac.Coe.f_i
ASP	7.59	2.5	8.62	2.6	5.44	2.2	-0.102
GLU	8.50	2.4	8.89	2.4	8.79	2.4	0.065
HIS	2.53	0.4	2.61	0.4	2.12	0.3	0.158
SER	6.00	2.4	6.32	2.5	5.71	2.3	0.015
ARG	12.53	0.3	11.53	0.4	14.49	0.6	-0.115
GLY	10.62	2.1	10.84	2.3	11.25	2.5	-0.099
THR	7.26	1.0	6.65	0.8	8.72	1.1	-0.129
ALA	7.76	1.3	7.70	1.3	8.72	1.4	-0.043
TYR	2.00	1.1	1.86	1.1	1.76	1.1	0.178
CYS	0.72	0.1	1.03	0.1	0.078	0.1	0.203
PRO	4.15	0.7	5.04	0.9	1.77	0.2	0.099
MET	1.55	1.0	1.10	1.0	1.37	1.0	0.134
VAL	6.01	1.7	5.44	1.6	6.45	1.8	-0.044
PHE	3.60	0.6	3.39	0.5	4.16	0.6	0.134
ILE	3.45	0.5	3.35	0.5	3.73	0.5	0.139
LEU	6.46	1.6	6.33	1.6	6.87	1.7	0.169
LYS	9.26	2.1	9.61	2.0	8.69	1.9	-0.086

3.3 Correlations between DI and Related Factors

To determine the biodegradation mechanism with related factors, we monitored three relevant ones as the known parameters. The correlation between DI and related factors can reflect the ammonification, nitrification and denitrification in the lake. Seasonality of Eh, temperature and microbial activity (FDA to represent) were shown in Tab. 4. These three factors have a similar trend which their contents are lower in winter and higher in summer. In Eh value, there was an obvious difference between two lakes after May while its value was no more than 200 mV rendering a reduction condition apparently. Contents of Eh and FDA in 'QNP' were slightly higher than 'QNC' in the same month. These suggested that photosynthesis was more susceptible and root parasitic in shallow water caused more microbial activity.

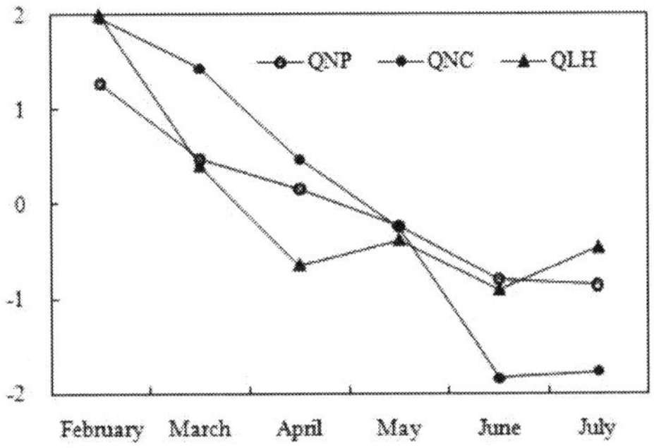

Fig 2 Seasonal dynamics of degradative index (DI) in two lakes in 6-month field study.

FDA had a positive correlation with THAA in two lakes indicating that aerobic bacteria adapts significantly to more oxygen and takes the majority in surface sediments. Greater seasonal fluctuations of glycine and glutamic acid in sediment can reflected this phenomenon. Glycine is the simplest amino acid and the primary product of metabolism for aerobic microorganisms in utility of sugar fermentation. Glutamate is the main structure amino acids and the intermediate products for aerobic metabolism. Thus, these positive coefficients reflected specific importance of microbial activity and sediment oxygen condition and hence other electron acceptors in surface sediments. From Tab. 4, correlation between three factors and DI in Lake Qingnian is different. That may revealed that out of the three related factors, phytoremediation plays is an important role in nutrient cycling because of plant utility of nutrients and its extended roots.

Table 4 Seasonal dynamics of Eh, temperature and FDA and their correlation to DI in two lakes.

	FDA (μmol/g)			Eh (mv)			Temperature (° C)		
	QNP	QNC	QLH	QNP	QNC	QLH	QNP	QNC	QLH
Feb.	0.38	0.36	0.77	52	46	32	4.1	4.1	2.3
Mar.	0.48	0.46	0.86	90	82	55	6.9	6.9	5.9
Apr.	0.68	0.62	0.97	102	99	62	13.2	13.2	11.2
May	0.86	0.81	1.01	121	115	87	21	21	20.4
June	0.94	0.88	1.08	148	107	110	24.4	24.4	22.9
July	0.91	0.88	1.02	136	103	114	27.9	27.9	26.9
Coe. f	0.8749	0.9494	0.9723	0.8689	0.7676	0.9813	0.9397	0.9720	0.9777

FDA and DI had an obvious negative correlation (R>0.95 in two lakes respectively). It indicated that biodegradation is prone react to the diagenesis. Meanwhile, it is reported that the activity of amination bacteria is increased with rising temperature in nature condition shown that temperature and DI had an obvious negative correlation in Lake Qingnian (R>0.95), but not in Lagoon Qilihai (R=0.77). The same condition carried out between Eh and DI (R≈0.8 in two lakes). Therefore, we can conclude that microbial activity in sediment is more important in these three factors for mineralization and diagenetic process. Changes of alanine, threonine and aspartate in the two lakes can be an explanation to the condition. Aspartate and alanine are synthetic materials for plant, microbial metabolism. Less in Lagoon Qilihai reflected greater biological activity there directly. And more threonine in Lagoon Qilihai revealed agriculture and fisheries are its major environmental factors. Because of threonine is the main component for feed additive. This conclusion is consistent with other studies. And the above all may be explained by the following reason: Although the redox potential and temperature have different situations, there always had some corresponding microbial which can still complete the degradation process for organic matter. The correlation parameters between DI and three factors for Lake Qingnian are greater than Lagoon Qilihai. It indicated that more oxygen, microbes and increasing temperature can promote the diagenesis in urban lake.

4 Conclusions

In the shallow eutrophic Lake Qingnian and Lagoon Qilihai, internal nitrogen release was significantly greater than the general level in sediment-water interface where DON takes the main contribution. The factors of microbes, Eh and temperature could affect the nitrogen releasing, degradation and diagenesis under oxidation conditions in lake sediment. Plants prosperity and microbial activity can relieve the hypoxia condition to inhibit biodegradation while temperature rising makes an exacerbation. And plant roots extension makes more loosed sediments could improve the oxygen conditions. In conclusion, endogenous organic nitrogen biodegrade accelerated in summer by the factors of temperature, phytoremediation and microbial conditions. Meanwhile, it can be quantitative decrypted the inhibition or promotion expressed by the form of DI and by their liner correlation. The activity of microorganisms in the sediments was primary factor than oxygen status or temperature. Seasonal variations for some amino acid indirectly confirmed this phenomenon. So that amino acid in mineralization and degradation is an important aspect to lakes ecological restoration.

Acknowledgements

This work was supported by "Study of foreign soil water retention, prevent salinization, heavy metal soil remediation, soil backfill and ecological restoration technique in land reclamation. (TKS150218).

 https://doi.org/10.1051/e3sconf/20183801027

References

1 A.D., T., G.E., M., and M.F., F. (2007) Distributions, cycling and recovery of amino acids in estuarine waters and sediments, Environmental Chemistry Letter 5, 7.

2 Duan, S., and Bianchi, T. S. (2007) Particulate and dissolved amino acids in the lower Mississippi and Pearl Rivers (USA), Marine Chemistry 107, 16.

3 Lomstein, B. A., Jørgensen, B. B., Schubert, C. J., and Niggemann, J. (2006) Amino acid biogeo- and stereochemistry in coastal Chilean sediments, Geochimica et Cosmochimica Acta 70, 20.

4 Dauwe, B., and Middelburg, J. J. (1998) Amino acids and hexosamines as indicators of organic matter degradation state in North Sea sediments, Limnology & Oceanography 43, 17.

5 Sattayatewa, C., Pagilla, K., Pitt, P., Selock, K., and Bruton, T. (2009) Organic nitrogen transformations in a 4-stage Bardenpho nitrogen removal plant and bioavailability/biodegradability of effluent DON, water research 43, 10.

6 Tappin, A. D., Millward, G. E., and Fitzsimons, M. F. (2007) Distributions, cycling and recovery of amino acids in estuarine waters and sediments, Environ Chem Lett 5, 7.

7 Jones, D. L., Shannon, D., Murphy, D. V., and Farrar, J. (2004) Role of dissolved organic nitrogen (DON) in soil N cycling in grassland soils, Soil Biology & Biochemistry 36, 8.

8 Badr, E.-S. A., Tappin, A. D., and Achterberg, E. P. (2008) Distributions and seasonal variability of dissolved organic nitrogen in two estuaries in SW England, Mar Chem 110, 12.

9 Jin, X. C., and Tu, Q. Y., (Eds.) (2000) Rules for investigation of lake eutrophication, China Environmental Science Press, Bei jing.

10 Jones, D. L., Shannon, D., Junvee-Fortune, T., and Farrar, J. F. (2005) Plant capture of free amino acids is maximized under high soil amino acid concentrations, Soil Biology & Biochemistry 37, 3.

11 Chong, J. C., and Fan, C. X. (2010) Environmental effect of sediment dredging in lake:Ⅳ.influences of dredging on microbial activity and functional diversity of microbial community in sediments and its significance, Journal of Lake Sciences 22, 8.

Linking Structural Equation Modelling with Bayesian Network and Coastal Phytoplankton Dynamics in Bohai Bay

Jiangtao Chu and *Yue Yang*

Laboratory of Waterway Environmental Protection Technology, Tianjin Water Transport Engineering Science Research Institute, Tianjin 300456, China

Abstract. Bayesian networks (BN) have many advantages over other methods in ecological modelling and have become an increasingly popular modelling tool. However, BN are flawed in regard to building models based on inadequate existing knowledge. To overcome this limitation, we propose a new method that links BN with structural equation modelling (SEM). In this method, SEM is used to improve the model structure for BN. This method was used to simulate coastal phytoplankton dynamics in Bohai Bay. We demonstrate that this hybrid approach minimizes the need for expert elicitation, generates more reasonable structures for BN models and increases the BN model's accuracy and reliability. These results suggest that the inclusion of SEM for testing and verifying the theoretical structure during the initial construction stage improves the effectiveness of BN models, especially for complex eco-environment systems. The results also demonstrate that in Bohai Bay, while phytoplankton biomass has the greatest influence on phytoplankton dynamics, the impact of nutrients on phytoplankton dynamics is larger than the influence of the physical environment in summer. Furthermore, despite the Redfield ratio indicating that phosphorus should be the primary nutrient limiting factor, our results indicate that silicate plays the most important role in regulating phytoplankton dynamics in Bohai Bay.

1 Introduction

The eco-environment is a complex system encompassing physical, chemical, and biological processes, but our understanding of these natural processes is limited. Modelling these complex systems faces various challenges: simulating all of these processes in one system with different spatial, temporal or functional scales, combining different sources of data and knowledge, properly addressing uncertainty and incomplete data sets, and so on. It is difficult to resolve these problems with traditional mechanistic models. Bayesian networks (BN), which utilize probabilistic expressions to describe relationships among variables, is an increasingly popular method for ecological modelling and environmental management [1-3]. This popularity should be attributed to its features: BN can be used to cope with uncertainty in a natural way, combine data with domain knowledge, provides greater prediction accuracy even with small sample sizes or incomplete datasets, and performs integrated ecological modelling [4-6].

In this study, we devised a method for linking BN with SEM and applied it to a study of phytoplankton dynamics in Bohai Bay, China to examine the effectiveness of using SEM in developing BN models for ecological and environmental research. This was carried out as follows: a) we built three SE models based on expert advice and previous studies to explore the causal relationship between environmental factors and phytoplankton dynamics in Bohai Bay; b) we then explored ways of linking SEM with BN; c) next, we evaluated the performance of the BN model when linked with SEM and d) finally, we examined the role of SEM in developing BN models and its effectiveness in modelling phytoplankton dynamics in Bohai Bay.

2 Methods

2.1. Study Area

Bohai Bay is located in the western region of the Bohai Sea in northern China, covering an area of 15,900 km2 at an average depth of 12 m (Fig. 1). It is a semi-closed, shallow water system, with the exchange of water between Bohai Bay and the outer sea being quite weak [7]; as such, the physical self-purification capacity of Bohai Bay is quite poor. Due to the rapid rates of economic development and population growth in the coastal zone of Bohai Bay, enormous quantities of pollutants from municipal wastewater, urban runoff, animal feeding operations and agricultural fertilizers are being discharged into Bohai Bay from Beijing, Tianjin and Hebei provinces, and most of these nutrients accumulate gradually. As a result, Bohai Bay has been experiencing severe eutrophication events in recent years, and excessive algal blooms and low levels of dissolved

* Corresponding author Yue Yang: lixin1609@tju.edu.cn

oxygen occur frequently, leading to declines in shellfish populations and an increasing number of fish kills [8]. Because of this, Bohai Bay has been declared as one of the most threatened marine systems in China.

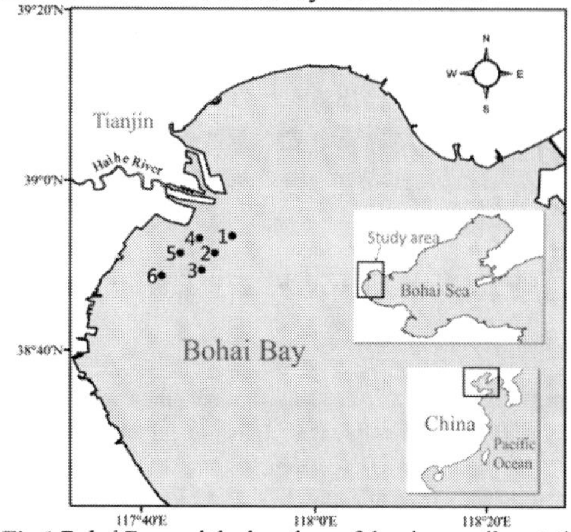

Fig 1 Bohai Bay and the locations of the six sampling stations.

2.2 Data description

The data used in this study were collected from 6 sampling stations in an ecological monitoring area of Bohai Bay, south of the estuary of the Haihe River (Fig. 1), with measurements of chemical, physical and biological processes. For chemical measurements, water samples were taken with Niskin water samplers and filtered immediately through pre-cleaned, 0.45 μ m pore-size acetate cellulose filters into a clean plastic tent, following which saturated HgCl2 solution was added to the filtrates. Nutrient (NO2-, NO3-, NH4+, PO43- and dissolved Si(OH)4 concentrations were quantified using the spectrophotometric methods previously reported in Liu et al. (2009). The analytical precision of NO2-, NO3-, NH4+, PO43- and Si(OH)4 were 0.06 μ mol L-1, 0.06 μ mol L-1, 0.09 μ mol L-1, 0.03 μ mol L-1, and 0.15 μ mol L-1, respectively. For physical measurements, salinity, pH and temperature were measured using YSI-6600 with an accuracy of 0.01 ppt, 0.01 and 0.01℃, respectively. For chlorophyll a measurement, the water samples were poured over Whatman GF/F glass microfiber filters. These filters were then immediately frozen under dark conditions until analysis in the laboratory using spectrophotometry with a Shimadzu UV-2550 [9].

3 Model Development

3.1 Relationship Exploration

In this study, SEM is used to provide a quantitative description of the relationship between phytoplankton dynamics and the coastal environment variables during the summer period in Bohai Bay. The work began by obtaining two competitive structural models, A and B (Fig. 2). Modelling processes described in previous literature asserted that the environmental factors only affect the dynamics of current phytoplankton [10], so the structural model A only considered the influence of the physical environment and nutrients on current phytoplankton dynamics, and did not include the lagged relationships (Fig. 2.a). However, the growth of phytoplankton is a two-step process: the first is to absorb nutrients, and the second to grow with the intercellular nutrient concentration. The response of phytoplankton dynamics may lag behind the effect of environmental factors. Thus, the structural model B evolves from the structural model A and is formulated based on the effects of the physical environment, nutrients and phytoplankton biomass on future phytoplankton dynamics (Fig. 2.b).

After completing the structural models, several sets of indicators should be identified that "best" describe each latent variable in the measurement models. Three observed variables – pH (pHt), salinity (St) and surface temperature (Tt) – were used in our models to measure the latent variable physical environment(the subscripted "t" denotes time and that the time step is 2 weeks).

It should be noted that chlorophyll- a was considered as a perfect measurement (i.e., no measurement error) of phytoplankton in this study. Thus, the latent variable current phytoplankton in the structural model A and structural model B was deputed by the concentration of current chlorophyll-a (Chlat), and the latent variable future phytoplankton in the structural model B was deputed by the concentration of chlorophyll-a two weeks later (Chlat+1). Two actual SE models were thus developed (Fig. 3a.b). In addition, considering that the effects on the total amount of future phytoplankton is different than the change rate of future phytoplankton caused by the environmental factors, the latent variable future phytoplankton was further set to be measured by the ratio of chlorophyll-a concentration after two weeks to current chlorophyll-a concentration (Relative Ratio), then the SE model B was developed to another actual SE model, called SE model C (Fig. 3c).

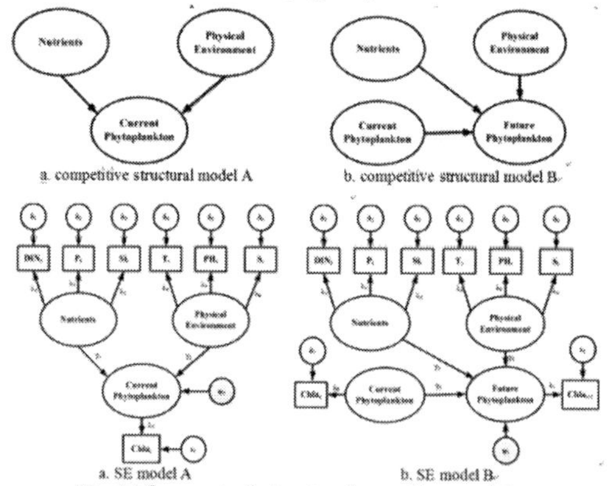

Fig 2 Three actual structural equation models

3.2 Models Definition

In general, there are four issues that need to be addressed when constructing a BN model: selecting the key factors (nodes), identifying the causal relationship between factors (nodes), discretizing the nodes and calculating the conditional probability tables (CPTs) associated with the nodes. Of these, the first two problems form the basis for the other issues of BN model, and they determine the accuracy and authenticity of the entire BN model. In this study, we used SEM to cope with these two issues. While the SE model has been shown to fit the observation data well, its measurement and structural models are considered to be realistic, which means that the selected observed variables are effective at measuring latent variables and the causal relationship represented by the structural model is reliable. Thus, in a BN model, the selection of key factors and the identification of the links can follow the measurement model and the structural model, respectively, of the SEM with desired results. In section 2.3, three satisfactory SE models were constructed. To compare and validate them, three BN models – A, B and C – were built based on the SE models A, B and C, respectively (Fig. 4).

For the discretization of nodes, the number of states and the cut-off values of the discrete states need to be defined. In general, nodes with fewer states may provide a more accurate prediction, although more states may provide greater precision. Thus, to balance accuracy and precision, the number of states was kept to four or fewer. In this study, all of the predictor variables were limited to 4 states and all the target variables were restricted to 3 states. Furthermore, to define the cut-off value of each discrete state, every state of the node should include similar amounts of observation data.

The most crucial problem of a BN model is determining the set of conditional probability tables (CPTs) underlying each node. The function of CPTs is to quantify the cause–effect relationship between variables. At the initial state, due to the analogous number of observation data in each state, all nodes were set to uniform probabilities and all CPTs were assigned to uniform values. Then, the BN models were updated automatically by using learning algorithm with sample data. Because the BN models include latent variables, the expectation maximization (EM) learning algorithm, which could cope with missing data in the case files and automatically calculate CPT values using case data, was used to update the BN models. Solving issues greatly improves the structure of BN models (Fig. 4).

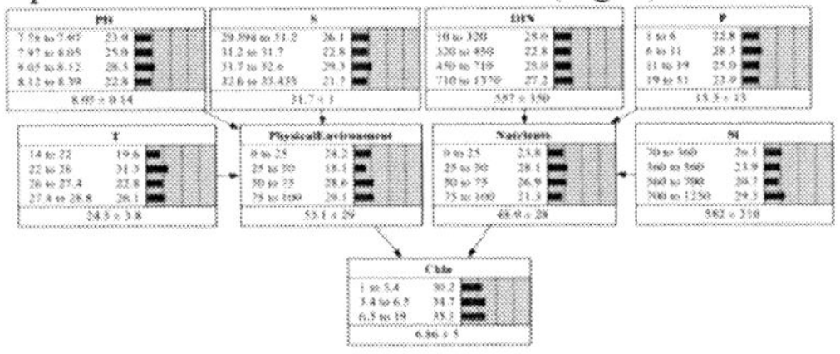

a. BN model A

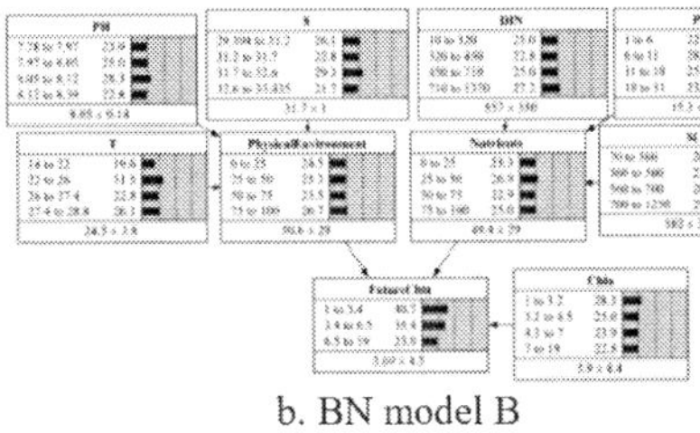

b. BN model B

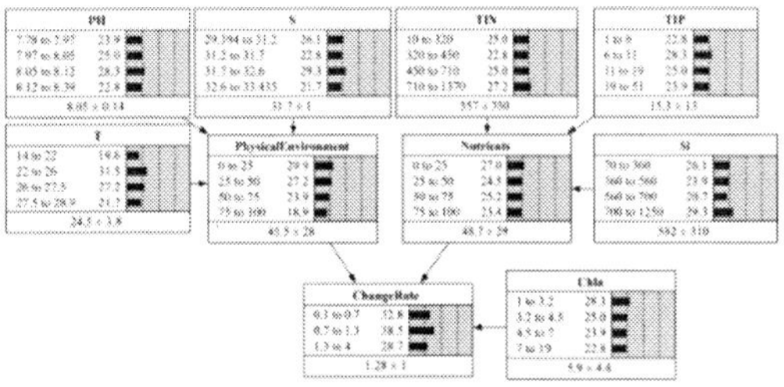

c. BN model C

Fig 3 The three non-informed BN models A, B and C.

BN models with different structure perform very differently, which was demonstrated by the error matrix (Table 2). For practical applications, the structure of a BN model is often difficult to specify, especially when a system is poorly understood and hard to characterize. Moreover, there is the possibility of spurious relationships in learning the structure. While these may limit the application of BNs in environmental research, the SEM provides a potential new approach for addressing such problems. Based on expert knowledge and the observation data, ecologists can use SEM to explore and validate causal relationships among ecological factors. After testing several pre-specified SE models, the model that provides meaningful interpretations and perfect goodness-of-fit is considered to represent the true "ecological causal web" of the key environmental correlates. Thus, SEM can provide more credible structure for and improve BN models during the process of structure building.

Compared with the previous method, linking SEM with BN models has several advantages:

a) It is suitable for relatively new and highly contentious issues, such as the initiation of harmful algal blooms in coastal areas, which are very common in ecological research. In general, the structure building mode of BN is often limited to expert-input and/or an initial review of the literature. However, if there is no widely accepted interpretation of the objective process, it would be inadequate for eliciting a prior structure, and may even result in the presence of spurious relationships. As an advanced statistical method, SEM combines expert knowledge and observations. With SEM, ecologists can hypothesize the relevance and influence of ecological variables with expert knowledge and test this causal relationship against observation data. After several specifications and with continual improvement, ecologists can obtain a truly ecological causal web and thereby improve the BN model structure via obtained causal relationships. Thus, SEM can help to minimize the need for expert elicitation and increase the application range of BN models.

b) It provides another reliable way to verify BN models. Verification is an important checkpoint in the construction of a BN model and a very necessary step for the model builder toper form. Generally speaking, verification is carried out to a certain extent according to some subjective criteria, whereas SEM provides an effective means of verifying BN models following objective criteria. Through different solution processes, SEM and BN models can both determine the absolute degree and the rank order of the influence of

https://doi.org/10.1051/e3sconf/20183801028

environmental factors or latent variables on outcome variables. Thus, by comparing the path coefficient of an SE model with the sensitivity analysis results of a BN model, the modeller can verify the BN model. Such an approach would be an important complement to the traditional verification process of BN models.

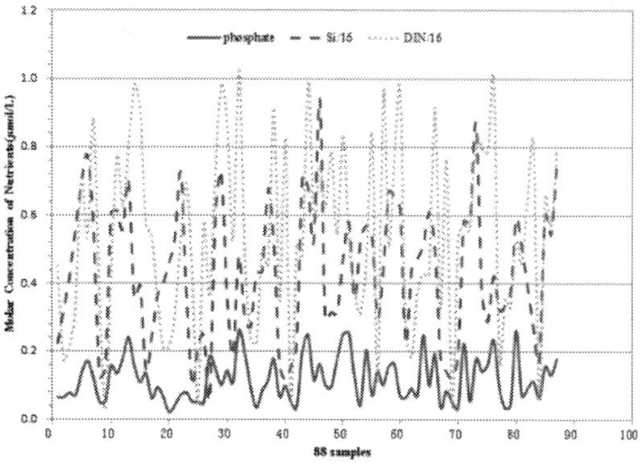

Fig 4 The molar concentration dynamic variation of the three nutrients.

4 Conclusions

In this study, a method of linking an SEM with a BN model is introduced as a means of overcoming the limitations of building BN model structure with inadequate existing knowledge. In the structural building step of a BN model, SEM was used to explore and validate the relationship between environmental factors, based on expert knowledge and observation data. The results indicate that SEM successfully selected the structure that was closer to the true causal relationship and minimizes the need of expert elicitation in the BN modelling process. In addition, SEM also increases the accuracy and reliability of the BN model. Thus, the application of a coupled SEM–BN model shows great promise as a tool for ecological research.

Our model was applied to Bohai Bay and some useful results were obtained: phytoplankton biomass has the largest influence on phytoplankton dynamics, the impact of nutrients on phytoplankton is larger than the physical environment and that controlling nutrient concentrations. To decrease the harmful algal blooms, controlling the amount of the nutrients would be the most effective means of reducing the frequency and magnitude of algal blooms. Finally, although of the three nutrients phosphorus is usually considered to be the limiting nutrient in Bohai Bay, the silicate is significantly correlated with phytoplankton dynamics.

Acknowledgements

This work was supported by "Study of foreign soil water retention, prevent salinization, heavy metal soil remediation, soil backfill and ecological restoration technique in land reclamation. (TKS150218).

References

1. Alameddine, I., Cha, Y., Reckhow, K.H., 2011. An evaluation of automated structure learning with Bayesian networks: An application to estuarine chlorophyll dynamics. Environmental Modelling & Software, 26(2), 163-172.
2. Anderson, R., 2004. Causal modeling alternatives in operations research: Overview and application. European Journal of Operational Research, 156(1), 92-109.
3. Arhonditsis, G., Paerl, H., Valdesweaver, L., Stow, C., Steinberg, L., Reckhow, K., 2007. Application of Bayesian structural equation modeling for examining phytoplankton dynamics in the Neuse River Estuary (North Carolina, USA). Estuarine, Coastal and Shelf Science, 72(1-2), 63-80.
4. Arhonditsis, G., Stow, C., Steinberg, L., Kenney, M., Lathrop, R., McBride, S., Reckhow, K., 2006. Exploring ecological patterns with structural equation modeling and Bayesian analysis. Ecological Modelling, 192(3-4), 385-409.
5. Barton, D., Saloranta, T., Moe, S., Eggestad, H., Kuikka, S., 2008. Bayesian belief networks as a meta-modelling tool in integrated river basin management — Pros and cons in evaluating nutrient abatement decisions under uncertainty in a Norwegian river basin. Ecological Economics, 66(1), 91-104.
6. Borsuk, M., 2004. A Bayesian network of eutrophication models for synthesis, prediction, and uncertainty analysis. Ecological Modelling, 173(2-3), 219-239.
7. Bromley, J., Jackson, N.A., Clymer, O.J., Giacomello, A.M., Jensen, F.V., 2005. The use of Hugin® to develop Bayesian networks as an aid to integrated water resource planning. Environmental Modelling & Software, 20(2), 231-242.
8. Brzezinski, M.A., 1985. THE Si: C: N Ratio of Marine Diatoms: Interspecific Variability and The Effect of Some Environmental Variables1. Journal of Phycology, 21(3), 347-357.
9. Castelletti, A., Soncinisessa, R., 2007. Bayesian Networks and participatory modelling in water resource management. Environmental Modelling & Software, 22(8), 1075-1088.
10. Chen, Y., Lin, L.-S., 2010. Structural equation-based latent growth curve modeling of watershed attribute-regulated stream sensitivity to reduced acidic deposition. Ecological Modelling, 221(17), 2086-2094.

An Unmanned Aerial Vehicle Cluster Network Cruise System for Monitor

Jirong Jiang[1], Jinpeng Tao[2], Guipeng Xin[3]

[1]Wuhan University of Technology,Wuhan, China
[2]Wuhan University of Technology,Wuhan, China

Abstract. The existing maritime cruising system mainly uses manned motorboats to monitor the quality of coastal water and patrol and maintenance of the navigation -aiding facility, which has the problems of high energy consumption, small range of cruise for monitoring, insufficient information control and low visualization. In recent years, the application of UAS in the maritime field has alleviated the phenomenon above to some extent. A cluster-based unmanned network monitoring cruise system designed in this project uses the floating small UAV self-powered launching platform as a carrier, applys the idea of cluster, and combines the strong controllability of the multi-rotor UAV and the capability to carry customized modules, constituting a unmanned, visualized and normalized monitoring cruise network to realize the functions of maritime cruise, maintenance of navigational-aiding and monitoring the quality of coastal water.

1 Introduction

With the development of the marine economy, the rapid development of the inland waterway transport, the distribution pattern of water transport and the risk of water traffic undergone tremendous changes, the task of safeguarding national sovereignty and maritime rights has been more prominent and the task of water safety management, marine life rescue and marine environment prevention and control has been getting more and more burdensome for the maritime departments. On May 31, 2017, Qinhuangdao Maritime Board organized UAV cruise.

operations to accurately grasp the actual distribution of aquaculture areas under administration, ship navigation in coastal waters and on-site construction operations, and help the maritime departments to grasp the information of the traffic conditions of vessels in key areas under administration and navigation dynamics more quickly and help the marine department effectively enhance the capability of warning, decision-making, command and disposal of emergency at the same time in the search and rescue of lives at sea .

At present, all maritime boards have gradually adopted sea cruising craft, unmanned aerial vehicles and shore-based integrated three-dimensional cruising system to step up maritime cruising and enhance the safety of navigation of ships. However, the existing cruise systems have the following problems.

(1) High energy consumption. The existing maritime cruising system mainly uses manned motorboats and large UAV to monitor the quality of coastal water and patrol and maintenance of the navigation-aiding facility with the problems of low autonomy of the weather, extremely large consumption of fossil fuel and high cost of maintenance and operating. To Guangdong Maritime board, for example, the annual fuel costs are 431,000 yuan and 204,800 yuan, the airport fees, insurance and other costs are 3.334 million yuan and 3.18 million yuan, the average cruise cost per unit of sea is 55.86 yuan and 221.55 yuan.

(2) Small cruise monitoring range. The cruising radius of manned motor boats and unmanned aerial vehicles are often delineated by half of their maximum range,because UAV needs the support of mother ships and shore-based charging stations, their cruising range is often limited by their mother ship or shore-based charging stations ,thus their cruise effect cannot be maximized.

(3) Informationcan not be obtained in time. Traditional methods for monitoring the water quality are the artificial tedious steps of assay after the manual sampling, with the problems for samples of small number, small scope, poor timeliness and so on, resulting in the monitoring of the sea not be centralized, real-time and large-scale.

In response to the increasingly severe maritime security situation, and existing cruise system problems, there is an urgent need for a new cluster unmanned network monitoring cruise system with a wide range of cruise monitoring, sufficient energy supply and timely information feedback so as to improve the monitoring efficiency.

2 Design principles

2.1. Design ideas

https://doi.org/10.1051/e3sconf/20183801029

We take designing a "cluster unmanned network monitoring cruise system" as the goal, first of all, design a capable of providing unmanned aerial floating platform, take the floating charging platform as the base and make reasonable arrangement in the offshore area to form a unmanned cluster network, leading to greatly extend of the UAVs cruise time and expanding of the cruising range of UAVs .By adding custom modules to unmanned aerial vehicles, realizing the unmanned monitoring of water quality, navigational aids, etc., and more timely information transmission can provide more accurate information on offshore maritime work.

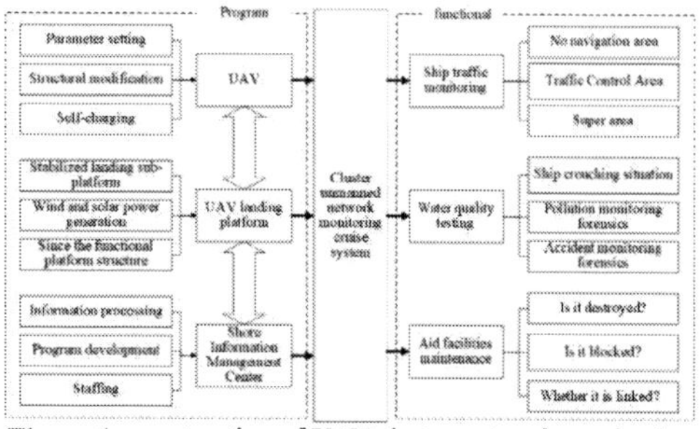

Figure 1. construction of UAV cluster network monitoring cruise system

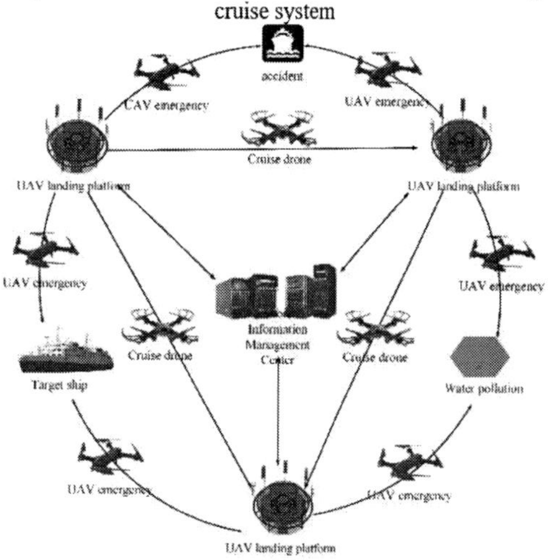

Figure 2. operation of UAV cluster network monitoring cruise system

2.2 The design of the whole program

The cruising UAV continuously collects video images of waterways in normal cruising and sends image information and self-status information to shore-based management center through the closer self-powered UAV landing platform. When the shore-based management center detects a ship accident, it sends an emergent order to the self-powered UAV landing platform which is closer to the accident site and the platform calls the emergent UAV to leave for the accidental waters to monitor and collect evidence, assisting search and rescue personnel in emergent handling of the accident, after that UAV continues to its scheduled conventional cruise. When there is a key ship entering the port and the maritime department needs to monitor it in real time, the shore-based management center send special monitoring instructions to the self-generating UAV Landing platform that closer to the ship route, the platform calls the emergent UAV to carry out follow-up monitoring of the whole relay, the real-time image information is fed back to the maritime regulatory authorities. Through the unmanned aerial vehicle equipped with maritime monitoring system, import the water image into the established neural network model to obtain the relevant data of the water body. When it is found that the water in the monitoring area is polluted, the system immediately sends an early warning to the shore-based staff, the relevant departments should immediately enter the emergent state and the shore-based management center shall call emergent UAV formation quickly activated and arrive in the area for on-site forensics and identification of sources of contamination. And feedback the type, degree, range and other information of the pollution timely.

2.3 The design structure of small floating self-generating UAV landing platform

The platform structure is composed of upper, middle and lower three layers as a whole, as shown in the figure, which is mainly composed of three modules of wind and solar power generating units, stable UAV landing platform, energy conversion module and etc.

The upper wind and solar power generation units include vertical axis wind power plants and solar power plants, and it has two vertical axis wind power plants. To ensure its stability, the bottom of it is connected to the bottom of the platform and go through the upper surface of the platform, symmetrically distributed on it, and the upper blade is used to capture wind power. The solar power generation device is mainly composed of three solar panels and a control device. Each solar panel is fixed by the pillar connection provided on the bottom of the landing platform and can be rotated and controlled around the axis of the contact point. The solar panel can automatically adjust the rotation angle through the photosensitive, maximizing the use of solar energy. In addition, the solar panel serves as the landing door of the landing platform, when the UAV needs charging from the power generation platform, the solar panel is sensed by the sensor and rotated to a position vertical to the upper surface to open the landing platform for the UAV .

The middle space, which is just below the solar panel door, is used to fix up the stable UAV landing and charging platform, including two cruise UAV landing and charging platforms and four emergent UAV landing and charging platforms, with the solar panel as the axis, arranged symmetrically, the specific distribution is shown as follow:

The lower space is the core component layer. This area has a sealed structure for storing energy storage, control circuit and information relay module to ensure the normal operation of the entire system.

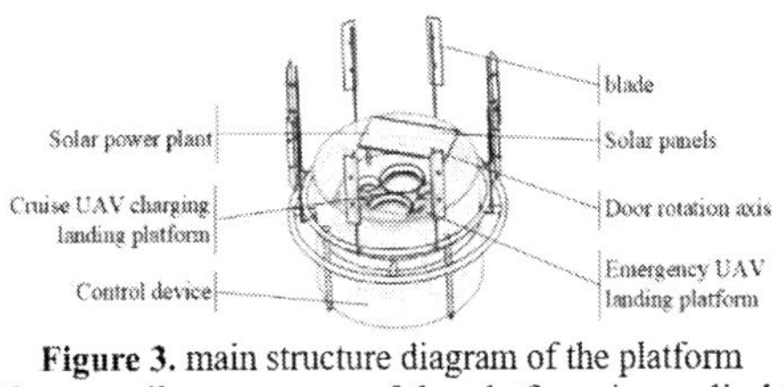

Figure 3. main structure diagram of the platform

The overall appearance of the platform is a cylindrical, with the bottom radius of 1.8 meters, height of 2 meters. In addition to the bottom part and the upper surface except by the three rotating solar panels outside the area, the platform is a closed compartment made of transparent materials, which can keep effectively off the wind and rain to protect the UAV.

2.4 UAV conversion design

Modified part mainly includes landing, charging device and protection if falling into the water of UAV in three parts.

1.UAV landing structure

According to the design of the stable UAV landing platform, caster is added to its four landing positions so that the UAV can slide down the circular slope of the platform within the allowable range and finally stop at the charging coil.

2. UAV charging device

In order to realize the wired charging of the UAV in the stable landing and charging platform, a wired charging device is specially designed at the bottom of the UAV. The wired charging device includes a positive charging annular groove, the size of the annular groove is coupled with the negative charging ring of the stable platform, and the negative charging contact can make contact with the positive electrode circular plate of the stable platform after the UAV lands to form a loop to start charging.

3. UAV protection device

UAV is expensive and often encounter stormy weather in actual cruise search and rescue process, UAV may fall into the water when landing on the platform or it breaks down. And UAV did not design watertight structure, so it will cause damage once it falls into the water. Therefore, we design the protective airbag at the bottom of the UAV, thus only the airbag on the bottom contacting with the water when the UAV falls into the water, then floats on the water waiting for manual recovery.

2.5 design module for remote control

UAV equipped with camera PTZ, water quality monitoring module, positioning module to help collect information of the status of navigation-aiding facility, images of water quality, its location or power and so on, and the information transforms long-distance transmission of signals to shore-based information management system through relay transmission from self-powered landing platform. Using computer to contact server, recept and run the remote monitoring software based on QT design processing , and analyses the data from the sensor on the self-powered landing platform. Then the shore-basing operator sends relevant instruction to control the operating of the self-powered landing platform and UAV to realize the long-distance control and management of the system and bidirectional transmission of information.

2.6 monitoring water quality and treatment of sudden cases of the ships for UAV

With the development of technology and the improvement of concept of environmental protection, environmental monitoring technology for UAV is inevitably one of the mainstream trends in the development of environmental protection technologies. In recent two decades, China has built the world's largest and most expensive online pollution emission monitoring system, costing tens of billions of yuan. However, according to the current situation, this system is difficult to prevent fraud and waste disposal, which is precisely one of the most important reasons for the deterioration of air and water environment in China. In the monitoring work of water quality in ports, the maritime departments use the UAV remote sensing image supervising site to supervise the scene to process evidence quickly, interprets aerial infrared thermal images, use spectral telemetry environmental monitoring technology to extract pollutant characteristics, provide high-resolution photos and interpretation data, capture other environmental violations and forensics timely.

Based on realizing the UAV cluster cruise, the system immediately sends an early warning to shore-based staff when it detects the presence of water quality pollution in the monitored area by carrying a marine monitoring system for water quality on the UAV, and the relevant departments should immediately enter the emergent state and the shore-based management center shall call emergent UAV formation quickly activated and arrived in the area for on-site forensics and identification of sources of contamination. And feedback the type, degree, range and other information of the pollution timely, carry out corresponding emergent measures according to specific water pollution, delineate the need to isolate the contaminated areas, thereby effectively avoiding the spread of the pollution and expansion in the accident site.

The cruising UAV continuously collects video images of channel waters in normal cruising and sends image information and self-status information to shore-based management center through the closer self-powered UAV landing platform. When the shore-based management center detects a ship accident, it sends an emergent order to the self-powered UAV landing platform that is closer to the accident site. The platform calls the emergent UAV to leave for the accidental waters to monitor and collect evidence, assisting search and rescue personnel in emergent handling of the accident, then the UAV continues its default routine cruise.

When there is a key ship entering the port and the maritime department needs to monitor it in real time, the shore-based management center send special monitoring instructions to the self-generating UAV Landing platform closer to the ship route, the platform calls the emergent UAV to carry out follow-up monitoring of the whole relay, and the real-time image information is fed back to the maritime regulatory authorities.

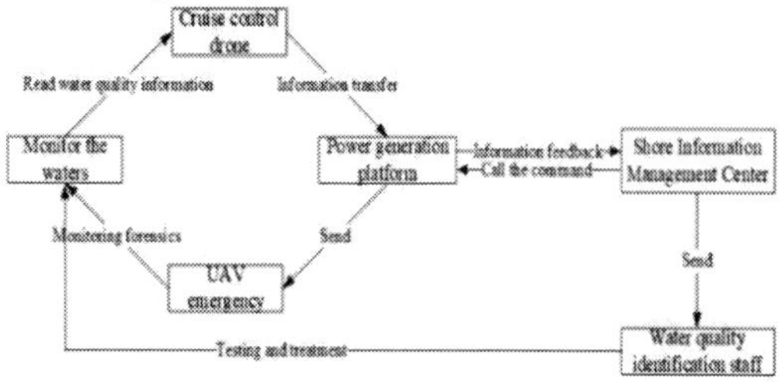

Figure 4. flow chart of emergency response to water pollution

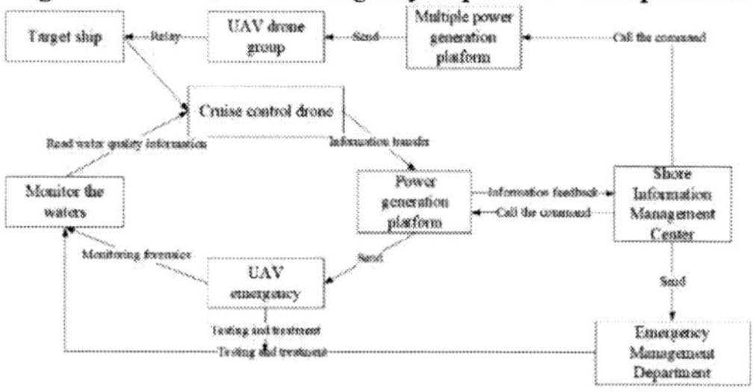

Figure 5. flow chart of emergency response

2.7 comparative analysis

As the platform is used in offshore waters, taking the various parameters of the Bohai Sea as an example, to calculate the capacity of the generation of solar-wind power generation platform. According to the monthly total solar radiation and the average wind speed in the Bohai Sea, and import the power consumption and power generation calculation formula, the calculated power generation and power consumption is compared as the Figure 6.

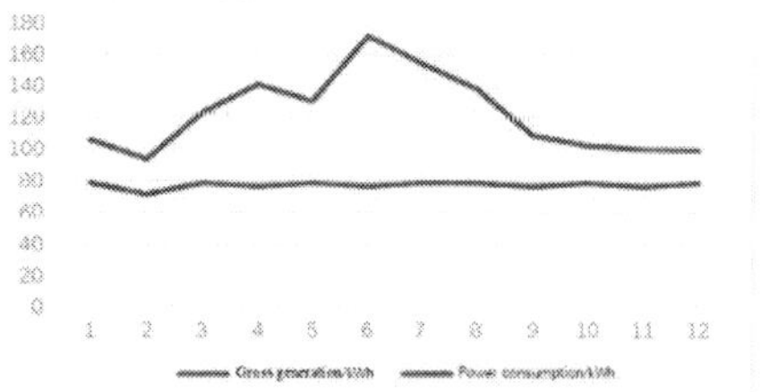

Figure 6. comparison chart of total power generation and power consumption

According to figure 8, the solar-wind power generation system can guarantee the power consumption of various equipment on the platform and provide enough power for the UAV lithium battery.

3 innovation and application prospects

3.1 innovation

1.The design of the small floating self-generating UAV landing platform. A multi-layer structure is designed to separate the stable UAV landing platform, power generation equipment and platform circuit components and to realize the full functionality of the platform on a limited area available. At the same time, the use of the platform effectively mitigated the short time of edurance of the multi-rotor UAV to impact its ability of cruising, making it possible for the medium-sized multi-rotor UAVs to be used in maritime cruising.

2. Integrated remote-control system. The information is relayed to the shore-basing information management system through relay transmission from the self-generation and landing platform. The remote monitoring software processes and analyzes the data from the sensors on the self-generation and landing platform to the shore-basing operators, the shore-basing operators can issue related commands to control the self-generation landing platforms and UAV to achieve remote control and management of the system and bidirectional transmission of information.

3. Monitoring program based on neural network model. Through the special surveillance system installed on the UAV, the water image is imported into the established neural network model, and the relevant data of the water body can be obtained after the operation of image cutting, feature extraction and other operations. The suspicious accidental waters may also be dispatched to supervise by the strength of surveillance.

4. Construction of a new type of UAV cluster network monitoring cruise system. Taking the floating platform of small self-powered UAV landing as the node, the UAV's trajectory is planned, the UAV cluster monitoring network is constructed, the visual supervision of offshore vessels' traffic and the continuous monitoring of the normalization of navigation- aiding facilities and offshore water quality are realized.

3.2 application prospects

With the development of science and technology and the unmanned and intelligent trend of future maritime cruising, the idea of clustering UAVs can be realized in the offshore monitoring cruise after the design of self-powered UAV landing platform solves the problem of UAV energy supplement. Through the separation of different watersheds and the rational design of patrol schemes, UAV stealth enforcement will become a notable feature of future maritime regulatory technology and will play an prominent role in future maritime for cruise monitoring.

References

1. Li Huawei. Research And Realization for Multi-UAV Cooperative Mission Planning[D] Xi'an University of Electronic Science and Technology,2014.
2. Yu Yongjun. Research on Key Technologies for High Altitude and Long-flight-time UAV Information Fusion Autonomous Navigation [D] Nanjing University of Aeronautics and Astronautics,2011.

 https://doi.org/10.1051/e3sconf/20183801029

3. Yu Sinan, Zhou Rui, Xia Jie, Che Jun. Decomposition and coverage of multi-UAV cooperative search area [J]. Journal of Beijing University of Aeronautics and Astronautics,2015,41(01):167-173.
4. Liu Fan, Shi Qianqing, Fei Xianyun. Object-oriented Remote Sensing for the Variation of Sea Area Utilization [J]. Journal of Huaihai Institute of Technology (Natural Science Edition),2015,24(02):82-86.
5. Liu Fuhu, Ma Xiaoping. Research for improving the endurance performance of minitype electric-powered UAV [J]. Flight Dynamics,2010,28(05):13-15.
6. Anonymous. ONR Awards Boeing $23 Million To Finish Free Electron Laser Design[J]. Defense Daily,2010,247(55).
7. Carlo Corradini,Lisa De Propris. Beyond local search: Bridging platforms and inter-sectoral technological integration[J]. Research Policy, 2016.
8. Paavo Ritala,Eelko Huizingh ,Argyro Almpanopoulou,Paul Wijbenga. Tensions in R&D networks: Implications for knowledge search and integration[J]. Technological Forecasting & Social Change, 2016.
9. Ma Minglu. Research on the Crucial Problems in Integration of Maritime Cruise Search and Rescue [J]. China Water Transport ,2016(7):36-38.

Experimental study of overland flow resistance coefficient model of grassland based on BP neural network

Peng Jiao[1,*], *Er* Yang[1], and *Yong Xin* Ni[1]

[1]Yellow River Institute of Hydraulic Research, Key Laboratory of Soil and Water Loss Process and Control on the Loess Plateau of Ministry of Water Resources, Zhengzhou 450003, China

Abstract. The overland flow resistance on grassland slope of 20° was studied by using simulated rainfall experiments. Model of overland flow resistance coefficient was established based on BP neural network. The input variations of model were rainfall intensity, flow velocity, water depth, and roughness of slope surface, and the output variations was overland flow resistance coefficient. Model was optimized by Genetic Algorithm. The results show that the model can be used to calculate overland flow resistance coefficient, and has high simulation accuracy. The average prediction error of the optimized model of test set is 8.02%, and the maximum prediction error was 18.34%.

1 Introduction

The overland flow resistance is a key parameter to reflect the hydraulic characteristics of the overland flow. Studying the law of overland flow resistance is of great significance for understanding the hydrodynamic mechanism of soil erosion process. Many scholars [1-7] carried out extensive and in-depth research on the law of overland flow resistance. In the past, the research mostly focused on the experimental analysis of the overland flow resistance. For example, Keli Zhang [3] thought that the resistance coefficient in the process of rill erosion was closely related to the flow conditions and slope morphology. Peiqing Xiao [7] thought that the average resistance coefficient of slope under different covering conditions was closely related to rainfall intensity, and decreased with the increase of rainfall intensity.

Due to the difference of study objects and calibration data, the formula of overland flow resistance coefficient is quite different in form and scope of application. In this study, the neural network theory was applied to the study of overland flow resistance coefficient. In order to provide a reference for improving the calculation method of overland flow resistance coefficient and the prediction accuracy of the erosion prediction model, a BP neural network model was established based on the simulated rainfall experiment data, the genetic algorithm (GA) was used to optimize the model parameters.

2 Experiment method

2.1 Experiment equipment

The experiments were carried out in the key laboratory of soil and water loss process and control on the loess plateau of ministry of water resources. The soil tank was 5.0m long, 1.0m wide and 0.6m deep. The slope was 20°. The representative plant alfalfa in the Loess Plateau was planted on the slope by bunch planting. The distance of bunch planting was 20cm. The experiments were carried out when the grass was covered by 60% - 70%. The simulated rainfall device was a side-jet artificial rainfall device. The simulated rainfall installation was 6.0m above the ground, and had a jet height of 1.5m. The values of rainfall uniformity measured before experiments were 86% -92%.

Soil was taken from the surface layer of loess slope of mountain Mang in Zhengzhou, and the particle composition was shown in Table 1. Loess was layered paving compaction to 50cm thick, and the soil moisture content was about 15%. The soil dry bulk density is about 1.05g/cm^3 above 20cm depth, and about 1.35g/cm^3 below 20cm depth. Samples taken from the upper, middle and lower part of the slope by cutting ring, are used to calculate the average bulk density and dry bulk density.

Table 1. Particle distribution of the experimental soil.

Particle Size (mm)	>1.000	1.000~0.250	0.250~0.050	0.050~0.010	0.010~0.005	0.005~0.001	<0.001
Proportion (%)	0	1.05	35.45	43.40	3.20	6.40	10.50

* Corresponding author: hnzzjp2010@126.com

2.2 Experiment scheme

According to the typical rainfall frequency characteristics in the hilly and gully regions of the Loess Plateau, three rainfall intensities of 45, 90 and 130mm/h were identified as the experiment conditions. Simulated rainfall of 30mm/h was applied to the experimental slope about 10 min on the day before the experiments to ensure that the soil moisture content of each experiment was consistent. Observation sections were set along the slope from top to bottom at an interval of 1m, and the 5th section was the exit section. In the measurement interval of 2min, the five sections of the experimental parameters were measured. The flow rate was determined by the dye method, and the flow width was the sum of the flow widths of each section. A runoff sample was taken at the outlet of the soil tank at each time interval. Take a sample of runoff sediment. After runoff occurred on the slope, the rainfall lasted for 60 min, and each experiment was repeated once. The overland flow resistance coefficient was calculated by the Darcy-Weisbach resistance coefficient formula.

3 Modeling method

3.1 BP neural network

BP neural network consists of input layer, hidden layer and output layer. Each layer has several neurons. Neurons in different layers are connected to each other with a certain weight. The hidden layer neurons have bias values and activation function, and the output layer neurons have bias values. The model input variables were simulated rainfall intensity, flow velocity, flow depth, slope roughness, and the model output variable was overland flow resistance coefficient. The model was a BP neural network with 3 layers of network topology, and the number of neurons in input layer, hidden layer and output layer are 4, 12, and 1, respectively. Hidden layer activation function is:

$$f(x)=\frac{2}{1+\exp(-2x)}-1 \tag{1}$$

3.2 Genetic algorithm

According to the topological structure and the characteristics of neuron activation function of BP neural network model, the initial weights and bias values of the model can be optimized by genetic algorithm. The weights and bias values optimization of BP neural network based on genetic algorithm, includes population initialization, construct fitness function, selection operation, cross operation, mutation operation.

Population initialization is to encode all the individuals in the population according to individual information. Each individual is a real number string, which is composed of input layer and hidden layer neuron connection weights, hidden layer neuron bias values, hidden layer and output layer neuron connection weights and output layer bias value.

To construct fitness function is to build a BP neural network model based on information of connection weights and bias values of the neurons carried by individual, to train BP neural network using training samples, and then to simulate and predict. The sum of the absolute value of the error between the predicted output and the expected output is taken as the fitness value F_i of the individual i.

$$F_i=k\left(\sum_{i=1}^{p}\text{abs}(y_i-t_i)\right) \tag{2}$$

F_i is the fitness value of individual i, k is a coefficient, p is the number of neurons in the output layer of the neural network, y_i is the expected output value of the i-th neuron in the output layer of the neural network; t_i is the predicted output value of the i-th neuron in the output layer of the neural network.

The selection operation refers to selecting an individual from a population. The method is roulette method based on the selection strategy of proportion of fitness, and the selection probability P_i of an individual i is expressed as:

$$f_i=\frac{k}{F_i} \tag{3}$$

$$P_i=\frac{f_i}{\sum_{j=1}^{N} f_j} \tag{4}$$

f_i is the inverse of the fitness value of individual I, N is the number of individuals in the population.

The cross operation refers to using real number cross method to process real-coded individuals. The cross operation expression of the k-th chromosome a_k and the l-th chromosome a_l in the j position is:

$$\begin{cases} a_{kj}=a_{kj}(1-b)+a_{lj}b \\ a_{lj}=a_{lj}(1-b)+a_{kj}b \end{cases} \tag{5}$$

a_{kj} is the j-bit coded value of chromosome a_k after the cross operation, a_{lj} is the j-bit coded value of chromosome a_l after the cross operation, b is a parameter of the cross operation.

The mutation operation refers to using a certain algorithm to make the gene fragments of selected individual mutation. The mutation operation expression of the i-th chromosome's j-th gene a_{ij} is

$$a_{ij}=\begin{cases} a_{ij}+(a_{ij}-a_{\max})f(g) & r>0.5 \\ a_{ij}+(a_{\min}-a_{ij})f(g) & r\le 0.5 \end{cases} \tag{6}$$

$$f(g)=r^{*}\left(1-\frac{g}{G_{\max}}\right)^2 \tag{7}$$

$a_{\max}$ is the upper bound of gene a_{ij}, $a_{\min}$ is the lower bound of gene a_{ij}, g is the current evolution generation; $G_{\max}$ is the largest evolution generation; r^* is a random number, r is a random number within [0,1], $f(g)$ is a function.

Considering the computational accuracy and computational requirements, set the population size to 20, the maximum evolutionary algebra to 50, the crossover probability to 0.7 and the mutation probability to 0.2.

* Corresponding author: hnzzjp2010@126.com

4 Results and discussions

The model of BP neural network optimized by genetic algorithm was BP-GA model. Because of the difference of the dimension and magnitude between the input and output variables can affect the accuracy of the model. Therefore, all data of the input and output matrices were transformed to the interval [-1, 1] using a normalization formula.

4.1 Model training results analysis

81 training samples of 3 simulated rainfall experiments under grassland conditions were put into BP-GA model for training. During the training of 10 BP-GA models, the model parameter optimization process of genetic algorithm was traced 5 times randomly, and the optimal fitness of all individuals in each generation was recorded. Optimization process of BP neural network model parameter using genetic algorithm was shown in Fig 1. Due to the random assignment of parameters during model initialization, the initial fitness of each population was different, the optimal fitness of the population decreased with the increase of the genetic generation, and the crossover and mutation of chromosomes leaded to the drastic changes of the optimal fitness, resulting in some inflection points in the optimal fitness curves. The optimal fitness of each population tended to be consistent at around 50 generations, and was stable at about 9.5.

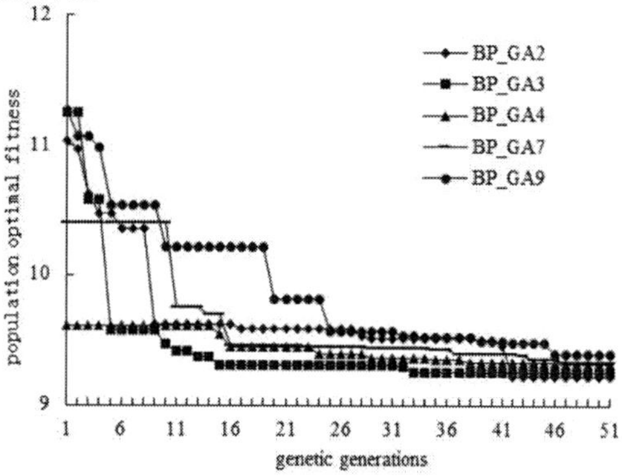

Fig. 1. Parameter optimization process of BP neural network by genetic algorithm.

The average prediction errors of 10 BP_GA models for the training samples were shown in Fig 2. When the simulated rainfall intensity is 45 mm/h, the average prediction error of all the training samples is 4.77%, and the maximum prediction error is 12.23%. When the simulated rainfall intensity is 90 mm/h, the average prediction error of all the training samples is 4.52%, and the maximum prediction error is 13.06%. When the simulated rainfall intensity is 130 mm/h, the model predicts the average error of all the training samples 6.52%, the maximum prediction error is 15.46%. The model had a high prediction accuracy. On the one hand, the grass can reduce the impact of raindrops on the surface of the slope, and on the other hand, it can increase the soil anti-corrosion ability and maintain the stability of the surface morphology.

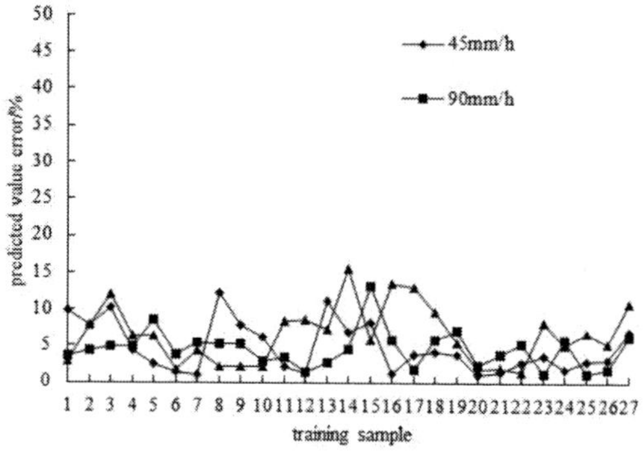

Fig. 2. Predicted value error of training samples of BP-GA model.

4.1 Model verification

9 groups of test samples of three simulated rainfall experiments under grassland conditions were respectively substituted into the BP-GA model for verification. The test samples were randomly selected in different stages of each experiment and had good representativeness. The average prediction error of the BP-GA model for the test sample set was 8.02%, and the maximum prediction error was 18.34%. The model had a high simulation accuracy and a good generalization ability.

Table 2. Model verification results of BP-GA model.

Number	Rainfall Intensity (mm/h)	Flow Velocity (cm/s)	Flow Depth (mm)	Roughness	Coverage	Measured Value	Predicted Value	Error (%)
1	45	0.025	0.5	0.1	0.5	18.30	17.26	5.68
2	45	0.036	0.5	0.1	0.5	16.00	16.41	2.56
3	45	0.034	0.8	0.1	0.5	15.60	14.78	5.26
4	90	0.064	2	0.2	0.63	12.70	13.12	3.31
5	90	0.067	2.8	0.2	0.63	10.40	12.14	16.73
6	90	0.063	2.6	0.2	0.63	12.90	13.63	5.66
7	130	0.075	3	0.3	0.9	9.62	8.37	12.99
8	130	0.077	3.1	0.3	0.9	10.98	10.80	1.64
9	130	0.093	3.9	0.3	0.9	8.56	10.13	18.34

The predicted value and the measured value calculated by BP-GA model were fitted and analyzed. The fitting relationship between the two was good, the fitting coefficient was 1.0045 and the coefficient of determination R^2 was 0.8588. Fig 3 showed the predicted and measured values have a good correlation and a one-to-one correspondence.

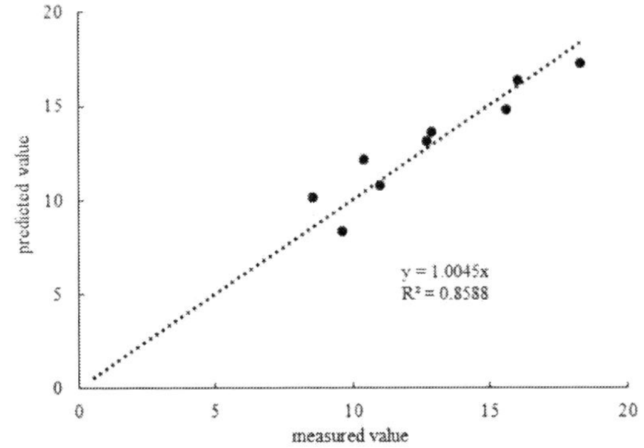

Fig. 3. Test set fitting relationship of BP-GA model.

5 Conclusions

According to the experiment data, model of overland flow resistance coefficient was established based on BP neural network optimized by genetic algorithm. The input variations of model were rainfall intensity, flow velocity, water depth, and roughness of slope surface, and the output variations was overland flow resistance coefficient. Genetic algorithm could improve the parameter optimization speed and simulation accuracy of BP neural network. The average prediction error and the maximum prediction error of the test set of the BP-GA model were respectively 8.02% and 18.34%. Because of the limitation of experiment conditions, the influences of different cover and slope on the overland flow resistance coefficient are not considered, and the prediction method of BP neural network model is only discussed by using simulated rainfall experiment data of grassland slope of 20°.

Acknowledgements

This paper is based on work supported by the National key research priorities program of China (2016YFC0402402) and the Central Nonprofit Research Institutions Basic Scientific Research Special Fund (HKY-JBYW-2014-19).

References

1. W Y. Yao. Journal of sediment research, **1** (1996): 74-82
2. A D. Abrahams, G. LI, A J. PARSON. Earth Surface Processes and Landforms, **21** (1996): 35-47
3. K L. Zhang. Yellow River, **20(8)** (1998): 13-15.
4. G H. Zhang. Advances in Water Science, **13(2)** (2002): 159-165
5. A. Knapen, J. Poesen, S. Baets de. Soil and Tillage Research, **94(2)** (2007): 425-440
6. M. Fatteta, Y. Fub, M. Ghestemc. Catena, **87(1)** (2011): 60-69
7. P Q. Xiao, W Y. Yao, L. Li, C X. Yang. Journal of sediment research, **3** (2013): 1-5

* Corresponding author: hnzzjp2010@126.com

Analysis of day rainfall characteristics of Zhengzhou

Zhenzhou SHEN, Wenyi YAO *, *Peiqing XIAO*, and *Xueqin YANG*

Key Laboratory of Soil and Water Loss Process and Control on the Loess Plateau of the MWR, Yellow River Institute of Hydraulic Research, Zhengzhou, 450003, Henan

Abstract. Raindrop characteristics, including speed and size of raindrops, in Zhengzhou city of Yellow River basin were analyzed through a natural rainfall on the loess slope. Results showed that the process of natural rainfall belonged to a parabola and counts, size and terminal velocity would increase with the rainfall intensity rising. Besides, the size and terminal velocity of natural raindrops were relatively scattered; In the process of individual rainfall, the terminal velocity and its peak value were mainly focused between 0.8~5m/s and 1m/s, respectively. Size of raindrops were mainly consisted of 0.125-0.5mm, among which the terminal velocity of raindrops with a size of 0.125mm, 0.25mm, 0.375mm, 0.5mm were primarily 0.8-3.4m/s, 0.6-3.4m/s, 0.8-1m/s, 1-1.4m/s, respectively.

1 Introduction

Loess Plateau, the region with extremely serious problems of loss of water and soils in the world, has been the main source of outflow sediments in Yellow River. Management of the sediments is the critical factor to curb sediment hazard[1-2]. However, due to complexity of soil and water loss and limitations in observation methods and procedures[3] for field research.For example, mechanism of watershed soil and water loss, erosion and sediment yieldis still unknown.,There are still lack of practical soil and water loss models for Loess Plateau[4], around the world [4-8]. These restrictions limit the comprehensive development of soil and water loss management[9-10].

Study of natural rainfall distribution [11] by combining the field observation and laboratory test are the focus of research to explore the mechanism of soil and water loss to develop the optimized model and to establish a mathematical model of soil and water loss on the Loess PlateauTherefore it is imperative to analyze the scale transformation and rainfall similarity and finally the raindrops characteristics in the soil erosion experiments must be solved [12-15]. This study aims to analyze the terminal velocity and gradation (sizes) of raindrops, combined with the eroded sediments resulted from different rainfall conditions in order to offer a technology support for dynamic rainfall simulation.

2 Experimental program

2.1. Experiment design of runoff plots

The experiment was carried out at the Zhengzhou test base of model Yellow River(34°45'36"N 113°40'27"E) of Yellow River Institute of Hydraulic Research. The length, width, depth and slope of runoff plots are 5.0 m, 1.0 m, 0.6 mand 20°, respectively. Structure of plots is brick-concrete and a water channel is at the outlet in each plots. Soil filled in each plots is surface loessal soil from the Mangshan Mountain in Zhengzhou, with the proportion of particle size of 0.05-0.01 mm, 0.02-0.05 mm being 43.4% and 35.45%, respectively. Soil bulk density is 1.20 g/cm^3.

2.2 Test methods

The 5 m long slope was divided into five fractural surface from top to bottom. The laser optical disdrometer was used here to collect the information about characteristics of raindrops, including intensity, size, terminal velocity and number under the same terminal velocity. Slope velocity radar gun and steel rule were used to record hydraulics parameters, including speed, width and depth of runoff. Sediments were collected by 1 minute after producing runoff and analyzed the parameters mentioned above. Erosion amounts and sediment concentration were calculated by the method of substitution or oven drying and calculated the parameters like infiltration of runoff according to rainfall and runoff yield.

* Corresponding author: 15981989220@139.com

Fig.1. Soil erosion model

Fig.2. The laser optical disdrometer

3 Results and discussion

In this paper, five typical rainfall time points in natural rainfall on April 23, 2016 (hereinafter referred to as 20160423) were selected. The length of each time point was 1min. The particle size of raindrops in the min length, the speed of raindrops, and the number of raindrops with the same raindrops were collected by LPM laser raindrop spectrometer. The raindrops characteristics were also analyzed. The results are as follows.

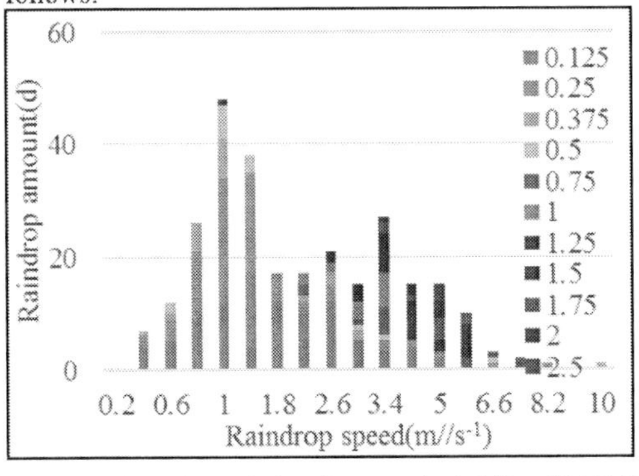

Fig.3. the first minute during the natural rainfall on 20160423

The distribution of the particle size and the final velocity of the raindrops in the first 1 minute during the natural rainfall on 20160423 is shown in Fig.3. It can be seen from the figure that the final velocity of the raindrops in the first 1min of the rainfall is mainly concentrated between 0.4-5.8m/s and the peak velocity is 1m/s, and the number of raindrops with the final velocity of the raindrop is about 48, accounting for about 17%; The rainfall is mainly composed of raindrops with particle size of 0.125-0.5mm. The final velocity of 0.125mm is mainly between 0.4-4.2m/s, and the final velocity of the raindrops with 0.25mm particle size is mainly concentrated on the range of 0.4-3.4m/s, 0.375mm diameter of the raindrops mainly concentrated on the 0.8-1m/s, the 0.5mm diameter of the raindrops mainly concentrated on the 1-1.4m/s.

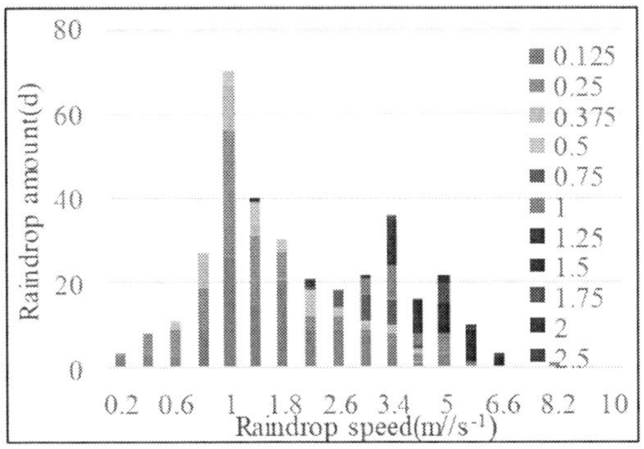

Fig.4 the second min during the natural rainfall on 20160423

According to the distribution of the raindrops and the final velocity of the raindrops in the second minute during the natural rainfall process (Figure 4), the final velocity of the raindrops in the second minute of the rainfall is mainly concentrated between 0.4-5.8m/s and the peak velocity is 1m/s, with the number of raindrops at the end of the raindrops is about 70, accounting for about 15%; We can also see the rainfall is mainly composed of 0.125-0.5mm diameter raindrops, the final velocity of 0.125mm particle size is mainly concentrated on the 0.8-3.4m/s, the final velocity of 0.25mm size is mainly concentrated in the 0.4-5m/s, the final velocity of 0.375mmparticle size is mainly concentrated in the 0.8-1.4m/s, the final velocity of 0.5mm particle size is mainly concentrated in the 1-2.6m/s.

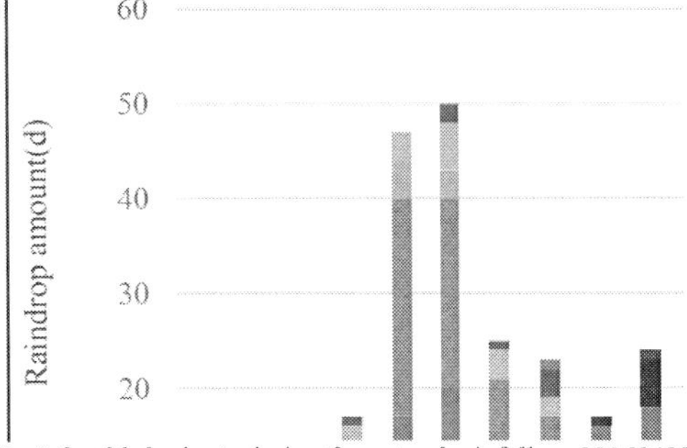

Fig.5 the third minute during the natural rainfall on 20160423

Fig.5 shows the distribution of the particle size and the final velocity of the raindrops in the third minute during the natural rainfall in 20160423, it can be seen from the figure that the final velocity of the rains in the third minute of the rainfall is mainly concentrated in the range of 0.4-6.6m/s, and the velocity of the raindrops in the third minute of the natural rainfall is analyzed. At the same time, the final velocity and the peak velocity was formed, the value is 1.4m/s and 3.4m/s respectively. That is, most of the raindrop speed is 1.4m/s and 3.4m/s or so, with the number of raindrops about 50 and 28 or so, accounting for about 30% or so; We can also see the rainfall is mainly composed of 0.125-1mm diameter raindrops, of which 0.125mm particle size of the end of the rain mainly concentrated in the 0.4-3.4m/s, 0.25mm diameter of the raindrops mainly concentrated in the 0.6-4.2m/s, the 0.375mm diameter of the raindrops mainly concentrated in the 0.6-1.4m/s, the 0.5mm diameter raindrops is mainly between 0.8-2.2m/s, the 0.75mm particle size of the raindrops mainly concentrated in the 1.4-2.6m/s, the 1mm particle size of the rain speed is mainly between 2.6-3.4m/s.

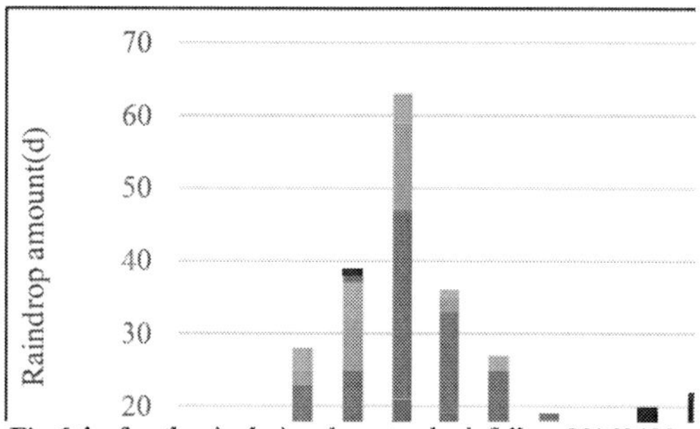

Fig.6 the fourth min during the natural rainfall on 20160423

Fig.6 shows the distribution of the particle size and the final velocity of the raindrops in the fourth minute during the natural rainfall in 20160423. It can be seen that final velocity of the raindrops in the fourth minute of the rainfall is mainly concentrated between 0.4-5m/s, forming two final velocity peaks, namely 1m/s and 4.2m/s respectively, that is, most of the raindrops is mainly 1m/s and 4.2m/s or so, with the number of raindrops is 63 and 30 or so, accounting for about 30%; We can also see the particle size of rainfall is mainly 0.125-0.5 mm raindrops. Of which 0.125, 0.25, 0.375, 0.5mm particle size of the rain speed is mainly concentrated in 0.4-3.4, 0.4-4.2, 0.8-1, 0.6-2.6 m/s.

Fig.7 shows the distribution of the particle size and the final velocity of the raindrops in the 5th minute during the natural rainfall, we can see that the final velocity of the raindrops in the first 5min of the rainfall is mainly concentrated between 0.4-6.6m /s, and the velocity of the raindrops in the 5th minute during the natural rainfall is analyzed. Forming a final velocity peak is 1m/s, that is, most of the raindrop speed is 1m/s, with the raindrop speed of the number of raindrops for about 54, accounting for about 17%; We can also see the rainfall is mainly composed of 0.125-0.5mm diameter raindrops, of which 0.125,0.25,0.375,0.5mm diameter of the raindrops mainly concentrated in 0.8-3.4,0.4-3.4,0.8-1.4,1-3.4m/s.

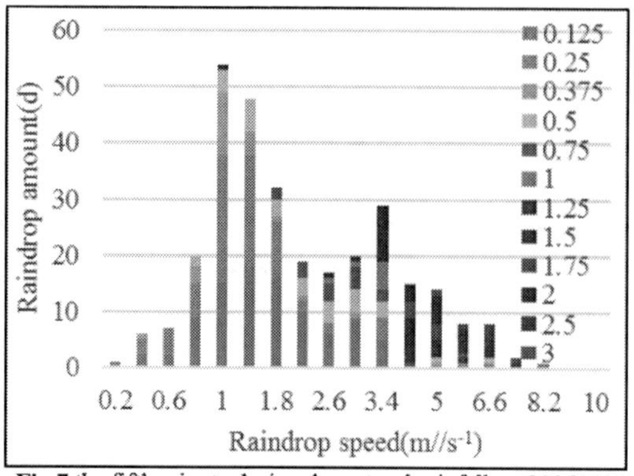

Fig.7 the fifth minute during the natural rainfall on 20160423

4 Conclusions

Through the analysis of natural raindrops, the following aspects were obtained:

(1) The natural rainfall process is a process of rainfall intensity changes.

(2) During the process of this rainfall, the final velocity of the rain is mainly concentrated between 0.8-5m/s, the peak of the final velocity is generally 1m/s; rainfall is mainly composed of the particle size of 0.125-0.5mm raindrops. The velocity of the raindrops of 0.125mm diameter is mainly between 0.8-3.4m/s, the final velocity of the raindrops of 0.25mm particle size is mainly concentrated between 0.6-3.4m/s, 0.375mm concentrated in the 0.8-1m/s, the 0.5mm particle size of the raindrops mainly concentrated in the 1-1.4m/s.

Acknowledgment

We greatly appreciate our group for useful discussions on this subject and all referees for their assistance and laboratory work. This research mainly supported by Key Laboratory of Soil and Water Loss Process and Control on the Loess Plateau of the MWR Foundation (2017005) ; National Key R&D Program of China (2017YFC0504503), National Key Technology R&D Program (2013BAC05B04) ; the National Key Basic Research Project(2011CB403303), ISTTCP of Henan Province (162101510004)

References

1. Tang K L 2004, Soil and water conservation in China [M]. Beijing: Science Press, 2004.
2. Chen L 2002, Soil and water conservation in China [J], 2002 (7): 4-6.
3. Li Z B 1991, Storm erosion test and small watershed sediment yield model of sloping land system in Loess Region [D], Doctoral Dissertation of Shaanxi Institute of Mechanical Engineering, 1991.
4. Moldenhauer W C 1965. Procedure for studying soil characteristics using disturbed samples and

simulated rainfall [J]. Transactions, American Society of Agricultural Engineer, 1965, 8(1):30-35.

5. Renard K G, Foster G R, Weesies G A, 1997. Predicting soil erosion by water: a guide to conservation planning with the Revised Universal Soil Loss Equation (RUSLE)[J]. Agriculture Handbook, 1997.
6. Foster G R and Lane L J 1987. User requirements, USDA-Water Erosion Prediction Project (WEPP) [R], NSERLReportNo1, West Lafayette: USDA-ARS National Soil Erosion Research Laboratory, 1987.
7. Nearing M A, Foster G R, Lane L J, 1989. A process-based soil erosion model for USDA-Water Erosion Prediction Project technology [J]. Trans. ASAE, 1989, 32(5): 1587-1593.
8. Laflen J M, Lwonard J L, Foster G R 1991. WEPP a new generation of erosion prediction technology[J]. J of Soil and Water Cons., 1991, 46(1): 34-38.
9. Yao W Y, Several design theories and applications of the Yellow River river channel physical simulation [D], Doctoral Dissertation of Hohai University, 2005.
10. Yao W Y , 2011, Review and Prospect of mathematical models for erosion and sediment yield in China [J]. Sediment Research, 2011, (2): 65-74.
11. Shen Z Z, Liu P L, Xie Y S 2007. Transformation of erosion types on loess slope by REE tracking [J], Journal of Rare Earths, 2007, 25 (4): 67-73.
12. Shen Z Z, Yao W Y, Li Z B, Li M, Xiao P Q, WU Y 2014. Transformation of Rainfall Characters on Soil Erosion Similar [J], Journal of Rare Earths, 2014, 32 (S): 130-133.
13. SHEN Z Z,YAO W Y, L I M,2008. Study on Correlativity Between Erosion and Seeping Under Different Underlying Horizon[J], Journal of Soil and Water Conservation, 2008, 22 (5): 43-46.
14. [1SHEN Z Z, YAO W Y, LI M, 2008. Runoff Sediment-Carrying Capacity and Critical Shear Force on Slope [J]. Journal of Tianjin University, 2008, 41(4):149-153.
15. Shen Z Z, Yao W Y, Li M 2009, Influence factors of runoff energy consumption and denudation amount on loess slope [J]. Science of Soil and Water Conservation, 2009, 7 (6): 9-13.

Effect of vegetation construction on runoff and sediment yield and runoff erosion ability on slope surface

Chun Xia YANG *, *PeiQing* XIAO, *Li* LI and *Peng* JIAO

Key Laboratory of the Loess Plateau Soil Erosion and Water Loss Process and Control of Ministry of Water Resources ,Yellow River Institute of Hydraulic Research, Zhengzhou,Henan 450003,China

Abstract. Land consolidation measures affected the underlying surface erosion environment during the early stage of vegetation construction, and then had an impact on rainfall infiltration, erosion and sediment yield. This paper adopted the field simulated rainfall experiments to analyze the function that pockets site preparation measures affected on rainfall infiltration, runoff sediment yield and runoff erosion ability. The results showed that, the measures can delay the rainfall runoff formation time of the slope by 3'17" and 1'04" respectively, Compared with the same condition of the bare land and natural grassland, The rainfall infiltration coefficient each increased by 76.47% and 14.49%, and infiltration rate increased by 0.26 mm/min and 0.11mm/min respectively; The amount of runoff and sediment yield were reduced because of the pockets site preparation , The amount of runoff reducing rate were 33.51% and 30.49%, and sediment reduction rate were 81.35% and 65.66%, The sediment concentration was decreased by 71.99% and 50.58%; Runoff velocity of bare slope and natural grassland slope decreased by 38.12% and 34.59% respectively after pockets site preparation . The runoff erosion rate decreased by 67.92% and 79.68% respectively. The results will have a great significance for recognizing the effect of water and sediment reduction about vegetation and the existence of its plowing measures at the early period of restoration.

1 Introduction

The site vegetation cover was greatly improved by forest and grass vegetation constructions since 2000a. It was well known that various vegetation measures played an important role in promoting rainfall infiltration, delaying runoff and reducing erosion [1-6]; Furthermore, it also had a certain effect to adjust the peak and bottom of the runoff and sediment yield process curve , which indicated that vegetation measures were one of the reasons for the phenomenon of sharply decrease in runoff and sediment production in recent years [7-10].But little attention has been taken to the effect of site preparation measures which existing in vegetation land on rainfall infiltration and runoff - sediment production. Pockets site preparation was an important and easy method for vegetation plantation and restoration, and it also played an important role in the soil erosion environment, So, It was great significance to study the influence which pockets site preparation affected on rainfall infiltration and runoff dynamics parameters. The study will have a great significance for exploring the causes of water and sediment reduction.

2 Materials and Methods

2.1 Site conditions

The research area was located at the slopes that belongs to LuoYuGou small watershed in Tianshui city. And some plots were built on the slope with 10 miles long and 2 miles wide, Study area distributed low liquid limit clay with hard texture. And covered with natural vegetation, The main species of vegetation were *Coronilla varia* and *Juncus effusus*, The plots site conditions were listed in Figure 1-2 and Table 1.

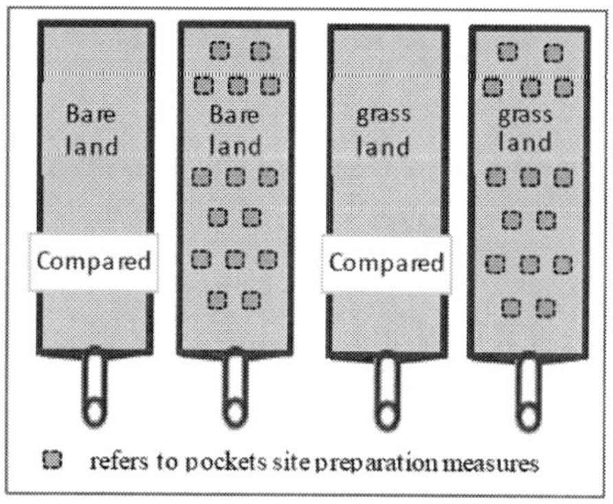

Fig. 1. Site conditions schematic diagram

The planting pockets site preparation diameter × depth was about 30cm × 30cm. And distribution of staggered on the slope. Each plot had 70 planting holes, all pockets site preparation area is about 4.95m2, accounting for 24.75% of the total area of the plot. The

* Corresponding author: yangchx312@126.com

original soil was dense, soil covered the planting pocket was soft. Amorpha fruticosa seedlings were planted about 5mm roots diameter, and cutted seedlings above ground surface, When experimenting, Amorpha fruticosa seedlings began to germinate and with the length of about 10-15cm only, In this article the role of Amorpha fruticosa plants was negligible.

Fig. 2. Experimental plots photograph

Table 1. Experimental plots basic site conditions

site conditions	Preparation methods	water content (%)	Rainfall intensity (mm/h)	pockets site preparation
Bare land	the surface grass were cutted off，no grass cover.	11.0	93.5	Compared
	the surface grass were cutted off，planted *Amorpha fruticosa L.* and distributed with triangle shape	15.8	100.9	Yes
Natural grassland	Cover with Coronilla varia and Juncus effusus , no treatment.	17	104.4	Compared
	Cover with Coronilla varia and Juncus effusus , planted *Amorpha fruticosa L.* and distributed with triangle shape	17.5	104.6	Yes

2.2 Simulated rainfall

QYJY-501 artificial rainfall system equipment was used to simulate rainfall, The portable device consists of master controller, water pump, rain shower, rain gauge and data download analysis software. There are five groups of rainfall sprinklers, each with three different aperture sprinklers, sprinklers can be used alone or in combination, to choice the different aperture sprinklers and pressure to achieve different rainfall intensity. And real-time online adjust the rainfall intensity through rainfall gauge data, and ensured the simulated rainfall was more than 80% similar to natural rainfall in rainfall intensity, drop velocity, and droplet size.

When start simulate rainfall and record the start time， and simulate rainfall lasted for one hour after the runoff occurred. During the experiment, observed runoff velocity and collected muddy sand samples.

3 Results and analysis

3.1. Pockets site preparation impacted on infiltration and runoff - sediment production

The calculation of rainfall infiltration rate was divided into two stages. Before runoff occurred, the rainfall infiltration rate was equal to the rainfall intensity. And after runoff generated, the infiltration rate was calculated according to the difference between rainfall and runoff [11]. runoff and sediment yield both were estimated based on the collected muddy water samples, Various types of runoff and sediment parameters were shown in Table 2.

The later runoff occurred indicated that the infiltration capacity of slope is stronger. Compared with the plots without pockets site preparation, The time of runoff occurred on that of slope with pockets site preparation was more later obviously， and rainfall infiltration rate was more higher too， The Runoff volume was significantly lower than that of the comparative slope without the pockets site preparation.

For the bare land, the time of runoff occurred was delayed from 1'38" (under rainfall intensity of 93.50mm/h) to 4'55" after conducted pockets site preparation. The infiltration coefficient increased from 0.34 to 0.60, with increments of 76.47%, and the infiltration rate increased from 0.31mm/min to 0.57mm/min. After the pockets site preparation, the infiltration rate increased by 0.26mm/min, resulted the total runoff volume decreasing from 1290.11L to 857.77L， Reduced by 33.51% .For the natural grassland, pockets site preparation measures can delay the time of runoff occurred from 2'24" to 5'19"， and it can increase the infiltration coefficient from 0.69 to 0.79, with increments of 14.49%, and increase the infiltration rate 0.66 mm/min to 0.76 mm/min, which increased by0.105 mm/min. Eventually, The measures of pockets site preparation can lead the total runoff volume decreasing from 714.74 L to 496.81 L , Reduced by 30.49% under natural grassland conditions. It can be seen that pockets site preparation measures can reduction runoff and sediment production obviously. This phenomenon can be clearly seen in Figure 3 and Figure 4 too.

About reducing sediment, pockets site preparation measures can cut down sediment reduction 81.35% and 65.66% corresponding to bare land and natural grassland, and sediment content reduced by 71.99% and 50.58% respectively.

Table 2. Comparison of runoff and sediment parameters

Item		Bare land		Natural grassland	
		No	Yes	No	Yes
Runoff generate	Observation(s)	1'38"	4'55"	2'24"	5'19"
	Prolongation(s)	/	3'17"	/	1'04"
Infiltration coefficient	calculation	0.34	0.6	0.69	0.79
	increased percentage	/	76.47	/	14.49
Infiltration rate	Calculation (mm/min)	0.31	0.57	0.66	0.76
	Increased infiltration rate (mm/min)	/	0.261	/	0.105
Runoff volume	Runoff total(L)	1290.11	857.77	714.74	496.81
	reduction rate(%)	/	33.51	/	30.49
Sediment quality	Sediment yield(kg)	28.69	5.35	9.29	3.19
	Sediment reduction rate/%	/	81.35	/	65.66
Sediment content	Calculation (kg/m^3)	22.24	6.23	12.99	6.42
	Reduced percentage (%)	/	71.99	/	50.58

Notes: "Yes"refers to"pockets site preparation"
"No"refers to "compared"

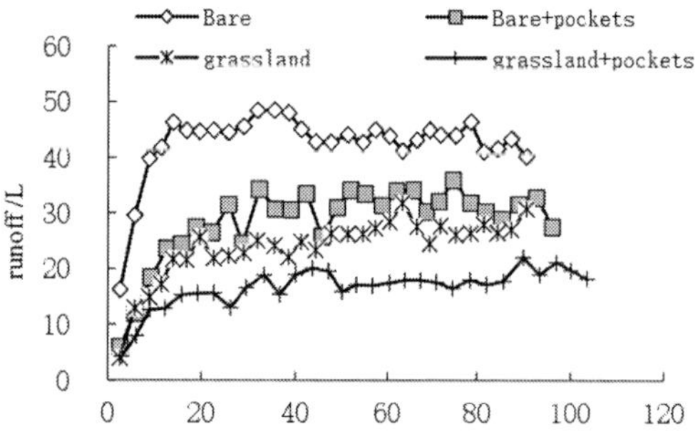

Fig. 3. Runoff process under simulate rainfall

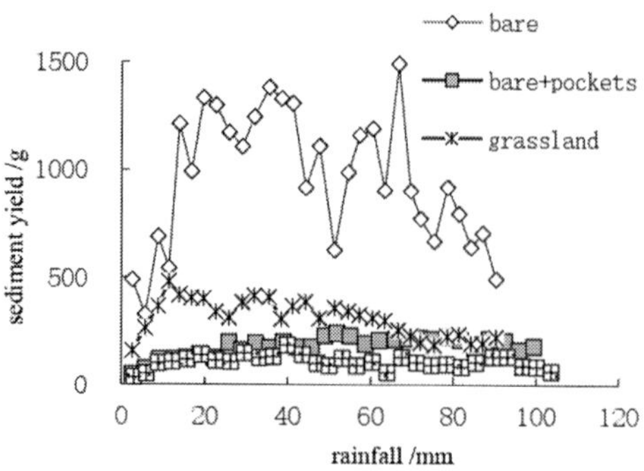

Fig.4. Sediment process under simulate rainfall

3.2 Influence of pockets site preparation affected on runoff erosion ability

Runoff velocity and soil detachment rate were used to indicate the strength of runoff erosion. Soil detachment rate [12] was that the quantity of runoff soil denuded and removed within a unit runoff area and a unit time. Runoff erosion ability parameters were calculated and shown in Table 3.

Table 3. Comparison of runoff erosion ability parameters

Item		Bare land		Natural grassland	
		No	Yes	No	Yes
Velocity	Observation (m/s)	0.27	0.17	0.24	0.16
	reduction rate (%)	/	-38.12	/	-34.59
Soil detachment rate	Observation (g/(s·m^2))	1.51	0.48	1.87	0.38
	reduction rate (%)	/	-67.92	/	-79.68

Notes: "Yes"refers to"pockets site preparation"
"No"refers to "compared"

Compared with the plots without pockets site preparation, the runoff velocity and soil detachment rate were all decreased after conducted pockets site preparation. For bare land the runoff velocity decreased from 0.27m/s to 0.17m/s, declined by 38.12%. And for natural grassland the runoff velocity decreased from 0.24m/s to 0.16m/s, declined by 34.59%. The influences of pockets site preparation measures affected on runoff velocity was evident in Figure 5. Under the similar conditions, the pockets site preparation measures reduced the runoff velocity throughout the experiment process.

For bare slope the soil detachment rate decreased from 1.51 g/(s·m^2) to 0.48 g/(s·m^2), declined by 67.92%. And for natural grassland the runoff velocity decreased from1.87 g/(s·m^2) to 0.38 g/(s·m^2), declined by 79.68%. It can be seen that the dynamic parameters of runoff erosion were obviously weakened after implementing pockets site preparation. and Figure 6 also intuitively reflected the difference of soil detachment rate under different site conditions.

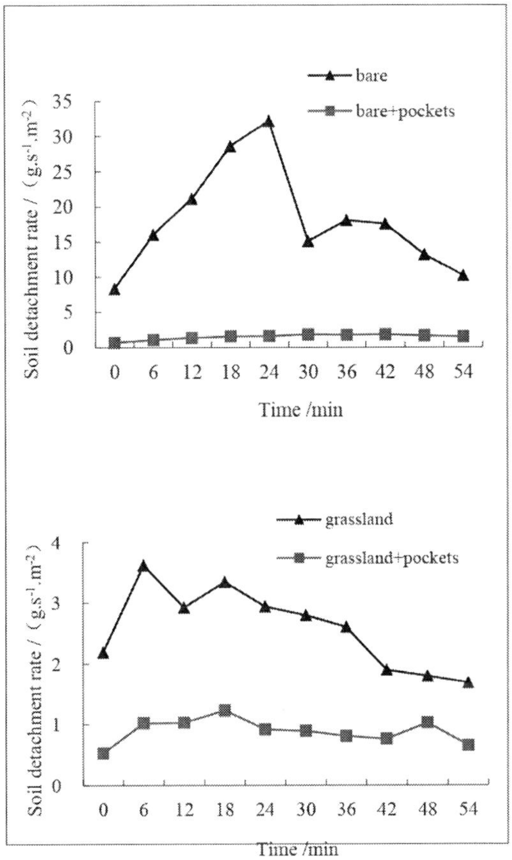

Fig.5. Runoff velocity process under different site conditions

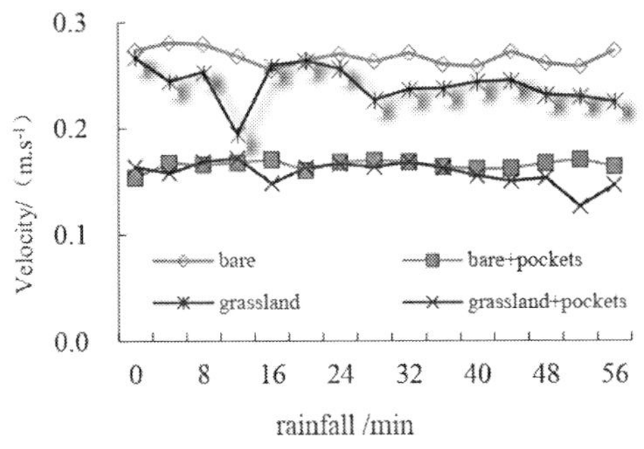

Fig.6. The difference of soil detachment rate under different site conditions

4 Conclusion and discussion

Based on the pockets site preparation plots and compared plots rainfall simulation experiment, some conclusions were made as follows:

(1) The pockets site preparation measures can delay the time of runoff generate on slope surface and increase the rainfall infiltration coefficient and infiltration rate; Reduce the runoff and sediment yield, Reduce the runoff sediment concentration.

(2) The pockets site preparation measures can significantly reduce the power of runoff erosion, and then reduces the sediment yield and runoff sediment concentration.

In the process of vegetation construction and restoration succession, the site preparation measures and the vegetation growth period all have certain impacts on the site conditions and the soil erosion environment in the initial stage of construction. As the vegetation grows, The impact of site preparation measures on the erosion environment will be weakened with the growth of vegetation and the extension of time, And then the adjustment of vegetation on runoff and sediment will be gradually highlighting with the extension of vegetation restoration period. It have great significance to study the both factors affected on soil erosion environment and the weakened or enhanced roles each other, More further researches must be taken for scientifically recognizing the impact of vegetation factors on water and sediment characteristics.

Acknowledgements

Foundation item: Under the auspices of National Science Foundation of China(No.41701326),Innovation Scientists and Technicians Troop Construction Projects of Henan Province(No.162101510004),Key Laboratory of Loess Plateau Soil Erosion and Water Loss Process and Control,MWR (201601).

References

1. Li Yi, Shao Ming an. Experimental study on influence factors of rainfall and infiltration under artificial grassland coverage[J].Transactions of the CSAE, 2007,23 (3) :18-23. (in Chinese with English abstract)
2. Chen Hong song, Shao Ming an, Zhang Xing chang, et al. Field Experiment on Hillslope Rainfall Infiltration and Runoff under Simulated Rainfall Conditions[J]. Journal of Soil and Water

Conservation，2005,19（2）：5-8(in Chinese with English abstract)

3. Xiao Pei qing, Yao Wen yi and Zhang Hai feng.Research Progress of the Role of Vegetable in Soil Conservation and Soil Erosion Reduction on the Loess Plateau[J].Research of soil and water conservation，2012,19（6）:282-286. (in Chinese with English abstract)
4. Yang Chun xia，Li Yong li，Li li，et al. Delay Effects of Ecological Restoration Grassland on the Runoff[J]. Yellow River，2014,36（8）：94-96. (in Chinese with English abstract)
5. Yang Chun xia，Xiao Pei qing, Zhen Bin, et al. Characteristics on Runoff and Sediment Yield in Field Prototype and different Coverage Slope[J].Journal of Soil and Water Conservation,2012,26(4):28-31. (in Chinese with English abstract)
6. Pan Cheng zhong，Shangguan Zhouping .Runoff hydraulic characteristics and sediment generation in sloped grass-plots under simulated rainfall conditions[J].J.Hydro.,2006,331:178-185.
7. [7] Yang Chun xia，Li Li, Zhen Bin, et al. Analysis on the Effects of Grass Coverage to the Runoff-Sediment Relationship[J]. Yellow River，2014,36（12）：91-93. (in Chinese with English abstract)
8. LI Xiao yu，LIU Xiao yan and LI Zhuo. Effects of rainfall and underlying surface on sediment yield in the main sediment-yielding area of the Yellow River[J].Journal of Hydraulic Engineering，2016,47（10）:1253-1268.(in Chinese with English abstract)
9. WU Shu fang，WU Pu te，SONG Wei xiu，et al. Study on the outflow processes of slope regulated by works and its effects on overland flow and sediment reduction [J].Journal of Hydraulic Engineering，2010 (7)：870-875.
10. QIN Wei，CAO Wen hong and ZUO Chang qing. Review on the coupling influences of vegetation and topography to soil erosion and sediment yieldJournal of Sediment Research，2015(3)：74-80.
11. Jiang Ding sheng.Soil and Water Loss and Management Models in the Loess Plateau[M]. Beijing ： China Water Conservancy and Hydropower Press，1997:34-35.
12. Wang Xuan, Li Zhanbin, Shang Baixiao, et al. Indoor simulation experiment of the relationship between soil detachment rate and water erosion factor[J]. Transactions of the CSAE, 2008,24(9)：22—26.(in Chinese with English abstract)

https://doi.org/10.1051/e3sconf/20183801033

Effects of water and soil conservation works on runoff and sediment variation in the areas with high and coarse sediment yield of the middle Yellow River

Wei Ying SUN, *Pan* ZHANG, *Li* LI, and *Jiang Nan* CHEN

Key Laboratory of Soil and Water Loss Process and Control on the Loess Plateau of Ministry of Water Resources, Yellow River Institute of Hydraulic Research, Zhengzhou 450003, China

Abstract. The areas with high and coarse sediment yield of the middle Yellow River is well known for its severe erosion, high sediment yields. Since 1982 when the 8 key soil and water conservation harnessing regions has been built, the ecological environment has been gradually improved and the amount of sediment and runoff entering the Yellow River has been reduced continuously. Some researchers considered that it was owing to the water and soil conservation works (WSCW), while others believed that it was caused by the rainfall variation, but this has not been quantified for the effect respectively. This paper deals with the effects of WSCW on runoff and sediment variation. The study has been carried out in the Sanchuanhe River watershed, where was listed as one of the 8 key soil and water conservation harnessing regions. The results show that the contribution rate of human activities was 80.2% after 1st harnessing stage (1970-1979), 43.0% after 2nd harnessing stage (1980-1989), in 3rd harnessing stage (1990-1996) it reached 98.4%, and was 44.8% after 4th harnessing stage (1997-2006). With regard to the influence on runoff reduction in the watershed, the contribution rate of human activities was 62.5% compared with the natural factors after 1st harnessing stage (1970-1979), 28.4% after 2nd harnessing stage (1980-1989), in 3rd harnessing stage (1990-1996) it reached 69.6%, and was 37.0% after 4th harnessing stage (1997-2006). The results revealed that human activities exerted the largest effects on the sediment reduction and explained 66.6% of the variation in the specific sediment yield. This study suggests that a combination of human activities and rainfall variation effectively reduces runoff and sediment delivery of the Loess Plateau. Generally The runoff reduction and contribution of rainfall variation to runoff reduction in this area were as large as human activities. After many years' harnessing the great benefit have been obtained in water and soil loss control in this watershed.

1. Introduction

Rainfall is the primary source of runoff and sediment yield in the middle Yellow River, and in recent several decades for soil erosion control and utilization of water resources, human activity intensity enhanced, the human activities may be another main cause for the variation in runoff and sediment besides the nature factors. In the past few years, with the progress of soil erosion control practices, the amount of sediment and runoff entering the Yellow River has been reduced continuously in areas with high and coarse sediment yield of the Loess Plateau. Some researchers attribute decreased soil erosion mainly to the WSCW, while others believed that it was caused by the rainfall variation. The reduction of runoff and sediment of the watershed cannot be done by simply the soil erosion control practices because the rainfall variation may also have played an important role in the sediment and runoff change. Liu [1] quantitatively analyzed the spatial-temporal evolution pattern of sediment yield in areas with high and coarse sediment yield of the Loess Plateau over the past 50 years, found that the decrease of the amount of sediment entering the Yellow River was caused by the changes of rainfall type in recent years. Ren [2] studied the runoff and sediment yield of Lianshui watershed in different stages by virtue of mathematical model, found that after the 3 soil and water conservation harnessing stages, the sediment yield has significantly decreased, and the annual runoff hasn't been changed along with the extension of the harnessing period. With the fast growth of WSCW-equipped areas, it becomes necessary to clarify the contribution of hydrological impacts soil erosion control to sediment and runoff reduction. The effect of rainfall variation and human activities (particularly WSCW) on runoff generation and soil erosion in the middle Yellow

https://doi.org/10.1051/e3sconf/20183801033

River has been an academic focus in recent years [3][4]. Thus, to better predict the impact that rainfall change or a future land-management scenario might have on land-use sustainability, an assessment of the effectiveness of WSCW used to control erosion processes is warranted. However, how to reasonably quantify the impacts of relevant factors on variations in runoff generation and to distinguish the contributions of climate change and human activities still remain as an unsolved theoretical and practical issue. Soil erosion and the resulting runoff and sediment export are caused by stochastic rainfall events and the combined effects of soil, topography, and human activities [5]. Nevertheless, it is necessary to quantify the effects of rainfall variation and human activities on runoff and sediment yield for effective watershed management.

Therefore, in this study the Sanchuanhehe River of Yellow River watershed was chosen as the case study area. The quantitative analysis model and monitoring data were used to explore the the influence of human activities and rainfall variation on sediment and runoff yield. The objectives of this study are the following: (i) to determine how the specific sediment and runoff yield is related to rainfall variation and human activities at the watershed scale; (ii) quantitatively calculate the contribution of long-term soil and water conservation harnessing measures to the runoff and sediment yield of this watershed in different stages and retain it for future reference

2. Study area

Sanchuanhe River watershed (36°55′-38°10′N, 110°33′-111°36′E) is located in the areas with high and coarse sediment yield (AHCS) of the Loess Plateau and covers 4161 km^2, in the reach from Hekouzhen to Longmen in the middle reaches of Yellow River (Fig.1). This watershed is representative of the AHCS in terms of its natural resources, land-use patterns, and population. Gauge records of the watershed's discharge and sediment yield have been collected since 1957. The area is influenced by dry and continental climates with an average annual precipitation of 506.3mm. 73% of this precipitation occurs in the period of May to September. In this area, the mean annual temperature is 7.3~10.5°C, with a minimum mean temperature of −9.5°C in January and a maximum of 24.3°C in July. In the summer, intense storms erode much soil; average annual sediment delivery in this area measured 10510 t/km^{-2} y^{-1}.

The topography of the watershed is characterized by mountain ranges, deep valleys, and loess hilly and gully region; and the elevation ranges from 650 to 1900 m. According to the Chinese soil classification system, the major soil types include brown loamy soil, loessial soil, grey cinnamonic soil, and meadow soil (National Soil Survey Office, 1998). The principal land cover type in this watershed is forest occupying 39.8% of the watershed area.

In the early 80s of last century, the soil erosion in local places was very serious, which resulted in riverbed sediment deposition and soil fertility decline and made the ecological environment very fragile. Since 1983 when Sanchuanhe River watershed was listed as one of the 8 key soil and water conservation harnessing regions, the state and local government have done a lot of hard work and effectively carried out many soil and water conservation measures such as soil and water conservation forest, closed forest, aerial tree planting, planting commercial fruit forest and grass, soil-retaining dam construction, etc. Those significant results have been achieved and the ecological environment has been gradually improved. During the period of 1970-2006, large-scale soil and water conservation harnessing was carried out in four stages in Sanchuanhe River watershed. namely, 1st stage: 1970-1979, 2nd stage:1980-1989, 3rd stage:1990-1996, 4th stage:1997-2006. By the end of 2006, the measures of soil and water conservation management area occupies 1735km^2.

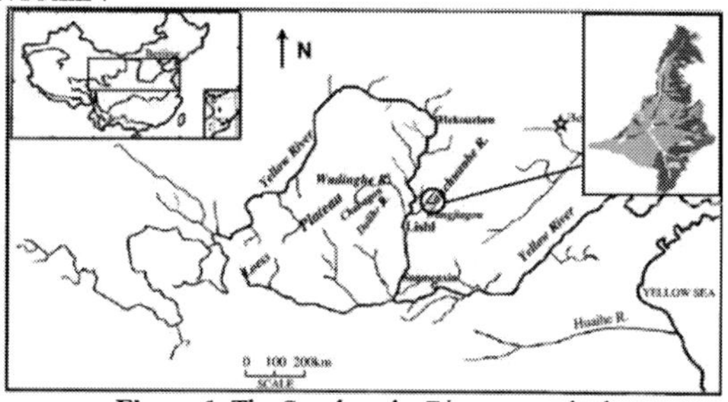

Figure 1. The Sanchuanhe River watershed

3. Results and discussions

3.1 The relationship between precipitation, runoff and sediment

The relationships between the runoff and the influence factors are constructed through stepwise regression for watersheds using the time series of data over the period of 1957–2006. Figure 2 to figure 4 is the diagram of relationship between precipitation, runoff and sediment in Sanchuanhe River watershed. From the diagram of years' precipitation to runoff (fig.2), precipitation to sediment discharge (fig.3) we can see that, the rainfall change is not consistent with the variation of runoff and sediment in the period of 1997–2006. The precipitation in recent years not has much difference with the precipitation before 1996, but the runoff and sediment discharge was greatly reduced. The relationship between runoff and sediment (Fig.4) is close. The appearance of this phenomenon indicates that the recent reduction in sediment and runoff not only due to the decrease of rainfall intensity, but also has a close relationship with the WSCW, humans consumption etc.

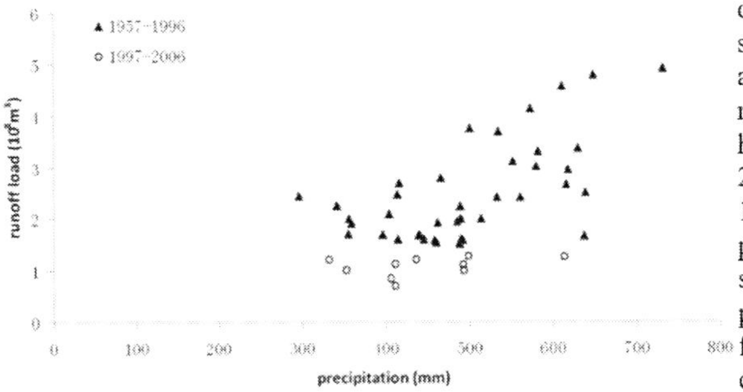

Figure 2. The relationship between the annual precipitation to runoff load

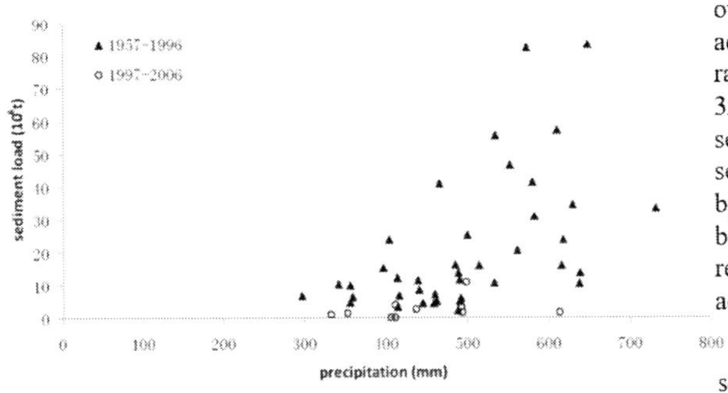

Figure 3. The relationship between the annual precipitation to sediment load

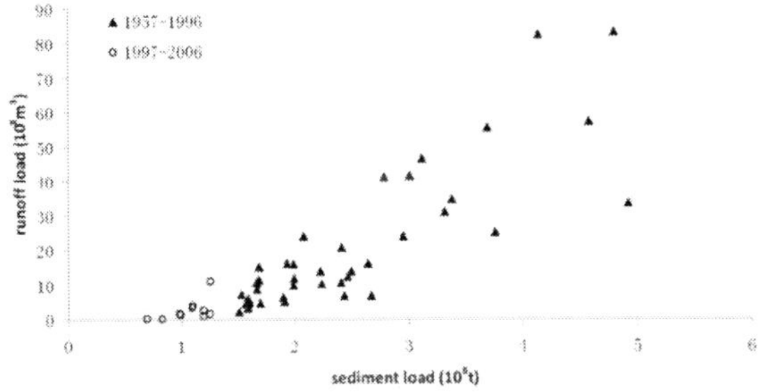

Figure 4. The relationship between the annual runoff load to sediment load

3.2 Effects of human and rainfall factors in the sediment reduction

Based on calculating model of Ran built above and the data of climate and water and soil conservation measures, computation is done on the runoff reduction due to soil erosion control practices and climate change. Table. 1 shows estimates of sediment reduction owing to human activities and rainfall factor for the Sanchuanhe River watershed. The actual monitoring data of the monitoring station in this watershed shows that great success has been obtained from the harnessing on soil erosion and the sediment yield has significantly declined. Based on the above results, we can divide the research into five stages now, namely, before the harnessing stage: 1959– 1969, harnessing of the 1st stage: 1970– 1979, harnessing of the 2nd stage: 1980– 1989, harnessing of the 3rd stage: 1990– 1996, harnessing of the 4th stage: 1997– 2006. The proportions of sediment reduction to the annual mean sediment over the period of 1970– 2006 and the percentages of contributions from rainfall variation and from human activities are given in Table 2. The contribution of different causes to the runoff deviation varied obviously over time. The runoff reduction by rainfall variation is the difference between deviation was over 50% in the 2nd stage and 4th stage. But human activities caused over 98% of the sediment reduction and rainfall variation contributed a proportion around 2% in the 3rd stage, the contribution of rainfall variation to the sediment deviation varied obviously over time. The sediment reduction by human activities was 2 times of that by rainfall variation according to their decade means. It can be seen that contribution of rainfall variation to sediment reduction in this area were generally lower than human activities.

Table 1. The proportion of human and rainfall factors in the sediment reduction of the Sanchuanhe River watershed during the stages

unit: 10^4t

The stage (year)	The measured value	The total sediment reduction	Contribution rate of human activities		Contribution rate of rainfall variation		The sediment reduction benefit (%)
			decrement	Proportion(%)	decrement	Proportion(%)	
1957~1969	3670						
1970~1979	1822	1848	1483	80.2	365	19.8	44.9
1980~1989	960	2710	1164	43.0	1546	57.0	54.8
1990~1996	1074	2596	2555	98.4	41	1.6	70.4
1997~2006	257	3413	1482	44.8	1931	55.2	83.6
1970~2006	1028	2642	1671	66.6	971	33.4	63.4

From the table 1, we can get the following conclusion: after the harnessing of the 1st stage, the human activities (namely, the soil and water conservation harnessing) are main factors influencing the sediment yield decreasing of the watershed by around 80.2% and the rainfall variation by around 19.8%. After the harnessing of the 2nd stage, the human activities contribution rate to the sediment yield decreased by around 43% and the rainfall factor by around 57%. After the harnessing of the 3rd stage, the human activities contribution rate amounts quickly to 98.4% meanwhile the rainfall variation accounted for about 1.6%. During the 4th stage from 1997 to 2006, water and soil conservation measures reduced sediment by about 14.82 million t annually for the Sanchuanhe River watershed. On the other hand, the rate of sediment reduction by rainfall variation was about 19.31 million t per year. Human activities contribution rate to the sediment yield decreased by around 44.8% and the rainfall factor by around 55.2%. The calculation results also indicate that the human factor can affect the sediment yield significantly, for long-term harnessing, the human factor can change even can decide the change trend of the sediment yield of the watershed. The above analysis of the relationship between runoff coefficient and influencing factors in the spatial dimension has excluded the impacts of water and soil conservation measures on runoff. The decrease of runoff modulus from the early to the later periods under a same climate condition declares that impacts of water and soil conservation measures on runoff has been growing.

3.3 Effects of human and rainfall factors in the runoff reduction

Table 2 shows the average runoff reduction due to human activities and rainfall factors in the drainage areas upstream of some hydrological stations. From the table 3, we can get the following conclusion: For different harnessing periods the runoff deviations and contributions of different causes were different. After the harnessing of the 1st stage, the human activities are still main factors influencing the runoff yield decreasing of the watershed and after the harnessing of the 2nd stage, runoff reduction associated with rainfall variation was much higher, rainfall variation caused over 70% of the runoff reduction and human activities contributed a proportion around 28%. Contribution rate to the sediment yield increased to around 71.6% and at the end of the harnessing of the 3rd stage, the human activities contribution to total runoff deviations was over 60%. During the 4th stage, influence of human activities and rainfall variation on the decrease of runoff in each half. It is clear that the decade means of runoff reduction by soil erosion control practices and climate change has a general increasing trend in Sanchuanhe River watershed. The runoff reduction and contribution of rainfall variation to runoff reduction in this area were generally as large as human activities. We find that the human activities can affect the sediment yield significantly, but the reduction benefit on sediment was smaller than runoff, and the contribution rate of human activities and rainfall variation had increased gradually before 2006 and the average runoff reduction benefit is about 44%.

In the period of 1970–2006, human activities reduced runoff by about 33 million m^3 annually for the Sanchuanhe River watershed, on the other hand, the rate of runoff reduction by rainfall variation was about 39 million m^3 per year (Table 2). The proportions of runoff reduction to the annual mean runoff over the period of 1970–2006 was 49.4% and the percentages of contributions from climate change was 50.6%. It can be seen that the contribution of rainfall variation to runoff reduction in this area were generally as large as human activities. Even though in the periods of 1971–1990 and 1991–2009 with intensive human activities, the natural factors had still a higher impact on the spatial distribution of runoff coefficient than human activities.

Clearly, although aimed at controlling soil erosion, soil conservation measures not only reduce sediment yield, but also cut down the runoff. Large scale soil conservation practices have been carried out on there in recent decades for reducing the sediment output from the Loess Plateau. Nevertheless, our results also show that the rainfall variation has a sizeable contribution to the runoff decrease in the study area. Similar cases have been reported for many worldwide rivers. In the whole middle Yellow River, climate change contributed over 40% of the runoff deviations, while water consumption induced also over 40% runoff reduction in the 1960s, 1970s and 1980s [6]. In Tigray, Northern Ethiopia, the variation in total rainfall amount explained 69% of the annual runoff volume variability [7].

Table 2. The proportion of human activities and rainfall variation in the runoff reduction of the Sanchuanhe River watershed during the stages

unit: 10^4m^3

The stage (year)	The measured value	The total runoff reduction	Contribution rate of human activities		Contribution rate of rainfall variation		The runoff reduction benefit(%)
			decrement	Proportion(%)	decrement	Proportion(%)	
1957~1969	11521						
1970~1979	6781	4740	2962	62.5	1778	37.5	30.4
1980~1989	4552	6969	1979	28.4	4990	71.6	30.3
1990~1996	4366	7155	4978	69.6	2177	30.4	53.3
1997~2006	1599	9962	3477	37.0	6485	63.0	62.7
1970~2006	4325	7206	3349	49.4	3857	50.6	44.2

4. Conclusions

This paper quantitative analyzed the contribution of WSCW harnessing to the runoff and sediment yield of Sanchuanhe River watershed in different stages by virtue of hydrological simulation model, thus to identify the influence on WSCW harnessing on runoff and sediment yield variation. The test for sediment and runoff prediction shows that the calculating model of Ran can be used to predict the annual runoff and sediment in Sanchuanhe River watershed. Analysis of Sanchuanhe River watershed indicates that runoff and sediment showed a clearly decreasing trend over the past 50 years. In the general decrease trend existed two abrupt changes around the years 1971 and 1981. The implementation of soil and water conservation measures in large scale has played great role to reduce the sediment and runoff yield of the watershed, and rainfall factors were of secondary importance. It can be seen that contribution of climate change to sediment reduction in this area were generally lower than human activities. The runoff reduction and contribution of rainfall change to runoff reduction in this area were generally as large as human activities. The contribution rate of these measures to the sediment and runoff yield of the watershed in the four stages from 1970 to 2006 gradually increased along with the spread of the implementation, the contribution rate rising from the initial 44.9% to 83.6%. The sediment trapping effect of soil and water conservation measures is further improved with the extension of the harnessing time, and the runoff yield capacity also changed greatly, the contribution rate rising from the initial 30.4% to 62.7%. The above-mentioned analysis shows that the large-scale soil erosion harnessing started in last century is rather proper on the principles of taking watershed as unit and harnessing by unit. After many years' harnessing the great benefit have been obtained in the reforestation and effective sediment trapping in this watershed.

Acknowledgments

This paper is based on work supported by an open project of the key laboratory (201801) and the Natural Science Foundation of China (51409110).

References

1. Liu, J.H., Wang, G.Q., Li, H.H., Gong, J.G., Han, J.Y., 2013. Water and sediment evolution in areas with highland coarse sediment yield of the Loess Plateau. International Journal of Sediment Research, 28, 448-457.
2. Ren, S. M., Liang, Y., Sun, B., 2011. Quantitative analysis on the influence of long-term soil and water conservation measures harnessing on runoff and sediment yield of watershed. Procedia Environmental Sciences, 10, 1732-1740.
3. Xu, Z.X., Zhao, F.F., Li, J.Y., 2009. Response of streamflow to climate change in the headwater catchment of the Yellow River basin. Quaternary International, 208 (2009), 62-75.
4. Xu, J.X., 2005. Temporal variation of river flow renewability in the middle yellow river and the influencing factors. Hydrological Processes, 19, 1871-1882.
5. Wei, W., Chen, L.D., Zhang, H.D., Chen, J.,2015. Effect of rainfall variation and landscape change on runoff and sediment yield from a loess hilly catchment in China. Environmental Earth Sciences, 73, 1005-1016.
6. Shi, C.X., Zhou, Y.Y., Fan, X.L., Shao, W.W., 2012. A study on the annual runoff change and its relationship with water and soil conservation practices and climate change in the middle Yellow River basin. Catena, 100, 31-41.
7. Girmay, G., Singh, B.R., Nyssen, J., Borrosen, T., 2009. Runoff and sediment-associated nutrient losses under different land uses in Tigray, Northern Ethiopia. Journal of Hydrology 376, 70–80.

Contrast Experiment on Advanced Treatment of Pharmaceutical and Paper-making Wastewater through Cinder Fenton-like Process

Jia Xiaofeng[1,2], *Chen Xiaoyu*[2], and *Mai Wenning*[3]

[1]Luohe Yinge Industrial Group Co., Ltd,462003, Luohe, Henan, China
[2]Zhengzhou University of Aeronautics,450000, Zhengzhou, China
[3]Zhengzhou University of Aeronautics,450000, Zhengzhou, China

Abstract. The Fenton-like process of catalyzing H_2O_2 with Fe^{2+} and cinder is adopted to subject pharmaceutical and paper-making wastewater to advanced treatment. The influence of each factor is determined using orthogonal experiment and single factor test. The optimal combination of influencing factors is 0.3mmol • L^{-1} of $FeSO_4$ • $7H_2O$, $[H_2O_2]:[Fe^{2+}]$=8:1 and 10g • L^{-1} of pyrites cinder. The reaction time in pharmaceutical wastewater and paper-making wastewater is 30min and 60min respectively, testifying to the fact that the reaction in pharmaceutical wastewater is faster than that in paper-making wastewater and the lower utilization rate of cinder in pharmaceutical wastewater. Under the optimal reaction condition, the COD removal rate of these two kinds of wastewater can reach as high as 65% and 72%. Characterized by simple operation and requiring less reagent dosage, this method does not have to regulate the pH of flooding water and allows the repeated usage of cinder.

1 Introduction

Fenton reaction[1] refers to the production of hydroxyl radical with strong oxidizing property by catalyzing H_2O_2 with Fe^{2+} that is used to oxidize and degrade contaminants. Despite its strong oxidizability, Fenton's reagent first requires the addition of acid and then alkali to adjust the pH to neutral. The complicated operation and frequent adjustment of pH will increase the operational costs and investment costs of infrastructure equipment[2].

With Fe content of 30~55% and Si, Ca, Mg, Mn, Cu and Zn, pyrites cinder refers to the waste residue from the production of sulfuric acid with pyrite[3]. Recent researches show that the pyrites cinder is highly active in catalyzing [4-5]the oxidation of H_2O_2 and effective in forming the new Fenton-like reaction with surface catalysis as the mechanism, which testifies to good treatment of Acid Red B, phenol and many other actual industrial wastewater.

The wastewater from a pharmaceutical company and a paper-making company in Henan are all subjected to the traditional anaerobic and aerobic treatment. The pharmaceutical wastewater, with complicated components, high concentration, high toxicity, dark color and strong smell, hardly degradable substances and inhibitory drugs[6-7], is difficult to be treated. With the implementation of *Discharge Standard of Water Pollutants for Pharmaceutical Industry Fermentation Products Category* (GB21903-2008), the single secondary biological treatment cannot satisfy the requirements. Characterized by high chromaticity and poor biodegradability, the paper-making wastewater usually contains chlorinated aromatic compounds[10] as a result of the usage of decolorizer, thus bringing harm to environment and human health[11]. For these reasons, it is a must to subject both the pharmaceutical and paper-making wastewater to advanced treatment. Under the condition of not regulating pH, the experiment of treating the secondary effluent from pharmaceutical and paper-making wastewater with H_2O_2 under the catalysis of pyrites cinder shows that there is no obvious decrease of COD. Therefore, the author intends to adopt the Fenton-like process of catalyzing H_2O_2 with pyrites cinder and Fe^{2+} to study the feasibility of the advanced treatment of pharmaceutical and paper-making wastewater.

2 Materials and methods

2.1 Test water quality

The water is from the aerobic treatment secondary effluent of a pharmaceutical and a paper-making company in Henan, whose quality is presented in Table 1.

Table 1. Water quality of secondary effluent

Item	ρ (CODcr)/ (mg·L^{-1})	ρ (BOD5)/ (mg·L^{-1})	ρ (SS)/ (mg·L^{-1})	pH

* Corresponding author: 191826850@qq.com

Pharmaceutical Wastewater	100~150	12~16	40~60	7.5~8.0
Paper-making Wastewater	240~270	20~40	30~50	7.5~8.5

For easier comparison, the paper-making wastewater is diluted with equal amount of water during experiment.

2.2 Experimental method

1000 mL wastewater, together with pyrites cinder and a certain amount of $FeSO_4 \cdot 7H_2O$ and H_2O_2(30%) is added to the 1000 mL beaker for reaction. After 20min of aeration and 30min of static settlement, the supernate is taken to test its COD.

2.3 Analysis item and methods

pH: glass electrode method; COD: potassium dichromate method; BOD: standard dilution method; SS: 103~105°C oven drying and weighing method[12].

3 Results and analysis

3.1 Orthogonal test

The four controllable parameters, namely, amount of pyrites cinder, $[H_2O_2]:[Fe^{2+}]$, dosage of $FeSO_4 \cdot 7H_2O$ and reaction time, are used as the influencing factor of orthogonal test. Three representative levels of each influencing factor is chosen and presented in Table 2.

Table 2. Influencing factor and level of orthogonal test

Level	Influencing Factor			
	Cinder Amount/(A) ($g \cdot L^{-1}$)	(B) $[H_2O_2]:[Fe^{2+}]$	$n(Fe^{2+})$ /(C) ($mmol \cdot L^{-1}$)	Reaction Time /(D) (min)
1	5	4:1	0.1	20
2	15	8:1	0.2	60
3	30	12:1	0.3	120

According to the determined factor, level and the L9(3^4) orthogonal table, the experiment is carried out with COD removal rate as the standard of evaluation, as shown in Table 3. The size of influencing factor determined through range is shown in Table 4.

Table 3. Orthogonal text results

Test No.	A	B	C	D	Removal Rate of COD in Pharmaceutical Wastewater%	Removal Rate of COD in Paper-making Wastewater%
1	5	4:1	0.1	20	18.6	25.7
2	5	8:1	0.2	60	57.3	67.5
3	5	12:1	0.3	120	61.1	70.2
4	15	4:1	0.2	120	39.2	43.7
5	15	8:1	0.3	20	64.7	68.6
6	15	12:1	0.1	60	42.6	54.7
7	30	4:1	0.3	60	48.3	44.3
8	30	8:1	0.1	120	41.8	48.7
9	30	12:1	0.2	20	55.3	63.5

It is shown in Table 3 and 4 that the influence of each influencing factor of pharmaceutical wastewater in descending order is C>B>D>A; the optimal combination is $A_2B_2C_3D_1$, which means the dosage of pyrites cinder of $15g \cdot L^{-1}$, dosage of $FeSO_4 \cdot 7H_2O$ of $0.3mmol \cdot L^{-1}$, $[H_2O_2]:[Fe^{2+}]$ (molar ratio) of 8:1 and reaction time of 20min. The influence of each influencing factor of pharmaceutical wastewater in descending order is B>C>A>D. The optimal combination is $A_1B_3C_3D_3$, which means the dosage of pyrites cinder of $5g \cdot L^{-1}$, dosage of $FeSO_4 \cdot 7H_2O$ of $0.3mmol \cdot L^{-1}$, $[H_2O_2]:[Fe^{2+}]$ (molar ratio) of 12:1 and reaction time of 120min. It shows that the degrading effect in pharmaceutical wastewater under the catalysis of H_2O_2 with cinder and Fe^{2+} is worse than that in paper-making wastewater; however, it takes longer time to treat the paper-making wastewater because cinder can obviously adsorb the chromaticity of paper-making wastewater. However, complete degradation is very slow.

Table 4. Range Analysis(a)

Analysis Item	Pharmaceutical Wastewater			
	A	B	C	D
K1	45.67	35.37	34.33	46.2
K2	48.83	54.6	50.6	49.4
K3	48.47	53	58.03	49.53
Range R	3.16	19.23	23.7	3.33

Table 4. Range Analysis(b)

Analysis Item	Paper-making Wastewater			
	A	B	C	D
K1	54.47	37.9	42.03	52.6
K2	55.67	61.6	58.23	54.5
K3	52.17	61.8	61.03	54.2
Range R	3.5	23.9	19	1.9

3.2 Single factor analysis

The single factor analysis is carried out by controlling the best level determined by orthogonal test.

3.2.1 Influence of molar ratio of $[H_2O_2]:[Fe^{2+}]$

To compare the treatment effect in the two different wastewater under the same condition, 10g pyrites cinder and 0.3mmol $FeSO_4 \cdot 7H_2O$ are added for 60min of reaction in order to study the influence of different H_2O_2 dosage on the removal rate of COD in these two different wastewater. The experimental results are shown in Fig. 1.

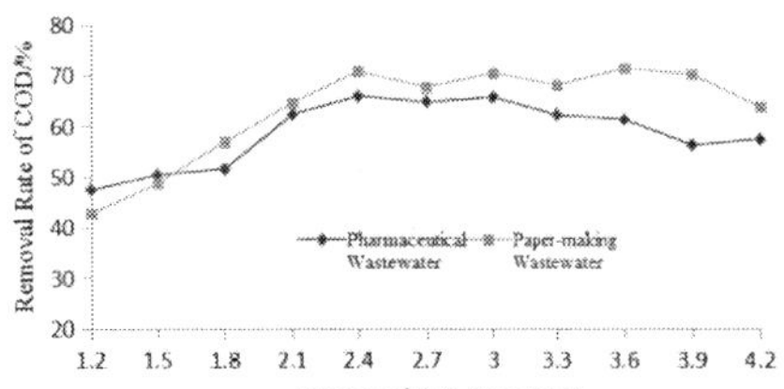

Fig. 1 The influence of H_2O_2 dosage on COD removal effect

Fig. 1 shows that along with the increasing dosage of H_2O_2, the removal rate of COD in pharmaceutical wastewater first increases significantly and then slowly, resulting in the decrease in the removal rate of COD. At lower dosage of H_2O_2, the production rate of OH· along with the increase in the initial concentration, thus significantly increasing the removal rate of COD. At higher dosage of H_2O_2, OH· will be captured by H_2O_2 and Fe^{2+} as it is produced at very fast rate. The ineffective decomposition of H_2O_2 is accelerated with Fe^{2+} oxidized into Fe^{3+}, which impedes catalysis and gives rise to the unobvious removal effect of COD. With excessive dosage of H_2O_2, Fe^{2+} will soon be oxidized into Fe^{3+} at the beginning of reaction. The excessive H_2O_2 will also soon be broken into H2O and O_2 and compete with organic matters for OH· and other free radicals, thus reducing the removal rate of COD. However, the COD removal rate in paper-making wastewater at the later stage decreases insignificantly because the cinder, with good adsorption effect, can thoroughly utilize the iron in the iron in the cinder. An overall consideration of the treatment requirements of wastewater and the important impact of H_2O_2 dosage on operation costs shows that 2.4mmol H_2O_2 can apply to these two wastewater; in other words, the molar ratio of $[H_2O_2]:[Fe^{2+}]$ should be 8:1.

3.2.2 The influence of the dosage of $FeSO_4·7H_2O$

Under the condition of dosage of cinder of 10g, $[H_2O_2]:[Fe^{2+}]$ of 8:1 and reaction time of 60 min, the dosage of $FeSO_4·7H_2O$ is adjusted. The influence of the dosage of $FeSO_4·7H_2O$ on COD removal rate is shown in Fig. 2.

Fig. 2 shows that COD removal rate rises along with the increase of Fe^{2+} dosage. Fe^{2+} is the essential condition for the production of hydroxyl radical. Under the catalysis of Fe^{2+}, the following reactions will take place[13-14]:

$$Fe^{2+}+H_2O_2 \rightarrow OH^-+\cdot OH+Fe^{3+} \quad (1)$$

$$Fe^{3+}+H_2O_2 \rightarrow Fe^{2+}+H^++\cdot O_2H \quad (2)$$

$$RH+\cdot OH \rightarrow R\cdot+H_2O \quad (3)$$

$$R\cdot+H_2O_2 \rightarrow ROH+\cdot OH \quad (4)$$

$$Fe^{2+}+\cdot OH \rightarrow OH^-+Fe^{3+} \quad (5)$$

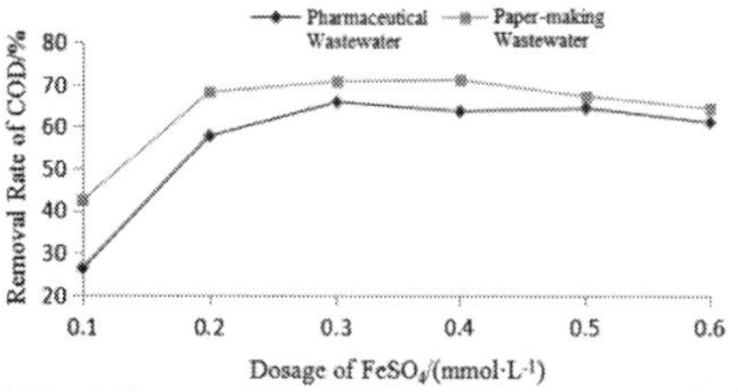

Fig. 2 The influence of $FeSO_4$ dosage on COD removal effect

Higher $[Fe^{2+}]$ content will promote the reaction in equation (1) and the high production rate of hydroxyl radicals. After the reaction, the Fe^{3+} produced is possessed of flocculation, which can remove some of COD after precipitation and filtration. However, the too high dosage of Fe^{2+} will also promote the reaction in equation (5) without significantly increasing the removal of COD. As a result, there will also be an increase in the dosage expense and sludge quantity, which will cause secondary pollution. For this reason, the dosage of $FeSO_4·7H_2O$ is determined as 0.3mmol·L^{-1}.

3.2.3 The influence of dosage of pyrites cinder

Under the condition of the dosage of $FeSO_4·7H_2O$ of 0.3mmol, $[H_2O_2]:[Fe^{2+}]$ of 8:1 and reaction time of 60min, the dosage of pyrites cinder is regulated, whose impact on COD removal rate is shown in Fig. 3.

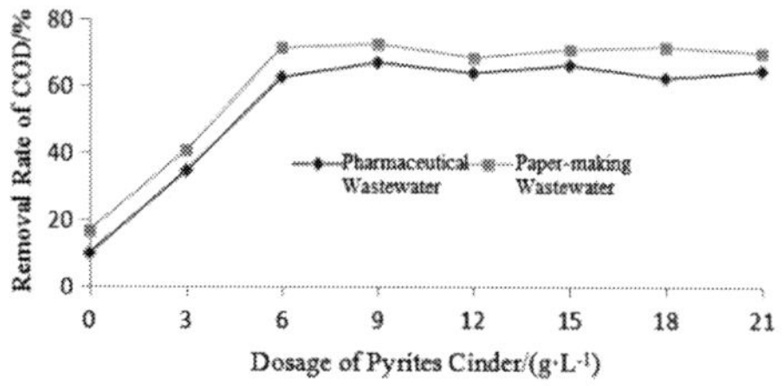

Fig. 3 The influence of dosage of pyrites cinder on COD removal effect

Fig. 3 shows that at lower content of cinder, the COD removal rate rises along with the increase in cinder dosage. After the dosage reaches a certain amount, there is a limited increase in COD removal rate. However, due to its large volume in the reactor and large pressure drop, the power expenses increase. Comparison shows that the optimal dosage of cinder in these two different wastewater is determined as 10g·L^{-1}.

3.2.4 The influence of reaction time

Under the condition of the dosage of $FeSO_4·7H_2O$ of 0.3mmol, 10g pyrites cinder and $[H_2O_2]:[Fe^{2+}]$ of 8:1, the influence of different reaction time on COD removal effect is investigated with experimental results shown in Fig. 4.

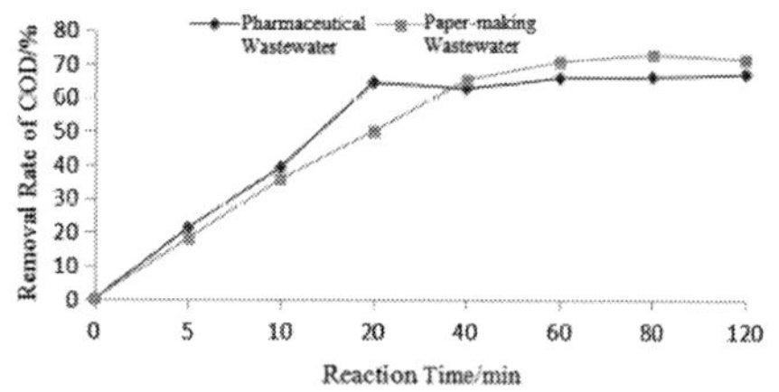

Fig. 4 The influence of reaction time on COD removal effect

It is presented in Fig. 4 that the pharmaceutical wastewater reacts faster than the paper-making wastewater. The COD removal rate basically reaches the highest after 20min in pharmaceutical wastewater and 60min in paper-making wastewater. As the reaction time prolongs, there is no obvious change in removal rate because at the beginning of the reaction, the large amount of H_2O_2 reacts with Fe^{2+} to produce abundant OH·, which immediately attacks organic matters and oxidizes the unreacted Fe^{2+}. As a result, at the beginning of the reaction, the degradation velocity of COD is very high. As the reaction continues, H_2O_2 and Fe^{2+} are gradually consumed. Although Fe^{3+} can still be reduced to Fe^{2+}, this reduction reaction is carried out at a lower rate. In this process, Fe^{3+} might be hydrolyzed, which will reduce the degradation rate of organic matters. For these reasons, the best reaction time of pharmaceutical wastewater is determined as 30min while that of paper-making wastewater as 60min.

3.2.5 Reutilization experiment of pyrites cinder.

The single factor experiment is conducted under the optimal condition, namely, dosage of $FeSO_4{\cdot}7H_2O$ of 0.3mmol, 10g pyrites cinder, $[H_2O_2]$:$[Fe^{2+}]$ of 8:1, 30min reaction in pharmaceutical wastewater and 60min reaction in paper-making wastewater. After each experiment, the wastewater is discarded and the cinder saved to investigate the influence of times of pyrites cinder usage on the COD removal rate in the wastewater with the results shown in Fig. 5.

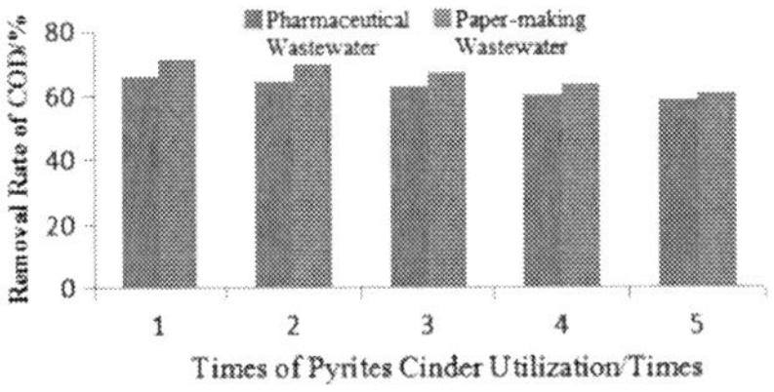

Fig. 5 The influence of the times of pyrites cinder utilization on COD removal rate

It can be seen from Fig. 5 that along with the increase of utilization times, the catalytic effect of cinder reduces but in an insignificant way. After five times, the COD removal rate in these two paper-making wastewater still remains at 60%, a fact indicating that the various elements in the cinder are basically not consumed but catalyze H_2O_2 together with Fe^{2+}. This expands the pH range of Fenton and reduces the dosage of Fe^{2+} and H_2O_2.

4 Conclusions

Through orthogonal test, the influence of each factor is presented in descending order as: pharmaceutical wastewater: $FeSO_4{\cdot}7H_2O$ dosage> $[H_2O_2]$: $[Fe^{2+}]$>reaction time>cinder dosage; paper-making wastewater: $[H_2O_2]$: $[Fe^{2+}]$> $FeSO_4{\cdot}7H_2O$ dosage>cinder dosage>reaction time. It shows that the utilization effect of cinder in pharmaceutical wastewater is less satisfactory than that in paper-making wastewater.

The optimal condition for the advanced treatment of pharmaceutical wastewater and paper-making wastewater by catalyzing H_2O_2 with cinder and Fe^{2+} is the dosage of pyrites cinder of $10g{\cdot}L^{-1}$, the dosage of $FeSO_4{\cdot}7H_2O$ of $0.3mmol{\cdot}L^{-1}$, $[H_2O_2]$:$[Fe^{2+}]$ of 8:1, the reaction time of pharmaceutical wastewater of 30min and reaction time of paper-making wastewater of 60min. Under the optimal condition, the removal rate of COD in pharmaceutical wastetwater and paper-making wastewater can reach 65% and 72% respectively. This method does not require the regulation of the pH of inflow or high dosage of reagents, which testifies to the simple operation.

The cinder still possesses certain concerted catalysis capacity even after five times of reutilization with a COD removal rate in the two wastewater of more than 60%.

Acknowledgements

Supported by Postdoctoral Science Foundation of Henan Province, China (NO.155889);
Supported by Programs for Science and Technology Development of Henan Province, China (NO.162102210084)

References

1. D. L. Zhang, Y.P. Huang, G.F. Luo, et al. *Environmental Chemistry*, **25** (2006).
2. L. P. Wang, H. Cai, Y.Z.Chen, et al. *China Water and Wastewater*, **26** (2010). .
3. X. S. Wang, D.H. Zhang, Y.F.Ren. *Environmental Engineering*, **17** (1999).
4. Tongji University. *Fenton-like Oxidation of Refractory Organic Contaminants in Waste Water Using Pyrite Cinder at Neutral pH* .P. China Patent: 101745197.2010-06-23.
5. D. L. Wu,D.Duan , L.M.Ma. *CIESC Journal(China)*, **61** (2010).
6. Z. P. Xing *Treatment of Antibiotic Fermentation Wastewater Using Combined Coagulation-Hydrolysis/Aerobic Moving Bed Biofilm Reactor-Fenton Process .D.* (2008).
7. Y. Q. Li, Y.Ma , G.E.Qian, . *Industrial Water Treatment*, **29** (2008).
8. X. Zhang, H. L.Jiang, *Industrial Water Treatment*, **27** (2007).
9. L. H. Mo,K. F .Chen, Z. X.Chen , *China Pulp & Paper*, **23**(2004).
10. Sealey J, Ragauskas A J. *Enzyme and Mierobial Technology*, **23**(1998).
11. D F Zeng, D Hu, J Cheng.*Journal of Water Chemistry and Technology*,**34**(2012).
12. *State Environmental Protection Administration. Detection and Analysis Method of Water and Wastewater (4th EDITION)*. (2002).
13. K Namgoo,S L Dong ,Y Jeyong . Chemosphere,**47**(2002).
14. C. H. Chen, B. Xie,Y. Ren , et al. J.E.S, **21**(2000)

The Location Analysis on Dry Port of the Inland in Jinan City

Jianmeng Sun[1,*], Haining Wang[2]

[1]Shandong Institute of Transportation Science, Jinan, Shandong 250023, China
[2]Shandong Hi-speed Group Co., Ltd, Jinan, Shandong 250013, China

Abstract. The port competition is more and more intense, starts to be the initial infrastructure and the operation ability competition, now is in the port economy hinterland competition. Dry Port has the functions of customs declaration, forwarder, inspection and quarantine, warehousing, loading and unloading, circulation processing etc. At present, the construction and layout of the inland dry port, the possibility and feasibility of construction are the focus of the research. The location of dry port in Jinan is fully analyzed on the basis of reference to the advanced operation mode and development experience of the excellent non-water port in the country and abroad. This paper first analyzed Location factors; Then center of gravity method were analyzed. Finally, the location of the dry port is determined.

1 The factor analysis of the location of dry port in Jinan

1.1 Proximity to market principles

The profitability of a dry port is usually dependent on the market. The choice of close to the market can effectively reduce the transport distance, reduce transshipment, avoid multiple loading and unloading, reduce freight, and timely replenishment. The price of land close to the market is higher than other places, so there are other factors that should be taken into consideration on dry port.

1.2 Close to the transportation hub principle

The transport facilities must be used for transport from the dry port to the port and from the port to the dry port. Fast and convenient transportation is also one of the factors to be considered.Site selection can quickly organize the delivery and purchase of goods in the more developed areas. It can also reduce the accumulation and retention of goods, improve the efficiency of logistics operation and improve the aging time.

1.3 Choose low land price principle

A dry port would normally choose a place with a lower land price. A dry port usually chooses a place close to the city. This can not only achieve the low cost of land, but also can make effective use of the city's developed traffic. It can comprehensively consider all aspects of the dry port to improve the efficiency.

1.4 The principle of complete infrastructure

The construction and operation of a dry port cannot be separated from the water port infrastructure, but also requires considerable manpower, material resources and financial resources. The infrastructure generally covers an area of large area, with difficulty in recovery, large investment, long payback period, and difficult to relocate.If you have existing infrastructure you can reduce the amount of money invested and increase the revenue.

2 The factor analysis of the location of dry port in Jinan

2.1 The relationship between the geographical location of dry port and city layout

According to the geographical location of the site, the distribution of resources, the structure of the region, the distribution of the consumption market and so on have a decisive influence on the development of the waterless port. Dry port, therefore, to comprehensively consider the site selection of regional traffic, surrounding the market situation, port and dry port distance, the city's industry layout, avoid benefit back against the effect.

2.2 The relationship between the construction address and transportation on the dry port

The development of a city cannot be separated from the traffic, and the construction and development of dry port cannot be separated from the traffic. The ideal effect of transportation goods is low freight, fast speed, and the choice of a well-developed area can achieve the desired effect. Meanwhile, multi-modal transport tools and methods can be selected flexibly. Dry port construction site should also consider the pressure of urban traffic.

* Corresponding author: mxr6465201@126.com

Choose the right location to avoid reducing the benefit of the dry port.

3 Model analysis of the center of gravity method

Gravity model:

$$MinTc=\sum ViRidi=\sum Cidi \quad (1)$$

In the formula:

T_c —Total transportation cost (yuan);

V_i—Transport capacity of I point (ton);

R_i—Transport rate of I point (yuan/ton. Km);

d_i—The distance from the location of the inland dry port to I point (km);

C_i—The transportation cost to I point (yuan/km).

The location of the lowest cost of transportation coordinates X and Y, gravity method using the formula is:

$$X=\sum dixVi/\sum Vi \quad (2)$$

$$Y=\sum diyVi/\sum Vi \quad (3)$$

In the formula:

$\overline{X}$ —The x-coordinate of the center of gravity;

$\overline{y}$ —The y-coordinate of the center of gravity;

d_{ix} —The x-coordinate of point I;

d_{iy} —The y-coordinate of point I;

V_i —The amount of goods shipped to or from I point.

The method of gravity is to calculate the location of the dry port according to the coordinates of each port. Other factors should be considered in the position of gravity. The actuality and candidate location of the land should be considered comprehensively. The position calculated according to the coordinates may be park, station and so on. So you have to think about other factors.

4 The development of the counties and districts in Jinan and the location of the dry port

4.1 The economic operation of various counties and districts in Jinan in 2016

The economic operation conditions of various counties and counties in Jinan in 2016 are shown in table 1.

As shown in table 1, in the various counties and districts of Jinan city in 2016, the total GDP of the Lixia District is the first. At the same time, in 2015, the region's GDP also ranked first. According to the GDP of dry port location, should be in lixia District, so that promotes the economic development of dry port, can provide affordable goods for dry port.

4.2 The import and export of various counties in Jinan in 2016

Evaluation In 2016, import and export situation of various prefectures in Jinan are shown in table 2.

As shown in table 2, in the various counties and districts of Jinan city, the total import and export volume of Lixia District is the first in 2016.Lixia District has an important position in the development of Jinan, control the economic lifeline of development of Jinan, only the total import and export in 2016 and the actual received foreign, the dry port should be constructed in the Lixia District.

4.3 The situation of industrial development and investment on each county area in Jinan

The industrial situation of each county in Jinan is shown in table 3.

As shown in table 3, in the various counties and districts of Jinan city in 2016, the Lixia District in investment in fixed assets is the first. Zhangqiu District ranks first in the major business income of the above industries. The total amount of industrial profits in Zhangqiu District is the first in the county districts of Jinan. It should be constructed in Zhangqiu District when considering industrial factors to construct the inland dry port in Jinan.

Table 1. The GDP of counties and districts in Jinan in 2016

Precedence	Counties	GDP (the hundred million)		The incremental (the hundred million)	Nominal growth (%)
		In 2016	In 2015		

	Jinan City	6536.12	6100.23	435.89	7.15
1	Lixia District	1837.1	1667.0	170.1	10.20
#	High-tech zone	650.3	596.3	54.0	9.06
2	Zhangqiu District	924.9	870.8	54.1	6.21
3	Licheng District	868.8	808.0	60.8	7.52
4	Shizhong District	826.9	751.3	75.6	10.06
5	Huaiyin District	431.9	387.1	44.8	11.57
6	Tianqiao District	418.0	384.9	33.1	8.60
7	Jiyang County	295.0	276.3	18.7	6.77
8	Changqing District	289.4	265.2	24.2	9.13
9	Pingyin County	245.2	227.0	18.2	8.02
10	Shanghe County	179.2	165.7	13.5	8.15
urban area		5597.0	5134.3	462.7	9.01
counties		719.4	669.0	50.4	7.53
total		6316.4	5803.3	513.1	8.84

Table 2. Import and export situation of counties in Jinan in 2016

Counties	Total retail sales of consumer goods(the hundred million)		Total import and export (the hundred million)		Actual to account foreign capital (the hundred million)	
	1-12 month	The cumulative year-on-year%	1-12 month	The cumulative year-on-year%	1-12 month	The cumulative year-on-year%
Total	3764.8	10.4	639.7	13.0	112.3	14.7
Lixia District	915.2	10.4	75.8	5.0	19.3	15.2
Shizhong District	484.6	10.2	31.7	17.7	15.4	12.7
Huaiyin District	479.2	10.4	33.3	15.4	3.5	57.5
Tianqiao District	408.7	10.5	26.8	8.5	2.3	19.1
Licheng District	519.6	9.5	106.4	-11.9	10.3	16.9
Changqing District	137.9	10.9	6.7	23.0	8.1	0.0
Pingyin County	96.3	11.2	45.2	5.1	1.5	42.1
Jiyang County	125.9	10.8	18.7	2.6	5.6	12.1
Shanghe County	88.8	11.2	7.3	9.1	3.1	12.7
Zhangqiu District	390.4	11.1	58.0	-2.1	14.9	10.5
High-tech zone	118.1	10.8	230.0	23.5	28.3	17.1

Table 3. The industrial condition of each county district in Jinan

Counties	fixed investments(the hundred million)	Scale above industrial added value	The main business income of the industries over the scale(the hundred million)	Total industrial profits above scale(the hundred million)

https://doi.org/10.1051/e3sconf/20183801035

	1-12 month	The cumulative year-on-year%	The cumulative year-on-year%	1-12 month	The cumulative year-on-year%	1-12 month	The cumulative year-on-year%
Total	3974.3	13.7	7.31	5722.6	7.31	368.6	12.23
Lixia District	248.9	14.3	5.84	289.3	-6.65	19.7	130.50
Shizhong District	217.2	12.3	16.37	430.5	12.84	14.1	43.82
Huaiyin District	268.7	4.9	4.35	164.8	1.54	14.0	28.86
Tianqiao District	169.4	-13.7	9.74	58.1	0.40	-4.43	-
Licheng District	594.8	16.6	0.75	591.0	3.14	32.38	-15.97
Changqing District	250.7	3.2	2.11	173.6	-7.54	-2.03	-
Pingyin County	248.3	7.6	11.56	357.5	8.72	40.7	14.61
Jiyang County	305.8	16.3	6.87	491.5	4.10	52.7	3.44
Shanghe County	127.0	10.9	6.00	178.8	-2.49	8.5	5.86
Zhangqiu District	676.2	20.0	11.50	1852.2	12.35	125.3	13.96
High-tech zone	684.8	18.0	8.31	1018.4	11.26	64.5	17.79

https://doi.org/10.1051/e3sconf/20183801035

5 Conclusion

The many factors influencing the development of dry port should be taken into consideration in the site selection. It is easy to have the benefit back effect and not get the best development benefit. In the long-term view of the development of dry port, a method of qualitative and quantitative combination should be adopted to locate the dry port. Considering the population, economy, transportation, industry and other conditions of various counties and districts in Jinan, it is suggested that dry port should be located in the district.

References

1. Roso.V., Philippe Leveque, Dry Port Concept for Seaport Inland Access with Inter-modal Solutions, [Master Thesis] for the Department of Transportation and Logistics, Sweden Chalmers University of Technology,2002.
2. Woxenius,J., Roso.V., The Dry Port Concept Connecting Seaports with Hinterland by Rail, the ICLSP Conference,China,Dalian,September,2004:22-26.
3. Rutten,B.C.M., The Design of Terminal Network for Intermodal Transport, Transport Logistics,1998,1(4):279-298.
4. Ballis.A.,Golias.J.,Comparative Evaluation of Existing and Innovative Rail Road Freight Transport Terminals, Transportation Research,2002,36:593-611.nagement: a state-ofthe-art literature review. International Journal of Management Reviews 2007;9(1):53–80.

Countermeasure Analysis on Internet Logistics

Shao Ying TENG[1,*],Xiao Jun LI[1] ,Zhi ZHAO[1] ,Peng Lei QIN[1] ,Ya Ya LU[1]

[1]Shandong Jiaotong University, Ji'nan, Shandong 250300, China

Abstract: The rapid development of Internet technology has caused a series of industrial revolution, which has provided strong impetus for economic development. The Internet + concept puts forward the deep integration between the Internet and traditional industries, which points out the direction for the development of various industries. For the logistics industry, "Internet +" provides a new way of transformation, and intelligent logistics, smart logistics and green logistics bring new business value to the logistics industry. This paper analyzes the current situation of the logistics industry in the context of Internet +, finds out the existing problems, and proposes corresponding solutions to provide the impetus for further development of the logistics industry.

1 Background

In other words, "the Internet" is "the traditional industries of the Internet," but this is not a simple combination of the two, but the use of information and communications technology and Internet platforms, so that the Internet and traditional industries are deeply integrated. Create a new development ecology. [1] the "Internet" of the logistics industry is not a mere addition of concepts, but a fusion of the two, and the Internet logistics industry, It is based on the application of Internet of things and Internet technology to realize the globalization, internationalization, information and intelligence of logistics, thus improving the level of productivity and market competitiveness. Optimize the allocation of social resources and promote the formation of a new social form of logistics development. Compared with traditional logistics, the "Internet" logistics can optimize logistics efficiency and become closer through the latest network technology. Connecting the integrated development model between logistics enterprises smoothly, reducing the waste of resources, achieving the goal of improving the efficiency of logistics management. Customers and suppliers are no longer isolated individuals. The online information platform formed by the "Internet" enables the third party logistics enterprises to realize the integration of resources, making vehicles and personnel, Physical and non-entity information implementation data, such as inventory, business processes, etc. Transformation and sharing [2]

At the third session of the 12 National people's Congress on the morning of March 5th 2015, Premier Li Keqiang put forward the "Internet" action plan for the first time in the government work report, and the "Internet" was included in the government work report. In the government work report, Li Keqiang put forward the "formulation of the" Internet "action plan. Promote the integration of mobile Internet, cloud computing, big data and Internet of things with modern manufacturing, and promote the healthy development of e-commerce, industrial Internet and Internet of Finance. It is proved that Internet logistics has become a major trend in the development of logistics industry in China in the future. Home will be the Internet logistics into the key planning, Internet logistics will take advantage of the popular entrepreneurship, innovation, rapid development. This is both an advantage and a challenge for the logistics industry. [3]

* Corresponding author:Email: 1300142749@qq.com

2 Industry Analysis

2.1 Low Efficiency of Logistics System and High Cost of Logistics Transportation

According to the data of the National Bureau of Statistics: in the first three quarters of 2017, the total amount of social logistics was 184.8 trillion, an increase of 6.9 percent over the same period last year. The total revenue of the logistics industry was 6.3 trillion yuan, up 13.3 percent from the same period last year. As shown in Figure 1 and 2 try, when the total amount of social logistics and transportation costs are so high,the production volume of transportation per 10,000 yuan gross product is 4,853 tons, while in the United States and Japan, the production volume of transportation per 10,000 yuan gross product is 870 tons and 700 tons, respectively. The gap is very big, and the operation efficiency of logistics system in our country is low, and the continuity of the link between each link is low. There is also unreasonable organization in the freight transportation link According to statistics, the logistics organization of road freight transport, the empty driving rate has remained at 50 for many years; railway, aviation, navigation, there are also a variety of problems.

 https://doi.org/10.1051/e3sconf/20183801036

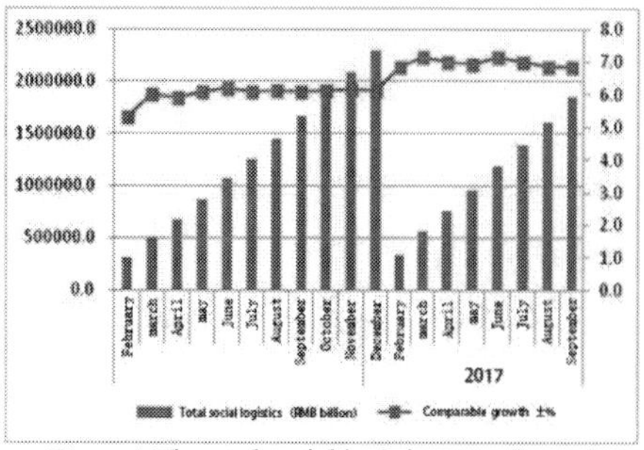

Figure1.The total social logistics growth trend

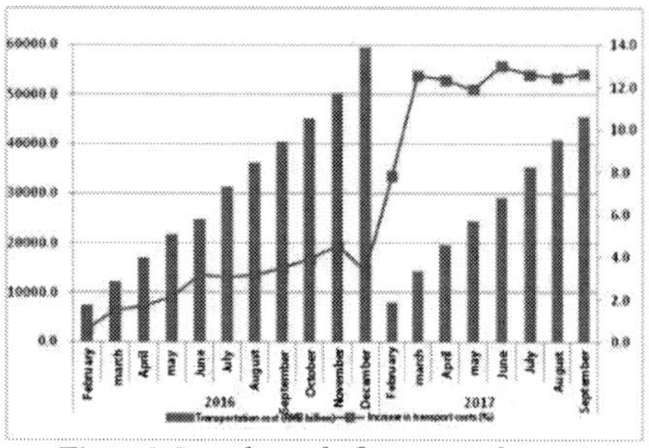

Figure2.Growth trend of transportation cost

2.2 Low Degree of Standardization in Logistics Enterprises

Although the state and related units have stipulated various indexes in the logistics process, there are still many problems. For example, there are two standards for domestic pallets and a variety of standards for containers. The standards for road and railway containers are also not uniform. There are different standards for transport equipment, handling equipment and storage equipment in different companies or regions. This is in transportation, storage, handling and handling. Packaging and circulation of processing and other aspects of the formation of different standards,logistics transportation also has a lot of inconvenience.

2.3 Insufficient Construction of Logistics Infrastructure

In 21th century, with the rapid development of traffic construction, although China made great achievements in railway and bridge construction, it was compared with other Asian countries, such as Europe, America and even India. There is still a certain gap. According to the land area and the density of transportation network, our country is only 1,344.48 km / km2, while the US is 6,869.3 km / 10000 km2, and Germany is 14680.4 km / 10,000 km2. India is 5403.9 km / 10 000 km2 [1]. Transportation network has great influence on logistics and transportation. From this point of view alone, the infrastructure construction of our country not only lags behind. There is also a significant gap between Dharma and developing countries such as India. In terms of building storage facilities, although the state has made great efforts to help and formulate relevant policies, it still has space efficiency and refrigeration. There are still many problems in the construction of warehouses such as air conditioners. As far as we know, among the warehouses owned by third-party logistics enterprises, bungalow warehouses account for 70%%, and there are cold storage and storage. Air conditioning and other functions of modern logistics warehouse accounts for a small proportion.

2.4 Low Degree of Automation in Logistics Enterprises

Since 2015, the state has vigorously developed automated logistics. According to the China Logistics Association, the penetration rate of warehousing automation in developed countries reached about 80% in 2017. In China, it is only about 20. And although domestic enterprises are vigorously developing automated warehousing and sorting, most small and medium-sized enterprises are still stagnant in automation development and have a lot of problems.The degree of automation also changes with the change of logistics price index (Figure 3).

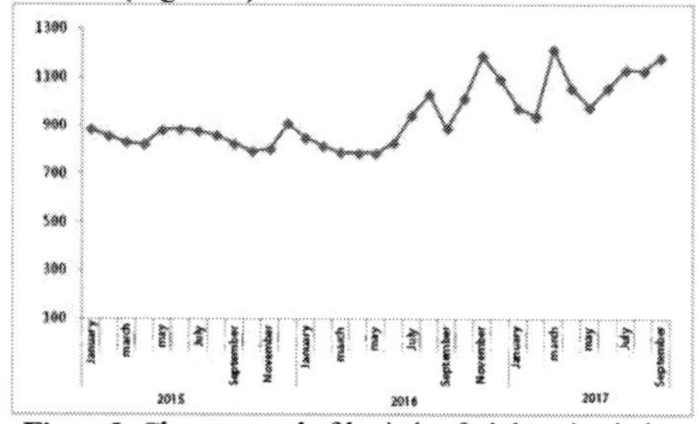

Figure3. Change trend of logistics freight price index

2.5 Logistics Talent Shortage

With the rapid development of logistics service industry in China, the demand for logistics talents is increasing day by day in the market, which requires a large amount of investment of logistics talents. According to a survey in 2017, due to the emergence of "people-oriented" social concept, Talent has become the first element for enterprises to seek survival and development. Our country now has a sharp shortage of middle and advanced logistics personnel, especially proficient in traditional transportation, warehousing and the effective use of modern information means. Familiar with logistics operators and experts. However, due to the recent emergence of logistics enterprises to vigorously train staff, as well as corporate system problems; and logistics professionals, most of the current high school students are not familiar with the situation of the enterprise, There are many problems in the formulation of enterprise goals and strategies, as well as in the decision of most programs, so high quality logistics personnel are very short.

2.6 Low Degree of Information in Logistics Enterprises

According to the information received, logistics enterprises plan, organize, lead, control and prevent all kinds of accidents from taking orders for transportation, warehousing, sorting and packaging, etc. However, the overall logistics planning capability of logistics enterprises in China is relatively low. The overall level of information application is relatively low, only a small number of enterprises have information systems that are limited to financial management, human resources management, customer relationship management, establishment of intranets, websites and web pages of enterprises, etc. Most businesses simply use computers as an operating tool to process simple documents and forms.

2.7 High Vehicle Vacancy Rate and Mismatch

There are a lot of problems in the logistics and transportation industry of our country, such as a large number of empty driving and unreasonable route, the empty driving rate of automobile is as high as 40%, and the mismatch of vehicle and cargo causes a lot of waste of resources. According to the recent survey data, The number of truck drivers in China has reached more than 50 million, carrying more than 75% goods in the country. Although there has been the emergence of software such as "Lala of goods", the transactions are still based on the introduction of family, fellow countrymen, and friends in the circle. Even transactions with logistics companies are based on acquaintance patterns. These phenomena have exacerbated the conflict that goods cannot be found and cars cannot be found. Drivers find the right goods on their way back It is very difficult, in order to avoid the return of empty driving, truck owners often through the intermediary to obtain source information, further increase costs, diluted profits.

3 Countermeasure Analysis

3.1 Strengthening the Leadership Function of the Government and Improving the Function of the Laws and Regulations of Logistics

With the rapid development of the Internet today, it is very necessary to actively establish a comprehensive coordination center centered on the unified planning of the government, and to plan the development of the logistics industry as a whole. We must take advantage of the beneficial side of the Internet for development. Perfecting the laws and regulations of the logistics industry and breaking down the barriers between different enterprises can not only maximize the utilization of resources among different enterprises, but also enable the enterprises to give full play to their respective duties and supervise each other. The role of standards in the industry.

3.2 Strengthen the Alliance of Small and Medium-Sized Logistics Enterprises, Perfect the Standardization and Automation of Enterprises

Small and medium-sized logistics enterprises are one of the biggest problems in the development of logistics process standardization and automation. What is necessary to develop Internet logistics is the standardization and automation of logistics process. It is more difficult to standardize and automate logistics. To strengthen the alliance between small and medium-sized logistics enterprises can improve the talents, funds, land area and other important obstacles among enterprises, which can make the resources of each enterprise to be maximized. For large enterprises, the problem of capital and industrial area is not a very important obstacle, which requires leaders to be conscious To gradually standardize and automate their logistics processes, such as the standardization of facilities and equipment in storage, handling and handling, and the automation of facilities and equipment in sorting and packaging, etc. Can effectively solve the small and medium-sized logistics enterprises in the development of Internet logistics process.

3.3 Improve the Infrastructure Construction in the Process of Logistics and Strengthen the Utilization Ratio of Space

According to the development trend of modern Internet plus logistics infrastructure, logistics process improvement is imperative. Perfect infrastructure can reduce space and time effective logistics process of each logistics enterprise. For example, the automation program of the enterprise will inevitably require to improve the infrastructure support, if infrastructure is still according to the original standard transport, packaging and other work, it is bound to the existence of middle Carlton, not only the high operation of logistics process, it is a waste of new resources, waste of the Internet. And every enterprise can use the Internet time allocation, to achieve win-win cooperation. In per capita land resources Nowadays, it is necessary to use the Internet to enhance the utilization of space in the process of logistics. The Internet can strengthen the cooperation among enterprises on a large scale and make the most effective use of space.

3.4 Actively Train Modern Logistics Talents

As a new concept of Internet logistics, the talents in the industry are scarce, and most of the talents in the enterprises are still using old knowledge and introducing new talents. Giving full play to the uniqueness of the new type of talents and making the new and old knowledge collide with the strongest force are the most important problems facing the enterprises today. The preservation of internal information in cooperation is also an aspect of identifying new talents.

3.5 Strengthening the Information Construction of Logistics

In the development of the Internet, the most important thing is the use of Internet information. The Internet is a shared platform. The information published and searched by each enterprise on the Internet will play an important role in the development of the enterprise. With the construction of logistics information, each logistics enterprise can not only see the information of other enterprises on the Internet. More can let their own information needs the enterprise and the talented person to see, each enterprise realizes the information sharing, the cooperation altogether wins the good desire.

3.6 Optimize the Transportation Route Reasonably and Improve the Transportation Efficiency

Using the Internet to understand information and rationally plan transportation routes can not only save time, but also save manpower, material resources, capital and other resources, reduce duplication of routes, and improve transportation efficiency. Optimize the transportation route, can make the enterprise internal resources reasonable use, efficient operation.

3.7 Strengthening the Construction of Network Platform

The network platform construction is the most important construction in the development of Internet, perfect network platform can not only be used as an advertising to promote their own business, more can let more enterprises and talents to see themselves in the Internet, increase the opportunities for cooperation. For a very strong momentum of development of enterprises in the Internet tide, perfect the network platform sharing platform can win more color and opportunities and better for themselves. The network platform is the foundation of a new era of logistics enterprise at the development of the network platform construction more perfect is the priority among priorities of each logistics enterprise.

4 Conclusion

With the continuous development of the "Internet" logistics model, the logistics industry will welcome a series of opportunities and challenges. Logistics industry should seize the opportunity to deal with the new development opportunities and problems brought by this model to the logistics industry with the correct mentality. This paper analyzes its own problems, thinks deeply according to the present situation, promotes its own strength, conforms to the trend of development of the times, and actively adapts to the new market demand, so that the industry is in an invincible position in the development trend of the times.

References

1. Interpretation: what is the "Internet +" in the report of li keqiang? HTTP: / / pa. Caijing. Com. Cn / 20150305/3832729. SHTML, 2015-03-05.
2. Wang xianqing, li zhengkun, liu fangdong, etc. Internet + logistics [M]. Beijing: people's post and telecommunications press, 2015.
3. Wang qifeng. Research on the value of logistics service in the "Internet plus" environment [J]. Logistics technology, 2016(11).
4. Xu chunyu. Analysis of logistics problems and solutions in small and medium-sized enterprises in China [J]. China science and technology expo, 2013, (12) : 23-27.
5. Zhang lifeng. Information research on logistics industry [M]. Dongying: petroleum university press, 2013:110-119.

Application of reverse osmosis in purifying drinking water

Lei Jiang, *Yue* Tu, *Xiangmin* Li, *Haixiang* Li*

College of Environmental Science and Engineering, Guilin University of Technology, Guilin 541006, China

Abstract. In view of the shortage of water resources and the pollution of water environment, the development of efficient water purification technology is one of the effective ways to ensure the safety of drinking water. The technology of reverse osmosis has attracted much attention in the application of drinking water security for its advantages of no phase change and simple equipment. This paper briefly introduces the research progress at home and abroad, basic principles, process flow diagram, advantages and disadvantages and development direction of the technology.

1 Introduction

The status of water resources is closely related to the improvement of economic and social development and people's living standard. Therefore, the utilization of water resources, rational use, effective protection and safety supply of water resources is of great significance for promoting economic and social development. At present, the outstanding problem of water supply industry in China is water quality. On the one hand, the water source is generally polluted, on the other hand, the standard of water quality has been continuously improved. The water quality of drinking water has become an important factor restricting the further development of the economy and the influence of social stability[1].

At the same time, drinking water is the basic demand for human survival. The safety of drinking water is closely related to the health of the people and the national security, so it has always been highly valued by our country. The Action Plan for Prevention and Treatment of Water Pollution, namely "water ten item", clearly pointed out that by 2020, the quality of water environment needs to be improved step by step, and the level of drinking water safety is continuously improving in 2015. However, the current situation of drinking water safety in China is still grim. Drinking water safety incidents often occur, such as groundwater pollution incidents in Xinhe County in 2015, and sudden water pollution incidents in Xinyu County, Jiangxi Province in 2016. The safety of drinking water is still one of the focus of public concern[2].

Compared with the conventional water purification process of increasing ozone and activated carbon adsorption equipment on the traditional water purification equipment, reverse osmosis water purification technology develop rapidly in recent years, the pressure driven water molecules flow through the dense reverse osmosis membrane to separate the solute and the solvent. This technology can improve efficiency, so it has made great strides, and at the same time, the development of reverse osmosis water purification technology has matured[3]. In theory, the reverse osmosis technology can not only remove the salts and other substances in the ion state, can also remove suspended solids , organic matter, colloids, bacteria and viruses[4], only water can through reverse osmosis membrane, so as to ensure the water from being polluted. Therefore, the drinking water purification technology which can maximum possibly remove the disinfection by-products and trace organic pollutants to be recognized by more and more people[5]. And it has become an effective way to ensure the safety of drinking water, and has been widely applied to improve drinking water quality, expand drinking water sources and ensure emergency water supply[2].

2 Domestic and foreign research

Research data show that scholars and researchers have detected 2221 kinds of organic matter in water all over the world, and there are 765 kinds of organic matter in drinking water, of which 107 species are carcinogenic, mutagenic, and distorted substances. Chlorination and other disinfection by-products were found after adding chlorine to the drinking water in the United States in 1974. Chloroform is the most likely to be carcinogenic in the water. The reverse osmosis membrane combined equipment effectively removes the residual chlorine and organic pollutants in tap water, and it can solve the key problem of the safety of drinking water[6].

The research and application of reverse osmosis technology started earlier in the US and Japan[7]. In the early 1980s, the United States overcome the defect of cellulose materials and developed a compound polyamide membrane with high water flux and high salt interception rate, which made reverse osmosis technology widely used in the industrial field[8]. Meanwhile, its self-designed reverse osmosis membrane emergency water supply desalination and purification equipment is easy to transport and install. It can be adapted to such as high

*Corresponding author's e-mail address: lihaixiang0627@163.com

turbidity water, seawater, nuclear polluted water, biochemical pollution water and so on[9].

In recent years, the reverse osmosis method for purifying drinking water has also been widely used in Russia. The small and medium-sized water reverse osmosis device designed by Russia can realize the automation of work. And it can be applied to the daily living conditions[4]. At the same time, in Russia, the reverse osmosis device is also widely used in the purification of well water and removing the fluoride, nitrate and nitrite ions which are the most difficult to remove[10].

G.R.Lashkaripour and others found that the water quality of desalination of Zahedan brackish groundwater with the salt content of 2697mg/L by reverse osmosis device can reach the standard of WHO, and the energy consumption is low. In Gaza area, the first brackish water reverse osmosis desalination plant was built in 1993. Now there are many equipments. So reverse osmosis desalination technology has become a mature technology in the local area.

The research of reverse osmosis membrane in China began in 1966 and it has been widely used in recent years. Reverse osmosis technology was initially applied to seawater desalination, and gradually expanded to brackish water desalination, food processing, medical and health, beverage purification, ultra pure water preparation and so on and resulting in high economic benefits[7]. In 2000, supported by the key technology research project of the Ministry of science and technology, "the production of 1000 tons of reverse osmosis seawater desalination system and engineering technology development", the 1000 t/d reverse osmosis seawater desalination demonstration project was built in Long Island, Shandong Province and Shengsi, Zhejiang Province[8].

3 Application progress

For a long time, reverse osmosis technology was considered less economical than other traditional methods of purification-coagulation and sedimentation, aeration, adsorption, etc. Most experts think this is a special and expensive method, and only limited to medicine, electronics, food and industry, etc. That is because of the high energy consumption (7.5~10kW/m^3), high investment of membrane components and pump equipment, and complex pretreatment system.

In recent years, because of its compact , simple and better effluent quality, the application of reverse osmosis has been considered as a good choice for small and medium-sized water treatment stations (50-1000m^3/d) in the selection of purifying surface water with toxic contaminant and underground hard water with higher iron content,.

Reverse osmosis technology can effectively remove water pollutants.At present, this technology can guarantee drinking water safety from many aspects such as water quality, water volume and emergency water supply[2].

3.1 Improve drinking water quality

In general, we believe that the membrane pore size of 1nm or less for reverse osmosis technology, it can effectively remove dissolved salts, colloids, microorganisms and organic compounds in water so as to ensure the safety of drinking water.

3.1.1 Removal of salt and hardness.

Salinity is an important component of healthy drinking water. But salty water, such as sea water in coastal areas and brackish water in inland areas, is not suitable for drinking. In addition, the high level of divalent ion concentration will cause water hardness. Hard water can easily cause calculi or other diseases. The toxic and harmful ions, such as fluorine and arsenic, can also lead to pathological changes and damage of human viscera. Reverse osmosis technology can effectively remove salt and hardness. Researches show that the removal rates of fluoride and arsenic in most of the commercial reverse osmosis membranes in high-fluoride water and high-arsenic water can reach more than 95%[11,12].

3.1.2 Removal of organics.

The concentration of organics in water is in a low level, but its composition is complex and it has potential harm to human health. Reverse osmosis technology helps to remove water traces of organic matter, thus ensuring the safety of drinking water. Studies have shown that reverse osmosis technology can successfully remove small molecule organic, such as emulsifiers[13], alkylphenol polyoxyethylene acid[14], etc. In addition, reverse osmosis technology is also quite complete for the removal of non-volatile dissolved organic carbon[15], total organic carbon[16] and assimilable organic carbon[17].

3.1.3 Removal of micro-organisms.

Micro-organisms are the common and widespread over-standard items in rural drinking water and secondary water supply in urban areas, mainly including protozoa and bacteria. Reverse osmosis membrane on a variety of microorganisms has a very high removal rate, up to 99.9% or more[18].

3.2 Increase the source of drinking water

3.2.1 Desalination of sea water.

The Earth's seawater reserves are about 1.35 billion km^3, providing a stable and reliable source of water for seawater desalination. Therefore, desalination is regarded as one of the important strategies to solve the global water shortage. Our country is a large marine country with a coastline of 18,000 km and provides convenient geographical conditions for desalination. In 2011, the largest seawater desalination project, using RO method, which independently designed and contracted by domestic enterprises was put into operation in Caofeidian, Hebei Province, marking the new stage of technological capability of membrane desalination project in China[19]. As of the end of 2015, the scale of nationwide reverse osmosis desalination water production reached 654, 535m^3/d[20], greatly alleviated the problem of insufficient drinking water in the coastal areas of our country. At present, the membrane desalination process with reverse osmosis technology as the core has made important progress. Among the installed capacity of seawater desalination, reverse osmosis accounts for 64.88%, accounting for the highest proportion[20].

3.2.2 Desalination of brackish water.

China has 1.6 million km^2 of land is brackish water distribution area, accounting for 16.7% of the country's land area, with a total of about 20 billion m^3 per year. However, there are only 1.066 billion m^3 of brackish water could be used for freshwater production each year in our country, accounting for only 4.99% of the total amount of brackish water that can be developed[21]. Water desalination of brackish water is an important supplement to freshwater resources and an effective way to alleviate the water crisis in inland areas of China. In view of the excellent desalination effect of reverse osmosis technology, it has gradually become the preferred technology for desalination of brackish water. In the brackish water treatment process design, reverse osmosis and other technologies can be combined flexibly, which is suitable for different salinities brackish water.

3.2.3 Reuse of rainwater.

The average annual total amount of rainwater in China exceeds 6 trillion m^3, most of the rainfall is lost and rainwater utilization is less than 1%. Rainwater, as an abundant freshwater resource, has simpler treatment process. It is of great practical significance to improve the water supply by rational exploitation and utilization. The rain water is collected by natural biological treatment at first and then treated by the membrane system, finally chlorinated and reused[22]. Among them, the reverse osmosis technology is mainly used for desalination, in order to keep the water with low salt. At the same time, it can be used as the depth of rain treatment process, for protecting the safety of rainwater utilization. In addition, reverse osmosis technology can ensure that the treated rainwater reaches the standard of living water[23].

3.2.4 Recycling of wastewater.

In recent years, the discharge of waste water which has reached 3.8 billion m^3 in 2015 in our country has continued to increase. Some rural non-point source wastewater and industrial wastewater were directly discharged without any treatment, which not only seriously polluted the environment but also caused waste of resources. It will open up a new stable source of water if the wastewater can be treated in depth. At present, due to the variety of waste water and complicated composition, the single membrane technology can not completely solve the problems of various waste water treatment. However, reverse osmosis technology has become the core technology of the advanced treatment technology of membrane regeneration water. There have corresponding research and engineering examples in the treatment of electroplating wastewater, printing[24] and dyeing[25] wastewater and papermaking wastewater, etc. It's both to reduce the discharge of a large number of waste water, and save a lot of groundwater, tap water, indirectly supporting the safety of drinking water.

3.2.5 Treatment of unconventional polluted water.

Unconventional pollutants are directly related to drinking water safety and public health and require more rigorous treatment. Reverse osmosis is considered as one of the most effective treatments[26]. An experimental study on the high natural radioactive content in drinking water compared the effects of ultrafiltration, reverse osmosis and electrodialysis on natural radioactivity[27]. The results showed that reverse osmosis has the best removal efficiency and it can effectively remove 95 % of total alpha activity and 93% of total beta activity.

3.3 Safeguard emergency water supply

The water supply system would be easily threatened by water pollution, pipeline rupture and sewage leakage when heavy natural disasters such as heavy rainfall happened, resulting in the excessive concentration of pollutants such as organic substances, heavy metals and microorganisms in water, seriously affecting the safety of drinking water. In recent years, sudden pollution incidents have occurred in urban water sources in China, so establishing and improving emergency water supply systems has important practical significance for ensuring drinking water safety and maintaining social stability. However, the emergency water supply is confronted with the requirements of complex water pollution components, urgent processing time and convenient equipment operation. Conventional water purification technology cannot meet the needs of emergency water supply. The application of reverse osmosis technology in emergency water supply has drawn more and more attention.

In view of the fact that domestic drinking water near residents in a nuclear power plant accident may be contaminated by radionuclides, the results[28] show that the reverse osmosis process puts up good stability in removing radionuclides from simulated wastewater.

For the problem of high concentrations of organic matter contaminated drinking water caused by sudden accidents, the results show[29] that the removal rate of trace benzene series and trichloroethylene by the osmotic membrane process is above 95%, and the removal rate of higher concentrations of organics is achieved more than 99%, effluent organic matter concentration is far lower than the drinking water quality standards limit, which can

https://doi.org/10.1051/e3sconf/20183801037

effectively ensure the safety of drinking water in sudden accidents.

4 Basic principles

Reverse osmosis technology is a membrane separation technology for pure water that makes use of reverse osmosis membrane to selectively pass through the solvent in the solution (usually water)[5]. It takes pressure as the driving force to overcome the osmotic pressure of water[8]. The process is a permeation process that is contrary to the natural infiltration[3].

In order to produce reverse osmosis pressure, a water pump is required to exert pressure on a saline solution or waste water[8]. When the pressure is equal to the natural osmotic pressure of the solution, the flow of the solvent will not occur. If the pressure is less than the natural osmotic pressure of the solution, the solvent flows from the dilute solution to the concentrated solution, and if not, the solvent will flow through the reverse osmosis membrane, so as to form dilute solution on the opposite side, and form a more concentrated solution on the side of pressure. All these will achieve solute separation and effectively remove salt, colloid, microorganism, heat source, organic matter and so on. In other words, the principle of desalination by using reverse osmosis membrane is to apply more pressure than natural osmosis pressure in water with salt (such as raw water). This will make the permeation go in the opposite direction, and press the water molecules in the raw water to the other side of the membrane, so as to achieve the purpose of removing the salt in the water. Reverse osmosis is also one of the most advanced membrane separation technologies in the world[3].

The principles are detailed in Figure 1[30].

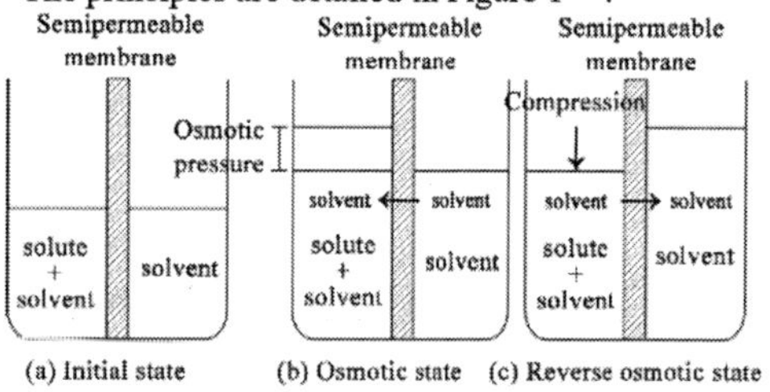

Figure 1. Principle of Reverse Osmosis

The process of reverse osmosis separation can remove the organic matter with molecular weight of more than 150, and the rate of desalination is over 95% one time. At the same time, as a special "filtration" process, microorganisms such as bacteria and viruses in water can be removed effectively to achieve aseptic water[5].

5 Process flow

The process flow is as follows:
Tap water→ raw water tank→ raw water pump→ mechanical filter→ activated carbon filter→ softener→ intermediate water tank→ booster pump→ 5μ security filter→ RO high-pressure pump → reverse osmosis system→ pure water tank→ pure water pump→ ozone generator→ 0.2μ fine filter→ outlet[5].

6 Advantages and disadvantages

6.1. Advantages

Reverse osmosis membrane technology is the most advanced and most energy-saving separation technology. It is a physical method. It has many advantages[31]:

- The operation of reverse osmosis only relies on pressure as a driving force. It has the lowest energy consumption compared with other traditional physical processing means, and it saves energy.
- Reverse osmosis can be carried out at room temperature without phase change.
- Reverse osmosis does not use chemical treatment reagents and it does not discharge chemical waste liquid. So it hardly pollute the environment.
- The space of operation equipment is small and the operation is simple.
- The water quality is excellent and the efficiency is high.

6.2. Disadvantages

In the process of continuous application, the shortcomings of reverse osmosis are increasingly revealed. For example, the operation and management is not strict. When the system is running, the pressure must be in the pressure range that the membrane can bear. In addition, the film materials and models can not be selected for the water quality of the original water[33].

On the other hand, the application of reverse osmosis device to improve well water and tap water has many advantages, but there is a serious contradiction between the purified water quality and the guidelines established by health institutions. Because the reverse osmosis device removes essential trace elements of human body at the same time[10].

7 Prospect of application

The application of reverse osmosis membrane technology in drinking water quality improvement plays an important role in solving the problems of imperfect water resources management system, poor drinking water quality, ecological environment investigation, ecological environment change, detailed management of water resources science and improvement of ecological environment. The technology of reverse osmosis membrane separation indicates the direction for the improvement of drinking water quality. Through a series of successful application in the water source, the water quality improvement project of drinking water has been pushed to a new height. In recent years, the technology has been implemented and operated frequently. So it is very mature now.

With the quickening of the pace of modernization, on the one hand, the water resources are lacking and the water sources with lower quality are constantly used. On the other hand, the pressure imposed by the government's environmental protection department and the public's demand for high quality drinking water ask the process to update and improve the degree of processing, this has brought great market potential and space for the membrane separation technology, especially the reverse osmosis membrane technology[8]. At the same time, from the development of membrane processing technology, it has a wide application prospect in environmental engineering, especially in the field of wastewater treatment and water reuse[32].

8 Direction of development

Nowadays, the world's water treatment industry is developing towards the dual goal of wastewater reuse and environmental protection. Because of its excellent separation performance, reverse osmosis membrane technology will be widely applied. In the future, the direction of its development is mainly focused on the following:

- Reverse osmosis membrane components of low energy consumption, anti pollution, high temperature resistance, high pressure and special separation.
- A polymer film which can remove small chlorinated organic molecules.
- Inorganic reverse osmosis membrane that can separate hydrocarbon mixtures[34].
- New membrane materials with acid alkali resistance, oxidation resistance and high water permeability.
- Combination of reverse osmosis membrane components and components of ultrafiltration, microfiltration, nanofiltration and EDI[8].

9 Conclusion

The application of reverse osmosis membrane separation technology in drinking water can effectively alleviate partial pressure caused by the pollution of water environment and the shortage of water resources. It also ensures the safety of drinking water and improves the quality of drinking water by its advantages. At present, in the process of applying the technology, some problems have also been encountered. However, the technology will become more and more mature with the promotion of technological innovation and the development of new components in the future. At the same time, it will applied to other fields gradually. This is of great significance to the sustainable utilization of water resources and the development of society.

References

1. Wenjie H. Technology to ensure the safety of drinking water Tianjin Construction *Science and Technology* J, **15**, 27-8(2005)
2. Li'an H, Haiyang Z, Xin G and Ming L. Application of reverse osmosis technology in ensuring drinking water safety in China *Water & Wastewater Engineering* J, **43**, 135-41(2017)
3. Shuai W. Application of reverse osmosis technology in purifying drinking water *Science & Technology Information* J, **27**, 781(2010)
4. Shoukai Z. RO for drinking water in Russia *Water & Wastewater Engineering* J **25**, 63-5+4(1999)
5. Weiyang W and Guichi L. Application of reverse osmosis technology in the manufacture of purified water *The Science Education Article Collects* J, **4**, 276(2009)
6. Mengchun H, Yongchun Z, Xiaoyan T and Haifeng S. Demonstration of deep purification technology for drinking water in reverse osmosis membrane in Yixing Dapu Water Treatment Plant *2007 Annual meeting of the Chinese Academy of Environmental Sciences*, 451-5(2007)
7. Shuxu W and Lirong W. Research status of reverse osmosis technology *Gansu Science and Technology* J, **21**, 172+94(2005)
8. Jun Z, Shaoming F, Hongzhong Z and Shaowei C. The progress of research on reverse osmosis membranes in water treatment *Journal of Filtration & Separation* J, **16**, 12-5(2006)
9. Shichang W and Shangbao Y. Emergency water supply technology for disaster relief Information of China *Construction.Water-Industry Market* J, **8**, 68-70(2011)
10. Application of reverse osmosis technology in purifying drinking water *China Science and Technology Information* J **15**, 39(2003)
11. Shen J and Schaefer A. Removal of fluoride and uranium by nanofiltration and reverse osmosis: a review *Chemosphere* J **117**, 679-91(2014)
12. Abejon A, Garea A and Irabien A. Arsenic removal from drinking water by reverse osmosis: Minimization of costs and energy consumption *Separation and Purification Technology* J, **114**, 46-53(2015)
13. Baudequin C, Mai Z, Rakib M, Deguerry I, Severac R, Pabon M, et al. Removal of fluorinated surfactants by reverse osmosis - Role of surfactants in membrane fouling *Journal of Membrane Science* J, **458**, 111-9(2014)
14. Fujita Y, Ding WH and Reinhard M. Identification of wastewater dissolved organic carbon characteristics in reclaimed wastewater and recharged groundwater *Water Environment Research* J, **68**,867-76(1996)
15. Juang L-C, Tseng D-H, Lin H-Y, Lee C-K and Liang T-M. Treatment of chemical mechanical polishing wastewater for water reuse by ultrafiltration and reverse osmosis separation *Environmental Engineering Science* J, **25**, 1091-8(2008)
16. Meckes MC, Haught RC, Kelty K, Blannon JC and Cmehil D. Impact on water distribution system biofilm densities from reverse osmosis membrane treatment of supply water *Journal of Environmental Engineering and Science* J, **6**, 449-54(2007)

17. Escobar IC, Hong SK and Randall AA. Removal of assimilable organic carbon and biodegradable dissolved organic carbon by reverse osmosis and nanofiltration membranes *Journal of Membrane Science* J **175** 1-17(2000)
18. Mi BX, Eaton CL, Kim JH, Colvin CK, Lozier JC and Marinas BJ. Removal of biological and non-biological viral surrogates by spiral-wound reverse osmosis membrane elements with intact and compromised integrity *Water research* J, **38**, 3821-32(2004)
19. Zhengfei F. The National Oceanic Administration issued the national report on the use of sea water in 2015,(2016)
20. Kai W and Ming Z. Problems of public engineering should be paid attention to in the project of large membrane seawater desalination: a case of Caofeidian 50 thousand / day desalination project *Silicon Valley* J, **6**, 96-7(2013)
21. Bingxin X, Ruyi L, Dongbo W, Yan Z and Yune C. Utilization status and research progress of brackish water *Journal of Anhui Agricultural Sciences* J, **41**, 13914-6+81(2013)
22. Kaijun W and Min Z. The comprehensive utilization of water resources in Sydney Olympic Games *Urban and Rural Development* J, **13**, 75-6(2004)
23. Dayong M. Improving domestic urban rainwater utilization rate by referring to foreign experience *Shanxi Architecture* J, **38**, 141-3(2012)
24. Weilei Z, Lixin P, Fengzhi Y, Shengping Z and Gang W. Application of RO technology to dyeing wastewater reuse *Industrial Water Treatment* J, **32**, 87-9(2012)
25. Xu Z and Junwei Z. Case study on the MBR+reverse osmosis process for treating reclaimed water in industrial parks *Industrial Water Treatment* J, **36**, 98-100(2016)
26. Jingjing L, Jian D and Weijin G Research on unconventional polluted water source *Science & Technology Information* J, **30**, 486-7(2013)
27. Montana M, Camacho A, Serrano I, Devesa R, Matia L and Valles I. Removal of radionuclides in drinking water by membrane treatment using ultrafiltration, reverse osmosis and electrodialysis reversal *Journal of Environmental Radioactivity* J, **125**, 86-92(2013)
28. Xingxing Z and Linping K. The treatment of emergency drinking water under severe nuclear power plant accident condition by reverse osmosis membrane *Technology of Water Treatment* J, **41**, 81-49(2015)
29. Zhong H, Zhiliang W, Shaogui Y and Cheng S. Safety Guarantee of Drinking Water on Basis of NF and RO Technology *Environmental Science and Technology* J, **29**, 29-31+7(2016)
30. Taotao N, Zhenyu L, Jiangen W and Xiao Y. Application of Reverse Osmosis in Wastewater Treatment *Guangzhou Environmental* Sciences J, **23**, 11-3+7(2008)
31. Introduction and Application Analysis of Reverse Osmosis Technology *Liaoning Chemical Industry* J, **43**, 308-9+11(2014)
32. Juan C, Jianping X, Ruiying Z and Junjun S. Application of Reverse osmosis membrane separation and processing technology in the improvement of drinking water quality in Huating county *Gansu Science and Technology* J, **32**, 96-9(2016)
33. Wei S. Experimental Study of Water Treatment Technology Using Reverse Osmosis *Dalian University of Technology*,(2006)
34. Xiaojie H and Baowei Y. The basic principle and development trend of reverse osmosis *Friend of Chemical Industry* J, **26**, 16-7(2007)

The Researches on Reasonable Well Spacing of Gas Wells in Deep and low Permeability Gas Reservoirs

Yu Bei Bei[1,*], *Li* Hui[1], and *Li* Dong Lin[2]

[1]Exploration and Development Research Institute of Dagang Oilfield, Binhai New area,Tianjin300280,China
[2]Oil & Gas Cooperation And Development Company, BHDC, Binhai New area,Tianjin300280,China

Abstract. This Gs64 gas reservoir is a condensate gas reservoir which is relatively integrated with low porosity and low permeability found in Dagang Oilfield in recent years. The condensate content is as high as 610g/m3. At present, there are few reports about the well spacing of similar gas reservoirs at home and abroad. Therefore, determining the reasonable well spacing of the gas reservoir is important for ensuring the optimal development effect and economic benefit of the gas field development. This paper discusses the reasonable well spacing of the deep and low permeability gas reservoir from the aspects of percolation mechanics, gas reservoir engineering and numerical simulation. considering there exist the start-up pressure gradient in percolation process of low permeability gas reservoir, this paper combined with productivity equation under starting pressure gradient, established the formula of gas well spacing with the formation pressure and start-up pressure gradient. The calculation formula of starting pressure gradient and well spacing of gas wells. Adopting various methods to calculate values of gas reservoir spacing are close to well testing' radius, so the calculation method is reliable, which is very important for the determination of reasonable well spacing in low permeability gas reservoirs.

1 Introduction

The determination of well spacing in low permeability gas reservoirs is the key parameter for its economic and reasonable development. Due to the small pore throat, poor pore connectivity, there are low permeability and high water saturation in low permeability gas reservoirs, Gas percolation is easy to represent unique percolation characteristics which is different from that of medium and high permeability gas reservoirs [1]. In low permeability reservoirs, gas percolation far away from bottom hole area appears low speed non-Darcy percolation, and how long distance away from bottom hole the gas can overcome starting pressure gradient to flow, that is the gas well spacing which we need to calculate. Many experts and scholars have their own opinions on the calculation method of well spacing. Li Shuang [2], Wang Xiuzhi [3] and others did not consider the influence of starting pressure gradient, and adopted economic limit control reserve method, single well area method and numerical simulation method to discuss the reasonable well spacing of low permeability gas reservoir. Although Guoqi [4] considered the influence of starting pressure gradient, deduced the formula for calculating reasonable well spacing of low permeability gas reservoir, but the calculation parameters are too many and the calculation is complicated. In this paper, starting from low speed non-Darcy seepage flow and combining with single well productivity equation, the control radius of single well considering starting pressure gradient is derived, combined with various methods of gas reservoir engineering and numerical simulation, The reasonable well spacing of low permeability gas reservoir is determined synthetically, the calculation is simple and the result is reliable.

2 Geological characteristics

The Gs64 block is located in the southern of Tangjiahe Oilfield, the descending plate of the gangdong fault. The main layers is Ban2 and Ban3 which are Low porosity and low permeability sandstone reservoir, reservoir depth is between 4200m and 4400m, the porosity is range from 11.5% to17.2%, the permeability is range from $3.7\times10^{-3}\mu\ m^2$ to $10.27\times10^{-3}\mu\ m^2$. The reservoir is in deep lake environment, with wide distribution, large thickness, deep burial, poor physical properties, low production degree and great difficulty in development.

3 Reasonable well spacing determination method

3.1 Seepage mechanics method

Domestic scholars' researches show that [5-7] the intact percolation of low permeability reservoir should be as shown in Figure 1: section I represents the low speed non-Darcy segment affected by the starting pressure gradient and threshold pressure effect, II represents the

* Yu Beibei: yu_beibei@foxmail.com

Darcy percolation stage; III means the high-speed non-Darcy stage under high yield conditions.

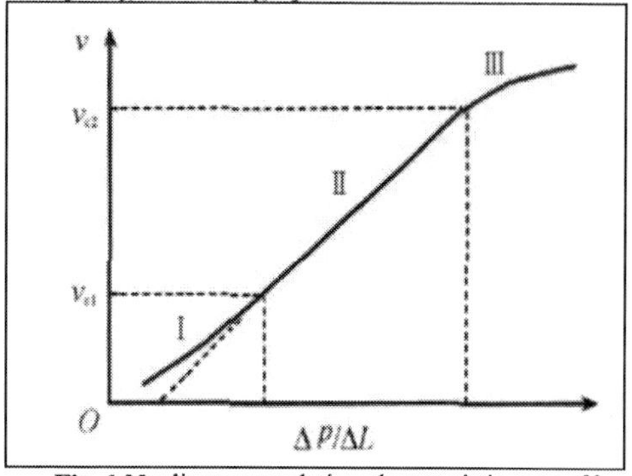

Fig. 1 Nonlinear percolation characteristic map of low permeability gas reservoir

When there is a starting pressure gradient, the linear relation with the origin between v and $\Delta P/\Delta L$ is not exist. Considering the starting pressure gradient, the equation which considered the starting pressure gradient in low velocity non-Darcy gas seepage flow is as follows:

$$\begin{cases} v = 0 & (\text{grad}p \le \lambda_B) \\ v = \frac{K}{\mu}(\text{grad}p - \lambda) & (\text{grad}p > \lambda_B) \end{cases} \quad (1)$$

The meaning represented by symbols in a formula:

v: velocity of gas movement in reservoir, m/s;

μ: gas viscosity, mPa.s;

k: absolute permeability, $10^{-3}\mu m^2$;

gradp: pressure gradient, MPa/m;

λ: starting pressure gradient, MPa/m.

The horizontal radial flow velocity of gas in the reservoir at r distance from the bottom of the well can be expressed by the underground flow rate:

$$v = \frac{Q}{2\pi rh} \quad (2)$$

The meaning represented by symbols in a formula:

Q: Underground flow in reservoir, m^3/s;

h: Reservoir thickness, m.

By using the equation of state, the velocity of gas flow can be expressed as the relation of surface production.

$$v = \frac{1}{2\pi rh}\frac{P_{sc}ZT}{Z_{sc}T_{sc}P_e}Q_{sc} \quad (3)$$

The meaning represented by symbols in a formula:

Q_{sc}: single well surface production, m^3/s;

T_{sc}: temperature in the standard state, K;

Z_{sc}: deviation factor in the standard state;

T: formation temperature, K;

Z: deviation factor;

P_{sc}: pressure in the standard state, MPa;

P_e: current formation pressure, MPa.

According to formula (1), when the pressure gradient of the resultant force of the gas at half of the well distance r is greater than the starting pressure gradient, the flow velocity of the gas is greater than 0, then the gas can flow. The formula of gas flow conditions at r can be obtained by combining the three equations (1) (2) and (3).

$$\frac{\mu}{K}\frac{1}{2\pi rh}\frac{P_{sc}ZT}{Z_{sc}T_{sc}P_e}Q_{sc} \ge 0 \quad (4)$$

Yihuaijian etc. [8] considering starting pressure gradient got the production equation formula.

$$Q_{sc} = \frac{\pi KhZ_{sc}T_{sc}[P_e^2 - P_{wf}^2 - 2\lambda P_e(r - r_w)]}{\mu ZTP_{sc}ln\frac{r}{r_w}} \quad (5)$$

The meaning represented by symbols in a formula:

P_{wf}: flowing pressure of bottom hole, MPa;

The control radius of a gas well can be obtained by taking formula (5) into formula(4)

$$r \le \frac{P^2{}_e - P^2{}_{wf}}{2P_e\lambda} + r_w \quad (6)$$

According to this formula, the control radius of gas well can be calculated by only knowing the formation pressure of gas reservoir, the flowing pressure of bottom hole and the starting pressure gradient of reservoir, and then the well spacing of gas reservoir can be obtained.

The original formation pressure of the Gs64 gas reservoir is 50.37MPa. the reasonable production pressure differential is 13MPa. the wellbore radius is 0.1m. the starting pressure gradient is 0.04 MPa/ m. taking these parameters into the formula(6) to calculate the control radius of single well is 283m, so the well spacing is 566m.

3.2 Gas reservoir engineering method

Only one well gs64 has been produced in GS64 gas reservoir, and the production is long and continuous. Therefore, the production dates of this well can be used to the decline curve analysis and the single well control reserves estimating, thus the control radius is obtained. Because the analytical method of decline curve is an empirical observation result of the well's production history, so the decline curve analysis is applicable in any case, mainly refers to the case of a single fluid or multiple fluids.

Using Blasingame Type curve Analysis method to fit the production data, figure 2, we calculate the well control reserves is $0.64\times10^8m^3$, the control area is $0.26\times km^2$, the radius is 290m, and the well spacing is 580m.

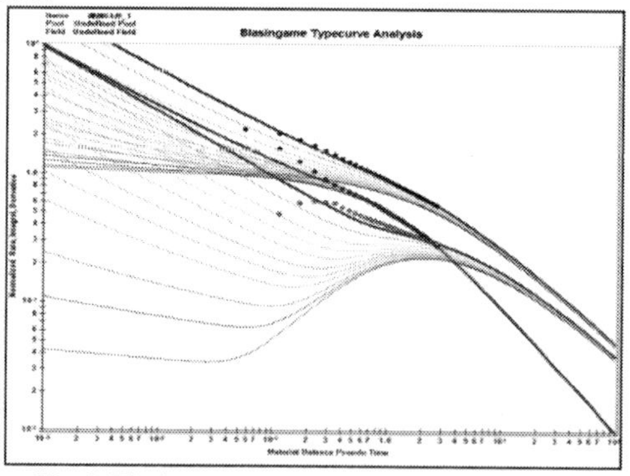

Fig. 2 Blasingame type curve analysis of well GS64

4 Comparative study on different well spacing schemes

According to the data of well spacing calculated by the former two methods, we designed three sets of well spacing schemes of 400-500m, 500-600m and 600-800m to predict and compare the project indexes.

4.1. Geological modeling and historical fitting

The geological model of gs64 fault block is established by using Petrol modeling software. The three dimensional geological modeling layer is Ban2, Ban3 oil group. The simulation flat grid step size is 25m × 25m, vertical grid step size is 0.5m, total grid number are 173×141×225= 5488425.

After the reserve fitting of the model, according to the actual data, we conducting the historical fitting about the pressure of the whole area and the single well production index, the permeability, pressure-sensitive parameters and permeability data of the gas reservoir are mainly adjusted. At the same time, considering the damage of drilling and operation to gas well productivity, the well point permeability and skin coefficient of gas well are corrected. The fitting rate of production index is more than 95%, which meets the standard of simulation and the model meets the requirement of project prediction.

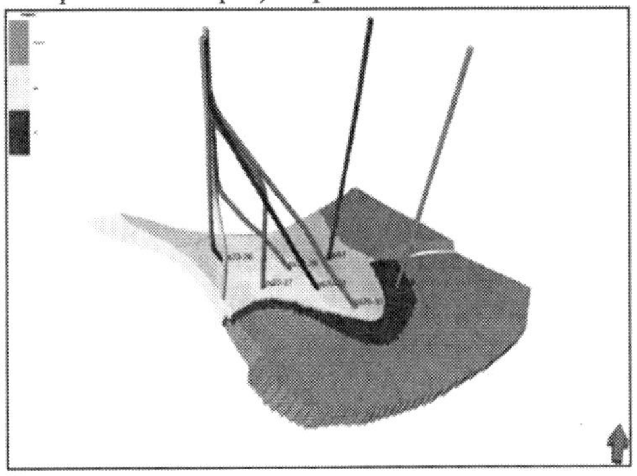

Fig. 3 Hydrocarbon model of GS64 reservoir

4.2 Comparison of different well spacing schemes

From the prediction results and curves of the three well spacing scheme indexes (Table 1, figure 4, figure 5), it can be seen that the oil and gas cumulative production of case one is close to that of case two, which are higher than that of case 3, in order to improve the efficiency of development, It is recommended that the case two with a fewer wells is the optimal scheme, and the optimum well spacing is between 500m and 600m.

The GS64 well has been done systematic well testing by using a high precision direct reading electronic pressure gauge. The well testing radius is 311m, so the well spacing calculated by the well testing is around 600m, which are close to the results calculated by Seepage mechanics method. It means the Seepage mechanics method which be deduced to calculate the well spacing is reliable, and the well spacing of the gas reservoir is about 600m. It accords with the production knowledge of low permeability gas reservoir.

Table 1 Forecast results of three program indicators

parameters	Case one	Case two	Case three
Well spacing (m)	400-500	500-600	600-800
Predicted number of years	10	10	10
Well number	8	7	6
Cumulative gas production	3.3	3.29	3.18
Cumulative oil production	3.4	3.38	3.25
FNPV (10^4RMB)	6114	6516	6232
payback period (year)	8.4	6.5	6.3

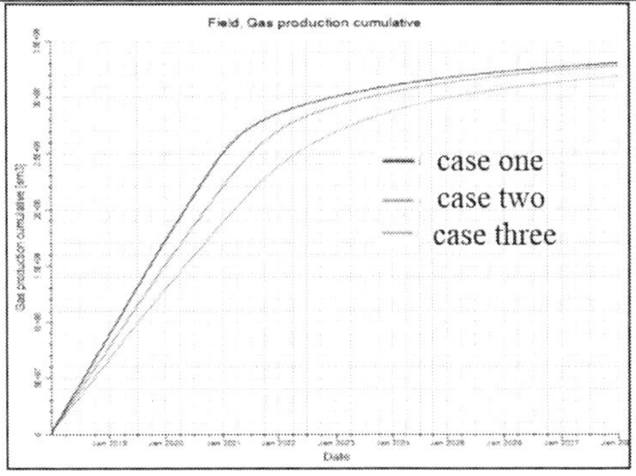

Fig. 4 Comparison of cumulative gas production of different schemes

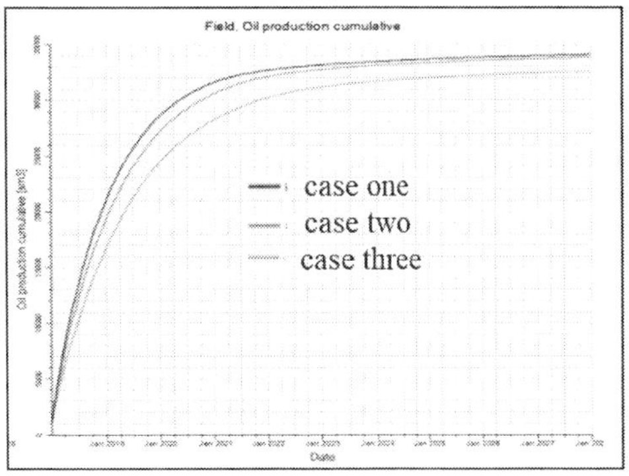

. **Fig. 5** Comparison of cumulative oil production of different schemes

5 Conclusion

(1) The determination of reasonable well spacing of low permeability gas reservoir is the key to ensure its reasonable and efficient development. According to its percolation characteristics, the influence of starting pressure gradient should be considered, and a variety of methods should be adopted to study and determine comprehensively.

(2) According to the flow conditions of reservoir fluid, the calculation method of reasonable well spacing for gas wells in low permeability gas fields is deduced in this paper. The calculation formula involves less parameters and is simple and convenient to calculate. The calculated results are compared with the well spacing calculated by well testing, The results of the new method are reasonable and can be used as a reference for well distribution in low permeability gas reservoirs.

(3) Through numerical simulation, the well spacing of GS64 low permeability gas reservoir is optimized, ranging from 500m to 600 m, which accords with the understanding of low permeability gas reservoir development.

References

1. Feng Wenguang. Non-Darcy Low-velocity Unsteady-State Seepage Flow of Natural *Gas.* J. *Natural Gas Industry*, 1986,6(3): 41-48
2. Li Shuang, Zhou Xinjia, Jin Hui, Jing Yuanshuai. Study on Rational Well Pattern and Well Spacing in Low Permeability Gas Field. J. *Special Oil&Gas Reservoirs*,2010, 17(5): 73-75
3. Wang Xiuzhi, Shi Zhiliang. Study on Economic and Reasonable Well Spacing of Changxing Formation Reservoir Development in Yuanba Gas Field. J. *Fault-Block Oil&Gas Field* 2012, 19(3): 340-342
4. Guo Qi, Li Zhen, Chen Kaiyuan. Well Spacing Optimization of Vertical Well in Low-Permeability Gas Reservoir. J. *Special Oil&Gas Reservoirs* 2016,23(4):109-111.
5. Zhou Keming, Li Ning, Yuan Xiaoling: Gas Percolation Mechanism of Low Permeability Reservoirs With Low Speed Under Residual Water Conditions. J. *Special Oil&Gas Reservoirs*, 2016,23(4):109-111
6. Wu Fan, Sun Lijuan, Qiao Guo'an. A Research on Gas Flow Property and Starting Pressure Phenomenon. J. *Natural Gas Industry*, 2001,21(1):82-84
7. Liu Xiaoxu, Hu Yong, Li Ning, Zhu Bin. Experimental Study and Analysis of the Special Percolation Mechanism in Low Permeability Sandstone Gas Reservoir. J. *Special Oil&Gas Reservoirs* 2007,14(1):80-83.
8. Yi Huaijian, Zhu Weiyao, Zhang Mingjiang. Analysis of Formation Pressure Distribution and Productivity of Low Permeability Tight Gas Reservoirs With Low-Velocity Non-Darcy Flow. J. *Special Oil&Gas Reservoirs*, 2012,19(1):70-73.

Analysis and discussion on the experimental data of electrolyte analyzer

XinYu DONG, JunJie JIANG, MengJun LIU, Weiwei LI

Tianjin Institute Of Metrological Supervision and Testing, Tianjin, 300192, China

Abstract. In the subsequent verification of electrolyte analyzer, we found that the instrument can achieve good repeatability and stability in repeated measurements with a short period of time, in line with the requirements of verification regulation of linear error and cross contamination rate, but the phenomenon of large indication error is very common, the measurement results of different manufacturers have great difference, in order to find and solve this problem, help enterprises to improve quality of product, to obtain accurate and reliable measurement data, we conducted the experimental evaluation of electrolyte analyzer, and the data were analyzed by statistical analysis.

1 Introduction

Electrolyte analyzer is commonly used in medical institutions at all levels. It is mainly used to detect human serum, plasma, whole blood and electrolyte ion in urine such as K^+, Na^+, Cl^-, Ca^{2+}, pH and carbon dioxide. The electrolyte analyzer uses the ion selective electrode measurement technology and the sensor technology, and uses microcomputers to realize rapid and convenient detection and data printing. The typical internal structure is shown in Figure 1.

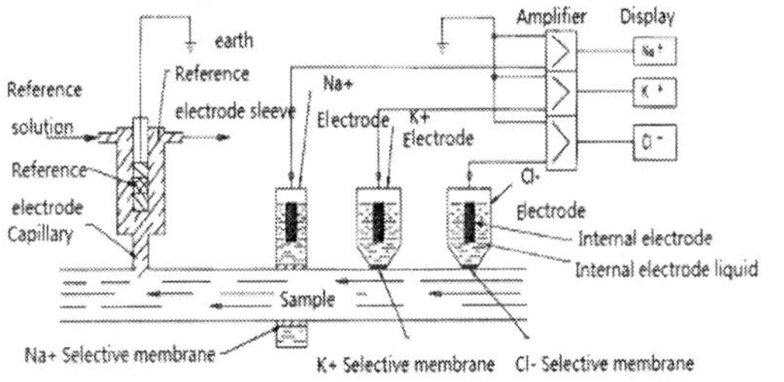

Fig. 1. The schematic diagram of the internal structure of the electrolyte analyzer

2 Measurement Principle and Development Status

2.1 Measurement principle of ion selective electrode

Ion selective electrode is a kind of electrochemical sensor. Its key component is the sensitive membrane which determines the potential response and selectivity. The sensitive membrane changes the potential by reacting with the corresponding ions in the sample. The membrane potential form a measuring cell through the liquid filled, internal reference electrode, by a wire conveying with the amplifier of high input impedance, obtained the concentration value of measured ion by calculating. The relationship between the potential of ion selective electrodes and the concentration of ions in the measured solution can be expressed by the Nernst equation, as shown in Formula 1:

$$E = E_0 \pm \frac{2.303\mathrm{RT}}{\mathrm{nF}} \ln\,(\alpha_i f_i) \qquad (1)$$

For a specific ion, at a certain temperature, 2.303RT/nF is a certain value, then Formula 1 can be simplified as:

$$E = E_0 + S\lg\alpha_i \qquad (2)$$

From the equations above, we know that under certain experimental conditions, the potential of ionic selective electrode (E) is linearly related to the logarithm of ionic activity (or concentration) in the measured solution. In practical applications, a comparative method is used to calculate the ionic concentration of the sample solution, and the standard liquid (A calibration solution and B calibration solution) of two known concentrations is measured first, with the two cell electromotive force of the battery, the slope of each electrode can be calculated and a calibration curve is set up in the instrument program, then the cell electromotive force of the unknown concentration sample is measured, and the concentration of the sample ion is obtained from the calibration curve established, as shown in Figure 2.

 https://doi.org/10.1051/e3sconf/20183801039

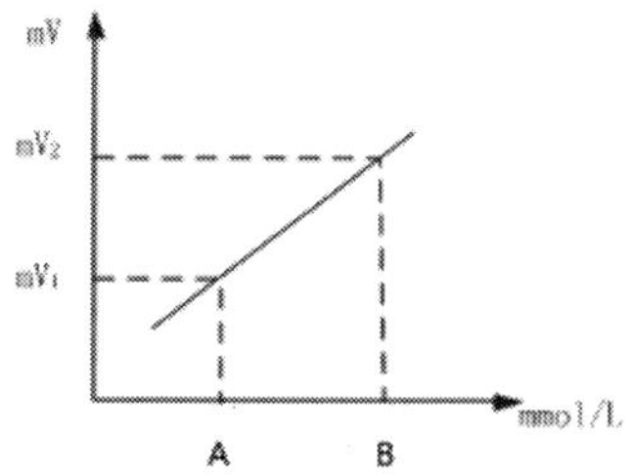

Fig. 2. Calibration curve

2.2 Development Status

The electrolyte analysis instrument used in China started late. In the early 1980s, a few units in China began to introduce the assembled clinical electrolyte analyzer. With the development of microelectronics and material science, the performance of electrolyte analyzer has been greatly improved, and it has been widely applied in various medical institutions and experimental institutes in China. According to incomplete statistics, there are as many as more than 30 manufacturers and companies that produce clinical electrolyte analyzer at present. The quality of products is also quite different. As a measuring instrument, it should have the metrological characteristics that conform to the standard requirements. In order to ensure the accuracy of the instrument, prevent the phenomenon that test results of patient in different hospitals, have obvious discrepancy, we need to do traceability of electrolyte analyzer, China promulgated verification regulation of electrolyte analyzer, which provide the technical basis for the quality control of the electrolyte analyzer.

3 Analysis of Test Results

3.1 Test evaluation experiment

The assessment experiment chooses three instruments , BK-2008E electrolyte analyzer made byBeijing Beiken, the latest model of URIT-910Plus electrolyte analyzer made by Guilin Uritest , the one made by German Abbott. The appearance identification of the three instruments are in line with the requirements, all of the fasteners are installed firmly, have good connection, no looseness; buttons and switches can work properly, which will not affect the normal operation of the instrument; sampling system and detection system are no leakage, no vibration and noise. This experiment using three concentrations of human serum inorganic component electrolyte standard material, the numbers are GBW (E) 090794, GBW (E) 090795, GBW (E) 090796, respectively. The potassium and sodium ions fixed values by means of atomic absorption, the chloride ion fixed values by means of ion chromatography.

3.2Analysis of test data

The instruments were tested by washing, calibration and other procedures. Six times of repeated measurements were done in a short time by the same operator using the three concentrations of standard material. The repeatability of the instrument can be calculated by formula 3 and formula 4. The test results were shown in Table 1.

$$r = s/\overline{x} \times 100\% \quad (3)$$

$$s = \sqrt{\frac{\sum_{i=1}^{n}(x_i - \overline{x})^2}{n-1}} \quad (4)$$

Table 1. Results of repeatability measurement

Model	Standard Material	Ion	x_1	x_2	x_3	x_4	x_5	x_6	$\overline{x}$	s	r (%)
BK-2008E	GBW (E) 090794	K^+	6.75	6.74	6.71	6.67	6.71	6.72	6.72	0.03	0.4
		Na^+	115.63	114.91	114.91	114.91	114.67	114.43	114.91	0.4	0.4
		Cl^-	82.84	82.43	82.43	82.84	82.43	82.43	82.57	0.21	0.3
	GBW (E) 090795	K^+	4.14	4.15	4.15	4.15	4.16	4.14	4.15	0.01	0.2
		Na^+	140.58	140.58	140.58	140.88	142.05	141.75	141.07	0.66	0.5
		Cl^-	107.71	107.71	107.71	107.71	105.6	106.12	107.09	0.97	0.9
	GBW (E) 090796	K^+	2.1	2.09	2.05	2.11	2.06	2.11	2.09	0.03	1.2
		Na^+	157.94	157.29	156.96	155.98	156.3	155.66	156.69	0.86	0.6
		Cl^-	140.75	142.15	143.57	143.57	144.99	144.28	143.22	1.53	1.1
URIT-910Plus	GBW (E) 090794	K^+	6.89	6.71	6.9	6.67	6.91	6.9	6.83	0.11	1.6
		Na^+	103.3	104.1	103.4	104.7	103.6	103.6	103.8	0.5	0.5
		Cl^-	91.31	92.02	91.66	92.19	91.49	91.84	91.75	0.33	0.4
	GBW (E) 090795	K^+	4.12	4.12	4.12	4.12	4.12	4.13	4.12	0	0.1
		Na^+	137.5	137.8	137.8	138	137.8	138.3	137.9	0.3	0.2
		Cl^-	105.8	105.8	106	106	106	106.3	106	0.2	0.2
	GBW (E)	K^+	1.97	1.94	1.94	1.95	1.95	1.95	1.95	0.01	0.6

	090796	Na^+	156.2	155.9	155.6	155.6	155.6	155.4	155.7	0.29	0.2
		Cl^-	129.5	129.3	129.3	129.3	129.5	129.5	129.4	0.11	0.1
Abbott	GBW (E) 090794	K^+	6.5	6.5	6.6	6.6	6.6	6.5	6.6	0.1	0.8
		Na^+	107	108	107	107	107	107	107.2	0.4	0.4
		Cl^-	100	101	101	102	101	102	101.2	0.8	0.7
	GBW (E) 090795	K^+	4.3	4.0	4.1	4.0	4.1	4.1	4.1	0.10	2.70
		Na^+	141	141	141	141	142	141	141.17	0.4	0.3
		Cl^-	108	108	108	109	109	109	108.5	0.5	0.5
	GBW (E) 090796	K^+	<2.0	<2.0	<2.0	<2.0	<2.0	<2.0	/	/	/
		Na^+	159	160	159	159	160	160	159.5	0.5	0.3
		Cl^-	131	131	132	130	132	132	131.3	0.8	0.6

By the repeatability measurement, the indication error of the instruments at three concentrations can be obtained. The indication error can be calculated by formula 5, and the results were shown in Table 2.

$$\delta = \frac{\bar{x} - D}{D} \times 100\% \qquad (5)$$

Table 2. Measurement results of the indication error

Model	Standard Material	Ion	Standard Value (mmol·L⁻¹)	Measured Value (mmol·L⁻¹)	Indication Error	
					Absolute Value (mmol·L⁻¹)	Relative Value (%)
BK-2008E	GBW (E) 090794	K^+	7.02	6.72	-0.30	-4.3%
		Na^+	110.7	114.91	4.21	3.8%
		Cl^-	99	82.57	-16.43	-16.6%
	GBW (E) 090795	K^+	4.41	4.15	-0.26	-5.9%
		Na^+	146.9	141.07	-5.83	-4.0%
		Cl^-	113.2	107.09	-6.11	-5.4%
	GBW (E) 090796	K^+	1.914	2.09	0.17	9.0%
		Na^+	161.7	156.69	-5.01	-3.1%
		Cl^-	133.5	143.22	9.72	7.2%
URIT-910Plus	GBW (E) 090794	K^+	7.02	6.83	-0.19	-2.7%
		Na^+	110.7	103.8	-6.9	-6.3%
		Cl^-	99	91.75	-7.25	-7.3%
	GBW (E) 090795	K^+	4.41	4.12	-0.29	-6.5%
		Na^+	146.9	137.9	-9	-6.2%
		Cl^-	113.2	106	-7.2	-6.4%
	GBW (E) 090796	K^+	1.914	1.95	0.04	1.9%
		Na^+	161.7	155.7	-6	-3.7%
		Cl^-	133.5	129.4	-4.1	-3.1%
Abbott (Germany)	GBW (E) 090794	K^+	7.02	6.6	-0.5	-6.7%
		Na^+	110.7	107.2	-3.5	-3.2%
		Cl^-	99	101.2	2.2	2.2%
	GBW (E) 090795	K^+	4.41	4.1	-0.3	-7.0%
		Na^+	146.9	141.17	-5.7	-3.9%
		Cl^-	113.2	108.5	-4.7	-4.2%
	GBW (E) 090796	K^+	1.914	/	/	/
		Na^+	161.7	159.5	-2.2	-1.4%
		Cl^-	133.5	131.3	-2.2	-1.6%

All of the three instruments have good repeatability, which conform to the requirements for the repeatability of the verification regulation of the electrolyte analyzer. The problem is under the same test conditions, the results of three instruments were quite different by the same batch of standard material. The indication errors were all significantly higher than the technical specifications of 4%.

Value error of measurement using the standard substances made by Shenzhen Kate biological medical

 https://doi.org/10.1051/e3sconf/20183801039

electronic technology co.LTD, which were numbered GBW (E) 090794, GBW (E) 090795, GBW (E) 090796 of three kinds of human serum inorganic electrolyte standard substances, whose characteristic value and uncertainty were shown as table 3.

Table 3. The characteristic and uncertainty of the human serum inorganic component electrolyte standard material

Name	Serial number	Standard value and uncertainty	Concentrations(mm/L)				
			K^+	Na^+	Cl^-	Li^+	iCa^{2+}
Human serum inorganic component electrolyte standard material	GBW(E)090794	Standard value	7.02	110.7	99.0	1.991	1.528
		uncertainty	0.16	3.1	2.8	0.060	0.064
	GBW(E)090795	Standard value	4.41	146.9	113.2	1.244	1.051
		uncertainty	0.11	2.6	3.2	0.036	0.025
	GBW(E)090796	Standard value	1.914	161.7	133.5	0.655	0.650
		uncertainty	0.056	3.2	3.7	0.025	0.020

MPE requirement of Verification regulation for K^+, Na^+ and Cl^- is ±4.0%, according to the above-mentioned conclusion $MPEV/U_{95}$ is less than 3, the influence of the measurement uncertainty of error value must be considered in the assessment. The eligibility criteria are:

$|\Delta| \leq MPEV - U_{95}$ Convicted for qualified

$|\Delta| \geq MPEV + U_{95}$ Disqualify

$MPEV - U_{95} \langle |\Delta| \langle MPEV + U_{95}$ Undetermined

It can be obtained by calculation which is shown in Table 4.

Table 4. Conformance determination of indication error

Evaluation instrument	Serial number	Indication error(mmol/L)		
		K^+	Na^+	Cl^-
BK-2008E	GBW(E)090794	-0.30	4.21	-16.43
	MPEV	0.28	4.43	3.96
	Conclusions	Undetermined	Undetermined	Disqualify
	GBW(E)090795	-0.26	-5.83	-6.11
	MPEV	0.18	5.88	4.53
	Conclusions	Undetermined	Undetermined	Undetermined
	GBW(E)090796	0.17	-5.01	9.72
	MPEV	0.077	6.47	5.34
	Conclusions	Undetermined	Undetermined	Disqualify
URIT-910Plus	GBW(E)090794	-0.19	-6.9	-7.25
	MPEV	0.28	4.43	3.96
	Conclusions	Undetermined	Undetermined	Disqualify
	GBW(E)090795	-0.29	-9.0	-7.2
	MPEV	0.18	5.88	4.53
	Conclusions	Disqualify	Disqualify	Undetermined
	GBW(E)090796	0.04	-6.0	-4.1
	MPEV	0.077	6.47	5.34
	Conclusions	Undetermined	Undetermined	Undetermined
Abbott(Germany)	GBW(E)090794	-0.5	-3.5	2.2
	MPEV	0.28	4.43	3.96
	Conclusions	Disqualify	Undetermined	Undetermined
	GBW(E)090795	-0.3	-5.7	-4.7
	MPEV	0.18	5.88	4.53
	Conclusions	Undetermined	Undetermined	Undetermined
	GBW(E)090796	/	-2.2	-2.2
	MPEV	0.077	6.47	5.34
	Conclusions	/	Qualified	Undetermined

Through the evaluation of uncertainty we found that the uncertainty caused by the inaccuracy of the fixed value is too large, even more than the maximum indication error of the instrument, has caused many test conclusion can only be "undetermined", which creates a big difficulty for quantity traceability of electrolyte analyzer

4 Discussions and Conclusions

Such problems are often encountered in daily life, the patient was treated in a hospital and tested, to another hospital for reexamination, the results of the tests are very different, it is also confirmed by the above experimental data. How to achieve the mutual recognition of different medical structures, improve the quality of instruments and equipment, save patients from repeated tests, and save medical costs for patients, which is the ultimate goal of this topic.

At present, the accuracy and reliability of clinical testing, traceability of clinical analysis of test results have gained wide attention in many countries, many countries have developed a large amount of clinical standard material, clinical laboratory reference system has also been enriched and developed, and gradually getting mature and perfect. According to the measurement principle of electrolyte analyzer, the system should be calibrated before the sample is measured, A, B standard solution are used as the calibration of ion selective electrodes with two kinds of known ionic concentration. It is assumed that the potential of A and B standard solution is EA and EB respectively, the ionic concentration is CA and CB, and the electrode slope is S. According to the measured potential value and the known ion concentration value, the actual slope of each electrode can be calculated by formula 6.

$$S = \frac{E_B - E_A}{\lg(C_B) - \lg(C_A)} = \frac{E_B - E_A}{\lg(C_B/C_A)} \quad (6)$$

Assuming that the potential of the sample is EX and the ionic concentration is Cx, the value of S remains unchanged under the same A and B standard solution, so that the ionic concentration Cx can be calculated according to the formula 7,8.

$$S = \frac{E_x - E_A}{\lg(C_x/C_A)} \quad (7)$$

$$C_X = 10^{\frac{E_X - E_A}{S}} C_A \quad (8)$$

However, each manufacturer use their own A, B standard solution and ion selective electrode, the value of the standard solution is not unified, the difference in concentration of standard solution and accuracy of indication will lead to different results. In addition, the performance gap of ion selective electrodes, which have different ability of Nernst response, commonly result in the difference of the measurement results by the various manufacturers, such as shown in figure 3.

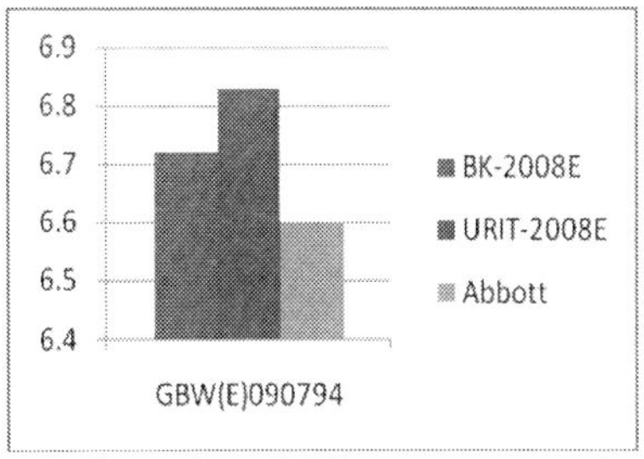

Figure. 3. Comparison of K^+ measurement results

In conclusion, we should develop clinical test standard substances with high quality as soon as possible, the value of the standard liquid used by the manufacturer should be traced to the unified value, manufacturer should also improve the performance of the ion selective electrode, quantity comparison can be carried out on a regular basis, in order to realize the unification of the quantity value, the quantity value of the electrolyte analyzer is ensured to be accurate and reliable.

References

1. Ding Haiming, Wu Guoqiang. "Electrolyte analysis and measurement techniques", Beijing: China Metrology Publishing House,2009: (128-131).
2. Ding Haiming, Wu Guoqiang. "JJG1051-2009 Verification Regulation for Electrolyte Analyzer".

 https://doi.org/10.1051/e3sconf/20183801040

Research on layout of multimodal transport center in Jinan City

Xue Qin ZHANG [1,*], Ying Ran CUI [1], Yuan LI [1] ,and Xiu Feng LIANG [1]

[1]Shandong Jiaotong University, Ji'nan, Shandong 250300, China

Abstract. Multimodal operation is an efficient form of transportation organization. By integrating various modes of transport, the advantages of each mode are fully exerted, and the efficiency and quality of transportation are improved to meet the demand for personalized transport. This article first analyzes the status of multimodal transport development in various regions of Shandong Province; through the analysis of the external environment in which Jinan City is located, it proposes the layout of the Jinan Multimodal Transport Center; finally, it proposes various solutions based on the actual development status.

1 Introduction

With the development of society, people pay more and more attention to the efficiency of transport. In this case, multimodal transport has emerged as the times require. It depends on the combination of sea and land, land and land, and so on, which greatly reduces the transportation time and saves manpower. Most multimodal transport today relies on large containers for loading and unloading of goods. Although China started early in multimodal transport, mainly because of the early system and facilities not perfect, missed a golden development period. Fortunately, the government has noticed the high efficiency and importance of multimodal transport, actively establish and improve the relevant legal system and facilities, so multimodal transport has been redeveloped at a high speed. With the development of our economy and the acceleration of the pace of opening to the outside world, the requirements of inland multimodal transport and international multimodal transport are becoming more and more intense. With the development of the times, logistics is developing rapidly. Multimodal transport as an important branch is concerned by people from all walks of life. Under the background of increasingly intense competition, "multimodal transport" has also been developing rapidly.

2 Analysis on Current Situation of Multimodal Transport in Shandong

2.1 Shandong Economic Development Analysis

In 2015, Shandong was China's economic province. Since the implementation of the "13th Five-Year Plan", its overall economic strength has been steadily improving. In 2017, the province's GDP reached 7.267818 trillion yuan, an increase of 7.4% over the previous year, as shown in Table 1.

Table.1 2011-2017Shandong Province GDP and industrial structure ratio

Years	GDP/ Billion	The proportion of three industries
2011	45429.2	8.8:52.9:38.3
2012	50013.2	8.6:51.4:40.0
2013	54684.3	8.7:50.1:41.2
2014	59426.6	8.1:48.4:43.5
2015	63002.3	7.9:46.8:45.3
2016	1604	7.3:45.4:47.3

(Source: China Statistical Yearbook)

2.2 Analysis on the Development Status of Multimodal Transport in Shandong Province

There are a total of 17 cities in Shandong Province. The development of multimodal transport in different cities is different. This paper selects Jinan, Qingdao and Weifang as the objects to analyze the development of multimodal transport in three cities with economic strength and geographical location.

(1) Jinan: Jinan is relying on the construction of railway logistics parks such as Shandong Ji'er Jiaozhou transport of railway containers. The Shandong Jiaozhou Jiaozhou Logistics Park Project was established solely by the Jinan Railway Bureau. It has five railway lines and a comprehensive modern logistics park centered on multimodal transportation of land, sea and air.

(2) Qingdao: In June 2016, the Qingdao "Belt and Road" cross-border container multimodal transport project was approved by the state and the only multimodal transport demonstration project in Shandong Province. The first multimodal customs supervision center in the coastal area of China—Qingdao Multimodal Customs Supervision Center achieved a total volume of shipments of more than 420,000 TEUs in 2017, an increase of 18.3% year-on-year, ranking fourth in the national railway container terminal. In 2017, the

* Corresponding author: 1454033921@qq.com

"One Belt and One Road" multimodal logistics bidirectional logistics hub centered on Qingdao Central Station was gradually formed. It was also the opening of two international trains of China-EU Classes and "ASEAN Special Lines".

(3) Weifang: The construction progress of expressway projects ranks in the forefront of the province, and the construction of "two-ring and six-shots" comprehensive transport corridors has been carried out in an orderly manner, forming the "One Belt and One Road" large-scale thoroughfare for full docking, vertical and horizontal development of the layout, and Luliao land and sea freight. Roll off the big transport lanes. It opened up the sea transportation channel for North China-Northeast commodity vehicles and agricultural equipment, and opened up the international major channel of "Liaoying Manchu and Europe".

The development of Shandong multimodal transport is steadily developing in recent years. Among them, the development of Qingdao is in a leading position in Shandong, and Jinan and Weifang are also steadily developing.

3 Analysis of the external environment of multimodal transport development in Jinan

3.1 Policy support for the development of multimodal transport

In the thirteenth Five-Year plan outline, the People's Republic of China clearly put forward the development plan of container multimodal transport. The key task is to improve the function of multimodal transport channels and strengthen the construction of comprehensive hubs. On June 3rd 2015 recently, the Department of Communications and Transport of Shandong Province ,the Department and Reform Commission of Shandong Province and other 24 departments jointly formulated the "Implementing the 18 sector notification of the Ministry of transportation and other departments to promote the implementation of multimodal transport in Shandong" .By the end of "the 13th Five-Year plan,in Shandong Province, a multimodal transport system with reasonable layout, advanced technology, perfect facilities, convenient and efficient, integrated linkage, green environmental protection, safe and orderly multimodal transport system has been established, and a modern multimodal transport system has basically been established.[1] Before end of 2017, the Provincial Bureau of Communications and Transportation and Jinan Railway Bureau signed the "Cooperation Agreement on Joint Promotion of Multimodal Transport Development Strategy", forming a cooperative propulsion mechanism.

3.2 The thrust of multimodal transport

Shandong Province vigorously promoted the construction of transportation infrastructure, while the economic strength of transportation continued to grow. The investment of natural assets in the year was 78 billion 300 million yuan, an increase of 21% , reaching a new high of history and exceeding the target.The total mileage of the highway is 265 thousand and 800 kilometers, and the road density is 169.6 km per 100 square kilometers. Shandong province business vehicles increased to 1 million 26 thousand vehicles, including 1 million trucksand 26 thousand passenger cars. The total transportation capacity of Shandong sea river is 19 million 80 thousand tons, and the average tonnage of a single vessel has increased to 1511 tons.

3.3 The economic power of multimodal transport in Jinan

Shandong's Gross Domestic Product (GDP) ranks third in the country, accounting for 1/9 of China's total GDP. In the past five years, Shandong's economy has continued to develop steadily, while the economy of Ji'nan city is in a steady state. In 2017, the total GDP of the city was 7 trillion and 267 billion 818 million yuan, an increase of 7.4 percent over the previous year. Jinan is the capital city of Shandong Province and the economic city of Shandong Province, which gathers a large number of professional technicians in there. Therefore, the advantages of the labor population in Jinan and the high quality of the labor provide a solid foundation for the development of multimodal transport in Jinan.

4 The layout of Jinan multimodal transport center

4.1 Analysis of internal function area of jinan multimodal transport center

(1) Storage area: used to store goods from the customer or port to the intermodal center, which can effectively organize the source of goods and increase the operational capacity of the center.

(2) Transit area: as a freight transit station between the transportation center and the customer, it can reduce the cost to a certain extent by reducing the inventory pressure and increasing the efficiency of transit.

(3) Loading and unloading handling area: the center is equipped with specialized handling

machinery to improve operation efficiency, reduce damage caused by operation and reduce operation cost.

(4) Delivery area: it is necessary to arrange delivery time and goods reasonably to avoid confusion and waste of time ,which caused by various goods.

(5) Customs declaration and inspection area: in order to meet the needs of multimodal transport, the center establishes customs declaration and inspection areas to declare goods inland, declare inspection, issue bills of lading, ship them directly to port, transport, lighten the pressure on the ports, and improve the throughput, reduce the retention time. To some extent, improve the income and reduce the cost.

(6)Information control center: build the information platform of transportation center, realize the information sharing of Anhydrous port, customers and shippers. Such as: cargo tracking system, information release, vehicle stowage and so on.

(7) Packaging area: according to the requirements of customers, we can make some necessary packing for the goods to reduce the wear and tear of the goods, increase the beauty and reduce the risk of transportation.

(8) Parking area: the parking area of the freight car and the two parts of the social vehicle parking area, which are used to park the freight vehicles in the park and temporary parking for customers and employees.

(9) Circulation processing zone: the goods can be processed according to the requirements of customers, w-hich can not only attract consumers' eyeballs, but also increase the profit of goods.

(10) Distribution and sorting area: the center can sort the goods and distribute the goods according to the customer's demand. When delivering the goods, we should fully consider the distribution route and so on, so as to deliver the goods to the customers with the minimum cost and the highest efficiency [2].

4.2 Layout of Jinan multimodal transport center

(1) Jinan External Highway Network

Jinan is located in the middle and western part of Shandong Province and has obvious location advantages. At the end of 2016, the highway mileage was 1,2730.2 kilometers. Jinan has a relatively developed transportation system. Jinan's external road network is shown in Figure 1. On December 28, 2016, Jidong Expressway was officially opened to traffic. It is one of the "two-ring, twelve-shot" high-speed road network routes in Jinan and is the "Five Vertical, Four Horizontal, One Ring, Eight Link" highway network in Shandong Province. The middle part is also an important channel connecting the Jinan metropolitan area in the strategic layout of the development of the highly efficient ecological economic zone in the Yellow River Delta.

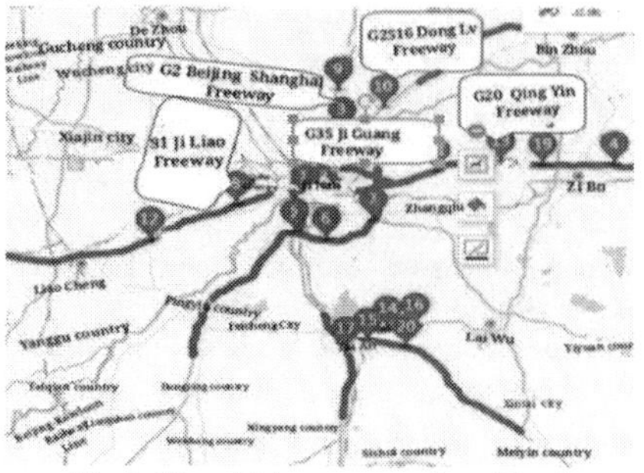

Fig.1. Ji'nan highway network ombined

(2) Railway network of Jinan

Jinan is in the key position through the four corners of the world. Jinan city, as the capital city of shandong province, has provided strong momentum for the steady economic growth of shandong.

Table.2 total imports and exports by category of jinan

Years	Total import and export	Total imports	Total export
2010	743776	338888	404888
2011	1041422	436966	604456
2012	913286	341844	571442
2013	957442	409126	548316
2014	1048867	442950	605817
2015	911424	311763	599661
2016	929391	592765	336626

(Source: Jinan Statistical Yearbook 2016.)

Foreign trade in Jinan City has increased. As shown in Table 2, Jinan City's import and export goods have increased, and Jinan's inland economic level has improved. This provides favorable conditions for the Jinan Inland Construction Intermodal Center. In addition, the railway network in Shandong Province includes the trunk lines in Shandong Province of the Beijing-Shanghai and Beijing-Kowloon Railways, the Blue Smoke Line, the Yanshi Line, the Jiaoxin Line, and the Jiaoji Line, and they are connected to the three major ports in Qingdao, Yantai and Rizhao respectively. Relatively speaking, the railway transportation is more developed and the Jinan railway network is shown in Figure 2.

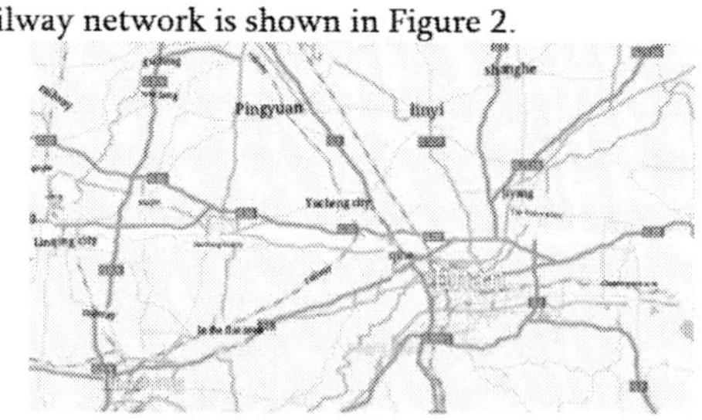

Fig.2. Ji'nan railway network

4.3 Development of logistics information platform for multimodal transport center in Jinan

In order to make the "Jinan multimodal transport cent er logistics information platform" smoothly exchanged ata with the user system, the development of interface must follow the principles of openness, extensibility, modularity, and standardization ,which ensure the exp ansion of the "information platform" to reduce the cost of ma-intenance.

(1) Customs system interface

For the convenience of the platform users to carry out the bonded warehouse, the supervision of the warehouse, customs declaration and other business, "i nformation platform" must establish EDI interface with customs system. Interface design must conform to cust oms data format requirements, and ensure effective co mmuni-cation with customs business system.

(2) Electronic data exchange (EDI) interface

EDI means to pass information through an electron ic data communication network in a set of common sta ndard formats specified by the protocol, which exchanges and processes automatically between the computersystems of the business and trade partne rs. The "information platform" needs to develop the de sign standard EDI in--terface.[3]

(3) SMS platform interface

At present, almost everyone has a mobile phone, short message platform can be used as an effective way to realize the basis of logistics information platform, including China Mobile, China Unicom, China Mobile MMS and other operators with SMS protocol.

In addition, a WAP interface is also reserved for th e SMS platform to provide support for the mobile Inter net business in the later stage, providing online mobile phone query service and other functions.

(4) Other business interfaces

Provides EDI interfaces with railway, aviation, port, fire, security, public security, insurance and financial systems to provide related business services to users.

5 Development Strategy of Multimodal Transport Center in Jinan City

5.1 Joint development and good investment opportunities

In the process of developing multimodal transport, each city in Shandong should break its oppositional attitude and form a joint situation. All kinds of transport should explore each other's joint additive effect and speed up the development of multimodal transport in Shandong Province. At the same time, we can focus on developing businesses in regions where economic development is relatively rapid and where there are potential bulk sources of supply. We can collect more cargoes and long-distance FCL cargoes on cargoes to increase economies of scale.[4]

5.2 Establish multimodal transport management information system

Through the collection of electronic information, the transmission of electronic information and the exchange of electronic data, not only multimodal transport operators, container liner companies, shipping agencies, truck companies, railway transport companies, container terminal companies, freight forwarding companies and inland container freight terminals. People and other transportation departments are linked to each other, and they combine the consignee and consignor as well as the "one customs and three inspections", the foreign trade departments such as banks and insurance, and the regulatory inspection and service departments, thereby realizing the container information system. Rapid flow increases the efficiency of the entire multimodal transport.

5.3 Straighten out the management system and remove obstacles

The seamless and efficient connection of different modes of transport is a fundamental feature of multimodal transport. Specifically, while carrying out traffic infrastructure planning and construction, it is necessary to fully consider the effective convergence between different infrastructures to meet the demand for the development of multimodal transport. In particular, it is necessary to reform the current system of multimodal transport, build an integrated traffic management system, avoid multi-headed management, simplify intermodal procedures among various departments, and truly achieve seamless convergence of different modes of transport. In addition, we should also improve China's multimodal transport laws and regulations and increase enforcement.[5]

5.4 Improve transport technology, innovate transport organization

Accelerate research on high-speed and heavy-loaded railway freight transport, and promote the development of double-decker container transportation. With flexible and diverse forms of transport organization, in the current situation where container traffic is insufficient in inland areas, ordinary trains can be opened first, and container trains can be opened as the market grows and develops. In the direction of larger traffic, container trains or double-decker container trains may be opened, while multimodal transport operators are encouraged to contract contracted container trains, and multimodal transporters are used for sole proprietorship or joint ventures with railways. Various forms of container train operations.

6 Conclusion

The construction of the multimodal transport center in Jinan City is of great significance to the development of the current logistics; the construction of the multimodal transport center can reduce the accumulation of port cargo in inland hinterland and can improve the operating efficiency of the port. The multimodal transport center serves as a connection to the outside world. The bridge has a great effect on the economic development of inland areas. Based on this, it can increase the speed of development in inland areas. Under the new situation, there are many opportunities for the Jinan Multimodal Transport Center and it faces many challenges.

References

1. Zhu Xiaoning. Container Transport and Multimodal Transport [M]. China Railway Publishing House,2010.11.
2. Tao Jinghui. Layout Planning and Operation of Logistics Park [M]. Beijing: China material Publishing House 2009.
3. Hu Yang.The Superiority of International Multimodal Transport [J]. Continental Bridge View, 2010, 08:58-59.
4. Zhang Zhihua. Development Countermeasures of International Multimodal Transport in China [M]. Water Transportation in China,2010(10).
5. Wang Wenjie. Brief discussion on the present situation and Development of China's International Container Multimodal Transport [M] .Entrepreneur 's World,2007(10).

Harbin 2020 R&D Personnel Demand Forecast Based on Manufacturing Green Innovation System

Xin Jiang[1,*], *Yu Ting* Duan[1], *Jun Yi* Shen[1], and *Dong Ying* Zhang[1]

[1]School of Economic and Management, Harbin University of Science and Technology, 150040 Harbin, China

Abstract. Because of the constraints of energy conservation and the impact on the environment, the manufacturing industry has adopted sustainable development as the goal, and a green manufacturing innovation system based on environmental protection has emerged. In order to provide R&D personnel support to manufacturing enterprises in Harbin, and in order to promote the construction of a green innovation system for manufacturing and the realization of the 13th Five-Year Plan, this article used the grey forecasting model and the univariate linear regression prediction to predict the number of R&D personnel in Harbin in 2020 based on the number of R&D personnel in 2010-2016, and the predicted values were 24,952 and 31,172 respectively. The results show that if Harbin continues to use its original development model, it will not be able to achieve the established development goals by 2020 because of the shortage of R&D personnel. Therefore, it is necessary to increase investment in R&D personnel so as to achieve the 13th Five-Year Plan of Harbin City and protect the ecological green development goals.

1 Introduction

In response to global climate change, energy conservation and emission reduction have been advocated by countries. As one of the major industries for energy consumption and pollutant emissions, manufacturing industry needs to assume specific tasks and responsibilities [1]. The manufacturing industry has become one of the major industries with high energy consumption, high pollution, and high emissions [2]. It is the municipal industry of Harbin which is the focus of development and plays an important role in achieving energy-saving emission reduction targets and green development strategies.

Today's society regards talent development as a strategic core. Talent has features that can be developed indefinitely. It is therefore seen as the main source of profit creation and has become the most important resource in social development. James M.W. Wong, Albert P.C.Chan, and Y.H.Chiang believe that the highest correlation with human resources demand forecast is economic indicators [3]. Chinese scholar Liu Fengchao once pointed out that when talent demand needed to be predicted, it is determined by economic development [4]. Therefore, based on the development goals of Harbin's 13th Five-Year Plan, the article forecasts the number of talents needed in Harbin in 2020. What are the requirements of the development goals in the 13th Five-Year Plan for the enterprises in the green innovation system of manufacturing industry in Harbin? Does the current development trend of the number of talents in Harbin City support the implementation of the corresponding 13th Five-Year Plan? These issues have become the main content of this study.

2 The Talent Types Needed by Enterprises in the Green Innovation System of Harbin Manufacturing

From the point of view of the development of the green innovation system for manufacturing industries in developed countries, most of them have high R&D strength and high R&D investment. According to the data released by the 2012 World Economic Forum, developed countries such as the United States and Japan rank top in R&D investment in companies (sixth and lth respectively), while China ranks 23rd in R&D investment rankings. Judging from this, R&D investment has a great influence on the ability of manufacturing green innovation.

R&D personnel is the main body of corporate R&D activities, and it is the most dynamic and the most important input factor in R&D activities of enterprises. It reflects the R&D input scale of the enterprise to some extent. Nowadays, green, low-carbon, and environmental protection are the themes advocated by industrial development in various countries. R&D personnel can carry out green technological innovation to achieve the goal of protecting the ecological environment. Therefore, this article will examine the needs of R&D personnel.

3 Forecasting Process

* Corresponding author: 507907432@qq.com

The R&D expenditure and R&D population table shown in Table 1 can be obtained by consulting the 2010-2016 Harbin Statistical Year book.

Table 1. Summary of R&D Expenditure and R&D Population.

Year	R&D Expenditure (Ten thousand)	R&D Population (Person)
2010	351282	15339
2011	404742	16997
2012	473101	16768
2013	548669	21314
2014	578513	19762
2015	585984	20156
2016	654493	20799

3.1 GM (1.1) Model

The grey forecasting model is based on the evolution of the system. Through grey forecasting is used, the evolution of the system can be accurately described, and valuable information can also be obtained through the analysis of known information.

The manufacturing green innovation system can be considered as a gray system because it includes not only known information but also unknown information. The grey forecasting model has three advantages. First, there is no special requirements or restrictions on the sample data. Second, human subjective assertions can be avoided because each person's experience, knowledge, and preferences are different. This method can avoid this error. Third, the growth trend of R&D in Harbin can be reflected. The steps to create a grey forecasting model are as follows:

Step1: Set the known historical load raw data sequence:

$$X^{(0)} = (x^{(0)}(1), x^{(0)}(2), x^{(0)}(3), \ldots, x^{(0)}(n))$$

Step2: Establishing the one-time AGO sequence:

Where, $X^{(1)}(k) = \sum_{i=1}^{k} x^{(0)}(i)$

Step3: Establishment of the following first order differential equation: Where, a,b is the unknown elements.

$$\frac{dx^{(1)}(t)}{dt} + ax^{(1)}(t) = b \quad (1)$$

Step4: Determine the data matrix:

$$\begin{bmatrix} x^{(0)}(2) \\ x^{(0)}(3) \\ \vdots \\ x^{(0)}(n) \end{bmatrix} = \begin{bmatrix} -\frac{1}{2}[x^{(1)}(1)+x^{(1)}(2)] & 1 \\ -\frac{1}{2}[x^{(1)}(2)+x^{(1)}(3)] & 1 \\ \vdots & \vdots \\ -\frac{1}{2}[x^{(1)}(n-1)+x^{(1)}(n)] & 1 \end{bmatrix} \begin{pmatrix} a \\ b \end{pmatrix}, \quad X_n = BN$$

Where, $X_n = (x^{(0)}(2), x^{(0)}(3), \ldots, x^{(0)}(n))^T$, $N = \begin{pmatrix} a \\ b \end{pmatrix}$

Step5: With the least square method, substituting a,b into the formula (1) and then solve the differential equation. The grey forecasting model available:

$$x^{(1)}(k+1) = \left[x^{(0)}(1) - \frac{b}{a}\right]e^{-ak} + \frac{b}{a} \quad (k = 0,1,2\cdots)$$

Step6: Predicting the value of . The predicted value of can be estimated as:

$$X^{(0)}(k+1) = (X^{(1)}(k+1) - X^{(1)}(k) \qquad k = 1,2,\ldots\ldots)$$

When, $k = 1,2,3,\ldots n-1$, the above equation yields the fitted value; when $k \geq n$, $x^{(0)}(k+1)$ is the predicted value.

Therefore, the steps using the grey forecasting model to predict the number of R&D people in Harbin in 2020 are as follows:

Step1: Set the known historical load raw data sequence:

$x^{(0)} = (15339, 16997, 16768, 21314, 19762, 20156, 20799)$

Step2: Using an accumulated generating series:

$x^{(1)} = (15339, 32336, 49104, 70418, 90180, 110336, 131135)$

Step3: Quasi-smoothness test for $X^{(0)}$:

$P(k) = X^{(0)}(k)/X^{(1)}(k-1)$, If $P(t) < 0.5$ then the quasi-smooth condition is satisfied. After the calculation, the following data is obtained, and when $k > 3$, a quasi-smooth test is passed.

$$p(2) = 1.11; p(3) = 0.52; p(4) = 0.43$$
$$p(5) = 0.28; p(6) = 0.22; p(7) = 0.19$$

Step4: Quasi-exponential law test on $x^{(1)}$:

$m^{(1)}(k) = X^{(1)}(k)/X^{(1)}(k-1)$, if $m^{(1)}(k) \in [1, 1.5]$, then the quasi-exponential law test is satisfied. After the calculation, the following data is obtained, and when $k > 3$, through the quasi-exponential law test.

$$m^{(1)}(2) = 2.11; m^{(1)}(3) = 1.52; m^{(1)}(4) = 1.43$$
$$m^{(1)}(5) = 1.28; m^{(1)}(6) = 1.22; m^{(1)}(7) = 1.19$$

Step5: If all of the above conditions are satisfied, a grey forecasting model can be established for $x^{(1)}$, otherwise it needs to continue accumulating. After calculation, a grey forecasting model is obtained:

x(k+1)=428336.324271exp(0.03988k)-412997.324271. Where, a=-0.03988,b=16470.423409

Comparing the actual values of the 2010-2016 R&D figures with the forecasted values, the calculation results shown in Table 2 and the predicted value of the R&D population in 2020 are obtained.

Table 2. Relative Errors and Forecasts

Year	Historical data (Person)	Predicted value (Person)	Residual	Residual
2010	15339	15339	0	0
2011	16997	17427	430	2.53
2012	16768	18136	1368	8.16
2013	21314	18874	-2440	11.45
2014	19762	19642	-120	0.61
2015	20156	20441	285	1.42
2016	20799	21273	474	2.28
2020		24952		

In order to ensure that the model has a good accuracy, the model must be tested. Table 3 is an accuracy check table.

Table 3. Accuracy check level reference table

Accuracy level	Index close value			
	Relative Error	Relational degree	Standard Deviation ratio	Small error possibility

	α	ε_0	C_0	P_0
First level	0.01	0.9	0.35	0.95
Second level	0.05	0.8	0.5	0.8
Third level	0.1	0.7	0.65	0.7
Fourth level	0.2	0.6	0.8	0.6

Where

$$S_1^2 = \frac{1}{n}\sum_{k=1}^{n}\left[x^{(0)}(k) - \bar{x}\right],$$

$$S_2^2 = \frac{1}{n}\sum_{k=1}^{n}\left[\varepsilon^{(0)}(k) - \bar{\varepsilon}\right]^2, C = \frac{S_2}{S_1}$$

$$p(k) = \left|\varepsilon^{(0)}(k) - \bar{\varepsilon}\right|$$

So: Raw sequence variance: S_1^2= 4629691.67

Residual sequence variance: S_2^2= 1190031.96

Standard Deviation ratio: C= 0.5

Small error possibility: P_0= 0.6475*S_1= 1393.21

As can be seen from Table 4, P(k) values are less than 1393.21 in all other years except 2013. So P=0.86.

Table 4. P(k) calculation result summary table

Year	P(k)
2010	0.43
2011	430.43
2012	1368.43
2013	2439.57
2014	119.57
2015	285.43
2016	474.43

Therefore, the prediction accuracy grade of this model is shown in Table 5.

Table 5. Accuracy level results

Index	Calculation results	Accuracy level
-a	0.03988	Medium and long term forecast
Relative Error	0.04	Second level
Relational degree	0.994	First level
Standard Deviation ratio	0.5	Second level
Small error possibility	0.86	Second level

-a is the coefficient of development and reflects the development of the system, indicating that the model is suitable for medium and long-term prediction. Other indicators are graded two, one, two, and two. So, the prediction accuracy of the model is acceptable. Therefore, the number of R&D personnel in Harbin in 2020 using the grey forecasting model is 24952.

3.2 Univariate Linear Regression Prediction

Taking The Thirteenth Five-Year Plan of Harbin as the research background, the development goal of the The Thirteenth Five-Year Plan of Harbin City has a guiding role for the number of R&D personnel required. Set the target as an independent variable, and sets the required number of R&D personnel as dependent variables, and establish a univariate regression prediction model to obtain the number of R&D personnel required for a given year and target. Therefore, it is particularly important to look for a variable that has a linear relationship with the number of R&D people. Cao Qinrun and other Chinese scholars have researched in this area. They used multiple linear regression method to obtain a significant positive correlation between R&D spending and R&D personnel turnover [5]. Therefore, the article sets R&D spending as an independent variable, and sets the required number of R&D personnel as the dependent variable, and establishes a univariate linear regression prediction, so as to obtain the number of R&D personnel required under the 2020 target.

The equations for R&D expenditures and R&D personnel numbers for Harbin manufacturing companies from 2010 to 2016 are established in Table 1. The results are as follows:

$Y=0.0195X+8724 \quad R^2=0.824$ (2)

In 2015, the GDP of Harbin was 575.12 billion yuan, with an average annual growth rate of 6.25%; The Thirteenth Five-Year Plan of Harbin Municipality requires that by 2020, the goal of the economic development of manufacturing enterprises is to regulate that the industrial added value exceeds 140 billion yuan, with an average annual increase of 7%; According to "Harbin Statistical Yearbook 2016", the total industrial output value of the manufacturing industry in 2015 was 328.308 billion yuan; The Thirteenth Five-Year Plan of Harbin requires that the total R&D input of the whole society up to 2.5% of GDP by 2020. Calculated from the above 4 information, Harbin's 2020 manufacturing R&D expenditure will be 11.517 billion yuan. Therefore, the R&D expenditure of RMB 11.5171 billion was included in formula(2), and the number of R&D personnel in manufacturing in 2020 totaled 31,172.

3.3 Analysis of Forecast Results

By using the grey forecasting model and the univariate linear regression prediction to predict the number of R&D personnel in 2020, the prediction results show that the number of R&D personnel in 2020 is not equal between the two prediction methods. The number of R&D personnel predicted by the univariate linear regression prediction based on R&D expenditure is 31,172, which is 6,220 more than the number of R&D personnel obtained from the grey forecasting model. The grey forecasting model predicts the number of R&D personnel required according to the historical development law, while the univariate linear regression prediction uses the historical data to establish the equation, and predicts the number of R&D personnel required under the requirements of future development goals. The predictive value is to achieve the future development goals, so it is more valuable.

R&D personnel is the main body of corporate R&D activities, and it is the most dynamic and the most important input factor in corporate R&D activities. The number of R&D personnel in the company reflects the importance of the company's R&D activities, and it also reflects the R&D investment scale of the company to some extent. According to the forecast of relevant agencies, R&D personnel is the most popular people currently and in the future [6]. The R&D level of a company embodies the competitiveness of a company. R&D personnel is the core strength of science and technology activities. Their current status and development trends can reflect the national level of scientific and technological competitiveness in a certain way. Obviously, the number of R&D personnel has played a key role in the green development of the city.

As Harbin's 13th Five-Year Plan puts forward requirements for green innovation,advocating vigorous efforts to cultivate green economy, promoting green consumption, focus on the development of green food, energy conservation and environmental protection and other green industries, and creating more "green wealth", it is necessary to in the ecological environment.

Under the requirements of the 13th Five-Year Green Development in Harbin, the effect of R&D investment on green technology innovation is positive and will continues to grow with time [7]. Large R&D investment will help improve the company's enthusiasm for innovation and willingness to develop new products, increase company productivity and energy conversion efficiency, and bring economic benefits to enterprises while reducing the amount of polluting waste and improving environmental conditions. Therefore, it is necessary to increase the input of R&D personnel to achieve the purpose of protecting the ecological environment.

To sum up, we should continue to increase investment in R&D personnel and enhance R&D personnel's important role in promoting green technology innovation so as to achieve the 13th Five-Year Plan of Harbin City and protect the ecological green development goals.

4 Suggestions for increasing the number of R&D personnel

Harbin is in a relatively backward area of economic and technological development. There is a phenomenon of insufficient input and loss of R&D personnel. Therefore, we must take active measures to cultivate R&D personnel, create opportunities for R&D personnel to display their talents, and increase their attractiveness so as to increase the input of R&D personnel.

4.1 Raising the income level of R&D personnel

The improvement of the disadvantages of Harbin's low level of development and poor technological environment can be achieved by appropriately raising the income level of R&D personnel. It can increase the government's investment in the basic construction of science and technology, widen the channels, and guide the entire society to provide basic hardware conditions for scientific and technological innovation so as to enhance the attractiveness of R&D personnel.

4.2 Perfecting R&D personnel incentive mechanism

Good incentive mechanisms and methods can increase the enthusiasm and passion of R&D personnel. The enterprises can attract R&D personnel by developing reasonable and appropriate incentive mechanisms for R&D personnel taking steps such as salary increase, promotion, continuing learning, and further education.

4.3 Increasing the number of R&D positions in companies, universities and research institutions

With the gradual implementation of Harbin's innovation strategy and policies, the market demand and social demand for science and technology activities will further increase, and the number of R&D employment positions will also increase year by year. Therefore, it is possible to increase the number of R&D positions in companies, universities, and research institutions. The proportion of R&D personnel in the labor force should be increased.

5 Conclusion

In order to forecast the number of R&D personnel in 2020, this paper is based on the green innovation system of manufacturing industry using the grey forecasting model and the univariate linear regression prediction according to the historical data of R&D personnel in Harbin from 2010 to 2016, the predicted values of R&D personnel in Harbin in 2020 are respectively 24952, 31172 .

Univariate linear regression prediction based on R&D expenditure is 6220 more than R&D personnel out of the grey forecasting model. It means that if Harbin will still be operating in accordance with the current development laws, the number of R&D personnel will not be sufficient to complete the corresponding development goals by 2020. At the same time, R&D personnel is required to carry out green technological innovation in order to achieve environmental protection and promote green development. This article proposes a total of three suggestions to increase the number of R&D personnel: Raising the income level of R&D personnel, perfecting R&D personnel incentive mechanism, and increasing the number of R&D positions in companies, universities and research institutions to provide personnel support for the implementation of the 13th Five-Year Plan of Harbin City and the protection of the environment.

References

1. Zhu, Q&Sarkis. An inter-sectoral comparison of green supply chain management in china: Drivers and practices[J]. Journal of Cleaner Production,

2006, 14(5): 472-486.

2. Cao H, DuY, ChenY. Exploring a new low-carbon development paradigm for China's future manufacturing sectors[J]. Journal of Science and Technology Policy in China, 2011, 2(2): 159-170.
3. James M. W. Wong, Albert. C. Chan, Y. H. Chiang. Construction Manpower Demand Forecasting: A Comparative Study of Univariate Time Series, Multiple Regression And Econometric Modeling Techniques[J]. Engineering Construction and Architectural Management, 2011, 18(1).
4. Liu Fengchao. Some Methodological Considerations on Personnel Needs and Research[J]. Prediction, 2001, (4): 11-13.
5. Cao Qinrun, Liang Laiyu, Xiao Yangyang. An Empirical Study on the Relationship between R&D Expenditure and R&D Personnel Change[J]. Science & Technology Progress and Policy, 2009, 26(8): 20-24.
6. Zhang Xu, Wang Yu. Influence of Environmental Regulation and R&D Investment on Green Technology Innovation[J]. Science & Technology Progress and Policy, 2017, 34(17): 111-119.
7. Zhu Changming, Chen Yongli, Deng Ying. Analysis on the Development Trend of R&D Resources in Hubei Province Based on Grey Prediction[J]. Pioneering With Science & Technology Monthly, 2012, (8): 14-17.

https://doi.org/10.1051/e3sconf/20183801042

Technique policy for concurrent development of natural and artificial water flooding of strong edge water reservoir: A case study of central layer Yu-III of Akshabulak oilfield

Zhang Angang, *Fan* Zifei, *Zhao* Lun, *Wang* Jincai, *Zhang* Xiangzhong, *Song* Heng, and *Hou* Qingying

PetroChina Research Institute of Petroleum Exploration & Development, Beijing, China

Abstract: The central layer Yu-III in Akshabulak oilfield is a sandstone reservoir with strong edge water, whose major development characteristics are high oil recovery rate and heterogeneous water invasion. Aiming at this problem, the development policy chart of concurrent displacement of natural water and injected water is established on the basis of material balance principle. Injection-production ratio and oil recovery rate are the main controlling factors for the concurrent displacement of natural water and injected water. Each injection-production ratio corresponds with only one rational oil recovery rate, and the rational oil recovery rate increases with the injection-production ratio. When the actual injection-production ratio of the central Yu-III reservoir is 0.9, the rational oil recovery rate should be 4%.

1 Introduction

During the development of strong edge water reservoir, how to maximize the usage of the driving energy of natural water and minimize the investment of artificial water injection becomes the urgent issue at present [1-2]. Therefore, for concurrent development of natural and artificial water flooding, it is significant to confirm the main influence factor and establish the correct technique policy [3-5]. By taking example of central layer Yu-III of Akshabulak oilfield, the development technique policy charts for concurrent development of natural and artificial water flooding of strong edge water reservoir are printed.

2 Geologic and production characteristics of central layer Yu-III

Akshabulak oilfield is located in Aryskum depression of South Turgai Basin, which is a NE structure controlled by faults. The oilfield is divided into three parts: the center, east and south part. Moreover, the center part is separated from the south part by EW fault. And the south part has more fault than the other two parts. As a whole, Akshabulak oilfield is a multi-layered anticline sandstone reservoir. The reservoir depth ranges from 1550m to 1950m. There are 7 oil layers vertically, which includes M-II-1, M-II-2, Yu-0-1, Yu-0-2, Yu-I, Yu-II and Yu-III (Fig. 1). The main rock types of central layer Yu-III are grit stone and medium sandstone. The average porosity is 25.8%, and the average permeability is 1320mD. Generally, the reservoir is a high porosity and super high permeability reservoir, but has strong heterogeneity. The central layer Yu-III was firstly put into production in October, 1996, and was transformed into out edge water flooding in November, 2002. The reservoir is developed by high oil recovery. The oil recovery rate was maintained at 3% since 2004, and the peak was high to 4.4%.

3 Technique policy for concurrent development of natural and artificial water flooding of strong edge water reservoir

3.1 Model establishment of reservoir engineering evaluation model

In order to carry out the concurrent development of natural and artificial water flooding efficiently, such influencing factors as water body energy, reserves, oil recovery rate and injection-production ratio are considered comprehensively (Fig. 2). Hence, how to define the production contribution ratio of natural water drive and artificial water flooding, becomes the urgent development issue at present.

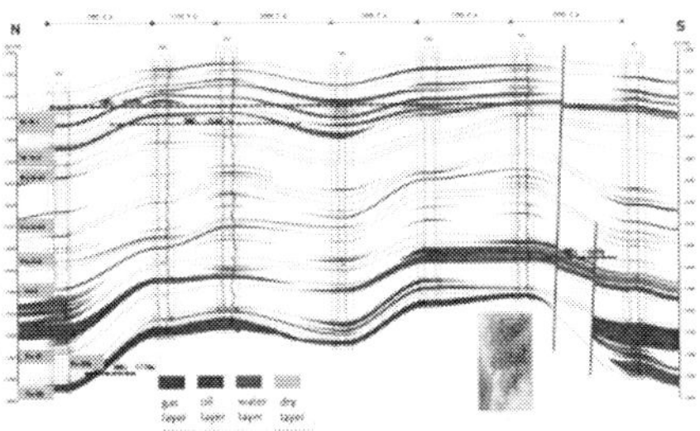

Fig .1 The north-south reservoir profile of Akshabulak oilfield

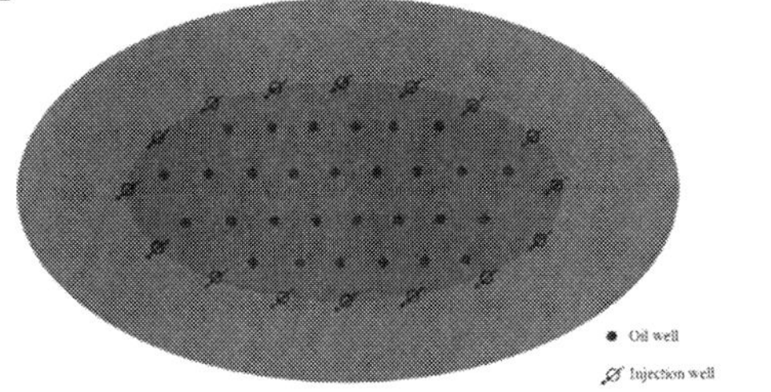

Fig. 2 Sketch map of out edge water flooding for strong edge water reservoir

According to material balance principle, the cumulative oil production in reservoir condition equals the fluid swelling capacity plus cumulative water injection [6-9].

$$N_p\left[B_o+\left(R_p-R_s\right)B_g\right]=NB_{oi}\left[\frac{B_o-B_{oi}+\left(R_{si}-R_s\right)B_g}{B_{oi}}+\frac{C_wS_{wc}+C_f}{1-S_{wc}}\Delta p\right] \quad (1)$$
$$+\left(W_e+W_i-W_p\right)B_w$$

Here, N_p is cumulative oil production at standard condition, m^3; B_o is oil volume factor at current condition, m^3/m^3; B_g is gas volume factor at current condition, m^3/m^3; B_{oi} is oil volume factor at initial condition, m^3/m^3; R_p is production gas oil ratio, m^3/m^3; R_s is dissolved gas oil ratio at current condition, m^3/m^3; R_{si} is dissolved gas oil ratio at initial condition, m^3/m^3; C_f is pore compressibility coefficient, MPa^{-1}; C_w is water compressibility coefficient, MPa^{-1}; B_w is water volume factor at current condition, m^3/m^3; S_{wc} is irreducible water saturation, fraction; Δp is reservoir pressure difference, MPa; W_e is cumulative water influx, m^3; W_i is cumulative water injection, m^3; W_p is cumulative water production, m^3.

In equation (1), cumulative water influx is obtained by Fetkovitch method [10-11]. The water influx equation about Fetkovitch method is shown as follows:

$$W_e=\frac{W_{ei}}{p_i}(p_i-p)\left(1-e^{-\frac{Jp_i}{W_{ei}}t}\right) \quad (2)$$

Hereinto

$$W_{ei}=V_w\left(C_w+C_f\right)p_i \qquad J=\frac{2\pi\varphi kha}{\mu_w(\ln\frac{r_e}{r_o}-\frac{3}{4})}$$

Here, W_{ei} is the maximum water flux potential, m^3; V_w is the volume of water body, m^3; p_i is initial reservoir pressure of water body, MPa; J is water influx coefficient, m^3/(d·MPa); φ is reservoir circumference coefficient, $\varphi=\theta/360$; θ is water influx angle, degree; k is formation permeability, μm^2 ; h is water formation thickness, m; a is conversion coefficient, a=86.4; μ_w is formation water viscosity, mPa·s; r_e and r_o are water body radius and reservoir radius, m; t is development time, d; p is the reservoir pressure at current condition, MPa.

During the reservoir development, the actual reservoir pressure changes with the development time. Thus, the water influx at the time of Δt_n can be expressed as:

$$\Delta W_{en}=\frac{W_{ei}}{p_i}\left(\bar{p}_{n-1}-p_n\right)\left(1-e^{-\frac{Jp_i}{W_{ei}}\Delta t_n}\right) \quad (3)$$

Then, the average pressure of water body $\bar{p}_{wn-1}$ can be expressed as:

$$\bar{p}_{n-1}=p_i(1-\frac{\sum_{j=1}^{n}\Delta W_{ej}}{W_{ei}}) \quad (4)$$

The cumulative water influx is the sum of water influx at several time slots.

$$W_e=\sum_{j=1}^{n}\Delta W_{ej} \quad (5)$$

From the above equation, the reservoir water influx is related to reservoir pressure closely. Hence, in order to predict the water influx at different oil recovery rate and injection-production ratio, the average reservoir pressure should be confirmed firstly. If the initial reservoir pressure, production history data and P.V.T parameters are given, the reservoir pressure at different time can be calculated by trial-and-error method. The detailed calculated process is shown as follows:

(1) Assuming the reservoir pressure is p', each part of material balance equation (1) is calculated. Hereinto, cumulative water influx W_e is obtained by the equation (3)-(5); Cumulative water production W_p is obtained by water-drive characteristic curve $\lg W_p=aN_p+b$;

(2) Judge the equality of material balance equation (1). If the equality is met, the current reservoir pressure is p'. If the equality is not met, a new pressure p'' is assumed and the above calculated process is repeated until the equality is met.

The cutoff condition of the trial-and-error method is that the reservoir pressure is less than the abandon pressure, or the development time reaches the end of contract term.

The production contribution ratio of natural water drive and artificial water flooding can be expressed as:

$$D_{we}=\frac{W_eB_w}{N_p\left[B_o+\left(R_p-R_s\right)B_g\right]+W_pB_w} \quad (6)$$

$$D_{wi}=\frac{W_iB_w}{N_p\left[B_o+\left(R_p-R_s\right)B_g\right]+W_pB_w} \quad (7)$$

Here, D_{we} and D_{wi} are the drive index of natural water drive and artificial water flooding, dimensionless.

By evaluating the production contribution ratio of natural water drive and artificial water flooding at different time, the synergistic effect of reservoir energy can be reflected to provide some evidences for the future reservoir development adjustment

3.2 Development technique policy charts

On the basis of the geologic and operational parameters,

the relevant reservoir engineering model is established. Such development indexes as reservoir pressure level, production contribution ratio of natural water drive and artificial water flooding, and oil recovery percent are studied by the above reservoir engineering model.

(1) Reservoir pressure level

Reservoir pressure variation law under different injection-production ratio and oil recovery rate is researched. Meantime, the charts of reservoir pressure level and cumulative water influx are printed (Fig. 3).

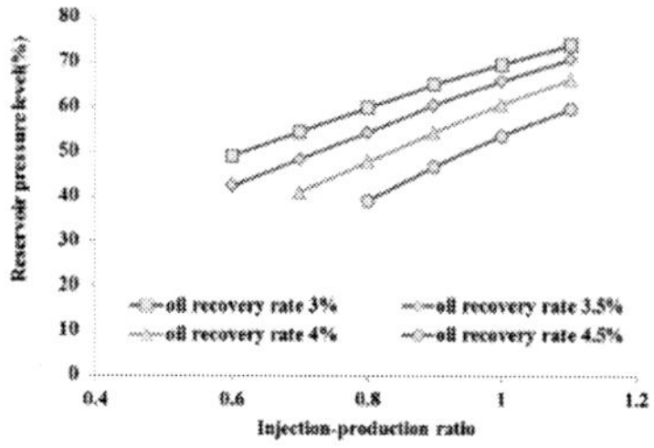

(a)Reservoir pressure level

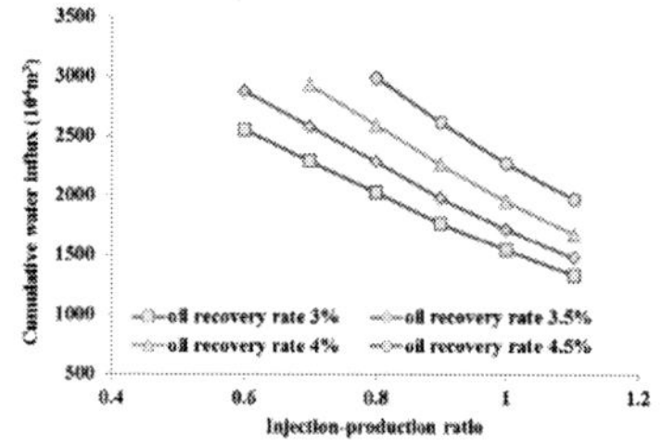

(b)Cumulative water influx

Fig. 3 Charts of reservoir pressure level and cumulative water influx under different injection-production ratio and oil recovery rate

Fig. 3 shows that when oil recovery rate is constant and the injection-production ratio increases, the cumulative water influx decreases and the reservoir pressure level increases. When the injection-production ratio is constant and oil recovery rate increases, the cumulative water influx increases and the reservoir pressure level decreases. In order to maintain the reservoir energy, injection-production ratio should be increased and oil recovery rate should be decreased.

(2) Production contribution ratio of natural water drive and artificial water flooding

Based on the reservoir engineering model, the charts of production contribution ratio of natural water drive and artificial water flooding are printed (Fig. 4).

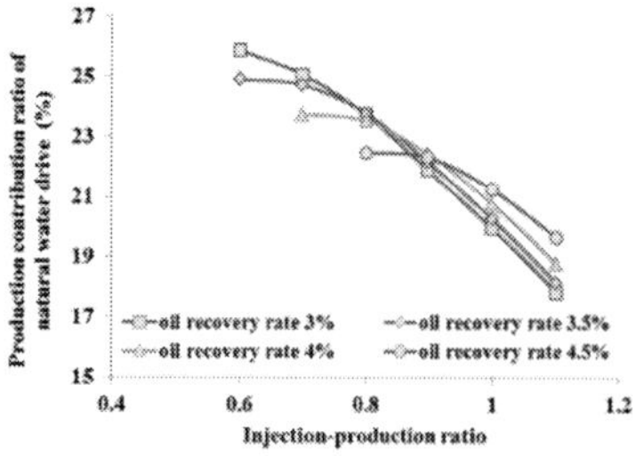

(a)Production contribution ratio of natural water drive

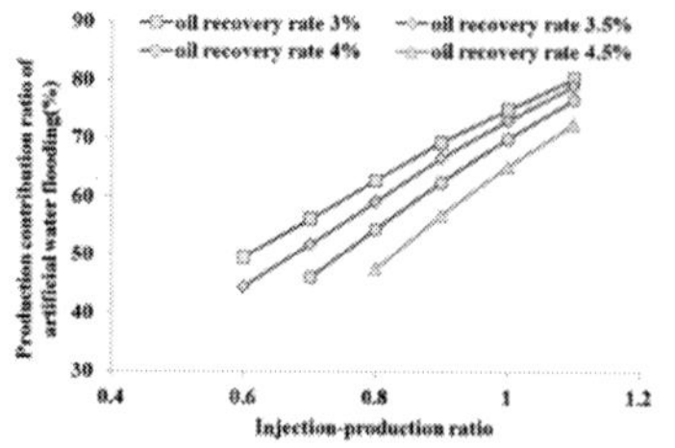

(b)Production contribution ratio of artificial water flooding

Fig. 4 Production contribution ratio of natural water drive and artificial water flooding under different injection-production ratio and oil recovery rate

Fig. 4 shows that when injection-production ratio is relatively small and oil recovery rate increases, the production contribution of natural water drive decreases; when injection-production ratio is relatively large and oil recovery rate increases, the production contribution of natural water drive increases. Whether oil recovery rate is small or large, the production contribution of natural water drive decreases with increasing injection-production ratio. On the other hand, the production contribution of artificial water flooding increases with injection-production ratio, and decreases with increasing oil recovery rate.

(3) Oil recovery percent

Considering the contract term of Akshabulak oilfield, the chart of oil recovery percent under different injection-production ratio and oil recovery rate is also printed (Fig. 5).

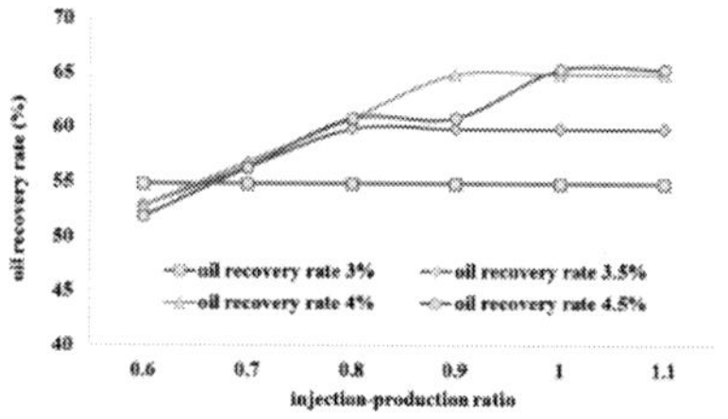

Fig. 5 Oil recovery percent under different oil recovery rate and injection-production ratio

Fig. 5 shows that oil recovery rate increases with injection-production ratio at the same oil recovery rate. However, when the oil recovery rate reaches to a certain level, the increase rate of oil recovery percent becomes slow. Each oil recovery rate refers to a rational injection-production ratio, and rational oil recovery rate increases with injection-production ratio (Fig. 6a).

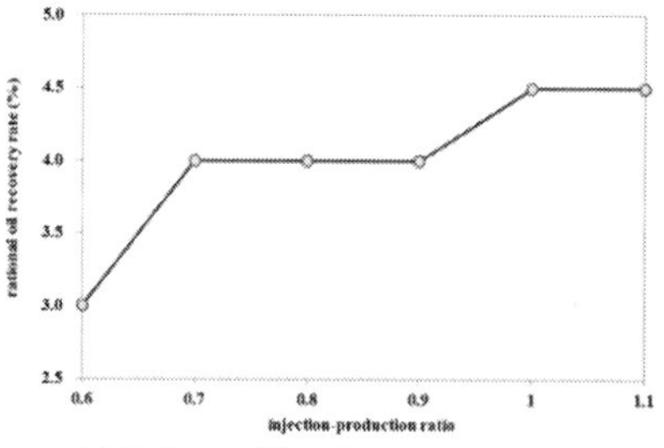

(a) Only consider oil recovery percent

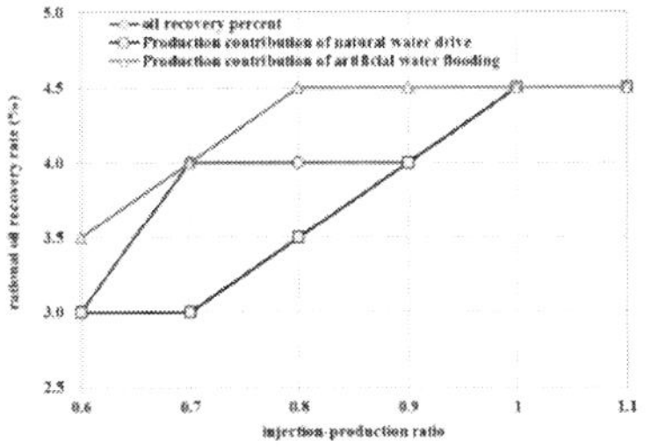

(b)consider oil recovery percent and production contribution ratio of natural water drive and artificial water flooding

Fig. 6 Relationship between rational oil recovery rate and injection-production ratio

For the water flooding of the oil reservoir with strong edge water energy, how to maximize the displacement of natural water and minimize the investment of artificial water injection becomes the urgent development issue at present. Thus, aimed at the development indexes of oil recovery percent, production contribution ratio of natural water drive and artificial water flooding, the relationship chart between oil recovery rate and injection-production ratio is established (Fig. 6b). For Fig. 6b, it can be seen that the best combinations of injection-production ratio and oil recovery rate are (1, 4.5%) and (1.1, 4.5%).

3.3 Validity verification of the technique policy charts

On the basis of the above reservoir engineering model, the change of drive index at different time is calculated. Since the reservoir is developed by water flooding in 2002, the drive index of natural water, dissolved gas and elastic energy decreased with increasing injection-production ratio, but the drive index of artificial water flooding increased with injection-production ratio (Fig. 7). In 2016, the cumulative water influx of central layer Yu-III is $1211\times10^4\,m^3$. Meantime, the drive index of natural water is 0.21, and the drive index of artificial water flooding is 0.72.

Assuming the reservoir injection-production ratio is constantly 0.9, the development effect at different oil recovery rate of 1%, 2%, 2.5%, 3%, 4% and 5% is studied by reservoir numerical simulation. The research results shows that different oil recovery cases have different oil recovery percent during the contract term. Furthermore, the oil recovery percent of the case of oil recovery rate 4% is 5.8% more than the case of oil recovery rate 1% (Fig. 8). For Fig. 6b, in order to maximize oil recovery percent and production contribution ratio of natural water drive, the best combination of injection-production ratio and oil recovery rate is also (0.9, 4%). Therefore, the numerical simulation results verify the effectiveness of the above reservoir engineering model.

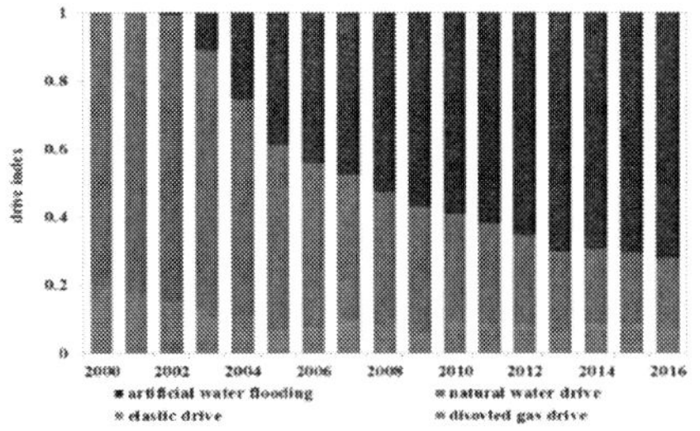

Fig. 7 Variation of drive index and injection-production ratio for central layer Yu-III

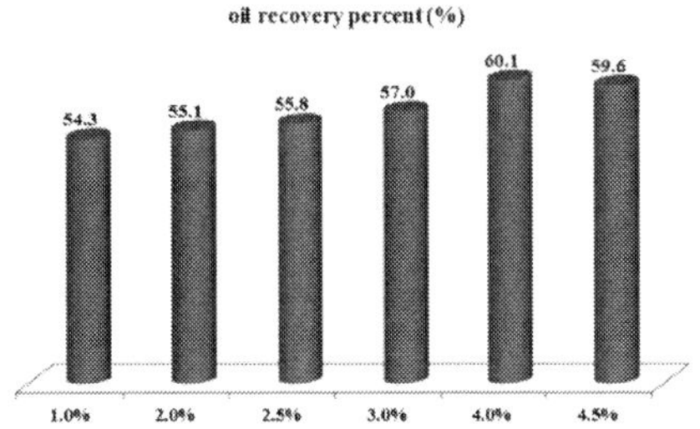

Fig. 8 Ultimate recovery factor and recovery percent of contract term under different oil recovery rate

4 Conclusion

Technique policy charts for concurrent development of natural and artificial water flooding of strong edge water reservoir are established, which indicates that injection-production ratio and oil recovery rate are the main influencing factors for the concurrent development. Each oil recovery rate refers to a rational injection-production ratio, and rational oil recovery rate increases with injection-production ratio. When the injection-production ratio of central layer Yu-III is 0.9, the rational oil recovery rates obtained by technique policy charts and reservoir numerical simulation are both 4%, which shows that the above technique policy charts is available.

References

1. Zhang Rui, Yu Qitai. An analysis of the development efficiency of a naturally water drive sandstone oil reservoir[J]. Petroleum Exploration and Development, 18(3): 46-54 (1991)
2. Mu Longxin, Wu Xianghong, Huang Qizhi. Development theory and technology of high pour point oil reservoir[M]. Beijing: Petroleum Industry Press, 108-115 (2015)
3. Jin Yong, Tang Jiandong, Zhao Juan, et al. Determining method for reasonable drawdown pressure of oil reservoir with edge and bottom water[J]. Acta Petrolei Sinica, 24(1): 68-72 (2003)
4. Hou Jian, Zhang Yanhui, Gong Ruxiang. Efficient development technique of high permeability sweet oil active edge water drive reservoir: case of Pai2 reservoir of Chunguang Oilfield[J].Petroleum Geology and Recovery Efficiency, 19(6): 82-86

(2012)

5. K. Ayeni, R. Wattenbarger, J. Maggard. Estimating water production behavior from edge water drive in a monocline reservoir[C]. SPE 114593-MS (2008).
6. Tang Ligen, Wang Jieming, Bai Fengjuan, et al. Inventory forecast in underground gas storage based on modified material balance equation[J]. Petroleum Exploration and Development, 41(4): 480-484 (2014)
7. Jiang Hanqiao, Yao Jun, Jiang Ruizhong. Principle and method of reservoir engineering[M]. Dongying: China University of Petroleum Press, 200-208 (2006)
8. Gao Chentai, Lu Tao, Gao Weixin, et al. Muti region material balance method and its use in the performance prediction and optimal well pattern design of edge water reservoir[J]. Petroleum Exploration and Development, 33(1):103-106 (2006)
9. Fan Zifei, Cheng Linsong, Song Heng, et al. Fluid interface moving for the concurrent production of gas cap and oil ring of gas cap reservoirs[J]. Petroleum Exploration and Development, 42(5):624-631 (2015)
10. M.J. Fetkovich. A simplified approach to water influx calculations-finite aquifer systems[J]. Journal of Petroleum Technology, 23(7): 814-828 (1971)
11. Lu Kefeng. Applicability analysis on classical empirical relation lnω=BlnR of water invasion prediction for water drive gas reservoirs[J].China Offshore Oil and Gas, 28(06): 40-45 (2016)

Discussion on runoff purification technology of highway bridge deck based on water quality safety

Sheng-guang TAN[1], *Xue-xin* LIU[2,*], *Guo-ping* ZOU[1], *Xin-zhu* XIONG[2], and *Shuang-cheng* TAO[2]

[1]Jiangxi Ganyue Expressway Co., Ltd, Nanchang, 330025, China
[2]China Academy of Transportation Sciences, Beijing, 100029, China

Abstract. Aiming at the actual problems existing, including a poor purification effect of highway bridge runoff collection and treatment system across sensitive water and necessary manual emergency operation, three kinds of technology, three pools system of bridge runoff purification, the integral pool of bridge runoff purification and ecological planting tank, are put forward by optimizing the structure of purification unit and system setting. At the same time, we come up with an emergency strategy for hazardous material leakage basing on automatic identification and remote control of traffic accidents. On the basis of combining these with the optimized pool structure, sensitive water safety can be guaranteed and water pollution, from directly discharging of bridge runoff, can be decreased. For making up for the shortages of green highway construction technology, the technique has important reference value.

1 Introduction

With the rapid development of highway construction, frequent traffic activities have raised more attention to the problem of road surface runoff pollution. Carrying large amounts of particulate matter, heavy metals, petroleum, nutrients and PAHs and other pollutants, road runoff put a negative impact on the ecological environment alongside the highway. The pollutants in road surface runoff mainly comes from the wear particles of vehicle tires and road materials, the leakage of hazardous goods in delivery, the particulate matter of snowmelt agent generated from auto parts, the pollutants in rainwater and the atmospheric dust-fall, and the erosion of road surface sediment by rainwater. [1]-[4]

The greatest potential danger to the water body lies in the serious pollution caused by the leakage from carriage of dangerous goods (CDG) due to accidents or some inner factors [5]-[6]. To prevent this kind of environmental risk, in *Notice on Strengthening Efforts on Environmental Impact Assessment of Highway Planning and construction (Doc. MEP [2007] No.184)*, requires that sedimentation tanks are required to set on both sides of the bridges over source water secondary protection area and quasi protection area and Class Ⅱ water area. Tanks are used for dealing with the bridge surface runoff after the pollution accident so as to ensure drinking water safety.

Since there is no specific design requirement for sedimentation tank, the design of bridge runoff sedimentation tank is mainly used to cope with the requirement of completion acceptance of environmental protection at present. As a result, the design is mostly too simple. Although some studies have optimized the tank structure [7], the overall pollutant precipitation grease effect is limited.

Drain valves were equipped and maintenance staff was required to regularly open the valve to empty runoff. But in fact, there was a long-term presence of sedimentation tank in the full state, resulting that tanks failed to play a role in the hazardous good leakage accident. In response to this kind of problem, researchers began to try to introduce some self-control measures into the bridge runoff emergency management system. Xu Jianping et al [8] proposed a technical system for emergency operation management based on water quality monitoring. However, it still needed to be further demonstrated whether the selected water quality indicators can cover all types of hazardous chemicals. Qin Haiwei et al. [9] proposed a selective collection system of runoff based on real-time monitoring according to different situations. But it lacks the corresponding remote automatic control mode in the disposal strategies.

In view of the above problems, three technologies are proposed to improve the water purification effect of pavement surface by adopting special measures based on smart transport means to ensure water quality and safety, aiming to provide technical support for highway water environment and water safety protection in environment sensitive areas.

2 The bridge runoff purification emergency three-tank system on different function

* Corresponding author: Xue-xin LIU, nkone@sina.com

The key point of the emergency treatment for hazardous chemical leakage runoff, in daily operations, is a better purifying function for collected runoff. In order to ensure sufficient accident response time during these two function switching processes, an intermediate buffer unit is also needed.

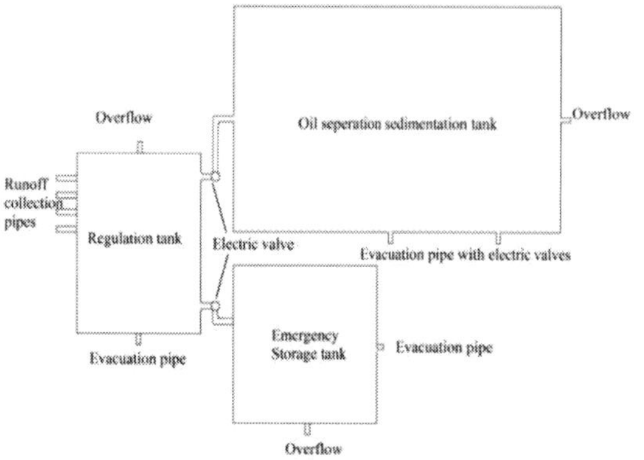

Fig. 1. Bridge runoff purification Emergency three-tank system.

Based on the considerations above, the bridge runoff purification emergency three-tank system was developed. It consists of a storage regulation tank, an oil separation sedimentation tank, an emergency storage tank and a monitoring control system. The surface runoff generated by rainfall firstly enters into the storage regulation tank, which plays a role as a buffer and provides the pretreatment for the bridge runoff.

When there is no leakage of hazardous chemicals, the runoff of bridge deck flows into the oil separation sedimentation tank through the electric butterfly valve which is normally opened. After sedimentation, filtered and purified runoff will be discharged into the surrounding water through the remote control opening the empty pipe valves of storage regulation tank and oil separation sedimentation tank.

When a traffic accident occurs on the bridge, it is first detected through the bridge head monitoring camera and the alarm telephone to determine whether there is any leakage of dangerous goods. After the leakage is confirmed, the monitoring center immediately starts the emergency control system, which orders to open the electric butterfly valve between storage regulation tank and the emergency storage tank, and close the electric butterfly valve between the storage regulation tank and the oil separation sedimentation tank at the same time. Then the leakage of dangerous goods and its dilution collected in storage regulation tank will flow directly into the emergency storage tank for post-treatment offered by the environmental protection department. This kind of technology has been applied to Jiangxi Chang Feng Expressway.

The oil separation sedimentation tank is equipped with baffle wall, pebble retaining wall and oil collection tank. Those kinds of structures are of sedimentation, oil separation and bio-degradation functions and have good removal effects on the primary pollutants in the initial rainfall runoff. Specifically, the vertical bend is set to keep folding flow and improve the sediment separation effect to stay the uniformity of water distribution.

A baffle plate is arranged at the bottom of the vertical baffle wall from top to bottom, which can improve the sedimentation effect of the system. The oil collecting tank is arranged inside the retaining wall. A set of cone mouth with a certain space is designed along the centerline of the bottom of the tank, inside which some nano-fiber oil absorption material is laid to ensure oil separation function of the oil collection tank. The pebble wall set up from the bottom is able to change the direction of water flow, and help with adsorption and biodegradation. The sand tank is set one each before and after the pebble retaining wall at the tank bottom and the blow-off pipe is emplaced one each on both sides of the pebble retaining wall. In addition, the use of electronic valve makes it possible to remote control the emissions when purification reaction time is done.

3 Coupling remote control of the bridge surface runoff purification and emergency integrated tank

Bridge runoff purification emergency three-tank system is able to perform the three functions independently, which includes regulation, purification and emergency disposal, and to protect the sensitive water quality to the fullest. However, the project quantity is relatively large and it is more difficult to implement especially when the space under short bridges is limited. So the three tanks mentioned above were combined to form a bridge runoff purification emergency integrated tank with all their functions and the remote control system was optimized.

The various control strategies for the emergency treatment of the bridge runoff are established on the basis of accurate identification of the hazardous chemicals in leakage accident. The accident identification of the three-tank system is worked through the manual confirmation by the camera on the bridge and the alarm telephone, and there is the possibility of

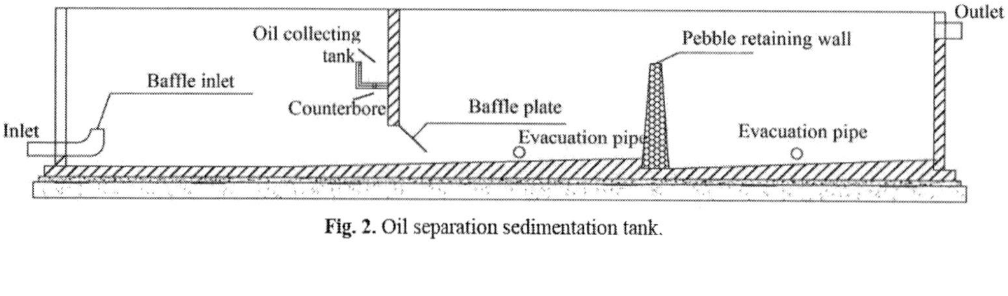

Fig. 2. Oil separation sedimentation tank.

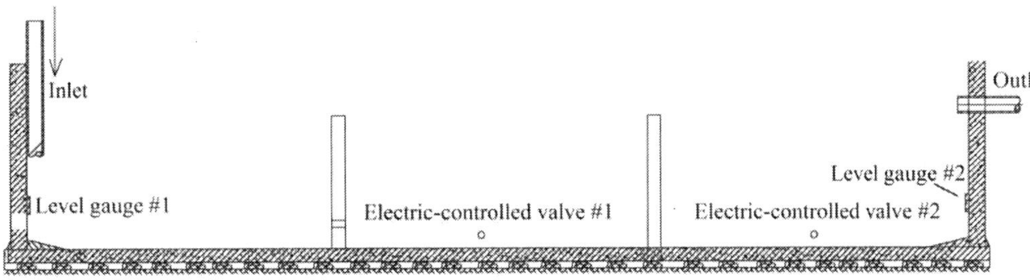

Fig. 3. Bridge runoff purification emergency integrated tank.

missing accidents. Therefore, through the bridge monitor in full coverage, the application of video-based traffic accident detection technology can achieve automatically identifying and issuing a warning to monitoring center. If manual determination ensures that the accident is involved with vehicles carrying hazardous chemicals, and then emergency operation will be confirmed to start, thus the automatic identification of accidents was optimized.

Compared with the oil separation sedimentation tank, the bridge surface runoff purification and emergency integrated tank is simpler in structure. At the same time it integrates the regulating, purifying and emergency disposal functions. There are two sets of retaining walls in the tank, dividing the tank body into three compartments. One plays a role in runoff regulation in water quality and quantity. when hazardous chemical leak accident occurs, the first two cell space with more than 100m3 area, can storage the chemicals and its flushing fluid, playing an emergency role. The water flows up and down in vertical baffle way in three cells. Normally, storm-water runoff flows into the tank from the upper part. The water level gradually increased in the first cell, and when the baffle height is reached, the runoff crosses the first baffle into the second cell. Cycling back and forth, the purpose of sedimentation purified layer by layer can be achieved. After the rainfall, the runoff purification is completed. Then the two electrically-controlled outlet drainage valves are remotely opened to realize the emptying function so as to ensure adequate storage space for the next leakage accident.

A group level gauge and electric control valve are set up one each inside the Cell 1 and Cell 3. After the rain stops, purification is also completed. According to the liquid level information, start the blow-off valve remotely to automatically discharge the treated bridge runoff. There are several overcurrent ports at the bottom of the retaining wall between Cell 1 and Cell 3. When the rainwater runoff is in a large amount, the flow passing through the overcurrent port is small. And the runoff will still pass through the first baffle and won't affect the baffle precipitation function. When the runoff in the tank settles, Cell 1 and Cell 2 will form a communication network. Therefore, only a group of level gauge and electronic controlled emptying valve are set up to save cost. This kind of technology has been applied to Heda Expressway in Jilin Province.

4 Bridge runoff purification emergency ecological planting groove Based on LID

Due to the extremely low probability of occurrence of hazardous chemicals leakage accidents, no matter the three-tank system or the purification and emergency integrated tank, they are all in masonry structures, which present a poor landscape effect even under the bridge. So they can only be a structure with temporary and limited function. At present, China is vigorously promoting the construction of "sponge city" with natural accumulation, natural infiltration and natural decontamination. Based on the concept of low impact development (LID), various ecological technologies including infiltration, retention, storage, purification, and reuse are integrated for application, thus playing a role in groundwater supplementary, and water cycle regulation [10]. Highway industry has carried out a lot of engineering practice that apply conventional treatment in the bridge runoff treatment and it has achieved initial success. But most of them fail to couple with the accident identification mechanism, so there is still the possibility of tanks full of rains when accident occurs. On the basis, guided by the LID conception, this paper puts forward the bridge runoff treatment process combining regulating tanks and ecological planting grooves, plus equipped with remote control module to achieve intelligent management.

The intelligence level of remote control technology has been further enhanced. It uses the camera with 1080P @ 30fps high-definition image full-frame detection technology. The real-time video image processing and system self-learning technology is carried to do real-time analysis of video images collected by the camera to identify road traffic accidents. After completing the initiative identification and track detection, it can initiatively remind staff through the sound and light alarming way to observe and deal with it. The alarming image will be immediately switched to the big screen, drawing attention to determine whether emergency operations should be carried out.

The treatment system is a combination of regulation ponds and ecological planting grooves. The pipeline between the regulation ponds and the ecological planting grooves are equipped with an electric-controlled connection valve, and electric-controlled emptying valves are arranged in the ecological planting grooves.

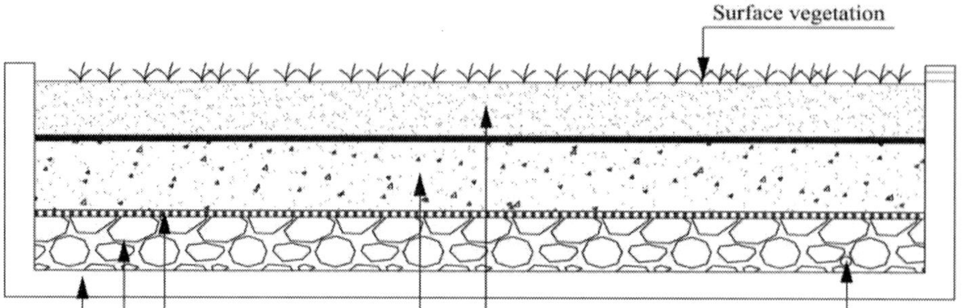

Fig. 4. Ecological planting groove unit.

When the hazardous chemical leakage occurs, staffs are required to shut down the valves remotely to fix the hazardous chemicals and their flushing liquid in the regulation tank with effective volume beyond 100m3. Under the extremely special circumstances of hazardous chemicals leakage happening in rainstorm conditions, it is possible that the regulation tank is overflowing. Then the electric-controlled connection valves are supposed to be opened while leaving the emptying valves closed. As a result, the ecological planting tanks are sacrificed in need to fully guarantee the water quality safety.

From top to bottom, each ecological planting groove unit consists of planting soil layer, adsorption purification layer, grille and non-woven fabric layer, aquifer layer and impermeable layer. Planting soil layer contains plants with good water purification effect and local soil, which needs allocation with a poor permeability. With multi-fillers of strong adsorbing and filtering ability for nitrogen and phosphorus removal, adsorption purification layer is the core unit of ecological planting groove. Grille and non-woven fabric, which plays a supporting role, separate the adsorption purification layer from aquifer layer. Filled with stones in a big size, the aquifer layer can make it possible to have reservoir of water, which can be provided for plants without rain. And anti-seepage geotextile is set in impermeable layer to prevent groundwater pollution from the leakage of hazardous contaminants.

This kind of structure has the integrated functions to deal with the bridge runoff, comprehensive utilization and accident emergency, which is an ecological technology with landscape effect. So it can meet the requirements of the green development and sponge city construction in the new era. This technology can also be applied to the road surface runoff purification emergency, which has been applied in Erhai East Lake Road.

5 Conclusions

1. The bridge surface runoff treatment three-tank system, the integrated tank and ecological planting grooves are all the new technologies of water protection especially for bridges over the sensitive water body. Compared to conventional technology, the surface runoff purification effect will be better. Due to the utilization of monitoring, accident identification, remote control and other intelligent means, emergency plan can be started in time, preventing the hazardous chemicals into the water body.

2. With the development of smart devices, accident identification and emergency response strategies adopted by the three technologies are still in the process of gradual improvement, and it can be interactively used under different situations. In the future, means of incidents identification should be further enriched to adopt video and microwave interaction technologies to further reduce the misjudgment rate, improving the accuracy of accident analysis.

3. Emergency control of the three technologies is based on accident identification. With the development of environmental monitoring technology, the three technologies mentioned in this paper could be improved with better measures, if the monitoring devices, which are of economic promotion value, could identify all kinds of hazardous chemicals and conventional bridge surface runoff.

4. From the perspective of water purification, the water purification effect increases in the order from the integrated tank, the oil separation sedimentation tank and ecological planting grooves. Besides, the discharge requirements and project cost could also be considered.

5. Under the guidance of green and innovation ideas, referring to new techniques such as sponge city and smart transport could accelerate the improvement of the technical system of water environment protection for bridge-crossing in China.

Acknowledgements

The research is supported by the science and technology project on transportation construction by Jiangxi Province (No. 2011C00005), the science and technology demonstration project of the Ministry of transport of the

People's Republic of China (No. 2016002) and the transportation science and technology project of Jilin Province (No. 2017ZDGC-2-8).

Reference

1. J.S. Wu, C.J. Allan. Characterization and Pollutant Loading Estimation for Highway Runoff. J. Environ. Eng. ASCE **124**(7): 584-592 (1998)
2. R.J. Charbeneau, M.E. Barrett. Evaluation of Methods for Estimating Stormwater Pollutant Loads. J. Water Environ. Res. **70**(7): 1295-1302 (1998)
3. J.K. Li, G.F. Du, H.E.Li. Purification Effects of Biological Filter on Urban Road Runoff. J. Soil Water Conserv. **26**(4): 1-6 (2012)
4. X. Zuo, D. Fu, H. Li. Variation Characteristics of Mercury in Speciation during Road Runoff for Different Rainfall Patterns. J. CLEAN–Soil, Air, Water **41**: 69-73 (2013)
5. F. Jia. The Protection Measures Study of Highway Operation Period on Water Environment Sensitive Points. J. Shanxi Sci. Tech. Commun. **5**: 94-96 (2015)
6. J.N. Liang, K. Liu, F. Pan. Impact and Protective Measures of Highway Construction in West Mountainous Area on Ecological. Environ. J. China Foreign Highway **30**(5): 20-24 (2010)
7. Y.H. Wang, Z.Q. Xiong, Y.T. Shan. Technical Research on Water Environmental Protection in Construction of Renhai-Chishui Expressway. J. Tech. Highway Transport **3**: 155-158 (2014)
8. J.P. Xu, S.G. Shao, W.S. Yu. Discussion of Integrated Design of Emergency Storage Facilities of Dangerous Chemicals in Expressway Transport Accidents and Treatment Facilities of Runoff on Bridge Surface. J. Highway Transport. Res. Dev. **31**(12): 139-144 (2014)
9. H.W. Qin, L.X. Zhu, X.F. Fu. Selective Collection System of Deck Runoff Based on Real-time Monitoring. J. Yellow River **31**(9): 24-25 (2009)
10. D. Zhou. The Application of the Sponge City in Modern City Construction. J. Anhui Agri. Sci. **43**(16): 174-175 (2015)

The Study of Sand-blocking Characteristic and Flow Field Simulation Analysis of the Ring-shaped Sand-barrier

Wei Qiang ZHENG, Wen Jun WEI, Ping Yi LIU *

China Agricultural University, Beijing100083, China

Abstract. Aiming at the complexity of wind direction and irregular sand flow in a desert area, a combinatorial ring-shaped sand barrier is used. Stokes law of inertia force and centrifugal force and gravity sedimentation are used. With CFD fluid software Fluent, laminar flow model Equation and κ-ε turbulence model, the wind speed of the sand in the sand-gas-solid two-phase flow passing through the circular sand barrier is studied at different distances and different altitudes after the sand barrier, the wind speeds before and after the sand barrier are compared and analyzed . The mean minimum wind speed behind the single sand barrier was reduced by 32.5% -49.4% compared with that before the sand barrier. The wind speed at different height of the composite sand barrier was reduced by 30% -58.3% compared with the inlet wind speed, which solved the problem of irregular wind and sand control in the desert wind direction.

1 Introduction

Sand particles forms aeolian flow under the action of wind farms, the aeolian flow aggravates the process of land desertification, study on the law of sand flow, master the law of sand flow,it can provide a theoretical basis for wind and sand stabilization.Wind sand liquidity is a complex multi-factor problem and a typical gas-solid two-phase flow in the desert[1].Sand-blocking process is a combination of a single sand macroscopical movement and macrovisual study[2,3].The sand barrier is one of the most effective sand fixation method, the sand barriers is mainly lattice sand barriers, the grass pane sand barries is used most on practical, most studies on sand barriers are also based on lattice sand barriers [4-8] .Some domestic scholars have also conducted some explorations and researches on the shape of sand barriers,such as Qu Jian Jun et al [9] studied the windbreak and sand fixation effects of six types of cotton sand barriers in the three structures of implantable,fenced and bundled structures,various structures and specifications of cotton stalk sand barriers can increase the surface roughness to a certain extent, reduce the near-surface wind speed, slow down the sand movement in the barrier and reduce the amount of sediment transported.Liu Jinhao et al [10] studied the windbreak and sand-fixing capabilities of the regular triangle,square and regular hexagonal grass barriers,the windbreak performance of a square grass sand barrier is obviously smaller than that of a regular triangle and a regular hexagon,the largest surface roughness is a regular triangular grass sand barrier.Wang Chengyu and Wang Yujie et al[11] conducted field trials to study the blocking effectiveness of positive semicircular,inverted semicircular and one-shaped sandbag sand barriers,the overall distance and sand erosion of sand barriers,the wind resistance benefits of obtaining a semi-circular sand barrier is best.Hayakawa et al[12]studied the flow around the cylinder and the influencing factors of wind speed variation in the flow field. This article proposes a combined ring-shaped sand-barrier,the fluent fluid software and the laminar flow model combined with κ-ε standard model [13] is used to study the sand-resistance characteristics of single and combined ring-shaped sand barriers and their effects on wind speed changes are analyzed, then the structure of the ring-shaped sand barrier is optimized, to get the best structure and arrangement.It can not only solve the problem of high experimental cost and long cycle in the study of windy and irregular regions,and also provide reference for windbreaks and sand fixing problems in areas with irregular wind directions

2 Sand particles separation principle

When sand particles move on the aeolian two-phase flow field, it can issubjected of the external forces, the external force including gravity, buoyancy, pressure gradient force, false mass force, basset force, mangus lift force and samffman lift force [14].In this paper, the study of sand-blocking characteristic of the ring-shaped sand-barrier is carried out, based on the dual principle of inertial force and centrifugal force (Fig.1).

2.1. Inertial force principle

When the sand passes through the ring-shaped sand-barrier windward side, due to due to the principle of the inertial force, a part of the sand particles is blocked when the sand particles knocks against the circular sand-barrier

* Corresponding author: liupingyi@cau.edu.cn

windward. If we consider the gravity sedimentation of sand particle obey the law stokes law in aeolian gas-solid two-phase flow, assuming that only gravity and drag act on the sand particle and the particle is in an initial stationary state, then the gravimetric momentum equation and its boundary conditions are when t = 0, the sand particle force equation of motion:

$$\sum F = G_O - F_d - m\frac{du}{dt} \qquad (1)$$

In the formula: ΣF is composition of forces（N）;G_0 is residual gravity (also known as effective gravity), gravity minus the buoyancy（N）;F_d is fluid resistance（N）;m is the quality of sand(kg); u is sand velocity(m/s).

Assume that the sand is spherical sand:

$$m = \frac{\pi}{6} d_p^{\ 3} \rho_p \qquad (2)$$

$$G_0 = \frac{\pi}{6} d_p^{\ 3} (\rho_p - \rho) g \qquad (3)$$

$$F_d = \zeta \frac{\pi d_p^{\ 2}}{4} \rho \frac{u^2}{2} \qquad (4)$$

In the formula: d_p-grit diameter（m）;ρ_p-grit density（kg/m^3）; ζ-resistance coefficient; ρ-gas-solid mixed fluid density（kg/m^3）.

Will G_0, F_d, m equivalent to (1) available:

$$\frac{d_u}{d_t} = g\frac{(\rho_p - \rho)}{\rho_p} - \zeta\frac{3}{4}\frac{u^2}{d^p}\frac{\rho}{\rho_p} \qquad (5)$$

It can be seen from the equation of motion, the acceleration of the sand depends on the residual gravity and air resistance in the static wind. G_0 is a constant for a certain size of the sand particle when the sand particle settles in the flowing air, fluid resistance increases with increasing velocity. If the gravity is greater than the buoyancy, the sand particle will accelerate to decline due to its own gravity at the beginning of settlement.

When suspended aeolian around the ring-shaped sand barrier, due to its inertia, the aeolian detaches from the fluid streamline, hits the windward surface of the ring-shaped sand barrier, inertial sedimentation occurs on the windward side of the ring-shaped sand barrier (Fig.1). The ring-shaped forms a large vortex, reaches full turbulence in turbulent state, flow rate is very big, the total resistance becomes smaller.

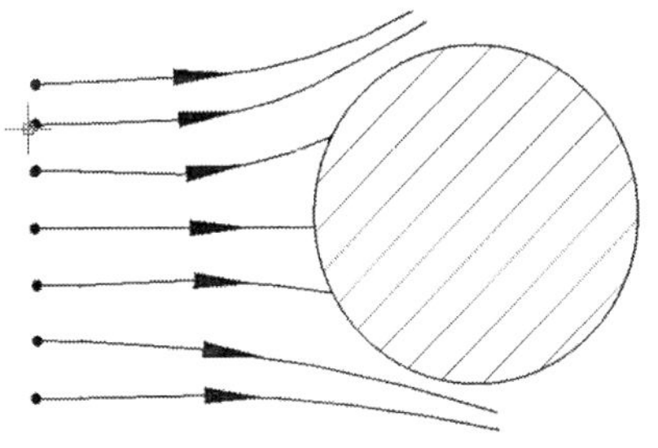

Fig.1. The inertia sedimentation schematicdiagram of sand particles

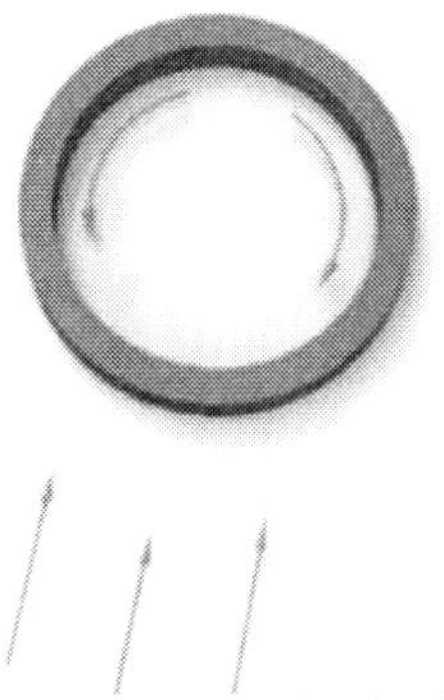

Fig. 2. The centrifugal sand resistance principle of ring-shaped sand-barrie

2.2. Centrifugal principle

When the aeolian passes through the ring-shaped sand barrier wall (Fig.2), under the action of centrifugal force, the aeolian rotates movement along the ring-shaped sand barrier wall, by the density differences of sand and fluid, according to the sedimentation coefficient of sand and buoyant density difference to separate sand. During the process of sedimentation, the movement direction and speed of sand particles have the following relations to the density, shape, size and intensity of sand particles.

(1)Centrifugal force of the sand particles is :

$$F_C = \frac{4}{3}\pi r^3 \rho \omega^2 R \qquad (6)$$

r- Sand particles radius（m）; ρ - sand particles density（kg/m^3）; ω-angular velocity (rad/s); R-airflow rotation radius.

(2)Settlement speed:

When the sand particles is $\frac{4}{3}\pi r^3 \rho$ **g**， sand buoyancy is $\frac{4}{3}\pi r^3 \rho'$ **g**:

$$V_c = \frac{2}{9}\frac{r^2(\rho - \rho')}{\eta}\omega^2 R \qquad (7)$$

When the sand particles diameter is d, in the centrifugal field velocity is:

$$V_c = \frac{d^2(\rho - \rho')}{18\eta}\omega^2 R \qquad (8)$$

Equation (8) below the unit of gravity, the velocity of the sand particles in the stokes theorem:

$$V_c = \frac{d^2(\rho - \rho')}{18\eta} g \qquad (9)$$

Above formula: sand sedimentation rate (V_c);the sand particle diameter is medium density (ρ'); gas-solid mixing medium viscosity(η).

Sand particles and air is separated under the action of centrifugal force, the greater the centrifugal force of sand, the higher the separation efficiency.

Analytical calculation: When the wind speed is 12m/s, R=4m, t=R-r=0.1m, h=1m is the best parameter of sand barrier.

 https://doi.org/10.1051/e3sconf/20183801044

3 The simulation of the ring-shaped sand-barrier sand-blocking characteristic

3.1. A single ring-shaped sand-barrier

To model and mesh a single ring-shaped sand barrier by GAMBIT of the fluent software preprocessing software(Fig.3),the size of three-dimensional calculation: the length is 20m,the width is 10m,the height is 1.5m;The diameter of ring-shaped sand-barrier is 8m,thickness is 0.1m, height is 1m;Mesh is divided into 0.2mm;Boundary conditions: ① Entrance:speed is12m/s; ② Export: Free flow; ③The component of each variable in the normal direction is zero;④ The remaining boundary conditions using slip-free solid wall boundary conditions [15-16].

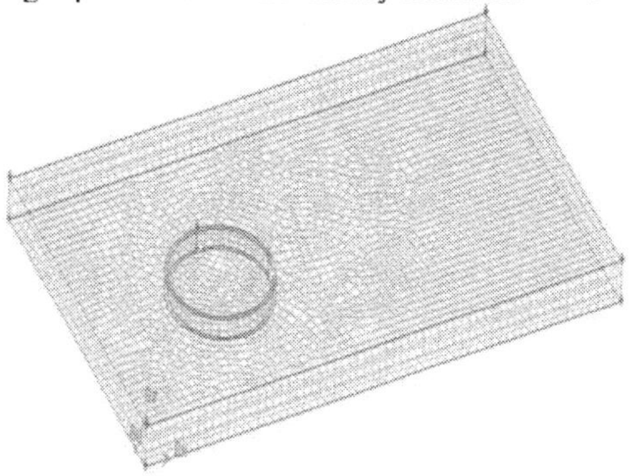

Fig. 3. The grid division graph of a single ring-shaped sand-barrier

The calculation area is modeled by GAMBIT preprocessing software, the model is read into by the FLUENT solver, checking the model file, choosing laminar calculation model, defining fluid physical properties, setting all the parameters, the simulation results are as follows:

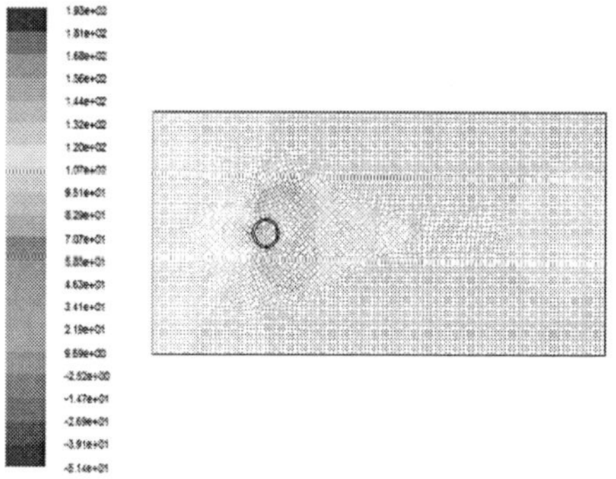

Fig. 4. The pressure nephogram of a single ring-shaped sand-barrier

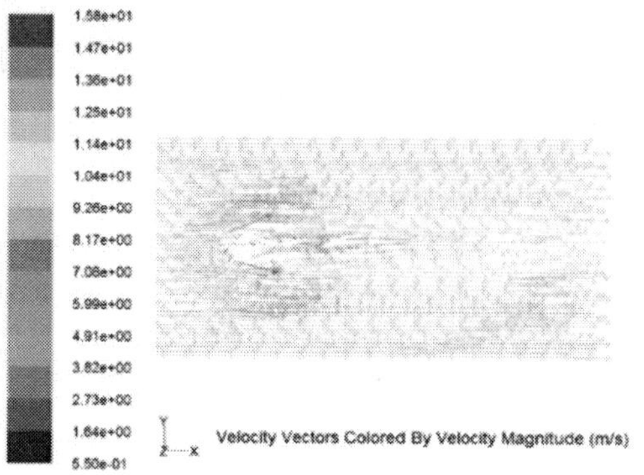

Fig. 5. The velocity vector of a single ring-shaped sand-barrier

In Fig4, the wind pressure is between-5.14e + 01 and 1.93e + 02, the wind pressure is the largest at the ring sand barrier windward and reduces first and then increase the trend along the radial to the sand barrier leeward in the model area. In Fig5, the speed of the aeolian passing the ring-shaped sand barrier is between 5.71e-02 and 1.77e + 01, the speed is the smallest in the ring-shaped sand barrier windward and in the area behind , the wind speed reaches the maximum along on both sides of the tangential direction of the ring-shaped sand barrier in the whole model area, along the radial to the ring-shaped sand barrier leeward,the wind speed increases first and then decreases and the wind pressure is exactly the opposite size. The wind speed gradually increases along the radial direction of the sand barrier surface, until the same with the mainstream wind speed outside, in the sand barrier windward surface the wind speed continues to decline due to sand barrier surface friction, but the main speed of the sand barrier boundary layer is gradually increasing, therefore wind speed will not decline. The prevailing wind speed gradually weakened in the sand barrier leeward, making the wind speed of the ring-shaped sand barrier surface reduced, because of getting the mainstream wind speed outside.

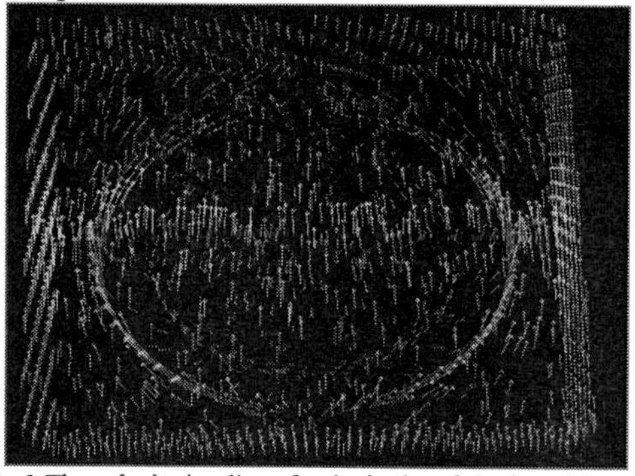

Fig.6. The velocity isocline of a single ring-shaped sand-barrier

 https://doi.org/10.1051/e3sconf/20183801044

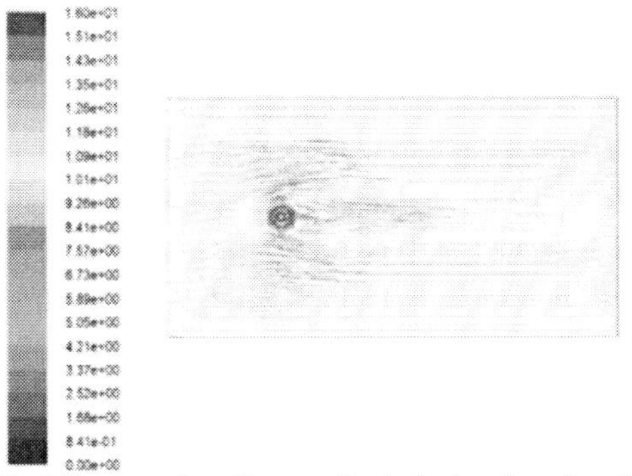

Fig.7.The streamline diagram of a single ring-shaped sand-barrier

It can be seen from Fig 6 that a part of the aeolian has straight the inner wall of the ring-shaped sand-barrier, a part of the aeolian moves along the direction of the center of the wall to move on both sides, due to the inertial force.It can be seen from the fig7 that there are two symmetrical whirling vortices behind the ring-shaped sand barrier, commonly known as Karman vortex,this process is turbulent and it is a steady-state process.

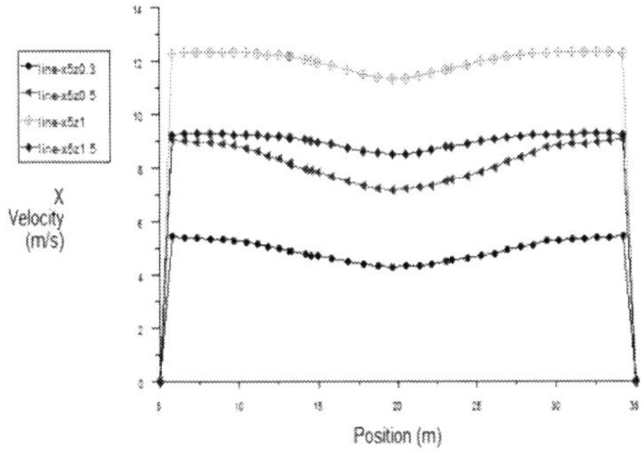

(a) x= 5m

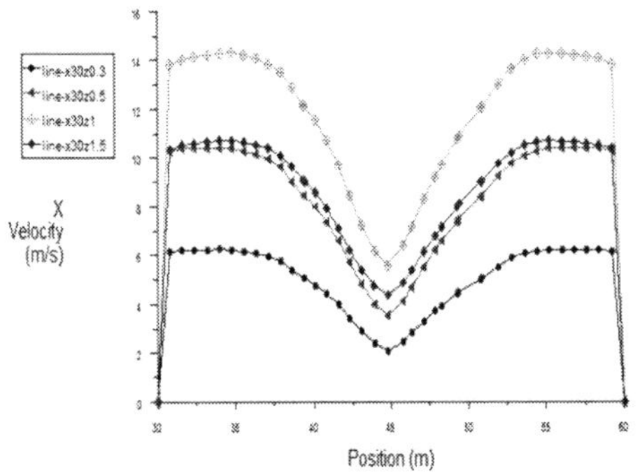

(b) x= 10 m

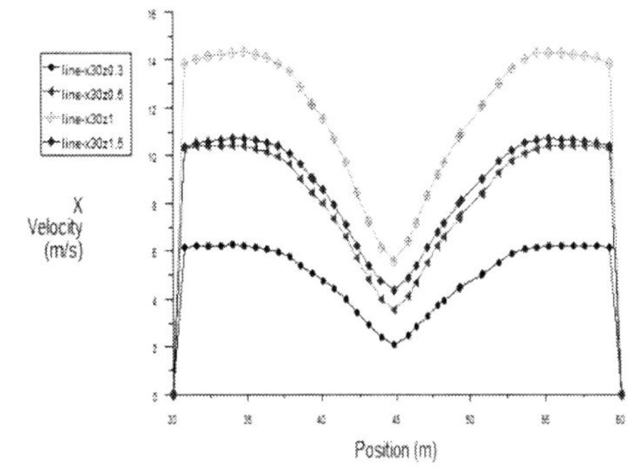

(c) x=20 m

Fig. 8. The wind speed before a single ring-shaped sand barrier

In Fig8(a), at 5 m before the single ring-shaped sand barrier, in the area just in front of the sand barrier, at 0.3m,0.5m,1m and 1.5m,the change of wind speed is about 1m /s, the height has little effect on wind speed changes. The height of the area just behind the sand barrier is 0.3m、0.5m、 1m and 1.5m, the change of wind speed is between 4m/s ~ 9m/s and 3.5m/s ~ 6.5m/s respectively, and the wind speed significantly reduces compared to the area before the sand barrier. The wind speed before and after the sand barrier increases with the increase of height, the closer to the ground or closer to the border the smaller the wind speed.

Table 1shows that the mean minimum wind speed at 10 meters after the sand barrier reduces by 49.4% than at 5 meters before the sand barrier, the average minimum wind speed at 20 meters after the sand barrier reduces to 32.5% than at 5 meters before the sand barrier.

Table 1 The sand barrier has different distance and high wind speed

Z	X=-5m (m/s)	X=-10m (m/s)	X=-20m (m/s)
0.5m	7	3.5	5.5
1m	11	6	7.2
1.5m	8.5	4.2	5
Average speed(m/s)	7.7	3.9	5.2

3.2. The combined ring-shaped sand barrier

The combined ring-shaped sand barrier arrangement is the sand barrier staggered arrangement of equal spacing in the desert, the distance between two adjacent sand barriers is greater than the outer diameter of sand barrier,scilice:d> 2（r+t）, three adjacent sand barrier center line was equilateral triangle(Fig.9).

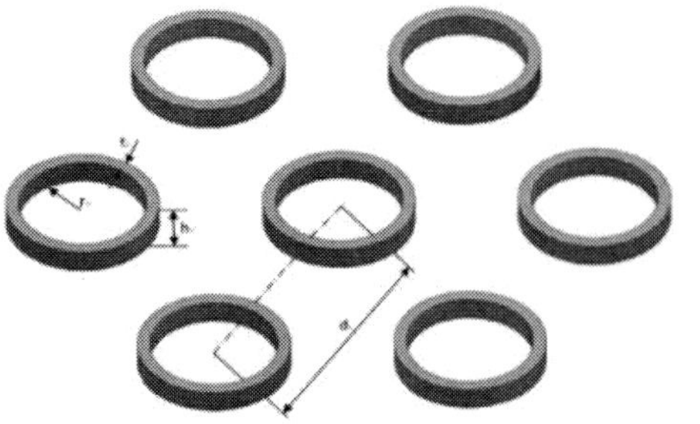

Fig. 9. 3d modeling diagram of combined ring-shaped sand barrier

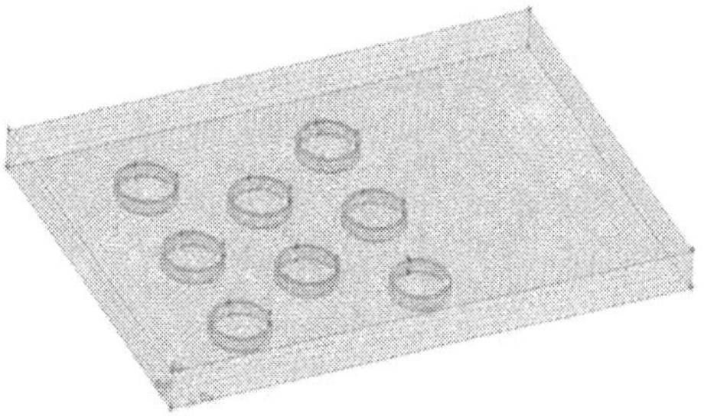

Fig.10 The combined ring-shapedsand barrier grid division map

The same as the study process of sand resistance of a single ring-shaped sand barrier, also use FLUENT software pre-processing software GAMBIT to model and mesh of combined ring-shaped sand barriers (Fig.10), its three-dimensional calculation area model size is: the length is 50m, the width is 35m, the height is 3m, the rest of the conditions are set the same as a single ring-shaped sand barrier, simulation results are as follows.

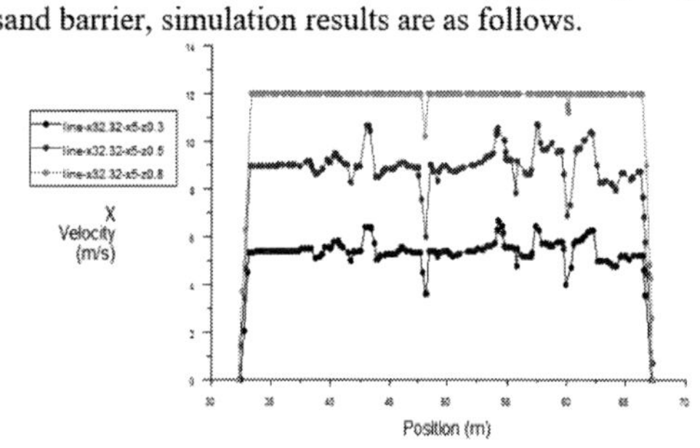

(a) x=5m

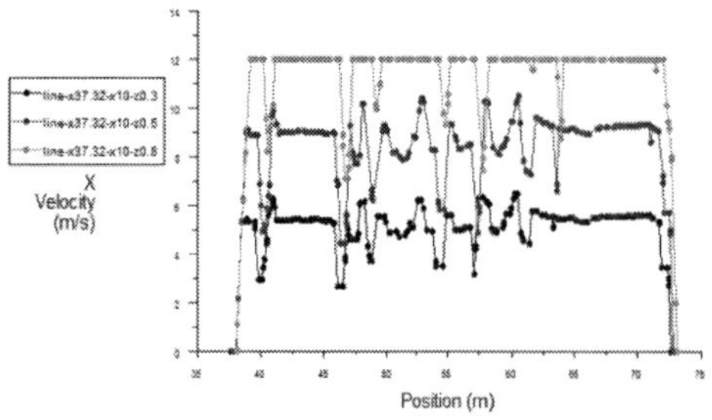

(b) x=10m

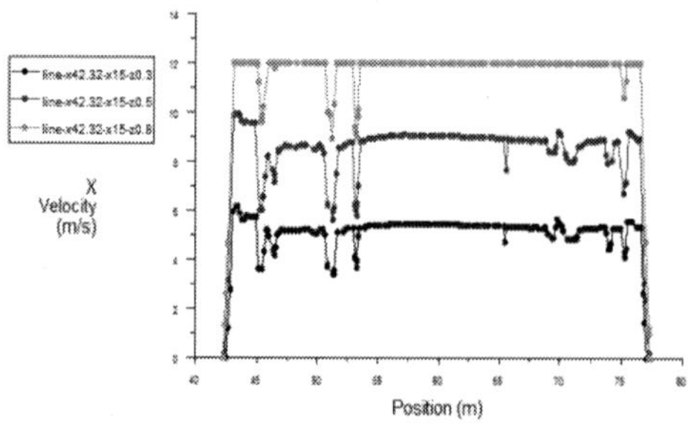

(c) x= 15m

Fig.11. Along x direction wind speed after the combined sand –barrier

In Fig.11, the lower the ground clearance the lower the speed and the higher the ground clearance the higher the speed after the sand barrier different distances, eventually approaching the inlet wind speed. At the height of 0.3 m, 0.5 m and 0.8 when the distance from the sand barrier is 5m, wind speed is about 5.5 m / s, 9 m / s and 12 m / s, the wind speed are basically the same at 10 m and 15 m away from the sand barrier and the wind speed at 5 m, but it has a greater frequency of change fter the sand barrier 10m, the speed fluctuations become smaller from the sand barrier near or far away ,more frequently in the sand barrier interception area.

Table 2 shows, at the same height, after the combining sand barrier at 5m, 10m, 15m, the average wind speed changes little, but the wind speed is increasing with the increase of the height after the combined sand barriers, with the increase of the distance, the wind speed basically remains unchanged at the same height.Wind speed decreased by 30% -58.3% at different distances and altitude after combined barrier compared with the inlet wind speed.

Table 2. The wind speed of pre and post combined sand –barrier at different height and distance

Z	X=5m (m/s)	X=-10m (m/s)	X=-20m (m/s)
0.3m	5.51	5.19	5.21
0.5m	9.17	8.53	8.64
0.8m	11.97	11.51	11.83

4 Conclusion

(1)Based on the dual principle of inertial force and centrifugal force, analysis and calculation of ring-shaped sand barrier: when wind speed is 12m/s, the radius of sand barrier is 4m, t=R-r=0.1m,h=1m,it is the best parameter for a single ring-shaped sand barrier; Ring-shaped sand barrier staggered arrangement of medium-spaced in the desert, adjacent two center distance is greater than the outer diameter of the sand barrier,it is the best layout for way that adjacent three centers from the connection was equilateral triangle.

(2)By using Fluent to simulate the wind speed of sand through the single and combinesand barriers, the average minimum wind speed at 10 meters after a single ring-shaped sand barrier decreased 49.4% compared with before the sand barrier, the average minimum wind speed at 20 meters after the sand barrier reduced by 32.5% compared with that before the sand barrier; Combined sand barrier wind speed at different heights than the inlet wind speed reduced by 30% -58.3%. The analysis shows that the average wind speed decreases less than before the sand barrier, the lower the ground clearance, the lower the wind speed, the higher the ground clearance is, the bigger the wind speed is.Compared with the current technology, with more effective anti-desertification effect, it can provide a reference for the actual sand control and sand fixation and be suitable for the desert environment in the complex control of the aeolian control project.

References

1. Wu sheng-zhi,Ren chun-yong.Numerical simulation of wind-blown sand based on the Eulerian model [J]. Journal of Lanzhou University(Natural Science),2012,48(1): 104-120.
2. S Fohanno,B Oesterle.Analysis of the Effect of Collisions on the Gravitational Motion of Large Particles in a vertical duct[J].International Journal of Multiphase Flow , 2000,26(2): 267-292
3. Salar Azizi , Dariush Mowla, Goodarz Ahmadi. Numerical evaluation of turbulence models for dense to dilute gas-solid flows in vertical conveyor [J]. Journal of particles, 2012, 553-561.
4. Sun Tao, Liu hu jun,Zhu guo -qing,et al.Timeliness of reducing wind and stabilizing sand functions of three mechanical sand barriers in arid region[J]. Journal of Soil and Water Conservation, 2012,26 (4) :12-16
5. Zhou na, Zhang chun-lai, Tian Jinlu et al. Flow field controlling the concave surface of the semi-buried checkerboards and its characterization by grain sizes of sediments[J]. Geographical Research,2014,33(11): 2145-2156
6. Huang qin qin.Research on Numerical Simulation and Feature Structure of Grass Sand Barrier for Mechanised Sand-fixing[D].Beijing forestry university,2014
7. Chen yong.Numerical simulation and optimal design of the vertical array sand trap and checkerboard sand-barrier[D]. Lanzhou University,2014
8. Ma xue-xi,Wang hai-feng,Li sheng yu et al. Comparison of Sand protecting efficiency and Terrain adapted of two checkerboard barriers [J]. Journal of Soil and Water Conservation, 2015,35(2):344-349
9. Ma rui,Wang ji-he, Qu jian jun et al. Study on protective effect of difference types of cotton haulm sand barriers[J]. Journal of Soil and Water Conservation, 2010,24(2):48-51
10. Sun hao,Liu jin-hao,Huang qing-qing et al. Research on the windproof efficiency of polygonal straw sand barrier[J]. Journal of BeijiuR- Forestry University, 2017, 39(10): 90-95
11. Wang chen-feng,Wang Yujie,Yuan li-min et al. Stability and Wind-break Effect of Sandbag Barriers of Different Shapes[J]. Journal of Soil and Water Conservation, 2015,29 (5) :7-12
12. Hayakawa M, Hussein A K M F. Turbulence structure in a cylinder wake. Advances in Turbulence[M].Springer-VerlagBerlin Heidelberg,1987:416-423
13. Yongxiang He, Haibo Zhao, Haoming Wang, et al. Differentially weighted direct simulation Monte Carlo method for particle collision in gas-solid flows[J]. Particuology, 2015(04): 135-145.
14. [14]Zhao Jiang. Numerical Simulation of Wind-Blown Sand under the Influence of Baffle[D]. Lanzhou University, 2013.
15. Wang fu-jun.Computational fluid dynamics analysis CFD software principles and applications [M]. Beijing: Tsinghua university press,2004
16. Shi F, Huang N. Computational simulation of blow sand fluxes over the surfaces of complex micro to pography[J].Environ mental Modeling & Soft ware, 2010,25: 362-367.

https://doi.org/10.1051/e3sconf/20183801045

Study on the Evaluation of Economic Support Index for Pig Breeding in Water Network Area of Middle Reaches of Yangtze River

Bi-Bin Leng, Qi -zhen Zhang, Wen-wei Lai and Xin-Fan Tang

Jiangxi Science &Technology Normal University, Nanchang, China

Abstract. The central region of China boasts a long history, abundant resources, convenient transportation, advanced economy, a strong industrial and agricultural foundation and a rapid development of modern service industry, with about 10.7% of the country's total land area, 26.5% of the country's population and a GDP of about 21.4% of the country's total . As the population center, transportation hub, economic hinterland and important market, it is the second echelon of China's economic development; the central region of China plays an important role in the division of labor in China. The middle reaches of the Yangtze River water network includes four neighboring provinces- Hubei, Hunan, Anhui and Jiangxi. Although all four provinces are located in the middle of the Yangtze River Basin, there are still quite some differences. Based on the 2017 Statistical Yearbook data, this paper studies the economic supportive index of pig breeding in four provinces located in the water network area of the middle reaches of the Yangtze River in Jiangxi, Hubei, Hunan and Anhui. The evaluation results show that among the four provinces, the socio-economic conditions of Hubei Province can most support the large-scale cultivation of live pigs, and the support of Jiangxi Province is the weakest.

1. Introduction

The Ministry of Finance, the Ministry of Environmental Protection, and the State Administration of Taxation issued a joint notice that the preparation for the collection and management of environmental protection tax was fully started and the Environmental Protection Tax Law was formally implemented on January 1, 2018. The implementation of the environmental tax law is a severe test for the breeding industry, which also means that the country pays much attention to the livestock breeding issue. Pig breeding is the top industry in animal husbandry. Inevitably, there will be a large number of "public hazards of livestock production", such as poultry excrement and aquaculture sewage which are arbitrarily dumped and discharged, and the environment is deteriorating in the concentrated areas where livestock farms are located, indirectly contaminating water sources and soil air.

The four provinces of Jiangxi, Hubei, Hunan, and Anhui are in the middle of the inland and coastal areas, and have become important links and bridges between inland and coastal cities. With the advantages and resources of the economic belt of the upper reaches of the Yangtze River and the economic belt of the lower reaches of the Yangtze River, these four provinces have become the "hearts" of China's economic geography. Based on the data of 2017 statistical yearbook, this paper uses the analytic hierarchy process to evaluate the economic support index of pig-scale aquaculture in the four provinces of the middle reaches of the Yangtze River.

2. Index determination and calculation

With reference to the existing literature, when the average GDP per capita, industrial enterprise research and experimental development expenditure equal to the ratio of regional GDP and total pork production are taken as evaluation indicators, the economic support of live pigs in six provinces in Central China is evaluated. According to the data of the 2017 statistical yearbook, the evaluation index of the economic support index for the scale pig breeding scale are calculated as Z1, Z2, and Z3, which are shown in Table 1:

Table 1. Calculation and Standardization of Economic Support Index of Live Pig Breeding in Four Provinces in the Middle Reaches of the Yangtze River in China

Province	Regional Product Value per capita with Dimensionless Process Z_1	R&D Expense of Industrial Enterprises Share of Regional GDP with Dimensionless Process Z_2	Pork Production with Dimensionless Process Z_3
Anhui	0.71	1.00	0.56
Jiangxi	0.72	0.63	0.56
Hubei	1.00	0.90	0.74
Hunan	0.83	0.81	1.00

Based on the table above, the descriptiveness of social and economic support indicators is further calculated and analyzed as shown in Table 2 below:

Table 2. Descriptive statistics

	N	Minimum	Maximum	Mean Value	Standard Deviation
Z1	4	.71	1.00	.8150	.13478
Z2	4	.63	1.00	.8350	.15716
Z3	4	.56	1.00	.7150	.20809
Valid N (the list state)	4				

Table 2 shows the descriptive statistics of the economic support indicators. Combining the data with that of table 1, we can see that the standard deviation of regional GDP per capita is small, and the degree of sample dispersion is low. The per capita GDP of Hubei is the highest; and the average per capita GDP of Anhui Province and Jiangxi Province are below the average value. As for the research expenditure of research and development of industrial enterprises share of regional GDP, Anhui owns the largest value, while Jiangxi owns the smallest one. And the indexes of Jiangxi and Hunan is below the average level. The standard deviation of total pork production is large, and the sample has a high degree of dispersion. Hunan has a maximum value while Jiangxi and Anhui own the smallest value. Besides, the indexes of these two provinces is below the average.

3. Evaluation on the economic support index of live pig breeding in four provinces in the middle reaches of the Yangtze River

After analyzing that factors analysis method for evaluation indexes of live pigs' scale aquaculture economic support index in six provinces in central China is applicable, we performed factor analysis on the indicators and obtained the factor variance contribution rate as shown in Table 3.

Table 3. Factor variance contribution rate

Component	Initial Egigenvalues			Extraction Sums of Squared Loadings			Rotation Sums of Squared Loadings		
	Summation	Of Variance %	Accumulation %	Summation	Of Variance %	Accumulation %	Summation	Of Variance %	Accumulation %
1	1.524	50.810	50.810	1.524	50.810	50.810	1.466	48.876	48.874
2	.995	33.153	83.963	.995	33.153	83.963	1.053	35.090	83.963
3	.481	16.037	100.000						

The factor variance after the extraction is very high, indicating that the extracted factors can best describe these three indicators. The variance contribution rate of factor 1 is 50.810%; the characteristic value is 1.524; the variance contribution rate of factor 2 is 33.153% and the characteristic value is 0.995. These two indexes explain 97.047% (more than 80%) of the variance, so we extracted the first two components as the first and second principal components. Therefore, we get Table 4 below:

https://doi.org/10.1051/e3sconf/20183801045

Table 4. Component Matrix[a]

	Component	
	1	2
Z1	.871	-.009
Z2	.371	.907
Z3	.793	-.415

As shown in Table 4, the first component has the maximum load on variables Z1 and Z3, and the correlation coefficients are 0.871 and 0.793 respectively; the second component has the maximum load on Z2 with the correlation coefficient as 0.907. It can be concluded that the variables Z1 and Z3 are mainly explained by the first principal component, and Z2 is mainly explained by the second principal component. The following factorial rotation factors are shown in Table 5 below:

Table 5. Rotating component matrix

	Component	
	1	2
Z2	.050	.979
Z1	.825	.280
Z3	.885	-.129

The results obtained after the rotation are shown in Table 5. The maximum load of component 1 on the variables Z1 and Z3 was 0.825 and 0.885 respectively. The maximum load of component 2 on Z2 was 0.979. Component 1 can be named as F1 and component 2 as F2. Therefore, the factor score coefficient matrix is as shown in Table 6.

Table 6 .Component Score Coefficient Matrix

	Component	
	1	2
Z1	.542	.181
Z2	-.072	.941
Z3	.629	-.222

Each factor expression is:

$F1 = 0.542 * Z1 - 0.072 * Z2 + 0.629 * Z3$

$F2 = 0.181 * Z1 + 0.941 * Z2 - 0.222 * Z3$

First two variance contribution rates (50.810%, 33.153%) were normalized (60.515%, 39.485%). Take the variance contribution rate of each factor was the weight, and the environmental pollution bearing index was obtained as follows:

$Y = 60.515\% * F1 + 39.485\% * F2$

Factor scores and energy system indexes are shown in Table 7 below.

Table 7. Social and Economic Support Index

Province	F1	F2	Y
Anhui	0.66506	0.94519	0.77567
Jiangxi	0.69712	0.59883	0.65831
Hubei	0.94266	0.86362	0.91145
Hunan	1.02054	0.69044	0.89020

The following figure shows the socio-economic support index Y in the Central Region:

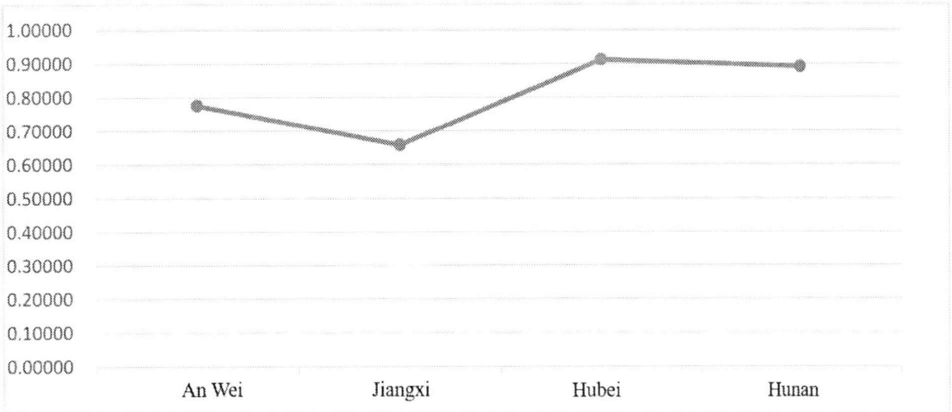

Chart 1. Line chart of environmental bearing capacity index of live pigs in four provinces in the middle reaches of the Yangtze River

As shown in the figure, Hubei has the highest socio-economic support index, followed by Hunan Province, Anhui Province, and Jiangxi Province. The lowest is Jiangxi Province. It shows that among the four provinces in the middle reaches of the Yangtze River, the social and economic conditions in Hubei Province can best support the scale breeding of live pigs, while Jiangxi Province has the weakest support.

4. Conclusion

With the intensification of market competition and people's increasing demand for quality of animal products, the scale of aquaculture has continued to increase. As a result, the pig breeding industry has become the top industry in animal husbandry. Inevitably, there will be a large number of "public hazards of livestock production". Animals and poultry excrement and aquaculture sewage are arbitrarily dumped and discharged, and the environment is deteriorating in the areas where livestock farms are concentrated, indirectly contaminating water sources and soil air.

The four provinces in the middle reaches of the Yangtze River water network - Jiangxi , Hubei , Hunan , and Anhui are in the middle of the inland and coastal areas, and have become important links and bridges between inland and coastal cities. With the advantages and resources of the economic belt of the upper reaches of the Yangtze River and the economic belt of the lower reaches of the Yangtze River, Hunan, Jiangxi, Hubei and Anhui have become the "hearts" of China's economic geography. Based on the data of 2017 statistical yearbook, this paper uses the analytic hierarchy process to evaluate the economic support index of pig-scale aquaculture in the four provinces of the middle reaches of the Yangtze River. The evaluation results show that Hubei Province has the highest index of economic support, followed by Hunan Province, Anhui Province, Henan Province and Jiangxi Province, and the lowest is Shanxi Province.

References

1. China Statistical Yearbook [M]. Beijing: China Statistics Press, 2017.
2. China Animal Husbandry Statistics Yearbook [M]. Beijing: China Agriculture Press, 2017.
3. National Pig Production Development Plan (2016-2020) [J]. China Agricultural Information, 2017(01):16-22.
4. Li Zhigang, Leng Bibin, Tu Guoping Comprehensive Evaluation of Environmental Bearing Capacity of Livestock Breeding Scale [J].Journal of Anhui Agricultural Sciences, 2014 ,42(27):9382-9385+9471.
5. Meng Xianghai, Zhang Juntao, Li Peng, Chen Xiaokun Summary of Environmental Pollution and Environmental Governance Policies in Animal Husbandry [J]. Journal of Ecology and Rural Environment,2014,30(01):1-8.

The Evaluation of Land Use Status in Mountainous Counties of Southwest China Based on Comprehensive Evaluation Models: A Case Study of Baoxing County, Sichuan Province

Mengtian Cao [1, *], *Jundi* Shen[1], *Zhehua* Chen[2]

[1]School of Tourism management, South China Normal University, Panyu district, No.378 Waihuan West Road, Guangzhou, 510006, Guangdong, P.R.China.
[2]Guangdong Forestry Survey and Planning Institute, Tianhe district, No.338 Guangshan 1st Road, Guangzhou, 510520, Guangdong, P.R.China.

Abstract. In mountainous areas of Southwest China, the land resources are scarce, and the ecological environment is fragile, so it is particularly important to carry out the evaluation of land use status for the sustainability of land development. Taking Baoxing County in Sichuan Province, the typical mountainous county in Southwest China, as an instance, this study refers to the existing research frameworks to establish the evaluation system of land use status. Meanwhile, the comprehensive evaluation models are used to evaluate land use status. As indicated from the results, in Baoxing County, the comprehensive evaluation score of the overall status of land use, the evaluation score of the development degree of land, the evaluation score of the intensive management degree of land and the evaluation score of the comprehensive benefits of land were 83.5, 108.24, 72.25 and 80.77, respectively. Land use status is generally at the relatively rational use stage, and the main problems are the lack of land investment and the low mechanization level of agricultural production. It is suggested to increase the financial investment in land and enhance the intensive degree and comprehensive benefits of land in the future.

1 Introduction

With an area of about 2/3 of China's territory area, China's mountainous areas are characterized by the complex geological environment, the scattered land resources and the outstanding contradictions between limited land resources and rapid urbanization. With the increase of population and the rapid improvement of urbanization rate, the industrialization degree is also being enhanced constantly. Under the pressure of these leading factors, the scales of industrial land and urban and rural construction land have been expanded gradually, which not only poses great challenges to the maintenance of limited land resources in mountainous areas but also can lead to the destruction of ecological environment in mountainous areas. Therefore, the reasonable development and use of the existing land resources in mountainous areas and the environmental management are crucial to ensuring the sustainable development of mountainous areas. However, the development and use of land in mountainous areas should be based on the evaluation of land use status, so as to carry out the overall planning and layout.

As a multi-level and multi-factor complex study, the evaluation of land use status is an analytical evaluation of characteristics, utilization structure and layout, use degree, utilization effect and existing problems of land resources in the research area, and the evaluation results are influenced by natural, human and other factors.The land evaluation frameworks are divided into three categories, namely the Goal-Estimate-Consequence(G-E-C) evaluation framework from "The Outline of Land Evaluation" and "The Outline of Sustainable Land Management Evaluation" (which analyzes the evaluation based on identifying the objectives, types and methods of land use), the Pressure-Status-Response (P-S-R) framework put forward by Organization for Economic Cooperation and Development (which emphasizes the evaluation of land based on the analysis on the correlations among human activities, social development and ecological environment) and the Nature -Economy-Society (N-E-S) framework (which analyzes the land use system as the natural, economic and social subsystems to measure the integrity of the land system and realize the dimensionality reduction of the complex system). The above frameworks play an important role in the overall evaluation of land, the quantification of relevant indicators and the environmental management.

At present, the statistical and econometric model analysis methods are mainly adopted for the land evaluation methods, and the relevant technical support is used. For example, overlap analysis method,

* Corresponding author: caomt@m.scnu.edu.cn (M. Cao) ;

multi-indicator decision model, vector-raster mixed data model, Markov model, neighborhood algorithm and GRNN model are used for land evaluation. The land is technically evaluated by GIS, RS, GPS and other means. Currently, there is the lack of the research on the evaluation of land use status in mountainous counties of Southwest China. As the giant panda habitat, Baoxing County belongs to the restricted development zone in the main functional areas, and the ecological environment is fragile, so the case selection is representative. Thus, drawing lessons from the previous research frameworks, this study uses the comprehensive evaluation models to evaluate the current status of land and analyzes the current status and future development direction of land use in Baoxing County.

2 Research region and data sources

2.1 General situations of research region

Baoxing County (102°28 - 103°02'E, 30°09' - 30°56'N) is located in the western edge of the Sichuan Basin and in the north of Ya'an City. Bordering Wenchuan County and Xiaojin County on the north and facing Tianquan County on the south, it is contiguous to Kangding County on the west and adjoins Lushan County on the east. The transportation is mainly based on highways, and provincial highway No. 210 runs through the whole territory. The county is 61 km wide from east to west and 81 km long from south to north, and the jurisdiction area is about 3114 km2.According to the second national land survey data, the total land area of Baoxing County is 311398.68 hm2, and the agricultural land, construction land and unused land in the county respectively account for 95%, 1% and 4% of the county's total land area.

2.2 Data sources

The latest DEM (1:50000 data), administrative division map, present land use map, river system map, fracture structure map, traffic road map, population and related socio-economic data were obtained in accordance with the MH tracking survey and CL development in Baoxing County from research groups between 2014 and 2016. The data originated from the Statistical Bureau, Meteorological Bureau, Land Bureau, archives, geological environment monitoring stations and other relevant government departments of Baoxing County.

3 Evaluation model calculation and design of evaluation indicator

3.1 Evaluation model calculation

(1) Single indicator score calculation

In this study, the comprehensive evaluation models are used to preprocess the selected indicator magnitude uniformly. Formula (a) is used for the calculation of the scores of positive indicators, while Formula (b) is adopted to calculate the scores of negative indicators. The specific formulas are as follows:

$$F_1 = CV/SV \times W_i \quad \text{(a)}$$

$$F_2 = SV/CV \times W_i \quad \text{(b)}$$

F1 and F2 refer to the score of the indicator; CV indicates the present value of the indicator (the current survey statistics); SV represents the standard value of the indicator (the target planning value in 2020); Wi denotes the weight.

(2) Evaluate the total score calculation

The calculation of the total score value of the evaluation is the linear weighting of the score of the individual evaluation indicator, and the formula can be expressed by:

$$LS = \sum W_i F_i (i = 1,2,...,n) \quad \text{(c)}$$

LS stands for the comprehensive level of land use; Wi refers to the weight of the indicator i; Fi represents the evaluation score of the indicator i; n denotes the number of indicators.

(3) Evaluation standard division

In order to deal with the data better and more intuitively, in this study, the total evaluation score is multiplied by 100, and land use status is divided into four stages, which are irrational use stage, general rational use stage, relatively rational use stage and sustainable use stage. The details are shown in Table1.

3.2 Design of evaluation indicators

Based on the actual characteristics of Baoxing County and the availability of data, the evaluation indicator system of land use status is established. Meanwhile, this study collects the data from the economic, environmental, transportation, agricultural and other departments in the county, the target planning value of the current overall land use planning and other relevant data, delimiting the standard value of each individual indicator. The details are presented in Table 2.

Table 1. Land use status evaluation score standard

Evaluation total score	<50	50-70	70-90	>90
Standard	Irrational use stage	General rational use stage	Relatively rational use stage	Sustainable use stage

Table 2. The Evaluation System and Standard Value of Land Use in Baoxing County

Target level	Criterion level (B)	Indicator level (C)	Calculation formula explanation	Standard
	Land	Land reclamation rate (%) (C_1)	Cultivated land area / Total land area·100	1.28

 https://doi.org/10.1051/e3sconf/20183801046

Status of land utilization (A)	development (B_1)	Land use rate (%) (C_2)	Used land area / Total land area·100	99.19
		Agricultural use rate of land (%) (C_3)	Agricultural land area / Total land area·100	98.65
		Land construction use rate (%) (C_4)	Construction land area / Total land area·100	0.55
		Multiple-crop indicator of cultivated land (%) (C_5)	Annual crop sown area / Total area of cultivated land·100	302.97
		Population density (%) (C_6)	Total population / Total land area·100	21.74
		Forest coverage rate (%) (C_7)	Woodland area / Total land area·100	73.67
		Per capita area of construction land (hm^2) (C_8)	Total area of construction land / Total population·100	2.51
	Intensive degree of land (B_2)	Fertilizer application rate of per-unit cultivated land(%) (C_9)	Fertilizer application amount / cultivated area·100	100.69
		Urbanization level (%) (C_{10})	Urban population / Total population·100	43.00
		Investment intensity of fixed assets in per-unit construction land (C_{11})	Fixed assets investment / Construction land	244.69
		Road network density (%) (C_{12})	Traffic land / Total land area·100	0.13
	Comprehensive benefit of land (B_3)	Output value of unit sown area (Ten thousand yuan) (C_{13})	Total agricultural output value / Crop sown area	1.86
		Output value of unit land area (Ten thousand yuan) (C_{14})	Total agricultural output value / Cultivated land area	9.85
		Total output value of per-unit agricultural land (Ten thousand yuan) (C_{15})	Total agricultural output value / Agricultural land area	0.13
		Land output rate (%) (C_{16})	GDP/ Land area·100	84.39
		Per capita area of cultivated land (hm^2) (C_{17})	Total area of cultivated land / total population	0.06

4 Evaluation result and analysis of land use status quo

4.1 Land use status evaluation weight calculation

The weights of the evaluation indicators of land use status are mainly determined by analytic hierarchy process (AHP), and the specific steps are as follows:(1)The establishment of the hierarchy judgment matrix;(2)The establishment of the related hierarchy sorting;(3)The consistency test of the judgment matrix;(4)The establishment of the related hierarchical level weight;In accordance with the above steps, the weights of various indicators are eventually calculated (Fig.1.).

4.2 Land use status evaluation score

The calculation results of the above weight values are substituted into Formula (c), and through calculation, it is concluded that the evaluation score of land use status in Baoxing County was 83.50. The scores and comprehensive scores of the main individual indicators are detailed in Table 3. From the overall evaluation standard delineated in Table 1, it is seen that land use status in Baoxing County was generally at the sustainable use stage.

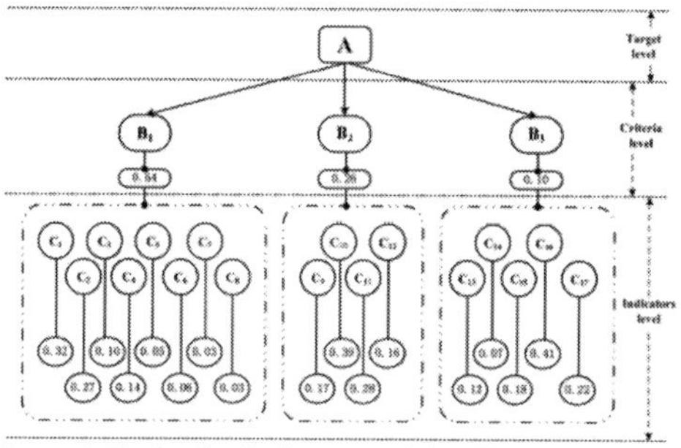

Figure 1. Evaluation indicators weight of land use status in Baoxing county

Table 3. Land use status evaluation score standard

Target level (A)	Criteria level (B)	indicator level (C)	Statistic	Single indicator score	Composite Score
A (83.50)		C_1	1.38	107.31	108.24
		C_2	95.88	96.66	

	B_1	C_3	95.09	96.39	
		C_4	0.79	144.51	
		C_5	279.40	92.22	
		C_6	18.92	87.01	
		C_7	76.77	104.21	
		C_8	4.17	166.08	
	B_2	C_9	134.69	74.76	72.25
		C_{10}	21.02	48.89	
		C_{11}	123.91	59.20	
		C_{12}	0.19	147.26	
	B_3	C_{13}	1.56	83.94	80.77
		C_{14}	11.53	44.28	
		C_{15}	0.17	49.30	
		C_{16}	64.97	76.99	
		C_{17}	0.07	123.32	

Compared with the target planning value in 2020, the comprehensive evaluation score of the development degree of land in this county was 108.24, showing that the development and use degree of land was high. Wherein, the scores of various indicators from large to small were as follows: per capita area of construction land > land construction use rate > land reclamation rate > forest coverage rate > land use rate > agricultural use rate of land > multiple-crop indicator of cultivated land > population density. The proportion of agricultural land was great, and agricultural land area was 296115.27 hm2, accounting for about 95% of the county's total land area. However, the land reclamation rate was only 1.37%. This is because the county is located in the mountainous areas and affected by the geomorphic environment, and the fragmentation degree of land is great. In addition, in recent years, most rural young and middle-aged laborers have gone out to work, and the rural left-behind residents are mostly the elderly and the sick, resulting in tremendous changes in agricultural farming methods. In the cultivation process, the rural left-behind residents strived for simplicity and convenience and did not attach importance to maintaining the fertility of land, leading to the decline in the quality of cultivated land and the ineffective exploration of land potential. The proportion of construction land in the county was only 1%, and the score of per capita area of construction land was 166.08, indicating that the industrial and mining land in urban areas was not enough; the sustained development of economy and society in the county has brought about a continuous increase in the demand for land, and the contradictions between the supply and demand of land will become more prominent and will restrict the overall development.

The evaluation score of the intensive management degree of land in the county was 72.25, and the intensive use level of land was lower than the target planning value. Wherein, the scores of various indicators from large to small were as follows: road network density > fertilizer application rate of per-unit cultivated land > investment intensity of the fixed assets of per-unit construction land > urbanization level. The score of urbanization was 48.89, indicating that the urbanization level was not high. The score of road network density was 147.26, suggesting that the infrastructure land in the county was constantly being improved. From 2005 to 2016, the decline in per capita area of cultivated land, the prominent contradiction between people and land and the increase in various indicators of non-agricultural construction land in the county have led to the serious problems of the protection of cultivated land and will result in the increasingly incisive contradictions of the allocation of land resources in various industries in the future. The total amount of reserved resources for cultivated land in the county was 13240.24 hm2, and the reserve resources for cultivated land suitable for development were relatively abundant. However, due to the insufficient development efforts, the situation of the sustainable use of land was not optimistic.

The evaluation score of the comprehensive benefits of land in the county was 80.77, showing that the comprehensive benefits of land were preferable. Wherein, the scores of various indicators from large to small were as follows: per capita area of cultivated land > output value of per-unit sown area > land output rate > total output value of per-unit agricultural land > output value of per-unit cultivated land area. The scores of the output value of per-unit sown area, the output value of per-unit cultivated land area and the total output value of per-unit agricultural land were 83.94, 44.28 and 49.30, respectively. This was related to the surface structure characteristics of Baoxing County, and the relative fragmentation of cultivated land resulted in the low yields of food crops. The land output rate was relatively high, with the score of 76.99, and the economic crops in the mountainous and hilly areas were developing well, so the total land output rate was higher. Due to the low population density and the large proportion of agricultural land in the county, per capita area of cultivated land had a higher score of 123.32.

5 Conclution and discussion

The overall characteristics of land use in Baoxing County are the diversification of land use types and the imbalanced spatial distribution of various types of lands, and the evaluation score of the overall status was 83.5. At present, land use is at the relatively rational use stage, and the main problems are the lack of land investment, the low mechanization level of agricultural production and the urgent need to improve the use and development degree, the intensive level and the comprehensive benefits of land.

After the county experienced the 5•12 Wenchuan earthquake in 2008 and the 4•20 Lushan earthquake in

2013, the number of hidden mountain disaster points has drastically increased, so the development of construction land is highly threatened by mountain disasters. Therefore, it is necessary to strengthen the existing financial investment in land, enhance the intensive degree of land and improve the mechanization level of agricultural production. At the same time, in the future process of use and development of land, there is a need to focus on the protection of ecological environment and the risks and management of mountain disasters, further carry out the suitability evaluation of land under the stress of mountain disasters and ensure the sustainable use of land resources.

Acknowledgments

We thank all funding support from the science and technology program in Guangdong province of China (Grant NO.2017A020220009), the 61st batch of Chinese postdoctoral program.(Grant NO.2017M612683),the south China normal university youth teacher scientific research cultivating fund program(Grant NO.2016JK102).

References

1. Chen G, Fang Y, Gao Y. China Mountain Development Report: New Developments and New Explorations of Mountainous Areas in China[M]. Commercial Press, 2010.
2. Wu C. Study of Land Use Evaluation in Urban Area and Analyzing its Potential: Taking Nanjing as an Example[D]. Naning Agricultural University, Naning, China, 2006.
3. Guo X, Qiu Y, Lian G, Liu K. The Progress and Prospect of Land Quality Indicators Based on"Press-State-Response" Framework[J].Progress in Geography, 2003(05): 479-489.
4. Feslm FAO. An international framework for evaluating sustainable land management[R]. World Soil Resources Report 73.1993, 1993.
5. Wang X, Bao Y. Study on the Methods of Land Use Dynamic Change Research[J]. Progress in Geography, 1999, 18(1): 81-87.
6. Malczewski J. Ordered weighted averaging with fuzzy quantifiers: GIS-based multicriteria evaluation for land-use suitability analysis[J]. International Journal of Applied Earth Observation and Geoinformation, 2006, 8(4): 270-277.
7. Chandio I A, Matori A N, Yusof K, et al. GIS-basedland suitability analysis of sustainable hillside development[J]. Procedia Engineering, 2014, 77: 87-94.
8. Promper C, Puissant A, Malet J P, et al. Analysis of land cover changes in the past and the future as contribution to landslide risk scenarios[J]. Applied Geography, 2014, 53: 11-19.
9. Ozdemir A, Altural T. A comparative study of frequency ratio, weights of evidence and logistic regression methods for landslide susceptibility mapping: Sultan Mountains, SW Turkey[J]. Journal of Asian Earth Sciences, 2013, 64: 180-197.
10. Koomen E, Diogo V, Dekkers J, et al. A utility-based suitability framework for integrated local-scale land-use modelling[J]. Computers, Environment and Urban Systems, 2015, 50: 1-14.
11. Wei H,Qin B,Peng J,Jin X. Evaluation on comprehensive exploitation suitability of low-slope hilly land based on GRNN model and neighborhood calculation:A case study of Wumeng Mountain continuous poverty-stricken region[J].Geographical Research, 2014, 33(5): 831-841.

Carbon Dioxide Emission Peak and Green Innovation-Driven —— Research of Escaping Middle Income Trap for China

Jianya Liu[1], *Zhenqing Sun*[1,*], *Zirui Lan* [1,2] *and Chunxiao Kou*[1]

[1]School of Economic and Management, Tianjin University of Science and Technology, Tianjin, 300222, China
[2]School of Economics and Business, Norwegian University of Life Science, As 1432, Norway

Abstract. This study explains and demonstrates whether China has the capability to avoid the Middle Income Trap. The 19th National Congress of CPC report points out: by 2035, China will become an international leader in innovation. At present, China is in the juncture of changing the mode of development, optimizing the economic structure and transforming the growth momentum. The juncture means that it is possible to be stagnation or retrogression of national economy as the national ability of innovation is insufficient, then sticking in the middle-income trap (MIC) for a long time. In this paper, we used the TFP to prove that the input-output ratio of different regions of China, with the per capita GDP and carbon emissions are fitting again, dividing China into optimized zones and non-optimized zones. It can be seen from the results that the optimized zones have achieved the peak of carbon emission and had relative advantages in green innovation. However, if China wants to achieve her overall carbon emissions and get out of the middle income trap, she must optimize the development zones to spillover technologies and talents.

1 Introduction

The climate issue has been becoming a focus of attention, and the Copenhagen conference in 2009 had written target that limit the 2 centigrade degrees rise into protocol. The 2015 Paris agreement is clearer about the temperature rise of 2 centigrade degrees, and as much as possible to meet the requirements of 1.5 centigrade degrees. In June 2015, China committed its goal of independent action to the United Nations: emissions of carbon dioxide would reach the peak by 2030 as soon as possible; the emissions of carbon dioxide per unit of gross domestic production would be decreased by 60% - 65% that would be compared with that in 2005. Under the aid of pressure of reaching peak, on the one hand, we should explore the system of driving innovation and development, and on the other hand, we should explore the path of crossing the "middle income trap" firstly.

China put forward the economic development strategy of Chinese Economic Reform and opens up since 1980, so the per capita GDP had been advanced from the end of the world to the middle income countries by 2007. China's economy has been entering a period of rapid development since 1980. But by the other countries' development experience, when a country's economy reaches middle income level, its economy is in a long-term stagnation due to various reasons and economic growth rate is slowing down obviously. In 30 years of rapid development by reform and opening up policy, China had successfully entered into Medium and high income countries. And the next goal is that China would enter the ranks of high-income countries. Although China's economy has been growing at a high speed for many years, the middle income trap is a problem that has to be faced.

2 Theories

When a country's economy has reached the level of middle income, her economy has been in a long-term stagnation and growth rate has slowed down as a variety of reasons, which is the most vulnerable stage for economic development. The World Bank (WB) calls this stage is a "Middle Income Trap (MIC)" [1]. In 2017, the latest WB income standard [2]: per capita GDP, 1006~12235 in current price of the US dollar is middle income, it means that high income is higher than 12235 US dollars, and 3955 US dollars is a boundary of middle upper and lower.

In 2008, China's per capita GDP firstly exceeded $3000, entering the level of middle-income countries. Whether the slowdown in economic growth has fallen into the middle income trap or not, which has attracted the attention of all walks of life. Kharas [3] believed that China would face the risks of falling into the middle-income trap. Hogo Erken [4] proposed to cross the middle-income trap by relying on total factor productivity (TFP), especially the productivity related to upgrading technologies. Barry Eichengree (2012) [5], based on Korean's experience, found that high income countries' economic growth was largely driven by TFP. The research found that the driving factor of TFP growth

* Corresponding author: sunzq@tust.edu.cn

was research and development (R&D), and the increase of R&D by 1% (the proportion of industrial added value) would increase the TFP growth rate of the corresponding industry by 0.13%. With the development of measurement technology and data collection, the research of TFP has expanded from macro field to industry and enterprise level, which is also fit for China [6]. Brandt etc (2012) [7] had found that China's TFP had been improved based on empirical analysis. The growth of China's economy over the past 30 years has not just depended on capital input.

Schumpeter (1912) is the first person to put forward the "innovation theory" in the perspective of technology and economy, emphasizing the important role of production technology innovation in the process of economic development. Nicholas Bloom (2013) [8] used a patent technology to measure the technological activity of every company in a period, and then used technology standard to measure the spillover effect of R&D.

3 Models

3.1 Calculations Model and Research Method

It is assumed that the production system has a number of n decision making units (DMU), which are represented as DMU_j (j = 1, 2, …, n). Respectively, x and y represents input and output variables, m and r is the number of input and output variables. s_i^- Represents the slack variable, which be denoted by input excess. Each DMU consumes m inputs of X_{ij} (i= 1, 2, …, m), production is the number of s_1 expected outputs Y^g and the number of s_2 undesirable outputs U^b, so these can be expressed as $x \in R^m$, $y^g \in R^{s1}$, $U^b \in R^{s2}$. Define the matrix X, Y^g, U^b are as follows:

$$X = [x_1, \ldots, x_n] \in R^{m \times n};$$

$$Y^g = [y_1^g, \ldots, y_n^g] \in R^{s_1 \times n};$$

$$U^b = [u_1^b, \ldots, u_n^b] \in R^{s_2 \times n};$$

$$X > 0,\ Y^g > 0,\ U^b > 0.$$

Under conditions of constant return to scale (CRS), the possible production that P-sets is defined as:

$$p = \{(x, yg, ub) \mid x \geq X\lambda,\ yg \leq Yg\lambda,\ ub \geq Ub\lambda,\ \lambda \geq 0\} \quad (1)$$

According to the SBM model proposed by Tone, it is obtained the SBM - Undesirable model, so the SBM model of undesired output under strong disposability is as follows:

$$\rho^* = \min \frac{1 - \frac{1}{m}\sum_{i=1}^{m}\frac{s_i^-}{x_{io}}}{1 + \frac{1}{s_1 + s_2}\left[\sum_{r=1}^{s_1}\frac{s_r^g}{y_{r0}^g} + \sum_{r=1}^{s_2}\frac{s_r^b}{u_{r0}^b}\right]} \quad (2)$$

$$\text{s.t.} \begin{cases} x_0 = X\lambda + S^- \\ y_0^g = Y^g\lambda + S^g \\ U_0 = U^b\lambda + S^b \\ s^- \geq 0;\ s^g \geq 0;\ s^b \geq 0;\ \lambda \geq 0 \end{cases} \quad (3)$$

In those formulas: s represents the slack variable of input and output; λ represents the weight vector;ρ* represents the objective function, which meets $0 \leq \rho^* \leq 1$, s^-, s^g and s^b are strongly decreasing. For specific DUMs, under the condition of that$\rho^* = 1$, $s^- = 0$, $s^g = 0$, $s^b = 0$, which are efficient ; if$\rho^* < 1$, it indicates that the DUMs are inefficient, which have the need of improving input and output.

3.2 Index selection and data source

As data of the regions of Tibet, Hongkong, Macao and Taiwan could not search , this study uses China's 30 provincial regional input-output panel data from 2005 to 2015 as the sample, which are from the year of 2006 to 2016 by Yearbooks including "China Statistical Yearbook", "Chinese City Statistical Yearbook", "China Statistical Yearbook on Science and Technology ", "Data of Gross Domestic Product of China(the year of 1952-2004)" and "Chinese Energy Statistical Yearbook" as the original data and the calculated results（Table1）.

Table1. The index of input and output.

index	Sub-index	Measurement index	unit
input	Capital input	Capital stock	Million yuan
	Labor input	R&D people number	per
		The number of people engaged in water conservancy, urban environment and public service.	per
	Resource input	The volume of water supply	Thousand stere
		Total electricity consumption	Million kilowatt hour
		The total gas supply of liquefied petroleum gas.	ton
	Energy input	Energy consumption	tce
output	Economic	GDP	Million yuan
	Technology output	New product sales revenue.	Thousand yuan
		Number of patents granted	piece
	Green output	Green land area	hectare
		The park area	hectare
	Environmental pollution output	Industrial SO2 emissions.	Million tons
		Industrial	ton

		wastewater discharge.	
		Industrial smoke (powder) dust emission.	ton
		CO2 emissions.	ton

The input-output indexes are treated with the maximum minimum value method, and the statistical description of input-output in China is obtained (Table 2).

Table 2. Statistical description of input-output in China.

	capital	labor	resource	energy	environmental pollution	economy	technology	green
Max	53486.38	337710.01	15493152.9	38179.53	917923.28	60891.28	49781620.5	234443.69
Min	1422.18	5942.80	337484.93	2124.91	2308.12	1762.27	22959.49	2824.30
Average	14540.94	106230.19	3415002.57	15597.29	354943.85	18955.35	5589284.33	43681.03
SD	11668.46	84232.94	3031614.96	8836.86	225389.72	14932.34	10263266.1	44451.15

3.3 Result

By the DEA-Solver Pro 5 software, the input-output indexes are calculated by using the SBM-DEA model. The statistical results of carbon dioxide emission efficiency were shown in Table 3.

Table 3. Carbon dioxide emission efficiency.

Area	CCR/SBM	Area	CCR/SBM	Area	CCR/SBM	Area	CCR/SBM	Area	CCR/SBM
Beijing	1/1	Liaoning	0.78/0.30	Zhejiang	0.96/0.73	Henan	0.93/0.58	Hainan	1/1
Tianjin	0.96/0.73	Jilin	0.88/0.43	Anhui	1/1	Hubei	0.82/0.59	Chongqing	1/1
Hebei	0.86/0.35	Heilongjiang	1/1	Fujian	11	Hunan	1/1	Sichuan	1/1
Shanxi	0.66/0.15	Shanghai	0.99/1.00	Jiangxi	1/1	Guangdong	1/1	Guizhou	0.81/.034
Neimenggu	0.66/0.08	Jiangsu	1/1	Shandong	0.97/0.96	Guangxi	1/1	Yunnan	1/1
Shaanxi	0.62/0.32	Gansu	0.57/0.15	Qinghai	0.73/0.43	Ningxia	1/1	Xinjiang	1/1

From table 3 we can see that the half of the Chinese efficiency value is 1, which indicated that Chinese serious carbon dioxide emissions of city, especially the economy of relatively developed cities, such as Beijing etc., have been under control and get certain effect. Under the condition of the serious environmental pollution, those cities utilize the approach of green innovation-driven that converts the input resources to the maximum utilization ratio of desired output resources. In some areas its green innovation ability is insufficient. On the one hand, such as Ningxia and Xinjiang, which have low emissions, less population, underdeveloped technology level, scarce talents, but nature own strong self-purification ability. On the other hand, the provinces such as Shanxi and Neimenggu, where is due to the waste of energy and the neglect of carbon dioxide emissions.

In all provinces, the values of unexpected output in CCR model are higher than SBM model. It is illustrated that the output of environmental pollution can boost the development of the economy, so it can't be ignored. Although some provinces' two efficiency values are very close, such as Beijing, Shanghai, Shandong and Tianjin, said that these provinces' economic development has been set the concept of green development into that, it can be said that the environment governance has a certain effect. But in the other provinces gap of two efficiency value are relatively large, such as Hebei, Shanxi, Neimenggu, etc. It can be clearly seen, the result of SBM model with unexpected output value are far lower than the efficiency without environmental pollution output value, indicating that these provinces' economic development have not yet the concept of green development among them, and the technical level in these areas is relatively low, meantime they are lack of green innovation.

4 The relationship between the carbon dioxide emission and across the middle income trap

China plans to peak its carbon dioxide emissions around 2030 and will take more measures to reach an early peak. For the optimized zones with demonstration effect, which have peaked in 2011 and gradually slowed down, the non-optimized zones was increasing year by year. For the three optimized areas, the carbon dioxide emissions are lower than the key development areas, and the time of the carbon dioxide emission peak is earlier than the key development area.

As far as the current carbon dioxide emissions are concerned, China must drive economic growth through green innovation in order to achieve the peak target of China. This is a path to reduce carbon dioxide emissions, to promote economic growth, and to jump out of the middle income trap. In order to intuitively discuss the relationship between peak and carbon emissions, carbon dioxide emissions are fitted with per capita GDP in MATLAB. The results are shown in Figure 1.

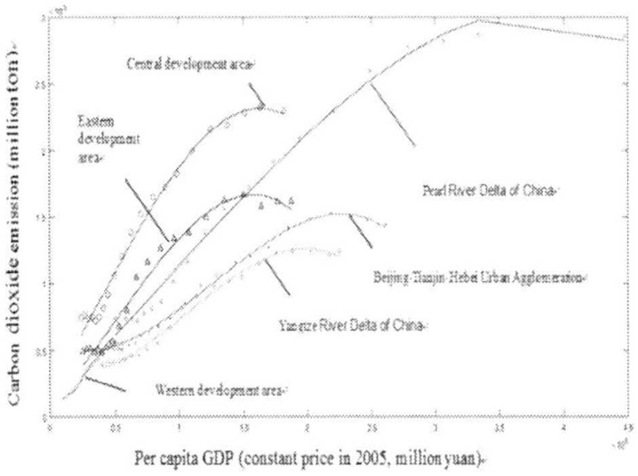

Fig. 1. Fitting of total carbon dioxide emissions with per capita GDP.

From Figure 1, we can see that with the development of economy and society, the results between carbon dioxide emissions and GDP exhibit the inverted U-shape pattern associated with the Environmental Kuznets Curve (EKC) hypothesis. When the per capita GDP reaches a certain value, carbon dioxide emissions will peak and then decrease. This trend is in line with the concept of green development whose economic growth is not at the cost of environmental pollution. China's economy is changing from high speed to "new normal", and the speed of economic growth is slowing down. The proportion of the general processing industry has declined, the proportion of technology intensive industries has increased, and especially the proportion of electrical machinery and equipment manufacturing, electronic and communication manufacturing industry has increased significantly.

In order to avoid the middle-income trap problems, it is necessary for China to optimize the regional development as the guide, and give demonstration to help other regions, in order to explain the correlation between China Carbon dioxide emissions' peak and jumping out of the middle-income trap. In order to achieve the peak goal, this paper demonstrates innovation ability from two perspectives. South Korea has been out of the middle-income trap which used the basis for the analysis of the possibility for China. Firstly, the tendency of patents number changed with time can be used to reflect the development of technology and to judge the innovation of technology. Secondly, the ratio of R&D to GDP is the basic method to measure the innovation ability of a region.

In terms of the number of patents[a], Taiwan, China now has reached the level of well-developed countries in terms of the number of patents, such as the United States and Japan, where China, Hong Kong, Korea and Singapore cannot compete. For China, the experience of Asian countries that jump out of middle income traps is the best example. In these areas, patent technology has become the leader of the world, which is attributed to excellent intellectual property protection mechanism, R&D expenditure support, and the transformation of technology imitation and import technology into independent technological innovation [9]. In the field of patent protection and application, China has issued a lot of strategic plans in 2017. The attention to the patent is obviously improved, and the Talent and Innovation-driven development strategy is realized.

The ratio of R&D[b]to GDP is the basic method to measure the innovation ability of a region. China's growth rate increased from 0.6% in 1995 to 1.3% in 2003, and has been growing since the 21st century. It has reached 2.1% in 2015. It is the second largest R&D investment country in the world behind the US, and also unique in middle-income countries. It is precisely because of the large amount of investment in research and development, the number of Chinese patents has also increased sharply, and the average growth rate of domestic innovation patents is also increasing. In 2015, the number of patents granted by China in the United States patent and trademark office (USPTO) exceeded 8000 pieces. It is far higher than other middle-income countries, and surpasses Israel who was called the "innovative country". In terms of the growth rate of patents, the growth rate in 2010 reached 64%, and the growth has slowed down since then.

[a] Source: USPTO report. https://www.uspto.gov/web/offices/ac/ido/oeip/taf/reports.htm
[b] Source: World Bank .the ratio of R & D data to GDP. http://date.worldbank.org.cn/indicator/GB.XPD.RSDV.GD.ZS

5 Conclusions

According to the current carbon dioxide emission status, it is concerned that the non-optimized zones have a great potential for reducing carbon dioxide emission. The reason is that the energy structure of non-optimized zones did not have completely converted. Firstly, using of non-renewable resources, such as coal and oil, is common. Secondly, it is Primary Industry and Second Industry, especially the Second Industry that plays a leading role in non-optimized zones' economic growth. For the optimized zones, they perform excellently in the two aspects. If we want to further reduce carbon dioxide emissions, we must start with green innovation.

There is great correlation between green innovation efforts and carbon emissions peak, reducing carbon dioxide emissions cannot be at the cost of economic development, and appear enormous environmental issues, while also face middle-income countries economic stagnation problem -- the middle-income trap. So the theory of green innovation is the key to solve the two problems. How to attain green innovation? On the one hand, the patent research and development should be considered energy conservation, environmental protection and green development. On the other hand, there is a need for policy support and strict enforcement. We should take the optimized zones as a model and learn its advanced low carbon dioxide technology. At the same time, it can also attribute other areas, so that China could solve the problem of carbon dioxide emissions peaking and jump out of the middle income trap as soon as possible.

Acknowledgements

The work was supported by the National Social and Scientific Fund of China (16AGL002), Major Program of the Ministry of Education (16JZD014). Special thanks are given to the anonymous reviewers for their incisive and comprehensive comments and suggestions.

References

1. A. P. Moniroth. The East Asian Economic Development Report. R., Washington D. C. . World Bank, (2006)
2. F. Gabriel Im. Middle-Income Traps. A Conceptual and Empirical Survey.R. Working Paper NO.6594, Washington D. C... World Bank, (2013)
3. H. Kharas. What Is the Middle Income Trap, Why do Countries Fall into It, and How Can It Be Avoided? J. Global Journal of Emerging Market Economies. 3(3): 281-289. (2011)
4. H. Erken. Why emerging economies are (un)successful in avoiding the middle income trap.

R. Rababank Raboresearch-Economic Research. (2017)

5. B. Eichengreen. From Miracle to Maturity: the growth of the Korean Economy. M.President and Fellows of Harvard College, (2012)
6. Scherngell. Effects of knowledge capital on total factor productivity in China: A spatial econometric perspective. J. China Economic Review. (29): 82-94. (2014)
7. Brandt. WTO Accession and Performance of Chinese Manufacturing Firms. R. CEPR Discussion Paper N. 9166. (2012)
8. N. Bloom. Identifying technology spillovers and product.J. Econometrica, 81(4): 1347-1393. (2013)
9. B. Eichengreen. When fast growing economies slow down: International evidence and implications for China. R. NBER Working Paper. No. 16919. (2011)

Biogas production of Chicken Manure by Two-stage fermentation process

Xin Yuan Liu, *Jing Jing* Wang, *Jia Min* Nie, *Nan* Wu, *Fang* Yang, *Ren Jie* Yang

College of Engineering and Technology, Tianjin Agricultural University, Tianjin 300384, China

Abstract. This paper performs a batch experiment for pre-acidification treatment and methane production from chicken manure by the two-stage anaerobic fermentation process. Results shows that the acetate was the main component in volatile fatty acids produced at the end of pre-acidification stage, accounting for 68% of the total amount. The daily biogas production experienced three peak period in methane production stage, and the methane content reached 60% in the second period and then slowly reduced to 44.5% in the third period. The cumulative methane production was fitted by modified Gompertz equation, and the kinetic parameters of the methane production potential, the maximum methane production rate and lag phase time were 345.2 ml, 0.948 ml/h and 343.5 h, respectively. The methane yield of 183 ml-CH_4/g-$VS_{removed}$ during the methane production stage and VS removal efficiency of 52.7% for the whole fermentation process were achieved.

1 Introduction

As the amount of the large scale livestock and poultry farms increased in China, the discharged manure increased accordingly and the environment pollution caused by manure became prominent gradually in recent years. Statistics suggested that the chemical oxygen demand (COD) and the ammonium nitrogen (NH_4^+-N) produced from livestock and poultry manure in China reached to 10.99 and 0.63 million ton, respectively, accounting for 45% and 25% of the summation discharged into environment nationwide, and also being equivalent to 95% and 78% of the summation produced by agricultural resources [1]. As one of the dominant manures in China, the chicken manure had the features of huge amount, complex component, high content of organic matters, and high concentration of nitrogen and phosphorus, etc. [2]. The centralized discharge of the chicken manure from large scale farms would directly get the surrounding soil, atmosphere and water polluted, and further bring down the local environment level in the rural area. However, the organic matters and nitrogen in chicken manure are the valuable biomass resource, and they could be reused and transferred into resources and energy materials by some recycle technologies. Anaerobic fermentation technology could produce biogas from the organic matters derived from manure by the metabolic activity of microorganisms, and it was recommended as the recycling method for manure treatment in China [3]. Besides, the NH_4^+-N in manure could be released greatly under the degradation of protein-based substrates, which would be benefit for increasing the fertilizer efficiency in the agricultural reuse.

However, the high content of free ammonia (NH3, main part of total NH_4^+-N) in the fermentation liquid would inhibit methanogens and reduce the methane production [4]. Among all kinds of livestock and poultry manure, chicken manure contains relatively high content of nitrogen as well as the low carbon to nitrogen ratio (C/N), which is more likely to cause severe inhibition effect as the fermentation substrate [5]. The simplest measure to reduce the reverse effects of NH_4^+-N on methanogens is diluting chicken manure with water by the application of wet anaerobic fermentation process. Two-stage fermentation process with pre-acidification stage and subsequence methane production stage was utilized in this study, and the component of volatile fatty acids (VFAs), methane yield and volatile solid (VS) removal efficiency were investigated to estimate the fermentation efficiency of chicken manure.

2 Materials and Methods

2.1. Seed sludge

The seed sludge was collected from sewage sludge digestion tank in Jingu wastewater treatment plant located in Tianjin, China. However, the seed sludge for pre-acidification was heat-treated at 100°C for 30 min to eliminate the methanogens and enrich acidate bacteria, and its VS/TS (total solid) was 44.9%. The collected seed sludge with VS/TS of 49.5% was directly used as methane-producing seed sludge.

2.2 Chicken Manure

* Corresponding author: liuxinyuan11@163.com

The chicken manure collected from household farms in the surrounding of west campus of Tianjin agricultural University located in Xiqing District of Tianjin, China. It was used as fermentation substrate and its main component was analyzed.

2.3 Experimental conditions

Batch experiments were conducted using 300 ml serum bottles filled with 150 ml of substrates. The substrate concentration was 10 g-VS/l in each bottle. Batch experiments were conducted in duplicate. The initial pH was 5.5 and the seed sludge for pre-acidification was 1.20 g-VS. Both bottles were purged by nitrogen gas for 3 min and then sealed by rubber stoppers. The bottles were put in a shaker with rotating speed of 120 rpm and the temperature of 37°C. Three day later, the methane-producing seed sludge with the VS weight of 0.8 g was added into each bottle. The pH of the substrate was adjusted to 7.0 and then the K_2HPO_4/KH_2PO_4 buffer solution (pH 7.0) was added. All the bottles were purged by nitrogen gas for 3 min again before sealing and then put in the incubator for methane fermentation at 37°C. The produced gas and methane content was measured periodically.

2.4 Analytical methods

Production of biogas was measured by a glass syringe. Methane content in biogas produced in methane production stage were analyzed by a gas chromatograph (Clarus 680 PerkinElmer, USA) equipped with a flame ionization detector and a capillary column (Elite-5, 30m×0.25mm×0.25μm). Nitrogen was used as the carrier gas at a flow rate of 2 ml/min. The operation temperatures of the injection port, oven and detector were 200°C, 150°C and 250°C, respectively. VFA including acetate, propionate, butyrate, *i*-butyrate and *i*-valerate in the mixed liquor were analyzed by another gas chromatograph (Agilent 7890B, USA) equipped with a flame ionization detector and a fused-silica capillary column (HP-5, 30mm×320μm×0.25 μm). Nitrogen was used as the carrier gas with a flow rate of 30 ml/min and the split ratio of 20:1. The temperature of injection port and detector were 250°C and 270°C, respectively. The temperature program for oven was as following: 80°C for 1 min, and then rise to 220°C with the rate of 20°C/min, 220°C for 1 min, further increased to 240°C with the rate of 20°C/min, finally maintained at 240°C for 5 min. The soluble carbohydrate was analyzed using anthrone-sulfuric acid method with glucose as standard [6], and the soluble protein was analyzed by Lowry method [7]. The samples for analysis of soluble carbohydrate and protein were filtrated by 0.45 μm membrane before detecting. TS, VS and NH_4^+-N were determined according to Standard Methods [8].

2.5 Calculation for cumulative methane production

In the batch experiment, the cumulative methane production was calculated described as the equation (1) [9].

$$V_2 = V_1 + V_G \cdot C_2 + V_0 \cdot (C_2 \cdot C_1) \qquad (1)$$

Where V_1, V_2 represents the methane production (ml) at the time of $t_1$and t_2, respectively; C_1, C_2 the methane content in the gas space of anaerobic bottle (%) at the time of $t_1$and t_2, respectively; V_0 the volume of the gas space of anaerobic bottle (ml); V_G the produced biogas (ml) during the time gap between $t_1$and t_2.

2.6. Kinetic analysis

The cumulative methane volume in batch experiments followed the modified Gompertz equation [10], as shown in Equation (2).

$$H = P \cdot \exp\{-\exp[\frac{R_m \cdot e}{P}(\lambda - t) + 1]\} \qquad (2)$$

where H represents the cumulative methane production (ml), the lag phase time (h), P the methane production potential (ml), and R_m the maximum methane production rate (ml/h). The values of P, R_m and λ were determined by best fitting the methane production data for Eq. (2) using Microsoft's software Excel 2010.

2.7 Calculation for VS removal efficiency

The VS removal efficiency in the overall fermentation process was calculated according to the equation (3), as following.

$$RE = (VS_f + VS_{S1} + VS_{S2} - VS_e)/(VS_f + VS_{S1} + VS_{S2}) \qquad (3)$$

Where *RE* represent the VS removal efficiencies in the whole process, respectively (%). VS_f, VS_{S1} and VS_{S2} were defined as the VS of initial feedstock, pre-acidification and methane-producing seed sludge (g). VS_e was the VS at the end of methane production (g).

3 Results and Discussion

3.1 Characteristic of chicken manure

The main component was shown in Table 1. Table 1 suggested that chicken manure contained 25.64% of TS in wet weight (ww) and 59.28% of TS was organic matters. The NH_4^+-N concentration was as high as 11.32 mg/g-ww in this study, which might be caused by the degradation of nitrogen-contain materials during the storage of chicken manure. Table 1 also shows that the sum of total carbohydrate and total protein account for 16.5% of VS, suggesting that great amount of other organic matters in chicken manure were not detected. The soluble component that was easily to be utilized by microorganisms was also measured. However, results show only 4.5% and 5.7% of total carbohydrate and protein were soluble, implying the chicken manure was hard to be hydrolyzed to some extent.

Table 1. Component of Chicken Manure based on wet weight (ww).

Program	Unit	Value
Total carbohydrate	mg/g-ww	18.0
Soluble carbohyrate	mg/g-ww	0.81
Total protein	mg/g-ww	7.17
Soluble protein	mg/g-ww	0.4
NH_4^+-N	mg/g-ww	11.32
TS	%	25.64
VS	%	15.20
VS/TS	%	59.28

3.2 VFAs produced in pre-acidification stage

The pre-acidification stage was performed for nearly three days. The VFAs including acetate, propionate, butyrate, i-butyrate, i-valerate were detectable in the digestate after pre-acidification stage, as shown in Figure 1. Due to the hydrolysis and acidification of the organic matters, a sum of 1588 mg-VFAs/l was produced in the end of pre-acidification stage. The acetate was the dominant among all kinds of VFAs, accounting for 68% of the total amount. The acetate could be directly utilized by the pathway of aceticlastic methanogenesis, which was the main pathway for *Methanosarcina*, *Methanosaeta*, and etc [11]. Followed by acetate, the propionate with the percentage of 19% was the second main component, which was not popular with methanogens and would inhibit the methanogens activity under the high concentrations.

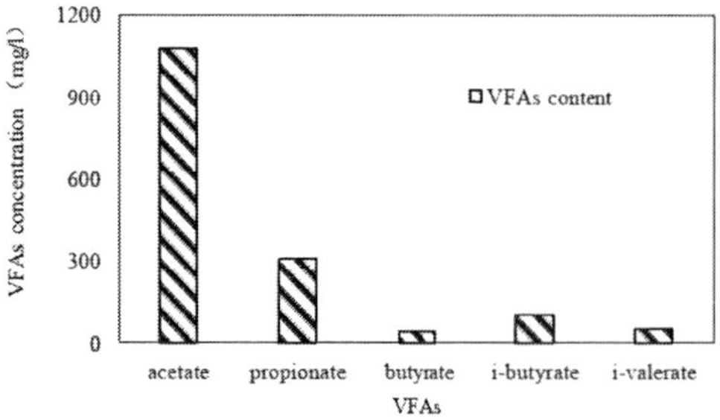

Fig. 1. VFAs component detected in the digestate in the end of pre-acidification stage.

3.3 Biogas production and methane content in methane production stage

After pre-acidification stage, the methane-producing seed was added into the anaerobic bottles, and pH was adjusted to 7.0, which is suitable for methane fermentation. The phosphate buffered saline was benefit for pH maintenance and the pH was 6.8-7.0 during the whole fermentation process. The daily biogas production and methane content in methane production stage was shown in Figure 2.

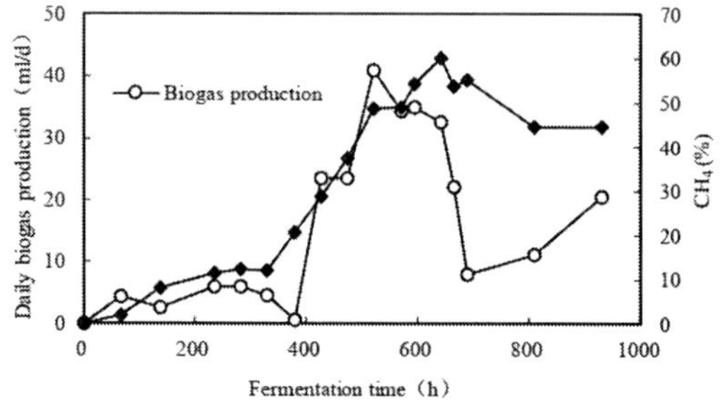

Fig. 2. The daily biogas production and methane content in biogas during the methane production stage.

As shown in Figure 2, the daily biogas production had experienced three peak periods during the total fermentation time of 930 h. In the first peak period (0-378.5 h), the daily biogas production was lower than 6 ml and even reduced to 1ml at 378.5 h, but the methane content increased smoothly to 20.7%. It could be speculated that a large percentage of the biogas produced by the biochemical reactions must be methane in the first peak period. In the second peak period (378.5-666.5 h), the daily biogas production increased greatly with the highest value of 40.8 ml/d and the average value of 32.9 ml/d. The methane content increased sharply to 60% with the increase rate of 3.3 ml/d. After the rapid increase stage, the daily biogas production slowly increased to 20.4 ml/d during the final peak period (666.5-930.5 h) and the methane content was slightly decreased to 44.5%. The tendency of the biogas production curve with three peak period during methane production stage might be caused by the insufficient of direct substrates for methanogens and the inhibition of metabolic products on the microorganisms.

3.4 Kinetic fitting of the cumulative methane production

Figure 3 shows the cumulative methane production calculated by Equation (1) by chicken manure digestion and the corresponding curves of Equation (2) using the best-fitted kinetic parameters. The results showed that the values of P, R_m and λ were 345.2 ml, 0.948 ml/h and 343.5 h, respectively.

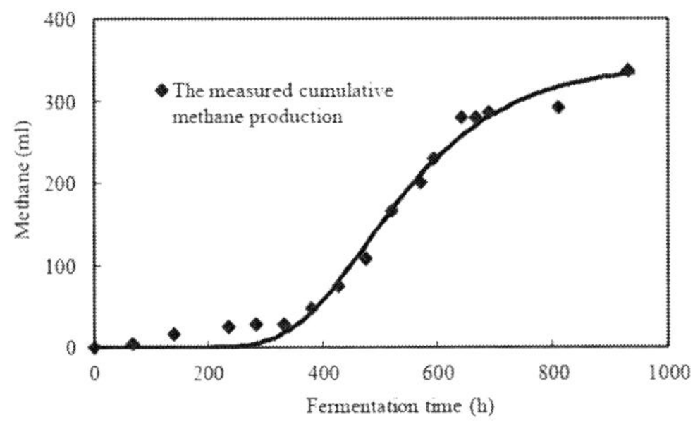

Fig. 3. The cumulative methane production in the methane production stage.

3.5 VS removal efficiency and methane yield

The VS removal efficiency during the whole fermentation process calculated according to equation (3) reached a value of 52.7%, implying that nearly half of the organic matters were converted into biogas. The methane yield was calculated through dividing the produced cumulative methane by the removed VS, and the value was 183 ml-CH_4/g-$VS_{removed}$.

4 Conclusion

The two-stage anaerobic digestion process with separated pre-acidification and methane production was applied in chicken manure treatment. Batch experiment results shows that the VFAs was produced in the pre-acidification stage with a total amount of 1588 mg-VFAs/l and the acetate was the main component. The methane yield reached 183 ml-CH_4/g-$VS_{removed}$ during the methane production stage and VS removal efficiency was 52.7% for the whole fermentation process.

Acknowledgments

Authors wishing to acknowledge the financial support by the Tianjin Research Program of Application Foundation and Advanced Technology, China (Program No.16JCQNJC08200) and the National Natural Science Funds for Distinguished Young Scholar of China (No. 21607114).

References

1. Y. Wang. China swine ind., **4** (2014)
2. J. Cao, B. Wu. Guangdong Feed, **26**, 11 (2017)
3. Decree of the State Council NO.643, *Regulations on the prevention and control of pollution by livestock and poultry*. (Decree of the State Council, Beijing, 2013)
4. H. Tian, N. Duan, C. Lin, et al. J. Biosci. Bioeng. **120**, 1 (2015)
5. C. Sun, W. Cao, C.J. Banks, et al. Bioresour. Technol. **218** (2016)
6. A. F. Gaudy. Industrial Water Wastes, 7, (1962)
7. O.H. Lowery, N.J. Rosebrough, A.L. Farr, et al. J. Biol. Chem. **193** (1951)
8. APHA. *Standard Methods for the Examination of Water and Wastewater*, 21st ed. (APHA, Washington DC, 2005)
9. T. Zhang. *Improved study on anaerobic fermentation of sewage sludge for volatile fatty acids by pretreatment with ultrasonic coupling technology*. (Suzhou University of Science and Technology, Suzhou, Jiangsu, China, 2015)
10. J.J. Lay, Y.Y. Li, T. Noike. J. Environ. Eng. **124**, (1998)
11. J.L. Garcia, B.K.C. Patel, B. Ollivier. Anaerobe, **6**, (2000)

https://doi.org/10.1051/e3sconf/20183801049

The design and implementation of automotive radar system based on MFSK waveform

Yong Kun SONG*, Yong Ji LIU, Zhan Long SONG.

School of Information and Communication, Guilin University of Electronic Technology, Guilin, 541004 , China

Abstract. As a driving aid system, automotive radar plays a more and more important role in reducing traffic accidents. In this paper, a simple and efficient automotive radar system based on MFSK (Multiple frequency shift keying) is designed, the system structure and implementation algorithm is presented. This system uses FPGA as the signal processing unit to ensure the response speed. In addition, the designed MFSK waveform algorithm achieved high precision and unambiguous measurement. The effectiveness of the system is verified by experiments and field measurements.

1 Introduction

The key technical challenge and problem of automotive radar system is how to get accuracy range, radial velocity and azimuth of multiple targets simultaneously, where the design of waveform plays an important role to solve those problems. The most commonly used LFM (Linear frequency modulation) waveform radar provides high range and velocity resolution, but it can not solve the problem of ghost targets in multiple targets situation. In addition, the radar system of FSK (Frequency shift keying) waveform has high velocity resolution and can avoid the ghost targets, but it is unable to measure the stationary target and multiple targets with the same radial velocity [1,2,3].

In order to solve these problems, many improved methods have been proposed. In [4,5], a variable-period trapezoidal wave radar system is studied, the system introduced a constant frequency on the basis of the traditional upper and lower sweep frequency, which can reduce the number of ghost targets. However, the complex range and velocity matching algorithm greatly increase the amount of calculation. Paper [6] proposed a period-varying sawtooth wave measurement method, but the method with minimum coupling range difference may loss the target with large Doppler shift and generate ghost targets. An adaptive beam-controlled phased array radar is designed to improve the measuring distance and precision, but the complicated hardware structure and high cost limit its application [7,8,9].

An automotive radar system with MFSK waveform is designed in this paper, which is implemented based on FPGA. This system overcomes the shortcoming of the LFM and FSK system, it combines the advantage of the two systems, provides high range and velocity resolution, short measurement time and also could avoids ghost targets [10]. Firstly, this paper analyzed the range and velocity measurement theory of MFSK radar system, which filled up the previous theoretical blank, then designed the corresponding system structure and FPGA implementation algorithm. The filed experiment results show that the system has the fast response time and high measurement accuracy.

2 Measurement Principle Analysis

This system uses MFSK waveform as the modulation signal, which is composed of LFM and FSK signals. The transmitted and received signals are shown in the Fig.1. A periodic transmitted signal consists of two alternate and stepwise modulation signals, A and B wave. The real line is transmitted signal and the dotted line is received signal.

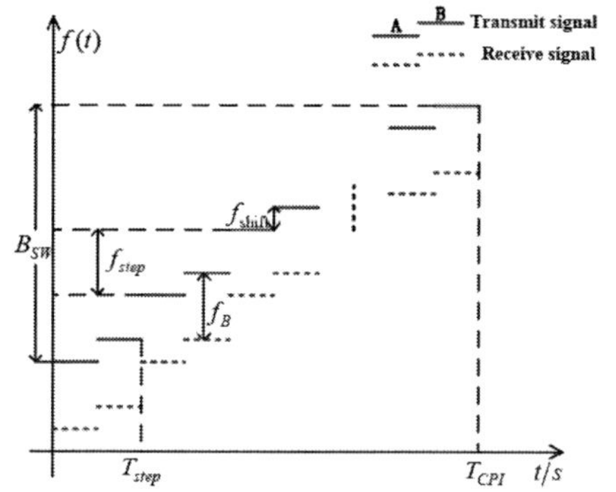

Fig. 1 MFSK transmitted and received waveform

Based on the characteristics of MFSK waveform, the radar system measurement principle can be analyzed.

Set N frequency transmitted signals as:

$$x_T(i) = A_T \cdot \exp\{-j2\pi(f_0 + i \cdot f_{step}) \cdot t\} \quad i = 1,2\ldots N-1 \quad (1)$$

The received signal is:

$$x_R(i) = A_R \cdot \exp\{-j2\pi(f_0 + i \cdot f_{step}) \cdot (t - \tau(i))\} \quad (2)$$

* Corresponding author: songyk_1993@163.com

After down-conversion, the difference frequency signal is simplified as follows:

$$x_s(i) = A_s \cdot \exp\left\{-j2\pi(f_0 + i \cdot f_{step}) \cdot \tau(i)\right\} \tag{3}$$

where $\tau(i) = 2(R + viT_{step}/2)/C$.

the N difference frequency signals are transformed by FFT, and it can be obtained that:

$$\begin{aligned} X(k) = A_s \cdot \exp\left(-j\frac{4\pi}{C} f_0 R\right) \cdot \exp\left(-j2\pi\frac{N-1}{2} \cdot F\right) \\ \cdot \sum_{i=0}^{N-1} \exp\left(\frac{-j4\pi v i^2 f_{step} T_{step}}{2C}\right) \cdot \frac{\sin(\pi N \cdot F)}{\sin(\pi \cdot F)} \end{aligned} \tag{4}$$

Where $F = (vf_0 T_{step} + 2Rf_{step})/C + k/N$, it can be known that $X(k)$ pick up the peak of the spectrum when $k = -N(vf_0 T_{step} + 2Rf_{step})/C$. Meanwhile $T_{step} = T_{CPI}/N$ and $f_{step} = B_{SW}/N$, so the beat frequency can be obtained.

$$f_B = \frac{k}{T_{CPI}} = -\frac{2 \cdot v}{\lambda} - \frac{2 \cdot R \cdot B_{SW}}{C \cdot T_{CPI}} \tag{5}$$

For equation (4), when $X(k)$ get to the peak, the phase of its peak is $\varphi = -4\pi f_0 R/C$, in addition, $R_B = R_A + v \cdot T_{step}/4$, which results that the phase difference is as follows:

$$\Delta\varphi = \varphi_B - \varphi_A = -\frac{\pi \cdot v}{(N-1) \cdot \Delta v} - 4 \cdot \pi \cdot R \cdot \frac{f_{shift}}{C} \tag{6}$$

With the two related equation (5) and (6), the target range and velocity can be resolved simultaneously. It can be seen that f_B and $\Delta\varphi$ are one-to-one correspondence, there is no cross aliasing. Therefore, the MFSK system can effectively avoids ghost target problems which exist in LFM system. Moreover, the MFSK waveform can get a large number of frequencies and a wide range of span, as it is composed of multiple stepwise FSK signals. Different targets are reflected by different frequency levels to obtain different IF(Intermediate Frequency) signals, which overcomes the defects that FSK system can not measure multiple targets with the same radial velocity since it only has two frequencies .

3 System Design

The hardware structure of MFSK automotive radar system consists of four parts is shown in Figure 2.

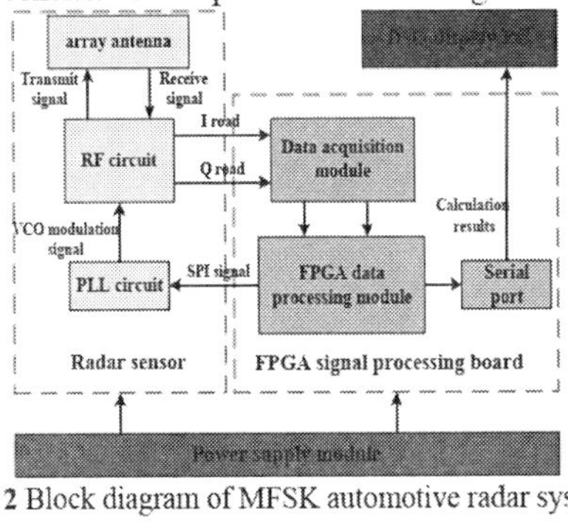

Fig. 2 Block diagram of MFSK automotive radar system

This system includes power supply module, radar sensor, FPGA signal processing board and data display PC. The power supply module provides the appropriate voltage and current for the system. The radar sensor includes 24GHz array antenna, RF (Radio Frequency) circuits and PLL (Phase-locked loop) circuit, it is responsible for signal emission and reception. The FPGA signal processing board includes data acquisition module, FPGA data processing module and serial port, which is mainly used for the signal processing and data transmission of the entire system. The data display PC is used to display the measurement results.

4 FPGA Signal Processing Algorithm

As the core of this system, FPGA controls the timing sequence, logic and signal processing of the whole system. The internal signal processing block diagram is shown in Figure 3 as below.

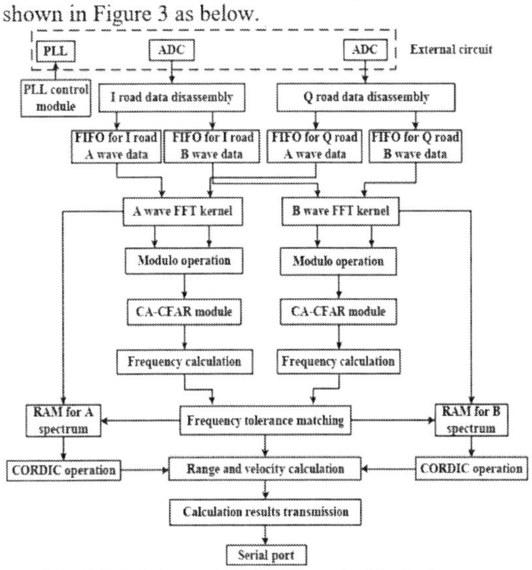

Fig. 3 FPGA internal signal processing block diagram

Firstly, the SPI(Serial Peripheral Interface) signal controls the PLL circuit of the radar sensor to generate the high precision MFSK signal, which could improve the modulation accuracy, then FPGA controls the ADCs to transform the orthogonal I and Q road signals into digital data. According to the measurement principle in section 2, the corresponding echo signals of A and B needs to be processed separately. The receiving orthogonal I and Q road data will be disassembled at the end of each short frequency step, after that the data are temporarily stored in the corresponding FIFO. Then the data are respectively transformed by FFT kernel, the transformed spectrum data is stored in RAM. At the same time, the absolute value of the spectrum data is obtained by modulo operation, then the data will be processed by the CA-CFAR module. This module can not only suppress system noise, but also effectively improve the detection efficiency of the spectrum peaks and reduce the system response time.

Then the data are respectively passed through the frequency calculation module to calculate the corresponding frequencies of the spectrum. Since the frequencies of the A and B wave which corresponding to the same target should be equal, due to the systematic error, the final calculation result may be little different. A frequency matching tolerance function is designed: $|f_{Ak} - f_{Bk}| \leq 2/T_{CPI}$, where is the modulation period of the MFSK signal. If the A and B frequency values satisfy this function, it means that the frequency corresponds to the same target, the target peak position and the frequency value can be obtained. According to the location of the peak points, the spectrum raw data of the corresponding location are extracted from A road RAM and B road RAM, then the data is sent to CORDIC (Coordinate Rotation Digital Computer) module to calculate the phase like equation (7).

$$\varphi = \arctan\left(\frac{I(t)}{Q(t)}\right) \tag{7}$$

In this system, an improved CORDIC algorithm is used to perform the inverse tangent operation, and the data compensation is carried out too. The floating point number is converted to the fixed point number to improve calculation precision. The phase difference is gained by subtracting the phase values of A and B wave, which is used to calculate the range and velocity of the target with the frequency obtained before. Finally, FPGA sends the results to external device through the serial port.

5 The FPGA Implementation And Field Experiment

The parameters of the system designed in this paper are as Table 1.

Table 1. System parameters

Parameter	Value
Center frequency	24GHz
Sweep bandwidth	150MHz
Modulation period	100Hz
Step number	100
Step frequency	1500KHz
Step time	0.1ms
Range resolution	1m
Velocity resolution	0.625m/s
Maximum range	80m

In this system, the PLL chip is ADF4158, which is a 6.1GHz fractional N frequency synthesizer with modulation and waveform generating capacity. Moreover, FPGA signal processing unit is EP4CE115F484. The system clock is 50MHz, and the sampling clock is 200KHz. 2048 points can be sampled in a modulation period, and the number of FFT point is 1024. In addition, the FPGA implementation language is Verilog HDL.

In order to test the performance of the system, some field experiments and matlab simulation are to be made, which measure the target of different motion state in the case of single target and multiple target. The whole system appearance and system filed test scene are shown in Fig.4, and the real time measurement results can be seen on the data display PC. It's like the results of the measurements shown in Fig.4, range is 15m, velocity is -2.5m/s.

Fig. 4 System field test scene

5.1 Single Target Field Measurement

The experiment set single target as measurement object, which includes the stationary target and the moving target. In order to see the process of signal processing more intuitively, the real-time data is extracted and processed by matlab with the algorithm designed in section 4. One of the real-time data processing results are shown in Fig.5, it is the frequency and amplitude spectrum of A and B wave. Meanwhile, the frequency and phase value of the peaks are marked in the figure. As we can see from Figure 6, the frequencies of the two signals of A and B are basically equal, which represent that they are the echo of the same target, but the phase value is different. So the beat frequency and phase difference can be obtained. Moreover, the range and velocity can be calculated simultaneously.

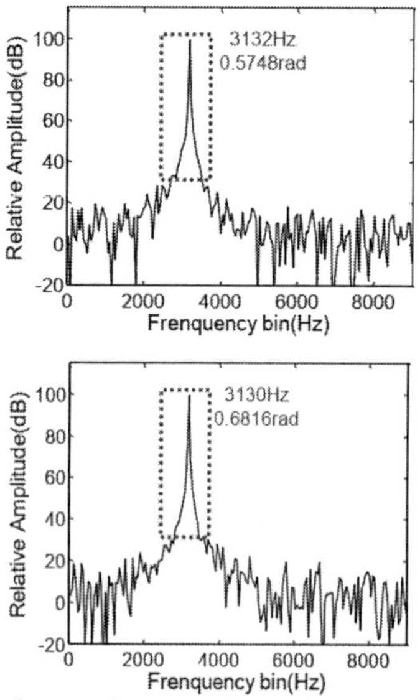

Fig.5 Single target data processing result, A wave(up), B wave(down)

Two hundred groups of experiments were done in total, and two hundred targets are measured, the measurement results are recorded from the data display PC, and the results are count as follow Table 2.

Table 2 Single target field measurement result statistics

Range RMSE	Velocity RMSE	Ghost target number	Lost target number	response time
2.5%	3%	0	0	30ms

5.2 Multiple Target Field Measurements

The multiple target measurement experiment to measure the multi-state target, which include stationary targets and moving targets, the targets with different velocity at the same range, the targets with different range at the same velocity, and the distinguishable target and non distinguishable target. Fig.6 shows one of the real-time data processing results, it can be seen that the three frequencies of the A and B wave correspond to each other, which represent three different targets respectively. After frequency matching and phase calculation, the three targets' beat frequency and phase difference can be obtained, and then get the target motion information.

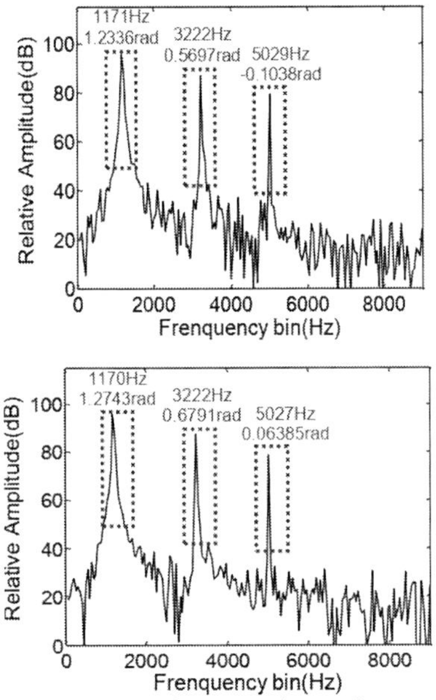

Fig.6 Multiple target data processing result, A wave(up), B wave(down)

The experiment also made two hundred groups field test, each group contains 2-4 targets, 500 targets are measured in total. The measurement result statistics as Table 3.

Table 3 Multiple target field measurement result statistics

Range RMSE	Velocity RMSE	Ghost target number	Lost target number	response time
3%	3.5%	0	0	35ms

5.3 Experimental results analysis

After massive experiments and field tests, the single target and multiple target measurement results are shown in Table II and table III. It illustrate that the system designed in this paper has high measurement accuracy, the measurement RMSE (Root Mean Square Error) of range and velocity is very small. Moreover, this system has little ghost targets in multi-target condition, and it can achieve unambiguous measurement even in the case of multiple targets with the same radial velocity. In addition, this system never lost target, and it also has fast response time.

6 Conclusion

In this paper, a FPGA-based MFSK automotive radar system is designed, the architecture of the system and the FPGA implementation algorithm are introduced in detail. At last, the system performance is proved by field test. The results show that MFSK method is better than traditional LFM and FSK methods, the FPGA implementation provides fast processing speed and high measurement accuracy. So the new system has good application prospect in real.

However, the automotive radar system designed in this paper has good measurement accuracy and fast response time under experimental conditions. There are no extra objects which affect the test, and the scene is relatively wide. In the case of real traffic road, the environment is more complex, and the system needs to be perfected continuously to reach the goal of adapting to more complex environment.

References

1. Patole S M, Torlak M, Wang D, et al. Automotive radars: A review of signal processing techniques[J]. IEEE Signal Processing Magazine, 2017, 34(2): 22-35.
2. Macaveiu A, Nafornita C, Isar A, et al. A method for building the range-Doppler map for multiple automotive radar targets[C]. Electronics and Telecommunications (ISETC), 2014 11th International Symposium on. IEEE, 2014: 1-6.
3. Rohling H, Moller C. Radar waveform for automotive radar systems and applications[C]. Radar Conference, 2008. RADAR'08. IEEE. IEEE, 2008: 1-4.
4. Ziran X, Dongchen Z, Xing J. An effective method for multi-target range and velocity measurement in FMCW radar[J]. Acta Electronica Sinica, 2016, 44(9): 2148-2157.
5. Geroleo F G, Brandt-Pearce M. Detection and estimation of LFMCW radar signals[J]. IEEE Transactions on Aerospace and Electronic Systems, 2012, 48(1): 405-418.

6. Kronauge M, Rohling H. New chirp sequence radar waveform[J]. IEEE Transactions on Aerospace and Electronic Systems, 2014, 50(4): 2870-2877.
7. Dudek M, Nasr I, Bozsik G, et al. System analysis of a phased-array radar applying adaptive beam-control for future automotive safety applications[J]. IEEE Transactions on Vehicular Technology, 2015, 64(1): 34-47.
8. Hu C, Liu Y, Meng H, et al. Randomized switched antenna array FMCW radar for automotive applications[J]. IEEE Transactions on Vehicular Technology, 2014, 63(8): 3624-3641.
9. Lee M S, Kim Y H. Design and performance of a 24-GHz switch-antenna array FMCW radar system for automotive applications[J]. IEEE Transactions on Vehicular Technology, 2010, 59(5): 2290-2297.
10. Rohling H. Development milestones in 24 GHz automotive radar systems[C]. Radar Symposium (IRS), 2010 11th International. IEEE, 2010: 1-5.

Optimization and research of primary frequency compensation strategy of the unit with supercritical forced circulation boiler

Ying-kun Han[1, 2], *Lin* Mou[3], *Wei-ming* Liu[2], *Peng-wei* Yang[2], *Xiang-Kun* Pang [1]

[1]State Grid Shandong Electric Power Research Institute, 250002 Jinan,Shandong, China
[2]State Grid East Inner Mongolia Electric Power Research Institute, 010070Hohhot, Inner Mongolia, China
[3]Shandong University Engineering Training Center, 250061 Jinan,Shandong , China

Abstract. The paper introduces the theory and application of Power Plant's primary frequency compensation(PFC) on the unit with the supercritical forced circulation boiler. First of all,the paper gives the typical parameters Setting and logic design of PFC. Then it explains some adverse factors which affect the PFC of the unit. With some problems in the operative process, the typical control strategy and some advices are given , which can improve PFC ability of Power Plant's PFC.

1 Introduction

With the development of new equipment and new technology, the power thermal units with supercritical forced circulation boiler will become the mainstream of technology due to the energy saving advantages. However, because of the technical characteristics on the reduction of heat storage capacity , the PFC capability of the units with the forced circulation boiler is weakened correspondingly. The PFC support to the power grid is affected. At the same time with the rapid development of the UHV grid and clean energy generation, more and more factors affect the power grid operation frequency. The large power grid safe and reliable operation with strong PFC ability is also more important. The units PFC ability has become an important technical support to the coordinated development on source-grid power system.

In order to improve the power quality and the control performance of the grid frequency in the operation, the coordination control system of the units must be optimized. The units are required to have the corresponding PFC capability. The paper will introduce the control principle and the typical technology solutions of PFC on the unit with the supercritical thermal unit. Combining the problems happened in the operative process, how to optimize PFC strategy can be illustrated. The correct control plan can avoid the adverse effect of PFC to the safe operation of the unit.

2 The overview of primary frequency compensation(PFC)

In the actual operation of the power grid, it is affected by the change of the user's consumption load, which causes the small and periodic changes of the power grid frequency. It is mainly based on the control system of the turbine generator to adjust the turbine valves automatically, which can complete the load compensation of the power grid and modify the fluctuation of the power grid frequency. This process is called PFC of the generator unit[1].

The PFC function in DEH system, speed deviation(Δn= rated speed-actual speed) is usually used instead of frequency deviation(Δf= rated frequency-actual frequency) , which is usually converted to power signal compensation or flow compensation. The control structure schematic diagram of PFC is shown in figure 1[2].

For example, the power grid rated frequency is 50Hz (the turbine rated speed is 3000 RPM) .The PFC function (FG) converts Δn (rated speed and turbine actual speed difference) to a PFC compensation factor. If PFC ON, the PFC compensation factor will be added with the load reference value to control the load of the unit. If PFC OFF,the PFC factor is set to zero, that not to participate in the system control.

* Corresponding author: hanyk8086@163.com

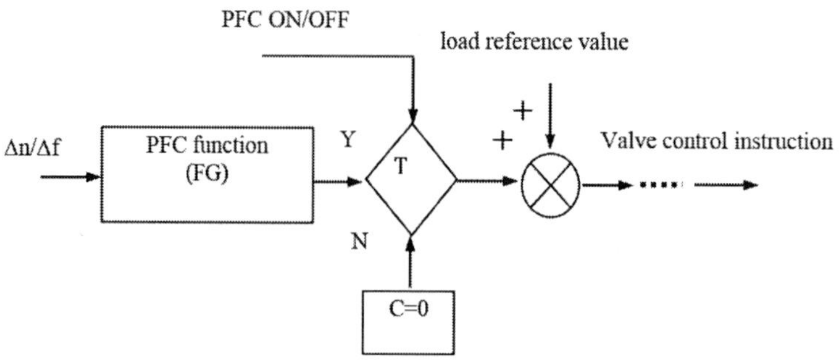

Figure 1 Control structure schematic diagram of PFC

3 Typical parameters setting of PFC function (FG) [3]

With the improvement of users' demand for power quality and frequency, the power grid has stringent requirements on PFC performance of grid-connected units. Therefore, the parameters setting of PFC is more important. The common PFC parameter setting methods are mainly used.

3.1 No dead zone, no limit setting mode

No dead zone, no limit setting mode. As long as the actual speed of the unit is deviated from the rated speed, the power grid frequency is adjusted. The control valves of the turbine are operated, which is unfavorable to the stable operation of the system. Limited to the capacity and stability performance requirements of China's grid, this setting mode is generally not applicable to the current generating sets.

3.2 With dead zone, no limit setting mode

With dead zone, no limit setting mode. The units can be set speed dead zone in this mode. When the Δn is in the range of the deviation dead zone, PFC OFF. When the Δn is beyond the deviation dead zone, PFC ON.

3.3 With dead zone, no changeδ,no limit setting mode

With dead zone, no change δ (speed governing droop),no limit setting mode. Compared with 3.2, this model can set the PFC dead zone. In the frequency difference of dead zone, the PFC factor has sudden change at the critical point, which will make the valves of the unit suddenly switch action. It will cause the changes of fuel, water supply, air supply and other parameters in the coordinated control system, which give rise to operation unstable of the system.

3.4 With dead zone, with limit setting mode

With dead zone, with limit setting mode. Compared with 3.3, this setting mode to increase the PFC factor limit setting function can effectively prevent the frequency difference signal from overloading. The parameters can also be modified to realize one-way limit function, which can meet the technical requirements of the power grid.

3.5 An instance of PFC function setting

For example, the unit with supercritical forced circulation boiler that rated capacity is 600MW.It's PFC control method use DEH+CCS. According to the related technical requirements of power grid, speed governing droop is set to 5%. At same time, speed dead zone is set to ±2RPM. The maximum compensation requirement of PFC is 6% of the rated load. The mode of the dead zone and limit setting is adopted in 3.4, and the set value of PFC command function is shown in table 1

Table 1 PFC function setting

signs	Δn（RPM） Δn= rated speed- actual speed	Load compensation command（MW）
2	-11	-36
3	-2	0
4	2	0
5	11	36
6	152	36

The setting curve of PFC function is shown in figure 2.

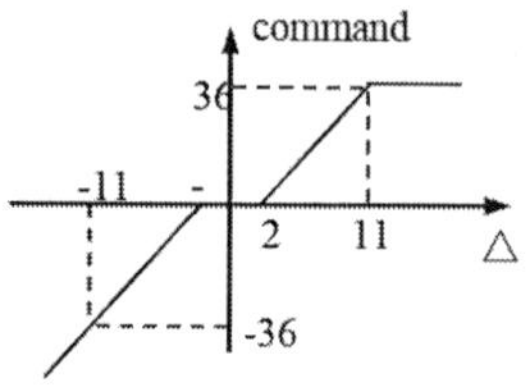

Figure 2 the setting curve of PFC function on

When |△n|>2, the controller will calculate the load compensation command according to the PFC function and △n, that add to the load reference value as the new command to control valve changing and complete unit's PFC function. When |△n|<=2, the compensation command is 0, PFC OFF.

4 Problems in the control logic of PFC

In order to ensure the safety and reliability of the unit in actual operation, the fixed range is generally used to set the PFC. When the frequency modulation happens, this mode ensure the unit can not exceed the stable operation range. However, due to the improper logic configuration in the load range limitation, some units can not provide the correct limit protection, but also adversely affect the safe operation of the unit. Here are some improper logical configuration methods.

4.1 Improper input conditions

For example, a 600MW unit is set as shown in figure 3. The unit's PFC ideal range is 300MW-600MW, and the PFC compensation load is set to -36MW-+36MW. The PFC will be OFF automatically when unit's load exceeds the stable operating range.

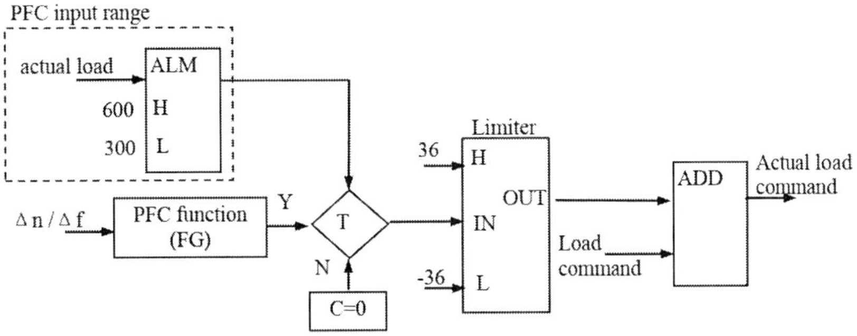

Figure 3 the diagram of improper input range on PFC

Figure 3 shows the logic configuration. When the unit's load is between 300MW and 600MW, the ALM module outputs is 1,PFC ON. When the unit's load is less than 300MW or the unit's load is greater than 600MW, the ALM module outputs is 0, PFC OFF.The fixed input range in the virtual frame is guaranteed PFC exit automatically when the unit's load exceeds the range of 300MW- 600MW. But at limiting points, such as at the 303MW load point, PFC command is -6MW. When the unit's load is less than 300MW, PFC command change to 0, so actual load command will be increased. When the unit's load is greater than 300MW, PFC IN again. The PFC command made load reduction and so on, that appear frequently ON or OFF on PFC. This configuration setting often causes the unit's load oscillate at the limit points, which is not conducive to the safe operation of the unit.

4.2 The PFC compensation command with improper limit setting logic

Figure 4 shows the logic configuration. A 600MW unit's configuration of PFC setting purpose is 36MW which is maximum value of PFC load compensation command. The unit's PFC range is 300MW to 600MW. At the limiting points, the controller will adjust PFC load compensation to the unit, which made operation load shall not exceed the limit value. Such as at the 590MW load point, if there is a PFC compensation command 20MW, the controller automatic computes the PFC command, that maximum is 10MW, up to the highest command for maximum load range 600MW, not 610MW. In order to ensure operation safety of the units, the load command can not exceed the maximum limit value.

Figure 4 the logic configuration calculates PFC compensation command, which select unit's actual load as comparing parameter. This mode will not affect the PFC adjustment under normal circumstances. When the load near the high or low load limit points, the PFC command may not meet the rating load, which affect the unit's PFC performance. For example, at the 590MW load point, the unit's command is 590MW before PFC action. The value will not change in PFC process. When PFC compensation command is 10MW, the unit's actual load commands become 600MW. The unit rise load immediately. However in this process, the controller calculates the PFC compensation command gradually decreases. The unit's actual load command decreases.

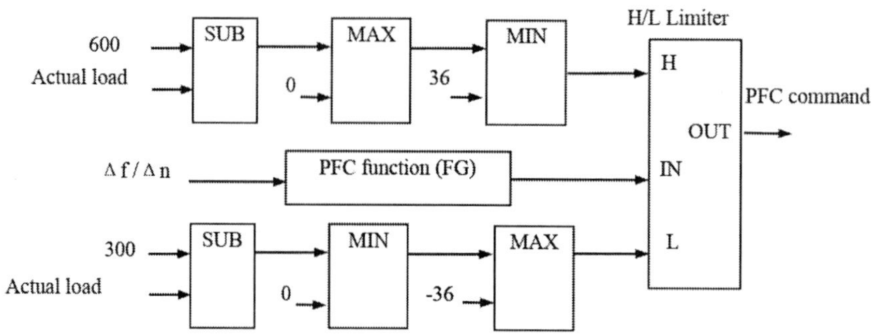

Figure 4 the diagram of PFC command logic

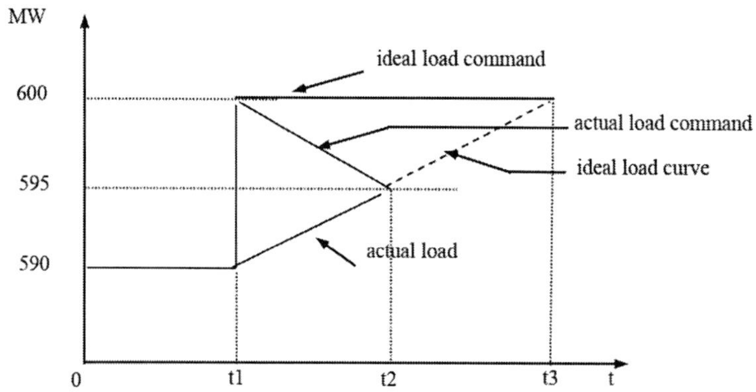

Figure 5 the response curve of PFC

The PFC process curve is shown in figure 5. If the PFC process is ideally, the unit's actual load command should equal to the ideal load command ,that the actual load both are 600MW finally, at the time point t3. Before time point t1, the load command and the actual load curve is overlapping, the values both are 590MW. The unit is operating stable. At time point t1,the PFC compensation load command is up to load 10MW because PFC happen, . The PFC compensation load command add with load command become the ideal load command as 600MW. From t1 to t2, PFC is operating. The load increase and PFC commands automatically reduce at the same time, due to the actual load command decreases. Finally the unit load stability in time point t2, at this time the unit's actual load command and the unit's load both are 595MW, which do not meet the technical index of PFC requirements.

5 Adverse factors affect the PFC

5.1 Low parameters operation

Considering the running efficiency, most of the unit adopts variable parameters at present, that is in different load, the parameters in different pressure. The units keep the turbine valves can be higher opening degree, in order to reduce the throttle of the adjustment and improve the economy. However, the units are operating at the higher opening degree, which has great impact on the PFC of the units. When the valves opening exceeds a certain value, its flow characteristic is poor. For example,the unit's PFC on, if increase load command, load compensation may be limited because the flow characteristic is poor. This mode will affect the unit's PFC capability.

5.2 Turbine following mode

When the unit is operating under the turbine following mode (TF mode), the control object of the turbine is the main steam pressure. When the frequency is adjusted, the load of the units is modulated according to the fluctuation of the power grid frequency. At the same time, the main steam pressure may be changed and modulated. When the controller modulates the main pressure in the TF mode, the process of the PFC is counterproductive. If the grid frequency is less than the rated frequency, the unit's PFC compensation load will rise, which make the main steam pressure drop. In order to adjust the main steam pressure, valves will be down, which will make the unit load drop. When the unit is running in TF mode that without special pressure correction scheme, the units PFC performance obviously can't meet the design requirements.

5.3 Control valves with poor linearity

When the control valves had been operating for long time, especially the unit valves after maintenance, the flow curve will drifted. If the false flow curve is set to the control valves, the control performance of the unit will get worse. The effect to PFC load compensation is not reliable at the same time.

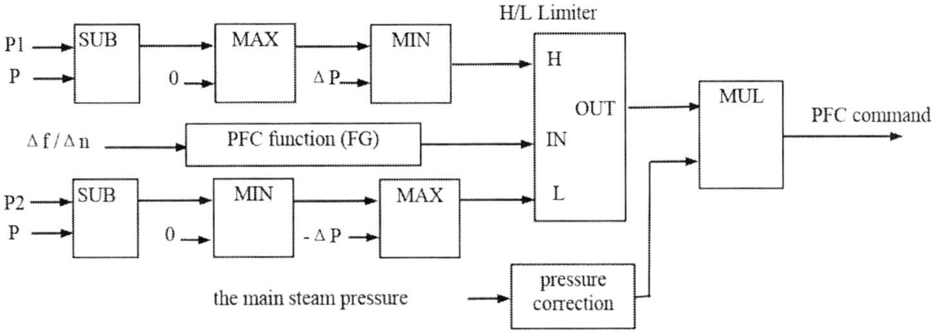

Figure 5 the diagram of an optimal design scheme

Notes: P: instantaneous power load command (no including PFC command)
P1: PFC working upper limit
P2: PFC working lower limit
Δf /Δn: frequency fluctuation difference/ the turbine speed fluctuation
ΔP: PFC ideal load command

6 An optimal design scheme for the PFC

The control logic configuration is shown in figure 6. Based on instantaneous power load command P, PFC upper limit P1 and lower limit P2, the controller automatically calculates the limiting values to set H/L limiter module on high limit value and low limit value. According to the grid frequency fluctuation difference Δ f (usually use the turbine speed fluctuation Δ n instead of Δ f) and PFC function (FG), the controller can calculate out the ideal PFC load command. Then the PFC ideal load command use H/L limiter module to guarantee system working in the normal adjustment range. The PFC compensation command is in range of -Δ P ~Δ P, which can prevent a large disturbance and guarantee the unit's safe and stable.

According to the real-time main steam pressure, the controller calculates a PFC correction coefficient by the pressure correction function block, which will adjust the PFC command. The correct PFC command can ensure the unit's PFC action under various operating mode. The unit's PFC can meet the design requirements, at same time it can ensure the quality of power supply.

7 Conclusion

As power system requirements to improve the performance of power-grid coordination, the power unit's PFC are usually using in the mode with DEH+CCS. IN DEH system, the PFC has a quickly response speed , that can be operated easily, and don't have big impact on the stability of the unit. IN the coordinated control system (CCS), the PFC adjust the load accurately. Using in the mode with DEH+CCS, the unit's PFC ability can be guaranteed. The units with the correct logic configuration of PFC not only can meet the requirements of the PFC performance, but also can meet the needs of the unit stable operating.

References

1. Xiao Z. DEH of the Steam Turbine [M]. Beijing: China Electric Power Press, 2005.
2. Wang F. Automatic Control and Protection on Power Units [M]. Beijing: China Electric Power Press, 2005.
3. Han Y, Meng X, Mou L. Application and Research on 1000MW Power Plant's Control Mode of Frequency adjustment in DEH[J]. East China Electric Power,2008,Vol36,No3:95-97

E3S Web of Conferences **38**, 01051 (2018) https://doi.org/10.1051/e3sconf/20183801051
ICEMEE 2018

Simulation of dissolved nutrient export from the Dongjiang river basin with a grid-based NEWS model

Qiangqiang Rong[1,2], *Meirong* Su[2], *Zhifeng* Yang[2,3], *Yanpeng* Cai[3], *Wencong* Yue[1,2], *Zhi* Dang[1,*]

[1]School of Environment and Energy, South China University of Technology, 510006 Guangzhou, China
[2]Research Center for Eco-environmental Engineering, Dongguan University of Technology, 523808 Dongguan, China
[3]State Key Laboratory of Water Environment Simulation, School of Environment, Beijing Normal University, 100875 Beijing, China

Abstract: In this research, a grid-based NEWS model was proposed through coupling the geographic information system (GIS) with the Global NEWS model framework. The model was then applied to the Dongjiang River basin to simulate the dissolved nutrient export from this area. The model results showed that the total amounts of the dissolved nitrogen and phosphorus exported from the Dongjiang River basin were approximately 27154.87 and 1389.33 t, respectively. 90 % of the two loads were inorganic forms (i.e. dissolved inorganic nitrogen and phosphorus, DIN and DIP). Also, the nutrient export loads did not evenly distributed in the basin. The main stream watershed of the Dongjiang River basin has the largest DIN and DIP export loads, while the largest dissolved organic nitrogen and phosphorus (DON and DOP) loads were observed in the middle and upper stream watersheds of the basin, respectively. As for the nutrient exported from each subbasin, different sources had different influences on the output of each nutrient form. For the DIN load in each subbasin, fertilization application, atmospheric deposition and biological fixation were the three main contributors, while eluviation was the most important source for DON. In terms of DIP load, fertilizer application and breeding wastewater were the main contributors, while eluviation and fertilizer application were the two main sources for DOP.

1 Introduction

Nutrient elements, such as nitrogen (N) and phosphorus (P), are the most important material basis for the development of aquatic ecosystem. The nutrient element contents of water body have great influence on the balance of aquatic ecosystem [1]. However, in recent years, under the combined impact of global change and human activities, nutrient element ratios of various aquatic ecosystems, especially the estuary ecosystem, has undergone great fluctuations [2]. Significant changes have taken place in estuarine nutrient cycle and flux in recent years [3]. Consequently, N and P contents of estuarine ecosystem loss the original balance, leading to eutrophication in estuary and coastal areas, which can further cause a series of ecological environment effects, such as harmful algal blooms, water hypoxia and acidification [4]. Thus, to prevent the ecological and environmental quality of estuarine area, it is very important to propose an effective method for nutrient export simulation in typical river estuary areas. Analyzing the N and P output from watershed to river estuary can provide a solid basis for aquatic environment regulation and water ecology protection. It will also satisfy the urgent requirement of watershed management and sustainable development.

In recent years, a number of scholars have carried out some research on the nutrient export model to analyze the nutrient output loads from various watersheds globally. Particularly, Global Nutrient Export from Watersheds Model, through integrating the migrating and transforming processes of multiple elements and forms, and establishing a unified simulation framework, has become an important tool for nutrient export simulation. For example, Mayorga et al. [5] proposed a Global NEWS model and provided a detailed model description. The model could be used as an effective tool to examine the impact of polices to reduce coastal eutrophication at regional to global scales. Strokal and Kroeze [6] used Global NEWS model and analyzed the past and future trends in river export of N and P to the coastal water of the Black Sea and assessed the associated potential for coastal eutrophication. At the same time, some researchers used Global NEWS Model to analyze the influence of social economy, farmland management and biogeophysical factors on the output of N and P. For instance, Yan et al. [7] estimated the inputs of nitrogen and exports of dissolved inorganic nitrogen from the Changjiang River to the estuary over the period 1970–2003. The result indicated that anthropogenically enhanced N inputs dominated river DIN yields under changing human pressures in the basin. Based on the Global NEWS model, Strokal et al. [8] developed a sub-basin model for the Pearl River basin over the period of 1970–2050 to analyze the trends in dissolved inorganic N and P and to identify the main sources of these nutrients and their locations. The results indicated that nutrient management in agriculture and sewage in down-stream areas was more effective in reducing coastal eutrophication than nutrient management in up- and middle-stream areas of the Pearl River basin. Thus, the Global NEWS model is an effective tool for nutrient export simulation on global and regional scales. However, most of the input data for NEWS model are spatially distributed at 0.5° resolutions (i.e. approximately 2890 km^2). The data source with such a low resolution would reduce the spatial difference to a certain extent. Concurrently, the N and P output is highly dependent on terrain and soil conditions within a watershed, especially when the watershed consists of a large area with heterogeneous terrain and soil types [9]. Thus, it is very difficult for the Global NEWS model to analyze the spatial distribution of N and P, when modeling the nutrient export on a watershed scale.

In this research, the Dongjiang River basin will be selected as the study area. Considering the nonuniformity of the underlying surface, a grid-based NEWS model will be first proposed, which improves the NEWS model through coupling the geographic information system (GIS) with the NEWS model framework. Then, the natural and social economic data of the Dongjiang River basin will be collected to establish the spatial and attribute database of the study area. Finally, the improved model will be applied to simulate the nutrient export in the Dongjiang River basin. Also, the spatial distribution will be analyzed to help provide a data basis for the NPS pollution management in the river estuary area. Thus, it is of great significance in both science and practice.

2 Materials and methods

2.1 Development of the grid-based NEWS model

NEWS model was developed to predict steady-state annual exports at the mouth of rivers for dissolved organic and inorganic nitrogen (DON and DIN); dissolved organic and inorganic phosphorus (DOP and DIP); and dissolved organic forms (DOC). It is a unified formulation of dissolved element-form sub-models [5].The export of each dissolved element form from watershed to river estuary can be calculated through the following equation:

$$\mathrm{Yld_F} = \mathrm{FE_{riv,F}} \cdot (\mathrm{RSpnt_F} + \mathrm{RSdif_F}) \tag{1}$$

where the subscript F means the nutrient form (i.e. DIN, DON, DIP and DOP); $\mathrm{Yld_F}$ means the general yield of each dissolved element form; $\mathrm{RSpnt_F}$ means F export through point sources (kg·km^{-2}·yr^{-1}); $\mathrm{RSdif_F}$ means F export through diffuse sources (kg·km^{-2}·yr^{-1}); $\mathrm{FE_{riv,F}}$ is a fraction (0~1) of nutrient form F inputs to rivers that is exported at the basin mouth, corresponding to retention within the river system $(1 - \mathrm{FE_{riv,F}})$.

Point and diffuse sources represent the human waste emission and watershed sources and sinks, respectively. The two sources can be described as follows:

$$\mathrm{RSpnt_F} = \mathrm{FE_{pnt,F}} \cdot \mathrm{RSpnt_E} = \mathrm{FE_{pnt,F}} \cdot (1 - \mathrm{hw_{frem,E}}) \cdot I \cdot \mathrm{WShw_E} \tag{2}$$

$$\begin{aligned}\mathrm{RSdif_F} &= (\mathrm{RSdif_{ant,F}} + \mathrm{RSdif_{nat,F}}) \\ &= \{[\mathrm{FE_{ws,F}} \bullet \mathrm{WSdif_{ant,E}} + \mathrm{Ag_{fr}} \bullet \mathrm{RSdif_{ec,F}}] \\ &\quad + [\mathrm{FE_{ws,nat,F}} \bullet \mathrm{WSdif_{nat,E}} \\ &\quad + (1 - \mathrm{Ag_{fr}}) \bullet \mathrm{RSdif_{ec,F}}]\}\end{aligned} \tag{3}$$

where $\mathrm{WShw_E}$ (kg·km^{-2}·yr^{-1}) represents a gross human-waste source to the watershed, $\mathrm{WShw_N} = \mathrm{WShwExc_N}$ and $\mathrm{WShw_P} = \mathrm{WShwExc_P} + \mathrm{WShwDet_P}$; I is the fraction of the population connected to a sewage system; and $\mathrm{hw_{frem,E}}$ is the fraction of E in sewage influent removed via wastewater treatment. $\mathrm{FE_{pnt,F}}$ is the fraction of $\mathrm{RSpnt_E}$ emitted as form F. $\mathrm{FE_{pnt,F}}$ for DIN is a linear, empirical function of $\mathrm{hw_{frem,N}}$, while for all other N and P forms it is equal to a calibrated or default constant, c_F. $\mathrm{RSdif_{ant,F}}$ and $\mathrm{RSdif_{nat,F}}$ mean the anthropogenic and natural inputs to watersheds, respectively, which include the net effect of land-based retention or removal (watershed sinks) of nutrients; $\mathrm{FE_{ws,F}}$ is a function of mean annual water runoff from land to streams ($\mathrm{R_{nat}}$): $\mathrm{FE_{ws,F}} = \mathrm{e_F} \cdot f_F(\mathrm{R_{nat}})$; $\mathrm{WSdif_{ant,E}}$ and $\mathrm{WSdif_{nat,E}}$ are the explicit budgets for N and P on agricultural (anthropogenic, ant) and natural areas of the watershed; $RS\mathrm{dif_{ec,F}}$ represents direct diffuse inputs to rivers; $\mathrm{Ag_{fr}}$ is the fraction of the basin covered by agricultural areas, and $\mathrm{FE_{ws,nat,F}} = \mathrm{FE_{ws,F}}$ except for DIN. Thus, the total load ($\mathrm{Ld_F}$, Mg yr^{-1}) of each form F exported to river mouth can be calculated as: $\mathrm{Ld_F} = \mathrm{Yld_F} \cdot \mathrm{A}$ [5].

When the model was used on a watershed scale, data with higher resolution should be inputted to improve the accuracy and rationality of the simulation results. To achieve this goal, a grid-based NEWS model is proposed in this research based on GIS. Through this method, the watershed can be divided into numerous grids. The size of each grid can be determined based on the area of the basin and the requirement of output accuracy. Through the GIS platform, the data related to population density, land use, livestock and poultry breeding, social economy, topography condition can be transformed into grid format. Then, all the input data for NEWS model can be obtained and a grid-based nutrient export simulation can be carried out on a watershed scale.

2.2 Research area and the data sources

The Dongjiang River basin is located in south China (N22°45′~25°20′, 113°30′~116°45′). Providing more than 40,000,000 people of Guangdong Province and Hong Kong, it is the third longest tributary of the Pearl River basin. This river is located in a subtropical monsoon climate area, with an annual average temperature of 20.4 °C. The precipitation is unevenly distributed during the year. The annual precipitation amount is 1500 ~ 2400 mm. The terrain of the Dongjiang River basin is very complex. Mountain, hills and plain approximately account for 34.77%, 38.58% and 13.01 % of the total area, respectively (Figure 1). Forestland, grassland, farmland, garden land, urban land and water area are the main land use types in this area [10].

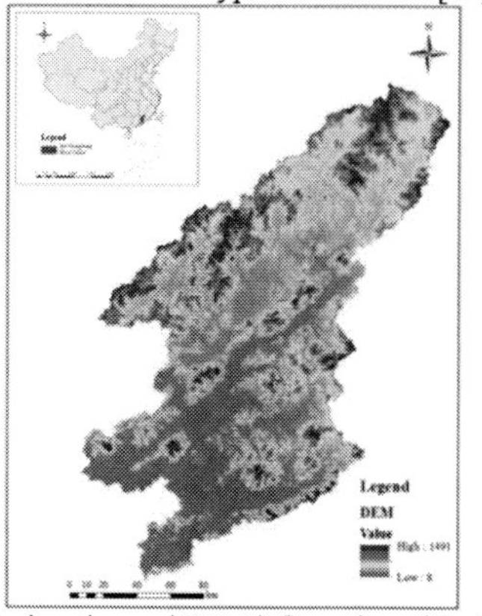

Figure 1 The location and DEM information of the Dongjiang River basin

In this research, the DEM was obtained from the Geospatial Data Cloud, with a resolution of 30 m × 30 m (http://www.gscloud.cn/). Based on the DEM, the subcatchment of the basin can be generated. The population data is from the National Earth System Science Data Sharing Infrastructure (http://www.geodata.cn/index.html), with a resolution of 1 km × 1 km. The landuse data was obtained from GlobalCover 2009 landuse map (http://due.esrin.esa.int/page_globcover.php), with a resolution of 300 m × 300 m. The agricultural information, including application amount of fertilizer, amount of livestock and poultry farming, crop area and yield, was obtained from County-level statistical yearbooks. All the input data was transformed in to raster format through GIS platform.

3 Results and Discussion

3.1 Overall description of the nutrient export loads in the Dongjiang River basin

The total amounts of the dissolved N and P exported from the Dongjiang River basin were approximately 27154.87 and 1389.33 t, respectively. Among these two loads, the inorganic N and P were the main forms, accounting for more than 90 % of the total loads. Specifically, the exported amounts of DIN and DON were 26206.29 and 948.58 t, respectively. Those of DIP and DOP were 1328.21 and 61.12 t, respectively. There are many sources that can lead to the nutrient export from a watershed. Different sources may have different influences on the output of each nutrient form. In terms of DIN load from the research area, fertilizer application, atmospheric deposition and biological fixation were the three main contributors. The DIN amounts exported from these three sources were 13611.12, 5514.62 and 3194.16 t, respectively. As for DON, the eluviation was the main contributor, accounting for more than 75 % of the total load. With regard to DIP, fertilizer application, breeding wastewater and domestic wastewater were the three largest contributors. The DIP amounts exported from these three sources were 580.24, 352.65 and 233.08 t, respectively. As for DOP, the influence of eluviation and fertilizer application to the DOP export amounts was much more significant than the other sources. These two sources contributed more than 90 % of the DOP load (Figure 2).

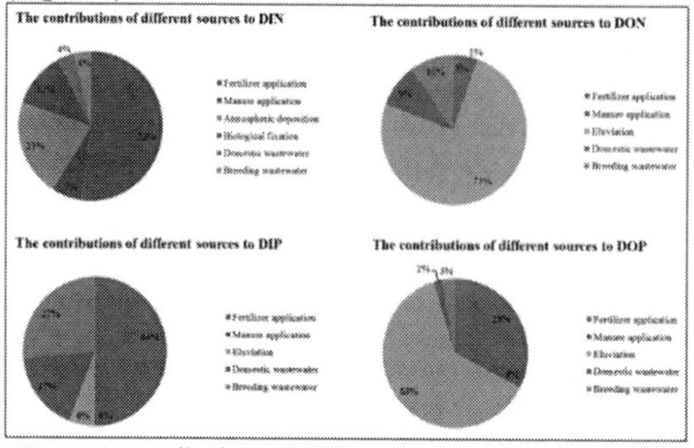

Figure 2 Contributions of different sources to the dissolved N and P in the Dongjiang River basin

3.2 The distribution characteristics of nutrient export in the Dongjiang River basin

The nutrient export loads did not evenly distributed in the area. Specifically, DIN and DIP export loads were largest in the main stream watershed of the Dongjiang River basin. Concurrently, the largest DON load was observed in the middle stream watershed of the basin (including Lijiang, Xinfeng and Qiuxiang River watersheds), while the largest DOP load was from the upper stream watershed of the basin (including Beiling and Xunwu River watersheds). The four largest loads from these four areas accounted for approximately 30 % of the nutrient loads exported from the entire basin. Although the middle stream watershed of the Dongjiang River basin has the largest area (i.e. 10035.46 km^2), all the nutrient export loads from this area were not the highest. On the contrary, with the smallest area (i.e. 4606.46 km^2), the upper stream watershed of the basin had a relatively higher dissolved P loads. The smallest loads of DIN, DIP were observed in the upper and middle stream watersheds of the basin respectively, while those of DON and DOP were from the downstream watershed (Figure 3).

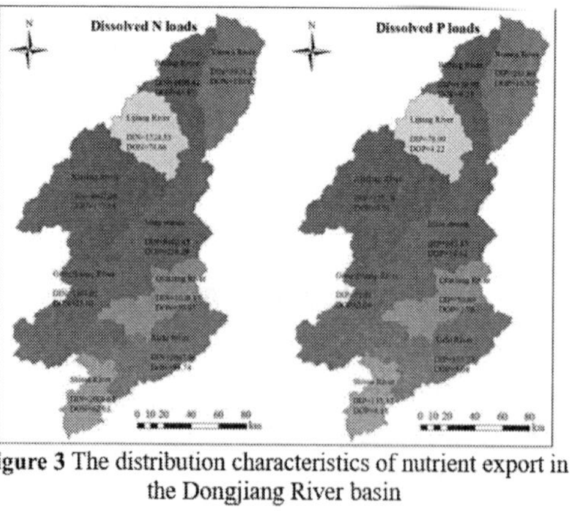

Figure 3 The distribution characteristics of nutrient export in the Dongjiang River basin

3.3 The contributions of different sources to the nutrient export in each subbasin

As for the DIN load exported from each subbasin, fertilization application was the biggest contributor. More than 40 % of the DIN load of each subbasin was from fertilization application. Besides, atmospheric deposition and biological fixation also had a significant influence on the DIN output. In terms of DON, eluviation was the most important source. More than 70 % of the DON loads from most of the subbasins were contributed by eluviation. However, for dissolved P loads, the influences of different sources varied significantly in different subbasins. Specifically, in the Xunwu, Beiling, Gongzhuang, Shima and the main stream watersheds, fertilizer application was the biggest contributor to the DIP export. However, breeding wastewater was the biggest source for the DIP exported from Lijiang, Xinfeng, Qiuxiang and Xizhi River watersheds. With regard to DOP, eluviation and fertilizer application were the two biggest sources, accounting for almost 90 % of the total DOP load in each subbasin (Figure 4).

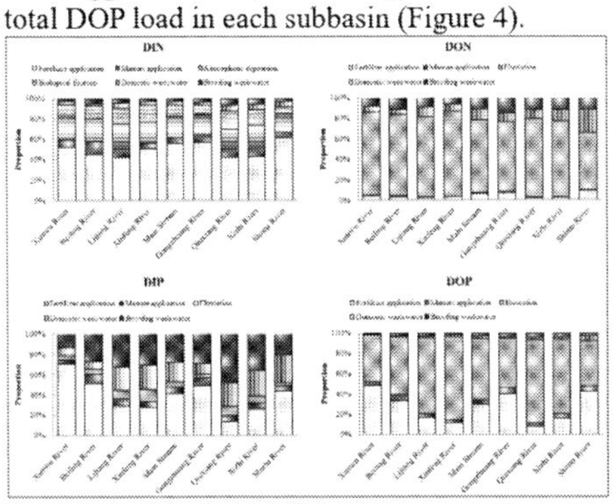

Figure 4 Contributions of different sources to the nutrient export in each subbasin

4 Conclusions

Under the influence of global change and human activities, plenty of nutrients have been discharged into the estuary area, leading to eutrophication in coastal areas. The ecological and environmental quality of estuarine area has obviously deteriorated. Thus, it is of great significance to propose an effective method for nutrient export simulation in typical river estuary areas. In this research, a grid-based NEWS model was proposed to simulate the nutrient export from the Dongjiang River basin. The results indicated that the total amounts of the dissolved N and P exported from the Dongjiang River basin were approximately 27154.87 and 1389.33 t, respectively. Among these two loads, the inorganic N and P were the main forms, accounting for more than 90 % of the total loads. The nutrient export loads did not evenly distributed in the area. DIN and DIP export loads were largest in the main stream watershed of the Dongjiang River basin. Concurrently, the largest DON and DOP loads were observed in the middle and upper stream watersheds of the basin, respectively. At the same time, the smallest loads of DIN, DIP were observed in the upper and middle stream watersheds of the basin respectively, while those of DON and DOP were from the downstream watershed. As for the contributions of different sources to the dissolved nutrient export, fertilization application, atmospheric deposition and biological fixation were the three main contributors for the DIN load in each subbasin, while eluviation was the most important source for DON. In terms of DIP load, fertilizer application and breeding wastewater were the main contributors for each subbasin, while for DOP, eluviation and fertilizer application were the main sources.

Acknowledgements

This work was financially supported by the National Key R & D Program of China (No.2017YFC0405900 and 2016YFC0502800), the Natural Science Foundation for Distinguished Young Scholars of Guangdong Province (No.2017A030306032),Guangdong Innovation Team Project for Colleges and Universities (No. 2016KCXTD023) , GDUPS (2017), the Scientific Research Foundation for High-level Talents and Innovation Team in Dongguan University of Technology (No. KCYKYQD2016001), and the Research Start-up Funds of DGUT (No. GC300501-16).

References

1. D.L. Stoliker, D.A. Repert, R.L. Smith, B. Song, D.R. Leblanc, T.D. Mccobb, C.H. Conaway, S.P. Hyun, D.C. Koh, H.S. Moon, Environ. Sci. Technol., **50**(7), 3649 (2016).
2. N. Peyman, S.B. Tavakoly Sany, M. Tajfard, R. Hashim, M. Rezayi, D.J. Karlen, Environ. Sci. Proc. Impacts, **19**(8), 1086 (2017).
3. Q.Q. Rong, J.T. Liu, Y.P. Cai, Z.H. Lu, Z.Z. Zhao, W.C. Yue, J.B. Xia, Ecol. Eng., **76**, 57-65 (2015).
4. H.Wang, M. Dai, J. Liu, S.J. Kao, Z. Chao, W.J. Cai, G. Wang, Q. Wei, M. Zhao, Z. Sun, Environ. Sci. Technol., **50**(5), 2255 (2016).
5. E. Mayorga, S.P. Seitzinger, J.A. Harrison, E. Dumont, A.H.W. Beusen, A.F. Bouwman, B.M. Fekete, C. Kroeze, G.V. Drecht, Environ. Modell. Softw., **25**(7), 837-853 (2010).
6. M. Strokal and C. Kroeze, Reg. Environ. Change, **13**(1), 179-192 (2013).
7. W.J.Yan, E. Mayorga, X.Y. Li, S.P. Seitzinger, A.F. Bouwman, Global Biogeochem. Cy., **24**(4) (2010).
8. M.Strokal, C. Kroeze, L. Li, S. Luan, H. Wang, S. Yang, Y. Zhang, Biogeochemistry, **125**(2), 221-242 (2015).
9. X.W. Ding, Z.Y. Shen, H. Qian, Z.F. Yang, W. Xing, R.M. Liu, J. Hydro., **383**(3–4), 233-244 (2010).
10. Q.Q. Rong, M.R. Su, Z.F. Yang, Y.P. Cai, W.C. Yue, Z. Dang. *3rd International Conference on Advances in Energy Resources and Environment Engineering*. (2018).

Granulation for Coking Wastewater Treatment in a Coupled Anaerobic-Aerobic Reactor

Chunjuan Dong[1,*], Bingnan Lv[2]

[1]Taiyuan University Department of Environmental Engineering Taiyuan, 030009, China
[2]Harbin Institute of Technology School of Municipal and Environmental Engineering Harbin, 150001, China

Abstract: A coupled anaerobic–aerobic granular bio-film reactor was employed with two operation stages: Stage Ⅰ, granular sludge was formed from digestion sludge using brewery wastewater, and Stage Ⅱ, granular sludge was acclimatized using coking wastewater. Two oxygenation methods (i.e. A and B) were employed to acclimatize the granules. For method A, dissolved O_2 was supplied through a continuous oxygenation way of 800-15000ml·min^{-1}. And for method B, dissolved O_2 was supplied of 800-15000ml·min^{-1} 18-12 times at 20-60min intervals, 1h each time. The experimental results showed that granules could quickly form in 10d in the EGSB reactor seeded with digestion sludge and little loose granules lack of nutrition, and it was the key factor for granules forming to add little loose granules. It took only about 6 months for granules acclimation using coking wastewater. Both oxygenation methods could run well when acclimatizing the granules. However, method A could have comparatively high and stable operation effect. The actual coking wastewater had distinct inhibition effect on the granules, but the supplement of some oxygen could promote the recovery of SMA, and $NaHCO_3$ supplement could also weaken the inhibition effect of the CWW. Method A had more strongly activity recovery ability than method B.

1 Introduction

The A^2/O biological treatment system is frequently used in order to treat coking wastewater for the toxic organic compounds and NH_3-N removal. But there are problems which are inherent to an anaerobic/aerobic treatment sequence (such as the A^2/O system) since the anaerobic transformation of toxic organic compounds is often incomplete[1-4]. The metabolites of anaerobic treatment may also be inhibitory to the methanogens themselves, resulting in declining anaerobic treatment efficiencies and adding to the load on the downstream aerobic treatment system [5-11]. Intimate contact of aerobes with anaerobes may reduce the accumulation of toxic intermediates as the aerobes would achieve an in situ removal of these anaerobic metabolites [12,13].

A simultaneous anaerobic/aerobic treatment methodology using granular sludge was recently described (example for the granule in the EGSB reactor) [14-17]. Faculative bacteria have been shown to be predominant in the peripheral layer while the methanogens are preferentially biodegradation of the recalcitrant xenobiotic compounds but also in the simultaneous removal of toxic organic compounds and NH_3-N in the coking wastewater. Therefore, integration of EGSB for the toxic organic pollutants removal and NH_3-N removal will be a powerful alternative for coking wastewater treatment.

However, the successfully culture and acclimation of the granule of anaerobic/aerobic coupled bioreactors are still a preliminary stage, In particular, forming and acclimation of the granular sludge treating coking wastewater were more important. Because the stably and highly efficiently granular sludge reactors used to treating industrial wastewater, particular for the coking wastewater were very less, and thus it is difficult and often impossible to obtain enough granular sludge to startup a large scale EGSB reactor in many countries, and other seed materials usually have to be chosen. The digestion sludge is considered to be available seed material because it has high methanogenic activity together with a complicated microorganism ecosystem that was suitable for treatment of many kinds of wastewater.

Brewery wastewater was usually used for the anaerobic granulation of the digestion sludge and only 46d was requisite for the forming of high activity granular sludge. But directly using coking wastewater to startup the granular sludge bioreactor was comparatively difficult. And even for the anaerobic SBR treating actual coking wastewater, it also took about 9 months to run well [18]. Thus it was necessary to search an available strategy for quickly start-up of the EGSB reactor (seeded with digestion sludge) treating actual coking wastewater. Two issues were needed to consider. One was how to quickly produce the granular sludge from digestion sludge. The other was how to quickly acclimatize the granules to highly efficiently treat the coking wastewater and meanwhile gain high COD and ammonia removal.

* Corresponding author: chunjuand@126.com

The investigation reported herein was tried to demonstrate the feasibility of using digestion sludge and little loose granules as the seed for quick startup of the EGSB reactor treating coking wastewater and also meanwhile contrastively analyze the influence of the continuous and intermittent oxygenation methods on the granular sludge acclimation process. In addition, batch bioassays were also performed to gain more insight into granular activity through assaying the SMA change.

2 Matericals and methods

The research was carried out in two lab-scale EGSB reactors. The 2.3m height EGSB reactor was an acrylic column with a conical-shaped bottom, a working volume of 12L, an internal diameter of 10cm The wastewater used as influent to the reactor was obtained one from synthetic wastewater using sucrose, beer and sodium acetate as the carbon source and NH_4Cl, KH_2PO_4, $Na_2S·9H_2O$, $FeCl_2·4H_2O$, $CoCl_2·6H_2O$, $NiCl_2·6H_2O$ as nutrients, and the other from the actual coking wastewater collected from the buffer tank of the first coking plant of Taiyuan Coal Gas and Chemical Stock Co., LTD. located in Shan'xi Province, North China with 818-978mg.L^{-1} COD, 12.8-49.3mg.L^{-1} phenol, 1.82-57.8mg.L^{-1} CN, 66.7-154.7mg.L^{-1} SCN and 6.5-8.1 pH. No KH_2PO_4 were added to the influent. But $NaHCO_3$ and some trace metals for Fe, Co, Ni were added. A schematic diagram of the experimental EGSB reactor used is presented in Fig.1.

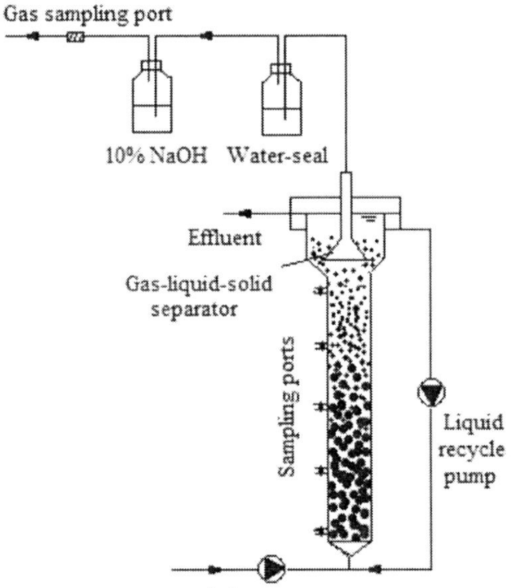

Fig.1 Schematic diagram of the experimental EGSB reactor

2.1 Seed sludge

The inoculum used in this experiment was municipal digestion sludge and little loose granules. The anaerobic digestion sludge was obtained from the dehydrated sludge from an anaerobic digester at YangJiaBu Wastewater Treatment Plant, TaiYuan (its activity was resumed before seeding into the EGSB reactor). And the granular sludge was obtained from a pilot-plant anaerobic EGSB reactor treating actual brewage wastewater for two years. Moreover, the granular sludge was deposited without any nutrients addition for two years under ambient temperature. Thus, the granules were loose. The EGSB reactor had 21.1gSS·L^{-1} biomass concentration and the VSS/SS was 0.51.

2.2 Analytical methods

Effluent was collected and centrifuged at 4000rpm for 5-6min with a centrifuge (80-2B, ANTING). And then the supernatant was used for further analysis. COD, ammonia, phenol and CN were measured according to the Chinese Standard Methods for Water and Wastewater Monitoring and analytical methods[19]. SCN measurement was performed with the American Standard Methods[20]. Specific methanogenic activity test was accomplished as described by Zhao et al [21].

2.3 Operating strategy

The operation process of anaerobic Granulation in EGSB reactor seeded with digestion sludge for treatment of actual coking wastewater was separated into two stages. First, at stageI, granular sludge was formed through using brewery wastewater as influence and seeding anaerobic digestion sludge in the EGSB reactor (meanwhile adding little granules, which were 1/7 of the total biomass in the EGSB reactor). And then at stageII, the granular sludge was acclimatized with the actual coking wastewater with about 900mg·L^{-1} COD concentration. And two oxygenation methods i.e. A and B were employed. For method A, the micro-aerophilic environments in the reactor A (RA) was generated through continuously supplying dissolved O_2 to the aeration column and then to the granule sludge bed in the EGSB reactor. For method B, the intermittent oxygenation way was used and different dissolved O_2 was supplied to the granule sludge bed in the reactor B (RB). An air pump supplied air through a porous stone diffuser, located at the bottom of the anaerobic column. The air flow rate was regulated by an air flow-meter. The oxygenation flow rate (air flow rate) increased from 800ml·min^{-1} to 15000ml·min^{-1} for the RA, and in a day, the RB was intermittently aerated with an air supply of 800-15000ml·min^{-1} 18-12 times at 20-60min intervals, 1h each time. Superficial flow rate and recycle ratio of 3.0m·h^{-1} and 22.3 were employed to offer a suitable hydraulic condition. The OLR was 1.1-2.4kgCOD·m^{-3}·$d^{-1}$with an influent flow of 1.0L·h^{-1} and a HRT of 12.0h. The process parameters such as COD, NH_3-N, phenol, cyanide and thiocyanate concentration and removal efficiency, ORP and dissolved O_2 concentration were regularly monitored.

Using $CaAc_2$ as substrate and taking granules samples from the RA and RB, the specific methanogenic activity (SMA) assays were performed for the varied operation stage: 1) before and after treating coking wastewater (CWW), 2) the anaerobic and microaerobic conditions after treating CWW, 3) SMA contrastive assay was also performed.

3 Results and discussions

The anaerobic granulation in RA (continuous oxygenation EGSB) and RB (intermittent oxygenation EGSB) seeded with anaerobic digestion sludge for treatment of actual coking wastewater was operated for about 7 months. First, the EGSB reactor was operated for approximately 1 month for the forming of granules from anaerobic digestion sludge. And then, the EGSB reactor was operated for the acclimating of the granules by actual coking wastewater for about 6 months (from days 29 to days 157 without the addition of $NaHCO_3$ and from days 158 to days 205 with the addition of $NaHCO_3$) . Fig.2 shows the influent COD concentration and COD removal efficiency changes for the intermittent and continuous oxygenation. And the ammonia removal efficiency variations for the RA and RB were presented in Fig.3 and Fig.4.

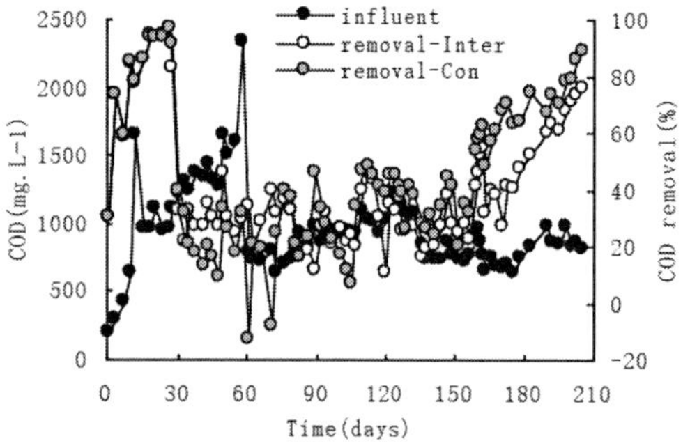

Fig.2. influent COD and COD removal for the intermittent and continuous oxygenation

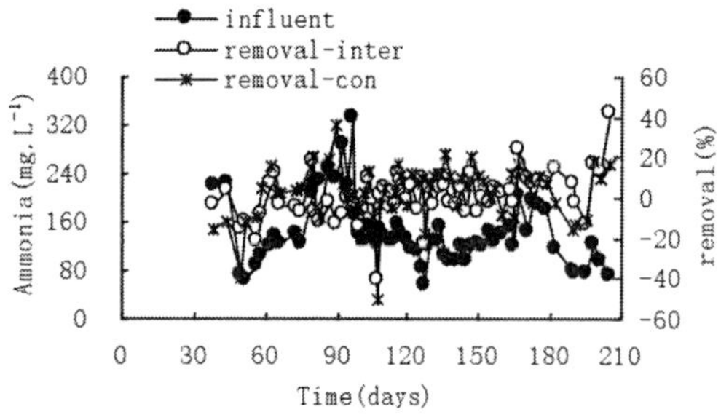

Fig.3. influent ammonia and ammonia removal for the intermittent and continuous oxygenation

StageI: granules forming. It took about only 10d for the forming of lots of granular sludge in the EGSB reactor seeded with digestion sludge and little loose granules. The COD removal efficiency had attained to 85%. However, at days 12, the EGSB reactor suffered a load shock, and the influent COD concentration abruptly increased from 630mg·L^{-1} to 1650mg·L^{-1}, meanwhile the organic load rates (OLR) also soon increased from 5.5kgCOD·m^{-3}·d^{-1} to 11.9kgCOD·m^{-3}·d^{-1}, thus the COD removal efficiency immediately decreased to 78%. However, only after 4 day, the COD removal efficiency resumed to 86%, and after 6 day increased to 94%. Which indicated that the granules cultured in the EGSB reactor through anaerobic digestion sludge had very strong supporting loads shock ability. In succession, through shorting HRT from 3.0h to 2.4h and increasing OLR from 7.5kgCOD·m^{-3}·d^{-1} to 18.0kgCOD·m^{-3}·d^{-1}, stably operated for about 10d, the EGSB reactor could gain a stable COD removal efficiency of about 94.8%. Thus the granulation process in the EGSB reactor seeded with digested sludge had accomplished, and meanwhile, biomass concentration of 19.6g·L^{-1}, MLVSS/MLSS of 0.64 and specific methanogenic activity of 0.75gCOD_{CH4}·gVSS^{-1}·d^{-1} were also attained in the EGSB reactor. The key factor for the rapid granulation in the EGSB reactor seeded with digestion sludge was not the use of brewery wastewater as substrate, but was the supplement of little granules which were loose and even decayed because of suffering the double disadvantage of lack of nutrients and low temperature.

StageII: granules acclimating (without the addition of $NaHCO_3$) (from days 29 to days 157). the RA and RB were operated for the acclimating of the granules by actual coking wastewater with about 1200mg.L^{-1} COD concentration (shorting HRT from 3.5h to 1.0h), and meanwhile 800ml/min air was supplied and no $NaHCO_3$ was added in the influent. Only after 10 day, the COD removal efficiency decreased from 97% to 9% in the RA, and 97% to 27% for the RB. However, at days 39, the COD removal efficiency in the RA and RB quickly ascended to 33% and 46% although with influent COD concentration increasing from 1200mg.L^{-1} to 1600mg.L^{-1}, which indicated that the granules had already adapted the coking wastewater quality. Subsequently, at 4 days after the 39th day, the influent quality was presented a great shift of increasing COD concentration abruptly from 1600mg.L^{-1} to 2330mg·L^{-1} and changing from golden and clear to black and ropy. The RA had a distinct COD removal decreasing of 33% to -13%, and only 46% to 34% for the RB, Which indicated that the intermittent oxygenation method (RB) had distinct advantage of supporting load shock compared with continuous oxygenation method (RA). But after 15 day, the COD removal efficiency resumed to 40% and 37% for the RA and RB, and then for about 2 months stable operation (from days 76 to days 157), the RA had an average COD removal of 32.7%, distinctly higher than the COD removal of 26.8% in the RB. But, subsequently though gradually increasing oxygen supplement from 1500ml.min^{-1} to 4000ml.min^{-1}, the maximal COD removal efficiencies in the RA and RB were still not exceeded 48% and 46%, and which might be caused by the presence of some toxic substances or thiocyanate(SCN^-). The SCN- removal of the RA was fluctuated between -12% and 8%, and the RB also only had an average SCN^- removal of -3.5% (-17.9%-3.6%). Moreover, the RA and RB only could have very low ammonia average removal efficiency of 2.4% and 2.3%, respectively. It has reported that ammonia oxidizing bacteria complete with thiocyanate-degrading bacteria for inorganic carbon and other essential nutrients [22]. And the competition of population growth between autotrophs for the SCN and CN degradation and heterotrophs for the organics degradation was also existed [23]. The actual coking wastewater had a BOD/COD value of 0.4-0.53, which perhaps indicated that the organics degradation heterotrophs was not restrained when organic carbon source was sufficient for the SCN and CN degradation bacteria. Thus, subsequently some $NaHCO_3$ was added to investigate the treatment performance of the EGSB reactor, such as the removal of the COD, ammonia phenol, CN^- and SCN^-.

StageIII: granules acclimating (with the addition of $NaHCO_3$) (from days 158 to days 205). At days 157, bicarbonate was added to feed wastewater (3.0g$NaHCO_3$/L) to be used as a source of inorganic carbon by autotrophic microorganisms, thus to favor the nitrification process or the thiocyanate degradation process. The lack of inorganic carbon may have caused incomplete degradation of thiocyanate in the aerobic zone, where thiocyanate-degrading bacteria complete with ammonia oxidizing bacteria for the carbon source [24]. For the RA and RB, the COD removal efficiency had a distinct improvement from 31.5% and 22.5% to 54.1% and 41.4%. Subsequently, at days 163, with an abrupt increasing of the oxygenation flow rate from 5000ml·min^{-1} to 15000 ml·min^{-1}, the COD removal efficiency had no increase instead of a distinct decrease from 54.1% and 41.4% to 48.4% and 31.6% for the RA and RB. But the RA (with continuous oxygenation way) could resume to 57.4% after 3days, and the RB (with an intermittent oxygenation way) still had a very low COD removal efficiency of 27.1% after 7 days. Subsequently, for the RA, the COD removal fleetly increased, and at days 203 attached to 86%, at days 205 high to 89%. But for the RB, the COD removal efficiency was comparatively low, and from days 170 to days 190, only had an increasing of 27%→41%→47%→52%→60%(68%→70%→64%→74%→67% for the RA), and even at days 203 and days 205, the COD removal efficiency was only 73% and 75%. In conclusion, the continuous oxygenation way was advantageous to the COD removal.

The supplement of $NaHCO_3$ was favorable for the removal of ammonia. When no $NaHCO_3$ supplied, the RA and RB only had a very low ammonia average removal efficiency of 2.4% and 2.3%. However, with 3.0g.L^{-1} $NaHCO_3$ addition, the ammonia average removal efficiency was increased to 2.8% and 8.9% for the RA and RB, respectively. It was very distinct that the RB (with intermittent oxygenation way) had higher ammonia removal of 41.9% than the RA of 16.7%. Moreover, the RB had more stronger supporting toxic substances (such as SCN) shock ability, and when influent SCN concentration having a fluctuation of 106.5mg.L^{-1}→82.9 mg.L^{-1}→121.3mg.L^{-1}→86.1mg.L^{-1} (presented in the Fig.3),the RB had an ammonia removal efficiency change of 7.7%→14.5%→7.2%→16.2%. however, the change of 7.3%→-2.0%→-15.8%→18.3% was for the RA. Thus, the intermittent oxygenation way (RB) was more suitable than the continuous oxygenation way (RA) for the ammonia removal.

The phenols removal efficiency variations for the RA and RB were presented in Fig.4.

From days 122 to days 137, with 95.3-97.6mg.L^{-1} influent phenol concentration, the RA could have 18.6%-53.1% phenol removal efficiency, but only -16.0%-24.2% for the RB. At days 138, the phenol removal efficiency fleetly increased to 93.1% for the RA, and only 34.8% for the RB. From days 140 to days 205, the RA could always keep very high phenols removal efficiency of 98.3%-100%, which is equivalent to an effluent phenol concentration level of 0.0-0.8mg•L^{-1}. Correspondingly, for the RB, the phenols removal efficiency began to increase from 34.8% (at days 138) to 99.2% (at days 164), and subsequently from days 164 to days 205, the RB had a stable phenols removal efficiency of 95.7-99.9%. For the RA, the phenol removal had no changes after adding $NaHCO_3$. But for the RB, the supplement of $NaHCO_3$ was very essential. Only with 3.0g.L^{-1} $NaHCO_3$ addition, the phenol removal fleetly increased from 27.4% to 70.2%. And then within 7 days (from days 157 to days 164), the phenol removal gradually rise to 99.2%.

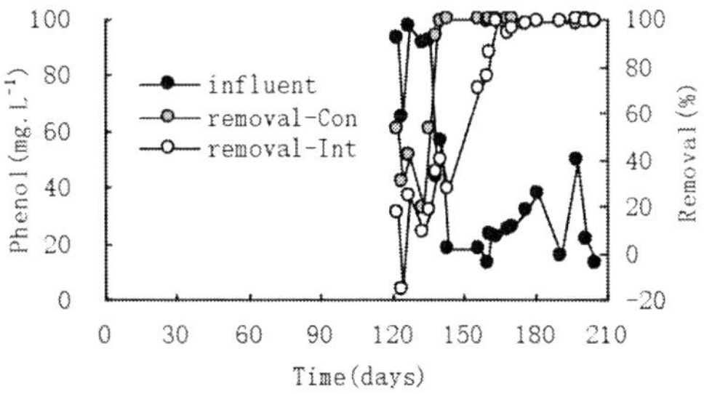

Fig.4. influent phenol and phenol removal for the intermittent and continuous oxygenation

Fig.5 presented the changes of influent cyanide (CN) concentration and cyanide (CN) removal efficiency for the RA and RB, and Fig.6 showed the variation of influent thiocyanate (SCN) concentration as well as the removal efficiency in the RA and RB.

From days 102 to days110, for the RA and RB, CN removal efficiency rapidly increased from 67.6% and 35.3% to 92.0% and 84.0%. Subsequently, the RA could keep stable high CN average removal efficiency of 95.7% (fluctuating between 89.2% and 100.0%). Correspondingly, the RB had a lag period for the high CN removal efficiency of 97.0% (at days 126) from 84.0% (at days 120). Subsequently, from days 126 to days181, the RB could attain a stable CN average removal efficiency of 95.1% (80.4%-100%). However, from days 190 to days 205, the CN average removal efficiency was abruptly decreased to 75.7% (71.4%-81.5%). Perhaps it was owing to the fluctuation of influent SCN concentration. But the CN removal in the RA was not decreased (94.5%-100%), which also predicated that the continuous oxygenation way (RA) was advantageous to the stable CN removal. Before and after adding $NaHCO_3$, the CN removal had no change for the RA and RB.

When no $NaHCO_3$ supplied, the RA and RB were always kept negative SCN average removal efficiencies of -1%(fluctuating between -12% and 8%) and -

https://doi.org/10.1051/e3sconf/20183801052

3.5%(fluctuating between -17.9% and 3.6%). However, with the addition of 3.0g.L^{-1} $NaHCO_3$ (at days 160), the RA had a distinct increasing of SCN removal efficiency from -6% to 21%, and after 20 days (at days 180), the SCN removal was attained to 100%. While for the RB, until to days 169(after 12 days for adding $NaHCO_3$), the SCN removal just attained to a positive value of 5.3% and then gradually increased to 44.0% (at days 181), and perhaps the shift of SCN removal from negative to positive was owing to the increasing of the oxygen supplement from 5000 ml min^{-1} to 15000ml min^{-1}. And which also indicated that sufficient oxygen supplement was very important for SCN removal, and compared with the continuous oxygenation way (RA), the RB with intermittent oxygenation way couldn't ensure enough oxygen for the thiocyanate-degrading bacteria and thus only had relatively low SCN removal efficiency. From days 181 to days 205, the RB had a stable SCN removal and the SCN average removal was 36.5% (27.5%-44.1%).

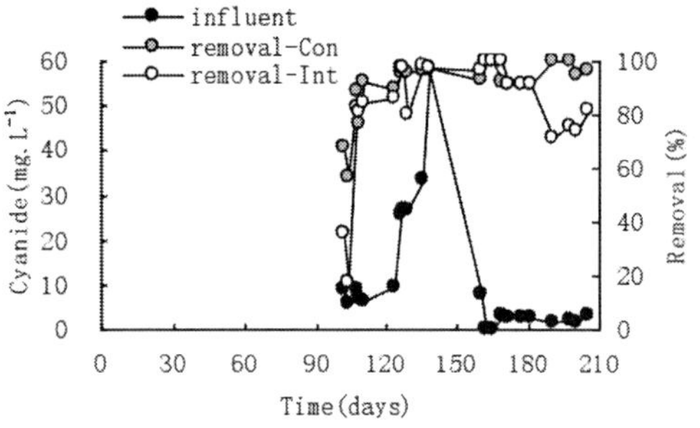

Fig.5. influent cyanide and cyanide removal for the intermittent and continuous oxygenation

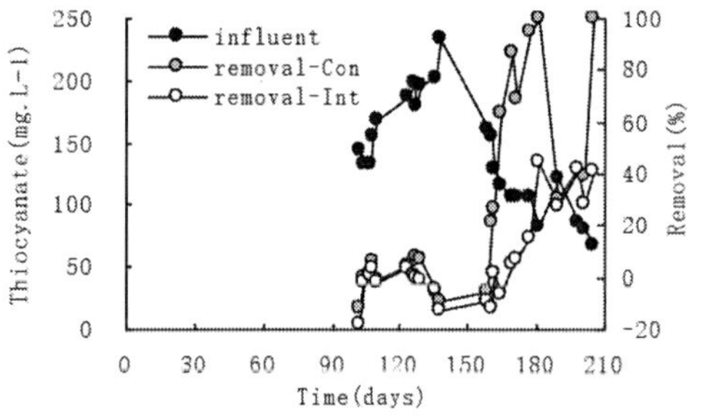

Fig.6. influent thiocyanate and thiocyanate removal for the intermittent and continuous oxygenation

Fig.7 presented the SMA changes of the granules samples taken from the RA and RB for varied operation stage: before treating CWW, after treating CWW for 33 days, 4 months. Not only the actual coking wastewater supplement but also the oxygen supplement could have influence on the SMA. Along with the treating of actual coking wastewater (with 1280-2200mg.L^{-1} COD and 1.0L/h influent flow) and oxygen supplement, the SMAs of the granules in the RA and RB were increased 14.3% and 5.3% (after 6 days, with 800ml.min^{-1} air flow rate), and 20.2% and 4.5% (after 33 days, with 1200 ml.min^{-1} flow rate), respectively. Some oxygen supplement was probably very important, and which could facilitate the removal of some toxic or inhibitory compounds in the coking wastewater and could weaken the inhibition of the toxic or inhibitory compounds on the methanogens. However, when treating coking wastewater for 4months, although already increasing air flow rate to 5000ml.min^{-1}, the SMA was not increased but reduced 26.5% and 34.0% compared with the parallel (before treating CWW), which indicated that the actual coking wastewater had distinct inhibition effect on the granules for the long term operation of the RA and RB. And what's the crucial factor influencing or restraining the SMA of the granules? The toxic or inhibitory compounds in the coking wastewater or the overmuch oxygen supplement (because with 1200ml.min^{-1} oxygen supplement, the SMA was not reduced but increased)?

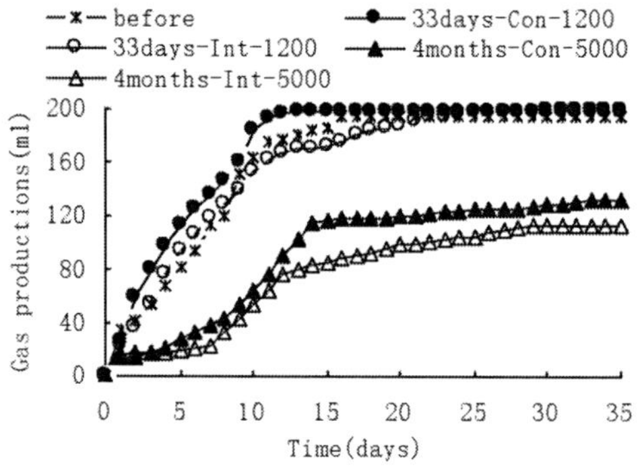

Fig.7. SMA for varied operation stage

We contrastively analyzed the SMA change at the two conditions of anaerobic and micro-aerobic with the same air flow rate of 5000ml.min^{-1} and 15000ml.min^{-1} and before and after adding $NaHCO_3$ (presented in Fig.8 and Fig.9).

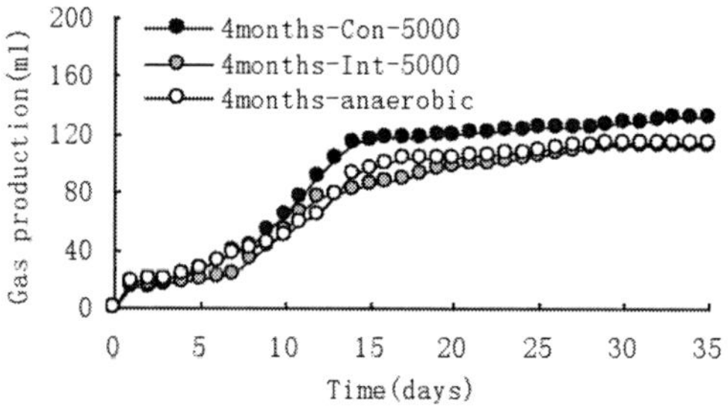

Fig.8. SMA for anaerobic and micro-aerobic (4 months)

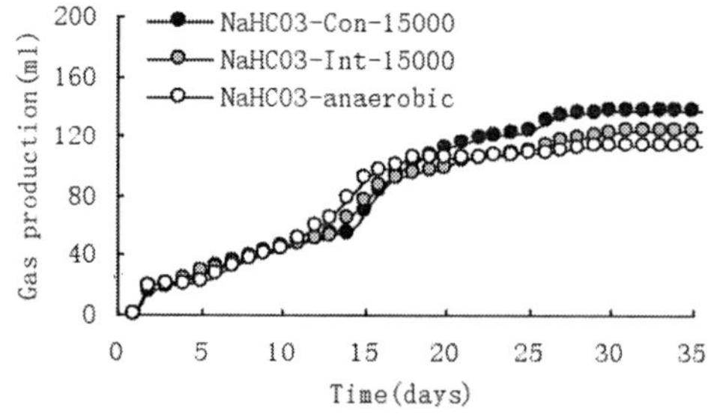

Fig.9. SMA for anaerobic and micro-aerobic (adding $NaHCO_3$)

Compared with anaerobic condition, for the RA and RB, supplement oxygen with 5000ml.min^{-1} air flow rate, the SMA was increased with 22.9% and 10.4%, respectively. Subsequently, enhancing the air flow rate to 15000ml.min^{-1} and meanwhile adding 3.0g.L^{-1} $NaHCO_3$, the SMA kept more distinct enhancing trend (35.4% and 17.7% for the RA and RB), and which indicated that the distinct reduction of SMA (treating CWW for 4 months) was not caused by the overmuch oxygen supplement, but caused by the toxic or inhibitory compounds in the coking wastewater. $NaHCO_3$ supplement could facilitate the degradation of SCN and some toxic compounds and thus weaken the inhibition effect of the CWW on the methanogens (or on the granules).

4 Conclusions

The micro-aerobic granulation process for actual coking wastewater treatment in a coupled anaerobic-aerobic reactor (micro-aerobic EGSB reactor) seeded with digestion sludge was divided into two stages. Stage one was to form granular sludge from digestion sludge using brewery wastewater at the anaerobic condition, and stage two was to acclimatize the granules using actual coking wastewater through micro-aerobic condition, and two oxygenation methods (intermittent oxygenation and continuous oxygenation) were employed.

The anaerobic granulation process was accomplished in 10d using brewery wastewater. And it was the key factor for the quickly forming of the granular sludge to add little loose granules.

It took only about 6 months for the successful micro-aerobic acclimating of the granular sludge by the actual coking wastewater. Oxygen supplement and inorganic carbon addition could distinctly improve the contaminant removal effect of the coking wastewater. With 15000ml.min air flow rate and 3g.L^{-1} $NaHCO_3$ addition, for the continuous oxygenation way, the COD, ammonia, phenol, CN and SCN removal efficiencies could attain to 89.0%, 16.7%, 99.3%, 95.7% and 100%, respectively, but only 75.0%, 41.9%, 96.8%, 89.9% and 44.0% for the intermittent oxygenation way.

The toxic or inhibitory compounds in the coking wastewater could distinctly restrain the granules and the methanogenic activity of the granules was reducing evidently. The actual coking wastewater had distinct inhibition effect on the granules, but the supplement of some oxygen could promote the recovery of SMA, and $NaHCO_3$ supplement could also weaken the inhibition effect of the CWW. The continuous oxygenation way had more strongly activity recovery ability than the intermittent oxygenation way.

Acknowledgements

This research was supported by grants from the Shan-Xi Province Key Technologies R & D Program of Shan Xi (NO.20120313008 and NO. 20150313002-1).

References

1. M. Zhang, J.H. Tay, Y. Qian, et al. Wat. Res. 32 (2):519-527(1998)
2. Y.M. Kim, D. Park, D.S. Lee, et al. J. Hazard. Mater. 152 (3): 915-921(2008)
3. Y.M. Li, G.W. Gu, J.F. Zhao, et al.. Chemosphere. 52:997-1005(2003)
4. Z. Wu, L. Zhu, J. Environ. Sci. 24 (2): 248-253(2012)
5. I. V'azquez, J. Rodr'ıguez, E. Mara˜n'on, et al. J. Hazard. Mater. B137 :1773-1780(2006)
6. Yong-Shik Jeong, Jong Shik Chung.Process Biochemistry, 41:1141-1147 (2006)
7. Corsino, S.F., Capodici, M., Morici, C.,et al. Wat.Res. 88: 329-336(2016)
8. Di Bella, G., Torregrossa, M., Bioresour. Technol. 142:706-713(2013)
9. A.M. Lotito, M. De Sanctis, C. Di Iaconi, G. Bergna, Wat. Res. 54:337-346(2014)
10. S.J. Lim, T.-H. Kim, Biomass Bioenergy, 60:189-202(2014)
11. K.Y. Show, D.J. Lee, J.H. Tay, Appl. Biochem. Biotechnol. 167:1622-1640(2012)
12. Li B, Sun Y L, Li Yu-ying. J Zhejiang Univ SCI. 6B (11):1115-1123(2005)
13. Q.Y. Gu, T.CH.Sun, G.Wu, et al., Bioresource Technology, 166:72-78(2014)
14. Chuan Chen, Aijie Wang, Nanqi Ren, et al. Process Biochemistry.45:1007-1010(2010)
15. R.-M.Wang, Y.Wang, G.-P.Ma, et al. Chemical Engineering Journal. 148:35-40 (2009)
16. F.Wang, Y.R.Hu, G.Chen, et al. Bioresource Technology,110:120-124(2012)
17. M. Goberna, M. Gadermaier, I.H. Franke-Whittle, et al. Biomass and Bioenergy, 75(5): 46-56(2015)
18. Li B, Sun Y L, Li Yu-ying. J Zhejiang Univ SCI. 6B (11):1115-1123(2005)
19. State environmental protection administration of china. Water and wastewater monitoring and analytical methods. Beijing: China Environmental Science Press, 2004.
20. APHA. Standard Methods for Examination of Water and Wastewater, 20th ed. American Public Health Association, American Water Works Association, Water Environment Federation,Washington, DC,1998
21. Zeng A P, Deckwer W D. Chemical Engineering Science, 51(10):2305-2314(1996)
22. Kim,Y.M., Park, D., Lee, D.S.,et al.J. Hazard. Master. 152,915-921(2008)
23. Vázquez, I., Rodríguez, J., Maranon, E., et al. J. Hazard. Master. 137, 1773-1780(2006)
24. Kumar MS,Vaidya AN, Shivaraman N, et al. Environ Eng. Sci. 17:221-226(2000)

Study on Activation Mechanism and Sustainable Development of Rural Human Settlements Based on Landscape Construction

Sui Xin[1], Lin Lin *[2(Correspondence author)], Sun Chaoyang [3]

[1]Dalian Nationalities University, No. 18, Liaohe West Road, Dalian Economic & Technical Development Zone, China, 116600
[2]Lecturer, School of Art and Design, Dalian Polytechnic Univ., Dalian, China, 116034, E-mail : guganmao@126.com
[3]Dalian Nationalities University, No. 18, Liaohe West Road, Dalian Economic & Technical Development Zone, China, 116600

Abstract. The quality of rural human settlement environment is directly related to the quality of urban human settlements, which is a big problem in the development of China's social economy. In the special period of social transformation and the key stage of building a well-off society comprehensively, it is of great practical significance to pay attention to the optimization of rural human settlement environment. China is in the transition from traditional rural landscape to modern rural landscape. However, how to realize the smooth transformation and the sustainable development of rural landscape is urgent and tough issues.

1 Introduction

The study of landscape construction based on rural settlements provides a framework and basis for rural landscape construction at the meso- and macro-levels. Based on the construction, landscape and planning disciplines, combined with the core spirit of cognitive development theory, synergetics, communication, psychology and other related theories, through combing the structure level and development thread of subject cognition, the author makes a detailed analysis of the villagers ' cognitive style and evolution characteristics in natural environment, social environment and built environment. The paper hope to play a certain practical guidance significance on the current rural settlement construction.

In the face of the regional cultural aphasia, the degradation of the ecological environment and the absence of root in the spatial form, this paper tries to extract some common meanings, features and laws in the village settlement construction in a way that is open and integrated. Its basic purpose is the cognitive considerations of villagers' subjects, "Seeing the big through the small" caused the rural settlements of the construction method of our thinking and improvement.

2 The theoretical significance of the study of rural human settlement environment

The research point of view of rural human settlements environment is the investigation and analysis of the laws and phenomena of rural natural ecological environment, traditional regional culture and regional human settlement space activities. The existing theoretical research on the one hand lacks systematicness, on the other hand, it lacks analysis and comparison under the microscopic perspective. Therefore, research on rural human settlements in underdeveloped areas based on surveys and studies of a large number of informative data is of great significance for the overall development of urban and rural areas and the improvement of the quality of rural development, as well as the improvement of the theoretical system of human settlements.

In the process of rapid urbanization, cities continue to expand into neighboring villages. The traditional rural settlement culture, personal relations, and community awareness are impacted by urban culture and gradually merged and replaced. The rural residential environment is in a state of disorder and transformation, and it urgently needs attention.

3 The realistic demand of activating the living environment in rural areas

Rural residential environment is a system of dwelling in rural area, that is, the organic combination of material and immaterial that farmers need to produce and live, is a dynamic complex giant system, its function transformation and evolvement have intrinsic law. The quality of rural human settlement environment is directly

* Corresponding author: guganmao@126.com

related to the quality of urban human settlements, which is a big problem in the development of China's social economy. In the special period of social transformation and the key stage of building a well-off society comprehensively, it is of great practical significance to pay attention to the optimization of rural human settlement environment. The new countryside construction provides the planning and the management policy basis for the rural residential area system optimization and the urban-rural coordinated development from the whole angle of view, is also an important way to realize the human settlements environment optimization.

4 Basic conditions for the study of rural human settlement environment

The rural human settlement environment and its location of the ecological environment, resources, social and economic has a very close material, energy exchange. These comprehensive environmental elements not only laid the foundation material platform for the development of rural settlements, but also affected the basic characteristics and internal laws of their distribution, form and combination, and finally contributed to the regional differences of rural settlements.

4.1. Social structure

With the development of rural economy and urbanization, the rural social structure gradually from the closed to open, from the homogeneous single society to the heterogeneous diversity of the transition and transformation of society, rural social structure began to transition from the traditional to the modern. The factors that promote the change of social structure in rural areas are also varied, which promotes the flow of large numbers of young and middle-aged labors to cities, the loss of rural Labor leads to the hollowing out of rural areas, the reform of production technology, and the related media, etc., which play a role in the process of rural social change

4.2. Folk culture

National culture refers to the culture that reflects the national characteristics of the nation in its own historical development process. National culture reflects a nation's historical development level and unique cultural temperament, including material culture, spiritual culture and institutional culture three levels.

4.3. Spatial structure of settlements

The rural settlement is influenced by many factors such as natural environment, mode of production and cultural difference, and forms a stable spatial structure pattern in the settlement space. According to the type of spatial structure, it can be divided into rural cluster villages, mostly in the area near the towns, and the linear distribution of rural areas. Most of them along the traffic arteries and linear arrangement. Pastoral diaspora is more distributed in grassland areas, rural homes or several households, or scattered distributed. (regiment, strip and scatter).

5 The development course of rural human settlement environment

For the development stage of rural construction, different scholars have roughly divided it into three to four stages according to the research emphases. Ren Qinguo thought that the course of rural construction in China can be divided into three stages of the construction experiment period, the founding to the reform and opening-up period and the later period. Li Jianqiao suggested that 1949-1978 is the stage of grain, 1979-2005 is the market stage, and 2006 is the stage of development in the new period. Lin Tao used its age as a node, which has made its profile four stages, 1920-1930,1950-1970,1980-2005 and 2005. Regardless of the definition, there are three important time nodes: The founding of 1949, 1978 years of reform and Opening-up and the fifth Plenary Session of the 16th CPC Central Committee in 2005. In the 30 years since the founding of the PRC, although the "goal of building a new socialist countryside" was clearly put forward in the three meetings of the 1956-year NPC, the agricultural auxiliary industry and the two-yuan pattern of the rural auxiliary cities made the gap between urban and rural areas widening, and the strict household registration system plus the two-element pattern makes the gap between the urban and rural areas widening, the strict household registration system and the restriction of the urban and rural product circulation, which leads to the problems of the three agricultural issues, and becomes the prominent feature of this stage. The reform and opening stage from the initial family contract to the well-off construction, fully mobilize the enthusiasm of the farmers at the same time the overall income also increased, the farmer's spiritual life is also gradually enriched. But in the late 90, as the market economy continued to increase the gap between urban and rural areas. The fifth Plenary Session of the 16th CPC Central Committee 2005 adopted the "Eleven-Five years outline proposal" proposed an important 20-word policy "production and development, well-off life, township wind civilization, tidy clean, management democracy", marking the new socialist rural construction into a fresh stage. These five requirements are the guiding ideology of the new rural construction, embody the organic unification of society, economy, culture and the construction of rural style, which are the coping strategies of the three agricultural problems under the background of harmonious society.

6 Characteristics of Rural Landscape

Since the reform and opening up, diversification trend has appeared in the rural industrial structure and social formation.With the development of economy, large-scale construction activities spread in the countryside spontaneously and consciously thus rural landscape has undergone significant changes which are reflected in the diversification of rural settlement types, the change of land use and material space in rural areas brought about by the change of industrial structure, as well as the change of people's way of life.

For a long time, agriculture, forestry, animal husbandry, fishery are the main body of rural landscape, as well as the leading factors of the industrial structure of rural economy. However, great changes have taken place in the industrial structure of the countryside since the reform and opening up. The drastic changes of the rural society and the gradually declined proportion of traditional agriculture are obvious among them. The traditional agricultural economy is an industrial activity characterized by family individual agriculture and manual work. Its function is mainly to provide the material foundation for the family which is the life guarantee.As the development of urbanization, the traditional agriculture has little significance to the villagers' life support due to the low economic benefits, and the industrial structure of many villages has been changing quietly.The type of village is divided from agriculture-based one to that based on non-agricultural industry. According to the experience of the world economic development, after the urbanization achieved, the industrial structure will enter the stage of accelerated differentiation.With the reform of rural economic system and the development of commodity economy, rural industrialization and urbanization in developed areas have entered the stage of accelerated development.Taking Zhejiang as an example, the level of urbanization has increased from 51% in 2002 to over 70% in 2009. The proportion of secondary industry and tertiary industry is increasing, while the proportion of primary industry is gradually decreasing. At present, rural settlement types can be divided into agricultural-based, industrial-based and tourism-based according to the features of its main industrial structure as different types displaying different landscape characteristics.

7 Space construction of rural human settlement environment

Space construction is a comprehensive embodiment of the ways, methods and technical means used by human beings to use natural environment, social economy, regional culture and productive life. Through the cognition and construction of the living environment, human beings constantly adapt to the social thought and demand, and gradually form an organic system of ecology, production and life. In the system, the elements of different types, different forms and different natures have their own development law and evolution process, and the human beings use these rules and processes to coordinate, two times cognition and function innovation to reflect different demands and the origin of human habitation.

7.1. Rural location and spatial pattern

The process of human settlement is influenced by multiple factors, showing different ways of settlement, and the role of space carrier of natural environment is especially prominent. Human beings in the use and enjoyment of the environment at the same time, constantly adapt to, change the environment. As the starting point of human settlement, rural settlement exudes the original taste, confirms the interdependence of human and space environment, and its location characteristics and spatial pattern embody human wisdom and courage.

7.2. Space elements

Influenced by the regional production life style and geographical conditions, the settlement form is mainly influenced by natural factors such as topography, landform and climate, and forms the typical steppe settlement form, which is composed of grassland, forest, river, dwelling, road and cattle pen. Due to the spatial dispersion and low-density population, the steppe settlements often appear scattered distribution characteristics, forming a point-like or flaky. In the process of formation and evolution of grassland settlements, religion, grassland and livestock become the typical elements of spatial form. With the change of production style and system, these typical spatial elements are also changing.

7.3. Mode of activity and construction

As a embodiment of human adaptation to the environment, architecture is the support of human production and life activities, mainly manifested in the environment, cost, technology and other aspects. The qualification of natural environment, the demand of function, the choice of material and the identity of culture can be said as the gene of construction entity, and it is reflected by human intelligence in the special regional building system.

8 Development trend of Rural Landscape Construction

"Poetic dwelling" is the ideal of people all the time. Instead of more and more similar cities constructed of concrete, the countryside,which allows people to enjoy the nature and carries Chinese traditional civilization,is more likely to be the ideal habitat. In terms of development orientation, the countryside can neither imitate the city's trajectory and lose its own characteristics, nor can it stand still and refuse to make progress. In the future, the countryside should complement the city's resources through the reconstruction of nature, production and life, so as to realize urban-rural integration. The possible trends are as follows.

8.1.A Settlement under the Concept of Slow Life

Through the promotion of the concept of slow life, during the construction and protection of the space of it, the countryside can be used as the balance point of rapid urban life. In economically developed areas, people's requirements for the quality of life are constantly raised in the process of rapid urban-rural integration and the proportion of tertiary industry serving the city is constantly increasing, and the "slow city", which can provide natural landscape as well as convenient urban life, is the most likely optimization direction of the rural areas, paying more attention to ecology and nature and the optimal integration of systems and efficient use of resources, which is more conducive to the long-term development of the economy and better to improve the quality of life of residents

8.2. Ecological Safeguard and Buffer between Cities

The model of German regional parks takes the construction of villages into the overall consideration of the whole region.Through the establishment of common regional sewage management institutions, the improvement of internal traffic conditions in regional parks, tourism services such as catering and accommodation, the construction of infrastructure and the development of markets, residents there have been provided with a special place for leisure and the entire regional economy and rural landscape development were promoted effectively. In the areas rich in landscape resources, it is of great significance for the development of urban and rural areas to integrate the villages along roads, rivers and scenic spots into a leisure land for people, as well as as ecological safeguard and buffer between cities.

8.3. Transformation of Rural Landscape Construction Method

The method of landscape construction should be applicable to social development demand. Since the 20th century, rural industrial structure, rural settlement types, land use and material space, as well as people's way of life have undergone profound changes and that of rural landscape are rapid and intense. In the context of the social transformation, to synthesize all elements in terms of ecology, production and life and to integrate multidisciplinary knowledge, we need to make decision in line with local conditions and adopt operational planning methods. In accordance with the requirements of sustainable development, some scholars also began to put forward some new ideas and concepts in landscape construction methods and put them into practice, such as ecologicalization, democratization, low carbonization and so on, which are summarized as follows.

8.4. Ecologicalization Direction

With the proposal of sustainable development, the study of ecological construction and ecological balance in pursuit of harmony between human and nature has been carried out in the construction of cities and villages. For example, Song Ku of Tsinghua University put forward a holistic view of ecological architecture based on some theories and concepts of ecology in his doctoral thesis "The study of Architectural Design based on Nature and Ecology." At the same time, the concepts of ecosystem structure framework and bioclimatic buffer layer are put forward, and the ecological architectural design strategy is discussed taking the design of agricultural house on Shuangshan Island in Zhangjiagang as an example and combined with the specific climate and environment conditions in southern Jiangsu.

8.5. Democratization Direction

The direction of democratization is an overall construction method aimed at the coordination and symbiosis of interests. With the deepening of rural practice and after the preliminary observation on the ecological pattern of landscape, people are more and more aware of the village, in the final analysis, belongs to the villagers and rural landscape is far from a matter of aesthetics or ecology, so people begin to pay more attention to the balance of interest pattern, and put forward the corresponding strategies and methods for the construction of rural landscape from the perspective of rural sociology.

Academician Wang Xiaodong of Xinjiang Architectural Survey and Design Institute, in the project of "Protection and renovation of Aho neighborhood in Luoshilaocheng District", holds that the protection and renovation of residential buildings not only involves to architecture and planning, but also involves to the whole social ecology and structural system. Therefore, it is not advisable to make great changes in construction and force emigration which may destroy the original handicraft, commercial and interpersonal structures. In terms of reconstruction mode, the planning and design should insist on maintaining the original overall style as much as possible, and break through the traditional mode, in which the managers and designers undertake the whole things. Instead, each household residents shall participate in the design. In the view of architects, the style and features of the old urban area is a combination of architecture, living habits, and national habits. It has a dynamic development process in which new ideas, new technologies and new ways are constantly mixed with the original set. And its success or not depends on the level of villagers' participation as well as the concept, attitude and working style of the managers and designers."

8.6. The Sustainable Development Direction of Low Carbonization

In Copenhagen climate change conference 2012, the Chinese government released the controlled emissions target of greenhouse gas that, by 2020, carbon dioxide emissions per unit of GDP are supposed to fall 40% to 45% year on year. Proposed by our government, low-carbon development model enters into people's vision and become a new hot spot. A key project of the National Natural Science Foundation of China, "Research on the Construction System of Low-Carbon Rural Residential Environment in the Yangtze River Delta", lead by Professor Wang Zhu of Zhejiang University from 2013 is in progress. The project faces the infiltration of the high-carbon behavior of urban construction into the rural areas and, from the perspective of "low carbon construction", aims at rectifying the misunderstanding of high carbon in rural construction in order to provide scientific support for the construction of low carbon rural human settlement environment based on the goal of "low carbon community" and guide the low-carbon transformation of communities.

9 Sustainable development of rural human settlement environment

9.1. From the strong as to the humble inaction

The people of the rural society in the same piece of land, born in here, grow in here and died in here. The agricultural mode of production determines the people for the climate, geographical location and other objective conditions of the great dependence. They have not enough strength to contend with the nature of the situation, took the way to adapt to nature, although at the time of the historical conditions is a kind of helpless, but at the same time is a wise choice, reflecting a kind of inaction of the construction concept. To some extent, the formation of this idea is closely related to the life style or cultural form of Chinese ancient farming.

The current rural construction is filled with too many rural lives outside of the "utilitarian as" heart, the rapid construction of many problems, the urgent need for a solution. In this context, the idea of inaction has important referential significance. The thought of inaction, which is characterized by strong interactive thinking, is another set of the diversity in the process of evolution. It provides a kind of reverse, transposition angle for people to think about problems. The road to the answer is not limited to the inherent routines, not only does not weaken the human initiative, but it can promote the evolution of thinking in a direction of innovation. In addition, the thought of inaction is a kind of thinking mode advocating nature to follow the natural law, which leads people to adopt a kind of attitude to the external environment in the process of recognizing nature and transforming nature, and thinks that objective condition is the vane of the subject consciousness, and any way of behaving against the natural law is the result of More importantly, the thought of inaction, as a complex way of thinking, can guide builders to transform contradictory opposing relations into a cooperative attitude. This kind of antagonism exists between the human development and the limited natural resources, also exists between the village builders and the villager main body, the inaction construction view through the compound various factors and the main body's way achieves to the contradictory difference softening. This way of thinking beyond the two-yuan opposition is also the essence of "inaction" wisdom. With the heart of inaction to integrate into the village construction, with tasteless heart to taste the local culture, with the use of useless for the big use, will itself from the strong controller into humble learners, is as the current rural construction participants should complete the self-examination.

9.2. From local spontaneous to cultural consciousness

As one of the most complete regions of Chinese regional culture, rural settlements have not only attracted extensive attention in the field of research but also become the prototype of design. The formation of this concern is not only the cultural value of the village settlement itself, but also the simple living form is disappearing and changing in an uncontrollable way. The traditional construction means, the cultural dissemination way as well as the external environment factor function make the local culture gradually form in a kind of spontaneous state. However, the tide of globalization, the advancement of urbanization and the acceleration of economic development make the original living form and the cognitive structure of the occupants' change, and the infiltration of the foreign culture has stimulated the appearance of the alienation of the rural settlement in a certain level. The emergence of these phenomena has already deviated from the original cultural attribute under the limit of existence, more is a manifestation of the concept level and the cognitive stratification, but it does represent a special regional phenomenon in the present. Therefore, only from the form to judge the merits of the phenomenon is clearly not comprehensive, instead, we should fully explain the rural construction in the new period, and analyze the relationship between cognition level and social culture from the main body of rural settlement construction and residence.

References

1. Wu Liangyong, an Introduction to human settlements environment Science, Beijing: China Construction Industry Press: 2001.38
2. Li Xiaofeng, vernacular architecture--interdisciplinary theory and method, Beijing: China Construction Industry Press, 2005
3. Wang Dong, communal construction of rural settlements and integration of architects, Times architecture, 2007 (04), 16-21
4. Wangzhu, Fan Liyang, Chen Zongyan, New Village "ecological human habitat" model--a case study of the rural areas of Jiangnan, China, 2011 (4) 22-26
5. ANSI B. AISC 360-10-Specification for Structural Steel Buildings[J]. Chicago AISC, 2010.
6. GB. 50010-2003, Code for design of steel structures[S]. Beijing: China Architecture & Building Press, 2003.

The Research and Application of Microbial Degradation Technology on Heavy Oil Reservoir in Huabei Oilfield

Guan *Wang* [1,*], Rui *Wang*[1], Yaxiu *Fu*[2], Lisha *Duan*[1], Xizhi *Yuan*[1], Ya *Zheng*[1], Ai *Wang*[1], Ran *Huo*[1], and Na *Su*[1]

[1]Engineering Technology Research Institute, Huabei Oilfield Company of PetroChina, Renqiu, He Bei Province, 062552, China
[2]The Erlian Filiate of Huabei Oilfield Company, Huabei Oilfield Company of PetroChina, Xilinhot, Inner Mongolia, 026000, China

Abstract. Mengulin sandstone reservoir in Huabei oilfield is low- temperature heavy oil reservoir. Recently, it is at later stage of waterflooding development. The producing degree of water flooding is poor, and it is difficult to keep yield stable. To improve oilfield development effect, according to the characteristics of reservoir geology, microbial enhanced oil recovery to improve oil displacement efficiency is researched. 2 microbial strains suitable for the reservoir conditions were screened indoor. The growth characteristics of strains, compatibility and function mechanism with crude oil were studied. Results show that the screened strains have very strong ability to utilize petroleum hydrocarbon to grow and metabolize, can achieve the purpose of reducing oil viscosity, and can also produce biological molecules with high surface activity to reduce the oil-water interfacial tension. 9 oil wells had been chosen to carry on the pilot test of microbial stimulation, of which 7 wells became effective with better experiment results. The measures effective rate is 77.8%, the increased oil is 1,093.5 tons and the valid is up to 190 days.

1 Introduction

Mengulin sandstone reservoir in Huabei oilfield is located in Erlian Basin. The average porosity is 22.4% and the permeability is between 334.8 to 749.7 mD. Under the reservoir temperature condition of 35 °C, the crude oil physical property is poor with high viscosity of 179.1 mPa•s, high oil-water viscosity ratio of 259, high colloid and asphalt quality score of 36.8% and high density of 0.9012 g/cm^3, being the typical heavy oil reservoir. Since entering the late waterflooding stage, the degree of water driving is poor and stable yield is difficult.

In order to solve the problem of lower waterflooding efficiency caused by huge crude oil viscosity, high oil-water viscosity ratio and poor crude oil liquidity, microbial degradation technology is researched. Through the microbe itself and its metabolites reacting with crude oil to improve product properties, the waterflooding displacement efficiency is improved [1-6].

The key for microbial enhanced oil recovery is to screen excellent strains. Based on the related theories of biological evolution [7-9], living organisms have innate ability of evolutionary and adaptation to the environment. So, selecting from the high colloid and asphalt content environment, through the enrichment and purification, 2 strains of HB-1 and HB-2 is eventually isolated.

2 Experiment

2.1 Test Materials

Activated medium (LB): peptone of 10.0g, yeast extract of 5g, NaCl of 10g, tap water of 1000ml, adjusting pH to 7.2.

Emulsified medium: yeast extract of 0.5g, NH_4Cl of 3g, KH_2PO_4 of 0.5g, glucose of 5g, peptone of 1.5g, formation water of 1000ml, crude oil of 50g.

Viscosity reduction medium: yeast extract of 0.5g, NH_4Cl of 3g, KH_2PO_4 of 0.5g, glucose of 5g, peptone of 1.5g, formation water of 1000ml. The volume ratio of nutrient solution, crude oil and bacteria solution is 1:1:1.

2.2 Evaluation Method

1. The adaptability of formation water and strains. LB medium is configured with formation water, and cultivated under formation temperature for 24 hours to observe the bacteria growth.

2. Emulsification test. The isolated bacteria is activated and cultured with LB activated medium for 24 hours and then inoculated on emulsion medium by the volume fraction of 5%. Under temperature of 35 ℃, on a table of 200r/min, it is cultivated for 3 days. Observe the emulsion phenomenon and measure surface tension using TX500C interface tensiometer.

3. Viscosity reduction test. The isolated bacteria is activated and cultured with LB activated medium for 24 hours, and then inoculated on viscosity reduction medium according to that the volume ratio of nutrient solution, crude oil and bacteria solution is 1:1:1. Under temperature of 35 ℃, on a table of 200r/min, it is cultivated for 7 days. Crude oil viscosity is measured before and after viscosity reduction using the Mars

* Corresponding author: cyy_wangg@petrochina.com.cn

rheometer.

3 Results and Discussions

3.1 Compatibility of Strains and Formation Water

From the water quality analysis of formation water, the formation water type of Mengulin is $NaHCO_3$, and the salinity is 1,301mg/L. The strains compatibility with formation water test results are shown in table 1. Under formation temperature, after joining strains of HB-1 and HB-2, the number of viable organism is higher than that before compatibility. It shows that strains HB-1 and HB-2 have well compatibility with formation water and can reproduce in the formation environment.

Table 1. Compatibility evaluation results.

Strain Number	Viable count before compatibility (cells/mL)	Viable count after compatibility (cells/mL)
HB-1	2.0×10^3	2.2×10^8
HB-2	2.0×10^3	1.1×10^8

3.2 Emulsion Performance

Take crude oil from Mengulin as samples, emulsification experiment was carried out. The emulsification effect is shown in table 2. The crude oil and bacteria liquid are separated after emulsification and determinate its interfacial tension, shown in table 3. After acting of strains HB-1 and HB-2, crude oil and water are emulsified completely. The strains fermentation liquor after oil emulsion has certain surface activity. Compared with the blank, interfacial tension reducing rate of strains fermentation liquor is as high as 77.91% to 91.05%, which illustrates that HB-1 and HB-2 can produce surface active substance after acting with crude oil to reduce the oil-water interfacial tension.

Table 2. Crude oil emulsion.

Strain Number	Well Meng 18-18	Well Meng 21-23
HB-1	+++++	+++++
HB-2	+++++	+++++

Note: emulsification effect will be divided into 5 levels, the methods are as follows:
Level 5: Oil and water can be completely miscible without oil-water boundary and will not layer standing for a long time.
Level 4: Oil and water can be mostly miscible and can mix basically by forcibly shake. Oil bead have better liquidity.
Level 3: Oil and water can be partly miscible and can mix partly by forcibly shake and then separate from each other quickly. Oil bead is droplets form.
Level 2: Oil and water can be a little miscible. The color of lower water is obviously stronger than the control sample. Oil bead is granular.
Level 1: Oil and water separate obviously. The lower water clears. It can't be mutually soluble after shocking forcedly.

3.3 Viscosity Reduction Property

Take crude oil from Mengulin as samples, viscosity reduction experiment was carried out. The results are shown in table 4. Strains HB-1 and HB-2 have better viscosity reduction effect. The viscosity reduction rate can reach from 39.8% to 49.5%.

Table 3. Interfacial tension measurement results.

Well	Interfacial tension of the blank ($mN\cdot m^{-1}$)	HB-1		HB-2	
		Interfacial tension ($mN\cdot m^{-1}$)	Reducing rate (%)	Interfacial tension ($mN\cdot m^{-1}$)	Reducing rate (%)
Well Meng 18-18	1.9765	0.338	82.9	0.4366	77.91
Well Meng 22-23	1.5224	0.1362	91.05	0.2069	86.41

Table 4. Viscosity reduction test results.

Well	Viscosity of the blank mPa·s	HB-1		HB-2	
		Viscosity mPa·s	viscosity reduction rate %	Viscosity mPa·s	viscosity reduction rate %
Well Meng 18-18	322.2	162.7	49.5	181.1	43.8
Well Meng 22-23	293.7	172.7	41.2	176.8	39.8

4 Microbial Mechanism Analysis

4.1 Crude Oil Component Analysis before and after Microbial Action

Selecting a set of test with best viscosity reduction effect (i.e. viscosity reduction test of strains HB-1 on well Meng 18-18), component analysis is carried out by gas chromatography. Chromatography conditions is weak polarity column of 30m * 0.32mm * 0.5μm, helium flow rate of 1.0ml/min, sample injection of 0.2μL and vaporization temperature of 260 ℃. On column temperature program, the initial temperature is 60 ℃. After 3 minutes, the temperature rises up to 260 ℃ by rate of 10℃/min. The analysis results are shown in Figure 1 and Figure 2.

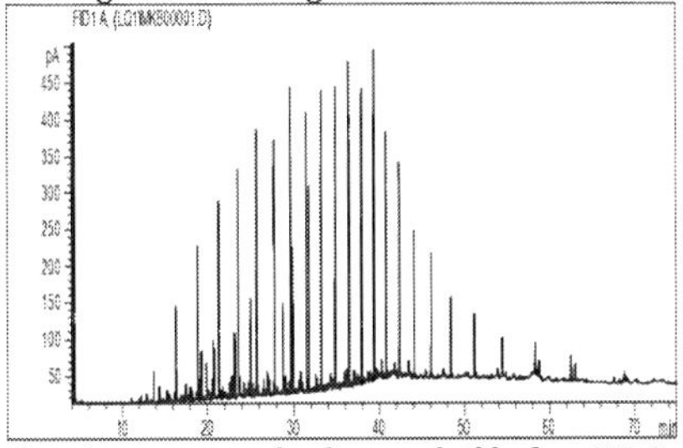

Fig. 1. Gas chromatography figure of oil before HB-1 action.

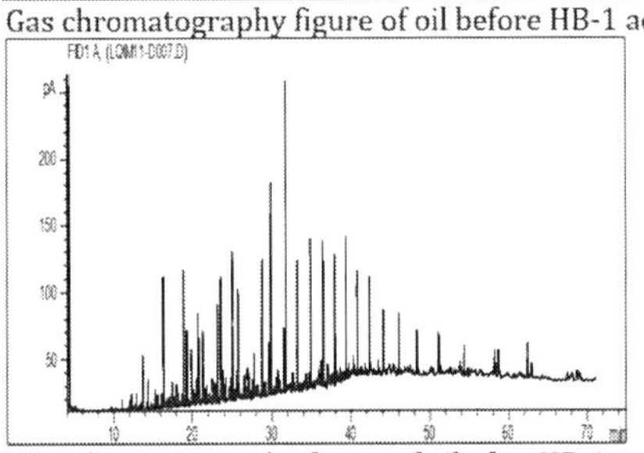

Fig. 2. Gas chromatography figure of oil after HB-1 action.

After microbial action, crude oil is decomposed, the peak of each composition obviously changes, and especially the characteristic peak of long chain hydrocarbon (C21-C50) begins to disappear. It shows that normal alkanes of different carbon chain length in crude oil are degraded by microbial. By chromatographic analysis statistics, pristine to C17 ratio of crude oil after HB-1 effect rises from 0.49% to 11.97% and phytane to C18 ratio rises from 0.86% to 22.97%. This suggests that the oil mobility is improved after microbial effect.

Through the paraffin change of crude oil (Figure 3), the mass fraction of different components in crude oil is changed after HB-1 effects. The mass fraction of heavy component significantly lowers and that of light component increases. It shows that HB-1 has a strong ability of growth and metabolism for petroleum hydrocarbon, achieving the purpose of reducing oil viscosity.

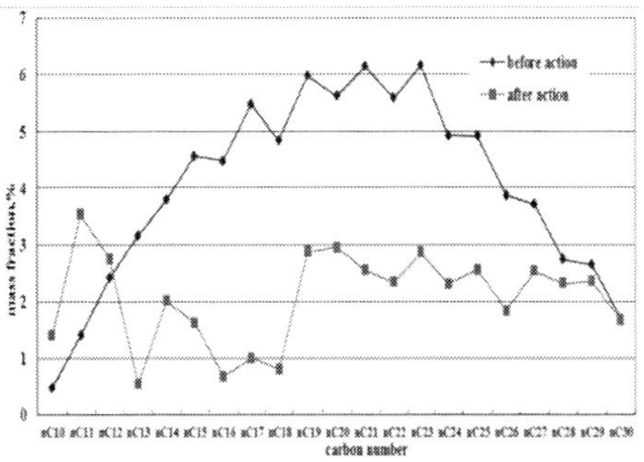

Fig. 3. Paraffin distribution of crude oil before and after HB-1 action.

4.2 Surface Active Material Analysis Produced by Microbial

Surface active substance could be produced by microbial metabolism. From FTIR (Figure 4), 3307cm^{-1} is NH contraction vibration caused by hydrogen bonding between molecular chain, 3070cm^{-1} is NH stretching band caused by intramolecular hydrogen bonding in N-H group, 1539cm^{-1} and 1648cm^{-1} is amide band Ⅰ and Ⅱ respectively. These characteristic absorption shows that the hydrophilic group of surfactant molecules is a peptide chain. 2856cm^{-1}- 2957cm^{-1} and 1387cm^{-1}-1463cm^{-1} is C-H stretching vibration of absorption aliphatic carbon chain. 1203cm^{-1} and 1735cm^{-1} are characteristic absorption of lactone. These show that the hydrophobic group of surfactant molecules is half-molecule fatty acid. Thus the surfactant molecule is inferred acyclic lipopeptide molecule. The microbe produces surfactant after acting with crude oil to reduce the oil-water interfacial tension and scatter oil with emulsifying ability.

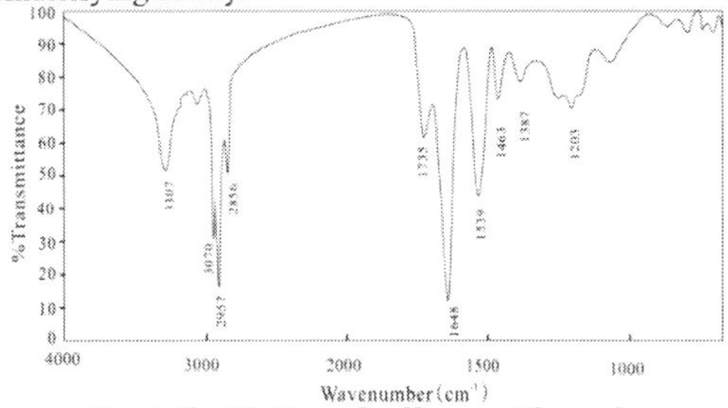

Fig. 4. The FT-IR graph of lipopeptid sample.

5 MEOR Field Applications

On May 2nd to April 27th 2013, in Mengulin oilfield, nine oil wells were chosen to carry on the pilot test of microbial stimulation, of which 7 wells became effective and the effective rate is 77.8%. In working wells, the average daily fluid rose from 81.34 tons to 103.96 tons, the average daily oil rose from 11.69 tons to 20.68 tons, and the average watercut decreased from 85.62% to 85.62%. The increased oil is 1,093.5 tons and the valid is up to 190 days.

https://doi.org/10.1051/e3sconf/20183801054

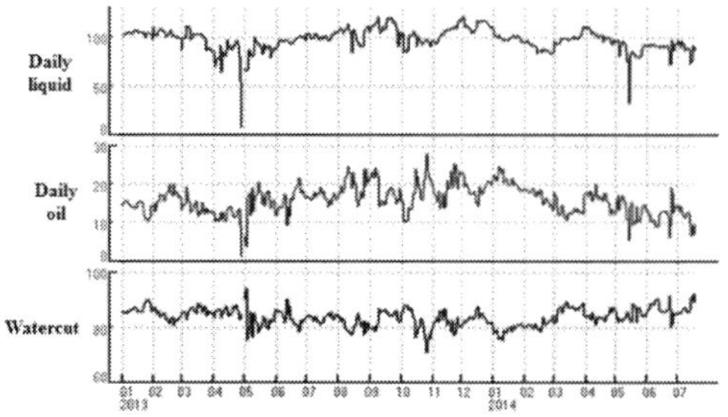

Fig. 5. Production situations of effective oil wells.

Table 5 The crude oil change before and after simulation

	Viscosity mPa·s	Interfacial tension (mN·m^{-1})	Wax mass score (%)	Colloid and asphalt mass score (%)
Before simulation	203.2	2.742	10.9	38.2
After simulation	111.5	0.689	6.6	30.1

From crude oil changes conditions before and after measures, the average viscosity of crude oil dropped from 203.2 mPa·s to 111.5 mPa·s, wax content of crude oil fell from 10.9% to 6.6%, colloid asphalt content fell from 38.2% to 30.1% and oil-water interfacial tension decreased from by 2.742 mN/m to 0.689 mN/m. The results show that the microbial degrades part of heavy components of formation crude oil, improves the liquidity of crude oil, and produces surface active substance.

6 conclusions

(1) According to the reservoir characteristics, 2 strains are optimized which can adapt to reservoir environment and reproduce massively.

(2) Microbe can improve crude oil properties after acting with formation crude oil. The light component increases and the heavy component decrease. The produced surfactants substances have emulsification ability to scattering oil, to reduce the oil-water interfacial tension and further improve the liquidity of crude oil.

(3) The microbial stimulation in Mengulin oilfield has apparent effective, showing that MEOR can improve oil displacement efficiency and recovery efficiency. It provides theoretical and practical support for implementation of microorganism to improve oil displacement efficiency of heavy oil reservoirs. MEOR can be the effective means to increase production at late stage of waterflooding development.

References

1. Wang C M, Li D P, Liu S G. Effects of Biophysiological and Biochemical Characteristics of a Strain of Microbacterium SP.on Properties of Thickened Oil[J]. Acta Petrolei Sinica, 2007, 28(5): 89-92.
2. Zhang Y S, Lan G Z, Deng L, et al. Experiments on Heaving Oil Degradation and Enhancing Oil Recovery by Microbial Treatments[J]. Acta Petrolei Sinica, 2001, 22(1): 54-57.
3. Yu J, Chen A H, Ma T L, et al. Analytic Mehtod of Oil Microbes and Their Metablizing Substances in the Third Oil Recovery[J]. Xinjiang Oil & Gas, 2008, 4(4): 88-92.
4. Li X M, Luan J Z, Xiao X M. Physical Simulation Situation of Microbe Oil Recovery Technology[J]. Oil Drilling & Production Technology, 2006, 28(1): 9-12.
5. Chen T, Yang S Z, MU B Z. Determination of Lipopeptide Surfactants in Cell-Free Broth[J]. Oilfield Chemistry, 2004, 21(4): 385-389.
6. Han S, Fu B F, Li C, et al. Characteristics and Field Application of Bio-Surfactant Produced by Bacillus subtilis BIT-BS002[J]. Journal of Microbiology, 2012, 32(4): 24-29.
7. Deng Z S, Ren G M, Zhang B C. Microorganism Screening for Petroleum Degradation and Their Degrading Capability in Shanbei Oil-field[J]. Journal of Yanan University(Natural Science Edition), 2005, 24(3): 65-67.
8. Luo J B, Wu W X,Zhang F, et al. Isolation and Metabolite Analysis of A Strain Producing Lipopeptide Biosurfactant[J]. Chemistry & Bioengineering, 2010, 27(2): 46-49.
9. Wen S C, Li X S, He S L, et al. Laboratory Evaluation of Bacterial Species for MEOR and Analysis of Their Field Results[J]. Oilfield Chemistry, 2009, 26(2): 195-198.

The Design and Implementation of Virtual Roaming in Yunnan Diqing Tibetan traditional Villages

Lucheng Cao, Wu Xu, Ke Li, Chunjie Jin, Ying Su, and Jin He*

School of Electronic and Information Engineering, Yunnan Minzu University, Kunming, China

Abstract. Traditional residence is the continuation of intangible cultural heritage and the primitive soil for development. At present, the protection and inheritance of traditional villages have been impacted by the process of modernization, and the phenomenon of assimilation is very serious. This article takes the above questions as the breakthrough point, and then analyzes why and how to use virtual reality technology to better solve the above problems, and take the Yunnan Diqing Tibetan traditional dwellings as the specific example to explore. First, using VR technology, with real images and sound, the paper simulate a near real virtual world. Secondly, we collect a large amount of real image information, and make the visualization model of building by using 3DMAX software platform, UV Mapping and Rendering optimization. Finally, the Vizard virtual reality development platform was used to establish the roaming system and realize the virtual interaction. The roaming system was posted online so that overcome the disadvantages of not intuitive and low capability of interaction, and these new ideas can give a whole new meaning in the protection projects of the cultural relic buildings. At the same time, visitors could enjoy the "Dian-style" architectural style and cultural connotation of dwelling house in Diqing Yunnan.

1 Introduction

With urbanization and development of the new countryside are proceeding rapidly, and traditional village of unified mode, this caused the emergence of homogeneity in villages, so that many ancient village assimilation phenomenon is very serious, particularly in Tibetan traditional villages in Shangri-La area of Yunnan Province. The architectural structure and layout of the Tibetan folk dwellings in Yunnan have changed considerably. The steel reinforced concrete and brick and tile has replaced the traditional timberwork structure, so that lost its original national characteristics and historical features .On November 1, 2016, experts and scholars from the first "Shangri-La for World" Rural Tourism Development Forum and "World Village in Yunnan" forum came to Wheelun in Shambhala to sign the "2016 Yunnan Traditional Ancient Village Protection and Development Shangri-La Initiative ", in order to protect Diqing Shangri-La Tibetan traditional village. In addition, the traditional residence has its particularity, it is the original soil of the existence and development of the intangible cultural heritage. Many intangible cultural heritages also disappear with the disappearance of residential houses. It can be said that the protection of a few traditional houses also protects the intangible cultural heritage. Therefore, the protection of national traditional houses has been urgent [1].

In recent years, with the rapid development of virtual reality technology, the combination of VR and network technology has brought the exhibition and protection of cultural relics to a brand-new stage. The virtual reality technology through acquiring image data to build 3D model library of traditional residential buildings, which preserved the original architectural style and spatial relations and a series of important data, moreover can achieve high precision of traditional houses, and permanent protection. It also can unify and integrate cultural relics resources through the Internet, whom to use the VR technology to display cultural relics more comprehensively, vividly and realistically through , in addition, cultural relics can get away from the fence of time and space, so that resources can be shared, and truly become a cultural heritage that all mankind can "own" [2].The use of virtual reality technology will promote the cultural industry to enter the information age more quickly and realize the modernization of cultural relics display and protection. This is also a strategic measure of social and cultural significance in the digital age.

2 Design of Virtual Roaming System

2.1. Construction of 3D Scene Model Library for Tibetan Dwellings

In the development of virtual reality roaming system, it is necessary to collect a large number of three views, architectural CAD drawings, and a large number of actual images of Tibetan houses in the field to build its three-dimensional model. Furthermore, the 3D model is the basis of the whole roaming system, and the quality of

* Corresponding author: hjmwh@sina.com

the model directly affects the running effect and the fidelity of the scene. The system uses 3DMAX software to build village spatial model, the software with powerful function, low price and relatively low requirement for hardware system, and has the advantages of simple operation and so on, which greatly reduces the threshold of use.

Because of the large amount of modeling, according to the layout of traditional residential buildings and the characteristics of the surrounding environment, this paper classifies the building and its internal layout as the main modeling object. The surrounding landscape such as trees, grasslands, hillsides and so on are regarded as secondary modeling objects, which can be imported from the model base, in a word, which can not only guarantee the quality of the main modeling objects, but also save a lot of time to finish the post-rendering.

The 3D model of Tibetan traditional residence and internal layout can be built by 3DMAX. In the process of constructing the model, we must pay attention to restoration of Tibetan blockhouse which includes carving, painting, coating and column beams, purlins, rafters and so on. Through modeling, making UV mapping, the basic model library of Tibetan houses has been completed. Figure 1 shows the hierarchical structure of the virtual scene of folk house [3].

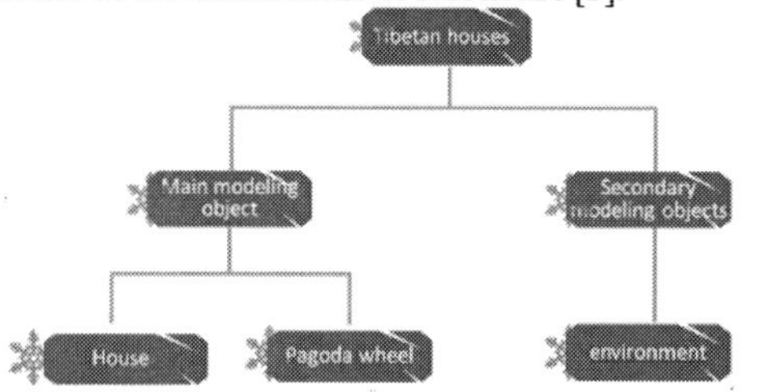

Fig. 1. Hierarchical Structure Diagram of Tibetan Residence Modeling

2.2. Rendering and Export of Models

The model rendering also uses 3DMAX software to set the lighting of the scene, thus forming the model rendering map, using the lighting, shadow effect to make the model more realistic, After rendering is complete, we need to install the plug-in OSGexplorer.exe, which can export OSGB file formats that can be recognized by Vizard. After the OSGB tilting photography format is exported, it is also convenient to follow up the operation processing on the Vizard virtual reality engine. In order to ensure that the Vizard can recognize all the information of the model, in this process, we should pay attention to the setting of the parameters when exporting. Since the subsequent virtual interaction is carried out on the virtual reality development platform, Vizard opened up the model built in 3DMAX directly and the modeling results are observed. Figure 2 shows the overall effect of Tibetan houses. Figure 3, Figure4 is part of the effect diagram, from which we can see all the information that Vizard can recognize the model. Accordingly, the virtual scene modeling is successful.

Fig. 2. Overall Effect of Diqing Tibetan Village

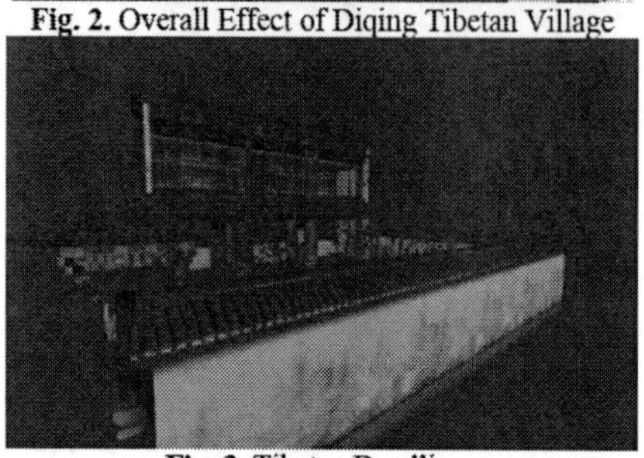

Fig. 3. Tibetan Dwellings

Fig. 4. Interior Painting and Engraving of Houses

3 Realization of virtual roaming system

3.1. the design process of the roaming system

The design of traditional Tibetan village roaming system can be divided into two processes: one is to construct 3D virtual scene of Tibetan traditional residence, the other is to realize virtual interaction. The main software used in this paper is 3D modeling software and virtual reality development platform software. According to the characteristics of this paper, and through comparative analysis, we select 3D Max and Vizard as main software.The specific process is shown in Figure 5.

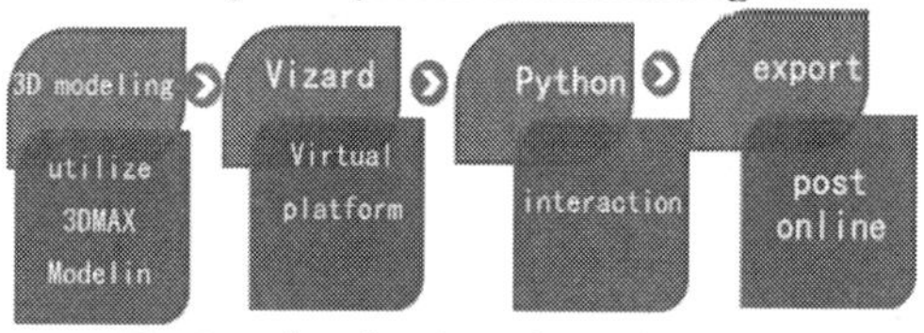

Fig. 5. Design Flowchart of Roaming System

3.2. Loading 3D Virtual Scene

The development of virtual roaming system of Tibetan village can not run without a feature-rich development platform, this paper adopts Vizard as the development platform. Compared to other virtual reality development platform, Vizard software provides abundant function modules, users can quickly develop a prototype system used in various occasions, moreover, it more easily to get started, and the programming is simple, the function is strong at the same time. The Vizard environment can use the Python language to load the successful 3D virtual model built in the 3DMAX. After loading, users can see the loading scene in the preview window. The Python statement to load the OSGB file is: cj= viz.addn.cj1.osgb. The basic process of loading a three-dimensional scene is shown in flowchart 6.

Fig. 6. Loading 3D Scene Flow Chart

3.3. Interaction of 3D Scenes

The realization of 3D scene interaction is the most important part of the development of roaming system, which is mainly divided into two parts: UI option interface and code design.

3.3.1. UI Option Interface

UI options interface plays an important role in the interaction of 3D scene, because a simple and full-featured interactive page will directly affect the visitor's experience, so the design of the option interface should be fully functional and easy to operate. In this paper, the option interface is divided into five main parts, respectively is general interface, residential interface 1, residential subinterface 1, residential interface 2 and residential subinterface 2, to guide the visitor to enter roaming system. options interface hierarchy diagram as shown in figure 7.

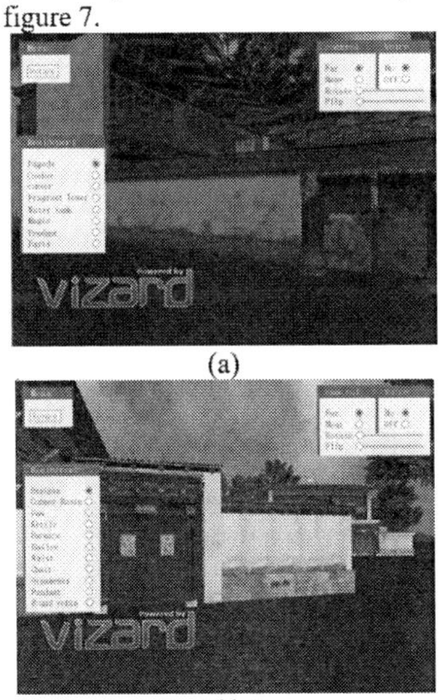

(a)

(b)

Fig. 7. Options Interface Hierarchy Diagram

3.3.2. Coding Design

The coding design can realize all the functions of the Tibetan residence roaming system, and make the roaming system have virtual interaction, and it is also an indispensable part in human-computer interaction. First of all, to determine the design goal, this paper designed a roaming system which can realize omnidirectional free roaming without dead-angle, and it can be controlled by the mouse angle at the same time. In order to increase the virtual interactive immersion of the roaming system, the audio editing software of Balabolka can be used to edit the sound, so that the audience can listen to the concrete explanation while the audience can enjoy it, so as to realize the function of the voice guide. Meanwhile, we can use Vizard through the Python language to write the content of the scene and achieve audio call [4].

The coding design process is as follows:

1)First, use Vizard to create a new Vizard file and name it as Tibetan dwellings.

2)Second, create a new blank scene, add the scene model (sky) to the blank scene, and adjust the background settings accordingly.

3)Third, establish UI option interface for scene, create UI material, and make main interface for interaction.

4)Then, load the establishing OSGB model, add the corresponding script, and build for the objects that need to interact by setting up the function. Finally realize the interaction of 3D scene, and call the audio at the same time.

5)Finally, test and compile run.

After the coding design is completed, the code is compiled and run successfully, the effect diagram is shown in Figure 8, Figure 9 [5].

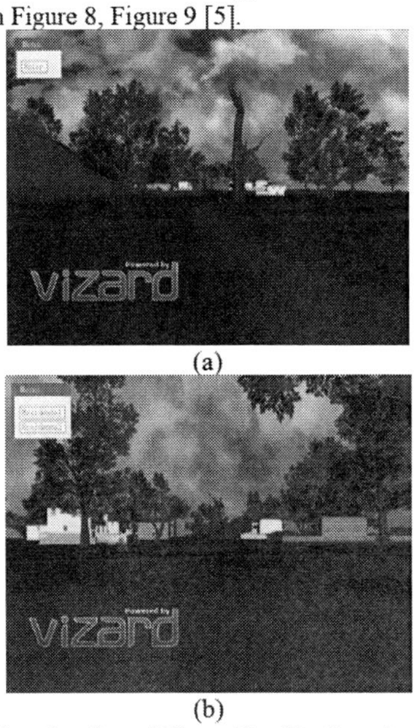

(a)

(b)

Fig. 8. Start Interface of Tibetan Dwelling Roaming System

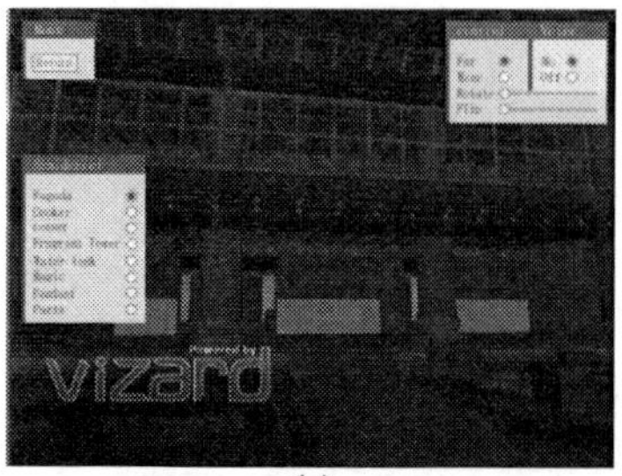

(a)

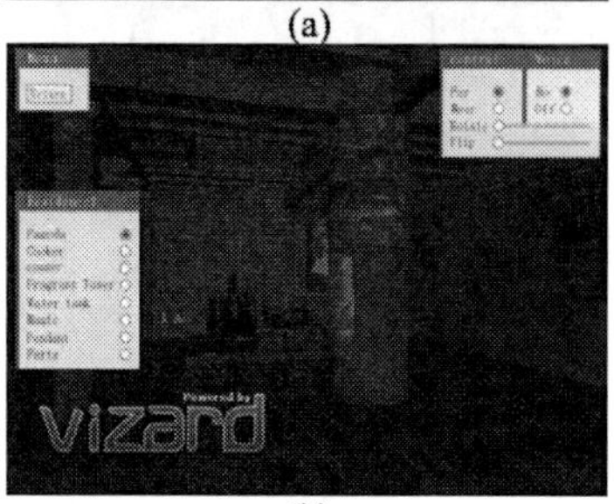

(b)

Fig. 9. Tibetan Dwellings Part1 Showing the Effect

Some of the key codes used are:

```
link =
viz.link(groupcm,viz.MainView,srcFlag=viz.ABS_GLOBAL)
mj1 =
modePanel.addItem(viz.addButtonLabel('Tibetan dwelling1'));
mj2 =
modePanel.addItem(viz.addButtonLabel('Tibetan dwelling2'));
vizact.onbuttondown(mj1,CMStart,'cm1');
vizact.onbuttondown(mj2,CMStart,'cm2');
diaoz_mode.endAction()
diaoz_ch.setPosition(00,00,00)
diaoz_mode.setAxisAngle([0.0, 0.0, 1.0, 0.0])
shipin_mode.endAction()
shipin_ch.setPosition(00,00,00)
shipin_mode.setAxisAngle([0.0, 0.0, 1.0, 0.0])
yizi_mode.endAction()
yizi_ch.setPosition(0,0,0)
yizi_mode.setAxisAngle([-0.0, 1.0, 0.0,
90.00000762939453])
```

3.3.3. Immersive Roaming

Vizard has strong device compatibility, supports helmet display and many other external input devices, so the use of stereoscopic head display device can achieve immersive roaming effect.This thesis has set Vizard in the code for full-screen display effect, specify the relevant helmet model and add tracking code in the software to track the helmet with the tracking code can achieve virtual roaming. The user can see a full view of the virtual village, and when you walk, you will see the picture will also move, so that the user is out of the real world, completely immersed in the virtual Tibetan dwelling world [6].

3.3.4. Promotion of Virtual Roaming System

Vizard can automatically use script the program to generate exe executable file, which means that it can roam on any computer, which provides a prerequisite for the promotion of roaming system. Therefore, this roaming system can post to the Internet, establish an independent website, and build a digital museum, meanwhile, realize the sharing of cultural resources, so that it let more people understand the Tibetan culture, and also help to protect and inherit the regional Tibetan dwelling culture.

4. Conclusion

This paper aims at the problem that the traditional villages of ethnic minorities are affected by urbanization and modernization, but lose their original national characteristics and historical heritage. This paper takes the Tibetan dwellings in Diqing Shangri-La, Yunnan Province as an example, puts forward the method of establishing virtual roaming system to protect traditional villages by using the VR technology, so that more people can view the architectural details and architectural characteristics of Tibetan houses at close range. And through the Internet, more people can cross the time and space constraints, enjoy Diqing Shangri-La Tibetan residence with "Dian-style" style anytime and anywhere, so that "cultural relics" really become the wealth of all mankind. The roaming system designed in this paper also provides a way of thinking for the digital protection of other endangered villages. At the same time, this paper also has some shortcomings, such as, due to time constraints, and the carvings of the dwellings are very complicated, the roaming system is still not able to fully depict the real delicate degree of them. But we still hope the research work can provide idea for the protection of cultural relics and contribute to the protection and inheritance of ancient buildings in China.

Acknowledgment

This research is supported by The National Natural Science Foundation of China: "contact + non-contact" mixed interaction interface design strategy and its key human factors problems (Grant: 61741206)

References

1. Zheng Xin. Study on the protection of traditional villages [D] .Beijing Construction University, 2014
2. Ren Ning-ning, Zhang Wei, Wu Hong-chuan, et al.Analysis of new forms of network display under 3D Internet [J] .computer knowledge and Technology, 31(2012)
3. Pan Xiuqiang.Design and implementation of virtual roaming system for ancient villages in Nanxi River based on OSG [J] .Journal of Zhejiang Institute of Technology and Industry, 15(2016)
4. Lin Chen, Lin xiaobin. Modeling of 3d virtual scene and its application in Vizard environment [J]. Journal of minjiang college, 36(2015)

5. Su Yang. Design and implementation of virtual tour of wuyuan huizhou architecture in jiangxi [D]. Beijing university of technology,2014
6. Zhao Wei, Xie Xiaofang.immersion roaming device based on Free Walking [J] .computer Engineering, 35(2009)

Study on Reuse Strategy of Abandoned Industrial Square – in the case of Jingxi Wang Ping Coal Mine

Xiaodan Li[1, *], *Zhiting* Chen[1], *Lijun* Jia[1], *Wei* Wu[1], *Hailiang* Zhang[1], *Tianyi* Ma[1], *Tao* Wang[1]

School of Mechanics and Civil Engineering, China University of Mining and Technology, Beijing100083, China.

Abstract. Wangping Coal Mine, whose industrial heritage is of great value, was one of the eight coal mines in Beijing. A large number of field surveys and analysis of the abandoned industrial facilities of Wangping Coal Mine were carried out in this paper. From the perspective of protecting industrial heritage culture and sustainable development, this paper studies the ideas and strategies for reusing the abandoned facilities of the Wangping Coal Mine. In order to protect its industrial heritage as much as possible, it is suggested to reuse the industrial square of Wangping Coal Mine as a community park.

1 Introduction

According to the instructions of *Responsibility Document for Resolving Overcapacity in the Coal Industry to Realize the Goal of Relief from Difficulty*, Wangping Coal Mine was asked to stop its production on August 16th, 2016. It was formally established in 1964 and suspended at the beginning of 1994. With 30 years of mining history, it has made outstanding contributions to the development of the coal industry of Beijing[1].

2 Overview of Wangping Coal Mine

2.1 Geographic location

Wangping Coal Mine is located in Wangping Town, which is in the west of Mentougou District in Beijing. The railway from Yongding Gate to Muchengjian passes through it. 109 National Road runs to the east of it. Tanwang Road passes through the middle of it (Fig. 2-1). The whole mining area is steep, because its north, west and south are surrounded by mountains.

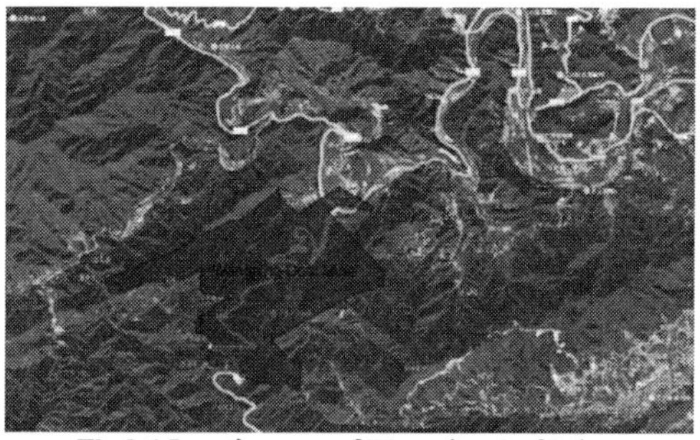

Fig.2-1 Location map of Wangping Coal Mine.

2.2 Historical evolution

Wangping Coal Mine has a long history (Table 2-1) and was one of the eight coal mines of Jingxi Mining Bureau[2].

Table 2-1 History of Wangping Coal Mine.

Age	Historical events
1919	Huazhao Wen opened Anjiatan Coal Mine around Anjiatan and Wangping. It was the predecessor of Wangping Coal Mine.
1951	The Beijing Public Security Bureau took over Yugong Mine and Shengli Mine and established Qinghe Mine in Anjiatan Village.
1953	In April, the local government took over Tongfeng Coal Mine in Anjiatan Village and all the surrounding coal mines.
1954	In March, Jingxi Mining Bureau took over them and established Anjiatan Coal Mine.
1958	In May, construction of Wangping shaft was started. In August, Jingxi Mining Bureau decided to build Huapogen footrill.

* Corresponding author: tjulxd@sina.com

1960	In April, Wangping shaft was put into operation.
1964	In March, Wangping Coal Mine was established.
1977	Wangping Coal Mine absorbed Dajiao Mine, which was part of Mentougou Coal Mine. In August, Jiulong footrill was started to be built.
1994	The production of Wangping Coal Mine was suspended. Prior to that, its annual output was 800,000 tons.

2.3 Small coal kilns

Before the foundation of New China, besides Tongfeng Coal Mine, there were five or six small coal kilns around Wangping Village. In 1951, Tongxian Military District, the Public Security Bureau of Beijing Municipal Public Security Bureau, and the National Coal Mine Trade Union and other units have successively established 16 small coal kilns in Anjiatan Village. In April 1953, The local government took over Tongfeng Coal Mine and all the surrounding small coal kilns [3].

Most of the small coal kilns in Wangping Coal Mine were located in Lujiapo, Jiaoling and Anjiatan. According to the government requirements, they were shut down successively in 2000, 2001 and 2009. Except for the comprehensive information on he small coal kilns closed in 2009, the data on that closed in 2000 and 2001 were barely preserved. (Table 2-2)

Table 2-2 List of closed coal kilns in Wangping Coal Mine, 2000-2009.

Age	List of closed coal kilns
2000	Xiwangping Coal Mine, Yumin Coal Mine, Huangqugou Coal Mine, Pingjiu Coal Mine, Houling Coal Mine, Menlingkou Coal Mine, Daqiao Coal Mine, Fuxing Coal Mine
2001	Dongwangping Coal Mine, Baoshun Coal Mine, Jiuyuan Coal Mine, Xingwang Coal Mine, Huanggang Coal Mine, Dongshan Coal Mine, Guacaodi Coal Mine, Shizidao Coal Mine, Wanda Coal Mine, Shanlindaxin Coal Mine
2009	Magezhuang Coal Mine (1), Nanjian Coal Mine (2), Ximage Coal Mine (3), Chunan Coal Mine (4), Jiaoling Xingda Coal Mine (5), Pingxi Coal Mine (6), Yunfutong Coal Mine (7), Nanwayao Coal Mine in Bai Daozi (8), Pingdong Coal Mine(9), Daqugou Coal Mine(10), Xinkou Coal Mine in Zhuanyaogou (11), Longshan Coal Mine (12), Changhong Coal Mine in Baidaozi (13), Suwa Coal Mine (14), Houtouyuan Coal Mine (15), Jinxiu Coal Mine (16), Jinyu Coal Mine (17), Qingtou Coal Mine (18), Shunmin Coal Mine in Bei Ling (19), Dongkou Coal Mine of Dahua (20), Zhaojiatai Coal Mine (21), Tanzhesi Coal Mine (22) (As shown on the right Fig.).

2009	

3 Security situation

3.1 Goafs and collapse sites

Most of the goafs in Wangping Coal Mine were located in the west and south of Anjiatan, Nanling, Beiling, Liujiafen, Shizidao and other villages. They covered about $2km^2$.

In August 2016, there were five small collapse pits on the hillside of wasteland in the west of Anjiatan. Based on many investigations and evaluations of the current geological environment of Wangping Coal Mine, the development of small collapse pits has been basically stable. [4] Under the current conditions, the mine geological environment has slight impacts on the human settlements and the environment around[5,6].

3.2 Groundwater

Related research showed that after the shutdown of Wangping coal mine, the groundwater level has picked up. The content and saturation of original water has been restored. The underground water in the mining area has no toxic or harmful substances and can be exploited and utilized [7,8].

4 Protective reuse strategy

It is found that the production equipments, office buildings, dormitory buildings, railroad tracks and others of Wangping Coal Mine industrial square are well preserved (Fig. 4-1). They are very valuable industrial heritage buildings.

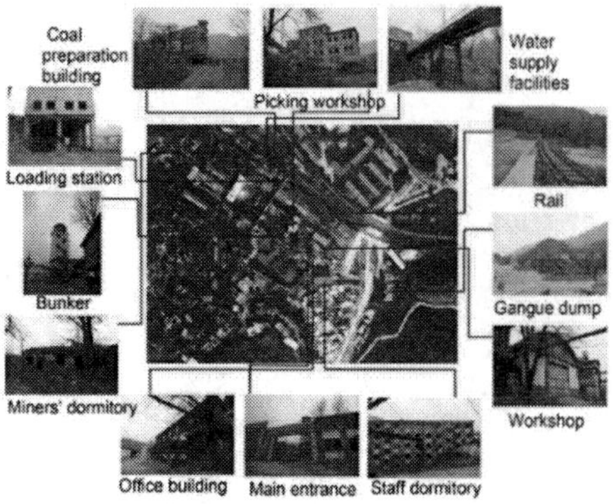

Fig.4-1 Current situation of industrial square facilities

4.1 Reuse principle

In order to take good use of the original appearance of the industrial square, it is suggested to take some reinforcement measures. And buildings that have nothing to save should be demolished, and some coal-industry-style facilities should be reserved.

4.2 Reuse strategy

Furthermore, the needs of local residents and urban tourists should be taken into account. Therefore, the industrial square of Wangping Coal Mine is suggested to be reused as a community park [9]. It consists of Pedestrian street, Industrial heritage park, Civic activity center, Railway park, Ecological experiment base (Fig.4-2) . This strategy creates more room for interaction and outdoor activities for people. The commercial pedestrian street at the entrance has promoted local economic development which also satisfies residents' shopping needs.

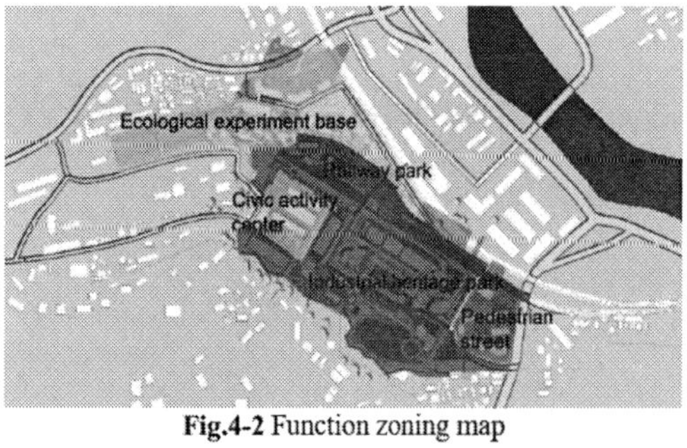

Fig.4-2 Function zoning map

The part of the coal mine ecological park includes the ring path system and the industrial park system. The design of the ring path, which allows visitors to overlook the Wangping Coal Mine, was inspired by the Acropolis' tour and the New York High Line Park. The design of the industrial park is based on the successful experience of the German Ruhr area.

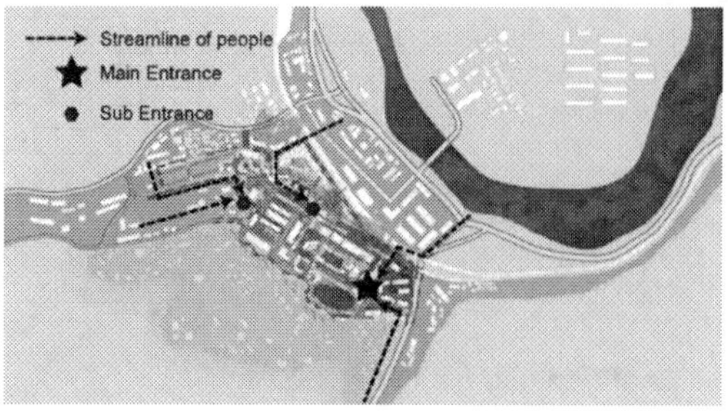

Fig.4-2 Streamline analysis diagram

Because of its high degree of advantage, the coal bunker has become a landmark of the entire mining area. It was reused as a watchtower connected to the circular path system. The Waterscape Park under the observation tower was transformed from the original collapsing pit, which echoes the northern wetland of Wangping Town.

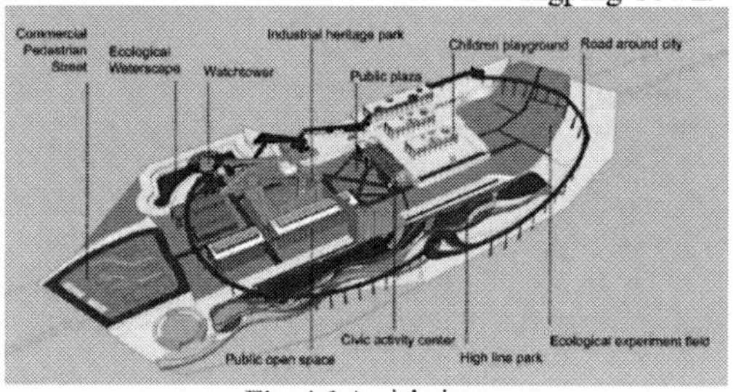

Fig. 4-3 Aerial view

5 Conclusion

Wangping Coal Mine have witnessed the profound impact of mining activities on Beijing. Its reuse is a respect for the history of cities and the culture of coal mines. At the same time it helps to enhance the local residents' sense of identity and social belonging. This study explores the reuse strategy of Wangping Coal Mine Industrial square and hopes to promote the reusing and sustainable development of Jingxi Coal Mine.

Acknowledgements

Thanks to Beijing Coal Group for providing a lot of information for this study. I would also like to thank my project assistants Lijun Jia and Wei WU, the anonymous reviewers and the Editor.

References

1. Li X, Yang H, Chen Z, et al. Evaluation system for prioritization tool to redevelop abandoned coal mine industry square based on DSR model.TCSAE, **34**,224(2018)
2. Closed Mine Report of Wangping Coal Mine, Beijing Jingmei Group Co., Ltd.
3. Liu Y, Sun X, Li J, et al. Restoration Effect Analysis of Abandoned Coal Mine Engineering of Mentougou in Beijing Based on Soil Physicochemical Properties. Chin. Sci. Bull.,**28**,246(2012)

4. Zhang H. Study on the Environmental management of Mentoukou Village Coal Mine in the Mentoukou District of Beijing. Urban Geology, **6**,54(2011)
5. Yan-Jun WU, Gen-Yi LI, et al. Analysis About Influence Factors and Collapse Disaster Characteristics of Small Coalmine Gob Areas in the West of Beijing Mountain Area. Journal of Geological Hazards & Environment Preservation, **28**,46(2017)
6. Cheng X Q, Zhao F Y. Quantitative classification and ordination of natural vegetation on abandoned mine land in Mentougou District, Beijing. J. Northeast For. Univ. (Chin. Ed.), **38**,71(2010)
7. LONG Jinghua, HU Zhenqi. Field Investigation and Reutilization Evaluation of Abandoned Mine Land in Mentougou District of Beijing.Coal Engineering, **47**,120(2015)
8. Zhang L, Zhang S, Huang Y, et al. Exploring an Ecologically Sustainable Scheme for Landscape Restoration of Abandoned Mine Land: Scenario-Based Simulation Integrated Linear Programming and CLUE-S Model. Int. J. Environ. Res. Public Health, **13**,354(2016)
9. Yi X. Problems and Redevelopment Strategy for Old Industrial Area. Building Resilient Cities in China: The Nexus between Planning and Science. Springer International Publishing, 2015:349-361.

Experimental study on laminated bamboo lumber column

Chenwei WANG[1, a], Huizhong ZHANG[1, a*], Chenjie ZHAO[1, a], Chenge ZHANG[1, a], Tongwei CAO[1, a], Haoran DONG[1, a], Jieying DING[1, a], Zhenhua XIONG[2,b], Xiaohiong XIONG[3,c], Wenfeng LIU[4,d], Kangwen WU[3,e], Rong Liu[1, a], Wen Ding[1, a], and Ziwen YAN[2,b]

[1] College of Civil Engineering, Nanjing Forestry University, Nanjing 210037, China
[2] Ganzhou Sentai Bamboo Company LTD, Ganzhou 341001, China
[3] Jiangxi Precious Bamboo Company LTD, Ganzhou 341001, China
[4] Jiangsu Fashion Safety Monitoring Consulting Co. LTD, Zhenjiang 212009, China

Abstract: This paper presents and discusses the experimental study on the mechanical properties of LBL column both under axial and eccentric compression. The results shows that the ultimate load for the eccentric compression specimens with the eccentricity values of 30 mm and 110 mm are 95.2 kN and 31.8 kN respectively. Eccentricity is one of the main influencing factors for the ultimate bearing capacity of the LBL columns. Because of the vulnerability of the mechanical connections or natural nodes to tensile stress and secondly, laminated bamboo is vulnerable to defects that has more detrimental influence on the tensile resistance of the material. The variation in strain for the laminated bamboo lumber column sections is linear throughout the loading process, following standard normal section bending theory which is similar as that for the beam.

1 Introduction

As a sustainable green building material, bamboo has attracted more and more researchers' attention [1-28]. Original bamboo culm has low rigidity and limited diameter. In order to solve these problems, industrial bamboo materials appeared in different forms such as parallel bamboo strand lumber (PBSL) [5-10], laminated bamboo lumber (LBL) [11-28] and so on. Laminated bamboo lumber (LBL) is made by laminating the thin flat bamboo strips together with adhesive under a certain treatment and the structural size could be controlled in the factory.

In order to apply laminated bamboo lumber to engineering area, many scientists [11-28] have done a lot of research on manufacturing technology and mechanical properties. Correal et al. [12] studied the adhesive bond performance in glue line shear and bending for glued laminated guadua bamboo. Verma et al. [13-14] have investigated the stiffness and strength of four layered laminate bamboo composite at macroscopic scale. Lee et al. [15] have studied the properties of laboratory-made laminated-bamboo lumber. Yeh et al. [16] has investigated finger joint performance of structural laminated bamboo member. Varela et al. [17] have investigated cyclic performance of glued laminated guadua bamboo-sheathed shear walls. Taheri et al. [18] have studied buckling response of glue-laminated columns reinforced with fiber-reinforced plastic sheets. Li et al. [19-21] have studied the flexural performance of laminated bamboo lumber beam and the ultimate load calculation formula for the LBL beams were proposed. Axial compression performance of LBL has been studied by Li et al. [22-24] and Su et al. [25]. Li et al. [26-28] also investigated the eccentric compression performance of LBL columns. Even though some researchers have studied the mechanical properties of laminated bamboo lumber column, the study for it is limit. There is still no standard for LBL columns under compression. More work needs to be done to investigate the mechanical performance for LBL columns.

This paper mainly studies mechanical properties for LBL columns based on the former research. In order to know the compression behaviour of LBL columns, the study will examine in detail the behaviour of structural members constructed from laminated bamboo lumber with a design dimension of 76 mm × 76 mm × 800 mm.

2 Materials Information and Test

Moso bamboo, harvested at the age of 3-4 years, was chosen for manufacturing laminated bamboo lumber columns. Phenol glue was used to manufacture the column specimens. Single layers were made first, and then they were pressed together to form the blocks (Figure 1). The pressing temperature was 140±5°C and the transverse compression was 1.82MPa for both the sheets and the blocks, and a confining pressure of 4.74MPa was used when manufacturing the sheets. Considering axial compression and eccentric compression with two

Corresponding author: Email: [a]310325816@qq.com, [b]792232771@qq.com, [c]1687699888@qq.com, [d]liu.wenfeng@free-sun.com.cn [e]767057291@qq.com, eccentricities of 30 mm, 110 mm, 5 column specimens (three for axial compression) were constructed with the cross-section of 76 mm× 76 mm and the length of 800 mm. The eccentric direction is tangential eccentric direction.

Figure 1 Laminated bamboo lumber (LBL)

Figure 2 Axial compression test

Figure 3 Eccentric compression test

Axial compression test photo and eccentric compression test photo can be seen from Figure 2 and Figure 3 respectively. Both the axial and lateral displacements were measured by Laser Displacement Sensors for all columns. Two main strain gauges along lateral and longitudinal directions were placed at the mid-height of specimens for axial compression tests. While for the columns under eccentric compression, besides two strain gauges were placed at the mid-height of four side surfaces, four more strain gauges were used on one side surface as can be seen from Figure 3. The loads were applied using a microcomputer-controlled electro-hydraulic servo universal testing machine with a capacity of 1000 kN with a TDS-530 Data Acquisition System.

The load was applied initially through load control in the elastic stage, and then was changed to displacement control before the proportional limit. The total loading time was controlled between 8-12 minutes. The test was continued at a certain displacement rate until the load reduced by 15% of the ultimate load and the specimen had sustained significant damage.

3 Test results and analysis

As for the columns under axial compression, the specimens behaved elastically at beginning and then buckling failure happened to them. The specimens bent to one direction suddenly. With the increasing of load and obvious deflection, cracks appeared. Finally the specimens crushed and the load values decreased. Typical final photos for the specimens under axial compression can be seen from Figure 4 and the ultimate load for three specimens are 274.7 kN, 270.2 kN, 275.0 kN respectively.

As for the specimens under eccentric compression, similar to the specimens under axial compression, the column specimens behaved elastically at beginning then followed by some plastic deformation with increasing load. The stiffness decreased in the plastic-elastic stage. With the increasing of load and obvious deflection, cracks (accompanied by a slight noise) appeared on the tensile surface D. The load application was stopped when the load was reduced by 15% of the ultimate load and the specimen sustained significant damage. Except for the compression surface B, cracks were visible on other three side surfaces. It was obvious that failure was triggered due to bending for all the two column specimens under eccentric compression. Typical final failure photo can be seen from Figure 5. The ultimate load for the eccentric compression specimens with the eccentricity values of 30 mm and 110 mm are 95.2 kN and 31.8 kN respectively. Compared with the axial compression results, eccentricity is one of the main influencing factors for the ultimate bearing capacity of the LBL columns. The ultimate load for the specimens with the eccentricity values of 30 mm and 110 mm decreased by 65.2% and 88.4% respectively.

(a) Face A

 https://doi.org/10.1051/e3sconf/20183802001

(b) Face B

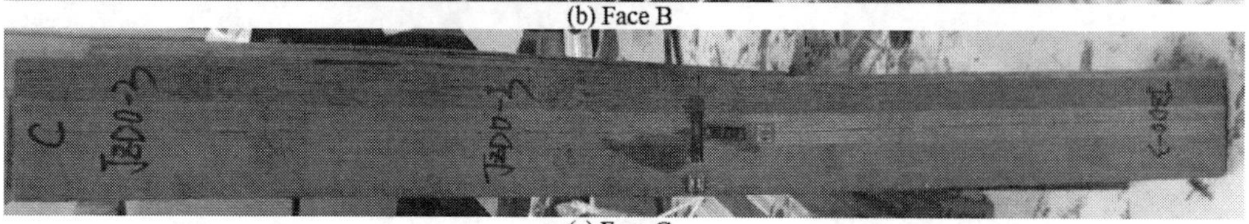

(c) Face C

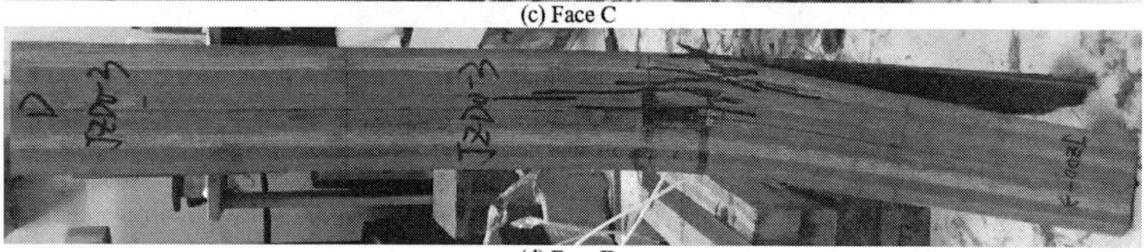

(d) Face D

Figure 4 Typical failure photos for axial compression specimen

(a) Face A

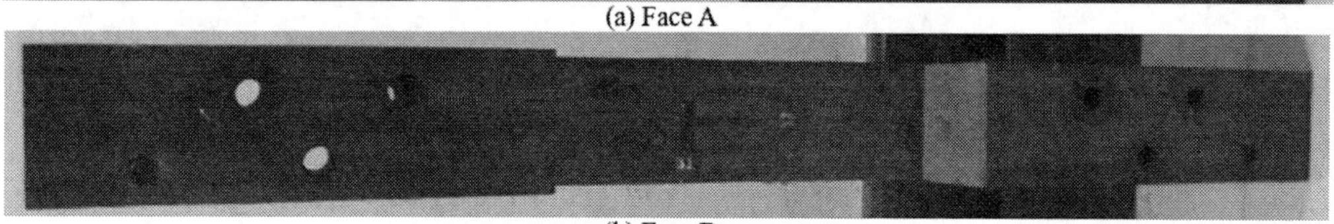

(b) Face B

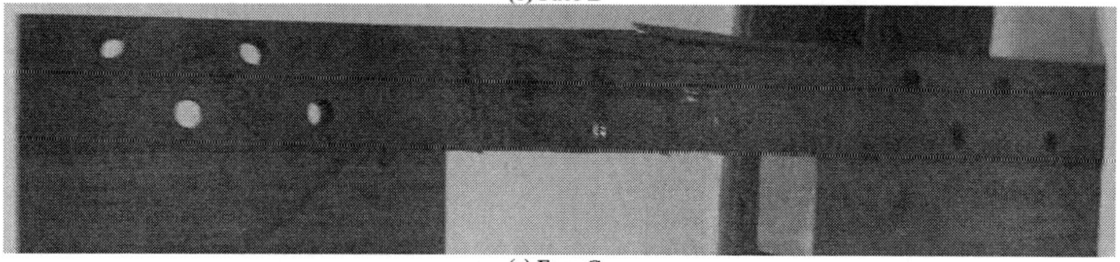

(c) Face C

Figure 5 Typical failure photos for eccentric compression specimen

4 Load vs strain relationship for LBL columns

Figure 6 shows the variations in longitudinal strains at the mid-heights of four side surfaces. As can be seen from the figures, all strains clearly displayed an initial elastic phase. The longitudinal strain values for face A and C were always consistent during the whole loading process, and with increasing eccentricity observed strains on faces A and C became insignificant. The mechanical behaviour of the columns with higher eccentricities resembled those observed for beams. Face B being subjected to the highest compression, always showed maximum deformations both in longitudinal and transverse direction. Although, failure was always triggered on the tension surface: firstly, because of the vulnerability of the mechanical connections or natural nodes to tensile stress and secondly, laminated bamboo is vulnerable to defects that has more detrimental influence on the tensile resistance of the material.

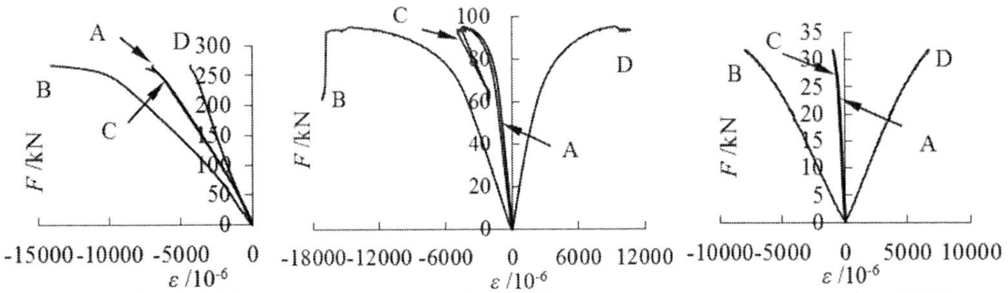

(a) JZD0-2 (b) JZD30 (c) JZD110

Figure 6 Load-longitudinal strain curves

5 Strain distribution over cross-sections

Fig. 7 shows the strain profile through the loading process for the mid-height cross-section of typical specimens for each group. The variation in strain for the laminated bamboo lumber column sections is linear throughout the loading process, following standard normal section bending theory which is similar as that for the beam (Li et al [19-21]).

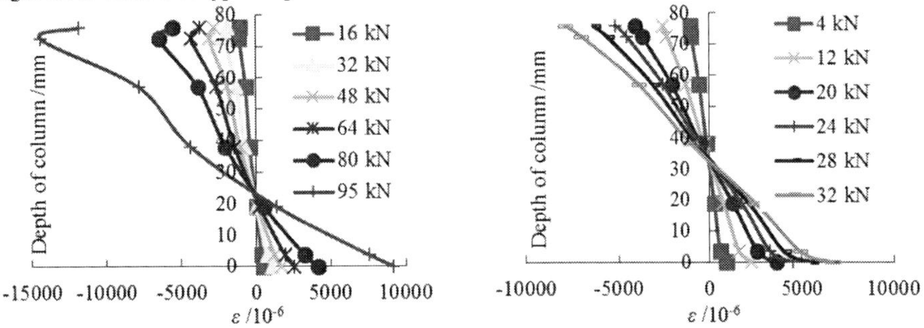

(a) JZD30 (b) JZD110

Figure 7 Typical strain profile development for the mid-span cross-section

6 Conclusion

This paper presents and discusses the experimental study on the mechanical properties of LBL column both under axial and eccentric compression. The results shows that the ultimate load for the eccentric compression specimens with the eccentricity values of 30 mm and 110 mm are 95.2 kN and 31.8 kN respectively. Eccentricity is one of the main influencing factors for the ultimate bearing capacity of the LBL columns. The ultimate load for the specimens with the eccentricity values of 30 mm and 110 mm decreased by 65.2% and 88.4% respectively. Because of the vulnerability of the mechanical connections or natural nodes to tensile stress and secondly, laminated bamboo is vulnerable to defects that has more detrimental influence on the tensile resistance of the material. The variation in strain for the laminated bamboo lumber column sections is linear throughout the loading process, following standard normal section bending theory which is similar as that for the beam.

Acknowledgements

This work was financially supported by the National University students practical and innovation training project (No. 201610298003Z, 201710298056Z), Postgraduate Research Innovation Project of Jiangsu Province (No. KYCX17_0831), the Project of Ministry of Housing and Urban-Rural Development of the People's Republic of China (No. 2014K4023), Project of the Housing and Urban-Rural Development Bureau of Jiang-su Province (No. JS2012ZD34), practical and innovation training project of Nanjing Forestry University (2017NFUSPITP180, 2017NFUSPITP182, 2017NFUSPITP183), and a Project Funded by the Priority Academic Program Development of Jiangsu Higher Education Institutions.

References

1. Li Y, Xu B, Zhang Q, et al. Present situation and the countermeasure analysis of bamboo timber processing industry in China. Journal of Forestry Engineering, 2016, 1(1):2-7.
2. Huang M, Zhang X, Yu W, et al. Mechanical properties and structure characterization of bamboo softened by high temperature steam. Journal of Forestry Engineering, 2016, 1(4):64-68.
3. Chen G, Zhou T, Li C, Zhang Q, Li H. Experimental study on the OSB webbed bamboo beams. Journal of Nanjing Forestry University (Natural Science Edition), 2016, 40(05): 121-125.
4. Yu Z, Wu Z, Su T. Microstructure analysis of the coating permeability of laminated bamboo panel for furniture[J]. Journal of Forestry Engineering, 2016, 1(2):130-134.

*Corresponding author: Email: [a]*310325816@qq.com, [b]792232771@qq.com, [c]1687699888@qq.com, [d]liu.wenfeng@free-sun.com.cn [e]767057291@qq.com,

5. Zhang W, Jiang W, Tang R. Study on short-term bending creep behavior and microstructure of bamboo scrimber[J]. Journal of Forestry Engineering, 2017, 2(3):33-37.
6. Li H, Su J, Wei D, Zhang Q, Chen G. Comparison study on parallel bamboo strand lumber under axial compression for different directions based on the large scale[J]. Journal of Zhengzhou University (Engineering Science), 2016, 37(2): 67-72.
7. Su J, Fan Wu, Li H, P Yang. Experimental research on parallel bamboo strand lumber column under axial compression[J]. China Sciencepaper, 2015, 10(1): 39-41.
8. Li H, Su J, AJ Deeks, Zhang Q, Wei D, Yuan C. Eccentric compression performance of parallel bamboo strand lumber column[J]. BioResources, 2015, 10(4): 7065-7080
9. Li H, Wei D, Su J, Yuan C, Chen G. Experimental study on PSBL under eccentric compression [J]. Journal of Building Materials, 2016, 19(3): 561-565.
10. Zhang S, Li C, Xiao Z, et al. Bending strength and loading simulation analysis of bamboo scrimber I-shaped beam[J]. Journal of Forestry Engineering, 2017, 2(1):125-129.
11. Li H, Zhang Q, Wu G, Xiong X, Li Y. Review on laminated bamboo lumber[J]. Journal of Forestry Engineering, 2016, 1(6): 110-116.
12. Correal J. F., Ramire L.F.. Adhesive bond performance in glue line shear and bending for glued laminated guadua bamboo[J]. Journal of Tropical Forest Science, 2010, 22(4): 433-439.
13. Verma C.S., Chariar V.M.. Stiffness and strength analysis of four layered laminate bamboo composite at macroscopic scale[J]. Composites part B-Engineering, 2013, 45(1): 369-376.
14. Verma C.S., Chariar V.M.. Development of layered laminate bamboo composite and their mechanical properties[J]. Composites part B-Engineering, 2012, 43(3): 1063-1069.
15. Lee AWC, Bai XS, Bangi AP. Selected properties of laboratory-made laminated-bamboo lumber[J]. Holzforschung, 1998, 52(2): 207-210.
16. Yeh M, Lin Y. Finger joint performance of structural laminated bamboo member[J]. Journal of Wood Science, 2012, 58(2): 120-127.
17. Varela S., Correal J., Yamin L., Ramirez F.. Cyclic Performance of Glued Laminated Guadua Bamboo-Sheathed Shear Walls[J]. Journal of Structural Engineering, 2013, 139(11): 2028-2037.
18. Taheri F., Nagaraj M., Khosravi P.. Buckling response of glue-laminated columns reinforced with fiber-reinforced plastic sheets[J]. Composite Structures, 2000, 50: 207–216.
19. Li H, Su J, Zhang Q, Chen G. Experimental study on mechanical performance of side pressure laminated bamboo beam[J]. Journal of Building Structures, 2015, 36(3): 121-126.
20. Li H, Deeks AJ, Zhang Q, Wu G. Flexural performance of laminated bamboo lumber Beam[J]. BioResources, 2016, 11(1): 929-943
21. Li H, Wu G, Zhang Q, Deeks AJ, Su J. Ultimate bending capacity evaluation of laminated bamboo lumber beams [J]. *Construction and Building Materials*, 2018, 160: 365-375
22. Li H, Zhang Q, Huang D, Deeks AJ. Compressive performance of laminated bamboo[J]. Composites Part B: Engineering, 2013, 54(1):319-328
23. Li H, Zhang Q, Wu G. Stress-strain model under compression for side pressure laminated bamboo[J]. Journal of Southeast University (Natural Science), 2015, 45(6): 1130-1134.
24. Li H, Su J, Zhang Q, Deeks AJ, Hui D. Mechanical performance of laminated bamboo column under axial compression[J]. Composites Part B: Engineering, 2015, 79: 374-382.
25. Su J, Li H, Yang P, Zhang Q, Chen G. Mechanical Performance Study on laminated bamboo lumber column pier under axial compression[J]. China Forestry Science and Technology, 2015, 29(5): 89-93.
26. Li H, Wu G, Zhang Q, Su J. Mechanical evaluation for laminated bamboo lumber along two eccentric compression directions[J]. Journal of wood science, 2016, 62(6):503-517
27. Li H, Chen G, Zhang Q, Ashraf M, Xu B, Li Y. Mechanical properties of laminated bamboo lumber column under radial eccentric compression[J]. Construction and Building Materials, 2016, 121: 644-652
28. Li H, Wu G, Zhang Q, Chen G. Experimental study on side pressure LBL under tangential eccentric compression[J]. Journal of Hunan University (Natural Science) , 2016, 43(5): 90-96.

Quality prediction modeling for sintered ores based on mechanism models of sintering and extreme learning machine based error compensation

Wu Tiebin, Liu Yunlian, Li Xinjun, Yu Yi, and Zhang Bin

Hunan University of Humanities, Science and Technology, Loudi, Hunan 41700, China

Abstract. Aiming at the difficulty in quality prediction of sintered ores, a hybrid prediction model is established based on mechanism models of sintering and time-weighted error compensation on the basis of the extreme learning machine (ELM). At first, mechanism models of drum index, total iron, and alkalinity are constructed according to the chemical reaction mechanism and conservation of matter in the sintering process. As the process is simplified in the mechanism models, these models are not able to describe high nonlinearity. Therefore, errors are inevitable. For this reason, the time-weighted ELM based error compensation model is established. Simulation results verify that the hybrid model has a high accuracy and can meet the requirement for industrial applications.

1. Introduction

Iron ores are raw materials in steel industry. After large sale mining for a long term, the reserves of natural rich ores have declined rapidly. In such context, poor ores have to be used to cope with the fact that the iron ore resources are increasingly scarce. However, poor ores cannot be directly put in blast furnaces, which otherwise would cause abnormal production under severe circumstances. Therefore, poor ores need to be agglomerated before being put into blast furnaces. In China, the agglomeration of more than 90% of poor ores is realized using sintering process **错误!未找到引用源。**. The sintering of iron ores refers to a process of roasting several types of iron ore powders, flux, and coke at high temperature to sinter and agglomerate the iron ore powders on incomplete melting conditions. The quality of sintered ores directly influences the yield and efficiency of blast furnaces, so accurately predicting technical indices including iron grade, alkalinity, and drum index of sintered ores is the precondition to effectively improve the quality of sintered ores. However, as sintering is a multi-variable, highly-nonlinear, and strongly-coupled complex process with large delay and coexistence of multiple reactions [2], accurately predicting the quality of sintered ores is always a challenge. According to the recondensation mechanism of water and the formation mechanism of over-wet zones, F. Patission established the mathematical model for the water migration in the sintering of iron ores. The model describes the relationship between the drum index and the liquid phases [3]. In previous research [4], the mechanism model for the sintering process was built, and then the programming-genetic algorithm was applied to identify the parameters therein. Finally, simulation proves that the model is effective. Literature [5] concretely elaborated the construction of sub-mechanism models including the combustion of coke, drying and condensation of fritted hearth, oxidation of iron ores, melting and solidification of minerals, and heat and mass transfer of fritted hearth. In addition, the models were verified to be reasonable based on the experimental data of sintering pots. As the pure mechanism modeling has to simplify the sintering process and make assumptions, it is inevitable that the mechanism models show certain errors with the actual condition. For this reason, mechanism models can hardly describe the coupling and high nonlinearity of the process. Aiming at this, the errors in the prediction of the mechanism models are adopted as the modeling objective to build the error compensation model based on time-weighted extreme learning machine (ELM). Then the errors predicted using the data models are compensated to the output of the mechanism models. The hybrid modeling ensures that the prediction model has strong generalization ability and accuracy.

2. Description of sintering process

In the sintering process, different iron ores and coke are blended at first and then mixed with some screened return fines, followed by burden distribution. Afterwards, the mixture is sintered at appropriate ignition temperature and moment, and the sintering duration is determined by the movement speed of pallets. The sintered ores are crushed, screened, and cooled to produce the finished products of sintered ores. The

* Corresponding author: *:Liu Yunlian, E-mail:liuyunlian85@163.com*

screened fine ores are mixed with the raw materials and sintered again.

There are many factors influencing the quality parameters (such as iron grade, alkalinity, and drum index) of sintered ores, including mixing ratio of raw materials, executing ratio of powdered coke, ignition temperature, bed depth, and movement speed of pallets. The mixing ratio of raw materials mainly involves FeO content, TFe (total iron) content, water content, and ratio of limestone.

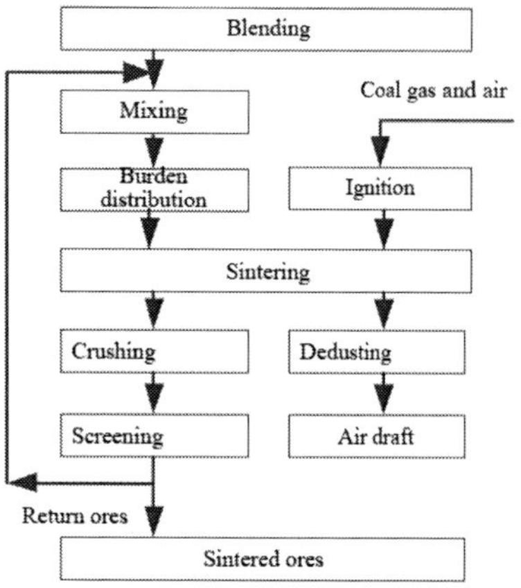

Figure 1 Flowchart of sintering process

3. Mechanism modeling of the sintering process

The quality of sintered ores greatly influences the yield and efficiency of blast furnaces, so accurately predicting technical indices of sintered ores including iron grade, alkalinity, and drum index is the premise of effectively improving the quality of sintered ores.

3.1 Mechanism modeling of drum index of sintered ores

Factors most greatly influencing the drum index include coke content, mass ratio of CaO to SiO_2, and percentage content of FeO in the mixture as well as movement speed of pallets. On this basis, the following formula is established:

$$Y_{drum} = \beta_1 + \beta_2 * x_c + \beta_3 * x_{CaSi} + \beta_4 * x_{FeO} + \beta_5 * x_v \tag{1}$$

Where Y_{drum} represents the drum index; x_c, x_{CaSi}, and x_{FeO} denote the percentage content of coke ($2\% < x_c < 7\%$), mass ratio of CaO to SiO_2 ($1.5 < x_{CaSi} < 3$), and the percentage content of FeO ($2\% < x_{FeO} < 7\%$), in the raw materials, respectively; and x_v refers to the movement speed of pallets.

3.2 Modeling of TFe(Total Ferrum) content in sintered ores

According to the conservation principle of matter, suppose that 100 kg of mixture is prepared and the percentage content of each type of iron concentrates is $x_i (i = 1, 2, \ldots, N)$, then the mass fraction of total iron in the mixture is expressed as Formula (2).

$$TFe_{mix} = \frac{\sum_{i=1}^{N}(100 * x_i * TFe_i)}{100} \tag{2}$$

Where TFe_N represents the iron grade of the mixture.

Production practice shows that the total iron in the mixture is basically consistent with that in the sintered ores. While, due to the consumption of all types of refined iron-mineral powders and powdered coke, the iron grade of sintered ores is higher than that of the mixture. The mechanism model for the total iron grade of the sintered ores is shown as Formula (3).

$$Y_{TFe} = \frac{TFe_{mix} \times 100}{100 - \sum_{i=1}^{n}(100 \times x_i \times d_i)}(\lambda_0 + \lambda_1 * x_c + \lambda_2 * x_c^2) \tag{3}$$

Where d_i represents the burning loss rate of each concentrate, and λ_0, λ_1, and λ_2 are coefficients. It means that the TFe content of sintered ores is not only related to the mixing ratio of concentrates in the mixture and the TFe content of the mixture, but also associated with the coke in the mixture.

3.3 Modeling of alkalinity of sintered ores

Suppose that 100 kg of mixture is prepared, the CaO and SiO_2 in the mixture basically remain consistent with those in the sintered ores. Owing to the consumption of all types of refined iron-mineral powders and coke in the sintering process, the mechanism model of alkalinity of the sintered ores is shown as Formula (4).

$$Y_{Alkali} = \frac{m_{cao} \times (\alpha_0 + \alpha_1 * x_c + \alpha_2 * x_c^2)}{m_{sio_2} \times (100 - \sum_{i=1}^{n}(100 \times x_i \times d_i))} \tag{4}$$

Where m_{cao} and m_{sio_2} denote the masses of CaO and SiO_2 in the mixture, respectively; and α_0, α_1, and α_2 are coefficients.

4. ELM based error compensation model

As mechanism prediction models for quality of sintered ores cannot describe the strong coupling and high nonlinearity, the errors in prediction of the models are applied as the modeling objective to establish the error compensation model based on ELM. Owing to equipment parameters slowly change with time in the sintering process of iron ores and the longer the time is, the smaller the reference value of samples, the sample with longest time is assigned with the least weight.

The semi-partial correlation analysis is carried out at first to obtain factors substantially influencing the quality of sintered ores. On this basis, the error compensation models of the mechanism models are constructed. The input variables of the ELM based compensation models include coke content, percentage contents of CaO, SiO_2, FeO, and each type of concentrates in the mixture, ignition temperature, movement speed of pallets, percentage content of coke, and final temperature. The output variables include compensation amounts of drum index, iron content, and alkalinity. The output of the hybrid model based on the mechanism models and the ELM based error compensation models is the sum of the output of the mechanism models and the compensation amounts of the compensation models.

Given a dataset (x_i, y_i) containing N samples, $x_i = [x_{i1}, x_{i2}, \cdots x_{in}]^T \in R^n$ and $y_i = [y_{i1}, y_{i2}, \cdots, y_{im}]^T \in R^m$ ($i = 1, 2, \cdots, N$). Assume that the hidden layer contains L nodes and the activation function is $g(x)$, then the ELM model can be expressed as [6]

$$\sum_{i=1}^{L} \beta_i g(a_i x_j + b_i) = y_i, \; j = 1, 2, \cdots, N \tag{5}$$

Where a_i and β_i represent connection weights and b_i refers to the offset. Formula (5) can be rewritten as

$$H\beta = Y \tag{6}$$

Where

$$H = \begin{bmatrix} g(a_1 \cdot x_1 + b_1) & \cdots & g(a_L \cdot x_1 + b_L) \\ \vdots & \vdots & \vdots \\ g(a_1 \cdot x_N + b_1) & \cdots & g(a_L \cdot x_N + b_L) \end{bmatrix}_{N \times L}$$

$$\beta = \begin{bmatrix} \beta_1^T \\ \vdots \\ \beta_L^T \end{bmatrix}_{L \times m}, \quad Y = \begin{bmatrix} y_1^T \\ \vdots \\ y_L^T \end{bmatrix}_{N \times m}$$

Where H represents the output matrix of the hidden layer of the neural network and the output weight can be calculated by solving Formula (7).

$$\hat{\beta} = H^{+} Y \tag{7}$$

Where H^{+} is the Moore-Penrose generalized inverse matrix of the output matrix H of hidden neurons.

In the training process of ELM, the samples trained for a longer time are assigned with smaller weights.

5. Simulation verification

The simulation results of the drum index, TFe content, and alkalinity based on the mechanism models and the hybrid model (based on the mechanism models and the ELM based compensation models) are shown in Figures 2, 3, and 4, respectively. In the figures, AV represents actual values, while MM and HM denote the simulation results based on the mechanism models and the hybrid model. Obviously, the accuracy of prediction results of the hybrid model is much higher than that when the mechanism models are used alone.

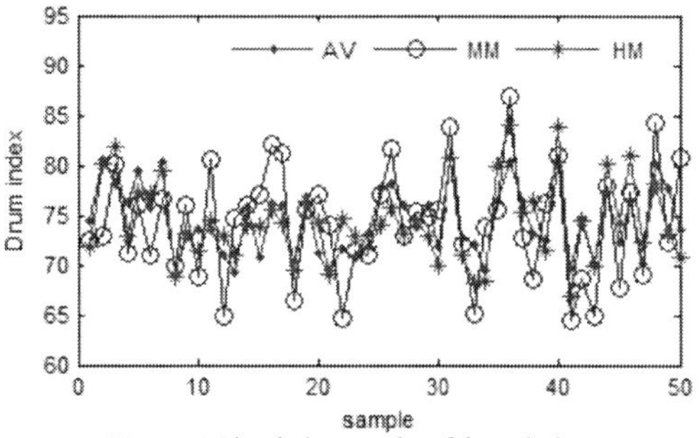

Figure 2 Simulation results of drum index

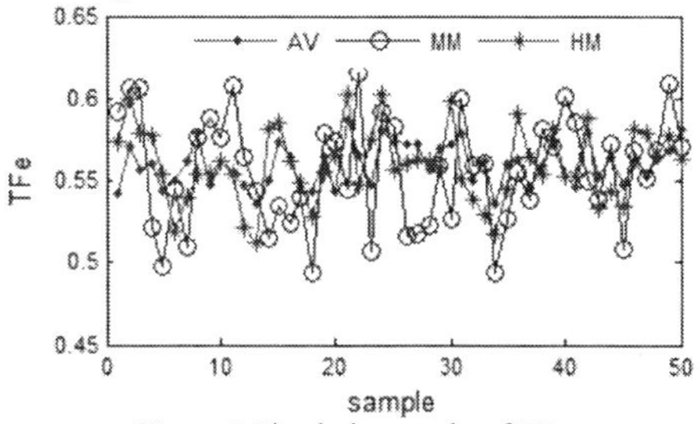

Figure 3 Simulation results of TFe

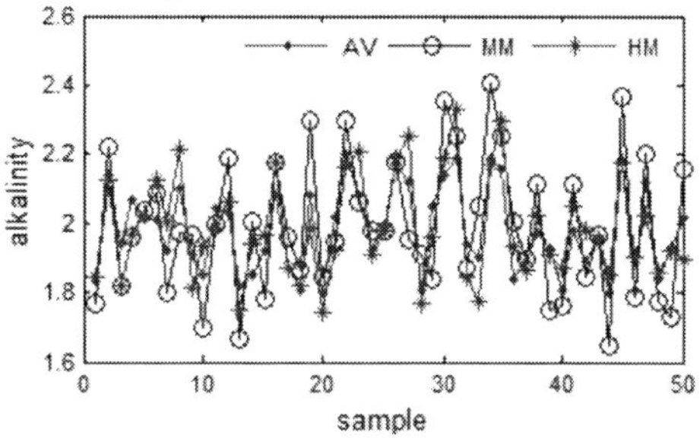

Figure 4 Simulation results of alkalinity

6. Conclusions

The research proposes the quality prediction model for sintered ores based on mechanism models and ELM based time-weighted error compensation. The simulation results verify that the hybrid model has much higher accuracy than that when mechanism models are used alone.

Acknowledgments

This work was supported by the Natural Science Foundation of Hunan Province (2016JJ3079), and supported by the Loudi Science and Technology Project(2017), and supported by the youth project of the Hunan Institute of Humanities and technology (2016QN18)

References

1. Liu Chengsong, Li Jing, Gao Ya Wei, Tang Haiyan. Optimization of ore matching for iron ore fines [J]. steel, 2013,48 (10): 6-11.
2. P. Lacroix, G. Dauwels, P.Dufresne, R. Godijn. High blast furnaces productivity operations with low coke rates in the European Union. LaRevue de Métallurgie-CIT, 2001, 3(1): 259-268
3. F.Patission.Study of Moister Transfer during the Strand Siniering Proeess. Metal Transacrions, 1990, 2(21B):37-47
4. Shang Xiu Qin. A new evolutionary computation method and its application in modeling and optimization of ironmaking and sintering process [D]. Hangzhou: Zhejiang University, 2010
5. numerical simulation and experimental verification of the sintering process of Zhao gape iron ore [D]. Hangzhou: Zhejiang University, 2012
6. Huang G B，Zhu Q Y，Siew C K.Extreme learning machine ： theory and applications[J]. Neurocomputing，2006，70：489-501.

Water-Sediment Partition of Polycyclic Aromatic Hydrocarbons (PAHs) in Nansi Lake

Guizhai Zhang, Youjiang Diao*

School of Environment Science and Engineering, Shandong Agriculture and Engineering University, Jinan, Shandong, 251100, P. R. China

Abstract: Based on field data of polycyclic aromatic hydrocarbons (PAHs) in water and sediment in Nansi Lake. The concentrations and the partitioning characteristic of PAHs in the water and sediment were studied. The lgKd of high molecular weight PAHs were higher than the low molecular weight PAHs. The most of PAHs Kd values were negligible correlated with TOC, soluble salt, clay and pH of the sediment in Nansi Lake.

1 Introduction

Polycyclic aromatic hydrocarbons (PAHs), containing two or more benzene rings, are widely distributed in the environment, for heir hydrophobicity, strong persistence and long-range transportation potential [1]. PAHs usually exist in the form of free, DOM binding, SPM binding and sediment combination in the aquatic environment. Investigation of the distribution of PAHs in an aquatic environment is great significance for strategy designing of pollution control. Partitioning in the two phases plays an important role in their behavior in the environment. The distribution of PAHs was mainly influenced by the characters of sediment (organic carbon content, particle composition) and environmental conditions (pH value, soluble salt content and so on), besides the physicochemical characters of PAHs and environmental conditions (temperature, precipitation) [2]. It is necessary to study the partitioning of PAHs in different aquatic environments, for the different partitioning behaviors of PAHs in different study areas.

Nansi Lake is the largest lake in Shandong province, the northern area of China, including Nanyang Lake, Dushan Lake, Zhaoyang Lake and Weishan Lake. It is an imporant buffer lake of the South-to-North Water Diversion Project (east route). Its water quality relates to the whole water quality conditions of South-to-North Water Diversion Project (east route) [3]. Some studies [1, 4, 5] assessed the PAHs residues in water, sediment and aquatic organisms of Nansi Lake. However, there is little information on water-sediment Partition of Polycyclic Aromatic Hydrocarbons (PAHs) in Nasi lake.

The objectives of this research are to survey concentrations of PAHs in the water and sediment, to evaluate water-sediment partitioning coefficient (K_d) of PAHs, and to assess the relationship of characteristics in sediments with K_d of PAHs in the Nansi Lake.

2 Materials and methods

2.1. Site location and sample collection

The locations of the sampling sites are shown in the paper by Zhang [1]. Water and surface sediment collected in May 2011.

2.2 Extraction procedure and chemical analysis

The Chemical reagents, extraction procedure, chemical analysis and Quality assurance are found in the literature [1].

2.3 The analysis of physicochemical character in surface sediment

The methods total organic carbon (TOC), mechanical analyses, soluble salt contents and pH of sediments were following the method of Zhang [3]. Statistical analysis software Spss16.0 was used in this study.

3 Results and discussion

3.1 Concentrations of PAHs in the water and sediment

The concentrations of PAHs in water and sediment samples are shown in Table 1. The results of PAHs in water and sediment samples came from the report by zhang [1]. The concentrations of total PAHs ranged from 27.52 ng/L to 55.06 ng/L in water, with the average value of 34.38±10.34 ng/L. The low molecular weight (2–3 rings) PAHs were dominant PAHs compounds, accounting for approximately 83% of total PAHs in water.

The concentrations of total PAHs ranged from 80.3 ng/g dw to 639.47 ng/g dw in sediment, with the average value of 225.18±206 ng/g dw. The high molecular weight (4–6 rings) PAHs were major compounds in sediment compared in water, accounting for approximately 56% of total PAHs in sediment. Those results were consistent with previous studies [6, 7].

Table 1 The concentrations of PAHs in water and sediment samples

PAHs	Water (n=6)	sediment(n=6)
	mean±SD	mean±SD
Acy	1.19±0.46	1.92±0.7
Ace	5.98±0.74	6.96±2.72
Fl	5.92±2.25	19.78±7.57
Ph	15.07±5.75	53.25±28.68
An	1.13±0.53	6.99±4.35
Flu	2.14±0.34	30.51±33.11
Pyr	2.29±0.36	27.74±25.7
Chr	1.03±0.54	13.22±16.75
BbF	0.25±0.08	15.75±17.95
BkF	0.077±0.02	nd
BaA	0.21±0.17	12.41±17.46
BaP	0.052±0.02	12.23±18.77
InP	0.038±0.01	12.33±16.05
DBA	0.025±0.01	2.82±3.88
BghiP	0.045±0.02	14.62±16.68
LMW (2-3 rings PAHs)	30.44±9.68	114.06±74.58
HMW (4-6 rings PAHs)	3.94±0.97	111.12±132.83
∑PAHs	34.38±10.34	225.18±206

nd not detected

3.2 Water-Sediment partitioning of PAHs in the Nansi Lake

The partitioning coefficient (K_d) of PAHs between sediment and water was calculated by the following formula: $K_d=C_s/C_w$ [8]. where C_s and C_w represent PAHs concentration in sediment (ng/g dw) and water (ng/L) samples, respectively. The lgK_d (L/kg) values at sediment-water interface from Nansi Lake are shown in table 2.

In this study, The Flu had higher lgK_d value among the low molecular weight (2–3 rings) PAHs. The lgK_d average values of the high molecular weight (4–6 rings) PAHs were in the range of 0.97-2.12 except the not detected compounds. The lgK_d values of InP, BaA, BaP, BbF were higher than the other high molecular weight PAHs. The InP, BaA, BaP and BbF were carcinogenic compounds, their average values of lgK_d in this study were lower than those in report [9].

This result shows that the low molecular weight PAHs are main compounds in water and sediment, but, the lgK_d values of high molecular weight PAHs were higher than the low molecular weight PAHs. This is mainly because the water solubility of PAHs is lower and the lipids are stronger with the number of benzene rings increasing. The higher octanol-water distribution coefficient of PAHs is stronger adsorbed in sediment.

Table 2 Lg Kd (L/Kg) at sediment-water interface from Nansi Lake

PAHs	S1	S2	S3	S4	S5	S6
Acy	0.22	0.02	0.08	0.36	0.15	0.39
Ace	0.18	0.08	0.17	0.33	0.1	0
Fl	0.64	0.28	0.5	0.24	0.64	0.81
Ph	0.47	0.29	0.55	0.29	0.69	0.86
An	0.76	0.55	0.74	0.38	1.01	1.13
Flu	0.97	0.96	0	0.46	0.96	1.61
Pyr	1.05	1.04	0.78	0.61	0.92	1.45

Chr	0.97	0.6	1.14	0.78	0.77	1.53
BbF	1.59	1.51	1.63	1.35	1.46	2.35
BkF	nd	nd	nd	nd	nd	nd
BaA	1.51	1.1	1.83	1.82	1.7	2.18
BaP	1.92	1.8	2.24	1.75	1.9	3.1
InP	2.27	2.23	2.2	2.03	2.12	nd
DBA	nd	1.91	nd	nd	nd	2.55
BghiP	nd	2.4	2.57	2.03	1.98	nd
LMW	0.53	0.35	0.4	0.22	0.63	1.02
HMW	1.28	1.16	1.29	0.96	1.15	1.9
∑PAHs	0.7	0.54	0.65	0.37	0.75	1.34

nd not detected

Table 3 The characteristics of samples sites in the Nansi Lake.

Sampling sites	pH	TOC(%)	Soluble salt(g/kg)	Clay(%)	Silt (%)	Sand(%)
S1	9.05	4.29	3.06	4.2	58.0	37.8
S2	9.58	6.06	4.66	4.9	70.3	24.8
S3	8.22	2.61	1.80	4.3	67.7	28.0
S4	8.38	5.19	4.26	3.6	52.9	43.5
S5	7.67	1.20	1.40	3.7	55.9	40.6
S6	7.60	1.73	3.34	3.4	72.1	24.5

3.3 The characteristics of surface sediment in the Nansi Lake

The results of characteristics of surface sediment are showed in Table 3. The value of TOC, soluble salt and pH ranged from 1.20 - 6.06%, 1.40- 4.66 g/kg and 7.60-9.58, respectively. Concentrations of clay was lower than silt and sand concentrations.

3.4 The relationship of characteristics in sediments with Kd of PAHs

Table 4 shows that the K_d value of InP is positively correlated with TOC; the K_d value of Ace is positively correlated with TOC, soluble salt and pH; the K_d value of BghiP is positively correlated with soluble salt and Clay. Previous reports found that organic carbon, soluble salt, Clay, pH in sediment could affect the partitioning of PAHs between sediment and water. In the present study, the analysis results indicate that TOC, soluble salt, clay and pH of the sediment was negligible correlated with the K_d, which was consistent with Zhang [10]. Further research is needed.

Table 4 Correlation of characteristics in sediments with Kd of PAHs in Nansi Lake

	pH	TOC	soluble salt	Clay
Acy	-0.512	0.231	-0.296	-0.861
Ace	0.199	0.824	0.524	-0.055
Fl	-0.594	-0.621	-0.89	-0.476
Ph	-0.802	-0.781	-0.922	-0.593
An	-0.661	-0.833	-0.929	-0.439
Flu	-0.15	-0.369	-0.471	-0.305
Pyr	-0.155	-0.52	-0.482	-0.135
Chr	-0.591	-0.339	-0.53	-0.519
BbF	-0.443	-0.473	-0.498	-0.389
BaA	-0.881	-0.247	-0.588	-0.866
BaP	-0.577	-0.513	-0.541	-0.469
InP	0.562	0.399	0.487	0.603
DBA	0.031	-0.36	-0.036	0.035
BghiP	0.16	-0.011	0.453	0.419

4 Conclusions

This study investigates the concentrations and the partitioning characteristic of PAHs in the water and sediment; evaluates the relationship of sediment characteristics with K_d of PAHs in the water and sediment of Nansi Lake. The results indicate that only K_d values of InP and Ace are positively correlated with TOC, the K_d of most PAHs was negligible correlated with TOC, soluble salt, clay and pH of the sediment in Nansi Lake.

Acknowledgments

This research was supported by the Science and Technology Development Project of Shandong Province (No. 2011YD10011) and the Science and Technology Project of Shandong Province (No. J17KB070).

References

1. Zhang GZ, Pan ZK, Wang XM, Mo XJ, Li XM *Distribution and Accumulation of Polycyclic Aromatic Hydrocarbons (PAHs) in Food Web of Nansi Lake, China*. Environ Monit Assess, 187:173(2015)
2. Liu J, Cui ZJ. Song XJ, Chen SF, Xu N, Lu F *Influences of soil characteristics on the PCBs distribution in soil of the modern Yellow River delta*. Ecology and Environmental Sciences, **19**(1):86-91(2010)
3. Zhang GZ, Pan ZK, Li J, Li XM *The Correlation of Sediments Characteristics with Organochlorine Pesticides (OCPs) in Surface Sediments of Nansi Lake, China*. Advanced Materials Research, **726-731**:926-930 (2013)
4. Li HL, Gao H, Zhu C, Li GG, Yang F, Gong ZY, Lian J *Spatial and temporal distribution of polycyclic aromatic hydrocarbons (PAHs) in sediments of the NansiLake, China*. Environmental Monitoring and Assessment, **154**:469–478(2009).
5. Liu Z, Zhang YS, Su ZZ, Zhang Y, Li TH, Yang X J *Distributions, sources and trends of polycyclic aromatic hydrocarbons in surface sediment in Nansihu lake. China*. Population Resource and Environment, **20**(6):136–140(2010)
6. Arias AH, Spetter CV, Freije RH, Marcovecchio JE *Polycyclic aromatic hydrocarbons in water, mussels (Brachidontes sp., Tagelus sp.) and fish (Odontesthes sp.) from Bahı 'a Blanca Estuary, Argentina*. Estuarine Coastal and Shelf Science, **85**, 67–81(2009)
7. Jiao WT, Wang TY, Khim JS, Luo W, Hu WY, Naile JE, Giesy JP, Lu YL *PAHs in surface sediments from coastal and estuarine areas of the northern Bohai and Yellow Seas, China*. Environmental Geochemistry and Health, **34**:445–456(2012)
8. Cao YX, Cao XZ, Wang H, Wan Y, Wang SL *Assessment on the distribution and partitioning of perfluorinated compounds in the water and sediment of Nansi Lake, China*. Environ Monit Assess, **187**: 611(2015)
9. Lan JC, Sun YC, Xiao SZ *Water-Sediment Partition of Polycyclic Aromatic Hydrocarbons in Karst Underground River*. Environmental Science, **36**(11):4081-4087(2015).
10. Zhang J, Cai LZ, Yuan DX *Distribution and sources of polynuclear aromatic hydrocarbons in Mangrove surficial sediments of Deep Bay, China*. Marine Pollution Bulletin, **49**(5-6): 479-486(2004)

Research on The Removal of Scale Ions from Circulating Cooling Wastewater by Chemical Coagulation Process

Song Du[1,2,*], Wenbiao Jin[1], FengDuan[2]

[1]Harbin Institute of Technology, Shenzhen Graduated School, Shenzhen518055, China.
[2]China Coal Research Institute, Beijing, China, 100013

Abstract: In this paper, the circulating cooling wastewater was treated by chemical coagulation process through adding $NaOH/Na_2CO_3$.The effect of NaOH and Na_2CO_3 dose on removal of scale ions, such as Ca^{2+}、Mg^{2+}、Ba^{2+}、Sr^{2+}、SiO_2, was studied and the removal mechanism was discussed. The results showed that the increase of NaOH dose was beneficial to the removal of above-mentioned scale ions. When NaOH was only added, the removal efficiency of Ca^{2+}、Mg^{2+}、Ba^{2+}、Sr^{2+}、SiO2 was 86.3%, 91.6%, 86.5%, 58.1%, 84.2%, respectively. When 680 mg/L of NaOH and 300 mg/L of Na2CO3 were added, and the effluent pH was above 11.2, the removal efficiency of Ca^{2+}、Mg^{2+} was 95.8% and 89.4%, respectively, and the concentration of Ca2+and Mg2+ was below 20 mg/L, which met the target of wastewater treatment. Finally the possible removal mechanism of Ca^{2+}, Mg^{2+}, Ba^{2+}, $Sr^{2+}$and SiO_2 was discussed.

1 Introduction

A power plant was designedto apply technology of reverse osmosis membrance to treat its circulating cooling water. And the treated water would be completely reused instead of being discharged. However the high hardness of water impacted the process of reverse osmosis, which had to be soft by a pre-treatment process. Conventional softening methods include chemical coagulation and ion exchange [1-3]. And as the most widely used method, chemical coagulation include several subentry, such as lime softening method, lime-sodium carbonate softening method, sodium hydroxide-sodium carbonate softening method[4-5]. Enterprise prefer sodium hydroxide soften method to lime softening, because it show less material-consumption, lower secondary-pollution and more simply operated [6].

The current literature about chemical coagulation soften methods mainly focus on removing conventional scaling ions such as Ca^{2+} and Mg^{2+}, and research on removal of unconventional scaling ions, such as Ba^{2+}, Sr^{2+} and $SiO_2$was not many[7-9]. Though applying "NaOH+Na2CO3" soften method to treat circulating cooling waterof the power plant, this paper studied the removal efficacy of scaling ions, such as Ca^{2+}, Mg^{2+}, Ba^{2+}, Sr^{2+} and SiO_2. The related reaction mechanism was deduced.

2 Material and methods

2.1 Quality of experimental water

Table 1 Circulating cooling water quality of the power plant unit: mg/L (except pH)

Item	Value
pH	8.5
Ca^+	254
Mg^{2+}	140
Ba^{2+}	0.267
Sr^{2+}	2.65
SiO_2	37.7
ALK	324
CODcr	59.5
NH_3-N	0.69

2.2 Experimental material and instrument

Experimental material: Sodium hydroxide, sodium carbonate, polymeric ferric sulfate (PFS), anionic polyacrylamide (PAM, molecular weight: 12 million), and all were solid state. Before being used, these materials would be configured to solution, the mass fraction respectively are 30%, 5%, 1% and 0.1%.

Experimental instrument: quadruple electric mixer (JJ-4A, Jiangsu Jintankexi instrument co., LTD), pH meter (FE20, Mettler Toledo).

2.3 Experimental approach

The soften coagulation experiment was carried out in a beaker [10]. To add NaOH (or NaOH and Na_2CO_3, 30% purity) into 500ml water to regulate pH with 200r/min stirring for 5 minutes. Then add the polymeric ferric

* Corresponding author: du@cct.org.cn

sulfate (PFS) with 200r/min stirring for 2 minutes. Next step is adding 1mg/L anionic polyacrylamide (PAM, molecular weight: 12 million) with 50r/min stirring for 20 minutes. After 30 minutes standing, the water was made into simple and analyzed. This paper treated concentration of Ca^{2+}, Mg^{2+}, Ba^{2+}, Sr^{2+} and SiO_2 as evaluation index to analyse the experimental result.

Fig.1 Experimental process

2.4 The analytical approach of water quality

Table 2 Analytical approach of water quality

Monitoring program	Monitoring Approach
Ca^{2+}, Mg^{2+}	Water quality-Determination of calcium and magnesium-Atomic absorption spectrophotometric method (GB 11905-89)
Ba^{2+}	Water quality-Determination of barium-graphite furnace atomic absorption spectrophotometry (HJ 602-2011)
Sr^{2+}	Water quality-Determination of 32 elements-Inductive coupled plasma optical emission spectrometry (HJ 776-2015)
SiO_2	Determination of silica (dissolved) (Hetropoly blue method)(SL 91.2-1994)
pH	Water quality-Determination of pH value – Glass electrode method

3 Results and discussion

3.1 The Influence of Dosing Quantity of NaOH on Removing Ca^{2+} and Mg^{2+}

Only adding NaOH without Na_2CO_3, the dosage of PFS was kept as 100mg/L with the gradual increase in NaOH ranging from 400 to 720 mg/L. The pH ranged from 10.5-11.3m, and the change about concentration of Ca2+ and Mg^{2+} was shown in Table 3, Figure 2and Figure 3.Table 3 demonstrated that the concentration of Ca2+trend down with increasing of pH, the concentration of $Ca^{2+}$of raw water was 254mg/L, when pH raise from 10.5 to 11.3, the concentration of Ca^{2+} declined from 56.2mg/L to 39.9mg/L, and the removal efficiency went up from 77.9% to 86%. Similarly, Table 2 shows that the concentration of Mg^{2+} gradually decreased with increasing of pH, the concentration of $Mg^{2+}$of raw water was 140mg/L, when pH raise from 10.5 to 11.3, the concentration of Mg^{2+} declined from 110mg/L to 11.7mg/L, and the removal efficiency went up from 21.4% to 91.6%. When pH was greater than 7.5, the removal efficiency of Ca^{2+} was greater than 75%. However, the removal efficiency of Mg^{2+} was able to exceed 75 when pH razing to 11.1. It states that comparing with Mg^{2+}, Ca^{2+} was able to be removed in lower alkalinity condition. The pH should be greater than 11 when Mg^{2+} forming to $Mg(OH)_2$[11].

Table 3 Change of Concentration of Ca^{2+}, Mg^{2+} with influence of dosage of NaOH

NaOH dosage(mg/L)	pH	Ca^{+} (mg/L)	Mg^{2+} (mg/L)	Removal efficiency of Ca^{2+}(%)	Removal efficiency of Mg^{2+}(%)
0(raw water)	8.5	254	140	-	-
400	10.5	56.2	110	77.9	21.4
480	10.6	59.2	95.2	76.7	32
520	10.7	53.4	86.5	79.0	38.2
560	10.8	62	78.5	75.6	43.9
600	11.0	35.5	38.1	86.0	72.8
640	11.1	55.3	29.2	78.2	79.1
680	11.2	34.8	20	86.3	85.7
720	11.3	39.9	11.7	84.3	91.6

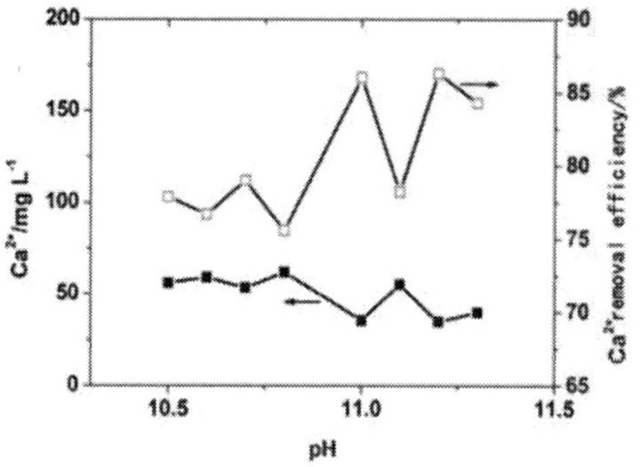

Fig. 2 Variations in Ca^{2+} removal at different pH

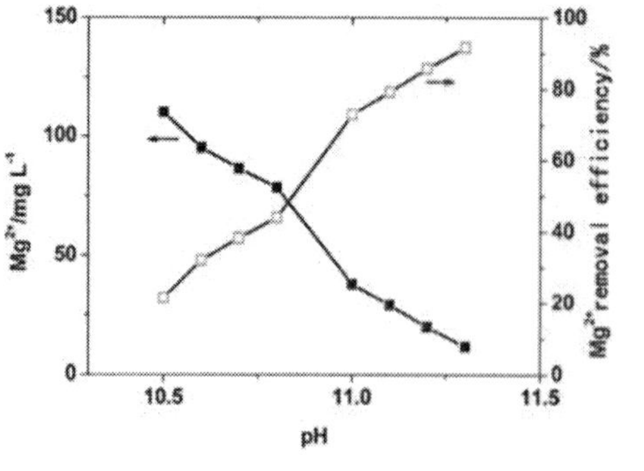

Fig. 3 Variations in Mg^{2+}removal at different pH

According to the literature of Wang [12], Ca^{2+} was mainly removed as CaCO3, and Mg^{2+} was mainly removed as $Mg(OH)_2$. It can be speculated that removing $Ca^{2+}$and Mg^{2+} according to following reactions:

$$Ca(HCO_3)_2+2NaOH \rightarrow CaCO_3\downarrow +Na_2CO_3+2H_2O \quad (1)$$

$$Mg(HCO_3)_2+4NaOH \rightarrow Mg(OH)_2\downarrow +2Na_2CO_3+2H2O(2)$$

$$MgCl_2+2NaOH \rightarrow Mg(OH)_2\downarrow +2NaCl \quad (3)$$

$$MgSO_4+2NaOH \rightarrow Mg(OH)_2\downarrow +2Na_2SO4 \quad (4)$$

3.2 The Influence of dosing quantity of Na_2CO_3 on removing Ca^{2+} and Mg^{2+}

As mentioned earlier, adding NaOH could lead high removal efficiency of Ca^{2+} and Mg^{2+}. For higher removal efficiency and more reasonable dosage of NaOH, "NaOH+Na_2CO_3"dual-alkali soften method was applied in this research. The dosage of PFS was kept as 100mg/L, while NaOH was 680mg/L and pH was about 11.2.To regulate the dosage of $Na_2CO_3$and the change about concentration of Ca2+ and Mg2+ was shown in Table 3, Fig.2 and Fig.3. Fig.4 shows when the dosage of Na_2CO_3 raised from 50mg/L to 300mg/L, Ca^{2+} decreased from 49.6mg/L to 10.6mg/L, and removal efficiency increased from 80.5% to 95.8%. Though adding the dosage of Na_2CO_3, Ca^{2+} was no longer change. It is because that $CO_3^{2-}$and Ca^{2+} reacted to form $CaCO_3$, which leading to decrease of Ca^{2+}. When CaCO3 reaches solubility equilibrium, the concentration of Ca^{2+} reaches its own equilibrium. Fig.5 states that when dosage of Na_2CO_3increased, the removal efficiency of Mg^{2+} went up slightly. It is because that Mg^{2+} was mainly removed as $Mg(OH)_2$, the addition of CO_3^{2-} made few contribution to pH.

Table 4 Change of concentration of Ca^{2+}, Mg^{2+} with influence of dosage of Na_2CO_3 (pH=11.2)

NaOH dosage(mg/L)	Na_2CO_3 dosage(mg/L)	Ca^{2+} (mg/L)	Mg^{2+} (mg/L)	Removal efficiency of Ca^{2+}(%)	Removal efficiency of Mg^{2+}(%)
0(raw water)	-	254	140	-	-
680	50	49.6	22.9	80.5	83.6
680	100	36.8	22.1	85.5	84.2
680	150	32.4	21.2	87.2	84.9
680	200	30.4	21.2	88.0	84.9
680	300	10.6	15.0	95.8	89.3
680	400	11.9	16.5	95.3	88.2
680	500	9.1	17.4	96.4	87.6
680	600	13.0	15.8	94.9	88.7
680	800	11.9	15.2	95.3	89.1
680	1000	10.8	14.8	95.8	89.4

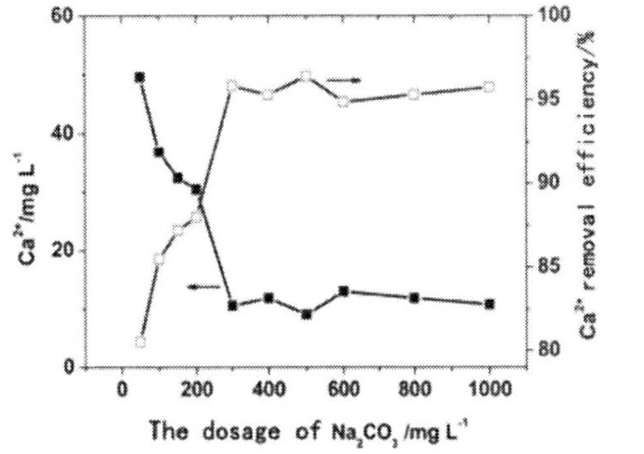

Fig. 4 Variations in Ca^{2+}removal at different Na_2CO_3

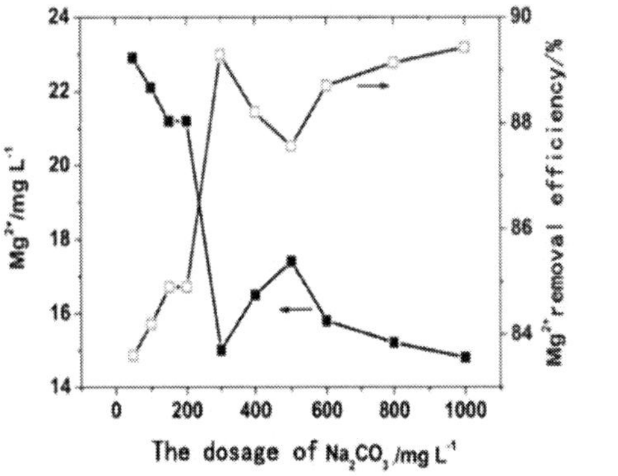

Fig. 5 Variations in Mg^{2+} removal at different Na_2CO_3

It can be speculated that "NaOH+Na2CO3" dual-alkali soften method not only involve in reaction (1) ~reaction (4), and there should be reaction (5):

$$CaCl2+Na2CO3 \rightarrow CaCO3\downarrow +2NaCl \quad (5)$$

$$CaSO4+Na2CO3 \rightarrow CaCO3\downarrow +Na2SO4 \quad (6)$$

Theoretical dosages of NaOH and Na_2CO_3 can be estimated by following reactions

$$NaOH=40\ (Hz+HMg+CO2+Fe+K+Ac) \quad (7)$$

$$Na2CO3=53\ (HCa-2\times Hz-CO2+\beta) \quad (8)$$

Hz—the hardness of carbonate in raw water, mmol/L;
HMg — the hardness of magnesium in raw water, mmol/L;
HCa—the hardness of calcium in raw water, mmol/L;
CO2 — the contentof carbon dioxide in raw water, nmmol/L;
Fe—the contentof iron in raw water, mmol/L;
K—the dosage of coagulant, 0.1-0.5mmol/L;
Ac—the excess alkalinity of NaOH, 0.2-0.4mmol/L;
β —the excess amount of Na2CO3, 1.0-1.4 mmol/L.

According to data of water quality, theoretical dosages of NaOH and Na_2CO_3 are 380mg/L and 46mg/L respectively. However, experimental dosages of NaOH and Na_2CO_3 are 680mg/L and 300mg/L respectively, greater than the theoretical value. The researcher Liu [14] agrees the point that the theoretical dosage is different from that of experiment. Therefore the theoretical value should only be intended as a reference, and the optimum dosage need to be ascertained by experiment.

Table 5 Change of concentration of Ba^{2+}, Sr^{2+}, SiO_2 with influence of dosage of NaOH

NaOH dosage(mg/L)	pH	Ba^{2+}(mg/L)	Sr^{2+}(mg/L)	SiO_2(mg/L)	Removal efficiency of Ba^{2+}(%)	Removal efficiency of Sr^{2+}(%)	Removal efficiency of SiO_2(%)
0(raw water)	8.5	0.267	2.65	37.7	-	-	-
400	10.5	0.09	2.09	27.6	66.3	21.1	26.8
480	10.6	-	2.27	17	-	14.3	54.9
520	10.7	0.052	1.43	5.74	80.5	46	84.8
560	10.8	0.036	1.11	5.96	86.5	58.1	84.2

3.3 The influence of dosing quantity of NaOH on removing Sr^{2+}, Mg^{2+} and Ba^{2+}

Only adding NaOH without Na_2CO_3, the dosage of PFS was kept as 100mg/L with the gradual increase in NaOH ranging from 400 to 560 mg/L. The pH ranged from 10.5-11.3m, and the change about concentrations of Ba^{2+}, Sr^{2+} and SiO_2 were shown in Table 5, Fig.6, Fig.7and Fig.8.The result demonstrated that the concentrations of Ba^{2+}, Sr^{2+} and SiO_2 were gradually decreased with increasing of pH, which consistent with the literature [6]. When pH raise from 10.5 to 11.3, the concentration of Ba2+ declined from 0.09mg/L to 0.036mg/L, and the removal efficiency went up from 66.3% to 86.5%; the concentration of Sr^{2+} decreased from 2.09mg/L to 1.11mg/L, and the removal efficiency increased from 21.1% to58.1%; the concentration of SiO_2descend from 27.6mg/L to 5.96mg/L, and the removal efficiency raised from 26.8% to 84.2%.

The increase of removal efficiency of Ba^{2+} and Sr^{2+} may be caused by rising pH. When pH kept rising, HCO3-continuously turned to CO_3^{2-}, which further made Ba^{2+} and Sr^{2+} translated to $BaCO_3$ and $SrCO_3$respectively (be removed as sediment). The rising pH also promoted the reaction that SiO_2 turned to SiO_3^{2-}, and then the SiO_3^{2-}reacting with Ca^{2+}(Mg^{2+}) to form $CaCO_3$ ($MgCO_3$) [15-16].

It can be speculated that using NaOH to remove Ba^{2+}, Sr^{2+} and SiO_2 should follow reactions below:

$$Ba(HCO_3)_2+2NaOH \rightarrow BaCO_3\downarrow+Na_2CO_3+2H_2O \quad (9)$$

$$Sr(HCO_3)_2+2NaOH \rightarrow SrCO_3\downarrow+Na_2CO_3+2H_2O \quad (10)$$

$$MgCl_2+SiO_2+2NaOH \rightarrow MgSiO_3\downarrow+2NaCl+H_2O \quad (11)$$

$$CaCl_2+SiO_2+2NaOH \rightarrow CaSiO_3\downarrow+2NaCl+H_2O \quad (12)$$

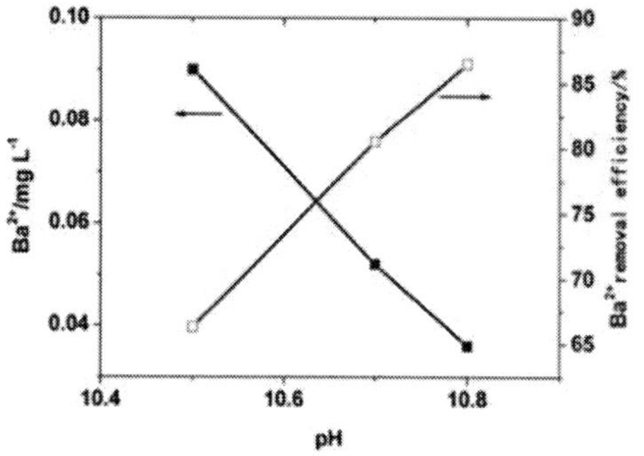

Fig.6 Variations in Ba^{2+} at different pH

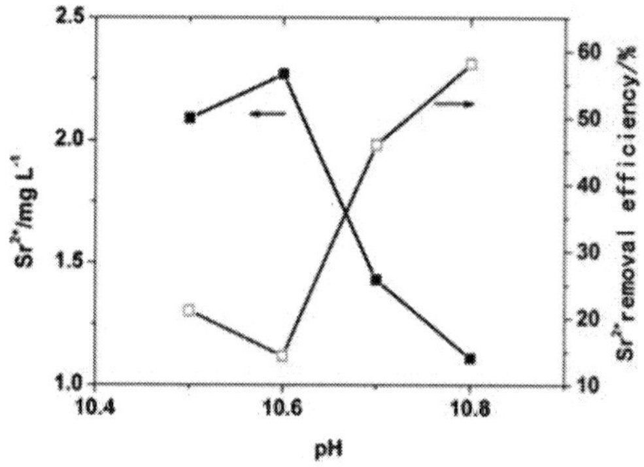

Fig. 7 Variations in Sr^{2+} at different pH

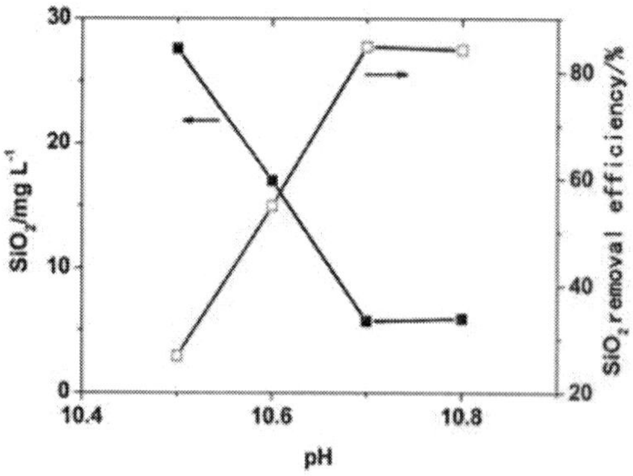

Fig. 8 Variations in SiO_2 at different pH

4 Conclusion

For preferable removal efficiency, the dosage of NaOH and Na_2CO_3 should be regulated to 680mg/L and 380mg/L respectively, and the pH should be no less than 11.2. The predicting outcomes are that Both Ca^{2+} and Mg^{2+} will be no less than 20mg/L, which meet the target of wastewater treatment.

This paper deducts the removal mechanism: scale ions are mainly removed by translated to sediment. Ca^{2+} is translated to $CaCO_3$; Mg^{2+} is turned to Mg(OH)2; Ba^{2+} and Sr^{2+} become $BaCO_3$ and $SrCO_3$ respectively; SiO_2 may be formed as $CaSiO_3$ and $MgSiO_3$.

References

1. S.U. Yan, X.U. Zhen, Z. Wang, J. Wang. Application of zero wastewater discharging system in a power plant with air-cooled units situated in northern china. Thermal Power Generation, 40(10), 74-77 (2011).
2. X.U. Yiming, H.E. Qingfeng, J. Huang , L.I. Pingli, Y. Chen , W.U. Haoyun, et al. Influence of pretreatment of reverse osmosis concentrate on performance of air gap membrane distillation process. Membrane Sci. Technol. 34(3): 86-91 (2014).
3. H. Zhang, L. Cui, B. Wu. Present situation of the research on water hardness removal. Ind. Water Treat. 31(12), 5-8 (2011).
4. J.T. Zhang, J. Dong. Comparative analysis of recycling technology of circulating sewage treatment of thermal power plant. Technol. Water Treat. 38(8), 124-127 (2012).
5. Z.Y. Xiao, Y. Chen, W.Q. Kang. Leakage test of rare earth tailings dam treated by lime softening - precipitation filtration - reverse osmosis. Technol. Water Treat. 38(9), 96-99 (2012).
6. R.Z. Hu, Y.L. Huang, S.Y. Yang. Comparison hardness removal of underground water resource by lime and sodium hydroxide. Water &Wastewater Eng. 42(7), 29-35 (2016)
7. L. Tian, Y. Qi, N. Yang. Research on the removal of scale ions from RO concentrated water by chemical coagulation process. Ind. Water Treat. 34(2), 43-46 (2014).
8. F.l. Gong, Y.Q. Wang, M.J. Shan, et al. To treat the hardness of reverse osmosis concentrate brine by lime softening method. Environ. Sci. & Technol., 37, 154-157 (2014).
9. L. L. Jing, W.U. Song, L. Wang, et al. Study on the Advanced Treatment of Groundwater with High Concentration of Hardness and Sulfate. J Chongqing Normal Univ. 31(6), 153-158 (2014).
10. F. Duan, W.G. Dong, L.F. Tian. Research on the removal of organic compounds from high salinity wastewater from coal chemical industry. J. Coal Processing & Comprehensive Utilization. (4), 21-24 (2016).
11. T.L. Ye, X.J. Wang, L. Reng. Advanced treatment and reuse of Printing RO concentrated water, Chinese J Environ. Engin. 9(2), 781-786 (2015).
12. Q.G. Wang, C. Le, P.F. Fu, W.L. Jiang, Y. Han, R.F. Zhuo. Study on the treatment of nanofiltration concentrated liquor of landfill leachate by the coagulation-coagulation-coagulating sedimentation-electrochemical oxidation process. Environ. Sci. & Technol. 27(3), 27-30 (2014).
13. East China architectural design institute co. LTD. Water supply and drainage design manual(Book 4: Industrial water treatment) 2nd edition. [M]. Beijing: China architecture &building press. (2002)
14. Yuhua L, Haiyang D. Study on the removal hardness of circulating cooling water of power plant. J. Refining & Chem. Ind. 23(12), 11-13 (2012).
15. Jing W, Zhengjiang W, Baohong Y. Experimental study on the pretreatment process of the terminal wastewater of coal-fired power plant. Thermal Power Generation. 39(10), 55-59 (2010).
16. R.F. Jin, D. Wang, J.T. Zhou, L.M. Teng, J. Su. Experimental study on the treatment of silicate wastewater by coagulation method. Ind. Water Treat. 23(1), 42-44 (2003).

Research progress on modification of activated carbon

Mengran Liu[1], and Cong Xiao[2*]

[1]Chengwu County Environmental Protection Bureau, He'ze, Shandong 274200, China
[2]Limited by Share Ltd Metallurgical Design Institute of Shandong Province, Jinan, Shandong 250101, China

Abstract. The paper overall the way and principle of how to modify activated carbon at home and abroad. Besides, the characteristics was given on physical, chemical modification respectively. The physical modification generally changed internal and external structure of the activated carbon, while chemical modification changed both its structure and surface properties. The advantages, disadvantages and the mechanism of the modifiers mentioned above are discussed, and their application are also prospected.

1 Introduction

Activated carbon is an amorphous carbon made from a special process. It has highly developed pore structure, huge specific surface area, varied surface chemical characteristics and high surface activity [1, 2]. It is reported that the annual consumption of activated carbon in the world is about 275000 t [3], which is used as an adsorbent for the removal of pollutants in the environment [4]. In addition, activated carbon can be used as catalyst carrier [5], electrode material [6], and can also be used for separation, concentration and decolorization [7]. According to the extensive study of activated charcoal by scholars at home and abroad, it is determined by two factors that activated carbon has strong adsorption properties: the surface structure characteristics of activated carbon (including specific surface area, pore structure and pore volume, etc.) and surface chemical properties (including surface acidity and basicity, functional groups and electrochemical properties) [1, 8-11]. Therefore, physical and chemical modification of activated carbon, making it have specific functional groups and changing pore structure to obtain multiple varieties, functional and high quality special activated carbon is urgent need to prepare.

2 Physical modification method

The principle of physical adsorption of activated carbon is that the pore diameter of activated carbon is much larger than or greater than or equal to the characteristics of adsorbate molecules, so that the molecules of adsorption can enter the pores of activated carbon surface, and then adsorb and remove the adsorbents. The common activated carbon has some disadvantages, such as high ash content, wide micropore distribution, small specific surface area and poor adsorption property [12]. In order to make active carbon have special properties and uses, it is usually modified to increase the specific surface area and pore size distribution.

2.1 Gas modification method

Activated carbon is modified by physical method. First, carbonization of raw materials is carried out to remove volatile matter, and then carbide is modified by gas. Pore structure is formed through opening, reaming and creating new holes. The typical modified gas has water vapor, CO_2 and so on. The main reactions occurred (1) ~ (2).

$$C+H_2O \rightarrow H_2+CO \quad (1)$$

$$C+CO_2 \rightarrow 2CO \quad (2)$$

Hakan Demira l[13] was used to modify the active carbon of olive residue at 700~900℃ with the flow rate of 0.08 mL/min. The results showed that the surface area increased from 523 m^2/g to 1106 m^2/g and the total pore volume increased from 0.2981cm^3/g to 0.6067 cm^3/g after the modification of activated carbon from olive residue. Su [14] used the water vapor and CO_2 modified 0.28~0.90 mm coconut shell activated carbon. After the reaction of 10~17 h, the activated carbon with a specific surface area of more than 2700 m^2/g was prepared.

2.2 Microwave modification method

The principle of microwave heating is to make use of the irregular movement of molecules inside the body to generate heat by friction, so as to rapidly and effectively heat the object, so that the surface functional groups of the activated carbon will react with the modifier adequately. The physical structure of activated carbon is adjusted by changing the microwave power and frequency, the time of radiation and the carrier gas. At the same time, the increase of the temperature of the system will further carbonization and change the pore structure [15]. Microwave heating has the advantages of high efficiency, low energy consumption, no pollution, energy saving and high efficiency, easy to control and so on.

*Corresponding author: xiaocong921101@163.com , sdliumengran@163.com

Jiang [16-18] investigated the effects of microwave power, activated carbon and carrier gas flow rate on the heating behavior of activated carbons in the microwave field, and the activated carbons with different particle sizes were modified. The analysis of finished products showed that the specific surface area of activated carbon changed little, the pore volume decreased slightly, the basic crystallites increased, the graphitization degree increased, the surface oxygen functional groups were decomposed in the form of COx, and the nitrogen content in pyrrole nitrogen form increased. Cao [19] made the pore channel of activated carbon more smooth by microwave modification of activated carbon, but at the same time it leads to the narrow channel and the contraction of carbon skeleton. Menendez [20] used microwave treated activated carbons showed that most of the oxygenated functional groups on the surface of activated carbons were removed while N_2 was protected, while pH values increased significantly.

2.3 Heating method

Using heating as a modification technology to modify activated carbon, changing the physical characteristics of activated carbon (specific surface area, pore volume and so on), improving its removal efficiency is also one of the research directions. Amina [21] treated the activated carbon at high temperature. After modification, the specific surface area of activated carbon increased (+ 7.2%, 400 ℃, + 6.6%, 600 ℃), and the total pore volume increased (+ 6.8%, 400 ℃, +8.6%, 600 ℃), but had no significant effect on the surface chemical properties of activated carbon. Rangel-Mendez [22] study of steam or steam and methane mixed gas under high temperature (1000 ℃) modification of activated carbon, the results showed that the activated carbon microporous and mesoporous volume increased 50% ~ 70%, 65% ~ 90%. The modified activated carbon can change the surface chemical properties of the activated carbon while changing the surface structure of the activated carbon. Heating treatment can destroy the unstable oxygen functional groups on the surface of activated carbon [21], and weaken the chelating ability of activated carbon and metal ions, thereby reducing the adsorption capacity of metal ions. However, high temperature is beneficial to the formation of alkaline groups on the surface of activated carbon, which is beneficial to the adsorption of organic matter in aqueous solution by activated carbon [8].

3 Chemical modification method

The chemical composition of activated carbon and the species and quantity of active functional groups on the surface have an important influence on the adsorption and catalysis. Therefore, it is of great significance to modify the activated carbon according to the different adsorbate. The main purpose of surface chemical modification is to change the surface acid and basicity of activated carbon, introduce or remove some surface functional groups, so that activated carbon has some special adsorption or catalytic properties.

3.1 Oxidation modification method

Oxidative modification is to adjust the appropriate temperature and react with the strong oxidants on the surface of activated carbon, so as to improve the content of oxygen-containing functional groups, such as carboxyl, phenolic hydroxyl, ester and carbonyl, and enhance the polarity and hydrophilicity of the surface [12]. The main oxidants commonly used are HCl, H_2SO_4, HNO_3, $HClO_3$, H_2O_2 and so on [23-26]. The types and quantities of oxygen functional groups on the surface were changed by the reaction of different oxidants with activated carbon. The higher the degree of oxidation, the more the content of oxygen functional groups. Gil [27] used HNO_3 and H_2O_2 to modify activated carbon, and the effect of temperature was discussed. The results showed that when HNO_3 was used as oxidant, the surface properties of activated carbon were greatly affected by temperature. When the temperature was increased to 363 K, the pore size increased gradually when the temperature was less than 333 K. But when the H_2O_2 is modified, the effect of temperature is much smaller. Liu [28] modified the mesoporous activated carbon by HNO_3, and then used it as a non-metallic catalyst to improve the conversion of propane. The results show that the oxygen containing functional groups on the surface of the modified activated carbon increase and can play an active role in the process of deoxidization.

3.2 Reductive modification method

The surface functional groups of activated carbon surface can be reduced and modified by reducing agents at proper temperature, which can improve the relative content of alkaline groups and enhance the non-polar surface, thereby improving the adsorption performance of activated carbon for non-polar substances. The methods of reducing modification are mainly concentrated on the high temperature treatment of activated carbon by inert gases such as H_2 and N_2. The surface reduction can also cause the change of the structure of the surface of the activated carbon.

Zhang [29] used NaOH to modify activated carbon. It was found that the surface oxygen containing functional groups decreased significantly, and the specific surface area and the concentration of pore volume increased with the increase of alkali concentration, which increased the adsorption capacity of methane. Gao [30] used H_2 modified activated carbon materials, change the pore properties of activated carbon materials after no significant change, but because of the surface oxygen groups, especially oxygen acid groups were significantly decreased, the adsorption capacity of the activated carbon for phenol increased nearly 2.5 times. Gu [31] used ammonia and aniline to modify the activated carbon, which partially eliminated the negative surface groups and enhanced the adsorption capacity of Au (III). Because when using concentrated ammonia and other polar aniline solution, the surface of the activated carbon acid reactive groups with ammonia and aniline, and then removed by heating, increasing the pore radius, for larger $AuCl^{4-}$ ions into the pore, also reduces the electrostatic negative electricity on the surface of the $AuCl^{4-}$ ion repulsion, thus improving the adsorption capacity of the activated carbon. Alessandro [32] used a NaOH impregnation ratio of 3: 1 to the reduction of Australian nut shell based activated carbon. The results show that there are a large number of basic functional groups on the surface of activated carbon after reduction and modification, and the specific surface area of BET is 1524 m^2/g. After 50 h reaction, the conversion rate of propane was increased from 22.4% to 86.6%.

3.3 Acid-base modified method

3.3.1 Acid modification

The acid modification technology of activated carbon is a typical wet oxidation technology. It means that using HNO_3, H_2O_2, HClO, H_2SO_4, CH_3COOH, citric acid and other oxidants on the activated carbon oxidation process under appropriate conditions, by changing the number and type of activated carbon surface oxygen functional groups, to improve the ability of removal of metal ions in aqueous solution. Meenakshi [33] studied the mechanism of Cu^{2+} adsorption on activated carbon. It was found that the unsaturated C=C on the surface of activated carbon was attacked by oxygen atom to form acidic structures such as carboxyl and lactone groups. In the aqueous solution, acid functional groups hydrolyze H^+, making the surface of the activated carbon with a negative contact site, which is combined with the positive metal ions in the solution. Badie [34] studied the method of pyrolysis modified by H_3PO_4 and the result showed that it can promote the formation of oxygen-containing functional groups on the surface of activated carbon, and the adsorption capacity for Pb^{2+} can reach 299 mg/g (about two times before modification).

3.3.2 Alkali modification

The alkali modification of activated carbon means using NaOH, ammonia (gaseous / liquid) and other basic reagents to modify activated carbon to improve the ability of reducing carbon and the adsorption capacity of organic and acidic gases [34-36]. Przepiórski [37] study showed that the adsorption capacity of activated carbon to phenol in aqueous solution increased by 29% after the modification of high temperature (400~800 ℃) in the gaseous ammonia environment. Mohammad[38] modified the activated carbon by oxidation pretreatment and high temperature (800 ℃) ammonia modification. After modification, the microporous structure of activated carbon is increased, the surface acidic oxygen functional groups are reduced, the surface alkaline of activated carbon is enhanced, and the adsorption capacity of CO_2 is increased. Valerie [39] used high temperature (700 ℃) and NH_3 modified method to treat activated carbon powder to improve its redox characteristics in microbial fuel cell. The results show that the redox ability of the modified activated carbon powder is enhanced under the neutral condition. The amount of oxygen functional groups on the surface decreased by 29% to 58%, the nitrogen content increased by 1.8%, and the surface alkalinity increased.

4 Load metal modification method

The surface load is modified by loading the metal ions in some soluble salt solutions to the surface of the activated carbon. Oxidizing redox reaction with the surrounding carbon atoms at high temperatures and it causes the loss of the surrounding carbon atoms to reach the purpose of enlarging the pores or metal ions themselves have a good ability to adsorb a particular substance. Qiao [40] were tested with Fe $(NO_3)_3$ and $Fe_2(SO_4)_3$ for activated carbon. The results show that at high temperature, the reaction of NOx and SOx to the pore channel of activated carbon is released by the decomposition of metal salt, and the pore size is increased. At the same time, the metal oxide metal oxide is reduced by activated carbon at high temperature, and the pore size is enlarged. In the presence of metal, the activation of the pore wall of activated carbon increases the mesoporous rate of activated carbon. Yang [41] used $FeCl_3$ as chemical additive and CO_2 as activator. The activated carbon was modified by chemical physical mixed activation method. The experimental results show that the pore structure and pore size distribution of activated carbon can be greatly improved by $FeCl_3$-CO_2 system. After modification, the average pore size of activated carbon decreased slightly, but the specific surface area and total pore volume increased by one times more than that of the raw carbon, and the pore size distribution was more uniform. Yu [42] loads Ag^+, Ni^{2+}, Cu^{2+}, Zn^{2+} and Fe^{2+} plasma on activated carbon. It is proved that the adsorption capacity of benzo thiophene containing sulfur organic compounds in oil can be obtained by loading different metal ions.

5 Conclusion

This study summarizes the modification methods of activated carbon and the characteristics of the modified methods. Acid modification is the most studied and most mature method of modification so far. Heating modification and alkali modification are beneficial to improve the adsorption of activated carbon to organic matter. When the load is modified, the chemical material can be loaded on the surface of the activated carbon to enhance its adsorption capacity to the target material. The special properties of activated carbon have been widely used in environmental treatment as an adsorbent. With the development of activated carbon modification technology, activated carbon can be specially modified according to the characteristics of pollutants in the environment, so as to realize the specialization of activated carbon in environmental treatment.

Reference

1. P. Chingombe, B.Saha, R.J. Wakeman. CARBON **43**, 3132-3143 (2005)

2. T.M. Alslaibi, I. Abustan, M.A. Ahmad, Eur. J CHEM TECHNOL BIOT **88**, 1183–1190 (2013)
3. B. Agarwal, P.K. Thakur, C. Balomajumder. CHEM ENG COMMUN **200**, 1278-1292 (2013)
4. O. Ioannidou, A. Zabaniotou. RENEW SUST ENERG REV **11**, 1966-2005 (2007)
5. M. Kubota, A. Hata , H. Matsuda. CARBON **47**, 2805-2811 (2009)
6. Y. Liu, Z.Hu, K.Xu, Eur. ACTA PHYS-CHIM SIN **24**, 1143-1148 (2008)
7. M.F.R. Pereira, S.F. Soares, J.J.M. Órfão, Eur. CARBON **4**, 811-821 (2003)
8. C.Y. Yin, M.K. Aroua, M.A.W.D. Wan.SEP PURIF TECHNOL**52**, 403-415 (2007)
9. Y. Han. TECHNIGUES AND EQUIPMENT FOR ENVIRO.POLL.CONT, (2003)
10. P.Wang, H.L. Zhang. CARBON TEC, (2003)
11. X.Y. Chen, X.L Lin, Q.H. Wei, Eur.SCI TECHN ENG, (2008)
12. X. Liang, X. Wang. TECHN WATER TREAT **37**, 1-6 (2011)
13. H. Demiral, D. Llknur. CHEM ENG RES DES **89**, 206-213 (2011)
14. W. Su, L. Zhou, Y.P. Zhou. Chem. Ind. Forest Prod **26**, 49-52 (2006)
15. L. Zhang, H.U. Jiapeng, S. Zhao, Eur. MATER REV, (2009)
16. W.J. Jiang, X. Jiang, X.F. Zhu, Eur. Chem. Ind. Forest Prod, **24**, 21-24, (2004)
17. W. Jiang, J. Xia, X. Zhu, Eur. Chem. Ind. Forest Prod, **24**, 39-42, (2003)
18. B. Yang, W. Jiang, Q. Chang. J LANZHOU JIAOTONG UNIV, 86, 46-50, (2006)
19. X.Q. Cao , X.M. Huang, S.R. Liu, Eur. J XIAN UNIV ARCH TECHNO, (2008)
20. J.A. Menéndez, E.M. Menéndez, M.J. Iglesias, Eur. CARBON **37**, 1115-1121 (1999)
21. A.A. Attia, W.E. Rashwan, S.A. Khedr. DYES PIGMENTS **69**, 128-136 (2006)
22. J.R. Rangel-Mendez, F.S. Cannon. CARBON **43**, 467-479 (2005)
23. P. Vinke, M.V.D. Eijk, M.Verbree, Eur. CARBON **32**, 675-686 (1994)
24. N.Wibowo, L. Setyadhi, D. Wibowo, Eur. J HAZARD MATER **146**, 237 (2007)
25. W. Huang, S.K. Sun, L.I. Yu-Jie, Eur. BIOMASS CHEM ENG, (2006)
26. J.L. Figueiredo, M.F.R. Pereira, M.M.A. Freitas, Eur. CARBON **37**, 1379-1389 (1999)
27. A. Gil, G.D.L. Puente, P. Grange. MICROPOR MESOPOR MAT **12**, 51-61 (1997)
28. L. Liu, Q.F. Deng, Y.P. Liu, Eur. CHEM COMMUN **16**, 81-85 (2011)
29. M.Z. Zhang, L. Li, J.X. Liu, Eur. J ENVIRON SCI-CHINA **34**, 39 (2013)
30. F.Gao, J.W. Patrick, A. Walker. COMPOS SCI TECHNOL **57**, 483-490 (1997)
31. J. Gu, D. Hu, W. Wang, Eur. BIOSENS BIOELECTRON **68**, 27-33 (2015)
32. A.C. Martins, O. Pezoti, A.L. Cazetta, Eur. CHEM ENG J **260**, 291-299 (2015)
33. M. Goyal, V.K. Rattan, D. Aggarwal, Eur. COLLOID SURFACE A **190**, 229-238 (2001)
34. B.S. Girgis, A.A. Attia, N.A. Fathy. COLLOID SURFACE A **299**, 79-87 (2007)
35. N.M. Haimour, S. Emeish. WASTE MANAGE **26**, 651-660 (2006)
36. 36. L. Monser, N. Adhoum. SEP PURIF TECHNOL 26, 137-146 (2002)
37. J. Przepiórski. J HAZARD MATER **135**, 453-456 (2006)
38. M.S. Shafeeyan, M.A.W.D. Wan, A. Houshmand, Eur. APPL SURF SCI **257**, 3936-3942 (2011)
39. V.J. Watson, C.N. Delgado, B.E. Logan. J POWER SOURCES **242**, 756-761 (2013)
40. W.M. Qiao, Y. Song, S.H. Yoon, Eur. NEW CARBON MATER **20**,198-204 (2005)
41. J.P. Yang, Y.H. Tian. J POWER SOURCES, (2005)
42. Y.U. Mo-Xin. J Appl Biomater Function Mater **37**, 1816-1818 (2006)

Effects of Different Material Total Solid on Biogas Production Characteristics

Yu-Ming Sun[1], Xiao-Mei Huang[1,*], Yin-Hu Kang[1]

[1]Chongqing University, Chongqing 400045, China

Abstract. In China, livestock manure emission has resulted in severe pollution to the environment and it is an efficient spreading agent of diseases. For this reason, the biogas has gotten a rapid development in the past few decades. As a kind of renewable and clean energy, many studies have indicated the prospect of biogas to replace fossil fuels in the future. However, the methane industrial production process is unstable due to various factors. Therefore, it is necessary to enhance the biogas fermentation efficiency. In this paper, the influences of the raw materials and the total solids (TS) concentration on biogas production characteristics are studied, where the utilization of raw materials can be reflected by the biogas production rate in the results. The results showed the anaerobic fermentation cycle is prolonged and biogas yield increases, but the utilization decreases with TS increases.

1 Introduction

With the exploration of world population and the rapid development of social economy, the earth's resources and environment are under great pressure in the 21st century [1]. Therefore, increasing interest and confidence have been paid to biomass, which is deemed as a valuable renewable energy from solar energy [2]. In China, biomass energy mainly derives from urban living garbage, forestry and timber waste, animal manure, industrial effluents, crop stalks, etc. All of them are featured with high carbon content and easy availability [3]. The development of biomass energy technique can not only reduce the ecological vicious cycle and greenhouse effect, but also substitute fossil fuels in the form of natural gas, coal, and petroleum. Thus it becomes an important pathway to relieve the energy crisis and environmental pollution [4].

As a new process converting the organic waste into usable energy, the biogas fermentation technology has received worldwide attention. In terms of technology, policy, economy and environment, Europe leads the utilization of biogas. It has the largest number of promotions, best efficiency, most mature technology of biogas anaerobic digestion in the world [5-6]. Among European countries Germany is the greatest biogas producer currently. Furthermore, Germany is transforming the raw material from livestock manure and crop straw into industrial scale production [7], such as organic crops (corn, straw, grass, etc.), municipal sewage sludge, livestock manure, industrial and food-processing enterprises waste, and urban household waste [8]. It also adopts different anaerobic digestion processes according to the engineering requirements. The United States started later until the 1970s to establish biogas facilities [9] because of the oil crisis. Therefore, biogas projects in the United States mainly concentrate upon farms. Most are used to generate electricity. In addition, since the 70's of the last century, Sweden has focused upon the development of biomass, proposing substituting renewable energy for fossil fuels before 2020. It has a great target to become the first country which doesn't rely on oil and coal throughout the world [10]. Besides, the raw materials of the biogas project in Sweden are mainly from the local industry, municipal wastewater and agricultural waste [11].

This paper shows that the total biogas yield increases and the anaerobic fermentation cycle prolongs but utilization efficiency of raw materials decreases with the increase in TS on the study of biogas production rate, biogas yield, production cycle and CH_4 content of the anaerobic fermentation through different TS on using different materials.

2 Materials and methods

2.1 Materials introduction

Fermentation materials: The cattle manure and wheat straw as the target fermentation materials and the characteristics of cattle manure and wheat straw listed in Table 1 which measured by a Muffle Furnace and an Elemental Analyzer.

Inoculum: The fermented substrate deriving from the laboratory after anaerobic digestion with dry cattle manure as the fermenting material and remaining in a stable state.

The wheat straw is minced into small segments with scales ranging from 30mm to 50mm in advance. The

* Corresponding author: hxm1980@cqu.edu.cn

pretreated wheat straw is used as material in a part of the experiment. The wheat straw is soaked in 8% $Ca(OH)_2$ solutions for 24 hours, soaked and then washed out to neutral before.

Table 1. The characteristics of cattle manure and wheat straw.

Characteristics	TS (%)	VS (%)	VS/TS	C (%)	N (%)	H (%)	S (%)	C/N
Cattle manure	85.04	72.27	0.85	31.29	4.50	1.11	0.26	6.95
Wheat straw	89.01	74.73	0.84	41.80	1.05	1.82	0.81	39.81

2.2 Experimental system of fermentation

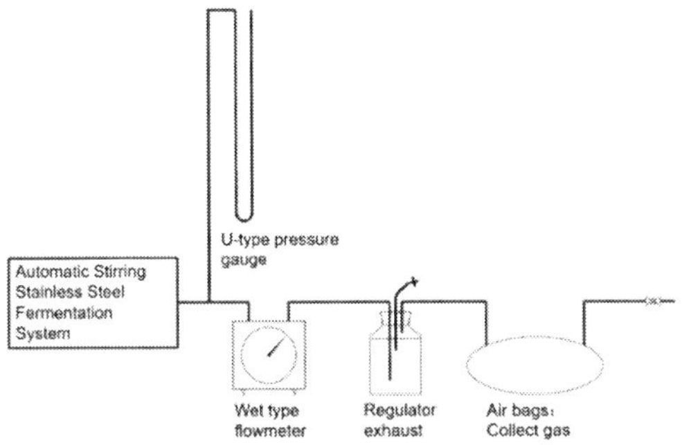

Fig. 1. Experimental system of fermentation.

The experimental system of anaerobic digestion is exhibited in Fig.1. The biogas, produced in the automatic mixing stainless steel fermentation system, passes wet flowmeter, regulator cylinder, and aluminum foil sampling bag in sequence. During the experiments, the internal pressure inside the reactor is kept constantly at 1600Pa.

2.3 Operating conditions

Six sets of experiments with different test materials and TS are involved in this paper, as Table 2 shows. The experiment mainly focuses on how Total Solid (TS) affects biogas anaerobic fermentation under different feed loads, raw materials, raw material processing and conditions as shown in Table 2.

Table 2. Operational conditions of biogas fermentation.

<table>
<tr><th>Groups</th><th>Materials</th><th>TS</th><th>Temperature</th><th>Tank pressure</th></tr>
<tr><td>I</td><td>Cattle manure</td><td rowspan="3">7.4%</td><td rowspan="6">35°C±5 °C</td><td rowspan="6">1600 Pa</td></tr>
<tr><td>II</td><td>Cattle manure + wheat straw</td></tr>
<tr><td>III</td><td>Cattle manure + pretreated wheat straw</td></tr>
<tr><td>IV</td><td>Cattle manure</td><td rowspan="3">10.0%</td></tr>
<tr><td>V</td><td>Cattle manure + wheat straw</td></tr>
<tr><td>VI</td><td>Cattle manure + pretreated wheat straw</td></tr>
</table>

3 Results and discussion

3.1 Effect of TS on anaerobic digestion characteristics

3.1.1 Biogas production rate

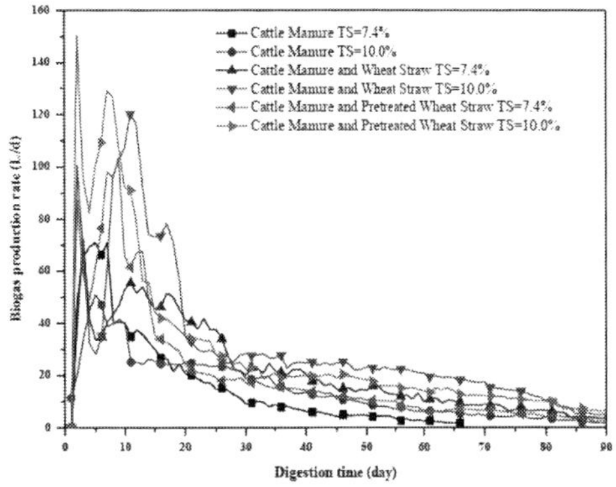

Fig. 2. Biogas production rate at different material and total Solid.

Although the raw materials and material ratios are varying, Fig. 2 reveals that the biogas production rates of the six sets of cases vary in a similar trend. The curved lines of biogas production rate reach the peak in the early days, and then the rate of daily biogas yield is damped. Compared with the line chart of the same kind of raw materials but different TS, it can be seen that the time of the peak daily biogas yield has been advanced and the daily biogas production rate has increased at TS=10.0%, only one raw material in different situations.

When using cattle manure and wheat straw as the fermentation raw material, compared with TS=7.4%, the biogas yield at TS=10.0% was one day ahead of time and increases by 38.7% at the first peak and by 117.2% at the second peak. In the middle and late stages of fermentation, the daily production rate at TS=10.0% is higher than that at TS=7.4%. By focusing on the set of cattle manure and pretreated wheat straw, the effect of TS on the biogas production rate can be examined. Compared with TS=7.4%, the daily biogas production rate at TS=10.0% is 72% higher at the first peak and 23% at the second. Furthermore, the biogas production rate reaches its peak value 2 days ahead of time. Under the same conditions, when single cattle manure is used as the fermentation raw material, the peak biogas production rate at TS=10.0% is 28.4% lower than that at TS=7.4%. The reason may be that after the capping, an anaerobic environment has not been formed in whole reactor, and the non-methanogens bacterium multiply quickly with the increase of organic matter content, which makes the accumulation of organic acid and the activity of methanogens inhibit in acidic environment.

3.1.2 Biogas yield

 https://doi.org/10.1051/e3sconf/20183802006

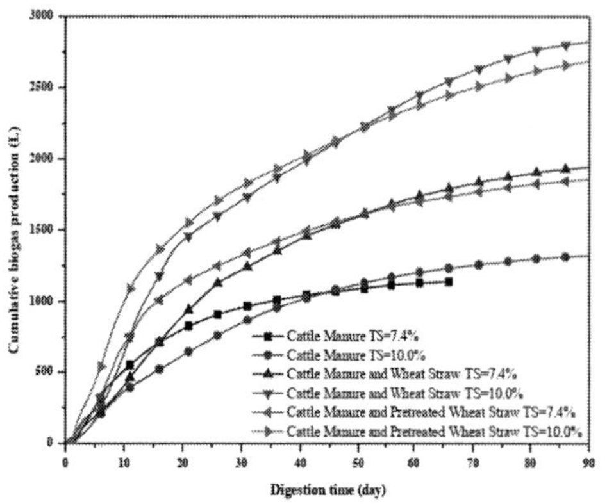

Fig. 3. The cumulative biogas production in different material and total solid.

It is observed that the total biogas production at TS=10.0% is higher than that at TS=7.4% in whole fermentation cycle.

When using single cattle manure as the raw material, in the total anaerobic fermentation cycle, it needs 12 and 38 days to achieve the cumulative biogas production of 50% and 90% at TS=7.4%, but 22 and 59 days at TS=10.0%. It can be seen that the biogas production cycle has a long delay at TS=10.0%. Under the same conditions, when cattle manure and wheat straw are used as the raw material at TS=7.4%, and reaches 50% and 90% of the total biogas production in the first 22 and 63 days. Compared with the former, the total biogas production increases by 45% over the entire reaction cycle, reaching 50% and 90% of the total biogas production on the 22nd and 65th day respectively at TS=10.0%. The total biogas production from cattle manure and pretreated wheat straw is 1853.36 L at TS=7.4%. It reaches 50% and 90% of the total biogas production in the first 14 and 57 days under the same conditions, and the cumulative biogas production in the first 30 days is 71% of the total biogas production. Being different from TS=7.4%, the total biogas production is 2684.54 L at TS=10.0%, although the biogas yield has increased, the time to reach 50% and 90% of total is delayed by 2 and 7 days, the cumulative biogas production in the first 30 days is 67% of the total biogas production.

3.1.3 CH_4 percentage

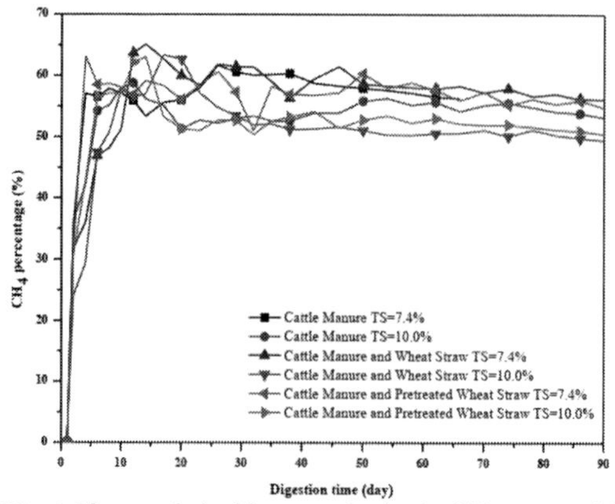

Fig. 4. The cumulative biogas production in different material and total solid.

It can be observed from the CH_4 percentage curve line of single cattle manure that the average percentage of CH_4 is 56.53% at TS=7.4%, the average percentage of CH_4 is 58.89% at TS=10.0%. As shown in Fig. 4, at TS=10.0%, the percentage of CH_4 is lower than that at TS=7.4% for most of time during the whole reaction cycle. In addition, the CH_4 percentage curve line of every groups show a similar conclusion, that the percentage of CH_4 at TS=10.0% is always lower than that at TS=7.4% after the early stage of reaction.

The main reason for this situation is that, in the early stage of reaction, the content of organic matter in the raw materials is more abundant and non-methanogens bacteria metabolize quickly and generates a large amount of acetic acid and CO_2. Then part of the CO_2 dissolves in the feed liquid and reduces the PH, which inhibits the methanogenic activity, and the percentage of CH_4 is low. In the middle and later stages of the reaction, the content of organic matter is sufficient, and the rate of non-methanogens bacterium degradation is faster than methanogens. More CO_2 surplus is overflowed with biogas, and the percentage of methane at TS=10.0% is lower than that at TS=7.4%.

3.2 Analysis of anaerobic digestion time and biogas production rate

3.2.1 Analysis of anaerobic digestion time

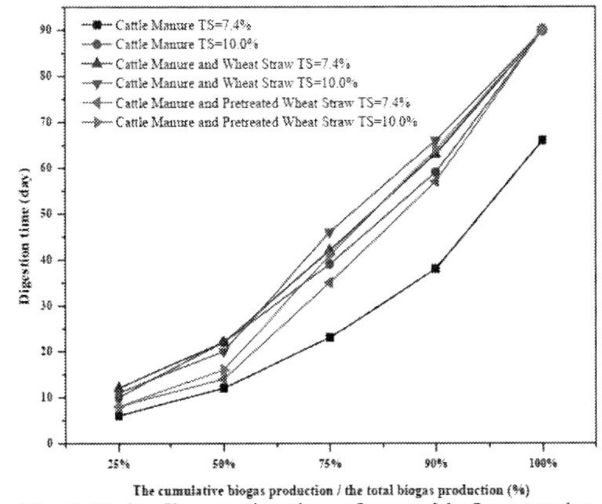

Fig. 5. Hydraulic retention time of anaerobic fermentation.

Fig. 5 shows that the times to reach 25%, 50%, 75% and 90% of the total biogas production are various. The group of single cattle manure is the shortest, the group of cattle manure and pretreated wheat straw is the next, and the cattle manure and wheat straw are the longest at TS=7.4%. It also shows that the time is prolonged by adding wheat straw into the cattle manure for anaerobic fermentation cause. Moreover, when wheat straw is pretreated, the time is less, indicating that the pretreatment process has shortened the biogas production cycle. Except the cases of single cattle manure at TS=7.4%, it takes 56-66 days for the cumulative biogas production to amount to 90% of the total biogas production, and 90 days to amount to 100% of the total production. Therefore, it is most efficient to stop the anaerobic fermentation process after 60 days from the beginning. This will not only significantly shorten the gas cycle, but also ensure that the raw materials are fully utilized.

3.2.1 Analysis of anaerobic digestion time

During anaerobic digestion process, methanogens and non-methanogens bacteria can only decompose to volatile solids content (VS) in the feed while ash cannot be degraded. Therefore, TS or VS biogas production rates can be used to determine the level of raw material utilization.

Table 3. The biogas production rate of TS and VS.

Nature	Group I	Group II	Group III	Group IV	Group V	Group VI
TS=	7.4%	7.4%	7.4%	10.0%	10.0%	10.0%
Cumulative biogas production in the first 30 days(m³)	0.954	1.218	1.322	0.847	1.707	1.805
Total biogas production(m³)	1.138	1.946	1.854	1.320	2.822	2.685
TS(kg)	6.609	7.145	7.145	8.564	9.501	9.501
VS(TS)	4.878	5.361	5.361	7.665	8.478	8.478
The relative percentage of the first 30 days (%)	83.9	62.6	71.3	64.1	60.5	67.2
TS biogas production rate(m³/kgTS)	0.188	0.272	0.259	0.154	0.297	0.283
VS biogas production rate(m³/kgVS)	0.233	0.363	0.346	0.172	0.333	0.317

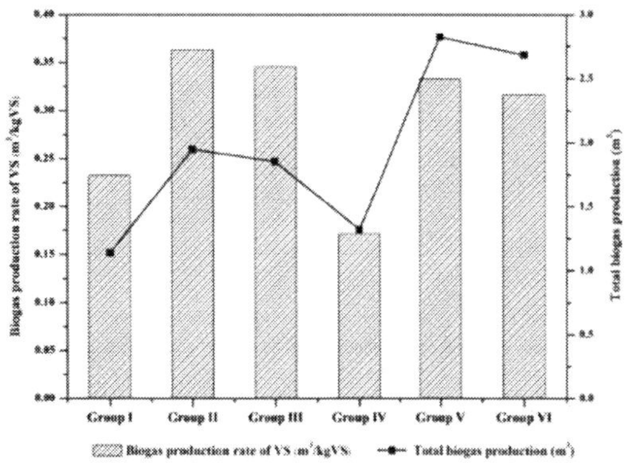

Fig. 6. The biogas production rate histogram of each group.

It can be seen from Table 3 that with the same raw materials, the utilization of raw materials at TS=7.4% is lower than that at TS=10.0% from biogas production rate of TS. However, the utilization of raw materials gets opposite results from biogas production rate of VS, because there is not volatile gravel in the cattle manure, so the biogas production rate of VS is used to reflect the level of raw material utilization in this paper. As shown in Fig. 6 and Table 3, the production rate ranking of the raw material biogas is Group I> Group IV> Group II> Group V> Group III> Group VI. The comparison of biogas yield is Group IV> Group I, Group V> Group II, Group VI> Group III. It is pointed out that under the same raw material and pretreatment conditions, when TS increases from 7.4% to 10.0%, biogas yield increases and the raw material utilization decreases.

4 Conclusions

1) In the same fermentation liquid, the total biogas yield increases with the growth of TS content, and the maximum cumulative biogas production occurs at TS=10.0% when cattle manure and wheat straw are used as the raw material.

2) The anaerobic fermentation of each group mainly occurs on the first 30 days. According to the ratio of the accumulated biogas production to the total biogas production in the first 30 days, it can be concluded that the utilization of the organic matter in the raw materials reduces with the increase of the total solid content. The optimal feeding utilization rate is found at TS=7.4% when cattle manure and pretreated wheat straw are used as the raw material.

3) Under the same conditions, when the TS increases from 7.4% to 10.0%, the anaerobic fermentation cycle is prolonged and the total biogas production increases, but the utilization rate of raw materials decreases and the environment of anaerobic fermentation microbial metabolic activity is more easily formed. In practical engineering, the fermentation cycle and the effect of total biogas production should be taken into consideration. In addition, the optimal total solid content

for different fermentation raw materials should also be determined to achieve the most efficient fermentation of biogas.

References

1. Zhang Dong, et al. *Treatment Technology of Livestock Manure Biogas Project.* Journal of Zhejiang agricultural sciences, **2**(2012): 223-227
2. Zhang Wudi, Song Hongchuan, Li Jianchang, et al. *Rural Energy for Sustainable Development of Agriculture; Biomass Energy.* Agriculture and Technology, **21**(2001): 8-12
3. Li Jingming. *Future: The Development of Biogas Industry in China.* China International Biomass Energy and Biomass Utilization Summit, (2014)
4. Liu Ronghou, Hao Yuanyuan, Ye Ziliang, et al. *Experimental Study on the Effect of Biogas Fermentation Process Parameters on Biogas and Biogas Slurry Composition.* Journal of Agricultural Engineering, **22**(2006): 85-88
5. C NMulligan. *An evaluation of the sustainability of the anaerobic digestion of manure.* Proceedings 10th World Congress AD, Montreal, Canada, (2004) 1786-17891.
6. Teodorita Al Seadi, et al. *Good Practice in Quality Management of AD Residue.*
7. Gorisch U , *Helm M. Biogas plants.* Stuttgart, Germany: Ulmer, (2006)
8. Lin Cong, Duan Na, Wang Yang, Zhang Youhong. *Development Status and Thinking of Biogas Engineering in Germany.* Swine Industry Science, **7**(2009)
9. Shou Yifeng. *Development Status and Trend of Biogas Industry in USA.* Scientific Research.
10. Tian Ying. *Energy Sustainable Development Strategy in Sweden.* http: //www. Isase /templates /News_50822. aspx. (2006)
11. LantzM, Svensson M, Bjrnsson L, Brjesson P. *The Prospects for an Expansion of Biogas Systems in Sweden - Incentives, Barriers and Potentials.* Energy Policy, **35** (2007): 1830 -1843.

Mechanism study and numerical simulation of Uranium nitriding induced by high energy laser

Yuan Zhu, Jingjing Xu, Yanwen Qi, Shengpeng Li, Hui Zhao

School of Mechatronic Engineering and Automation, Shanghai University, Shanghai 200072, China

Abstract. The gradients of interfacial tension induced by local heating led to Marangoni convection, which had a significant effect on surface formation and the process of mass transport in the laser nitriding of uranium. An experimental observation of the underlying processes was very difficult. In present study, the Marangoni convection was considered and the computational fluid dynamic (CFD) analysis technique of FLUENT program was performed to determine the physical processes such as heat transfer and mass transport. The progress of gas-liquid falling film desorption was presented by combining phase-change model with fluid volume function (VOF) model. The time-dependent distribution of the temperature had been derived. Moreover, the concentration and distribution of nitrogen across the laser spot are calculated. The simulation results matched with the experimental data. The numerical resolution method provided a better approach to know the physical processes and dependencies of the coating formation.

1 Introduction

Since uranium metal readily corrodes under the ambient environment, many technologies were developed to improve its corrosion resistance [1,2], where nitriding and carbonation of the surface are well known for improving the tribological properties of different metal compounds. Pulsed laser nitriding has shown that the metal-nitride layer produced by a high-power pulsed laser greatly improves the surface hardness and the corrosion resistance [3]. Due to its high power and flexibility in its temporal shaping, this type of laser could be the appropriate tool. The nitrogen transport and corresponding coating properties are determined by the time of treatment, for which diffusion will be assisted by the Marangoni convection for extended time regimes. Because time-dependent occurrences were inaccessible in experiment, many researches have done using numerical simulation method. Daniel Höche et al [4,5] established the N-Ti two-dimensional (2-D) model in Comsol, which was under the influence of the Marangoni effect on the formation of molten pool. Fuzhen Xuan [6] established the coupling model of the transient temperature field and the concentration field. The model was based on the unsteady state Fourier heat conduction equation and the Fick law of modified diffusion equation, the effect of a moving heat source on the formation of temperature gradient, which lead to nitrogen diffusion in titanium, was taken into account. However, as of now, there has been no in depth research conducted on the nitriding of actinide metal. Meanwhile, in the previous study, the factors in the process of laser nitriding for different problems are not considered comprehensively. In this paper, the effects of Marangoni convection, temperature gradient and fluid flow are investigated using the computational fluid dynamic analysis technique of FLUENT. The transient N-U two-dimensional model is established, and the two-film theory model is applied. Absorption of nitrogen into a laminar falling film on the interface is presented using the volume of fluid (VOF) model. Consequently, the mechanism of heat and mass transfer were established, and the simulation results agree with the experimental results. In addition, the relationship between the parameters and the properties of nitride coating was revealed. This finally provides references for theory analysis and allows an optimization of the coating synthesis.

2 Numerical Simulation Model

2.1 Physical Model

Uranium sheets (20 mm in diameter, 2mm in thickness) were placed in a chamber evacuated and filled with 99.99% pure nitrogen gas. In order to treat extended areas, the vacuum chamber was placed on a two-dimensional monitored linear stage. The laser scanned along the Y direction with speeds ranging from 0.1to 1.28mm/s. After scanning a line in the Z direction, the vacuum chamber moved 1mm in X direction [7,8]. In Daniel Höche and Sven Müller's research, a feasible two-dimensional simplified computational model was constructed, and the simulation results matched experimental results very well. The scan velocity, which the model is based on, is much lower than the melt flow velocity. Here, we assumed that the present case

* Corresponding author: zhuyuan1993@163.com

matched with Daniel Höche's, so a transient and two-dimensional model (plane XY) is used to describe the nitriding of uranium. The 5×4 mm2 simplified plane model was established in the simulation.

Figure 1 shows the different domains (N_2 and uranium) and several boundaries. The temperature condition of the thin uranium sheet is given by T_0=300K. The PLANE13 2-D 4-node quadrilateral mesh was applied in simulation, which contains 50000 elements with 50451 nodes. Elements were linear in this case of heat transfer and diffusion. At the boundaries of the model (walls), the velocity was set to be zero (no slip). On the gaseous side (outlet), a pressure-outlet condition was employed. The initial temperature was set to 300K, the equivalent laser heat source is setted at the surface of uranium.

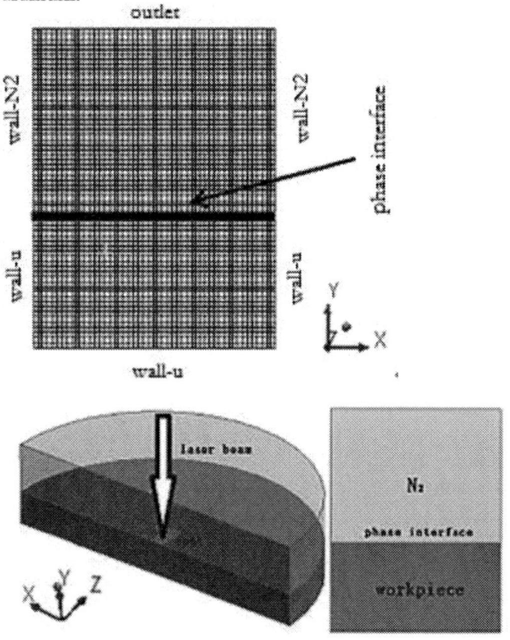

Fig. 1. Geometry and Grid

The laser beam was focused on the sample using a convex lens with spot size of 1×2 mm^2, which was a temporal pulse structure in a continuous wave mode. The beam was approximately 1mm in width at cross-section.

In the simulation, in order to better use numerical simulation methods to study nitriding process, the millisecond time of uranium treatment is applied. The form of the laser-pulse intensity I adopted in our simulation will be taken as "rectangle":

$$\begin{cases} I(x,t) = (1 - R_{s,l})\dfrac{p}{A}, 0 \le t \le \tau, -0.0005 \le x \le 0.0005 \\ I(x,t) = 0, \tau \le t \end{cases} \quad (1)$$

2.2 Heating Transport

The effect of atomic fission was ignored because of the short interaction time and insufficient energy of the laser. We made use of the heat flow equation in two-dimensional (2-D) space to describe the heat flow. The mathematical description of phase changes is a difficulty formally known as the Stefan problem. To solve the problem the smoothed Heaviside function (T, B) was used. B is the width of the temperature region where (T, B) changes from 0 to 1 and was assumed to be 100K.

$$G(T) = G_{sol} - (G_{sol} + G_{liq})\Theta(T,B) \quad (2)$$

Here, the energy conservation equation is established in the solid phase and liquid phase region, respectively, the temperature T as the only variable.

$$\beta(T)C_p(T)\frac{\partial T}{\partial t} = \nabla(\kappa(T)\nabla T) + S - v_t\rho(T)C_p(T)\nabla T \quad (3)$$

Where T, $c_p(T)$, and $\kappa(T)$ are, respectively , temperature, specific heat and thermal conductivity and S is heat source term given by Eq.(4). The quantities c, ρ, and κ will be subsequently assumed to be temperature and space independent. S can be written using UDF in Fluent.

Where T_0 is the ambient temperature (T_0=300K).

Table 1 and table 2 give the material physical properties used in the present modeling work. Additionally, the external absorption parameter α_{ext} was defined to account for heat losses due to the optics.

Table 1. Physical Properties of uranium Used in Simulations

Melting temperature/K	T_L	1405
Boiling temperature at 1 atm/K	T_b	4091
Latent heat of fusion /vaporization /(kJ·kg^{-1})	L_m/L_{ev}	1159.2/ 38364
Solid/ Liquid phase density /(kg·m^{-3})	ρ_s/ρ_l	19100/ 17300
Molar mass /(g·mol^{-1})	M_u	238
Viscosity at 1405K /mpa·s	η_u	6.55
Surface tension /(N·m^{-1})	σ	1.55
Gradient of surface tension/ (N·m^{-1}·K^{-1})	$\partial\sigma/\partial T$	-1.4× 10^{-4}

Table 2. Physical Properties of Nitrogen Used in Simulations.

Density/(kg·m^{-3})	ρ_{N_2}	1.138
Thermal conductivity/ (W·m^{-1}·K^{-1})	κ_{N_2}	0.026
Specific heat ratio/ (J·kg^{-1}·K^{-1})	c_{PN_2}	1040
Viscosity/μpa·s	μ_{N_2}	3.2

2.3 Mass Transport

With the laser irradiated, nitrogen gas absorbed energy, forming nitrogen atoms, which were diffuser into a molten pool. This process occurred at the gas-liquid interface, and was in accord with the falling film desorption. Interphase mass transport always exists if only the bulk concentration of gas-liquid does not reach equilibrium. The process consists of a gas-liquid two-phase stratified flow. In order to model surface tracking, we applied the VOF interface tracking technology. VOF can accurately simulate the unstable of liquid film at free surface, and is especially suitable for flow filed, which has obvious interface.

The process includes the mass conservation Eq. (5), momentum conservation Eq. (6).

$$\frac{\partial}{\partial t}(\rho)+\nabla\left(\rho\vec{v}\right)=0 \quad (5)$$

$$\frac{\partial}{\partial t}(\rho\vec{\mu})+\nabla\left(\rho\vec{v}\vec{v}\right)=-\nabla p+\nabla\left[\mu\left(\nabla\vec{v}+\vec{v}^{T}\right)\right]+\rho\vec{g}+\vec{F} \quad (6)$$

$\vec{v}$ is the expansion velocity in the molten pool, ρ is the sum of the different phase density, and $\vec{v}$ the momentum source.

When the molten pool was formed, there was obvious gas-liquid interface; combined with double film theory and penetration theory, mass transport equation is as follows:

$$S_{GL}^{k}=\frac{\partial\alpha_q\rho_q w_q^k}{\partial t}+\nabla\left(\alpha_q\rho_q\vec{\mu}w_q^k-\alpha_q\Gamma_q^k\nabla w_q^k\right);q=L,G;k=1,\dots,r \quad (7)$$

The mass transfer source S_{GL}^{k}, which is the speed of species k in one phase transport to another phase ; w_q^k is the mass fraction of species k in phase q; Γ_q^k is the coefficient of mass transfer; α_q is the volume fraction of phase q, and the gas-liquid two phase is in state of laminar during the mass transport. In this simulation, the rate of nitrogen decomposition into nitrogen atoms is constant.

$$\Gamma_q^k=D_q^k\rho_q \quad (8)$$

Due to the heating, the phase transition and flow in molten pool occurred, which led to fluctuation at the phase interface. VOF method obtains the position of the phase interface by calculating and tracing the distribution of volume fraction α_q. In a computing unit, when α_q=0, phase q is not shown in this unit; If α_q=1, the phase q is filled in this unit; when $0<\alpha_q<1$, phase q and other phases are contained in this unit. Therefore, the distribution of interface can be calculated by the following.

$$\frac{\partial\alpha_q}{\partial t}+\vec{v}\nabla\alpha_q=0 \quad (9)$$

$$\sum_{q=L,G}\alpha_q=1 \quad (10)$$

The physical properties in the control equation are determined by each phase in the system. Thus, ρ and v in the gas-liquid two phase flow can be described by Eq. (11) (12).

$$\rho=\alpha_L\rho_L+\left(1-\alpha_L\right)\rho_G \quad (11)$$

$$v=\alpha_L\rho_L+\left(1-\alpha_L\right)v_G \quad (12)$$

The mass source S_{GL}^{k} can be obtained based on the two film theory.

$$S_{GL}^{k}=(mx_k-y_k)\left(\frac{m}{k_L a_e M_k C_{LT}}+\frac{1}{k_G a_e M_k C_{GT}}\right) \quad (13)$$

Where x_k、y_k is the mole fraction of component k in gas and liquid at phase interface. a_e is the effective mass ratio of the surface area ($m^2.m^{-3}$), M_k is the relative atomic mass of the transfer element, and C_{LT} and C_{GT} are the mole concentrations in liquid and gas phases ($kmol.m^{-3}$).

S_{GL}^{k} can be written using UDF in Fluent. Here, mass transfer conformed to penetration theory.

$$k_{G,L}=2\sqrt{\frac{D_{G,L}}{\pi T}} \quad (14)$$

Where T is the contact time, and D_L、D_G are the diffusion coefficients of species k in liquid and gas phases (cm^2/s).

So the melting and solidification model, VOF model, the mass transfer model, and so on in Fluent were applied to simulate the process of laser nitriding. The source term of energy, mass transport, mass and momentum were written using in Fluent.

3 Resulting and Discussion

3.1 The Temperature Field and Flow in Molten Pool

From the numerical calculations, many details of the process were obtained. For the energy density was $8.5\times10^8 W/m^2$, the distributions of temperature field, the flow in molten pool, and the change in size of molten pool at different times are shown in Figure 2. This figure shows the temperature field (color scale), the flow direction of the melt (indicated by arrows), and the melt pool shape (surrounded by the red line) in a cross-sectional view. The time-dependent development of a track during irradiation was calculated to be 5ms. The region of molten pool grew wider and deeper with continued irradiating.

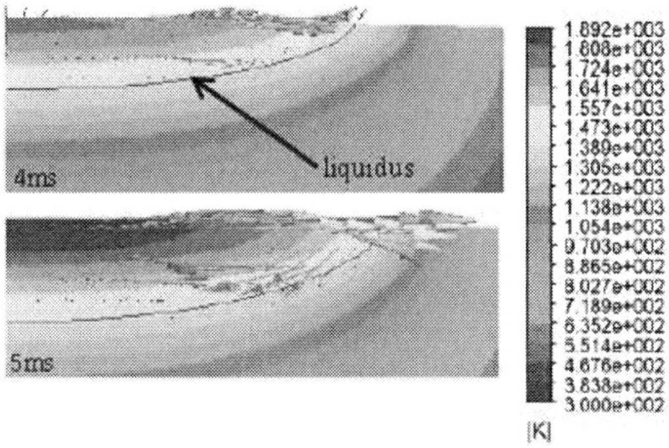

Fig. 2. Variation of molten pool shape and temperature field

The profiles of surface temperature versus time at different density are shown in Figure 3. The temperature did not reach the boiling point in simulation because of the limiting treatment parameters (laser intensity and interaction time). Consequently, keyhole formation and evaporation could be avoided or minimized, which is a main requirement for cladding or alloying at laser material processing. With the sample is irradiated with the laser, the temperature on the surface increases as shown in Figure 3. Increasing the energy density caused the maximum temperature of the surface to increase and the flow in the pool to become accelerated. The

 https://doi.org/10.1051/e3sconf/20183802007

temperature and the flow in the melting pool affect the behavior of heat transfer.

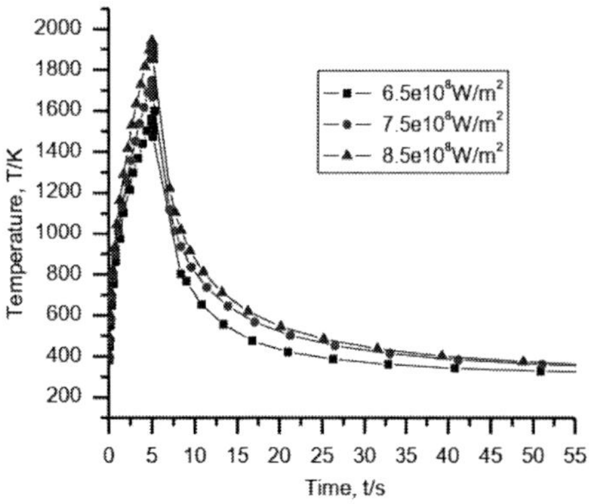

Fig. 3. variation of surface temperature with time

3.2 The Concentration of Nitrogen in Melt Pool

the higher temperature, the greater diffusion coefficient, means the nitrogen spread faster in uranium. At the same time, the greater gradient of the temperature provided thicker nitride layer. The pressure was at 0.5MPa, and the energy density varied from 6.5×10^8 W/m^2 to 8.5×10^8 W/m^2; experimental and simulation results are shown Figure 4. At the same pressure as in ambient, the concentration of nitrogen in melt pool rose with increasing energy density. The trend of simulation agreed with the experiment. Meanwhile, the decomposition of nitrogen gas was assumed to be constant in the numerical simulation, and the nitrogen in uranium would form nitride. The correction coefficient was applied to correct the differences.

$$[R]_e = \beta[R]_s \tag{15}$$

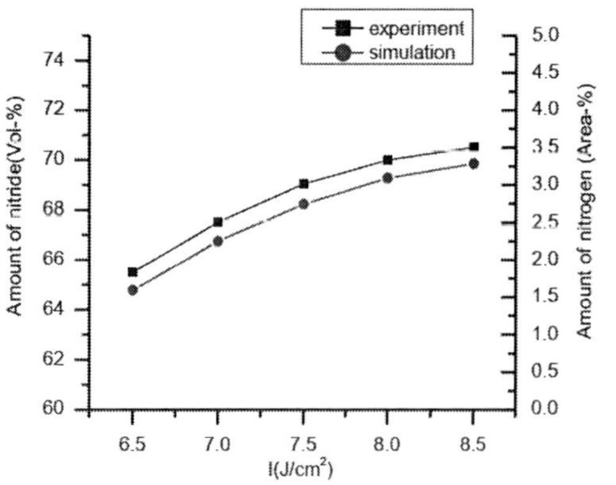

Fig. 4. Dependence of nitrogen concentration on nitrogen laser energy

According to the results of experiment and simulation above, the nitriding model is feasible. At the same time, the decomposition rate of nitrogen gas varied with temperature, and the nitrogen in the material will form nitride. All of these need to be further studied.

The influence factors on the diffusion and distribution of nitrogen can be acquired based on this nitriding model. The results were obtained to provide guides for production.

3.3 The Distribution of Nitrogen

Due to the strong convective flow, the nitrogen transport is a combination of diffusion and convection. Nitrogen surface profiles were measured for irradiated samples. The energy varied from 6.5×10^8 W/m^2 to 8.5×10^8 W/m^2 for one pulse. In Figure 5, when the energy is 6.5×10^8 W/m^2, only a small region in the central part is being nitrided. When the energy was 8.5×10^8 W/m^2, a region with broadest nitriding area appeared, covering the entire laser area. It was also observed that the edge of the nitriding region had higher concentration of nitrogen than the central region. It indicated that with the increasing of energy density, the temperature increased. Although the diffusive mass transport enhanced, the convective mass transport occupied the dominant position in the process of mass transport.

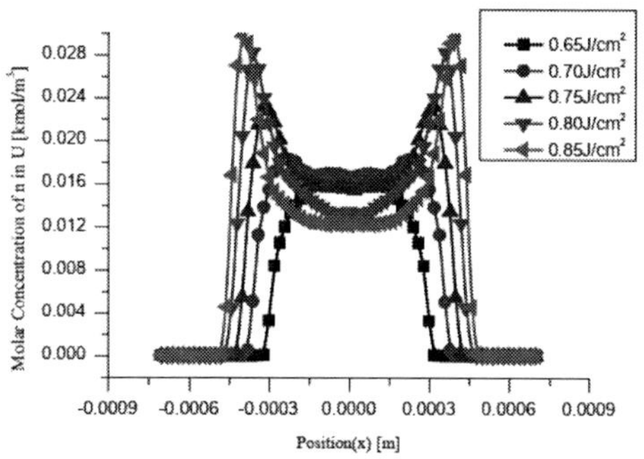

Fig. 4. Distribution of nitrogen at surface

4 Conclusion

This paper established a numerical model for simulating the laser nitriding of uranium in an ambient nitrogen atmosphere. The result showed that the energy density of the laser had the main influence on the temperature field of laser nitriding. The higher the laser density, the greater the gradient of temperature; it led to strong Marangoni convection in the molten pool. Due to the supplemental convective, the aspect ratio of the melt pool was very high. However, the higher temperature would lead to decrease in solidification rate. As a result, a coarse solidification microstructure would be developed.

The density of energy not only determines the size of nitriding area, but also determines the distribution of nitrogen in the molten pool; it affects the formation of the nitride layer. The higher energy density would not lead to ideal distribution of nitrogen at the surface. Meanwhile, the lower energy density only provided a narrow nitriding region. In order to get the uniform nitriding layer, reasonable laser parameters should be taken into consideration. In this simulation, the ideal threshold laser energy is around 7.0×10^{8} W/m^{2}. The calculated results are consistent with the experiments.

References

1. D. M. Mattox, R. D. Bland. Journal of Nuclear Materials **21**, 3 (1967):349-352.
2. R. Arkush, M. H. Mintz, N Shamir, & R Arkush. Journal of Nuclear Material **281**, 2(2000): 182-190.
3. Peter Schaaf. Progress in Materials Science **47**, (2002): 1-161.
4. D. Höche, S. Müller, G. Rapin, M. Shinn, E. Remdt & M. Gubisch et al. Metallurgical & Materials Transactions B **40**, 4(2009): 497-507.
5. Höche D, Rapin G & Schaaf P. Applied Surface Science **254**, 4(2007): 888–892.
6. FuZhen, Xuan, LiQin, Cao, Zhengdong, Wang et al. Computational Materials Science **49**, 1(2010):104-111.
7. Yongbin Zhang, Daqiao Meng Youshou Zhang. Journal of Nuclear Materials **397**, (2010): 31–35.
8. ZhiLei Chen, Yongbin Zhang. Acta Phys. – Chim.Sin. (Suppl.), 2015, 0001-0009.

Application Research on Stabilizing Treatment of Dredged Sediment

Shi Yannan[1], Zhang Chaojie[1,2,3], Yu Jiongqi[1,2,3], Tang Mingli[1]

[1] Zhejiang Guang Chuan Engineering Consulting Co., LTD, Hangzhou 310020, China;
[2] Zhejiang Institute of Hydraulics & Estuary, Hangzhou 310020, China;
[3] Zhejiang Provincial Key Laboratory of Hydraulic Disaster Prevention and Mitigation, Hangzhou 310020, China.

Abstract. In order to improve water quality, to ensure the capacity of normal flood discharge and river transport, it carried out a lot of dredging work across the country recently. For harmful sediment's second pollution and recycling use, this study selected five kinds of sediment. And point A was determined as the research object by using the geo-accumulation index to evaluate the heavy metal. Then the sediment was mixed with lime, fly ash, kaolin clay, sodium silicate, sodium carbonate and polyvinyl alcohol as mud stabilizing materials. Experimental research shows that, copper loss of stabilized soil's toxicity leaching agent reaches more than 95%, and the permeability coefficient was 10-5 cm/s orders of magnitude. In this paper, it ensures GW7 as the best choice of plants experiments, which are 5% fly ash, 1% lime, 3% kaolin clay, 3% sodium silicate, 3% sodium carbonate and 1% polyvinyl alcohol, through the ways of permeability, water retention, SEM and XRD. The scheme effectively stabilizes copper, and keeps the plants as better form than others. But it has different effect on different plants. Thus it should consider planting varieties optimization in actual applications.

1 Introduction

With the deepening of Zhejiang Province "five water treatment", it appeared a lot of dredged sediment in a short time. Because dredged sediment had high moisture content and large volume, it needed a large amount of land for stacking. In addition, a big part of dredged sediment is polluted by different degree, especially the urban river sediment, it contains a lot of organic matter, nitrogen and phosphorus nutrition, and some harmful heavy metals, pathogenic bacteria, virus microorganism, and toxic organic compounds, and so on. It make the mud easy to blacken and rot [1]. If untreated direct stacking, it would cause secondary pollution, especially in cultivated land. Therefore, the dredging sediment's stabilization and harmlessness disposal has become one of the problems to be solved urgently.

Stabilization refers to the transformation of harmful substances into substances with low solubility, low toxicity and low mobility, that reduces the pollution potential of waste materials [2]. Stable technology is generally implemented by curing, that the curing materials change the soil mechanics properties, and reduce pollutants dissolved out[3]. The techniques mainly include cement, thermoplastic materials, hot and hard materials, lime, large encapsulation, self-cementing, and glass solidification, etc. [4-5].

According to the literature, lime, cement, fly ash can improve the soil pH, and reduce the microorganism and organic matter content in the sediment, while steady heavy metals form of cadmium and chromium. Because microbial decomposition and metabolism constantly changes sediment chemical conditions, it will affect the form of heavy metals [6]. Kaolin is basically neutral and has fine pores on the surface, which offers good adsorption and desorption properties to nitrogen, phosphorus, potassium and organic carbon [7]. And Iron powder, sodium carbonate and sodium silicate can stabilize copper [8].

In 2016, Zhejiang Province had dredged more than 100 million cubic meters sediment. At present, its utilization mode was simplex, that mainly was backfilled. In this, the use of agricultural fields is accounted for 22%, and backfill is at 33%, contained low-lying areas, mines and reclamation areas backfill. And landscaping is at 7%. But agricultural fields and landscaping are affected by mud properties. Additionally, the search for land is increasingly difficult as the backfill area decreases. Although the concept of "dredging sediment is an available resource" has been gradually recognized, it is necessary to carry out systematic research to translate this understanding into practice[9].

Based on the problems of dredged sediment's high water content, diversified pollution, and it has difficulty in storage yard, this paper put forward to use mud as planting soil through solidifying single pollution factor silt, which will provide more resource utilization possibility for polluted sediment.

* Corresponding author: syn.0419@163.com

2 Screening of Sediment

2.1 Evaluation Standard of Geo-accumulation Index Method

The geo-accumulation index (Igeo) is a quantitative index of heavy metal pollution in water environment. Because it not only considers the man-made pollution factors, environmental geochemical background values, etc., but also in consideration of the dynamics of background value by the natural rock forming effect. Therefore, the application is widely distributed[10]. Evaluation standard of geo-accumulation index is shown in table 1.

Table 1. Evaluation Criterion of Geo-accumulation Index and Pollution Series

Level	I_{geo}	Degree of Contamination
0	(-∞, 0)	No
1	(0,1]	Mild
2	(1,2]	Less Moderate
3	(2,3]	Moderate
4	(3,4]	Less Heavy
5	(4,5]	Heavy
6	(5,+∞)	Serious

2.2 The Calculation Formula of Geo-accumulation Index

Computational formula of geo-accumulation index is shown in formula (1). The background value of heavy metals in soil of Zhejiang Province [13] is shown in table 2.

$$I_{geo} = \log_2[C_n/(K * B_n)] \quad (1)$$

Where Cn is the content of heavy metal n in sediment, Bn is the background value of heavy metals in soil of Zhejiang Province, K is correction factor, and usually K is 1.5.

Table 2. Soil element background values of heavy metals in Zhejiang Province

heavy metal	background values
Cu	22.63
Zn	83.06
Pb	23
Cd	0.17
Ni	23.93
Hg	0.17
As	6.88

2.3 The Evaluation Results of Geo-accumulation Index

This study screened five different sediments as research object. The results of Igeo were shown in table 3. The results showed that the Igeo of Cu in sample A was less heavy pollution, and the factor was relatively simple. Therefore, the test chose sample A, and its physical properties were shown in table 4.

Table 3. Geo-accumulation index(Igeo) calculation of heavy metal in sediment

The sample of Sediment	I_{geo} Cu	I_{geo} Zn	I_{geo} Pb	I_{geo} Cd	I_{geo} Ni	I_{geo} Hg	I_{geo} As
A	3.15	0.74	1.59	0.08	0.33	0.61	0.83
	Less Heavy	Mild	Less Moderate	Mild	Mild	Mild	Mild
B	1.7	0.97	1.15	1.85	0.51	0.25	0.5
	Less Moderate	Mild	Less Moderate	Less Moderate	Mild	Mild	Mild
C	-0.63	-0.22	-0.06	-1.67	-1.58	-0.41	0.29
	No	No	No	No	No	No	Mild
D	-0.56	-0.02	-0.06	-0.28	0.03	-1.21	-0.27
	No	No	No	No	Mild	No	No
E	0.34	0.25	-0.56	-0.21	-0.92	-2.21	0.05
	Mild	Mild	No	No	No	No	Mild

Table 4. The statistical table of point A sediment's physical property

The sample of Sediment	A
moisture content (%)	51.8
wet density (g/cm³)	1.67
dry density (g/cm³)	1.1
void ratio (%)	1.486
liquid limit (%)	54.6
plastic limit (%)	31.2
the specific gravity of solids	2.74
coefficient of vertical permeability (cm/s)	4.4×10^{-7}

3 the Stabilization Experimental Scheme on Sediment

Firstly, it took 5% fly ash + 1% lime +3% kaolin as the main curing agent, because the Igeo of Cu in sample A was less heavy pollution. The mixing amount of curing agent was in percentage of dried mud quality. And the specific solutions were determined by the adjustment of pH. They were shown in table 5. In this experiment, it studied the reduction rate of copper concentration, water retention and permeability.

Table 5. Curing ratio test scheme

Sample number	the mixing amount of curing agent /%					pH
	Na_2SiO_3	Na_2CO_3	$MgSO_4$	$Fe_2(SO_4)_3$	PVA	
GW 0	/	/	/	/	/	6.9

Sample number	the mixing amount of curing agent /%					pH
	Na_2SiO_3	Na_2CO_3	$MgSO_4$	$Fe_2(SO_4)_3$	PVA	
GW 6	3	/	/	3	1	7.3
GW 7	/	3	/	3	1	8.6
GW 8	1	1	1	3	1	6.2
GW 9	/	/	3	3	1	5.4
GW 10	/	/	6	/	1	7.9

4 Results Analysis

4.1 Heavy Metal Leaching Test

Heavy metals are mostly stabilized in the form of metal hydrates and metal hydroxides in solidified bodies [3]. Heavy metal morphology are change by pH of soil. Thus, it affects the bio-availability of heavy metals[11]. On the condition of acidic, H^+ can destroy the hydration products and obstruct the settlement of metal compounds, or reduce calcium hydroxide concentration reacted with OH^-, which promoting the dissolution of hydration products and accelerating the leaching of heavy metals [12].

The test took acetic acid solution (pH = 4.93 + 0.05) as leach liquor. It simulated acid condition, then study all kinds of solidified sediment's stability. Refer to "soil environmental quality standard" (GB15618-1995), it tested eight heavy metals' content and leaching concentration of copper, zinc, nickel, lead, cadmium, chromium, mercury and arsenic. As shown in table 6, according to "surface water environment quality standard" (GB 3838-2002), leaching liquid of original sample GW0's copper, lead and cadmium were beyond standard of the class V water quality. Solidified sediments of GW6-GW10's toxic leaching results were much better, and they meet standard of the class I ~ III water quality.

Table 6. The heavy metals table of solidified sediment leachate

Testing items	GW0	GW6	GW7	GW8	GW9	GW10
Cu (mg/L)	4.29	0.02	ND	0.12	0.18	0
Zn (mg/L)	0.55	0.42	0.39	0.38	0.21	0.45
Ni (mg/L)	0.23	0.17	0.06	0.27	0.28	0.14
Pb (mg/L)	0.15	ND	ND	ND	7.7×10^{-3}	ND
Cd (mg/L)	2.0×10^{-2}	5.5×10^{-4}	ND	1.5×10^{-3}	3.0×10^{-3}	4.5×10^{-4}
Cr (mg/L)	0.03	ND	ND	0.04	0.05	0.04
Hg (mg/L)	8.0×10^{-4}	ND	ND	ND	6.0×10^{-5}	5.0×10^{-5}
As (mg/L)	1.0×10^{-3}	ND	ND	4×10^{-4}	1.0×10^{-4}	ND
surface water environment quality standard	Beyond class V	I~III	I~III	I~III	I~III	I~III

Copper leaching reduced rate before planting (α_n) and copper leaching changed rate before and after planting (Δx_n) were calculated by formula (2) and (3).

$$\alpha_n = \frac{\rho_0 - \rho_n}{\rho_0} \quad (2)$$

$$\Delta x_n = \frac{\rho_n - \rho_n'}{\rho_n'} \quad (3)$$

Where ρ_0 is original sample's copper leaching concentration before planting (mg/L), ρ_n is solidified sediment of n's copper leaching concentration before planting (mg/L), ρ_n' is solidified sediment of n's copper leaching concentration after planting (mg/L).

From figure 1, copper leaching reduced rate before planting reached more than 95%. It proves that the group of Na_2SiO_3, Na_2CO_3, $MgSO_4$ and $Fe_2(SO_4)_3$ can stable copper ions effectively.

As shown in the figure 2, solidified sediment GW7's copper leaching concentration was minimum after planting kale, and it was basically the same as before curing. This shows that GW7 is the most effective for curing copper, and there is no soluble free copper ion in solidified sediment.

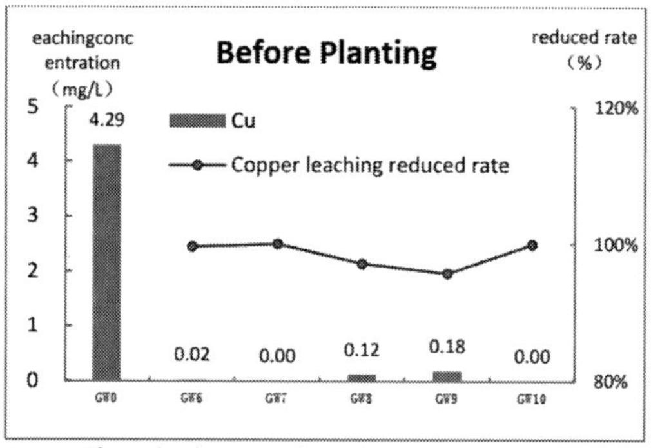

Fig. 1. The reduction rate of preplant stabilized soil's copper leaching

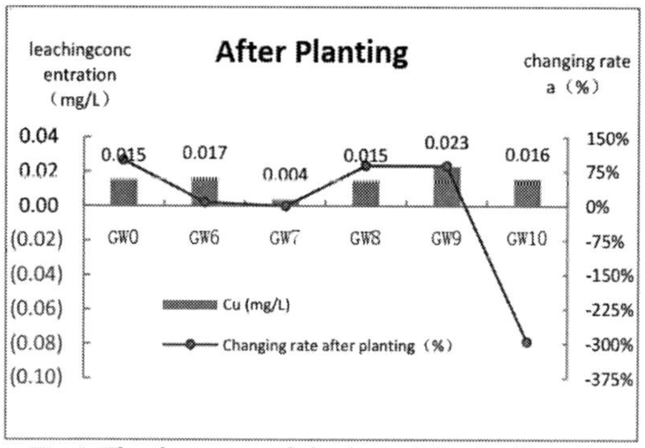

Fig. 2. The change rate of planting stabilized soil's copper leaching

4.2 Percolation Test

According to "Rules of Geotechnical Testing" [14], infiltration is the phenomenon of water moving in porous media. It tested the seepage rate of solidified sediments under 20kPa vacuum pressure in order to understand the influence of different dosage of curing stabilizer.

As shown in the figure 3, it used quantifiable multiunit vacuum suction-filter device by Independent R&D[15]. It recorded the relationship between the suction filter water and time at 20℃, and drawed the relation curve of water filter and time, and measure the mud sample thickness, then got the permeability coefficient of each solidified soil. It was calculated by constant head penetration test formula.

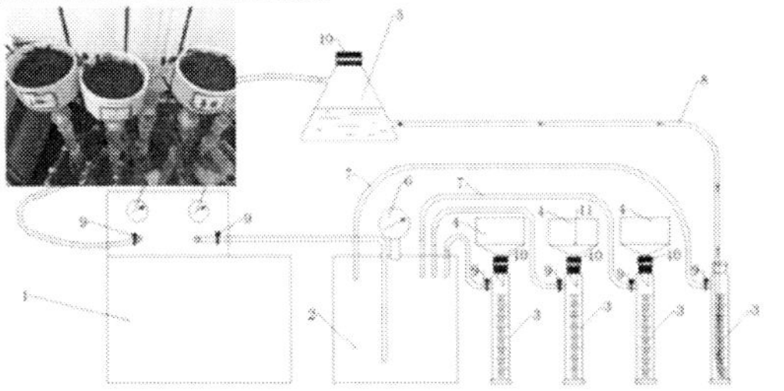

Fig. 3. The change rate of planting stabilized soil's copper leaching

From figure 4 and table 7, under the constant vacuum of 20kPa, the osmotic coefficient of GW10 is 10^{-4}cm/s, and the rest is about 10^{-5}cm/s. The ranking of infiltration is GW10＞GW8＞GW9＞GW0＞GW6＞GW7. At the same time, it shows that the seepage rate increases with time, because there is a lot of fiber in stabilized soil. It provides water seepage channels to improve late permeability.

Table 7. The table of solidified sediment permeability coefficient

vacuum degree	Sample number	water penetration rate mL/s	permeability coefficient cm/s
20kPa	GW0	0.88	6.07×10^{-5}
	GW6	0.73	5.01×10^{-5}
	GW7	0.45	3.11×10^{-5}
	GW8	1.24	8.59×10^{-5}
	GW9	1.14	7.88×10^{-5}
	GW10	8.57	5.92×10^{-4}

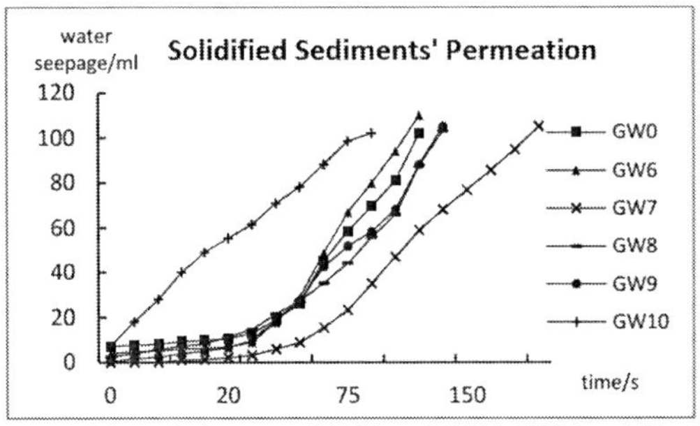

Fig. 4. The emission rate of GW0 and GW6-GW10 stabilized soil

4.3 Water Retention Test

It was helpful to research plant planting experiments through study soil water retention. Put the mud into beaker after stir well, add water to the saturation, and took out and put it in a funnel for 24 hours, then weigh the soil and calculate the initial water content. After that, weigh the soil every 24 hours, observed about 30 days, and calculated soil moisture content.

As shown in the figure 5, the ranking of water retention is GW10＜GW8＜GW6＜GW9＜GW0＜GW7. It is mainly consistent with the law of permeability.

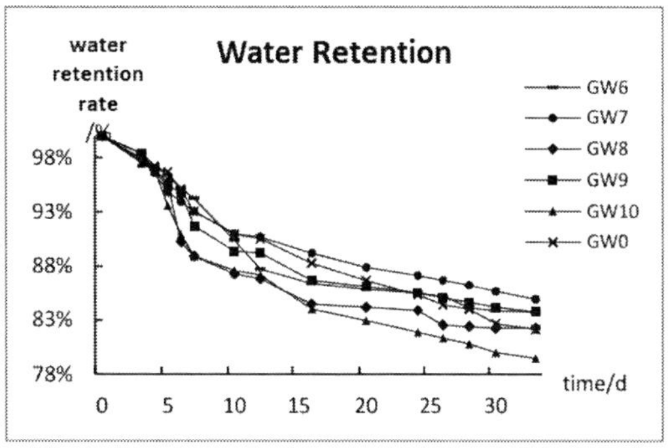

Fig. 5. The water-retention rate of GW0 and GW6-GW10 stabilized soil

4.4 Microscopic Test

From figure 6 and table 8, it had stronger peak and larger area near 26.72°. It proves that solidified sediments consists of quartz primarily. From figure 7, there is a lot of fiber in GW7. It is ensured that plants can breathe normally by the pore and seepage diameter. Therefore, the cultivation of kale grows best in the solidified soil of GW7.

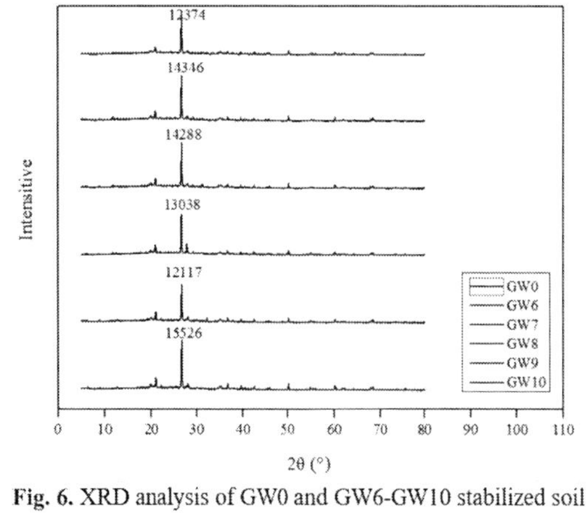

Fig. 6. XRD analysis of GW0 and GW6-GW10 stabilized soil

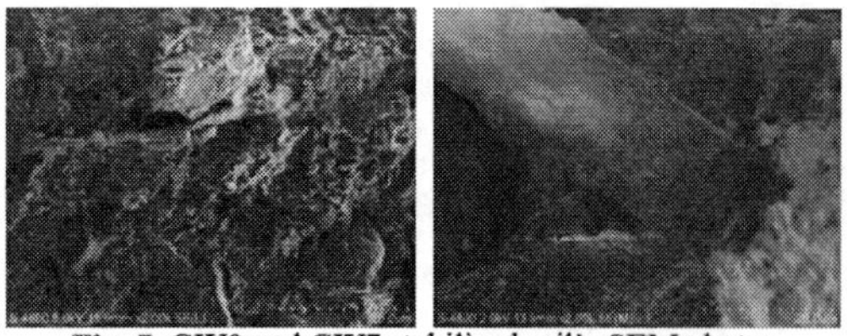

Fig. 7. GW0 and GW7 stabilized soil's SEM photos

Table 8. The eigenvalue of GW7 stabilized soil's XRD

Sample number	2θ (°)	d (Å)	Intensitive
GW7	20.92	4.24	3999
	26.72	3.33	13038
	27.86	3.20	4158
	50.22	1.81	2254

5 Conclusion

This paper presents and research stabilizing treatment of dredged sediment. The result shows that copper loss of stabilized soil's toxicity leaching agent reaches more than 95%, and the permeability coefficient was 10-5 cm/s orders of magnitude. And it ensures GW7 as the best choice of plants experiments, which are 5% fly ash, 1% lime, 3% kaolin clay, 3% sodium silicate, 3% sodium carbonate and 1% polyvinyl alcohol, through the ways of permeability, water retention, SEM and XRD. The scheme effectively stabilizes copper, and keeps the plants as better form than others. However, it has different effect on different plants. Thus it should consider planting varieties optimization in actual applications.

Acknowledgements

This work was financially supported by Zhejiang Province Science and Technology Agency Key Research and Development Project (2017C03008), Zhejiang Province Science and Technology Agency Research Institutes Special Project (2016F50003), Zhejiang Institute of Hydraulics & Estuary Dean's Fund Project(ZIHE20160l1),Zhejiang Province Science and Technology Agency Innovative Methods for Technical Projects (2016IM020100-4), Zhejiang Province Water Resources Department Science and Technology Planning Project(RB1614), Zhejiang Province Water Resources Department Science and Technology Planning Project(RC1733).

References

1. Jia Yunhui. Technology of Repair and treatment on Polluted river sediment[J]. Science & Technology Information, 2014, (4): 226-227.
2. Wang Zhiyong. Analysis of solidification/stabilization on urban life sludge[J]. Urban and rural construction, 2012, (18): 55-58.
3. Malviya R, Chaudhary R. Leaching behavior and immobilization of heavy metals in solidified/stabilized products[J]. Journal of hazardous materials, 2006, 137(1): 207-217
4. Jiang Jianguo, Wang Wei. The history, present status and development of solidification/stabilization technology in treating with hazardous waste[J]. Advances in Environmental Science, 1998, 6(1): 55-62.
5. Wang Wei, Yuan Guangjue. The state and development of solid waste disposal in China [J]. Environmental Science, 1997, 18(2): 87-90.
6. Ji Shunjian. Effects of solidification/stabilization of sludge on the morphology of heavy metal[D], Hohai University, 2008.
7. Liu Xiumei, Zhang Fudao, Zhang Shuqing, etc. Study on adsorption and desorption properties of nano-kaoline to nitrogen, phosphorus, postash and organic carbom[J]. Scientia Agricultura Sinica, 2005, 38(1): 102-109.
8. Chen C L, LOSL, Kuan Wenhui, etal, Stabilization of Cu in Acid extracted Industrial Sludge Using a Microwave Process[J]. Journal of Hazardous Materials, 2005, 123(13): 256-261.
9. Fu Gui, Zhao Dezhao, Cheng Haifeng. Comparison and analysis of comprehensive utilization of dredged materials at home and abroad[J]. Port & Waterway ngineering, 2011, 3(451): 90-96.
10. Li Gongzhen, Wang Xiao. Study on the pollution and release of heavy metals in the main canal of the Beijing-Hangzhou Grand Canal (Subei Section) [M]. Jiangsu Xuzhou: China University of Mining and Technology , 2013:55-58.
11. Ding Jianghua, Wen yanmao, Shu Qiang. Fraction transformation of cadmium and Zinc in soils[J]. Urban environment & Urban ecology, 2001,14(2): 47-49.
12. Chen Q Y, Tyrer M, Hills C D, et al. Immobilization of heavy metal in cement-based solidification/stabilisation:A review[J]. Waste Management, 2009, 29(1): 390-403
13. Fan Yunhui, Wang Yanqing. Background characteristics of soil elements in four plants of Zhejiang Province[J]. Geophysical and Geochemical Exploration, 2009, (02):132-134.
14. SL 237-1999, Test methods of soils[S]. Beijing: Ministry of Water Resources of the People's Republic of China,1999.
15. Shi Yannan, Zhang Chaojie, Yu Jiongqi, etc. A quantifiable multi-link vacuum extraction stabilizer[P]. China: ZL201620060109.2, 2016.

Corrosion resistance of the microarc oxidation coatings prepared on magnesium alloy

Ying Lv, *Jun Gang* Li*, *Ming Zhong* Wu, *Zhen* Ma, *Jing Qiang* Zhang, and *Le Le* Wang

School of Materials Science and Engineering, Jiamusi University, Jiamusi154007, China

Abstract. Ceramic coatings were prepared on the surface of AZ91D magnesium alloy by microarc oxidation technology. The effects of different voltages on morphology, phase composition and thickness of the coatings were characterized by SEM and XRD. The corrosion resistance of the coatings was measured by electrochemical workstation. Results indicated that the microarc oxidation coatings prepared in sodium silicate electrolyte exhibited porous surface and mainly comprised MgO, Mg_2SiO_4 and a small amount of $MgAl_2O_4$. The thickness of the oxide coatings increased rapidly with the increase of voltage. The coating prepared at 400V voltage had good electrochemical corrosion resistance in 3.5wt% NaCl solution.

1 Introduction

In light metal structural materials, magnesium alloys possess many excellent properties such as high specific strength, good electric conductivity and thermal conductivity, vibration damping, electromagnetic shielding, easy processing molding and recycling use [1]. Therefore, they have wide prospect in the fields of electronic industry, aerospace, automobile, biomedicine, etc. However, magnesium alloys have poor corrosion resistance and high-temperature resistance. The standard electrode potential of pure magnesium is only -2.37V, which is the lowest in all the structural metals[2]. The oxide film formed on the magnesium surface in the air has porous structure, resulting in the poor protection for the substrate[3]. So magnesium alloys are not suitable for most corrosive environments, which has restricted their widespread applications.

The use of surface treatment technology can improve the properties of magnesium alloys[4]. Microarc oxidation (MAO) is a new nonferrous metal surface modification technique developed on the basis of anodic oxidation[5,6]. Under the action of a strong electric field, the gas in the electrolyte is ionized, and the microarc discharge is produced[7]. As a result, the ceramic oxide coating in-situ forms on the surface of nonferrous metal such as Al, Ti and Mg. MAO has the advantages of simple process, high efficiency and low pollution, which has became the most potential surface treatment technology for magnesium alloys[8]. In the present study, ceramic coatings were obtained on the surface of AZ91D magnesium alloy by the MAO technology. The effects of different voltages on morphology, microstructure and thickness of the MAO coatings were studied, and the corrosion resistance of the coatings was also investigated.

2 Experimental

The substrate material was AZ91D magnesium alloy ingot which was cut into the samples with a dimension of 20mm×15mm×3mm by line cutting device. Every sample was drilled a hole with 2mm in diameter, then ground with 200, 400, 600 and 800 mesh SiC sandpapers step by step, finally immersed into the acetone and cleaned 3min by ultrasonic washing machine.

An alkaline electrolyte was prepared from the deionized water solution of $Na_2SiO_3 \cdot 9H_2O$ (20g/L), NaOH (5g/L) and $C_3H_8O_3$ (10mL/L). A bipolar pulsed power supply produced by Harbin Institute of Technology was used to fabricate the MAO coating. The stainless steel electrobath was as the cathode. The samples, suspended with aluminum wire, were immersed in the electrolyte as the anode. The experiments were carried out under constant voltage mode with frequency of 400Hz and duty ratio of 8%. The voltage was adjusted to 350V, 400V, and 450V, respectively. The temperature of the electrolyte was controlled under 38°C. After the samples were oxidized for 15min, closed the power, then removed the samples. The preparation process of the MAO coatings was shown in Fig.1.

3 Results and discussion

3.1 Phase composition of the MAO coating

XRD pattern of the MAO coating fabricated at 400V voltage was shown in Fig.2. The coating was mainly composed of MgO, Mg_2SiO_4 and a small amount of and $MgAl_2O_4$. In the process of MAO, the oxidation reaction of the magnesium substrate and oxygen produced MgO under the action of arc, then Mg_2SiO_4 was generated owing to the participation of the electrolyte.

* Corresponding author: 8921107ljg@163.com

 https://doi.org/10.1051/e3sconf/20183802009

Fig. 1. The preparation process of the MAO coatings.

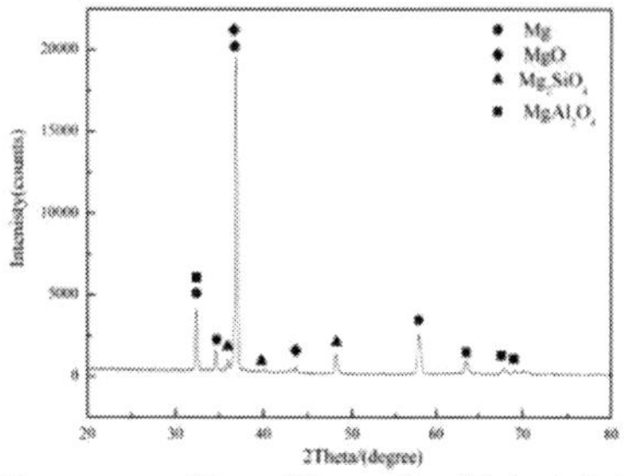

Fig. 2. Phase composition of the coating fabricated at 400V.

3.2 Morphology of the MAO coating

The morphology of the MAO coating fabricated at different voltages was represented in Fig.3~5. It was found that the surface layer of the MAO coatings had porous structure with irregular micropores. When the voltage was 350V, the arc was so weak that electric breakdown occurred in the same position for a long time. As a result, the layer surface was uneven and distributed 2~15μm diameter micropores. There were some protuberances around the micropores, as shown in Fig. 3. When the voltage was 400V, the arc starting was stable and fine, so the layer surface was smooth with uniform micropores of 2~5μm diameter, as shown in Fig. 4. When the voltage was raised to 450V, the arc starting was acute and the arc became thicker. Consequently, the molten substances that sprayed out along the arc channel were on the increase, which made the surface rougher, as shown in Fig. 5.

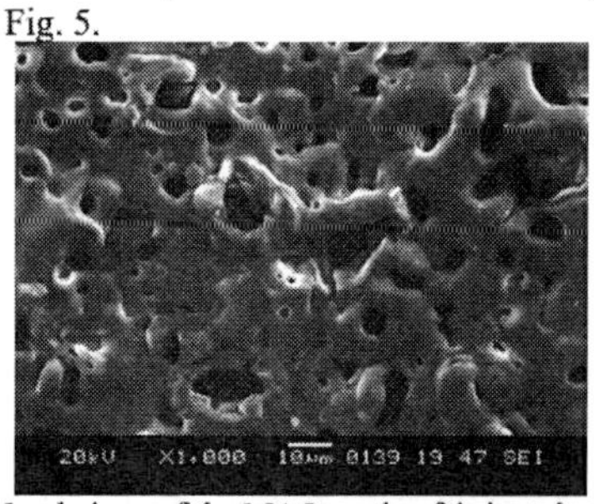

Fig. 3. Morphology of the MAO coating fabricated at 350V

The thickness of the MAO coatings was illustrated in table.1. It could be seen that the thickness of the coatings increased with the voltage. When the voltage was 450V, the coating was the thickest with 16.4μm.

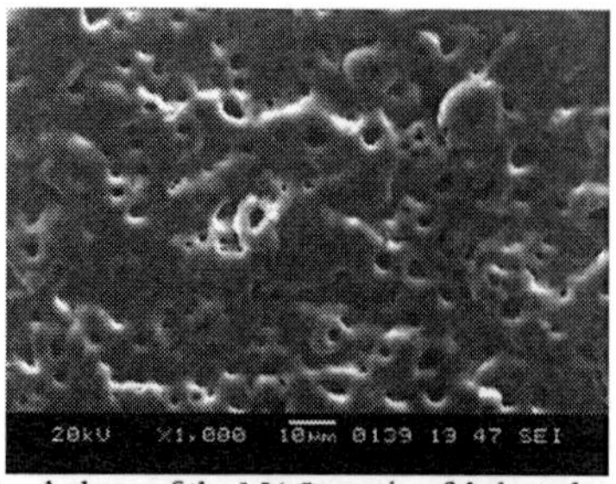

Fig. 4. Morphology of the MAO coating fabricated at 400V.

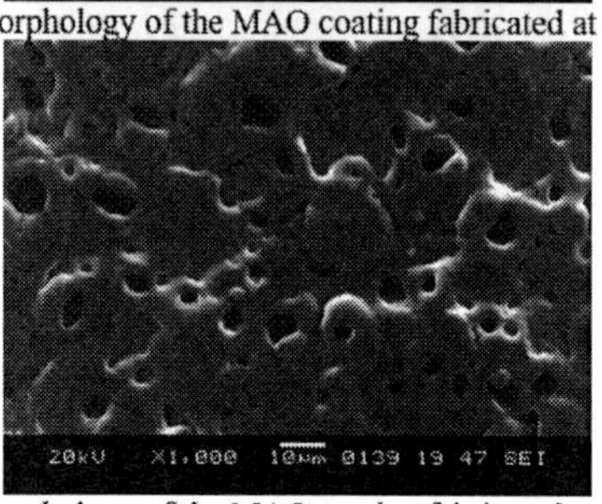

Fig. 5. Morphology of the MAO coating fabricated at 450V.

Table 1. The thickness of the MAO coating.

Voltage(V)	350	400	450
Thickness(μm)	10.1	14.2	16.4

3.3 Formation process of the MAO coating

After the AZ91D alloy was placed in the electrolyte and electrified, a thin insulated oxide film formed immediately on the substrate surface. Once the voltage applied on the samples exceeded a certain critical value, some weak portions on the oxide film were broken down and microarc discharge occurred. The impact of arc on the substrate surface resulted in a severe oxidation reaction, releasing a large amount of gas. The reactions generated in the electrolyte were as follows:

$$Mg \rightarrow Mg^{2+} + 2e \quad (1)$$

$$Al \rightarrow Al^{3+} + 3e \quad (2)$$

$$O^2 + 4e \rightarrow 2O^{2-} \quad (3)$$

$$Mg^{2+} + O^{2-} \rightarrow MgO \quad (4)$$

$$Mg^{2+} + H_2O \rightarrow MgO + H_2\uparrow \quad (5)$$

$$2Al^{3+} + 3H_2O \rightarrow Al_2O_3 + 3H_2\uparrow \quad (6)$$

$$2Mg^{2+} + SiO_3^{2-} + O^{2-} \rightarrow Mg_2SiO_4 \quad (7)$$

$$Mg^{2+} + Al_2O_3 + O^{2-} \rightarrow MgAl_2O_4 \quad (8)$$

Under the action of the arc, the molten MgO, $MgAl_2O_4$ and a small amount of $MgAl_2O_4$ were mixed together. Due to high temperature, large amounts of H2 in the arc channel produced high pressure which helped above mixture to eject from the arc channel and accumulate around the channel. Ultimately, owing to the rapid cooling of the electrolyte, molten mixture solidified and produced circular protuberances covering on the AZ91D alloy surface.

In the MAO process, the thickness of the coatings changed constantly. When the voltage was 350V, the coatings became thicken, which caused arc starting to be weaker. Arc was only generated at local position for a long time, therefore many large diameter arc channel appeared. When the voltage was 400V, the arc starting was uniform and the arc position changed frequently. Fine arc could transfer rapidly in all the positions, so the arc channel formed had relatively small diameter. When the voltage was 450V, the arc became thicker and its duration prolonged, causing the increase of the ejected mixture and the height of the protuberances around the arc channel. Ultimately, the oxide coating was formed on the surface of AZ91D alloy under the action of arc. With the increase of oxidation time, the oxide coatings became thicker and had good bonding with the substrate.

3.4 Electrochemical corrosion analysis of the MAO coating

Fig. 6 presented the dynamic polarization curve of the MAO coatings prepared at different voltages in 3.5wt%NaCl solution. The corrosion potential and corrosion current of the coatings were shown in Table 2.

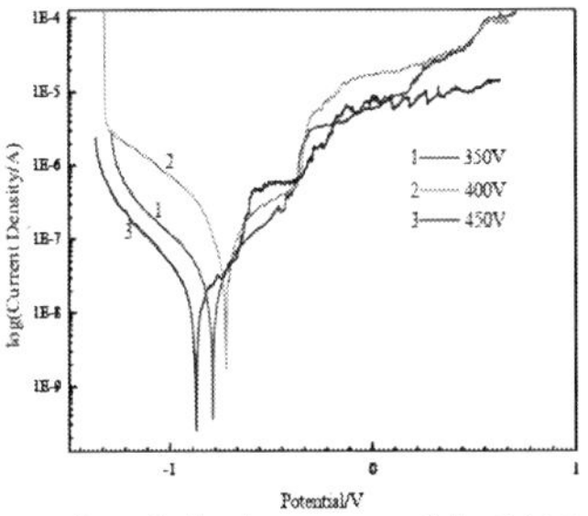

Fig.6. Dynamic polarization curves of the MAO coating.

Table 2. Electrochemical corrosion current and potential.

Voltage (V)	350	400	450
Corrosion potential (mV)	-780.755	-734.353	-852.568
Corrosion current (mA)	8.348 $\times 10^{-5}$	2.385 $\times 10^{-4}$	1.173 $\times 10^{-4}$

During the electrochemical corrosion, the cathode polarization mainly involved hydrogen evolution process in which the H^+ in the solution had been replaced by Mg. Anodic polarization referred to the active dissolution of Mg alloy, that is, Mg^{2+} and Al^{3+} transferred from the substrate to the NaCl solution and produced hydrated cations. According to Fig. 6, a buffer platform of corrosion current density appeared in curve 2 and curve 3 in the anodic polarization process, and this anodic passivation phenomenon was benefit for the improvement of the corrosion resistance. When the voltage was 400V, the MAO coating exhibited the lowest corrosion current and the highest corrosion potential, indicating that its corrosion resistance was best of all the MAO coatings.

4 Conclusions

The MAO coatings prepared on AZ91D alloy exhibited porous surface and mainly comprised MgO, Mg_2SiO4 and a small amount of $MgAl_2O4$.When the voltage was 400V, the coating had the best surface quality with small and uniform micropores. The thickness of the oxide coatings increased rapidly with the increase of voltage. The thickness of the coating was 10.1μm at 350V and 16.4μmm at 450V. When the voltage was 400V, the MAO coating obtained the lowest corrosion current and the highest corrosion potential, indicating that its corrosion resistance was best of all the MAO coatings.

Acknowledgements

This work was supported by the Basic Scientific Research Foundation of Heilongjiang Provincial Education Department (2016-KYYWF-0552), Key Scientific Research Foundation of Jiamusi University (Lz2013-013) and Open Foundation of University Key Laboratory of Biomaterial (YB12001). The authors thank the Engineering Center for Wear Resistant Materials and Surface Technology of Ministry of Education (Jiamusi University).

References

1. G. Wu, K.C. Chan, L.L. Zhu, L. Sun, J Lu, Nature, 545 (2017)
2. A. Pardo, M.C. Merino, M. Mohedano, P. Casajús, A.E. Coy, R. Arraba, Surf. Coat. Technol. 203 (2009)
3. Y.H. Sun, R.C. Wang, C.Q. Peng, Y. Feng, M. Yang, T. Nonferr. Metal. Soc. 27 (2017)
4. Y.J. Zhang, C.L. Zhao, S. Tao, S.G. Jia, Z.C. Liu, Appl. Mech. Mater. 713 (2015)
5. X.J. Cui, X. Dai, B.Y. Zheng, Y. Zhang, J. Chin. Soc. Corros. Prot. 37 (2017)
6. L.Y. Cui, S.D. Gao, P.P. Li,R.C. Zeng, F. Zhang, Corros. Sci. 118(2017)
7. B. Zou, G.H. Lu, G.L. Zhang, Y.Y. Tian, T. Nonferr. Metal. Soc. 25 (2015)
8. M. Tang, Z. Feng, G. Li, Z. Zhang, R. Zhang. Surf. Coat. Technol. 264(2015)

Optimization of extraction of chitin from procambarus clarkia shell by Box-Behnken design

Fang Dong, *Hailong* Qiu, *Shaoqian* Jia, *Cuiping* Dai, *Qingxin* Kong, *Changliang* Xu

Jiangsu Food & Pharmaceutical Science College, 223003, Huai'an, China

Abstract. This paper investigated the optimizing extraction processing of chitin from procambarus clarkia shell by Box-Behnken design. Firstly, four independent variables were explored in single factor experiments, namely, concentration of hydrochloric acid, soaking time, concentration of sodium hydroxide and reaction time. Then, based on the results of the above experiments, four factors and three levels experiments were planned by Box-Behnken design. According to the experimental results, we harvested a second-order polynomial equation using multiple regression analysis. In addition, the optimum extraction process of chitin of the model was obtained: concentration of HCl solution 1.54mol/L, soaking time 19.87h, concentration of NaOH solution 2.9mol/L and reaction time 3.54h. For proving the accuracy of the model, we finished the verification experiment under the following conditions: concentration of hydrochloric acid 1.5mol/L, soaking time 20h, concentration of sodium hydroxide 3mol/L and reaction time 3.5h. The actual yield of chitin reached 18.76%, which was very close to the predicted yield (18.66%) of the model. The result indicated that the optimum extraction processing of chitin was feasible and practical.

1 Introduction

Chitin owns excellent biological efficacy and diverse applications in different fields, such as textile industry, feed industry, medical materials, environmental protection, and so on [1]. For example, chitin is utilized for treating wastewater as adsorbents and flocculants [2,3], and is vital for artificial skin, artificial dialysis membrane and artificial blood vessels in medical materials field [4,5]. The shell of arthropod and fungus are rich in chitin. Such as prawns and crabs, the content of chitin is up to 58%~85%. Procambarus clarkia is one of the most popular aquatic products in China, but the shell of procambarus clarkia is usually discarded as waste of the food processing industry. Herein, the extraction of chitin from the shell of procambarus clarkia has dual significance for saving resource and protecting environment. In order to achieve the above purpose, this paper extracted chitin from the shell of procambarus clarkia by acid-alkali method, explored the vital factors influencing extraction by single factor investigation and selected the optimum processing by Box-Behnken design, one of the response surface methods.

2 Experimental section

2.1 Experimental methods

The certain weight of powder made from clean dried rocambarus clarkia shell was soaked in HCl solution with different concentrations at room temperature for certain soaking time to remove minerals. Then, HCl solution was discarded and the precipitate was washed pH to neutral by water. After that, the precipitate was reacted in NaOH solution with different concentrations in 95°C water-bath for different reaction time to remove protein. After dumping NaOH solution and washing the precipitate to neutral by water, the crude chitin was obtained.

The crude chitin was soaked in 3g/L KMnO4 solution for 1.5h. Then, the chitin was filtered and washed to remove KMnO4. 5g/L oxalate solution mixed with chitin in 65°C water-bath to decolor until chitin was turned white. Finally, the refined chitin dried at a lower temperature for 24h was obtained after removing oxalate solution and washing pH to neutral [6].

Yield of chitin (%) = (Weight of chitin/Weight of dried powder of rocambarus clarkia shell (30g))×100%.

2.2 Single factor investigation

The main factors affecting extraction of chitin were concentration of HCl solution (mol/L), soaking time (h), concentration of NaOH solution (mol/L) and reaction time (h). Therefore, the four factors were explored respectively through single factor experiment to select the optimal conditions of each influence factor.

2.3 Optimization of Box-Behnken design

Based on the single factor test results, the four factors (A, concentration of HCl solution; B, soaking time; C, concentration of NaOH solution; D, reaction time) at

* Corresponding author: dong1234fang@126.com

https://doi.org/10.1051/e3sconf/20183802010

three levels was performed. According to the principle of Box-Behnken design, the low, medium and high levels of each factor as independent variable was coded -1, 0, and +1, respectively. Meanwhile, the yield of chitin was the response value (Y) as dependent variable. The factors and their levels are shown in Table 1.

Table 1 Factors and levels of the extraction of chitin process

Levels	Factors			
	A (mol/L)	B (h)	C (mol/L)	D (h)
-1	1	15	1.5	3
0	1.5	20	2.5	3.5
+1	2	25	3.5	4

2.4 Statistical analysis

Design-Expert software (Version 7.0) was employed in planning Box-Behnken design and analyzing the following data. All extraction experiments were implemented in triplicate.

3 Result sections

3.1 Single factor assays

Fig.1 showed the results of single factor investigation of extraction of chitin. Fig. 1(a) indicated that the yield of chitin rapidly increased when the concentration of HCl solution reached 1.5mol/L(other factor levels: soaking time 10h, concentration of NaOH solution 0.5mol/L, reaction time 2h), but the yield of chitin did not obviously increase while concentration of HCl solution reached 2mol/L and 2.5mol/L. Similarly, Fig.1(b) showed the yield of chitin signficantly increased at soaking time 20h, while slowly increased at soaking time 25h and 30h(other factor levels: concentration of HCl solution 1.5mol/L, concentration of NaOH solution 0.5mol/L, reaction time 2h). The reason for these results was that the rocambarus clarkia shell was fully remove minerals when concentration of HCl solution reached 1.5mol/L and soaking time was 20h. The higher concentration of HCl solution and the longer soaking time could not obviously promote the yield of chitin. Fig. 1(c) illustrated that the yield of chitin reached a maximum value at concentration of NaOH solution 2.5mol/L(other factor levels: concentration of HCl solution 1.5mol/L, soaking time 20h, reaction time 2h). Fig.1(d) showed that the yield of chitin evidently increased when the reaction time was 3.5h. The reason was that the higher concentration of NaOH solution and the longer reaction time could decompose the chitin. Herein, concentration of HCl solution 1~2mol/L, soaking time 15~20h, concentration of NaOH solution 1.5~3.5mol/L and reaction time 3~4h were considered to be optimal levels in the Box-Behnken design.

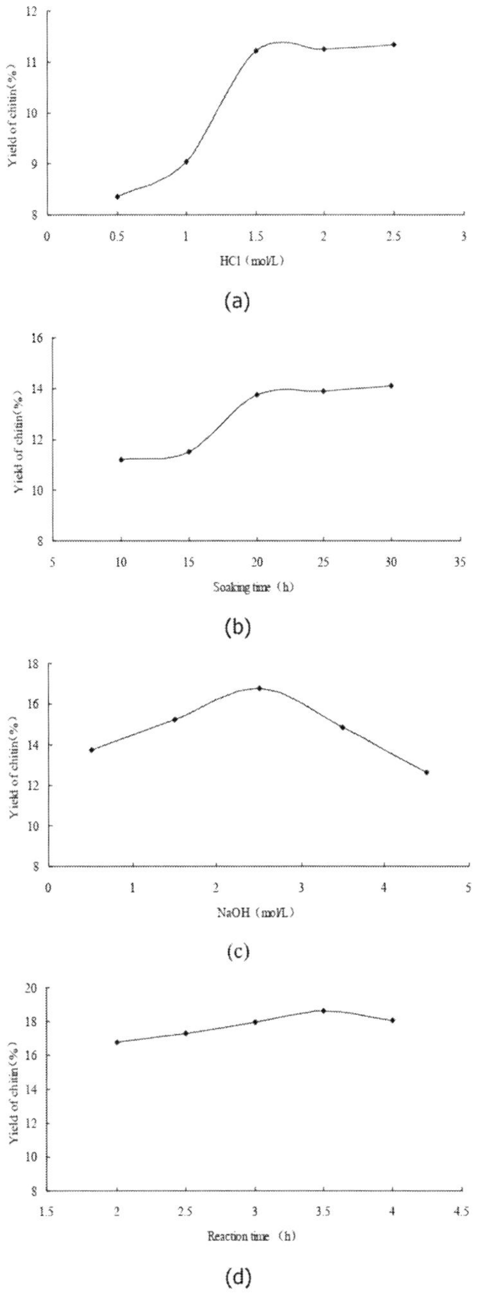

Fig. 1. Effect of four single factors on the extraction of chitin

3.2 Experimental results of the yield of chitin by Box-Behnken design

Twenty-nine experiments planned by Box-Behnken design and their results of the yield of chitin were shown in Table 2. No.1~24 was the factorial experiment, and No.25~29 was the central experiment. It could be seen from Table 2, the yield of chitin ranged from 16.18% to 18.79%.

Table 2 Box-Behnken design and the results of yield of chitin

No.	Factors and levels				Yield of chitin (%)
	A (mol/L)	B (h)	C (mol/L)	D (h)	
1	-1	-1	0	0	16.32
2	1	-1	0	0	16.67
3	-1	1	0	0	16.52
4	1	1	0	0	16.18
5	0	0	-1	-1	17.21
6	0	0	1	-1	17.83
7	0	0	-1	1	17.16
8	0	0	1	1	17.65
9	-1	0	0	-1	17.28
10	1	0	0	-1	16.68
11	-1	0	0	1	16.89
12	1	0	0	1	17.71
13	0	-1	-1	0	16.64
14	0	1	-1	0	17.72
15	0	-1	1	0	17.95
16	0	1	1	0	17.64
17	-1	0	-1	0	16.27
18	1	0	-1	0	16.25
19	-1	0	1	0	17.08
20	1	0	1	0	17.93
21	0	-1	0	-1	17.52
22	0	1	0	-1	17.63
23	0	-1	0	1	17.33
24	0	1	0	1	17.98
25	0	0	0	0	18.55
26	0	0	0	0	18.79
27	0	0	0	0	18.48
28	0	0	0	0	18.69
29	0	0	0	0	18.37

3.3 Model fitting and statistical significance analysis

According to the experimental results of Table 2, a second-order polynomial regression equation of extraction of chitin was obtained by Design-Expert 7 software analysis and fitting.

$$Y=18.49+0.15A+0.094B+0.53C+0.12D-0.022AB-0.038AC+0.21AD-0.048BC+0.11BD-7.500E-003CD -1.23A^2-0.80B^2-0.69C^2-0.22 D^2$$

In the above formula, Y indicated the yield of chitin, A was the concentration of HCl solution, B was soaking time, C was the concentration of NaOH solution, and D was the reaction time.

Table 3 showed that yield of chitin was extremely significantly affected by linear parameters (C), quadratic parameters (A^2, B^2, C^2) ($P<0.01$), and significantly affected by interactive parameters (AD,BC) and quadratic parameters(D^2) ($P<0.05$). Other parameters showed no significant effect ($P>0.05$).In addition, the R2 of chitin extraction model was 0.9157, the Adeq Precision was more than 4, and coefficient variation (C.V.) was 0.0128. These indicated that the model obtained by optimization experiment had the high degree of precision. The simulation of four factors and three levels could be used to optimize the experiment, and the accuracy of the experiment was higher.

Table 3 Analysis of optimal process regression model for extraction of chitin

Parameter	Sum of Squares	Df	Mean Square	Standard Deviation	F Value	P Value
Model	15.24	14	1.09	-	10.86	<0.0001**
A	0.094	1	0.094	0.091	0.93	0.3502
B	0.13	1	0.13	0.091	1.28	0.2772
C	1.94	1	1.94	0.091	19.39	0.0006**
D	0.027	1	0.027	0.091	0.27	0.6114
AB	0.12	1	0.12	0.16	1.19	0.2943
AC	0.19	1	0.19	0.16	1.89	0.1911
AD	0.50	1	0.50	0.16	5.03	0.0416*
BC	0.48	1	0.48	0.16	4.82	0.0455*
BD	0.073	1	0.073	0.16	0.73	0.4081
CD	4.225E-003	1	4.225E-003	0.16	0.042	0.8403
A^2	9.87	1	9.87	0.12	98.51	<0.0001**
B^2	3.12	1	3.12	0.12	31.15	<0.0001**
C^2	1.89	1	1.89	0.12	18.87	0.0007**
D^2	0.78	1	0.78	0.12	7.82	0.0143*
Residual	1.40	14	0.10	-	-	-
Lack of Fit	1.29	10	0.13	-	4.65	0.0759
Pure Error	0.11	4	0.028	-	-	-
Cor Total	16.65	28	-	-	-	-

R^2 =0.9157, Adjusted R^2=0.8314, C.V.%=1.82, Adeq Precision=10.125,$^*P < 0.05$, $^{**}P < 0.01$.

3.4 Optimization of extraction processing of chitin

On the basis of the regression equation obtained by Design-Expert 7.0 software, the response surface plots of the yield of chitin affected by A, B, C and D were shown in Fig.2. The contour plots represented the intensity of the interaction actions. According to Table 3 and Figure 2, the interaction between concentration of HCl solution and reaction time was significant ($P<0.05$). Likewise, the interaction between soaking time and concentration of NaOH solution aslo showed the significance ($P<0.05$).

Meanwhile, the predictive optimum processing of extraction of chitin was harvested from the response surface analysis. The was shown as follows: concentration of HCl solution 1.54mol/L, soaking time 19.87h, concentration of NaOH solution 2.9mol/L and reaction time 3.54h. The maximum predicted yield of chitin was 18.66%.

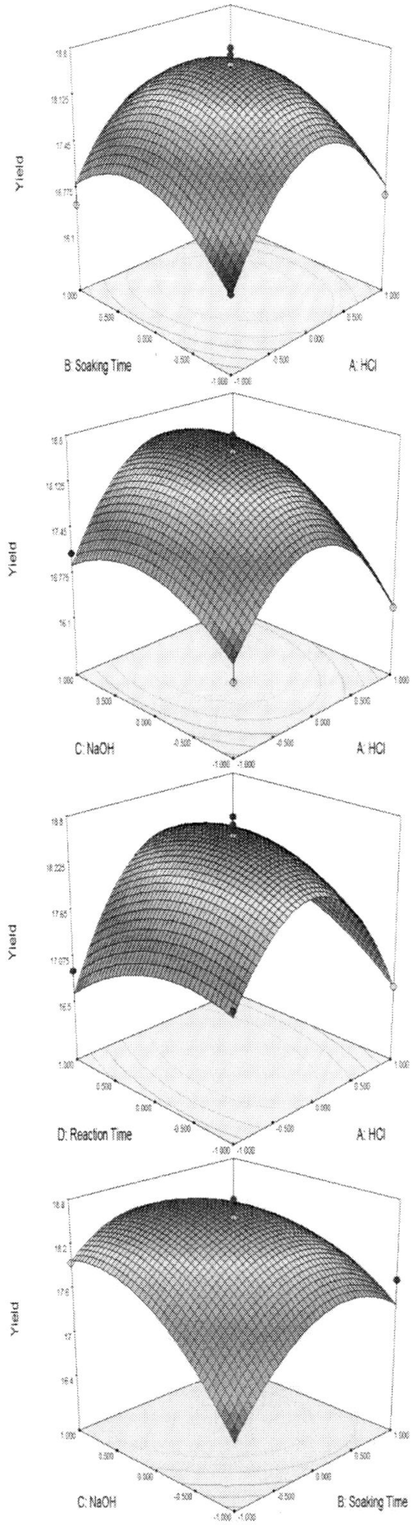

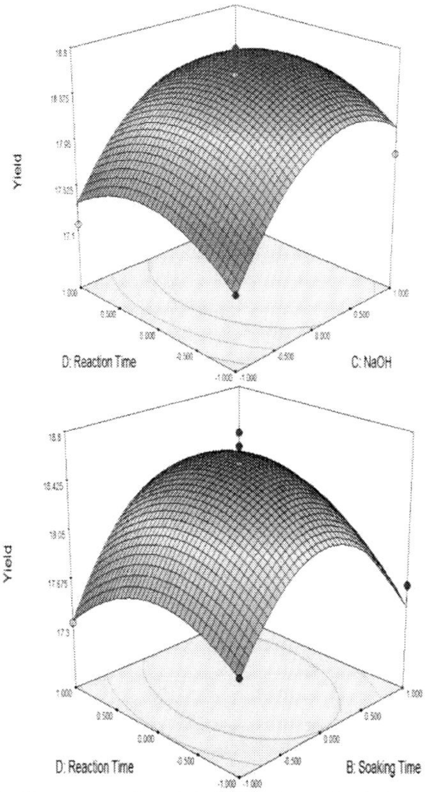

Fig. 2. Response surface plots (3D) of Interaction between the four factors to the yield of chitin

3.5 Model verification

In order to verify the accuracy and reliability of the predictive optimum processing of extraction of chitin, moreover, considering the convenience of the actual operation, the optimum processing of extraction of chitin were implemented as follows: concentration of HCl solution 1.5mol/L, soaking time 20h, concentration of NaOH solution 3mol/L, reaction time 3.5h. As Table 4 showed, the actual result was close to the predictive result.

Table 4 Predictive and actual results of extraction of chitin

	Concentration of HCl solution (mol/L)	Soaking time (h)	Concentration of NaOH solution (mol/L)	Reaction time (h)	Yield of chitin (%)
Predictive	1.54	19.87	2.9	3.54	18.66
Actual	1.5	20	3	3.5	18.76

4 Conclusion

Procambarus clarkia shell is a new source of chitin. How to make full use of procambarus clarkia shell has great significance to environmental protection and new resources development. This paper investigated the optimizing extraction processing of chitin from procambarus clarkia shell by Box-Behnken design. Through single factor investigation, we found the optimal levels of four factors (concentration of HCl solution, soaking time, concentration of NaOH solution and reaction time). According to the single factor assays, Box-Behnken design planned 29 experiments to optimize of the extraction processing of chitin. The actual processing was shown as follows: concentration of HCl solution 1.5mol/L, soaking time 20h, concentration of NaOH solution 3mol/L, reaction time 3.5h. The actual yield of chitin (18.76%) was close to the predictive value (18.66%). It indicated that the model was accurate and the optimum extraction processing of chitin was feasible and practical.

References

1. E.Tabata, A. Kashimura, S. Wakita, et al, Int. J. Mol. Sci **19**, 14(2018)
2. M Osada, C. Miura, Y.S. Nakagawa, et al, Carbohydr. Polym. **134**, 8(2015).
3. Y.Q. Li, C.Yang, X.Q. Gong, et al. Nat. Prod. Res. Dev. **28**, 4(2016)
4. Y.L. Li. Cott. Textile Technol, 3**8**, 4(2010)
5. K. Brodaczewska, K. Donskow-Lysoniewska, M. Doligalska, Acta Parasitol. **60**, 8(2015)
6. Q. L. Xu, Q. Z. Zeng. China Food Addit. 6 (2013)

Modelling for Arsenic's transferring process in Jiaozhou Bay

Dongfang Yang[1, 2, *], *Xiaolong* Zhang[1], *Qi* Wang[1], *Yuan* Zhang[1], and *Haixia* Li[1]

[1]Accountancy Shool, Xijing University, Xian 710123, China
[2]North China Sea Environmental Monitoring Center, SOA, Qingdao 266033, China

Abstract. This paper quantified the horizontal and vertical transferring processes of As in Jiaozhou Bay using investigation data in April and August 1988. Results showed that the horizontal absolute loss amounts of As contents in surface waters were 0.10-1.42 μg L^{-1}, and the horizontal relative loss amounts in surface waters were 7.04%-25.00%, respectively. The horizontal absolute loss amounts of As contents in bottom waters were 0.04-1.00 μg L^{-1}, and the horizontal relative loss amounts in bottom waters were 2.85%-41.66%, respectively. The vertical absolute dilution amounts of As contents were 0.02-0.28 μg L^{-1}, and the vertical relative dilution amounts were 1.78%-16.66%. The vertical absolute accumulation amounts of Pb contents were 0.10-0.86 μg L^{-1}, and the vertical relative accumulation amounts were 7.35%-35.83%. During the transferring processes through the bay mouth, no matter from the internal waters to the open water, or from the open waters to the internal waters, As contents in both surface and bottom layers were decreasing in a certain degree. The changes of As contents in surface and bottom waters were determined by the transferring distance of As in surface waters.

1 Introduction

The excessive existence of As in the environment is harmful to organism and ecosystem since As is high toxic ([1-2]). Many marine bays have been polluted by As due to the rapid development of industrial economic and the increase of population size ([3-6]). Understanding the transferring processes of As in marine bays is essential to pollution control. Jiaozhou Bay is a semi-closed bay located in Shandong Province China, and has been polluted by various pollutants including As after the rapid increasing of industry the past three decades ([1-2]). This paper quantified the horizontal and vertical transferring processes of As in Jiaozhou Bay using investigation data in April and August 1988. The aim of this paper was to better understand the transporting processes of As in marine bay, and provide basis for scientific research and environment remediation.

2 Materials and method

2.1 Study area

Jiaozhou Bay is located in the south of Shandong Province, eastern China (35°55'-36°18' N, 120°04'-120° 23' E). The total area and average water depth are 446 km^2 and 7 m, respectively. The bay mouth is very narrow (3 km), and is connected to the Yellow Sea in the south (Fig. 1). There are a dozen of rivers including Dagu River, Haibo Rriver, Licun Rriver, and Loushan Rriver etc., all of which are seasonal rivers ([7-8]).

2.2 Data collection

In April 1988, As's contents in surface and bottom waters were measured in Site D5 in the east of bay, in Site A5 in the bay mouth, and in Site A1 in the open waters, respectively (Fig. 1). In August 1988, As's contents in surface and bottom waters were measured in Site B5 in north of the bay, in Site A5 in the bay mouth, and in Site A1 in the open waters, respectively (Fig. 1). As in waters was sampled and monitored follow by National Specification for Marine Monitoring ([9]).

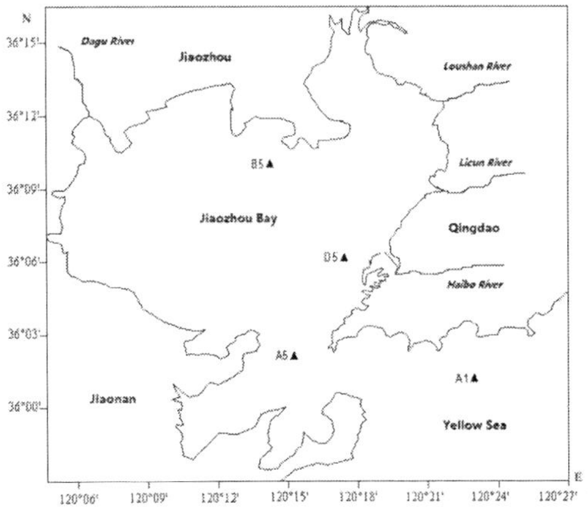

Fig. 1. Geographic location and monitoring sites in Jiaozhou Bay.

* Corresponding author: dfyang_dfyang@126.com

2.3 Modelling for horizontal transferring processes of As

By means of the water exchange via the bay mouth, substance's contents in waters in Jiaozhou Bay were changing continuously ([4-6], [10]). Supposed that certain substance's contents in surface and bottom waters in the bay are A and a, in the bay mouth are B and b, and in the open waters are C and c, respectively.

In surface waters, and from the inner of the bay to the open waters, substance's contents are changing, and the calculation formula for transferring process is:

$$D=A\text{-}B,\quad E=(100\times|A\text{-}B|/\max(A, B))\% \quad (1)$$

where, D is the horizontal absolute loss amount in surface waters from the inner of the bay to the open waters, E is the corresponding horizontal relative loss amount in surface waters.

In surface waters, and from the open waters to the inner of the bay, substance's contents are changing, and the calculation formula for transferring process is:

$$F=B\text{-}C,\quad G=(100\times|B\text{-}C|/\max(B, C))\% \quad (2)$$

where, F is the horizontal absolute loss amount in surface waters from open waters to the inner of the bay, G is the corresponding horizontal relative loss amount in surface waters.

In bottom waters, and from the inner of the bay to the open waters, substance's contents are changing, and the calculation formula for transferring process is:

$$d=a\text{-}b,\quad e=(100\times|a\text{-}b|/\max(a, b))\% \quad (3)$$

where, d is the horizontal absolute loss amount in bottom waters from the inner of the bay to the open waters, e is the corresponding horizontal relative loss amount in bottom waters.

In bottom waters, and from the open waters to the inner of the bay, substance's contents are changing, and the calculation formula for transferring process is:

$$f=b\text{-}c,\quad g=(100\times|b\text{-}c|/\max(b, c))\% \quad (4)$$

where, f is the horizontal absolute loss amount in bottom waters from open waters to the inner of the bay, g is the corresponding horizontal relative loss amount in bottom waters.

2.4 Modelling for vertical transferring processes of As

Supposed that certain substance's contents in surface and bottom waters in Site n in the bay are A and a, respectively. From surface waters to bottom waters, the calculation formula for this transferring process is:

$$V_{na}=A\text{-}a,\quad V_{nr}=(100\times|A\text{-}a|/\max(A, a))\% \quad (5)$$

where, V_{na} is the horizontal absolute dilution amount from surface waters to bottom waters, V_{nr} is the corresponding horizontal relative dilution amount. While from bottom waters to surface waters, V_{na} refers to the horizontal absolute accumulation amount, and V_{nr} refers to the corresponding horizontal relative accumulation amount.

3 Results

3.1 Horizontal loss of As in surface and bottom waters

In April 1988, As's contents in surface and bottom waters were measured in Site D5 in the bay, in Site A5 in the bay mouth, and in Site A1 in the open waters, respectively. In August 1988, As's contents in surface and bottom waters were measured in Site B5 in the bay, in Site A5 in the bay mouth, and in Site A1 in the open waters, respectively. In according to Eq. (1) and (2), the horizontal loss of As in surface waters were calculated listed in Table 1. In according to Eq. (3) and (4), the horizontal loss of As in bottom waters were calculated and listed in Table 2.

Table 1. Horizontal loss of As in surface waters in Jiaozhou Bay 1988

	From D5 to A5		Form A1 to A5	
Month	D/μg L^{-1}	E/%	F/μg L^{-1}	G/%
April	-0.14	8.33	1.42	25.00
August	0.10	7.04	0.20	15.15

Table 2. Horizontal loss of As in bottom waters in Jiaozhou Bay 1988

	From B5 to A5		Form A1 to A5	
Month	d/μg L^{-1}	e/%	f/μg L^{-1}	g/%
April	1.00	41.66	0.04	2.85
August	0.36	20.00	0.34	23.61

3.2 Horizontal loss of As in surface and bottom waters

In April 1988, vertical loss of As's contents in waters were calculated in Site D5 in the bay, Site A5 in the bay mouth, and in Site A1 in the open waters, respectively. In August 1988, vertical loss of As's contents in waters were calculated in Site B5 in the bay, Site A5 in the bay mouth, and in Site A1 in the open waters, respectively. In according to Eq. (5), the vertical loss of As in waters were calculated and listed in Table 3.

Table 3. Vertical loss of As in waters in Jiaozhou Bay 1988

Time	Sampling Site	V_{na}/μg L-1	V_{nr}/%
April	D5	-0.86	35.83
	A5	0.28	16.66
	A1	-0.10	7.35
August	B5	-0.38	21.11
	A5	-0.12	8.33
	A1	0.02	1.78

4 Discussion

4.1 Horizontal and vertical changes of As

In April 1988, the horizontal absolute loss amount in surface waters from the bay mouth to the open waters was as high as 25.00%, while the horizontal absolute loss amount in bottom waters from the inner of the bay to the bay mouth was as high as 41.66% (Fig. 2). There were accumulation processes in the inner of the bay and the open waters, and the relative accumulation amounts were 35.83% and 7.35%, respectively, while in the bay mouth there was dilution process, with a relative dilution amount of 16.66% (Fig. 2).

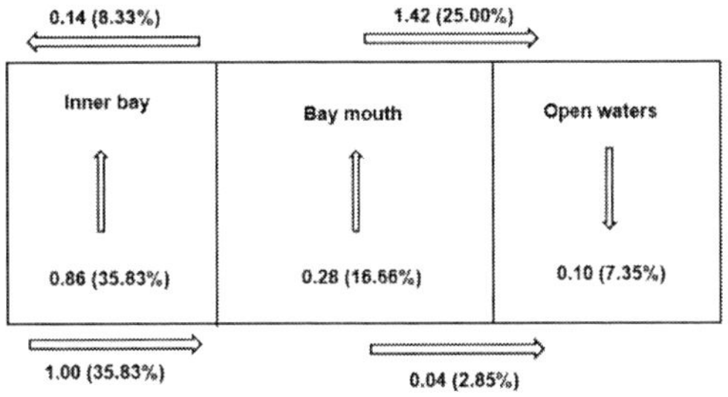

Fig. 2. Block diagram model for horizontal-vertical changes of As in Jiaozhou Bay in April 1988.

In August 1988, the horizontal absolute loss amount in surface waters from the bay mouth to the open waters was as high as 15.15%, and the horizontal absolute loss amount in bottom waters from the bay mouth to the open waters was as high as 23.61% (Fig. 3). There were accumulation processes in the inner of the bay and the bay mouth, and the relative accumulation amounts were 21.11% and 8.33%, respectively, while in the open waters there was dilution process, with a relative dilution amount of 1.78% (Fig. 3).

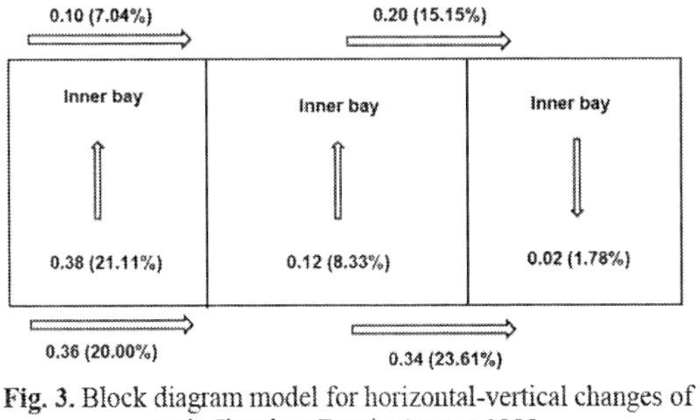

Fig. 3. Block diagram model for horizontal-vertical changes of As in Jiaozhou Bay in August 1988.

In surface waters in April and August 1988 in Jiaozhou Bay, the horizontal absolute loss amounts of As ' s contents were 0.10-0.42 μ g L^{-1}, and the corresponding relative loss amounts were 7.04%-25.00%. In bottom waters, the horizontal absolute loss amounts of As ' s contents were 0.04-1.00 μ g L^{-1}, and the corresponding relative loss amounts were 2.85%-41.66%. The absolute dilution amounts of As's contents were 0.02-0.28 μg L-1, and the corresponding relative dilution amounts were 1.78%-16.66%. The absolute acumulation amounts of As's contents were 0.10-0.86 μg L^{-1}, and the corresponding relative dilution amounts were 7.35%-35.83%.

In general, by means of the modelling processes (i.e., Eq. (1) to Eq. (5)), the changes of As's contents in transferring processes in Jiaozhou Bay were quantified ([4-6]). Furthermore, the block diagram models were demonstrating the horizontal and vertical changes of As' s contents visually.

4.2 Horizontal loss of As in the bay mouth

In April 1988, from the inner of the bay to the bay mouth and then to the open waters, the horizontal relative loss amount of As's contents in surface waters were 8.33%-25.00%, while in bottom waters were 2.85%-41.66% (Fig. 2). It could be found that in transferring process from the inner of the bay to the open waters, the horizontal loss amounts in surface waters were relative low, while in bottom waters were relative high.

In August 1988, from the inner of the bay to the bay mouth and then to the open waters, the horizontal relative loss amount of As's contents in surface waters were 8.04%-15.15%, while in bottom waters were 20.00%-23.61% (Fig. 3). It could be found that in transferring process from the inner of the bay to the open waters, the horizontal loss amounts in surface waters waters were relative low, while in bottom waters were relative high.

As a whole, during the transferring processes through the bay mouth, no matter from the internal waters to the open water, or from the open waters to the internal waters, As contents in both surface and bottom layers were decreasing in a certain degree. The changes of As contents in surface and bottom waters were determined by the transferring distance of As in surface waters.

5 Conclusions

The changes of As's contents in transferring processes in Jiaozhou Bay were quantified, and block diagram models were demonstrating the horizontal and vertical changes of As's contents visually. In surface waters in April and August 1988 in Jiaozhou Bay, the horizontal absolute loss amounts of As's contents were 0.10-0.42 μ g L^{-1}, and the corresponding relative loss amounts were 7.04%-25.00%. In bottom waters, the horizontal absolute loss amounts of As's contents were 0.04-1.00 μg L^{-1}, and the corresponding relative loss amounts were 2.85%-41.66%. The absolute dilution amounts of As's contents were 0.02-0.28 μg L^{-1}, and the corresponding relative dilution amounts were 1.78%-16.66%. The absolute acumulation amounts of As's contents were 0.10-0.86 μg L^{-1}, and the corresponding relative dilution amounts were 7.35%-35.83%. During the transferring processes through the bay mouth, no matter from the internal waters to the open water, or from the open waters to the internal waters, As contents in both surface and bottom layers were decreasing in a certain degree. The changes of As contents in surface and bottom waters were determined by the transferring distance of As in surface waters.

Acknowledgments

This research was sponsored by Research Projects of Guizhou Nationalities University ([2014]02), Research Projects of Guizhou Province Ministry of Education (KY [2014] 266), Research Projects of Guizhou Province Ministry of Science and Technology (LH [2014] 7376).

References

1. D.F. Yang, W.P. Song, S.T. Chen, J.H. Guo, L. Zheng, Coastal Engineering, **31**, 4(2012)
2. D.F. Yang, Y.H. Zhao, Z.G. Piao, Ocean Development and Management, **31**, (2014)
3. D.F. Yang, S.X. Zhu, F.Y. Wang, S.T. Chen, Z.J. Xu, S.X. Zhu, Y.J. Wu, Meterological and Environmental Research, **5**, (2014)
4. D.F. Yang, F.Y. Wang, H.Z. He, Y.J. Wu, S.X. Zhu, Proceedings of the 2015 international symposium on computers and informatics, (2015)
5. D.F. Yang, F.Y. Wang, X.L. ZhaoY.J. Wu, S.X. Zhu, Sustainable Energy and Enviroment Protection, (2015)
6. D.F. Yang, F.Y. Wang, X.Q. Yang, Y.J. Wu, S.X. Zhu, Advances in Computer Science Research, **3252**, (2015)
7. D.F. Yang, Y. Chen, Z.H. Gao, J. Zhang, F. Wang, Chinese Journal of Oceanology and Limnology, **23**, 1(2005)
8. D.F. Yang, F.Y. Wang, Z.H. Gao, W.L. Cui, S.X. Huo, Marine Science, **28**, 6(2004)
9. China's State Oceanic Administration, *The specification for marine monitoring* (Ocean Press, Beijiang, 1991)
10. Yang DF, Miao ZQ, Xu HZ, Y. Chen, J.Y. Sun, Marine Environmental Science, **32**, 3(2013)

Design and Application of Integrated Assembly Technology of FRG in Residential Ceiling

Xiuyun Li[1,*], *Changyong* Yu[2]

[1]Department of Architectural Engineering, Guangzhou Nanyang Polytechnic, Guangzhou, Guangdong, 510925, China
[2]Guangdong Meisui Industrial Development CO., LTD. Guangzhou, Guangdong, 510550, China

Abstract. FRG material is a new environmentally friendly indoor decoration materials and popular in prefabricated construction, the paper introduces the performance and design of materials, and takes FRG in the residential ceiling integrated assembly process into a demonstration project, which showed that FRG in the prefabricated modules integrated ceiling of the whole template scheme has a great artistry and application effect. Meanwhile it provides reference for the integrated ceiling assembly modular process design of similar indoor decoration.

1 Introduction

In recent years, the development of the world integrated ceiling has focused on China. With the promotion of fabricated buildings, China's integrated ceiling has occupied a large international market, and has displayed distinctive cultural characteristics in the residential ceiling style. From the plane design to the three-dimensional development, the integrated ceiling will turn into a new, green and integrated field which promoting industry to upgrade.

2 The Material Characteristics of Green Environmental Innovative FRG

During the gradual attention and development of residential integrated ceiling in the past more than 10 years, more and more attention has been paid to functional, artistic and environmental protection in construction applications. The aluminum buckles as the initial material for the integrated ceiling, then from GRG to FRG, and there are different features among them.

FRG is a typical inorganic cementitious material, English Name: Change the nature of the ontology by liquid phase method fiber-reinforced gypsum complete mould for decoration member. It is made up of a gypsum decorative art component composed of a liquid phase modification of a alpha natural gypsum, a green fiber and a photocatalyst [1].

As a large common construction decoration fixture of environmentally friendly building materials, GRG (Glass Fiber Reinforced Gypsum) has arbitrary shape, class A1 fire prevention, no dehiscence and non deformation, excellent performance, strong acoustic comfort effect. Compared to GRG, FRG core performance of the integrated ceiling indoor decoration reflected in the whole mould customization, lightweight and high strength, absorbing aldehyde, weathering and corrosion resistance, mechanical properties in bending strength can reach above Mpa15-25 (ASTMD790-2002 test); tensile strength can reach above Mpa4-15 (ASTMD256-2002 test), the thermal properties also reflects the low coefficient of thermal expansion and dry shrinkage rate of less than 0.01%, the product of FRG under the environment of cold and hot, dry and wet deformation, showing performance is extremely stable. In the integrated ceiling material selection, FRG is an innovative trend of green building materials.

3 General Situation of Integrated Assembly of Residential Ceiling

The case of Mr. Tong's integrated ceiling wants household ceiling design that has exclusive and comfortable effect which locates in the city of Heyuan, Guangdong, China. The project by the Guangdong Meisui industrial development CO., LTD. is responsible for the design and construction, and the owner offered the ceiling artwork for the bold black lines of Figure 1, the original design decorative ceiling intends to use block board and gypsum board, because of complicated shape, block board failed to meet the needs of the beauty of the shape, also face a long period and high labor cost.

Some problems proposed and need attention in examination of working drawing. On account of shaped plate curvature, arch height and angle accuracy, particularly accuracy in manual operation of the traditional process of custom-made block board is easy to produce error. Coefficient of variation of block board is a little large, the transportation and storage is inconvenient due to large shape, also facing installation and construction technology of high working, edge closing, maintenance series of technological problems.

* Corresponding author: sueone921@163.com

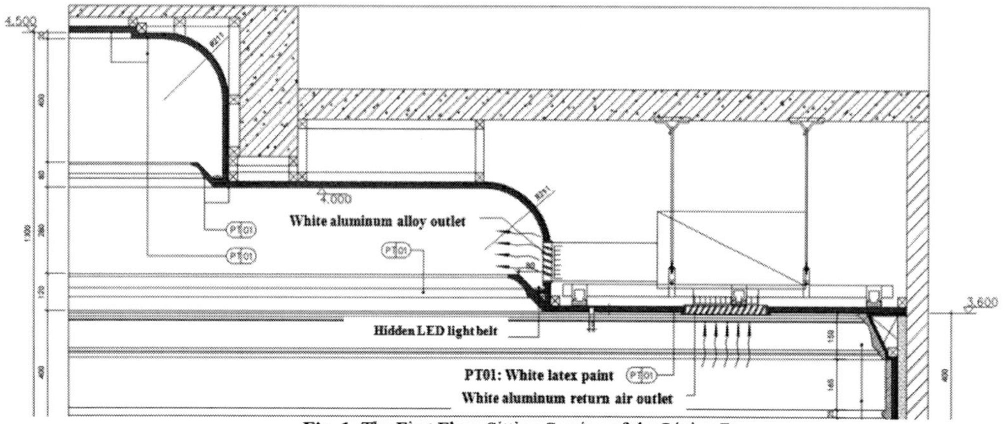

Fig. 1. The First Floor Sitting Section of the Living Room

4 Design idea of FRG whole forming mold

One of the core performance of FRG is the whole design of module, the design of the whole mold has a strong overall sense, which avoided disorderly seam fracture phenomenon, showing clean and elegant. In the process of the design of the whole forming mold, several difficulties are solved.

(1)Precision: the project process using FRG whole module integration designed to replace handmade block board, FRG as the main material, making troffer into design integration, in the design process in advance of having orifice, hole, and integrated silicone mold into molding, which solves the continuity of special-shaped wood angle for precision and complexity; for getting higher data, 3D data collection of actual space is required. The first drawing is in CAD, second is deepen in Rhinoceros software, and the establishment of 1:1 model in the process of the production, then imported CNC engraving and carving basementmembrane, through accurate mold, it can produce a variety of FRG blocks, various FRG is equipped with the mould to ensure the accuracy of whole design.

(2)Seamless forming: The sense of integrity of FRG is embodied in seamless connection. The gap between the end plate and the other plate is filled with GRG powder at the connection joint, make effective use of FRG powder material consistent seams to fill the gap and the putty powder is used for polishing. In order to avoid color aberration when condensation generating, with white latex spraying whitewash, the final effect of continuous of the whole color, like nature itself. In closing, to solve the hidden peril of cracking in splicing installation process, or multiple plates influence the roughness and other issues, this can be accurate reservation for a custom lamp box. Or late for opening, it is advisable to use opening machine to avoid excessive cracking, the gap edge necessary reserved 5mm width, and collocation of aluminum components, plus the joints the same treatment method, effectively achieve seamless effect.

(3)Installation: FRG integrated ceiling has the characteristics of assembly construction, in the approach of integrated ceiling materials, preparation of varieties, specifications, size and appearance inspection should be Check in order to improve the efficiency of installation , all materials in the transport, handling, storage, installation take protective measures to prevent extrusion impact, damp, deformation and damage the surface of the plate and corner. Considering the safety in the process of installation, the scaffolding is set up to meet the construction needs of the working face.

5 Application of Integrated Ceiling Assembly

5.1 Prefabricated Modularization Scheme for FRG Integrated Ceiling

This case takes into account the traditional ceiling method with high cost and long construction period, the FRG ceiling design is adopted, and the whole living ceiling design of FRG is designed as shown in Figure 2. The integral ceiling is made up of a customized lamp slot component and the same material inorganic high crystal composite board. The product is fixed by the light steel keel lock. The seams among the block boards by using the same material FRG powder mixed with white latex and filling sealer belt, so that the top surface does not appear gap, at the same time to solve the cracking problem of joints. The material use FRG silicone rubber mould integrated molding, the complex and changeable modeling of the pattern is well solved; auxiliary parts used in light steel joist, hanger rod, self tapping screws, explosion screws, FRG powder, sealing tape, white latex.

In the process of design deepening, we should give full consideration to the reserved holes at the ends of lamps and other devices, and divide them into integral modules as much as possible on the curved surfaces and irregular corners of the block board, so as to facilitate installation and beauty. During the production process, mold engraving, feeding, pouring, controlling size, demoulding, drying and other processes are strictly in

* Corresponding author: sueone921@163.com

 https://doi.org/10.1051/e3sconf/20183802012

accordance with the acceptance criteria, ensuring silicone mold integrated molding.

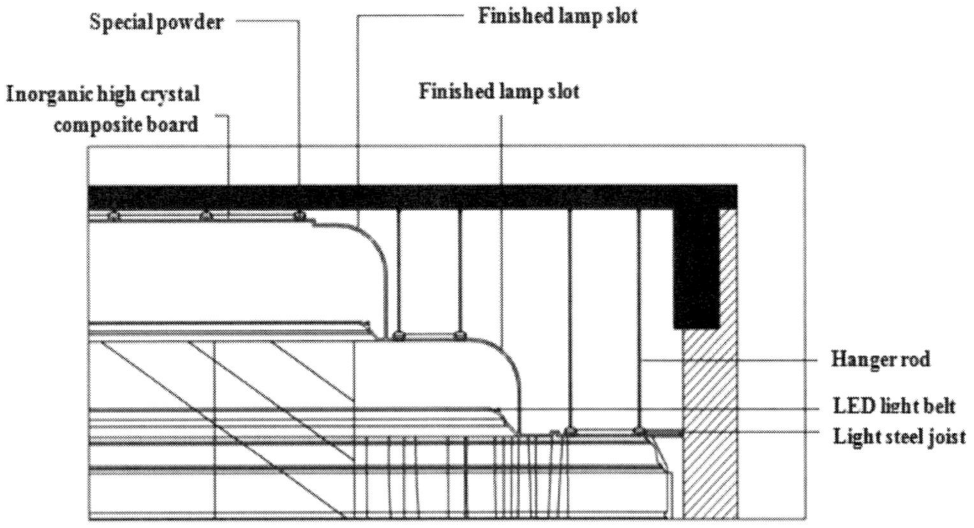

Fig. 2. The First Floor Sitting Section of the Living Room

4.2 Construction Technology Process

Projectile line positioning→light steel joist installation→ FRG product installation→product crevice treatment→ surface layer spraying white latex → engineering acceptance

Before the construction of the ceiling works, we need to be familiar with the drawings and worksite conditions, make sure that the equipments and pipes in the ceiling have been completed. The incoming materials are inspected according to the construction standards, and the construction personnel are aware of the technical disclosure.

(1)Projectile line positioning

Before installation, strictly according to the drawing line, ensure the large area of finished curve surface to install and make convenient for the second and third level of the light steel joist[2]. By using the CAD diagram, the top surface of the ceiling is divided into a proper vertical partition to form a plane (Figure 3):, a facade and a profile, and positioning of Actinobacillus is carried out.

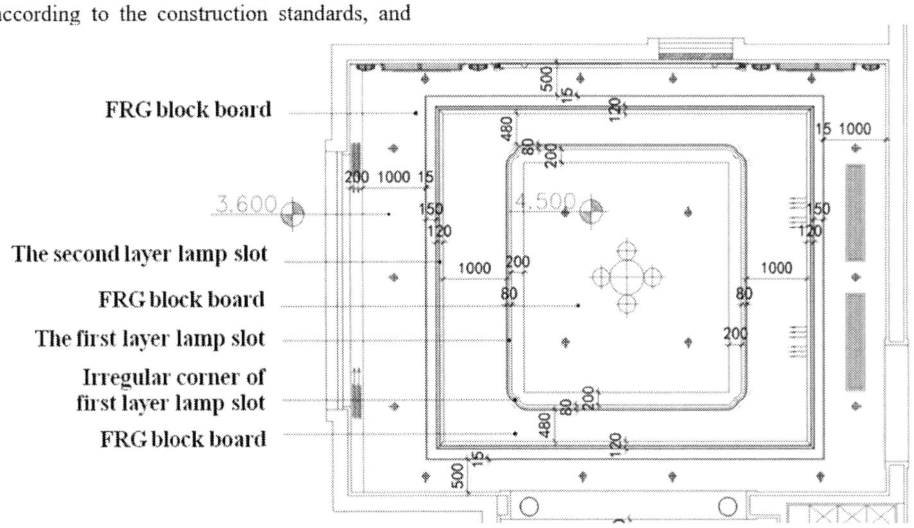

Fig. 3. First Floor Living Room FRG Drawing

(2) Preliminary installation

After placing the positioning line, the expansion screw, the hanger rod, and the light steel joist are laid according to the point position, and the light steel joist frame is adjusted according to the national standard.

(3) Product installation

Supporting material of ceiling use light steel joist. and according to the requirement of positioning line to prepare the hanger rod and auxiliary accessories, and change the traditional manual wooden cutting and lumber fixed keel, which improves the overall sense of flow lines and smoothness, and principal position according to the construction drawings, then determine to install and check.

(4)Gap treatment

Crack is a common disease of all decorative surfaces, and the influencing factors are many and complex. Especially after cracking, it is very difficult to repair[3]. Therefore, the first treatment is particularly important. In order to better solve the noise reduction function, the living room ceiling can also produce good acoustics. When processing, use of FRG powder grout in the gap, On the one hand, make sure the block board and powder are made up with the same material, for reducing the influence of the coefficient of variation; on the other hand, the strength and the density of the block board are reaching unanimity. Particularly in the process of solidification, the chromatic aberration caused by successively drying can not be too obvious. At the same time, slots reserved for a certain width can make the FRG powder bond better, filled the weak part of the pore, and reduced the late cracking and reduced the background noise. Further, in order to bond more firmly, after making up the surface smooth, paste the sealing belt on the gap, or the elastic putty powder is added, which can scrape the powder in horizontal or vertical direction, and usually also need to polish and clean the surface in advance.

(5) Surface treatment

General to create a sense of beauty, in the surface layer evenly coated with white latex, note the sagging or color chromatic aberration, according to the needs in the product surface layer it can also apply with three-dimensional relief texture, colored painting, metal finishes to a variety of realistic decorative effect, which are making the overall integrated ceiling appear seamless, continuous, smooth and comfortable.

(6) Acceptance of work

The engineering case of integrated ceiling acceptance according to JGJ 345-2014 "*Technical specification for ceiling engineering of public building*", CECS 255-2009 "*Technical specification for installation of ceiling systems in buildings*", GB50325-2010 "*Code for indoor environmental pollution control of civil building engineering*", GB50210-2001 "*Code for construction quality acceptance of building decoration*" of the relevant provisions of approval, the whole integrated ceiling in this case design by FRG prefabricated modularization has been verified for acceptance by the customer satisfaction(Figure 4).

Fig. 4. First Floor Living Room FRG Real View Diagram

5 Conclusion

FRG integrated ceiling has excellent decorative effect. Even damaged, it can immediately maintenance and leave no trace, but it also needs to avoid heavy weight, avoid contact with gasoline and diluent, and also pay attention to make it soak in water for a long time while water leakage. FRG is lightweight, can be personalized customization of the whole, also is a green innovative constructive materials in 2017. With the modular design and assembling design, integrated ceiling has a commendable performance. This case successfully achieve the project acceptance, in the following 3-5 years, FRG integrated ceiling is expected has a broad market space, and will be favored, usher in a huge development space.

References

1. Wang wei. Green and environment-friendly innovation material – FRG [EB/OL]. ECONOMIC DAILY-www.ce.cn, 2017-11-25. 17:52
2. Li Zhen. Anti cracking construction technology for installation of special shape GRG ornament panel in large area [J]. City & House 2017 (11):107-108
3. Tan Zebin,Yi Hui. The construction organizations and technical points of Gansu grand theatre audience hall GRG acoustic peg board [J]. Environmental Engineering, 2012 (s1):25-27

Research on the dynamic buckling of functionally graded material plates under conditions of one edge fixed and three edges simply supported

Wang XIN, Han Zhi JUN, Wu Ya LI

College of Mechanics, Taiyuan University Of Technology Taiyuan 030024, China

Abstract: Based on the strain assumption and linear mixing rate of Vogit, the physical property parameter expression of functionally graded material plates is obtained. According to the theory of small deformation and Hamilton principle, the dynamic buckling governing equation of functionally graded material plates under longitudinal load is obtained. Using the method of trial function, the analytical expression of critical load and the buckling solution of the functionally graded material plate under conditions of one edge fixed and three edges simply supported is obtained. The analytical expression of critical load is numerically calculated by METLAB. The influence of geometric size, gradient index, modal order and material composition on critical load is discussed. The results show that the critical buckling load decreases exponentially with the increase of critical length, decreases with the increase of gradient index k, increases with the increase of modal order, and the elastic modulus of constituent materials has significant effect on the critical load. The higher-order buckling modes of functionally graded material plates are prone to occur under the condition of high longitudinal load.

1 Introduction

Functionally graded materials are non-homogenous new composite materials that show smooth and continuous characteristics among different materials. Functionally graded materials can effectively avoid or reduce stress concentration without sudden changes in physical properties. Due to its good mechanical properties, functionally graded materials have become research hot spot[1,2]. It has been widely used in aerospace, mechanical engineering, biomedical engineering and other fields. Researchers on the dynamic buckling of functionally graded materials plate are explored Constantly[3,4]. Feldman and Aboudi[5] studied the linear dynamic buckling of functionally graded plates under in-plane loading. Shen [6] used the quadratic perturbation method to solve the nonlinear bending problem of simply-supported functionally graded material plates under the coupling condition of thermo and mechanical. The results show that the changes of volume fraction index and temperature have significant influence on the nonlinear bending behavior of the plate. Najafizadeh and Eslami[7,8] discussed the thermal buckling of circular plate in the in-plane load under simply supported and clamped supported conditions.

In this paper, Vogit's strain assumption, linear mixing rate and Kirchhoff sheet theory are used to derive the governing equations with Hamilton's principle. The expression of critical load for the functionally graded material plate is obtained. The expressions are numerically calculated by METLAB software programming. The influences of critical length, gradient index, material properties and boundary conditions for critical load are discussed.

2 The governing equation of functionally graded material plates

* Corresponding author email:wx2015tyut@126.com

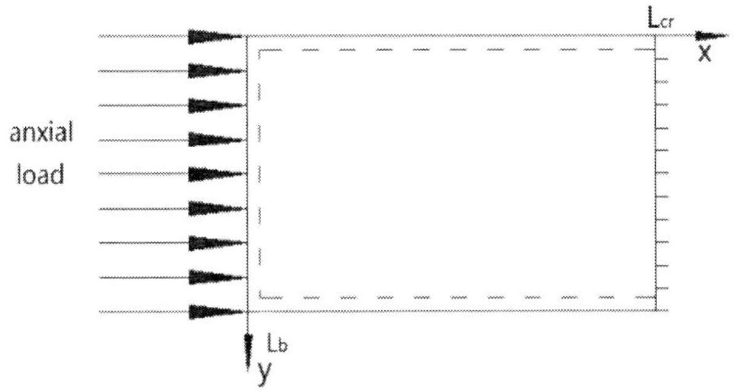

Figure 1 The plate with the length Lcr, width Lb

The boundary conditions of one edge fixed and three edges simply supported can be written as

$$\begin{cases} w(0)=w_{xx}(0)=w(L_{cr})=w_{x}(L_{cr})=0 \\ w(0)=w_{yy}(0)=w(L_{b})=w_{yy}(L_{b})=0 \end{cases} \tag{1}$$

The material properties of functionally graded material along the z-direction can be written as

$$\begin{cases} E_{(z)}=(E_c-E_m)\left(\dfrac{2z+h}{2h}\right)^k+E_m \\ \rho_{(z)}=(\rho_c-\rho_m)\left(\dfrac{2z+h}{2h}\right)^k+\rho_m \\ \mu_{(z)}=(\mu_c-\mu_m)\left(\dfrac{2z+h}{2h}\right)^k+\mu_m \end{cases} \tag{2}$$

The constitutive equation of functionally graded material plates is

$$\begin{Bmatrix} N_x \\ N_y \\ N_{xy} \\ M_x \\ M_y \\ M_{xy} \end{Bmatrix}=\begin{bmatrix} A_{11} & A_{12} & A_{16} & B_{11} & B_{12} & B_{16} \\ A_{12} & A_{22} & A_{26} & B_{12} & B_{22} & B_{26} \\ A_{16} & A_{26} & A_{66} & B_{16} & B_{26} & B_{66} \\ B_{11} & B_{12} & B_{16} & D_{11} & D_{12} & D_{16} \\ B_{12} & B_{22} & B_{26} & D_{12} & D_{22} & D_{26} \\ B_{16} & B_{26} & B_{66} & D_{16} & D_{26} & D_{66} \end{bmatrix}\begin{bmatrix} \varepsilon_x^0 \\ \varepsilon_y^0 \\ \gamma_{xy}^0 \\ k_x \\ k_y \\ k_{xy} \end{bmatrix} \tag{3}$$

Kinetic energy with moment of inertia is

$$\begin{aligned} T&=\frac{1}{2}\int_{-\frac{h}{2}}^{\frac{h}{2}}\int_0^{L_{cr}}\int_0^{L_b}\rho_{(z)}\left[\left(\frac{\partial u}{\partial t}\right)^2+\left(\frac{\partial v}{\partial t}\right)^2+\left(\frac{\partial w}{\partial t}\right)^2\right]dxdydz \\ &=\frac{1}{2}\int_0^{L_{cr}}\int_0^{L_b}\left[I_0\left(\frac{\partial u_0}{\partial t}\right)^2+I_0\left(\frac{\partial v_0}{\partial t}\right)^2+I_0\left(\frac{\partial w_0}{\partial t}\right)^2-2I_1\frac{\partial u_0}{\partial t}\frac{\partial^2 w_0}{\partial x\partial t}-2I_1\frac{\partial v_0}{\partial t}\frac{\partial^2 w_0}{\partial y\partial t}\right. \\ &\quad\left.+I_2\left(\frac{\partial^2 w_0}{\partial x\partial t}\right)^2+I_2\left(\frac{\partial^2 w_0}{\partial y\partial t}\right)^2\right]dxdy \end{aligned} \tag{4}$$

The deformation energy is

$$U=\frac{1}{2}\int_0^{L_{cr}}\int_0^{L_b}\left(N_x\varepsilon_x^0+N_y\varepsilon_y^0+N_{xy}\gamma_{xy}^0+M_xk_x+M_yk_y+M_{xy}k_{xy}\right)dxdy \tag{5}$$

The energy of external power is

$$W=\frac{1}{2}\int_0^{L_{cr}}\int_0^{L_b}N_{(t)}\left(\frac{\partial w_0}{\partial x}\right)^2dxdy \tag{6}$$

The coefficients are

$$A_{11}=A_{22}=\int_{-\frac{h}{2}}^{\frac{h}{2}}\frac{E_{(z)}}{1-\mu_{(z)}^2}dz \quad A_{12}=A_{21}=\int_{-\frac{h}{2}}^{\frac{h}{2}}\frac{\mu_{(z)}E_{(z)}}{1-\mu_{(z)}^2}dz \quad A_{66}=\int_{-\frac{h}{2}}^{\frac{h}{2}}\frac{E_{(z)}}{2(1+\mu_{(z)})}dz$$

$$B_{11}=B_{22}=\int_{-\frac{h}{2}}^{\frac{h}{2}}\frac{zE_{(z)}}{1-\mu_{(z)}^2}dz \quad B_{12}=B_{21}=\int_{-\frac{h}{2}}^{\frac{h}{2}}\frac{z\mu_{(z)}E_{(z)}}{1-\mu_{(z)}^2}dz \quad B_{66}=\int_{-\frac{h}{2}}^{\frac{h}{2}}\frac{zE_{(z)}}{2(1+\mu_{(z)})}dz$$

$$D_{11}=D_{22}=\int_{-\frac{h}{2}}^{\frac{h}{2}}\frac{z^2E_{(z)}}{1-\mu_{(z)}^2}dz \quad D_{12}=D_{21}=\int_{-\frac{h}{2}}^{\frac{h}{2}}\frac{z^2\mu_{(z)}E_{(z)}}{1-\mu_{(z)}^2}dz \quad D_{66}=\int_{-\frac{h}{2}}^{\frac{h}{2}}\frac{z^2E_{(z)}}{2(1+\mu_{(z)})}dz$$

Functionally graded material plates have the isotropic mechanical properties

$$A_{16} = A_{26} = B_{16} = B_{26} = D_{16} = D_{26} = 0$$

Generalized inertia is defined in Eq. 4:

$$(I_0, I_1, I_2) = \int_{-\frac{h}{2}}^{\frac{h}{2}} \rho_{(z)}(1, z, z^2)dz \quad (7)$$

Substituting Eq.3, Eq.4, Eq.5, Eq.6 and Eq.7 into Hamilton's principle $\delta\int_{t_2}^{t_1}(T - U + W)dt = 0$,the governing equation of functionally graded material plates can be written in the following form through calculation and simplification

$$D_{11}w_{xxxx} + (2D_{22} + 4D_{66})w_{xxyy} + D_{11}w_{yyyy} + N_{(t)}w_{xx} - I_0\ddot{w} + I_2\ddot{w}_{xx} + I_2\ddot{w}_{yy} = 0 \quad (8)$$

3 Dynamic buckling analysis of functionally graded material plates

In y direction, set the buckling mode of functionally graded material plates is $\sin\frac{m\pi y}{l_b}$ [9], set

$$w = X(x)T(t)\sin\frac{m\pi y}{L_b} \quad (9)$$

Substituting Eq. 9 into Eq. 8, the following equations can be obtained:

$$D_{11}X_{xxxx}T - (2D_{12} + 4D_{66})X_{xx}T\frac{m^2\pi^2}{L_b^2} + D_{11}XT\frac{m^4\pi^4}{L_b^4} + N_{(t)}X_{xx}T - I_0XT_{tt} + I_2X_{xx}T_{tt} - I_2XT_{tt}\frac{m^2\pi^2}{L_b^2} = 0 \quad (10)$$

Separating the variables, Eq.10 can be transformed into:

$$\begin{cases} X_{xxxx} + \alpha X_{xx} + \beta X + \lambda\xi X - \lambda\eta X_{xx} = 0 \\ T_{tt} + \lambda T = 0 \end{cases} \quad (11)$$

The coefficients of Eq.11 are

$$\alpha = \frac{N_{(t)}}{D_{11}} - \frac{(2D_{22} + 4D_{66})m^2\pi^2}{D_{11}L_b^2} \qquad \beta = \frac{m^4\pi^4}{L_b^4}$$

$$\xi = \frac{I_2}{D_{11}} \qquad \eta = \frac{I_0}{D_{11}} + \frac{I_2m^2\pi^2}{D_{11}L_b^2} \quad (12)$$

The solution of Eq. 11 is divergent when λ < 0 ,the system is not stable and the dynamic buckling will occurs[10]. So the Eq. 11 has the dynamic buckling solution, the parameter λ of Eq. 11 must satisfy the following conditions

$$\lambda < 0, (\alpha - \lambda\xi)^2 - 4(\beta - \lambda\zeta) > 0 \quad (13)$$

The dynamic buckling solution to Eq. 13 is

$$X(x) = C_1\sin k_1x + C_2\cos k_1x + C_3\sin k_2x + C_4\cos k_2x \quad (14)$$

The coefficients of Eq.14 are

$$\begin{cases} k_1 = \sqrt{\dfrac{\alpha - \lambda\xi + \sqrt{(\alpha - \lambda\xi)^2 - 4(\beta + \lambda\eta)}}{2}} \\ k_2 = \sqrt{\dfrac{\alpha - \lambda\xi - \sqrt{(\alpha - \lambda\xi)^2 - 4(\beta + \lambda\eta)}}{2}} \end{cases}$$

Substituting Eq. 14 into Eq. 1, the linear equation group can be obtained

$$\begin{bmatrix} 0 & 1 & 0 & 1 \\ 0 & -k_1^2 & 0 & -k_2^2 \\ \sin k_1L_{cr} & \cos k_1L_{cr} & \sin k_2L_{cr} & \cos k_2L_{cr} \\ k_1\cos k_1L_{cr} & -k_1\sin k_1L_{cr} & k_2\cos k_2L_{cr} & -k_2\sin k_2L_{cr} \end{bmatrix} \begin{bmatrix} C_1 \\ C_2 \\ C_3 \\ C_4 \end{bmatrix} = 0 \quad (15)$$

The conditions of Eq. 15 with non-zero solution are $k_1l = n_1\pi$, $k_2l = n_2\pi$ [10],where $n_1 = 1,2,3(1,2,3\cdots), n_2 = n_1 + 1$

The following forms can be obtained

$$\begin{cases} k_1^2 + k_2^2 = \dfrac{\pi^2(n_1^2 + n_2^2)}{L_{cr}^2} = \partial - \lambda\xi \\ k_1^2k_2^2 = \dfrac{n_1^2n_2^2\pi^4}{L_{cr}^4} = \beta + \lambda\eta \end{cases} \quad (16)$$

Simplifying Eq.16, the expression of critical dynamic buckling load N_{cr} is

$$N_{cr} = \frac{D_{11}\pi^2(n_1^2 + n_2^2)}{L_{cr}^2} + \frac{D_{11}\pi^4n_1^2n_2^2I_2}{L_{cr}^4(I_0L_b^2 + I_2m^2\pi^2)} - \frac{D_{22}m^4\pi^4I_2}{L_b^2(I_0L_b^2 + I_2m^2\pi^2)} + (2D_{12} + 4D_{66})\frac{m^2\pi^2}{L_b^2} \quad (17)$$

Using the boundary conditions and the constraints conditions Eq.1 and Eq.14 the buckling mode can be obtained

$$w = T_{(t)}C_1\left[\left(\frac{n_2\sin\frac{n_1\pi}{L_{cr}} - n_1\sin\frac{n_2\pi}{L_{cr}}}{n_2\cos\frac{n_2\pi}{L_{cr}} - n_1\cos\frac{n_1\pi}{L_{cr}}}\right)\left(\cos\frac{n_1\pi x}{L_{cr}} - \cos\frac{n_2\pi x}{L_{cr}}\right) + \sin\frac{n_1\pi x}{L_{cr}} - \frac{n_1}{n_2}\sin\frac{n_2\pi x}{L_{cr}}\right]\sin\frac{m\pi y}{L_b} \quad (18)$$

4 Numerical results and discussion

Table 1 Material parameters of functionally gradient materials plates

materials	Elasticity Modulus $E(GPa)$	Density $\rho(g/cm^3)$	Poisson's Ratio μ
Ceramics	385	3.96	0.23
Titanium	108.5	4.54	0.41
Iron	155	7.86	0.291
Copper	119	8.96	0.326

The relationship between the critical load *Lcr* and critical length *Ncr* in different situations can be acquired and different orders of the buckling modes can be obtained by MATLAB.

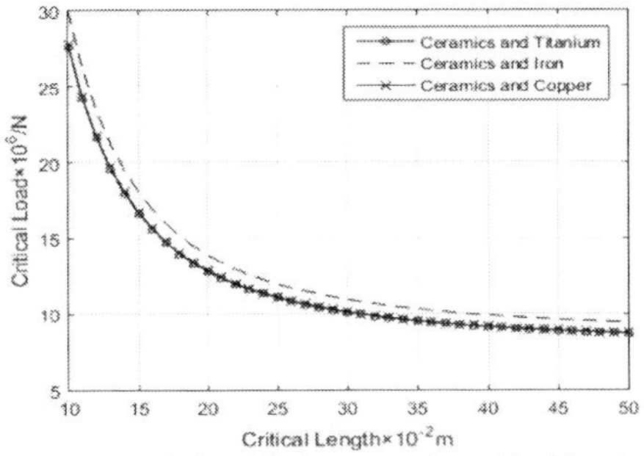

Figure 1. Relationship between the critical load and critical length when material is different

The critical load *Ncr* decreases exponentially with the increase of the critical length *Lcr*. Fig. 1 shows with Table 1 that the change of the elastic modulus of the material has obvious influence on the critical load of dynamic buckling of the functionally graded material plate, and the influence of Poisson's ratio is relatively small.

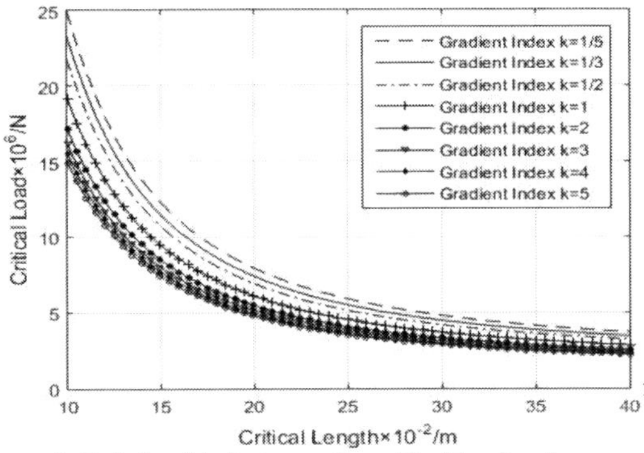

Figure 2. Relationship between the critical load and length when gradient index is different

The critical load *Ncr* decreases with the increase of gradient index *k*. The relationship between *Ncr* and *Lcr* changed obviously when the value of *k* is small and the change of *k* value in the range of (0-1) has great impact on *Ncr*. The critical load curve corresponding to different *k* values tends to level with the increase of *Lcr*.

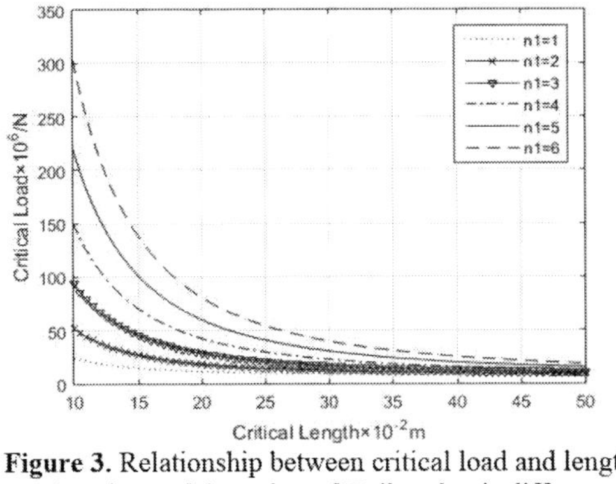

Figure 3. Relationship between critical load and length when the modal number of X direction is different

The critical load *Ncr* increases with the increase of modal number *n1*. The change of *Ncr* with *Lcr is* more obvious in the condition of high value of *n1*

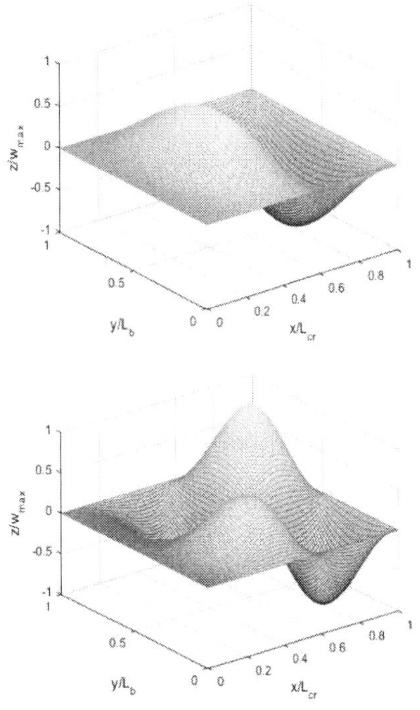

$(a)\, n_1 = 2, m = 1$

$(b)\, n_1 = 2, m = 2$

Figure 4. Different orders of the buckling modes

The amplitude modes w of functionally graded material plates increasing positively in X direction.

5 Conclusion

1. The analytical expression of dynamic buckling critical load (17) and buckling modal expression (18) of functionally graded material plates are obtained. The buckling modes are shown in Fig.3.

2. Calculate Eqs. (17) and (18) by MATLAB, and discuss the influence of critical length, gradient index, material property and order of buckling modes on the critical buckling load and buckling mode. The results show that the critical load of dynamic buckling decreases exponentially with the increase of length, and the critical load of buckling decreases with the increase of gradient index *k*. The gradient index *k* have great influence on buckling critical load in conditions that the gradient index *k* changes in the range of (0-1). The buckling critical load is greatly affected by the elastic modulus and the boundary conditions have great influence on the buckling mode of the functionally graded material plates.

Acknowledgements

The project is supported by Science Foundation for

Young Scholars of Shanxi Province of China (NO.201601D021127) and the Graduate Educational Reformation of Shanxi Province of China (NO.2015JG41)

References

1. Na K S, Kim J H. Three-dimensional thermomechanical buckling analysis for functionally graded composite plates[J]. Composite Structures, 2006, 73(4):413-422.
2. Huang X L, Shen H S. Nonlinear vibration and dynamic response of functionally graded plates in thermal environments[J]. International Journal of Solids & Structures, 2005, 41(9–10):2403-2427.
3. Huang X L, Shen H S. Nonlinear vibration and dynamic response of functionally graded plates in thermal environments[J]. International Journal of Solids & Structures, 2005, 41(9–10):2403-2427.
4. Shen H. Bending, buckling and vibration of functionally graded plates and shells[J]. Advances in Mechanics, 2004, 34(01):53-60.
5. Feldman E, Aboudi J. Buckling analysis of functionally graded plates subjected to uniaxial loading[J]. Composite Structures, 1997, 38(1–4):29-36.
6. Shen H S. Nonlinear bending response of functionally graded plates subjected to transverse loads and in thermal environments[J]. International Journal of Mechanical Sciences, 2002, 44(3):561-584.
7. Najafizadeh M M, Eslami M R. Buckling analysis of circular plates of functionally graded materials under uniform radial compression[J]. International Journal of Mechanical Sciences, 2002, 44(12):2479-2493.
8. Najafizadeh M M, Eslami M R. First-Order-Theory-Based Thermoelastic Stability of Functionally Graded Material Circular Plates[J]. Aiaa Journal, 2002, 40(7):1444-1450.
9. Leissa A W, Kang J H. Exact solutions for vibration and buckling of an SS-C-SS-C rectangular plate loaded by linearly varying in-plane stresses[J]. International Journal of Mechanical Sciences, 2002, 44(9):1925-1945
10. HAN Zhi-jun, Experimental and theoretic studies on the impact buckling of columns and the effect of stress wave [D].Taiyuan University of Technology, (2005).

 https://doi.org/10.1051/e3sconf/20183802014

Enhanced photocatalytic activity of wool-ball-like TiO_2 microspheres on carbon fabric and FTO substrates

Yu Zhang[a*], Jian Gu[b,c], Mengqi Zhang[a]

[a]School of materials science and engineering, Hubei University of Automotive Technology, Shiyan, 442002, P.R. China.
[b]College of Engineering and Applied Sciences, Nanjing Univerisy, Nanjing 210093, China.
[c]Institute of Materials Engineering, Nanjing University, Nantong 226019, Jiangsu, P. R. China.

Abstract: The wool-ball-like TiO_2 microspheres on carbon fabric (TiO_2-CF) and FTO substrates (TiO_2-FTO) have been synthesized by a facile hydrothermal method in alkali environment, using commercial TiO_2 (P25) as precursors. The XRD results indicate that the as-prepared TiO_2 have good crystallinity. And the SEM images show that the wool-ball-like TiO_2 microspheres with a diameter of 2-3 μm are composed of TiO_2 nanowires, which have a diameter of ~50 nm. The photocatalytic behavior of the wool-ball-like TiO_2 microspheres, TiO_2-CF and TiO_2-FTO under ultraviolet light was investigated by a pseudo first-order kinetic model, using methyl orange (MO) as pollutant. The wool-ball-like TiO_2 microspheres obtained a degradation rate constant (K_{ap}) of 6.91×10^{-3} min^{-1}. The K_{ap} values of TiO_2-FTO and TiO_2-CF reach 13.97×10^{-3} min^{-1} and 11.80×10^{-3} min^{-1}, which are 2.0 and 1.7 times higher than that of pristine wool-ball-like TiO_2 microspheres due to the "sum effect" between TiO_2 and substrates. This study offers a facile hydrothermal method to prepare wool-ball-like TiO_2 microspheres on CF and FTO substrates, which will improve the recyclability of phtocatalysts and can be extended to other fields.

1 Introduction

With the increase of environment pollution, photocatalytic degradation of environmental pollutants has been the focus in the field of environmental applications. Titanium dioxide (TiO_2), as a well-known good photocatalyst in semiconductor nanomaterials, has been intensively investigated in photocatalysis due to its low cost, high biocompatibility and chemical inertness. Semiconductor nanomaterials are morphology dependent physicochemical properties [1]. By accurate control of size, shape, crystallinity, and degree of exposure of reactive crystal facets, one can tune the properties of nanomaterials.[2-4] Recently, the synthesis of (001) facet-dominated anatase TiO_2 based on powdered shape has attracted intensive attention in photocatalysis due to the higher reactive facets of (001) to effectively enhance surface properties.[3, 5-8] [9, 10]

Compared with powdered TiO_2 nanomaterials, one-dimensional TiO_2 nanomaterials have higher photocatalytic activity due to reduced agglomeration and increased surface area.[11] Moreover, the unique structure of one-dimensional TiO_2 nanomaterials provide longitudinal transmission channel for charge transport and inhibit the recombination of charge carriers.[12] Hence, many efforts have been devoted to synthesis of one-dimensional TiO_2 nanomaterials, including sol-gel method [13], hydrothermal method [14] and electrochemical anodic oxidation[15]. In order to obtain one-dimensional TiO_2 nanomaterials, various templates are developed [15-17]. Template-free synthesis of one-dimensional TiO_2 nanomaterials also gained great attention [18, 19]. In addition, these traditional methods of synthesizing one-dimensional TiO_2 nanomaterials often used titanium compounds (titanium alkoxide, titanium halide, et.al) as Ti precursors. Recently, TiO_2 as a starting material to prepare nanostructured TiO_2 with template-free attracted great interest of researchers [20, 21]. Li et.al used commercial P25 as precursor dissolved in a 10 M NaOH alkaline solution and gained TiO_2 nanowires after 24h hydrothermal treatment at 269°C in a Teflon-lined stainless steel autoclave.[11]

However, recycling these one-dimensional TiO_2 nanomaterials from aqueous solution after photocatalytic reactions are not easy and these photocatalysts cause secondary pollution. [5, 22-24] To address this problem, growing these one-dimensional TiO_2 nanomaterials on certain substrates is an ideal way.[25] In this work, we demonstrated a facile synthesis of wool-ball-like TiO_2 microspheres composed of one-dimensional TiO_2 nanowires on fluorine doped SnO_2 (FTO) and carbon fabric using P25 as precursor by a modified hydrothermal method. With the "sum effect" between TiO_2 and substates on decomposition of methylene

* Corresponding author email: zhangy@huat.edu.cn

orange (MO), TiO_2 growing on substrates exhibits better photocatalytic efficiency than that of wool-ball-like TiO_2 microspheres.

2 Experimental Procedures

2.1Material preparation

In a typical synthesis process of wool-ball-like TiO_2 microspheres on substrates, 1.4 g of P25 (Degussa Industries) was dissolved in a 10 M NaOH alkaline solution; after the solution had been stirred for 1 h, the fully mixed solution was transferred into a Teflon-lined stainless steel autoclave. Next, the treated substrates (FTO, CF) were immersed into the mixed solution and the hydrothermal reaction was carried out under 120℃ for 24 h. Finally, the product was obtained after being washed by MilliQ water (resistivity: 17 MΩ*cm) for several times and dried in an oven at 80℃ overnight. To get Anatase TiO_2, the dried product was calcined at 550℃ for 5 h.

2.2 Characterizations and measurements

The phase and microstructure of as-prepared samples were examined using powder X-ray diffraction (XRD) on a PANalytical X-ray diffractometer (Cu K_α radiation, 45kV, 40mA) and Field Emission Scanning Electron Microscopy (FESEM, JSM-7100F, JEOL). The ultraviolet light photocatalytic activity of as-prepared samples was investigated by the photodegradation of methylene orange (MO) in aqueous solution. In a typical experiment[11], 50 mg of photocatalysts were dispersed into 50 mL of MO solution (10 mg/L). Next, the mixed solution was kept in the dark at room temperature under stirring for 1 h to get the balance of adsorbing-desorbing. Photocatalytic activity was estimated in the presence or absence of ultraviolet light irradiation provided by XPA photochemical reaction apparatus equipped with 100 W mercury lamp. During the photoreaction process, 5 mL of solution was collected every 10 minutes. The collected solution was centrifuged at 8000 rpm to remove remaining photocatalysts and the concentration of MO was determined by UV-Vis spectroscopy at 465 nm using a standardized calibration curve.

3 Results and discussion

3.1 Characterization of crystal structure and morphology

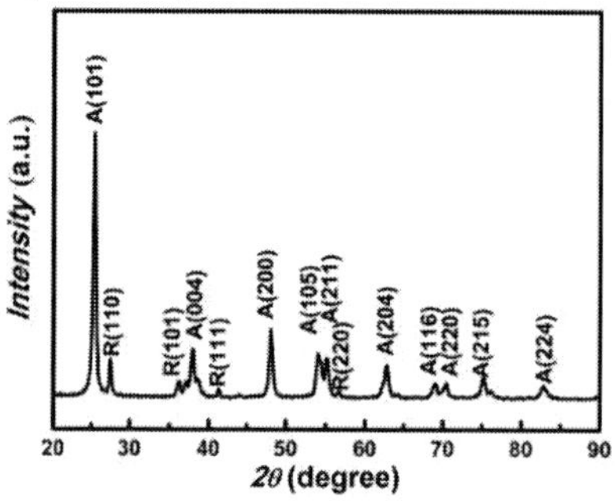

Fig. 1. X-ray diffraction spectra of as-prepared wool-ball-like TiO_2 microspheres.

Fig. 1 shows the X-ray diffraction (XRD) patterns of as-prepared wool-ball-like TiO_2 microspheres, indicating that the main phase is Anatase. The 2θ values of 25.3°, 37.9°, 48.0°, 54.0°, 55.1°, 62.8°, 68.8°, 70.3°, 75.2° and 82.7° are characteristic of Anatase (JCPDS Card no. 21-1272) and correspond to (101), (004), (200), (105), (211), (204), (116), (220), (215) and (224) crystal planes[11, 26]. It is noted that the 2θ value of 25.3° has the highest peak in all diffraction peaks, which is considered an index of good crystallinity for Anatase. The low-energy (101) crystal facets, approximately 0.44 J m^{-2},[27] leads to the highest peak at 2θ = 25.3°. Additionally, small amounts of rutile phase are observed. For example, Rutile (JCPDS Card no. 21-1276) peaks are found at 2θ values of 27.5°, 36.0°, 41.1° and 56.8°, which correspond to crystal planes of (110), (101), (111), and (220). [11, 26]

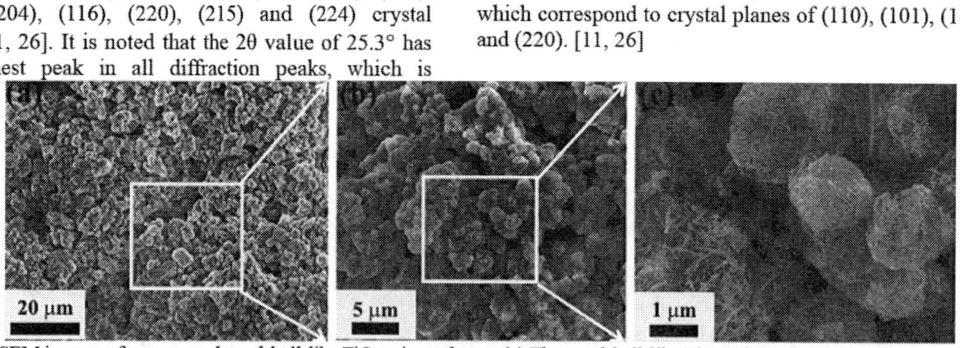

Fig. 2 SEM images of as-prepared wool-ball-like TiO_2 microspheres. (a) The wool-ball-like TiO_2 microspheres; (b) the magnification image of specific area in (a); (c) the magnification image of specific area in (b), the wool-ball-like TiO_2 microspheres are composed of TiO_2 nanowires.

Fig. 2 depicts the scanning electron microscopy (SEM) morphologies of the wool-ball-like TiO_2 microspheres. As shown in Fig. 2a, the large-scale synthesis of wool-ball-like TiO_2 microspheres can be realized by this facile hydrothermal method. Fig. 2b represents the wool-ball-like TiO_2 microspheres that have a diameter of 2-3 μm. These wool-ball-like TiO_2 microspheres are composed of one-dimensional TiO_2 nanowires with a diameter of ~50 nm (as shown in Fig. 2c), which increase the specific surface area. Due to the large specific surface area and multiple active sites provided by this 3D hierarchical structure, these wool-ball-like TiO_2 microspheres would enhance their photocatalytic properties[28].

3.2 Evaluation of photocatalytic activity

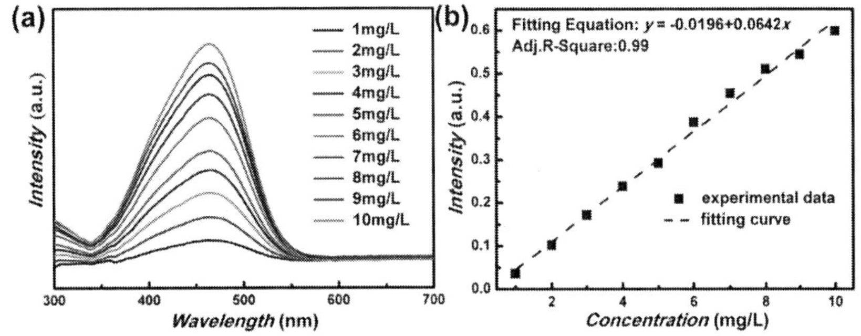

Fig. 3 (a) The absorbance curve of different MO concentration; (b) the calibration curve for MO concentration.

The ultraviolet light photocatalytic activity of all samples was studied using MO as pollutant. As shown in Fig.3a, the absorbance intensity at 465 nm, which is the characteristic absorption wavelength of MO, increases systematically with increasing MO concentration. Based on the data provided by Fig.3a, the calibration curve for the MO concentration was generated (as shown in Fig.3b).

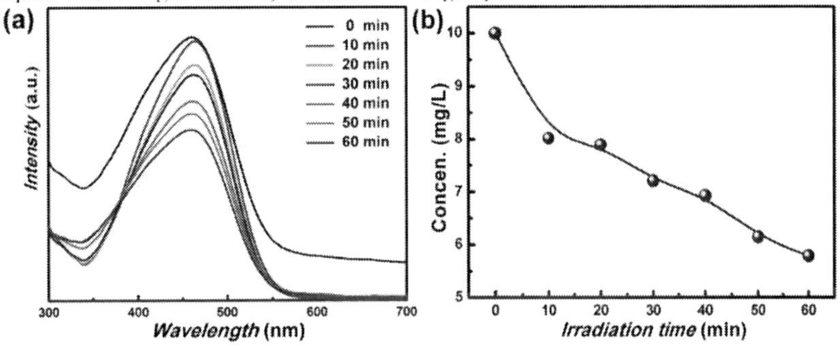

Fig. 4 (a) The absorbance curve of the wool-ball-like TiO_2 microspheres after different irradiation time; (b) MO concentration in TiO_2 microspheres solution as a function of irradiation time, in which the MO concentration was estimated by the calibration curve.

Fig. 4a shows the absorbance curve of the wool-ball-like TiO_2 microspheres after different irradiation time from 0 min to 60 min. The intensity of peak at ~ 465 nm decreases with prolonging radiation, implying the photocatalytic behavior under ultraviolet light. As shown in Fig. 4b, the MO concentration in TiO_2 microspheres solution decreases nearly linearly with increasing radiation time, indicating the good photocatalytic activity.

However, the recyclability of these TiO_2 microspheres is not very good [5, 22-24]. Growing these TiO_2 nanomaterials on certain substrates, such as FTO and CF [27], is an ideal way to address this problem.[25]

3.3 Optimization of photocatalytic activity

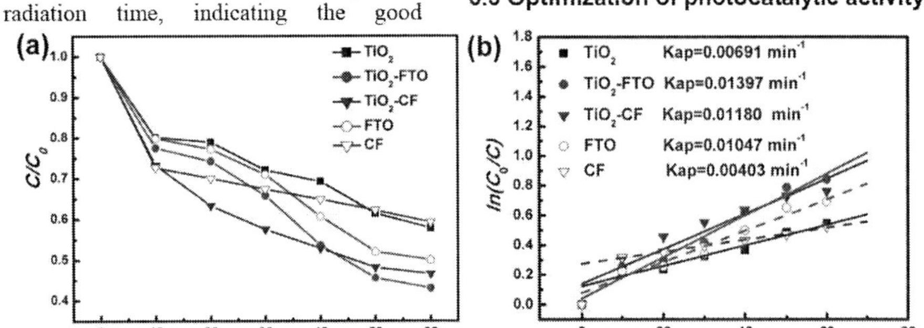

Fig.5 The photocatalytic kinetics of TiO_2, TiO_2-FTO, TiO_2-CF, FTO and CF.

In this work, the wool-ball-like TiO_2 microspheres were grew on FTO and CF substrates, respectively. Fig. 5 shows the photocatalytic kinetics of TiO_2, TiO_2-FTO, TiO_2-CF, FTO and CF. As shown in Fig. 5a, 42.2%, 57.0%, 53.4%, 50.0% and 40.8% of MO are decomposed by TiO_2, TiO_2-FTO, TiO_2-CF, FTO and CF after irradiation for 60 min, respectively. Compared with TiO_2, TiO_2-FTO and TiO_2-CF exhibit better degrading efficiency. The apparent reaction rate constant (K_{ap}) is applied to quantitatively evaluate the photocatalytic activity of these photocatalyts[11]. The apparent reaction rate constant is described by the apparent pseudo-first-order equation $\ln(C_0/C)=K_{ap}t$, where K_{ap} is the apparent rate constant, C_0 is the concentration of MO after darkness adsorption for 1 h and C is the concentration of MO after irradiation. Fig. 5b shows the linear relation of $\ln(C_0/C)$ versus irradiation time for degradation of MO using different photocatalysts. The determined rate constants (K_{ap}) and linear regression coefficient (R^2) from these curves are presented in Fig. 5b. All the kinetic data agrees well with the pseudo-first-order model because the R^2 values are higher than 0.95. The K_{ap} values of TiO_2-FTO and TiO_2-CF for the photodegradation of MO reaches 13.97×10^{-3} min^{-1} and 11.80×10^{-3} min^{-1}, which are 2.0 and 1.7 times higher than that of TiO_2, respectively. This result indicates that growing TiO_2 on FTO and CF can enhance photocatalytic activity. In order to get a better understanding of the enhanced photocatalytic activity, the photocatalytic behavior of (FTO and CF) were also quantitatively estimated by the apparent reaction rate constant (as shown in Fig. 5b). It is observed that the K_{ap} value of TiO_2-FTO is almost the sum of that of TiO_2 and FTO. This "sum effect" is also occurred for TiO_2-CF. These results indicate that the "sum effect" between TiO_2 and substrates can improve the photocatalytic activity.

4 Conclusions

The wool-ball-like TiO_2 microspheres on carbon fabric (TiO_2-CF) and FTO substrates (TiO_2-FTO) were successfully synthesized by a facile hydrothermal method. These obtained 3D hierarchical microspheres have a diameter of 2-3 μm, which are composed of TiO_2 nanowires with a diameter of ~50 nm. The photocatalytic behavior of the wool-ball-like TiO_2 microspheres, TiO_2-CF and TiO_2-FTO were estimated by a pseudo first-order kinetic model. Due to the "sum effect" between TiO_2 and substrates, the photocatalytic activity of TiO_2-CF and TiO_2-FTO was significantly enhanced. The K_{ap} value of TiO_2-CF and TiO_2-FTO were 2.0 and 1.7 times higher than that of the wool-ball-like TiO_2 microspheres. This work not only offers a facile hydrothermal method to prepare wool-ball-like TiO_2 microspheres on CF and FTO substrates, but also provides an 3D hierarchical structure based on TiO_2, which can be extended to other fields (hydrogen evolution, energy storage, et.al).

Acknowledgements

This work was financially supported in part by the National Natural Science Foundation of China (No. 51701092), the Natural Science Foundation of Hubei Province (Grant Nos. 2014CFB631, 2015CFB498, 2017CFB740), the Foundation of Hubei Educational Committee (No.Q20141802), the Doctoral Scientific Research Foundation of Hubei University of Automotive Technology (BK201503), the Natural Science Foundation of Jiangsu Province (No. BK20160419), the Science and Technology Plan of Nantong City (No. GY12016048) and the Fundamental Research Funds for the Central Universities (No. 021314380053). The authors gratefully acknowledge financial supported by Hubei Provincial Collaborative Innovation Center for Optoelectronics.

References

1. Y. Sun, Y. Xia, Science 298 (2002) 2176-2179.
2. H.G. Yang, C.H. Sun, S.Z. Qiao, J. Zou, G. Liu, S.C. Smith, H.M. Cheng, G.Q. Lu, Nature 453 (2008) 638.
3. X. Han, Q. Kuang, M. Jin, Z. Xie, L. Zheng, J. Am. Chem. Soc. 131 (2009) 3152-3153.
4. S. Sun, C.B. Murray, D. Weller, L. Folks, A. Moser, science 287 (2000) 1989-1992.
5. S. Liu, J. Yu, M. Jaroniec, Chem. Mater. 23 (2011) 4085-4093.
6. Z.Y. Jiang, Q. Kuang, Z.X. Xie, L.S. Zheng, Adv. Funct. Mater. 20 (2010) 3634-3645.
7. T. Tachikawa, S. Yamashita, T. Majima, J. Am. Chem. Soc. 133 (2011) 7197-7204.
8. Y. Dai, C.M. Cobley, J. Zeng, Y. Sun, Y. Xia, Nano Lett. 9 (2009) 2455-2459.
9. A.-W. Xu, Y. Gao, H.-Q. Liu, J. Catal. 207 (2002) 151-157.
10. H. Li, D. Wang, P. Wang, H. Fan, T. Xie, Chemistry-A European Journal 15 (2009) 12521-12527.
11. W. Li, R. Liang, A. Hu, Z. Huang, Y.N. Zhou, RSC Advances 4 (2014) 36959-36966.
12. S. Ikeda, N. Sugiyama, S.-y. Murakami, H. Kominami, Y. Kera, H. Noguchi, K. Uosaki, T. Torimoto, B. Ohtani, PCCP 5 (2003) 778-783.
13. Z. Miao, D. Xu, J. Ouyang, G. Guo, X. Zhao, Y. Tang, Nano Lett. 2 (2002) 717-720.
14. T. Cao, Y. Li, C. Wang, C. Shao, Y. Liu, Langmuir 27 (2011) 2946-2952.
15. M. Paulose, K. Shankar, S. Yoriya, H.E. Prakasam, O.K. Varghese, G.K. Mor, T.A. Latempa, A. Fitzgerald, C.A. Grimes, The Journal of Physical Chemistry B 110 (2006) 16179-16184.
16. J.-H. Lee, I.-C. Leu, M.-C. Hsu, Y.-W. Chung, M.-H. Hon, The Journal of Physical Chemistry B 109 (2005) 13056-13059.
17. H. Yu, J. Yu, B. Cheng, S. Liu, Nanotechnology 18 (2007) 065604.

18. Q. Wang, Z. Wen, J. Li, Inorg. Chem. 45 (2006) 6944-6949.
19. F. Shao, J. Sun, L. Gao, S. Yang, J. Luo, ACS applied materials & interfaces 3 (2011) 2148-2153.
20. H. Yin, Y. Wada, T. Kitamura, S. Kambe, S. Murasawa, H. Mori, T. Sakata, S. Yanagida, J. Mater. Chem. 11 (2001) 1694-1703.
21. T. Kasuga, M. Hiramatsu, A. Hoson, T. Sekino, K. Niihara, Langmuir 14 (1998) 3160-3163.
22. P. Chen, L. Gu, X. Xue, M. Li, X. Cao, Chem. Commun. 46 (2010) 5906-5908.
23. B. Wang, R. Karthikeyan, X.-Y. Lu, J. Xuan, M.K. Leung, J. Hazard. Mater. 263 (2013) 659-669.
24. X. Jin, J. Xu, X. Wang, Z. Xie, Z. Liu, B. Liang, D. Chen, G. Shen, RSC Advances 4 (2014) 12640-12648.
25. P. Liu, S.C. Mao, L.H. Wang, X.D. Han, Z. Zhang, Scripta Mater. 64 (2011) 343-346.
26. A. Hu, X. Zhang, D. Luong, K. Oakes, M. Servos, R. Liang, S. Kurdi, P. Peng, Y. Zhou, Waste and Biomass Valorization 3 (2012) 443-449.
27. W. Guo, F. Zhang, C. Lin, Z.L. Wang, Adv. Mater. 24 (2012) 4761-4764.
28. L. Mi, W. Wei, Z. Zheng, Y. Gao, Y. Liu, W. Chen, X. Guan, Nanoscale 5 (2013) 6589-6598.

 https://doi.org/10.1051/e3sconf/20183802015

Failure warning of hydrous sandstone based on electroencephalogram technique

Kai Tao, and *Wei Zheng* *

College of Optoelectronic Engineering, Chongqing University, Chongqing, 400044, China

Abstract. Sandstone is a type of rock mass that widely exists in nature. Moisture is an important factor that leads to sandstone structural failure. The major failure assessment methods of hydrous sandstone at present cannot satisfy real-time and portability requirements, especially lacks of warning function. In this study, acoustic emission (AE) and computed tomography (CT) techniques are combined for real-time failure assessment of hydrous sandstone. Eight visual colors for warning are screened according to different failure states, and an electroencephalogram (EEG) experiment is conducted to demonstrate their diverse excitations of the human brain's concentration.

1 Introduction

Acoustic emission (AE) of rocks is a phenomenon in which cracks inside a rock develop high energy under local stress and release energy to achieve a stable state by generating transient elastic waves [1]. Sandstone, which is mainly composed of quartz and feldspar, is a common rock type in geotechnical engineering. Moisture can change the physical properties of sandstone particles and soften loading resistance. Stability is also affected, and this scenario leads to different levels of failure [2-3].

In this study, Failure assessment was completed in real time when AE signals were collected. A visual warning mechanism of assessment results was established based on EEG.

2 Mapping rule of AE and moisture through the SVM algorithm

Training samples with five types of moisture contents were obtained through a weighing method. The moisture contents were 0%, 25%, 50%, 75%, and 100%. The AE signals of all samples were collected, and the energy and duration parameters were extracted afterward. The mapping rule of AE and moisture was obtained using an SVM classifier. The original energy and duration parameters are shown in Figure 1(a) and (b), where the horizontal axis is a random sample number and the vertical axis is the value of a characteristic parameter. Of the 100 samples for each moisture content, 70 were used as training groups, and the rest (30 samples) were utilized as test groups. The C-SVC structure network and linear kernel function were used to verify the SVM classifier's identification of the effect of different moisture contents. The identification result is shown in Figure 1(c).

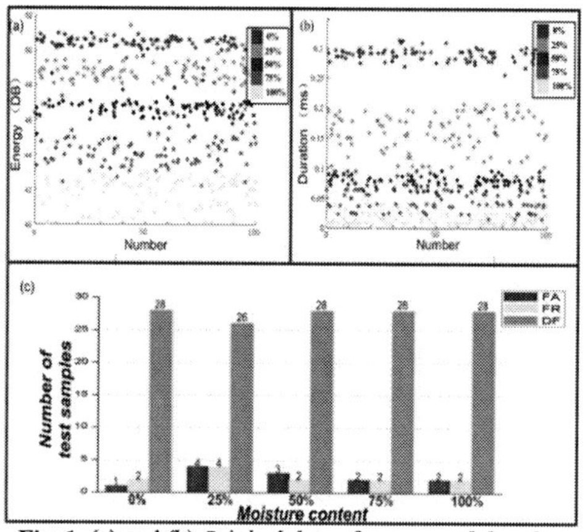

Fig. 1. (a) and (b) Original data of energy and duration parameters. (c)Moisture content identification result of SVM

According to the numbers of false acceptance (FA; classified into the same level but the actual moisture is different), false rejection (FR; not classified into the same level but the actual moisture is the same), and discriminant formula (DF; classified into the same level and the actual moisture is the same), which are displayed in the diagram, 138 were classified correctly, and the overall identification accuracy of SVM was 92%. Aside from the efficacy of the network structure, the significant difference in AE parameters with diverse moisture was also an important factor for the high identification accuracy. This result indicates the feasibility and rationality of this method.

3 Mapping rule of moisture and failure level through CT technique

* Wei Zheng: zw3475@163.com

3.1 Comprehensive index failure assessment of sandstone

(1) Morphologic quantification of cracks

The morphologic information (such as area, length, and width) of cracks can be quantificated on the CT image after the connected domain-quantifying filtering [4-5]. The number of elements, which are incorporated into the slit region, can be calculated as the area A_c of this region. Skeleton extraction can be employed to obtain the eigenvalue of the crack image and the number of pixels L_c in the skeleton diagrams is the crack length. The average width of a crack can be calculated using Eq. 1.

$$W_c = A_c / L_c \quad (1)$$

The above crack's morphological parameters are aimed toward a single crack. If the rock surface has many cracks or a bifurcation phenomenon exists, then the cracks should be numbered, the area, length, and width of a single crack should be calculated, and all crack parameters should be summed up. We assume that the number of sandstone cracks is n, and the area, length, and width of a single crack are A_{c_i}', L_{c_i}', and W_{c_i}', respectively, where $i \in (1,n)$. For the sample crack, $A_c = \sum_{i=1}^{n} A_{c_i}'$, $L_c = \sum_{i=1}^{n} L_{c_i}'$, $W_c = \frac{1}{n}\sum_{i=1}^{n} W_{c_i}'$.

(2) Comprehensive failure index

To present the failure information of rock mass in many aspects, the following three failure morphological indexes are defined based on reference [6].

$$\alpha = A_c / A_R \text{ , } \beta = L_c^2 / A_c \text{ , } \gamma = W_c^2 / A_c \quad (2)$$

where A_R is the area of the CT image; A_c, L_c, and W_c are the area, length, and width of the crack, respectively. Parameter α reflects the overall failure degree of the rock mass, β reflects the longitudinal expansion of the crack, and γ reflects the transverse expansion of the crack.

Given the different physical properties of materials, the diversity of crack expansion is obvious. To describe the rock failure scientifically, a comprehensive index is provided as follows:

$$\phi = \omega_1 * \varepsilon_1 * \alpha + \omega_2 * \varepsilon_2 * \beta + \omega_3 * \varepsilon_3 * \gamma \quad (3)$$

where ω_1, ω_2, and ω_3 are the weight of the three morphological failure indexes, and $\omega_1 + \omega_2 + \omega_3 = 1$. The specific value reflects the diversity of crack expansion tendency and varies according to different materials. Significant diversity exists in the trend of crack expansion due to the difference in the rock particles' physical properties. For example, the consistency of sandstone particles is obvious, and the expansion degree of cracks is uniform in each direction. Thus, the difference in crack length and width is small among the three weights. However, the longitudinal trend of crack is highly obvious due to the lamellar joints. Hence, the crack is long, and parameter is large. The magnitudes of the three indexes are different, but all of the indexes require the same magnitude. Thus, magnitude adjustment coefficients ε_1, ε_2, and ε_3 were set to adjust the magnitudes into the same interval. The specific values were determined by the rock crack propagation characteristics.

3.2 Failure assessment rule table

The failure level of rocks is defined from one to four grades, and the failure intensifies with the increase in value. Moisture reduces the compression capability of sandstone and increases the expansion trend of failure cracks [7]. Therefore, the comprehensive failure index of the saturated and extremely failured sample was used as a basis. This value is divided into four equal parts, which are regarded as a judging rule for failure assessment. The weight of the three morphological failure indexes was 0.33, and magnitude adjustment coefficients ε_1, ε_2, and ε_3 were set to 100, 0.1, and 100, respectively. The calculated value of ϕ is 9.91, and the failure assessment rule table is shown in table 1.

Table 1. Failure assessment rule table.

Failure level	comprehensive failure index interval
First level	[0,2.48)
Second level	[2.48,4.96)
Third level	[4.96,7.44)
Fourth level	[7.44,9.91]

3.3 Mapping rule of moisture and failure level

By calculating the morphological parameters and comprehensive index of the five crack images, the failure level of each moisture was classified according to the rule table of failure assessment, as shown in table 2.

Table 2. Failure classification of different moisture samples.

Moisture content	α	β	γ	ϕ	Failure level
0%	0	0	0	0	First level
25%	0.014	27.57	0.036	2.56	Second level
50%	0.034	67.73	0.015	3.85	Second level
75%	0.051	105.21	0.0095	5.47	Third level
100%	0.047	116.41	0.086	8.23	Fourth level

4 Real-time failure assessment function

The external environment in actual geotechnical engineering is complicated. The moisture and loading state of sandstone may change simultaneously. The synergistic effect of moisture and accumulative loading should be considered to obtain the exact damage state. A real-time damage assessment function of hydrous sandstone was proposed by considering two pieces of information: the assessment level, which varies with the time of single AE, and the superposition effect of many AE events. The function at time t during the sandstone failure expression is as follows:

$$Q(t)=\sum_{i=1}^{n}Q_i=\sum_{i=1}^{n}\chi_i * d \quad (4)$$

where n is the number of AE events up to time t and Q_i is the failure increment caused by AE event i. Parameter d is the unit of failure increment, which can be simplified to 1 in the actual application of failure assessment. $\chi_i \in \{1,2,3,4\}$, which is the same value as the failure level decided by AE event i. According to the mapping rules between moisture content and failure state, the expression of Q_i is as follows:

$$Q_i=\begin{cases} d & C_i=0\% \\ 2d & C_i=25\% \,\|\, C_i=50\% \\ 3d & C_i=75\% \\ 4d & C_i=100\% \end{cases} \quad (5)$$

where C_i is the moisture content that corresponds to AE event i.

5 Real-time visual warning based on EEG technique

The Beta wave (13–30 Hz) in the human forehead is the main reflection of the human mental state. This rhythm becomes highly obvious when humans are excited or focused; thus, the Beta wave is the major reference for human concentration assessment [8]. The stimuli provided by different colors can directly affect the Beta wave in EEG and bring diverse emotional expressions to the human brain, such as tension and relaxation [9-10]. To acquire the most scientific visual warning according to the failure level, red, orange, royal blue, rose red, green, yellow, violet, and blue were used as alternative colors to assess the Beta wave in EEG. The order of the colors was determined according to the Beta wave value under the stimulation of the different colors. Assuming that the color order of damage level from low to high is $Col_1, Col_2, ..., Col_8$, thresholds $\theta_1, \theta_2, ..., \theta_7$ are the color switching basis. Five volunteers with normal intelligence and no color blindness were recruited. The EEG signal under the stimulation of the eight colors was collected, and the Beta wave was extracted to obtain different excitations of the concentration of the human brain. To avoid sample errors caused by the emotional state of volunteers, three experiments were conducted on each volunteer, and the most stable data were selected. The acquisition device was from NeuroSky Company, and data transmission was completed via a PC with Bluetooth capability. The typical original EEG waveforms under different color stimulations are shown in Figure 2(a).

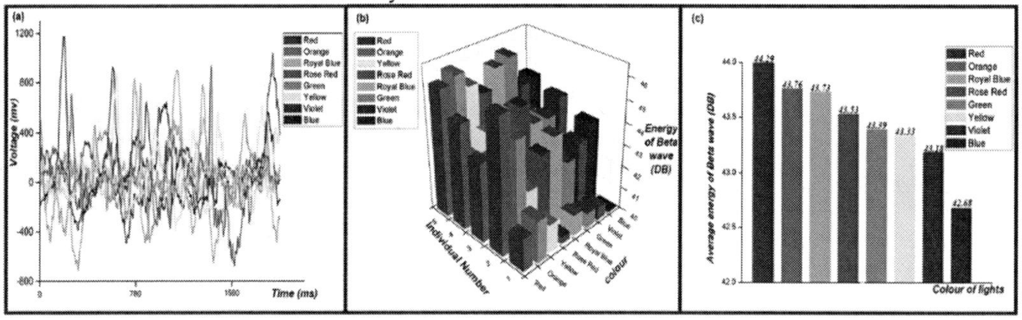

Fig. 2. (a) Typical original EEG waveforms under different color stimulations. (b) Beta wave energy in the EEG of five volunteers. (c) Average Beta wave energy of five volunteers

The original waveform shows that the morphological difference of EEG under diverse color stimulations is obvious. Hence, the concentration information of the human brain is difficult to observe directly. The Beta wave energy in the EEG of the five volunteers was extracted and is shown in Figure 2(b). The excitation regulation of the different colors on concentration was not identical due to the individual differences. However, warm colors, such as red and orange, were more likely to excite the Beta wave than cold colors, such as violet and blue. The average Beta wave energy under different color stimulations is shown in Figure 2(c). Red exerted the strongest excitation effect on the Beta wave in the EEG, which can lead to the maximum degree of concentration. Blue had the weakest effect. Hence, the visual warning order is blue, violet, yellow, green, rose red, royal blue, orange, and red according to the Beta wave excitation intensity.

6 Application experiment

To verify the feasibility of this real-time damage assessment method, an application test was conducted. First, slow-change uniaxial loading was applied to a dry sandstone sample. Its AE signal was collected, and the real-time damage state was assessed. Second, the moisture content was controlled up to 25%, and slow-change uniaxial loading was exerted on the sample without withdrawing the previous loading. The AE signal was collected, and the real-time damage state was assessed again. This step was repeated until the damage state of the 100% moisture sample was achieved. The visual warning color changed in real time during this process. The AE signals within 0-5 milliseconds were shown in Figure 3(a). The energy and duration parameters of the AE signal were extracted, and the

SVM classifier was used to realize AE–moisture mapping. The damage state in each of the different periods was calculated according to the moisture–damage level mapping rule and the real-time damage assessment function. The color switching failure thresholds were 10, 20, 30, 40, 50, 60, and 70. The failure state and visual warning color are shown in Figure 3(b).

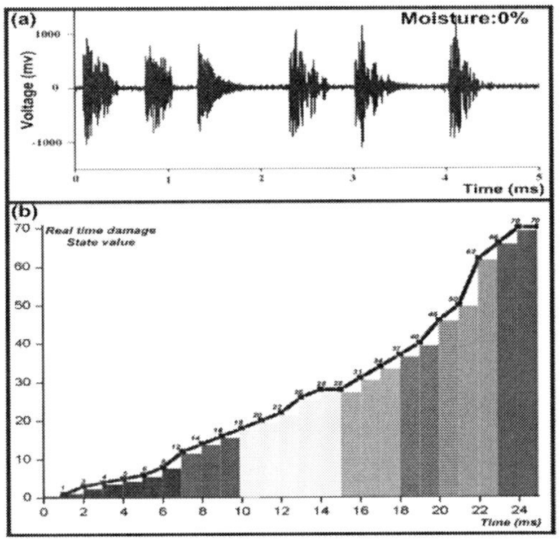

Fig. 3. (a) Original AE signals. (b) Real-time failure state and visual warning color

As shown in the figure, the damage state under the dual effect of moisture and loading can be obtained by the real-time damage assessment function, and the colors can be adjusted in real time. In actual geotechnical engineering, the change in moisture and loading can be very slow. Thus, the monitoring period is longer than that in this application test.

7 Conclusion

Hydrous sandstone failure plays an important role in geotechnical engineering safety. Currently, quantitative assessment of hydrous sandstone demonstrates poor real-time performance and insufficient portability, especially lacks of warning function. In this study, a failure assessment function of hydrous sandstone was proposed to assess the failure state of hydrous sandstone in real time. Consequently, real time and portability requirements in structural monitoring can be met. The visual warning color order was screened via an EEG experiment and was determined to be reasonable for warning geotechnical engineering staff.

Acknowledgements

The authors would like to thank the project supported by the National Natural Science Foundation of China (Grant No. 61573073) and the Chongqing City frontier general and applied basic research projects (Project No. cstc2015jcyjA40008).

References

1. Scruby CB. An introduction to acoustic emission. Journal of Physics E Scientific Instruments. 1987;20:946.
2. Sun Q. The influence of moisture content on the acoustic emission at threshold of rock destruction. Acta Geodynamica Et Geomaterialia. 2015:279-87.
3. L. Girard, S. Gruber, S. Weber, J. Beutel. Environmental controls of frost cracking revealed through in situ acoustic emission measurements in steep bedrock. Geophysical Research Letters. 40 (2013) 1748–53.
4. S. Chen, L. Liang, Y. Tian. The number of connected components in a graph associated with a rectangular (0,1)-matrix. Linear Algebra & Its Applications. 487 (2015) 74-85.
5. E.S.A.D. Silva, H. Pedrini. Connected-component labeling based on hypercubes for memory constrained scenarios. Expert Systems with Applications. 61 (2016) 272-81.
6. Suzuki T, Shiotani T, Ohtsu M. Evaluation of cracking damage in freeze-thawed concrete using acoustic emission and X-ray CT image. Construction & Building Materials. 2017;136:619-26.
7. Wan LH, Ping C, Huang YH, Wang YX. Study of subcritical crack growth of rocks and threshold values in different environments. Rock & Soil Mechanics. 2010;31:2737-42.
8. Costa ET, Costa EQ. Changes in EEG amplitude (Alpha and Beta waves) with Thermal environment. 2016;83:page. 87-93.
9. Münch M, Plomp G, Thunell E, Kawasaki A, Scartezzini JL, Herzog MH. Different colors of light lead to different adaptation and activation as determined by high-density EEG. Neuroimage. 2014;101:547.
10. Alharbi ET, Alharbi ET, Alharbi ET. SINGLE TRIAL CLASSIFICATION OF EVOKED EEG SIGNALS DUE TO RGB COLORS. Brain Broad Research in Artificial Intelligence & Neuroscience. 2017.

Lodging markedly reduced the biomass of sweet sorghum via decreasing photosynthesis in saline-alkali field

Jian Rong Guo, *Hai* Fan, *and Bao Shan* Wang*

Key Lab of Plant Stress Research, College of Life Sciences, Shandong Normal University, Ji'nan, Shandong, 250014, P.R. China

Abstract. Lodging is a serious problem in plant growth, especially in crops growth of the natural habitat. In order to determine the influence of lodging on the growth characters of sweet sorghum, plants grown in natural saline-alkali environment were used to investigate the fresh weight, dry weight, sugar content in the stalks and the photosynthesis index of salt tolerant crop sweet sorghum. Results showed that lodging significantly reduced the growth of sweet sorghum, the fresh weight and dry weight was only 28.3% and 22.5% of the normal plants when lodging occurred after 49 days. Lodging also reduced the stalks sugar content of sweet sorghum, the stalk sugar content of lodged plants was only 45.4% of that in the normal plants, when lodging occurred for 49 days. Lodging reduced the growth and sugar content by reducing the photosynthesis parameters of sweet sorghum grown in the saline-alkali field, thus, affected the accumulation of photosynthate. Interestingly, with the extension of the lodging time, lodging led to a decrease in photosynthetic rate of sweet sorghum mainly due to non-stomatal factors.

1 Introduction

Soil salinization is an increasing problem throughout the world. Nearly 10% of the soils were affected by salt in the world and there are 50% in irrigated land [1, 2]. Furthermore, the seedling growth and the yield of salt sensitive crops were significantly inhibited when grown in such environments [3], while the salt-tolerant plants could grow well in such conditions [4], and salt-tolerant crops cultivated in saline-alkali soils could be considered as the most economical and effective way to use and improve saline-alkali land. Therefore, revealing the cultivation techniques of salt-tolerant crops on saline-alkali soil is necessary for the better utilization of saline-alkali soil.

High crop yields must have a good and complete growth cycle from seedling to harvest. Lodging is one of the important factors which limit crop growth and yield in the growth period [5]. There are two types of lodging, stem lodging and root lodging [6]. Lodging will be affected by plant's own and environmental factors, high plant density and high level of nitrogen fertilization could directly influence the growth of crops, thereby reducing the lodging resistance of crops [7, 8], the lodging resistance is also related to the developmental stages of crops [7]. The environmental factors, such as rain, wind and together could cause crop lodging [9], therefore, crop lodging mainly occurred in the rainy season with wind. When lodging occurs, the absorptive capacity of plants would be destroyed, water and minerals transportation was blocked, as while as the capacity of photosynthetic rate and the dry mass were reduced, thus leading to a decline in the yield and quality of plants [10]. For example, lodging reduced the biomass production and cane quality in sugarcane, so lowered the productivity of sugarcane [11]. In rice, lodging at one day before the grain-filling stage reduced the yield, decreased the seed-setting percentage, reduced the 1000-grain weight, subsequently lower the eating quality [12]. Lodging at different development stage may have different impact on the growth of plants. Artificial lodging experiments for oilseed rape indicated that, lodging during flowering, the yield of oilseed rape could reduce by 50%, approximately, while the yield lost was smaller when the lodging occurred during seed filling, but the maximum decrease in seed oil content when lodging occurred at early flowering [13]. However, it remains unknown that whether or not the growth and the yield of sweet sorghum also decreased by lodging when grown in the saline-alkali field.

Sweet sorghum [*Sorghum bicolor* (L.) Moench] is a C_4 crop that can grow in saline-alkali environment, and it is an important raw material employed as ethanol production and fodder [14]. Additionally, sweet sorghum has high salt tolerance in seed germination and seedling growth, and it can complete the life cycle under saline-alkali environment. And the tolerance of sweet sorghum to saline environment seems to be related to the ability to avoid the excessive accumulation of harmful ions of Na^+ and Cl^-, and keeping a certain concentration of K^+, especially in shoot [15]. Compared with ordinary sorghum crops, sweet sorghum can be used in a wide range of applications, because of the characteristics of high biomass and high sugar content in its stalks [16]. There are a lot of saline-alkali lands in the Yellow River Delta of Shandong, and it has high economic values for

* Corresponding author: bswang@sdnu.edu.cn

using and developing saline-alkali soils by planting sweet sorghum in such environment. Therefore, it is necessity to investigate the cultivation technology of sweet sorghum in saline-alkali soils. The lodging of sweet sorghum was found during the cultivation after the heavy rain in the summer, so the changes and the reasons in physiological indicators of sweet sorghum after lodging were analyzed.

Based on the above analysis, the objective of this study was therefore to determine the effect of nature lodging in sweet sorghum on the growth, sugar content and photosynthesis when grown in saline-alkali soil, thus may help to find the suitable cultivation techniques for planting sweet sorghum on saline-alkali fields.

2 MATERIALS AND METHODS

2.1 Plant materials and growth conditions

Sweet sorghum (*Sorghum bicolor* L. Moench) seeds were sown in the field at Dongying (N 37°27'; E 118°30'), province of Shandong (north China). The field experiments were conducted during 2016/06-2016/11 in the growing seasons. The salinity of the soil was 0.34-0.63%. Rainfall from sowing to maturity was 430-460 mm mainly concentrated in July and August, the lodging of sweet sorghum occurred at August 26 after intense rainfall and wind at 91 days after sown, and the lodging rate was 52.1% approximately.

2.2 Growth assessment

To check if lodging affects the growth and dry mass material accumulation of sweet sorghum, at different growth time after lodging, the above-ground parts of six replicate sweet sorghum plants from normal and lodging, respectively, were harvested and put to the laboratory. The plants were washed with distilled water. After the distilled water on the plant was dried by filter paper, the fresh weight (FW) was measured. The dry weight (DW) was measured after the fresh material was dried in an oven at 105 °C for 10 min then turned to 80 °C for 72 h.

2.3 Stalk sugar content determination

In order to determine whether the sugar content in the stalk of sweet sorghum was influenced by lodging, at different growth time after lodging, six replicates of sweet sorghum plants from normal and lodging, respectively, were selected. The same section, No. 7 counted from the bottom, was harvested, and the juice was extracted by squeezing. Then, the total sugar content in the juice was determined by using a hand sugar meter (WYT-5, China) according to [14].

2.4 Photosynthesis determination

In order to determine whether the photosynthesis of sweet sorghum was influenced by lodging, at different growth time after lodging, thirty replicates of sweet sorghum plants from normal and lodging, respectively, were selected. Net photosynthetic rate, stomatal conductance, intercellular CO_2 concentration and transpiration rate of the functional leaves were measured with a portable photosynthesis system (CIRAS-3, PPS, USA).

2.5 Statistical analysis

The statistical results are described as means ± s.d., where n is the number of replicates. The data were analysed using the statistical software SPSS ver. 17.0 (SPSS Inc.) and one-way ANOVA software packages. Different letters in the table and figures indicate significant difference among the means (at 0.05) according to Duncan's test.

3 Results

3.1 Lodging reduced the fresh weight and dry weight of sweet sorghum

In order to analyze the effect of lodging on the growth and the stalk yield of sweet sorghum grown in the saline-alkali field, the whole plant above ground was harvested, the fresh and dry weight of the individual sweet sorghum plant was determined at different times (September 2, September 23 and October 14) after lodging at August 26 due to heavy rainfall and wind. The results showed that, after lodging occurred, the growth of sweet sorghum was inhibited when compared with the normal plants, and the inhibitory effect becomes more pronounced as the lodging time increases. Compared with the normal plants, when lodging for 7 days, the fresh weight and dry weight of sweet sorghum decreased significantly, which were 82.2% and 55.2% of the normal plants, respectively. When the lodging time was 28 days, the fresh weight and dry weight were 40.9% and 38.2% of the normal plants, respectively. But when the lodging time increased to 49 days, the fresh weight and dry weight of sweet sorghum plants decreased extremely significantly, which only were 28.3 % and 22.5% of the normal plants (Fig. 1). This result indicated that lodging markedly affected the biomass of sweet sorghum plants.

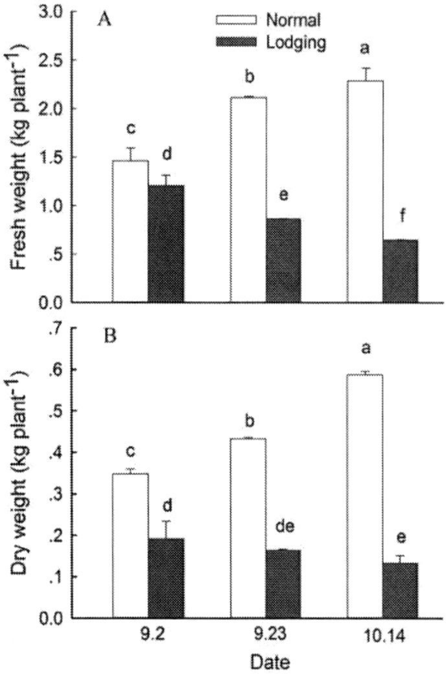

Fig. 1. Fresh weight (A) and dry weight (B) of sweet sorghum of the normal and lodging plants in the field, at different harvest date in the year of 2016. The data are described as the means ± s.d. of six replicate plants. Different letters indicate significant differences at the level of 0.05.

3.2 Lodging reduced the sugar content of sweet sorghum

Lodging reduced the accumulation of sweet sorghum material, whether did lodging also affect the sugar content of sweet sorghum stalks, so the stalk sugar content was determined at different times (September 2, September 23 and October 14) after lodging at August 26. The results showed that, lodging significantly reduced the content of sugar in the stalk of sweet sorghum, and the sugar content in the stalk decreased more significantly with the extension of lodging time. Compared with the normal plants, when lodging for 7 days, the sugar content in the stalk of lodging plants was 81.1% of the normal plants. When the lodging time increased to 49 days, the sugar content in the stalk of lodging plants was only 45.4% of the normal plants (Fig. 2). The results indicated that the lodging may affect the accumulation of photosynthate, thus reduce the sugar content of sweet sorghum.

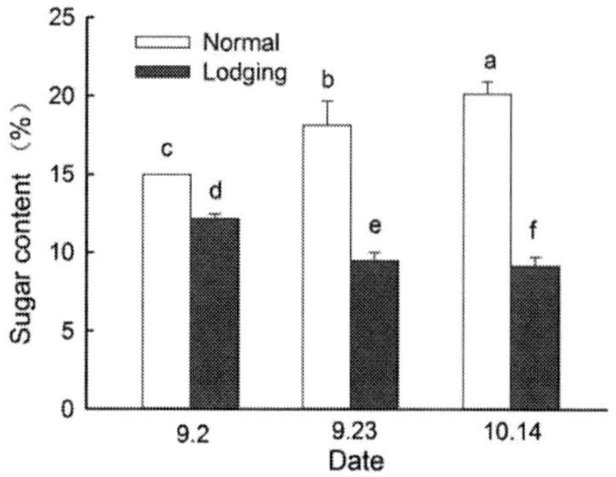

Fig. 2. Sugar content in the stalk of the No. 7 section, counted from bottom of sweet sorghum from the normal and lodging plants in the field, at different harvest date in the year of 2016. The data are described as the means ± s.d. of six replicate plants. Different letters indicate significant differences at the level of 0.05.

3.3 Lodging reduced the net photosynthetic rate and stomatal conductance of sweet sorghum

In order to analyze the effect of lodging on the photosynthesis parameters of sweet sorghum grown in the saline-alkali field, the net photosynthetic rate and the stomatal conductance of the functional leaves of sweet sorghum plant was determined at different time (September 2 and September 23) after lodging at August 26. The results showed that lodging significantly reduced the net photosynthetic rate of sweet sorghum leaves (Fig. 3A). Compared with the normal plants, when lodging for 7 days, the net photosynthetic rate of sweet sorghum leaves was 60.6% of the normal plants. When the lodging time increased to 28 days, the net photosynthetic rate of the leaves of the lodging plants was only 13.4% of normal plants. The results from the stomatal conductance of sweet sorghum leaves were consistent with the net photosynthetic rate results. When lodging for 7 days, the stomatal conductance of sweet sorghum leaves was 66.9% of the normal plants, while the stomatal conductance was 41.0% of the normal plants, when the lodging time increased to 28 days (Fig. 3B). The result of transpiration rate was consistent with stomatal conductance (Fig. 3C). Interestingly, with the extension of the lodging time, the decline degree of intercellular CO_2 concentration was significantly less than the degree of photosynthesis, it was only 70.7% of the normal plants at the day 28 after lodging, when the lodging time increased to 28 days (Fig. 3D). Lodging led to a decrease in photosynthetic rate of sweet sorghum due to stomatal factors in the early stage, and non-stomatal factors in the later stage after lodging.

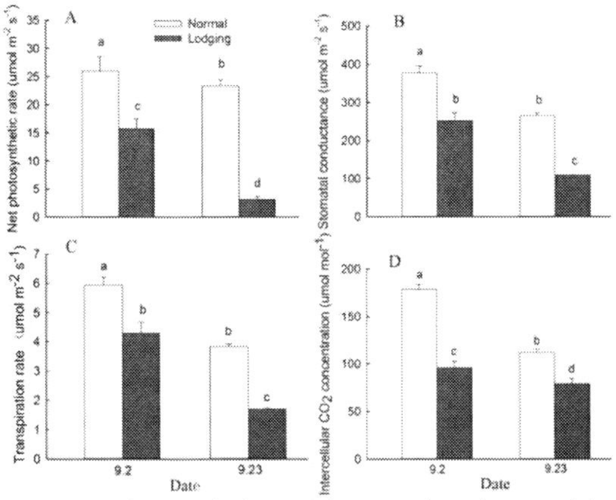

Fig. 3. Net photosynthetic rate (A), stomatal conductance (B), transpiration rate (C) and intercellular CO_2 concentration (D) of sweet sorghum from the normal and lodging plants in the field, at different harvest date in the year of 2016. The data are described as the means ± s.d. of thirty replicate plants. Different letters indicate significant differences at the level of 0.05.

4 Discussion

Lodging had a consistently negative effect on crop growth [17], sugar accumulation [11] and seed yield [18]. In the present paper, for the first time, the effect of lodging on the plant growth, sugar accumulation and photosynthetic parameter of sweet sorghum when cultivated in the saline-alkali soil was investigated. Sweet sorghum has high salt tolerance and can be grown in saline-alkali habitat, and can be used as fodder and fuel ethanol [14]. Therefore, it has a crucial significance for the utilization of saline-alkali land by cultivating sweet sorghum in saline-alkali soil. While, high percentage of lodging often occurred because of heavy rainfall in the experimental field of Dongying during the growth period of sweet sorghum. Thus, it is very important to understand the lodging effect on sweet sorghum when planting on saline-alkali soils.

Biomass yield of crops will be reduced by lodging [17, 19]. In the present study, the fresh weight and dry weight of individual sweet sorghum plant was significantly decreased after lodging (Fig. 1A), and which was gradually decreased with the increase of lodging time when compared with the normal plants, thus reduces the harvest index and total yield of sweet sorghum, and which is unfavorable for fodder and fuel ethanol production. Just like the growth in wheatgrass, lodging resulted in a severe of crop loss [20].

Stalk sugar content is an important indicator of sweet sorghum use as bioenergy feedstock, with the growth progresses, large amounts of sugars accumulated in sweet sorghum stalks, especially in saline-alkali habitats [21]. But the sugar content in sweet sorghum stalks significantly reduced when lodging occurred for 7 days (Fig. 2), and with the increase of lodging time, the sugar content in stalks decreased more significantly. The sugar and dry matter accumulation of plant comes from the photosynthesis of plants, lodging reduced the sugar and dry weight of individual sweet sorghum plant, so, what about the photosynthetic index of sweet sorghum plants after lodging. From the detection of the net photosynthesis of the lodging plants, we found that the net photosynthesis of the lodged plants was significantly reduced (Fig. 3), as while as the stomatal conductance. As showed in rice, lodging reduced the yield of rice by the reductions in photosynthesis [22]. At 7 days after lodging, the net photosynthetic rate of sweet sorghum leaves decreased in a similar as that of the stomatal conductance, while, with the extension of the lodging time, the decrease of net photosynthetic rate was more significantly than that of the stomatal conductance and the intercellular CO_2 concentration. Therefore, at early lodging, the decline of the net photosynthetic rate may be due to stomatal factors, with the lodging time is prolonged, the decline of the net photosynthetic rate of lodged sweet sorghum plants may mainly be because of non-stomatal limitation factors. Long period of lodging maybe resulted in the damage of the photosynthetic system, reduced the using efficiency of carbon dioxide, thus reduced the net photosynthetic rate.

5 Conclusions

Sweet sorghum can be used as feedstock for fuel ethanol and livestock husbandry. Sweet sorghum has strong salt resistance and can grow in the saline-alkali environment, so it is a good material for the utilization and development of saline-alkali soil. Lodging markedly reduced the fresh weight, dry weight and the sugar content in the stalks of sweet sorghum, and the decrease of matter accumulation mainly due to the decline of photosynthesis. So, it is important to improve the lodging resistance of sweet sorghum in the field, future studies should focus on the fertilizer supply and hormone regulation in nature growth conditions.

Acknowledgements

This work was supported by the National Natural Science Foundation of China (31570251 and 31770288), the independent innovation and achievement transformation of special major key technical plans of Shandong Province (2015ZDJS03002 and 2017CXGC0313), the Natural Science Research Foundation of Shandong Province (ZR2017MC003), and the Higher Educational Science and Technology Program of Shandong Province (J17KA136).

References

1. J. Song, W. Shi, R. Liu, Y. Xu, N. Sui, J. Zhou, Plant Spec. Bio. **32**, 2 (2016)
2. J. Guo, S. Suo, B. Wang, Seed Sci. Res. **25**, 3 (2015)
3. L. Kotula, H.A. Khan, J. Quealy, N.C. Turner, V. Vadez, K.H. Siddique, P.L. Clode, T.D. Colmer, Plant Cell Environ. **38**, (2015)
4. J. Guo, Y. Li, G. Han, J. Song, B. Wang, Funct. Plant Biol. **45**, 3(2018)

5. A.S. Peake, K.L. Bell, P.S. Carberry, N.Poole, S.R. Raine, Crop and Pasture Sci. **67**, 9 (2017)
6. M.M. Sposaro, P.M. Berry, M. Sterling, A.J. Hall, C.A. Chimenti, Field Crops Res. **119**, 1 (2010)
7. M.M. Sposaro, C.A. Chimenti, A.J. Hall, Field Crops Res. **106**, 2 (2008)
8. M. Podgórska-Lesiak, P. Sobkowicz, Field Crops Res. **149**, (2013)
9. P. Martinez-Vazquez. Agri. Forest Meteorol. **228–229**, (2016)
10. T. Kashiwagi, H. Sasaki, K. Ishimaru, Plant Prod. Sci. **8**, 2 (2005)
11. P.D.R. van Heerden, A. Singels, A. Paraskevopoulos, R. Rossler, Field Crops Res. **180**, (2015)
12. Y.Z. Lang, X.D. Yang, M.E. Wang, Q.S. Zhu, Rice Sci. **19**, 4 (2012)
13. S.L. Kendall, H. Holmes, C.A. White, S.M. Clarke, P.M. Berry, Field Crops Res. **211**, (2017)
14. S. Audilakshmi, A.K. Mall, M. Swarnalatha, N. Seetharama, Biomass & Bioenergy **34**, 6 (2010)
15. A. Almodares, M.R. Hadi, B. Kholdebarin, B. Samedani, Z.A. Kharazian, J. Environ. Biol. **35**, 4 (2014)
16. A. Goshadrou, K. Karimi, M.J. Taherzadeh, Ind. Crop Prod. **34**, 1 (2011)
17. G. Singh, S.G. Chapman, P.A. Jackson, R.J. Lawn, Aust. J. Agr. Res. **53**, 11 (2002)
18. Y. Xiao, J. Liu, H. Li, X. Cao, X. Xia, Z. He, Frontiers Agr. Sci. Engin. **2**, 2 (2015)
19. J.A. Rueda, E. Ortega-Jiménez, A. Hernández-Garay, J.F. Enríquez-Quiroz, J.D. Guerrero-Rodríguez, A.R. Quero-Carrillo, Biomass and Bioenerg. **88**, (2016)
20. M.M. Acreche, G.A. Slafer. Field Crops Res. **122**, (2011)
21. N. Sui, Z. Yang, M. Liu, B. Wang, BMC genomics **16**, 1 (2015)
22. T.L. Setter, E.V. Laureles, A.M. Mazaredo, Field Crops Res. **49**, (2-3) (1997)

Application of BIM technology in green building material management system

Tong Zhineng

Jiangxi Normal University of science and technology, Nanchang,Jiangxi,China

Abstract. The current green building materials management system in China's construction industry is not perfect, and there are still many shortcomings. Active construction of green building materials management system based on BIM technology, combined with the characteristics of green building materials and its relationship with BIM technology application, is urgently needed to better realize the scientific management of green building materials.

1 Introduction

In the current social development and construction, the establishment of environment-friendly and resource-saving society is of great significance. In this process, the development of green building is essential. The prerequisite for the realization of green building development is to strengthen the management of green building materials, and to ensure that building materials are comfortable, economical, healthy, energy saving, environmental friendly and other green features. At present, China's green building materials for the development and research in the "four sections of an environmental protection" stage, the green building design and construction also put forward higher requirements. However, there are still some problems and shortcomings in the construction of green building materials management system.

2 Management characteristics of green building materials

First of all, manufacturers of green building materials should clearly recognize the standards of green building materials, and strictly manage the green degree in the process of production. At the same time, in accordance with the actual needs of different green buildings, green building materials with different green levels can be supplied, and the management of inventory, sales volume and production capacity shall be strengthened. Then, as green building materials come in many different categories and projects, it will involve many different material suppliers and manufacturers.

So in the green building materials management, the traditional management mode of building materials are too extensive, have been difficult to meet the current needs of [1]. management in this regard, take the fine management mode of management of green building materials, it is necessary to. Finally, in the actual construction of green building, according to the actual requirements of architectural design, different types of green building materials should be chosen. All parties involved in the construction of the project and their upstream and downstream enterprises should pay attention to strengthening coordination and communication. After the completion of the project delivery, shall provide the corresponding information of green materials and green building information to the property side, so that helps the property side in the days after the use of green materials for maintenance or replacement and maintenance.

3 Relationship between BIM technology and green building material management characteristics

3.1 Modeltraceability of information

In the actual use of green building, combined with the specific requirements of functional use, you can adjust, replace and maintain the corresponding green building materials. Therefore, in the process of operation and management of green building, it is very important to the construction of green building materials, application materials, material production function parameter information such as the history of engineering information[2]. And the construction process of green building in the design, BIM technology can accurately record the green building materials information and detailed engineering information. At the same time, it can also record the occurrence of engineering changes in detail, and accurately evaluate the green grade of green buildings.

3.2 Collaborative management

* Corresponding author: tongzhineng2008@sina.com

In the process of the construction of green projects, green building materials manufacturers and suppliers, construction units, design units, the owners should be collaborative management of project construction material, which can form an industrial chain. Materials manufacturers, suppliers, construction units, the owners shall actively participate in the green building design stage, the value of individual demand of reasonable formulation, so as to clear the production and green building materials, green building construction scheme and provide technical measures. The application of green building materials in green building puts forward higher cooperative management requirements for all parties involved in the project.

3.3 Fine management

Green building materials in construction, maintenance, operation, construction, design and other stages, all need to build a fine library of information. In this way, the parties involved in the green building project can be more accurate production, use, ordering and query of green building materials, resulting in the use and management of green building materials fine practical application in the process of. In the process of building the information base of green building materials, BIM technology can play a great role in the process of maintenance, use and production of green building materials, and realize fine management. In addition, the information breach of green building materials management will be broken, and the parties involved in the project can share the information better.

3.4 3D visualization

In the process of green building design and construction, through the establishment of three-dimensional model, can provide visual simulation to the customer, make it more intuitive understanding and understanding of the project. The 3D visualization model characterized by its intuitive, in green materials and other site management, construction design, green material selection, green building apartment layout design and other aspects can, play a great role in the green building project, the establishment of three-dimensional model of green building materials and construction, also need to BIM technical support.

4 Construction of green building material management system based on BIM Technology

4.1 Composition of green building material management system

4.1.1 Application management layer

In the three-dimensional model based on BIM technology, the application of management role is to manage green building materials, construction and design, green building projects and other aspects of the model. And in the use and maintenance of green building on the management, use the corresponding maintenance information and historical information traceability. In the green building life cycle, through the three-dimensional model based on BIM technology, the project participants can parameter information for green building materials for accurate extraction, in order to better communication and information sharing, unified and effective integration of management resources, developing better project management decision.

4.1.2 Data exchange layer

In the whole life cycle of green building, green building materials to achieve collaborative management through the BIM technology, the network information transmission platform and cloud services platform is built, which can effectively manage and share information. In addition, in the BIM technology data exchange process, the handset, the notebook and so on the terminal equipment provided the good hardware support for it.

4.1.3 BIM model layer

Before the construction of green building construction, construction units, design units, the owners shall advance the purchase of green building materials, and to make a comparison between the different simulation models were completed, with the development of construction technology and construction scheme. Therefore, in the application of BIM model in green building material management system, more attention has been paid to the construction of technical model, material model, material database and so on. According to the actual demand of green building materials management, in the model layer, the core of BIM model were set up, which makes the whole life cycle of green building, green building materials are able to achieve the perfect management.

4.1.4 Basic data layer

In the process of the management of green building materials, it concerns the function and production information of green building materials, architectural framework design model information, green building materials evaluation information and so on. Therefore, in the construction of green building materials management system based on BIM technology, IFC format should be used to convert and store data, and to manage BIM basic data uniformly in the same platform. After the completion and acceptance of the project, the operation and maintenance of the later stage needs the BIM information database to provide relevant information, so as to form a cycle feedback, and provide relevant reference for other construction projects. So rich, huge BIM database, for the state, industry, enterprises in the green building materials management level has a great role in promoting.

4.2 Implementation of green building material management system

4.2.1 Production and transportation of materials

The RFID tag will be attached to the green building materials that have just been shipped. The green building materials will be scanned through the reader and then uploaded to the BIM to achieve the management of green building materials. Thus, in the management of green building materials, material suppliers and manufacturers can achieve zero inventory and zero defects. In the process of material transportation, the order, route and train number of transportation are optimized on the basis of visualization model and storage area.

4.2.2 Material construction

Storage and lifting of materials, RFID tags can be used for tracking and monitoring, management of materials access construction. Through the cloud platform for the transmission of these information, to avoid the artificial input of information arising from negligence or error. For example, in the process of material access check, you can use RFID reader to automatically collect data to manage the material.

4.2.3 Material operation

The RFID tag also contains material maintenance information, which allows rapid location of material and storage of maintenance records in the RFID tag. In addition, it can also integrate maintenance information in the BIM property management system, thus more intuitive management and maintenance of green building materials.

4.2.4 Material removal

When the construction project reaches the life cycle, the green building materials can be recycled and used at the time of demolition, and the BIM database and RFID tags can be used to judge. This can better realize the recycling management green building materials, play a role in saving resources and reducing consumption, so as to better realize the green development of construction projects.

5 Conclusion

In the current society, people pay more and more attention to the green environmental protection of buildings. Green building material is the important condition and foundation to realize the greening of building project. However, China's current management of green building materials is not perfect, there are still many deficiencies. In this regard, we should actively build green building materials management system based on BIM technology, combined with the characteristics of green building materials and its relationship with BIM technology, so as to better realize the management of green building materials.

References

1. Cheng Yuan cheng. Application Research of green building design based on BIM technology [D]. Hunan University, 2013.
2. Guo Yunpeng. Study on the application strategy of BIM technology in green building life cycle [D]. Harbin Institute of Technology, 2013.
3. Kang Ming, Zhang Hao. Discussion on the application of environmental protection and energy saving design in green building [J]. Building materials and decoration (early part). Marketing, 2010.4 (10): 23 - 24.
4. Wang Zhenshuang, Wang Liguo, Gaoping, Chen Xiaobo. Research on green building materials management system based on BIM technology [J]. building economy, 2015,04: 83 -86.

Anti-wear Mechanism Analysis of Nano-$CaCO_3$ Additives

Zhen-Xu[1], Junfeng-SUN[2]

[1]Faculty of Mechanical and Electrical Engineering,Yunnan Open University,Kunming 650500,China
[2]Faculty of Mechanical and Electrical Engineering, Kunming University of Science and Technology, Kunming 650504, China

Abstract: In this paper, the wear test was carried on with cylinder piston by the wear test device, receiving the results of the piston ring wear and abrasive characteristics by monitoring the wear process, the thesis analysis and put forward the nano-$CaCO_3$ lubricating oil additive anti wear mechanism by the ferrography analysis technology, and provide the technical reference for the relevant measures to reduce wear and the friction, and provide reference value for further study on the related theories of reducing wear and reducing friction.

1 Preface

Internal combustion engine is the most important power machinery, which is widely used in all walks of life, such as aviation, railways, machinery[1].With the rapid development of production practice, the wear of internal combustion engines is diversified and the degree of wear is getting deeper and deeper. with the rapid development of nanotechnology, nonmaterial have the characteristics of small size, large specific surface area, quantum effect, quantum tunneling effect and interface effect[2].The application of nonmaterial in tribology has become an important research topic of nanoscale tribology[3].And we can analyze the characteristics of abrasive grain by the ferrography analysis technology. It has strong explanatory for the fault diagnosis of lubrication and wear. It is an important method for the study of lubrication theory [4].

2 The Wear Test

Taking the piston-cylinder produced by a combustion engine factory as the test object (hereinafter referred to as piston-cylinder,model 4100QB).The domestic Kunlun lubricating oil was used as the lubricating condition,we can analyze the characteristics of abrasive grain and the wear of piston in the three wear period by the ferrography analysis technology and analyze the influence of nano-$CaCO^3$ lubricant additive on wear from macroscopic and microscopic perspectives.

3 The Analysis of Wear

3.1 The Wear Test without Adding Nano-$CaCO_3$ Additive

The specifications of the lubricant oil used in the test are shown in Table 1.

Table 1 the condition during the wear of cylinder

Load and speed	Lubricant oil	the type of lubricant oil
1Kg and 270r/min	CD 20W-50	CD 20W-50

In the case of fixed test conditions, the changes of the wear of piston changing with time are shown in Table 2.

Table 2 the Wear of Piston

time（min）	30	60	100	200	300	400	600	800	1000	1200
piston(mg)	2.0	4.0	8.7	11.4	11.1	11.6	19.3	23.2	30.0	38.0

3.2 The Wear Test with Adding Nano-$CaCO_3$ Additive

The specifications of the lubricant oil used in the test are shown in Table 3.

*Corresponding author email:250242124@qq.com; junfengsun@139.com

Table 3 the Condition during the Wear of Cylinder

Load and speed	Lubricant oil	the type of lubricant oil
1Kg and 270r/min	CD 20W-50+1%CaCO3	CD 20W-50

In the case of fixed test conditions, the changes of the wear of piston changing with time are shown in Table 4.

Table 4 the Wear of Piston

time（min）	30	60	100	200	300	400	600	800	1000	1200
piston(mg）	0.9	1.1	1.5	2.1	2.2	2.3	2.4	2.5	3.0	4.5

3.3 Comparatively analyze the curve of wear under two kinds of lubrication conditions

Figure 1 shows the time-wear curve under the Kunlun engine oil without adding additive and adding $CaCO_3$.From the figure1, the two curves far away, the wear station is far difference. From the wear period, the first 60 minutes is the run-in stage under the condition with adding $CaCO_3$, froming 100 minutes is the stable wear period, and after 1000 minutes is the serious wear period. Under without additive, the first 100 minutes is the run-in period, stable wear period is shorter from 200 minutes to 600 minutes. From the analysis of the amount of wear, the addition of nano-$CaCO_3$ greatly changes the amount of wear, which is about 1/6 of that of the non-additive.

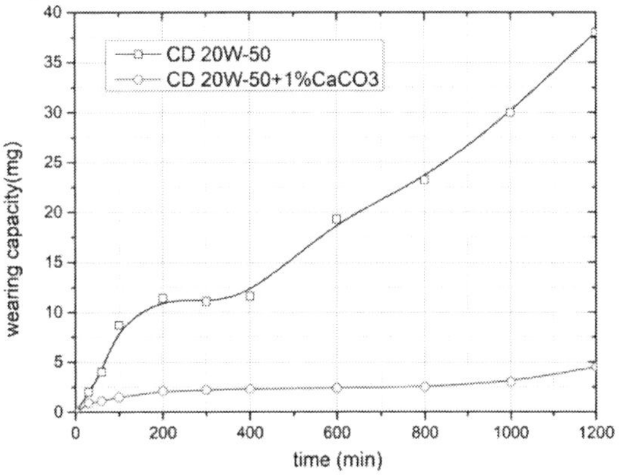

Figure1 the time-wear curves under the conditions of CD 20W-50 and CD 20W-50+1%$CaCO_3$

4 Ferrography analysis of adding Nano-$CaCO_3$

4.1 The ferrography analysis in run-in period

The following figures are the fields of abrasive grains in run-in period which under the condition of adding nano-$CaCO_3$. In the following figures, the abrasive grains are evenly distributed along the lines of magnetic force and catenary, and the red pellets Fe_2O_3 and the black abrasive grains Fe_3O_4 appear.

By comparing the characteristics of wear particles without adding additive and adding $CaCO_3$ additive, it can be seen that the magnetic lines of force with $CaCO_3$ additive are less than those without additive, as shown in Figure 2 and Figure 4. The typical grain size is also about 5μm smaller than the typical grain size without $CaCO_3$ additive.

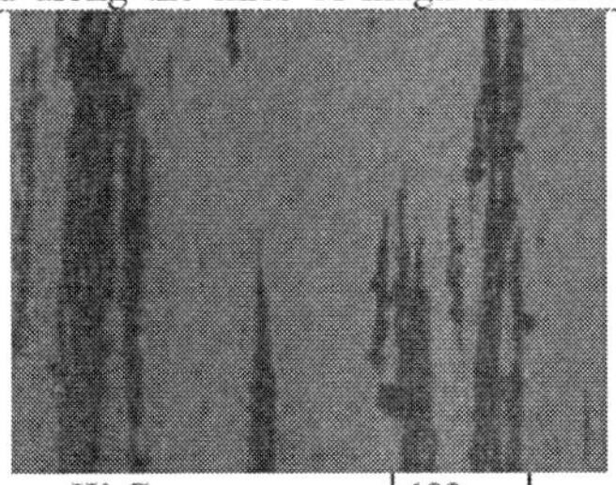

W-G |-100μm-|

Figure 2 CD 20W-50 60min

G-G |-50μm-|

Figure 3 CD 20W-50 100min

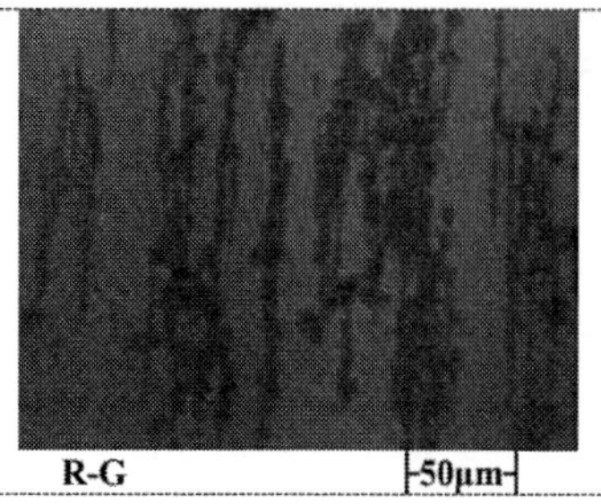

Figure 4 CD 20W-50+1%$CaCO^3$ 60min

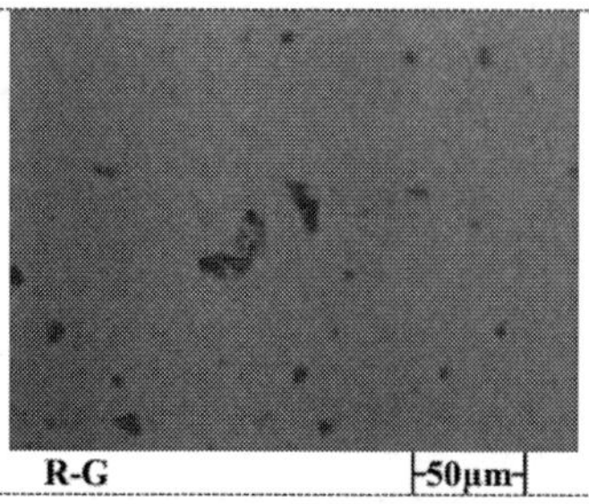

Figure 5 CD 20W-50 +1%$CaCO^3$ 30min

4.2 The ferrography analysis in stable wear period

Figure 6 shows the field of abrasive grains without additive at 200 minutes. In Figure 7 and Figure 8, typical large abrasive grains can be clearly seen along the lines of magnetic force. The abrasive grain surface is smooth. In Figure 7, the large abrasive grain size is about 10μm and the edge of the abrasive grain is irregular. Figure 8 shows the test results for 400 minutes. In the Figure 8, the large abrasive grain size is about 15μm.The large abrasive grain has a straight edge and the other edges are irregular. Figure 9 shows the 800 minutes test. The abrasive grains accumulate more chains, and the amount of friction polymer increases.

The entire stable wear period, uniform and slender magnetic lines, abrasive grains sizes range are not large, there are only a few large abrasive grains. By comparison, it can be found that the abrasive grains with adding additive all need to be more than 200 times in order to observe the large clear field of abrasive grains. And the typical large abrasive grains can be observed after 500 times the field. During this period, the large abrasive grains sizes with additive were about 10μm-15μm while the typical large abrasive grains with no additive were about 15μm.

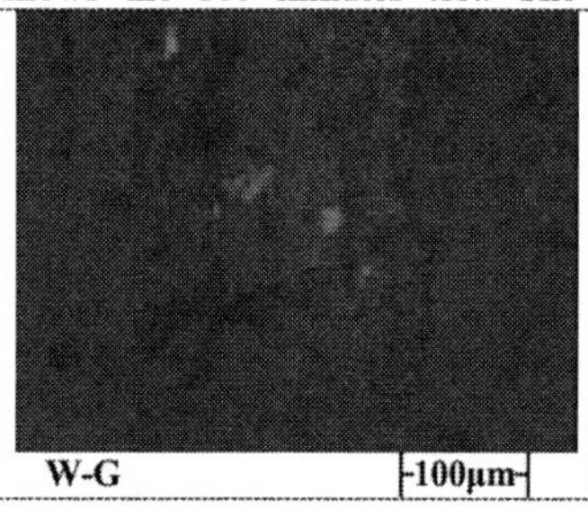

Figure 6 CD 20W-50 200min

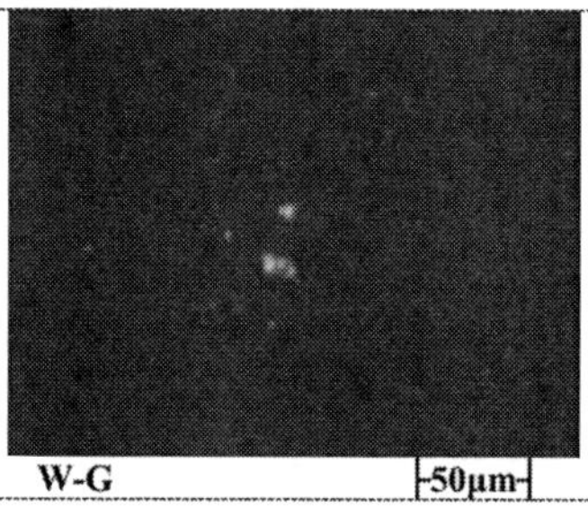

Figure 7CD 20W-50+1%$CaCO^3$ 300min

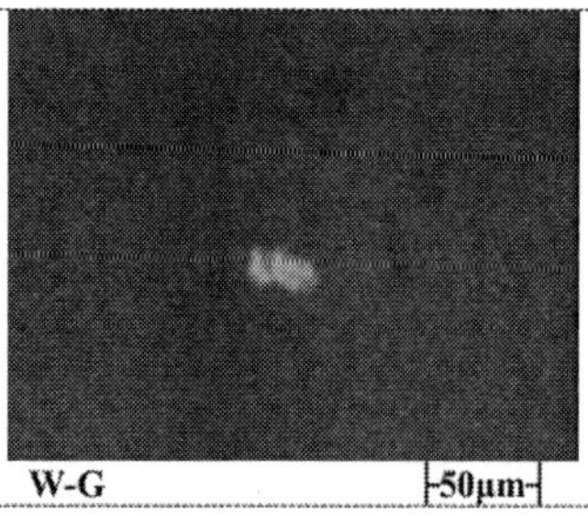

Figure 8 CD20W-50+1%$CaCO^3$ 400min

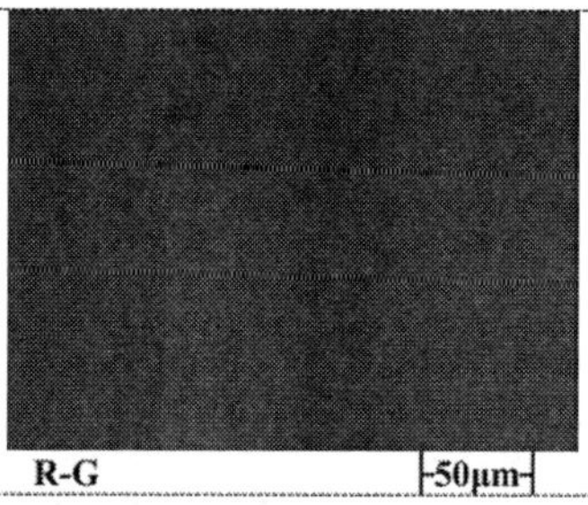

Figure 9 CD 20W-50+1%$CaCO^3$ 800min

4.3 The ferrography analysis in serious wear period

Figure 10 is a spectrogram of 200 times field of adding nano-$CaCO^3$ additive to Kunlun oil.It can be clearly seen that there are many friction polymers, and there are many chains of abrasive grains, their surfaces are smooth. And large abrasive grains appear more.

At 1200 minutes of testing, as shown in Figure 10, a 200x field clearly shows that the chains of abrasive grains grow larger and larger, the sizes of the larger abrasive grains increase, and the presence of friction polymer occurs. Fatigue spalling grains appear in the magnetic flux as shown in Figure 11.The surface of the abrasive grains is relatively flat and thick. Fatigue spalling grains are mainly caused by pitting corrosion on the friction surface, causing the metal material to be spalled. Initially, only the microscopic peeling, with the continuance of wear, one after another macroscopic peeling until the wear failure. Figure 12 is the serious sliding wear abrasive grains, the size of them are larger than 15μm, the surfaces of abrasive grains have serious scratches, and have temper stain.

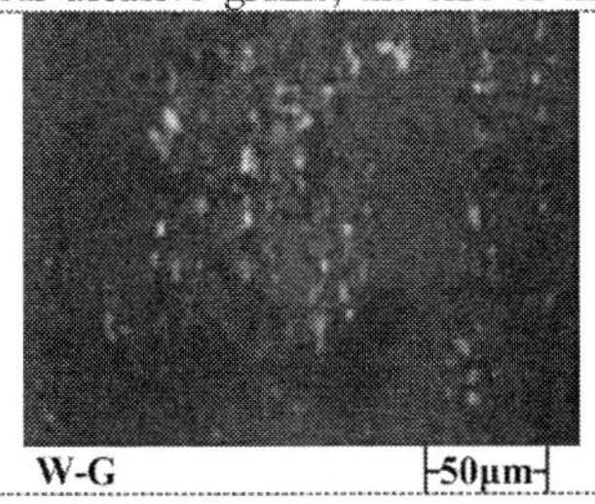

Figure10 CD 20W-50+1%CaCO3 1200min

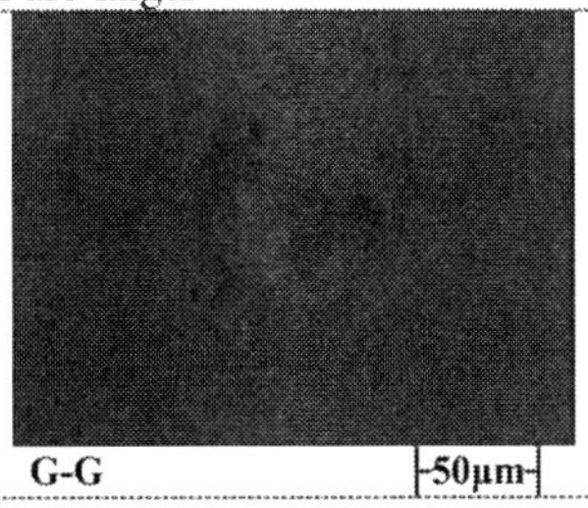

Figure11 CD 20W-50+1%CaCO3 1200min

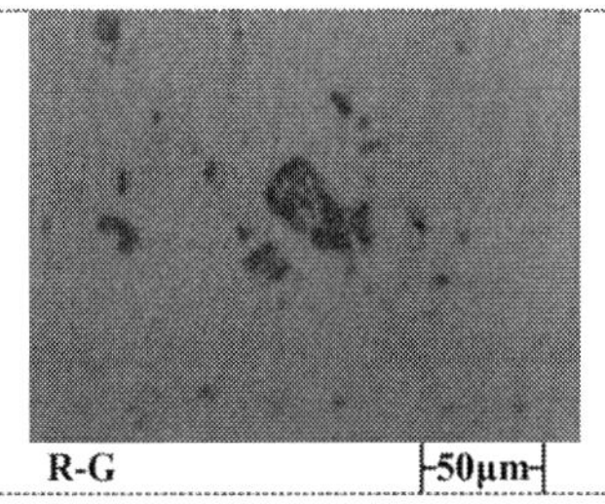

Figure12 CD 20W-50+1%$CaCO_3$ 1200min

5 Anti-wear Mechanism Analysis of Nano-$CaCO_3$ Additive

During the process of friction, the nano-$CaCO_3$ in the lubricating oil are affected by two factors resulting in their migration of the frictional contact surface [5].First, the local high temperature generated in the process of friction makes the molecular weight greater than that of the bulk phase, and these strong molecular fluctuations occur non-directional migration, thus they increase the chances of nano-$CaCO_3$ migrating to the frictional contact surface. Second, the frictional contact surface produces an interfacial electric field such as an escape electron that enhances the frictional surface, and the enhanced interface electric field produce enhanced magnetic field. Due to the presence of weak electromagnetic field and the adsorption of nano-$CaCO_3$ surface extremely strong, so the nano-$CaCO_3$ more easily enriched in the bad contact surface of liner - piston, shown in Figure 13.Due to its small particle size and high surface energy [6], nano-$CaCO_3$ form a layer of physical adsorption film on the frictional contact surface just before the friction begins. The adsorption layer is oriented perpendicularly to the frictional contact surface, and has strong adhesion.

As the local friction temperature increases slowly, due to the nano-$CaCO_3$ has a very high pressure, even if aboving adsorption temperature of the adsorption layer will not be destroyed. When the temperature continues to rise to a certain temperature, the elements in the nano-$CaCO_3$ penetrate into the friction surface or the friction sub-surface and form a solid solution with the liner of the liner-piston, and at the same time chemical reaction forms a thin wear-resistant chemical reaction layer, Figure 14.Thus they improve the anti-wear and anti-wear lubricating oil performance.

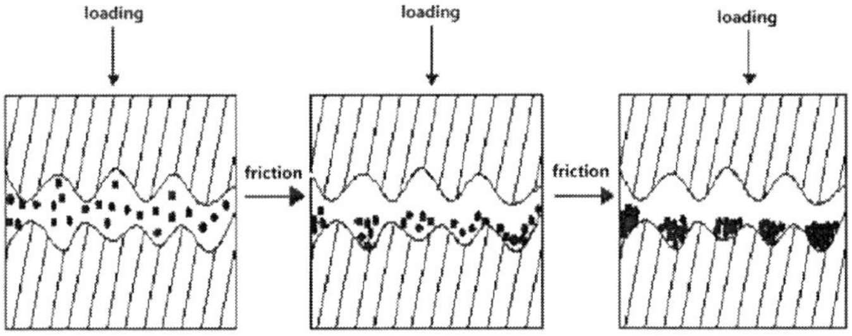

Figure13 an example of enrichment effect of nano-$CaCO_3$ on the friction surface

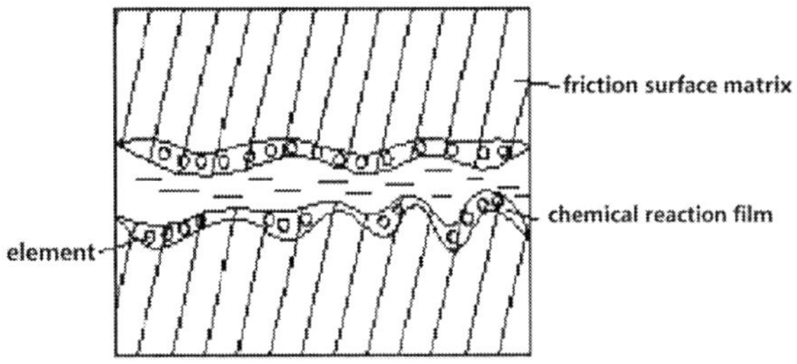

Figure14 an example of the effect of nano-CaCO3 formation chemical reaction membrane

 https://doi.org/10.1051/e3sconf/20183802018

6 Conclusion

Based on the wear test of nano-$CaCO_3$ as a lubricating oil additive,based on the nano-material science and tribology, the anti-wear mechanism of the nano-$CaCO_3$ lubricant additive was studied and analyzed based on the characteristics of the abrasive particles.The results show that compared with the lubricating oil without nano-$CaCO_3$,the lubricity of nano-$CaCO_3$ is greatly improved and the wear loss is significantly reduced,which greatly improves the lubrication condition of the cylinder-piston ring.

Acknowledgements

The work described in this paper was fully supported by the Youth Project of Science and Technology Department of Yunnan Province (No.2017FD132) and the Analysis Test Fund of Kunming University of Science and Technology (No.2016T20100075). We are very grateful to the organizations concerned.

References

1. ZL Cui. ZK Zhang.Nanostenet.Mater.1996.7 (3):355-361
2. Kaufman R E.Remaining Usethl Life Measuremeuts of Diesel Engine Oils, Automotive Engine Oils, Hydraulic Fluids, and Greases Using Cyclic Vohammetrie Methods[J].Lub Eng,995:51(3)
3. Kaufman R E,Rhine W E.Assessment of a Remaining Lubricant Life[R].Report No.AFWALTR-86-2024,Nevember,1986
4. Xiaohan Liang et al. Ferrography and its application in mechanical monitoring and diagnosis.Beijing: China Communications Press, 1993.10
5. Zhang E et al. Ferrography and its industrial application. Xi'an: Xi'an Jiaotong University Press, 2001.10
6. L Rapoport,M Lvovsky,I Lapsker,et a1.Friction and wear of bronze powder composites including fullerene-like WS2 nanoparticles.Wear,2001,249:150-15706.(in Chinese)

Structures and Properties of Self - crosslinking Silicone Resin

*Yong-xin Zhao , and Ying-qiang Zhang**

Shanghai Institute of Technology, School of Materials Science and Engineering, Shanghai, 201418

Abstract. Highly transparent silicone resin with self-crosslinking structure was prepared using phenyltrimethoxysilane, diphenyldimethoxysilane, 1,3,5,7-cyclotetra(methyl siloxane) and bisvinyltetramethyldisiloxane as main raw materials. The structure of silicone resin was determined by Fourier Transform Infrared Spectroscopy (FT-IR). The light transmittance was measured by UV-Vis spectroscopy. Thermogravimetric analysis (TGA) was used to study the thermal decomposition process. The microstructure of cured self-crosslinking silicone resin is more uniform, resulting in better light transmittance up to 100% in the range of 400nm ~ 800nm. The cured has relatively good heat resistance, the initial thermal decomposition temperature of the cured could be up to 315.8 ℃. SEM observations show that the self-crosslinking silicone has a uniform, textured structure, higher transparency compared with the existing condensation silicone material, and can be used as advanced architectural translucent materials and optics packaging materials.

1 Introduction

Silicone resin has the advantages of high light transmittance, high refractive index, high thermal stability, high yellowing resistance, low temperature resistance, etc. It is currently the most widely used optical packaging material, which is widely used in the field of construction sealing, the packaging of solar energy, high power LEDs and optical devices[1-3]. In recent years, with the rapid development of high-rise buildings, and research of connecting materials and new construction seals shows active, there are new and higher requirements for the light transmittance and heat resistance of packaging materials. The research in this field of technology is very promising [4].

At present, double-packed condensation silicones sealing materials, such as dow corning's GP300, GP800, 1-2577 and shin-etsu's KE445 and KR251, and other double-packed addition types used in the construction lighting industry. However, there are some problems in double-package silicone adhesives: Mixed unevenness is almost present in all types of two-component silicone adhesives. For instance, when the refractive index difference between vinyl phenyl silicone resin and cross-linking agent is large, the difference in the molecular structure between the two is large, so the compatibility becomes worse. When the two are mixed and cured, it is difficult to achieve uniform microscopic mixing and full curing, resulting in a decrease in the light transmission performance of the sealant. So it is difficult to meet the development needs of high-end light transmissive materials. In order to overcome the above problems, the author has recently tried a preparation technique of a self-crosslinking silicone resin (which integrates Si-H bonds and Si-vinyl groups into one molecular structure through chemical synthesis), That is, through the redesign and integration of the curing groups in the molecular structures of methylphenyl vinyl silicone resins and methylphenyl hydrogen-containing silicone oils, a highly light-transmitting encapsulating material is prepared by utilizing the chemical structural compatibility[5-6]. This technology adjusts the amount and proportion of monomers, selects the appropriate feeding sequence and feeding method, and controls the appropriate reaction temperature and polymerization time. The prepared packaging adhesive has adjustable viscosity, uniform composition, high light transmittance and heat resistance. It is expected to be used as an encapsulation material for advanced architectural light transmission materials and optical devices. Self-crosslinking silicone resin was prepared in this paper, and the structures and properties of its cured product were analyzed.

2 Experimental

2.1 Materials

Phenyltrimethoxysilane, diphenyldimethoxysilane, 1,3,5,7-cyclotetra (methyl siloxane) and bisvinyltetramethyldisiloxane, industrial grade, commercial; toluene, analytical grade, Shanghai Titan Technology Co., Ltd.

2.2 Preparation and Curing of Self-crosslinking Silicone Resins

* Yingqiang Zhang: sh_yqzhang@sina.com

According to stoichiometry, phenyltrimethoxysilane, diphenyldimethoxysilane, 1,3,5,7-cyclotetra(methyl siloxane) and bisvinyltetramethyldisiloxane, distilled water, toluene and concentrated sulfuric acid were weighed into a 500 ml four-necked flask with a spherical condenser tube. The temperature was raised to 65°C, stirred at a constant temperature for 3 hours, and then heated to 100°C. Stirring was continued at a constant temperature for 5 hours. Heating was stopped, and the mixture was poured into a separatory funnel. The lower layer liquid was removed out, then added sodium carbonate solution, and washed to neutral. The lower layer liquid was distilled to obtain the product. The synthesis route is shown in Scheme 1. The curing reaction of the self-crosslinking silicone resin is shown in Scheme 2.

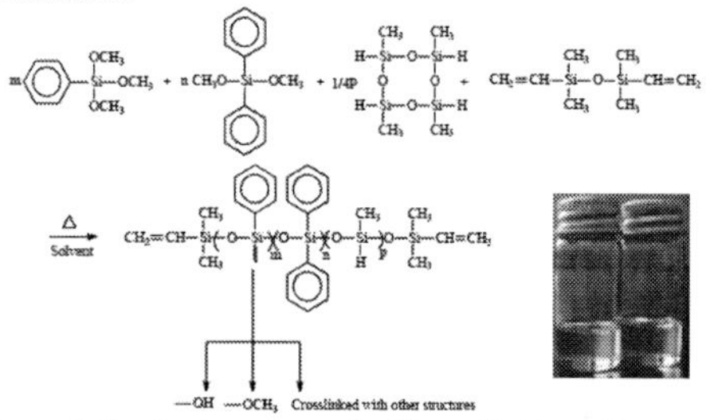

Scheme 1 Synthesis process of self-crosslinking silicone resin

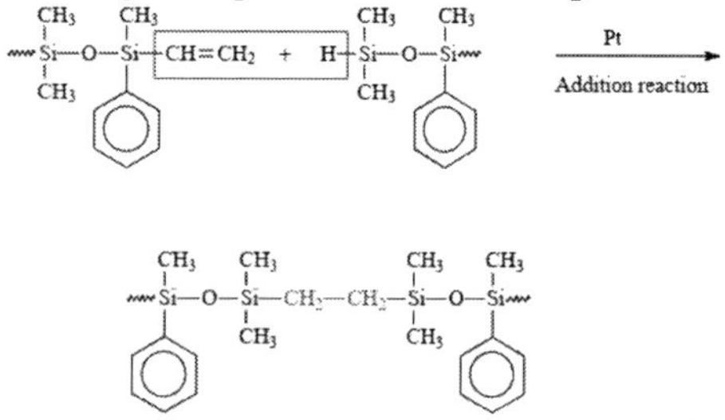

Scheme 2 Curing crosslinking reaction of silicone resin

2.3 Test equipment and methods

A self-crosslinking silicone resin was added with a 3% platinum complexed curing agent and cured at 150°C for a certain period of time to prepare a sample with a thickness of 1 mm. The air was used as a reference and UV757CRT UV-visible spectrometer was used for analyzing the transmittance of sample, testing wavelength range was from 200 nm to 800 nm. The thermal decomposition analysis of the cured product was carried out using a German TG synchronous thermal analyzer. The heating range was from 25°C to 300°C and the heating rate was 10°C/min. The morphology of the material was tested using a Quanta 200 FEG scanning electron microscope (SEM) from FEI, the Netherlands.

3 Results and discussion

3.1 FT-IR analysis

The FT-IR spectroscopy of self-crosslinking silicone resin of before and after curing is shown in Fig.1.

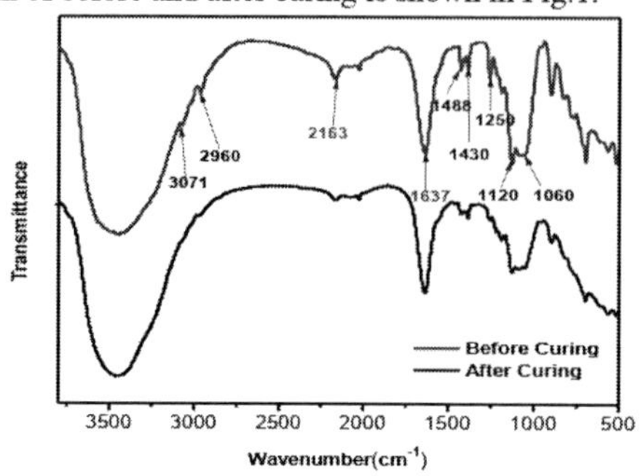

Fig.1 The infrared spectra (FT-IR) were compared with before and after curing

FT-IR (500-4000cm^{-1}, v): 3071cm^{-1} (Ph-CH2=CH-); 2960cm^{-1} (-CH3); 2163cm^{-1} (-Si-H); 1637cm^{-1} (-CH=CH_2); 1488cm^{-1} (-Ph); 1430cm^{-1}and 1120cm^{-1} (-Si-Ph); 1250cm^{-1} (-Si-CH_3); 1060 cm^{-1} (Si-O-Si). The FT-IR data indicate the synthetic silicone resin consisting of -Si-H, -CH=CH_2, -Si-Ph and other structures. And the skeleton vibration peak of Si-O-Si is a strong broad peak, indicating that the Si-O-Si structure is a chain which was the structure of the copolymerization formed in the condensation reaction. After curing, the signal for 2163cm^{-1} (-Si-H) and 1637cm^{-1} (-CH=CH_2) got weakened, which indicated that hydrosilylation addition reaction was conducted while curing.

3.2 Transmittance analysis

As can be seen from Fig. 2, the light transmittance of the product is almost 100% in the range of 400 nm to 800 nm. There are two main reasons for the higher light transmittance of the self-crosslinking silicone resin. First, in order to increase the light transmission properties of the cured product, the density of crosslink bonds is appropriately reduced in the design of its molecular structure. The molecular structure of the resin contains both vinyl groups, Si-H bonds, which do not require the addition of a cross-linking agent when it is cured. Under the action of the catalyst, the vinyl and Si-H bonds of its own molecular structure mutually reacts with each other. An addition reaction can occur, that is, without involving a conventional two-pack silicone resin mixing process, the curing structure of the silicone resin becomes relatively simple, so that the cured product achieves the uniformity of the microstructure.

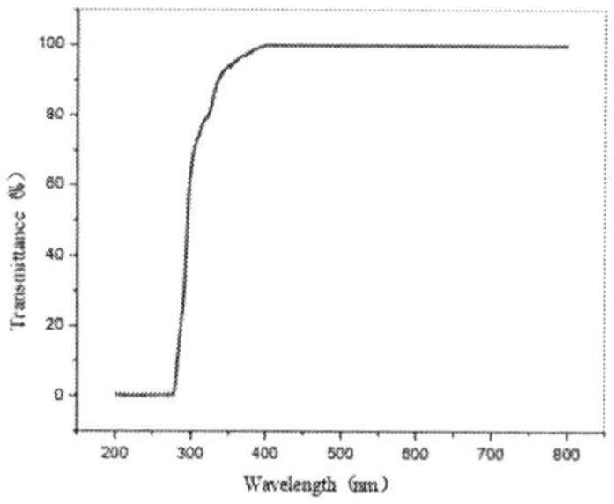

Fig.2 Transmittance curves of sample

3.3 TG-DTG analysis

From Fig. 3, it can be seen that the cure reaction of self-crosslinking silicone resin is relatively complete, and its relatively perfect cross-linking structure of the cured product makes it have a higher thermal decomposition temperature. From Fig. 3, it can be seen that the cured product has a high initial thermal decomposition temperature of 315.8°C. The thermal decomposition process is a one-stage reaction. During the thermal decomposition process, when the temperature rises as high as 500°C, the silicon resin backbone-Si-O-Si- is largely degraded by the ring, and the bonds break into rings of various sizes, and the weight loss rapidly increases. After that, the system reaches the maximum degradation rate.

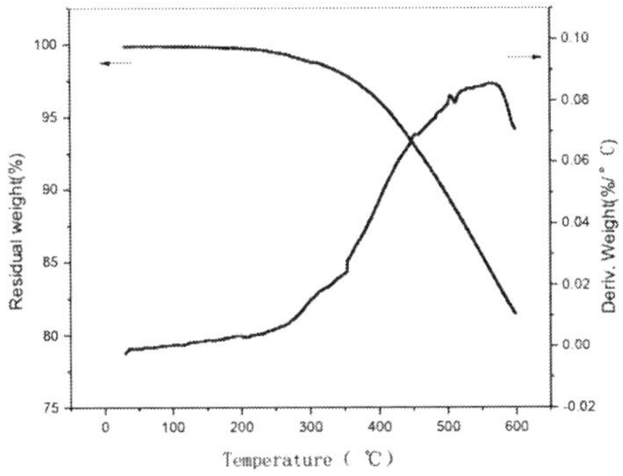

Fig. 3 TG-DTG curves of different samples

3.3 SEM analysis

SEM image of the cured product surface with different cure times were shown in Fig. 4. Comparing Fig. 4 (a), (b), (c) and (d), it can be seen that the four products have similar cured structures and all exhibit a uniform and textured structure. With the prolongation of the curing time, the texture structure appears more, which indicates that the heat resistance is getting better and the light transmission is slightly reduced.

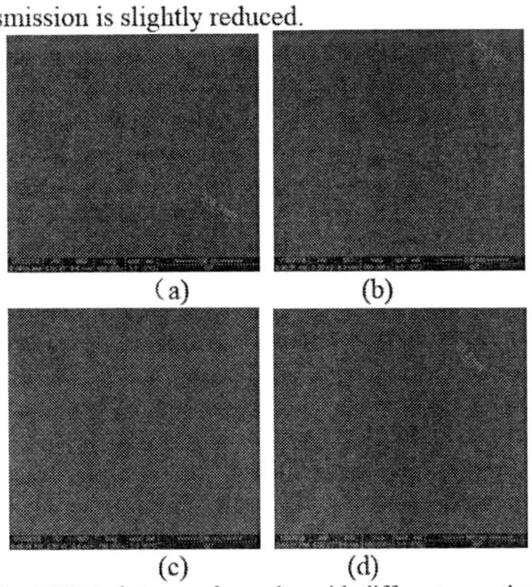

Fig. 4 SEM picture s of samples with different cure times
(a) 10min; (b)20min; (c) 30min; (d) 60min

4 Conclusion

Different proportions of self-crosslinking silicone resins were prepared and their light transmission and thermal properties were investigated. The conclusions are as follows: The structure of silicone resin was determined by Fourier Transform Infrared Spectroscopy (FT-IR). Self-crosslinking silicone resins have a uniform microstructure with good light transmission. Transmittance in the range of 400~800nm can reach 100%. The cured product of the self-crosslinking silicone resin has a relatively good heat resistance, and the cured product has a high initial thermal decomposition temperature of 315.8°C. SEM observations show that the self-crosslinking silicone has a uniform, textured structure

References

1. JANG S J, YUN H D, LEE M S. Effects of stiffening sealant thickness on the structural performance of structural silicone glazing (SSG) sealantconnections in curtain wall systems[J]. Archives of Civil and Mechanical Engineering, 2017, 17(1): 65-74.
2. BUYL F D. Silicone sealants and structural adhesives. International Journal of Adhesion and Adhesives, 21 (5) (2001), pp. 411-422.
3. PANTALEO A, ROMA D, PELLERANO A. Influence of wood substrate on bonding joint with structural silicone sealants for wood frames applications[J]. International Journal of Adhesion and Adhesives, 2012,37:121-128.
4. PEREZ G G, BURILLO G. Modification of silicone sealant to improve gamma radiation resistance, by addition of protective agents[J]. Radiation Physics and Chemistry, 2013, 90(9): 98-103.
5. KIM D S, PARK Y S. 10、Zhang Yingqiang, Zhao Yongxin, Li Ye. Single-package silicone rubber encapsulant for high-power white light LED and preparation method thereof[P]. ZL2014104712166, 2016.08.24.
6. Zhang Yingqiang, Zhang Linlin, Wu Zhen. High-viscosity self-crosslinking LED encapsulant resin and its preparation method[P]. ZL2014104712753, 2017.06.23.

Rigid-flexible Coupling Analysis of Backhoe Hydraulic Excavator Working Device

Yongliang Yuan

School of Mechanical Engineering, Dalian University of Technology, Dalian 116024, China

Abstract. This paper develops a rigid-flexible coupling approach to predict the dynamic performance of the excavator, including the stress distribution and stress diagram of curves. In order to study the dynamic performance of the excavator during the mining process, the rigid-flexible coupling model of a excavator is established and carried out for dynamics simulation. The results show that the strength of excavator meets the expected requirement and there is a large optimization space. In order to reduce the mass of the excavator's boom, this paper optimizes the design of the boom. Based on the results of the optimization design, the rigid-flexible coupling simulation of the excavator was performed again, which verified that the strength of the boom is sufficient. Simulation results show that the virtual prototype software can effectively predict the dynamic performance of the excavator. This paper provides a solid foundation for further study of the lightweight design of the whole excavator component.

1 Introduction

Excavators are widely used as primary production equipment in surface mining operations, the dynamic response of the excavator reflects its performance. In order to improve the dynamic performance of the excavator, it is necessary to build the excavator model based on the virtual prototype technology just as to build a physical prototype. Li et al [1]. conducted an excavator model confirmed its feasibility with the virtual prototype software and obtained its kinematics principle. K.Awuah-Offei et al [2]. proposed the numerical method and obtained the lifting speed and thickness of cutting relational curves. Ma used Matlab/Simulink to establish the model and carry out the simulation [3]. He et al. [4] used the rigid rod to replace the steel wire rope to establish the virtual prototype model, and carried out the simulation. From the above literature, it can be seen that scholars' researches on excavators mainly focus on the study of movement laws and stress distribution, and there is less research on the light weighting of excavators. In order to improve the dynamic performance of the excavator, this paper optimizes the design of the excavator's boom. The rest parts of this paper are organized as follows: Section 2 discusses the simulation process of virtual prototype technology. Section 3 discusses the simulation results. The conclusion is presented in Section 4.

2 Simulation process of virtual prototype technology

Conventional design process uses a laboratory test to simulate the physical prototype to evaluate the performance of the excavator. Adams simulation of virtual prototype technology in product design has gradually become a major tool. Modern design processes in virtual simulation environments enable the testing of the excavator performance prior to final design and practical application. The virtual prototype technology not only reduces costs the product development cycle, but also predicts the performance of products and provides the basis for further optimizing product. Technology of virtual prototype simulation process is shown in Fig 1.

Step 1 Build the 3D model by CAD

In view of the fact that the ADAMS software establishes the excavator model is very difficult and has many limitations. The three-dimensional (3D) dynamic model is established by the third-party CAD software. In this study, the 3D dynamic model is saved as parasolid (.x_t) format and imported into the software of Automatic Dynamic Analysis of Mechanical Systems (ADAMS), and completed the rigid body model the establishment.

Step 2 Add the material property and constraints

The material property of the parts are added, according to the actual working condition. In accordance with the system actual working state to add restrictions and drive components.

Step 3 The establishment of a flexible body

In order to carry on the rigid-flexible coupling simulation, in this study, the flexible body is obtained by using the finite element software for modal analysis of the model. Using the ADAMS/Flex body module connection to import the file (.mnf) into ADAMS, finally we obtain the system rigid-flexible coupling model.

Step 4 Select the solver of ADAMS

* Corresponding author: yuan-yong-liang@163.com

https://doi.org/10.1051/e3sconf/20183802020

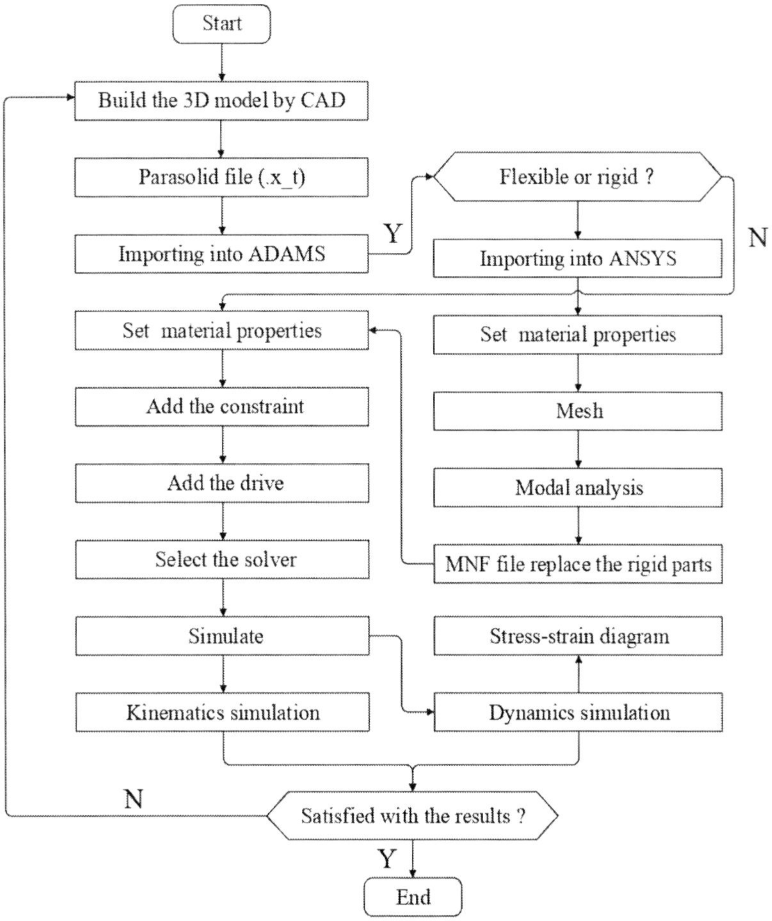

Fig. 1. The Simulation Process of Virtual Prototype.

In view of the fact that the ADAMS software has many solvers, in order to obtain the accurate results, it becomes very important to select the appropriate solver and the integration method. To solve the differential equations of the motion, we also choose the appropriate step length of the coupled simulation.

Step 5 Simulation Results & Analysis

After the simulation, using the ADAMS post-processing module to carry on the simulation result, we can obtain the motion law of system and the dynamics curve. Finally, we can grasp its dynamics performance.

3 Rigid-flexible coupling simulation analysis

3.1. Pre-optimization simulation

The mechanism used in this case has the same characteristics as previously, except that the boom and dipper handle. The boom and dipper handle are saved as the need neutral file (.mnf) and import into the Adams to obtain the rigid-flexible coupling model of excavator. In order to ensure the authenticity of the excavator model, the model of the material, constraints, and parameters are set according to the actual situation[5-8]. This paper chooses the integration tolerance is 0.001, the simulation time is 16s and the step size is 0.02s.

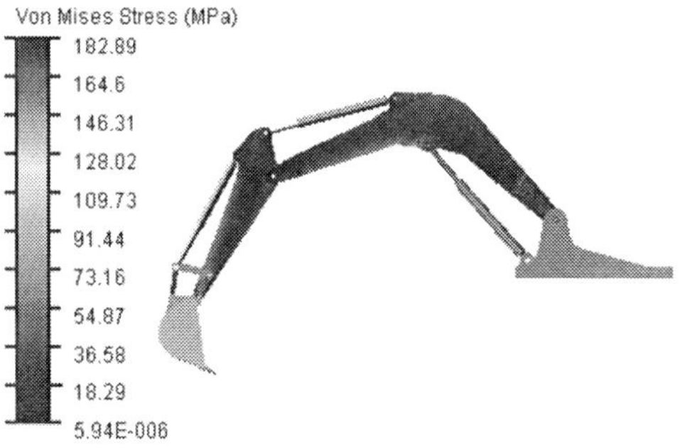

Fig. 2. The Stress Distribution of Excavator.

As can be seen from Figure 2, the maximum stress of the excavator is 182.89 Mpa, and the maximum stress stress occurs at the hinge hole of the boom of the excavator. The main reason for this is the uneven force exerted by the excavator during excavation. From the value of the maximum stress, it can be known that the strength of the excavator can meet the design requirements and the large arm has a large optimization space.

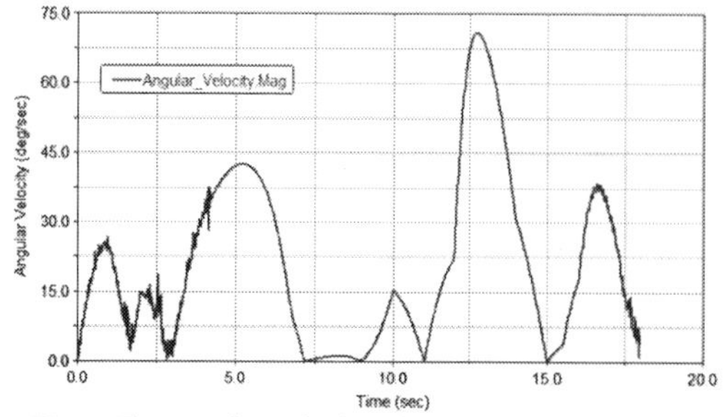

Fig. 3. The Angular Velocity of Arm and Lifting Cylinder Hinge.

As shown in Figure 3, the angular velocity of arm and lifting cylinder hinge is 72.8deg/s. It can be seen from the figure that the excavator does not work smoothly during the work, but there is some vibration. As the excavation material increases, the vibration of the excavator decreases and gradually stabilizes.

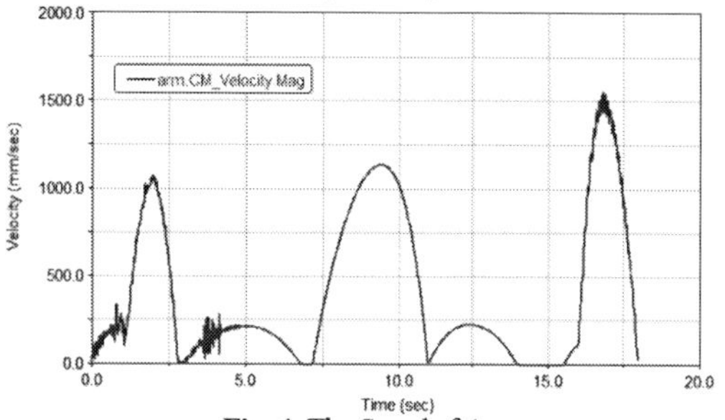

Fig. 4. The Speed of Arm

Figure 4 illustrates that there is also some vibration in the whole work process of the arm, which is similar to the reason of Figure 3. The maximum value is 1547.2mm/s. After the excavator dumped the material, the excavator began to vibrate again. In order to reduce the vibration of the excavator, reduce the speed of the hydraulic cylinder as much as possible while satisfying the excavation requirements.

3.2. Optimization simulation

The analysis of section2 shows that the strength of the arm is sufficient. In order to reduce the weight of the excavator, this paper optimizes the lightweight design of the boom. Combined with the optimized results, the dynamics simulation was performed again. By comparing with the previous results, the influence of the dynamic performance of the excavator was obtained.

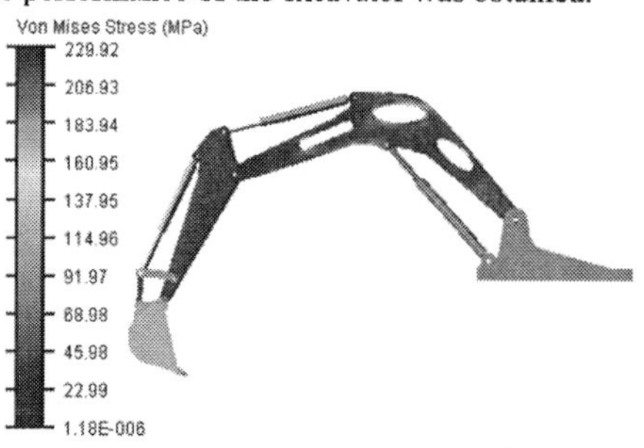

Fig. 5. Optimized Stress Distribution.

Figure 5 shows that the stress of the excavator after the optimization of the boom is greater than the stress before the optimization, mainly because the weight of the boom becomes smaller. The maximum stress after optimization is 229.92MPa, which is 47.03MPa larger than before optimization. Although the optimized stress is larger, it still meets the strength requirements. So we can continue to optimize the design based on this result.

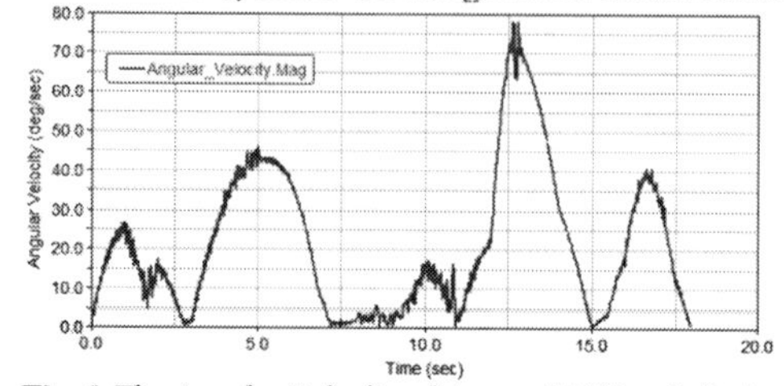

Fig. 6. The Angular Velocity of Arm and Lifting Cylinder Hinge of Optimized Result.

By comparing Figure 3 and Figure 6, it can be seen that the optimized maximum value is 72.8deg/s. Figure 6 shows that the optimized vibration is more pronounced than before the optimization, which is mainly due to the lower weight of the big arm and the shift of the centroid. In order to reduce the vibration of the excavator, reduce the speed of the hydraulic cylinder as much as possible while satisfying the excavation requirements.

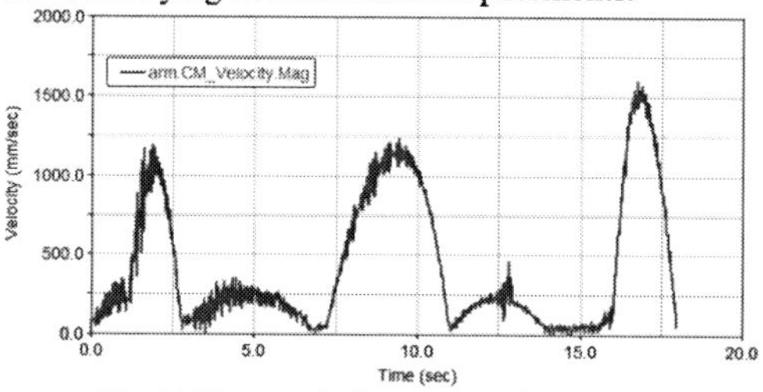

Fig. 7. The Speed of Arm of Optimized Result.

Figure 7 illustrates that there is also some vibration in the whole work process of the arm, which is similar to the reason of Figure 6. The maximum value is 1631.9mm/s. After the excavator dumped the material, the excavator began to vibrate again. Through the optimization analysis of the boom of the excavator, it can be seen that as the mass of the boom decreases, the vibration of the excavator increases significantly. In order to reduce the vibration of the excavator, it should be optimized from the whole to avoid resonance between the components and the excavator.

4 Conclusion

This paper uses virtual prototyping to verify the necessity of rigid-flexible coupled simulation of excavator. First, we establish the model which is the rigid body dynamics simulation of excavator by UG. The modal analysis is carried out on the boom and the dipper handle and the modal neutral file is got at the same time. Second, the stress distributions of the boom and dipper handle are obtained, including the pre-optimization simulation results and the optimized simulation. Third, by comparing the results before and after optimization, the results show that the virtual prototype software can effectively predict the dynamic performance of the excavator. In later studies, in order to make the performance of excavator to be the best, it' s necessary that we should consider the boom and dipper handle are flexible body at the same time.

References

1. Y. Li, S. Frimpong. Int. J. Adv. Manuf. Tech. **37**, 5 (2008).
2. K. Awuah-Offei, S. Frimpong. Mech. Mach. Theory. **42**, 8, (2007)
3. M.H. Ma. Jilin University. (2009)
4. B. He, G. Zhou, S. Hou, et al. Int. J. Adv. Manuf. Tech. **89**, 9, (2017)
5. Y.L. Yuan, L. Du, W. Sun, et al. International Conference on Mechanical Design, (2017)
6. X.L. Luo, A.T. Lewandowski, D. Berlin, et al. The First Int Colloq Microfluid, Shenyang, (2007)
7. W. Lin, G. Zhu, Y. Tang, et al. P. I. Mech. Eng. C-J. Mec. **229**, 2, (2015)
8. S. Šalinić, G. Bošković, M. Nikolić. Automat. Constr. **44**, (2014)

Regeneration of paint sludge and reuse in cement concrete

Enqi Feng[1], *Jitao* Sun[2], *and Liming* Feng[3,*]

[1] Xiamen University,College of Chemistry and Chemical Engineering, 361005Ximen,Fujian, China
[2]Tai'an LEBANG Environmental Technology Co., Ltd, 271208Tai'an,Shandong, China
[3]Shandong Jianzhu University, School of Materials Science and Engineering, 250101 Ji'nan,Shandong, China

Abstract. Paint Sludge (PS) is a hazardous waste. Inappropriate disposal of PS might be harmful to public health and the environment. Various size of Paint Sludge Solid Powder (PSSP) particles have been produced by automatic processing equipment via dewatering, crushing, screening removing Volatile Organic Compounds (VOCs), and etc. Meanwhile, the test results show that PSSP is not a hazardous waste. Both flexural and compressive strength are increased by adding PSSP of polyurethane to cement concrete at a level of below 10% of cement weight. However, the strength has a significant reduction at a level of above 15% of cement weight. The reason for the increase of strength is probably due to a slow coagulation and copolymerization of PSSP and cement. The reduction is likely due to the self-reunion of PSSP.

1 Introduction

The raw PS produced by automotive industry by using Sediment Scraper usually contains 75-90wt.% of water and 10wt.% of volatile organic compounds(VOCs). Therefore, removing water and VOCs from the raw PS is very important for recycling PS. However, it is difficult to reduce the water below 40wt.% of PS by mechanical dehydration, besides, the dewatered PS is easy to form a soft lump quickly. This also increases the difficulty of recycling. E. M. James designed a Vacuum Filtration System for dewatering PS [1]. It could enhance dewatering in vacuum chamber by using negative pressure. Then, remove residual water by using hot pressurized air. R. Elangovan also designed PS recovery and reuse converting machine [2]. First, the water and solvent in raw PS turn into steam (vapor and VOCs) by heating the PS. Then, the mixed steam was collected through cooling unit, collection unit, wet scrubber unit, and ect. This machine could recover 34% of reusable solvent, 48% of stabilized solid residue and 18% of water. Although these equipment can effectively dewater the PS, they couldn't dewater enough if there was no crushing device. Therefore, the reprocessed PS is difficult to reuse directly. Recently, a comprehensive processing and utilization equipment for PS was designed by LEBANG Co., Ltd in Tai'an, China [3]. All processing units such as dewatering, drying, crushing, and etc. are integrated together. So it is easy to produce PSSP with 50-1000 mesh and less than 2% of water. Especially, the treating process will not cause environmental pollution because it has a treatment system for waste water, waste gases, and dedusting.

Recycling and reusing the PS has been an important topic in recent years. The primary study focuses on construction and road pavement materials. Lightweight construction materials have been obtained using PS of acrylic resin as a partial replacement for sand in cement concrete [4]. The results have demonstrated that these materials can be used in residential construction. Furthermore, as modifier replaced partial asphalt, satisfactory results have been achieved in road pavements [5,6].

In this study, we produced PSSP of polyurethane which was regenerated from the automatic equipment for recycling PS made by LEBANG Co., Ltd. The purpose is to add PSSP of polyurethane into the concrete product as a filler and investigate the effects of different additives and percentages of PSSP on flexural and compressive strength.

2 Materials and Methods

Materials for this study include cement, sand, water, and regenerated PSSP. The raw PS was provided from automotive industries of SINOTRUK Co., Ltd in Ji'nan, China. The cement is the commercially available type P.O32.5（GB175-2007）. Sand was acquired from an unknown source.

The experiments include two steps. First, the raw PS was processed into PSSP with 200 mesh size. Then, this PSSP was mixed with cement, water and sand in certain proportions. The standard size of molds was 40mm×40mm×160mm cubes. The preliminary assessment for all specimens was performed at 7 days of aging .

2.1 Preparation of PSSP

According to PS processing flow chart in Figure1, the raw PS of polyurethane (Figure2) was regenerated into

* Corresponding author: flm@sdjzu.edu.cn

 https://doi.org/10.1051/e3sconf/20183802021

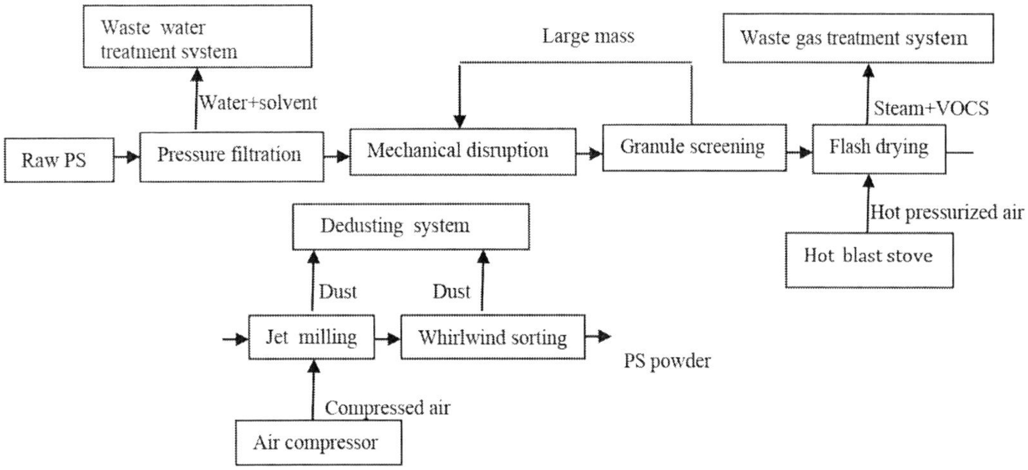

Fig. 1. PS process flow chart.

PSSP with 200 mesh (Figure3). The chemical composition and leaching toxicity before and after treatment were shown in Table1.

Fig.2 .The raw PS.

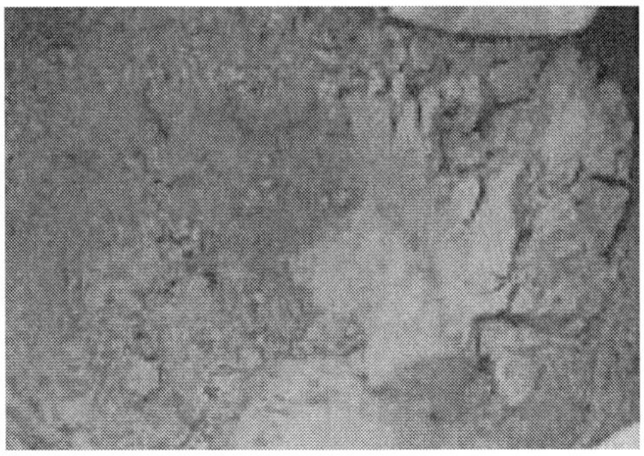

Fig.3. PSSP with 200 mesh.

2.2 Preparation of the PSSP-concrete specimens

According to the test of water absorption rate of PSSP and method of cement mortar test in ISO679:1989, the water was increased by one third of PSSP accordingly.

This study investigated the effects of additives and PS content on mechanical properties of PSSP-concrete.

Bubbles and interfaces may be created when PSSP is mixed with concrete because the PSSP is mainly composed of polymer resin and pigment. Therefore, it is necessary to modify it using additives such as surfactants (sodium polyphosphate), defoamers (DF460) at a level 2‰ and 2.5‰ of total water, respectively. The exact compositions with different additives are shown in Table2. Furthermore, the compositions with same additives, PSSP contents of 5,10,15,20wt.% of cement are shown in Table3.

2.3 Mechanical Properties

Flexural and compressive tests were performed by using Full-automatic cement pressure testing machine (DYH-300B) whose precision is grade1 and maximum stroke is 100mm.

3 Results and Discussion

3.1 Results of PSSP regeneration treatment

According to the results in Table 1, the regenerated PSSP is no longer hazardous waste because leaching toxicity is far lower than the national standard values of hazardous waste[7]. In addition, the water and VOCs level of PS are significantly decreased. The data indicates that water content from 31.5 to 1.7wt.% with the removal efficiency at 94.6 %.

3.2 Effects of additives on strength of PSSP-concrete

Figure 4 and Figure 5 show the results of flexural and compressive test for PSSP-concrete respectively. As can be seen, when PSSP accounts for 5% of cement weight, flexural and compressive strength of specimens are increased by 5.76%,12.46% with single surfactant, 3.36%, 12.69% with single defoamer, 1.4%, 16.8% with both surfactant and defoamer, respectively. In addition, when PSSP accounts for 10 % of cement weight, flexural strength is decreased by 8.6%, 2.16%, compressive strength is increased by 10.8%、14.8% with single surfactant and defoamer. As expected, flexural and compressive strength of specimens are increased by

Table 1 .Changes in chemical composition of PS before and after treatment.

Chemical composition	Text method	Level in before treatment	Level in after treatment	National standard
Water weight (%)	CJ/T221	31.5	1.7	
Petroleum solvent (mg/kg)	GB5085.6	——	1.54×10^3	
Isophorone(mg/kg)	GB5085.3	——	<5	
acetone(mg/kg)	GB5085.3	588	<20	
benzene(mg/kg)	GB5085.3	<0.05	<0.05	
toluene(mg/kg)	GB5085.3	2.92×10^3	1.33	
p-xylene(mg/kg)	GB5085.3	1.29×10^4	11.0	
styrol(mg/kg)	GB5085.3	12.4	0.17	
o-xylene(mg/kg)	GB5085.3	7.05×10^3	8.52	
chlorobenzene(mg/kg)	GB5085.3	<0.05	0.29	
1,3-dichlorobenzene(mg/kg)	GB5085.3	<0.05	<0.05	
1,4-dichlorobenzene(mg/kg)	GB5085.3	<0.05	<0.05	
1,2-dichlorobenzene(mg/kg)	GB5085.3	<0.05	<0.05	
1,2,4-trichlorobenzene(mg/kg)	GB5085.3	<0.05	<0.05	
1,2,3-trichlorobenzene(mg/kg)	GB5085.3	<0.05	<0.05	
PO(mg/kg)	GB5085.3	<25	<25	
MTBE(mg/kg)	GB5085.3	<0.5	<0.5	
Epoxyethane(mg/kg)	GB5085.3	<10	<10	
Epichlorohydrin(mg/kg)	GB5085.3	<50	<50	
TBA(mg/kg)	GB5085.3	<10	<10	
ETO(mg/kg)	GB5085.3	<1	<1	
n-butyl alcohol(mg/kg)	GB5085.3	<10	<10	
2-butanol(mg/kg)	GB5085.3	<50	<50	
MEK(mg/kg)	GB5085.3	<20	<20	
MIBK(mg/kg)	GB5085.3	<20	<20	
Isobutanol(mg/kg)	GB5085.3	<50	<50	
Acrolein(mg/kg)	GB5085.3	<10	<10	
2-nitropropane(mg/kg)	GB5085.3	<10	<10	
Acetonitrile(mg/kg)	GB5085.3	<5	<5	
δ-chlorotoluene(mg/kg)	GB5085.3	<1	<1	
EAC	USEPA 8260C	<0.1		
n-butyl acetate	USEPA 8260C	61.5		
Total vocs (mg/kg)	GB5085.3		<6.6	
leaching toxicity				
Ba(mg/L)	GB5085.3		0.06	≤100
Cd(mg/L)	GB5085.3		<0.001	≤1
Total Cr(mg/L)	GB5085.3		0.07	≤15
Hg(mg/L)	GB5085.3		<0.005	≤0.1
Pb(mg/L)	GB5085.3		<0.01	≤5
Zn(mg/L)	GB5085.3		1.35	≤100
As(mg/L)	GB5085.3		<0.05	≤5
Cr^{6+}(mg/L)	GB/T 15555.4		0.067	≤5
DOP(mg/L)	GB5085.3		<0.0025	≤3
benzene(mg/L)	GB5085.3		<0.0500	≤1
toluene(mg/L)	GB5085.3		<0.0500	≤1
xylene (mg/L)	GB5085.3		<0.100	≤4

 https://doi.org/10.1051/e3sconf/20183802021

1.4%、16.8% with both surfactant and defoamer. Unfortunately, both flexural and compressive strength of specimens declined with increasing PSSP to top 15 % of cement weight in either group. It is evident that surfactant and defoamer have similar effect when PSSP content at 5%. However, the defoamer had more significant effect when PSSP content is more than 10%. This may be because sodium polyphosphate not only

Table 2. Compositions with different additives.

Series	Cement (g)	Sand (g)	PSSP (g)	Water (ml)	Surfactants (g)	Defoamer (ml)	Flexural (MPa)	Compressive (MPa)
1	450	1350	0	225			4.17	30.10
2	450	1350	22.5	232.5	0.465		4.41	33.85
3	450	1350	22.5	232.5		0.57	4.31	33.92
4	450	1350	22.5	232.5	0.465	0.57	4.30	35.13
5	450	1350	45	240	0.48		3.81	33.36
6	450	1350	45	240		0.6	4.08	34.55
7	450	1350	45	240	0.48	0.6	4.23	35.16
8	450	1350	67.5	247.5	0.495		3.11	28.28
9	450	1350	67.5	247.5		0.619	3.86	31.60
10	450	1350	67.5	247.5	0.495	0.619	3.76	34.46

Table 3. Compositions with the same additives , different percentages of PSSP.

Series	Cement (g)	Sand (g)	PSSP (g)	Water (ml)	Surfactants (g)	Defoamer (ml)	Flexural (MPa)	Compressive (MPa)
1	450	1350	0	225			4.17	35.10
2	450	1350	22.5	232.5	0.465	0.57	4.30	35.13
3	450	1350	45	240	0.48	0.6	4.23	35.16
4	450	1350	67.5	247.5	0.495	0.619	3.76	34.46
5	450	1350	90	255	0.51	0.64	3.50	27.44

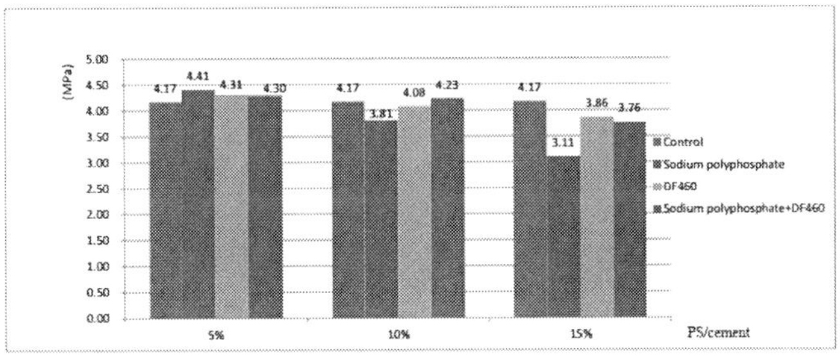

Fig.4. Effects of different additives on flexural strength.

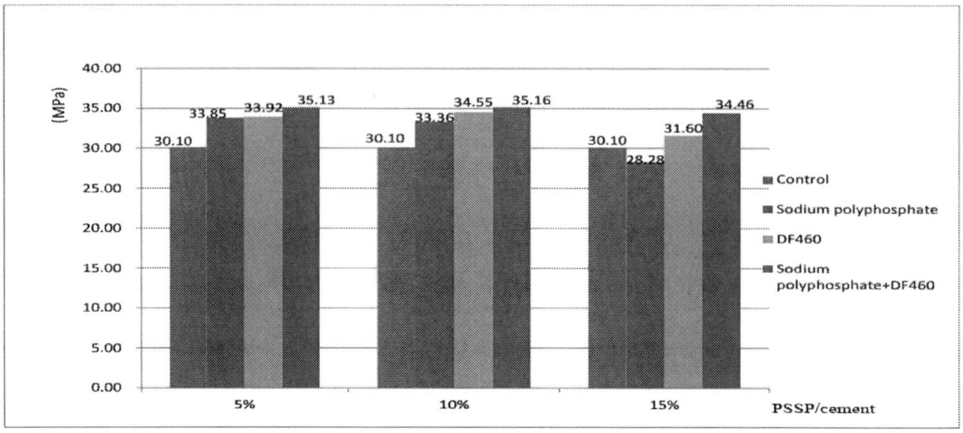

Fig. 5. Effects of different additives on compressive strength.

increases wettability but also forms complex compound with Ca^{2+}, Mg^{2+} in PSSP and cement as a bridge. Moreover, DF460 is mainly to eliminate bubbles. Usually, the PSSP contains small amount of residual flocculant, more bubbles come into mortar mixture with an increase of PSSP content. Apparently, the bubbles become the main factors. Using sodium polyphosphate and DF460 together had synergistic effect to increase strength of specimens.

3.3 Effects of PSSP Content on Strength of PSSP-Concrete

The strength tests have been performed to determine the effects of PSSP content using sodium polyphosphate and DF460 together. Figure 6 shows the result of strength test for PSSP-concrete. As seen in figures, when PSSP content below 10%, the flexural and compressive strength have been greatly improved. In here, the maximum compressive strength increased by 16.7% compared with the control sample. However, the strength began to decline when the PSSP reaches to 15%. Unfortunately, the strength is far lower than control sample when PSSP reaches to 20%. These regularities are probably due to small amount of residual isocyanate and hydroxyl in PSSP can cause copolymerization with the hydroxyl in cement. So the degree of cross-linking is improved. In the mean time, PSSP may alleviate hydration of cement and reduce stress between sand and cement as filler to increase strength. Of course, PSSP can be agglomerated when the dosage is too high, which could increase the interface between powders and cement paste to decrease strength.

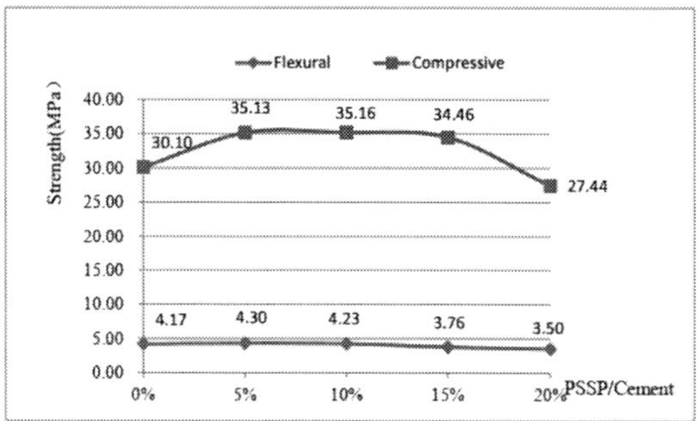

Fig. 6. Relationships between the strength and wt.% PSSP.

4 Conclusion

In this study, we focused on the regeneration of PS and reuse in cement concrete. We produced nontoxic PSSP. Based on the results from PSSM-concrete, changes of mechanical property have been observed.

PSSP with different particle size range can be obtained via high efficiency dehydration, crushing and screening by using comprehensive processing and utilization equipment for PS. More importantly, the PSSP is no longer hazardous waste.

The flexural and compressive strength of specimens are improved by adding PSSP below 10% of cement content in concrete. The flexural and compressive strength of specimens are significantly declined by adding PSSP above 15% of cement content in concrete.

References

1. E. M . James. US. pat. No.20040031739A1 (2004)
2. R.Elangovan, TAB, BARATHI. WO. pat. No. 2014037954A1 (2014)
3. L. M. Feng, F. Q. He, et al. CHN. pat.No.2017106782839(2017)
4. H.Avci, H.Ghorbanpoor, etal. J. Environ. Chem. Eng. 5(1),861-869 (2017)

https://doi.org/10.1051/e3sconf/20183802021

5. D. Dalmazzo, A .Vercelli ,E.Santagata ,B. Ruffino, et al. Mater. Struct. 50 (1) ,1-14(2017)
6. M. C. Zanetti, B. Ruffino, et al. CEST2017, Rhodes, Greece.
7. GB5085.3-2007, SEPA and AQSIQ , Beijing , China

https://doi.org/10.1051/e3sconf/20183802022

Research on Extrusion of Rubber Composites Reinforced by Short Fibers Orientation Based on FEA

Dewei Zhang[1, 2], Chuansheng Wang[1, 2*], Bo Shen[1, 2], Shaoming Li[1, 2] and Huiguang Bian[1, 2]

[1]Shandong Provincial Key Laboratory of Polymer Material Advanced Manufacturing Technology Qingdao University of Science & Technology, 266061, Qingdao, China
[2]College of Electromechanical Engineering, Qingdao University of Science & Technology, 266061, Qingdao, China

Abstract: In recent years, rubber composites reinforced by short fibers has been researched deeply, because of its good performances such as higher wear resistance, higher cut resistance and so on. Some research results indicated that if short fibers get orientation in rubber composites, the performances of rubber products could be promoted greatly. But how to make short fibers get orientation in rubber matrix during extrusion is still a real problem. And there are many parameters affect the short fibers orientation. So, in this paper, the effects of die structure including expansion-die and dam-expansion-die on extrusion flow field of short fiber and rubber composite material during extrusion process has been researched by Polyflow. And the FEA results about the pressure field, velocity field and the velocity vector of the rubber composites flow field indicate that, comparing with expansion-die and the dam-expansion-die, the latter one is better for the extrusion process of rubber composites and making short fibers get radial orientation in rubber matrix.

1 Introduction

Rubber composites which could be reinforced by short fibers has been started to be researched from 1970's. Because it combines the rigidity of short fiber and elasticity of rubber together, the rubber composites reinforces by short fibers have became a new kind of high polymer composite material[1-4]. As a result, the rubber composites has been researched deeply and used widely in recent years[5-12]. Moreover, some research results showed that if short fibers get orientation in the rubber composites material, the rubber products using properties could be promoted. For example, if short fibers get radial orientation tire tread products which means the orientation direction of short fibers is at the radial direction of tiers, or at the thickness direction of tire tread, the wear resistance ability, rolling resistance ability, tearing resistance ability, etc of tire tread could be promoted greatly. During the manufacturing process of rubber composites, the die structure is one of the most important factors that impacts short fibers radial orientation during extrusion process. Therefore, in this paper, the effects of two kinds of different die structure which is dam-die, expansion-die on extrusion flow field of rubber composites have been researched by FEA method using Polyflow. And the analyzed flow fields were pressure field, velocity field and velocity vector.

2 FEA models

2.1 Physical model and FEM model

The physical model and FEM model must to be established before FEA. The physical model is the material that fulls of the die channel, which have been shown in the Fig.1. And the FEM model is the results of net dividing, which have been shown in the Fig.2.

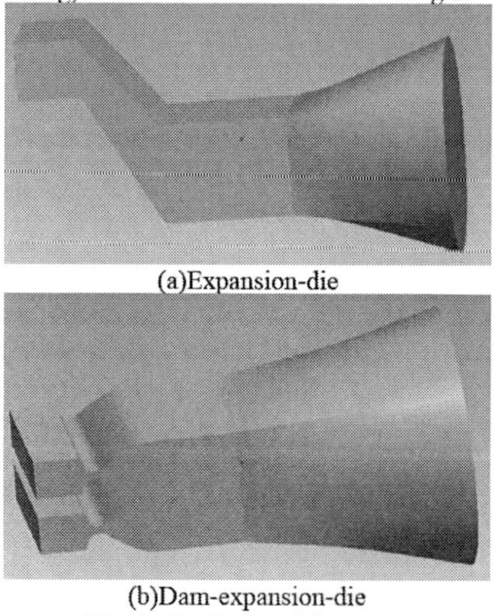

(a)Expansion-die

(b)Dam-expansion-die

Fig. 1. Physical model of FEA

*Corresponding author: zhangdewei1421@163.com, wcsmta@qust.edu.cn

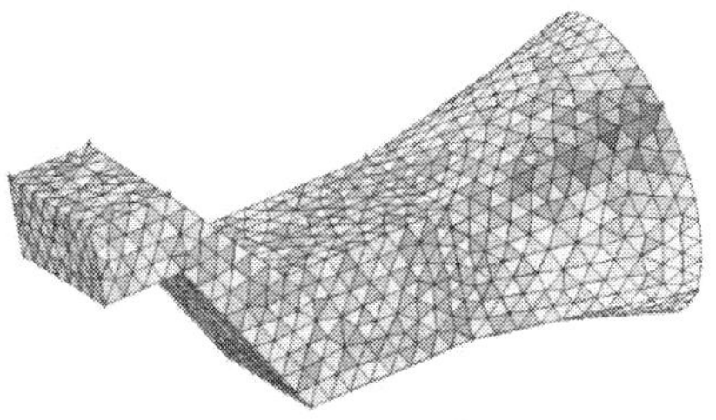

(a)Expansion-die

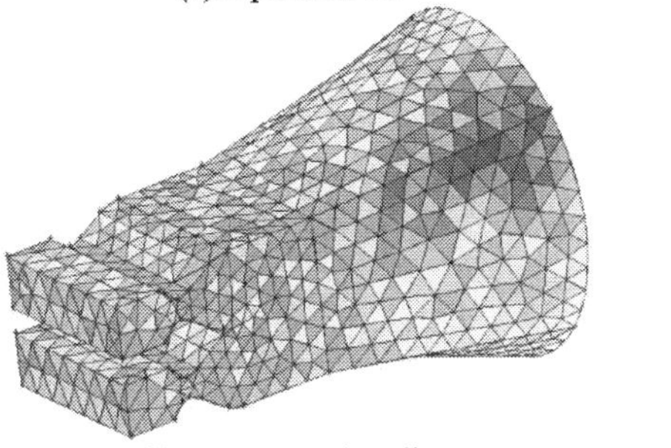

(b)Dam-expansion-die

Fig. 2. FEM model of FEA

2.2 Boundary conditions and physical parameters

The boundary conditions and physical parameters that have been set for the FEA process have been shown in the Table 1 and Table 2.

Table 1 Dies boundary conditions

Die flow amount (mm^3/s)	Velocity in the channel face(m/s)
40	0

Table 2 Physical parameters for FEA

Density (Kg/m^3)	Milv Index	Material Viscosity ($Pa \cdot S^n$)	Material Viscosity ($Pa \cdot S^n$)	Relax time (s)
1066	0.23	1000000	10	10

3. Analysis results and discussion

3.1 Pressure field

The effects of the two kinds of structure which were expansion-die and dam-expansion-die on extrusion process pressure field of rubber composites have been shown in the Fig. 3.

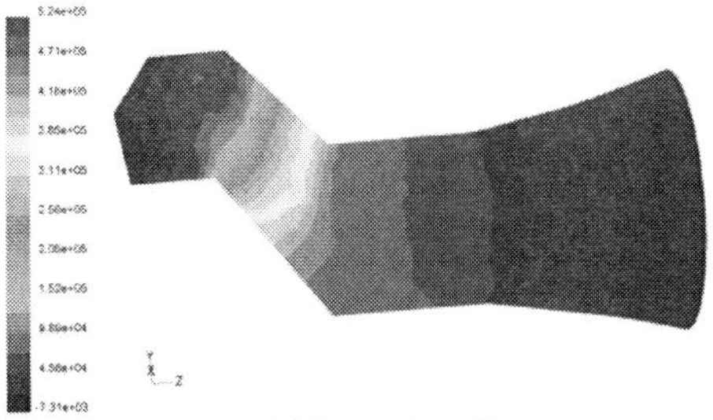

(a)Expansion-die

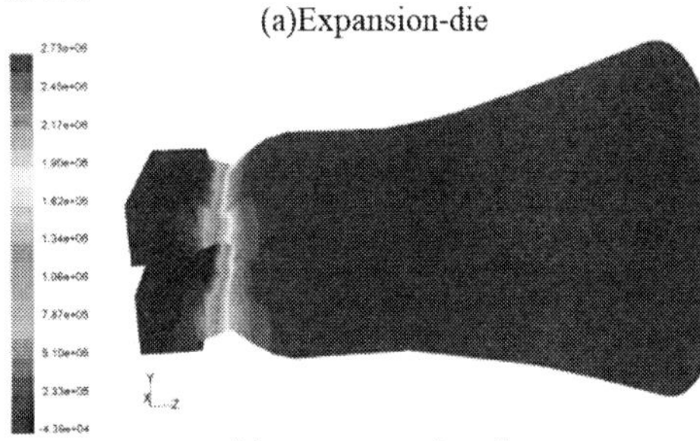

(b)Dam-expansion-die

Fig. 3. FEA Results of Pressure Field

As vividly shown in the Figure 3, at the same extrusion conditions, it is obviously that the pressure field of dam-expansion-die is changing more intensely than that of expansion-die. And the maximum pressure value of dam-expansion-die is about 2.73MPa, while the maximum pressure of expansion-die is 0.52Mpa.

As a result, at the same extrusion conditions, on one hand, the dam-expansion-die could establish a higher die pressure for extrusion of rubber composites. Due to it is necessary to make rubber material with short fibers get compressed for short fibers get radial orientation in rubber matrix, so a higher die pressure is also needed. On the other hand, the dam-expansion-die pressure change is intense, which indicate that the rubber material could expand as fast as possible after passing the dam structure, as a result, short fibers would turn to a bigger angle with the expanding of rubber material than the situation of rubber material expanding slowly. Therefore, the dam-expansion-die is could establish a better pressure field for radial orientation of short fiber than expansion-die.

3.2 Velocity field

The effects of the two kinds of structure which were expansion-die and dam-expansion-die on extrusion process velocity field of rubber composites have been shown in the Fig. 4.

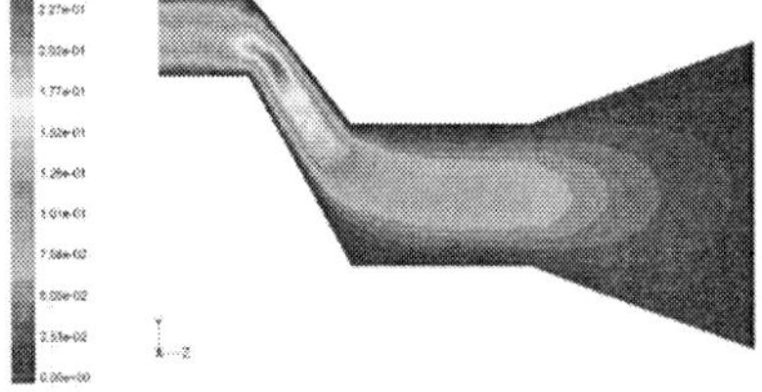

(a)Expansion-die

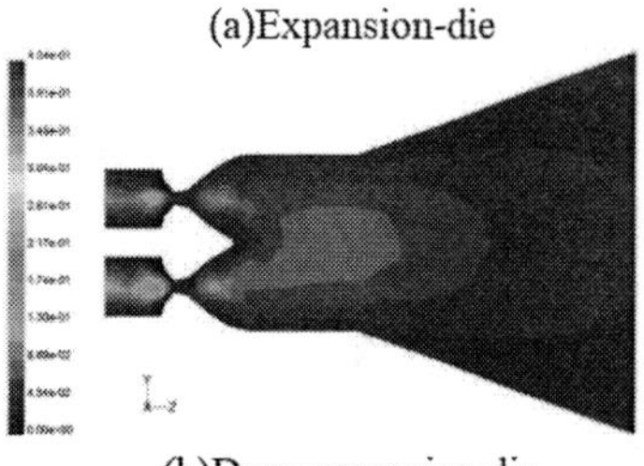

(b)Dam-expansion-die

Fig. 4. FEA Results of Velocity Field

As what have been shown in the Figure 4, at the same extrusion conditions, the velocity field of dam-expansion-die is steadier than that of expansion-die, which means the change of velocity value is not intensely. And the maximum velocity value of dam-expansion-die is about 43.4mm/s, while the maximum velocity value of expansion-die is 25.3mm/s.

As a result, at the same extrusion conditions, not only could the dam-expansion-die have a bigger extrusion speed, which does not cut down the production ability of extruder, but also the velocity field of the dam-expansion-die is steadier than that of expansion-die. If the changing of velocity value is intense, such as the velocity field of expansion-die which has been shown in the Figure 4, the flow of rubber material with short fibers would generate turbulence, this would impact radial orientation of short fibers in the rubber composites during the extrusion process. So, the dam-expansion-die has a bigger velocity value and a steadier velocity field than those of the expansion-die. Therefore, the dam-expansion-die is better for the extrusion of rubber composites to make short fibers get radial orientation.

3.3 Velocity vector

Considering velocity is a vector, and the Figure 4 just only showed the velocity value of the velocity field which generated during extrusion process of rubber composite. And there was no velocity direction. So the velocity direction of the two kinds of structure which were expansion-die and dam-expansion-die has also been analyzed based on FEA, and the results have been shown in the Fig. 5.

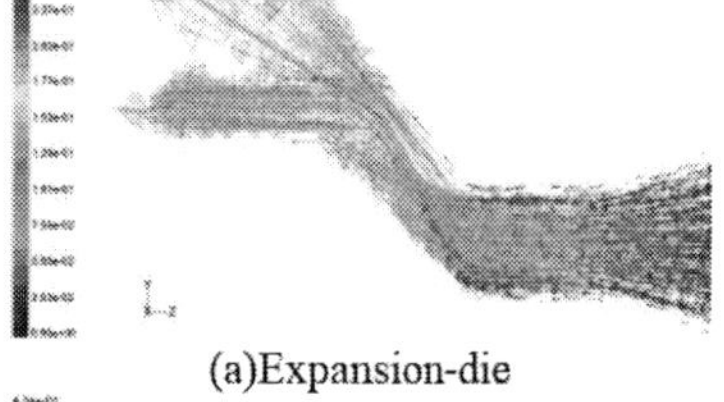

(a)Expansion-die

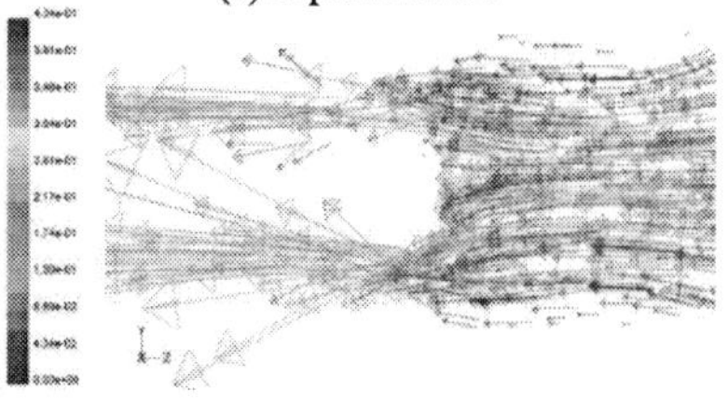

(b)Dam-expansion-die

Fig. 5. FEA Results of Velocity Vector

As what have been vividly shown in the Figure 5, the velocity direction could be seen directly together with the velocity value. Only considering the velocity direction, at the same extrusion conditions, the direction angle of dam-expansion-die is bigger than that of expansion-die, which means the short fibers turn direction of dam-expansion-die or dam-die is better than that of expansion-die, this also means the dam-expansion-die could make short fibers get better radial orientation during the extrusion process of rubber composites.

4. Conclusions

The conclusions about effects of the two kind of different structure which were dam-expansion-die and expansion-die on extrusion flow fields of rubber composites during extrusion process could be drawn as following,

1) At the same extrusion conditions, the dam-expansion-die could establish a bigger pressure field that changes intensely and a steadier velocity field that has a bigger velocity value than the expansion-die. Moreover, the velocity direction angle of dam-expansion-die is bigger than that of expansion-die.

2) The effects of die structure on extrusion field of rubber composites could be researched easily and directly and vividly without experiments, avoiding the bad effects of some experimental factors. Also, the design of die structure can be changed according to the FEA results. As a result, the design efficiency and quality can be promoted.

Acknowledgements

The authors would like to thank the National Natural Science Foundation of China (NO. 50775116), Project of Shandong Province Science and Technology Development Program (NO. 2013TD16006), Project of Shandong Province Higher Educational Science and Technology Program (J15LB73) for finance support.

References

1. Li H T. Development of Short Fiber-rubber Composite Material[J]. Modern Rubber Technology, 2006, 32(4):5-12
2. Wang C S, Liu C J, Bian H G. Radial Orientation Mechanism and Experimental Research of Short Fiber in Tread Compound[J]. Journal of Donghua University, 2009, 26(6):666-672.
3. Wang C S, Zhang D W, Bian H G. Effects of Mixing Parameters on Properties of Short Fiber and Rubber Composite Material[J]. Journal of Functional Materials, 2011, 42(8): 1448-1452.

4. Nuttapong H, Sombat T and Taweechai A. Improving the mechanical properties of short pineapple leaf fiber reinforced natural rubber by blending with acrylonitrile butadiene rubber[J]. Polymer Testing, 2017, 57:94-100
5. Pitchapa P, Sombat T and Taweechai A Comparative study of natural rubber and acrylonitrile rubber reinforced with aligned short aramid fiber[J]. Polymer Testing, 2017, 64:109-116
6. Zhang D W, Wang C S, Lin G Y. Effects of Rotor Speed on Manufacturing Process of Short Fiber-rubber Composite Material[J]. Advanced Materials Research, 2011, 211(1): 350-355.
7. Zhang D W, Wang C S, Lin G Y, etc. Effect of Die Structure on Radial Orientation of Short Fiber in Tire Tread and its Performance[J]. Key Engineering Materials, 2012, 501(1): 247-252.
8. Li Li, Wang Chuansheng, Zhang Dewei, Effects of Radial-Orientation-Die Structure Parameters on Properties of Short Fiber and Rubber Composite Material, Journal of Donghua University (Eng. Ed), 2014, 31(3): 249-255
9. L. Liu, L.Q. Zhang, Z.X. Feng, Development of Green Tires, J. China Rubber Industry. 46(1999)245-248.
10. C.S. Wang, C.J. Liu, H.G. Bian, Radial Orientation Mechanism and Experimental Research of Short Fiber in Tread Compound, J. Journal of Donghua University. 6(2009)666-672.
11. Rijpkema B. The use of short fiber to reduce the rolling resistance of tires[J], Plastomere Elastomere Duromere, 1994, 47(10): 748～752
12. Walker L A, Pans W, Harber John B et al.. Santoweb fiber reinforcement of ruber compounds[J] , Kautachuk Gummi Kunststoffe, 1985, 38(6): 494～612
13. Datta R N, Peters M. Approaches to Improve Cut/Chip/Chunk Resistance in Truck and Off-the-road Tread Compounds [C]. Symposiums of International Rubber Conference, 2004, 158-165.
14. Li G F, Yang P. Application of Short Fiber-rubber Composites in Rubber Pipes[J]. Ningxia Chemicals, 1994, 1: 1-4.
15. Liu Z J. Application of Nylon Fiber/CR/EPDM Composite Material in Rubber Dam[J]. China Synthetic Rubber Industry, 1997, 20(5): 311-312.
16. Liu J W. Application of Polyester Fiber in Rubber Belts[J]. Special Purpose Rubber Products, 2003, 24(1): 25-40

The Removal of Perfluorinated Carboxylic Acids (PFCAs) on Zinc Anode by Electrocoagulation and the Effects of Chemical Flocculants

Lei Xu[1,2], *Jinhua* Wu[1,*]

[1]Key Laboratory of Pollution Control and Ecosystem Restoration in Industry Clusters, Ministry of Education, School of Environment and Energy, South China University of Technology, Guangzhou 510006, China
[2] Research Center for Eco-Environmental Engineering, Dongguan University of Technology, Dongguan 523808, China

Abstract. This paper focus on the removal of PFCAs (C4-C10) in aquatic solution by electrocoagulation using Zn as anode. The effects of carbon chain length on the removal efficiencies of PFCAs were studied, in which the removal priorities of PFCA with different carbon chain length as well as the change of concentration during the electrocoagulation process were evaluated in the mixed solution of PFCAs (C4-C10). The addition of chemical flocculants was carried out to improve the electrocoagulation performances of PFCAs. The results shows that PFCAs with longer carbon chain length (C8-C10) have better removal efficiencies (almost 100%), and the longer one has the removal priorities in the mixed solution containing PFCAs (C4-C10). The addition of gelatin, APAM, and CPAM have different effects on the removal performances of PFCAs (C4-C7), in which C4 and C5 have no improve removal with any chemical flocculant due to their small molecular weight. APAM can improve the removal of C6 and C7 obviously, because of the larger floc generated by the electrostatic attraction between of APAM and $Zn(OH)_2$ floc.

1 Introduction

Perfluorinated carboxylic acids (PFCAs) are a kind of typical perfluorinated compounds (PFCs), which have been widely used in a variety of industries, such as fire-fighting foams, textiles and semiconductor industry [1]. PFCAs are so stable that they are difficult to be biodegraded and can exist in various environmental media for a long time. Besides, they can accumulate in plants, animals, and human beings through biological enrichment and amplification, resulting serious threats to environment and human health [2]. Metal hydroxide flocs, formed during coagulation, can strongly sorb certain contaminants and then remove them from aquatic solution [3]. It is reported that electrocoagulation is efficient in the contaminants removal process by coagulation, considering the *in-situ* formed metal hydroxide flocs [4]. Electrocoagulation is also a combined process consisted of coagulation, electroflotation, and electrooxidation/reduction, in which coagulation is the main process for removing the contaminants. Thus the electrocoagualtion mechanism is partly similar with chemical coagulation process [5]. In the industrial wastewater treatment by coagulation, addition of chemical flocculants, such as gelatin, polyacrylamide (PAM), and activated hydrated silica, is helpful to the removal efficiency. In this study, we mainly discussed the removal of PFCAs in water by electrocoagulation and chemical flocculants will also be added in the process to evaluate their effects.

2 Experimental Sections

2.1. Materials

All chemicals used in the experiments were reagent grade and used as received. Perfluorobutyric acid (PFBA, C4, 99%), perfluoropentanoic acid (PFPeA, C5, 98%), perfluorohexanoate (PFHxA, C6, 98%), perfluoroheptanoic acid (PFHpA, C7, 98%), perfluorooctanoic acid (PFOA, C8, 98%), perfluorononanoic acid (PFNA, C9, 98%), and perfluorodecanoic acid (PFDA, C10, 98%) were bought from Sigma-Aldrich Chemical Co., Ltd. (St. Louis, MO, USA). Deionized water with conductance of 18.2 MΩ cm at 25±1 ℃ used in all experiments was prepared by a Millipore water system in our laboratory. Gelatin, anionic PAM (APAM, >90%), and cationic PAM (CPAM, >88%) were purchased from Sinopharm (Beijing, China).

2.2 Experimental Setup and Procedures

The electrocoagulation reactor system (Fig. 1) was consisted with direct-current power supply, magnetic stirring, and electrolytic cell, in which zinc plate was used as anode and titanium plate was used as cathode. The effective size of electrodes was 8 cm × 4 cm × 0.2

* Corresponding author: jinhuawu@scut.edu.cn

cm, and the distance of anode and cathode was 5 cm. The total volume of aqueous solution for each treating was 100 mL. The initial concentrations of PFCAs (C4-C10) were all 100 ppm with NaCl (20 mM) was the electrolyte and the electrolysis time was 20 min. The current density and stirring speed were 2 mA/cm^2 and 500 rpm, respectively, for all experiments.

Fig. 1 The electrocoagulation system

2.3 Analyses

The concentrations of PFACs (C4-C10) were analyzed by high performance liquid charomatography (HPLC-UV, Dionex U3000, USA) with WpH C18 (4.6 mm × 250 mm, 5 μm) as column and AD-20 (210 nm, 30 ℃) as UV detector. The operating parameters were following: (1) isocratic elution with acetonitrile/20 mM Na_2HPO_4 (50/50. v/v), (2) flow rate of 1 mL/min, (3) each sample injection with 20 μL.

3 Results and Discussion

3.1 Removal of PFCAs (C4-C10)

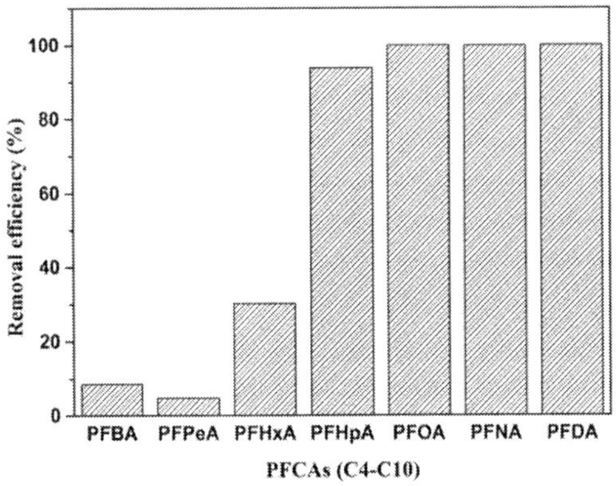

Fig. 1 Removal efficiency of PFCAs (C4-C10) by electrocoagulation

Fig. 1 shows the results of removal efficiencies of PFCAs (C4-C10) after 20 min by electrocoagulation, which clearly indicates that the removal efficiency is improved with the increase of the carbon chain length. Electrocoagulation can remove PFOA, PFNA, and PFDA (C8-C10) thoroughly with their removal efficiencies are all almost 100%. As for PFCAs (C4-C7), especially for PFCAs (C4-C6), the removal efficiencies are lower than those of PFCAs (C8-C10). The value of PFBA and PFPeA are only 8.6% and 4.7%, respectively. In the mixed solution containing PFCAs (C4-C10), the concentration of which were all 100 mg/L, their removal performances are studied further. As shown in Fig. 2, PFOA, PFNA, and PFDA are all almost removed thoroughly in 10 min, meanwhile the PFCA with longer carbon chain possess removal priority. PFOA, PFDA, and PFNA are all have worse performances than PFCAs (C8-C10) in electrocoagulation process, which are similar with the results in Fig. 1. Thus we will evaluate the effects of chemical flocculants, such as gelatin, APAM, and CPAM, to identify whether they could improve the removal efficiency of PFCAs (C4-C7) during the process of electorcoagulation under the above conditions.

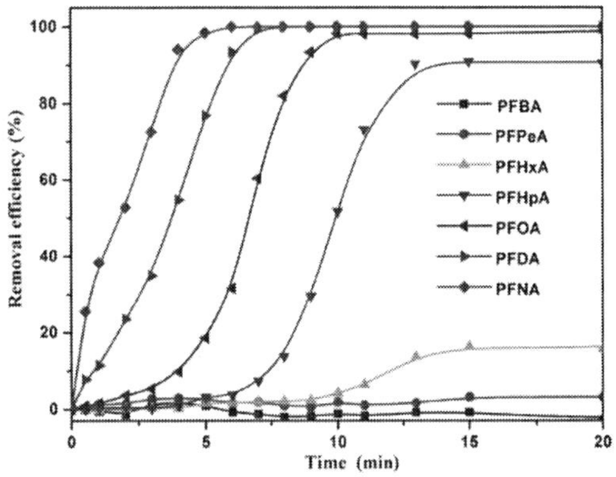

Fig. 2 Removal efficiency of mixed PFCAs (C4-C10) by electrocoagulation

3.2 Effect of gelatin

Gelatin is a polypeptide compound, mainly consists of amino acid, which has the properties of low viscosity, high affinity, and dispersion stability. It is always used as organic polymeric flocculant for treating Cr^{6+} [6] and PFOA [7]. In this work, the addition amount of gelatin was 6 mg/L, meanwhile the other conditions were all same with the content in **3.1**.

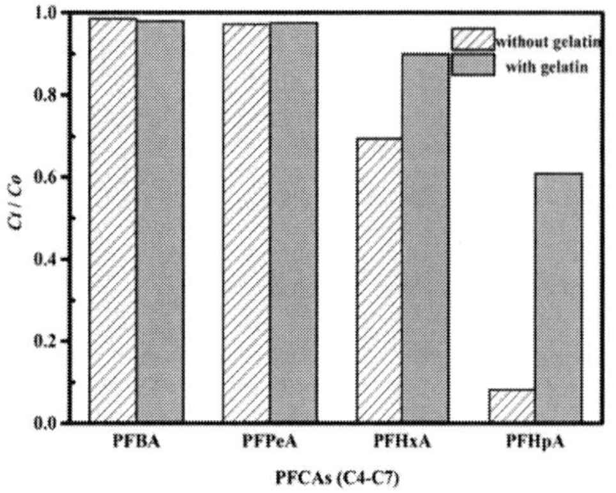

Fig. 3 The effect of gelatin on removal efficiency of PFCAs (C4-C7) by electrocoagulation

It can be seen in Fig. 3 that the removal performances of PFBA and PFPeA have no differences under the conditions with/without gelatin, which both show very low removal efficiencies or show no removal. It is another case for PFHxA and PFHpA that their removal efficiencies are sharply decreased after the addition of gelatin, in which the removal ratios are changed from 69.4% and 92.3% to 10% and 40.2%, respectively. Thus the addition of gelatin is not helpful to improve the removal efficiency of PFCAs (C4-C7), to the contrary it can inhibits the removal by eectrociagulation. It is known that there are many carboxyl and amine groups in the molecular structure of gelatin, which could partly increase the size of floc as well as combination degree. But the carboxyl groups show competitive adsorption with PFCAs on the surface of generated $Zn(OH)_2$ floc during electrocoagulation using Zn as anode. Then the adsorption of PFCAs, such as PFHxA and PFHpA, are weakened under the competition.

3.3 Effect of PAM

PAM is a frequently used organic flocculation aid in the industrial wastewater treatment, which can be divided into APAM and CPAM. APAM has high water solubility, and exists in the form of anion after dissolution in the water solution. APAM is often used in the flocculent precipitate and clarification during the treatment of industrial wastewater as well as in the purification of drinking water. There are many active groups in the structure of CPAM, and they can sorb contaminants by the effect of hydrogen bond. In particularly, the colloids with positive charge are often used as the flocculant in removing organic matters in wastewater. In treating dye wastewater by flocculation, $Al(OH)_3$-PAM shows higher COD removal and decolourization ratio than iron ion [8]. In this part, we will evaluate that whether CPAM or APAM can improve the removal efficiencies of PFCAs during electrocoagulation. The addition amounts of APAM and CPAM were both 10 ppm.

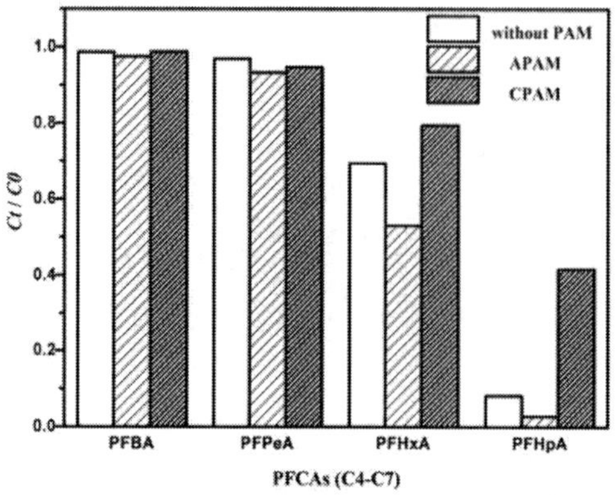

Fig. 4 Effect of PAM on the removal efficiency of PFCAs (C4-C7) by electrocoagulation

As shown in Fig. 4, the addition of APAM and CPAM has different effects on the removal performances of PFCAs (C4-C7). The electrocoagulation performances of PFCAs (C4-C7) are partly improved with the presence of APAM, e.g., the removal efficiencies of PFBA, PFPeA, PFHxA, and PFHpA are changed from 1.6%, 2.8%, 30.6%, and 91.8% to 2.5%, 6.7%, 47%, and 97.3%, respectively. However, the addition of CPAM is not useful to the electrocoagulation performance improvement of PFCAs (C4-C7). For the shorter carbon chain length, e.g., PFBA and PFPeA, the addition of PAM has no effect to improve their removal efficiencies substantially, which is the same as the addition of gelatin. The improved removal performances of PFHxA and PFHpA after the addition of APAM in electrocoagulation can be explained by the adsorption of APAM on $Zn(OH)_2$ floc. Because of the electrostatic attraction, APAM can be combined with $Zn(OH)_2$ floc, which carries positive charge, then forming lager floc through the action of bridge connection. CPAM carries negative charge and cannot be useful to from lager floc.

4 Conclusion

This paper presents and discusses the removal of PFCAs (C4-C10) by electrocoagulation using Zn as anode. The results show that the longer carbon chain length PFCAs, such as C8-C10, can be removed effectively in 10 min. The addition of chemical flocculants has different effects on the removal of PFCAs (C4-C7) in electrocoagulation process. Using APAM as chemical flocculant, the electrocoagulation performances of PFCAs (C6-C7) can be improved obviously. The addition of gelatin and PAM (both CPAM and APAM) cannot change the removal performances of PFCAs (C4-C5) effectively due to their short carbon chain length or their small molecular weight.

References

1. Z. Wang, I. T. Cousins, M. Scheringer, Environ. Int., **60(5)**, 242-248 (2013).

 https://doi.org/10.1051/e3sconf/20183802023

2. J. C. D'Eon, S. A. Mabury, Environ. Sci. & Technol., **41(13)**, 4799-4805 (2007).
3. M. Y. A. Mollah, P. Morkovsky, J. A. G. Gomes, J. Hazard. Mater., **114(1)**, 199-210 (2014).
4. P. Sakthisharmila, P. N. Palanisamy, P. Manikandan, J. Clean. Prod., **172**, 2206-2215 (2018).
5. J. Sun, C. Hu, T. Tong, Environ. Sci. & Technol., **51(15)**, 8544-8551 (2017).
6. Y. Wang, M. Liu, N. Liu, Journal of Northeast Normal University, **43(1)**, 104-108 (2011).
7. H. Lin, J.F. Niu, S.Y. Ding, L.L. Zhang, Water Res., **46(7)**, 2281-2289 (2012).
8. Y. Al-Ani, Y. Li, J. Taiwan Inst. Chem. E., **43(6)**, 942-947 (2012).

Superhydrophobic Surface Fabricated on Carbon Steel Substrate by Zinc Electrodeposition for Prevention of Corrosion

Liping Wang[1,2,*], *Lei* Zhuang[2,3], *Gang* Chen[2], *Dong* Leng[2,3], *Jian* wang[2,3], *Li* Cheng[2,3], *Wei* Liu[2,3], *Jianbo* Wang[2] and *Baorong* Hou[1]

[1]Key Laboratory of Marine Environmental Corrosion and Bio-fouling, Institute of Oceanology, Chinese Academy of Sciences, Qingdao 266071, China
[2]Rizhao Port Group Co., Ltd, Rizhao 276826, China
[3]Shandong Harbour Engineering Group Co., ltd, Rizhao 276826, China

Abstract. A facile, low cost, and relatively environmental friendly method was presented for the fabrication of superhydrophobic surface on carbon steel substrate. The superhydrophobic surface was obtained by zinc electrodeposition and chemical modification with ethanolic stearic acid. The wettability of the superhydrophobic surface was measured by a water contact angle (WCA) with a highest value of 155.7°. The morphology of the fabricated film was characterized by scanning electron microscopy (SEM). The chemical composition of the fabricated surface was analysed by energy dispersive spectrometer (EDS) and fourier transform infrared spectroscopy (FTIR), respectively. The corrosion resistance of the carbon steel substrate was measured by potentiodynamic polarization test in 3.5 wt% NaCl solution. The electrochemical measurements show excellent corrosion resistance, which may have promising applications in anticorrosion of carbon steel.

1 Introduction

Carbon steel, a common kind of metallic material, has been widely used in many applications such as architectural engineering, bridges, railways, vehicles, ships and various machinery manufacturing industries and other areas owing to its outstanding mechanical performance, low cost, perfect electrical conductivity and so on [1, 2]. Due to the high chemical reactivity, carbon steel was easy to be corroded in a wet environment, causing lots of economic losses and security issues [3]. To reduce these problems, zinc was usually used to protect steel from corrosion by electrodeposition [4, 5]. At present, the galvanizing process is still one of the most economic and efficient technology for carbon steel protection. However, chromic acid salt was usually used for passivation of zinc to improve the performance of corrosion resistance, which may cause critical problems in environment [6].

In recent years, superhydrophobic surface technique have been extensively developed and applied in the self-cleaning [7-9], oil-water separation [10-13] and anti-corrosion [14-16]. Generally, the superhydrophobic surface possess a water contact angle (CA) more than 150° and a sliding angle (SA) lower than 10°, which can be commonly found from lotus leaf, butterfly wings and mosquito eyes in nature [17-19]. These superhydrophobic states grant excellent anti-corrosion properties to the metal material that has attracted our great interest. In this work, superhydrophobic surface was fabricated on carbon steel via chemical modification after zinc electrodeposition. This kind of method was simple and did not require complex instruments. Moreover, the fabricated film shows excellent anticorrosion and the entire process of fabrication was relatively environmental friendly without any toxic substances. Therefore, the developed method will be one of the best substitute technologies for the traditional passivation of galvanized zinc.

2 Experiment

Carbon steel (Q235), purchased from Baosteel Group Corp., and was clipped into pieces of 200mm×150mm × 2 mm as samples. Zinc chloride, potassium chloride, boric acid and stearic acid with analytical grade were obtained from Sinopharm Chemical Reagent Co.,Ltd).

To remove the iron rust and other oxide film, the substrates were polished by sand paper with different mesh number of 400, 800, 1200, and 2000 successively and cleaned with acetone, ethanol, deionized water orderly and then air dried. The carbon steel and a 300mm×20mm×2mm pure Zinc has been chosen as cathode and counter electrode, and 120g/L zinc chloride, 60g/L potassium chloride, 15g/L boric acid aqueous solution has been chosen as electrolyte. In the electrodeposition, a direct current power was used to supply a constant current with the bath voltage of 4.0V at room temperature for 20 min. After the electrodeposition, the treated carbon steel substrates

* Corresponding author: wanglp88@126.com

were cleaned with deionized water and dried with cold air, followed by immersion into 0.05 mol/L stearic acid. Then the pieces were dried at 105 °C for 1h, the superhydrophobic Zn surface on carbon steel was obtained.

The wettability properties of the prepared superhydrophobic substrates were measured by a contact angle analyser (JCY-2, Shanghai FangRui Instrument CO,.LTD) with 5μL water droplet at room temperature. All the results were taken from the mean of the five different measurements. The morphologies of the superhydrophobic film were characterized by scanning electron microscope (SEM, Carl Zeiss sigma300, Germany) with an energy dispersive spectrometer (EDS). The chemical composition of the surface was characterized by a Fourier transform infrared spectroscopy (FTIR, Nexu Thermo Fisher Scientific Inc. USA). The corrosion resistance of the samples was investigated by potentiodynamic polarization curves at a scan rate of 1 mV/s and potential from -250 mV to +250 mV vs Open Circuit Potential, which were carried out using by Gamry Reference1000 (USA). All the electrochemical experiments were measured in 3.5 wt% NaCl aqueous solution at a room temperature.

3 Result and discussion

3.1. The morphology of the fabricated film

As is well known, the excellent superhydrophobic properties of the substrate depend on its special microstructure and compositions, corresponding to the phenomena of "Capillarity" and "Cushion effect" [20]. Figs. 1a-1c shows the SEM images of bare carbon steel, carbon steel with electrodeposited Zn and superhydrophobic Zn surface.

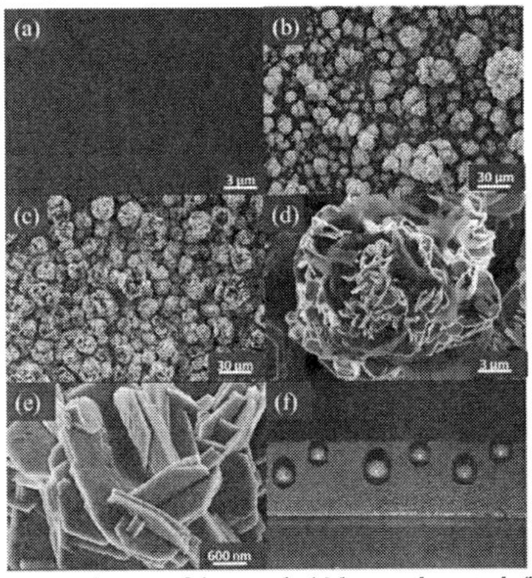

Figure 1. SEM images of the sample (a) bare carbon steel, (b) carbon steel after Zn electrodeposition , (c, d, e) carbon steel after Zn electrodeposition and modification, (f) Photograph of water droplets on the superhydrophobic film.

An approximately smooth flat surface has been shown in Fig.1a. Compare to Fig.1a, it was clear seen from Fig.1b and Fig.1c that a rough surface has been fabricated in the substrate on account of large amounts of electrodeposited protrusive Zn microparticles. Thus, we may find several micron sized hills and valleys in Fig.1b and Fig.1c. Along with the amplification, it may be found from Fig.1d and Fig.1e that multiple nanosized embossments exist on the surface. The fabricated miro-nano structures allow air to be trapped within the surface, which can effectively prevent water drop from the substrate. Fig.1f shows the photograph of carbon steel after Zn electrodeposition and modification with steric acid. The water on the surface are all of sphere shapes, that revels the successful superhydrophobic fabricated on the carbon steel substrate.

3.2. The composition of the fabricated film

The composition of the pure carbon steel and samples after electrodeposition and modification was analysed by EDS and FTIR, and results were shown in Fig.2 and Fig.3. It may be seen from Fig.2a that the carbon steel was mainly composed of element C and Fe with small amount of O, which may be owing to the steel oxidation. Fig.2b shows that the carbon steel was completely covered by Zn after electrodeposition. Compared to Fig.2b, the Zn surface after modification in Fig.2c contained the additional element C and O, revealing that the stearic acid has successfully assembled to the surface. Fig.3a and Fig.3b show the FTIR spectra of carbon steel after Zn electrodeposition and carbon steel after Zn electrodeposition and modification with stearic acid in the range of 4000 to 500 cm^{-1}, respectively. We may find an approximate smooth curve in Fig.3a with unnoticeable absorption peak. Compared to Fig.3a, an obvious series of absorption peaks were detected and shown in Fig.3b. In the high frequency region, the peak at around 2916 and 2847 cm^{-1} correspond to the methyl and methylene stretching vibration absorption peaks of stearic acid, respectively[21]. The significant peaks present at 1537 and 1468cm^{-1} was also observed, assigned to COO− groups of $CH_3(CH_2)COO$- on the carbon steel after Zn electrodeposition and modification with stearic acid. It is further indicated that the stearic acid had successfully assembled on this Zn coating surface, which contributed to the decrease in the surface energy of this coating.

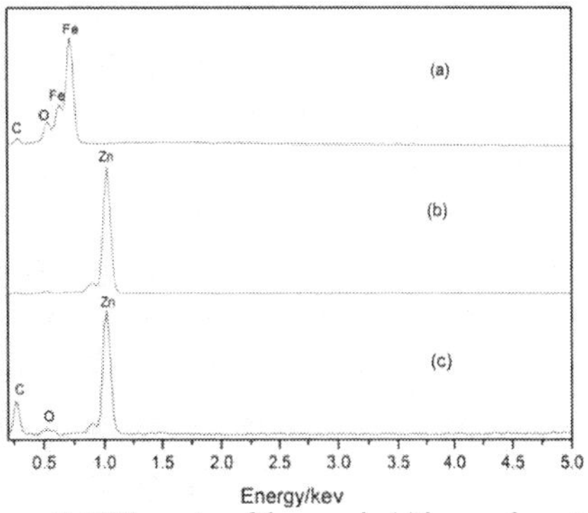

Figure 2. EDS spectra of the sample (a) bare carbon steel, (b) carbon steel after Zn electrodeposition , (c) carbon steel after Zn electrodeposition and modification.

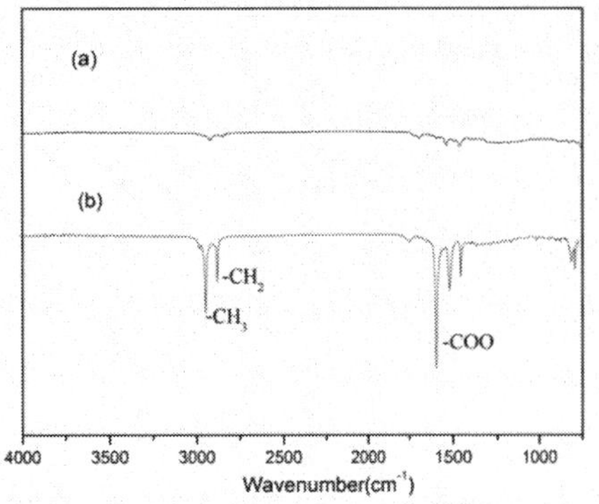

Figure 3. FTIR spectra of the sample (a) carbon steel after Zn electrodeposition, (b) carbon steel after Zn electrodeposition and modification

3.3. The effect of current density on the wettability of the fabricated film

In order to study the effect of current density on the wettability of the superhydrophobic Zn surface, the carbon steel was electrodeposited with a constant bath voltage of 4.0V for 20 min under different current density, and all the results were shown in Fig.4. It can be seen from Fig.4 that the water contact angle of the fabricated surface increases significantly with the enhance of current density, and reaches the highest value of 155.7° when the current density added up to 5 A/dm^2. After that, the water contact angle decreases slowly with the growing current density. It can be concluded that the electrodeposition current density was an important factor that affects the superhydrophobic performance.

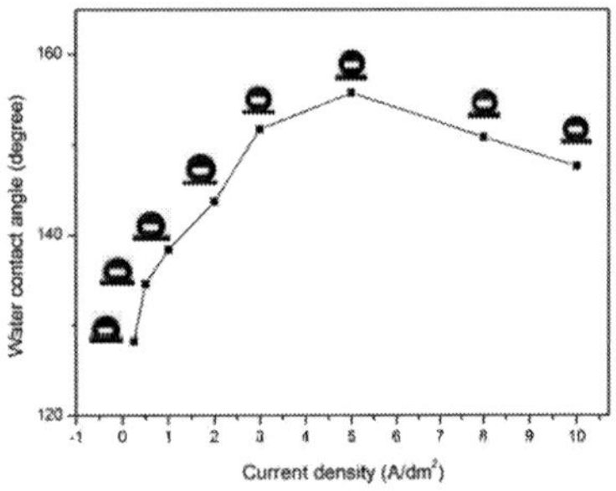

Figure 4. Contact angle of the superhydrophobic Zn surface with different current densities.

3.4. The corrosion property of the fabricated film

In order to evaluate the corrosion property of the prepared samples, the potentiodynamic polarization tests on bare carbon steel, carbon steel after Zn electrodeposition and superhydrophobic Zn surface were investigated in 3.5wt% NaCl aqueous solution. All the experimental data were plotted in Fig.5. The corrosion potential (E_{corr}), the corrosion current intensity (I_{corr}) and the corrosion rate (v_{corr}) were obtained from the polarization curves in Fig.5 according to Tafel extrapolation. The protection efficiency (η) of carbon steel after Zn electrodeposition and superhydrophobic Zn surface were calculated by the equation[22]:

$$\eta = (I_{corr} - I'_{corr}) / I_{corr} \times 100\% \quad (1)$$

Where I_{corr} was the corrosion current of the bare carbon steel, I'_{corr} presents the corrosion current of carbon steel after Zn electrodeposition or superhydrophobic Zn surface. In addition, the polarization resistances (R) was calculated using by Stern-Geary equation[23]:

$$R = (b_a \cdot b_c)/2.303/(b_a + b_c)/I_{corr} \quad (2)$$

Where b_a and b_c were the slopes ($\Delta E/\log I$) of the anodic and cathodic by Tafel fitting. All the values were listed in Table 1.

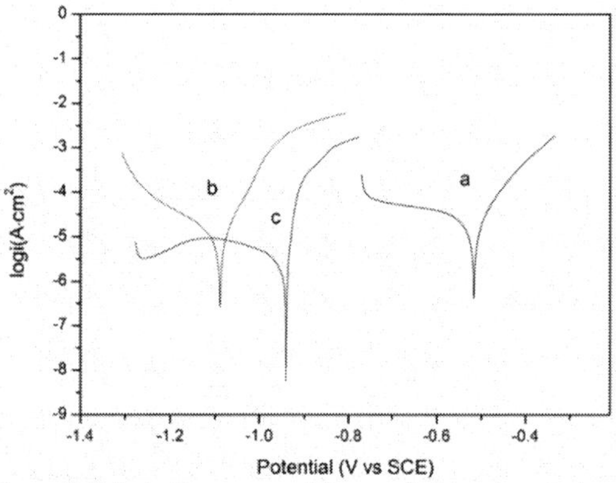

Figure 5. Polarization curves of the sample (a) carbon steel after Zn electrodeposition, (b) carbon steel after Zn electrodeposition and modification.

Table 1. Comparison of E_{corr}, I_{corr}, v_{corr}, η and R of different samples

Samples	E_{corr}(mv)	I_{corr}(A/cm^2)	v_{corr} (mm/a)	b_a(V/dec)	b_c(V/dec)	η(%)	R(kΩ·cm^2)
Bare carbon steel	-517	3.57×10^{-5}	16.32	0.1014	0.7694		1.09
Electrodeposited Zn surface	-1090	1.35×10^{-5}	6.19	0.0600	0.1577	62.18	1.40
Superhydrophobic Zn surface	-941	3.64×10^{-7}	0.17	0.0063	0.0106	98.98	4.71

From Table 1 we may find that the corrosion potential of carbon steel after Zn electrodeposition was lower than bare carbon steel, which means an easier corrosion for Zn. The result verifies that carbon steel can be protected by sacrificial anodes of Zn. The corrosion potential of electrodeposited Zn was increased from -1090 to -941 mv after modification with stearic acid, which means a fewer degree of corrosion of superhydrophobic Zn surface. The value of corrosion currency intensity and the corrosion rate of the fabricated superhydrophobic surface was 3.64×10^{-7} A/cm^2 and 0.17 mm/a, which was much lower than electrodeposited Zn surface (1.35×10^{-5} A/cm^2, 6.19 mm/a) and pure carbon steel (3.57×10^{-5} A/cm^2, 16.32 mm/a). The calculated protection efficiency of superhydrophobic Zn surface was 98.98% compared with the value of 62.18% of the electrodeposited Zn surface without modification. It can be confirmed the fact that the superhydrophobic Zn surface owns the best anticorrosion properties. Moreover, the polarization resistance of the superhydrophobic Zn surface was as high as 4.71 kΩ·cm^2, which was nearly 4.3 times of the bare carbon steel and 3.3 times of the electrodeposited Zn surface without modification. From the above, the superhydrophobic Zn surface has the best corrosion protective properties from various corrosion parameters, which has great potential applications in the field of carbon steel protection.

4 Conclusion

In this work, a method of fabrication of superhydrophobic surface on carbon steel substrate by Zn electrodeposition and chemical modification with stearic acid has been presented. SEM results show that the superhydrophobic property was caused by the fabrication of mico-nano structure. The asprepared superhydrophobic Zn surface can exhibit excellent reversible wettability controlled by changing electrodeposited current density. In addition, potentiodynamic polarization test shows the superhydrophobic Zn surface has the best anticorrosion property compared with bare carbon steel and the electrodeposited Zn surface without modification. The established method was facile, low cost, and relatively environmental friendly, and can be suitable for the protection of carbon steel.

Acknowledgments

This work was financially supported by Shandong Postdoctoral Fund (2017157).

References

1. A. Fihri, E. Bovero, A. Al-Shahrani, A. Al-Ghamdi, G. Alabedi, Colloids and Surfaces a-Physicochemical and Engineering Aspects, **520** (2017) 378-390.
2. H. Gao, S. Lu, W. Xu, S. Szunerits, R. Boukherroub, Rsc Advances, **5** (2015) 40657-40667.
3. G.N. Kirby, Chemical Engineering, **86** (1979) 72-84.
4. K. Alipour, F. Nasirpouri, Rsc Advances, **7** (2017) 51879-51887.
5. O.S.I. Fayomi, A.P.I. Popoola, Surface Engineering and Applied Electrochemistry, **51** (2015) 76-84.
6. M. Dong, X. Xue, A. Kumar, H. Yang, M.I. Sayyed, S. Liu, E. Bu, Journal of Hazardous Materials, **344** (2018) 602-614.
7. H. Bagheri, M. Aliofkhazraei, H.M. Forooshani, A.S., Applied Surface Science, **436** (2018) 1134-1146.
8. J. Cremaldi, B. Bhushan, Journal of colloid and interface science, **518** (2018) 284-297.
9. Y. Xing, Y. Xue, J. Song, Y. Sun, L. Huang, X. Liu, J. Sun, Applied Surface Science, **436** (2018) 865-872.
10. Z. Guo, F. Yang, , Fabrication and Application, (2017) 245-272.
11. R.K. Gupta, G.J. Dunderdale, M.W. England, A. Hozumi, , Journal of Materials Chemistry A, **5** (2017).
12. Y. Xiang, Y. Si, Y. Xin, Z. Guo , Chemistry Letters, **46** (2017) 1393-1395.
13. X. Zhu, Z. Zhang, B. Ge, X. Men, X. Zhou, Q. Xue, Journal of colloid and interface science, **432** (2014) 105-108.
14. C. Lv, H. Wang, Z. Liu, W. Zhang, C. Wang, R. Tao, M. Li, Y. Zhu, Applied Surface Science, **435** (2018) 903-913.
15. C.-W. Yao, D. Sebastian, I. Lian, O. Gunaydin-Sen, R. Clarke, K. Clayton, Coatings, **8** (2018).
16. S. Zhou, X. Zhu, Q. Yan, Surface and Interface Analysis, **50** (2018) 290-296.
17. Y. Qing, C. Yang, C. Hu, Y. Zheng, C. Liu, Applied Surface Science, **326** (2015) 48-54.
18. L.-K. Wu, X.-F. Zhang, J.-M. Hu, **85** (2014) 482-487.
19. D. Yu, J. Tian, J. Dai, X. Wang, Electrochimica Acta, **97** (2013) 409-419.
20. Y.J. Cho, H. Jang, K.-S. Lee, D.R. Kim, Applied Surface Science, **340** (2015) 96-101.
21. Y. Liu, S. Li, J. Zhang, Y. Wang, Z. Han, L. Ren, Chemical Engineering Journal, **248** (2014) 440-447.
22. B.V.A. Rao, M.Y. Iqbal, B. Sreedhar, Electrochimica Acta, **55** (2010) 620-631.
23. J. Tan, J. Hao, Z. An, C. Liu, International Journal of Electrochemical Science, **12** (2017) 40-49.

Study on environment-frientdly UV-led ceramic ink transfer film

Jingru Jia[1,*], *Jinhai* Zhang[2]

[1]Beijing Institute Of Graphic Communication, 102600 Beijing, China
[2]Beijing Institute Of Graphic Communication, 102600 Beijing, China

Abstract. UV polyurethane acrylate (PUA-1) oligomer was synthesized and used to prepare the UV-LED ceramic ink transfer film. The effects of active thinner (or active monomer), photosensitizer type, photo-sen-sitizer percentage, photosensitizer consumption, etc. on the ceramic ink transfer film performance was studied. The results showed that the prepared PUA prepolymer, single functional group thinner, and the UV-LED curing system consisting of 2,4,6-Trimethyl Benzoyl Diphenyl Phosphine Oxide (TPO), 2,4-Diethyl-9H-Thioxanthen-9-one(DETX) and Reactive Amine (P115) could meet the technical re-quire-ments of transfer film.

1 Intruduction

The ceramic ink transfer film traditionally contains sol-vent-type acrylic resin as bonding agent, and is thermally cured by infrared ray. Organic solvent volatilization during curing causes serious air pollution, so the use of ceramic ink transfer film is restricted. UV curing technique rapidly develops because of quick curing, no solvent, environment-friendliness, etc. However, traditional UV curing techniques generally have high energy consumption, high temperature, ozone release[1], etc., due to use of high pressure mercury lamp or metal halide lamp, etc., and therefore their development is adversely affected. With technology progress, UV-LED curing technique can eliminate the defects of traditional UV curing technique, including high energy consumption, high heat and ozone release[2,3], and therefore represents the photo curing development trend.

As an optoelectronic device for semiconductor, light emitting diode (LED) converts electric energy to luminous radiation. Compared with traditional light source, LED has the following features: (1) emission of monochromatic light, with very small spectral line width, 5~20 nm; (2) its light output of nearly 100%; (3) low energy consumption; (4) no ozone production; (5) low heat release; and (6) long service life. In the industrial production sector, the LED used for ultraviolet light curing is called UV-LED. UV-LED technique has good development prospect in the light-curing field because of its advancement, economy, environment-friendliness, etc.

UV-LED curing system mainly consists of prepolymer, active thinner, photoinitiator, and aids. Under UV-lED radiation, the photoinitiator in this system absorbs the ray at a wavelength to initiate polymerization of oligomer and active thinner before final crosslinking into a net structure resulting in curing.

Although UV curing technique is convenient, quick, energy-saving, and environment-friendly, the cured film has high hardness, and low flexibility and ductibility. Ceramic ink transfer film is different from ordinary films. During transfer, irregular ceramic surface requires bending, stretching, etc. Thus, the ceramic ink transfer film must have good hardness, flexibility and ductibility. Very flexible polyurethane acrylate (PUA) shall be selected from UV prepolymers. General polyurethane acrylate still fails to reach the requirement of ceramic ink transfer film for flexibility.

In this study, the flexibility and ductibility of ceramic ink transfer film were adjusted by (1) synthesizing the more flexible PUA prepolymer[4]. According to the data, hydroxy terminated polybutadiene (HTPB) (relative molecular weight: 2000) and polydiol (relative molecular weight: 2000) were used as the skeleton to synthesize polyurethane acrylates. It was found during performance comparison that the PUA with HTPB as skeleton had good flexibility, ductibility and low temperature resistance; (2) selecting an appropriate active thinner to reduce the crosslinking density and enhance the flexibility and ductibility; and (3) selecting an appropriate photosensitizer composition and consumption to control the crosslinking density and curing rate.

2 Experiment

Equipment: UV-LED plane light source PT-75, wavelength of 395nm, made by IGEtec; DRK101B tensile testing machine, made by Shangdong Drick; IMT-NZ01 folding endreanc tester, made by Dongguan City International Material Tester Co., Ltd.

2.1 Flexible Polyurethane Acrylate Synthesis and Performance

* Corresponding author: hellojiajingru@163.com

2.1.1 Main Chemical Materials

Hydroxy terminated polybutadiene(HTPB) : technically pure, imported; Toluene diisocynate (TDI): technically pure, made by Mitsubishi Group; Dibutyltin dilaurate (DBT): chemically pure, made by Beijing Chemical Factory 3; 2-Hydroxyethyl acrylate: chemically pure, made by Beijing Dongfang Chemical Factory of Beijing Chemical Industry Group Co., Ltd.; paradioxybenzene: chemically pure, made by Shanghai Chemical Reagent Factory 6; Commercial PUA prepolymer samples: SM6206 from Jiangsu Sanmu Group; CR3210 from Huizhou City Changchang Electronic Materials Co., Ltd.; DH316 from Shenzhen Dinghao PhotoChemical Technology Co., Ltd.; Leveling agent JZL-033: technically pure, home-made; Anti foamer: technically pure, home-made.

2.1.2 Experiment Procedure

HTPB was added to a three mouth bottle with a stirrer, thermometer and condenser, and then dibutyltin dilaurate (DBT) (consumption: 0.1%) was added. TDI was dropwise added. The mixture of reactants was heated under nitrogen supply until the reflux occurred. The reaction lasted for about 4 h. The system was cooled to 60°C, HEA was dropwise added, and paradioxybenzene polymerization inhibitor (400 ppm) was added. The reaction lasted for 3-4 h at 60-65°C. The free isocyanic acid group content was determined by di-n-butylamine method. The reaction was terminated when the free isocyanic acid group content is below 0.5%. The resulting product was PUA (PUA-1).

2.1.3 Photosensitizer Selection

Most of the photosensitizers in the market have strong absorbance at 300~370 nm, and poor absorbance above 370 nm. Hence, the emission spectrum of UV-LED curing system does not match with the absorption spectrum of traditional photosensitizers, resulting in a decrease in initiation efficiency of photosensitizers. This problem was solved by selecting many photosensitizers and mixing some of them in this study.

The spectral line of UV-LED had very narrow width, only 5~20 nm, and the light from UV-LED had a long wavelength, generally 395~405 nm. 395 nm was the common wavelength used for UV-LED curing[5]. Thus, few commercial photosensitizers applied to UV-LED curing. The wavelength used by UV-LED curing machine in this study was 395 nm. The followings were the optional photosensitizers and their maxim absorption wavelength at 395 nm: Isopropyl Thioxanthone (ITX): 382 nm; phenyl bis(2,4,6-trimethylbenzoyl)-phosphine oxide (819): 370 nm; TP0:393 nm; Photoinitiator 651(DMPA): 390 nm; DETX: 384 nm.

According to the data analysis and experimental verification[6], the cured DMPA film seriously yellowed, and DMPA photolysis products gave out odor, so DMPA was excluded. However, the absorption peak of 819 was far from 395 nm, and the light source at the wavelength of 395 nm had low initiation efficiency. The experiment also verified this presumption. Thus, 819 was excluded.

As a cracking (NorrishI) efficient photosensitizer, TPO especially applied to coloured system and thick-film curing system. However, the use of TPO alone would lead to bad surface curing because oxygen inhibited polymerization; as hydrogen abstracting (NorrishII) photosensitizers, ITX and DETX could avoid polymerization inhibition by oxygen, and were used along with amine potentiating agent, such as P115, etc., to greatly enhance the initiation efficiency. In addition, ITX always yellowed. According to various factors and experimental verification, TPO, DETX, and P115 finally were selected as the photosensitizer composition in this paper.

According to the following formulation, the transfer film was prepared: 15 g of PUA-1, 5 g of Tri(propylene glycol) diacrylate (Diluent TPGDA), 0.2 g of anti foamer + Leveling agent, and 2.2 g of photosensitizer. The reactants were printed on the transfer printing paper by the 100 mesh screen process press, and cured by the UV-LED curing machine. No sign after hand pressure represented complete curing. The curing times were recorded and listed in the table below:

Table 1. Effect of Photosensitizer Composition on the Curing Time.

No.	1	2	3	4	5	6	7
PUA-1, g	15.0	15.0	15.0	15.0	15.0	15.0	15.0
TPGDA, g	5.0	5.0	5.0	5.0	5.0	5.0	5.0
Anti-foamer +Leveling agent, g	0.2	0.2	0.2	0.2	0.2	0.2	0.2
TPO , g	2.2	0.0	1.2	1.0	1.2	1.6	1.8
DETX, g	0.0	2.2	1.0	0.8	0.6	0.4	0.3
P115, g	0.0	0.0	0.0	0.4	0.4	0.2	0.1
Curing time, s	60	60	40	30	12	25	60

The experimental results review: No.1, Bad surface curing; No.2, Bad surface curing; No.3, Bad surface curing; No.4, Good curing; No.5, Good curing; No.6, Good curing; No.7, Bad surface curing.

According to the experimental results, No.6 is the best formula, the photosensitizer consisting of 55% TPO, 27% DETX and 18% P115 was finally selected.

According to the photosensitizer formulation, the effects of photosensitizer content on curing time were studied and shown in Figure 1:

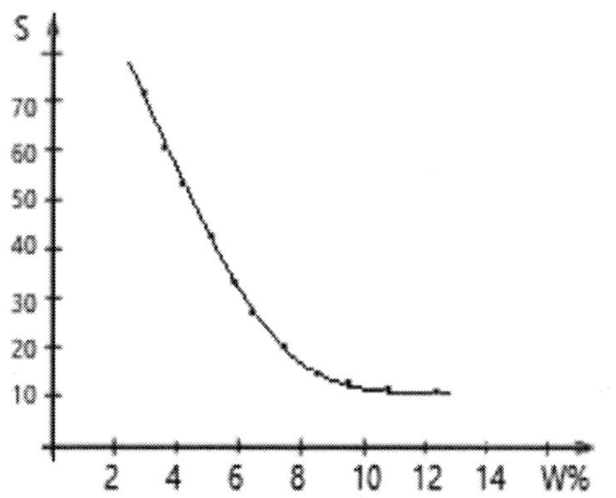

Figure. 1. Photosensitizer Content (w%) Vs.Curing Time(s).

The results showed that different photosensitizer contents led to different curing times of ceramic ink transfer film. With increasing photosensitizer content below 9%, the curing time gradually decreased; with increasing photosensitizer content, the curing time did not change significantly. Thus, the photosensitizer content of 8%~9% was selected.

2.1.4 PUA Performance Comparison

The transfer film formulation: 67% PUA; 24% TPGDA; 8.8% photosensitizer; 0.1% leveling agent; and 0.1% anti foamer. The reactants were printed on the transfer printing paper by the 100 mesh screen process press, and cured by the UV-LED curing machine. No sign after hand pressure represented complete curing.

The dried transfer film was tested on its pencil hardness, elongation at break, and folding endurance. The results were listed in the table below. The self-prepared PUA-1 had higher flexibility than the control sample.

Table 2 Effect of Prepolymer Type on Transfer Film Performance.

Sample	PUA -1	SM -6206	CR -3210	DH -316
Transfer film thickness, μm	35	35	35	35
Curing time , s	12	8	10	8
Hardness	H	3H	2H	2H
Folding endurance (at 25°C), times	180	60	80	50
Elongation at break, %	20	5	10	10

2.2 Effect of Thinner Type on Ceramic Ink Transfer Film Performance

At the constant consumption of main resin, PUA, active monomers with varying functionality are generally used to adjust the flexibility of transfer film. The active monomer with higher functionality will provide higher cross linking density, smaller distance between molecular chains, higher hardness and lower flexibility of transfer film, and higher curing rate because of its more functional groups, whereas the active monomer with mono functional group will provide lower cross linking density, bigger intermolecular distance, higher flexibility and lower strength of transfer film, lower curing rate because of few functional groups, and lower industrial production efficiency. Active monomers with two functional groups are used to enhance the curing rate and strength and those with mono functional group is used to adjust flexibility in order to try to reach both specified flexibility and curing rate considering all the factors, including strength, flexibility, curing rate, etc.

2.2.1 Main chemical materials

PUA-1; bifunctional group thinner (technically pure): TPGDA, 1,6-Hexanediol diacrylate (HDDA); single functional group thinner (technically pure): isodecyl acrylate (IDA), and 2-ethylhexyl acrylate (2-EHA), Isobornyl acrylate (IBOA).

2.2.2 Preparation and Test

Ceramic ink transfer film formulation: 67% PUA-1; 24% thinner; 8.8% photosensitizer; 0.1% leveling agent; and 0.1% anti foamer. HDDA, TPGDA, IDA, 2-EHA and IDOA, were used to prepare five samples.

According to the formulation, five different diluents were used to prepare five ceramic ink transfer films, which were cured, respectively. The curing times were recorded and the ceramic ink transfer films were tested on their pencil hardness, elongation at break, and folding endurance.

Table 3. Effect of Thinner Type on Transfer Film Performance.

Thinner	Curing time	Hardness	Folding endurance , times	Elongation At break , %
HDDA	8	H	130	20
TPGDA	12	H	180	18
IDA	23	2B	600	45
2-EHA	25	2B	500	42
IBOA	16	H	300	30

It could be seen from the above table that bifunctional group thinner had higher curing rate, but low ductibility and flexibility, while single functional group thinner had the best ductibility and flexibility, but lower curing rate. It was recommended that the single functional group, especially IDA, should be selected in practice considering the special requirements of ceramic transfer printing ink for ductibility and flexibility.

3 Ceramic printing decoration process

3.1 Water transfer printing paper

The water transfer printing is a decoration process in which a pattern is printed on a piece of special paper to which watersoluble adhesive is applied, and the paper is soaked and wetted in water before the pattern is transferred to the surface of an article. Water transfer printing paper is mainly used for decoration of irregular surfaces of ceramic, glass, metal, etc. Water transfer printing paper consists of base paper as the lower layer, watersoluble adhesive as medium layer, pattern as the upper layer, and transfer film ink on surface.

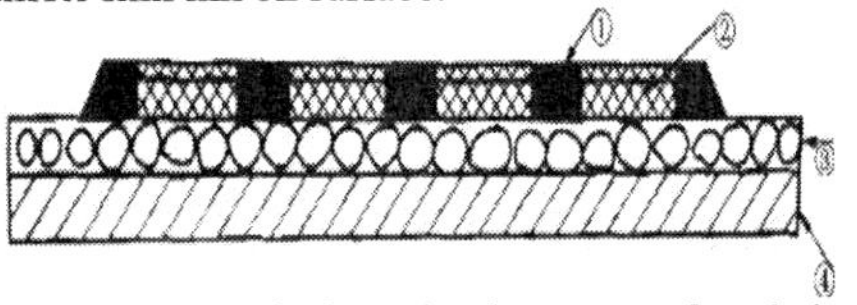

Figure. 2. Structural schematic of water transfer printing paper.

①transfer film②printed pattern ③watersoluble adhesive layer ④absorbent wood paper

A layer of watersoluble adhesive is applied to the surface of 120-180 g/m2 very absorbent raw paper. Then a pattern is printed on the watersoluble adhesive layer. Finally, a layer of cover ink is applied to it by screen printing, and dried to form a transfer film, which is called water transfer printing paper.

Water transfer printing paper manufacturing process: Base paper → Pattern printing → UV-LED drying → printing by transfer film ink → UV-LED drying → finished product.

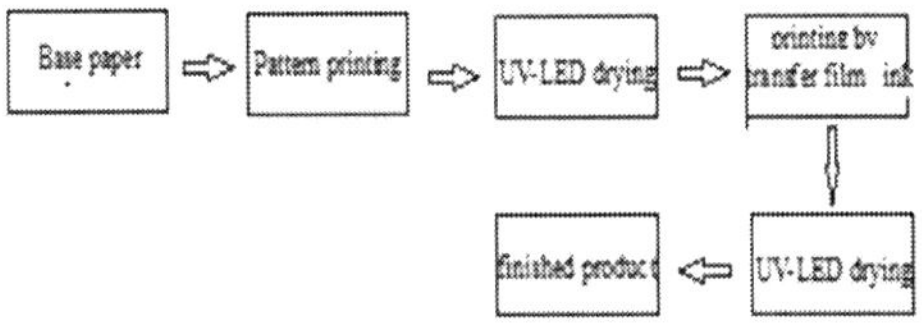

Figure. 3. Water transfer paper production process

Water transfer printing paper use method: The water transfer printing paper is cut along the edges of the pattern, and soaked in water until the adhesive is dissolved in water and it is separated from the base paper. It is applied to green ceramic, biscuit or ceramic glaze, flattened by a rubber scraper, naturally aired, and fired at 800~1350°C until the transfer film decomposes and volatilizes at high temperature. Finally, the ceramic ink develops into a color pattern.

Water transfer printing paper quality criteria: (1) During transfer, the cover film has ideal ductibility, high strength and ideal firing effect;(2) The water transfer printing paper has short release time, good gliding property, and no residue after release;(3) No pop and film trace during firing at high temperature. (Pop refers to crack or pinhole in the pattern after firing; film trace refers to residual cover film.)

3.2 Testing of transfer film property

The transfer film ink is made according to the following formulation: PUA-1 67%; HDDA 6%; IDA 18%; photosensitizer 8.8%; leveling agent 0.1%; anti foamer 0.1%.

The release performance and ductibility of the cover film (80 mesh screen printing, 20 μm thick):

Table 4. Release time and elongation at break.

NO	Temperature, °C	Release time, s	Elongation at break (at 25°C), %
1	20	0	
2	25	3	
3	30	42	28
4	35	83	
5	40	90	

The base paper with pattern is printed by cover ink in the 80 mesh screen printing machine, cured by UV, released with water, transferred to ceramic plate surface, naturally aired, kept at constant temperature in a tubular furnace at varying temperature for 40 min, tested on the weight loss percentage at varying temperature, and observed under a magnifier for printing to check crack or pinhole. For the results, see the table below.

Table 5. Volatility of cover film at varying temperature.

Temperature, ℃	350	400	450	500	550	600
Volatility, %	3	42	83	90	100	100
Pop	No	No	No	No	No	No

The resulting transfer film has outstanding ductibility, ease to release and aging resistance, and completely volatilizes below 600°C.

4 Conclusion

The self-prepared PUA-I, i.e. polyurethane acrylate oligomer, was used to prepare the ceramic transfer printing ink for UV-LED curing in order to study the photosensitizer composition and consumption and the effects of active thinner type on the cured film ductibility and flexibility. The followings were concluded from the results:

(1) The optimal photosensitizer composition for UV-LED curing at the wavelength of 395 nm consisted of 55% TPO, 27% DETX and 18% P115;

(2) The optimal photosensitizer consumption was 8~9% of total ink mass. With increasing photosensitizer content above 8~9%, the curing rate did not significantly increase;

(3) Among the thinners, multi-functional group thinners had high curing rate, but low ductibility and flexibility. Among single functional group thinners, IDA had the best flexibility and ductibility, but lower curing rate. Therefore, the combination of IDA and HDDA should be used, and the effect is better.

(4) The resulting transfer film has outstanding ductibility, ease to release and aging resistance.

References

1. R.W. Stowe, Key Factors in the UV Curing Process-the Relationship of Exposure Conditions and Measurement in UV Process Design and Process Control partII: Lamp Systems.J. Metal Finishing, 4,50(2006)
2. S.F. Zhang, Y.S. Ma, W.W Huang, Application of UV-led Technology in number Ink.J. Label Technology, 5,30(2015)
3. D.L. Ji , The application and development direction of led in uv printing.J. Printing Field, 5,56(2010)

4. Wu S, Deng J R, Study on high flexibility polyurethane acrylate synthesis and application.J. Exploration and Research, 6,1(2004)
5. L. Zhang, Development Trend of UV-led Curing Techniques.J. Electronic Components and Materials, 6,99 (2015)
6. K. Jian, D.B. Li, C.Q. Hu, J.L. Yang, Research Progress in Photoinitiator for LED-Curable System.J. Paint & Coatings Industry, 10,75 (2016)

The Effect of Different Bags on Chlorophyll and Carotenoids in the Peel of Chinhuang Mango

MA Zhi Ling, *WEI* Chang Bin*

Key laboratory of Tropical fruit Biology, Ministry of Agriculture, South Subtropical Crop Research Institute, Chinese Academy of Tropical Agricultural Sciences, Zhan Jiang, Guangzhou, 524091, China

Abstract. The effect of different fruit bags on the content of chlorophyll and carotenoids in the fruit peel of Chinhuang mango during normal storage and after-ripening process was studied. The results showed that the content of chlorophyll and carotenoid in the bag was significantly lower than that of the control fruit. The skin color of the white paper bag fruit and the no bagged bag fruit was green during the green ripening process, while the outer yellow inner black paper bag fruit showed a yellow green halo. When ripe, the control fruit is yellowish green and orange red. The white paper bag fruit is greenish orange yellow, and the outside yellow inside black composite paper bag fruit is golden yellow.

1 EXPERIMENTAL MATERIALS AND METHODS

1.1 Experimental Materials, Instruments and Reagents

1.1.1 Experimental Materials

The experiment was carried out at the demonstration orchard of tiandong Mango of CAAS in 2017. The test species was Chinhuang Mango. There are 3 bagging treatments. A is treated as white paper bag; B processing is the outer yellow inner black composite paper bag; CK is the control, not bagging, all the specifications are 23cm x 36cm.The bags are all provided by fruit bag factory of the CAAS. The bagging time was carried out after the second times of physiological fruit drop (April 20th). Before the bagging, the whole fruit was set and fixed, Spray pesticides and nutrient solutions, then the bag was single bagged. Each treatment was treated with 5 repetitions, which were Randomly arranged, and the bag was removed when harvesting. Take out 20 fruits per treatment and placed indoors, Let the fruit mature naturally. The representative fruit samples were quickly separated the peel from (about 0.8 ~ 1 mm) the pulp by a scalpel at the same time and frozen in liquid nitrogen. The content of chlorophyll and carotenoid in fruit peel was measured every two days for a total of five measurements.

1.1.2 Instruments and Reagents

UV-Vis Spectrophotometer (Shimadzu UV-2700), Balance (Mettler Toly Electronic Balance AB204-S\A).

Extract: acetone 80 mL and plus distilled water to 100 mL.

1.2 Sample Preparation Method

Accurately weigh 0.50 g representative skin samples in a mortar, A few quartz sand and CaCO3 and 3 mL extracts were lapping evenly and transferred into the 50 mL bottle, and the mortar was washed into the capacity bottle and fixed all by the extractant. The dark condition was extracted at room temperature for 24 h, and the supernatant was used for analysis.

1.3 Instrument Analysis Methods

The absorbance of the sample solution was measured respectively at 665 nm, 649 nm, and 470 nm.

Calculate the concentration of chlorophyll a and b, carotenoids according to the formula:

Chlorophyll a concentration: C_a=13.95×A_{665}-6.88×A_{649}, unit mg/L;

Chlorophyll b concentration: C_b=24.96 ×A_{649}-7.32×A_{665}, unit mg/L;

Carotenoid concentration: $C_{x.c}$ = (1000 x A_{470}-2.05 x C_a-114.8 x C_b)/245 unit mg/L.

The content of each pigment component in the sample is calculated based on the concentration of each pigment: X=C×V×B/1000/M

X —the content of each pigment, mg/g; C—the concentration of each pigment, mg/L; V—the volume of the extract, mL; B—the multiple of the supernatant when analyzed (as the case may be); M—sample quality, g.

2 RESULTS AND ANALYSIS

* Corresponding author: mbelle@163.com

 https://doi.org/10.1051/e3sconf/20183802026

2.1 Effect of Different Fruit Bags on Chlorophyll Content in Peel of Chinhuang Mango

After the Chinhuang mango was picked and unbagged, the content of chlorophyll a in the control group was reduced first, and then reached the highest level by the 5th day (Figure 1)), and then decreased. The content of chlorophyll a in the peel of white single layer paper bags increased continuously until the fruit ripening, and then decreased. The content of chlorophyll a in the outer yellow and black composite paper bag was decreased first, and reached the minimum at fifth days and then slightly increased. The content of chlorophyll a in the pouch fruit peel was higher than that of white single paper bag, and that was the lowest in the outer yellow inner black paper bag fruit in the different bagging treatment.

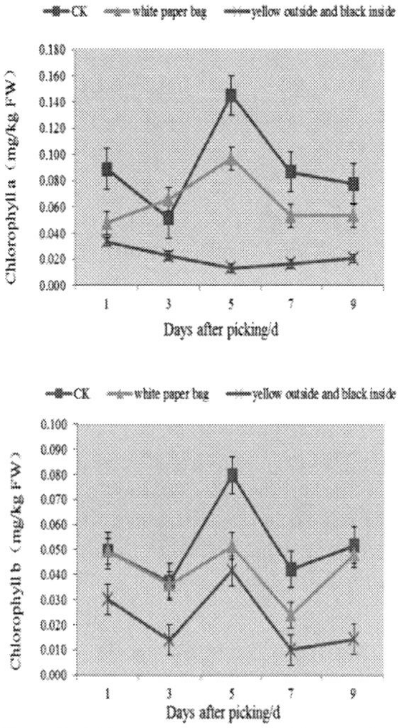

Fig. 1. Effect of Different Fruit Bags on chlorophyll Content in Peel of Chinhuang Mango in the process of aging

The overall trend of chlorophyll b content is reduced first and then reached the highest when it reached fifth days after Chinhuang mango was picked and unbagged. It fell to a minimum on the seventh days and increased slightly by the ninth day (Figure 1). After reaching the ripeness of the bagging treatment, the content of chlorophyll b in the control peel was higher than white single-layer paper bag, which was the lowest in outer yellow black composite paper bag.

2.2 Effect of Different Fruit Bags on Carotenoid Content in Peel of Chinhuang Mango

The overall trend of carotenoid content in control and white single-layer paper bags was consistent after the Chinhuang mango was picked and unbagged. It first rosed to the maximum on the fifth day of ripening, and then decreased slightly on the next seventh days and slightly increased on the ninth day. However, the content of carotenoids in the peels of the outer yellow inner black composite paper bag increased first, and then reached the lowest level on the fifth day, then reached the maximum on the next ninth days. Between different bagging treatments, the carotenoid content of unbagged fruit was slightly higher than that of the white single-layer paper bag, and the outer yellow inner black paper bag was the lowest (Figure 2).

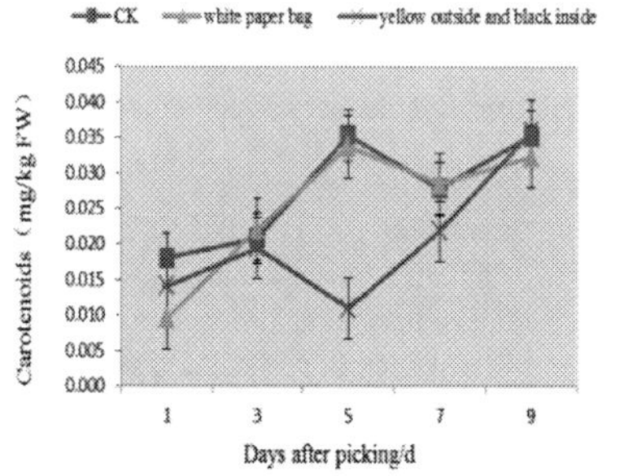

Fig. 2. Effect of Different Fruit Bags on Carotenoid Content in Peel of Chinhuang Mango in the process of aging

3 DISCUSSION

Bagging the outer yellow inner black composite paper bag was significantly reduced the content of chlorophyll and carotenoids of Chinhuang mango peel in its process of aging. The content of chlorophyll and carotenoid in the peel was significantly higher than that of the outer yellow and black composite paper bag, and the white single layer paper bag had great differences in color compared to the un-jacketed fruit.The shading rate of the outer yellow inner black paper bag was higher than that of the white single paper bag. The inhibition of the chlorophyll and carotenoid in the peel was more obvious than that of the white single paper bag and the un-jacketed fruit.

The skin color is the result of interaction of various related pigments [1]. The color of the peel is the comprehensive performance of chlorophyll, carotenoids, flavonoids, and anthocyanins. such as after using different types of bags can change partial light and temperature condition of the fruit development, effect the metabolism of skin pigment, so as to improve the color and quality of fruit[2]. A large number of studies have shown that bagging can promote the synthesis speed and accumulation of anthocyanins in fruit peel of apple at maturity, while reduces the content of chlorophyll in fruit peel, changes the color background of fruit, and makes the fruit colorful [3]. The content of the photosensitive pigment was increased after the bagging, and the chlorophyll content was decreased. The pear surface was colored rapidly after picking the bag. At the same time, the anthocyanin color background was improved, and the bagged pear was fresh red. In this study, the color change of the fruit of Chinhuang Mango without control was more obvious (Figure 3). The bagging obviously inhibited the synthesis of chlorophyll and carotenoid in the peel, but the change of anthocyanin content in the peel had to be further studied.

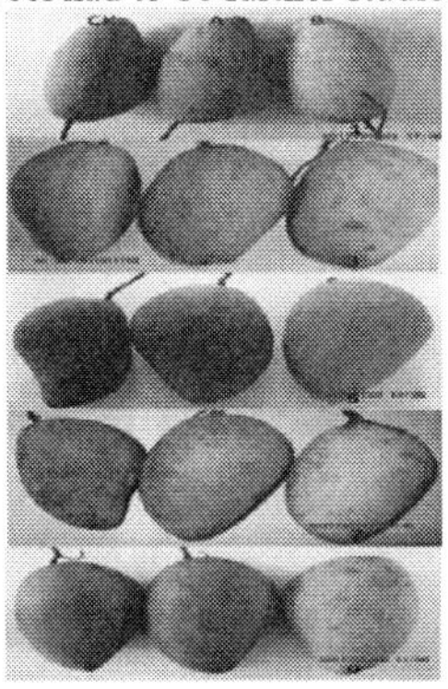

Fig. 3. The effect of different fruit bags on the pigmentation of the fruit peel in the ripening process of Chinhuang mango.

In addition, through observation, it was found that different bags can significantly improve the appearance of Chinhuang mango, and the skin has a high degree of s smooth finish. Because bagging avoids permanent scars on the peel and wound stream juice corrosion peel due to mechanical friction caused by external forces[4]. In addition, bagging can reduce the number of spraying pesticides, thus reducing production costs and producing pollution-free green fruits [5].

References

1. Zeng k F, Jiang M. Effects of bagging on the quality and post-harvest of mango fruit [J]. Food science, 2007(12):507-511.
2. Jiang J H, Tang S H. Comparative experimental study on different bagging materials of mangol in late mature [J]. Modern agricultural science and technology, 2015(06):80-81.
3. Li Y J, Xie D F. Research on the influence of bagging on mango nutrition [J]. Tropical agricultural science, 2015, 35(04):11-14.
4. Wu H X, Ma W H, wang pansong, luo deqiang. Effects of bagging on Irwin mango [J]. Journal of tropical crops, 2005(04):1-5.
5. Ma X W, Xing S S, Wang S B, Yao S, Zhan R L, Wu H X, Ma W H. Effects of Bagging on the Pericarpial Coloration of 'Irwin' Mango During Fruit Development[J]. Chinese Journal of Tropical Crops, 2011, 32(07): 1206-1208.

 https://doi.org/10.1051/e3sconf/20183802027

Tool Efficiency Analysis model research in SEMI industry

MA Lei*, ZHANG Nana, ZHANG Zhongqiu

Northwestern Polytechnical University Ming De College, Xi'an, China

Abstract: One of the key goals in SEMI industry is to improve equipment through put and ensure equipment production efficiency maximization. This paper is based on SEMI standards in semiconductor equipment control, defines the transaction rules between different tool states, and presents a TEA system model which is to analysis tool performance automatically based on finite state machine. The system was applied to fab tools and verified its effectiveness successfully, and obtained the parameter values used to measure the equipment performance, also including the advices of improvement.

1. Introduction

In semiconductor manufacturing industry, production tools are the most important part of the whole fab, usually they are very expensive and have higher human maintenance costs. So the biggest concern of semiconductor factory is how to improve fab tools performance, how to improve the productivity and yield, and how to reduce scrap.

Most of fab in global wide use automation application which are mainly focus on SPC and Run to Run control in recent years. Both of SPC and R2R are working during production process, they can monitor and forecast product quality by statistical analysis of recipe parameter. But from the perspective of equipment, how to analysis tools' performance and find out the reason why production efficiency is not high, how to realize tools' fullest utilization are still in an primary stage, so they could be a key point of semiconductor manufacturing industry's development direction in the future.

This paper is on the basis of SEMI international standard, presents an automation model to analysis production equipment's performance by finite state machine and verify its efficiency, finally analysis the operation result of this model and come to a conclusion.

2. Basic knowledge

2.1 Basic states of tool

Before analysis tool's performance and utilization, there should define tool states in each stage. In SEMI E10 standard, they defined 6 basic states for tools.

(1) Non-Scheduled State; (2) Unscheduled Down State; (3) Scheduled Down State;

(4) Engineering State; (5) Standby State; (6) Productive State.

Figure 1 and Figure 2 describe the six basic equipment states defined in SEMI E10.

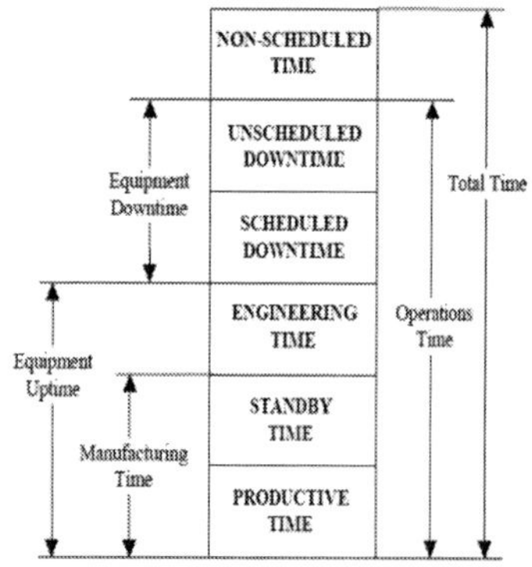

Figure 1. SEMI E10 Equipment States Stack Chart

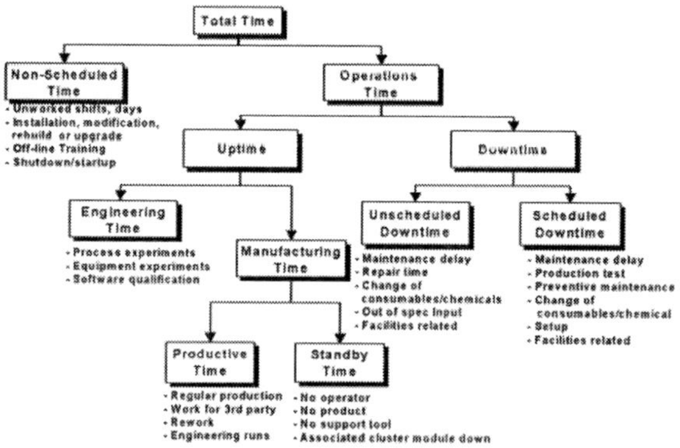

Figure2. SEMI E10 Summary of Time

*Corresponding Author. E-mail: 357023814@qq.com

2.2 Equipment productivity measurement

An OEE (Overall Equipment Effectiveness) concept is given in E79, the calculation method is listed below:

Overall Equipment Efficiency (OEE) = (Availability Efficiency) X (Performance Efficiency) X (Quality Efficiency).

Availability Efficiency = (Equipment Uptime) ÷ (Total Time);

Performance Efficiency = (Operational Efficiency) X (Rate Efficiency);

Operational Efficiency = (Production Time) ÷ (Equipment Uptime);

Rate Efficiency = (Theoretical Production Time for Actual Units) ÷ (Production Time);

Theoretical Production Time for Actual Units =I (Actual Units of Recipe I x THTI);

Theoretical Production Time for Effective Units =I (Effective Units of Recipe I x THTI);

THTI = theoretical production time per unit of recipeI;

Quality Efficiency = (Theoretical Production Time for Effective Units)÷ (Theoretical Production Time for Actual Units);

In order to calculate tool OEE value accurately to judge its performance, TEA (Tool Efficiency Analysis) model is presented in the following paper. The advantages of this model are to calculate each parameter of tool performance and find out the reason which affect tool performance through data analysis.

3. Model introduction

3.1 TEA model framework

TEA model includes ADC (Automation Data Collection) and DA (Data Analysis) two parts.

(1) Automation Data Collection Framework

ADC is divided into three modules: event trigger, event mapping module, and states change engine, as shown in figure 3.

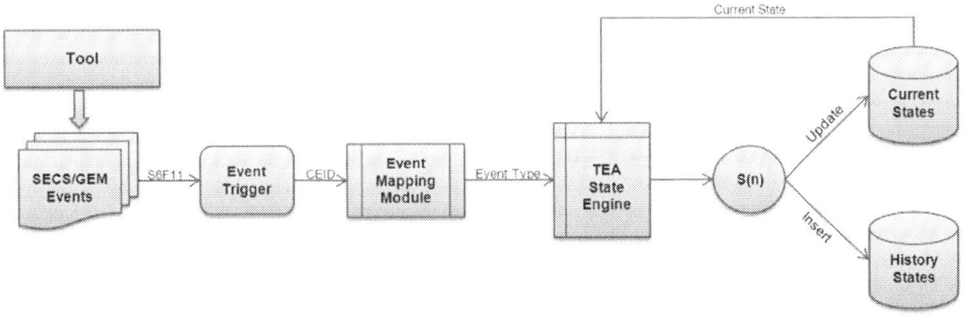

Figure 3. TEA data collection framework

Event Trigger is responsible for capturing S6F11 message reported by tool and retrieving collection event ID (CEID) value. SEMI communication standard uses SECS/GEM, which defined the communication rules between tools and host, tools and tools. S6F11 is one of message format defined in SECS/GEM, it has been used for sending event from tool to host.

Event Mapping Module is responsible for relationship mapping between event CEID captured by event trigger and pre-defined event in TEA system and return this captured event's type. The relationship mapping table is based on event definition in tool manual, mapping example is like below:

Table 1. TEA event mapping example

CEID	Event Type
500	Material Placed
1000	Wafer Start
1001	Wafer End
600	Material Removed
....	....

TEA State Engine defined tools' all states, including each parent state and sub state. It records tool current state according to the event type, and changes tool current state when certain conditions are met and saved the result into database. The detail is below:

Define a finite state machine $A = (Q, \sum, \delta, s, F)$:

- Q is non-empty finite state set, including all sub state, as shown in table 2;
- $\sum$ is non-empty finite character set, which defined all the event type;
- δ is transition function, $\delta\ (S_c\ ,\ \sigma) = S_d, (S_c\ , S_d \epsilon\ Q\ ,\ \sigma \epsilon \sum)$;
- s is initial state, $s\ \epsilon\ Q$;
- F is accepted state set, $F \leq Q$.

S_d is destination state, S_c is current state, their mapping relationship when certain event type $\sigma(\sigma \epsilon \sum)$received is defined in δ transition function. As shown in table 3.

 https://doi.org/10.1051/e3sconf/20183802027

Table 2. State of the machine definition

Parent State	Sub State
Productive	Productive/Processing
	Productive/Processing/Step1
	Productive/Processing/Step2
	....
Standby	Standby/Door_Opened
	Standby/Material_Loaded
	Standby/ReadyForUnload
	....
Engineering	Engineering
Scheduled_Down	Scheduled Down Time
Unscheduled_Down	Unscheduled Down Time
Non_Scheduled	Non-Scheduled Time

Table 3. State transition rules

Current State	Destination State	Event Type
Standby	Productive/Processing	Wafer Start
Productive/Processing	Productive/Processing/Step1	Step Change
Productive/Processing	Standby	Wafer End
......	......	......

Data Analysis Framework

Data Analysis is divided into two modules: DSM (Data Statistics Module) and DCM (Data Calculation Module). Its main function is to analysis data retrieved by data collector, and calculates the OEE value. As shown in Figure 4.

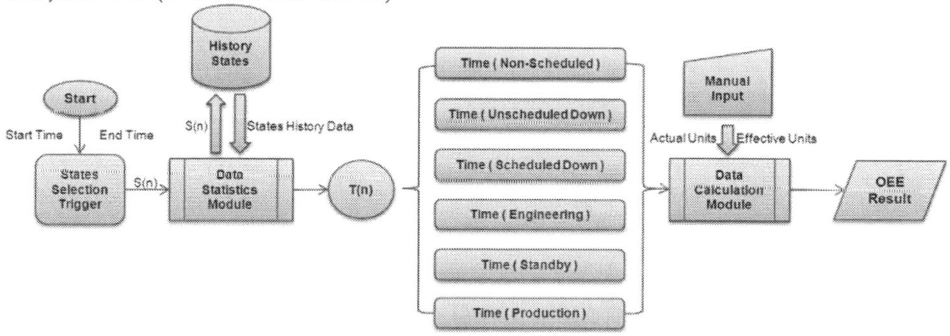

Figure4. Data Analysis Framework

Data Analysis model filters valid data based on time range selected by user, and calculates six states' time separately in this time range. The data source is in History States table in Data Statistics Module. Data Calculation

Module calculates OEE value based on production time per unit and total time which according to the six states time, actual units and effective units.

4. Experiment, Example calculations

The calculations in this section are based on a seven-day period sample dataretrieved by Data Collection module, and calculate the following states data by Data Analysis module:

- Non-Scheduled Time 0.00 hours
- Unscheduled Downtime 4.00 hours
- Scheduled Downtime 8.00 hours
- Engineering Time 3.00 hours
- Standby Time 6.00 hours
- Production Time 147.00 hours
- Total Time 168.00 hours

Recipe	Theoretical Production Time (hr/wafer)	Theoretical Production Number(wafer/ hr)	Actual ProductionNumber	Effective Production Number
A	0.03333	30	1420	1400
B	0.04000	25	600	600
C	0.05000	20	800	800
D	0.06667	15	500	480

Data Calculation Module calculates like below:

- *Equipment Uptime*
  = (Production Time) + (Standby Time) + (Engineering Time)
  = (147.00 hours) + (6.00 hours) + (3.00 hours)
  = 156.00 hours
- *Production Time (given)*
  Theoretical Production Time for Actual Units
  = i *(Actual Units of Recipe i X THTi)*
  = [(1420 units X 0.03333 hr/unit)
  + (600 units X 0.04000 hr/unit)
  + (800 units X 0.05000 hr/unit)
  + (500 units X 0 .06667 hr/unit)]
  = 144.66 hours
- *Theoretical Production Time for Effective Units*
  = i *(Effective Units of Recipe i X THTi)*
  = [(1400 units X 0.03333 hr/unit)
  + (600 units X 0.04000 hr/unit)
  + (800 units X 0.05000 hr/unit)
  + (480 units X 0.06667 hr/unit)]
  = 142.67 hours
- *Availability Efficiency*
  = (Equipment Uptime) ÷ (Total Time)
  = (156.00 hours) ÷ (168.00 hours)
  = 0.9286
- *Operational Efficiency*
  = (Production Time) ÷ (Equipment Uptime)
  = (147.00 hours) ÷ (156.00 hours)
  = 0.9423
- *Rate Efficiency*
  = (Theoretical Production Time for Actual Units)
  ÷ (Production Time)
  = (144.66 hours) ÷ (147.00 hours)
  = 0.9840
- *Performance Efficiency*
  = (Operational Efficiency) x (Rate Efficiency)
  = (0.9423) X (0.9840)
  = 0.9272
- *Quality Efficiency*
  = (Theoretical Production Time for Effective Units)
  ÷ (Theoretical Production Time for Actual Units)
  = (142.67 hours) ÷ (144.66 hours)
  = 0.9862

- *Overall Equipment Efficiency (OEE)*

  = (Theoretical Production Time for Effective Units)

  ÷ (Total Time)

  = (142.67 hours) ÷ (168.00 hours)

  = 0.8492

OEE is the key indicator of measuring tool performance. According to the OEE value fab customer can find out how to improve tool efficiency and performance, as shown in Figure 5. This experiment calculates the OEE value of the test tool and verify the validation of the TEA model finally.

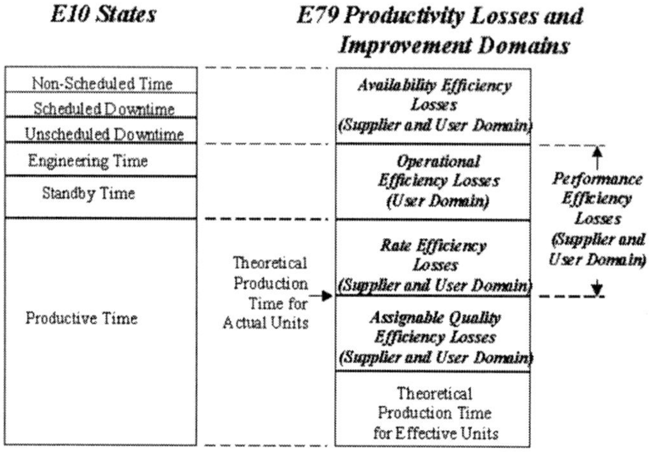

Figure 5. E10 and E79 mapping relationship

5. Conclusion

In recent years, the semiconductor manufacturing industry has high development, which makes more demands to the automation software technology. Currently, much more related researches are focused on improvements of manufacturing process. This paper discusses production equipment itself, presents TEA (Tool Efficiency Analysis) model which is based on SEMI standard, and also considers equipment performance and efficiency, finally verifies its effectiveness. This model would be useful to monitor tool behavior and analyzed tool performance automatically.

Reference

1. SEMI Standards: E10, E79. .
2. Generic Model For Communications and Control Of Manufacturing Equipment(GEM). SEMI E30-1000 .
3. Stability analysis of double EWMA run-to-run control with metrology delay. Good R,Qin S J. Proc. of American Control Conference . 2002
4. "FACTORYworks System Architecture Overview"Version2.5. Brooks Automation. . 2002
5. 《SEMI Equipment Communication Standard 1 Message Transfer》. Global Information and Control Committee. . 1999
6. 《High-Speed SECS Message Services (HSMS) Generic Services》. Global Information and Control Committee. . 2002
7. 《SEMI Equipment Communication Standard 2 Message Content》. Global Information and Control Committee. . 2000
8. A Drumea C. Popescu "Finite state machines and their applications in software for industrial control[C]" ,Electronics Technology: Meeting the Challenges of Electronics Technology Progress 2004. 27th International Spring Seminar on, vol. 1 pp. 25-29 2004.
9. V. Estivill-Castro R. Hexel "Logic labelled finite-state machines and control/status pull technology for model-driven engineering of robotic behaviours" ,Proc. 26th InternationalConference on Software & Systems Engineering and their Applications, May 2015.
10. D. Bresolin, K. El-Fakih, T. Villa, N. Yevtushenko, "Deterministic Timed Finite State Machines: Equivalence Checking and Expressive Power", Proceedings of the 7th InternationalSymposium on Games Automata Logics and Formal Verification, pp. 203-216, 2014.

Experimental study on flame propagation characteristics of Hydrogen premixed gas in gas pipeline

Danzhu Ma[1*], *Zhuang Li*[1], *Fengrui Jia*[1], *Zhou Li*[1]

[1]Liaoning Shihua University, Fushun, Liaoning 113001, China

Abstract. Hydrogen is the cleanest high-energy gas fuel, and also is the main industrial material. However, hydrogen is more explosive and more powerful than conventional gas fuels, which restricts its application. In particular, the expansion of premixed combustion under a strong constraint is more complicated, the reaction spreads faster. The flame propagation characteristics of premixed hydrogen/air were investigated by experiment. The mechanism of reaction acceleration is discussed, and then the speed of the flame propagation and the reaction pressure were tested and analysed.

1 Introduction

Flammable gas refers to the gas that can cause combustion or explosion and release a large amount of energy under the ignition energy excitation within the environment full of combustion improver. The flammable gases play a very important role in the fields of petrochemical industry, metallurgical industry and gas production and supply industry, such as methane and hydrogen. By far, hydrogen is the cleanest high-energy gas fuel, and also is the main industrial material. However, hydrogen is more explosive and more powerful than conventional gas fuels due to the lower ignition energy, the wider explosion limitation and the larger laminar burning and diffusion velocity. The hydrogen explosion accidents occur frequently. The research on the combustion and flame propagation characteristics of hydrogen premixed system has a great practical significance. The propagation of gas explosion wave, the kinetic theory of laminar flame and the constant flame theory were discussed in the early 1950s [1-2]. The propagations of gas explosion wave in the closed system were also focused as follow [3-6]. But the investigations about the hydrogen premixed system are not enough. It has very important practical significance and is the key of whether hydrogen can be used as a conventional energy.

In this paper, the deflagrating characteristic of premixed hydrogen was concerned, based on the theory of flame dynamics and pressure change. The speed of flame propagation and the pressure fluctuation were studied and analysed.

2 Hydrogen premixed combustion and its mechanism in the constrained space

The combustion and explosion of premixed gas needs the conditions, such as the suitable flammable gas concentration, enough oxygen and ignition source. It is a process of chemical reaction flow. The typical process of premixed flame propagation in the constrained space includes the ignition, the laminar combustion, the fire perturbation, the transformation from laminar to turbulent combustion and the final detonation. At the beginning of the reaction, the laminar flame is unstable. The fold appears caused by the instability of the thermal diffusion perturbation. The vortex generated to increase the disturbance of the flame front and to prompt the turbulent combustion, when the explosion space is constrained or there are obstacles. Then, if the flame continues to accelerate, the explosion occurs and even transforms to detonation. Figure 1 gives a sketch of the propagation process of the reaction.

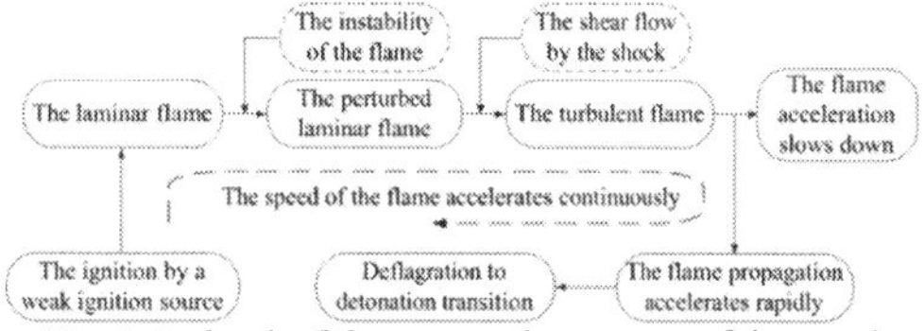

Fig. 1. A sketch of the propagation process of the reaction.

The following combustion chemical reaction occurs, when premixed hydrogen is ignited.

$$H_2+0.5O_2+1.88N_2 \rightarrow H_2O+1.88N_2$$

In this paper, four different hydrogen concentrations were considered for test, they are 13.5%, 18%, 24%, 28.5% and 38.5%.

3 Experimental system

The experimental setup is shown as in figure 2, which consists of an automatically gas supply system, a high explosive pressure test system, a high-speed photography system, a high-voltage ignition system and a pressure relief channel. The simulated pipe used in the

* Corresponding author: danzhuma@163.com

experiment was made by polymathic methacrylate tube. The inner diameter of the tube was 108 mm, with a uniform length of 2 m. Two relief channels are set on the side of the pipeline to ensure the safety. The ignitions were charged by high voltage spark discharge. The reaction pressures and the flame propagation were recorded by the pressure sensors and high-speed camera.

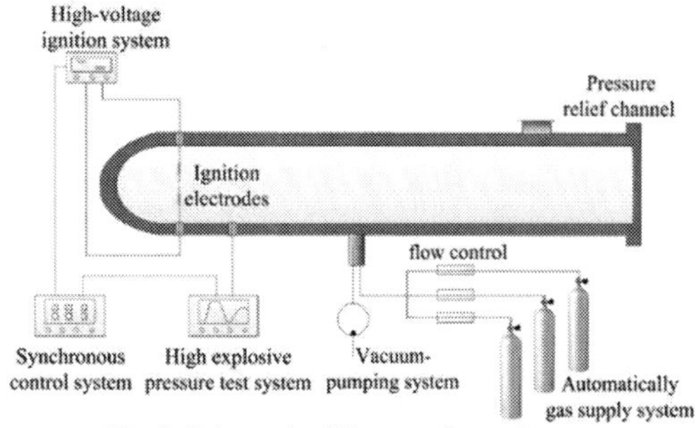

Fig. 2. Schematic of the experimental setup.

4 Results and discussion

4.1 The speed of flame propagation of premixed hydrogen in the pipeline

The reaction wave fronts were recorded by the high-speed camera, and then the speed of the flame propagation could be calculated by the image processing method. The position of flame front is calculated and the speed of propagation is the ratio between the distance and the time difference between two positions, as shown in figure 3. In this paper, the speed of flame propagation under different hydrogen volume fraction is calculated by using the above method.

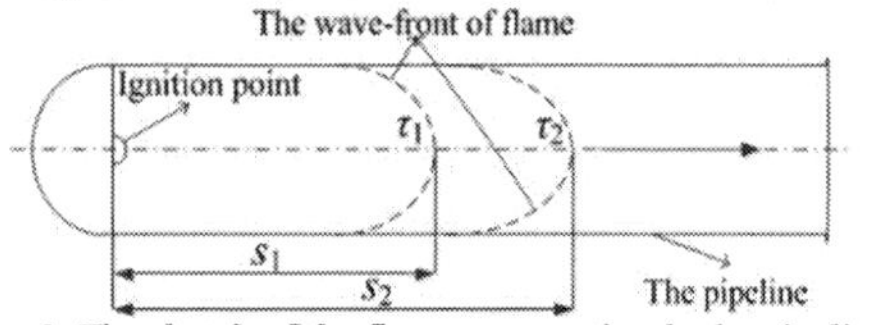

Fig. 3. The sketch of the flame propagation in the pipeline.

The effect of different hydrogen volume fractions on the speed of flame propagation was discussed. The results are shown in figure. 4. The propagation speed of the flame in the pipe increased with the flame front propelling, and then decreased after the peak for $X_{H_2} = 13.5\%$ and $X_{H_2} = 18\%$. The speed of flame propagation decreased with the irregular flame front. But the decrease of the speed was not tested in the higher hydrogen content.

The peak speed of flame propagation and the location of the peak are all related to the hydrogen concentration. The higher the concentration of hydrogen, the more violent the reaction is, the faster the flame spread. The propagation speed of the premixed gas in the restricted pipeline is closely related to the shape of the flame front. In the early stage, the flame is not restrained by the wall, it is semi-spherical and rapidly evolves into a fingertip flame, the surface area of the flame is accelerated by exponential growth. When the flame comes into contact with the pipe wall, the acceleration is terminated into the plane flame phase, the propagation speed of the flame front is beginning to decline. Then the flame forward reverses, gradually forming the tulip flame, which increases the surface area of the flame, the flame forward speed increases again. It can be seen from figure 4 that the tulip flame appears at $X_{H_2} = 18\%$. But, at the higher hydrogen concentration, the premixed flame is in the accelerated propagation period, there is no feature of tulip flame in the length of the experimental pipeline.

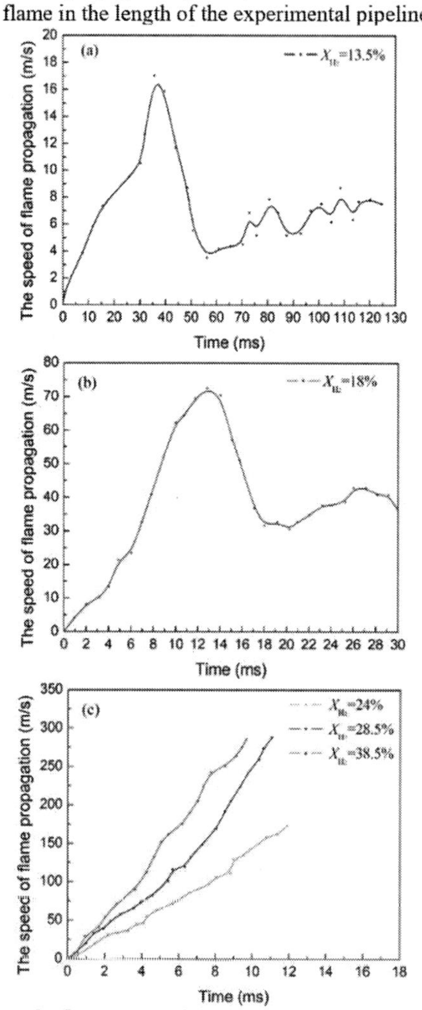

Fig. 4. The flame propagation speed curves of pipelines with different hydrogen concentration.

4.2 The reaction pressure

The reaction pressure was recorded by the pressure sensor, which was lie in the front of the pipe. The typical curves were shown in figure 5. As shown in the figure, the reaction pressure of premixed hydrogen-air has obvious pressure oscillation and appears negative pressure. At the beginning of the reaction, the pressure increases rapidly with the reaction progresses. Then the film on the other side of the pipeline breaks, the pipe becomes to a semi-enclosed space, part of the premixed

https://doi.org/10.1051/e3sconf/20183802028

gas washes out of the pipe. At the same time, the gas in the pipeline continues to participate in the reaction and consumed, the pressure drops in the pipe, resulting the negative pressure. The negative pressure in the tube causes the ejected gas to be sucked back into the reaction, causing the pressure rise again, thus causing the pressure oscillation. With the constant consumption of unburned gas in the tube, the absolute value of the pressure oscillation decreases and finally returns to the initial value.

The results of the pressure test show that the pressure propagation of hydrogen-air in the pipeline is divided into three stages. (1) After the input of the ignition energy, the premixed gas is ignited, the hydrogen-air continue to react, the pressure in the pipe continuously accumulates to reach the first crest value. (2) Due to the high temperature of the premixed combustion, when the water vapor which generated by the reaction contact the lower temperature environment to condense into droplets. This process causes the negative pressure in the pipe. And then, the unreacted hydrogen-air are sucked back to the pipe to join in the reaction, the pressure goes up again. This causes the pressure shock in the pipeline. (3) The pressure amplitude reduces to the initial value with the complete reaction of the premixed gas. The peak pressure and the intensity of the pressure oscillation increase with the increasing of hydrogen content, while the time of peak pressure decrease.

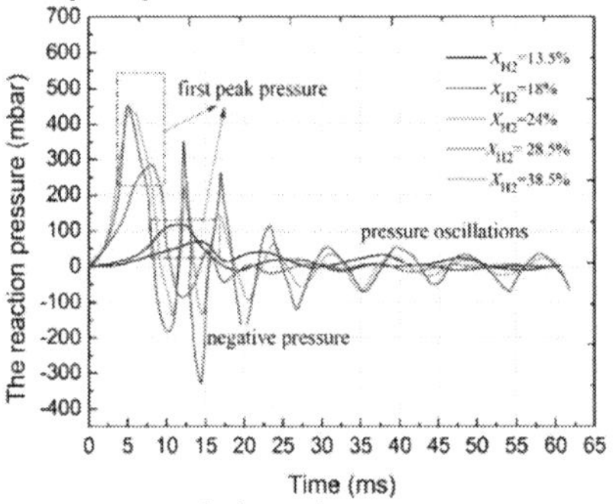

Fig. 5. Typical reaction pressure curve.

5 Summary

The combustion characteristics of premixed hydrogen is the foundation of its safe application as a high-energy fuel. The experiment results show that the peak speed of flame propagation and the location of the peak depend on the concentration of hydrogen. The higher hydrogen concentration, the more violent the reaction is, and the faster the flame spread. The reaction pressure develops through three stages: the pressure rises to the first peak; the pressure continues to oscillate and the pressure reduces to the initial value.

Acknowledgements

This work was financially supported by the Scientific Research Project of the Department of Education of Liaoning Province (L2015303) and the Breeding program of Liaoning Shihua University (2016[177]-19).

References

1. W. Bartknecht, *Explosions: Course, Prevention, Protection* (Berlin, Springer, 1981)
2. D. Bjerketvedt, J. R. Bakke, K. Wingerden, *Gas Explosion Handbook* (Journal of Hazardous Materials, 1997)
3. K. K. Kuo, *Principles of Combustion* (Wiley-Interscience, Switzerland, 2005)
4. Mustafa Ilbas, Serhat Karyeyen, Ismail Ozdemir, Int. J Hydrogen energy, E **26**, 1 (2015)
5. Xianzhong Hu, Qingbo Yu, Nan Sun, Qin Qin, Int. J Hydrogen energy, E **39**, 9527 (2014)
6. Hongfa Huo, Xingjian Wang, Vigor Yang, Combustion and Flame, E **161**, 3040 (2014)

Research on Closed Residential Area Based on Balanced Distribution Theory

Si LAN, Ni FANG, Hai Peng LIN, Shi Qi YE

School of Information Science and Technology, Jinan University, Guangzhou, Guangdong 510000, China

Abstract: With the promotion of the street system, residential quarters and units of the compound gradually open. In this paper, the relationship between traffic flow and traffic flow is established for external roads, and the road resistance model is established by internal roads. We propose a balanced distribution model from the two aspects of road opening conditions and traffic flow inside and outside the district, and quantitatively analyze the impact of the opening and closing on the surrounding roads. Finally, it puts forward feasible suggestions to improve the traffic situation and optimize the network structure.

1 Introduction

February 21st, 2016, the State Council issues the "opinions" on further strengthening the administration of city planning and construction, saying views aroused widespread concern and discussion, in principle no longer closing residential construction, opening built-residential area and the unit compound gradually.

Although the open area may lead to security issues, the focus of the discussion is whether open area can achieve the optimization of network structure, improve the road capacity and traffic conditions. One view is that the enclosed residential cell destroys the urban road network structure, which is easy to cause traffic jams while the open cell will reduce the density of the road network.

2 Model establishing

This paper analyzes the influence of open area on the surrounding road traffic. The mathematical model of vehicle traffic includes the opening condition of the road and the balance to reach the driving time of the traffic inside and outside the residential cell.

2.1. Cost - minimized Goal Planning Model

It is indicated that the parameter k is the total number of openings in the residential cell while K is the maximum acceptable number, which satisfies the condition $k \le K$. Because there is a certain amount of expenditure to be used for every opening, we should make the number of openings as few as possible.

Regarding the cost of per road section as the same, the total length of all sections which will be built should be as short as possible in terms of economy. Therefore, the goal is

$$\min \sum_{i=1}^{k} \sum_{j=i+1}^{k} L_{ij} \tag{1}$$

In which L_{ij} represents the length of the road from the ith opening to the jth opening of the residential cell.

What's more, when the judging condition $Q_{\text{out}} \ge Q_{\max}$ is true, in which Q_{out} represents the traffic outside the residential cell and Q_{out} represents the total carrying capacity of the lane, it is necessary to increase the number of openings to realize the diversion of vehicles.

2.2 Time - balanced Distribution Model of Road Network

Traffic assignment is to allocate the traffic demand to the traffic network according to a certain routing principle, so the traffic flow is got. If the time outside the residential cell is less than that through it, it is of little significance to open the cell. On the contrary, if most of the vehicles have chosen to pass through the cell after opening, then the cell will be in congestion and the traffic flow will be diverted into the outer road. Finally, all vehicles are randomly selecting paths, which can achieve the equilibrium state. At this time, the time spent on the road either inside or outside the cell is same.

According to the principle of time balance, a traffic assignment model of road network is established

*Corresponding author: Email: ojwlg@jnu.edu.cn

$$\frac{x_{ij}}{v_1}=\frac{y_{ij}}{v_2} \quad (2)$$

In which x_{ij} means the length of the road outside from i^{th} cell to j^{th} cell; y_{ij} means the length of the road inside; v_1 represents the driving speed outside the cell while v_2 represents the speed inside, both of which will be obtained in the following models.

2.3 Speed-flow Relationship Model of Traffic Flow

The traffic capacity of the external road is determined by the flow of the road. According to Greenshields theory, we can establish the speed-flow relationship model, as shown in figure 1.

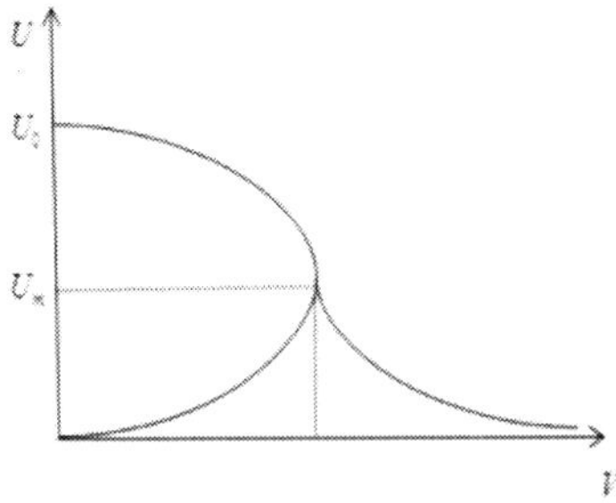

Figure 1 Schematic Diagram of Greenshields Theory

The average speed of the vehicle arriving in the unit time is

$$v_1=U_0/(1+\frac{V}{C}) \quad (3)$$

In which U_0 represents the average speed of vehicles when the traffic volume is zero and $\frac{V}{C}$ is a ratio about the actual traffic volume and the capacity of the road. In addition, according to the relationship between the whole and the part, we can get the following equation

$$V/C=\sum V_i L_i/\sum C_i L_i \quad (4)$$

In which V_i means the distribution of traffic in section i of the road network, C_i means the traffic capacity of the section i and L_i means the length of section i.

However, the formula is obtained under the assumption that the vehicle uniformly dissipates, so two correction coefficients α and β are quoted to obtain the traffic impedance model

$$v_1=\frac{U_0}{1+\alpha\left(\frac{V}{C}\right)^{\beta}} \quad (5)$$

To obtain a more accurate prediction result, we make $V/C<0.8$ and $\gamma=1$. As a result, we get

$$v_1=\alpha_0 U_s\left(1+\gamma\left(\frac{V}{C}\right)^{\beta}\right)^{-1} \quad (6)$$

In the model, U_s is modified into the design speed and β is a nonlinear function to V/C, which is expressed as

$$\beta=\alpha_2+\alpha_3\left(\frac{V}{C}\right)^3 \quad (7)$$

Therefore, the speed-flow relationship model at any arbitrary level and traffic load is

$$\begin{cases} v_1=\dfrac{\alpha_1 U_s}{1+\left(\frac{V}{C}\right)^{\beta}} \\ \beta=\alpha_2+\alpha_3\left(\frac{V}{C}\right)^3 \end{cases} \quad (8)$$

2.4 Road Resistance Function Model

According to the traffic flow theory, the relationship $Q_{in}=\rho\cdot v$ will be obtained when the road facilities are in certain circumstances, in which Q_{in} is traffic flow inside the cell, v is driving speed and ρ is traffic density.

Under the assumption of continuous flow in the absence of interference, the model of road resistance function is

$$\begin{cases} t=\dfrac{L}{v} \\ v=\dfrac{v_m}{2}+\sqrt{\left(\dfrac{v_m}{2}\right)^2-\dfrac{Q_{in}v_m}{1000\rho_m}} \end{cases} \quad (9)$$

In which L, v_m, ρ_m mean the length of the road, the smooth speed and the blocking density respectively.

In specific practical applications, the road traffic is usually intermittent, so we fix the processing model from four aspects of the interference, including intersection spacing, lane width, non-motor vehicles and pedestrians.

So we learned that $v_m=\mu_1\mu_2\mu_3\mu_4 U_s$, in which μ_1, μ_2, μ_3, μ_4 are four correction coefficients of intersection of spacing influence, lane width influence, bike impact, pedestrian interference respectively.

Set the number of lanes to n, the average body length to l, the average block from the car to l_0. As the blocking density meets the condition $\rho_m=\frac{n}{l+l_0}$, the corrected one can be transferred into

$$\rho_m=\frac{\mu_1\mu_2\mu_3\mu_4 n}{l+l_0} \quad (10)$$

According to the speed-flow relationship model and road resistance function model based on the traffic flow theory, we get the traffic balance inside and outside the district formula

$$Q_{in} = Q_{out} \tag{11}$$

3 A Comparison of Roads in Two Exact Opening Conditions

According to the classification of the cell 5.3.1, two types of cells closure are mainly discussed.

Figure 2 shows one residential cell whose opening level is not high in a medium-sized city. Assuming the external trunk road is two-way four-lane, it is known that the road designed speed is 40km / h; capacity is 900pcu / h; width of a lane is 3.75m. As for the cell we design, we make the two-lane road speed 40km / h, capacity 900pcu / h, width of a lane 3.75m.

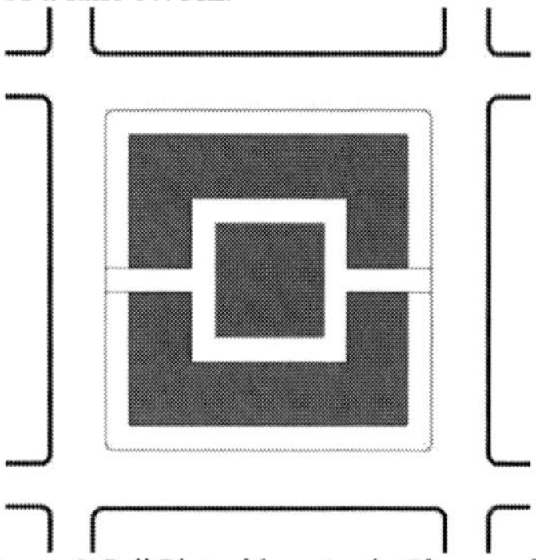

Figure 2 Cell Plot with an Angle Closure of 2

Base on the actual data of a residential cell in Changsha, the plot covers an area of a square, side length 500m, area 250000, the resident population of about 15000 people. Two import and export corresponding to the external road traffic is 2000pcu/h while the other two external roads' traffic is 800pcu/h. The two import and export flow data are shown in Table 1:

Table 1 Inlet and Outlet Flow Meter

Models	Outlet (pcu/h)	Inlet (pcu/h)
Non-Motor Vehicle	136	66
Bicycle	224	59
Pedestrian	920	360

Substituting data above into the model with MATLAB, and the result shows below:

$v_1 \approx 36.67\text{km}/\text{h}$, $v_2 \approx 27.40\text{km}/\text{h}$, $V = 1142.86\text{pcu}/\text{h}$, $\Delta Q = 857.14\text{pcu}/\text{h}$, $U_1 = 0.9$, $\beta = 13.34$, $V_m = 28.35\text{km}/\text{h}$, $\rho_m = 0.1357$

We can see that after opening the cell, the density of the trunk road network increases and the external roads of the cell share the traffic flow of 857pcu / h for the external ones, making the internal and external road load are about 1.2. In terms of speed, the open cell reduces the degree of peripheral lane congestion and improves the traffic speed. Although the speed of the internal lane is limited by the crowd and the width of the road, it reaches the requirements of the equilibrium theory in general, the speed of internal and external traffic can make the entire road network inside and outside keep relative balance, so it is reasonable and effective under the conditions of such a cell.

Figure 3 shows one cell whose opening level is high in a general city. Assuming the external trunk road is two-way six-lane, access to information that the road designs speed of 60km / h, capacity of 1100pcu / h, width of a lane of 3.75m. As for the cell we design, we make it the two-way four-lane road speed of 40km / h, capacity of 900pcu / h, width of a lane of 3.75m.

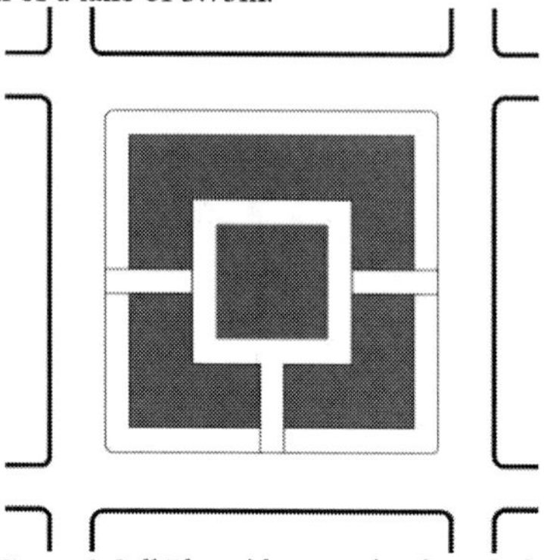

Figure 3 Cell Plot with an Angle Closure of 3

Substituting data above into the model with MATLAB, and the result shows below:

$v_1 \approx 46.31\text{km}/\text{h}$, $v_2 \approx 34.52\text{km}/\text{h}$, $V = 1152.32\text{pcu}/\text{h}$, $\Delta Q = 1047.38\text{pcu}/\text{h}$, $U_1 = 0.863$, $\beta = 17.99$, $V_m = 35.76\text{km}/\text{h}$, $\rho_m = 0.1987$

Base on the actual data of a cell in Changsha, the plot covers an area of a square, side length 600m, area 360000 m^2, the resident population of about 20000 people. Three import and export corresponding to the main external road traffic is 2500pcu/h while the other road's traffic is 1200pcu/h. The data about three import and export flow are shown in Table 2:

Table 2 Inlet and Outlet Flow Meter

Models	Outlet (pcu/h)	Inlet (pcu/h)
Non-Motor Vehicle	164	85
Bicycle	240	67
Pedestrian	980	390

We can see that after opening the cell, the density of the trunk road network increases and the external roads of the cell share the traffic flow of 1047pcu / h for the external ones, making the internal and external road load are about 1.3 and 1.1 respectively. In terms of speed, the internal and speed almost reaches the requirements after the cell opens. The degree of congestion in the outer lane has alleviated and the traffic speed has increased, but still far less than the road design speed. Therefore, the external road is still congested and cannot make the road network balanced. That is to say, such an open residential cell settings cannot meet the requirements of the road network.

Compared the two types of cell models, we can conclude that the first one is a cell model whose opening degree is relatively low, roads within the cell are narrow, pedestrians are more and traffic capacity is poor, which is suitable for the condition that the traffic density is not too high; the second one is a cell model whose opening degree is relatively high, roads within the cell are wide and traffic density is not too large. However, the second cell model of the blocking density $\rho_m = 0.1987 > 0.1357$, which shows the role of road traffic is not as well as the first category. Therefore, the selection of the cell model needs to be controlled on the external traffic flow within a certain range.

After analyzing the changes of different types of residential cells in different conditions, it is concluded that when the external traffic density does not exceed the limit, the effect of the open cell is positively correlated with the width of the road and the density of the road while it is negatively correlated with the population density and the number of non-motor vehicles. The open residential area can reduce the road load and blocking density, playing a role of sharing traffic flow, improving the road capacity and the traffic congestion.

4 Conclusions

When the external traffic density does not exceed the limit, the effect of the open cell is positively correlated with the road width and the trunk density while it is negatively correlated with the population density and the number of non - motor vehicles. Open residential area can reduce the road load and blocking density, share traffic flow, increase the road capacity and improve the traffic congestion.

References

1. Zhu F L. Research on Index System of Urban Traffic Congestion Measures[D]. Southeast University, 2006: 14-15.
2. Liu D, Lu F, Ma S Q. Screening and Determination of Evaluation Indexes for Comprehensive Evaluation System of Road Traffic Safety[J]. Journal of Chinese People's Public Security University, 2005, 1: 42.
3. Yang G J, Zhao C Q, Zhang D H. Discussion on Evaluation Method of Urban Road Network Structure[J]. Journal of Zibo University, 2002, 4(1): 43-46.
4. Hu C, Zou Z Y, Mei Y N. Study on Evaluation Index System of Urban Road Network Planning[J]. Journal of Huazhong University of Science and Technology, 2006, 23: 98-99.
5. Liu M. Urban Transportation Continuous Equilibrium Network Design Based on Travel time Budget[D]. Southwest Jiao tong University, 2006: 44.
6. Wang W. Practical Speed-flow Relationship Model of Highway Traffic-flow[J]. Journal of Southwest University, 2003, 33(4): 489-490.
7. Chang Y L, Wang X L, Zhang P. Speed-flow Relationship Model Based on the Upstream Traffic Volume of the Cosine Function Road[J]. Science Technology and Engineering, 2016, 16(11): 258.
8. Wang W, Zhang G H. Study on Road Resistance Function of Urban Road[J]. Journal of Southwest University for Nationalities, 1992, 11(3): 85-91.
9. Huang Y, Zhou W, Li J. Equilibrium of Traffic Flow in Transportation Network[J]. Journal of Southwest University for Nationalities, 2006, 32(2): 213-241.
10. Li X P. Urban Traffic Congestion Countermeasures-Opening Research of Enclosed Type District Traffic[D]. Henan University, 2014, 32(2): 46.
11. Qin G. Study of the Compatibility between Enclosed Residential Areas and Developing Transportation in Urban Areas[D]. Chang'an University Press, 2010: 51.
12. Gao Y. Research on Optimization Strategy of Residential Street Space Based on Traffic Microcirculation--for Example in Chengdu[J]. Southwest Jiao tong University, 2015: 20-21.

Theoretical Analysis for Local Buckling of Corrugated Steel Plate

Bai Jian LI, Liang Sheng ZHU, Xin Sha FU

School of Civil Engineering and Transportation, South China University of Technology, Guangzhou, Guangdong, 510640, China

Abstract: To study local buckling of Corrugated Steel Plate under concentrated loads. Through experimental tests and theorical analysis, bearing capacity and failure form of Corrugated Steel Plate were discussed. Bearing capacity of Corrugated Steel Plate associated with local buckling, which can be assumed to be composed of three parts: buckling of plane rigid frame caused by concentrated loads, buckling of roof and web caused by bending stress. These three parts were unified by buckling relevant equations, then local buckling calculation formula was obtained. Comparing with experimental results, the loads obtained by local buckling calculation formula agree with test results very well. Since the buckling calculation is independent of the material strength, the calculation formula of local buckling is reliable, it can be used to evaluate local buckling of Corrugated Steel Plate.

1 Introduction

Corrugated Steel Plate (CSP) is pressed or bent corrugated by steel plate, which has high bearing capacity and high stability, has been widely used in bridge and culvert engineering [1-3] and PC box girder with corrugated steel webs [4-11]. In recent years, researchers have developed a new type of structure on the basis of the application of steel-concrete composite deck called corrugated steel-concrete composite bridge deck. This type of structure is first found in a bridge reinforcement project located in Pingnan County, Xunjiang River. The corrugated steel bridge deck was used to improve the shear bearing capacity [12]. Experiments of the corrugated steel-concrete composite bridge deck by Prof. Xu Haiyan and his team obtained load-displacement curves and stress-strain state, which indicated that the structure had characteristics of high bearing capacity and good ductility. Bearing capacity could be calculated by composite structure theory [13]. Prof. Su Qingtian also conducted laboratory experiments, confirming that the corrugated steel-concrete composite bridge deck had the characteristics of light weight, high capacity and high cost; compared with the steel deck, it had the characteristics of low cost, easy to pave, high fatigue strength and high weight; compared with the ordinary steel plate-concrete deck, it had the characteristics of low weight, high shear strength. It is a feasible option in the bridge deck of medium span bridges [14-15].

In summary, the application of corrugated steel plate in multiple fields was studied deeply, but there were more problems need to be solved, such as local stability of the corrugated steel-concrete composite deck, which affects the mechanical properties of composite deck during construction, the present study mainly focus on the shear buckling performance of corrugated steel webs and the buckling of corrugated steel plate arch. There is no research on local buckling of CSP. Therefore, this paper focuses on solving local buckling problem of CSP.

2 Structural Form and Experimental Test

2.1 Structural Form

Compared with the steel platforms, CSP platform will reduce the number of supporting ribs, greatly increase the span of the structure, and reduce the number of steel reinforced beams. The ends of CSP are usually supported on steel beams or concrete beams with setting connecting bars.

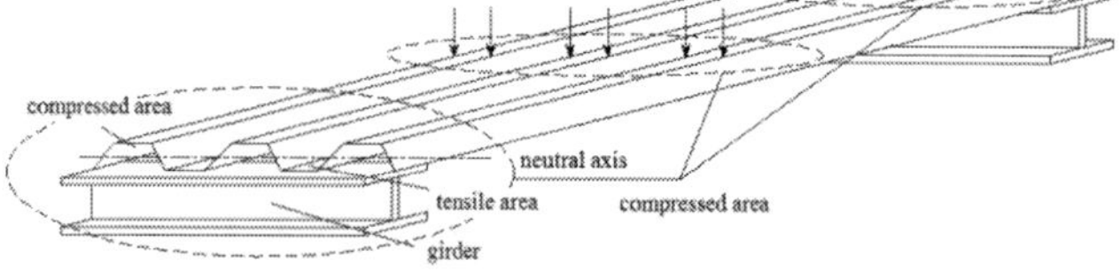

Figure 1 CSP Platform Structure

*Corresponding author: Email: BJian_LI@163.com

This kind of platform is mainly applied in the architectural engineering called Composite Floor, the thickness of steel plate used in Composite Floor is less than 1.5mm, but the thickness of CSP is usually from 3mm to 10mm, maybe thicker. The increasing of the thickness will supply higher bearing capacity.

2.2 Experimental Test

In this paper, the experimental test of CSP platforms will be conducted with the following three kinds of dimensions: 400×50×3mm (Width, Height and Thickness), 500×50×5mm and 500×86.6×5mm. The span of the structure is 2390mm, the distance of the support of the distribution beam is 800mm, and CSP section characteristics are shown in Figure 2 and table 1.

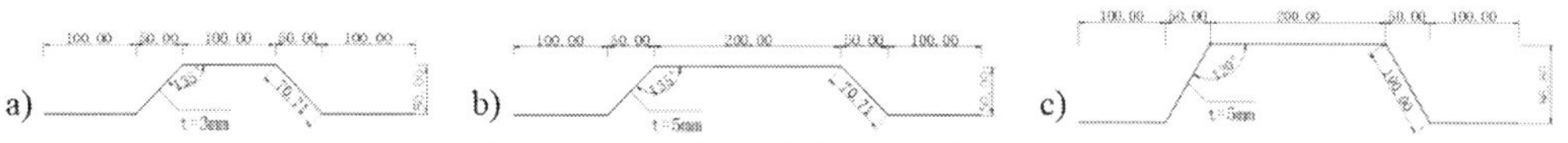

Figure 2 Graphic Presentations of CSP: a) CSP1; b) CSP2; c) CSP3

Table 1 Section Properties of CSP

Components	Moment of inertia I/mm^4	Area A/mm^2	plastic modulus S/mm^3	Height of centroid y/mm
CSP1	611676	1324	26471.9	20
CSP2	1043717	2707	58875.3	27.5
CSP3	4382812	3000	108189.7	45.8

Testing loads are controlled by the variation of load-displacement law, CSP1: the load step is 3kN, and break for 1 minute at the end of each load step, until reaching the ultimate bearing capacity; CSP2, CSP3: the load step is 5kN, and break for 1 minute at the end of each load step, until reaching the ultimate bearing capacity.

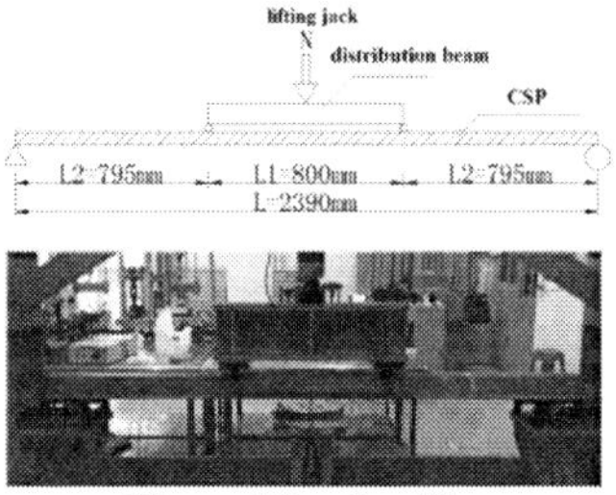

Figure 3 Loading Plan

3 Result Analysis and Discussion

The load-displacement curves obtained by experimental tests are drawn in Figure 4. According to the bearing capacity curve of Figure 4, it is seen that the ultimate bearing capacity of CSP1 is 30kN, CSP2 is 67.1kN and CSP3 is 125kN. All CSPs showed good elastic performance, load-displacement curve is basically linear before yielding.

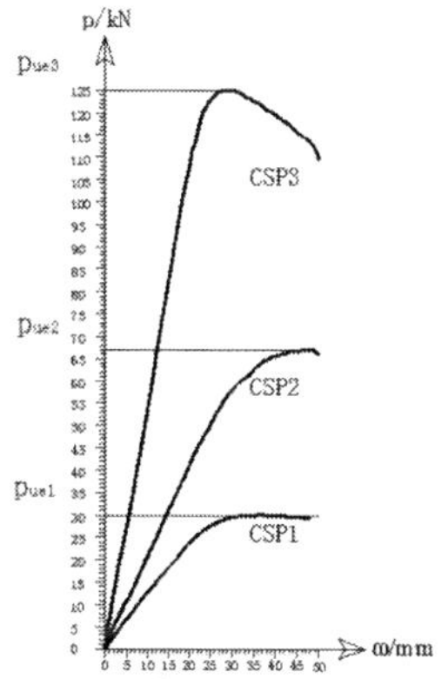

Figure 4 Load - Displacement Curves

The deformations of the loading position are shown in Figure 5, both CSP1 and CSP3 show as elastic-plastic buckling, which occurred after yielding with unrecoverable deformation; CSP2 shows as elastic buckling, the plate deformed in the process of loading and deformation recovered when the load was removed. The deformations of the loading point illustrates that the ultimate bearing capacity of the components is relevant to the local buckling.

Figure 5 Deformation of Loading Position: a) CSP1; b) CSP2; c) CSP3

4 Analysis of Local Buckling

The mechanical behavior of local compressive position of CSP is very complicated. It contains vertical concentrated loads and bending stress in the roof and web. Furthermore, since the web skews cross the roof, the concentrated loads will be decomposed into two component forces along the direction of the web and roof. In addition, the relationship between the roof and the web buckling must be considered. Combining these factors above, local buckling is very difficult to be analyzed. The calculation diagram is shown in Figure 6.

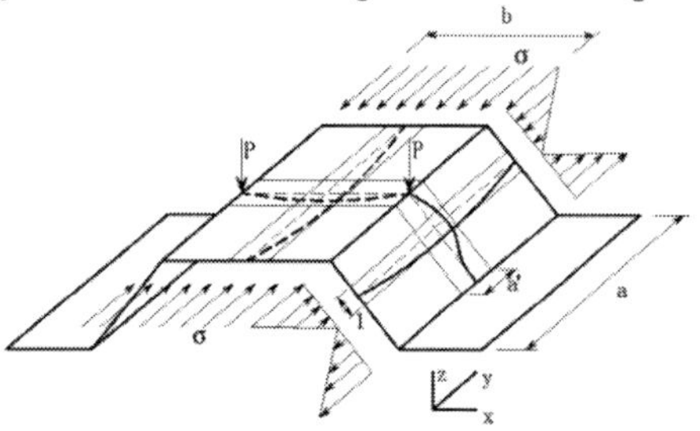

Figure 6 Analysis Diagram of Local Buckling

The bending stress σ in Figure 6 can be solved easily according to bending theory. The parameters a and b both are half wavelength, which should be calculated by the differential equation of plate buckling [17].

4.1 Half Wavelength

Due to the skew between the web and the roof, the horizontal component force of the X direction will be generated in the roof under the vertical load P, and the roof will be in a state of biaxial compression. In which, p_y is produced by the bending stress of the roof, and p_x is produced by the horizontal component force of P.

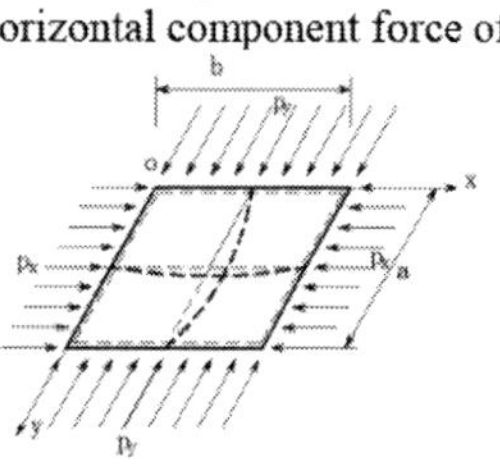

Figure 7 Simply Supported Plate of Two-way Compression

The solutions of buckling modes of biaxial compression can be used to solve the half wavelength a and b.

The equilibrium equation is:

$$D\left(\frac{\partial^4\omega}{\partial x^4}+2\frac{\partial^4\omega}{\partial x^2\partial y^2}+\frac{\partial^4\omega}{\partial y^4}\right)+p_x\frac{\partial^2\omega}{\partial x^2}+p_y\frac{\partial^2\omega}{\partial y^2}=0$$

Assuming the deflection surface equation is a trigonometric series, the following equation is obtained.

$$\omega=\sum_{m=1}^{\infty}\sum_{n=1}^{\infty}A_{mn}\sin\frac{m\pi x}{a}\sin\frac{n\pi y}{b}$$

Where: m, n - the half wave number of X and Y direction; A_{mn} - undetermined constant.

The relationship between buckling load p_x and p_y can be obtained by assuming the half wave numbers of two directions are 1, and meet with the condition of both A_{mn} and the sine term value is not zero.

$$\frac{p_x}{a^2}+\frac{p_y}{b^2}=\pi^2D\left(\frac{1}{a^2}+\frac{1}{b^2}\right)^2$$

Make λ=a/b, the formula above becomes as:

$$p_x+\lambda^2p_y=\frac{\pi^2D}{b^2}\left(\frac{1}{\lambda}+\lambda\right)^2$$

When λ=1 (that is a=b), the formula above will achieve the minimum, and buckling will occur easily, in this time:

$$p_{crx}+p_{cry}=\frac{4\pi^2D}{b^2}$$

Therefore, when the simple supported plate is compressed by two directions, buckling modes will generate with the same value of half wave numbers in two directions, that is a=b. In this view, this problem can be solved only when a or b is known. Here, the value of b is just equal to the width of roof, so buckling will produce a square grid region with length b, this agrees with the square caused by local buckling in figure 5.

4.2 Analysis of Local Buckling

It is complicated to analyze the buckling mode shown in Figure 6, and the buckling differential equation cannot be used to solve this problem directly. Therefore, the buckling mode of CSP is decomposed along the X direction and the Y direction, in this way, the three-dimensional buckling problem is transformed into two-dimension plane buckling problem. In X direction, the buckling problem is considered as the buckling of plane rigid frame, and the influence of Y direction is considered as the constraint of X direction. Considering the composite buckling of the roof and web, this problem can be solved successfully.

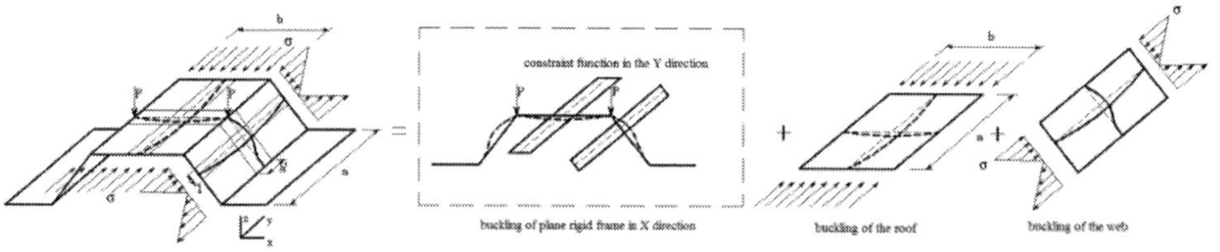

Figure 8 Analysis Diagram of Local Buckling

1) Buckling of plane rigid frame in X direction

The stability analysis is carried out by using buckling theory of rigid frame [18].

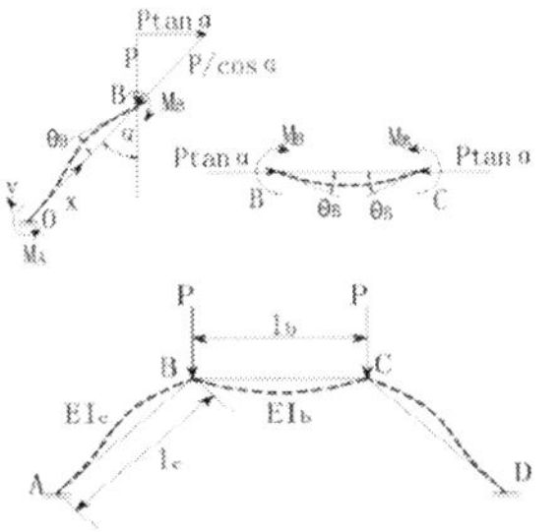

Figure 9 Buckling of Plane Rigid Frame

The load P is decomposed along the direction of the inclined web and the roof, the component force along the web P/cosα and along the roof Ptanα can be obtained. Buckling of the inclined web and roof will take place under the action of axial compressive loads, and the buckling equation is established according to the equal rotation of the point B, see Figure 9.

The differential equation of inclined web is:

$$y_c'' + k_c^2 y_c = \frac{M_B - M_A}{EI_c l_c} x_c + \frac{M_A}{EI_c} \qquad k_c^2 = \frac{P}{EI_c \cos\alpha}$$

The following formula will be obtained by solving differential equations above.

$$y_c = -\frac{M_B - M_A}{Pk_c l_c}\cos\alpha \sin k_c x_c - \frac{M_A}{P}\cos\alpha \cos k_c x_c + \frac{M_B - M_A}{Pl_c} x_c \cos\alpha + \frac{M_A \cos\alpha}{P}$$

$$M_A = \frac{\sin k_c x_c - k_c l_c}{\sin k_c x_c - k_c l_c \cos k_c l_c} M_B$$

$$\theta_B = \frac{2k_c l_c - 2k_c l_c \cos k_c l_c - (k_c l_c)^2 \sin k_c l_c}{\sin k_c l_c - k_c l_c \cos k_c l_c} \times \frac{M_B \cos\alpha}{Pl_c}$$

The differential equation of the roof is:

$$y_b'' + k_b^2 y_b = \frac{M_B}{EI_b} \qquad k_b^2 = \frac{P\tan\alpha}{EI_b}$$

The following formula will be obtained by solving differential equations above.

$$y_b = \frac{M_B + M_B \cos k_b l_b}{P\tan\alpha \sin k_b l_b} \sin k_b x_b - \frac{M_B}{P\tan\alpha} \cos k_b x_b + \frac{M_B}{P\tan\alpha}$$

$$y_b' = \frac{M_B + M_B \cos k_b l_b}{P\tan\alpha \sin k_b l_b} k_b \cos k_b x_b + \frac{M_B}{P\tan\alpha} k_b \sin k_b x_b$$

$$y_b'(0) = \frac{M_B + M_B \cos k_b l_b}{P\tan\alpha \sin k_b l_b} k_b$$

The correlation equation between the roof and the inclined web is established according to the equal rotation of point B.

$$\frac{M_B + M_B \cos k_b l_b}{P\tan\alpha \sin k_b l_b} k_b = \frac{2k_c l_c - 2k_c l_c \cos k_c l_c - (k_c l_c)^2 \sin k_c l_c}{\sin k_c l_c - k_c l_c \cos k_c l_c} \times \frac{M_B \cos\alpha}{Pl_c}$$

$$\frac{1 + \cos k_b l_b}{\tan\alpha \sin k_b l_b} k_b = \frac{2k_c l_c - 2k_c l_c \cos k_c l_c - (k_c l_c)^2 \sin k_c l_c}{\sin k_c l_c - k_c l_c \cos k_c l_c} \times \frac{\cos\alpha}{l_c}$$

Simplifying the formula above, obtain:

$$\tan\left(\frac{k_b l_b}{2}\right) = \frac{2 - 2\cos k_c l_c - (k_c l_c)\sin k_c l_c}{\sin k_c l_c - k_c l_c \cos k_c l_c} \sqrt{\sin\alpha} \qquad (1)$$

The tangent value of the left side of the equation is 0~∞. When the left side is 0, the right side will become the following:

$$\frac{2 - 2\cos k_c l_c - (k_c l_c)\sin k_c l_c}{\sin k_c l_c - k_c l_c \cos k_c l_c} = 0$$

$k_c l_c = \pi/0.5$ is obtained by solving the equation above.

When the left side is ∞, the right side will become the following:

$$\frac{2 - 2\cos k_c l_c - (k_c l_c)\sin k_c l_c}{\sin k_c l_c - k_c l_c \cos k_c l_c} = \infty$$

$k_c l_c = \pi/0.7$ is obtained by solving the equation above.

Make $\pi/\mu = k_c l_c$, μ will be change from 0.5 to 0.7 with different tangent values.

It is difficult to solve the equation above exactly, which needs to be constantly repeated to get the buckling load P.

2) Influence of Y Direction

The buckling analysis above is based on the considerations of buckling of plane frame in X direction, but the slab in Y direction will produce restraint to the buckling in X direction, which will increase the buckling loads. The constraint function in the Y direction can be solved by bending differential equation of plate with 4 sides simply supported [19].

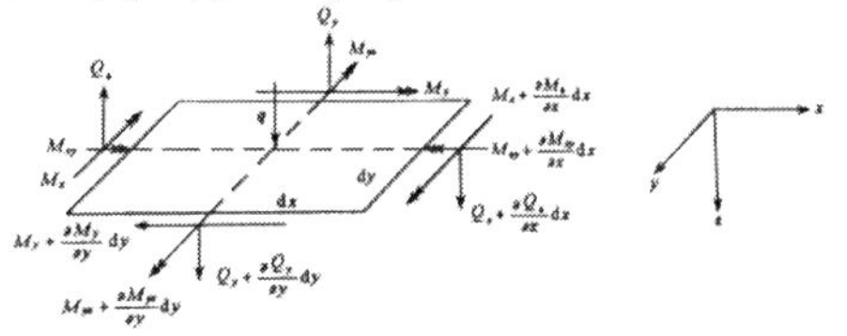

Figure 10 Infinitesimal Body of Plate

The differential equation corresponding to figure 10 is:

$$D\left(\frac{\partial^4 \omega}{\partial x^4} + 2\frac{\partial^4 \omega}{\partial x^2 \partial y^2} + \frac{\partial^4 \omega}{\partial y^4}\right) = q$$

The solution of ω is the key to calculating the internal force of the plate. With the solution of ω obtained above, the constraint problem in Y direction has been solved too. Fortunately, ω has been solved.

When there is no constraint in Y direction, the internal force of the plate is transmitted along the X direction, in this way, the 3D space problem can be reduced to a 2D problem. When the constraint of Y direction is considered, the internal force problem becomes a 4-side simply supported slab problem. The structural stress state of the two cases mentioned above can be converted by the deflection ratio.

Supposing that P_1 is the buckling load without considering the constraint of Y direction, while P_2 is the buckling load with the constraint of Y direction, ω_1 is the deflection of the plate simply supported by both sides caused with unit force, ω_2 is the deflection of the plate simply supported by 4 sides caused with unit force, then the relationship between P_1 and P_2 can be expressed as:

$$\frac{P_1}{P_2} = \frac{\omega_2}{\omega_1}$$

The ratio of the deflection ω of the center point is associated with the ratio of the lengths of the two directions of simple supported plate, which is given by Table 2 [20].

Table 2 Load Conversion Coefficient of Buckling Considering Constrains in y Direction

l_1/l_2	0.5	0.55	0.6	0.65	0.7	0.75	0.8	0.85	0.9	0.95	1.0

ω_2/ω_1	0.778	0.722	0.666	0.611	0.558	0.509	0.463	0.42	0.381	0.345	0.312

Where: l_1- length of short edge; l_2- length of long edge; when l_1/l_2 is less than 0.5, the situation can be considered as without constraints in Y direction.

3) Width of Plane Rigid Frame in X Direction.

When the buckling problem of plane rigid frame and the constraint of Y direction is solved, the stiffness of the frame I_b and I_c, which are relevant to the width of plane rigid frame, needs to be solved, the key parameter to solve I_b and I_c is a', shown in Figure 6 and Figure 8.

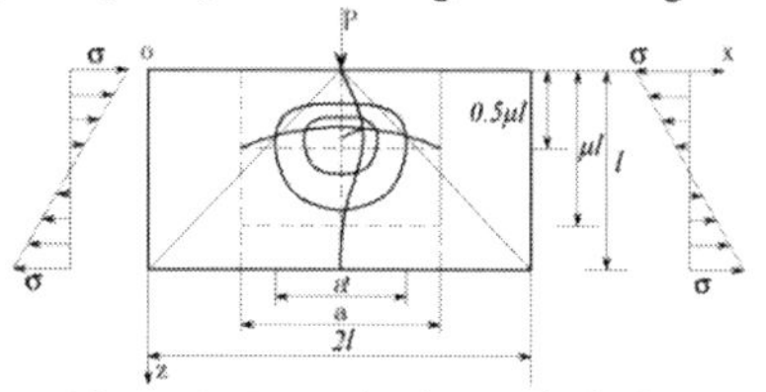

Figure 11 Determination Method of a'

According to "Code for Design of Steel Structures", when the load is applied to the top of the web, it will diffuse along the web with 45 degree angle, and the stress is gradually decreasing along the structural height direction.

If the height of web is l, the distribution width of the bottom will be $2l$. Assuming the width with l or $2l$ in the plane rigid frame is not suitable obviously, because the stress distribution is not uniform. So, a relatively concentrated area need to be found, and assuming the stress distribution is uniform in this area, load P is mainly borne by the width of the strip in this area.

In the theory of elastic stability, μl is used to calculate the length of the compression column according to the different supporting conditions. The largest deformation and the maximum internal force is in $0.5\mu l$ position of the component. So, the stress distribution width of the web in $0.5\mu l$ position (Figure 11) is $a'=\mu l$ (when $\mu l<a$; if $\mu l>a$, $a'=a$). With the assumption that the load P is uniformly distributing along the width, see Figure 11.

4) Buckling relevant equation

When buckling load of the plane rigid frame is calculated, buckling of the roof and web (Figure 8) needs to be considered, which affect the buckling of local area. Buckling of the roof and web can be solved by the methods mentioned in reference [21].

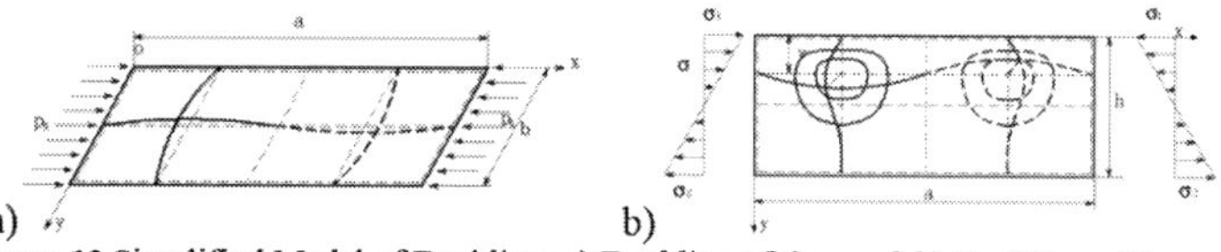

Figure 12 Simplified Model of Buckling; a) Buckling of the roof; b) Buckling of the web

Table 3 Calculation Formula of Local Buckling

Category	Critical stress	k	χ
Buckling of the roof	$\sigma_{crx}=\frac{\chi k}{12(1-\upsilon^2)}\times\frac{\pi^2E}{(b/t)^2}$	4	1.3
Buckling of the web	$\sigma_{crx1}=\frac{\chi k}{12(1-\upsilon^2)}\times\frac{\pi^2E}{(h/t)^2}$	23.9	1.0

When the three parts decomposed in Figure 8 are solved. The buckling relevant equation is used to estimate whether buckling happens in local area. If $P>P_y$, it is elastic-plastic buckling, otherwise, it's elastic buckling.

The buckling relevant equation can be expressed as:

$$\frac{P}{P_{cr}}+\frac{\sigma_x}{\sigma_{crx}}+\frac{\sigma_{x1}}{\sigma_{crx1}}=1 \quad (2)$$

The following equation will be obtained after all expression introducing into equation (2).

$$\frac{P}{\frac{\pi^2EI_c}{(\mu l_c)^2}}+\frac{\frac{M}{I}y_c}{\frac{\chi_1k_1\pi^2E}{12(1-\nu^2)(b/t)^2}}+\frac{\frac{M}{I}y_c}{\frac{\chi_2k_2\pi^2E}{12(1-\nu^2)(h/t)^2}}=1 \quad (3)$$

Notice: the relationship between P in equation (3) and P in load-displacement curve in Figure 4 is that the value of P in equation (3) is equal to 1/4 of the value of P in Figure 4, which is the distribution load of the lifting jack. So to make these two unified, the equation (3) becomes the following form:

$$\frac{\frac{P}{4}}{\frac{\pi^2EI_c}{(\mu h)^2}}+\frac{\frac{PL_2}{2I}y_c}{\frac{\chi_1k_1\pi^2E}{12(1-\nu^2)(b/t)^2}}+\frac{\frac{PL_2}{2I}y_c}{\frac{\chi_2k_2\pi^2E}{12(1-\nu^2)(h/t)^2}}=1 \quad (4)$$

Where: l_c - the height of the wed, l_c=h.

The equation (4) is the formula to calculate the local buckling under the loads shown in Figure 3. $P_{cr}=\pi^2EI_c/(\mu h)^2$, P_{cr} can be obtained if the value of μh/b agree with table 2 or not.

Where: I_c- moment of inertia of the web with width a',$a't^3/12$; P- concentrated load applied by lifting jack; E-elastic modulus of steel; μ- calculating factor of length; h- height of CSP; ν- Poisson's ratio; χ_1, χ_2- the elastic restraint coefficient; b- width of the roof; t- thickness of the plate; k_1, k_2- local buckling coefficient; I- moment of inertia of CSP; y_c- distance from centroid to the roof; L_2- distance from loading point to support point.

4.3 Method validation and discussion

The calculation results of CSP1, CSP2 and CSP3 are summarized in Table 4.

Table 4 Results of Buckling Calculation

component	μ=0.5	μ=0.7	P_{ue}	conclusions
CSP1	37.6kN	36.6kN	30kN	yielding

CSP2	67.8kN	67.08kN	67.1kN	buckling
CSP3	144.7kN	140kN	125kN	yielding

P_{cr} calculated by formula (4) can only give the distribution range, when the stiffness of the roof from 0 to ∞, which correspond to the situation of μ=0.5 and μ=0.7, it is very difficult to give an exact result. According to table 4, the buckling load range of CSP1 is 36.6kN - 37.6kN, which is greater than the experimental load, the ultimate bearing capacity is controlled by the bending strength; The buckling load range of CSP2 is 67.08kN - 67.8kN, the experimental load is between the range, the ultimate bearing capacity is controlled by the buckling strength of the structure; The buckling load range of CSP3 is 140kN - 144.7kN, which is greater than the experimental load, the ultimate strength is controlled by the bending strength. The results above agree with the experimental results, which illustrate the formula (4) is reasonable. For checking general buckling under concentrated loads, formula (3) is required to be changed to establish the relevant equations of the rigid frame, the roof and the web.

Thickness ratio of the roof of CSP1 is 100/3=33.3, and the web is 70.71/3=23.57, which are close to or less than b/t=30. Thickness ratio of the roof of CSP2 is 200/5=40, and the web is 70.71/5=14.142, which are greater or less than b/t=30; thickness ratio of the roof of CSP3 is 200/5=40, and the web is 100/5=20, which are greater or less than b/t=30. It is not accurate to judge if buckling of CSP takes place or not according to thickness ratio of the plate. The "code for design of steel structures" [21] stipulates that: when the width-thickness ratio b/t is less than or equal to 30, local buckling won't take place, experimental phenomena does not conform to the provisions in the code obviously, so the width-thickness ratio can't be used to determine whether buckling of CSP take place or not.

According to the analysis above, the buckling load calculated by formula (4) is reliable. Since the buckling calculation is only relevant to the stiffness of the components and not relevant to the strength of materials, which does not need to be obtained by the material stress-strain relationship. The buckling capacity calculated by the stiffness agrees with the experimental phenomena and results very well, so the formula is reasonable.

The boundary condition of the plane rigid frame is simplified to fixedly connect with the foundations, but the actual boundary condition of the structure is elastic boundary. The bottom plate is partly tensile and partly compressive, the tensile stress will increase the buckling capacity in compressive direction. The concentrated force acts on the roof, which will be transferred to the bottom plate, the transferred pressure is less than that of the roof. The buckling of the web will only take place near the roof. When these factors were considered, the fixed connected boundary condition of plane rigid frame is reasonable.

Experiment used one waveform of CSP, which is not fully reflect the actual structure, but due to the bottom plate of CSP is pulled and is not prone to buckling, so the whole section of actual structure can be divided into a number of single waveform by bottom plates. Buckling loads of actual structure and one waveform of CSP are similar in this paper, so the analysis method and the experimental method are reasonable.

5 Conclusion

The following conclusions can be drawn by the analysis above:

Load-displacement curve of CSP is basic linear before yielding which shows these structures have good elastic properties; the final failure mode has two kinds: one is yielding with the plastic deformation; the other is the elastic buckling with the instable elastic deformation.

Checking if local buckling takes place at the loading point or not, formula (4) can be used. In general cases, formula (3) can be used. The formula includes three parts: buckling of the simplified plane rigid frame, buckling of the roof and the web. Buckling loads calculated by the formula agree with the experimental results very well, which can be used to determine the local buckling loads.

References

1. Minkwan Ju, Hongseob Oh. Static and fatigue performance of the bolt-connected structural jointed of deep corrugated steel plate member[J]. Advances in Structural Engineering, 2016, 19 (9): 1435-1445.
2. Damian BEBEN, Adam STRYCZEK. NUMERICAL ANALYSIS OF CORRUGATED STEEL PLATE BRIDGE WITH REINFORCED CONCRETE RELIEVING SLAB[J]. JOURNAL OF CIVIL ENGINEERING AND MANAGEMENT, 2016, 22 (5): 585-596.
3. LI Bai-jian, FU Xin-sha, JIANG Xiao-li. Technology of Corrugated steel plate - concrete composite beam used in bridge reinforcement [J]. Highway, 2016, 10: 77-80.
4. CHEN Shui-sheng, ZHANG Xiaop-guang, GUI Shui-rong. Parametric Analysis of Dynamic Characteristics of Multi-Cell Single-Box Composite Box Girder Bridge with Corrugated Steel Webs[J]. Bridge Construction, 2014, 44 (6): 40-45.
5. LI Jie, WU Hai-peng, CHEN Huai. Analysis of Steel Web Shear Capacity of Variable Section Continuous System Composite Box Beam Bridge with Corrugated Steel Webs[J]. Bridge Construction, 2015, 45 (1):79-84.
6. Nie Jianguo, Zhu Li, Tang Liang. Shear strength of trapezoidal corrugated steel webs[J]. CHINA CIVIL ENGINEERING JOURNAL, 2013, 46 (6): 97-109
7. Tong Guo, Richard Sause. Analysis of local elastic shear buckling of trapezoidal corrugated steel webs[J]. Journal of Constructional Steel Research, 2014, 102: 59-71.
8. LI Ming-hong, YANG Bing-wen, WANG Yi-li etc. Design Method of Corrugated Steel Webs in Prestressed Concrete Composite Girder with the

Webs [J]. Journal of Highway and Transportation Research and Development, 2013, 30 (6): 40-44.

9. Zhang Feng, Li Shucai, Ye Jianshu etc. Shear strength of corrugated steel web partially encased with concrete[J]. JOURNAL OF SOUTHEAST UNIVERSITY (Natural Science Edition), 2016, 46 (6): 1264-1270.
10. Yang Le, Zheng Kaifeng. Research on the dynamic properties of steel bottom slab and corrugated steel web bridge[J]. Gummi, Fasern, Kunststoffe, 2016, 69 (13): 218-224.
11. Moon Jiho, Ko Hee-Jung, Sung Ik Hyun etc. Natural frequency of a composite girder with corrugated steel web [J].Steel and Composite Structures, 2015, 18 (1): 255-271.
12. ZHAO Jun-hao. Corrugated steel-reinforced concrete combination bridge panel capacity experiment [D]. Nanchang: East China Jiaotong University, 2012.
13. XU Jia-qi. Corrugated steel-reinforced concrete combination bridge panel capacity experiment [D]. Nanchang: East China Jiaotong University, 2012.
14. SU Qing-tian, TIAN Le, ZENG Ming-gen etc. RESEARCH ON BASIC PERFORMANCE OF COMPOSITE BRIDGE DECKS WITH ORTHOTROPIC CORRUGATED STEEL PLATES AND CONCRETE [J]. ENGINEERING MECHANICS, 2016, 33(suppl): 138-142.
15. HU Su, SU Qing-tian, WU Chong. Study on the Section Optimization of the Orthotropic Steel-composite Bridge Deck [J]. Structural Engineers. 2015, 31 (2): 131-137.
16. WANG Xin-min. ANSYS Numerical Analysis of Engineering structures [M]. Beijing: China Communication Press, 2007.
17. Timoshenko S P, Gere J M. Theory of Elastic Stability, 2nd, Ed. McGraw-Hill, 1961, 180.
18. Chen W F, Liu E M. Structural Stability-Theory and Implementation, Elsevier, New York, 1987.
19. Han Qiang. Higher plate and shell theory [M]. Science Press Ltd. 2002. 01
20. Shu Shi-lin. Design of Reinforced Concrete Structures (Third edition) [M]. Zhejiang University Press. 2011.
21. GB. 50017-2014, Code for design of steel structure[S]. Beijing: China Architecture & Building Press, 2014.

Comparison of Standards for Accessible Design between America and China

Ying HAN, Xu Feng TAO

Jinling Institute of Technology, Jiangsu Nanjing 211169, China

Abstract: Aiming at the accessibility design and environmental constructions of China, comparing the differences of codes for accessibility design between domestic and abroad. It's including four aspects: the difference in audience of accessible design, content of the facilities, quantitative indexes and vision and hearing accessible design. It analyzed that the early stage of our country is backward in terms of accessibility and construction, mainly based on differences in values and professional education. In the end it put forward three suggestions to improve the construction of barrier-free environment in China.

1 Introduction

Accessible design begins in 1950s when people start to realize the issues of the disabled in Japan, Europe and America. They found the way to get rid of the various obstacles in the surroundings of the disabled by accessible design. Sweden and Denmark have built facilities dedicated to the disabled back then. UN released international documents like Declaration on the Rights of Disabled Persons, Convention on the Rights of Persons with Disabilities. Those documents emphasized the importance of accessible design. At that time, accessible design targeted the elder and the disabled but using a way of quarantine and special treatments. With the development of humanism design concepts, accessible design seeped in our society gradually, it has became compulsory criteria in architecture and public environment design field. Hundreds of countries and districts have made law related to the disabled also regulatory and technical standards of accessible design. Nowadays the governments of all countries are still exploring and extending their connotation while building and reforming barrier-free environment. They want to extend the meaning of accessible design to make it a universal design idea which suitable for more people [1]. Even those two concepts can exchange in some standards [2].

China has entered an aging society in 21st century; while the disabled and the elder have diversified demand compared the general population. It is architects' career requirement and social duty to figure out how to fulfill their need in city constructions and architectural environment.

2 The current situation of accessibility legislation and environmental construction in China

The progress about public building accessible environment constructions was slow from Reform and Opening-up to year 2000.The government mainly focus on making welfare system better. But after 2000, especially after the 2008 Beijing Olympic Games and the 2010 Shanghai EXPO were held, the government promoted the development of accessible design in several big cities of my country. And it has formed the situation that national standards are the dominant factor, but also refined by local standards. We issued Code for design on urban roads and buildings for disabled persons in 1989 first, the code revised as Codes for design on accessibility of urban roads and buildings in 2001, and it revised again as Codes for accessibility design in 2012 [3]. The codes covers the construction of accessible environment more comprehensively, quantification degree was higher with supporting atlas. It belongs to the national mandatory standards of constructions, and it went through the test of more than 10 years of construction practice. According to the MOHURD's experts, Many local governments made local standards including Hebei Province, Liaoning Province, Hubei Province, Guangdong Province, 5 provinces of the Inner Mongolia Autonomous Region, Beijing, Shanghai, Tianjin, Nanjing, Dalian, Suzhou and other 14 city and the District of Shenzhen [4]. However, the requirements of accessible design in public area, accessible streamline, vision and hearing aid equipment in public area etc. are incomplete. Our current accessible design standards made for specific types of public buildings includes Technical standards for airport passenger terminal

*Corresponding author: Email: chelsea_han@jit.edu.cn

facilities with accessibility by individuals with disabilities and Code for design on accessibility of railway passenger station buildings. Locations which have detailed requirements of accessible design relatively referred in national and local standards mainly includes urban road, squares, bridges, tunnels, residential areas, public parking lot, public toilets, leisure buildings for the aged etc., but we don't have specific rules for other public which have large numbers of people and wide services like offices, cultural and display public buildings.

Compared to the legislation and practice of public building accessible design in western developed countries, our country's accessible design at present stage still has a lot to catch. The gap reflected in the results of accessible environment construction on the surface, but it essentially reflected in construction legislation like standards, codes and so on which related to the hardware. The accessible design system has a deep and permanent reflection on our country's level of accessible design environment for it is a significant component of standard system of construction. There is a big disparity between Chinese and Western Developed Countries' legislation and standard of accessible design from the aspects of constitution system of legislation and standard, the depth of clauses and the expression characteristics.

3 The comparison of accessible design between Chinese and American

This paper compared our country's standard (i.e., Codes for accessibility design published in 2012) with western developed countries (e.g., American 2010 ADA Standards for Accessible Design short for "America standard" below). There are four main differences between those two standards.

3.1 The difference in audience of accessible design

Our code's general provisions 1.0.1 points out the accessible design is and we quote "urban road and buildings that guarantee the handicapped can use conveniently and safely", although "the handicapped" should include the people who have eyesight obstacles, the aged and so on, but our country's codes mainly focus on the accessible design for people who are physically handicapped rather than people who are vision and hearing handicapped.

American standards, by contrast, fully considered human primary disability types and needs in compiling concept. For example, it clearly states that "If providing accessibility in conformance with this section to individuals with certain disabilities (e.g., those who use wheelchairs) would be structurally impracticable, accessibility shall nonetheless be ensured to persons with other types of disabilities (e.g., those who use crutches or who have sight, hearing, or mental impairments) in accordance with this section." It also takes the different body size between adults and children in to considerations. Normal Principles of General Design Policy [5] of Japan noted that "until now, the measure to achieve accessible society is focus on removing the obstacles of the aged and the handicapped rather than considering the diversity of users. We should take measures to remove the obstacles of the aged and the widespread handicapped which not considered before including the retards or people who have mental disorder, the foreigner, the children and the adults with children. " The differences in the purpose of the standards and the scope of applications reflected the gaps in accessible design idea and practice between our country's inland cities and the developed countries.

3.2 The difference in facility types of accessible design

Our country's codes of accessible design was written by the sequence of building's composition, the clause's content is: 1. General Provisions; 2. Terms; 3. Design Requirements of Accessible Facilities; 4. Urban Road; 5. Urban Square; 6. Urban Green Space; 7. Residential District, Residential Building; 8. Public Buildings; 9. Barrier-free Renovation of Historic Buildings. The code is applicable for various types of residential buildings and public buildings but not with a strong pertinence, the clause and design atlas are separate. Also, it lacks of guidance details. For now, only a few traffic public buildings like civil aviation terminals, railway stations etc. and buildings for the aged have specific codes for accessible design.

All types of public buildings in the United States refer to all construction facilities that are newly built, constructed, renovated, purchased, occupied or rented by the State, local and individual investments, including 20 sets or more or 20 sets of temporary occupancy [6]. Including the office of scientific research, business services, catering, health care, culture memorial, play sports, hotels, transport facilities, schools, justice, and discipline, detention and punishment, houses, gardens, public land, entertainment facilities, etc.. There is corresponding administration section and codes for accessible design for almost each kind of buildings like colleges, courts, prisons and so on. It has formed the overall accessible design environment that has law to abide by. Also, American standards have the applicable scope of mandatory provisions cleared, it didn't just define the categories of applicable buildings but also includes the applicable scope of individual buildings, for instance, when it comes to accessible design of buildings' main functional area, it said "the path of travel to the altered area, including the rest rooms, telephones, and drinking fountains serving the altered area ; Alterations made to provide an accessible path of travel to the altered area will be deemed disproportionate to the overall alteration when the cost exceeds 20% of the cost of the alteration to the primary function area. "

American standards mainly composed by the need and the routes people choose, it also covers a lot and has a lot of details. Basically, it covers from new buildings to traffic and residential buildings, from outdoor places to buildings' component and indoor equipment, from block planning to special rooms and the space and details of accessible design.

Another important difference is western developed countries emphasize the importance of accessible routes. The example is that American standards have regulations about accessible routes indoor and outdoor. It said "A 'path of travel' includes a continuous, unobstructed way of pedestrian passage by means of which the altered area may be approached, entered, and exited, and which connects the altered area with an exterior approach (including sidewalks, streets, and parking areas), an entrance to the facility, and other parts of the facility. " The code has illustrations of word descriptions and dimension markings to avoid the ambiguity caused by word expression. While there is no specific regulations of accessible route and the reach range in our standards. The differences are not just reflecting on ideas, it appears that the accessibility facilities constructed by the code can't form a complete accessible route which leads to the imperfection of public accessibility environment. The legal disputes caused by the imperfection will be more complex than others. The most appropriate example nowadays is the enormous blind tracks which are discontinues and unavailable for the blind constructed in many big cities according to our country's current code.

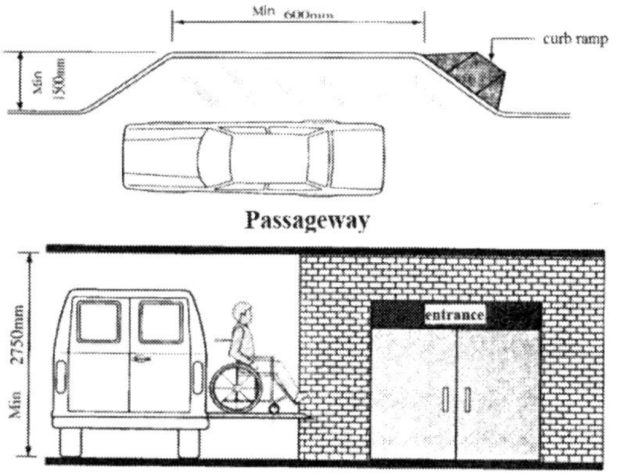

The clearance height of the passenger car area
Figure 1 Parking space access aisle and vehicle parking spaces

3.3 The difference in quantitative indexes

Compare to foreign standards, the biggest difference is we have less content of quantitative indexes in code for accessible design and standard atlas. While almost each American standards' clause has clear quantitative index or data, it also has illustrations where may cause ambiguities, it not only gives the minimum requirements but also the advisory requirements.

For example, we don't have specific rules for rugs, but in practice, rugs often become obstacles. American standards have clear regulations for the thickness of rug, Carpet or carpet tile shall have a level loop, textured loop, level cut pile, or level cut/uncut pile texture. Pile height shall be ½ inch (13 mm) maximum. Exposed edges of carpet shall be fastened to floor surfaces and shall have trim on the entire length of the exposed edge.

For another example, we only set standards for plan rather than the need of space when it comes to accessibility parking lot design. But the American standards have clear regulations, it said the passageway area includes at least 2750 mm of clearance along the way the vehicle approaches the route or exit line and the height of the parking lot within the specified space (Fig. 1).

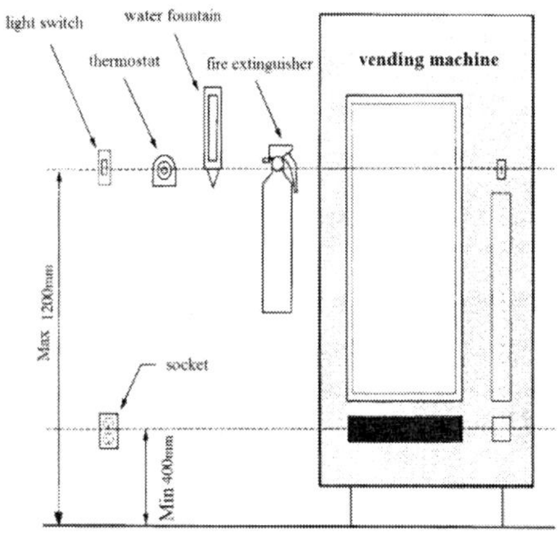

Figure 2 Height of switches

The control switches are massive and indispensable auxiliary facilities in modern architecture, it reflects the level of accessible design directly. But we don't have unified regulations about this, many important and frequently-used equipments are installed too high to reach. It is inconvenient not only for the man who use wheelchairs but also inconvenient for children and the aged. American standards put this part in the first place, they made regulations about control switch, door handle and door lock, window handle, faucet, electrical socket and switch, thermostat, pull fire alarm station and alarm start device. The standard said in the vicinity of the operation control space should be provided at least a 750mm × 1200mm flat barrier-free area to facilitate the front and back and both sides close to the control switch; control switch axis height should be set from the ground 400mm ~ 1200mm position. The control switch button should have a bulge mark or voice message (Fig. 2), indicating the function and the position of the control button.

3.4 The difference in vision and hearing accessible design

Our code about accessible design focus on building structure, it stresses the regulations about the handicapped that use wheelchairs but ignore the regulations of hardware like illumination, color contrast, hearing aid system. America standards have clear requirements for hardware about vision and hearing aid in public buildings. For instance, the hearing aid system is clearly defined. This clause applies to areas where hearing is the main mode of communication, such as concert halls, lecture halls, theaters, theaters, conference rooms, etc. Such a public space must be fitted with a qualified listening system, if it can accommodate at least 50 people, or they have an audio amplifier system and have a fixed seat. .American standards refined several different hearing aid systems as applied range of necessary accessibility facilities in buildings. Take museum for example, the hearing aid system installed includes FM broadcast, infrared ray, traditional induction circuit, 3-D loop system and so on.

4 The reason for those gaps

Based on the comprehensive analysis of the content of domestic and abroad accessible design, I think the reason why our country is behind the American and European countries in aspect of accessible design standards and constructions mostly is the difference based on values and professional education.

4.1 The differences in values

The formation of western countries' values can be traced back to the Renaissance whose guiding ideology idea is humanism. The humanism advocates individual as center, preach individualism and suggest to develop people's ego and self- expression. The disabled and the aged in western countries don't want to be separated from society and treated differently. That require society environment and architecture environment can satisfy their need to exercise alone. In addition, children of America and England usually become fully independent in economical and life aspects. Identically, old parents would not become burden, they choose to live in nursing home and supported by the society. Therefore, the European and American countries embodies the civil rights which means the government serves the public in aspects of laws and welfare policy.

Due to our culture of group and tradition of respecting the old and cherishing the young, we tend to provide special services to the disabled, the aged, and the children and so on. But we usually ignore that the obstacles made by the architecture environment itself can bother the disadvantaged when they participate in society activities individually. In fact, the expense of the construction of accessible design only a little part of construction cost. The ratio of the construction of accessible environment cost to the total cost is approximately 1%, there is no need to add more space to many facilities in the construction if we take the necessary width of the door, the opening direction, the suitable height of working face of various equipment in kitchen, the threshold, the handrail in the bathroom into considerations during early process of design. Otherwise, it will cost a lot to reform the building to suit the requirement of accessible design, and it will also cause the architect and owner to regret the decision. We should see that the improvement of accessibility environment may cost some fund, but it can reduce the fee of social service. Accessibility environment construction has considerable economic benefits from the long, social and general aspects.

4.2 The differences in professional education

Nowadays, the education of accessible design in our country is relatively weak, the pertinent researches start and we have few facilities related. In China, except several universities like Tongji University who offered relevant courses, most universities usually referred the practice of accessible design and relevant codes just in curriculum design practice. The depth and popularization are not enough. Our first institute which targets the construction and development of accessibility facilities named as "Joint research center of accessibility construction project" established until May 16, 2011. By contrast, the architecture of EU, America and Japan think highly of accessible design, their research in the field of accessible design has been in the leading position in the world. Many majors in Danish universities offer curriculum related to accessible design, they also hold design competition on the subject of accessible design to promote students learn relevant knowledge further [7]. New York State University started the research about accessible design in 1970s, they have specialized lab and teaching group. There is 70% of Japanese universities offer curriculum about accessible design [8].

5 Conclusions

We should perfect the corresponding codes, standards, and atlas and so on, including dimension markings and atlas illustrated clearly to avoid the ambiguity caused by word expression. The codes should clear the applicable scope of mandatory provisions and main accessibility area in various public buildings. The codes should emphasize the completion of accessible routes to make sure the accessibility environment built is continuous systematic and efficient.

We should increase the investment of new technology and equipment which is helpful to remove the obstacles of vision and hearing. The development of accessibility products should complement current accessibility facilities to remove the obstacles people meet in society when they walking, watching and hearing as far as possible. We should help the handicapped to walk outdoor and join the society.

Government should establish monitoring system about the construction of accessibility environment, including the implement of accessible design, the evaluation and feedback of the environment built already and the maintenance and management measures. European and American public buildings not only have clear and mature technical support but also have the guarantee of stable legislation, judicature and law enforcement system. So, it is important to build excellent accessibility environment judicature guarantee system and carry out the generality of built environment for administration.

References

1. Han Ying, Wang Yan, Wang Qiang. Evaluation on Indoor Barrier-free Stream Line of Museum [J]. Architecture and Culture. 2013(10): 81.
2. Premier's Council on the Status of Disabled Persons. Improving Universal Design Requirements in the New Brunswick Building Code [D], 2007.
3. Ministry of Housing and Urban-Rural Development of the People's Republic of China. Code for Accessibility Design[S].Beijing: China Architecture & Building Press, 2012.
4. Hu Chuanhai. The Establishment and Implementation of Technical Standards System for Accessibility Construction in China [DB/OL].http://www.doc88.com, 2013-4.
5. Li Xiuying. An Analysis of Japan 's Aging Society and Its Problems [J].Japan Studies.1989(3): 36.
6. Department of Justice. 2010 ADA Standards for Accessible Design [S]. 2010.
7. Naoto Tanaka, Kunio Hoshiba.Accessible design: design method of five stimulate ways [M].Beijing: China architecture & building press, 2013: 170.
8. Ecns. Chinese first Joint research center of accessibility construction project established at Tongji University.[EB/OL]http://edu.ifeng.com/gundong/detail_2011_05/16/6426685_0.shtml, 2011-05-16.

 https://doi.org/10.1051/e3sconf/20183803004

Application of Digital Technology and PDA+APPS Mobile Terminal in Historical Architecture Culture Protection and Development

Zhi Feng LIU, Ying WANG, Dong Hui HUANG

Architectural Engineering Institute, Jin ling Institute of Technology, Nanjing, Jiangsu 210069, China

Abstract: In the wake of big data and Internet plus era, continuous infiltration of digital technology has been happening in various fields of social and economic development. As the most important material carrier of historical culture, the historical value of historical buildings is produced and accumulated in its historical evolution, and it can only be protected from being created again. Based on the background of digitization of cultural resources, this paper summarizes the relevant digital technologies for the digital translation of information on buildings of historical and cultural heritage, as a means to promote the spread of the PDA+APPS mobile terminal, so as to achieve the purpose of preservation, protection, management and publicity. Meanwhile, this paper analyzes the application of digital technology in this field and the prospect of its function.

1 Research Background

With the advent of big data and Internet plus era, continuous infiltration of digital technology has been happening in various fields of social and economic development, thus mutual cooperation among traditional industry and information industry is playing a key role in promoting the development of industrial transformation and upgrading. Historical buildings, as the most important material carrier of historical culture, the historical value of whom is produced and accumulated in its historical evolution, and can only be protected or developed protectively from being created again. Once lost, it will be lost forever.

Using digital technology to protect and develop historical architectural culture is a product of the development of information technology and computer technology. The relevant information of historical architecture culture heritage is digitally translated through computer network and software technology, by means of intelligent mobile terminals, and eventually achieve the purpose of preservation, protection and promotion[1].

The specific advantages are listed as follows:

(1) The digital acquisition and storage technology provides guarantee for the complete protection of historical architectural cultural resources, and the digital restoration and reproduction technology provides support for its effective inheritance;

(2) Virtual reality technology expands space for the development and utilization of these cultural resources;

(3) PDA+ APPS mobile terminal provides a wide publicity platform for the display and dissemination of the technology, broadening the horizon of the audience.

Compared with earlier studies of ancient books, dance and other intangible cultural resources, digital development and utilization of historic architecture is a relatively new topic. At present, those remaining historic buildings in the city, which are still in use and of sound quality, have a distinctive characteristic of the times and the local traditional culture. In addition, the use of new technology to preserve these precious intangible heritage information development and its utilization, will also become the inevitable trend in this field.

2 Main Technical Means

With the rapid advancement of computer and network technology, the depth and breadth of the research work about digital protection and development of historic architectural culture are also developing rapidly, and gradually formed an intensive, intelligent and systematic stage of technology on the basis of three-dimensional scanning technology, information modeling technology and virtual reality (VR).

PDA includes consumer goods and industrial grade, mainly refers to the intelligent mobile phone, tablet computer, POS machine, RFID reader and other handheld mobile intelligent terminal; APPS is an intelligent software rooted in certain operating system, known as the "application"(APP). PDA+APPS, which is based on Internet, equips those current mainstream like QQ, We Chat, micro-blog and other social network media on the intelligent mobile phone, tablet computer, portable computers and other mobile intelligent terminal

*Corresponding author: Email: wangying1212@jit.edu.cn

for publicity and promotion.

The main technologies of digital protection and development of historical buildings include raw data acquisition and measurement technology, three-dimensional modeling technology and three-dimensional simulator technology.

2.1. Raw Data Acquisition and Measurement Technology

The raw data collection of historical buildings is the basis of digitization. The original data is usually obtained by field surveying, and its accuracy directly affects the accuracy of building rehabilitation design. The establishment of the database is to archive and protect the historical building information, and also to provide basis for the rehabilitation design.

In the past, people mainly used historical drawings, tables, text and other forms of recording to archive and preserve historical buildings, which cannot reflect the real effect of the building in addition to large quantity of data; While the use of photos, videos and other image archive is intuitive, it cannot bring about accurate physical data and structural relation. Thus, it is difficult to accurately protect those buildings or carry out scientific research, and the reconstruction, imitation and restoration work will be hard to access to. Therefore, GPS, total station, measurement robot and photogrammetry have been the representative of three-dimensional modern building surveying and mapping technology, which is not only gradually replacing the traditional two-dimensional measurement method but also improving and enriching the modern means of collecting original data. Among them, the more advanced digital mapping methods can be combined with 3D scanners to facilitate point sampling on different scales of real objects and obtain cloud data of different precision.

Through the processing, analysis and recognition algorithms of point cloud and grid, 3D surface model is constructed, and the 3D entity information is taken as the form of mapping results directly[2]. Compared with the traditional database which is made up of two-dimensional information, this kind of database which contains a lot of 3D building information is more comprehensive, intuitive and easy to check.

2.2 Three-dimensional Modeling Technology

On the basis of collecting original data through surveying and mapping, the two-dimensional data is transformed into 3D information model by using professional software. Modern computer technology has been able to overcome geographical restrictions to achieve real, rapid and accurate visual display, and can even interact with the environment model. In this way, it can not only reduce the man-made damage to the physical object, but also expand the publicity and bring economic benefits.

In addition to the three-dimensional scanning model mentioned above, the building modeling in the current industry is mainly based on the two-dimensional data obtained by surveying and mapping (such as space size, component size, graticule modeling photos, etc.), using universal professional modeling software (such as AutoCAD, 3DMax, Revit, SketchUp etc.), to establish certain three-dimensional modelt which can meet the requirements of surveying and mapping accuracy and scale representation. In the meantime, A large number of architectural models can form model libraries to provide protection for the construction of virtual simulation and preservation of architectural information.

2.3 Three-dimensional Simulator Technology

With the advancement of three-dimensional modeling technology, rendering technology and computer aided design (CAB) technology, three-dimensional simulator technology has become more and more widely used in various construction work. Through digital technology, professional tools are used to measure the plane and elevation of the existing historical buildings, then the original data are collected, processed, drawn and digitally modeled to form a digital resource library.

Through the intuitive image formed by digital software and media, virtual reality technology is used to reproduce the scene, and then realize the virtual simulation of historical buildings. The application of 3D digital technology is of great practical significance for the protection, renovation, data preservation, excavation of historical buildings and promotion of ancient architecture culture[3]. Among them, VR (Virtual Reality), also known as virtual environment (Virtual Environments), simulator technology (Simulator Technology) and so on, creates a 3D image world for the user in real-time to reflect the changes and interactions of the entities. In the realistic experience of seeing, listening, touching and smelling, the ordinary audience can explore and directly experience the historical building environment in the virtual, as if in the historical process of the building.

3 Main Technical Applications

At present, the digital protection and development of historic buildings include single building maintenance and renovation, data acquisition of historical blocks during their preservation and renovation, data analysis and processing, the effect of virtual display, forming an intensive resource sharing platform to facilitate the unified management of the government. In this way, the intelligent mobile terminal is used to spread the information, broadening the horizon of audience and improving data utilization level as well as the public awareness of protection. Figure 1 is a framework of digital technical protection and development of historical buildings.

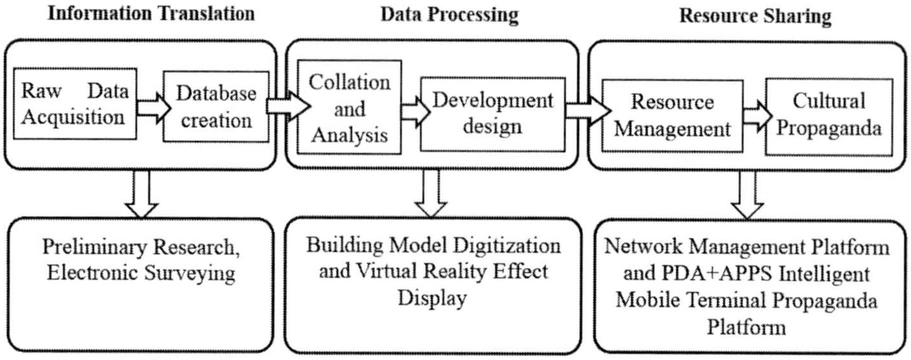

Figure 1 Framework of Digital Technical Protection and Development of Historical Buildings

3.1 Protection and Renewal of Historical Buildings

Under the background of rapid construction of modernization in the city, many historic buildings which are not included in the security unit have been frequently demolished or rebuilt, facing the danger of disappearing forever. Due to the complexity of the present situation, the protection and renewal of the historical buildings is also tend to be diversified. Especially for those badly damaged and badly preserved urban historical buildings, they should be restored at the site according to its original features. At the same time, the internal layout or structure should be properly reformed to meet the requirements of modern utilization[4].

Digital technology, on the one hand, is significant for the upcoming demolition, the overall relocation or rehabilitation of historical buildings, as detailed preservation of three-dimensional data will become an important basis for the restoration of its original shape. On the other hand, by means of digital technology and the accurate three-dimensional digital model, the engineering data, material data and process data of the historical buildings which have disappeared or are disappearing can be originally identified in the early stage of building information collection. In the meantime, the construction quality and feasibility of repair can be evaluated on this basis. Besides, the combination of digital 3D building model and virtual reality technology is also used for tourism development and cultural promotion.

3.2 Preservation and Renovation of Historical Blocks

As an important carrier of urban cultural heritage, historical blocks are also important elements of urban historical environment and characteristics of the times. Therefore, when it comes to the protection of cultural heritage, we should avoid the misunderstanding that only protection units instead of general houses and alleys should be paid attention to. We can never develop and construct our urban at the cost of losing historical blocks and urban traditional features. We should keep it firmly in mind that preservation is combined with utilization and development. With the advancement of digital technology, it plays an increasingly more important role in the whole renovation of historical blocks.

For example, located in the downtown of Nanjing, the modern historic district Meiyuan Village, who covers an area of about 45 thousand square meters, is one of the residential areas directly promoted by government during the period of Republic of China. From the perspective of construction features, in addition to the civilian housing built by the national government, there are also self-designed houses for middle and high class, and the forms of construction, land use area and living standards also show obvious diversification. Since Liberation, due to the complex ownership of property rights and other reasons, in those residential areas there still exist varying degrees of damage and other issues such as serious renovation and reconstruction, disunity of overall features and so on[5]. Recently, in order to meet the requirements of the cultural relics census and the protection and renovation works of historical blocks, a large-scale investigation has been carried out within the neighbourhood. In the course of the investigation, digital technology has also been applied in many places, mainly embodied in:

(1) By means of photogrammetry and three-dimensional measurements, relatively accurate architectural mapping data of nearly 60 modern houses within the neighbourhood was collected in the case of not or less disturbing local residents. Through AutoCAD, SketchUp and other model software, models that can reflect the shape, space, material of the building have been created as the basic data. At the same time, each residential monomer was numbered, a small database was formed based on location index, history status, elevation mapping, model data and other information, so as to provide effective resources for the overall style and environment improvement of the block .

(2) Classify and sort out the building information in the established database. On the one hand, according to the location, blocks was divided into three zones- the Meiyuan, Yongyuan and Taoyuan, where modern houses are more concentrated, so that it is beneficial to patch location management. On the other hand, those buildings

were divided into three types: the western style villa, the detached house and the platoon apartment according to the architectural form and the living standard, and then we classified the existing modern houses separately. Through these methods, the system and retrieval of database are further strengthened.

(3) According to the current situation of blocks in the database, a unified renovation scheme was formulated for most of those modern residence (especially residence along the street). Through digital software and media, an intuitive image was formed, and a static display of the restored part, the facade shape, the exterior wall material and the color of the restored single building was made .The use of 3D simulation technology makes people feel personally on the scene of the effect of renovation , and certain publicity was made through the network platform.

3.3 Create a platform for digital management and publicity

At present, in the process of digital construction of national cultural resources, low level construction and low sharing degree are common. Therefore, it is necessary to establish an effective resource management and publicity sharing platform through effectively organizing and standardizing the description of resources. Digital management and publicity platform mainly refers to the implementation of information integrated management and publicity system based on the established database by means of Web and other network technologies. Through this platform, the relevant written materials, pictures, multimedia and video of the historical buildings in the database can be digitally processed and edited, and many types of mainstream social network media is equipped with a variety of mobile terminal in the form of "PDA+APPS". In this way, the intensive means of promotion as well as the sharing of historic buildings resources can be realized.

Through the creation of digital platform, first of all, it is beneficial for relevant departments to improve the management of historical buildings and ensure the current management more rigorous and efficient; Secondly, the network sharing of digital cultural resources also provides technical support for the protection and development of historical buildings. In addition, the digital platform in the future might help the government realize the information management of historic buildings's preservation and redevelopment, including the implementation of relevant policies and regulations, broadening the investment channels and other functions.

For example, the scope of platform management can involve government, developers, masses, historical construction operators and historical construction managers an so on Among them, the government can publish policies through the platform, mainly focuses on management and guidance. Owners of property rights, managers and social masses can also obtain all the information about historical buildings in a timely manner through this platform, which can provide services for tourism development, cultural exchanges and educational research,etc. Ordinary users are also end-users, they can search, check, download certain information about historical building resources as well as participate in the feedback of these resources object[6].

4 Conclusion and Outlook

At present, in some specific domestic cases where the digital development of historic buildings are mainly concentrated in , there are still lacks in established system and uniform standards. Besides, the integrity and accuracy of the database, the innovative application of its digital function still need further study. At the same time, the digital resources of historical buildings are mostly limited to the internal sharing of the research institutions, and have not been effectively developed and exchanged in a wider scope. Therefore, the digital technology and PDA+APPS mobile terminal is a creative combination of modern information technology and mobile communication technology, it can effectively translate, preserve and protect the data of historical architecture culture. And through digital management and publicity platform it is feasible to realize intensive management, utilization, publicity and sharing. It is believed that with the advancement of technology and the continuous improvement of policies and measures, digital development will become a new starting point for the protection and cultural heritage of modern historical buildings.

References

1. Zhang Hui, et. Extended research on the function of digital application of public cultural resources [J]. information and communication, 2014.10 (142): 78-79.
2. Fan Zhang Wei, Xing Yu. Research on Ancient Architecture Conservation Based on Digital Technology [J]. Beijing surveying and mapping, 2010 (3): 18-21.
3. Wang Haiying, Liu Zhifeng. Conservation and Reuse of Digital Technology in the Minguo Architectural Landscape of Culture [J]. architecture and culture, 2015.08 (137): 113-114.
4. Wang Ying, Wang Yongping. Protection and Renovation of Historical Architecture in Block Reconstruction: Taking Two Republican — Style Buildings in Shengzhou Road in Nanjing [J]. Hundred Schools in Arts, 2012, 28 (8): 148-152.
5. Wang Ying, Wang Yongping. A Preliminary Exploration on the Residential Area Built Before 1949 in Meiyuanxincun, Nanjing [J]. Hundred Schools in Arts, 2009 (04): 39-45.
6. Li Yan. Suggestionon the Development of Digitization Management Platform for Haikou Historic Architecture [J]. Journal of Hainan University, Natural Science Edition,.2009.6 (27): 173-175.

CF60 Concrete Composition Design and Application on Fudiankou Xijiang Super Large Bridge

Yi Mei QIU [1, a], Sen Yuan WEN [2,b] and Jun Xiang CHEN [3,c]

[1]Guangxi Vocational and Technical College of communications, Nanning, China
[2]Guangxi road & bridge engineering group co., LTD, Nanning, China
[3]Guangxi Transportation Research & Consulting co., LTD, Nanning, China

Abstract: Guangxi Wuzhou City Ring Road Fudiankou Xijiang super large bridge CF60 concrete is a new multi-phase composite high-performance concrete, this paper for the Fudiankou Xijiang bridge structure and characteristics of the project, in accordance with the principle of local materials and technical specification requirements, combined with the site conditions of CF60 engineering high performance concrete component materials, proportion and the technical performance, quantify the main physical and mechanical performance index. Analysis main influencing factors of the technical indicators, reasonable adjustment of concrete mix design parameters, and the use of technical means of admixture and multi-function composite admixture of concrete, obtain the optimal proportion of good work, process, mechanical properties stability and durability of engineering properties, recommend and verification of concrete mix; to explore the CF60 high performance concrete Soil in the Fudiankou Xijiang bridge application technology, detection and tracking the quality of concrete construction, concrete structure during the construction of the key technology and control points is proposed, evaluation of CF60 high performance concrete in the actual engineering application effect and benefit to ensure engineering quality of bridge structure and service life, and super long span bridge engineering construction to provide basis and reference.

1 Introduction

Guangxi wuzhou of ring road Fudiankou Xijiang super large bridge is from 1 # main bridge [145 m + 270 m + 145 m of short tower cable-stayed bridge] + 2 # main bridge [131 m + 198 m + 131 m continuous rigid frame bridge] + approach bridge [11 × 40 m first simply supported then the continuous T girder bridge], total length of 1474 m bridge. The integral deck a total width of 28.5 meters, cantilever construction is used super wide and multiple trusses hanging basket, the largest cantilever segmental weight of 667 tons, wide and segment of hanging basket are domestic rare. This bridge uses CF60 high performance concrete, which is a new multi-phase composite high-quality concrete, is made of high-strength concrete, high-flow concrete is optimized, requires a good workability, process, mechanics, stability and durability and other engineering properties, the quality of its design and casting quality is particularly important. This paper will focus elaborate CF60 high performance concrete mix optimization design and application technology.

2 CF60 concrete raw materials optimization

Fudiankou Xijiang super large bridge CF60 high performance concrete compared with ordinary concrete, it has more raw material variety, its uniformity, compactability to good , technical performance has improved significantly. Experience has shown that the most important to obtain high performance concrete technology is using the appropriate admixture and multifunctional composite admixture.

CF60 concrete according to the principle of local materials and the requirements of technical specification, for a variety of raw materials, such as cement, coarse aggregate, fine aggregate, water, admixtures and admixtures etc., physical, chemical and mechanical properties of the test, thus reasonable choice.

2.1. Cement

Cement as concrete cementation material, direct impact on the key technical performance of concrete structures. Through contrast test, CF60 concrete using in Guangxi Huayun (Pingnan) company limited P.II52.5 silicate

*Corresponding author: Email: [a]gx5635604@sina.com, [b]jychxb@sina.com, [c]chenjunxiangde@126.com

cement, quality is stable, the intensity fluctuation is small. Its main physical, chemical and mechanical properties of technical indicators test results as shown in table 1 and table 2.

2.2. Coarse aggregate

Coarse aggregate using Shiqiao Ma An Shan stone field (5-10) mm, (10-15) mm, (15-25) mm mixed macadam, Blending proportion:(5-10)mm:(10-15)mm:(15-25)mm=10%:30%: 60%.Physical and mechanical properties of its main technical indicators test results such as listed in table 3 and table 4.

Table 1 P. II52.5 physical and mechanical properties of cement

Strength grade	Density (kg/m³)	Specific surface area (m³/kg)	Normal cosistency (%)	Stability C-A value (mm)	Condensation Time (min)		Compressive strength (MPa)		Bending strength (Mpa)	
					Initial	Fina l	3d	28d	3d	28d
52.5	3160	373	26.70	2.0	154	207	32.3	60.3	6.0	9.0
Conclusion	The are all items conform to "General Portland Cement " GB175-2007 strength grade standard P.II 52.5 of the technical requirements									

Table 2 P.II52.5 Cement chemical composition analysis

Project \ Chemical composition	CaO (%)	MgO (%)	SO_3 (%)	Ignition Loss (%)	chloride ion (%)	Alkali Content (%)	Insolubles (%)	Conclusion
Determination value	0.64	2.17	2.47	2.06	0.009	0.58	0.67	Qualified
Predetermined value	/	≤5.0	≤3.5	≤5.0	≤0.6	≤0.6	≤1.5	

Table 3 Physical and mechanical properties of coarse aggregate index

Large st grain size (mm)	Apparent density (kg/m³)	Packing density (kg/m³)	Tap density (kg/m³)	Accumulation void fraction (%)	Vibrated void fraction (%)	Needle flake content(%)	Crush value (%)	Silt content (%)	Clay content(%)	Absorption rate (%)
25	2639	1510	1580	42.8	40.1	5.9	7.5	0.4	0	0.1
Conclusion	The sample inspection by project meet "The technical specification for construction of highway bridge" JTG/T F50-2011 requirements for high performance concrete with coarse aggregate.									

Table 4 Coarse aggregate grading

Mesh size (mm)	31.5	26.5	19	16	9.5	4.75	2.36	<2.36
Cumulative sieve residual rate (%)	0	2.8	38.2	61.0	82.8	94.4	99.8	0.2
Conclusion	The sample inspection by project meet "The technical specification for construction of highway bridge" JTG/T F50-2011 requirements for high performance concrete with medium sand distribution.							

The mechanical strength of the coarse aggregate, particle shape, surface characteristics, grading, impurity content, water absorption, etc have important influence on the performance of the CF60 concrete. The selection of coarse aggregate is very important for the efficient concrete, high quality material can be used to make high performance special concrete. Therefore, choose texture hard, clean, good grain shape, good gradation, maximum nominal size appropriate gravel, can ensure the performance of concrete construction, minimizing water usage and slurry volume, improve the strength and durability of concrete.

2.3. Fine aggregate

Fine aggregate used in Tengxian Huang Huahe natural sand, medium sand, Particle clean, hard texture, the main physical and mechanical performance index test results as shown in Table 5, table 6.

Table 5 Physical and mechanical properties of fine aggregate index

Apparent density (kg/m³)	Packing density (kg/m³)	Tightly packed density (kg/m³)	Accumulation of void fraction (%)	Tightly packed void fraction (%)	Silt content (%)	Clay content (%)	Organic matter content	Fineness modulus (Fm)
2631	1589	1688	39.6	38.0	1.0	0.2	Qualified	2.76

*Corresponding author: Email: [a]gx5635604@sina.com, [b]jychxb@sina.com, [c]chenjunxiangde@126.com

Conclusion	The sample inspection by project meet "The technical specification for construction of highway bridge" JTG/T F50-2011 requirements for high performance concrete with fine aggregate.

Table 6 Fine aggregate grading

Mesh size (mm)	9.5	4.75	2.36	1.18	0.6	0.3	0.15	0.075
Cumulative sieve residual rate (%)	0	4.0	16.7	38.8	54.6	80.0	95.6	98.4
Conclusion	The sample inspection by project meet "The technical specification for construction of highway bridge" JTG/T F50-2011 requirements for high performance concrete with medium sand distribution.							

The effect of fine aggregate on the mixing of high grade concrete is larger than the rough aggregate. Should choose grading good river sand. If the sand is too thin, satisfies the requirement of the same workability, will increase the dosage of cement. This not only increases the cost of concrete, but also causes cracks in the shrinkage, which can affect the technical performance of concrete, especially durability. But the sand also should not be too coarse, the fineness modulus is at more than 3.3, it is easy to cause the new mixed concrete to be segregation in the transport pouring process, protect the water ability is poor, which can affect the quality of the concrete.

2.4. Water

Concrete mixing water, natural water or clean drinking water, conform to the content of chloride ion in water is not more than 200 mg/L, sulfate content at SO_3 is not more than 500 mg/L.

2.5. Admixtures

The CF60 concrete adopts Guangxi Huangteng chemical building materials company limited production of KL - HPC retarding type high performance water reducing agent, Due to the larger dosage of cement, low water cement ratio, high strength requirement, concrete mixing fluidity big, so the design of concrete and construction put forward higher request, in order to meet the requirements of the performance of concrete and construction, at the same time reduce the dosage of cement, reduce the engineering cost, the choice of admixtures is particularly important. Admixtures is chosen mainly considered from the following aspects: delay the initial setting time of concrete, improve the early strength of concrete, decrease the loss of concrete slump, its compatibility with cement, admixtures stability and reduce concrete hydration heat, etc.

Through contrast test, choose KL - HPC retarding type high performance water reducing agent, its main technical indicators performance test results as shown in table 7.

Table 7 KL - HPC retarding type high performance water reducing agent of quality inspection result

Technical indicators	Solid content (%)	Density (g/ml)	PH	Total alkali content (%)	Chloride ion content (%)	Fluidity of net cement slurry (mm)	Mortar water reducing rate (%)
Determination value	22.5	1.070	6	1.4	0.01	260	26
Predetermined value	22±1	1.070±0.002	4~6	<10.0	<0.02	>240	>25.0
Conclusion	According "Concrete admixture"GB8076-2008 standard, the product the project reach the retarded high performance water reducing agent grade a index requirements.						

2.6. Fly ash

Fly ash contains the active ingredient slower than the C_3S and C_2S cement hydration speed, a secondary hydration can effectively fills the hole in the concrete internal structure, makes concrete internal more dense, so as to improve the technical performance of concrete. Selecting high quality fly ash can effectively improve the working and durability of concrete, save cement and reduce cost.

The concrete useIgrade fly ash is provided by Zhuhai Yuezu environmental protection science and technology development co., its main technical indicators performance test results as shown in table 8.

Table 8 Igrade fly ash quality inspection results

Technical indicators	Fineness (%) (45um Square hole sieve)	Water requirement than (%)	Ignition loss (%)	Sulphur trioxide (%)	Water content (%)
Determination value	10.1	93	2.21	1.75	0.30

Predetermined value	≤12.0	≤95	≤5	≤3	≤1
Conclusion	The sample inspection by project meet "Used in cement and fly ash in concrete" GB/T1596-2005 Igrade fly ash of the technical requirements.				

2.7. Steel fiber

CF60 high-performance concrete mixed incorporation a certain amount different distribution of end hook steel fiber, can effectively hinder the expansion of microcracks in concrete and the formation of macroscopic fissures, significantly improve the concrete tensile, bending, impact resistance and fatigue resistance, and have good ductility.

The selection of steel fiber concrete, and its main technical indicators performance test results as shown in table 9.

Table 9 End hook steel fiber products quality inspection results

Inspection items	Length deviation (%)	Equivalent diameter deviation (%)	Length to diameter ratio deviation (%)	Tensile strength (MPa)	Bending performance	Conclusion
Determination value	−1.2	-4.3	+1.8	640	No fracture	Qualified
Predetermined value	±10	±10	±15	>600～1000	No fracture	

2.8. Polyacrylonitrile fiber

The polyacrylonitrile fiber and concrete has good mixed sex, CF60 high-performance concrete mixed with a certain amount of polyacrylonitrile (12 mm), which can effectively improve the shock resistance, tensile strength, crack resistance and durability, the more fiber, the smaller the distance, the stronger the anti-cracking ability, resistance to early plastic better and brittleness smaller.

The main performance and technical indexes of polyacrylonitrile fiber selected in this test are shown in table 10.

Table 10 Polyacrylonitrile fiber quality inspection result

Inspection items	Tensile strength (MPa)	Elongation at break (%)	Modulus of elasticity (MPa)	Melting point (°C)	Diameter (um)	Conclusion
Determination value	991	20	17936	245	18	Qualified
Predetermined value	>910	>15	>17100	≥220	/	

3 CF60 concrete optimization design

CF60 high performance concrete should have good workability, manufacturability, mechanical properties, stability and durability of engineering characteristics. Must be in accordance with the requirements of technical specification and engineering design, based on "The common concrete mixture ratio design regulation" (JGJ55-2011) to the high performance concrete CF60 composition design, determine the various ingredients of satisfying the engineering properties of concrete unit dosage, conduct performance test study, comprehensive evaluation of its rationality and economy.

3.1. Determine benchmark concrete mixture ratio

The design requirements. The strength grade of concrete CF60, mixing content slump of 160 ~ 200 mm, sand ratio of 35 ~ 45%, the initial setting time is not less than 12 h.

Design principle. Concrete mixture ratio of main design parameters are: the unit water consumption, water, cement ratio, sand ratio, can choose three different water-binder ratio (±0.02～0.03) and the sand ratio (±1% ~ 2%) accordingly combination, through contrast test, the determination of different mixing proportion of concrete slump and compressive strength, etc, the main technical indicators, thus determine the optimal dosage of the unit of the composition of the material proportion, that is the benchmark concrete mixture ratio.

3.2. Recommended concrete construction mix ratio.

Through a large number of experiments, the optimum mixture ratio design and field verification, recommended concrete construction mix ratio, its test results as shown in table 11.

3.3. The workability of concrete

The workability of the newly mixed high-flow concrete determines whether the construction of concrete site is smoothly, whether or not the quality is guaranteed, its best effect should be to control the loss of big slump and little slump; Not secrete water, segregation resistance, good uniformity; Good pumpability, no blocking tube, no detonation tube; Good filling and tightness. Therefore, in the design of mix proportion, must focus on three main contradictions of fresh concrete mixture, that is the contradiction between big slump and little slump loss, the contradiction between deformation ability and segregation resistance, the contradiction between the liquidity and adhesiveness.

The experimental results show that the main method of solving the above contradictions effectively is: choosing the proper type of cement, the strength grade, the amount of cement, the unit water use; The variety and quantity of blended materials; The dosage of admixture and its applicability to cement; Raise sand rate, increase water retention and suitable environmental conditions, etc. The good workability of concrete is a prerequisite for the orderly construction of concrete.

Table11 Recommend CF60 concrete mixture ratio test results

Raw materials	Specifications and instructions	Relative mixing proportion	Unit consumption (kg)	Trial mix material dosage (kg)	Note
Cement	Guangxi Huayun (Pingnan) co., LTD., P. II52.5 silicate cement	1.000	449	21.10	1. Sand and stone are in dry state as a benchmark, the coarse aggregate and fine aggregate moisture content are all 0%, construction according to the measured the moisture content use conversion. 2. This mixture is according to the selected material design, engineering parts use limited to regulation, Including the bridge piers, cable tower, box girder. 3. The concrete is suitable for use in the environment category for general, its effect is C grade.
Sand	Tengxian Huang Huahe natural sand, medium sand	1.423	639	30.04	
Stone	Shiqiao Ma An Shan stone field (5-10) mm, (10-15) mm, (15-25) mm mixed macadam	2.347	1054	49.54	
Water	Drinking water	0.345	155	7.28	
Admixture	Zhuhai Yuezu environmental protection science and technology development co.I grade fly ash	0.176	79	3.72	
Water reducing agent	Guangxi Huangteng chemical building materials co., LTD. Production of KL - HPC retarding type high performance water reducing agent	0.015	6.864	0.32	
Steel fiber	Steel fiber co., LTD. Chongqing city port (End hook type)	0.174	78	3.66	
Polyacrylonitrile fiber	Taian wisdom let engineering materials co., LTD (12mm)	0.002	0.8	0.04	
Test mixing water than rubber	0.29		Dosage of admixture	1.5%	
Sand ratio	38%	Mixture ratio (Mass ratio)	449:639:1054:155:79:6.864:78:0.8		
Mixture density	2440kg/m^3				
Initial condensation time(h)	15	Final condensation time (h)	17		
Slump	180mm	The measured compressive strength (Mpa)	3d	/	
Design strength	60 MPa		7d	65.9	
Try match strength	69.9 MPa		28d	74.7	
Conclusion	The concrete that using each performance index of all kinds of materials in accordance with the relevant specification requirements, recommend mixture ratio each performance index conforms to "Technical specification for construction of highway bridge" JTG/T F50-2011 requirements.				

3.4. The mechanical properties of concrete

The mechanical properties of high performance concrete mainly depends on three aspects, one is the mechanical strength of the gelled material, the second is cementing material and aggregate interface cementation strength, The third is to aggregate their strength. When under external load, load can only be borne by the solid part of the concrete, porosity can't bear any load, micro cracks so will continue to expand, interpenetration lap as crack network; Interface transition zone also influence its role in aggregate and passed, become the best way to crack propagation and well versed in, thus significantly reduce the strength of concrete. Therefore, concrete construction should be uniform mixing, vibrating compaction, should immediately after forming the suitable temperature and humidity in curing, prevent moisture loss and the generation of micro defects.

4 Construction technology and quality inspection of concrete structure

4.1. Concrete structure construction technology

4.1.1 Construction organization design

In bridge construction organization design, combined with the scene of the topography, geology, soil quality, hydrology, climate and other natural conditions, making reasonable construction scheme, the choice of the reasonable construction method, especially the key project, key parts and important link, such as bridge foundation, bridge span structure, etc.

4.1.2 Construction technology key points

Configuration choice for raw materials of high performance concrete and its mixing proportion are have high requirements, but to ensure the high performance concrete meet the requirements of the technical performance and the use of engineering structures, main technical points of the construction and quality control is very important. Specific as follows:

Control the quality of the raw materials. All raw material quality inspection including four areas: in the former, after the play, before use and in use process, according to regulations frequency to sampling inspection.

Raw material ingredients accurately. Dosage should be automatic measurement of all kinds of materials, coarse and fine aggregate of aggregate control error within 2%, other material error control within 1% accurate configuration.

Mixture mixing evenly. Concrete mixing in forced mixing equipment, and strictly control the mixing time of 60 ~ 180 s, colour and lustre is uniform.

Mixing content should have good workability and manufacturability. Using the pumping construction technology, to ensure that meet the requirements of the construction of concrete slump, at the same time, pay special attention to slump by the problem of loss too fast, and when it is necessary to take effective measures, to ensure the liquidity and the pumpability of concrete.

Reasonable control site construction process. According to the concrete mixing content setting time, reasonable control site construction process; Concrete to uniform vibrating close-grained, template shall be installed in accordance with the requirements and reasonable control the time to dismantle.

Protect wet curing. Concrete open the template should be timely watering after moisture maintenance, and maintenance of not less than 14 days.

Inspection and evaluation. By using the method of scientific detection technology and concrete construction of every working procedure quality inspection and evaluation, and constantly improve the quality awareness and strengthen quality management.

Establish a perfect quality assurance system. A comprehensive, whole link, the whole process of quality supervision, inspection and management.

4.2. The engineering quality inspection and evaluation

The engineering quality inspection including process quality inspection and quality acceptance of the completion of. Quality inspection process is more important than the quality acceptance of the completion. Engineering quality test results show that the optimized design of concrete mixture ratio of high performance concrete with good workability, pumpability, mechanics, volume stability; Structure of the concrete surface color is consistent, without the voids and pits, the actual strength to meet the requirements of the design strength grade of concrete. Built the appearance quality of concrete structure and the intrinsic quality of overall in good condition. But in a complex environment, the stress conditions and construction control and so on the many kinds of damage factors, make concrete structure cracking phenomenon still is inevitable.

5 Conclusion

Research and practice show that CF60 high performance concrete has suitable working performance, good mechanical properties, long-term stability and excellent durability. Guangxi Wuzhou City Ring Road Fudiankou Xijiang Bridge is a long, large span bridge structure, CF60 high performance concrete on its application, effectively enhance the ability of concrete to resist the long-term damage factors, delay and reduce the early shrinkage and the emergence of cracks ,improve the engineering quality and service life of the structure, and can obtain the good investment efficiency, has good popularization value and broad prospects for development.

References

1. GB175-2007. General Portland Cement [S].
2. JTG E30-2005. Highway engineering cement and concrete test procedures [S].
3. JTG/T F50-2011. Technical specification for construction of highway bridge [S].
4. GB8076-2008, Concrete admixture [S].
5. JGJ55-2011. Code for design of concrete mix proportion [S].
6. Qiu Yi-mei, Wang Yu-hong, Chen Jun-xiang: C50 self-compacting high performance concrete mix proportion design and engineering application[J].World Bridge, 2011, 4:61-64.
7. QIU Yi-mei, CHEN Xiao-bing:Application and popularization of highway steel fiber concrete pavement [J].Highway traffic science and technology,2011,3:70-72.

Mechanical Performance Test of Rubber-Powder Modified Concrete

Yan Cong ZHANG [1,a], Ling Ling GAO [2, b]

[1] Key Laboratory of Highway Construction & Maintenance in Loess Region, Shanxi Research Transportation Institute, Taiyuan 030006, China
[2] Shanxi Conservancy Technical College, Yuncheng, 044004, China

Abstract: A number of rubber cement concrete specimens that rubber powder dosage different were obtained using same cement, water and fine aggregates, by adjusting the dosage of rubber powder. Then it was used to research the influence of rubber powder dosage on performance of cement concrete by measuring its liquidity, strength and toughness. The results show that: when water-cement ratio was equal and rubber powder replacing the same volume sand, the fluidity of cement concrete almost linear increased with rubber powder dosage increasing. With dosage of rubber powder increasing, compressive strength and flexural strength reduced, but toughness linear growth trend when dosage of rubber powder less 30%.

1 Introduction

Waste rubber powder was a renewable resource. It was widely used in road construction technology at home and abroad, not only improved road performance but also significantly reduced noise, prolong its life [1]- [3].

Therefore, scholars in domestic and foreign do more research on rubber powder cement concrete or concrete [4]-[7]. Most of researchers replaced rubber particles with coarse aggregate, and then researched the properties of rubber concrete [5]-[18]. So, study on rubber powder cement concrete that replaced rubber powder with fine aggregate was less. This paper obtained a number of rubber cement dosage specimens that rubber powder dosage different to study effect of rubber powder ash on cement concrete fluidity, strength and toughness.

2 Raw Material

2.1. Cement

Lafarge P.O42.5 cement made in Beijing was used in the testing. The physical and mechanical properties of cement were shown in Table 1.

Table 1 Performance index of cement

Fineness / (m^2/kg)	Setting time/min		Compressive strength/MPa		Flexural strength/MPa	
	Initial setting	Final setting	3d	28d	3d	28d
350	180	275	5.6	49.8	6.1	8.9

2.2 Fine Aggregate

Fine aggregate was river sand choosing from Qinhuangdao. It belonged to zone II. The fineness modulus was 2.8, and apparent density was 2.6 g/cm^3. The fineness modulus of ultra-fine sand was 0.8. The grading of ultra-fine sand and ordinary sand was fine.

2.3 Coarse Aggregate

Coarse aggregate was choosed from Beijing. The apparent density was 2.7 g/cm^3.

*Corresponding author: Email: [a]568223103@qq.com, [b]380294235@qq.com

2.4 Rubber Powder

80 mesh rubber powder was used in this testing. The density was 1.2 g/cm^3 and its bulk density was 0.3 g/cm^3.

3 Testing Methods

Single factor method was used in the testing to determine the effect of rubber powder dosage on liquidity, compressive strength, flexible strength and toughness by adjusting the dosage of rubber powder. Meanwhile the other raw material was same.

Workability, strength test referenced to the guide that "cement and concrete testing procedures for highway" (JTG E30-2005). Rubber concrete toughness was determined by flexural load-deformation curve and trabecular specimen testing. Specimen size and loading position was equal to flexural strength testing. The curve that flexural load-deformation of rubber concrete trabecular was measured using MTS-810 at the rate of 0.5mm/min.

When workability was measured, select three kinds of rubber powder dosage replace sand with equal volume rates were as follows: 10%, 20%, and 30%. The basis amount of cement was 600g, sand was 1500g, aggregate was 3100g, and water-cement ratio was 0.5.

When strength was measured, select ten kinds of rubber powder dosage replace sand with equal volume rates were as follows: 1%, 3%, 5%, 7%, 9%, 10%, 15%, 20%, 25% and 30%. The basis amount of cement was 600g, aggregate was 3100g, sand was 1500g, and water-cement ratio was 0.5.

When toughness was measured, select three kinds of rubber powder dosage replace sand with equal volume rates were as follows: 10%, 20% and 30%. The basis amount of cement was 600g, sand was 1500g, and water-cement ratio was 0.5. The basis specimen was marked 3-0, and the other was 3-1, 3.2, and 3-3.

4 Testing Results and Analysis

By test method described above, the influence of rubber powder dosage on performance of cement concrete by measuring its liquidity, strength and toughness was shown as follows.

4.1 Effect of rubber powder dosage on workability

The workability and liquidity testing results of cement concrete mixed with different rubber powder dosage was shown in Figure 1 and Figure 2.

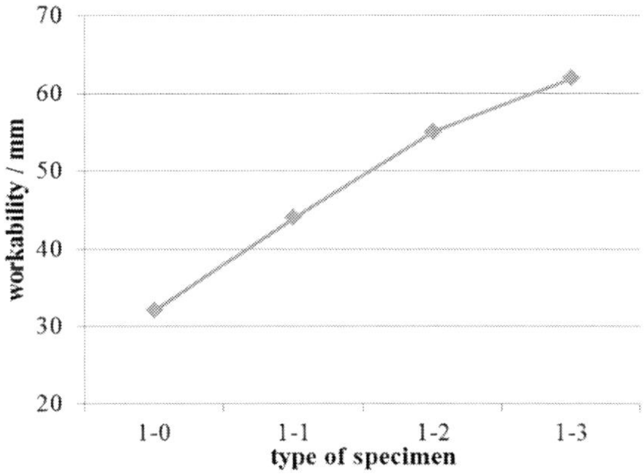

Figure 1 Rubber powder dosage and workability

As can be seen from the Figures, when cement ratio was same, workability of the cement concrete linearly increased with rubber powder dosage increased. Crumb replaced by fine sand dosage volume ratio, due to less dense rubber powder was loose state, resulting in lesser quality rubber powder instead of the larger mass of fine sand, in the water-cement ratio under the same conditions, with the significantly crumb content increased, the corresponding amount of water absorption of the cement concrete has been reduced, an increase in free water. While the reduction of fine sand, causing concrete cement particles and aggregate particles sliding frictional resistance reduction. Therefore, the workability of the cement concrete increased with increasing of rubber powder dosage.

 https://doi.org/10.1051/e3sconf/20183803006

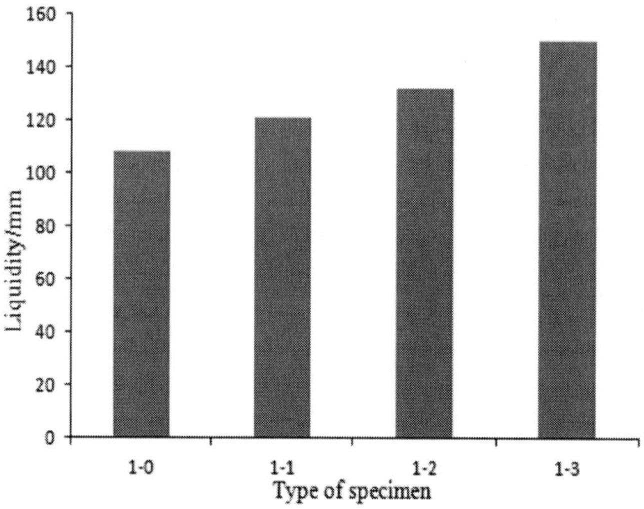

Figure 2 Rubber powder dosage and liquidity

4.2 Effect of rubber powder dosage on cement concrete strength

The compress strength and flexible strength testing results of cement concrete mixed with different rubber powder dosage was shown in Figure 3.

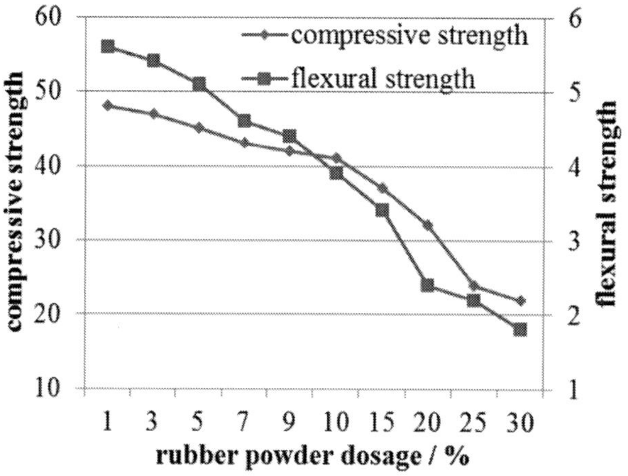

Figure 3 Rubber powder dosage and strength

It can be get following conclusions from Figure 2: ① With the dosage of rubber powder increasing, 28 days compressive strength and flexural strength of rubber powder significantly decreased. Compress strength decreased by 43%, while flexural strength dropped 32%. The decreasing amplitude of flexural strength less than compress strength. ② The ratio of flexural strength and compress strength increases gradually with rubber powder dosage increasing.

4.3 Effect of rubber powder dosage on toughness

The toughness testing results of cement concrete mixed with different rubber powder dosage was shown in Figure 4.

Toughness increased with rubber powder dosage increasing. It was relevant with that rubber powder was low modulus and elastic material, the elastic modulus of the cement concrete material greatly concerned. Incorporated into rubber powder in the cement concrete, the deformation small capacity concrete material was rubber powder instead of the deformation capacity. So, when load was equal, the greater dosage of the rubber powder, the greater the deformation capacity of the material. Stress-strain curves belonged to specimen that rubber powder dosage was different was close to a straight line. Rubber powder improved the elasticity of cement concrete material, even larger strain, the material remains at elastic state.

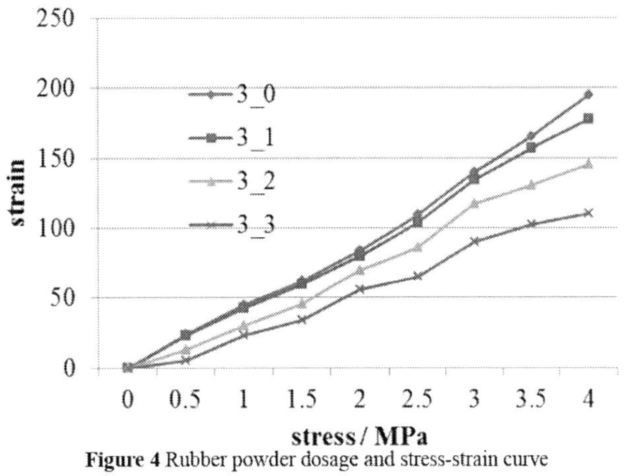

Figure 4 Rubber powder dosage and stress-strain curve

5 Conclusion

A number of rubber cement concrete specimens that rubber powder dosage different were obtained using same cement, water and fine aggregates, by adjusting the dosage of rubber powder. Then it was used to research the influence of rubber powder dosage on performance of cement concrete by measuring its liquidity, strength and toughness.

The result shows that: 1, When water-cement ratio was equal, and rubber powder replaced sand by equivalent volume amount, the workability of cement concrete linearly increased with rubber powder dosage increasing. 2, Compress strength and flexural strength of cement concrete gradually reduced with dosage of rubber powder increased. When dosage was greater than 10%, strength decreased sharply with dosage increasing. Meanwhile, the ratio of flexible strength and compress strength increased. 3, When water-cement ratio was equal, rubber powder replaced sand by equivalent volume amount, and rubber powder dosage was less than 30%, toughness of cement concrete increased with dosage of rubber powder increasing. 4, Rubber powder improved brittleness of cement concrete. It maintained concrete keeping flexible at a large strain range.

Acknowledgements

This research was supported by Shanxi natural science foundation project (2015021115), and Shanxi province communications department Science and Technology Project (2015-1-26, 2017-1-17).

References

1. SHEN Junmin, LI Yinbang. The damage survey and evaluation of old cement concrete pavement under heavy-load [J]. Shanxi science & technology of communication, 2012,1:15-17.
2. F. Hernandez - Olivares, G. Barluenga. Fire performance of recycled rubber-filled high strength concrete [J]. Cement and Concrete Research, 2004, 34: 109-117.
3. Li Lijuan, Chen Zhize, Xie Weifeng. Experimental study of performance of high s trength concrete modified by recycled rubber powder [J]. Concrete. 2007 (05): 60-63.
4. Jeong Jin-Hoon, Dan G, Zollinger P E. Finite-element modeling and calibration of temperature prediction of hydrating Portland cement concrete pavements [J]. Journal of Materials in Civil Engineering, ASCE, 2006, 18(3): 317-324.
5. William G. D, Joe P. M. Experimental Verification of Rigid Pavement Joint Load Transfer Modeling with Ever FE [J]. Transportation Research Record: Journal of the Transportation Research Board.2007, 1684/1999: 81-89.
6. HU Xinmin. Effects of transition layer on cement concrete pavement slab structure [J]. Journal of Chang'an University (Natural Science Edition).2010, 30 (3):37-40.
7. Zhang Yan-Cong, Gao Ling-Ling. Influence of rubber powder dosage on performance of cement mortar[C]. 3th international conference on civil engineer, architecture and building materials, Jinan, 2013.
8. Jeong Jin-Hoon, Dan G, Zollinger P E. Finite-element modeling and calibration of temperature prediction of hydrating Portland cement concrete pavements [J]. Journal of Materials in Civil Engineering, ASCE, 2006, 18(3): 317-324.
9. Ioannides A M, Alexander D R, Hammons M I, et al. Application of artificial neural networks to concrete pavement joint evaluation[R].Washington D C:Transportation Research Board,1996:54-64.
10. Jiang Yingjun, Dai Jingliang. Contact stresses at interfaces between dowels and surrounding concrete [J]. China Journal of Highway and Transport, 2007, 20(2): 29-34.(in Chinese)
11. ASTM Committee. Standard Test Method for Airport Pavement Condition Index Surveys[S]. ASTM Committee, 2004.7

12. AASHTO. Pavement Management Guide: Executive Summary Report [S]. America Association of State Highway and Transportation Officials, 2001
13. Tournon, G. Segregation of Non-cohereents Mixtures and Concretes [J]. Rilem Bulletin, 1995, 24:15-31.
14. Menkhoff, H. Density Measurement with Radioactive Isotope [J]. Der Bauingenieur, 1958, 12:465.
15. William G. D, Joe P. M. Experimental Verification of Rigid Pavement Joint Load Transfer Modeling with Ever FE [J]. Transportation Research Record: Journal of the Transportation Research Board.2007, 1684/1999: 81-89.
16. ZHANG Ruizhuo, LING Tianqing, YUAN Ming. Influence of Asphalt Semi-rigid Base Modules on Pavement Structural Stress [J]. Journal of Chongqing Jiaotong University (Nature Science). 2011, 30 (4):755-758. (in Chinese)
17. YAO Jialiang, HU Keyi, YUAN Jianbo. Mechanical Performance of Interface between Cement Concrete Pavement and Lean Concrete Base Treated with Different Bond-breaking Layers [J]. Journal of Highway and Transportation Research and Development. 2012, 29 (2):7-12.
18. HU Xinmin. Effects of transition layer on cement concrete pavement slab structure [J]. Journal of Chang'an University (Natural Science Edition .2010, 30 (3):37-40.

Experimental Study on Load-Displacement Relationship of Prestressed Cable in Red Clay Stratum

Yong Xing JI[1, 2, a], Yuan Jie XIANG[1, 2, b], Xiao Yong ZHAO[1, 2, c]

[1] Guizhou Construction Science Research and Design Institute of CSCEC, Guiyang 550000, China.
[2] Inspection Centre of Construction Science Research of Guizhou Province, Guiyang 550000, China.

Abstract: The P-S curves of prestressed anchor cable are obtained by field pull-out tests in the red-clay stratum in Guiyang area. The P-S curves of different theoretical models are discussed base on the theoretical analysis. The elastic and plastic displacement of prestressed anchor cable are analyzed. The results shows that, the anchoring effect are mainly effected by the red-clay mechanic properties in the red-clay stratum; the P-S curve of anchor bolt can be fitted well by the exponential model and conformed to the actual; the plastic displacement is used to anchor's damage index are more truthfulness in the red-clay stratum.

1 Introduction

Prestressed anchors as a safe, reliable and cost-effective geotechnical reinforcement measures have been widely used in slope reinforcement, dam engineering, foundation pit engineering and tunnel and underground chamber engineering, the outer anchorage section provides tension, Section anchors and anchors and anchors and rock mass interface will be transmitted to the rock mass [1]. The failure modes of the prestressed anchor are mainly in the following forms: the cable strand breaks; the slippage between the strand and the anchoring agent; the slippage between the anchoring agent and the wall of the hole; the anchored rock Weakened; between the strand and the hole between the anchoring agent weakened. Engineering practice, the third form of the most common destruction [2].

The research on the strengthening mechanism of the prestressed anchor cable under different soil conditions is not deep, and the anisotropy of the rock and soil itself .It is very necessary to study the mechanism of prestressed anchor cable under certain soil conditions. Based on this, the ultimate drawing test of the prestressed anchor cable is carried out, and the comparative analysis and research on the ultimate pull out force and failure mechanism of the prestressed anchor cable in the red clay stratum of Guizhou are carried out.

2 Prestressed Anchor Cable Load-Displacement Curve Function Model

2.1 Double-broken Line Model

When the interface of the anchor layer is in the elastic phase, the relationship between load and displacement of the anchor cable under the pulling force is given by the following formula: [3]

$$u = P\left[\frac{L_1}{E_s A_s} + \frac{4Coth\left(\beta L_b\right)}{\pi D^2 E_a \beta}\right] \qquad (1)$$

When the anchoring interface is in the plastic stage, the relationship between the load and displacement of the anchor cable under the pulling force is given by the following formula:

$$u = \frac{4\tau_1}{\beta^2 DE_a} - \frac{2\tau_2}{DE_a}L_p^2 + \frac{4P}{\pi D^2 E_a}L_p + \frac{PL_1}{A_s E_s} \qquad (2)$$

In the formula (2), E_a is the elastic modulus of the anchor, E_s is the elastic modulus of the anchor, D is the diameter of the anchor , L_1 is the length of the free section, L_b is the length of the anchorage section , A_s is the cross-sectional area of the anchor, β is a constant , τ_1 is the ultimate resistance Shear strength, τ_2 is the residual shear strength, u is anchor head displacement.

2.2 Exponential Model

Using the displacement solution of the Boussinesp problem with the vertical force acting on the Semi-infinite boundary, the formula for deriving the axial force distribution of the anchor anchorage section is shown below. [4]

$$P = P_u\left(1 - e^{aS}\right) \qquad (3)$$

In the formula (3), P_u is the ultimate bearing capacity of the Anchor Cable, a is the settlement attenuation factor, P_u and a can be determined by analysis of the pull-out test data.

Email: [a]17866631@qq.com, [b]hunanxyj@qq.com, [c]Corresponding author: mrzhao@vip.126.com

2.3 Hyperbolic Model

Using the load transfer function method, the shear stress of the anchorage section is analyzed and solved, and the distribution curve of shear stress along the axial direction of the anchorage section is simulated. [4]

$$P = \frac{S}{aS + b} \quad (4)$$

In the formula (4), P is the anchor load (kN), S is the anchor displacement (mm), a and b are the fitting parameter.

3 Prestressed Anchor Cable Field Test

3.1 Project Overview

The anchor cable pull-out test is done at the xingyunjiadiancheng foundation pit project in Guiyang. The soil layer of the foundation pit is a red clay stratum in the south area, and the stratum is more stable and has no bad geological phenomenon. The tensile strength design value f_{py} = 1320N/ mm² is used for the 1860 grade steel wire with the specification Φ15.2. The design value of the prestressed design is 100kN and the single cable anchor is 110kN.

3.2 Pull-out Test Process

The maximum test load of the single test cable is 150 kN (1.5 times the design value of the prestress), the cable tensioning step:

(1) the overall pre-tensioning, pulling the maximum test load of 10% to eliminate the inelastic deformation of the anchor cable;

(2) The cable was subjected to a seven-stage pull-out test. Each stage of the tensile load, respectively, take the maximum test load of 20%, 20%,20%,10%, 10%, 10% and 10%.

(3) When the following requirements are met, the acceptance shall be judged:

After loading to the maximum value of the test load plan, the deformation is stable.

① The total displacement measured under the maximum test load should exceed 80% of the theoretical elastic elongation of the free length of the anchor under the load and less than the theoretical extension of the length of the free section of the anchor Long value.

② Under the last load of 1 minute to 10 minutes, the anchor head is stable and the displacement is less than 1 mm. When not satisfied, the continuous load 60 minutes, bolt displacement should be less than 2mm.

(4) An attack in the cable test may be considered as damage when one of the following conditions occurs.

① Anchor head displacement does not converge, the anchor from the rock layer pull out the anchor or pull out from the anchor;

② The total displacement of the anchor head exceeds the design allowable value;

③ The increase of the anchor head displacement in the post-level load of the soil anchor test is more than twice the displacement of the upper load.

Fig.1 Manometer in the experiment

4 Comparison of Test Results

4.1 Test to Determine the Results

Field tests randomly selected 10 prestressed anchors in the site. According to the criteria described in 3.2, the test of the 10 anchor to determine the results shown in Table 1.

Tab.1 The testing results of the prestressed anchors

Specimen	L (m)	Anchor cable displacement (mm)			Test result
		(ΔL_1)	(ΔL_2)	(ΔL)	
N13-2	26.5	66.96	112.52	99.41	Qualified
N15-1	33.5	113.07	163.51	143.35	Qualified
N18-1	33.9	114.85	165.73	147.13	Qualified
2-10-1	23.6	50.11	92.57	90.63	Qualified
2-13-2	23.6	59.86	99.49	93.34	Qualified
2-15-1	31.8	54.98	99.22	98.48	Qualified

2-17-1	27.4	67.84	113.91	105.11	Qualified
2-18-1	26.9	64.3	109.19	91.9	Qualified
2-19-1	27.7	56.76	107.81	97.85	Qualified
2-20-1	27.3	66.51	112.8	89.91	Qualified

L- length of prestress anchor cable (m)

ΔL_1- 80% of the theoretical elongation value of the free-segment length.

ΔL_2- The theoretical elongation value of the sum of the free-segment length and 1/2 anchorage length.

ΔL- Maximum actual elongation of anchor under maximum experimental load.

Three of the test data were selected for processing and the *P-S* curve was drawn, as shown in Figure 2.

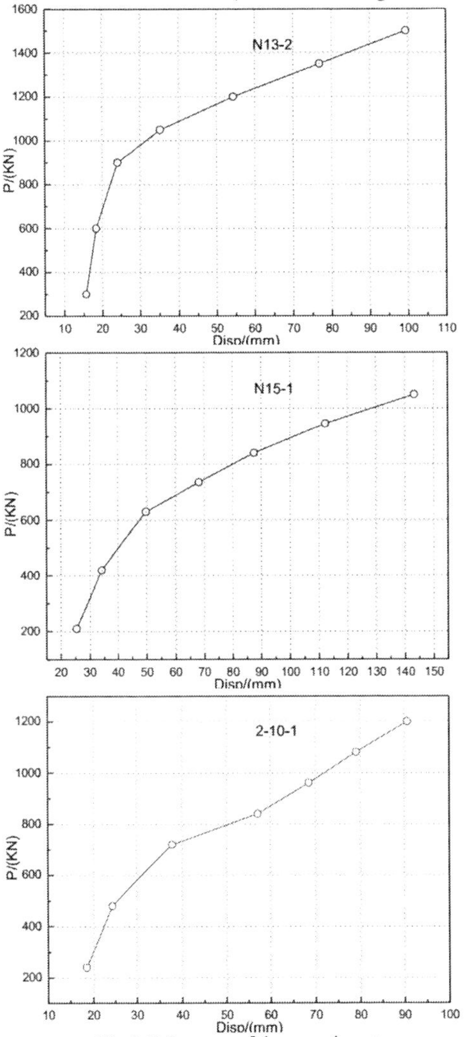

Fig.2 *P-S* curves of the experiment

4.2 Analysis of Anchoring Effect

From the analysis of Fig. 2, the pull force and the displacement of the cable are basically linear. When the anchor displacement is small, the interface between the anchor and the anchor does not produce shear yield point, that is, the interface between the anchor and the anchor has not yet slipped or destroyed.

After the displacement of the anchor is obviously increased, the shear yield point is just generated at the interface between the anchor and the anchor, that is, the interface between the anchor and the anchoring agent is just sliding or breaking. After the displacement increases, the tension remains constant.

In the analysis of N13-2 in Fig. 2, when the pull force reaches 1000KN and the displacement reaches 35mm, the anchoring section of the anchor section is partially sheared and the *P-S* curve is became flat.

When the displacement of the N15-1 cable reaches 700KN and the displacement reaches 55mm, the *P-S* curve enters the plastic growth stage, which indicates that some shear failure occurs between the grout and the red clay.

Compared with the anchor cable number 2-10-1 and the anchor number N15-1 in Fig. 2, it can be seen that the displacement of the anchor cable increases with the increase of the pulling force in the elastic range, and the mechanical properties are the same.

The anchors 2-10-1 are pressure dispersible. When the anchorage length reaches a certain value, and then continue to increase the length of the anchor, can only lead to the potential of the cable material waste, while the load capacity to help little.

4.3 *P-S* Curve Analysis

According to the different load displacement theory model, the relevant parameters are substituted into the *P-S* curve of each model, as shown in Fig.3

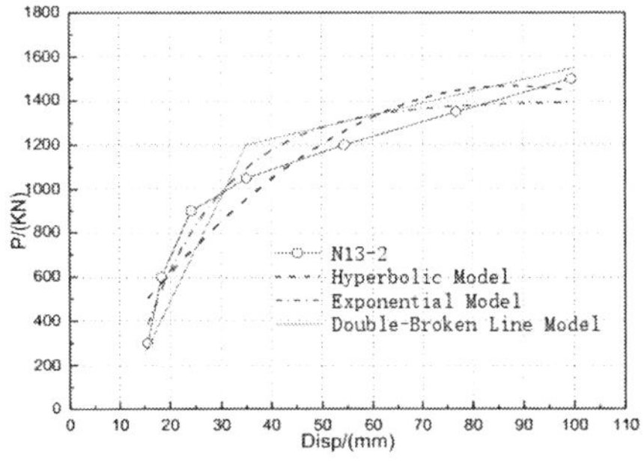

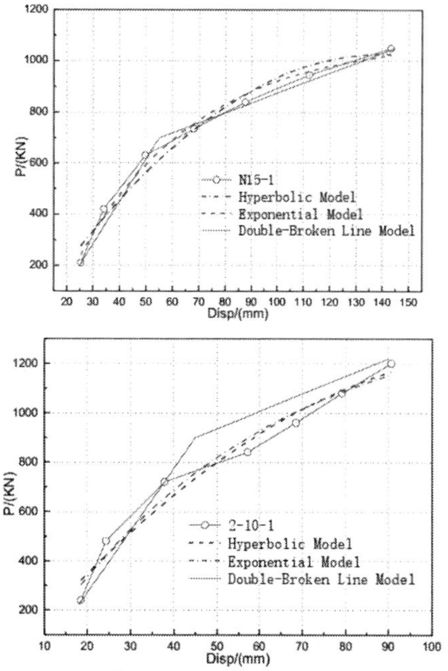

Fig.3 *P-S* fitting curves of the experiment

From Figure 3 three kinds of prestressed anchor cable *P-S* curve comparison shows that the displacement of the anchor head is not generated, and the distribution of the hyperbolic model and the exponential model is almost the same in the three cases.

In contrast, the accuracy of the fitting curve obtained by using the exponential model is high, and the obtained prediction bearing capacity is more reasonable, and the error between the measured values is very small.

Under the three cases, the prediction model of the ultimate bearing capacity of the anchor cable and the prediction accuracy of the double-broken line model at each stage load point are slightly worse.

Hyperbolic model and exponential model, the prediction accuracy of the load at all levels has a good fit with the measured *P-S* curve. The accuracy of the load-displacement correlation error of the exponential model at each stage load is higher.

From the analysis of the *P-S* curve with three kinds of anchor numbers, it is found that the damage form of the anchor cable in the red clay strata is mainly the relative displacement between the anchor and the soil when the anchor body has enough strength. That is, the plastic displacement, so the use of plastic in the red clay layer as the anchor of the destruction of the standard is more realistic.

The exponential model has better adaptability in the prediction of the ultimate bearing capacity of the red clay strata, and can be used as the index of prestressing control.

5 Conclusion

Through the pull-out test and the *P-S* curve fitting model analysis of prestressed anchor cable in red clay stratum in Guiyang, the following conclusions can be drawn:

(1) In the red clay strata, the damage of the anchor body occurs mainly on the bonding surface of the slurry and the soil, and the anchor anchorage force is mainly affected by the mechanical properties of the red clay.

(3) Increasing the anchorage length of the anchor cable can not effectively improve the ultimate bearing capacity of the anchor cable. The anchoring length is not linearly related to the ultimate drawing force.

(3) In the red clay strata, the exponential model of the *P-S* curve fitting of the prestressed anchor cable has better adaptability, which can provide the control standard for the prediction of the prestressing anchor cable performance index.

Acknowledgments

The authors gratefully acknowledge the support from Guizhou Provincial Science and Technology Department (QKH-SY[2015]3055).The authors also gratefully acknowledge the support provided through the research project of science and technology in China Construction Fourth Engineering Division Corporation Ltd (No. CSCEC4B-2015-KT-03).

References

1. You Chunan, Zhan Yubao. Distributing characters and analysis of stresses in prestressed cables. Chinese Journal of Rock Mechanics and engineering, 2005; 24(6):925-928
2. Fan Yujie, Zheng Qizhen, Wei Lin. Research on failure mechanism and ultimate load carrying capacity of prestressed cable. Chinese Journal of Rock Mechanics and engineering, 2005;15:2765-2769
3. Li Jinhua, Yuan Bing. Research on relationship between anchor pulling capacity and displacement in soil. Coal Science and Technology, 2010;(03):32-34,38
4. Ying Zhimin, Zhang jie, Shang yuequan. Exponential model for simulating load-displacement curve of anchor rod.Rock and Soil Mechanics, 2005;(08):1331-1334

Effect of Processing Parameters on 3D Printing of Cement - based Materials

Jia Chao LIN[1], Jun WANG[1], Xiong WU[1], Wen YANG[1] , Ri Xu ZHAO[1], Ming BAO[1]
National Enterprise Technology Center, China State Construction Ready Mix Concrete CO., LTD, Wuhan, China

Abstract: 3D printing is a new study direction of building method in recent years. The applicability of 3D printing equipment and cement based materials is analyzed, and the influence of 3D printing operation parameters on the printing effect is explored in this paper. Results showed that the appropriate range of 3D printing operation parameters: print height/nozzle diameter is between 0.4 to 0.6, the printing speed 4-8 cm/s with pumpage 9 * 10^{-2} m^3/ h.

1 Introduction

3D printing, as a rapid prototyping technology, is called "industrial revolution with the manufacturing technology." Compared with traditional manufacturing technology, Through a relatively simple process 3D printing can make the complex shape of the product without mold[1-3]. Because of the characteristics of materials saving and energy saving, 3D printing has the advantages of traditional construction methods in the construction industry, while its low energy consumption, high automation of the building model will greatly expand the diversity of construction methods, an also points the direction of the development for China's construction industry.

2 Analysis of Printing Equipment and Material Performance

2.1 The Status of 3D printing Equipment

According to the different forming principle, 3D printing technology can be divided into FDM, SLS, SLA, 3DP, LOM, etc. [4] Due to different forming principles and printing process, 3D printing equipment cannot be universal in different areas, but the equipment operating principle is relatively consistent. Through the installation of x, y, z three axis to achieve the core components of three-dimensional space movement. According to the computer data, the 3D printer creates a single plane layer by layer until the entire three-dimensional model is created. To the construction plane as a benchmark, 3D printers can be divided into:

(1) Mobile construction plane structure: the construction plane with the work platform in one or two axis direction of movement, with the nozzle work plane to complete the process.

(2) Fixed construction plane structure: transmission mechanism for the X, Y-axis linkage, Z-axis running alone, the construction plane in the entire printing process does not shift.

2.2 Performance of Cement - based Materials

Different from the traditional construction mode, 3D printing puts a more special requirement on the new performance of cement-based materials (rheological properties, plasticity, self-condensation performance). Therefore, 3D printing concrete as a new type of concrete, its performance should meet the following requirements [5,6]:

(1) Good extrudability: Cement-based materials can be formed in the filament after extrusion, cross-section was oval.

(2) Excellent constructability: The deformation of the filamentous concrete in the subsequent stacking process should be minimal.

(3) Controllable working time: The lower concrete quickly harden to provide adequate support strength for the upper layer.

3 Raw materials and experimental methods

3.1 Raw materials

P·O42.5 cement is produced by Hubei Yadong Cement Plant, S95 mineral powder is produced in Xinzhou, Wuhan. And microspheres is produced by Tianjin ZHUXIN New Material Technology Co., Ltd. Their basic performance indexes and particle size distribution are shown in Table 1.

*Corresponding author: Email: Sillybear@yeah.net

Table 1. Physical Properties of Cement and Mineral Admixtures

Raw Materials	Density/g·cm^{3}	Specific Surface Area/m^{2}·kg^{-1}	Water Requirement/%	Fluidity Ratio/%	28d Compressive Strength Ratio/%
Cement	3.20	350	100	100	100.0
S95 Slag	2.80	425	105	100	90.1
Microsphere	2.73	710	95	105	83.5

3.2 Test Methods

Mortar fluidity test is performed according to the standard for methods of testing uniformity of concrete admixture (GB8077-2012), 3D printing test is performed according to the construction process of project developed equipment with the component quality of the mortar, Cement: Microsphere: Slag Powder: Sand: Water=40:9:5:74:15. 3D printing equipment is based on the method of FDM, using the fixed construction plane structure. The equipment includes the storage device, the main support frame, the transmission device, the print device, the numerical control device, etc. as shown in Figure 1. And the device parameters are as follows: output displacement of the screw pump 9 * $10^{-2}m^{3}$ / h, φ32mm high pressure rubber tube, φ30mm nozzle.

The main support frame **Transportation device** **Control device**

Fig.1 3D printing equipment

4 Experimental Results and Discussion

4.1 Effect of Material Properties on Process

By printing 13mm layer height closed graphics with running speed 4cm/s, the number of stable layers using mortar mix ratio in the chapter 2.2(the fluidity range 150~290 mm) is studied.

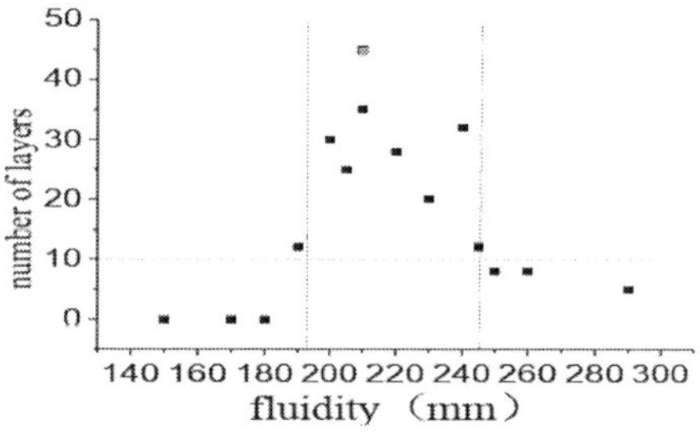

Fig.2 Effect of Material Properties on Process

The results are shown in Fig.2, with the increase of the admixture content, the fluidity of the mortar increase gradually, and the number of 3D stable layers rise first and then decreased. When the fluidity is less than 190mm, it is prone to extrusion difficulties and intermittent extrusion phenomenon, or even blocking, as shown in Fig.3. When the fluidity is greater than 245mm, the stable number of layers is not more than 12 layers, as shown in Fig.4. Therefore, when the fluidity between 190-245mm, the mortar has a certain degree of 3D printing performance, as shown in Fig.5. To a certain extent, the degree of fluidity can restrict the scope of the 3D print performance. However, it is to be noted that there is no strong linear relationship between the printability and the flow ability of the mortar. With the same fluidity range, the thickener group (green dot in Fig. 1) enhances the number of stable printing layers with lower single-layer stacking deformation. From this, we can see the 3D print mortar should be appropriate to improve the cohesion of getting better 3D printing performance under the premise of a certain fluidity.

Fig.3 Effect of dry mortar

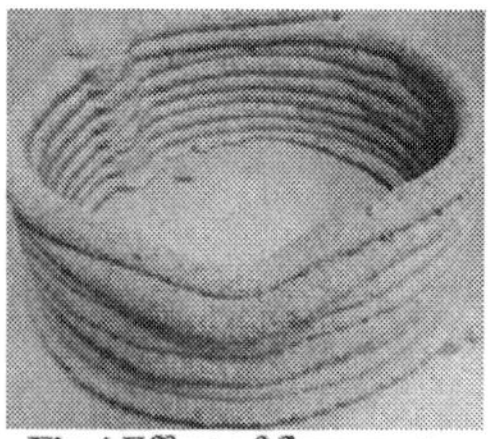

Fig.4 Effect of flow mortar

Fig.5 Effect of suitable mortar

 https://doi.org/10.1051/e3sconf/20183803008

4.2 The effect of the print layer on Process

By printing closed graphics, test study single layer height, width and apparent effect using mortar mix ratio in the chapter 2.2(the flow performance range 210±20 mm).

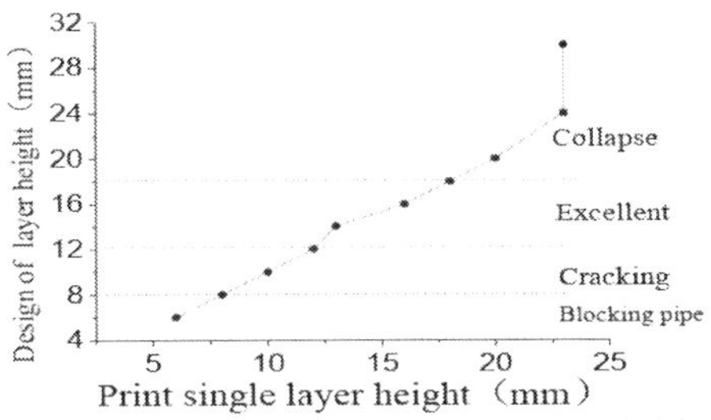

Fig.5 Effect of single layer height

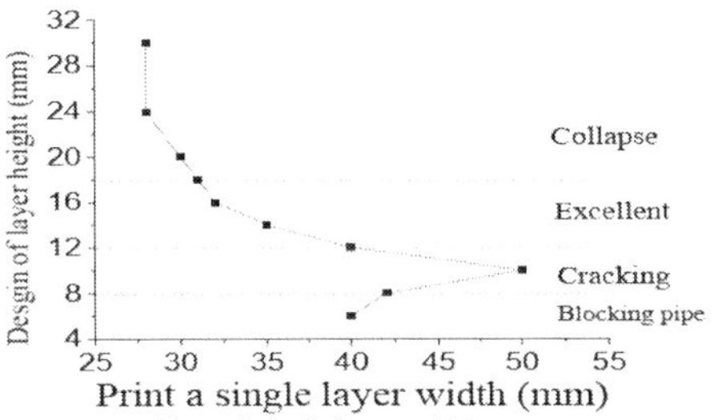

Fig.6 Effect of single layer width

According to Fig.5, we can see that the layer height have a more obvious effect on the 3D printing effect. When the single layer height is more than 20mm, because of the deformation of the slurry causing the structural deformation to collapse, the mortar cannot pile up according to the design trajectory. As shown in Fig.6, if the single layer height is lower than 12mm, the mortar extrusion volume is greater than the single layer limit. This will increase the single layer width, and the edges of the print layer are cracked. When the single layer height is lower than 8mm, the mortar cannot be discharged. And around the nozzle, the mortar accumulates together to form a closed structure, and even cause the phenomenon of blocking, as shown in Fig.7. But when single layer height is between 12-18mm, the nozzle extrusion pressure concentration problem is not significant. The printing process is relatively smooth. Therefore, in the test conditions, the best single-layer height and nozzle diameter ratio is between 0.4-0.6.

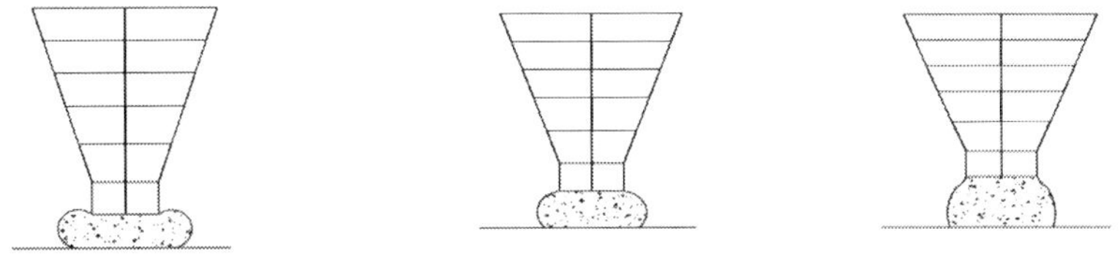
Lower layer height Suitable layer height Higher layer height

Fig.7 3D print effect of different single layer high

4.2 The Effect of Printing Speed on Process

By printing 13mm layer height closed graphics, test study single layer height, width and apparent effect using mortar mix ratio in the chapter 2.2(the flow performance range 210±20 mm).

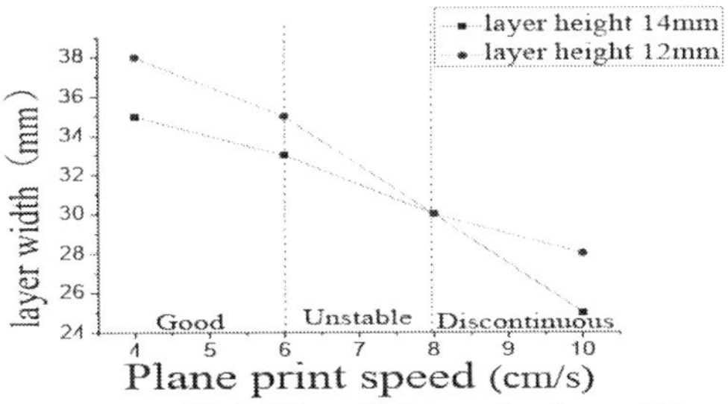

Fig.8 Effect of Print speed on layer width

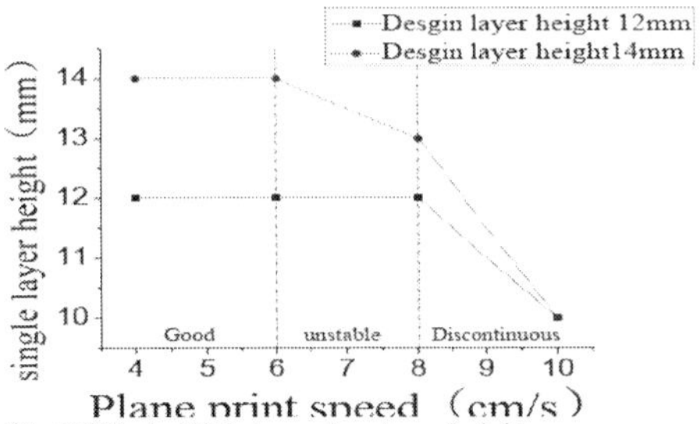

Fig.9 Effect of Print speed on layer height

According to the fig.8, we can see that under the same layer height and print volume, different printing speed directly affects the effect of component molding. In the range of suitable layer height, the excessive speed will have an impact on single layer stack stability. When the plane running speed is more than 6cm / s, the width of the single layer is biased. When the speed exceeds 8cm / s, the extrusion speed of the mortar cannot reach the moving speed. And in this test case, the most appropriate flat print speed should be between 4-8cm / s.

4.3 The Effect of Pipe length, arrangement and delivery equipment on Process

By printing 14mm layer height closed graphics with 2m, 3m, 4m of φ32mm high pressure rubber tube, test study single layer height, width and apparent effect using mortar mix ratio in the chapter 2.2(the flow performance range 210±20 mm).

Table 2 The Effect of Pipe length, arrangement

Pipe length (m)	The initial fluidity (mm)	The export fluidity (mm)		Suspension pipe
		Straight pipe	Crooked pipe (<90°)	
2	220	205	190	Discontinuous
3	230	200	180	Discontinuous
4	230	195	175	Discontinuous

The mortar material is pumping from the screw pump to the 3D printer. Due to the limited diameter of the screw pump output, the diameter of the rubber tube can only be selected as 32mm. During the printing test, the performance of the mortar is reduced by pumping, and the mortar performance decreases with the length of the pipe. At the same time, when the rubber tube is inevitably bending, the load of the pump is obviously increased, and the performance of the extruded mortar is also significantly reduced compared with the initial state, as shown in fig.10. And when the rubber tube is suspended, the mortar with fluidity between 210 ± 20mm cannot be continuously printed. Therefore, in the screw pump and rubber tube conditions, 3D printing should be based on construction requirements to select the appropriate length of the pipeline, and should ensure that the pipeline straight to avoid the pipe bending and suspension, it will help reduce the risk of blocking.

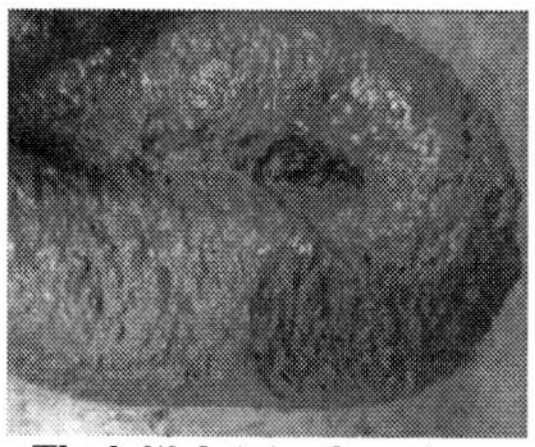

The initial state of mortar

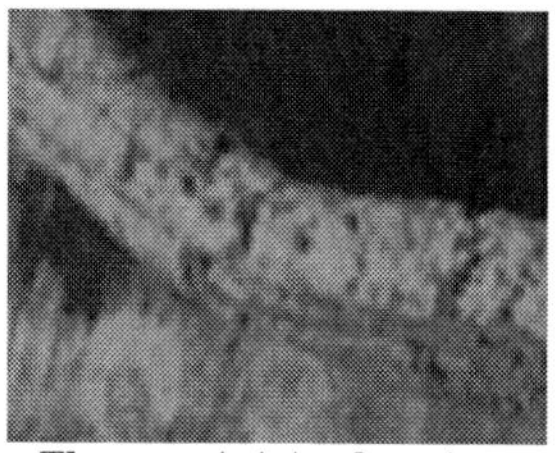

The export state of mortar

Fig.10 The effect of the conveying pipe

At the same time, it is found that the metal screw and rubber bushing of the screw pump produced some frictional heat during operation. If the extrusion rate is matched with the screw rotation rate, the heat was taken away from the mortar without affecting its pumping efficiency. However, if the material performance is degraded or the local pipe is slightly clogged, the heat generated by the screw pump will be insufficient for rapid dissipation, which will cause further deterioration of the mortar performance, resulting in pipe plugging. According to the experience of the test, the impact of printing equipment on the success of 3D print is shown in table 3:

Table 3 The impact of printing equipment on the success of 3D print

Influencing factors	Equipment performance degradation	Pipe length	Distribution method	Sealability
	Extremely	Middle	High	Extremely
Control measures	To remove hard materials in time	According to the component size	Try to remain straight	Keep fine airtightness

5 Conclusion

In the 3D printing with cement-based materials, the material properties and process parameters have a great impact on the actual printing process. In the test conditions, the results show that the best single-layer height and nozzle diameter ratio are between 0.4-0.6. The best parameters for the equipment are: the output displacement is 9 * $10^{-2}m^3$ / h, the printing speed should be controlled in 4-8cm / s; At the same time, 3D printing should try to choose a shorter length of the pipeline, and should ensure that the pipeline straight to avoid the pipe bending and suspension, it will help reduce the risk of blocking.

Acknowledgements

This paper is supported by Cscec Fund Project (cscec-2014-z-33).

References

1. LI Dichen, HE Jiankang,TIAN Xiaoyong,etc.

Additive Manufacturing: Integrated Fabrication of Macro/Microstructures [J]. Journal of Mechanical Engineering, 2013,06:129-135.

2. WANG Zhonghong, LI Yang, ZHANG Manying. The Present Situation and Development of China's 3D Printing Industry [J]. Economic Review,2013,28(1):90—93.
3. HUANG weidong. How to rationally view the material manufacturing (3D printing) technology [J].Adbanced Materials Industry,2013(8):9-12.
4. LIU Yuanyuan, ZHANG Fuhua, CHEN Weihua, etc. CAD/CAM System and Experimental Study of Biological 3D Printing Composite [J]. Journal of Mechanical Engineering, 2014,50 (15):147-154.
5. T.T.Le,S. A. Austin,S.Lim, R.A.Buswell, A.G.F.Gibb, T.Thorpe. Mix design and fresh properties for high-performance printing concrete[J].Materials and tructures, 2012, 45:1221-1232.
6. JIANG Jianing, GAO Yuxin,WU Xiong, etc. Research status and problem analysis of 3D printing concrete [J]. Concrete, 2015, 05: 62-65.

One-dimensional soil temperature assimilation experiment based on unscented particle filter and Common Land Model

Xiao Lei FU[1], Bao Ming JIN[1], Xiao Lei JIANG[2], Cheng CHEN[1]

[1] College of Civil Engineering, Fuzhou University, Fuzhou 350116, China
[2] State Key Laboratory of Hydrology-Water Resources and Hydraulic Engineering, Hohai University, Nanjing 210098, China

Abstract: Data assimilation is an efficient way to improve the simulation/prediction accuracy in many fields of geosciences especially in meteorological and hydrological applications. This study takes unscented particle filter (UPF) as an example to test its performance at different two probability distribution, Gaussian and Uniform distributions with two different assimilation frequencies experiments (1) assimilating hourly in situ soil surface temperature, (2) assimilating the original Moderate Resolution Imaging Spectroradiometer (MODIS) Land Surface Temperature (LST) once per day. The numerical experiment results show that the filter performs better when increasing the assimilation frequency. In addition, UPF is efficient for improving the soil variables (e.g., soil temperature) simulation/prediction accuracy, though it is not sensitive to the probability distribution for observation error in soil temperature assimilation.

1 Introduction

Soil temperature is an important variable in land surface processes, as it not only influences the energy partitioning [1-2], but also affects the water cycle of land-atmosphere system [3-4]. However, it is difficult to obtain accurate soil temperature information at different soil depths at large scales through *in situ* observation, land surface model, or remote sensing. It is impractical to conduct intensive *in situ* observations over large areas although its accuracy is high [5]. Using a land surface model, one can simulate soil temperature at the large scales even for different soil layers, but the accuracy of such simulation is variable and can be quite low [6-7]. Through satellite remote sensing, one can derive soil surface temperature information at continental and global scales, but satellite retrieval of soil temperature is also limited to the soil surface at a relative large sacale, and is not error free [8-10].

Data assimilation techniques were introduced to geosciences such as weather forecasting and hydrology as an efficient way to incorporate observations into model states to improve the accuracy of simulated state variables [11-13]. In the past few decades, numerous data assimilation methods have been developed for different applications [5,14]. For example, Evensen (1994) proposed ensemble Kalman filter to solve the high-order nonlinear problems, and has been used in hydrologic applications [15]. Van der Merwe [16] introduced unscented particle filter, and its performance was tested by Han and Li [17] through soil temperature assimilation experiment. Yu et al [2,5] introduced ensemble particle filter, compared it to ensemble Kalman filter and particle filter through soil temperature simulation, and used it to improve the soil temperature profile by assimilating MODIS LST. This study takes unscented particle filter as an example, and aims to test its performance with different assimilation frequencies and probability distributions of observation error in land surface state application.

2 Study area and data description

Walnut Gulch Experimental Watershed (WGEW), located in southeastern Arizona and with an area of approximately 149 km^2 [18], is selected as the study area (Fig. 1), given the large amount of in situ and remote sensing data available in the watershed. The data used in the study include the meteorological forcing data, soil hydrological properties, and MODIS LST. Detailed information about the study area can be found in Keefer et al. [18], hence not repeated here. The observed data in KENMet site are used to evaluate the filter performance.

3 Methodology

3.1. Model operator

The model used in this study is the Common Land Model (CLM), and details about CLM can be found in Oleson et al. [19]. The CLM can have up to 10 layers in the soil column and up to five layers of snow depending on the snow depth [19]. In this study, the soil column is discretized into five layers. Only the one-dimensional

*Corresponding author: Email: fuxiaolei518@163.com

heat diffusion equation for soil temperature in CLM describes here briefly as follows:

$$c\frac{\partial T}{\partial t} = -\frac{\partial F}{\partial z} \quad (1)$$

$$F = -\lambda\frac{\partial T}{\partial z} \quad (2)$$

Where c is volumetric soil heat capacity ($Jm^{-3}K^{-1}$), T is soil temperature (K), t is time (h), F is heat flux (Wm^{-2}), λ is thermal conductivity ($Wm^{-1}K^{-1}$), and z is vertical depth (m).

3.2. Unscented particle filter (UPF)

The unscented particle filter is proposed by Van der Merwe [16], using the unscented Kalman filter (UKF, which can be found in Han and Li [17]) to obtain the mean and covariance of the proposal distribution for each particle. The details of UPF can be found in Han and Li [17] and not list in this study to avoid redundancy.

In addition, to evaluate the performance of the filter, the root mean square error (RMSE) is used to be the only evaluation metric for soil temperature.

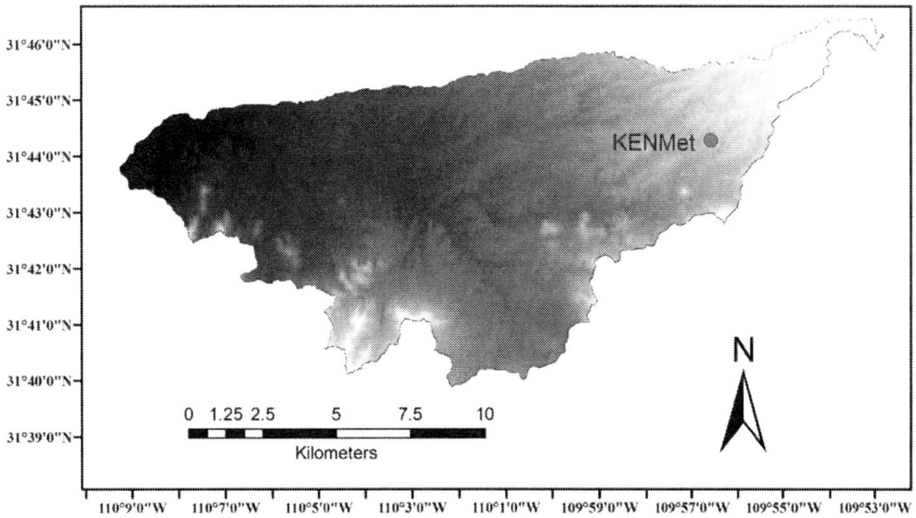

Fig. 1. The location of KENMet site which is selected to verify the filters performance at WGEW.

4 Results and discussions

In the numerical experiments, the observed soil surface temperature, MODIS LST from 2 October (day 276) to 21 October (day 295), 2004 were assimilated into CLM. To show the comparison results clearly, the last 5-day (from day 286 to day 290) simulated and assimilated results are used to verify the performance of the assimilation methods.

4.1. Assimilating hourly in situ soil surface layer temperature into the model

The RMSE between the simulated/assimilated results and observed data in KENMet site of the 5-day is shown in table 1. From the table, it is noted that UPF performs well for improving soil temperature profile except the last soil layer. In addition, it can be found that the filter is not sensitive to the probability distribution in soil temperature simulation. Therefore, only the average simulated and assimilated results of the 5-day with Gaussian distribution for hourly observations assimilation shown in Fig. 2.

From Fig. 2, it is noted that the averaged open loop (simulated results without assimilation) is far from the observed data at each layer, especially the open loop underestimated soil temperature significantly when soil temperature reached to its peak value for the first three layers. The reason may be attributed to the uncertainty of input and model.

Fig. 2 shows that the assimilated results of the filter are closer to the observed data than the simulated data in the first four layers. But for the fifth layer (e.g., the bottom layer), the performance of the filter is poor, this may be because only the soil surface temperature was assimilated, the soil heat flux is assumed to zero at the under boundary in the model and soil temperature at this layer changes in-obvious. The figure shows that the assimilated results underestimated the soil temperature at some time as the open loop underestimated it at the same time. However, the filter performed much better when soil temperature reached to its peak value for the first three layers. In addition, it is noted that the amplitude of the simulated/assimilated soil temperature is smaller than observed at these layers.

Table 1. The 5-day soil temperature RMSE (K) of simulation and UPF.

Depth (cm)	Simulation	Hourly Gaussian	Hourly uniform	LST Gaussian	LST uniform
0	2.2024	1.5575	1.5396	2.0537	2.0763
5	1.6029	1.0079	1.0038	1.1670	1.1774
15	0.9673	0.6585	0.6504	0.8619	0.8857
50	0.4277	0.2810	0.3019	0.2820	0.3127
100	0.1801	0.2597	0.3134	0.3846	0.3701

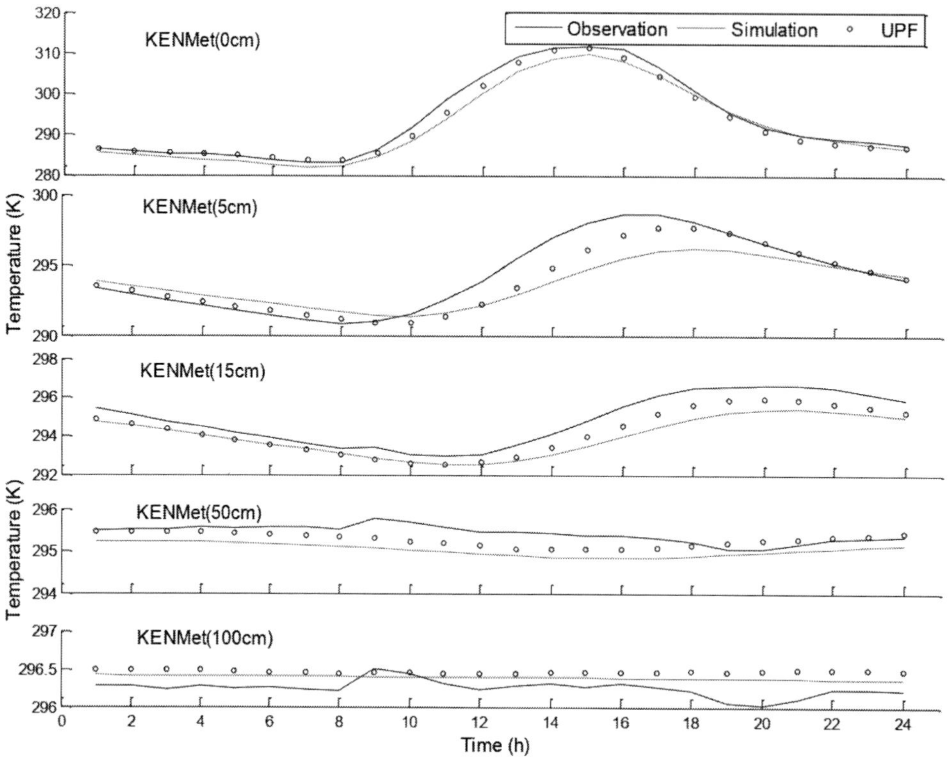

Fig. 2. The 5-day average results of assimilating hourly in situ soil surface observations using EnKF and UPF with Gaussian distribution.

4.2. Assimilating the original MODIS LST once per day

In this experiment, we assimilate the original MODIS LST into the model once per day as there is only one MODIS LST of the WGEW every day when the MODIS sensor passes over the area about from 10:30am to 11:00am. Compared with assimilation of hourly in situ observations in section 4.1, the assimilation frequency in this experiment is 1/24 of that in above experiment. Similarly, the 5-day (from day 286 to day 290) results are also used to verify the filter performance.

Similarly, only the average results of the 5-day with Gaussian distribution for MODIS LST assimilation are shown in Fig. 3. The results in this figure are similar to that in Fig. 2. However, it is noted that the filter assimilated results are closer to the observed data at the first four layers in Fig. 3, but not significant like that in Fig. 2. The filter performs poorly in the fifth layer, which is consistent to the results in section 4.1.

Comparing the assimilating hourly in situ soil surface temperature and assimilating MODIS LST experiments, it is noted that the main difference of the two experiments is the assimilation frequency, and the filter performs better with increasing the assimilated observations. Comparing the experiment with Gaussian distribution to that with uniform distribution for observation, it can be found that the difference of the filter performance is small. This demonstrates that the filter is not sensitive to error distribution for observation in some state variables (e.g., soil temperature) simulation without considering the assimilation frequency, this may be because the variables are relative stable and not varied sharply in spatiotemporal.

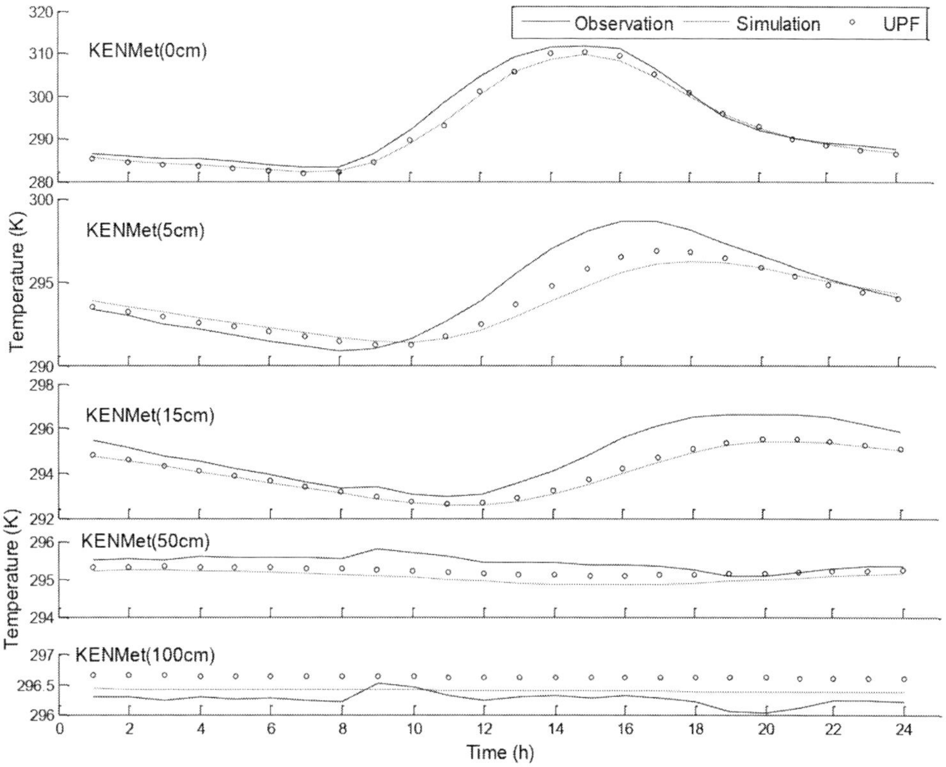

Fig. 3. The 5-day average results of assimilating MODIS LST using EnKF and UPF with Gaussian distribution. The MODIS sensor passes over the area about from 10:30am to 11:00am.

5 Conclusion

In this study, we evaluate UPF performance based on different assimilation frequencies and different probability distributions of observation error through soil temperature simulation at Walnut Gulch Experimental Watershed (WGEW) in Arizona, United States. The numerical experiment shows that the filter performs well in soil temperature assimilation at upper several soil layers. When assimilation frequency is low (e.g., once per day), it can improve the simulation accuracy, although the improved accuracy is not significant. Furthermore, it performs well significantly as increased the assimilation frequency (e.g., once per day to hourly). In addition, the filter is not sensitive to error distribution for observation in soil temperature assimilation without considering assimilation frequency.

References

1. Yu Z, Carlson T N, Barron E J, et al. On evaluating the spatial-temporal variation of soil moisture in the Susquehanna River Basin [J]. Water Resources Research, 2001, 34: 1313-1326.
2. Yu Z, Fu X, Lü H, et al. Evaluating ensemble Kalman, particle and ensemble particle filters through soil temperature prediction [J]. Journal of Hyrologic Engineering, 2014, 19: 04014027.
3. Yu Z, Barron E J, Yarnal B, et al. Evaluation of basin scale hydrologic response to a multi storm simulation [J]. Journal of Hydrology, 2002, 257: 212-225.
4. Huang C, Li X, Lu L. Retrieving soil temperature profile by assimilating MODIS LST products with ensemble Kalman filter [J]. Remote Sensing of Environment, 2008, 112(4): 1320-1336.
5. Yu Z, Fu X, Luo L. One-dimensional soil temperature simulation with Common Land Model by assimilating in situ observations and MODIS LST with the ensemble particle filter [J]. Water Resources Research, 2014, 50: 6950-6965.
6. Mihalakakou G. On estimating soil surface temperature profiles [J]. Energy and Buildings, 2002, 34(3): 251-259.
7. Wu C L, Chau K W, Huang J S. Modeling coupled water and heat transport in a soil-mulch-plant-atmosphere continuum (SMPAC) system [J]. Applied Mathematical Modeling, 2007, 31 (2): 152-169.
8. Wan Z. Dozier J. A generalized split-window algorithm for retrieving land-surface temperature from space [J]. IEEE Transactions on Geoscience and Remote Sensing, 1996, 34(4): 892-905.
9. Njoku E G, Li L. Retrieval of land surface parameters using passive microwave measurements

at 6-18GHz [J]. IEEE Transactions on Geoscience and Remote Sensing, 1999, 37(1): 79-93.

10. Owe M, Van de Griend A A. On the relationship between thermodynamic surface temperature and high frequency (37GHz) vertical polarization brightness temperature under semi-arid conditions [J]. International Journal of Remote Sensing, 2001, 22(17): 3521-3532.
11. Kumar S V, Reichle R H, Peters-Lidard C D, et al. A land surface data assimilation framework using the land information system: Description and applications [J]. dvances in Water Resources, 2008, 31(11): 1419–1432.
12. McLaughlin D. An integrated approach to hydrologic data assimilation: interpolation, smoothing, and filtering [J]. Advances in Water Resources, 2002, 25(8-12):1275–1286.
13. Reichle R H. Data assimilation methods in the earth sciences [J]. Advances in Water Resources, 2008, 31(11): 1411–1418.
14. Fu X, Yu Z, Luo L, et al. Investigating soil moisture sensitivity to precipitation and evapotranspiration errors using SiB2 model and ensemble Kalman filter [J]. Stochastic Environmental Research and Risk Assessment, 2014, 28: 681-693.
15. McLaughlin D. Recent developments in hydrologic data assimilation [J]. Reviews of Geophysics, 1995, 33: 977-984.
16. Van der Merwe R. Sigma-Point Kalman Filters for Probalistic Inference in Dynamic State-Space Models [D]. Oregon Health and Science University, 2004.
17. Han X, Li X. An evaluation of the nonlinear/non-Gaussian filters for the sequential data assimilation [J]. Remote Sensing of Environment, 2008, 112(4): 1434–1449.
18. Keefer T O, Moran M S, Paige G B. Long-term meteorological and soil hydrology database, Walnut Gulch Experimental Watershed, Arizona, United States [J]. Water Resources Research, 2008, 44: W05S07.
19. Oleson K, Dai Y, Bonan G, et al. Technical description of the Community Land Model (CLM) [M]. NCAR Technical Note NCAR/TN-461+STR, National Center for Atmospheric Research, Boulder, Colo, 2004.

Analysis of the Psychological Conflict of Contemporary Street Aesthetics in China from the Views of Media Communications

Chen FENG, Hua Wei XU

Faculty of Architecture and Urban Planning of Chongqing University, Chongqing, China

Abstract: New media patterns generate more complicated and diversified space forms. Spatial experiences that are instant, fragmented, reality-virtuality interlaced are ubiquitous. Such interlacing brings an enormous impact on traditional Chinese streetscapes, increases the risks of the fragmentation of streetscapes, and leads to the disorder of visual aesthetics. Moreover, it necessarily imposes a great impact on traditional street aesthetics, and causes various psychological conflicts. This study aims to describe this phenomenon and try to explore the reason behind it in new view.

1 Introduction

The physical environment, electronic information and activities of Chinese urban street spaces in modern times are interacting with each other under the influences of the new media, demonstrating a highly interactive situation, namely, the "mediated quasi-interaction" [1]proposed by John B. Thompson. Physical space is being gradually included as a part of media, acting as a role that disseminates information. Streetscape is enclosed by spatial media of various forms from top to bottom, constructing the appearance of the city with a variety of ways and affecting how people recognize urban space psychologically.

2 Defamiliarization of Street Visions Caused by Foreign Scenes

The boundaries of connection in a traditional urban street space place people in a continuous and enclosed realm where the spatial experiences are homogeneous and linear. People's spatial perceptions change along with the patterns and speeds of directions and movements, and people perceive spatial order and structure according to such progressive continuity. The features of visual communication in the age of new media diversify the public interface that was used to be plain, leading to frequent changes of the visual images of street space.

2.1. The Unconventional "Foreign Landscapes"

"The worship of differences is built on the loss of differences."[2] With the rapid dissemination of a massive number of image information, disseminators choose to use images with a stronger visual impact while disseminating information in order to prevent them from being devoured. Because visual information that is of easier format and clearer theme can be more easily recognized and accepted in a visual way in terms of avoiding visual fatigue, and it can easily generate visual memory. Architectural forms and street scenes start "disintegrating", bringing an enormous impact on the visual image of street space.

The unconventional "foreign landscapes" are increasingly becoming an integral part of urban public landscapes, breaking the homogeneity of space and forming "singularities" one after another. Endless newfangled buildings seize the eyeballs of pedestrians with their exaggerated and powerful visual appeal. Visual spectacles are now common in street visual images.

2.2. The Fragmentation and Collage of Streetscapes

The visual spectacles of streets and surroundings are broken off, splitting the street space, and the saltant image icons seriously damage the connections and integrity of the convey of streetscape traits. Street space is filled with interrupted fragments, fractures and cracks, gradually disproportionating the balance of traditional landscape order, and making the "fragmentation of streetscapes" an increasingly serious problem.

The visual state of being "collage" and disorder make people become used to fragmented thinking. The fission of the mass communication in the urban street public space is now conveying information to smaller groups. The diverse representations in the streets are conducting free collage, and visual methods, such as "incompleteness, fragment, subversion, delusion, absurdity, and eeriness" are like "noisy points in images", breaking off with the surroundings and disproportionating the traditional balance in terms of

*Corresponding author: Email: 19535411@qq.com

visual images.

2.3. The Visual Defamiliarization of Street Space

We may borrow the term, "noisy point", in electronic imaging, as a way to conclude such phenomenon existed in the visual images of Chinese urban streets. The aesthetic concepts that are harmonious and complete are now being deconstructed and reshaped due to information differentiation of urban street space. The entire image is now disappearing, and those visual orders and rules of previous street space start collapsing, and the streetscapes that were used to be visible to people now no longer have integral clues to be found.

The fragmented thinking caused by such "visual mosaic and excessive symbolic elements", however, covers the differences among symbol vision. The street aesthetics, in general, demonstrate the state of "heteroglossia"[3]and the trend of "visual defamiliarization", quoted Bakhtin. The interface of streets resembles a paper of beautifully written words, which is visible yet unable to interpret the general implication. Foreign landscapes generate powerful visual defamiliarization. Despite of its efforts in innovation and striving for changes, such defamiliarization effects still pose challenges against traditional classic aesthetics. It questions about some of the basic aesthetic principles of urban design, and thoroughly overturns the traditional constituent principle of design, which is "symmetry, proportions, and sequence".

3 The Tearing of Street Interface Brought by the Trend of Labeling

Architectural forms may contend for public recognition, stimulate public awareness, and construct the value standard of users with the help of system symbols. Charles Sanders Peirce and Ferdinand de Saussure have proposed a modern academic thought, "symbology", in earlier days. With the use of the basic theories of "symbology", we may regard and interpret modern society in a more clarified way. This provides research basis for us to better understand various symbols.

3.1. The Unconventional "Foreign Landscapes"

With the development of new media technologies and modes of transmission, enterprises now have more and more powerful control over visual communications, and their discursive right in terms of space expression has been continuously enhanced. However, spatial visual images with clarified image perception may completely represent unique corporate images, and construct the enterprise's visual image in the city in a labeling way, bringing enormous economic values to the enterprise. Moreover, the commercial architectural interface, which is better in demonstrating hi-tech packaging and the fineness of decoration, becomes the best representation of aesthetic experiences, showing a trend of labeling that is more recognizable and memorable.

Under such influence, designers also believe that "in the construction process of complicated complex of commercial media, a conspicuous information transfer system shall be adopted to replace those inconspicuous ways of express"[4]. Lots of commercial buildings often come across as striking with their supernormal shapes, colors and textures, and they are of intense traits of "sculpture" or "painting", giving people incredible visual association, and becoming "image features" as well as unique "commercial symbols".

3.2. Exaggerated Symbols that are Different from "Official" Landmarks

Unlike the "official" background of normal urban landmarks, commercial buildings that are affected by pan-aesthetic orientation are more instantaneous and dramatic. They instinctively cater to the public psychology of cultural consumption, and satisfy people's pursuit of fashion and pleasures. They release the tension of the pursuit of the absolute "official" landmarks. Because their familiar cognitive impression of the public, they achieve a short moment of relaxation psychologically while enhancing visual attraction. Various graphic symbols that are "created" try their uttermost to meet people's pursuit of fashion and pleasures, making them filled with instantaneous and dramatic effects, catering to the public psychology of cultural consumption, quickly occupying our daily vision, and bringing deep "capital branding" on urban public space.

3.3. The Extreme Simplification of Street Images

In a chaotic street space where fragmented shapes are scattered around, diverse information symbols are concentrated, and simplistic as well as primitive forms will undoubtedly arouse visual attraction. Under such influence, modern architecture starts pursuing concise "spirit of geometry" again. Streets, as a result, become the arena of commodity devices, symbols and information, and street images quickly become flattened and symbolized.

Concise space images start becoming brand symbols after they have achieved tremendous commercial values, which can be seen from the color tones of buildings to their surfaces, their volumes, and even urban space. Such "labeling" phenomenon of street space has been penetrating ubiquitously, imposing an enormous negative impact on the visual order of urban streets. The labeled buildings resemble symbolic scenes in cities, and bringing negative impacts such as the tearing of urban street interface and the loss of spatial cognition.

4 The Overload of Street Visual Information Resulted from Image Sticking

David Shenk discovered that "developed media technologies have brought people from the age where information was insufficient to the age of information redundancy"[5]. Excessive hands-down information has been interfering how people estimate and view authentic space environment. People spend lots of time and energies on information processing, and overloading information even makes information anxiety disorder a new issue of the urban development promoted by media.

4.1. The Explosive Growth of the Total Amount of Street Information

The simplicity of digital reproduction leads to the explosive growth of the total amount of street information, and people are able to access information of which the amount is many times higher than that in the past in every unit period of time. The life pace of modern society does not allow people to meticulously focus on one object. Therefore, people have to rush through when the street space is filled with overloading visual information, and there is no more viewing of perspective and contemplation. Instead, we have to accept visual objects by means of viewing or glancing. An excessive amount of information leads to the dilemma of information receiving, and we are put in a predicament of insufficient information.

When there is an excessive amount of information, people are interfered so much that they can hardly determine, and their attention becomes the thing being "gnawed". To arouse people's attention better, scenes of stimulation have to be constantly changed in order to promote people's attention shift. Therefore, urban street space has officially entered a space where the competition for attention is extremely fierce.

4.2. The Collapse of the Hierarchy of Street Information

Urban streets are gradually evolving into a world of symbols made of information under the influence of new media. Various kinds of information exchange, radiation, spreading and transfer gather here rapidly, and media of communications appear to be highly saturated. The interactive experiences of streetscapes also accelerate the speed of information communications. The openness of structure, the overlapping of functions, and diversity of forms all provide new possibilities for the richness of urban streets. Excessive symbolism and abduction lead to massive waste of urban street space information resources since information is often replaced before it is expressed, and electronic footprints and bitty data collections are scattered everywhere in the street space.

The communications of image information are often ignored and drowned, so epidermidalized method with a strong visual impact is chosen for the expression. This image information that is created in different times is being continuously accumulated on the skins of buildings, gradually leading to overflow and overload. The visual system of streets is manifested as information surplus, overflow and the collapse of information hierarchy.

4.3. The Influence of Image Saturation on the Delivery of the Street Visual Information

As the mark of the age of media, overflowing surplus visual information is invading the limited visual space. The rapid information flow of new media has been impacting people's working load limit fiercely. Such surplus information gradually become the so-called "photochemical garbage", generating a sense of pressure on our vision. The bright and noisy urban visual background interfered people's biological rhythms, and affecting the general health levels of people.

The excessive visual image sticking and saturation have completely surpassed the basic reading of the audience towards space sceneries. Various kinds of information are polluting our vision and invading our visual space, thus causing the loss of valuable information, affecting the effective delivery of visual information, causing the fatigue of attention, leading to excessive consumption of attention, and making people unable to extract useful information out of a complicated environment. Such phenomenon of information overload, incapability of filtering and evaluating information is referred to as the "AIDS (anti-information deficiency syndrome)"[6] , according to Boltzmann.

5 The Space-Time Disorder Resulted from Multi-Dimensional Concatenation

Various virtual media are creating phantasmagoric reality for cities, and a feeling of space-time tension is generated by the virtual space squeezing the real space. Media arts brought us the idea of "de-distantiation", which eliminates the "aesthetic distance", a point of aesthetic view of authority that has been advocated by traditional aesthetic theories. Media led to the information grafting in the age of globalization, and the random redirection between images that has no inner logics leads to the space-time condensation of street visual scenes.

5.1. The Dissipation of the Sense of Dimensions of the Street's Real Space-Time

Urban street space is no longer a pure material and substantial space under the influence of electronic media. Instead, it is changed into a composite state where information elements and virtual space are intermingled, isomorphic co-existing. Therefore, the space boundaries and rules that were used to be well-defined are now promiscuous and confused. "The things that are not supposed to be together and shown to all are now shown." [7]

The "parts and fragments" in street space are collaged, and forms of different histories, cultures and geographies show themselves simultaneously in the same scene, "reflecting the sense of human nature and the historical connotations of the base in the scene"[8]with "real analogs" that extend public memory and feelings. People cannot precisely locate in space dimensions easily, and we are no longer able to determine the environment and define our directions by mobilizing our sensory cognition as we are in it. Gradually, we lost the sense of dimensions of the real space and time.

5.2. The Montage-Style Reconstruction of Street Conception Brought by Space-Time Condensation

Traditional space-time view has gone through fundamental changes in the age of information. The urban street space becomes smaller and smaller, and the interface is no longer continuous and complete. However, the rapid development of science and technology makes the collage of any images easy. As a result, the fragments of city images come in flocks under the background of "space-time condensation", and they change quickly. Street scenes are dissevered into countless non-continuous fragments.

"The non-linear narrative scene" has completely deconstructed and overturned logical narration and time narration. Such "fragmented" scenery fracture facilitates even larger and more zones of fracture in people's daily life like a "butterfly effect". The order between the past, present and future in space becomes confused. Also, the fracture of time and the highlighting of "present" make everything lose their inner connections.

5.3. The Streetscape Delusion Has Altered the Real Experiences of Space

The powerful media technologies have altered the way the landscapes are created, allowing the audience to immerse themselves in a world of virtual simulacrums. As a result, the streets are filled with a massive number of exaggerated and implicitly metaphorical "landscape delusion". Such "virtual speed" has breached the boundaries between the real and virtual world, and the extremely surreal world overwhelms us. So many circulating image signals have altered the real experiences of people's urban life. Unknowingly, the recognition of the city's geographical, historical and cultural depth is weakened, and the cultural space is dissevered and overturned.

6 Conclusion

This study interprets the aesthetic phenomena of urban streets and aesthetic psychological conflicts, which is different from the studies that traditionally focus on "whether a form is good or bad". It is not limited to the aesthetic study of a certain attribute such as materials, structure and color. Instead, the aesthetic environment and the shift of people's aesthetic psychology are discussed from the perspective of media, and a special study of various psychological conflicts that may take place during the process is carried out.

References

1. John B. T. Ideology and Modern Culture [M]. Trans. by Xian G. et al. Nanjing: Yilin Press, 2012.
2. Jean B. The Consumer Society [M]. Trans. by Chengfu L., Zhigang Q., Nanjing: Nanjing University Press, 2000.
3. Jean F. L. The Condition of Postmodernity [M]. Trans. by Daozi, Changsha: Hunan Arts Press, 1996.
4. Robert V. Learning from Las Vegas [M]. Trans. by Yifang X., Jian W., China WaterPower Press, 2006.
5. Shenk D. Data Smog: Surviving in the Information Glut [M]. Jiangxi Education Press, 2001.
6. Neil P. Technopoly – The Surrender of Culture to Technology [M]. Trans. by Daokuan H., Peking University Press, 2007.
7. Joshua M. No Sense of Place: The Impact of Electronic Media on Social Behavior [M]. Trans. by Zhijun X. Beijing: Tsinghua University Press, 2002.
8. Colin R. Collage City [M]. Trans. by Tong Ming, Beijing: China Architecture & Building Press, 2003.

Study on load test of 100m cross-reinforced deck type concrete box arch bridge

Jing Xian SHI[1], Ying Jie CHENG[2]

[1]Oxbridge College, Kunming University of Science and Technology,KunMing 650106 Yunnan, China;
[2]Yunnan Aerospace Engineering Geophysical Detecting Co.Ltd, Kunming 650217, Yunnan, China

Abstract: Found in the routine quality inspection of highway bridge that many vertical fractures on the main beam (10mT beam) of the steel reinforced concrete arch bridge near the hydropower station. In order to grasp the bearing capacity of this bridge under working conditions with cracks, the static load and dynamic load test of box arch bridge are carried out. The Midas civil theory is calculated by using the special plate trailer - 300 as the calculation load, and the deflection and stress of the critical section are tested by the equivalent cloth load in the test vehicle. The pulsation test, obstacles and no obstacle driving test were carried out. Experimental results show that the bridge under the condition of the test loads is in safe condition, main bearing component of the strength and stiffness meet the design requirements, the crack width does not increase, in the process of loading bridge overall work performance is good.

1 Introduction

Arch bridge as an application example of arch structure, also become one of the longest bridge in the history of mankind, its light structure, beautiful shape, has the strong compression capability, widely used in highway construction, especially in the mountainous area in Yunnan province.

This article analyzes the object on the secondary highway of a mountain in Yunnan province. On the highway bridge during commissioning acceptance, routine inspection found in the 10m T beam bridge upper structure, there are more vertical cracks, especially the second and ninth on arch, the bridge in the case of fracture of carrying capacity should be further identified. The bridge load test is the most effective and direct method to understand the performance parameters of the bridge and analyze its working condition. In order to study the carrying capacity of this bridge, load test includes static load should be used.

2 Engineering Overview

The structure of the bridge is the upper bearing reinforced concrete box arch bridge (figure 1), and the main arch ring consists of 6 arch boxes, the arch axis coefficient is 2.1, the vector ratio is 1/5. The main span is 100 meters, the vertical column and the guide hole are the steel reinforced concrete T beam with an average span of 10m, and the two-lane two-way driving. Auto load level for highway- II ; The calculation load is special plate trailer -300; Seismic fortification for the VII; The peak acceleration was 0.10 g.

Figure 1 Panorama of bridge

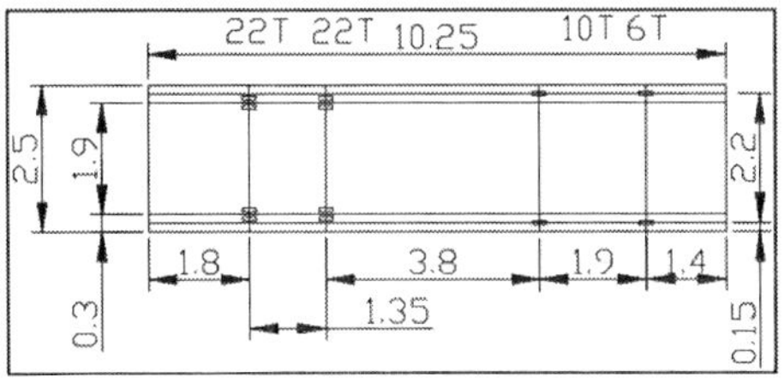

Figure 2 Loading planar size of the vehicle(unit/m)

*Corresponding author: Email: sara_shivip@163.com mail: 532965722@qq.com

The prefabricated arch, the current roof and the arch are all C40 concrete. The precast T-beam, the frame and the wall of the arch, the wall of the arch, the bridge pier, the cover girder is C30 concrete; The base class adopts C25 concrete. The diameter of reinforcement is greater than or equal to 12 mm, and HRB335 is adopted.

3 Load test pattern

According to the principle of internal force equivalent, the internal force of the structure under the action of the test load should be equal or close to the internal force of the structure under the design load. The strain (stress) and deflection of the control cross section under test load are analyzed, and the results of the test are compared with the calculation results.

3.1 Statical test

In order to meet the requirements of bridge bearing capacity test, there are five working conditions of the most adverse stress state of bridge structure were selected. Respectively are: working condition1-load test of maximum pressure stress in the top section of the main arch; working condition2-load test of the maximum compressive resistance in the cross section of the 9th midspan of T beam; working condition3---load test of the maximum compressive resistance in the cross section of the 11th midspan of T beam; working condition4-1#load test of the maximum axial force of the frame section; working condition5-9#load test of the maximum axial force of the frame section. The loading vehicle calculation diagram is shown in figure 2.

The deformation measurement points are arranged on the deck, with a total of 4 sections (figure3) and 3 points of each section (figure 4). At the corresponding points of the bridge, the height mark nails are set, and the relative elevation of the various working conditions is tested by the water level, and the deformation (displacement) of the structure is determined.

The strain tests in five control sections are shown in figure 6.The lateral arrangement is shown in figure 5,7 and 8, and the triangle symbol in figure 7 is the dynamic strain point. The static strain test system is used to test the resistance strain gauge. The static strain test system is used to test each cross section with resistance strain gauge.

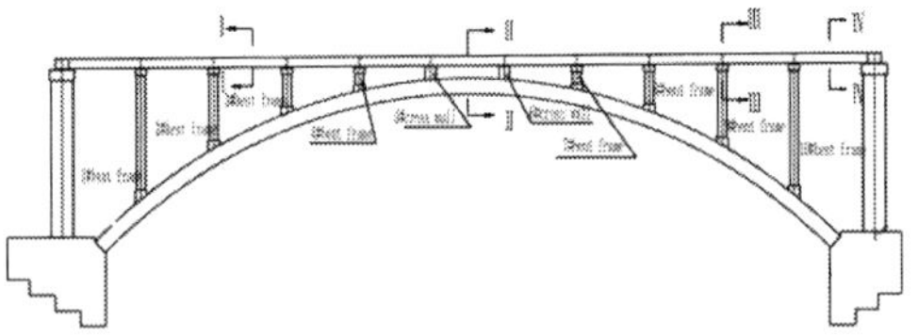

Figure 3 Vertical layout of of deformation measuring points

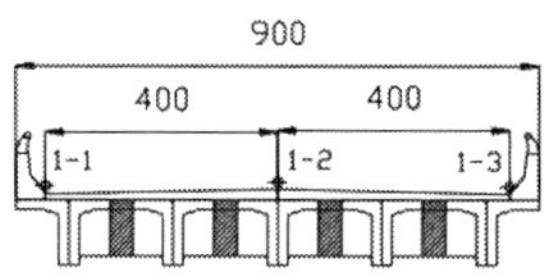

Figure 4 Deformation measuring point of 1-1 section

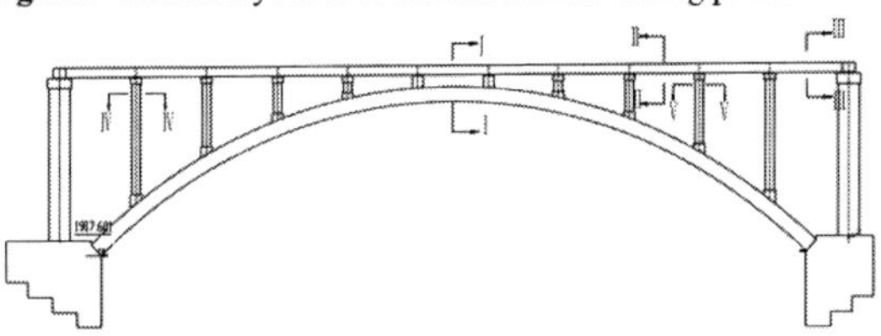

Figure 6 Vertical layout of Strain test points

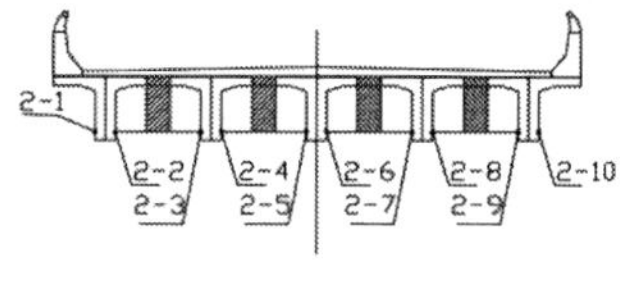

Figure 5 Strain test points of II - II section

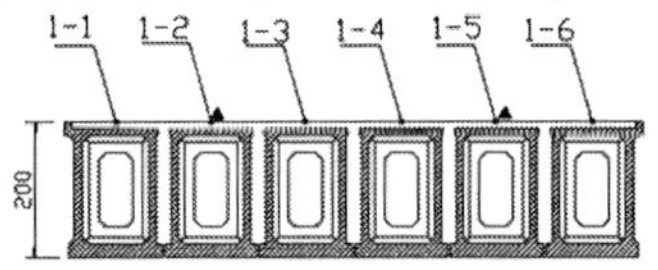

Figure 7 Strain test points of 1-1 section

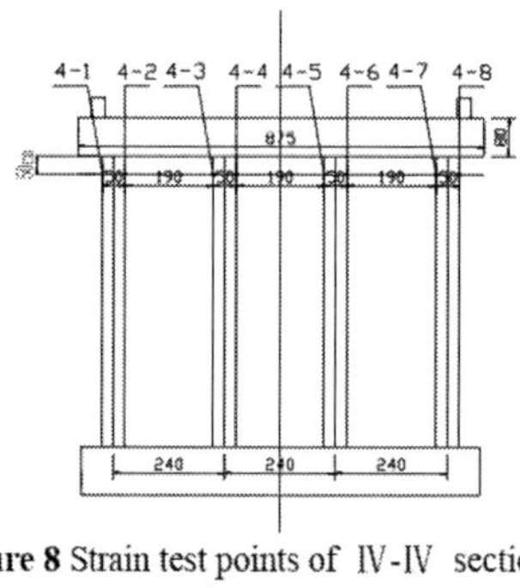

Figure 8 Strain test points of IV-IV section

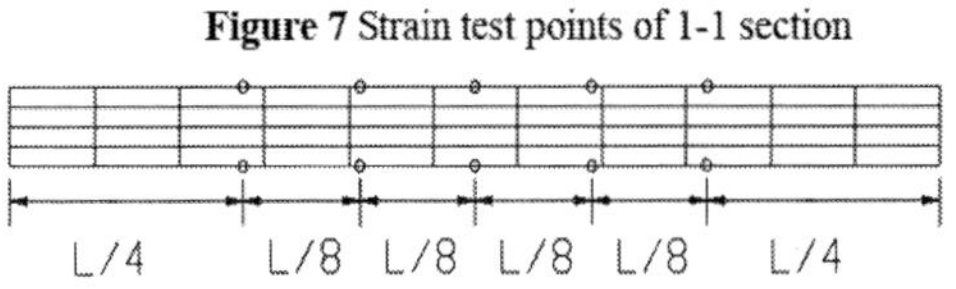

Figure 9 Measuring points of pulsating test

3.2 Dynamic test

3.2.1 Pulsating test

The vibration frequency, vibration mode and damping ratio are the inherent characteristics of the bridge, which

*Corresponding author: Email: sara_shivip@163.com mail: 532965722@qq.com

are related to the span, structure and material of the bridge. The self-vibration characteristics of the bridge structure are obtained through the pulsation experiment (the heavy vehicle excitation is adopted when necessary). The pulsation test point is arranged on the deck (figure 9) to test the acceleration of the free vibration of the bridge.

3.2.2 Barrier-free driving test

Barrier-free driving test, for testing: the dynamic response of the bridge structure under the load of the moving vehicle when the bridge deck pavement is in perfect condition, then simulate normal driving condition, using one 30 tons truck and on 5, 10, 20, 30, 40 km/h speed commute through the bridge, measuring the dynamic response of the bridge, the bridge test and calculate the actual impact coefficient.

3.2.3 Obstacles driving test

Simulating the condition of local damage of deck pavement, the dynamic response of bridge span structure under the action of moving vehicle load. The simulated obstacle height is about 5 - 10cm, with 1 30 ton trucks passing through the bridge at the speed of 5km/h.

4 Theoretical calculation

By using the bridge structure analysis software Midas Civil, the static dynamic calculation model of the bridge space is established (only the test span is modeled and calculated). According to the regulations of the bridge design, considering the actual bridge operation load and calculate the indicators in the design criterion. The calculation results provide a reference basis for the detailed plan of the load test.

The finite element calculation model is shown in Figure 10.The contents and results of the theoretical calculation are follows:

1) Calculate the load efficiency factor based on test load(test internal force / design internal force). According to the influence line to determine the test load size and position to ensure that the load efficiency coefficient is within 0.85 ~ 1.05 range.

The working conditions1:1-1~1-3 measuring point load test efficiency were 0.97, 1.01, 1.01; condition2:2-1~2-3 measuring point load test efficiency were 0.87, 0.91, 880; condition3:3-1~4-3 measuring point load test efficiency were 0.88, 0.98, 0.93; condition4:4-1 and 4-2 condition measuring point load test efficiency were 0.86, 0.88 and condition5: 5-1,5-2 condition measuring point load test efficiency were 0.87, 0.85.

2) Calculate the magnitude of each experiment result under the test loading, including stress, strain and displacement. The loading deformation at working condition 1-3 is shown in figure 11.

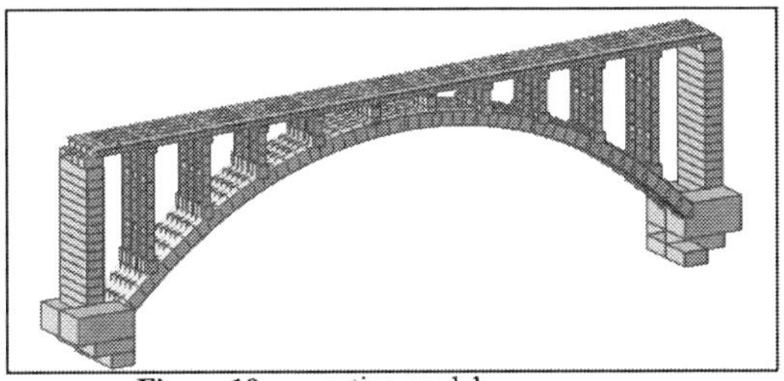

Figure 10 computing model

(a) Working condition 1

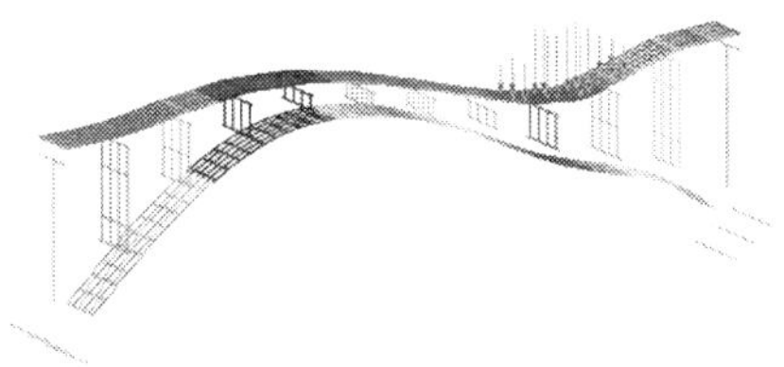

(b) Working condition 2

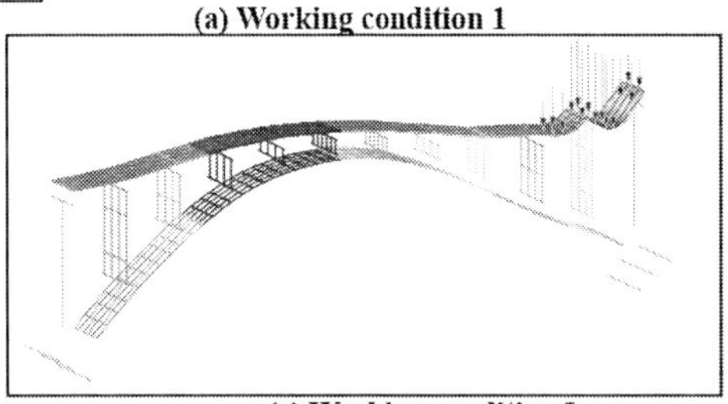

(c) Working condition 3

Figure 11 loading graph

3) Dynamic calculation results

The first 5 orders of frequencies are 1.800 HZ, 2.695 HZ, 4.841 HZ, 5.121 HZ and 5.750 HZ, respectively. The first and second modes are shown in the figure 12.

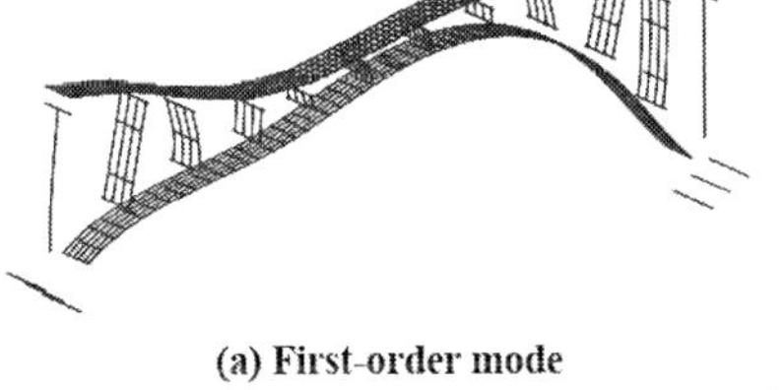

(a) First-order mode

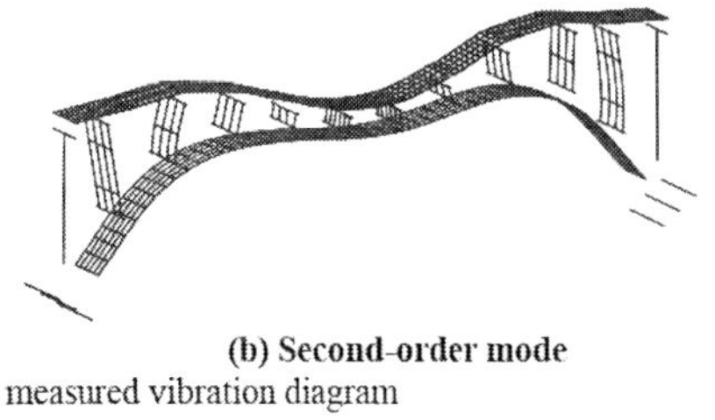

(b) Second-order mode

Figure12 Modal measured vibration diagram

5 The results of static test

5.1 Deformation test result and testing coefficients

The deformation test results of working condition 1-5 and the residual deformation after unloading are shown in Figure 13. The data in each cross section are disturbed and the points that the instrument can not collect are eliminated.

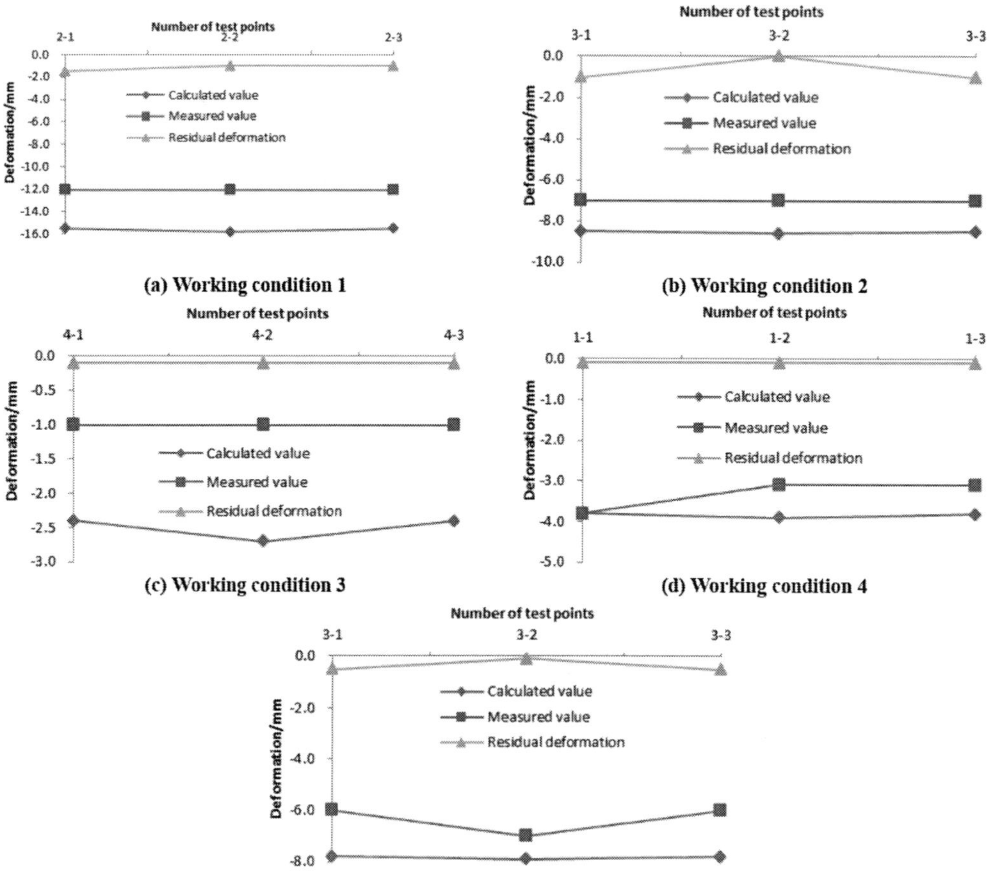

(a) Working condition 1

(b) Working condition 2

(c) Working condition 3

(d) Working condition 4

(e) Working condition 5

Figure 13 Deformation test results

5.2 Stress test and calibration coefficients

The stress condition of 1-5 (the measured strain is converted) test results and the residual strain after

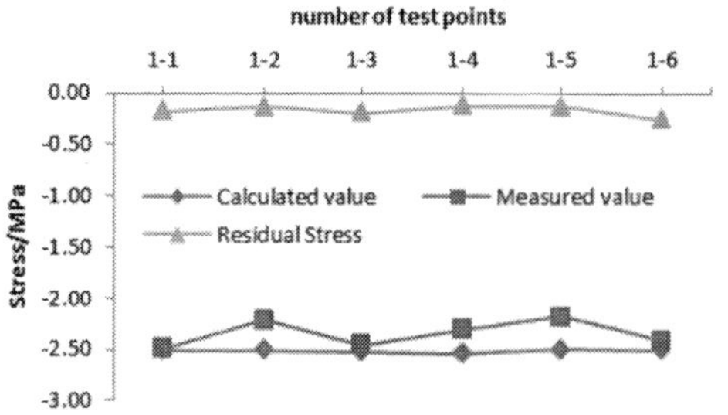

(a) Working condition 1

unloading values are shown in Figure 14, it can be show that the calculated values and test values, the residual stress value after unloading is between 1%-11% (the standard limit value is 20%)

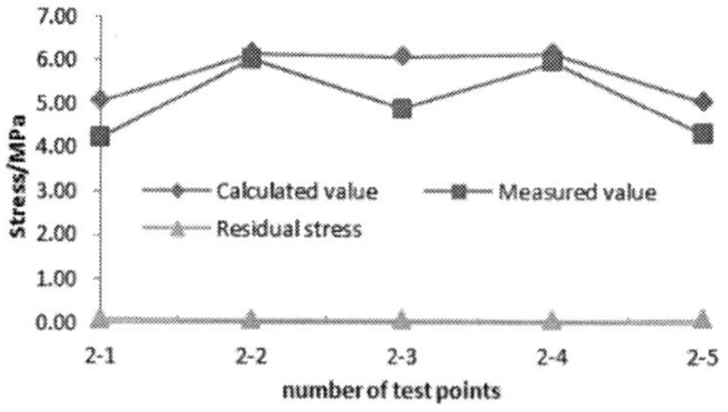

(b) Working condition 2

https://doi.org/10.1051/e3sconf/20183803011

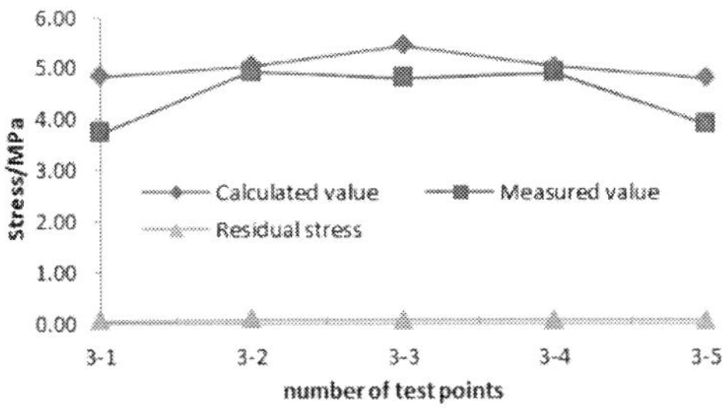

(c) **Working condition 3**

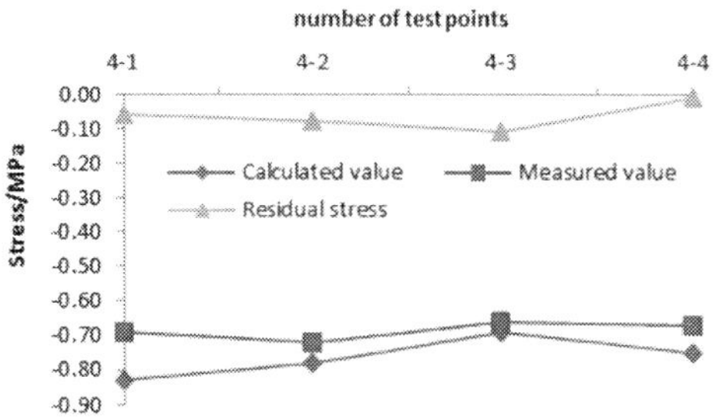

(d) **Working condition 4**

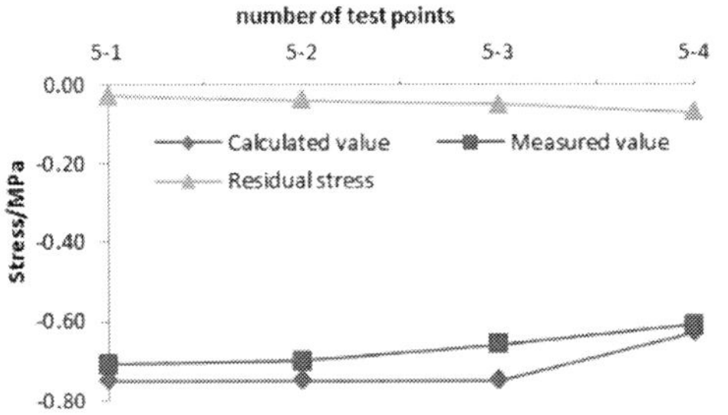

(e) **Working condition 5**

Figure 14 Stress test results

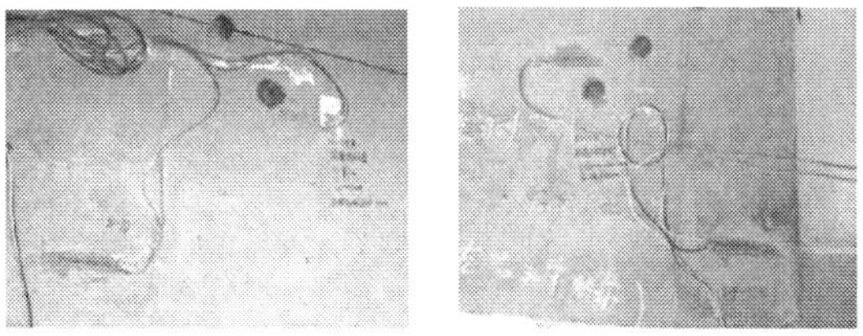

Figure 15 Site placement diagram

Table 1 Calibration coefficients of each test position

Position	Calibration coefficient	Position	Calibration coefficient	Position	Calibration coefficient	Position	Calibration coefficient	Position	Calibration coefficient
1-1	0.99	1-6	0.96	2-5	0.86	3-5	0.81	5-1	0.95
1-2	0.88	2-1	0.84	3-1	0.77	4-1	0.83	5-2	0.93
1-3	0.97	2-2	0.98	3-2	0.98	4-2	0.92	5-3	0.88
1-4	0.91	2-3	0.80	3-3	0.88	4-3	0.96	5-4	0.97
1-5	0.87	2-4	0.97	3-4	0.98	4-4	0.89		

5.3 Fracture stress test results

During the load test, the cracks observation point was arranged in the T beam of sixth and eleventh span of the main arch. In the crack position, the cross-seam patch is carried out (figure 15). The crack position is located on the second test section and third test section, Beam3-3, Beam3-5, Beam5-2, Beam5-5 left and right sides. The position diagram of the patch is shown in figure 15 and the test results is shown in table 2.

Table 2 Test results of crack deformation

Measuring position	Crack initial width/mm	Increased crack width/mm	Whether to return to zero after unloading	Measuring position	Crack initial width/mm	Increased crack width/mm	Whether to return to zero after unloading
Beam3-3-L	0.1mm	0.063	return to zero	Beam5-2-L	0.1mm	0.056	return to zero
Beam3-3-R	0.1mm	0.060	return to zero	Beam5-2-R	0.13mm	0.055	return to zero
Beam3-5-L	0.15mm	0.064	return to zero	Beam5-5-L	0.1mm	0.054	return to zero
Beam3-5-R	0.05mm	0.030	return to zero	Beam5-5-R	0.13mm	0.053	return to zero

Note: 3-3- L indicates the direction of the road forward, third spans, third T beams, and left side of the beam.

6 The results of Dynamic test

6.1 Free - vibration characteristic test

The modal shapes are shown in Figure 16. After the data analysis and processing, the fundamental frequency of the structural free - vibration is obtained, the measured values and theoretical values are shown in table 3.

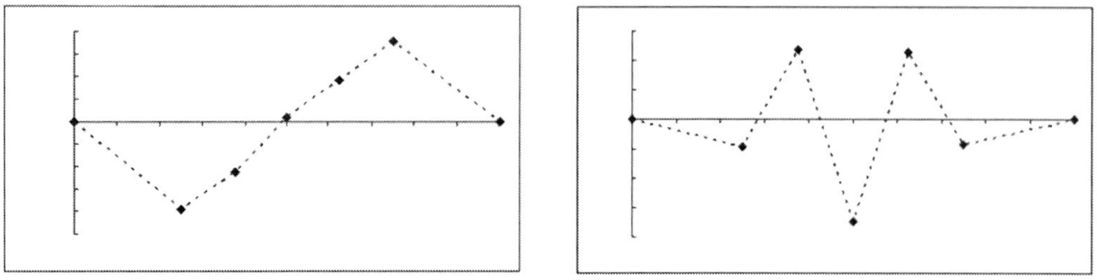

(a) **First-order mode** (b) **Second-order mode**

Figure16 Modal measured vibration diagram

Table 3 Natural frequency and damping test value

Measured order	Test value (Hz)	Calculated value (Hz)	ζ	Main vibration direction
First order	1.981	1.800	0.018	Vertical bending
Second order	2.854	2.695	0.015	Vertical bending

6.2 Impact coefficient test results

Through test, the typical dynamic response curve at different speed is tested. After processing the data, the impact coefficient of the bridge at different speed is shown in table 4 (The design specification calculates $1+\mu = 1.23$).

Table 4 Test value of impact coefficient

Measuring point	Speed/km/h	Maximum stress/MPa	Impact coefficient	Measuring point	Speed/km/h	Maximum stress/MPa	Impact coefficient
Dynamic strain measuring point 1 (1-2)	5	2.33	----	Dynamic strain measuring point 2 (1-5)	5	2.26	----
	10	2.38	1.018		10	2.29	1.010
	20	2.37	1.014		20	2.30	1.014
	30	2.38	1.018		30	2.31	1.018
	40	2.40	1.027		40	2.31	1.020
	5(Vehicle bu mping)	2.47	1.058		5(Vehicle b umping)	2.42	1.071

7 Conclusion

The test vehicle is conducted on the equivalent load according to the design load, and the results of static and dynamic tests show that the bridge has a good bearing capacity, and the bridge is in a safe state under the test load condition.

1) The strain and deformation of the bridge under the static load conditions show that the bridge is under the design load and is in the elastic working stage. After unloading, the zeroing condition is normal and no new cracks occur, which proves the strength of the main load-bearing members of the bridge meets the design requirements.

2) The test stress calibration factor of the bridge under static load is between 0.85~1.01, The measured maximum displacement is 12mm (less than the design limit), It shows that the bridge structure and materials meet the design requirements and the overall performance is good.

3) The bridge dynamic test results show that the first vertical free-vibration frequency is 1.981Hz, the damping ratio is 0.018, the impact coefficient is 1.010-1.071, the test mode and the theoretical modal shows good agreement and stiffness of bridge structure meets the design.

References

1. Wang Hai-peng, Study on the Bearing Capacity Estimation of Deck Type Concrete Box Arch Bridge Based on Load Test [J],CONST R UCTION TECHNOLOGY,2015(09):33-35.
2. HU Yong, TAN Dong-lian. Large-Tonnage Static Load Test and Simulation Analysison the Approach Way Bridge of a Hydropower Station[J]. esearch & Explore.2010(07).
3. Shuang Yang Zhang, Qian Hui Pu, Ren Da Zhao, Zhou Shi. Load Test Design and Stability Analysis of Basket Handle Arch Bridge with Reinforced Concrete[J]. Advanced Materials Research,2014,2837(838).
4. Cristopher D. Moen, Elaine E. Shapiro, Julia Hart. Structural Analysis and Load Test of a

Nineteenth-Century Iron Bowstring Arch-Truss Bridge[J]. Journal of Bridge Engineering,2013,18(3).

5. Devin K. Harris, John M. Civitillo, Amir Gheitasi. Performance and Behavior of Hybrid Composite Beam Bridge in Virginia: Live Load Testing[J]. Journal of Bridge Engineering,2016.
6. Łukasz Filar, Jerzy Kałuża, Marek Wazowski. Bridge Load Tests in Poland Today and Tomorrow – The Standard and the New Ways in Measuring and Research to Ensure Transport Safety[J]. Procedia Engineering,2017,192
7. Shervan Ataei, Meysam Jahangiri Alikamar, Vahid Kazemiashtiani. Evaluation of Axle Load Increasing on a Monumental Masonry Arch Bridge Based on Field Load Testing[J]. Construction and Building Materials,2016.
8. JTG/T J21-2011, Specification for Inspection and Evaluation of Load-bearing Capacity of Highway Bridges [S].
9. JTG / T J21-01-2015, Load Test Methods for Highway Bridges [S].
10. WU Xiao-guang, BAI Qing-xia, LEI Zi-xue. Application Calculation Example of Specifications for Strengthening Design of Highway Bridges [M]. Beijing: China Communications Press, 2011.

A Review of Research on Human-induced Vibration Serviceability and Dynamic Properties of Long-span Floor

Li Lin CAO[1, a], Cheng QIAN[1], Lan Zhe CHU[2],Guo Jun YU[1]

[1]Faculty of Civil Engineering and Mechanics, Jiangsu University, Zhenjiang 212013, China.
[2] Construction Engineering Quality Supervision Station, Zhenjiang 212000, China.

Abstract: Studies on the dynamic properties of long-span floor is an essential basis for a solution to the problem of human-induced vibration serviceability. According to the recent research, this review introduces the application of numerical simulation in the research of dynamic properties and summarizes the modeling method of sophisticated finite element model. The reliability of the finite element model is verified by the field test and model test. Comparing the application of studies on dynamic properties among numerical simulation and model test, the differences between numerical simulation and model test are summarized. Finally, some problems that need further study are proposed.

1 Introduction

In recent years, people began to pay attention to the comfort issue of the floor in human-induced vibrations. Composite floor is widely used in new buildings makes the structure with larger span and lower damping. Aiming at the vibration comfort issue of long-span floor, many scholars put forward the feasible solution, as shown in table 1. At present, the commonly used vibration control methods include structural modification, active control, passive control, semi-active control and hybrid control. In order to avoid the comfort issue of long-span floors, the design needs to consider the most unfavorable conditions and analyzes the response of floor vibration. Dynamic characteristic parameters of the floor including natural frequency, vibration mode and damping ratio. Long-span floor vibration control method can be reduced to two types: 1) Change the floor vibration frequency to avoid overlapping with pedestrian excitation frequency, so as to avoid resonance. 2) Improve structural damping to reduce vibration frequency and vibration amplitude. According to the dynamic characteristics of floor vibration comfort analysis, and summarizing existing research.

Table1 Typical Long-span Floor Structures

Name	Structure Type	Span/(m)	Issues	Solution
Beijing South Railway Station	H-shaped solid steel beam	40.5	Emergency evacuation	Considering the structural vibration caused by the evacuation, the structure of the crowd should be adjusted and optimized
Shenyang Olympic Center	Long-span prestressed concrete frame beam slab structure	41.5	It is difficult to determine the dynamic characteristics	Finite element analysis and field test
Changsha Railway Station	Profiled steel sheet concrete composite floor	49	Peak acceleration exceeds the limit	Installing tuned mass damper, altering the high span ration of the supporting steel trusses
National Fitness Center in EjinHoro Banner	Large-span and cantilevered steel floor	40	Effect of exercise motivation and floor decoration on the floor vibration	Field test

2 Serviceability assessment index of human-induced vibration

*Corresponding author: Email: cll@ujs.edu.cn

At present, many study the dynamic characteristics of floor by the combination of numerical simulation and field measurement. According to the study of different scholars, different vibration comfort evaluation criteria are established in different countries, and the commonly used evaluation indexes.

It is difficult to satisfy the requirement of vibration comfort for single control index. The international organization for Standardization (ISO) in the 1989 edition of the ISO 2631-2 standard first used frequency and peak acceleration to evaluate human-induced vibration, and frequency weighting function for vertical vibrations as shown in Figure 1 [1]. The American Institute of steel construction (AISC) and the Canadian Standards Association (CSA) subsequently proposed a similar evaluation program. In the revision process of domestic relevant standards, the frequency and vertical acceleration limits are also added to the assessment of vibrations. The peak acceleration is the maximum value of the acceleration in the process of excitation. In the whole process, the structure will not always be in the peak of vibration [2]. Therefore, ISO 2361-2:2003 cancels the frequency-weighting function for vertical vibrations and the peak acceleration limit value. In the measurement and evaluation of the impact of vibration on the human body, ISO 2361-2:2003 recommend to use the frequency weighting method to deal with complex vibration signals, using the w_m curve shown in Figure 2 for frequency weighting [3].

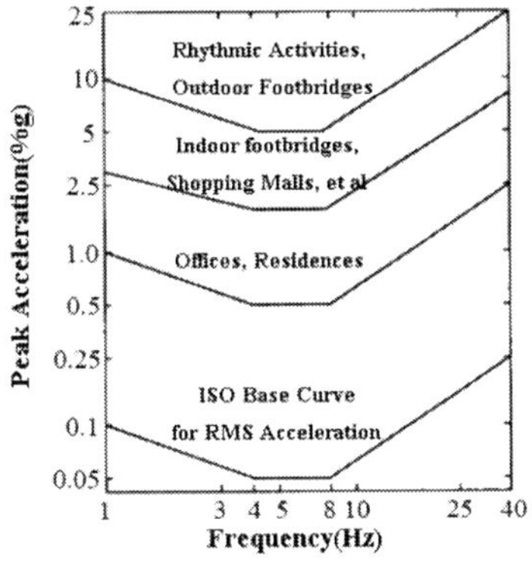

Figure 1 ISO Standards

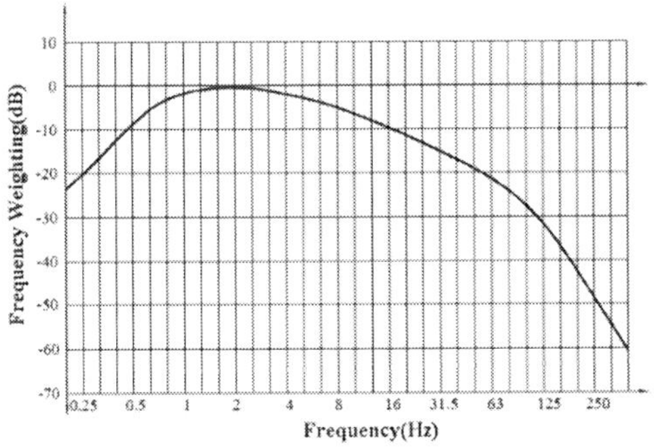

Figure 2 ISO Frequency weighting

3 The Numerical Simulation of Floor

Establishing suitable finite element model of long-span floor is the first step of numerical simulation. Due to the limitation of computer processing ability, rational simplification of finite element model can accelerate the analysis speed. However, the unreasonable simplification of the model, such as the wrong boundary conditions and the over simplification of the components, makes the finite element calculation results not consistent with the actual situation. In the actual numerical simulation, it is necessary to pay attention to the analysis speed and the calculation error of the finite element. And the finite element model should be modified by the results of field test and model test.

3.1 Refinement of Finite Element Model

In the process of establishing the finite element model, there are many uncertain factors, such as the determination of the boundary conditions, the selection of the material properties, the modeling of the non-structural parts and so on. Especially in the field of civil engineering, there are non-structural stiffness and elastic modulus of concrete and concrete cracking stiffness. These factors are difficult to determine and uncertainties must be expressed as parameters[4]. So the uncertainty can be quantitatively assessed. Many scholars for the study of the sophisticated finite element model is mainly reflected in the treatment of boundary conditions and the impact of non-structural components on the dynamic characteristics.

The essence of the finite element calculation can be reduced to the solution of the differential equation. And the boundary condition is an additional condition for the solution of the differential equation, it is very important to the finite element results. In the actual structure, there are usually loose connections between components. It is very difficult for the finite element to simulate these connections without additional stiffness in the modeling process[5]. Floor vibration could not be delivered to the adjacent floor under human-induced vibration, columns can be extended to the adjacent floors and the ends are clamped [6].

People do not consider the non-structural effects on the dynamic characteristics of the floor under the strong vibration condition, but the vibration comfort analysis is in weak vibration condition. Reynolds et al. believes that in the weak vibration conditions, the quality and the stiffness of non-structural components will affect the dynamic characteristics of the floor[6]. For example, the filling wall of the common office area will affect the structural damping of the floor. Ma et al. shows that the increase of the thickness of the floor will reduce the vertical acceleration and the natural frequency of the structure[7]. According to the principle of equivalent stiffness, the influence of the non-structural parts such as the floor decorative layer on the dynamic characteristics is simulated by increasing the floor thickness [8]. For the comfort analysis of human induced vibration, connections between filling wall and floor are usually not damaged, and the openings of filling wall will reduce the

 https://doi.org/10.1051/e3sconf/20183803012

stiffness of the structure, and the modeling should consider the influence of filling wall and its openings on the dynamic characteristics[9].

It is worth mentioning that long-span floor is not a complete rectangle, because of the stairs or elevator, it often requires opening. In the finite element analysis of many scholars, the opening of floor structure is not taken into account. Liu studied the effect of floor openings on the dynamic characteristics, opening rate compared to the opening shape had greater impact, and the natural frequency increasing first and then decreasing Stairs and elevators as the part of long-span public buildings are indispensable, the finite element model should be modified for floor openings[10].

3.2 The Establishment of Finite Element Model

For long-span floor structure, the refinement of the slab structure is the main part of the sophisticated finite element model, the establishment process is as follows: 1) Based on the calculation model of ultimate capacity, the structural floor slab is introduced; 2) Considering the influence of the quality and stiffness of the non-structural parts (floor decorative layer, such as marble, wood floors, etc.), adjusting concrete slab boundary combine with construction process; 3) The influence of live load is considered in the form of added mass, and the combination coefficient of live load is adjusted; 4) Adjusting the floor parameters related to dynamic characteristics (mainly damping ratio) [11].

4 Model Test

The numerical simulation method has been widely used in floor dynamic analysis. With the development of computer technology and the improvement of the finite element model, the error between the calculated results and the actual results is reduced. However, it is difficult to select the relevant parameters of the finite element and determine the boundary conditions because of the new materials, new structures and new construction technology. The dynamic test is also dependent on the construction of long-span structures, which makes the validity of the measured work is not satisfactory. Therefore, model test based on the similarity theory is an important part of structural dynamic characteristics research. The test model according to the proportion can be divided into geometric model and scale model.

Varela and Battista built a 9.2m*9.2m square geometric model of composite floor to carry out free vibration tests. It obtains some of the dynamic characteristics of the model floor, such as the first three natural frequencies as well as the corresponding mode shapes and modal damping ratios[12]. Carrying out single and crowd excitation test after determining the parameters of dynamic characteristics of the floor, the test process is shown in Figure 3, the experimental results verify the effectiveness of floor seismic design. Chen et al. proposed the floor acceleration response spectrum under bounce load, and the model test was carried out to verify the rationality of the spectrum by a 10m*6m prestressed floor model[13], the experiment process is shown in figure 4. Yang et al. designed the assembled monolithic steel 1/5 scale model of 30m * 30m concrete hollow floor, the structural dynamic characteristics and load-carrying behavior were studied through scaled model test and finite element analysis [14]. The deflection test got floor structure and natural frequency, comparing with the numerical simulation results to verify the reliability of finite element model. That provides the basis for the design of such structures.

Figure 3 Model Test under Human-induced Vibration

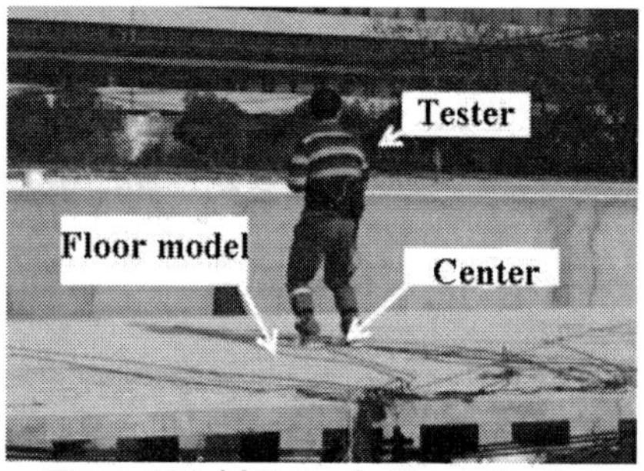

Figure 4 Model Test under Bounce Load

Compared with the traditional numerical simulation method, the same material and construction technology are used in the model test. And the numerical simulation are easy to overlook some factors have significant impact on the floor structure vibration peak acceleration, such as: walking path, walking route. When using the finite element software to analyze the time history, due to the complexity of the pedestrian load, the assumption is usually introduced to improve the computational efficiency, such as not considering the differences between the individual pedestrians, ignoring the moving characteristics of pedestrian loads, etc. [15]. Jiang et al. shows that the walking path, walking frequency and other factors have significant influence on the vertical acceleration of long-span floor, thus ignoring the mobile characteristics of the pedestrian load will bring obvious errors[16]. In the numerical simulation, the constraint condition of the finite element model is not the same as the actual connection mode. The inevitable defects in the finite element analysis are solved in the model test. When floor system occur comfort issues, structural modification is generally considered, such as increasing the additional column to increase the stiffness of the structure. In addition, the appropriate control scheme can be selected (active control, passive control, semi-active control, etc.). Compared to the high cost of repair of the two solutions, the model test takes into account both the accuracy of the data and the rationality of the cost.

5 Conclusion

Based on the review of long-span floor dynamic characteristics research in recent years, we put forward the following summary and suggestions on the research on dynamic characteristics of long-span floor:

(1) The finite element model in numerical simulation is improved continuously, from simple beam model to complex finite element model, from the simplification of model structure to the sophisticated FEM model with non-structural components. There are few research about the influence of the dynamic characteristics of non-structural components, further studies still need to be refined model.

(2) Considering the influence of walking route on vibration comfort, the route for the random walk is a straight line through the geometric center of the floor, not carrying out analysis of the actual walking path of the population in the structure. The different functions of the long-span structures and the different situations that may occur different conditions, such as the structure of the station and the commercial square, the normal use and emergency evacuation, will have an impact on the pedestrian route. In the analysis of human induced vibration comfort, it is necessary to consider the actual situation of the possible route of walking, the corresponding numerical simulation or model test should be conducted.

Acknowledgements

This study is financially supported by the National Natural Science Foundation of China (Grant Nos.51408267,51508237) and the Natural Science Foundation of the Jiangsu Higher Education Institutions of China (Grant No.14KJB560005) and Energy saving projects in Colleges and universities of Jiangsu Province (Grant No.JSSNZH2016301) .

References

1. IS02631-2(1989). Evaluation of human exposure to whole-body vibration-Part 2 : Continuous and shock-induced vibration in buildings(1 to 80Hz)[S]. In: IS02631/2-1989.
2. Zhou X, Cao L, Chen Y F, et al. Experimental and analytical studies on the vibration serviceability of pre-stressed cable RC truss floor systems[J]. Journal of Sound and Vibration, 2016,361:130-147.
3. ISO10137, Bases for design of structure-serviceability of buildings and walkways against vibrations[S]. Geneva: International Organization for Standardization, 2007.
4. Brownjohn J M W, Xia P Q. Dynamic Assessment of Curved Cable-Stayed Bridge by Model Updating[J]. Journal of Structural Engineering, 2000,126(2):252-260.
5. Racic V, Pavic A, Brownjohn J M W. Experimental identification and analytical modeling of human walking forces: Literature review[J]. Journal of Sound and Vibration, 2009,326(1-2):1-49
6. Reynolds P, Pavic A. Effects of false floors on vibration serviceability of building floors. I: Modal properties[J]. Journal of Performance of Constructed Facilities, 2003,17(17):75-86.
7. MA Bo-tao, YANG Guo-li, PANG Yu-tao, et al. Study on comfort analysis of sport field structure with large-span and cantilevered steel floor [J]. Journal of Building Structures, 2012,33(4):95-103.
8. XIE Wei-ping, MA Zhao-xia, HE Wei. Study on the relation of natural frequency and vibration serviceability of large-span floor structure under crowd load [J]. Journal of Wuhan University of Technology, 2012(04): 96-101.
9. XIE Wei-ping, ZHAO Na, HE Wei, et al. Analysis on vibration serviceability of over-track buildings [J]. China Civil Engineering Journal, 2013(06):90-96.
10. LIU Zhi-guo. Analysis and Study on the comfort of long-span flat slab with opening vibration excitation load [D]. Xian University of Architecture and Technology, 2013.
11. HE Wei, XIE Wei-ping. Study on sophisticated calculation model of large-span railway station structures based on vibration serviceability evaluation [J]. China Civil Engineering Journal, 2014(01):13-23.
12. Varela W D, Battista R C. Control of vibrations induced by people walking on large span composite floor decks[J]. Engineering Structures, 2011,33(9):2485-2494.
13. CHEN Jun, WANG Lei, LOU Jia-yue, et al. Acceleration response spectrum for predicting floor vibration due to single human bounce load [J]. Journal of vibration and shock, 2015(05):14-19.
14. YANG Qi-zhu, MA Ke-jian, HU Lan, et al. Experimental study on load-carrying behavior of long-span assembled monolithic steel-concrete composite open-web plate [J]. Journal of Building Structures, 2013(10): 32-40.
15. XU Ruo-tian, CHEN Jun, LIU Wei. Optimal design of multiple tuned mass dampers applied in reducing floor vibration induced by pedestrian's walking loads [J]. Journal of vibration and shock, 2014(09):100-106.
16. JIANG Lan, ZHAN Hua-gang, YUAN Bo, et al. Research on vibration response of long-span reinforced concrete open-web sandwich plate structures under walking excitation [J]. Sichuan Building Science, 2012(01): 9-13.

Study on Human-structure Dynamic Interaction in Civil Engineering

Feng GAO[1], Li Lin CAO[2,*], Xing Hua LI[3]

[1]Department of State-owned Assets, Jiangsu University, Zhenjiang 212013, China.
[2]Faculty of Civil Engineering and Mechanics, Jiangsu University, Zhenjiang 212013, China.
[3]Jiangsu University Library, Zhenjiang 212013, China.

Abstract: The research of human-structure dynamic interaction are reviewed. Firstly, the influence of the crowd load on structural dynamic characteristics is introduced and the advantages and disadvantages of different crowd load models are analyzed. Then, discussing the influence of structural vibration on the human-induced load, especially the influence of different stiffness structures on the crowd load. Finally, questions about human-structure interaction that require further study are presented.

1 Introduction

In recent years, with the use of structural design, high-strength lightweight materials and construction technology, the structure of the building is gradually developing towards soft, long-span and low damping. The bearing capacity and deformation of the static performance can generally meet the requirement in the traditional design, but its dynamic performance, especially the excessive vibration of crowd excitation, may become a key problem restricting its application and development. The vibration caused by human activities is low frequency vibration, when the vertical natural frequency of this kind of structure is close to the frequency of the human activity, it is easy to cause resonance. This kind of vibration may exceed the tolerance limit of human comfort, cause panic in the psychology and seriously affect the performance of the structure. In the field of civil engineering, the perception of the human-structure interaction (HSI) was first derived from some phenomena found in some structural modal tests. Many studies show that people stay in long-span structure will affect the dynamic characteristics of the structure, and the vibration of long-span structure will have an impact on the excitation of crowd load, human-structure interaction is a new interdisciplinary scientific problem.

2 The influence of crowd load on structural dynamic characteristics

The crowd that stay in the building structure not only cause structural vibration, but also may change the modal characteristics. Therefore, how to consider the crowd stay in the structure is an urgent problem. Early on, Ebrahimpour regarded the crowd that stay in the structure as additional quality to consider its influence on the dynamic characteristics [1]. ATC design guide also uses a similar idea considering the influence on structural modal characteristics [2]. Obviously, the result of this treatment leads to a decrease in the natural frequency of the structure. However, Pimentel's research shows that presence of the crowd in remarkably crease structural damping [3]. Moreover, Brownjohn found that the single person that stay in concrete hollowed slab absorb more energy than structure itself[4]. Canada design code also consider the potential beneficial effects of crowd load on structural vibration response, it is proposed to increase the viscous damping ratio of the structure to 12% in the condition that crowd will stay in the structure[2]. The presence of people only regarded as the additional quality obviously cannot explain the research results given above.

In order to solve this problem, Rainer and Pernica pointed out the damped dynamic model of the crowd, this model can explain the phenomena that human occupants significantly increase the damping of the structure [5]. However, the additional mass model is widely accepted in a period, researchers believe that it can accurately predict the influence that human occupants change structural vibration frequency. Lenzing found that keep the single person weight over wood slab weight of more than 2 times, the frequency of the structure is not reduced, but a slight increase [6]. On 1997, Ellis and Ji carried out field dynamic characteristic test under two condition, the first was no load, the second considered the human occupants [7]. The second condition found a new mode of the stadium stands, and assume that the crowd as powered system with mass

*Corresponding author: Email: cll@ujs.edu.cn

spring damper can better explain test result[7]. In 2002, Sachse et al. pointed out mathematical model of human occupants confirmed this assumption [8].In short, human occupants can increase structural damping, reduce fundamental frequency and add new modal mode. To accurately predict the effects of human occupants is of great significance for the economic design and safety evaluating of long-span structure. Therefore, a reasonable crowd load dynamic model must be established. In 1998, Griffin et al. used electro-hydraulic servo vibration table to test the contact force and displacement between the human body and the table of standing posture and sitting posture random vibration excitation, then obtained the basic frequency and damping ratio of the 60 tested human body[9]. In 2006, Liang et al. analyzed and compared the different degrees of freedom model based on the experimental data of vertical vibration of European and American sitting posture [10]. In 2011, Gao et al obtained the basic parameters of human vibration based on sitting posture of few Chinese people [11]. The above results are the basic human body parameters biomechanics field, mainly used in automobile, machinery and other fields of medicine, cannot directly apply to the civil engineering structure model of the human body, but useful research ideas for other researchers.

In the field of civil engineering, the simplified model of human body mainly consists of model with damping and without damping, and the model without damping is divided into discrete model and continuous model [12] (shown in Figure 1). Lenzing and Hothan studied the human single degree of freedom discrete model without damping, the lumped mass m_H of the simplified human model was assumed to be equal to the total mass $m_T$of the human body, and given the values of the stiffness and the natural frequency parameters of the human body model [13]. Later, Ji assumed that the upper and lower parts of the human body had different stiffness and mass, Falati assumed the stiffness and mass of the human body along the height distribution and established a continuous model without damping [14]. Foschi first studies the single degree of freedom damping model (Figure 2), obtained dynamic response based on the measured foot displacement time history curve and the calculation of single degree of freedom human body impact floor considering the human-structure interaction, the stiffness k_H of the simplified kinetic model of human body is 40kN/m and the viscous damping c_H is 1.25kNs/m, but the premise of the study is that the lumped mass m_H is equal to the total mass m_T[15].The values of the natural frequencies (f_H) and the damping ratio (ξ_H) of the human body in the model are different from different scholars, and specific results are presented in table1.

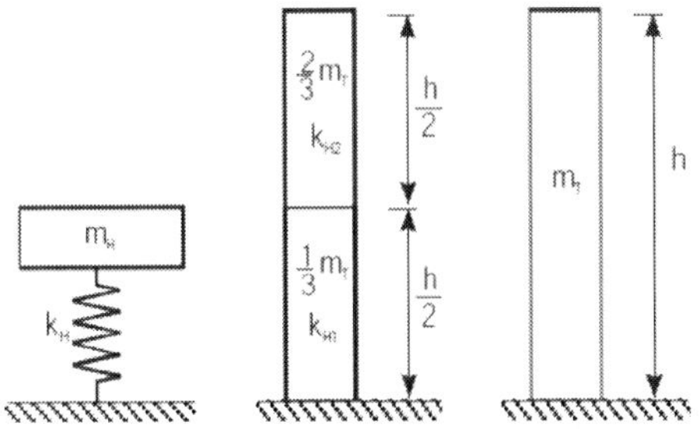

Figure 1 Simplified Model without Damping

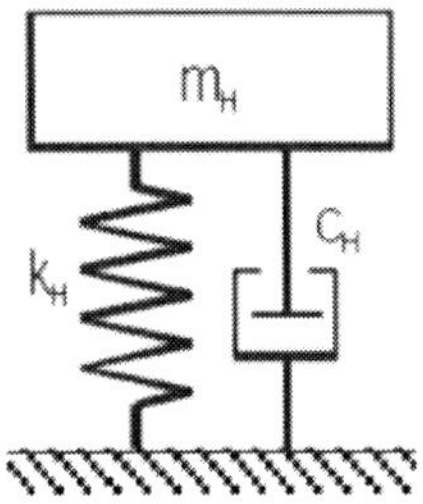

Figure 2 Single Degree of Freedom Model with Damping

Table 1 Comparison Between Numerical and Experimental Results

Author	The natural frequency of human body f_H(Hz)	The damping ratio of human body ξ_H	Lumped mass value
Falatiet al.	10.43	45%-55%	$m_H = m_T/3$
Zhengand Brownjohn	5.24	39%	$m_H = m_T$

After the human body model is determined, the combined dynamic system of human-structure interaction can be analyzed. Sachse has established the system of two degrees of freedom dynamic human-structure coupling model (Figure 3a), and the natural frequencies, mode shapes, modal mass and damping ratio of the coupled system are analyzed in detail, the results of analysis showed that damped discrete model of the human-structure system with two degrees of freedom can better explain changes across the dynamic characteristics of structures with human occupants [16]. Zhou et al. uses a continuous model without damping to analyze the dynamic characteristics of a coupling system of human-structure interaction [17] (shown in Fig. 3b).

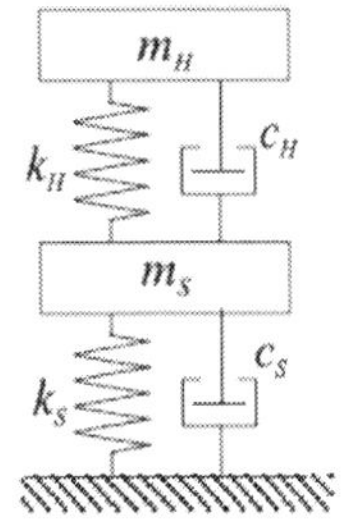

(a) Two Degrees of Freedom Discrete Model with Damping

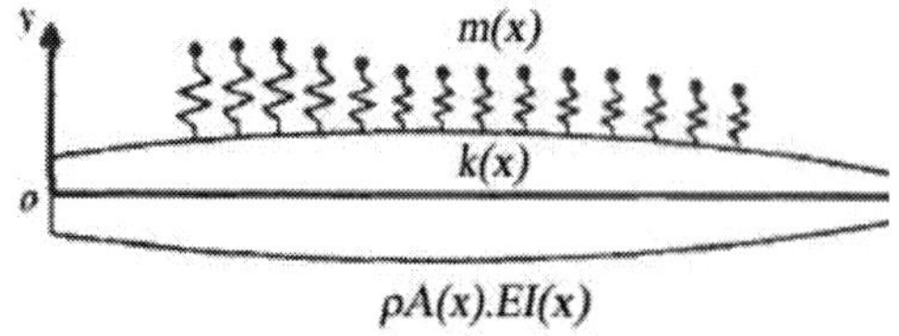

(b) Continuous Model without Damping
Figure 3 Human-structure Interaction Analysis Model

3 Influence of structural vibration on human-induced load

Effects of structural dynamic characteristics and vibration levels on crowd loads will happen when the vibration response of the structure is too large, the pedestrian adjust the pace and phase in order to improve walking comfort and synchronize with the structural vibration. In 1982, Ohlsson found that there were some differences in the measured walking load between the rigid ground and the long-span soft slab [18]. In 1988, Baumann and Bachmann indicated that the dynamic load coefficient measured on the rigid ground was increased by 10% compared to the 19m long prestressed beam[19]. In 2002, Yao et al. pointed out that the dynamic load of jumping and running on the long-span floor is smaller than the rigid ground [20]. All the above studies show that the vibration of structure affects human-induced load, but there is some difference between the human activities in experiment and the actual structure, and the influence of individuals among the crowd is not considered. Therefore, it is of great significance to study the influence of long-span structure vibration with synchronous excitation on the human-induced load.

In recent years, the study of dynamic load of crowd excitation has attracted much attention. Research shows that the dynamic load generated by crowd excitation is much larger than that of single excitation, but the number of people and human-induced load is not a simple linear relationship. In some cases, people on the structure will do conscious or unconsciously synchronous activities. For example, the crowd in a rock ballroom will consciously synchronize with musical rhythm, which will have a greater impact on structural vibration. The involuntary, synchronous activity of the crowd may cause a violent vibration, which makes the structure uncomfortable and even insecure. It is an important research direction of the human-structure interaction that structural vibration is caused by the involuntary synchronization of the crowd. When the vibration of long-span structure reaches a certain extent, the crowd will adjust the walking pace and keep pace with the structural vibration. Grundmann et al. studied the synchronous modulation probability of vertical vibration between the crowd and structure, and proposed a method for calculating the synchronous modulation probability and vertical vibration acceleration [21]. So far, fewer research about synchronous modulation probability can be retrieved, it has important practical significance to deeply study synchronous modulation probability in the actual civil engineering structure.

4 Conclusion and Prospect

By reviewing the research of human-structure interaction, the following suggestions are put forward for the study of structural dynamic characteristics:

(1) Carrying out structural dynamic measurement. According to the experimental data, the parameters such as the lumped mass, the natural frequency of the body and the damping ratio of the human body are corrected and the crowd load model is refined.

(2) Considering the influence between individuals, developing the method for calculating vertical acceleration of structures based on synchronous modulation probability.

(3) At present, the study of human-structure interaction is more focused on the vertical vibration of structures, and the horizontal vibration of structures should be included in further study.

Acknowledgements

This study is financially supported by the National Natural Science Foundation of China (Grant No.51408267) and the Natural Science Foundation of the Jiangsu Higher Education Institutions of China (Grant No.14KJB560005) and Energy saving projects in Colleges and universities of Jiangsu Province (Grant No. JSSNZH2016301).

References

1. Ebrahimpour A, Sack R L, Vankleek P D. Computing crowd loads using a nonlinear equation of motion[C]. Proceeding of the 4th International Conference on Civil and Structural Engineering, 1989, 2: 47-52.
2. Allen, D E, Onysko D M, Murray T M. ATC Design guide 1: Minimizing floor vibration[S].1999, Redwood City, Canada, Applied Technology Council (ATC).
3. Pimentel R L, Waldron P. Validation of the numerical analysis of a pedestrian bridge for vibration serviceability applications[C]. Identification in Engineering Systems, Swansea, UK, 1996, Cromwell Press. 648-657.
4. Brownjohn, J M W. Energy dissipation from vibrating floor slabs due to human-structure interaction [J]. Shock and Vibration, 2001, 8(6):315-23.
5. Rainer J H, Pernica G. Vibration characteristics of a floor sample[C]. Ottawa, Canada: NRCC, Division of Building Research, 1985, NRCC-24298.
6. Lenzing H. Durch Menschen induzierteSchwingungen[D]. University Hannover, Hanover, Germany, 1988.
7. Ellis B R, Ji T. Human-structure interaction in vertical vibrations[C]. Proceedings of the ICE: Structures and Buildings, 1997, 122(1):1-9.

8. Sachse R. The influence of human occupants on the dynamic properties of slender structures[D]. University of Sheffield, Sheffield, UK, 2002. [9] ANSI B. AISC 360-10-Specification for Structural Steel Buildings[J]. Chicago AISC, 2010.
9. Wei L, Griffin M J. Mathematical models for the apparent mass of the seated human body exposed to vertical vibration[J]. Journal of Sound and Vibration, 1998, 212: 855-74.
10. Liang C C, Chiang C F. A study on biodynamic models of seated human subjects exposed to vertical vibration[J]. International Journal of Industrial Ergonomics, 2006, 36(10): 869-890.
11. Gao J H, Hou Z C, He L, et al. Vertical vibration characteristics of seated human bodies and a biodynamic model with two degrees of freedom. SciChina Tech Sci, 2011, 54: 2776-2784
12. Ji, T. A continuous model for the vertical vibration of the human body in a standing position[C]. UK Informal Group Meeting on Human Response to Vibration, Silsoe, UK, 1995, 18-20.
13. Lenzing H. Durch Menschen induzierteSchwingungen[D]. Universität Hannover, Hanover, Germany, 1988.
14. Falati S. The contribution of non-structural components to the overall dynamic behaviour of concrete floor slabs[D]. University of Oxford, Oxford, UK, 1999.
15. Foschi R O, Neumann G A, Yao F, et al. Floor vibration due to occupant and reliability-based design guidelines[J]. Canadian Journal of Civil Engineering, 1995, 22(3):471-79.
16. Sachse R, Pavic A, Reynolds P. Parametric study of modal properties of damped two-degree-of-freedom crowd–structure dynamic systems[J]. Journal of Sound and Vibration, 2004, 274(3): 461-480.
17. Zhou D, Ji T. Free vibration of rectangular plates with continuously distributed spring-mass[J]. International journal of solids and structures, 2006, 43(21): 6502-6520.
18. Ohlsson S. Floor vibrations and human discomfort[D]. Chalmers University of Technology, Gothenburg, Sweden, 1982.
19. Baumann K, Bachmann H. Man-induced dynamic loads and their influence on beam structures[R]. Zürich, Switzerland: Institute of Structural Engineering (IBK), 1988.
20. Yao S, Wright J, Pavic A, et al. Forces generated when bouncing or jumping on a flexible structure[C]. Proceedings of the International Conference on Noise and Vibration, Leuven, Belgium, 2002, 563-572.
21. Grundmann H, Kreuzinger H, Schneider M. Dynamic calculations of footbridges[J]. Bauingenieur, 1993, 68:215-225.

Parametric Study on Dynamic Tension Behaviour of Offshore Pipeline for Deepwater S-lay Operation

Pu XU[1], Shun Feng GONG[2], Ting ZHANG[1], Cheng CHEN[1]

[1] College of Civil Engineering, Fuzhou University, Fuzhou 350116, China
[2] Institute of Structural Engineering, Zhejiang University, Hangzhou 310058, China

Abstract: In recent years, higher demands for offshore pipeline design of deepwater installation are put forward. This paper mainly aims to investigate the influence of pipelay parameters on dynamic tension behaviour of deepwater S-lay pipeline to optimize the design of pipe structure and tensioner. A full finite element model for deepwater S-lay systems is developed to simulate the dynamic response of offshore pipeline from the pipelay vessel via stinger to the seabed. The influences of water depth, outer diameter and submerged self-weight of the pipe, and stinger length on the dynamic tension behaviour of the pipeline have been quantitatively studied. A strong relevance between the axial tension of offshore pipeline and pipelay parameters is observed, which offers very intuitive evidences for the design of pipe structure and tensioners.

1 Introduction

With the exploration of offshore oil-gas resources expanding into deep waters, pipelines play an indispensable role of large-scale transportation for crude oil and natural gas. S-lay technique is widely applied to install the pipeline from sea level to the seafloor due to its advantages of higher pipelay efficiency and better workability [1]. In the S-lay method, the pipeline passes through the tensioners and slides over the stinger at the stern of the vessel via a sloping ramp before reaching the seabed, as shown in Figure 1. The tensioners mounted on the lay vessel are functioned to apply a horizontal tension to the pipe at the top end and control the pipeline configuration in the sagbend. The axial tension response of the pipeline in the deepwater S-lay process is very significant and variable, and directly dominates the design of the pipe structure and tensioners. Analytical and numerical solutions were developed to carry out the dynamic response analysis of offshore pipeline for S-lay operation [2-4]. Mattiazzo et al. [5] proposed a numerical model to analyze the performance of the tensioner used in the process of pipeline installation. Yue et al. [6] presented an analytical model to predict the tension behavior of the flexible pipeline in shallow water. Lately, Gong et al. [7-8] developed a full element model to simulate the dynamic response of deepwater S-lay pipeline. The objective of the present study is to explore the influence of pipelay parameters on the axial tension behavior of offshore pipeline for deepwater S-lay operation, and promote the optimal design of pipe structure and tensioners.

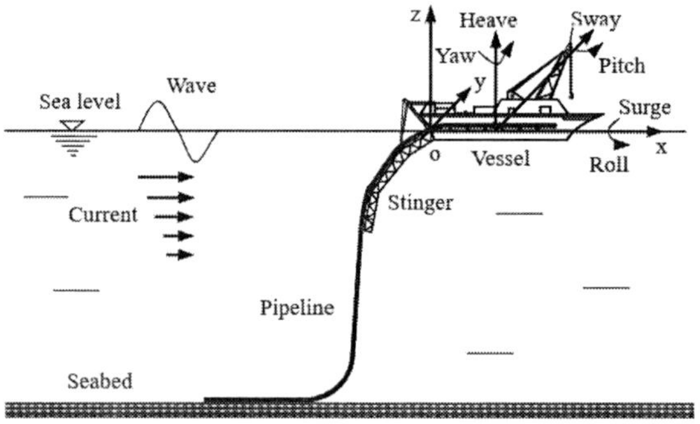

Figure 1 Deepwater S-lay pipeline

2 Theory Models

2.1. Analytical Model of Pipe Tension Behaviour

*Corresponding author: Email: puxu@fzu.edu.cn

A comprehensive finite model for deepwater S-lay system is established to simulate the tension behavior of offshore pipeline based upon the commercial commercial computer code OrcaFlex [9]. The effectiveness and practicality of the model is adequately validated through a practical engineering example of a 6-inch pipeline laying to the LW3-1 gas filed in the South China Sea [7]. In this model, the pipeline is divided into a series of line segments which are modelled by straight massless model segments with a node at each end. A representative single mid-line node and line segments either side of it are selected to establish the local xyz-frame frames of reference, as illustrated in Figure 2.

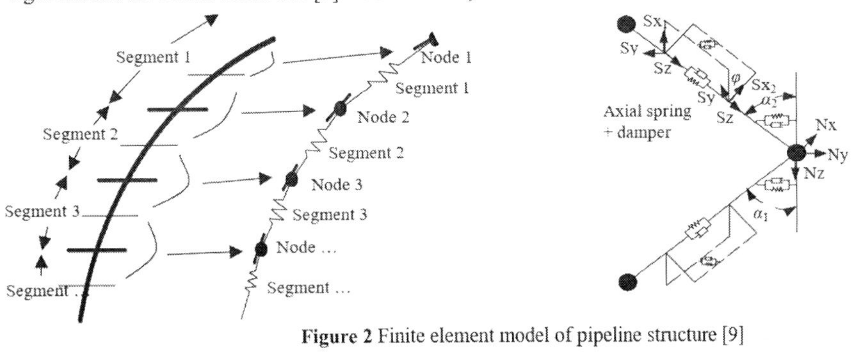

Figure 2 Finite element model of pipeline structure [9]

An axial spring and a damper at the center of line segment is used to simulate the axial stiffness and damping of the pipeline, and applies an equal and opposite effective tension force to the nodes at each end of line segment. Meanwhile, it is assumed that the line segment is subjected to external and internal fluid pressure, as shown in Figure 3, the effective tension at the mid-point of line segment can be expressed as

$$T_e = T_w(\varepsilon) + (1-2\nu)\cdot(P_o A_o - P_i A_i) + EA_{nom}\cdot\xi\cdot(dL/dt)/L_0 \qquad (1)$$

Where T_w is the function relating axial strain to wall tension, P_o and P_i are external and internal pressure, A_o and A_i are external and internal cross-section areas of the stress annulus, ν is the Poisson's ratio, EA_{nom} is the nominal axial stiffness which is defined to be the axial stiffness at zero strain, L and L_o are the instantaneous and unstretched length of line segment, ξ is the axial damping coefficient.

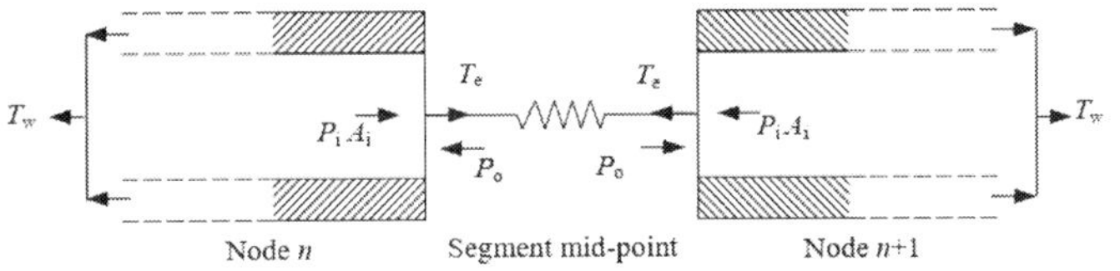

Figure 3 Tension analysis of the pipe at the mid-point of line segment [9]

The non-linear stress-strain relationship of the steel pipe is characterized by Ramberg-Osgood model [10], which is expressed as

$$\varepsilon(\sigma) = \sigma / E + B(\sigma/\sigma_y)^n \qquad (2)$$

Where B and n are the coefficient and power exponent constants that depend on the steel material grade.

2.2 Interaction Models

At the top end, the pipeline is usually supported by the tensioner and multiple discrete rollers on the main work line of the pipelay vessel, as shown in Figure 4. The tension winch element attached to the vessel and the pipe is adopted to simulate the tensioner, ensuring that the effective tension fed from the tensioner is applied to the top end of the pipeline. The parts of the pipeline in the overbend are supported by the roller boxes on the stinger hinged off the back of the HYSY 201 pipelay vessel. The S-lay stinger is made up of three sections of truss structures mutually connected via hinges, and on which 10 roller boxes are regularly-spaced and mounted. Each roller in the roller box should be examined to identify whether the pipeline is in contact with the roller. If the pipeline is confirmed to be touching the roller, the support reaction is then calculated and applied to the pipe and the roller.

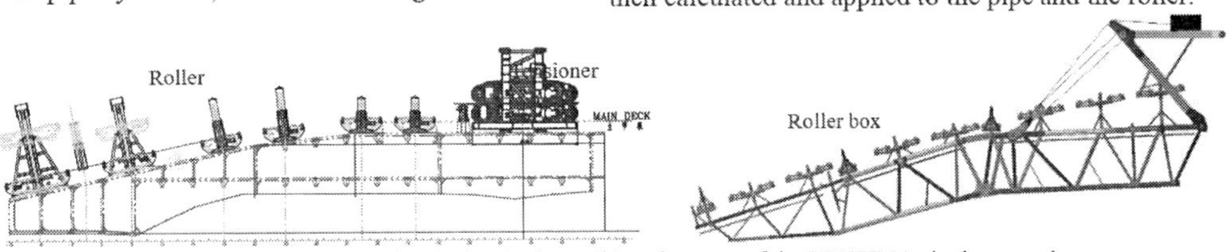

Figure 4 Main work line (a) and articulated stinger (b) at the stern of the HYSY201 pipelay vessel

At the bottom, the pipeline is gradually laid on the seabed during deepwater installation. The pipe-seabed vertical interaction under cyclic motions is modelled by a non-linear seabed model based on a hyperbolic secant stiffness formulation [11], as shown in Figure 5. In this model, four penetration modes, namely not-in-contact, initial penetration, uplift and repenetration, are used to represent the variation of the pipeline motion on the seabed. In each penetration mode, the seabed normal resistance per unit length, $P(z)$, is modelled by an analytic function of the non-dimensionalized penetration, z/D, where z is the penetration and D is the pipe diameter. On the other side, the pipe-seabed lateral interaction is simulated by the modified Coulomb friction model, as illustrated in Figure 6. Based upon the seabed soil type, the lateral seabed resistance in the model is represented as the product of the vertical seabed reaction force per unit length $P(z)$ and a soil friction coefficient u. The friction force is assumed to be linear variation from $-uP(z)$ to $+uP(z)$ as the deflection ranges from $-y_{breakout}$ to $+y_{breakout}$, and given by $F_y = -k_s Ay$ to a magnitude of not more than $uP(z)$, where y is the displacement from the un-sheared position, k_s is the seabed shear stiffness, and A is the contact area.

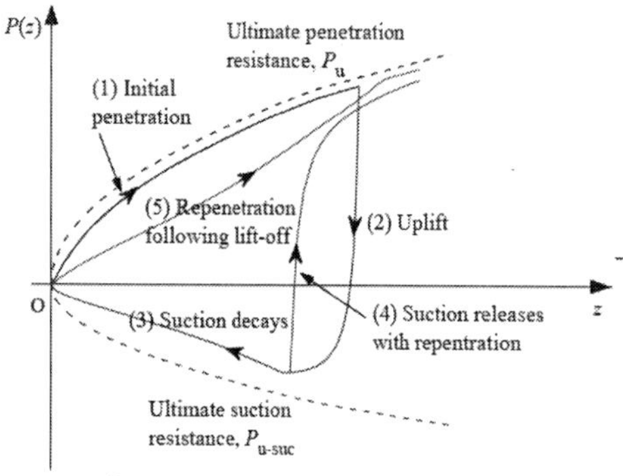

Figure 5 Non-linear hysteretic soil model [11]

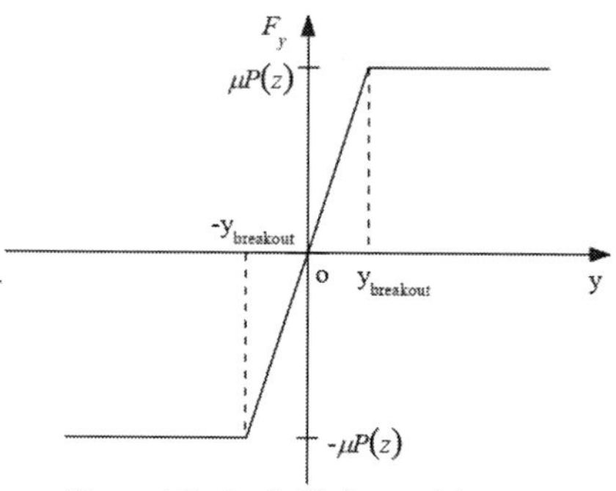

Figure 6 Coulomb friction model

2.3 Vessel Motions

The vessel model for deepwater S-lay pipeline is based upon the HYSY 201 pipelay vessel with 204.65 m in length and 39.2 m in breadth. The vessel generally has 6 degrees of freedom at the center of gravity, i.e. 3 translations (surge, sway, heave) and 3 rotations (roll, pitch, yaw), as illustrated in Figure 1. The vessel motions induced by surface waves can be calculated by the displacement response amplitude operators (RAOs). Each displacement RAO is composed of a pair of numbers that define the vessel response to one particular degree of freedom for one particular wave direction and period.

3 Numerical Implementation

3.1 Metocean Conditions

Based upon the marine environmental statistics in the gas field of the South China Sea, the standard wave spectrum JONSWAP is selected to represent the sea state with significant wave height H_s=2.0 m and peak period T_p=8.7 s. The equal energy method is adopted to discretize the wave spectrum, and one hundred linear wave components are then generated to construct an irregular random wave train by linear superposition, as illustrated in Figure 7. In order to cove the time when the maximum wave height 3.76 m occurs, the simulation time origin of the pipelay system is shifted to 8600 s with the specified simulation duration 500 s. The vertical profile curve of current velocity based on the practical measurement in the gas field is illustrated in Figure 8 with surface current speed 1.0 m/s. The current direction is assumed to be coincident with the lay vessel heading.

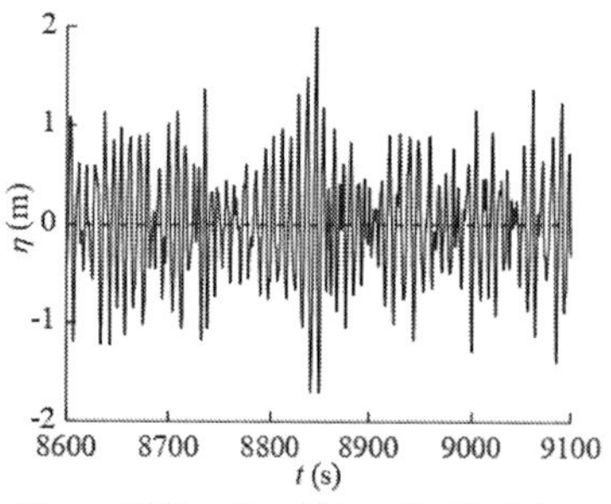

Figure 7 Wave time history for Hs=2.0 m

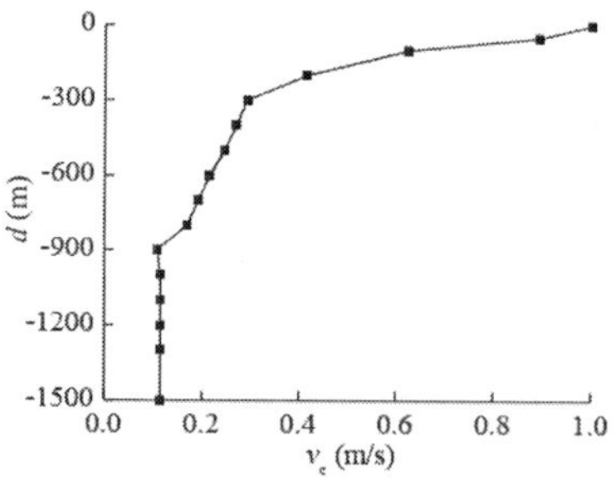

Figure 8 Current velocity distribution

3.2 Calculation Methods

The whole calculation analysis for dynamic response of deepwater S-lay pipelines contains two steps, the first step is to conduct static analysis based upon catenary solution and determine the equilibrium position of pipelay system under weight, buoyancy, hydrodynamic drag, etc, the full static calculation is subsequently implemented particularly considering the effects of bending stiffness and interaction with solid shapes. The second step is to carry out dynamic analysis starting from the static equilibrium position. The explicit integration scheme with a constant time step is adopted to simulate the dynamic response of offshore pipeline, which sufficiently considers various geometric non-linearities and the spatial variation of both wave loads and contact loads. The axial numerical damping in a pipe segment consists of critical damping and target damping, which is given by $\xi = \xi_c \cdot \xi_t$. The axial critical damping for a line segment is expressed as $\xi_c = (2 \cdot m_s \cdot L_0 / EA_{nom})^{1/2}$, and the axial target damping is specified the value of 10%. The numerical damping is capable of effectively damping down the high frequency noise and has litter influence on the calculation results of the pipelay systems.

4 Result and Discussion

4.1 Tension Varied with Water Depth

According to the practical engineering case in the LW3-1 gas field in the South China Sea, the pipeline need to laid with diameter 168.3 mm, wall thickness 15.9 mm, and a submerged weight 361.0 N/m. The response analyses of offshore pipeline in the process of S-lay operation are carried out under water depths being 500, 1000, 1500, 2000, and 2500 m, respectively. As shown in Figure 9, with the increase of water depth, the axial tension of the entire pipeline significantly increases. A dynamic amplification factor (DAF) is defined as the dynamic response of the pipeline relative to the static response. The maximum axial tensions of the pipeline for both static and dynamic analysis linearly increases, whereas the corresponding DAF gradually drops. It is evidently illustrated from these results that the water depth has a great influence on axial tension of the pipeline, and the increase of water depth weakens the dynamic lay effect on maximum axial tension of the pipeline to some extent.

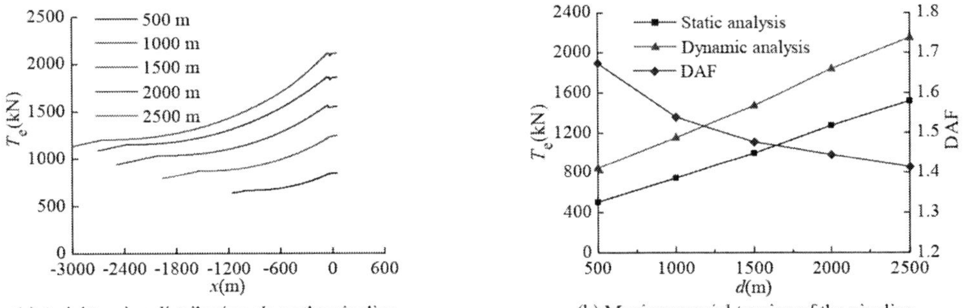

(a) Axial tension distribution along the pipeline (b) Maximum axial tension of the pipeline

Figure 9 Axial tension response of the pipeline varied with water depth

4.2 Tension Varied with Stinger Length

Based upon the geometrical characteristic of the articulated stinger specially designed for HYSY 201 pipelay vessel, the stinger lengths of 75, 84, and 93 m are selected to conduct the dynamic response analysis of the 6-inch. pipeline laid into a water depth of 2500 m. As shown in Figure 10, as the stinger length increases, the axial tension of the entire pipeline has a slight increase, the maximum axial tension of the pipeline for static and dynamic analysis and its DAF mildly increase. In summary, the stinger length has little influence on the axial tension, and the increase of stinger length slightly enhances the dynamic lay effect on maximum axial tension of the pipeline.

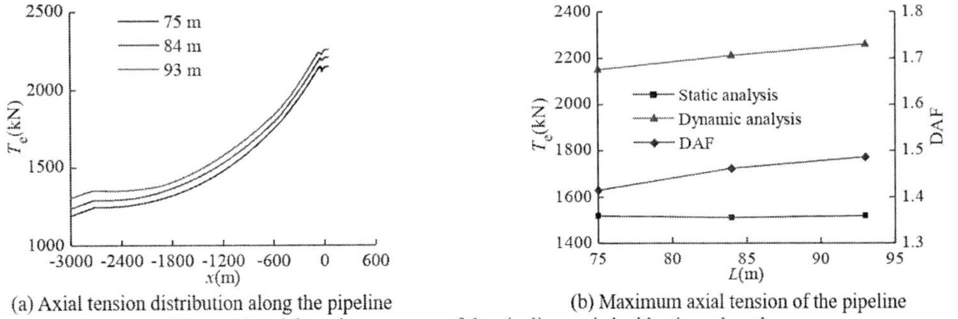

(a) Axial tension distribution along the pipeline (b) Maximum axial tension of the pipeline

Figure 10 Axial tension response of the pipeline varied with stinger length

4.3 Tension Varied with Pipe Diameter

In order to explore the axial tension response of the pipeline varied with pipe diameter, a group of steel pipes with outer diameters 8.625, 10.75, 12.75, 14.00 and 16.00 inch. are selected in terms of the API standard [12]. The corresponding wall thicknesses are 22.2, 20.6, 20.6, 20.6 and 22.2 mm, and the submerged self-weights of the pipe per unit length are nearly the same and equal to 0.676, 0.667, 0.680, 0.668 and 0.756 kN/m, respectively. The axial tension responses of offshore pipeline laid into a water depth of 1500 m are shown in Figure 11. As the pipe diameter increases, the axial tension of the pipeline becomes apparently larger, the maximum axial tension of the pipeline for static analysis changes very small, while considering the dynamic lay effect, the maximum axial tension of the pipeline for dynamic analysis and its DAF remarkably raise. These results adequately demonstrated that the pipe diameter strongly influences the axial tension of the pipeline, and the increase of pipe diameter obviously enhances the dynamic lay effect on the maximum axial tension.

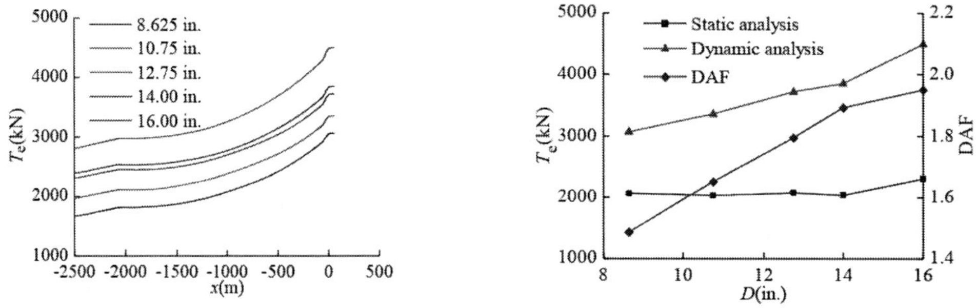

(a) Axial tension distribution along the pipeline (b) Maximum axial tension of the pipeline

Figure 11 Axial tension response of the pipeline varied with pipe diameter

4.4 Tension Varied with Submerged Self-weight of the Pipe

To investigate the tension response of offshore pipeline varied with submerged self-weight of the pipe, a group of steel pipes with outer diameter 323.9 mm, and wall thickness 19.1, 20.6, 23.8, 27.0 and 31.8 mm are selected in accordance with the standard API [12]. The corresponding submerged self-weights of the pipe per unit length are 0.58, 0.68, 0.90, 1.11 and 1.41 kN/m, respectively. As shown in Figure 12, the axial tension of the pipeline significantly increases with the increase of submerged self-weight of the pipe, the maximum axial tension of the pipeline for both static and dynamic analysis linearly increases, whereas the DAF rapidly drops. In summary, the submerged self-weight of the pipe has a strong influence on axial tension of the pipeline, and the increase of submerged self-weight of the pipe apparently weakens the dynamic lay effect on the maximum axial tension.

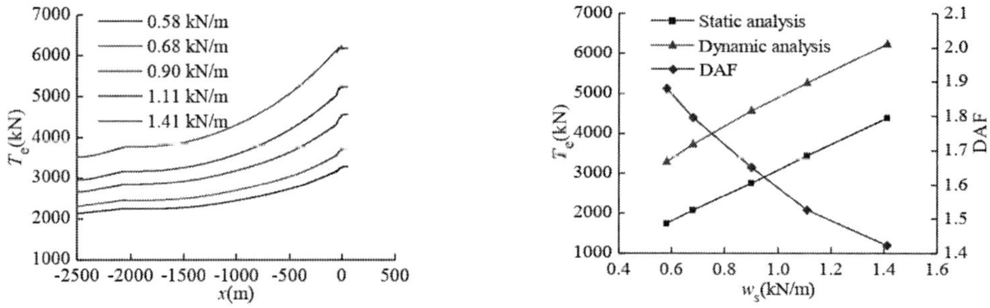

(a) Axial tension distribution along the pipeline (b) Maximum axial tension of the pipeline

Figure 12 Axial tension response of the pipeline varied with submerged self-weight of the pipe

5 Conclusion

This paper presents a full finite element model for deepwater S-lay system to explore the dynamic tension behaviour of offshore pipeline varied with various pipelay parameters. A strong relevance between pipelay parameters and axial tension of the pipeline is observed. These results show that the axial tension responses of the pipeline are greatly influenced by water depth, pipe diameter, and submerged self-weight of the pipe. Therefore, the design of the tensioners should sufficiently take into account the variations of the pipe tension under different pipelay parameters.

Acknowledgements

The authors express their gratitude to the National Natural Science Foundation of China (Grant nos. 51479176, 51679042), Youth Teacher Education Scientific Research Project in the Education Department of Fujian Province (Grant no. JAT170064), and Talent Introduction Scientific Start-up Foundation of Fuzhou University (Grant no. 510434) for the financial support to this study.

References

1. Li Z G, Wang C, He N, et al. An overview of deepwater pipeline laying technology[J]. China Ocean Engineering, 2008, 22(3): 521-532.
2. Clauss G F, Weede H, Riekert T. Offshore pipe laying operations—Interaction of vessel motions and pipeline dynamic stresses[J]. Applied Ocean Research, 1992, 14(3): 175-190.
3. Yun H D, Peek R R, Paslay P R, et al. Loading history effects for deepwater S-lay of pipelines[J]. Journal of Offshore Mechanics and Arctic Engineering, 2004, 126(2): 156-163.
4. O'Grady R, Harte A. Localised assessment of pipeline integrity during ultra-deep S-lay installation[J]. Ocean Engineering, 2013, 68: 27-37.
5. Mattiazzo G, Mauro S, Guinzio P S. A tensioner simulator for use in a pipelaying design tool[J]. Mechatronics, 2009, 19(8):1280-1285.
6. Yue Q J, Lu Q Z, Yan J, et al. Tension behavior prediction of flexible pipelines in shallow water[J]. Ocean Engineering, 2013, 58(2):201-207.
7. Gong S F, Xu P, Bao S, et al. Numerical modelling on dynamic behaviour of deepwater S-lay pipeline[J]. Ocean Engineering, 2014, 88: 393-408.
8. Gong S F, Xu P. The influence of sea state on dynamic behaviour of offshore pipelines for deepwater S-lay[J]. Ocean Engineering, 2016, 111: 398-413.
9. Orcina. OrcaFlex User Manual, Version 9.7a[M]. Cumbria, UK: 2014.
10. Ramberg W, Osgood W R. Description of stress-strain curves by three parameters[C]. National Advisory Committee for Aeronautics (NACA) Technical Note, No.902, NACA, Washington, 1943.
11. Randolph M F, Quiggin P. Non-linear hysteretic seabed model for catenary pipeline contact[C]. In: Proceedings of the International Conference on Offshore Mechanics and Arctic Engineering, OMAE2009-79259, Hawaii, USA, 2009.
12. American Petroleum Institute (API). API Specification 5L: Specification for Line Pipe [S]. Washington: API Publishing Services, 2013.

 https://doi.org/10.1051/e3sconf/20183803015

Application of Tessellation in Architectural Geometry Design

Wei CHANG

School of Civil Engineering, Tangshan College, Tangshan, Hebei 063000, China

Abstract: Tessellation plays a significant role in architectural geometry design, which is widely used both through history of architecture and in modern architectural design with the help of computer technology. Tessellation has been found since the birth of civilization. In terms of dimensions, there are two- dimensional tessellations and three-dimensional tessellations; in terms of symmetry, there are periodic tessellations and aperiodic tessellations. Besides, some special types of tessellations such as Voronoi Tessellation and Delaunay Triangles are also included. Both Geometry and Crystallography, the latter of which is the basic theory of three-dimensional tessellations, need to be studied. In history, tessellation was applied into skins or decorations in architecture. The development of Computer technology enables tessellation to be more powerful, as seen in surface control, surface display and structure design, etc. Therefore, research on the application of tessellation in architectural geometry design is of great necessity in architecture studies.

1 Introduction

Architectural geometry design is an important aspect of architectural design from ancient times to the present. Tessellation, or, Tilling, has been widely used in the field of architectural geometry design, which mostly deals with topological and fractals. A tessellation of a flat surface is the tiling of a plane using one or more geometric shapes (tiles), with no overlaps and no gaps[1]. Also, tessellations can be generalized to higher dimensions and a variety of geometries. There are many types of tessellations, among which the most basic and common ones are grid, hexagon tessellation, triangle tessellation, Penrose tessellation, Kelvin tessellation and so on. It has many advantages: figures in tessellations seem simple, but it can be shaped into extremely complex graphics; tessellation generates according to precise rules and orders, conducive to the standardization of production components; with spatial tessellation design, load distribution can be more reasonable in structure, and the damage of single component will not lead to the collapse of the whole structure; besides, tessellation gets strong adaptability to be connected into various irregular shapes.

Tessellation is of great significance to contemporary architectural geometry design. How this ancient art can solve the present problems in architectural design, and how it can inspire architects to create more rational and complicated forms should be studied and stressed, and space in architecture can be well understood through this procedure. This paper will summarize mathematical rules in tessellations, present the application of tessellation in architecture through history, analyze typical cases in contemporary architectural design with application of tessellation, and explore the feasibility and significance of application of tessellation in architectural geometry design.

2 Rules of Tessellation

Defined as tilling of a plane or of space with on overlaps and no gaps, tessellations can be classified as blow:

2.1. Two-Dimensional Tessellation

Tessellation in two-dimensional space refers to the tessellation of a plane or of two-dimensional surfaces, and can be simplified into some basic geometries, which are called "prototypes".

[1] https://en.wikipedia.org/wiki/Tessellation

*Corresponding author: Email: successC_W@163.com

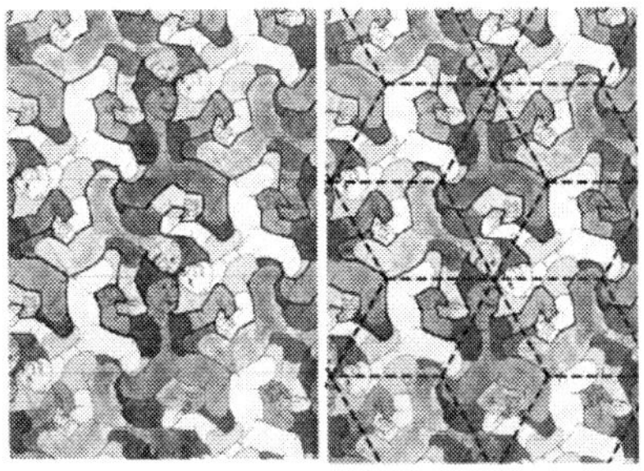

Figure 1 Parts of M. C. Escher Pictures and Its Tessellation

2.1.1. Symmetry and periodic tessellation

Many basic tessellations have the attributes of certain symmetries. Symmetry refers to the phenomenon when under some transformation conditions (e.g., around the line for plane rotation, reflection, etc.), an object or a figure in the same part are regular repeated. The symmetry modes of tessellations include translation symmetry, rotation symmetry, reflection symmetry and sliding reflection symmetry[2].

Tessellation with translational symmetry is called periodic tessellation, such as the regular hexagon tessellation. Periodic tessellation can be represented by lattice, with simple expressions: tessellation systems = elements * lattices. In addition to the common attribute of translational symmetry, plane periodic tessellation tends to have attributes of rotational symmetry, reaction symmetry sliding symmetry.

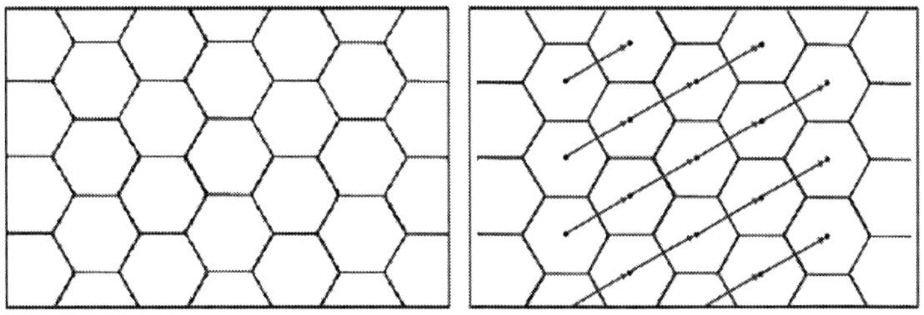

a b

Figure 2 Translational Symmetry Attribute of Regular Hexagon Tessellation

2.1.2. Aperiodic Tessellation

Aperiodic Tessellation refers to the translational tessellation without translational symmetry. As shown in Figure3, the figure in the tessellation has rotationally symmetry but no translational symmetry. Similar tessellations include Penrose aperiodic tessellation, Pinwheel aperiodic tessellation, Robinson aperiodic tessellation and so on.

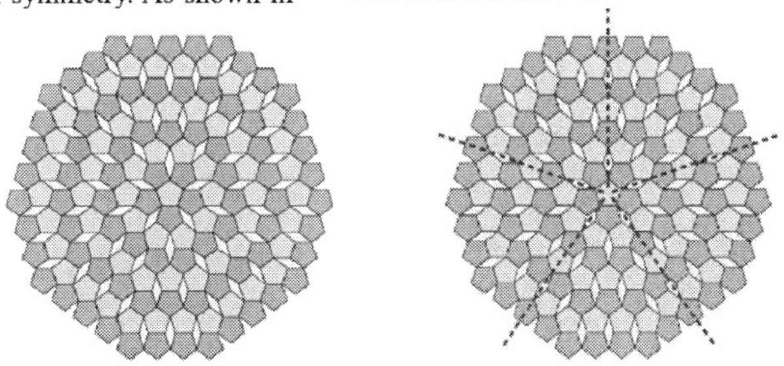

a b

Figure 3 Example of Aperiodic Tessellation

2.1.3. Fractal Tessellation

Fractals and tessellations are generally considered as two different fields of mathematics, but they can be combined to form a variety of forms with visually attractivities, concluding both fractal and tessellation features. Such form is called fractal tessellation[3].

[2] Branko Grunbaum, Tilings and Patterns. W. H. Freeman, 1987, p26.

*Corresponding author: Email: successC_W@163.com

[3] Robert W. Fathauer, Fractal tilings base on v-shaped prototiles, Computers & Graphics 26, 2002, p635-643.

 https://doi.org/10.1051/e3sconf/20183803015

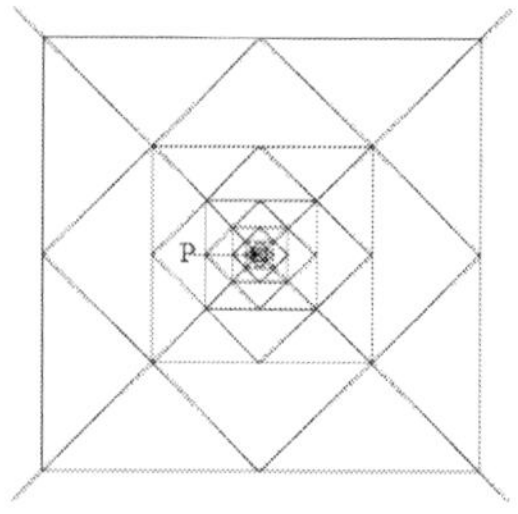

Figure 4 Fractal Tessellation with Isosceles Right Triangle

2.1.4. Radial Tessellation and Spiral Tessellation

Another way to create aperiodic tessellation is to select prototypes which can form concentric circles. Such tessellation is called radial tessellation. If parts of the radial tessellation are shifted, a swirl of new type of tessellation can be produced, as called a spiral tessellation.

Figure 5 Radial Tessellation and Spiral Tessellation with Isosceles Triangle

2.2. Three-Dimensional Tessellation

Three-dimensional tessellation, also called spatial tessellation, has more dimensional variations than two-dimensional ones, thus generating more types of geometries.

2.2.1. Crystal structure and Quasicrystal structure

Similar with two-dimensional tessellations, translational symmetry is still the basis criterion of periodicity for three-dimensional tessellations. Much attention has been focused on the field of crystal morphology in the study of three-dimensional periodic tessellation since crystals have internal structures arranged according to certain geometric rules, while internal structures of non-crystals are arranged without regular rules or even randomly. Therefore, the internal structure of crystal can be regarded as a periodic spatial tessellation, which can fill the space closely by convex polyhedrons of the same shape and size in ideal conditions.

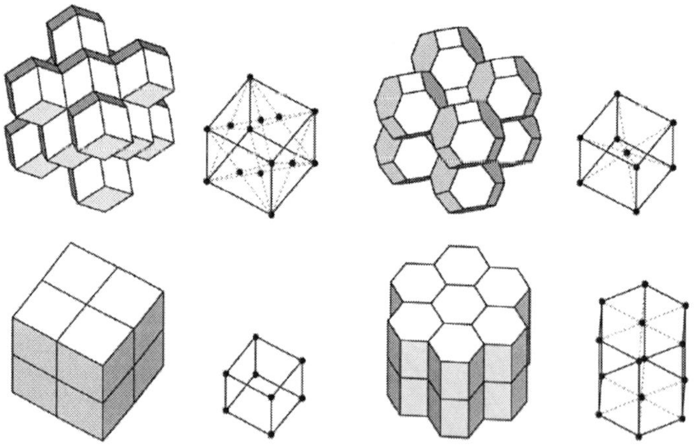

Figure 6 Examples of Combination of Convex Polyhedrons in Crystal Structures

Quasicrystal structure does not own the attribute of translational symmetry, but can repeats in certain rules in various positions and directions, following Bravais laws to generate geometric polyhedrons spontaneously. For example, Rhombic Hexecontahedron can be composed of 20 long rhombohedral bodies, while Rhombic triacontahedron can be composed of 10 "long rhombus" and 10 "flat rhombus". Combination of these two polyhedrons generates on typical aperiodic three-dimensional tessellation.

2.2.2. Other Types of Three-Dimensional Tessellation (E.G.)

2.2.2.1. WP structure

The WP structure consists of two irregular polyhedrons of equal size but of different shapes. This tessellation gives great answers to the famous question: how to divide space into equal volume units with the most economical structure segmentation and the least amount of interface areas.

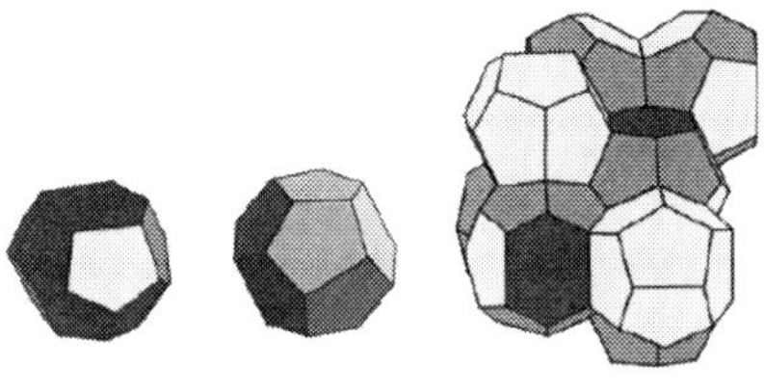

Figure 7 WP Structure Unit

2.2.2.2. Danzer Tiling

The Danzer Tiling consists of four different tetrahedra A, B, C, and K, and the unfolded diagrams of which can be regarded as the different arrangements of 4 different triangles. Each tetrahedron can be split into another Danzer Tiling with smaller proportions tetrahedron mosaic, and its expansion rate is exactly the golden number π. Different combinations of these tetrahedra can form a variety of rotationally symmetric three-dimensional aperiodic tessellation.

Figure 8 Tetrahedra and Expansion Rate in Danzer Tiling

3 The Application of Tessellation in Architectural Geometry Design

3.1. Application of Tessellation in Architecture Through History

Application of tessellation in architecture can be dated back to the beginning of civilization. It is widely found in many civilizations of the world, including ancient Egypt, Moors, Romans, Persia, Greece, Byzantium, Arabia, Japan and China, existing throughout the history of architecture development.

The ancient Romans liked to use mosaics, as one of the most common tessellations to decorate their floors and ceilings. These decorations varied according to different spaces inside the building. This style also influenced Islamic architecture greatly, using ceramics, marble, glass, glazed pottery to complete mosaic decoration. While in Gothic cathedral, radial tessellation was designed as rose windows, and arrangement of arch and its complicated structures were also typical examples of application of tessellation in architectural design.

Figure 9 Ground Floor Pavement of Rome Villa in Italy

https://doi.org/10.1051/e3sconf/20183803015

Figure 10 Tessellation in Alhambra Palace

In Chinese traditional architecture, flower windows and floor tiles are often decorated with geometric patterns to form regular, rhythmic tessellations.

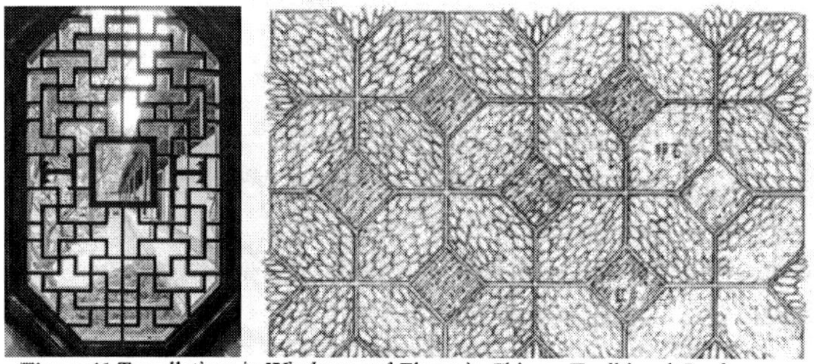

Figure 11 Tessellations in Windows and Floors in Chinese Traditional Gardens.

3.2. Application of Tessellation in Modern Architecture

Corbusier' promotion of modular design after his theoretical research and analysis of human body and its related mathematics bought in new generation of modern architecture. Standard modular system in architecture is encouraged world-widely, including basic modulus and enlarged modulus. It is a three-dimensional modular grid structure of certain rules, which can be regarded as one type of three-dimensional tessellations. Therefore, this tessellation provides the basis for the form, structure and space design of the building.

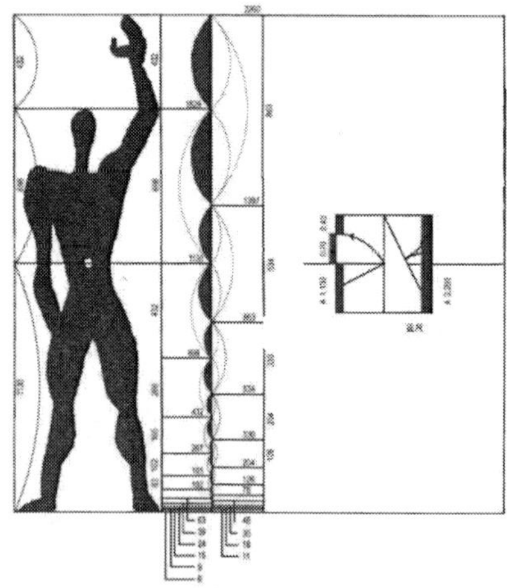

Figure 12 Human Body Modular Analysis by Corbusier

3.3. Application of Tessellation in Computer Geometric Technology

Geometric tessellation is one of the foundations of computer graphics. The application of geometric tessellation in computer technology is based on the algorithm of surface, from the input of the surface to the output of the geometry.

3.3.1. Surface Control in Modeling

Computer surface modeling can use a variety of methods, such as NURBS curve mesh modeling, mesh (Polygon Mesh) modeling, polygon mesh subdivision modeling (Subdiv). All these methods need the control and editing of the surface mesh, which can never be achieved without the application of tessellation.

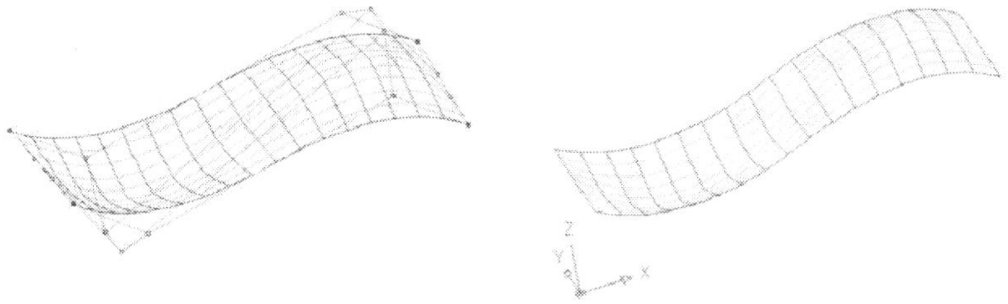

Figure 13 NURBS Control

3.3.2. Subdivision of Surfaces in Display

In the aspect of computer graphics display, surface subdivision technology is the technology of managing graphic data by embedding graphics, and subdividing the graph to make it easy to render. It is called "mosaic (one classical tessellation) processing technology". Computers can not directly generate curves, nor can they directly generate surfaces. The surface displayed on a computer screen is actually made up of a number of dots of polygonal patches. Computer surface subdivision is to refine the original rough mesh model by using proper topologies and geometric rules, so as to generate smooth surfaces.

4 Modern Cases

4.1. 0-14 Tower

The 0-14 Tower in Dubai was designed by Reiser + Umemoto in 2007. Rather than a glass curtain wall which is visible everywhere in Dubai, the facade uses concrete pouring to form a shell structure. Large and small circular windows composed its unique facade. These windows were generated accoding to various factors based on rectangular grid. Approximate circular window arrays were formed by inner offset and chamfer of each unit in the oblique rectangular grid. The amount of offseting was related with parameters such as indoor lighting conditions and structure forces. Moreover, the mutation of several units in the oblique rectangle grid provides better outside views for the inside.

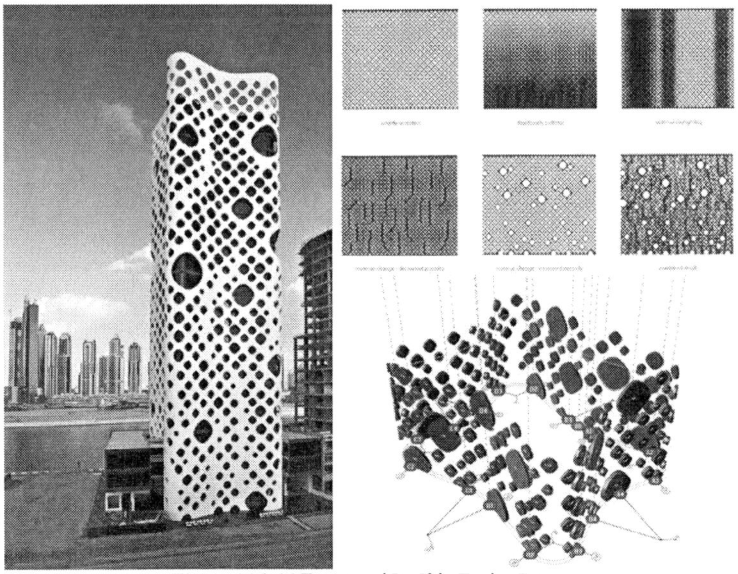

Figure 14 0-14 Tower and Its Skin Design Process

4.2. Water Cube

Designed by CSCEC, CCDI, PTW and ARUP, the steel structure of the "Water Cube" (National Aquatics Center of the Beijing Olympic Games) is a three-dimensional tessellation structure made of WP units. In the design process WP units were firstly used to produce a tessellation structure construction larger than the volume of "Water Cube", after which the structure was rotated by 60 degrees based on vector axis, from which a 176.5 m x 176.5m x 29.4 m cubic was cutting out as the building's gerometry. Spaces of different funtions were later generated from inside, thus forming a spatial grid steel structure. Since the cutting direction is not parallel or orthogonal to its translation direction, the facade is not smooth. However, in fact it still keeps the translational periodicity as WP tessellation does.

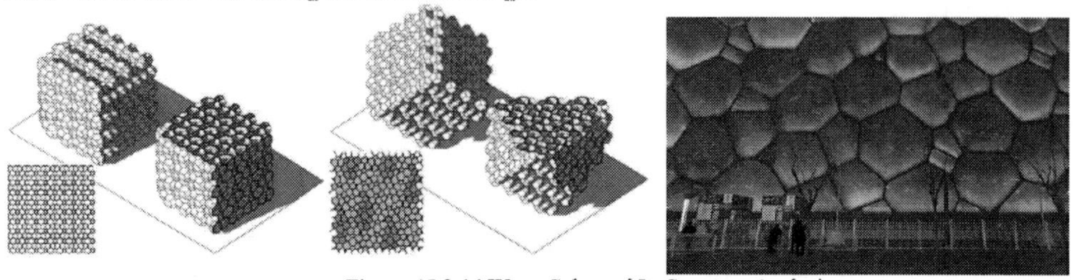

Figure 15 0-14 Water Cube and Its Structure Analysis

4.3. Hybrid Bio Structures

Hybrid Bio Structures was designed by Riyad Joucka in his master's degree project in AA (U.K.). This project revealed another application of Voronoi tessellation. Voronoi tessellation get features of "field", so in the conditon of membrane structure with a plurality of support points, the grid generated through Voronoi algorithm reflected the gravity of maximal deformation in the membrane structure. Voronoi tessellation was generated after setting "seeds" (based on indoor function, lighting environment, etc.) positions. Roof of complex membrane structure was formed through point stretching, smoothing and other movements of the units. This architecture is a pavalion with multi-functions, and the Voronoi grid remained through the tour route for visitors.

Figure 16 The Component Unit of Hybrid Biostructure

5 Conclusion

As significant influential factor of architecture, tessellation runs through the whole architecture history of the East and the West. At present, it is also the basis of graphic technologies in the development of digital technology. It acts as a link between architectural design and digital technology, widely used in contemporary architectural design, especially architectural geometry design.

Since tessellation geometry is usually a combination of homogeneous units, it has outstanding advantages in building skin, construction ability, structure property and so on. In addition, standing at the source position in the computer surface technology, tessellation contributes to the easy set-up and control of surface design in architectural geometry. With the application of tessellation, architectural geometry can be more flexible and contains various forms of different compositions, being more adopted in dealing with different problems. Therefore, such application in architectural geometry design will be of great benefit and potential with thoughtful consideration.

References

1. Branko Grunbaum, Tilings and Patterns[M]. W.H. Freeman, 1987.
2. Lynn G. Animate Form[M]. Princeton Architectural Press, 1999.
3. Farshid Moussavi, The Function Of Form[M]. Harvard University Graduate School of Design,2009.
4. Kostas Terzidis, Algorithmic Architecture[M]. Architectural Press, 2006.
5. Michael Meredith, From Control to Design: Parametric/Algorithmic Architecture[M]. Actar/Barcelona Regional, 2008.
6. Paul Jackson, Folding Techniques for Designers: From Sheet to Form[M]. Laurence King, 2011.
7. Eric Gjerde, Origami Tessellations Awe-Inspiring

Geometric Designs[M]. A K Peters,2008.
8. Michael Fox, Interactive Architecture[M]. Princeton Architectural Press, 2009
9. Jingzhong Chen, Basic properties of quasicrystals[J]. Proceedings of China University of Geosciences, 1993-S1.
10. Xianguang Chen, Study on the Shape Finding of Tensegrity Units [J]. Space Structure, 2003-01.
11. Robert W.Fathauer, Fractal tilings based on v-shaped prototiles[J]. Computers & Graphics, 2002.
12. Danzer L., Three-dimensional analogs of the planar penrose tilings and quasicrystals[J]. Discrete Mathematics, 1989.

https://doi.org/10.1051/e3sconf/20183803016

Research on Bidding Decision-making of International Public-Private Partnership Projects

Zhen Yu HU, Shui Bo ZHANG, Xin Yan LIU

College of Management and Economic, Tianjin University, Tianjin 300072, China

Abstract: In order to select the optimal quasi-bidding project for an investment enterprise, a bidding decision-making model for international PPP projects was established in this paper. Firstly, the literature frequency statistics method was adopted to screen out the bidding decision-making indexes, and accordingly the bidding decision-making index system for international PPP projects was constructed. Then, the group decision-making characteristic root method, the entropy weight method, and the optimization model based on least square method were used to set the decision-making index weights. The optimal quasi-bidding project was thus determined by calculating the consistent effect measure of each decision-making index value and the comprehensive effect measure of each quasi-bidding project. Finally, the bidding decision-making model for international PPP projects was further illustrated by a hypothetical case. This model can effectively serve as a theoretical foundation and technical support for the bidding decision-making of international PPP projects.

1 Introduction

PPP (Public-Private Partnership), a new type of project construction model, is widely used in infrastructure projects. And as there are more and more international projects with the participants from multiple countries, the number of PPP projects has shown a rapid growth trend in the international construction market [1]. Some investment enterprises want to bid for international PPP projects to get the opportunity to undertake them and then obtain high revenue. But compared with general projects, international PPP projects take more time and funds in the bidding, because of the massive investment and high project complexity of these international PPP projects, together with the huge difference in institutional environments. If an improper international PPP project is selected, the investment enterprise will not only suffer from a waste of resources and the miss of better project bidding opportunities, but also face enormous risks that can cause project failure in the implementation of the project [2]. Therefore, the investment enterprises that intend to invest in the international PPP projects must adopt scientific bidding decision-making methods to make rational bidding decisions in the bidding of international PPP projects.

At present, some scholars have conducted researches on bidding decision-making methods. Han [3] used a cross-impact analysis method (CIA) to develop a risk-based bidding decision-making model applied to general projects. Based on the project portfolio management theory, Li [4] adopted the fuzzy comprehensive evaluation method through analyzing project profitability to make bidding decisions of general projects. Liu [5] proposed a multi-objective bidding decision-making method based on the fuzzy set theory to solve the bidding decision-making problem in the bidding of general projects. However, these studies were mainly focused on general projects, and there has been relatively little scientific research on the bidding decision-making of international PPP projects. Therefore, this paper aims to establish a bidding decision-making model for international PPP projects to select the optimal project from the quasi-bidding projects.

2 Bidding decision-making index system for international PPP projects

Based on the characteristics of the international PPP projects, this study searched for the literature related to the bidding decision-making of international PPP projects through the Google Scholar. A total of 84 key papers were screened out and analyzed by adopting the literature frequency statistics method [6]. Then the bidding decision-making indexes which appear in at least 10 papers, were selected and used to construct the bidding decision-making index system for international PPP projects as shown in Fig 1.

*Corresponding author: Email: huzhenyugg@tju.edu.cn

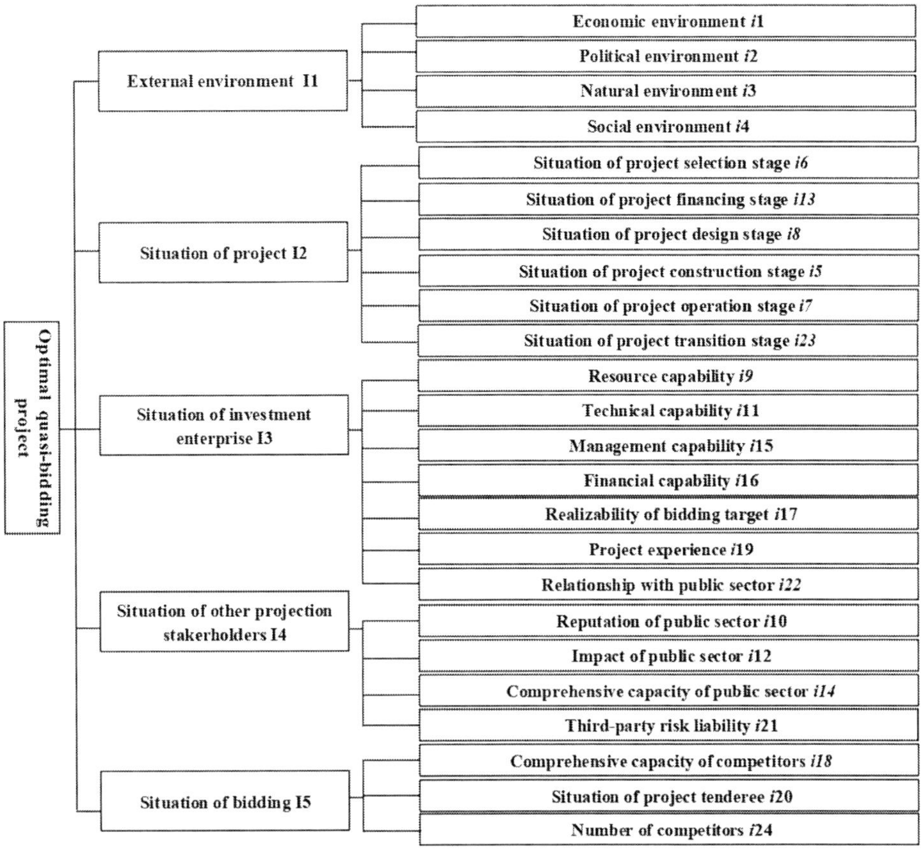

Figure 1 Bidding decision-making index system for international PPP projects

3 Bidding decision-making model for international PPP projects

It is assumed that the number of the quasi-bidding projects is l, and the number of bidding decision-making indexes is m. In the bidding decision-making model, y_{ih} represents the index value of the project $h\ (h=1,2,\text{L}\ ,l)$ under the bidding decision-making index $i\ (i=1,2,\text{L}\ ,m)$, and $Y=(y_{ih})_{m\times l}$ refers to the matrix of the decision-making index values. As the above bidding decision-making index system for international PPP projects is adopted in this study, the value of m in this study equals to 24. Among the bidding decision-making indexes in the bidding decision-making index system, only the bidding decision-making indexes of 'Number of competitors' is a quantitative bidding decision-making index, and the other 23 bidding decision-making indexes are all qualitative bidding decision-making indexes. The value of a quantitative decision-making index is determined by collecting empirical data, while the experts' evaluation scores are used to determine the value of a qualitative bidding decision-making index.

3.1 Setting decision-making index weights

In the proposed bidding decision-making model, the group decision-making characteristic root method [7] is used to set the subjective weights of the bidding decision-making indexes, which can be combined with the objective weights of the bidding decision-making indexes set by the entropy weight method [8] to obtain the comprehensive weights of the bidding decision-making indexes as the decision-making index weights. Setting the comprehensive weights of the bidding decision-making indexes can both make full use of the knowledge and experience of experts, and effectively make up for the shortcomings of subjective judgments, thus making the bidding decisions more scientific and reasonable.

1) Determination of subjective weights of bidding decision-making indexes through group decision-making characteristic root method

First of all, experts are invited to score the relative weights of the first-level bidding decision-making indexes under the general objective, and the relative weights of the second-level bidding decision-making indexes under their first-level bidding decision-making indexes. For example, the expert group determines the score matrix of $Q=(q_{ij})_{m\times n}$ representing the relative weights of certain second-level bidding decision-making indexes under their first-level bidding decision-making index, and q_{ij} refers to the relative weight of the bidding decision index i judged by the expert j $(j=1,2,\ldots,n)$. The multiplication of the scoring matrix Q with its transpose is denoted as matrix F, that is $F=QQ^T$. The expert group's judgment vector of the relative weights of the bidding decision-making indexes is a positive eigenvector corresponding to the maximum positive eigenvalue of the matrix F [7]. So after obtaining and normalizing this positive eigenvector, the subjective relative weights of bidding decision-making indexes can be determined. And then according to the hierarchical structure of the bidding decision-making index system, the subjective comprehensive weights of bidding decision-making indexes represented by $\omega=(\omega_1,\omega_2,\ldots,\omega_m)^T$ can be obtained.

2) Determination of objective weights of bidding decision-making indexes through entropy weight method

The greater the variability of a decision-making index value is, the smaller its entropy but the larger its weight is, thus the entropy weight method can be used to determine the objective weights of the bidding decision-making indexes by analyzing the degree of variation of each decision-making index value [8].

At first, the matrix $Y=(y_{ih})_{m\times l}$ is normalized to obtain the matrix $R=(r_{ih})_{m\times l}$. Among the 24 bidding decision-making indexes used in this study, only the bidding decision-making indexes of "Impact of public sector", "Third-party risk liability", "Comprehensive capacity of competitors" and "Number of competitors" are cost-type bidding decision-making indexes, whose values are better when they become smaller, while the remaining bidding decision-making indexes are benefit-type bidding decision-making indexes, whose values are better when they become larger.

The normalized equation for the benefit-type bidding decision-making indexes is [8]:

$$r_{ih}=\frac{y_{ih}-\min\limits_{1\le h\le l}\{y_{ih}\}}{\max\limits_{1\le h\le l}\{y_{ih}\}-\min\limits_{1\le h\le l}\{y_{ih}\}} \tag{1}$$

Similarly, the normalized equation for the cost-type bidding decision-making indexes is [8]:

$$r_{ih}=\frac{\max\limits_{1\le h\le l}\{y_{ih}\}-y_{ih}}{\max\limits_{1\le h\le l}\{y_{ih}\}-\min\limits_{1\le h\le l}\{y_{ih}\}} \tag{2}$$

The equation used to calculate the entropy of the bidding decision-making indexes is [8]:

$$E_i=-k\sum_{h=1}^{l}p_{ih}\ln p_{ih},\quad i=1,2,\ldots,m \tag{3}$$

$$p_{ih}=\frac{r_{ih}}{\sum_{h=1}^{l}r_{ih}}\ (\text{or } r_{ih}=0 \text{ with } p_{ih}\ln p_{ih}=0),\quad k=\frac{1}{\ln l},$$

$$0\le E_i\le 1$$

The equation to obtain the objective weights of the bidding decision-making indexes is [8]:

$$\mu_i=\frac{1-E_i}{m-\sum_{i=1}^{m}E_i},\quad i=1,2,\ldots,m \tag{4}$$

$$0\le\mu_i\le 1,\ \sum_{i=1}^{m}\mu_i=1$$

3) Determination of comprehensive weights of bidding decision-making indexes through optimization model based on least square method

It is assumed that $w_i(i=1,2,\ldots,m)$ is the comprehensive weight of the bidding decision-making index i and $\sum_{i=1}^{m}w_i=1$, so $W=(w_1,w_2,\ldots,w_m)^T$ represents the comprehensive weight vector. To combine the subjective weights and the objective weights of the bidding decision-making indexes, the optimization model based on least square method can be expressed by the equation as [9]:

$$\begin{cases}\min I(W)=\sum_{h=1}^{l}\sum_{i=1}^{m}\{[(\omega_i-w_i)r_{ih}]^2+[(\mu_i-w_i)r_{ih}]^2\}\\ s.t.\sum_{i=1}^{m}w_i=1\\ 0\le w_i\le 1, i=1,2,\ldots,m\end{cases} \tag{5}$$

It is has been proved that the exact solution of this equation, i.e. the comprehensive weight, is equivalent to the arithmetic mean of the objective weight and the subjective weight, so the comprehensive weights of the bidding decision indexes can be obtained by the equation expressed as [9]:

$$w_i=\frac{\omega_i+\mu_i}{2},\quad i=1,2,\ldots,m \tag{6}$$

3.2 Evaluating quasi-bidding projects

1) Transformation of decision-making index values into consistent effect measure

Because of the various dimensions and heterogeneous nature of the decision-making index values, consistent effect measure function can be constructed to transform the decision-making index value y_{ih} into consistent effect measure $d_h^{(i)}\in[-1,1]$ $(h=1,2,\ldots,l;i=1,2,\ldots,m)$, to make these decision-making index values comparable. The consistent effect measure matrix $D^{(i)}=(d_h^{(i)})$ can be obtained accordingly [10].

At first, the critical value y_{ih_0} of the bidding decision-making index i is determined by averaging the critical values of this bidding decision-making index set by each expert. Then the consistent effect measure is calculated based on the following two equations. The larger the value of $d_h^{(i)}$ is, the more satisfactory the decision-making index value y_{ih} is. '$d_h^{(i)} \in [0,1]$' means the decision-making index value of the project h under the decision-making index i meets the requirements, but '$d_h^{(i)} \in [-1,0)$' has the opposite meaning.

If the satisfactory range of a benefit-type decision-making index value is $y_{ih} \in [y_{ih_0}, \max y_{ih}]$, the consistent effect measure function for the benefit-type decision-making indexes can be proposed as [10]:

$$d_h^{(i)} = \frac{y_{ih} - y_{ih_0}}{\max y_{ih} - y_{ih_0}} \quad (7)$$

Similarly, if the satisfactory range of a cost-type decision-making index value is $y_{ih} \in [\min y_{ih}, y_{ih_0}]$, the consistent effect measure function for the cost-type decision-making indexes can be proposed as [10]:

$$d_h^{(i)} = \frac{y_{ih_0} - y_{ih}}{y_{ih_0} - \min y_{ih}} \quad (8)$$

2) Calculation of comprehensive effect measure of quasi-bidding projects

According to the following equation, the comprehensive effect measure $d_h \in [-1,1]$ of the quasi-bidding project h can be calculated by synthesizing the weights of each bidding decision-making index [10]. The larger the value of d_h is, the more satisfactory the quasi-bidding project h is. '$d_h \in [0,1]$' means the quasi-bidding project h meets the requirements of the investment enterprise, but '$d_h \in [-1,0)$' means oppositely. Then the comprehensive effect measure matrix $D = (d_h)$ can be obtained, based on which the optimal quasi-bidding project can be chosen from the satisfactory quasi-bidding projects by comparing the comprehensive effect measure.

$$d_h = \sum_{i=1}^{m} w_i \bullet d_h^{(i)} \quad (9)$$

If '$\max_{1 \le h \le l} \{d_h\} = d_{h_0}$' exists, the quasi-bidding project h_0 is the optimal quasi-bidding project.

4 Hypothetical case

An investment enterprise intends to bid for an international PPP project which is chosen from Project h_1, Project h_2 or Project h_3, and a expert group with five members is set up to analyze the bidding decision to select the optimal quasi-bidding project.

4.1 Determination of decision-making index values

The expert group uses the Likert five-component method to score the 23 qualitative decision-making index values of each quasi-bidding project. The score increases from 1 to 5, indicating the gradual increment of the qualitative decision-making index value. The scores of the five experts are averaged to be used as the qualitative decision-making index values. As for the only quantitative bidding decision-making index "Number of competitors", it is found there are 8, 12, 10 quasi bidders in addition to this investment enterprise for Project h_1, Project h_2 and Project h_3 respectively through a survey. Therefore, all decision-making index values of the three quasi-bidding projects are obtained and shown in Table 1.

Table 1 Decision-making index values and weights

Indexes	Decision-making index values			Decision-making index weights		
	Project h_1	Project h_2	Project h_3	Subjective comprehensive weights	Objective weights	Comprehensive weights
i1	3.2	3.8	3.4	0.067	0.041	0.05
i2	3.4	4	3.6	0.064	0.041	0.05
i3	4.2	3.2	3	0.061	0.052	0.06
i4	4	3.6	3.4	0.058	0.041	0.05
i5	3.2	3.2	2.2	0.048	0.031	0.04
i6	3.8	3.2	4.2	0.042	0.033	0.04
i7	3	3.8	3.8	0.050	0.031	0.04
i8	3.4	3.4	3.2	0.038	0.031	0.03

i9	3.8	4.2	3.8	0.034	0.083	0.06
i10	3.8	3.8	4.4	0.043	0.083	0.06
i11	3.2	3.8	4.2	0.034	0.033	0.03
i12	3.2	4.2	3.4	0.042	0.031	0.04
i13	3	3	3.6	0.035	0.083	0.06
i14	3.6	3	3.6	0.041	0.031	0.04
i15	3.4	3.2	4	0.031	0.045	0.04
i16	4.2	4	2.6	0.039	0.031	0.03
i17	3.4	3.2	3	0.034	0.035	0.03
i18	3.2	3.8	3	0.041	0.031	0.04
i19	3.2	3.6	3.8	0.031	0.032	0.03
i20	4.2	3	3.8	0.042	0.032	0.04
i21	4.2	3.4	4.4	0.032	0.049	0.04
i22	3	3.2	3.2	0.029	0.031	0.03
i23	3.8	3.6	3.4	0.025	0.035	0.03
i24	8	12	10	0.036	0.035	0.04

4.2 Determination of decision-making index weights

The expert group also adopts the Likert five-component method to score the relative weights of the bidding decision-making indexes, and the score ranges from 1 to 5 representing the continuum from the end of the minimum relative weight to the end of the maximum relative weight. For example, the expert group determines the score matrix of the subjective relative weights of the first-level bidding decision-making indexes under the general objective, which is $Q_1=\begin{bmatrix}5&5&4&5&4\\4&5&5&4&4\\4&5&4&5&3\\2&3&3&4&2\\1&2&2&3&3\end{bmatrix}$. Then a positive eigenvector corresponding to the maximum positive eigenvalue of this matrix is obtained by using MATLAB 7.0, which is $(0.542,0.518,0.504,0.342,0.259)^T$. This positive eigenvector is normalized and the subjective relative weight vector of the first-level bidding decision-making indexes under the general objective is obtained, which is $\omega_1^{'}=(0.250,0.239,0.233,0.158,0.120)^T$. Similarly, according to the experts' score matrixes of the subjective relative weights of the second-level bidding decision-making indexes under their first-level bidding decision-making indexes, which are $Q_2=\begin{bmatrix}5&5&4&4&5\\5&4&4&5&4\\4&5&4&4&4\\4&4&5&4&3\end{bmatrix}$, $Q_3=\begin{bmatrix}4&5&3&4&4\\3&3&4&4&3\\4&4&3&4&3\\5&4&5&5&4\\5&5&5&4&5\\2&3&2&3&2\end{bmatrix}$, $Q_4=\begin{bmatrix}4&5&3&4&4\\4&3&4&4&5\\3&4&5&3&3\\5&5&4&5&4\\4&3&4&5&4\\3&4&3&4&4\\3&3&4&4&3\end{bmatrix}$, $Q_5=\begin{bmatrix}4&5&5&5&4\\4&4&5&4&5\\4&5&4&4&5\\3&4&3&3&4\end{bmatrix}$, $Q_6=\begin{bmatrix}5&4&5&4&4\\5&5&3&4&5\\4&4&4&3&4\end{bmatrix}$, the subjective relative weight vectors of the second-level bidding decision-making indexes under their first-level bidding decision-making indexes are obtained by calculating and normalizing with MATLAB 7.0, which are $\omega_2'=(0.268,0.256,0.243,0.233)^T$, $\omega_3'=(0.177,0.148,0.159,0.201,0.209,0.106)^T$, $\omega_4'=(0.147,0.147,0.132,0.169,0.148,0.132,0.125)^T$, $\omega_5'=(0.275,0.263,0.260,0.202)^T$, $\omega_6'=(0.345,0.354,0.301)^T$. And then according to the hierarchical structure of the decision-making index system, the subjective comprehensive weights of the bidding decision-making indexes are obtained and shown in Table 1.

Furthermore, the decision-making index values are normalized according to the equation (1), equation (2) and equation (3). Then the objective weights of bidding decision indexes are calculated according to the equation (4) and shown in Table 1. Finally, according to the equation (6), the comprehensive weights of the bidding decision-making indexes are obtained and also shown in Table 1.

4.3 Determination of optimal quasi-bidding

project

The critical values of the bidding decision-making indexes are determined by averaging the critical values of these bidding decision-making indexes judged by each expert, and the vector of these critical values is obtained as $Y_0 = [3,3,3,3,2.5,3,3,3,3,3,3,4.5,3,3,3,3,3,4,3,3,4.5,3,3,10]^T$, and the consistent effect measure of each bidding decision-making index satisfies the condition of $d_h^{(i)} \in [-1,1]$. After the consistent effect measure matrix $D^{(i)}$ is obtained according to the equation (7) and equation (8), the comprehensive effect measure matrix is obtained according to the equation (9), that is $D = W \bullet D^{(i)} = [0.64, 0.51, 0.59]$.

The result shows $d_1 > 0$, $d_2 > 0$, $d_3 > 0$, so the three quasi-bidding projects all meet the requirements of the investment enterprise. But $\max_{1 \le h \le 3}\{d_h\} = d_1 = 0.64$, so the Project h_1 is the optimal quasi-bidding project for the investment enterprise.

5 Conclusion

Through the above case analysis, the feasibility and applicability of the proposed bidding decision-making model for international PPP projects have been verified. Although the model can be applied to the bidding decision-making of international PPP projects, some suggestions are still put forward to strengthen its validity further.

Firstly, the developed model has integrated sufficient macro-level factors of the bidding decision-making of international PPP projects, such as the 'Economic environment' and 'Political environment', but some micro-level psychological factors that may affect the decision behavior of the decision maker, such as the decision maker's risk preference, were not taken into account. Future research can further improve the model by resolving the issue to make it more practical and effective.

Secondly, only a hypothetical case was used in this study to verify the proposed bidding decision-making model for international PPP projects. Future research can conduct empirical studies to collect real-life project data to further validate this model, and to provide practical proof of the conclusion of this study. Such rich empirical investigation would contribute to the further improvement of the bidding decision-making model for international PPP projects.

Acknowledgements

We gratefully acknowledge the National Natural Science Foundation of China (NSFC) for its financial support for this research work (Grant Nos. 71572124).

References

1. Grimsey D, Lewis MK. Evaluating the risks of public private partnerships for infrastructure projects[J]. International Journal of Project Management, 2002, 20(2): 107-118.
2. Zhang XQ. Critical success factors for Public–Private Partnerships in infrastructure development[J]. Journal of Construction Engineering and Management, 2005, 131(1): 3-14.
3. Han SH, Diekmann J. Making a risk-based bid decision for overseas construction projects[J]. Construction Management and Economics, 2001, 19(8): 765-776.
4. Li Hong-bing, Li Hong, Chen Yu. Application of Portfolio Management to Bidding Decision-making[J]. Journal of Wuhan University of Technology, 2008, 30(6): 170-173.
5. Liu Er-lie, Wang Jian, Luo Gang. Project bidding decision based on fuzzy logic[J]. China Civil Engineering Journal, 2003, 36(3): 57-63.
6. Lindsay R K, Gordon M D. Literature-based discovery by lexical statistics[J]. Journal of the Association for Information Science and Technology, 1999, 50(7): 574-587.
7. Chen Wei, Kang Xin, Feng Zhi-jun, Tian Shi-hai. Identification of evaluation indexes for intellectual property's exploitation of high-tech enterprises based on group decision-making characteristic root method[J]. Science & Technology Progress and Policy, 2011, 28(11): 116-119.
8. Hong Wei-min, Liu Hong-mei, Wang Zhuo-fu. Project transaction mode decision based on entropy weight fuzzy comprehensive evaluation method[J]. Science and Technology Management Research, 2010(3): 122-125.
9. Xu Ze-shui. Improvement and Extension of a Decision Making Method[J]. Operations Research and Management Science, 2000, 9(4): 46-51.
10. Liu Si-feng, Yuan Wen-feng, Sheng Ke-qin. Multi-attribute intelligent grey target decision model[J]. Control and Decision, 2010, 25(8): 1159-1163.

 https://doi.org/10.1051/e3sconf/20183803017

Research on Green Construction Technology Applied at Guangzhou Hongding Building Project

Yong Zhong LOU

School of Architecture & Civil Engineering, Huizhou University, Huizhou, Guangdong 516007, China

Abstract: The green construction technology is the embodiment of sustainable development strategy in the construction industry, and it is a new construction mode which requires a higher environmental protection. Based on the Hongding building project, this paper describes the application and innovation of technical in the process of implementing green construction in the project, as well as the difficulties and characteristics in the specific practice; .The economic and social benefits of green construction are compared to the traditional construction model; .The achievements and experience of the green construction technology are summarized in the project; The ideas and methods in the process of implementing green construction are abstracted; some suggestions are put forward for the development of green construction.

1 Background of Implementing Green Constructions

With the fast development of Chinese urban construction scale, a large amount of materials and energy are consumed in the construction process. The current situation of the polluted environment goes against the concept of Energy-saving & Low-carbon and Green & Environmental-Friendliness currently promoted by China. Therefore, developing green construction technology becomes the urgent requirement for the current construction industry and the inevitable path for Chinese construction industry to achieve sustainable development [1]. Responding to the call of China and following the requirements of green construction's guidelines and assessment standards, Guangzhou Hongding Building Project actively constructs green construction sites, promotes the effective measures of "four conservancies and one environmental-protection" and achieves the aim of "low energy consumption, low pollution, recycled use of resources and the harmony between humans and the nature" on the precondition of guaranteeing constructions' quality, safety and schedule.

2 Pre-phase Planning of Green Constructions

2.1. Green Construction Management System

Guangzhou Hongding Building Project is to build one high-rise building complex. The architectural planning is: five floors for commercial use and 30 floors for flats, with an architectural area of 17,000 square meters and overall construction area of 140,000 square meters. The building's structure form is frame—core tube structure.

* Corresponding author: Email: 58307085@qq.com

To promote the green construction management comprehensively, it is necessary to set up a green construction leading team first, which is equipped with green construction departments and teams that assign different work to each personnel; Next, formulating regulations and systems to implement systematic guarantee, such as *Green Construction Index Management System*, *Green Construction Assessment Responsibility System*, *Energy Use and Management System*, *Environmental Protection and Management System*, *Green Construction Assessment and Evaluation System* [2].

2.2 Construction Management

The department should carry out dynamic management of the whole implementation process, manage and supervise each link from construction planning, material procurement, on-site construction and project acceptance.

It is advised to put up placards in the living zone and construction zone on the construction site, thereby creating the atmosphere of green constructions and enhance employees' awareness of green constructions.

2.3 Assessment Management

Together with proprietors and supervisors, the project department examines the implementation of green constructions every month, grades and assesses the implementation in accordance with GB/T50640-2010 in *Construction Project Green Construction Assessment Standards* to achieve sustainable improvement; .The green construction leading team randomly examines green constructions and comprehensively assesses the implementation of the project's green constructions until the project is accepted and delivered to proprietors for use [3].

3 Measures of Green Constructions

3.1 Environmental Protection

3.1.1 Control of Dust Pollution

It is necessary to carry out hardening and greening of the on-site construction's primary and secondary roads, the material processing zone, the office zone and the living zone in accordance with requirements. Meanwhile, pour water on the primary and secondary roads to reduce dust regularly.

Service-racks, dampness-absorbing cushion and high-pressure water guns are arranged at the entrance and exit, which ensure vehicles' sanitation and prevent vehicles' tires from bringing mud out of the construction site and making outside roads dirty.

It is advised to adopt measures to cover vehicles that may easily cause dust when delivering earthwork, concrete and other things. It is prohibited from overload to avoid spillage.

Under the weather with wind bigger than Level-4, carrying out earthwork refilling, transshipment and other construction operations that may cause dust pollution will not be allowed.

In the structure construction phase, premixed concrete and mortar will be adopted.

Professional examination unit regularly tests the TSP concentration in positions with the same height of the shield near the construction site and achieves constant dynamic supervision. The testing results all meet normal requirements.

3.1.2 Control the Emission of Harmful Gases

It is prohibited from burning various wastes on the construction site. And transportation vehicles must meet the local region's exhaust emission standards. Vehicles failing to meet standards are prevented from entering the construction site.

The welding of steel components is carried out by carbon dioxide gas protection welding and other welding techniques that produce fewer waste gases. The smoke emission of electrical welding should meet the regulations of national standard—*Comprehensive Emission Standards for Atmospheric Pollutants*.

3.1.3 Control of Construction Wastes

Steel bars, templates and other architectural materials are blanked by professional personnel that have rich experience, so as to avoid unnecessary rework that increases construction wastes; improve the qualities of templates' seams and the seams of steel bar truss steel plates, so as to avoid or reduce spillage.

The department should enhance the recycled and reuse management of construction wastes: Setting up closed waste separation station on the site and process construction wastes according to different categories; smash gravels, broken bricks, concrete and other basic root that are paved for temporary roads into gradations for refilling foundation pits; the recycling of steel bar truss plate's leftover materials is carried out by a professional company; the wood material and plats are cycled by plywood mill and paper mills; short steel bars are used to fix electrical switches and siding wall's position bars.

The dust bins can be divided into recycled use, non-recycled use and toxic & harmful wastes, which are cleared and transported regularly.

Construction materials' packages are centrally piled before being uniformly recycled by manufacturers, and the rate of recycling packages reaches 100%.

Printer's oil boxes, toner cartridges, cells and other toxic & harmful wastes are recycled in closed box. Manufacturers adopt the service of exchanging old with new ones and recycled them for reuse.

3.1.4 Control of water and land pollution

The sewage should be directly discharged into the municipal sewage pipe network. It should be first discharged into the sedimentation tank. After undergoing twice sedimentation & filtration of cement paste and mortar, it is used to spray on the hardened floor within the construction site to reduce dust or discharged into the municipal sewage pipeline.

The special sewage pipelines are arranged in the toilets and canteens in the living zone. Meanwhile, link them to the septic tank, oil separator and other facilities to process life sewage; set filtration webs in the sewage discharge pipelines of toilet and canteens, which are cleared by professional personnel regularly.

The site's sewage discharge entrusts qualified units to carry out sewage water quality testing regularly. The sewage discharge meets the requirements of GB20425-2006 in Sewage Discharge Comprehensive Standards.

The place for storing chemical products and other toxic materials has strict water-proof designing to prevent oil and liquid from leaking.

3.1.5 Control of light pollution

It is prohibited from using lightening equipment that has no cover and protection in the construction site. The lights for lightening at night are energy-conserving and have a set cover, with the lightening direction concentrated in the construction range to avoid influencing the common life of surrounding residents. And it should reasonably formulate construction operation plans and try to avoid have construction between 22:00~6:00 a.m.

It is suggested adopting LED light in the basement of the construction site and for floor lightening. It is prohibited from using incandescent light bulbs. LED lights not only have good outcomes but also save electricity.

3.1.6 Noise and Vibration Control

The department should promote civilized constructions, enhance management of human noises and reduce humans' uproars. The schedule plan should be reasonably arranged and avoid construction operations that have huge noises at night, such as concrete pouring.

The mechanic equipment that produce strong noises should be kept away from the office zone and the living zone of the construction site, as well as surrounding living residential zones, so as to prevent the construction from disturbing residents.

The environmental-friendly and low-noise vibrators and other machines are used for constructions. When vibrating, it is prevented from contacting modules and steel bars to prevent the vibrator from running idly and avoid or reduce construction noises. These machines are maintained by professional personnel regularly.

Establishing noise supervision sites surrounding the construction site and arrange professional personnel for dynamic supervision.

3.2 Measures of saving and utilizing materials

The department implement quota material-picking, formulate and implement measures of controlling crucial points with targets, so as to improve the rate of saving materials. And the material procurement plan, time and sequence of entering the site reasonably should be formulate reasonably to reduce the stock and avoid overstock or waste.

The detachable and environmentally-friendly walls that can be used for many times are adopted; the protection parts of adjacent edges and holes should use canalized, tooled and standardized wall that is detachable and used in recycled ways. The results are shown in Fig 1 and Fig 2.

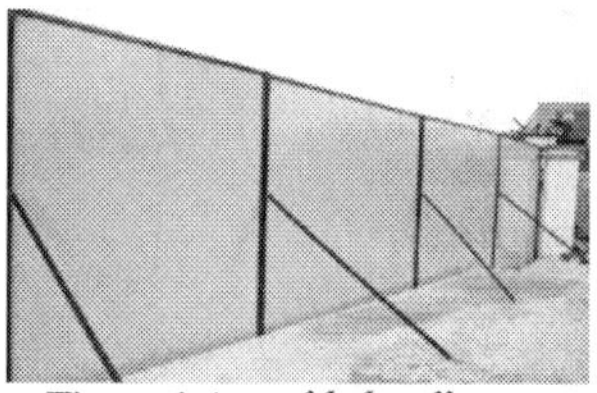

Figure 1 Assembled wall

Figure 2 Detachable 360°-turning protection rail

3.3 Measures for conserving water and utilizing water resources

The pipe's diameter is reasonable and the pipeline is easy and convenient, which reduce the leakage and loss caused by the pipe network and water-using tool. When signing labor distribution sub-contract and participating unit's cooperation contact, incorporate the water conserving index into the contract's clauses and determine water conserving indexes, making it convenient to control water and fully utilize water resources, as well as carry out measurement and assessment by professional personnel in each month.

It calculate water according to different regions and phases, count and manage waster used for the project and living. It adopt advanced water-conserving construction procedures:

① The concrete is maintained by covering straw bags and thin films;

② Cover the independent column concrete with plastic thin film for maintenance;

③ Spray concrete maintenance protection liquid on the surface of walls to save water used for constructions.

3.4 Measures of Saving Energy and Utilizing Energy

The energy-conserving & efficient electric appliance and construction equipment recommended by the country and industry are utilized; it is recommended to utilize invertor air-conditioners. On the site, utilize inverter high-voltage water pump to save energy. It is recommended to fully utilize renewable energy such as solar energy. Table of economic analysis before and after using energy-conserving equipment is shown in Table 1.

Table 1: Economic analysis before and after using energy-conserving equipment

Part of using energy conserving equipment	Electricity before using/month (kilowatt)	Electricity before using/month(kilowatt)	Saved electricity/month (kilowatt)	Saved electrical fee/month (yuan)
54 dormitories (use USB socket and LED light)	13,500	8,100	5,400	7,020

Tower cranes (four LED lights are used)	1,920	480	1,440	1,872
Human &cargo elevator (use frequency converting equipment)	3,010	1,600	1,410	1,833
Construction elevator zone (use LED light belt)	720	360	360	468
Dormitory zone (use air energy water heater)	2,160	600	1,560	2,028
Sum	21,310	11,140	10,170	13,221

3.5 Measures of Saving Land and Protecting Construction Land

The on-site place is arranged reasonably and compactly. In the precondition of meeting the requirements on the environment, occupational health, security and civilized construction, the discarded place and blind corners are reduced. The construction on-site reduces the arrangement of temporary roads following the principle of combining permanent roads and temporary roads;

The construction plan of deep foundation pit is optimized, which reduces the earth excavation and refilling amount, reduces the disturbance to land to the largest extent and protects the surrounding natural and ecological environment.

The office zone and workers' living zone are two-floor assembled houses, which increase the rate of using land. The office zone and living zone are separated from the production zones and equipped with standard separation facilities.

4 Conclusion

In conclusion, this project has obtained significant economic benefits and social benefits by implementing green construction plans, which proves the green construction's feasibility and provides reference significance to promoting and implementing green construction projects in various regions in the future.

Acknowledgments

Supported by Huizhou Municipal Natural Science Foundation (No.2013B020015004)

References

1. Zhang Xiqian, Lin Lin, Wang Jun. The Present Situation and Prospect for Green Construction and Green Building. Construction Technology, 40:339 (2011), 1-7.
2. Ye Haowen, Zou Jun, Sun Hui, Yang Wei. The Key Green Construction Technologies of Guangzhou Zhoudafu Finance Center. Optimization of Capital Construction, 44:10(2015), 28-32.
3. Zhu Jun, Zhang Haibin. Construction Management for Green Construction Demonstration Project. Construction Technology, 30:12(2015), 44-49.

https://doi.org/10.1051/e3sconf/20183803018

Experimental Study on Impermeability of Recycled Concrete

Shao Zhen WANG, Jian Gong YANG, Lu WEI

Tianjin College, University of Science and Technology Beijing, Tianjin, 301830, China

Abstract: Recycled concrete is a kind of concrete which is constructed by crushing and removing the building waste and concrete blocks and mixing them according to a certain proportion after grading. In this study, the applicability of recycled concrete is studied only in terms of impermeability.

1 Introduction

Under the background of protecting environment, saving resources and harmonious development of human and nature, recycled concrete is bred. Recycled concrete is the construction waste to be broken, classified according to a certain proportion, used for coarse aggregate of concrete, so as to replace the original natural aggregate. There are many factors affecting the concrete, the applicability of recycled concrete is studied only in the aspect of impermeability.

2 Experimental ideas

The micro factor that affects the impermeability of concrete is cracks in the concrete. To enhance the impermeability of recycled concrete, it is necessary to reduce the internal cracks. The influencing factors of the impermeability of recycled concrete can be studied by means of the control variable method, such as sand ratio, replacement rate, water cement ratio, etc. A gradient test designed for influencing factors under the conditions of the specification.

Calculation formula of impermeability grade:

$$S = 10H - 1$$

S: Impermeability grade of recycled concrete;

H: The water pressure at the top of the recycled concrete specimen when it has seepage。

Notes: 1、Concrete impermeability grade is divided into $S_4, S_6, S_8, S_{10}, S_{12}$, If pressurized to 1.2MPa, After 8 hours, it is still impervious to water, then stop the test, the impermeability of the test piece is indicated by S_{12}.

2、According to the calculation of anti permeability grade, if the maximum water pressure value H is even, then the calculated S value is odd, at this time S-1 is the impermeability grade, for example: H=0.8MPa, S=7, then the impermeability grade is S_6.

The anti permeability level take calculation formula of impermeability coefficient:

$$K = aD^2 / (2 \times T \times H)$$

K: relative impermeability coefficient, unit cm/s;

D: average water penetration height, unit cm;

H: the pressure of water is expressed as the height of water, unit cm;

T: constant pressure time, unit s;

a: water absorption rate of recycled concrete.

Water absorption formula:

$a = (m_0 / m) \times 100\%$

m_0: The quality of non absorbent, unit g;

m: quality after water absorption unit g.

Notes: Impermeability grade and impermeability coefficient are different, impermeability coefficient is used in hydraulic concrete, relative permeability coefficient is calculated by reference seepage height. In this test, the anti - permeability instrument is a step-up anti-seepage instrument, so the formula of the impermeability grade is selected.

3 Experimental instruments

*Corresponding author: Email: wangshaozhen724@sina.com

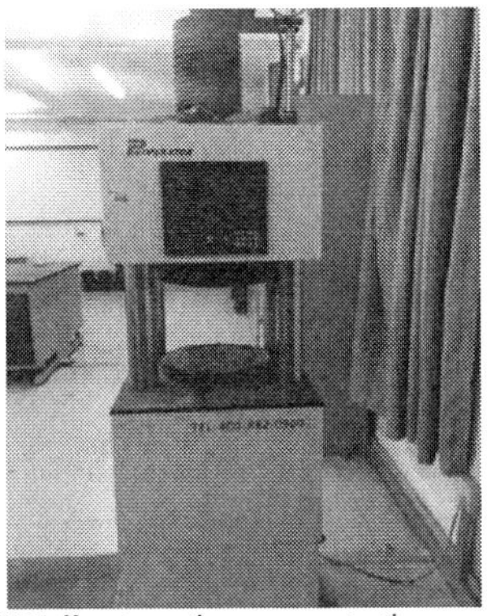

DY-2008MC Full automatic pressure testing machine

HP-4.0 Full automatic compression concrete impermeability tester

4 Materials selection

Choose suitable construction waste as the test coarse aggregate, because the waste concrete requires the machine to break it, artificial fragmentation can not be achieved. In addition, the manufacturing of asphalt materials and sintered brick materials of concrete, under the same condition, the impermeability of sintered brick materials better than asphalt brick materials, also the strength of concrete and asphalt material is shortage, so this experiment selects sintered brick as coarse aggregate of recycled concrete.

5 experimental dates

The recycled concrete strength grade in this experiment is C30, the pressure increase v be equal to 0.0125Mpa/h, initial pressure 0.3MPa, the impermeability grade is S_4, a total of five specimens of each test group, each group of 3 specimens, each gradient experiments were performed two times. Test results are shown in table 1 and table 2. Comparison of the results of the two tests is shown in figure 1-figure 4.

Table 1 Water cement ratio and sand rate, impermeability gradient test results

Specimen	Water cement ratio	Average pressure (Mpa)		Sand rate (%)	Average pressure (Mpa)	
		The first time	The second time		The first time	The second time
The 1 group	0.4	0.538	0.546	30	0.488	0.483
The 2 group	0.44	0.467	0.446	32	0.538	0.521
The 3 group	0.5	0.433	0.446	33	0.479	0.467
The 4 group	0.56	0.429	0.434	34	0.480	0.471
The 5 group	0.58	0.442	0.438	35	0.475	0.47

Table 2 Test results of substitution rate and content of fly ash against seepage gradient

Specimen	Substitution rate (%)	Average pressure (Mpa)		Fly ash content (%)	Average pressure (Mpa)	
		The first time	The second time		The first time	The second time
The 1 group	100	0.454	0.463	0	0.51	0.525
The 2 group	75	0.463	0.454	15	0.504	0.492
The 3 group	50	0.559	0.559	30	0.567	0.563

The 4 group	25	0.554	0.542	45	0.525	0.525
The 5 group	0	0.471	0.463	60	0.463	0.450

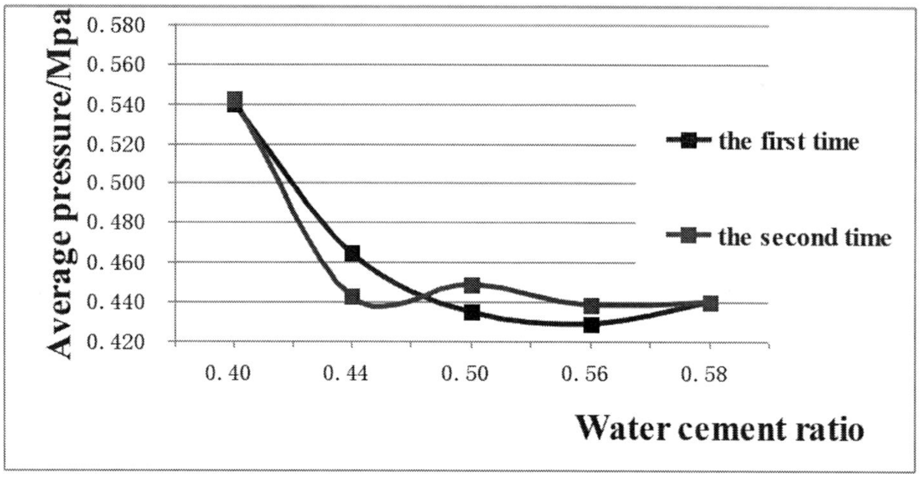

Figure 1 Comparison of gradient test results of water cement ratio

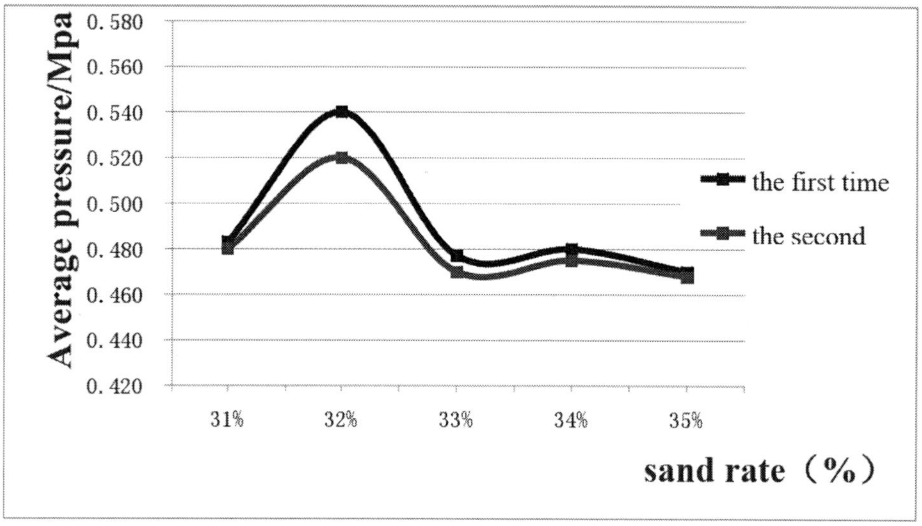

Figure 2 sand rate test results of water cement ratio

https://doi.org/10.1051/e3sconf/20183803018

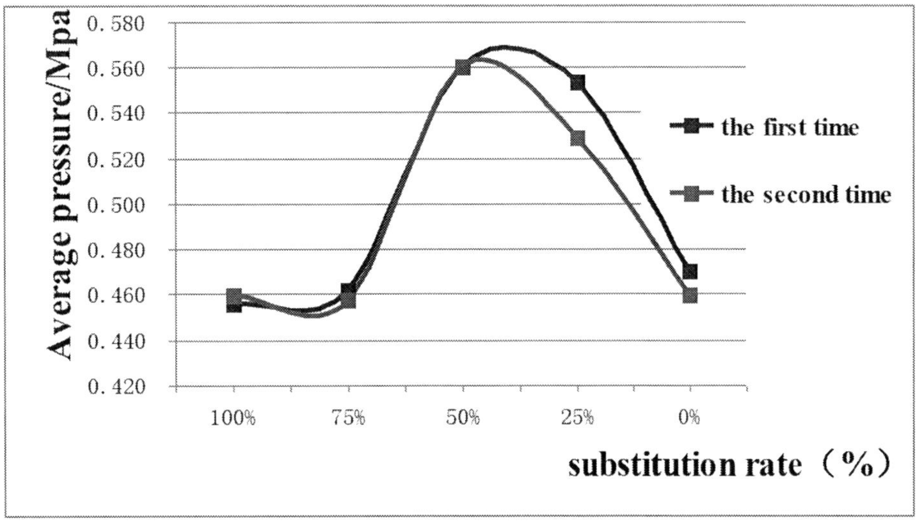

Figure 3 substitution rate results of water cement ratio

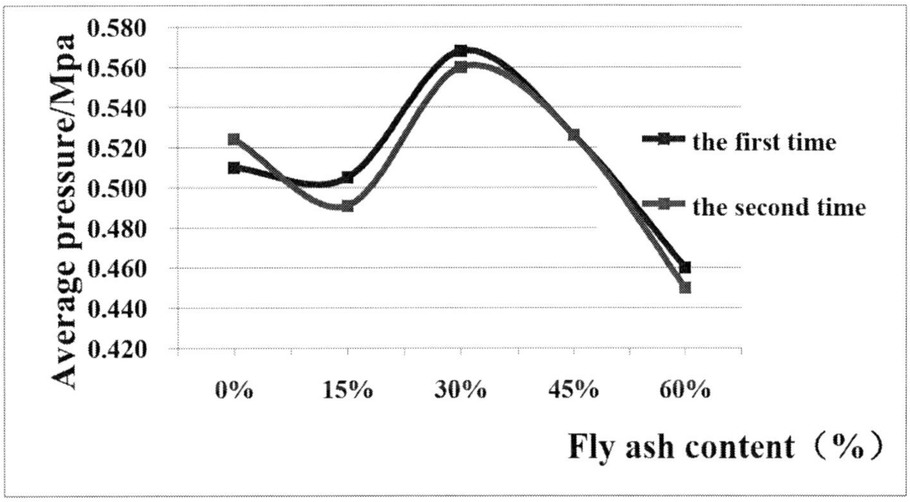

Figure 4 fly ash Volume test results of water cement ratio

6 Conclusions

1. From the water cement ratio data, it can be seen that when the other conditions are unchanged, when the water cement ratio is 0.4, the seepage resistance pressure is the largest. With the increase of water cement ratio, the resistance to seepage pressure decreases continuously. Therefore, the water cement ratio is controlled at about 0.4, which can improve the impermeability of recycled concrete.

2. From the sand rate data, it is found that when the sand rate is about 32%, the impermeability pressure of recycled concrete is the maximum.

3. From the data of substitution rate, it can be seen that when the substitution rate is about 50%, the maximum impermeability pressure is observed. Therefore, adding 50% natural aggregate to the waste aggregate can not only improve the impermeability of recycled concrete, but also save the cost effectively.

4. From the data of the fly ash, it can be seen that the pressure decreases when the fly ash is greater than 30%. Therefore, when the content of fly ash is about 30%, the impermeability of recycled concrete can be appropriately improved.

The test shows that the impermeability of recycled concrete can be improved by controlling the water cement ratio, sand rate, replacement rate of recycled aggregate and adding admixtures. But because of the particularity of recycled concrete itself, the crack inside the aggregate can not be avoided. Therefore, it is suggested that from the macroscopic and microcosmic two angles, it is a new direction to study how to remedy the cracks produced by recycled aggregate effectively.

https://doi.org/10.1051/e3sconf/20183803018

References

1. Zhang Zhen, Lou Zongke, Zhang Yingxue. Study on permeability resistance of recycled aggregate concrete based on orthogonal experiment. China rural water conservancy and Hydropower 2016.10.15.
2. Zhang Haobo, Ren Huichao, Kou Jialiang. Fly ash on the experimental study on the compressive strength and durability of recycled concrete. Journal of Xi'an University of technology.2016.12.30.
3. Wei Chengjuan, Gao Hui, Xue Pengfei, Dong Huifang. Experimental study on durability of recycled aggregate concrete. Construction technology.2015.12.28.
4. Yue Gongbing, Li Qiuyi, Gao Song. The quality of recycled fine aggregate replacement rate and effect on the frost resistance of concrete. Journal of natural disasters. 2015.10.15.
5. Chen Yungang; Xiao Jian Zhuang; Sun Zhenping. The recycled coarse aggregate concrete durability test. Journal of Anhui University of Technology (NATURAL SCIENCE EDITION).2013.7.15.
6. Geng Jian; Sun Jiaying; Mo Liwei; Zhang Guoliang. Microstructure characteristics of recycled fine aggregate and its concrete. Civil engineering and environmental engineering.2013.4.15.
7. Jing Xinpei; Shu Qian; Dong Yunting. Rich brick grain test the impermeability of recycled concrete. Journal of Lanzhou University of Technology.2017.3.3.
8. Qin Li, Ding Jingnan, Zhu Jinsong. Test on impermeability and frost resistance of high volume fly ash and slag high strength concrete. Chinese Journal of agricultural engineering.2017.3.23.

Design and development of Building energy simulation Software for prefabricated cabin type of industrial building (PCES)

Jun ZHANG [1,2], Ri Yi LI [1]

[1] South China University of Technology, State Key Laboratory of Subtropical Building Science, Guangdong Guangzhou 510640, China;
[2] China Energy Engineering Group Guangdong Electric Power Design Institute Co., Ltd., Guangzhou 510663, China

Abstract: Building energy simulation is an important supporting tool for green building design and building energy consumption assessment. At present, Building energy simulation software can't meet the needs of energy consumption analysis and cabinet level micro environment control design of prefabricated building. thermal physical model of prefabricated building is proposed in this paper, based on the physical model, the energy consumption calculation software of prefabricated cabin building(PCES) is developed. we can achieve building parameter setting, energy consumption simulation and building thermal process and energy consumption analysis by PCES.

1 Introduction

The background of PCES design and development of prefabricated compartment industrial building energy consumption simulation software mainly includes the following three aspects:

1) Prefabricated cabin buildings are faced with unprecedented opportunities for development: with the development of smart grid, as a center of power dispatching and control, intelligent substation will usher in a new wave of construction. Rapid growth in the number and size of substation integrated automation system. From this, prefabricated, assembled, compact and many other advantages of prefabricated cabin architecture came into being.

In practical application, prefabricated building module not only exist in the power industry, such as container data center, communications industry, mobile base station are prefabricated building module category, the common features for precast monolithic construction, lightweight enclosure structure, heat rejection to heat equipment fever; the difference of power equipment the tank body is not equipped with a, so the energy density is different.

2) The high energy consumption of industrial buildings has become a bottleneck restricting the development of prefabricated industrial buildings. Taking the power grid industry as an example, Only Guangzhou, Shenzhen and Dongguan three places, the number of 110kV and above substations as many as 771, although the power consumption of industrial building process huge base, substation building power consumption (including energy consumption of Technology) still accounted for the total power consumption of 26%[1]. In the prefabricated class of buildings widely used in the communications industry, the situation is not optimistic, the entire telecommunications industry's annual electricity consumption has more than 20 billion kwh [2], of which the base station power consumption accounted for 73%[3] of total electricity consumption.

High energy consumption, and the present situation of the project construction of prefabricated cabin supporting structure and most engineering application without such design, there is a lack of optimization in energy consumption.

3) The current building energy simulation software is not specially designed for prefabricated cabin industrial buildings, and has limitations. It is difficult to meet the needs of energy consumption analysis and cabinet level micro environment control design of prefabricated cabin buildings.

In the middle of 60s of last century, Some scholars used dynamic simulation method to analyze the heat transfer characteristics of building envelope. with the rapid development of computer technology, a great deal of complex computing could be put into practice. Therefore, this short presentation of a number of building energy simulation software, including the United States' BLAST[5], DOE-2[6], Europe's ESP-r[4], China's DeST[7] and Japan's HASP, etc.. But the above research is to comfort people for the moderate target of environment control of civil construction type, and more to the building function zoning thermal environment as the research subject, the core operation simulation of cabinet level micro environment almost no parts of equipment.

*Corresponding author: Email: zhangjun@gedi.com.cn

 https://doi.org/10.1051/e3sconf/20183803019

2 Establishment of simplified calculation model of prefabricated building energy consumption

2.1. The foundation of building thermal physics model

The thermal process disturbance of prefabricated cabin building has the following characteristics:

1) Periodic double outer disturbances: solar radiation and outdoor air temperature changes periodically;

2) Stability single interference: Prefabricated cabin type industrial buildings are unattended operation mode, no human body heat dissipation, in addition, cabin electrical equipment function for signal acquisition and relay protection of power system, industrial control. Under the condition of signal acquisition, data points and control and protection action, the internal heat disturbance of equipment is negligible and stable.

3) No room interference: Prefabricated cabin independent layout, no adjacent room;

2.2. Establishment of building thermal physics model

The sketch of the thermal physics model of the prefabricated building is shown in figure 1.

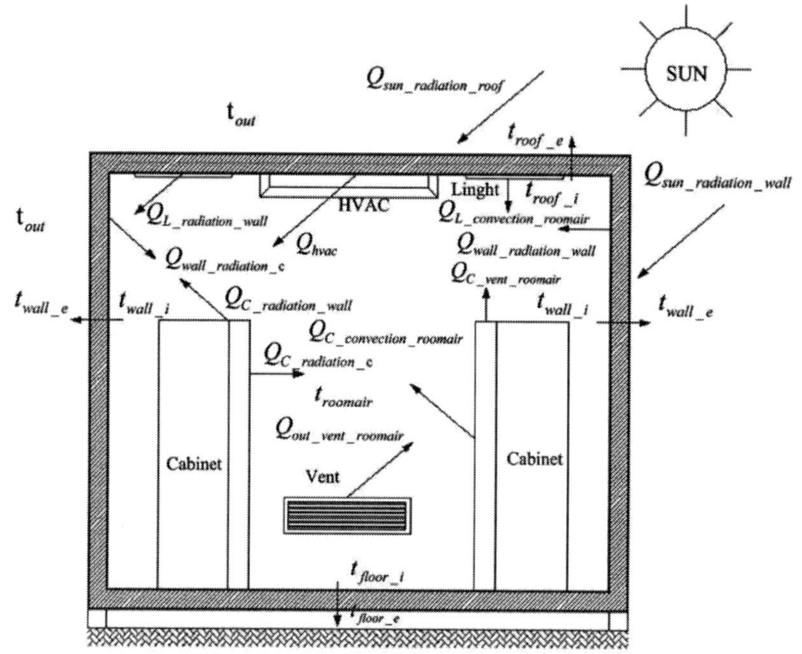

Figure 1 A physical model of building thermal processes

The meaning of each symbol in Figure 1 is shown in table 1.

Table 1 Various parameters of building thermal process

t_{out}	Outdoor dry bulb temperature	$t_{roomair}$	Indoor air temperature
t_{wall-e}	Exterior surface temperature of enclosure structure	$t_{wall-air}$	Internal surface temperature of enclosure structure
t_{roof-e}	Roof surface temperature	t_{roof-i}	Roof inner surface temperature
$t_{floor-e}$	Surface temperature of ground surface	$t_{floor-i}$	Surface temperature of ground
$Q_{l-radiation-roomair}$	Thermal radiation disturbance of lamps	$Q_{l-convection-roomair}$	Heat transfer disturbance of lamps
$Q_{sun-radiation-wall}$	Solar radiation heat disturbance in wall	$Q_{sun-radiation-roof}$	Solar radiation heat disturbance of roof
$Q_{c-convection-roomair}$	Heat disturbance of Cabinet convection heat transfer	$Q_{c-vent-roomair}$	Heat disturbance of Cabinet ventilation heat transfer
$Q_{c-radiation-c}$	Long wave radiative heat transfer of cabinet wall to other cabinet walls	$Q_{c-radiation-wall}$	Long wave radiative heat disturbance of enclosure wall by cabinet
$Q_{wall-radiation-wall}$	Long wave radiative heat transfer of enclosure wall facing other walls	$Q_{wall-radiation-c}$	Long wave radiation heat shock of enclosure wall facing the wall of cabinet
Q_{hvac}	Room air conditioning cooling capacity	$Q_{out-vent-roomair}$	Outdoor ventilation heat disturbance

The building envelope heat balance system including outer surface heat balance of surface thermal model, heat transfer model, inner balance model; air heat balance system including heat balance model, heat balance model, cabinet air heat balance model of the air chamber of the equilibrium model to discuss below.

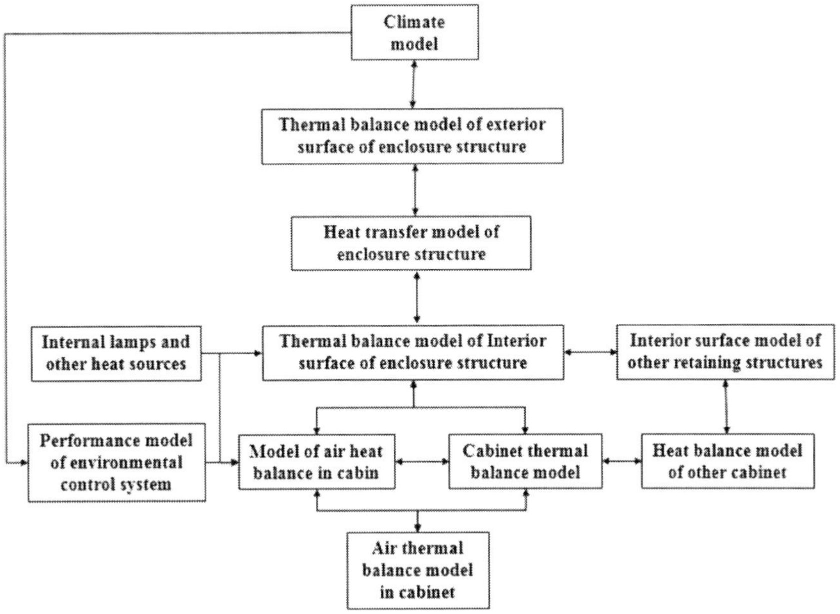

Figure2 Thermal equilibrium model system of prefabricated building

A heat balance equation can be established for the exterior surface of the enclosure:

$$q_e = \rho I + h_{c,e}(t_0 - t_e) + h_{r-e,0}(t_0 - t_e) + h_{r-e,sky}(t_{sky} - t_e) + h_{r-e,g}(t_g - t_e) \quad (1)$$

$q_e$is the exterior wall heat flux, W/m^2; ρ is the outer surface of the absorption rate of shortwave radiation; I is the total solar radiation intensity, the outer surface of W/m^2; $h_{c,e}$ is the exterior surface of the convective heat transfer coefficient ,$W/(m^2 \cdot K)$; t_0 is the outdoor dry bulb temperature, K; t_e is the external surface temperature of K; t_{sky} is K; t_g is the sky temperature, surface temperature, K; $h_{r-e,0}$ is the exterior surface long wave radiation between the outdoor air and the heat transfer coefficient, $W/(m^2 \cdot K)$; $h_{r-e,sky}$ is the exterior surface and space day long wave radiation heat transfer coefficient, $W/(m^2 \cdot K)$; $h_{r-e,g}$is longwave radiation and surface between the wall surface heat transfer coefficient, $W/(m^2 \cdot K)$.

The heat balance equation of the enclosure structure is:

$$q_\lambda = \frac{t_e - t_i}{\sum R} \quad (2)$$

$$\sum R = \frac{d}{\lambda} \quad (3)$$

$q_\lambda$is a heat retaining structure of heat flux density, W/m^2; $\sum R$ is thermal resistance of building envelope, $m^2 \cdot K/W$; $t_i$is the surface temperature of the outer walls, K.

The heat balance equation can be established for the inner surface of the enclosure :

$$q_i = h_{c,i}(t_r - t_i) + \sum_{k=1}^{N_i} h_{r-ik}(t_{i-k} - t_i) + \sum_{k=1}^{N_c} h_{r-ik}(t_{ce-k} - t_i) + \frac{Q_L(1 - C_1)}{F_{tol}} \quad (4)$$

$q_i$is through the building envelope, the heat flux into the room by room outside, W/m^2; $h_{c,i}$ is the inner surface of the convective heat transfer coefficient$W/(m^2 \cdot K)$; $t_i$is the surface temperature of the enclosure structure, $t_r$is the indoor air temperature, K; t_{i-k} is the first k wall temperature, K; t_{c-k} is the k cabinet surface temperature, K; h_{r-ik} is the heat transfer coefficient of the K wall on the wall surface radiation, $W/(m^2 \cdot K)$; h_{r-ic} is the k cabinet wall on the wall of the radiation heat transfer coefficient,.

The equation of heat balance can be established for the air in the cabin:

$$c_p \rho_e V_{ra} \frac{dt_{ra}}{d_\tau} = \sum_{k=1}^{N_i} h_{c,i} F_{i-k}(t_{i-k} - t_{ra}) + c_p \rho_e V_e (t_o - t_{ra}) + Q_L C_1 + \sum_{k=1}^{N_c} F_{ce-k} h_{c,c}(t_{ce-k} - t_{ra}) + \sum_{k=1}^{N_c} c_p \rho_e V_{ca-k}(t_{ca} - t_{ra}) + Q_{ac} \quad (5)$$

$c_p\rho_eV_{ra}\frac{dt_{ra}}{d\tau}$is incremental sensible heat room air per unit time , W; $F_{i-k}$is the k wall surface area, m^2; t_{ra} is the air temperature of the cabin, K; c_p is the specific heat capacity at constant pressure air, $J/(kg\cdot K)$; ρ_e is the air density, kg/m^3; V_e is indoor ventilation, m^3; $h_{c,c}$ is a cabinet wall facing the flow heat transfer coefficient; t_{ce-k} is the k cabinet surface temperature, K; F_{ce-k} is the k cabinet the outer wall of the surface area, m^2 ; V_{ca-k} is the k cabinet ventilation air exchange, m^3; Q_{ac} is the air conditioning system in heat sensible heat, W.

The heat balance equation can be established by taking a single cabinet as an isolator:

$$\frac{dQ_C}{d\tau}=Q_C+c_p\rho_eV_{ca}(t_{ra}-t_{ca})+F_{ce}h_{c,c}(t_{ra}-t_{ce})+\sum_{k=1}^{N_i}h_{r-ci}F_{ce}(t_{i-k}-t_{ce})+\sum_{k=1}^{N_{ce}}F_{ce}h_{r-cc}(t_{ce-k}-t_{ce}) \quad (6)$$

For the air in the cabinet, a heat balance equation can be established:

$$c_p\rho_eV_{ra}\frac{dt_{ca}}{d\tau}=F_{ce}h_{c,c}(t_{c,i}-t_{ca})+F_{ce}h_{c,c}(t_{c,e}-t_{ca})+c_p\rho_eV_{ca}(t_{ra}-t_{ca}) \quad (7)$$

The building heat balance equations are composed of the envelope structure, the indoor air, the cabinet and the heat balance equations of the air in the cabinet, so as to establish the building heat balance equations.

3 Prefabricated compartment industrial building energy consumption simulation software development

3.1. Requirements analysis and interface design

The target users of the prefabricated building energy consumption calculation software PCES include the prefab, the designer and the operation manager, and the software should operate under the Windows operating system, taking into account the user's needs. The main functions of the software include parameter setting, energy consumption simulation, building thermal process and energy consumption analysis.

Building energy simulation: under the given climatic parameters and environmental control objectives, the energy consumption of each part of the building is solved. building thermal process analysis: the temperature of bulkhead surface, air temperature in cabin and temperature distribution in cabin are solved under the given boundary conditions such as climate parameter and environment control target or energy consumption target.

3.2. Software process

The software calculation process is shown below:

 https://doi.org/10.1051/e3sconf/20183803019

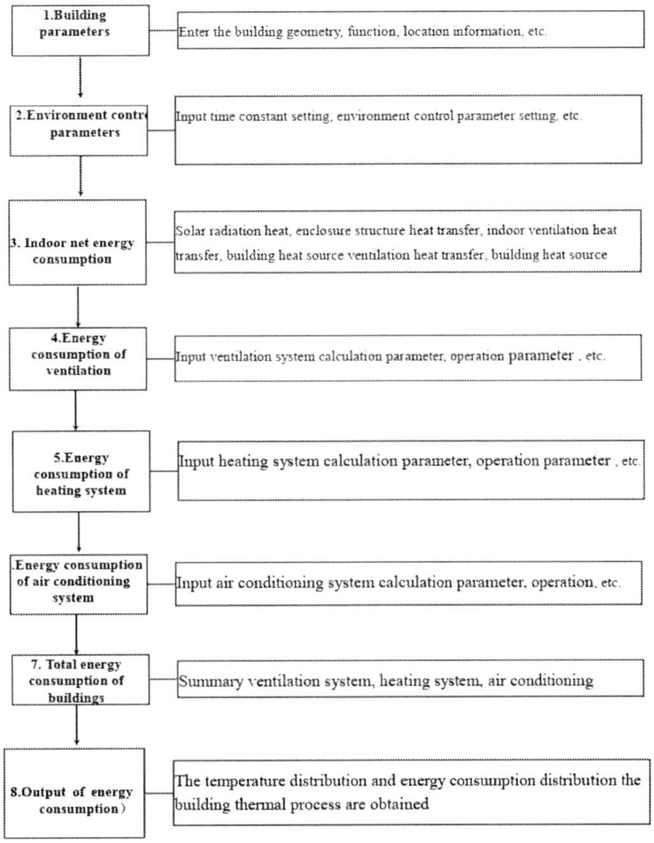

Figure 3 Processing flow of PCES

4 Conclusion

Based on the heat transfer of building construction prefabricated cabin ideal building thermal physical model theory, and based on the physical model of the development of prefabricated cabin building energy consumption calculation software PCES, analysis of setting, energy consumption simulation and building thermal process and energy consumption to achieve the construction parameters, accuracy finally through the calculation results of the model test method is verified.

Prefabricated cabin class building energy consumption calculation software PCES can be achieved by solving the prefabricated cabin building envelope, cabin heat and cabinet level thermal environmental conditions, and the results are accurate, suitable for engineering design and energy saving analysis.

References

1. Li Tienan, National Center for industrial energy saving and emission reduction technology promotion center. Progress report on China's industrial energy saving. 2013: medium term progress in 12th Five-Year, [M]., China Quality Inspection press, China Standard Press, 2014.
2. Qin ting. Energy consumption analysis of communication base station [J]. Journal of Xi'an University of Posts and telecommunications, 2011, 16 (2): 76-78.
3. Liao Xiaofang, Xiang Bing, Xie Fengzhu, et al. Energy consumption analysis and integrated energy saving solution for mobile communication base station [J]. telecommunications engineering technology and standardization, 2010, 23 (4): 53-57.
4. Hittle D C. The Building Loads Analysis and System Thermodynamics (BLAST) program, version 2.0. Users manual, volume 1[J]. 1979.
5. Yan D, Xia J, Tang W, et al. DeST — An integrated building simulation toolkit Part I: Fundamentals[J]. Building Simulation, 2008, 1(2):95-110.
6. Michael J, Witte , RH Henninger , J Glazer . TESTING AND VALIDATION OF A NEW BUILDING ENERGY SIMULATION PROGRAM [J]. Seventh International IBPSA Conference,2001: 353-359.
7. Yan D, Xia J, Tang W, et al. DeST — An integrated building simulation toolkit Part I: Fundamentals[J]. Building Simulation, 2008, 1(2):95-110.

Development of Simulation Program for Temperature Field of Mass Concrete Structures

Zheng SI, Qian ZHANG, Ling Zhi HUANG, Dan YANG

Xi'an University of Technology, Xi'an, Shanxi 710048, China

Abstract: Most existing temperature field calculation programs have relative defects. In the present paper, based on merits of ANSYS platform, a temperature field calculation program of mass concrete structure is developed and demonstrated. According to actual pouring progress and thermodynamic parameters, a roller-compacted concrete dam is simulated. The difference of maximum temperature between calculated and measured values of measuring points is less than 1.8°C. Furthermore, the relative difference is -5%–5%. This result shows that the calculation program developed based on ANSYS platform can simulate and calculate the temperature field of mass concrete structure.

1 Introduction

For finite element calculation of mass concrete structure temperature field, many research institutes have developed temperature field calculation program. Invariably, the development mode uses finite element program compiled using FORTRAN programming language, whereas the visual interface uses Visual Basic programming language. This development mode has many defects in practice, such as non-intuitive pre-treatment, non-direct post-processing, limited element types, and high requirement of entity model, element shape, and so on. In the current paper, secondary development is achieved based on the large-scale commercial finite element software ANSYS. Existing resources are reformed and extended to adapt temperature field calculation to all complex shape mass concrete structure, which could enable the simulation program of mass concrete structure temperature field to become well-known and universally used.

2 Merits of ANSYS Software

ANSYS is a commercial finite element software developed by the ANSYS Corporation (US). It is the first software to achieve ISO9001 certification in the FEA industry, thereby leading the development of the world's finite element technology. ANSYS software is capable of performing combined analyses, such as structural, thermal, fluid, magnetic field, and sound field analyses, and can also be used to analyze coupling of multi-physics. Thus, it is widely used in research and design in various industries, such as nuclear, astronautics, national defense and military industry, civil engineering, hydraulic engineering, and others [1–2]. The main features of ANSYS software are as follows:

(1) Complete pre-treatment functions: ANSYS software is a powerful solid modeling and meshing tool for conveniently establishing mathematical and finite element models that provides nearly 200 element types. Engineers and technicians are able to easily and accurately establish the simulation model to reflect the actual engineering structures using the solid modeling, meshing tools, and abundant element types.

(2) Powerful solver: ANSYS software is capable of analyzing multiple physical fields. Its analysis calculation module includes structure, thermal, fluid, magnetic field, sound field, and coupling of multi-physics. Analysis types include linear and nonlinear analyses. Moreover, the ANSYS software has the functions of interaction of multi-physics mediums, sensitivity, and optimization analyses.

(3) Convenient post-processor: The post-processing of ANSYS software includes common post-processing module POST1 and time history post-processing module POST26. Using the post-processor, results or outputs can be displayed as charts, curves, animations, and other formats. Results can be presented in various ways, such as color cloud map, contour map, vector diagram, particle flow track, slice map, time history curve, and so on.

(4) Flexibility: Aside from the complete analysis functions, the ANSYS software has flexibility. It offers many secondary development tools for users, such as user interface design language, ANSYS parametric design language, user program features, and so on. With these tools, users are able to develop special programs according to their needs. In the present paper, the author makes full use of the ANSYS secondary development features to develop the mass concrete structure temperature field calculation program.

*Corresponding author: Email: sz123hlz@163.com

3 Feasibility of Concrete Structure Temperature Field Calculation by ANSYS

3.1 Thermal Analysis Module

When calculating concrete structure temperature field, ANSYS thermal analysis module is generally be used. The temperature of every node can be calculated using heat balance equations and the finite element method (because space is limited, readers are referred to Reference [3] for the temperature field finite element equations). ANSYS thermal analysis includes steady-state heat transfer and transient heat transfer. When calculating the temperature field of concrete structures, transient heat transfer is used.

3.2 Simulation of Concrete Pouring Process

The "element birth and death" function of ANSYS is used when simulating the concrete pouring process. "Kill" does not mean to delete the "killed" elements from the whole model, but to multiply the stiffness matrix of "killed" elements by an extremely small factor (default value is 1×10^{-6}). The element load of "dead" elements is 0, thus, these elements have no effect on load vector. Similarly, "birth" does not mean to add these elements to the whole model, but to reactivate them. The secondary development commands are as follows:

Ekill,whole_structure_element(element component) !Kill the whole structure element

Ealive,layer%i%_element

!Activate one layer concrete elements

In finite element analysis, a three-dimensional finite element model of a complete concrete structure and bedrock should be established initially before all concrete elements are "killed." During the concrete pouring, the corresponding concrete elements are then activated.

3.3 Element

The eight-node hexahedron element SOLID70 of ANSYS thermal analysis module possesses heat transfer capacity in three directions. Each node has only one temperature degree of freedom. SOLID70 element can be used in three-dimensional steady state or transient thermal analysis. SOLID70 element has the "element birth and death" function, thus, it can be activated according to the pouring progress. The "DO" loop statement is then used to apply hydration heat and convection boundary conditions to each layer of concrete. In addition, in stress analysis, SOLID70 element can be easily switched to structural element.

3.4 Impose Boundary Conditions

When calculating temperature field, all concrete elements are first "killed". The initial foundation temperature can be applied according to the temperature of measured point along the depth direction. The middle regions between two measured points should be plotted by linear interpolation. In analyzing the unsteady temperature field of concrete structure (e.g., concrete gravity dam), the four sides and bottom surface of the foundation are taken as adiabatic boundary, whereas the top surface is taken as the third boundary condition. The dam transverse joint surfaces are regarded as adiabatic boundary, whereas the upstream surface, downstream surface, and concrete storehouse surface are regarded as the heat dissipation boundary (i.e., the third boundary condition). After impoundment, the upstream surface of the dam below water level is the first boundary condition, whereas the upstream surface above the water level and downstream surface of the dam are the third boundary condition [4-5]. The secondary development commands are as follows:

Sfa,all,1,hflux,0 !Impose adiabatic boundaries on the foundation side and bottom surfaces

Sfa,all,,conv, heat_transfer_coefficient,external temperature !Impose the third boundary on surfaces

Sf,all,conv, heat_transfer_coefficient,external temperature !Impose the third boundary on nodes

3.5 Impose Temperature Control Measures

3.5.1 Simulation of Water Pipe Cooling

The ANSYS heat transfer analysis module provides a uniform model of heat source and heat sink. The effect of pipe cooling can be considered as imposing concrete hydration heat [6]. The temperature increase per unit volume of concrete in unit time is assume to need quantity of heat Q ,

$$Q = Q_1 - Q_2 \qquad (1)$$

Where Q_1 is the concrete hydration heat and Q_2 is the heat absorption of pipe cooling.

$$Q_1(\tau) = c\rho \frac{\partial \theta(\tau)}{\partial \tau} \qquad (2)$$

$$Q_2(\tau) = c\rho \frac{\partial \phi(t)}{\partial t} \qquad (3)$$

Where c 、 ρ and θ are the concrete specific heat, density, and hydration heat temperature rise, respectively; τ is the concrete age; ϕ is the temperature drop of concrete caused by the cooling water pipes; and t is the time of water cooling. The expressions $\theta(\tau)$ and $\phi(t)$ are discussed in Reference [3]. As hydration heat of concrete is imposed, the cooling effect of water pipes can be embedded simultaneously.

3.5.2 Simulation of Superficial Thermal Insulation, Watering, and Spray

Using the equivalent surface heat transfer coefficient method in Reference [3], the surface heat transfer coefficient after insulation can be calculated. If watering on the concrete surface for curing, the surface boundary condition is the first category, and the concrete surface temperature is the water temperature. If spray is present on the surface, the concrete surface is the third boundary condition, and the environmental temperature should be decreased according to actual effect of the spray.

4 Comparative Analyses between Calculated Values by ANSYS and Measured Values

The maximum height of a roller-compacted concrete dam is 130.0 m. The pouring concrete is composed of two-gradation and three-gradation concrete. The calculation model is shown in Figure 1. The calculation range of dam foundation along the depth direction and in the upstream and downstream directions extends 130.0 m. The coordinate origin of the overall model is at the left heel of the dam. The X-direction of the coordinate is along the dam axis, with positive direction pointing to the right bank. The Y-direction is along the flow direction, with downstream direction being positive. The Z-direction is the vertical direction, and with upward direction being positive.

The concrete and bedrock calculation parameters are shown in Table 1.

The roller-compacted concrete dam temperature field is calculated by ANSYS according to the actual construction schedule, pouring temperature, material partition, surface heat transfer conditions, and the actual impoundment process. Calculation time step of construction period is 0.25 d, and time step of operation period is 0.5–30 d of the variable step size.

The Locations of the thermometers embedded in the dam are shown in Figure 2. Results of the comparative analysis of calculated temperature vs. measured temperature values are described below.

Table 1 Thermodynamic parameters of dam concrete and bedrock

Sequence Number	Item	Roller-Compacted Concrete		Bedrock
		Two-gradation Concrete	Three-gradation Concrete	
1	Thermal Conductivity ($kJ \cdot m^{-1} \cdot h^{-1} \cdot °C^{-1}$)	7.668	7.668	6.87
2	Thermal Diffusivity ($m^2 \cdot h^{-1}$)	0.003039	0.003039	0.00319
3	Specific Heat ($kJ \cdot kg^{-1} \cdot °C^{-1}$)	0.94	0.94	0.77
4	Expansion Coefficient ($10^{-6} \cdot °C^{-1}$)	5.823	6.744	7.0
5	Heat Transfer Coefficient ($kJ \cdot m^{-2} \cdot h^{-1} \cdot °C^{-1}$)	42	42	42
6	Bulk Density ($kN \cdot m^{-3}$)	24.53	24.53	28.00
7	Expression of Adiabatic Temperature Rise	$T_\tau = \frac{23.33\tau}{5.07+\tau}$	$T_\tau = \frac{17.5\tau}{4.03+\tau}$	—

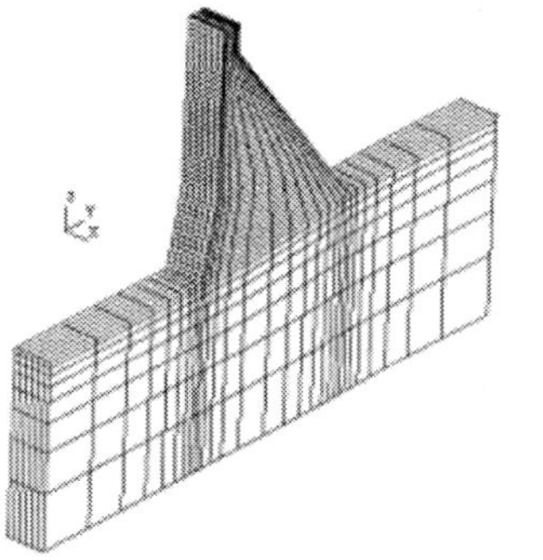
Figure 1 Calculation model chart of the dam

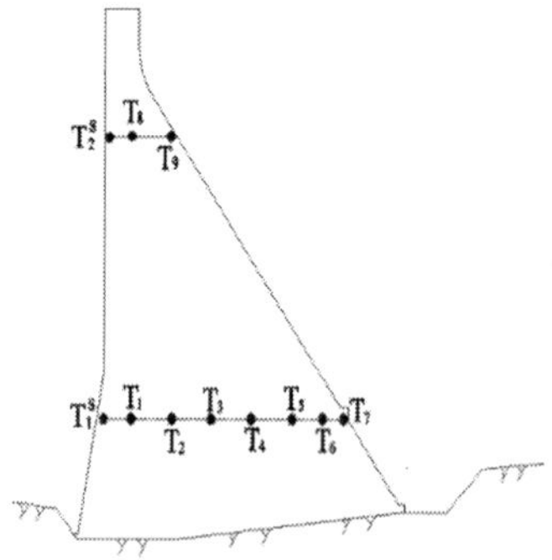

Figure 2 Location of thermometers embedded in the dam

(1) Comparative analysis of the dam surface point temperature

Figure 3 shows the temperature history curves of calculated values and measured values of measuring point T_1^S in the upstream of the dam, 29.0 m from the foundation. The surface point temperature generally

https://doi.org/10.1051/e3sconf/20183803020

involves three stages: 1) the initial temperature rise caused by hydration heat, 2) temperature decrease as heat dissipates to the outside, and 3) periodic temperature changes according to environmental temperature. The calculated values and the measured values of measuring point T_1^S accord with the general law. The measured maximum temperature, calculated maximum temperature, temperature difference, and relative difference are 30.6 °C, 29.3 °C, 1.3 °C, and 4.25%, respectively, which shows the calculated values are in compatible with the measured values.

(2) Comparative analysis of the central point temperature

Figure 4 shows the temperature history curves of the calculated and measured values of measuring point T_3, in the center of the dam, 29.0 m from the foundation. After the concrete is poured, the temperature rises because of the hydration heat. With the large amount of fly ash mixed in the concrete, the roller compacted concrete heat dissipation rate is slower, temperature reaches the maximum value approximately two months after pouring. Measuring point T_3 is in the center of the dam concrete, the condition of heat dissipation to outside is worse, therefore, the temperature of measuring point T_3 drops gradually. Figure 4 shows similar variation with time of the measured and calculated temperature at the measuring point T_3. The measured maximum temperature, calculated maximum temperature, temperature difference, and relative difference are 37.6 °C, 36.2 °C, 1.4 °C, and 3.72%, respectively, which also shows the calculated values are in compatible with the measured values.

Table 2 shows the measured and calculated temperature values of each measuring point. The change laws of measured values and calculated values are nearly the same, and the maximum temperature difference is 1.8 °C. The relative difference is limited to ±5.00%, which proves that the calculation program developed using the secondary development based on ANSYS platform can simulate and calculate the temperature field of mass concrete structure with high accuracy.

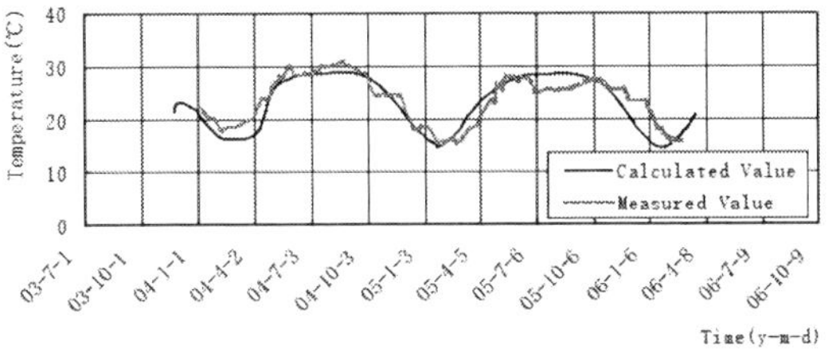

Figure 3 Temperature history curve of measuring point T_1^S

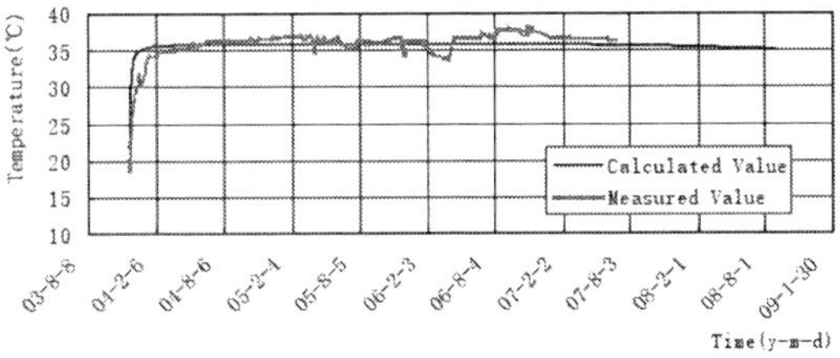

Figure 4 Temperature history curve of measurement point T_3

Table 2 Comparison of calculated and measured values

Measuring Point	Maximum Value of Temperature (°C)		Difference	
	Measured Value	Calculated Value	Temperature Difference (°C)	Relative Difference (%)
T_1^S	30.6	29.3	1.3	4.25
T_1	38.1	36.6	1.5	3.94
T_2	37.1	36.0	1.1	2.96
T_3	37.6	36.2	1.4	3.72
T_4	37.6	36.4	1.2	3.19
T_5	37.6	36.0	1.6	4.26
T_6	33.1	34.2	-1.1	-3.32
T_7	32.1	33.3	-1.2	-3.74

https://doi.org/10.1051/e3sconf/20183803020

T_2^S	37.1	35.9	1.2	3.23
T_8	37.8	36.7	1.1	2.91
T_9	37.9	36.1	1.8	4.75

5 Conclusion

In the present paper, a temperature field calculation program is developed using secondary development based on ANSYS platform in order to simulate the actual construction schedule of mass concrete structure with complex shape and boundary conditions. An actual project is simulated and calculated by the developed program. Comparing the temperature values of measuring points, the calculated temperature values are very close to the measured temperature values, and the relative difference is limited to ±5.00%. These results indicate that the developed program is able to simulate and accurately calculate the temperature field of mass concrete structure.

Acknowledgments

This study was financially supported by National Natural Science Foundation of China (51409207, 51609200), the Key Research Projects of Shaanxi Provincial Loess Mechanics and Engineering Key Laboratory (No.09JS103), and Shaanxi provincial key innovative research team (2013KCT-15).

References

1. Zhu Xiaohua, Yu Zhixiang. Finite Element Analysis Choiceness Example of Advanced Engineering by ANSYS [M]. Beijing: Publishing House of Electronics Industry, 2004.
2. Kan Qianhua, Tan Changjian. Application Example of Advanced Engineering and Secondary Development of ANSYS [M]. Beijing: Publishing House of Electronics Industry, 2006
3. Zhu Bofang. Thermal stresses and temperature control of mass concrete [M]. Beijing: China electric press, 1999.3
4. Shouyi Li, Jinkai Zhang, Ting Wang. Effect of Pouring Temperature on Temperature Stress of RCC Gravity Dam[C], Fourth International Conference on Natural Computation 2008, 2008:355–359
5. Zhang Xiaofei, Li Shouyi, Chen Yaolong, Chai JunRui. The development and verification of relocating mesh method for the computation of temperature field of RCC dam [J]. Advances in Engineering Software, 2009, 40(11):1119–1123
6. Jin Feng, Zhou Zhidan, Zhou Yuande, Wang Jinting. Development and application of thermal stress computation procedure of concrete dams on the ABAQUS platform [J]. Journal of Hydroelectric Engineering, 2006, 25(4):75–78

Research on the Method of Setting Waiting Area for Non-motor Vehicle at Signal Control Intersection

Yun Xia WANG, Dong Bo LIU

Traffic management research institute of the ministry of public security, Wuxi 214151

Abstract: Electric bicycle has become an indispensable important component of the transportation system. The fact is that traffic organization and channelizing design of signal control intersection is not intensive, which cannot adapt to the current traffic demand of non-motor vehicle, such as unclear traffic rules and poor visibility, thus the traffic safety of non-motor vehicle is not optimistic. Therefore, it is necessary to study on traffic organization method based on the demand of non-motor vehicle, which can provide certain theoretical basis for traffic administrative department to make policy and traffic design. This article focuses on the method of setting waiting area for non-motor vehicle at signal control intersection, including the advantages, disadvantages and the applicable conditions.

1 Introduction

With the rapid development of non-motor vehicle, the traffic order, traffic efficiency and traffic safety at signal control intersection is not optimistic. From the perspective of intersection traffic management, traffic organization and channelization design is extensive, without consideration of the characteristics of electric bicycle. How to give full play to the advantages of electric bicycles, and regulate the traffic order, improve the traffic efficiency and traffic safety level has become a major problem for traffic managers. Therefore, it is necessary to analyze the characteristics of electric bicycle, and propose the method of traffic space optimization, which can adapt to the current non-motor vehicle traffic.

2 Analysis of traffic characteristics of electric bicycle

2.1 Physical Characteristics

According to the actual measurement data, the average body width of electric bicycles is about 70cm, and the average body width of bicycles is about 60cm.

2.2 Queuing Characteristics

During the red light, bicycles slow down and queue before the stop line. But electric bicycles tend to staggered together, which is different from motor vehicles. According to the actual observation, it is very common for electric bicycles to line up across the crosswalk.

Figure 1 Queue of electric bicycles at signal intersection

*Corresponding author: Email: wangyunxiahc@126.com

2.3 Passing through the Intersection

Electric bicycle is very flexible, and its instantaneous starting speed is faster than motor vehicles. According to the actual survey, the speed of most electric bicycles through the intersection is relatively close and moderate. The average starting time of the first line of a queue is less than 0.5 seconds. The average speed of straight line electric bicycles is about 14km/h, and the average speed of left-turn electric bicycles is about 13km/h.

When the signal is green, electric bicycles spread to the intersection gradually just like a fluid. Then they expand to both sides in order to pass through the intersection quickly, taking up the space of motor vehicles and pedestrian. With the gradual increase of electric bicycles, the expansion width increases, forcing the vehicle to slow down, which brings the reduction of traffic capacity as well as the increase of traffic accident. According to the actual survey, the average width of the electric bicycle crossing the street is about 1.4 times compared with the width of the queue.

Figure 2 Expansibility of bicycles

2.4 Traffic Violation Characteristics

As the electric bicycle is very flexible, start faster than motor vehicle, its illegal behavior at signal control intersection is very common, such as turning off the priority order, do not take non-motor vehicle lanes, reverse driving, and so on.

Based on the traffic characteristics of electric bicycle at signal control intersection, it is practical to study on the traffic organization and management of the non-motor traffic. It is necessary to optimize the design of intersection, the timing of the signal and the supporting management facilities to protect the safety of electric bicycle at signal control intersection, and regulate the traffic order. Based on the characteristics of electric bicycle, the "blank area" which is not used can be used to set waiting space for non-motor vehicle in order to increase its waiting space, to reduce the traffic conflict between motor vehicle and non-motor vehicle, and to improve the intersection of traffic efficiency. According to the location, the waiting space can mainly be divided into the three types as following.

3 Waiting Area in Front of Stop Line for Non-motor Vehicle

3.1 Setting Method

The stop line of motor vehicle can be stepped back, then the stop line of non-motor vehicle can be set in front. The distance between the motor vehicle and non-motor vehicle stop line is generally determined by the traffic volume of the non-motor vehicles and motor vehicles, as well as the geometry of the intersection, as well as signal control factors. During the red signal phase, the non-motor vehicles are waiting in front of motor vehicles. And once the green signal is turned up, the non-motor vehicles can drive into the intersection immediately, which reduces the interference between main flow non-motor vehicles and motor vehicles. The method of step back of motor vehicle stop line can improve the traffic operation order. In addition, the color pavements (such as green) can be applied to make the non-motor vehicle waiting area more visible[1].

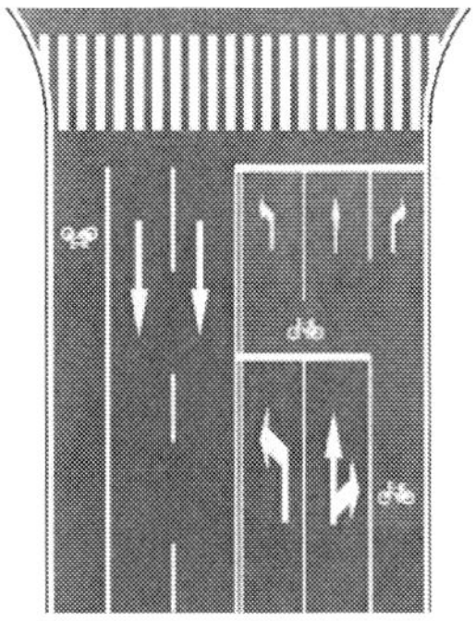

Figure 3 Setting illustration of waiting space in front of motor vehicle stop line for non-motor vehicle

3.2 Advantage and Disadvantage

The non-motor vehicle queuing waiting space can be greatly increased, then the efficiency of non-motor vehicles can be improved. While the non-motor vehicles waiting during the red light tend to flock into the intersection. If the traffic volume of non-motor vehicles is large, due to its expansion and lateral movement characteristics, some non-motor vehicles will occupy motorway, which reduces the speed of motor vehicles, as well as the saturated flow and capacity.

3.3 Applicable Conditions

This method is fit for the two-phase signal control intersections with 1-2 entrance lanes[2][3]. The successful implement of this method requires enhanced education management of non-motor vehicle drivers, which interpret the reasonable and normative waiting behavior to the drivers. In addition, if the traffic volume of non-motor vehicle is very big, it is required to set specialized left-turn phase for non-motor vehicle in order to reduce the impact of non-motor vehicles on motor vehicles.

4 "Non-signal Control" Waiting Area for Non-motor Vehicle

4.1 Setting Method

With the conditions of high non-motor vehicles volume and enough intersection space, the non-motor vehicle waiting area can be set at the "black area" to reduce the crossing time of non-motor vehicles. The "black area" is the intersection space which is not occupied by motor vehicles. When the high non-motor vehicle volume leads to the interference of normal operation between straight and left-turn flow, the non-motor vehicle waiting area can be divided into straight waiting area and left-turn waiting area. In addition, the color pavements and text makeup can be applied to make the non-motor vehicle waiting area more visible[4].

4.2. Advantage and Disadvantage

The non-motor vehicle queuing waiting space can be greatly increased, then the efficiency of non-motor vehicles can be improved. However, when the non-motor vehicles enter the pending area, they are possible to intersect with pedestrians who are crossing the street.

4.3 Applicable Conditions

This design is suitable for large intersections. The successful implement of this method requires enhanced education management of non-motor vehicle drivers, which interpret the reasonable and normative waiting behavior to the drivers.

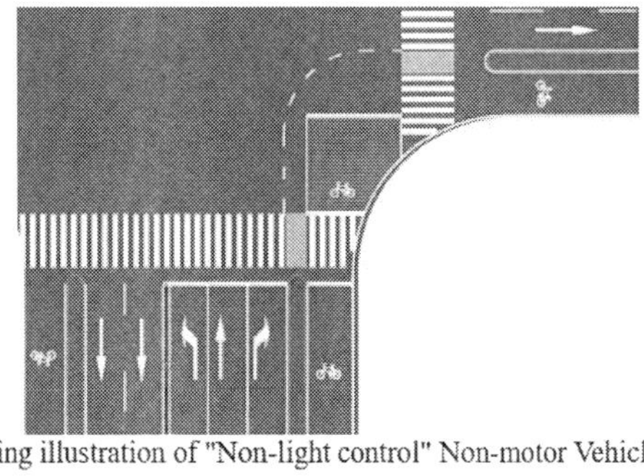

Figure 4 Setting illustration of "Non-light control" Non-motor Vehicle waiting Area

 https://doi.org/10.1051/e3sconf/20183803021

5 "Signal Control"Waiting Area for Non-motor Vehicle

5.1 Setting Method

With the conditions of high non-motor vehicles volume and enough intersection space, the non-motor vehicle waiting area can be set ahead of crosswalk lines. When the high non-motor vehicle volume leads to the interference of normal operation between straight and left-turn flow, the non-motor vehicle waiting area can be divided into straight waiting area and left-turn waiting area.

It is required to use anti-clockwise unilateral release, when releasing the sraight motor vehicles, the left-turn non-motor vehicles should be prohibited, and when the non-motor vehicles are entering the waiting area, the right-turn motor vehicles should be prohibited. Meanwhile, the LED display screens or traffic signs should be laid out to guide the non-motor vehicles to drive into the waiting areas. For example, when the signal control intersection indicates left-turn phase starts with the end of straight phase, the drivers should be guide into the waiting areas during the left-turn phase of the crossing road. In addition, the color pavements (such as green) can be applied to make the waiting area much more visible.

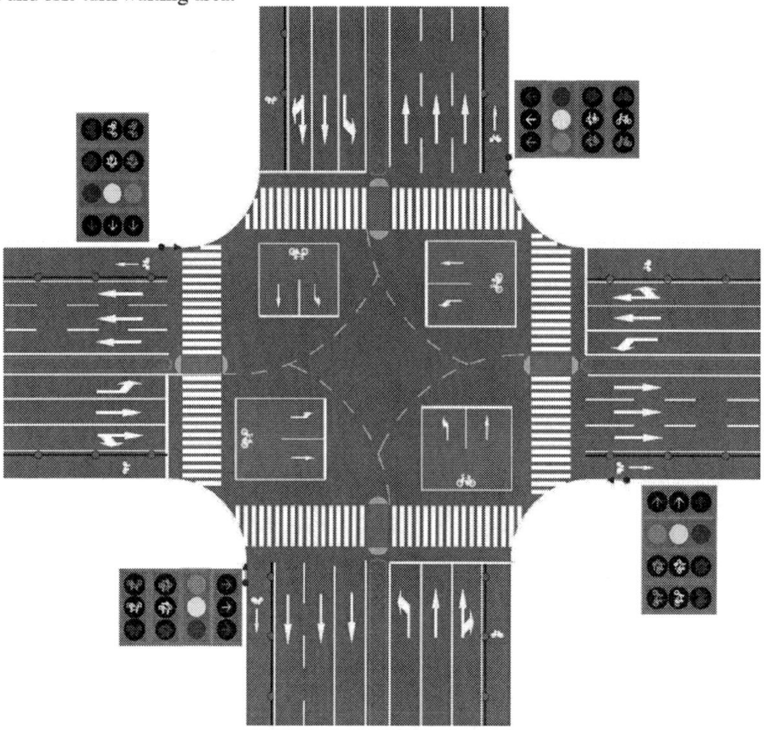

Figure 5 Setting illustration of "light control" Non-motor Vehicle waiting Area

5.2 Advantage and Disadvantage

This design makes reasonable use of road resources at signal control intersection, increases the non-motor vehicle queuing waiting area and the green light time, shortens the running cycle time, reduces the intersection signal control delay as well as the conflict between motor vehicle and non-motor vehicle, then the traffic capacity of the intersection is significantly improved. However, it is not easy to implement.

5.3 Applicable Conditions

This design is suitable for large intersections (entrance lanes should be at least 3), the size of the area should be able to store the non-motor vehicles. At the same time, it requires non- motorcyclists to take the initiative to comply with the rules.

6 Typical Case Analysis

6.1 Channelization State and Existing Problem

The case intersection is a very important intersection, which connects the east and west district of the city. The traffic volume of non-motor vehicle is very large, and there is serious conflict between motor vehicles and non-motor vehicles which seriously affects the traffic efficiency of the intersection. To improve the situation, the intersection "blank area" is used to set the "signal control" waiting area for non-motor vehicle, with four directions

https://doi.org/10.1051/e3sconf/20183803021

unilateral clockwise release to eliminate the traffic conflict. Specific measures are shown in Fig.5.

6.2 Analysis of Program Implementation Effect

Before the improvement, the motor vehicles and non-motor vehicles are in the same signal control phase, and non-motor vehicles often occupy the motorway, which leads to serious traffic safety risks. After the improvement, on one hand, road space resources are made full use, and non-motor vehicles are concentrated in the waiting area to be released before motor vehicles, which brings the reduction of the traffic conflict and regulates traffic order of the intersection; on the other hand, motor vehicles and non-motor vehicle traffic efficiency has been improved significantly, for example, both morning and evening peak motor vehicle capacity are increased by about 7%.

Figure 6 Before the Improvement

Figure 7 After the Improvement

7 Conclusions

Based on the analysis of typical traffic characteristics of non-motor vehicles, the space optimization method for non-motor vehicles of typical signalized intersections is proposed in this research, which aims to support the traffic channelization design of signal controlled intersections and improve the operation environment of non-motor vehicles. The future research will focus on the space and time integrated optimization method which fit for the operation demand of non-motor vehicle crossing.

Acknowledgement

The authors thanks for the sponsor of National Science and Technology Support Program with Grant No. 2014BAG03B01.

References

1. Junjie Lu. Technology and application of colored pavement [J]. Fujian Construction Science & Technology, 2013, (5): 65-67.
2. Xiaoguang Yang. Traffic Design [M]. Beijing: China Communications Press, 2010.
3. Xiaoguang Yang. Manual of urban traffic design [M]. Beijing: China Communications Press, 2003.
4. National standards "Road traffic signs and markings-Part3: Road traffic markings" (GB5768.3-2009)

 https://doi.org/10.1051/e3sconf/20183803022

Cyclic Flexural Behavior of Reinforced Concrete Beams

Prafulla B MALLA[1], Hong ZHOU[2], Yi CHE[1]

[1] State Key Laboratory of Costal and Offshore Engineering, Dalian University of Technology, Dalian, Liaoning 116024, China
[2] Department of Architectural Engineering, Shandong Water Conservancy Vocational College, Rizhao, Shandong Code, China

Abstract: The present study aims at investigating the cyclic flexural behavior of reinforced concrete beams with varying depths. Five reinforced concrete beams with beam depth ranging from 250 mm to 750 mm were tested under reversed cyclic loading and the influence of beam depth on the flexural strength and ductility of reinforced concrete beams was investigated. In addition, OpenSees was used to model the test specimens and the analytical results were compared with the experimental reuslts. It is shown that there is no apparent size effect on the normalized ultimate flexural strength of the tested beams, while for the displacement ductility factor, a significant size effect is observed. Load-deflection hysteric curves of test specimens obtained by the fiber-based element of OpenSees with Concrete03 and Hysteric models are in good agreement with those from experimental tests.

1 Introduction

The size effect has been experimentally evidenced for most brittle types of failures of plain and reinforced concrete members, for example, shear and torsional failure of beams, punching shear failure of slabs and pullout failure of bars etc. [1, 2]. In addition, the effect of size on flexural behavior of reinforced concrete beams was also investigated [3-9]. Early work on flexural behavior of reinforced concrete beams indicated that there was no apparent size effect on flexural strength and rotational capacity [3, 4]. Appa et al. [5] and Yi et al. [6] reported from their studies that size effect is significant for both flexural strength and ductility. Most of the research on under-reinforced concrete beams confirmed the existence of size effect on rotational capacity [7-9]. Some researchers also investigated behavior of over-reinforced and prestressed concrete beams failing due to concrete compressive failure and found that significant size effect on rotational capacity or ductility of beams exists [10-12].

However, most of the studies reported in the literature are concerned with the size effect of flexural behavior of reinforced concrete beams that contain only tensile reinforcement and are subjected to monotonic loading. In reinforced concrete moment-resistance frames, the cross section of the beam end to resist the positive and negative moment induced by lateral earthquake and vertical loads, locates in the potential plastic hinge region where sufficient ductility and rotational capacity is required. Therefore, a better understanding of size effect on flexural behavior of doubly reinforced concrete beams subjected to cyclic loading is necessary to ensure the safety of reinforced concrete structures.

In view of this, the influence of beam depth on the cyclic flexural behavior of RC beams was investigated experimentally in this study. Besides, nonlinear analysis of test specimens was carried out using OpenSees models.

2 Experimental Program

2.1 Test Specimens and Material Properties

In total, five specimens with beam depth ranging from 250 mm to 750 mm were tested under reversed cyclic loading. The specimens used in the experimental program consisted of two beam elements with identical section cast integrally with a stub as the column element at mid-span. The details of specimens are shown in figure 1 and table 1.

All specimens were cast from the same batch of concrete supplied by a local ready-mix plant. Concrete cubes of 150×150×150 mm^3 were casted along with the specimens, and compressive strength tests were performed at time of cyclic loading tests. Hot-rolled deformed bars with diameters of 12, 20, 22 and 25 mm were used as longitudinal reinforcement giving longitudinal reinforcement ratios of 1.08 % and 0.69%, respectively. Hot-rolled plain bars with 6, 8 and 10 mm diameters were used as vertical stirrups, with a constant transverse reinforcement ratio equal to 0.52%. Properties of steel bars are presented in table 2.

*Corresponding author: Email: cheyi@dlut.edu.cn

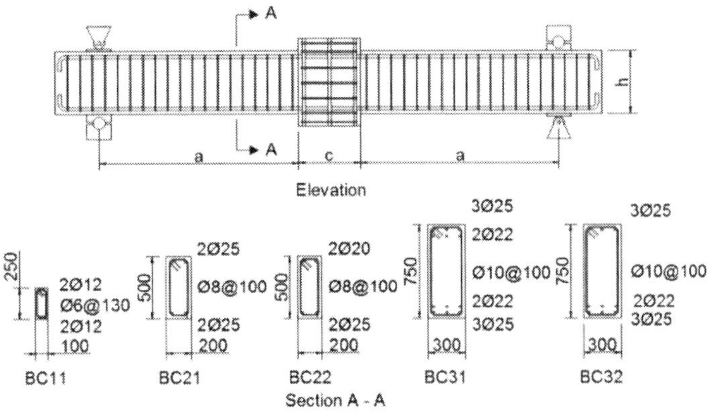

Figure 1 Details of test specimens (mm)

Table 1 Parameters of specimens

Specimen	Dimension (mm)				Concrete cubic strength f_{cu} (MPa)	Longitudinal rebar ratio (%)		Stirrups ratio ρ_{sv} (%)
	Width b	Depth d	Effective depth d_0	Span a		Top bar ρ_s	Bottom bar ρ_s'	
BC11	100	250	210	3	73.6	1.08	1.08	0.52
BC21	200	500	456	3	73.8	1.08	1.08	0.52
BC31	300	750	688	3	72.0	1.08	1.08	0.52
BC22	200	500	456	3	73.8	1.08	0.69	0.52
BC32	300	750	688	3	72.0	1.08	0.69	0.52

Table 2 Properties of steel bars

Diameter of steel bar (mm)	Yield strength (MPa)	Tensile strength (MPa)
6	407	578
8	467	507
10	384	535
12	420	618
22	423	598
25	470	660

2.2 Loading Protocol and Test Setup

Figure 2 shows the planned loading protocol for the reversed cyclic loading test. Each specimen was subjected to progressively increasing load cycles. After yielding of tensile reinforcement, progressively increasing displacement cycles were applied until the specimen failed completely. For each displacement step, three fully reversed cycles were applied. The test setup is shown in figure 3.

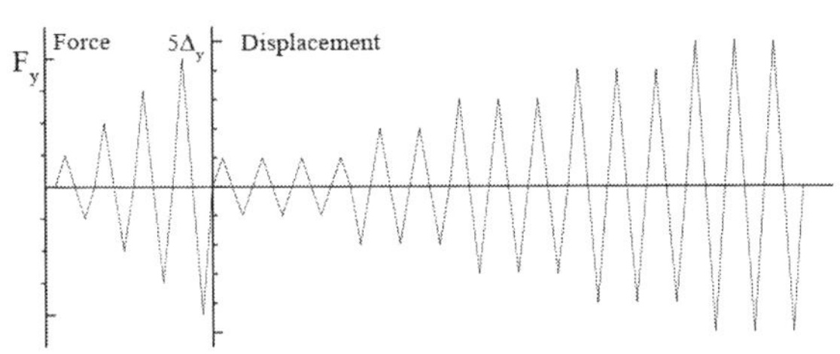

Figure 2 Loading protocol

Figure 3 Test setup

3 Test results and Size Effect

3.1 General Behavior

 https://doi.org/10.1051/e3sconf/20183803022

Similar failure patterns were also observed in other specimens regardless of specimen sizes. Flexural cracks were firstly observed at sections adjacent to the central stub. The failure of the specimen resulted from the yielding of tensile reinforcement followed by the crushing of concrete. Figure 4 shows the typical failure patterns of Specimens BC11, BC21 and BC31 at the end of the test.

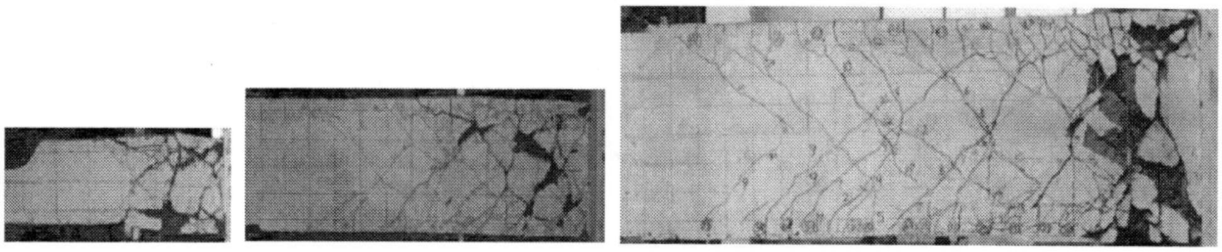

(a) Specimen BC11 **(b) Specimen BC21** **(c) Specimen BC31**

Figure 4 Failure patterns of specimens

3.2 Test Results

Test results are listed in table 3, in which the ductility factor for each specimen is calculated by

$$\mu = \Delta u / \Delta y \quad (1)$$

Where μ is the displacement ductility factor, and Δy and Δu are displacements at yield load and ultimate load, respectively.

Table 3 Experimental results

Specimen	Loading direction	At yield		At ultimate		Ductility factor μ
		Moment My [kN.m]	Displacement Δy [mm]	Moment Mu [kN.m]	Displacement Δu [mm]	
BC11	+	17.46	3.32	20.12	29.23	8.80
	-	17.42	3.00	20.29	-	-
BC21	+	194.14	7.55	235.48	43.46	5.76
	-	192.30	7.62	223.43	-	-
BC31	+	656.43	13.01	754.49	65.11	5.00
	-	652.31	19.03	752.41	-	-
BC22	+	190.05	8.97	236.00	46.83	5.22
	-	114.70	4.66	137.68	-	-
BC32	+	662.22	12.24	779.62	59.16	4.83
	-	481.04	9.40	533.01	-	-

3.3 Size Effect

The normalized flexural ultimate strength, m, for each specimen is calculated by

$$m = \frac{M_u}{f_{cu} bd} \quad (2)$$

Figure 5 shows the variation of normalized flexural strength with beam depth. As shown, there is no apparent size effect on the normalized ultimate flexural strength of the tested beams. Figure 6 plots the ductility factor varying with the depth of beams. For beams with symmetric longitudinal reinforcement, i.e. Specimens BC11, BC21 and BC31, the ductility factor decreases with increase of the depth of beams exhibiting a significant size effect. However, for beams with unsymmetric longitudinal reinforcement, i.e. Specimens BC22 and BC32, the size effect cannot be confirmed due to the insufficiency of test data.

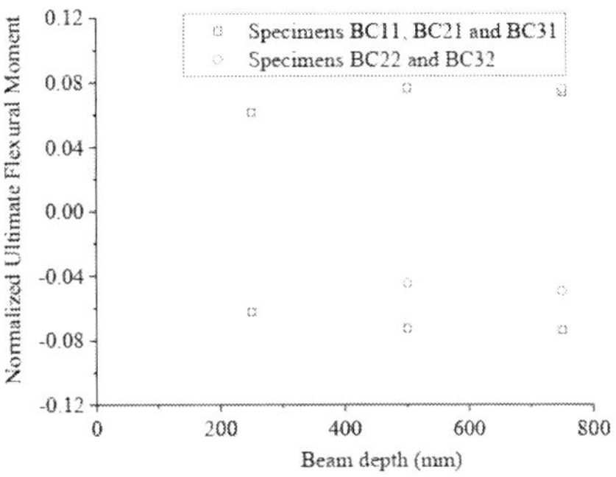

Figure 5 Normalized ultimate flexural moment

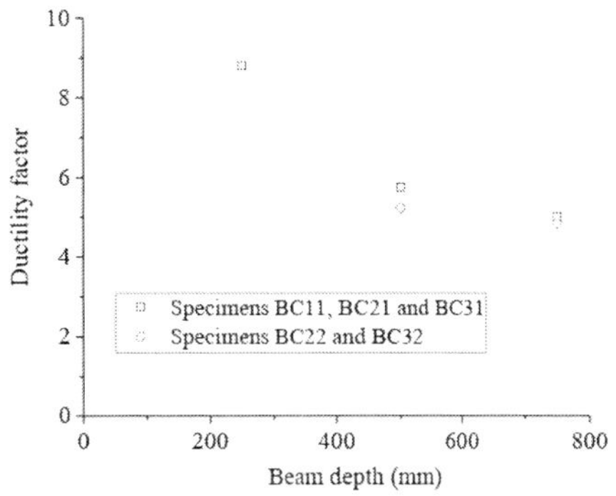

Figure 6 Ductility factor

4 OpenSees Modeling

4.1 The FE Model

Nonlinear analysis of the cyclic experiments was conducted by using OpenSees software which is based on fiber based finite element analysis. In the analysis, the influence of shear deformation was considered, whereas the slippage of reinforcement was ignored.

The uniaxial material concrete03 for concrete and the hysteretic model for reinforcement provided in OpenSees, as shown in figures 7 and 8, were used to simulate the nonlinear behavior of concrete and reinforcement, respectively. Detailed information of the FE models and the analysis of test specimens can be found elsewhere [13, 14].

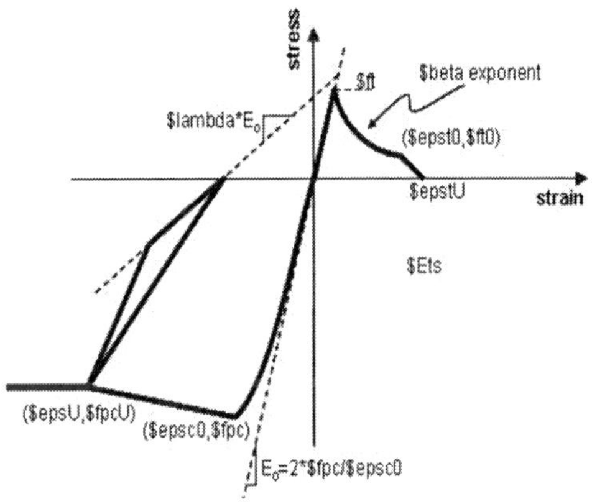

Figure 7 Concrete 02 model

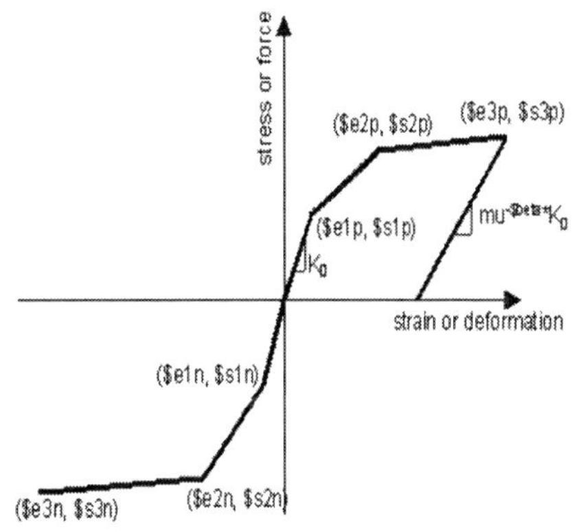

Figure 8 Hysteretic model

4.2 Analytical Results and Comparison

Figure 9 shows the load-deflection hysteric curves of Specimens BC11, BC21 and BC31obtained from the experiments and analysis by OpenSees. As can be seen, the results of the analysis are in good agreement with the experimental results, and the pinching effects of experimental curves are well captured by the FE analysis. The fiber-based element of OpenSees with Concrete03 model and Hysteric steel bar model is suitable to simulate cyclic behavior of RC beams and is able to consider the pinching effect.

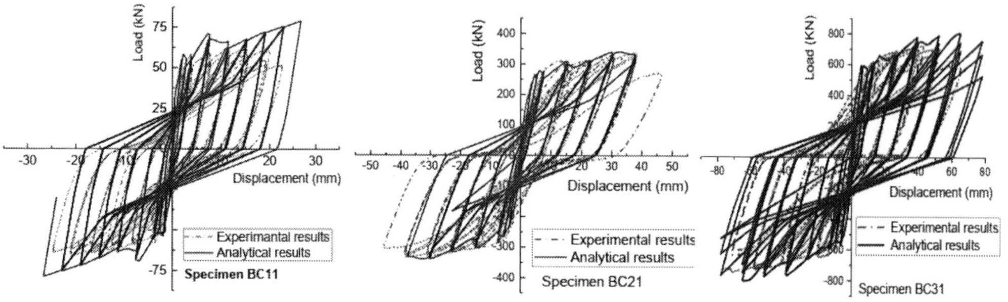

Figure 9 Comparison of load-deflection hysteric curves

5 Conclusion

Based on the present study, the following conclusions may be drawn:

(1) Test results show that there is no apparent size effect on the normalized ultimate flexural strength of the tested beams.

(2) For the displacement ductility factor, a significant size effect is observed experimentally in beams with symmetric reinforcement.

(3) The load-deflection hysteric curves obtained by the OpenSees analysis are in good agreement with those from experimental tests. It is found that the fiber-based element of OpenSees with Concrete03 and Hysteric models is suitable to simulate cyclic behavior of RC beams and is able to consider the pinching effect.

Acknowledgement

This research was financially supported by the Science Foundation of the State Key Laboratory of Costal and Offshore Engineering.

References

1. Bazant Z P and Planas J. Fracture and size effect in concrete and other quasibrittle materials [M]. 1st ed., CRC Press, London, 1998.
2. Kim J K, Yi S T and Kim J H J. Effect of specimen sizes on flexural compressive strength of concrete [J]. ACI Structural Journal, 2001, 98 (3): 416-424.
3. Corley W G. Rotational capacity of reinforced concrete beams, Journal of Structural Division [J]. 1966, 92 ST5: 121-146.
4. Alca N, Alexander S D B, and MacGregor J G. Effect of size on flexural behavior of high-strength concrete beams [J], ACI Structural Journal, 1997, 94 (1): 59-67.
5. Appa R G, Vijayanand I and Eligehausen R. Studies on ductility and evaluation of minimum flexural reinforcement in RC beams [J]. Materials and Structures, 2008, 41: 759-771.
6. Yi S T, Kim M S, Kim J K and Kim J H J. Effect of specimen size on flexural compressive strength of reinforced concrete members [J]. Cement and Concrete Composites, 2007, 29 (3): 230-40.
7. Bigaj A and Walraven J. Size effect in plastic hinges of reinforced concrete members [J]. Heron, 2002, 47 (1): 53-75.
8. Carpinteri A, Corrado M, Mancini G and Pagi M. Size-scale effect on plastic rotational capacity of reinforced concrete beams [J]. ACI Structural Journal, 2009, 106 (6): 887-895.
9. Khatieb M D. Experimental investigation of scale influence on plastic rotational capacity of reinforced concrete beams [J]. Engineering, 2016, 8:196-203.
10. Ozbolt J, Mestrovic D, Li Y and Eligehausen R. Compression failure of beams made of different concrete types and sizes [J]. Journal of Structural Engineering, 2000, 126 (2): 200-209.
11. Belgin C M and Şener S. Size effect on failure of overreinforced concrete beams [J]. Engineering Fracture Mechanics, 2008, 75(8): 2308-2319.
12. Vorel J, Gattu M and Bazant Z P. Size effect in flexure of prestressed concrete beams failing by compression softening [J]. Journal of Structural Engineering, 2014, 140 (10): 04014068.
13. S. Mazzoni, F. Mckenna, M. H. Scott, et al. OpenSees command language manual [R]. Pacific Earthquake Engineering Research (PEER) Center, 2007.
14. P. B. Malla P B. Nonlincar analysis of reinforced concrete based on Abaqus and OpenSees [D]. Dalian University of Technology, Dalian, China, 2016.

Exploratory Research on Bearing Characteristics of Confined Stabilized Soil

Shuai Shuai WU, Zheng Guo GAO, Shi Yang LI, Wen Bo CUI, Xin HUANG

School of Transportation Science and Engineering, Beihang University, Beijing 100191, China

Abstract: The performance of a new kind of confined stabilized soil (CSS) was investigated which was constructed by filling the stabilized soil, which was made by mixing soil with a binder containing a high content of expansive component, into an engineering plastic pipe. Cube compressive strength of the stabilized soil formed with constraint and axial compression performance of stabilized soil cylinders confined with the constraint pipe were measured. The results indicated that combining the constraint pipe and the binder containing expansion component could achieve such effects: higher production of expansive hydrates could be adopted so as to fill more voids in the stabilized soil and improve its strength; at the same time compressive prestress built on the core stabilized soil, combined of which hoop constraint provided effective radial compressive force on the core stabilized soil. These effects made the CSS acquire plastic failure mode and more than twice bearing capacity of ordinary stabilized soil with the same binder content.

1 Introduction

A confined stabilized soil pile (CSSP) is proposed in this paper as a new kind of foundation treatment method. It is constructed by filling the stabilized soil, which is made by mixing soil with a binder containing a high content of expansive component, into a constraint pipe with enough tensile modulus and strength fixed in ground.

It is supposed that combining the constraint of the pipe with the binder containing expansion component would bring several unique effects. 1) Volume expansion produced by expansive component in the binder can reduce voids in the stabilized soil so as to significantly enhance the strength of stabilized soil [1] [2]. Under the constraint conditions, higher content of the expansive component could be adopted so as to further improve strength of the stabilized soil. 2) Hoop constraint of confined concrete can enhance the bearing capacity (BC) obviously [3]; similar to this, hoop constraint of CSS should also enhance the BC greatly. 3) Bringing prestress on confined concrete can enhance its BC further [4]; according to this principle, under the constraint, the expansion produced by the binder might build prestress upon the core stabilized soil, which would enhance the radial confining force on the core stabilized soil and further enhance the BC when bearing load.

Combining these benefits together is expected to provide the CSSP with rather high BC and make it become an effective foundation treatment method. But whether this idea is feasible or not should be verified. No literature about research on the CSS has been found by the authors. Although the fact that BC of confined concrete is much higher than that of unconfined concrete have been proved, whether the result fit the CSS or not is unconfirmed. Therefore the performance of the CSS was investigated in this exploratory research.

*Corresponding author: Email: hx2000@buaa.edu.cn

2 Experiment Schemes and Methods

2.1. Experiment Scheme

1) Experiment one: cube compressive strength of the stabilized soil. The binder content of 15 and the mass ratio of cementing component and expansive component (C/E) of 7/3, 6/4 and 5/5 were chosen in this experiment. For each C/E, stabilized soil samples formed with constraint (experimental group) and those formed without constraint (control group) were prepared; in addition, an expansion group was also included in which five faces of the cubic stabilized soil samples were confined and one face was left free for measuring the expansion quantity during curing time.

2) Experiment two： axial compression performance of the confined stabilized soil cylinder(CSSC). The effects of confinement stiffness (CS, product of tensile modulus and thickness of the constraint pipe) and density of the stabilized soil on bearing behavior of the CSSC were considered. C/E = 6/4 and binder content =15% were chosen, with which the stabilized soil samples formed with constraint reached the highest strength according to the results of experiment one. Polyvinyl chloride (PVC) pipe was adopted as the constraint pipe, whose CS was adjusted by changing its wall thickness. The density of the stabilized soil was adjusted by changing water content of the soil; a plastic state soil with optimum water content and a plastic-flow state soil with water content higher than

https://doi.org/10.1051/e3sconf/20183803023

liquid limit were chosen respectively. Two groups of stabilized soil cylinders were prepared, and both groups were formed with the constraint pipes, but the group of CSSC bore load with the constraint pipe while the group of unconfined ones bore load without the constraint pipe. The scheme of Experiment two is shown in Table 1.

Table 1 Scheme of Experiment Two

Number	Water content (%)	PVC pipe wall thickness(mm)
C13-1/2/3	13	2.2/2.5/3.2
C30-1/2/3	30	2.2/2.5/3.2

2.2 Experimental Materials

Physical indexes of the soil were shown in Table 2.

Table 2 Major Physical Indexes of the Experimental Soil

Natural water content (ω/%)	Specific gravity of soil particle (Gs)	Natural porosity (e)	Liquid limit (ωL/%)	Plastic limit (ωP/%)
20.4	2.61	0.66	28.9	12.7

The cementing component of the binder, which would mainly produce calcium silicate hydrate (CSH), consisted of Portland cement, slag, fluidized-bed combustion fly ash and hydrated lime, and the expansive component that would mainly produce calcium sulfoaluminate hydrate (AFt) was made of aluminate cement, gypsum and hydrated lime.

Elastic modulus and strength of the PVC pipe was 2753 and 31.89MPa respectively. The external diameter of the pipe was 11 centimeters, and its length was 20 centimeters. In order to eliminate vertical bearing function of the pipe, it was divided into 3 equal sections, and between each section there was an about 2mm-wide seam.

2.3 Experiment Methods

The method for preparing stabilized soil (according to JGJT 233) was generally as follows. Manually mix all kinds of components of the binder with water, and the water binder ratio was 0.5. Then mix the mixture and soil sample by Type NJ-160 cement paste mixer. Slowly stir (62 r/min) for 1 minute and then quickly stir (125 r/min) twice for 2 minutes sequentially.

For the cubic samples, divide the mixture into three parts then push and hammer them in turn into a steel mold (5cm*5cm*5cm), after that compact the mixture on the vibrating table for 2 minutes. As for preparation of the cylinders, divide the mixture into ten parts and push and hammer each part in turn into the constraint pipe. Finally cover the exposed face of the samples with plastic sheeting.

For the experimental group in Experiment one, the cubic stabilized soil samples were cured in room with average temperature of 20°C and average humidity of 42% during the first 14 days. Enough weights were pressed on the top of the steel mold to prevent expansion of the samples. During this period, cover them with cloth. At the 15th day, unwrap the molds and put the samples into the standard curing box with temperature of 20±2°C and humidity of 95±2% for another 14 days. Cubic samples of the control group and the expansion group were cured in the same condition with those of the experimental group, while they were unmolded at the second day and the 15th day respectively.

Vertically placed in water, the stabilized soil cylinders were cured in the room for 28 days, meanwhile press weights on the top of the cylinders.

In order to measure the expansion quantity, a dial gauge was placed vertically on the free face of the cubic stabilized soil samples during the first 14 days of the curing period. The strength of cubic samples were measured by pressed at the speed of 2.4mm/min with the Pavement Material Strength Experimenter (0~30KN). The cylinders were pressed at the speed of 0.1kN/s by the microcomputer controlled electro hydraulic servo universal testing machine. Damage load, load-displacement curves and the failure modes were recorded.

3 Results and Analysis

The strength and the expansion data of experiment one were shown respectively in Figure 1. BC (damage load) and load-displacement curves of the cylinders were shown in Figure 2. And failure modes of some cylinders were shown in Figure 3.

 https://doi.org/10.1051/e3sconf/20183803023

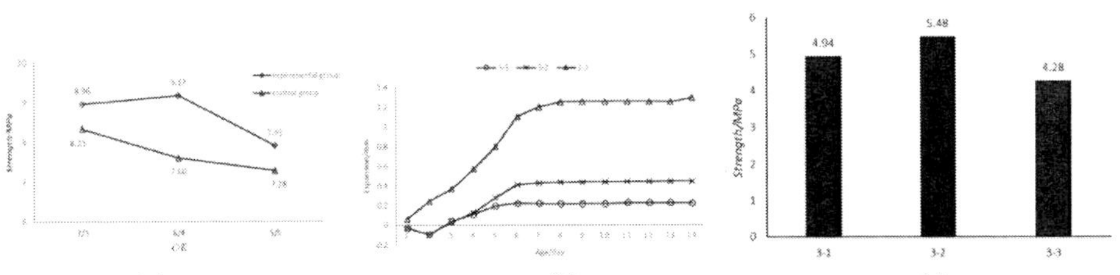

(a) (b) (c)

Figure 1 The Cubic Strength and Expansion Data of Experiment One

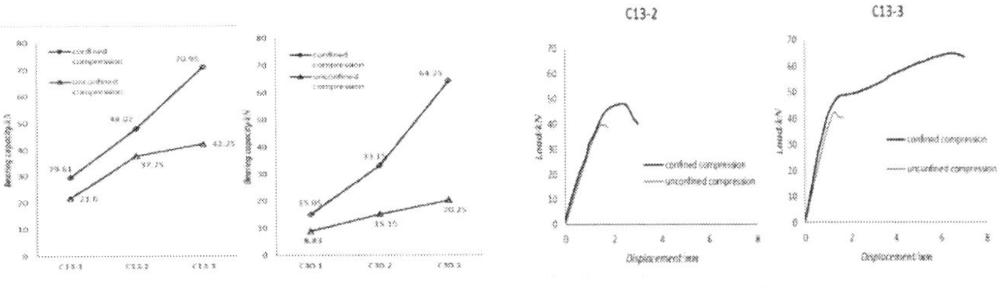

(a) (b)

Figure 2 BC and Load-displacement Curves of Cylinders

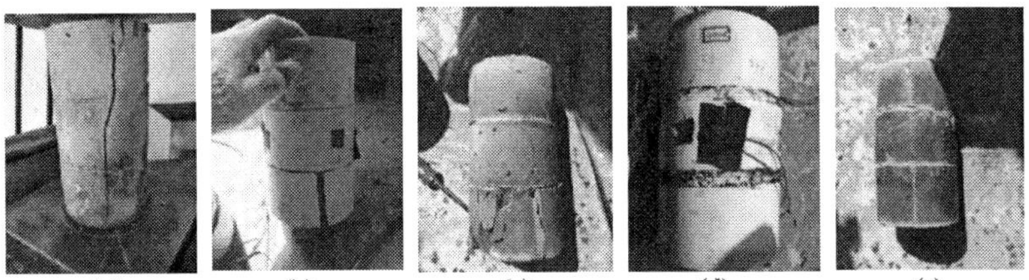

(a) (b) (c) (d) (e)

Figure 3 Failure Modes of Some Cylinders

3.1 Compressive Strength of the Stabilized Soil Formed with Constraint

Figure 1(a) showed that: strength of the stabilized soil formed without constraint reached its highest at C/E = 7/3, and declined with C/E reducing. While, as C/E dropped strength of the stabilized soil formed with constraint increased at first then decreased, and reached the highest strength 9.17MPa at C/E = 6/4, which was higher than the highest strength of the stabilized soil formed without constraint, 8.33MPa. Figure 1(b) showed that: apparent volume expansion emerged during the stabilized soil structure forming process in consequence of AFt developed. The expansion increased with the decrease of C/E. The strength of the corresponding stabilized soil, as shown in Figure 1(c), also increased first then decreased.

AFt generated volume expansion during the stabilized soil structure forming process. When the stabilized soil formed with constraint, instead of expanding outwards, AFt volume expansion would squeeze the internal space of the stabilized soil so as to decrease its porosity and increase the strength within a certain range of C/E. But it might also damage the stabilized soil structure formed by the cementation of CSH, which led to the decrease of strength of the stabilized soil. The decrease of C/E would lead the quantity of CSH to decrease but that of AFt to increase. When C/E was less than a certain value, the quantity of CSH was too less to form sufficient cementation and too much expansion increased damage on the stabilized soil structure, which led to the decrease of strength of the stabilized soil. So the strength of sample 1-3 with C/E = 5/5 greatly decreased.

As shown in Figure 2(a), BC of the unconfined stabilized soil cylinders (USSC) were related to the CS of the constraint pipe. Within a certain range of the CS, BC of the cylinders increased along with the increase of the CS.

If equal amount of the same binder was used, quantity of AFt volume expansion would be equivalent, and the expansion force put on the constraint pipe should be the same too. Hoop tensile deformation of the constraint pipe would generate under the expansion force. When the CS was small, a large proportion of AFt volume expansion expended outwards while only a little part of it left to squeeze the internal space of the stabilized soil. On this condition, the compactness and strength of the stabilized soil increased just in a limited range, such as C13-1 and C30-1 shown in Figure 2(a). With the increase of CS, the hoop tensile deformation of the constraint pipe decreased and volume of the cylinders grew a little under the same expansion effect. So a large part of the volume expansion contributed to squeeze the pores in the stabilized soil. Therefore, the compactness and strength of the stabilized soil increased significantly. These results indicated that in order to improve strength of stabilized soil CS should be high enough.

3.2 Axial Compression Performance of CSSC

As shown in Figure 2, BC of the CSSC was apparently higher than that of the unconfined ones, and about once higher could achieved as shown in C13-3 and C30-3 when strength of the stabilized soil of each pair of the cylinders were roughly the same. The load-displacement curves of the USSC were almost linear increasing before reaching the peak load then abruptly dropping, which presented a brittle failure mode. While the load-displacement curves of the CSSC could be roughly divided into two stages, the first was a linear stage, after which the deformation increased rapidly with load increasing until the curves reached the peak load and presented a plastic failure mode.

In the first stage the core stabilized soils were intact cemented bodies. The linear relationship between load and deformation of the cylinders, which were similar to that of unconfined ones, indicated that there were no apparent changes of properties of the stabilized soil. It could be deduced that in the stage the load was born by the core stabilized soils which kept intact state. As the load increased, radial strain of the core stabilized soil gradually reached its limit tensile strain and the core stabilized soil began cracking. In the second stage, as its integrality lost gradually, the core stabilized soil expanded horizontally and was compacted under the load, as a consequence of which considerable compressive deformation was presented and the constraint pipe bore increasing tension. It was the confining force of the constraint pipe that made the CSS bear load further. With the increase of the vertical load, tension stress in the constraint pipe increased constantly, until the pipe was broken as shown in Figure 3(b), or yielded as shown in Figure 3(d). Finally the CSSC lost bearing ability due to failure of the constraint pipe, and BC of the CSSC was determined by strength of the constraint.

3.3 Compressive Prestress Building on the Confined Stabilized Soil

AFt volume expansion under the constraint could be described in two stages as shown in Figure 6. Assuming there was no lateral constraint on the stabilized soil, during its structure forming expansion stress σ_1 made the diameter of stabilized soil samples expand freely from L_0 to L_1; then a radial pressure σ_2 was applied to compress the samples to the final diameter L_2. Fig.6 implied that if a CSS sample was prepared by using a binder containing expansive components, considerable confining force would be applied to the core stabilized soil during its structure forming, which indicated that there existed already radial compressive prestress on the stabilized soil before it bore vertical load.

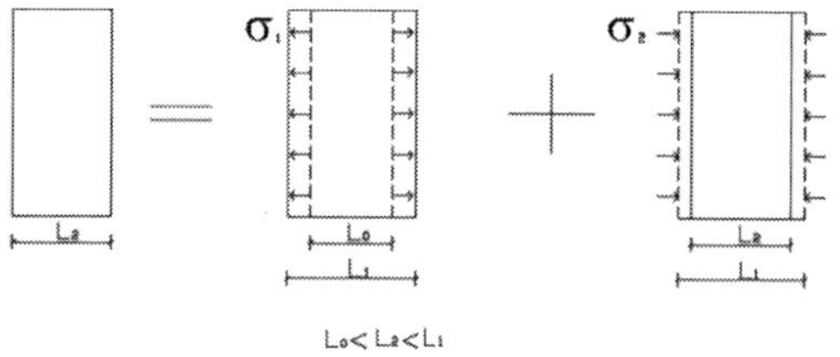

Figure 4 Section Sketch Drawing of Prestress Generation

The radial compressive prestress could enhance the radial confinement effect of the constraint pipe on the core stabilized soil when it bore vertical load, so as to further increase its BC. Without prestress, there was little radial compressive stress on the core stabilized soil in the early stage of loading. Only when the stabilized soil produced considerable lateral dilation deformation, could the radial compressive stress be produced passively. So the constraint pipe might hardly provide effective radial confining force to the core stabilized soil before it damaged. However, when bearing radial compressive prestress, the stabilized soil was in triaxial compressive stress state from the beginning of loading. As the existence of the radial pressure hindered the radial dilation of the stabilized soil, higher vertical pressure was needed to make the stabilized soil reach its ultimate tensile strain. Above analysis was supported by the facts in Figure 4 in which the linear stage of the CSSC lasted longer than that of the unconfined ones. Obviously, the greater the radial compressive prestress acted on the stabilized soil, the longer the linear stage lasted. Also the prestress improved the stress state of the core stabilized soil in the second stage of the load-displacement curves, which also improved the bearing performance and deformation characteristics of the CSSC.

3.4 Influence of CS and Soil Density

As shown in Figure 2(a), the BC increment of the CSSC was not significant as the thickness of the constraint pipe increased from 2.2mm to 2.5mm, but it increased obviously when the thickness increased from 2.5mm to 3.2mm. As shown in Figure 4, BC of the USSC of C13-2 and C13-3 were roughly same. When the thickness of the constraint pipe was 2.5mm, BC and deformation quantity of the confined cylinders didn't increase significantly, compared with that of the unconfined ones; while, as the thickness of constraint pipe increased to 3.2mm, BC and deformation quantity of the confined cylinders increased greatly.

The USSC was destroyed due to the vertical crack produced by the transverse tension stress as shown in Figure 3(a). C13-2 was destroyed due to the fracture of the confining pipe caused by transverse tension as shown in Figure 3(b), and failure mode of the core stabilized soil of C13-2 shown in Figure 3(c), was similar to that in Figure 3(a). C13-3 bulged outwards obviously before failure as shown in Figure 3(d). And the core stabilized soil of corrupted C13-3 only bulged without vertical crack as shown in Figure 3(e).

These results revealed the failure of all of the CSSC were due to the constraint pipe losing efficacy. Only the constraint pipe possessing sufficient CS, could the CSSC enhance effectively their BC and take favorable failure mode.

As shown in Figure 2(a), the bearing characteristics of samples C13 were respectively similar to those of samples C30 although density difference of the two core stabilized soil (shown in water content) was rather large. These results demonstrated that whether the CSSC made of plastic state soil or those made of plastic-flow state soil, which represented the range of soil used in engineering practice, all could improve BC effectively.

3.5 Prospective Use of Confined Stabilized Soil

The stabilized soil pile is a widely used foundation treatment method, which can make full use of the soft soil in situ and need not transport outward the abandoned soil as well as need carry much less pile materials into the site than other piles. But the low BC of the stabilized soil pile limited its application scope. Based on above results it was expected that the CSS technique might provide an effective foundation treatment method which could retain the advantages of the existing stabilized soil technique and also obtain rather high BC.

Salt content of the saline soil around the Qarhan Salt Lake in Qinghai Province of China is even as high as 40 wt. %. All kinds of pile foundation and foundation treatment methods containing Portland cement could not be used in the area for severe corrosion caused by the saline soil. The research of our team showed that strength of the stabilized saline soil stabilized with a special binder could increase to 11MPa and keep stable in the saline soil [5], combining which with the technique of the CSSP is expected to provide saline soil area with a high BC foundation treatment methods.

4 Conclusion

In the paper axial compression performance of the CSSC was investigated, following conclusions were gained. 1) The combination of the constraint pipe and the binder containing expansion component provided the stabilized soil with high strength. Made with same binder content, the CSSC could reach more than twice BC of the unconfined ones, also the CSSC showed rather larger plastic deformation. 2) The improvement of the CSS may be due to three effects: ① with the constraint pipe, higher production of AFt could be adopted to increase the void-filling rate so as to further improve the strength of the core stabilized soil; ② hoop constraint provided radial compressive force on the core stabilized soil so as to enhance the BC; ③ with the constraint pipe AFt expansion produced radial compressive prestress on the core stabilized soil. 3) The failure of CSSC was almost all due to the constraint pipe losing efficacy. Only the constraint pipe possessing sufficient CS could the CSSC enhance effectively their BC and take favorable failure mode.

Acknowledgments

The research is funded by the National Natural Science Foundation of China (No. 51578033).

References

1. Huang X, Li Zhanguo, Ning Jianguo, and Xu Sheng. Principle and Method of Optimization Design for Soft Soil binder[J]. Journal of Wuhan University of Technology-Mater, 2009, 24(1): 154-160.
2. Li Z G, Ning J G, and Huang X. Discussion on the Reasons of Different Curing Effects of Cement Expansive Composite Soft Soil[J]. China Civil Engineering Journal, 2011, 33(3): 420-426.
3. Heon-Soo Chung, Keun-Hyeok Yang, Young-Ho Lee, and Hee-Chang Eun. Stress-strain curve of laterally confined concrete[J]. Engineering Structures, 2002, 24(9): 1153-1163.
4. Ciupala M.A., Pilakoutas K., Mortazavi A.A., and Taranu N. RC Columns Lateral Prestressing with Composites[J]. 2th International Conference on Advanced Polymer Composites for Structural Applications in Construction, 2004.
5. Pan Yang, Xin Huang, Jin Pi, Jianwei Yang. Study on new cementitious materials used for pile and stabilized soil in super saline soil[J]. Japanese Geotechnical Society Special Publication, 2016, 2(62): 2111-2114

Prediction of Surface Subsidence Caused by Rectangular Pipe Jacking Construction

Chao ZHANG[1], Qi Jun JIA[1], Ri Qing XU[2], Su Yang FENG[2*]

[1]China Road and Bridge Corporation, Dongcheng, Beijing 100011, China
[2]College of Civil Engineering and Architecture, Zhejiang University, Hangzhou, Zhejiang 310058, China

Abstract: Based on the development of rectangular pipe jacking, the characteristics of rectangular pipe jacking are introduced. The difference between surface settlements caused by rectangular pipe jacking is calculated by stochastic medium method and peck formula method. The influence of section shape on rectangular pipe jacking is analyzed. The results show that the calculation results of the stochastic medium method are more reasonable.

1 Introduction

Pipe jacking is a technique used for installing pipes under the ground using a tunnel boring machine and hydraulic jacks located inside a jacking pit. A shield is inserted into the ground through a drive shaft or reception point and cuts a bore. As the shield advances forward, excavating the earth in its path, sections of pipe are jacked into place directly behind it using powerful hydraulic jacks. As a means of trenchless construction, the pipe jacking technology has been fully developed in the world. It has accumulated a lot of experience in terms of the large diameter, long distance and curve jacking construction [1-3].

The cross-section of the jacking is mostly circular. In the early 1970s, the rectangular pipe jacking technology was used in Japan for the first time. While in 1999, the first rectangular pipe jacking was completed in the underground pavement project of entrance 5# in Lujiazui station, line 2, Shanghai Metro, China[4].

Since most rectangular pipe jacking projects are located in the bustling area, there is a high demand for surface subsidence control. The existing calculation methods of ground settlement include Peck formula method [5] and stochastic medium method [6] [7]. Peck formula method can be applied to circular tunnel and jacking, but the rectangular pipe jacking remains to be verified. Nevertheless, stochastic medium method can be used for a variety of shapes of tunnels or pipe jacking. Wang [7] calculated the surface subsidence induced by ground loss under rectangular pipe jacking construction. However, it was assumed that the soil in the rectangular pipe jacking was uniform convergence, and lacked in-depth study. In this paper, stochastic medium method is used to calculate the surface subsidence caused by rectangular pipe jacking under uniform and inhomogeneous convergence modes. What's more, compared with measured values, the applicability of Peck formula in rectangular pipe jacking is discussed.

2 The characteristics of rectangular pipe jacking

This paper collects some representative pipe jacking projects around the world. As can be seen from Table 1, the cross-section of the rectangular pipe jacking is too large, as the equivalent radius is often greater than 1.8m and it can be regarded as large diameter pipe jacking according to the code for circular one. It also indicates that the depth is shallow because the ratio of depth to the equivalent radius is usually less than 3. Besides, the jacking path is shorter. Rectangular pipe jacking is often used for subway entrances; the ground outside is mostly trunk lines and dense buildings, so a stricter ground subsidence control standard is required.

Table 1 Partial geometric parameters of rectangular pipe jacking projects

Project	Buried Depth /m	Width×height	z/R'	Jacking Path /m	Number of Jacking
Underpass tunnel in Zhongzhou Avenue, Zhengzhou, China	8	9.0m×6.1m	1.91	110	4
Entrance 5# in Culture Park Station,	6.9	6m×4.3m	2.41	52	1

* Corresponding author: fsy@zju.edu.cn

https://doi.org/10.1051/e3sconf/20183803024

line 6, Guangzhou Metro, China					
Entrance 4# in Wangjiadun Station, line 2, Wuhan Metro, China	7.9	6m×4m	2.86	62	1
Entrance 3# in Pudian Road Station, line 6, Shanghai Metro, China	7.9	6.2m×4.4m	2.68	42	1
Channel from Huaqiang North Station to Huaxin Station, Shenzhen, China	5.95	6.9m×4.9m	1.81	41	3
Sewer tunnel in Kikuta-gawaroute 2, Narashino, Chiba-ken, Japan	5.0	4.4m×4.0m	2.11		2
Box-type pipe jacking pedestrian tunnel in Kota, Indonesia	6.65	6m×4m	2.41	45	1

3 Stochastic medium theory

The whole excavation is decomposed into infinitely small excavations from a statistical point of view, assuming that the soil is not condensed, the density does not change, and the rock mass is incompressible. The effect of the upper rock mass should be equal to the sum of the effects of infinitely small excavations. That is what we call stochastic medium theory. Under the plane strain condition, the ground traverse settlement trough caused by excavating the micro-unit at deep z is:

$$s_e(x) = \frac{1}{r(z)} \exp(-\frac{\pi}{r(z)^2} x^2) d\xi d\eta$$

$$r(z) = z / \tan\beta$$

Where $r(z)$is the main influence radius on the horizontal plane, otherwise known as the main scope of influence, β is the main influencing angle of upper tunnel surrounding rock. The relationship between β and the width of settlement trough i can be obtained by comparing Peck formula method and stochastic medium method:

$$\tan\beta = \frac{z_0}{\sqrt{2\pi} i}$$

The calculation method of i value in stochastic medium method includes Knothe Formula and the method proposed by Han [8].

(1) Knothe Formula

$$i = \frac{z_0}{\sqrt{2\pi} \tan\left(45° - \frac{\varphi}{2}\right)}$$

Where φ is internal friction angle of soil.

(2) Han's method

$$i = K z_0$$

$$\tan\beta = \frac{20}{50 - \varphi}$$

Where K is based on the experiences of different types of soil.

As Knothe Formula is mainly applied to all types of rock, while pipe jacking technology is usually used in soft soil areas, the second method is used to calculate the influence angle β in this paper.

The calculation of the surface settlement horizontally caused by ground loss can be obtained by integrating the settlement area:

$$s(x) = \iint_{\Omega} \frac{1}{r(z)} \exp(-\frac{\pi}{r(z)^2} x^2) d\xi d\eta$$

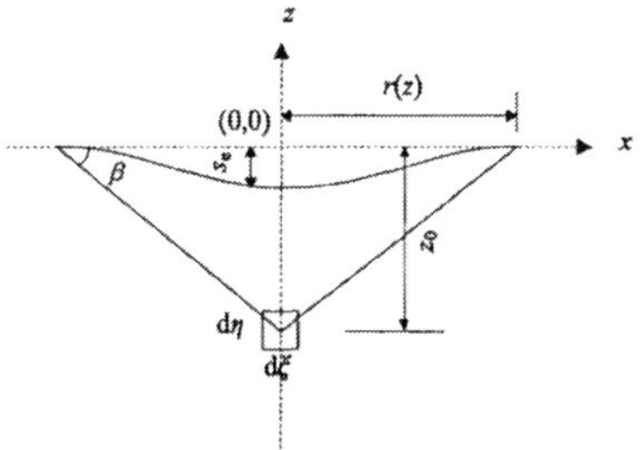

Figure 1 The excavation of unit under the plane strain condition

With reference to the engineering experiences of circular pipe jacking [8], there are two kinds of convergence modes for rectangular pipe jacking, uniform convergence and non-uniform convergence. In the uniform convergence mode, the pipe jack is suspended in the slurry jacket, and the gap between the top pipe and the soil is equal to the bottom clearance. In the non-uniform convergence mode, by contrast, the jacking is sinking as a whole and the bottom clearance is zero.

When tunneling machine advancing, the angle between the advancing direction and the horizontal direction is greater than zero during rectangular pipe jacking. And the cross section of rectangular pipe jacking is generally large; the bottom of the pipe is in direct contact with the soil due to its own weight. Therefore non-uniform convergence mode is more in line with the actual situation.

Assuming $\Delta A = k\Delta B$, ΔB can be inferred by the rate of ground loss:

$$\Delta B = \frac{1}{4k}[-(A + kB) + \sqrt{(A + kB)^2 + 4\frac{kAB\varepsilon}{1 - \varepsilon}}]$$

Where ε is the rate of ground loss, that is, the amount of ground loss is divided by the actual excavation area of the pipe jacking.

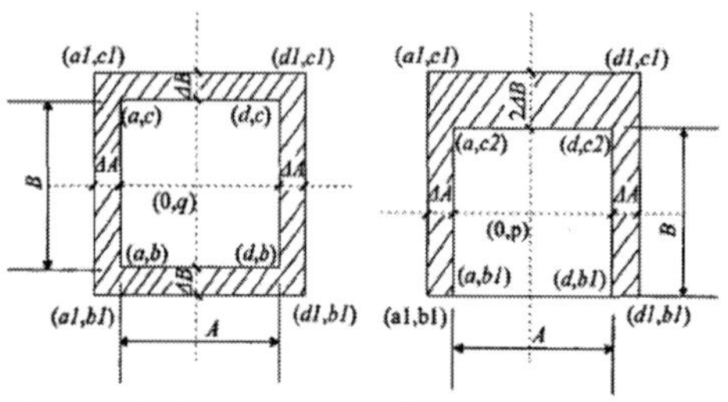

Figure 2 Uniform and non-uniform convergence mode

Take the vertical axis of the surface z=0, the calculation formula of soil settlement in two modes can be obtained:

(1) Uniform convergence mode

$$s(x)=\int_{b1}^{c1}\int_{a1}^{d1}\frac{\tan\beta}{\eta}\exp[-\frac{\pi\tan^2\beta}{\eta^2}(x-\xi)^2]\mathrm{d}\xi\mathrm{d}\eta-\int_{b}^{c}\int_{a}^{d}\frac{\tan\beta}{\eta}\exp[-\frac{\pi\tan^2\beta}{\eta^2}(x-\xi)^2]\mathrm{d}\xi\mathrm{d}\eta$$

(2) Non-uniform convergence mode

$$s(x)=\int_{b1}^{c1}\int_{a1}^{d1}\frac{\tan\beta}{\eta}\exp[-\frac{\pi\tan^2\beta}{\eta^2}(x-\xi)^2]\mathrm{d}\xi\mathrm{d}\eta-\int_{b1}^{c2}\int_{a}^{d}\frac{\tan\beta}{\eta}\exp[-\frac{\pi\tan^2\beta}{\eta^2}(x-\xi)^2]\mathrm{d}\xi\mathrm{d}\eta$$

4 Comparison of stochastic medium method and Peck formula method

In order to estimate the settlement, Peck put forward the empirical formula for the shield tunnel or circular pipe jacking, that is, Peck formula.

$$S(x)=S_{\max}e^{-\frac{x^2}{2i^2}}$$

$$S_{\max}=\frac{V_{\text{loss}}}{i\sqrt{2\pi}}$$

Where $S(x)$ is the amount of ground subsidence, x is the transverse horizontal distance from the axis of the tunnel, $S_{\max}$ is the maximum settlement above the tunnel axis, V_{loss} is the amount of ground loss per unit length of the tunnel, i is the width of settlement trough.

Peck formula was summarized on the basis of a large number of measured data, most of that was deep circular tunnel [9], while rectangle pipe jacking was mainly shallow buried. As a result, the accuracy of Peck in predicting ground settlement trough of a rectangular pipe is still to be studied. In order to compare the differences between the stochastic medium method and Peck formula, the two types of methods are used to calculate the surface subsidence of rectangular pipe jacking under different burial conditions. Supposing that A=5m, B=5m, the top clearance ΔA is equal to the bottom clearance ΔB, and the rate of ground loss is 1%. In addition, the stochastic medium method is divided into two modes: uniform convergence and non-uniform convergence.

(1) Comparison of uniform convergence and non-uniform convergence

As shown in the figure, there is a big difference between the uniform convergence mode and the non-uniform convergence mode. The maximum settlement value of the uniform convergence mode is always smaller than that of the non-uniform convergence mode. The latter settlement trough is narrow and deep, while the former one is wide and shallow. As the burial depth decreases, the difference between the two modes is increasing.

With respect to the uniform convergence mode, the bottom clearance is moved to the top in the non-uniform convergence mode. This shows the influence of the location of ground loss on surface subsidence. The amount of ground loss in the same condition, the more shallow the location of the loss, the greater the impact on the surface.

(2) The differences between stochastic medium method and Peck formula

Han [10] theorized that if the depth was deep enough or the excavation unit was small enough, the results of the two methods should be consistent. To verify his theory, two methods were used to calculate the settlement of the ground caused by circular tunnel construction. It turned out that the calculated results of Peck formula were between the results of the uniform and non-uniform convergence modes when using the stochastic medium method. With the increase of the buried depth, the gap between the three decreased gradually, but not completely consistent.

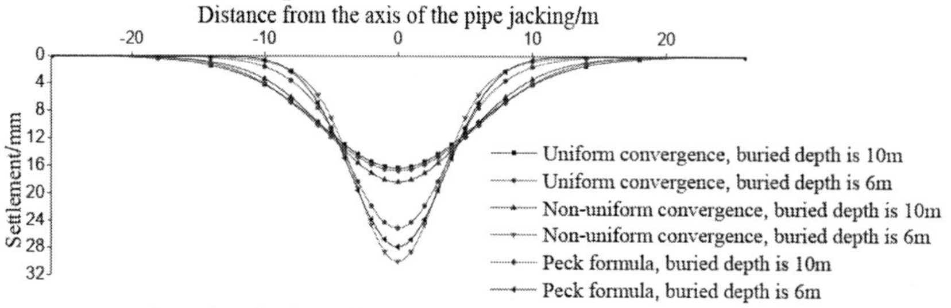

Figure 3 Comparison of stochastic medium method and Peck formula under shallow buried condition

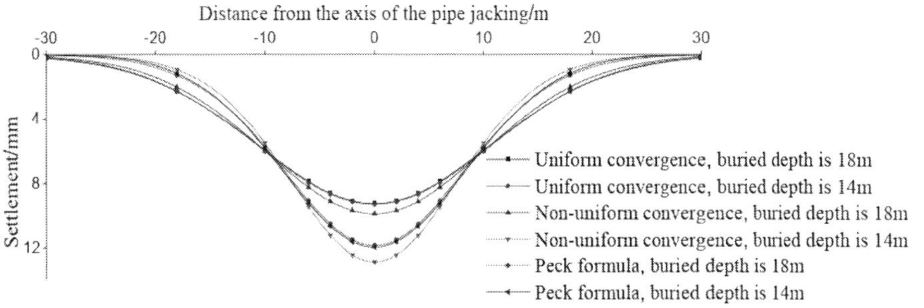

Figure 4 Comparison of stochastic medium method and Peck formula under deep buried condition

As shown in Fig. 3 and Fig. 4, the calculated results are consistent with those of Han's [10].The calculated results of Peck formula were between the results of the uniform and non-uniform convergence modes when using the stochastic medium method. With the increase of the buried depth, the results of the uniform model are consistent with those of Peck formula. However, there are always some differences between the calculation results of the non-uniform convergence model and those of Peck formula. It demonstrates that Peck formula is more suitable for predicting the subsidence of deep buried tunnels under uniform convergence mode. For shallow buried pipes, the results when using Peck formula to predict and actual results often are put in certain error.

(1) Comparison of round and rectangular pipe jacking

In order to compare the differences between the circular and the rectangular pipe jacking, a circular pipe jacking model is established to calculate the surface subsidence under non-uniform convergence conditions. The pipe jacking area and the stratum loss rate are set the same as the previous example.

Compared with the calculation results of the two types of pipe jacking, the shape of the two settlement trough is not very different on the condition that both have the same height and width. When the depth is shallow, the maximum settlement value of circular pipe is slightly smaller than the rectangular one. While when the depth is deep, the settlement regularities of circular and rectangular pipe are basically the same.

5 The effect of the cross-sectional shape

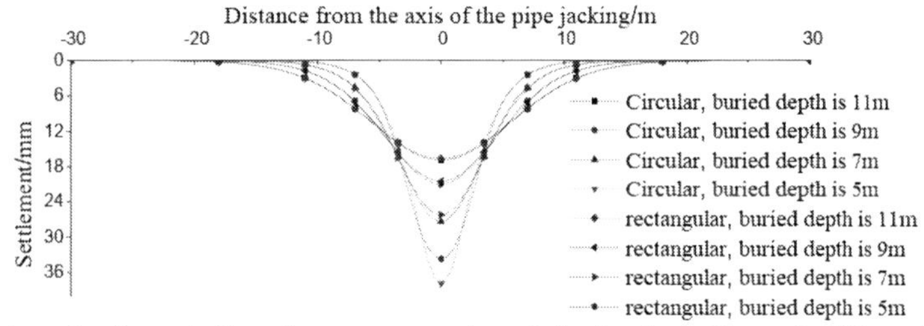

Figure 5 Comparison of surface subsidence between rectangular and circular pipe jacking under different buried depth

(2) The influence of pipe width on surface subsidence

Due to the particularity of rectangular pipe jacking, the width of the rectangular pipe jacking is often greater than the height; the aspect ratio is up to 2.0 in actual projects. In order to study the influence of the aspect ratio on the surface subsidence, this paper changes the width of the pipe to obtain a different aspect ratio. Meanwhile, the height and the stratum loss rate are consistent with the previous example, and the depth of the pipe is set to -8m. The results are shown in Fig. 6.

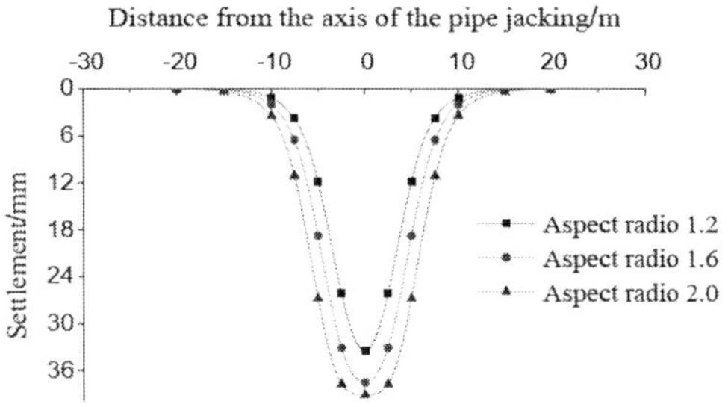

Figure 6 Surface subsidence curves under different aspect ratios

With the increase of the aspect ratio, the actual excavation area increases and the total loss of stratum increases correspondingly. Moreover, the settlement trough curve moves down as a whole, and the width of settlement trough also increases. Take the settlement of 18mm corresponding to the abscissa i_{18}, and the width of settlement trough is quantitatively analyzed with i_{18} as the standard. As shown in Fig. 7 and Fig. 8, the width of settlement trough is linearly related to the aspect ratio, and the maximum settlement shows an exponential relation with the aspect ratio.

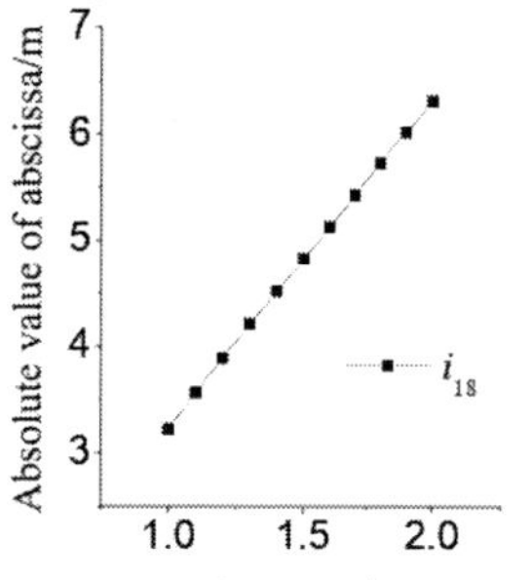

Figure 7 The relationship between i_{18} and the aspect ratio

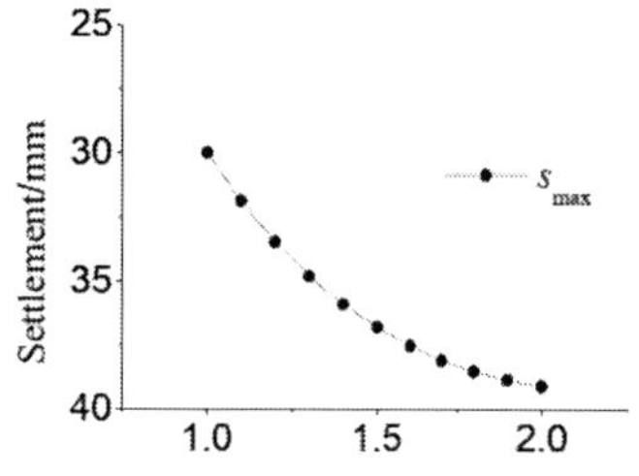

Figure 8 The relationship between the maximum settlement and the aspect ratio

6 Influence of k

According to the gap parameters proposed by Lee[11], the gap g between the soil and the pipe jacking is mainly composed of three parts:

$$g = G_p + U_{3D} + \omega$$

Where G_p is the physical gap, usually refers to the difference between the external dimensions of the roadheader and the pipe; U_{3D} is the three-dimensional elastic-plastic deformation of the front part of the roadheader; ω is the construction factors (including shield steering adjustment, throwing, retreat, buckle, etc.). For the rectangular pipe jacking, the physical gap between the top and the bottom is greater than or equal to the physical gap on both sides, such as channel from Shenzhen Huaqiang North Station to Huaxin Station, the physical gap on both sides is one-tenth that between the top and the bottom. And due to the existence of hydrostatic pressure and lateral earth pressure, the mud supporting role on both sides is greater than the top. Therefore, the ratio k of ΔA to ΔB should be less than 1.

In this paper, the influence of the physical gap upon the surface subsidence on both sides is studied by changing the ratio of ΔA to ΔB. Meanwhile, the size of pipe jacking and the amount of stratum loss are consistent with the previous example. The results are shown in Fig. 9. The smaller the proportion of stratum loss on both sides of the total, the deeper and the narrower the settlement trough is. What's more, the curve passes through (± 3.5, -16.6) regardless of how the k value changes. In other words, other things being equal, the settling curve always passes through the point where the settlement value is about half of the maximum settlement value.

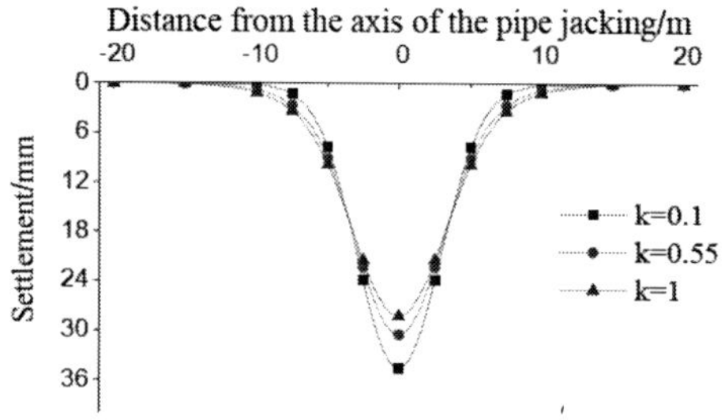

Figure 9 Influence of k value on surface subsidence

7 Example

A slurry pressure balanced rectangular pipe jacking machine is used for construction in the exit of a subway station. The length of the pipe is 43.5m, the slope is 0.5% and the average covering thickness is 5.2m.The external dimension of the pipe is 6.0m×4.2m.The surrounding soil is silt fine sand and silty clay, where the internal friction angle φ ranges from 10° to 16°.According to the literature and the data, we take tanβ as 0.59.By calculating the clearance ratio k = 0.1. In order to better compare with the measured values, suppose that the stratum loss rate ε is 0.77% in Peck formula, 0.62% in the stochastic medium method, and the convergence mode is assumed to be non-uniform.

As shown in Fig. 10, Peck formula and the stochastic medium method can achieve good results. However, the curve obtained by the Peck formula is relatively wide. While the curve obtained by the stochastic medium method is closer to the measured value whose shape is narrow and deep. It is more reasonable to use the stochastic medium method to calculate the settlement of the rectangular pipe jacking.

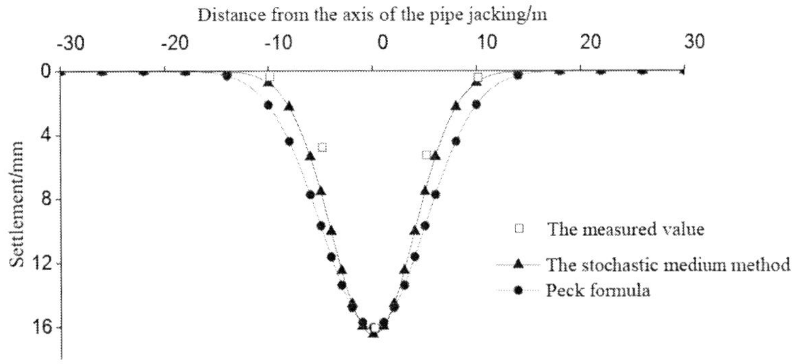

Figure 10 Comparison of the measured and calculated values

8 Conclusions

In this paper, the stochastic medium method is used to calculate the surface settlement caused by rectangular pipe jacking. Compared with the calculation results of Peck formula, the results show that:

(1) The maximum settlement of the uniform convergence mode is smaller than that of the non-uniform convergence mode. The settlement trough is wide and shallow under the uniform convergence mode, while narrow and deep under the non-uniform convergence mode. As the burial depth decreases, the difference between the two modes is increasing.

(2) The calculated results of Peck formula are between the results of the uniform and non-uniform convergence modes when using the stochastic medium method. With the increase of the buried depth, the results of the uniform model are consistent with those of Peck formula, while the results under the non-uniform convergence mode and those of Peck formula always have a certain difference.

(3) With the increase of the buried depth, the settlement trough of the circular and rectangular pipe jacking tends to be the same. The width of settlement trough increases linearly with increase of the aspect ratio.

(4) The increase in the clearance ratio of the settlement trough tends to be wide and shallow, but always through the fixed point.

(5) Compared with the measured results, the calculation results of the stochastic medium method are more reasonable.

References

1. ZHANG Yu-sheng, ZHOU Ming-yi. Jacking process of a pipe with large diameter through a long distance[J]. Chinese Journal of Geotechnical Engineering, 2000,04:506-508.
2. ZHU He-hua, WU Jiang-bin, PAN Tong-yan. Theoretical analysis of three-dimensional mechanical model of curved pipe jacking and its application[J]. Chinese Journal of Geotechnical Engineering, 2003,04:492-495.
3. WEI Gang, XU Ri-qing, HUANG Bin. Analysis of stability failure for pipeline during long distance pipe jacking[J]. Chinese Journal of Rock Mechanics and Engineering, 2005,08:1427-1432.
4. LV Jian-zhong, YANG Lei. Jacking construction method for soil pressure balance stylerectangle jacking-pipe (YJGF15-2000)[J]. Construction Technology, 2002,09:43-45.
5. LIN Qiang-qiang. Analysis and control of the surface deformation caused by rectangular pipe jacking construction[D].Tongji University,2008.
6. WANG Ri-dong. Analysis of additional load imposed on underground pipelines nearby induced by the construction of rectangular pipejacking[J].Low Temperature Architecture Technology,2014,09:124-127.
7. WANG Ri-dong. Research of computing method of soil deformation induced by the construction of rectangular pipejacking[J]. Journal of Jamusi University, 2014,05:711-714+722.
8. Dingwen Zhang, Bo Liu, Yujun Qin. Construction of a large-section long pedestrian underpass using pipe jacking in muddy silty clay: A case study[J]. Tunnelling and Underground Space Technology, 2016,60:151-164.
9. PECK R B. Deep excavations and tunneling in soft ground[C]. Proceedings of 7th International Conference on Soil Mechanics and Foundation Engineering, Mexico City: State of the Art Report, 1969: 225-290.
10. HAN Xuan, LI Ning. Comparative analysis of strata prediction models for ground movement induced by tunnel construction[J]. Journal of Rock Mechanics and Engineering, 2007,03:594-600.
11. Lee K M, Rowe R K, Lo K Y. Subsidence owing to tunneling I: Estimating the gap parameter[J].Canadian Geotechnical Journal,1992,29:929-940.

Monitoring calculation of closure change of Extradosed Cable-stayed Bridge

Jing Xian SHI[1], Zhi Hong RAN[2]

[1]Oxbridge College, Kunming University of Science and Technology,KunMing 650106 Yunnan, China;
[2]School of Architecture and Urban Planning, Yunnan University, Kunming 650091, Yunnan, China

Abstract: During the construction of extradosed cable-stayed bridge in Yunnan province, China, the construction unit has made certain changes in the construction process of the closure section due to environmental restrictions: remove the hanging basket after the closure, the sling shall not be provided in closure section, the function of the sling is realized by the hanging basket on the 16th beam. In case of this change, the bridge has been constructed to section 15th. In order to ensure the smooth and orderly progress of each stage in the closure phase, this article is arranged according to the construction plan, appropriate adjustment of related procedures, checking the bridge safety at all stages of construction, the stress and force of the main girder are compared to ensure the safety of the construction after closure changes. Adjust the height of the beam of the 16th and 17th to adapt the new construction plan, and the bridge closure smoothly.

1 Engineering Overview

The bridge analyzed in this paper is located in Yunnan province in China, main spans of extradosed cable-stayed bridge is 2*85m, bridge width is 27m; the grade of Urban trunk road is III, the motor vehicle is two-way and four-lane; the design speed is 30km/h; design load for highway is levelII, the throng is 3.5kN/m^2; design useful life of 100 years. The design elevation of the bridge is 454.589-451.389m.

The facade layout of the main bridge is shown in figure1, and the structure of the cable-stayed bridge is adopted. The main beam adopts single box three-compartment large cantilever variable cross-section PC continuous box beam (shown in figure2), the main pier is thin wall type pier, high 28m, and the deck thickness is 5m. The pier beam adopts the consolidation form. The bridge model material is summarized in table 1.

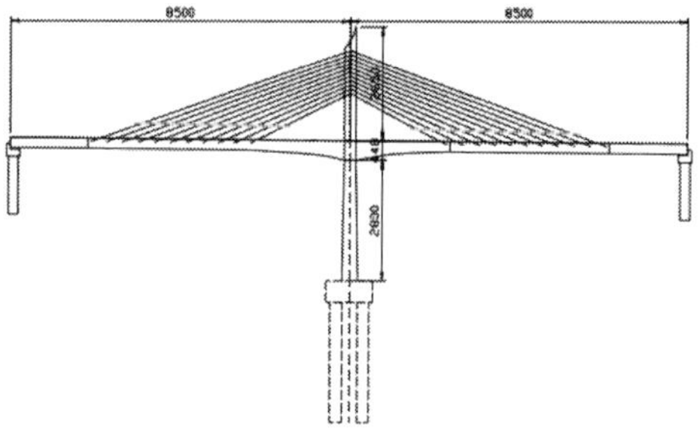

Figure 1 Main Bridge Elevation Layout/ cm

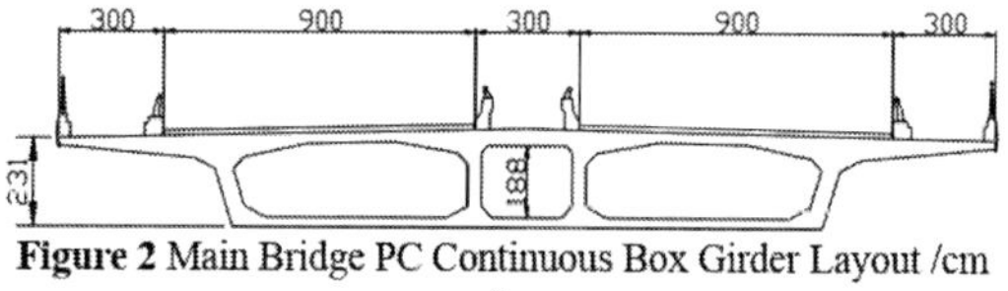

Figure 2 Main Bridge PC Continuous Box Girder Layout /cm

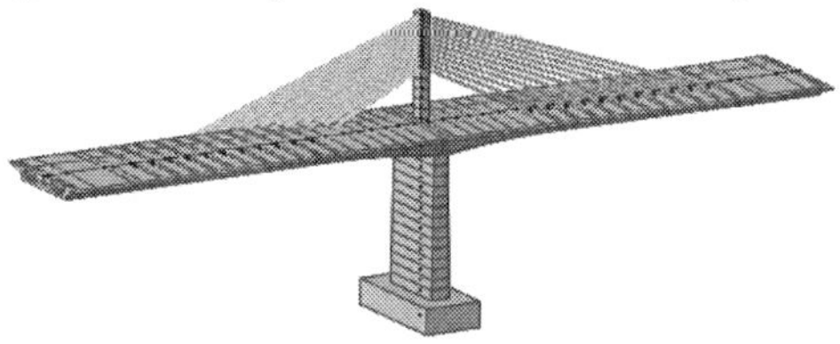

Figure 3 The overall calculation model of structure

Table 1 Material parameters Summary of calculation model

Material code	Unit type	Material type	Elastic Modulus (MPa)	Volumetric weight (kN/m3)	Coefficient of thermal expansion (1/°C)

* Corresponding author: sara_shivip@163.com; 1973941@qq.com

1	Bridge pier, main beam, main tower	Concrete50	3.45E+04	26	1.00E-05
2	Stay Cables	High strength steel wire	1.95E+05	78.5	1.20E-05
3	Prestressed	High strength steel wire	1.95E+05	78.5	1.20E-05

This monitoring calculation adopts the MIDAS Civil8.32 analysis program. The structure of the structure is analyzed by the structure of the space bar system. The structure consists of bridge pier, main beam, main tower and lasso. The structural analysis model is shown in figure 3. The total bridge has 146 units and 172 nodes. The boundary conditions in the analysis are: the bottom end consolidation of the cable tower, the transition pier and the auxiliary pier are simulated by the movable hinge support, and the lower end of the cast-in-place bracket is fixed hinge support.

2 Construction change requirement

According to the requirements of construction sequence diagram in the original bridge design document, the cantilever of closure strap and its formwork system are 32tons heavy, remove the hanging basket before closure stage, at the end of the closure, the weight is 58tons. But the construction unit proposed: remove the hanging basket after the closure, the sling shall not be provided in closure section, the function of the sling is realized by the hanging basket on the 16th beam, that meansthe former lifting point is located on the 17th beam segment and is 58cm from the starting point, to realize the function of the dragon sling (number of beam segment is shown in Figure 4). The concrete wet weight of all the hedrons is borne by the cantilever end, so the closure weight is increased from 58 to 105tons.

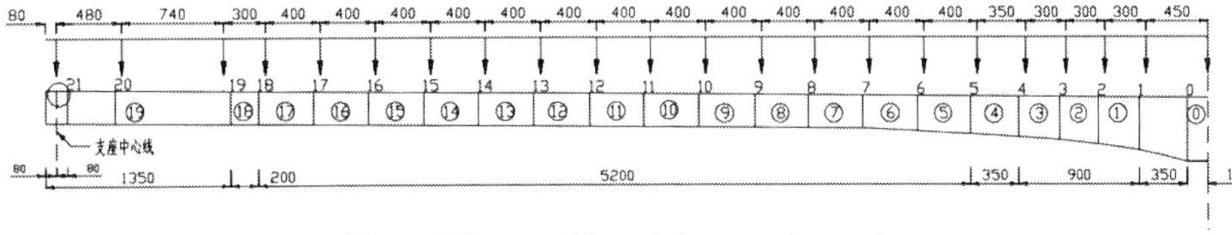

Figure 4 Diagram of the main beam section numbers

When this change is proposed, the bridge has been constructed to 15th beam segment, in order to ensure the successful completion of the sub stages in the closure section, the relevant procedures are adjusted properly in accordance with the construction plan. The hanging basket shall not be removed until it is closed and only moved forward; the weight of the cantilever end is 105tons; after repeated trial calculation, in order to make the bridge state as consistent as possible with the original design, it is necessary to adjust the single tension of 11th cable to 415tons, and make two adjustments after the completion of the bridge, reduce the single tension of 11th cable to 395tons. The closure conditions of the two models are shown in table 2.

Table 2 Comparison table of condition adjustment

Name of construction component	Description of the original model	Described conditions of the adjustment model
16th section of Beam	Tension cable-stayed	Cable Tension of 11th adjustment to 415t
17th section of Beam	Move forward Hanging basket、Pouring concrete、Tensioning prestress	
Side-span	With full framing construction	
Closure Section	Tear down hanging basket、Put 32t in final closure at both ends of the closure	Move forward Hanging basket (The front hanging point is located on the 17th beam section and from the starting point of 0.58m)
	Pressure 58t in the end of the closure	Pressure105t in the end of the closure
	Installation of closure and steel skeleton、Unloading and pouring concrete、Waiting for age then tensioning prestress	
	Tear down Hanger	Tear down hanging basket
Side-span	Tear down full framing	
	Stage II load	Stage II load, The second adjustment of 11th cable tension to 385t
	Into the bridge conditions	

 https://doi.org/10.1051/e3sconf/20183803025

3 Safety checking calculation in construction stage

The stress calculation results of the main girder in each construction stage are shown in figure5 and 6.It can be seen from the picture, the stress change is small compared with the original model and the adjusted model, but they are all within safe limits. The results of pier strength calculation are shown in table 3. As you can see from the table, the capacity of the maximum cantilever stage (cast concrete in the concrete) meets the requirements.

《Guidelines for Design of Highway Cable-Stayed Bridge》(JTG/T D65-1-2007) Specification 3.4.2 stipulate: Under construction the safety factor of the cable-stayed cable should greater than 2.0. The calculations results of the force of the bridge are shown in table 4, it can be seen from the table that all the forces of the cable meet the standard requirements.

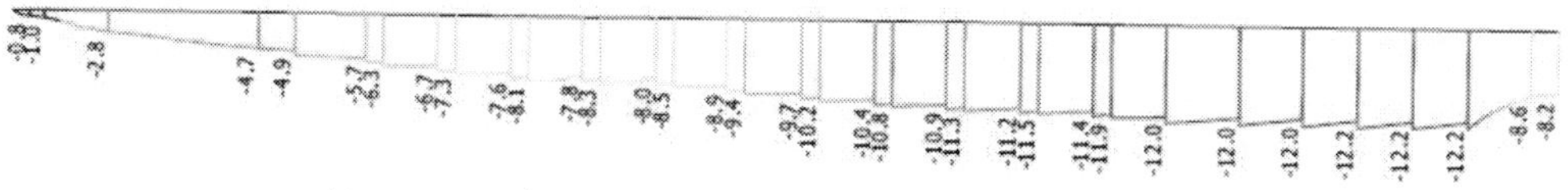

(a) Compressive stress envelope of the upper edge of the main beam

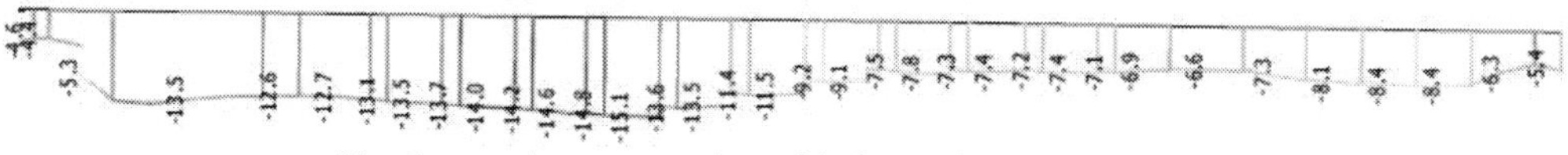

(b) Compressive stress envelope of the lower edge of the main beam

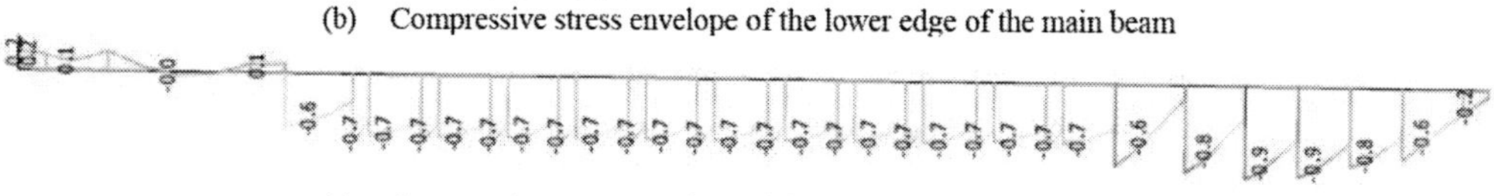

(c) Compressive stress envelope of the lower edge of the main beam

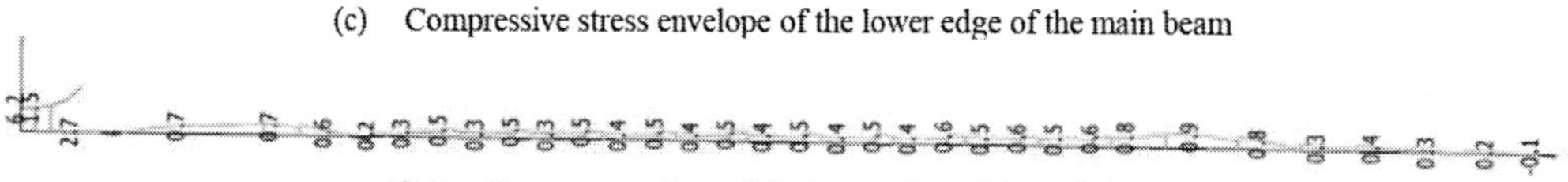

(d) Tensile stress envelope of the lower edge of the main beam

Figure 5 Calculation results of main beam stress in construction stage the original model / MPa

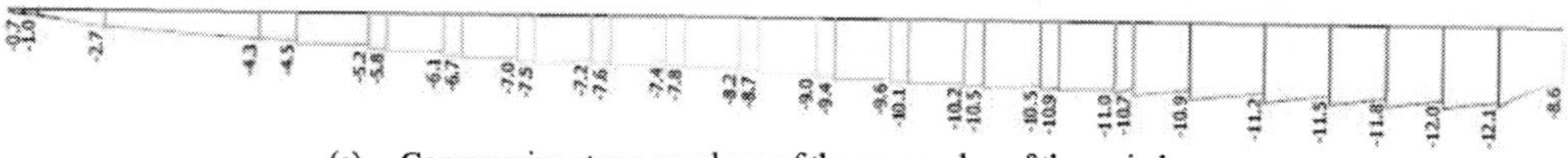

(a) Compressive stress envelope of the upper edge of the main beam

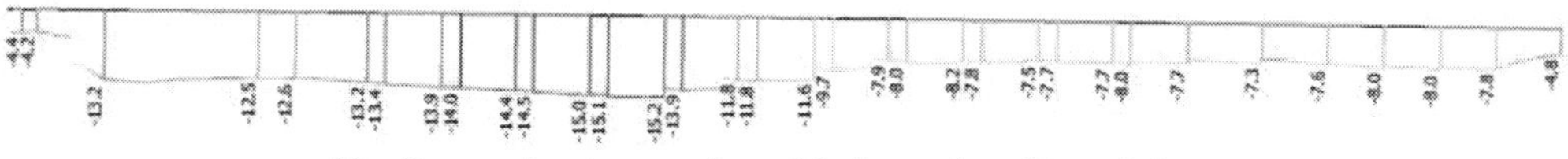

(b) Compressive stress envelope of the lower edge of the main beam

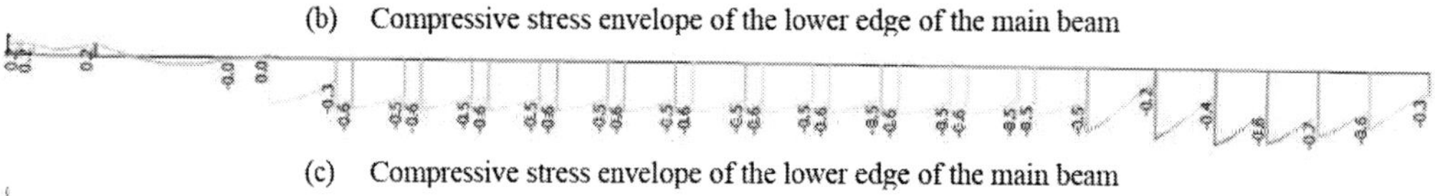

(c) Compressive stress envelope of the lower edge of the main beam

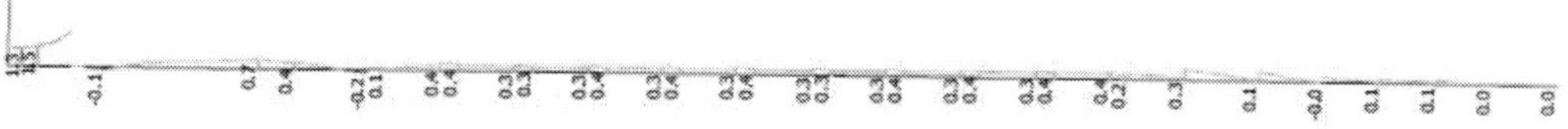

(d) Tensile stress envelope of the lower edge of the main beam

Figure 6 Calculation results of main beam stress in construction stage the adjusted model / MPa

Table 3 Calculation Results of Bridge pier and tower Strength in Construction Stage

	Load Conditions	Largest Cantilever Stage (Original model)	Largest Cantilever Stage (Adjusted model)
Pounding Bottom Section	Axial Force/kN	-130460	-132177
	Forward Bending Moment/kN.m	522	536
	Section Resistance/kN	1480000	1450000
	Safety coefficient	11.65	10.97

The Tower Bottom Section	Axial Force/kN	-55713	-57407
	Forward Bending Moment/kN.m	421	434
	Section Resistance/kN	143000	141000
	Safety coefficient	2.6	2.45

Table 4 The maximum cable force calculation results in the construction stage

Serial number	The original model/kN			The adjusted model/kN			Safe coefficient	
	Main beam side	Bridge tower side	Average value	Main beam side	Bridge tower side	Average value	The original model	The adjusted model
1	3266	3270	3268	3266	3270	3268	2.47	2.47
2	3283	3287	3285	3283	3287	3285	2.46	2.46
3	3317	3321	3319	3317	3321	3319	2.43	2.43
4	3614	3620	3617	3614	3620	3617	2.45	2.45
5	3648	3654	3651	3648	3654	3651	2.43	2.43
6	3680	3686	3683	3680	3686	3683	2.4	2.4
7	3980	3987	3984	3980	3987	3984	2.42	2.42
8	4006	4013	4010	4006	4013	4010	2.4	2.4
9	4028	4036	4032	4028	4036	4032	2.39	2.39
10	4047	4055	4051	4047	4055	4051	2.38	2.38
11	4062	4071	4067	4720	4729	4724	2.37	2.04

4 The comparison of stress of the main beam and the cable force

The upper margin and lower edge stress of the original model and the adjusted model are shown in figure 7 and 8. It can be seen from the picture, the maximum value of upper margin compressive stress is 0.3 MPa, the lower margin stress difference is 0.8 MPa. The lower edge of the bridge produces tensile stress due to the automobile live load, therefore; the 0.8 MPa compressive stress of the modified model is beneficial to the structure. Table5 is a comparison of the calculation results for the bridge stage. The maximum difference was 2.4% after adjustment of cable force. Table 8.10.4-1in Terms 8.10.4 of 《Standard for quality inspection and assessment of highway engineering》 (JTG F80/1-2004) stipulate: The measured force and the design of the force extreme error are 10%, therefore, the original model and adjusted model of the force difference of 2.4% can not affect the load-carrying capacity of the bridge.

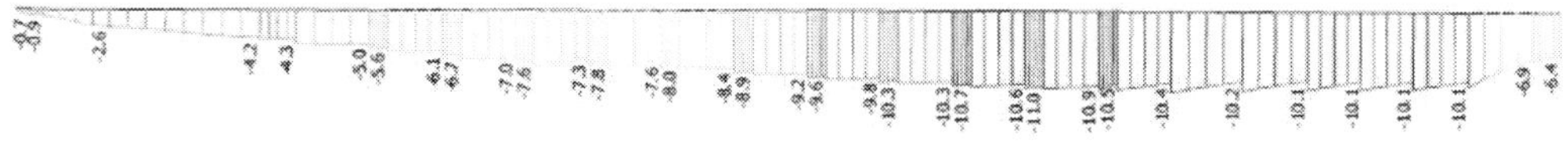

(a) The stress value of the upper edge

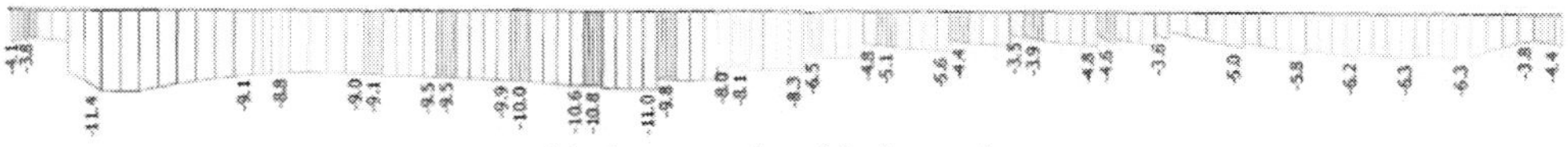

(b) The stress value of the lower edge

Figure 7 The stress map of the original model when end of construction /MPa

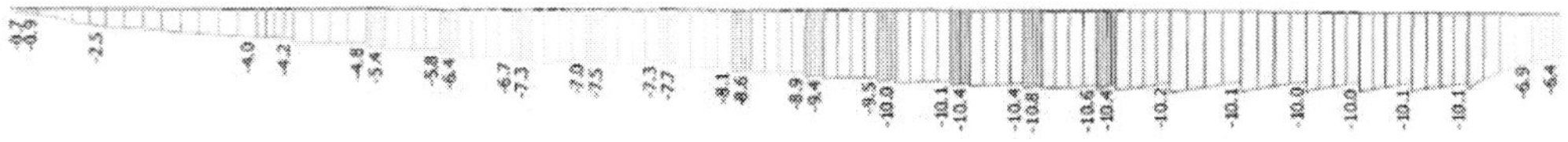

(a) The stress value of the upper edge

(b) The stress value of the lower edge

Figure 8 The stress map of the adjusted model when end of construction /MPa

Table 5 The calculation results of cable force in the end of the construction stage

Serial	Main	Bridge	Average	Main	Bridge	Average	The difference of cable	Relative

number	beam side	tower side	value	beam side	tower side	value	force between original and adjustment model/kN	difference (%)
1	2624	2628	2626	2637	2642	2640	14	0.5
2	2658	2663	2660	2677	2682	2680	19	0.7
3	2699	2704	2702	2725	2729	2727	25	0.9
4	2965	2971	2968	3000	3005	3003	35	1.2
5	3028	3034	3031	3070	3076	3073	42	1.4
6	3092	3099	3095	3142	3148	3145	49	1.6
7	3415	3422	3418	3477	3484	3481	62	1.8
8	3522	3530	3526	3593	3601	3597	71	2
9	3631	3639	3635	3710	3718	3714	80	2.2
10	3735	3743	3739	3823	3831	3827	88	2.4
11	3834	3843	3838	3915	3924	3920	81	2.1

5 Adjustments Calculate of shuttering elevation

Due to the change of construction sequence and plan, the height of the bridge cannot reach the design value under the completed shuttering elevation. Detailed calculations revealed that the maximum difference between the predicted primary bridge and the original design is 17.5mm. Therefore, the beam 16th and 17th sections should be adjusted, and found by the adjusted model, the folding angle of the 17.5mm is generated on the beam 15th-16th section. To mitigate this situation, the following are dealt with:

1) In order to closure, the 17th beam is not changed;

2) Raise the shuttering elevation of the 16th beam segment to 8mm.

At the time of closure,altitude difference is 15mm between west of final closure and side span,altitude difference is 16mm in East side. However, based on 0.78% design data, the theoretical height difference should be 15.6mm,then successfully complete the bridge construction.The table 6 can be show that the calculation of completed bridge after the adjustment of elevation of the formwork.

Table 6 Calculation result of completed bridge Linetype

		Deflection at the stage of pouring		Difference analysis with design				
Beam section number	Shuttering elevation/m	Placing concrete to make the front of the casting beam downwarping/ mm	The hanging basket produces a downwarping/ mm	Finished bridge elevation/ m	Design elevation/ m	Altitude difference between Design elevation and bottom of beam	The difference between the predicted primary bridge and the design elevation	Elevation at closure/ m
West 17th	451.893	-103	-30	451.735	454.425	2.69	0	451.715
West 16th	451.824	-93	-30	451.709	454.399	2.69	0	451.688
East 16th	450.802	-93	-30	450.687	453.377	2.69	0	450.666
East 17th	450.809	-103	-30	450.651	453.341	2.69	0	450.631

Note: The design elevation is the elevation value of the construction drawing + aftershrinkage creep+ Live load displacement /2

6 Conclusion

Through the calculation of the stress of the main girder and the cable force at each stage of construction and the calculation results of the elevation of the vertical mold, it can be seen that the bridge is in a safe state after the adjustment of the formwork.

1) The stress of the adjusted formwork is within the safe range, and the bearing capacity meets the requirement during the maximum cantilever stage, and the bridge is in a safe state.

2) The calculation result of cable force satisfies the requirement that the safety factor of stay cable is more than 2.

3) After adjustment, there is a small gap between the cable forces, and the maximum gap is 88ton in cable 10th. Compared with the original model, the difference of 2.4% does not affect the carrying capacity of the bridge.

4) After adjusting the elevation of the formwork of the beam section, the finished bridge shall conform to the design line according to the changed working procedure.

References

1. JTG/T D65-1-2007.Guidelines for Design of Highway Cable-Stayed Bridge [S].
2. JTG F80/1-2004.Standard for quality inspection and assessment of highway engineering[S].
3. Shi Jun,Zhao Yue. Closure Technology of Huiqing Yellow River Bridge[J]. Construction Technology.2011(05):44-48.
4. WEI Zuo-ming,MENG Fang-cheng,CHEN Wen. Linear Sensitivity Analysis at Construction Phase of Extradossed Cab1e-stayed Bridge[J]. Western China communication Science and Technology.Bridge and Tunnel engineering.2014(06):65-68.
5. ZHAO Fu-xing. The analysis and research on construction control of low-pylon cable-stayed bridge[J]. HeiLongJiang jiaotong keji.2016(02):80-81.
6. Sun Fuyang. Construction Monitoring of Asymmetric Extradosed Railway Cable-stayed Bridge Tower Beam [J].Railway Construction Technology 2016 (04):15-19.
7. Ming Yuan,Dong Huang Yan. Optimization on Closure Scheme of Multi-Span Prestressed Concrete Box-Girder Bridge[J]. Advanced Materials Research,2011,1067(163).
8. Zhan Yong Yao,Xiao Meng Zhang,Shuai Li,Lian Fu Li,Xiang Hong Pan,Kai Yao. Influence of Closure Sequences on the Rigid Frame-Continuous Girder Bridge Closure[J]. Applied Mechanics and Materials,2013,2547(361).
9. Jun Dong,Tian Liang,Jiang Xiao Feng. Investigation on Optimization of Cable Tension for Extradosed Cable-Stayed Bridge under the Rational Completion Stage[J]. Advanced Materials Research,2014,3456(1020).
10. Shao-min Jia. Geometric Closure Control of the Main Channel Girder of Jiaxing-Shaoxing Bridge[J]. Journal of Highway and Transportation Research and Development (English Edition),2015,9(1).

Using BIM Technology to Optimize the Traditional Interior Design Work Mode

Ning Ke ZHU

Beijing University of Civil Engineering and Architecture, Beijing, 100044, China

Abstract: the development of BIM technology and application in the field of architecture design has produced results, but BIM technology and application in the field of interior design is still immaturity because of construction and decoration engineering separation. The article analyzes the problems that BIM technology lead to the interior design work mode optimization, from the 3D visualization work environment, real-time collaborative design mode, physical analysis design mode, information integration design mode state the application in interior design.

1. Introduction

The Building Information Modeling technique first appeared in the United States. The concept of BIM undergo nearly 30 years. in 1975, Dr Charlie Eastman , Carnegie Mellon University initiated the "Building the Description System" (architecture Description System) working prototype, in 1984, the Hungarian Graphisoft company researched and developed architectural design software named AtchiCAD, in 1986, Robert Ash in GMW company proposed "Building Modeling" conception, the definition of Building Information Model has been updated and improved. In 2002, With the aid of continuous application of computer aided design in the construction industry and the continuous exploration in architecture engineer of the software development company, Autodesk company puts forward the concept of BIM (Building Information Modeling).

BIM technology is based on 3D digital technology, the engineering data model integrate of all kinds of information related to construction engineering project, it provided information for the design and construction with mutual coordination, internally consistent , operation. People can use BIM technology at architecture design, structure design, MEP design, construction engineer and operations management and associate them, it make a designer to focus on the design better, free designer from tedious drawing works, it help budget company work more accurate, it help construction organization work more precise and greatly improve the project progress, it help the operation management people manage and maintain the whole life cycle of construction more convenient and accurate.

In the design stage, The application of BIM technology is aimed at architectural design mainly. Several major design software, such as Autodesk Revit, Bentley, ArchiCAD, etc, have a lot of information building block libraries which is easy to adjust architecture building information modelling by means of parametric design model, and they have effectively integrated many block libraries, such as structure, water supply and drainage, electrical, HVAC and other profession, it is convenient to carry out various profession collaborative design.

In 2003, SOM applied BIM technology to the building design of the world trade center in New York City, which had a great success. Countries such as the UK, Singapore and South Korea have also applied the technology to the design and construction of buildings in large Numbers. Countries have also compiled and published BIM technical standards. In recent years, many large design units in our country have also used BIM technology in architectural design, such as table 1, and have written relevant standards.

Table 1. projects analysis using BIM technology

Project	Description	Design Company
Shanghai Tower	Super high architecture design, curved surfce design, various profession coordination	Gensler (USA), Tongji Architecture Design (Group) Co.,Ltd
Tianjin 117 Building	The structure of the building is 596.5 meters high, realizing the integration of professional model information and business information, saving cost,	The P&T Group, ARUP, Shanghai Xian Dai Architecture Design (Group) Co., Ltd

* Corresponding author:zhuningke@bucea.edu.cn

	improving management and standard construction.	
Zhuhai Grand Theater	BIM technology is widely used in the design of thin-wall curved surface and theater interior design	China Reconstruct Design

Traditional Interior Design Mode

Interior design is the continuation and deepening of architecture design, which is the re-creation of interior space and environmental design. Interior design aims for the people who live in or use the space. According to the natural functional quality of the building, environment and the corresponding standards, using the material technical means and building design principle, create reasonable function, comfortable indoor environment and satisfy people's material and spiritual need. This space environment has both the use value to satisfy functional requirements and reflects the spiritual factors such as historical context, architectural style and environmental atmosphere.

2. Question

BIM technology has been used more and more in architecture design stage, but the study of using BIM technology in interior design has seldom been seen although the interior design is a important part of architecture design. The domestic BIM technology project also seldom has the case of combining construction engineering and decoration engineering completely. The reasons are as follows.

(1) Construction engineering and decoration engineering are separated. Usually the developer is responsible for the construction work, the work contents generally include the building main body and the exterior decoration, the internal decoration is usually very simple and isn't a detail project. As the user of the building is often uncertain when construction is completed, it is impossible to determine the internal function of the building and its decorative contents.

(2) While in construction projects, the BIM technology has been applied from time to tome, but in the next phase of the project, both the project owner and designer are not very interest in using BIM technology in indoor decoration and haven't strict requirements on that. Interior design units usually use traditional design patterns which are more comfortable and familiar. It also eliminates the manpower and time to learn and promote BIM technology.

(3) Interior design also involves multiple professional coordination problems, decoration, water supply and drainage, electrical, HVAC and even construction, etc. In the design and construction phrase, decoration design, construction engineering, electrical and other professions cross-work, the branch itemized project is more, the process is complex, it often lead to the three-side project because of catching time.

3. Optimization of interior design work mode brought by BIM technology

3.1 D visualization working environment

Visualization is a necessary result of interior design work mode, converting interior design into visual graphics can facilitate design project better and also conduce to better communication between people. In the design process, sketch, three views, and 3D model are all commonly used visualizations, each with its characteristics, and table 2.

Table 2. visualizations of sketch, three views, 3D model

Sketche	Three views (Plane, elevation, section)	3D model
	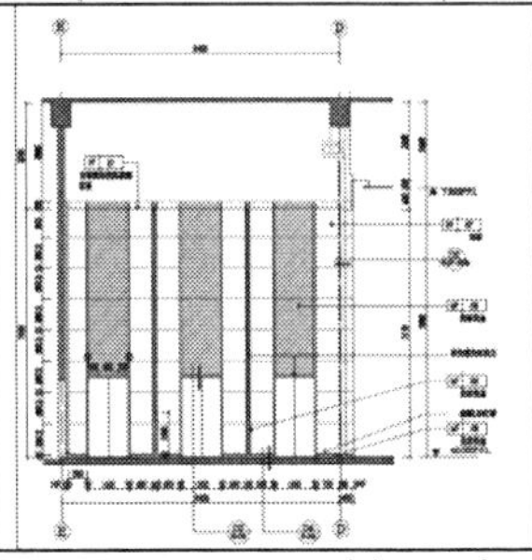	

(1) Schetch

Advantages: the fastest and most intuitive way to record designers' inspiration , strong and expressive.

Disadvantages: it is difficult to be very accurate, the perspective is more random, no dimension mark, no accurate data, it is difficult to achieve the lighting, material and other effects. It is hard to achieve what you see is what you get.

(2) Three views

The three views is the most widely used visual design model, which has been used since ancient times. Three views separate a three dimensional space into different angles of view to express such as plane, elevation, section, the viewers use comprehensive analysis ability of the brain to combine different drawings and get a full recognition about the space .The advantage of the three

 https://doi.org/10.1051/e3sconf/20183803026

views is to display the spatial shape on the two-dimensional plane, which is easy to draw and label the dimensions, and the dimensions are accurate, it also can annotate materials and other relevant information. The disadvantage is that both the drawing staff and viewers usually need professional training. The drawing person need to divide the three dimensional space in his mind into three views to draw separately, so the plane, elevation and sections will not matched often. It is necessary for viewers to combine the three kinds of drawings into a three-dimensional space to understand, it has some difficulty on spatial imagination. Due to the increasingly complex space in modern engineering, it is also inevitable to make mistakes whether in the process of segregation or the process of integration,.

(3) 3D Model

The 3D model can be divided into solid 3D model and the virtual 3D model of computer simulation. Solid 3D model is usually made in a reduced scale, so that the observer can have a bird's eye view, and the effect is real and intuitive. In interior design process, solid 3D models are used less because the its scale is small, the observer can't look at inside and also can't experience indoor space feeling and effect.

With the development of computer technology, virtual 3D models are increasingly applied to interior design. 3ds Max, Sketchup, Autodesk Revit and other softwares can accurately model and render, so that the designer and the viewer can visually observe the interior space, materials and details from different angles.

BIM technology is more closely designed with "real" 3D visualization mode. The designer directly constructs the model in 3D space, and the model has real space information, construction information, material information and cost information. Designers not only can observe the indoor space from any Angle, but also can simulate sunlight and material effects at any time, realizing real-time design, real-time modification and real-time vision.

3.2 Real-time Collaborative Design Mode

BIM model is an open, Shared, update design work platform that provides a platform for real-time collaborative design. According to the characteristics of the interior design, interior design collaborative process which based on building information model can be divided into: internal work in interior design, external work in different profession, interior design and related products. All three collaborative work is based on the web, but they are in different ways. Internal collaborative work mainly based on working set, external collaborative work mainly based on links, interior design and related product design is accomplished by information model data library.

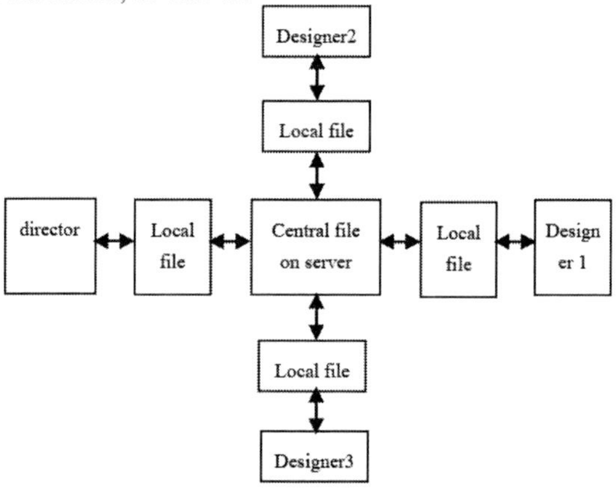

Figure 1. Building Information Model Set

The work set can divide the permissions between team members and save the information model to local computers and servers to realize the relative independent work of team members on the same model. As shown in figure 1, both director and ordinary designers, their design information are passed through the central file which ensures that all people are working on the same working model, and "graph combining" between each member is highly automated.

The link can be used to combine interior design model with other equipment models (water supply and drainage, electrical, HVAC) into the same information model. Various professional designers can be easily read other professional design file and check the collision conflict problems, in this stage, interior designer is usually responsible for the whole coordination, form the final results.

The information resource library provided by the indoor building information model connects the up industry and down industry, including interior designers, product designers, manufacturers and so on. The information library can include appearance, material, price, service life, etc.

3.3 Physical Analysis Design Mode

In past, physical analyses of Interior space environmental in the interior design are often ignored, designers usually just do it according to the experiences. for example, about intensity of illumination of indoor lighting, designer usually set indoor lights and supply power usually according to the illuminance standard in different space which caused a lot of energy waste, and didn't meet the need of highlighting key space and key lighting. In recent years, with the gradual deterioration of the ecological environment, indoor decoration has caused a quality of kinds of diseases and more and more people has been caring about it. Aiming at this kind of phenomenon, the new revision of 《The Green Building Evaluation Standard》 (GB/T 50378-2014) was implemented on January 1, 2015, in《The Green Building Evaluation Standard》, indoor environment quality has a separate section, it give specific requirements on acoustic environment, light environment, thermal environment and air quality in interior space. Obviously, it is difficult to achieve these requirements by rough experiences, it requires the application of a professional computer-aided design tool. BIM technology is an effective means to realize simulation and verification which provides a platform to cooperate with related professional software.

In light environment simulation, the domestic main lighting calculation software are DIALux, AGi32, etc, the BIM model can be imported to lighting design software, after illumination calculation and setting lights, it can quickly get illumination calculation results of working face and show it in the graphic. As figure 2.

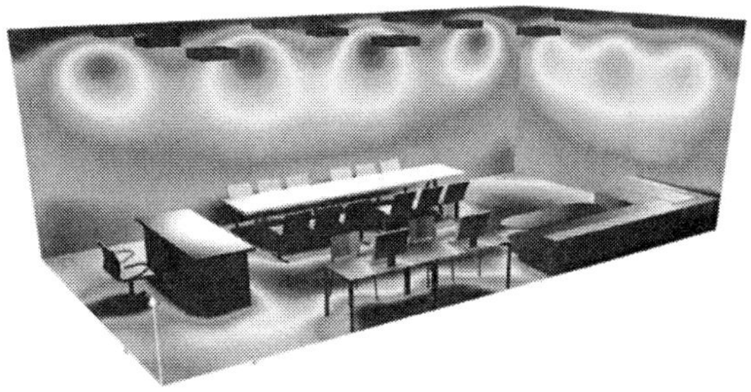

Figuire 2. Illumination calculation in meeting room

In acoustic environment simulation, acoustic calculation (mainly reverberation time) mainly depends on the type of decoration materials, sound absorption coefficient and area, which requires a large amount of data calculation. BIM model can be convenient to count data of decorate material and output forms. Base on this form, by writing acoustic formula in Excel spreadsheet, the calculation and adjustment of reverberation time in interior space becomes extremely fast.

In thermal environment simulation, the BIM model can accurately simulate the location of the building and simulate the irradiation of natural light. The model can also be imported into the relevant software to simulate air flow and calculate it, which can achieve the comfort of the space environment and reduce the energy consumption.

Indoor air quality optimization refers to natural ventilation and controlling of harmful substances in decoration materials, especially formaldehyde, benzene, ammonia, radon, etc. While establishing BIM model, it builds the decoration material data library which records the relevant physical data of the material. using the statistical function, BIM software can calculate the content of the harmful substances such as VOC in each room. Using BIM technology, the air quality control can be advanced to the design construction stage, instead of putting activated carbon or green planting into space when the decoration project has been finished.

3.4 Information integration design model

The building information model, by using BIM technology, is definitely not a simple geometric solid, but is a set of components with data attributes. On the platform of building information model, all kinds of information rely on each other. Interior designers constantly optimize their design project by this model, and pass the information model to the entire building life cycle, which can realize to reuse the information model in different professions and different stage, and constantly improve it.

In the BIM model, the model is not a simple geometry solid, but many components with attributes and constraints. As shown in figure 3, the door has its own attribute definition, elevation, etc. The width, height and position of the gate could be adjusted at any time in the design. The wall attached to it will be adjusted automatically, as well as the plan and other views. In the same way, the wall in the BIM model is divided into structure, substrate and surface layer, and the thickness of each layer can be defined. The building information model established by this method is not only an information model for interior designers, but it can be transmitted and Shared among different professions. By putting the model into the building physical environment analysis software, the indoor physical environment of the space can be analyzed. The designer can optimize the project further base on the acoustic, light, heat analysis results. The collision error can be checked through the comprehensive model of water, heat, electrical engineering. This model is used for view switching, and to generate 2D drawings, 3D models, animation, etc.

In the whole life cycle of Building information model, different participants, such as the owner, design units, government departments, supervision unit and construction unit, material suppliers, operations management, etc. will work on it. Various departments communicate information through the same building information model which can greatly save energy, reduce waste and improve efficiency.

4 Conclusion

If BIM technology has been used effectively in architectural design, the design institute and design company that applied BIM technology for interior design is still in the beginning. The visual design model of BIM technology is more comprehensive, more straight and more efficient, which provides new design thinking and design methods for interior designers. As an important part in the whole life cycle, interior design can't avoid BIM reform. Fortunately, BIM technology is becoming more and more humanized, and the technical threshold is getting lower and lower. But at the same time, the design institute and design company should adjust the business model and train a group of designers who master BIM technology and promote BIM technology in interior design in the whole industry chain.

References

1. Luo Heping. The Application of Bim Technology In Interior Design Field [J]. Shanxi Architecture, 2016 (4): 257-258.
2. Luo Lan, Zhong Fan. Application Research of Bim Technology Based On Sketchup [J]. Journal of Information Technology in Civil Engineering and Architecture, 2015 (4): 37-42.
3. Luo Lan, Wang Fang. A Project Decoration Project Is Based On Revit BIM Technology Application Research [J]. Journal of Information Technology in Civil Engineering and Architecture, 2014 (8): 55-61.
4. Xiong Peng. Interior Design Based On Building Information Model (BIM) [J].Popular Science, 2016 (10): 150-151.
5. Liu Liejun. Interior Design In Information Age - Building Information Model And Its Application In Interior Design [J]. Art & Design. 2010. 2:125-126.
6. Jia Shuang, Shen Xiaowei, Zhang Baogang. Application Analysis Of Green Building Design Based On BIM Technology [J]. Shanxi Architecture, 2016 (4): 209-210.
7. Asojo A O. An instructional design for Building Informa-tion Modeling (BIM) and Revit in interior design curriculum[J].Art, Design & Communication in Higher Education.2013, 11(2) :p143-154.

Influence of Soil Structure on the Mechanical Response of Soft Soil

Bin Bin XU

Tianjin Port Engineering Institute, Key Lab. of Geotechnical Engineering of Tianjin, Key Lab. of Geotechnical Engineering, Ministry of Communication, Tianjin, 300222, China

Abstract: Usually the natural sedimentary soils possess structure more or less, which makes their mechanical response much different from the fully remolded soils. In this paper, the influence of soil structure on the mechanical response such as compressibility, shear, permeability is literately reviewed. It is found that the compressibility and consolidation behavior of structured and remolded soils can be divided clearly before or after the structural yield stress. The stress-strain relationship can be divided into two segments before and after the structural yield stress. Before the yield stress, the curve is elevating and after the yield stress the curve is decreasing. The increasing rate of pore water pressure increases after the soil reached yield stress.

1 Background

As the development of the economic, many large engineering such as port, breakwater, storage tank, airport, railway and express highway have been built on the natural structured ground. If there is no comprehensive and deep acquaintance to the natural sedimentary clay, it is impossible to forecast the mechanical response. For example, the Kansai international airport is built by the land reclamation in the sea to form an artificial island. There has been over 12m settlement since its operation, which gives significant influence on the following maintenance. Now in the Hongkong-Zhuhai-Macau bridge engineering, there are also two artificial islands in the sea to connect the bridge passage and the immersed tube tunnel, its post-construction settlement is bound to be very important for the engineers.

As the development of boring technology, the recognition to the natural structured soil is more and more thoroughly. Due to the structure, the mechanical response of natural soil is quite different from the corresponding fully remolded soil. For the geo-materials, the structure displays as the shape of soil particles and void ratio, the pattern of arrangement, the mechanical action among particles. When the structured soil is loaded, its structure varies accordingly, which results in the corresponding variation of physical and mechanical properties of soft soil. From the aspects of practical engineering, there may be two problems. If we ignore the structure of natural soil and only predict its response based on the fully remolded samples or disturbed samples, the stiffness and strength of natural soils may be underestimated so that the design may be relatively safe. On the other side, if we only recognize the low compressibility of natural soil under low stress level but neglect the significant compressibility when the stress is over the yield stress, the stiffness of natural soil would be overestimated, which may threaten the safety of the construction. In this paper, the influence of soil structure on the mechanical response of ground is reviewed in detail to help the engineer to comprehend its importance.

2 Conception of Soil Structure

Leroueil et al. [1] pointed that the structure of soil is the expression of sedimentary, load experience. For this view, any soil at any state should be more or less structured. Also he proposed four different state of soil corresponding to the structures: undisturbed state, destructured state, remolded state, resedimented state. The undisturbed state is the natural sedimentary state which is a complex proceeding. The destructured state corresponds to the variation of structure due to volumetrical or shearing deformation. The remolded state means that the soil is loaded enough so that the structure is fully destroyed. The resedimented state is the slurry state of remolded soil and it is depended on the mineral constituent, particle size, sedimentary rate and other environment.

The remolded state can reflect the mineral constituent and sedimentary environment of natural soil. The variation from the undisturbed state to the destructured state until the remolded state represents the gradual loss of structure. The remolded soil represents the state that the combination among the soil particles is destroyed and the large void is eliminated, where the soil is an even configuration at the macroscopic view. Therefore, the mechanical properties of remolded soil are the most

* Corresponding author: xubinbin@tpei.com.cn

essential characteristics. Leroueil and Vaughan [2] defined the concept of structured permitted space, as shown in Figure 1, to demonstrate the relationship of three states. The influence of soil structure has been recognized in the early stage. Leroueil et al. [2, 3] proposed the concept of structure by comparing the natural soil and soft rock. The mechanical response of nature soil and remolded soil can be distinguished when considering the structure. Therefore, the structure of soil is as important as the initial void ratio, stress history to determine the mechanical property.

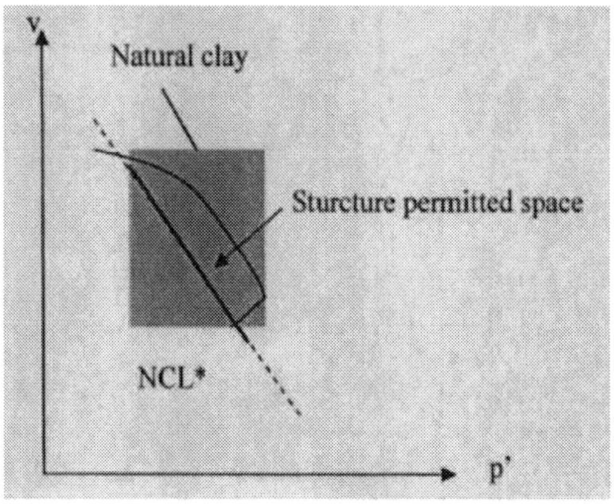

Figure 1 Sketch of structured permitted space of nature clay

3 Influence on Mechanical Property

3.1 Influence on Compressibility

The compressibility of soil is always the highlight in the research and it is the foundation for solving many geotechnical engineering problems, especially for the settlement. Mesri [4], Zhang [5], Zhou [6], Wang [7] and other scholars have researched the structure of soil in different areas including Mexico, Zhanjiang, Hangzhou, Wenzhou etc. The research results show that the compression curve of the structured soil is quite different from that of the remolded soil. Because of the soil structure, there is obvious turning point in the compression curve of the undisturbed soil. The stress corresponding to the turning point on the compression curve is usually called structural yield stress or apparent pre-consolidation pressure, which is usually larger than the overburden effective stress. When the current stress is smaller than the structural yield stress, the compressibility of the structured soil is relatively low. If the current stress is larger than the structural yield stress, the compressibility is much high. Finally, the compression curve of the structured soil is gradually close to the compression curve of the remolded soil as the stress increases. Usually the void ratio of the structure soil is larger than that of the remolded soil for the same stress level.

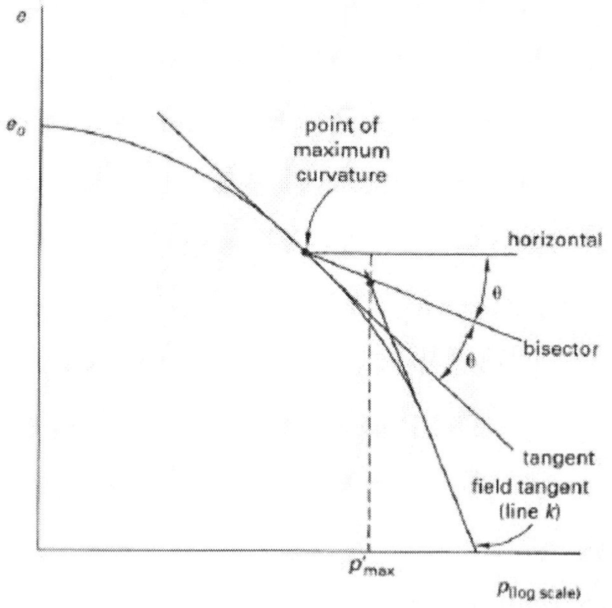

Figure 2 Casagrande's method to determine pre-consolidation pressure

Usually the response of undisturbed soil is used to represent the response of the in-situ natural soil considering that the in-situ soil specimen is different to obtain. However, the undisturbed soil in the laboratory is more or less disturbed such as the unloading influence of taking specimen, the mechanical disturbance of core sampler or the disturbance during the sampling, which may make the compression curve of undisturbed specimen quite different from the in-situ compression curve. Many scholars have researched the method to predict the in-situ compression curve according to the compression curve of the undisturbed specimen. The most common method to determine the pre-consolidation pressure is the Casagrande's method [8], as shown in figure 2. Figures 3 and 4 give Schmertmann and Nagaraj methods [9-10] to determine the in-situ compression curve based on the undisturbed compression curve. Detailed explanation and operation can be found in relative references.

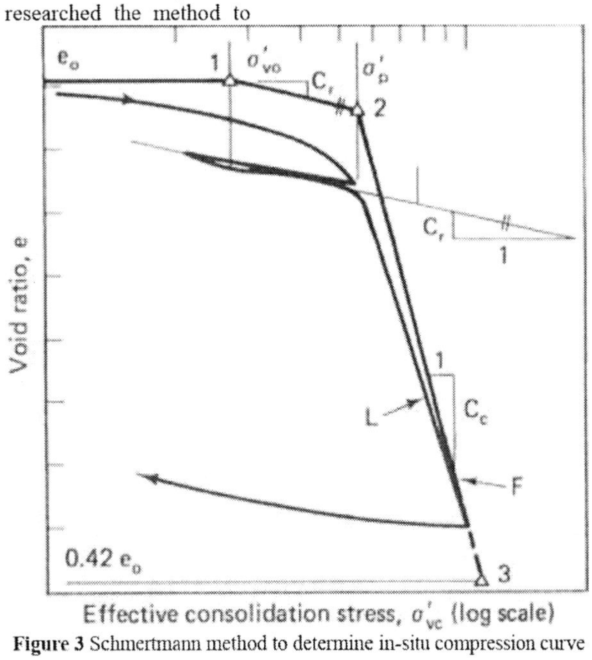

Figure 3 Schmertmann method to determine in-situ compression curve

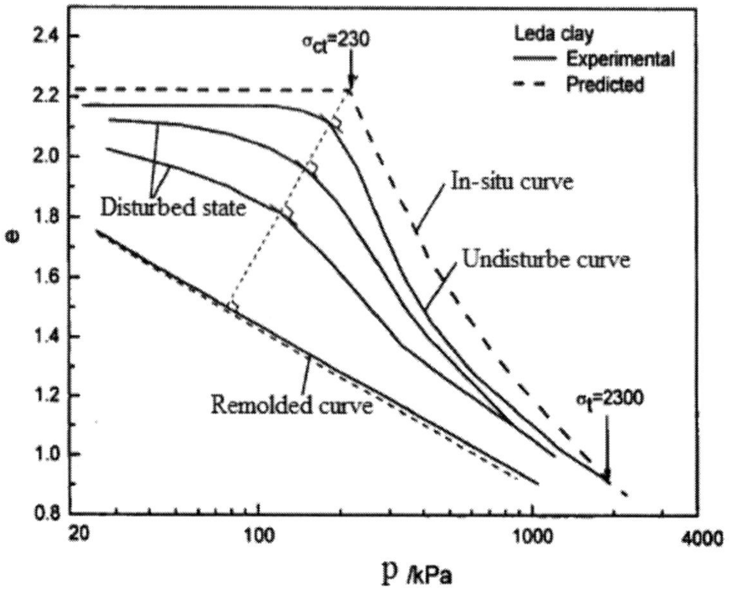

Figure 4 Nagaraj method to determine in-situ compression curve

3.2 Influence on the Compression Index and Secondary Consolidation Coefficient

Shen [11] believed that the soil structure would influence the consolidation coefficient. When the current stress is lower than the structural yield stress the consolidation coefficient of structured soil is larger than that of the remolded soil. When the current stress is larger than the structural yield stress, the consolidation coefficient becomes close, as shown in figures 5 and 6.

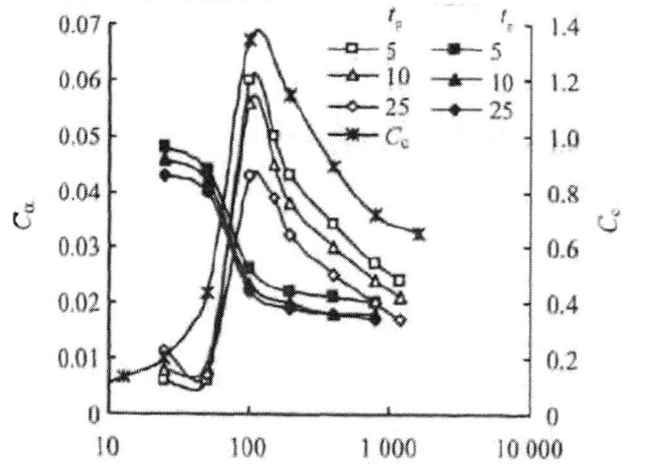

Figure 5 Relationship between compression index, consolidation coefficient and stress in Tianjin

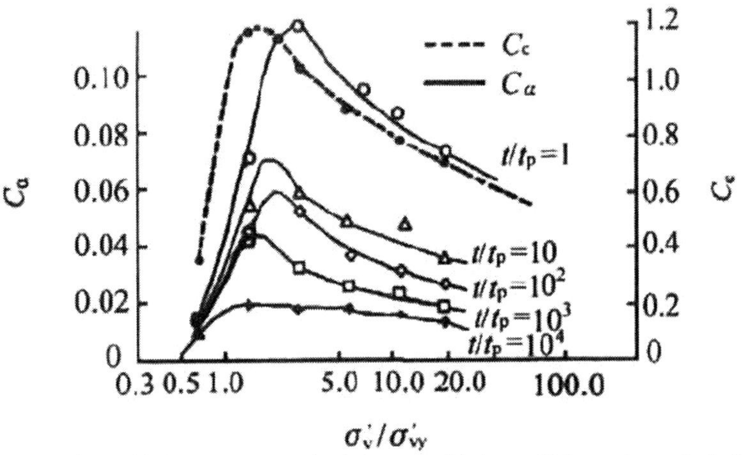

Figure 6 Relationship between compression index, consolidation coefficient and stress in Osaka

3.3 Influence on Stress-strain Relationship

In the triaxial consolidation drained compression test, for the structured natural clay the stress-strain relationship shows strain softening behavior when the consolidation stress is lower than the structural yield stress. When the consolidation stress is larger than the structural yield stress the stress-strain curve presents strain hardening behavior. The triaxial undrained compression tests of typical structural clay also show that the stress-strain relationship can be divided into two segments before and after the structural yield stress. Before the yield stress, the curve is elevating and after the yield stress the curve is decreasing. If the structured soil is disturbed, the structural strength decreases and peak strength also decreases, which means that the existence of structure can enlarge the stiffness of the soil. In the initial stage, the stress-strain relationship is under elastic state and then the percentage of plastic increases. When it reaches the peak strength, the soil structure is destroyed significantly so that the deformation increases greatly.

3.4 Influence on Pore Pressure

Li [12] found that in the undrained shear test the relationship between pore water pressure and structural yield stress differs with the variation of consolidation stress, and sometimes the Skempton coefficient is larger than 1. Lei [13] and Fang [14] investigated the undrained shear behavior of marine soft clay in Tianjin and it is found that after the soil reached yield stress the increasing rate of pore water pressure increases which proved that the pore water increases after the structural yield stress.

4 Conclusion

The influence of soil structure on the mechanical behavior has been reviewed literately and the conclusions are as follows:

1) There is obvious turning point in the compression curve of the undisturbed soil and before or after the turning point the compressibility of structured soil is quite different.

2) When the current stress is lower than the structural yield stress the consolidation coefficient of structured soil is larger than that of the remolded soil, after which the behavior becomes close.

3) The stress-strain relationship can be divided into two segments before and after the structural yield stress. Before the yield stress, the curve is elevating and after the yield stress the curve is decreasing.

4) The increasing rate of pore water pressure increases after the soil reached yield stress.

Acknowledgement

This work is supported by the Natural Science Foundation of Tianjin(16JCZDJC38800).

References

1. [1] S. Leroueil, F. Tavenas, J. Locat, Discussion on: Correlations between index tests and the properties of remoulded clays – Carrier W.D and Beckman J.F. [J]. Geotechnique, 35(2), (1985), p.223-226.
2. S. Leroueil, P.R. Vaughan, The general and congruent effects of structure in natural soils and weak rocks [J]. Geotechnique, 40(3): (1990), 467-488.
3. S. Leroueil, Compressibility of clays: Fundamental and practical aspects – closure [J]. Journal of Geotechnical and Geoenvironmental Engineering, 123(9): p. 895-897.
4. Mesri G., Roskhsar A., Bohor B.F., Composition and compressibility of typical samples of Mexico City clay. Geotechnique, 25(3): (1975), p.527-554.
5. C. H Zhang. Geotechnical properties of two structural soils. [J] Hydro-Science and Engineering, 4, (1983) 65-71.
6. J. A. Zhou, X. N. Gong. Experimental study on soft clay structure property [J]. Industrial Construction, 32(3), (2002), p.35-37.

https://doi.org/10.1051/e3sconf/20183803027

7. L. Z. Wang et al. Study on compressibility of structured soft soil [J]. Chinese Journal of Geotechnical Engineering, 4: (2004), p.46-53.
8. Casagrande A., The structure of clay and its importance in foundation engineering [J]. J. Boston Soc. Civ. Engrs., 19(4): 168-209.
9. Schmertmann J.H. The undisturbed consolidation behavior of clay. Transaction of ASCE, 120(2): 1201-1226, 1955.
10. Nagarj T.S., et al. Analysis of compressibility of sensitive soils. Journal of Geotechnical Engineering, ASCE, 116(1): 105-118, 1990.
11. Z. J. Shen, An elasto-plastic damage model for cemented clay [J]. Journal of Geotechnical Engineering, 15(3): (1993), p.21-28. Chinese
12. Li Z.Q. Mechanical properties of underconsolidated soil with structural strength [J]. Journal of Geotechnical Engineering, 4(1): (1982), p.34-45. Chinese
13. LeiHuayang . Experimental studyon the structuralmodel of marine soft soil and its application to foundation pit excavation[D]. Changchun: Jilin University, 2001.
14. Fang H.G. et al. Structural strength of tianjin marine sofi soil and its influence on mechanical property [J]. Journal of Jilin University (earth science edition), 32(1), (2002), p. 73-76.

Study on the Old Girders in the Widening Hollow Slab Girder Bridge

Ying WANG, Li Fang ZHANG, Hai Yan MA

Department of Civil Engineering, Nanjing University of Aeronautics and Astronautics, Nanjing, Jiangsu 210016, China

Abstract: Taking the bridge widening project of Shanghai-Hangzhou-Ningbo expressway widening construction project (China) as the background in this paper, the variation law of the internal force of the old bridge in the widening hollow slab girder bridge under vehicle load is studied, which is under the condition of different span lengths and different widening widths. Three different span lengths of the pre-tensioned prestressed hollow slab girder bridges are selected, the spatial finite element models of both the old bridge and the whole structure of widening bridge are established and calculated respectively by Midas/Civil software. The influences of widening and load increasing on the old bridges under the vehicle load are compared and analyzed. In addition, the authors also analyze the influences of different widening widths on the force state of old bridges under the condition of widening the same number of lane. Moreover , the effects on the old bridges that are caused by the uneven foundation settlement of widening bridge structure are also studied in this paper. This paper can provide some references for widening design of hollow slab bridges.

1 Introduction

The vast majority of expressways built last century in China are two-way four lanes expressways due to the restrictions of many factors. However, with increasing traffic volume in recent years, most of the expressways in China have been unable to meet the current traffic demands, so it is necessary to broaden and rebuild them to improve service levels. The majority of expressway widening and reconstruction projects involve the widening and reconstruction of bridges, and hollow slab bridge is the common bridge type in China, so the key technology to widen and rebuild hollow slab bridges has been a research focus.

There are currently three ways to widen the hollow slab bridges in China: neither the superstructure nor the substructure of the old and new bridges being connected; both the superstructure and the substructure of the old and new bridges being connected; the superstructure of the old and new bridges being connected while the substructure being separated 1. Considering the adverse effect of differential foundation settlement between the old and new structures on widening bridges, the third method of widening the hollow slab bridges is generally adopted in China. At present, there are many engineering practices at connection method of widening hollow slab bridges 2, and a suit of relatively mature construction technologies and processes has been formed.

Most of relevant literatures mainly focus on the effects of different connection methods on bridges, or analyze a bridge engineering example. But in this paper, it is systematically studied that the influence of widening on the main girders in old bridges which are the same type of bridge on the same line under the condition of different spans, different widening widths and load increasing.

2 General Situation of the Project

Shanghai-Hangzhou-Ningbo expressway widening construction project (China) is taken as the background in this paper, three different span lengths of the pretensioned prestressed hollow slab girder bridges are selected to be calculated and analyzed, in order to find out the variation law of the internal force of the old bridge in the widening hollow slab girder bridge under vehicle load. Three span lengths of the bridges are 10m, 13m and 16m respectively, and the heights of the main girders are 40cm, 55cm and 70cm. These old bridges are made up of 12 prestressed concrete hollow slabs, and the widening structures consist of 7 prestressed concrete hollow slabs. The cross- section of superstructure is illustrated as Figure 1 (Only half bridge is shown).

* Corresponding author: crystaling_wy@163.com

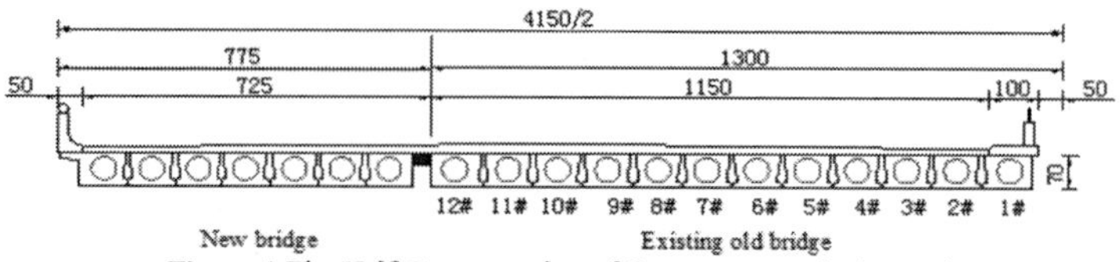

Figure 1 The Half Cross- section of Superstructure (unit = cm)

3 Calculation Mode

3.1 Design Codes

The old bridges were built on the basis of Code for Design of Highway Reinforced Concrete and Prestressed Concrete Bridges and Culverts JTJ023-85 (Old Standard for short), and the designing load of the old bridges was Truck-Load over 20. For bridge widening construction project, current Code for Design of Highway Reinforced Concrete and Prestressed Concrete Bridges and Culverts JTG D62-2004 (New Code for short) stipulates that the standard value of design load shall be the first grade loading of roadway. In this paper, the standard value of the first grade loading of roadway is determined by General Code for Design of Highway Bridges and Culverts JTG D60-2015.

The main girders of old bridges adopt C40 concrete according to the Old Standard, and the main girders of new bridges adopt C40 concrete due to the New Code.

3.2 Finite Element Modeling

Based on Grillage Theory, the spatial finite element models of old bridges and the overall bridge superstructures with span 10m, 13m, 16m are established and calculated respectively by Midas/Civil 8.32. In order to gain the vehicle load effects on these old bridges, Old Standard load and New Code load are both considered.

Each piece of prestressed hollow slab is modeled as one longitudinal beam, the transverse connections between longitudinal beams are simulated by rigid arms, which are perpendicular to the longitudinal beams. Each rigid arm between longitudinal beams is divided into two units, and the middle node is simulated by rigid joint or hinge joint according to the actual joint form between prestressed hollow slabs. The splicing joint between the new structure and the existing old bridge is rigidly connected, while the joints between the other main girders are hinged. Hinge joints are simulated by releasing beam end restraints. The space grillage model of superstructure is illustrated in Figure 2.

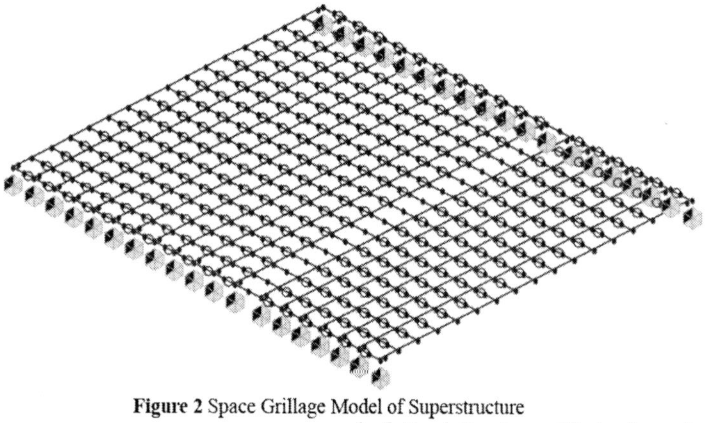

Figure 2 Space Grillage Model of Superstructure

4 Internal Force Analysis

4.1 Influence of Widening on the Old Hollow Slab Girder Bridge with Different Spans

On the basis of the articulated slab girder theory 8, the Transverse Distribution Factor of each main girder will decrease with the increase of the number of hollow slab, so the internal forces of the girders will decrease too, which shows that widening is beneficial to the main girders of the old bridges. Figure 3 shows the relationship between the load reduction rate of old bridges and the span length of bridges, in which the vehicle load on bridges before and after the broadening are both Truck Load over 20. As shown in Figure 3, after broadening, the maximum mid-span bending moment of vehicle load effect (VMM-bending for short) of main girders in old bridges decreases, and the decrease amplitude increases with the increase of the span length. Meanwhile, it can be seen that the decrease amplitude of VMM-bending of main girders in old bridges near the broaden splicing side is more than which far from broaden splicing side.

Bridge widening and reconstruction is often accompanied with load increasing, but load increasing is unfavorable to existing old bridges. As illustrated in Table 1, the old bridges before widening are calculated on the basis of Old Standard while the whole bridge superstructures after widening are calculated based on New Code. We can find that the VMM-bending of main girders in old bridges increases after widening and load increasing. Moreover, as the span length of bridges increased, the increase amplitude has a decreasing trend.

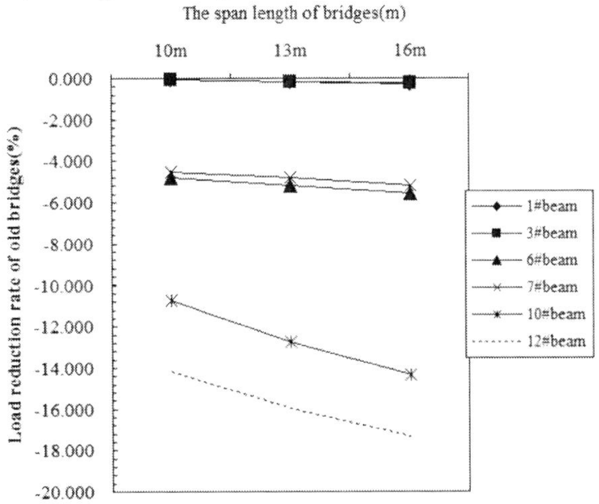

Figure 3 Relationship Between Load Reduction Rate of Old Bridges and Span Length of Bridges

Table 1 Comparison of VMM-bending of Main Girders in Old Bridge Between Existing Bridges and Whole Superstructures After Widening

Beam Number \ Span Length		10m M (kN·m)	13m M (kN·m)	16m M (kN·m)
1#beam	$M_{before\text{-}Old}$	166.1	218.2	314.2
	$M_{after\text{-}New}$	224.6	287.9	399.5
	Change Rate (%)	35.22	31.94	27.15
3#beam	$M_{before\text{-}Old}$	191.8	247.0	339.6
	$M_{after\text{-}New}$	269.6	337.3	448.2
	Change Rate (%)	40.56	36.56	31.98
6#beam	$M_{before\text{-}Old}$	182.9	236.4	330.4
	$M_{after\text{-}New}$	242.9	301.0	403.3
	Change Rate (%)	32.81	27.33	22.01
7#beam	$M_{before\text{-}Old}$	211.6	269.4	360.2
	$M_{after\text{-}New}$	294.7	364.3	483.8
	Change Rate (%)	39.27	35.23	34.31
10#beam	$M_{before\text{-}Old}$	229.2	297.3	402.4
	$M_{after\text{-}New}$	299.0	377.6	489.7
	Change Rate (%)	30.45	27.01	21.69
12#beam	$M_{before\text{-}Old}$	221.1	292.3	401.6
	$M_{after\text{-}New}$	277.5	354.9	460.3
	Change Rate (%)	25.51	21.42	14.62

Note: 1. $M_{before\text{-}Old}$ =VMM-bending of Main Girders in Old Bridge before widening based on Old Standard
2. $M_{after\text{-}New}$ =VMM-bending of Main Girders in Old Bridge after widening based on New Code
3. Change Rate = $(M_{after\text{-}New} - M_{before\text{-}Old}) / M_{before\text{-}Old} \times 100\%$

4.2 Influence of Widening Width On Old Bridge of the Widening Hollow Slab Girder Bridge

The hollow slab girder bridge with span 16m is taken as an analysis object in this paper, which consists of 12 main girders. To widen the old bridge from two lanes to four lanes, there are five kinds of widening width as follows:

1) Widening width is 5.75m: consisting of 5 main girders;

2) Widening width is 6.75m: consisting of 6 main girders;

3) Widening width is 7.75m: consisting of 7 main girders;

4) Widening width is 8.75m: consisting of 8 main girders;

5) Widening width is 9.75m: consisting of 9 main girders.

The splicing joint between the new structure and the existing old bridge is rigidly connected, while the joints between the other main girders are hinged. According to the most dangerous state of the old bridge, the distribution method of vehicle load is to asymmetrically load four lanes near the original inner side beam of old bridge as shown in Figure 4. The finite element models are established by Midas/Civil, and the vehicle load on bridge is Truck-Load over 20. The VMM-bending of main girders in old bridge under the condition of different widening widths is shown in Table 2.

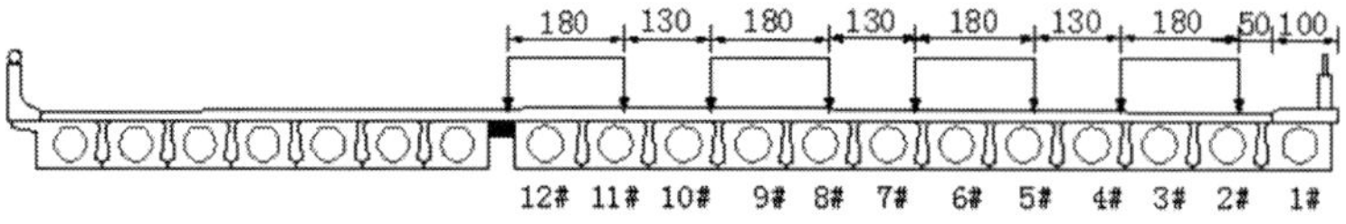

Figure 4 Distribution Method of Vehicle Load (unit = cm)

Table 2 The VMM-bending of Main Girders in Old Bridge under The Condition of Different Widening Widths

Widening Width \ Beam Number		VMM-bending of Main Girders in Old Bridge（M/kN·m） 1#beam	6#beam	10#beam	12#beam
0m	$M_{before\text{-}Old}$	314.2	330.4	402.4	407.6
Widening width: 5.75m	$M_{after\text{-}Old}$	313.4	312.2	345.7	334.6
	Change Rate (%)	-0.25	-5.51	-14.09	-17.91
Widening width: 6.75m	$M_{after\text{-}Old}$	313.4	312.1	345.0	332.9
	Change Rate (%)	-0.25	-5.54	-14.26	-18.33
Widening width: 7.75m	$M_{after\text{-}Old}$	313.3	312.1	344.7	332.1
	Change Rate (%)	-0.29	-5.54	-14.34	-18.52
Widening width:8.75m	$M_{after\text{-}Old}$	313.3	312.1	344.5	331.8
	Change Rate (%)	-0.29	-5.54	-14.39	-18.60
Widening width: 9.75m	$M_{after\text{-}Old}$	313.3	312.0	344.2	331.6
	Change Rate (%)	-0.29	-5.57	-14.46	-18.65

Note: 1. $M_{before\text{-}Old}$ =VMM-bending of Main Girders in Old Bridge before widening based on Old Standard
2. $M_{after\text{-}Old}$ =VMM-bending of Main Girders in Old Bridge after widening based on Old Standard
3. Change Rate = $(M_{after\text{-}Old} - M_{before\text{-}Old}) / M_{before\text{-}Old} \times 100\%$

As shown in Table 2, the decrease amplitude of the VMM-bending of main girders in old bridge has a slow increasing trend with the increase of widening width with the condition of broadening the old bridge to four lanes. Moreover, the increase of widening width has almost no influence on the main girders far from the broaden splicing side.

4.3. Influence of Foundation Settlement of Widening Structure on Old Bridge

In general, when the old bridge was built and used for many years, its foundation settlement has been basically completed, but the foundation settlement of the new bridge has not been completed. So there is an uneven vertical foundation settlement between the old and new bridges 7. To analyze the influence of foundation settlement of widening structure on old bridge, the foundation settlement of widening structure is illustrated in Figure 5. Considering the engineering practice and the restriction of old bridge on widening structure, the foundation of the old bridge is considered having no settlement, and the foundation settlement of the main girder of the new bridge near the old bridge is adopted as 2mm, the foundation settlement of the outside girder of the widening bridge structure is adopted as 5mm, the foundation settlement of the other main girders of the widening bridge structure is calculated by linear interpolation simultaneously. The distribution method of vehicle load is shown in Figure 4, and the vehicle load on bridge is Truck-Load over 20. The VMM-bending of main girders in old bridge with the influence of uneven foundation settlement of widening structure is shown in Table 3.

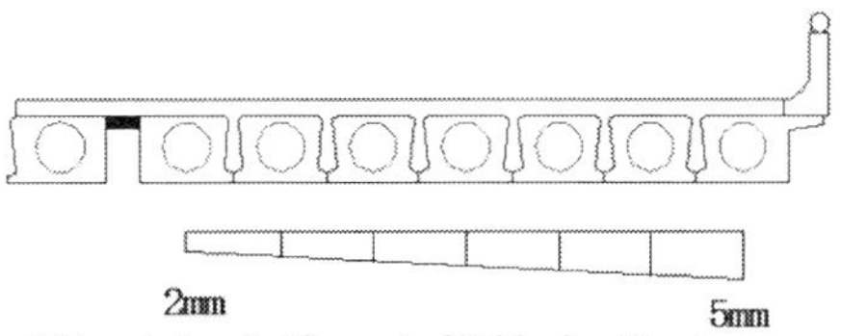

Figure 5 Foundation Settlement of Widening Structure

Table 3 The VMM-bending of Main Girders in Old Bridge with the Influence of Foundation Settlement of Widening Structure

Beam Number \ Span length		VMM-bending of Main Girders in Old Bridge (M/kN·m)		
		10m	13m	16m
8#beam	$M_{(without\ settlement)}$	204.5	261.4	347.9
	$M_{(with\ settlement)}$	204.5	261.4	347.9
	Change Rate (%)	0.000	0.000	0.000
9#beam	$M_{(without\ settlement)}$	173.9	221.5	306.3
	$M_{(with\ settlement)}$	173.9	221.5	306.3
	Change Rate (%)	0.000	0.000	0.033
10#beam	$M_{(without\ settlement)}$	204.6	259.4	344.7
	$M_{(with\ settlement)}$	204.6	259.5	344.8
	Change Rate (%)	0.000	0.039	0.029
11#beam	$M_{(without\ settlement)}$	211.9	271.5	358.3
	$M_{(with\ settlement)}$	212.1	271.6	358.5
	Change Rate (%)	0.094	0.037	0.056
12#beam	$M_{(without\ settlement)}$	189.8	245.6	332.1
	$M_{(with\ settlement)}$	190.1	245.8	332.4
	Change Rate (%)	0.158	0.081	0.090

Note: Change Rate = $(M_{(with\ settlement)} - M_{(without\ settlement)}) / M_{(without\ settlement)} \times 100\%$

As shown in Table 3, the VMM-bending of main girders in old bridge near the broaden splicing side increases with the influence of the foundation settlement of widening structure, but the changes are very small (The change rates of the VMM-bending of main girders in old bridge are less than o.2%). Besides, the foundation settlement of widening structure has almost no influence on the main girders of old bridge which are far from the broaden splicing side.

5 Conclusion

Through tabular approaches to the study of the old girders in the widening hollow slab girder bridge, we can come to the following conclusions:

When the vehicle load on bridges before and after the broadening are both Truck-Load over 20, the VMM-bending of main girders in old bridge reduces, and the reduction of main girders of old bridge near the broaden splicing side can reach about 18%, which shows that widening can play an obvious unloading effect on the old bridges. Meanwhile, we can find that the decrease amplitude has an increasing trend with the increase of the span length of bridges, and decrease amplitude of VMM-bending of main girders in old bridge near the broaden splicing side is more than which far from broaden splicing side.

Besides, when the vehicle load on bridges after the broadening is the first grade loading of roadway, the VMM-bending of main girders in old bridge is greatly improved than which before widening. It indicates that load increasing is unfavourable for the old bridge. Therefore, it is necessary to check and analyze the state of old bridges during bridge widening design, and the old bridges should be reinforced if necessary. What's more, according to analysis, the influence of the widening width on the old bridge can be be neglected under the condition of widening the same number of lane because the changes are very small. And the foundation settlement of widening structure has little influence on the old bridge under vehicle load.

Acknowledgments

 https://doi.org/10.1051/e3sconf/20183803028

The writers gratefully acknowledge the support provided by Youth Foundation Project of Jiangsu Natural Science Foundation of China (Grant No.BK20150750).

References

1. FAN Xiao-jiang. Finite Element Analysis of Widening Plan for Voided-Slab Bridge[J]. Shanxi Science & Technology of Communications, 2008, (02): 49-51+75.
2. LIAN Jun, JIA Lei, XU Yan-ling. The research on longitudinal connection in the widening and rehabilitation designing of the existing bridges[J]. Shanxi Architecture, 2007, (29): 313-314.
3. LI Qun, YE Jianlong, YU Maofeng. Technology for Splicing between New and Old Bridges in Open-to-Traffic State in Widening Project of Shanghai-Hangzhou-Ningbo Expressway[J]. Technology of Highway and Transport, 2011, (02): 74-78.
4. LIANG Zhi-guang, WANG Ping, WANG J ia-chen, SONG Shen-you. Connection of New and Old Structures of Bridges on Guang-Fo Expressway[J]. China Municipal Engineering, 2006, (01): 30-31+34+89.
5. Liu Xiaojun, Liu Zhongtian. Rebuilding Design of Sujiahu Expressway South Section Extension Project[J]. Urban Roads Bridges & Flood Control 2007, (04): 113-115+18.
6. SHEN Xiaolei, YE Jianlong, GUO Binqiang. Study on Broadening Design of Simply Supported Hollow Slab Bridge in Old Bridge Reconstruction[J]. Technology of Highway and Transport, 2016, (04): 101-104.
7. HAO Fu-jun, WU Dian-kun. Study on effect of foundation settlement on widening continuous hollow slab[J]. Journal of Henan University of Urban Construction, 2011, (05): 11-15.
8. JTG D60-2015, General Code for Design of Highway Bridges and Culverts[S]. Beijing: People Communications Press.
9. JTG D62-2004, Code for Design of Highway Reinforced Concrete and Prestressed Concrete Bridges and Culverts[S]. Beijing: People Communications Press.
10. JTJ021-89, General Code for Design of Highway Bridges and Culverts[S]. Beijing: People Communications Press.
11. JTJ023-85, Code for Design of Highway Reinforced Concrete and Prestressed Concrete Bridges and Culverts[S]. Beijing: People Communications Press.

Efficient web opening shape and stiffener for large web opening

Xiao Ying WU[1], Ling Xu LI[2], Dong Hua ZHOU[3]

[1]Faculty of Architectural Engineering, Dianchi College of YunnanUniversity, Kunming, Yunnan650000, China
[2]Faculty of Architectural Engineering, Kunming University of Science andTechnology, Kunming, Yunnan 65000, China

Abstract: to save material circle and hexagon are often used shapes for web opening in praxis; horizontal plates are often welded on web for stiffening large web opening. These common used shape of opening and stiffening way for large opening are proved as inefficient by finite element analyze. There are new shape of opening and new type of stiffener for large opening, which proved to have the best load bearing behavior respectively. More details will be introduced in following.

1 Introduction

Bending moment and shearing force in a beam will be carried by different part of the beam. For an I-beam, for example, the web takes mainly the shearing force up, the flanges carry the tensile and compressive force due to the bending moment. The bigger the flanges are, the bigger the web is needed. When more cross sectional area of web is required, the depth of web will be generally increased in stead of raising its width. Because the increasing depth of web makes the distance between flanges greater, as the result the internal level arm and moment bearing capacity become also greater. Now someone would ask whether the web of a beam has always an appropriate dimension in accordance with the flanges. The answer is no. Generally a web is dimensioned with its ratio of depth to width under consideration of buckling aspect. In respect to cross section of flange the web is usually over dimensioned, that is, in some cases the material strength of web can not be fully utilized and thus some material of web can be saved, for instance, with web openings. Which shapes of web openings (circles, rectangles or others) should be used that the beam has a better load carrying capacity? The design of steel beams with large rectangular openings is an important structural and practical problem which is caused by the need to pass service ducts through the structural zone of floor systems. Careful stiffening of these openings in the beam webs can minimize their adverse effects on the shear and the bending resistances of steel beams. How to stiffen opening and what kind of stiffener would have the best load bearing behavior? The answer to theses questions all mentioned above is the main content of this paper.

2 Study of beams with different web opening shapes using nonlinear FEM

We can be inspired from struts of a trussed framework. Beside struts or post there are only triangular hollow areas. If we choose triangular shape as web opening on an I-beam, what would happen, will this have a better load bearing behavior than those very commonly used, such as circle or hexagon? To answer this question a wide-flange beam IPB450 with following three different web openings is computed with ANSYS program:

1) Triangular openings
2) Circular openings
3) Hexagonal openings

In odder to comparison from each other all openings have the same hollow area, it means, all beams have also the same web area in elevation. In the analyzing the nonlinear material behavior and the geometric symmetry in direction of breadth are taken into account. That is, only one half beam is modeled in direction of width. The beam and its cross section as well as the stress-strain relationship used in analyses are shown in Figure 1.

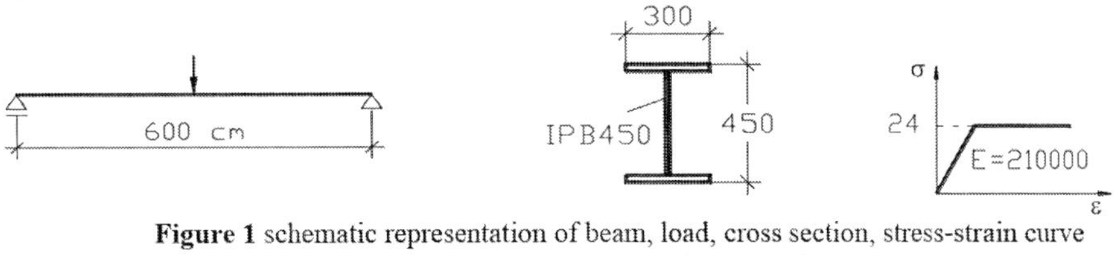

Figure 1 schematic representation of beam, load, cross section, stress-strain curve

From the computed results shown in Figure 2 we can see that beam in case 1 has the greatest load bearing capacity (266 kN), which is same as that per hand computed (see Eqs. 2) . Beam in case 2 and 3 have a

* Corresponding author:758975451@qq.com

lower ultimate failure loads.

Plastic moment of flange is obtained by Eqs. (1):

$$M_{pl} = 15 \times 2.6 \times 24 \times (45 - 2.6) = 39686.4\ kNcm \quad (1)$$

Then the bearing load due to M_{pl} would be

$$P = \frac{4 \times M_{pl}}{L} = \frac{4 \times 39686.4}{600} = 264.6\ kN \quad (2)$$

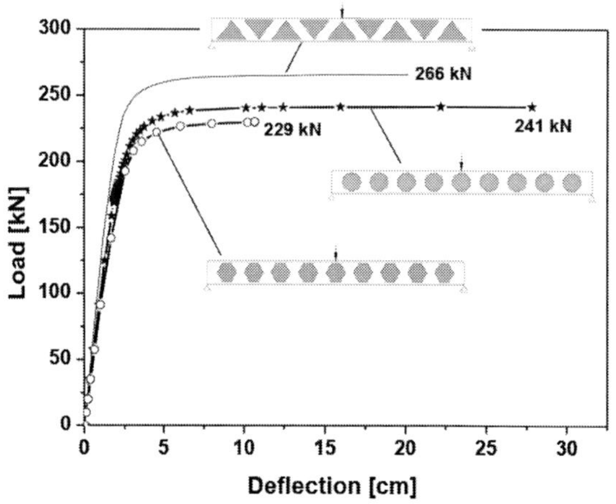

Figure 2 load-deflection curves

The reason for the different responses shown in Figure 2 can be clarified as the following:

1. The approximate resultant of all principle stresses in a web for case 1 as shown in Figure 3 run into a intersection point at flange and no flange is thereby subjected to vertical shearing force, while other two cases show a opposite behavior, that is, there is no intersection point of the approximate resultant of all principle stresses, the part of flange will subjected to local bending.

2. In each web strut of case 1 there is only one single stress state, namely one principle stress state, while in tow other cases there is a combined stress state in each web strut, that are the tensile and compressive principle stresses. As we know, a combined stress with different sign leads to lower stress failure criteria than a single stress state.

3. The principle stresses in web of case 1 are fairly uniformly distributed and this is not so in case 2 and 3.

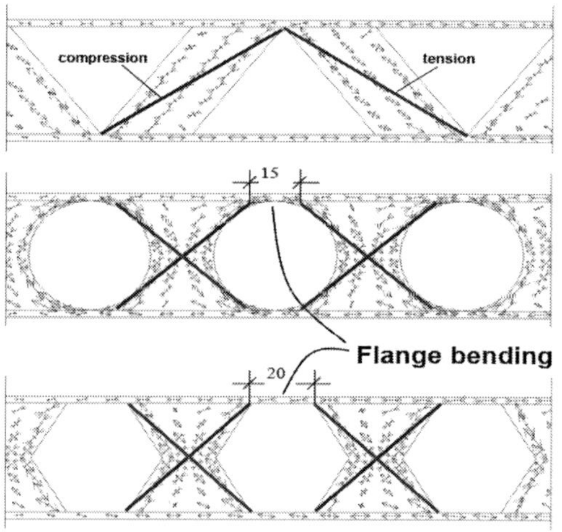

Figure 3 principle stress

3 Study of beams with different stiffeners for large web opening using nonlinear FEM

How most efficiently stiffening large Opening is another key point of this paper. To make this clear, beams with different stiffener for web opening are investigated and one beam with the same web opening but without any stiffener is also computed for comparison. Four cases being investigated are shown in Figure 4:

https://doi.org/10.1051/e3sconf/20183803029

1) Opening with two slanting plates as struts
2) Opening with a semicircle as arc
3) Opening with two horizontal plates
4) Opening without any stiffener

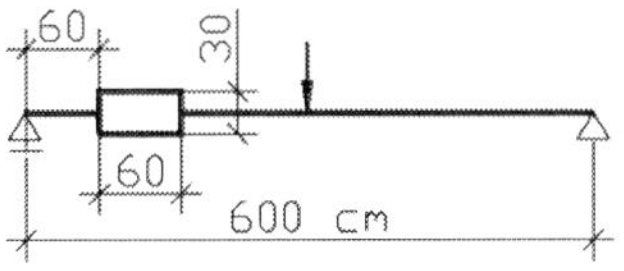

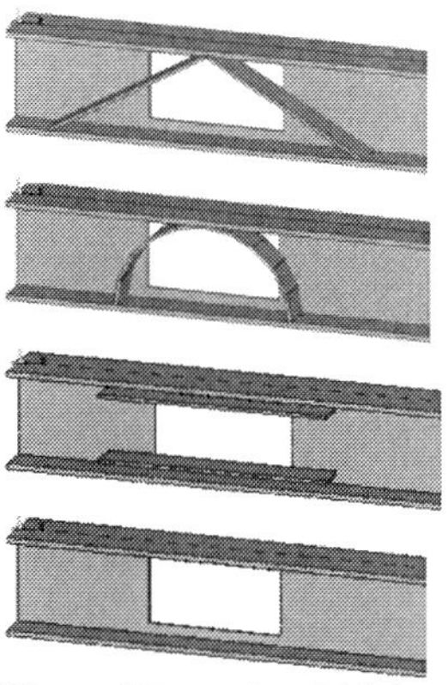

Figure 4 schematic representation of beam (link) and stiffener oblique view (right)

In the analyses the same steel wide-flange shape and the same stress-strain relationship used in preceding point 2 are adopted. The computed load-deflection curves shown in Figure 5 indicate that the stiffener in case 1 and 2 are most efficient, with which beams have the largest failure load and behave ductile. Although the beam with traditional stiffener in case 3 has a considerable higher failure load than beam of case 4 without any stiffening, but lower than beam of case 1 and 2, besides the beam of case 3 behaves very brittle.

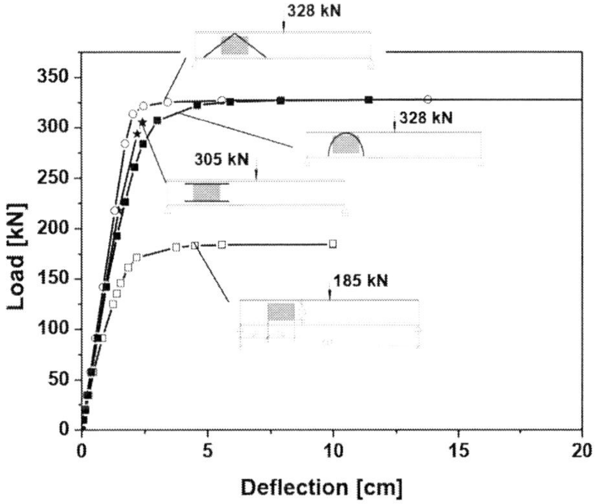

Figure 5 load-deflection curves

From additional results as shown in Figure 6 and 7 we can see more deeply what has happened in the beams. In Figure 6 and 7 the stress ratio and von Mises plastic strain are represented respectively at their own failure load levels. The stress ratio means the ratio of equivalent stress to yield stress. Through these computed results the following special features can be summarized:

1) Beams of case 1 and 2 fail to exploitation of material strength at midpoint of span, while in case 3 and 4 the material fail at opening instead of at center of span.

2) The efficient grade of stiffening of opening is decreased in order of case 1 to case 3. With this decreased grade the failure places transfer from center of span to the region of opening.

 https://doi.org/10.1051/e3sconf/20183803029

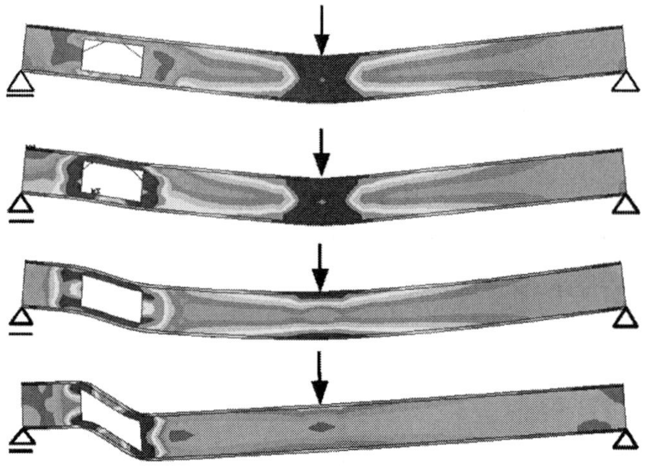

Figure 6 stress ratio (the ratio of equivalent stress to yield stress)

From von Mises plastic strain shown in Figure 7 it can be confirmed the feature described above, namely exploitation of material in the center of span in case 1 and 2 and in the region of opening in case 3 and 4.

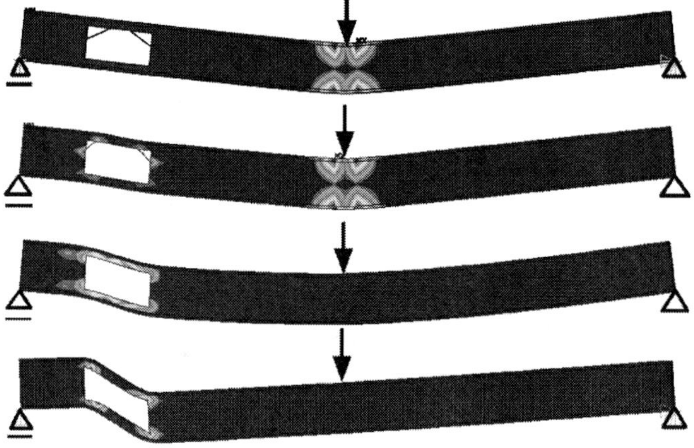

Figure 7 von Mises plastic strain

It is interesting to look at region of opening of case 3 and 4 closely; they both fail at rest web material over and under opening, but the causes led to failure are quite different. The strength of rest webs over and under opening in case 3 is exploited by shear stress, and therefore the normal stress due to bending moment is there very small. In case 4 it shows an opposite behavior, the rest webs failed to normal stress due to bending moment instead of shearing force.

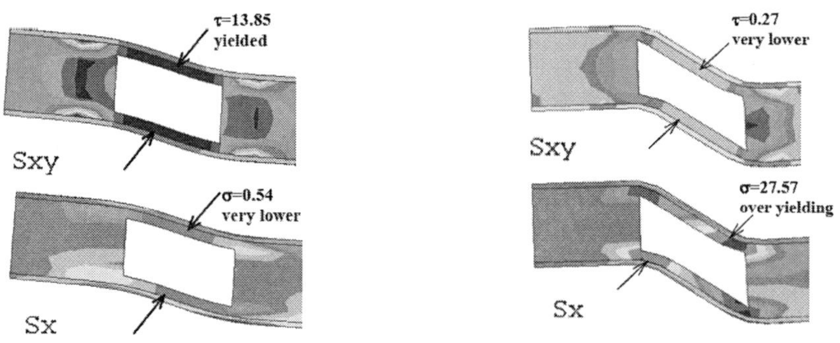

Figure 7 shear stress (Sxy) and normal stress (Sx) in case 3 (link) and 4 (right)

4 Conclusions

We can very well inspired form load bearing mechanism of the strut of trussed framework by choosing opening shapes or stiffener for large opening. Through the analyses carried out in this paper the following important features can be summarized:

1. Opening with triangular shape provides a better bearing behavior than those with traditional opening shapes like circle and hexagon.
2. Triangular stiffener over the opening is the best way to stiffen the large web opening.
3. Beams with the triangular opening shape or beams with triangular stiffeners behave just like slanting struts of trussed framework and present a very good load bearing behavior, high ultimate failure load and ductile deformation.
4. Traditional opening shape like circle and hexagon as well as traditional stiffening way like using horizontal plates has their disadvantage and they should therefore replaced by those mentioned above.

Acknowledgments

This work was financially supported by the National Natural Science Foundation (51668027 and 51468026).

References

1. Bode, H., Ramm, W. und Zhou, D.: Physikalisch nichtlineare Berechnungen von Verbundträgern mit Stegöffnungen mit dem Programmpaket ANSYS. Verbundkonstruktionen Neues aus Forschung, Entwicklung und Normung Beiträge zur Fachtagung an der Universität Kaiserslautern (Juli 1997)
2. Bode, H., Stengel, J. und Zhou, D.: Composite Beam Test For A New High-Rise Building, Engineering Foundation Conference, Composite Construction III, Irsee, Germany, June 1996.
3. Donghua Zhou: Tragverhalten von Einfeldverbundträgern mit Stegöffnungen, Stahlbau 3/2004)
4. Donghua Zhou: Besonderheiten von Durchlaufträgern mit Stegöffnungen, Stahlbau 5/2004)

The Researches on Damage Detection Method for Truss Structures

Meng Hong WANG, Xiao Nan CAO

School of Civil and Transportation Engineering, Beijing University of Civil Engineering and Architecture, Beijing, China

Abstract: This paper presents an effective method to detect damage in truss structures. Numerical simulation and experimental analysis were carried out on a damaged truss structure under instantaneous excitation. The ideal excitation point and appropriate hammering method were determined to extract time domain signals under two working conditions. The frequency response function and principal component analysis were used for data processing, and the angle between the frequency response function vectors was selected as a damage index to ascertain the location of a damaged bar in the truss structure. In the numerical simulation, the time domain signal of all nodes was extracted to determine the location of the damaged bar. In the experimental analysis, the time domain signal of a portion of the nodes was extracted on the basis of an optimal sensor placement method based on the node strain energy coefficient. The results of the numerical simulation and experimental analysis showed that the damage detection method based on the frequency response function and principal component analysis could locate the damaged bar accurately.

1 Introduction

With the development of prefabrication and the mechanized construction of truss structures, more and more buildings are adopting a truss structure. However, the weight of equipment is increasing. Under the long-term action of moving loads, bolts loosen, rods bend and trusses fracture due to local damage, which can easily lead to a catastrophic accident. It is difficult to locate damage owing to the large number of structural members. Therefore, it is very important to identify the location of damage in a truss structure accurately.

Adams [1] and Cowley have detected damage in structures by collecting parameters related to frequency, accurately determining the damage location by using the ratio of the corresponding frequency changes before and after the structure was damaged. Numerical simulation of a simple cable has been carried out by Zheng Minggang et al. [2], frequency response function (FRF) curvature has been used and who found that after a damage index had been selected, FRF curvature could determine the damage location and damage degree sensitively. Yang and Wu [3] detected damage in a simple model using the acceleration frequency response function. The analysis of a combined T-shape beam was carried out by Zhang [4], who used the difference between the matrices formed by the frequency response function of the structure before and after damage as the basic parameter. At the same time, principal component analysis (PCA) was used for dimensionality reduction. Numerical analysis of the cantilever beam model was carried out by Li and Yu [5], who detected damage using the frequency response function of the structure and achieved satisfactory accuracy of damage identification.

There has been many research on structural damage detection based on the frequency response function, but most of the research is aimed at the simply-supported beam, cantilever beam and bridge structures. Research on damage detection in truss structures is scarce. Therefore, in this paper, a method of damage detection in truss structures will be researched by means of numerical simulation and experimental analysis.

2 Basic Theory

2.1 Frequency Response Function

The transfer function is the ratio between the output signal and the input signal in a structure and is one of the functions used most widely to describe the vibration characteristics of a structure.

After Laplace transformation, the multi-freedom vibration equation can be simplified as:

$$Y(s) = H(s)F(s) \tag{1}$$

In the equation, $Y(s)$ is the output signal; $F(s)$ is the input signal.

After Fourier transformation, the above equation can be written as follows:

$$H(\omega) = \frac{Y(\omega)}{F(\omega)} \tag{2}$$

* Corresponding author: wangmh@bucea.edu.cn

Where $H(\omega)$ is the frequency response function.

2.2 Principal Component Analysis(PCA)

Assuming that there are n samples, and each sample contains m indexes, the original data matrix $X_{n\times m}$ can be obtained. The new matrix $Y_{n\times m}$ will be obtained after the original data matrix $X_{n\times m}$ is standardized. Subsequently, for the n samples given, the covariance matrix is obtained for the new matrix $Y_{n\times m}$; each element of the covariance matrix is represented by its corresponding covariance. The covariance matrix is a real symmetric matrix; therefore, it has m eigenvalues. Assuming that the eigenvalues are λ_1 、 λ_2 、 Λ 、 λ_m ($\lambda_1 \ge \lambda_2 \ge \lambda_3 \ge \Lambda \ge \lambda_m$) the normalized feature vectors corresponding to the m eigenvalues are u_1、u_2、Λ 、u_m. Finally, these feature vectors are used as the coefficient vectors of the principal components, then, combined with the standard matrix $Y_{n\times m}$, each principal component Z_1、Z_2、Λ 、Z_m of the original data can be obtained:

$$\begin{cases} Z_1 = u_{11}Y_1 + u_{12}Y_2 + \Lambda + u_{1m}Y_m \\ Z_2 = u_{21}Y_1 + u_{22}Y_2 + \Lambda + u_{2m}Y_m \\ \Lambda\ \Lambda\ \Lambda\ \Lambda \\ Z_m = u_{m1}Y_1 + u_{m2}Y_2 + \Lambda + u_{mm}Y_m \end{cases} \quad (3)$$

The cumulative contribution rate of the principal components Z_1、Z_2、Λ 、Z_m should exceed 85%:

$$\sum_{i=1}^{p} Z_i \Big/ \sum_{i=1}^{m} Z_i \ge 85\% \quad (4)$$

3 Numerical Simulation

3.1 Numerical Simulation Model and Working Condition Setting

Parameters of the model are as follows: The span of the truss structure is 3m×5 m, the truss structure contains 60 nodes and 174 rods and the size of all member sections is Φ42mm×2.5 mm. Q235 steel was selected, the elastic modulus is E=200GPa. In the numerical simulation, the self-weight of the rod was transferred into the equivalent node load without considering dust, snow or wind loads. Two conditions are defined in this paper. Condition I: no damage to any members; Condition II: damage has just occurred in rod 34; the ends of the rod are node 7 and node 16, as shown in Figure 1.

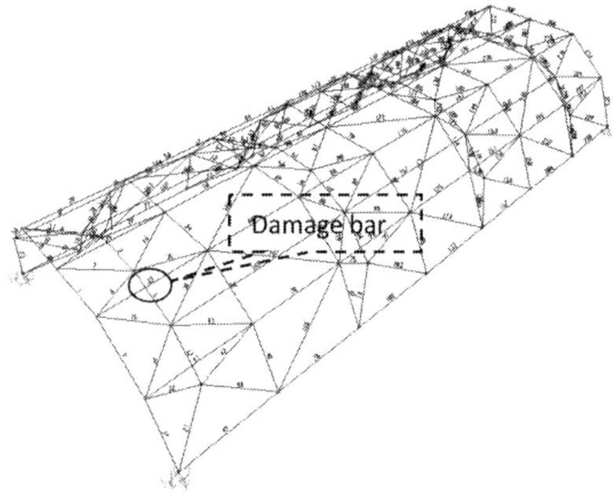

Figure 1 Damage Location

3.2 Acquiring the Time Domain Signal and Matrix Dimensionality Reduction

In this paper, the finite element analysis software SAP2000 was used for numerical analysis. The sampling frequency was 512 Hz and the sampling time was 2 s. Using MATLAB, frequency domain signals were derived from the time domain signals based on the Fourier transform. A column vector was made up of the frequency domain signal of each node, and the frequency response function matrix was composed of all of the column vectors.

The frequency response function matrix was input to the statistical analysis software SPSS, in which principal component analysis (PCA) was applied to reduce the dimension of the matrix. The results of the dimensionality reduction under the two working conditions are shown in Tables 1 and 2, respectively.

Table 1 Result of Dimensionality Reduction Under Condition I

Total Variance Explained	
Initial Eigenvalues	Extraction Sums of Squared and Loading

 https://doi.org/10.1051/e3sconf/20183803030

Component	Total	% of Variance	Cumulative %	Total	% of Variance	Cumulative %
1	717.269	69.977	69.977	717.269	69.977	69.977
2	178.833	17.447	87.425	178.833	17.447	87.425
3	74.087	7.228	94.653	74.087	7.228	94.653
4	39.358	3.840	98.492	39.358	3.840	98.492
5	8.969	.875	99.367	8.969	.875	99.367
6	3.428	.334	99.702	3.428	.334	99.702

Table 2 Result of Dimensionality Reduction Under Condition II

Total Variance Explained						
	Initial Eigenvalues			Extraction Sums of Squared and Loading		
Component	Total	% of Variance	Cumulative %	Total	% of Variance	Cumulative %
1	680.221	66.363	66.363	680.221	66.363	66.363
2	177.148	17.283	83.646	177.148	17.283	83.646
3	60.005	5.854	89.500	60.005	5.854	89.500
4	34.181	3.335	92.835	34.181	3.335	92.835
5	25.535	2.491	95.326	25.535	2.491	95.326
6	15.042	1.467	96.793	15.042	1.467	96.793
7	9.811	.957	97.750	9.811	.957	97.750
8	6.711	.655	98.405	6.711	.655	98.405

From Table 1 and Table 2, we can see that the cumulative variance contribution rate of the third matrix component under the two working conditions has exceeded 85%. Therefore, extraction of the first three principal components meets the requirement for dimensionality reduction. In SPSS, using the method of dimensionality reduction by principal component analysis, the first three principal component coefficients can be derived automatically, and then, the principal components Z_1、Z_2、Λ 、Z_P can be calculated.

3.3 Damage Index

According to the results of the numerical simulation, the frequency domain signals of each node can be regarded as a three dimensional vector before and after the damage. The angle between the two vectors before and after the damage was selected as the damage index. In the analysis of the results, if the angle between the two vectors before and after the damage of a node takes a maximum value, it can be concluded that the damage occurred in the vicinity of the node.

To make the simulation results clearer, the difference between the FRF vector angle and the upper confidence limit was used to judge damage location in this paper. If the calculated result was negative, then the difference was taken to be zero.

3.4 Numerical Simulation Analysis

The difference between the FRF vector angle and the upper confidence limit is shown in Figure 2. As can be seen in Figure 2, the difference between the FRF vector angle and the upper confidence limit took its maximum values at node 7 and node 16, namely, damage occurred in the vicinity of node 7 and node 16. Therefore, it could be estimated that rod 34, which connected the two nodes, had been damaged.

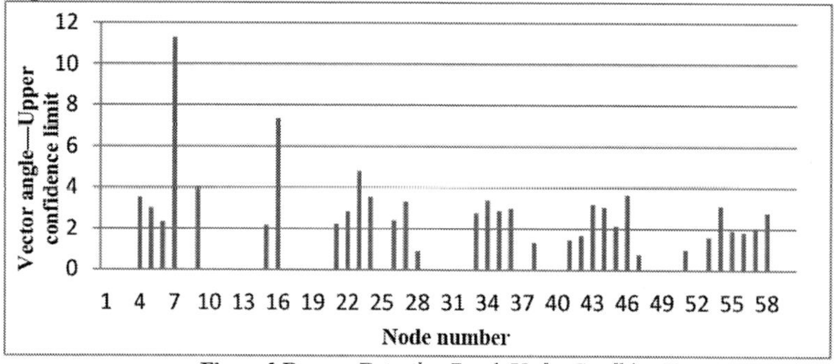

Figure 2 Damage Detection Result Under Condition II

4 Experimental Analysis

4.1 Experimental Procedure

In this paper, the working conditions and the size of the experimental model was the same as that of the finite element numerical simulation model. In this experiment, damage detection on the truss structure was divided into two steps: First, with the sensors arranged as shown in Figure 4, the time domain signal of 20 nodes was obtained by the hammering method. After data processing, we found the suspected damage nodes through the identification result obtained in the first step. Second, based on the suspected damage nodes from the first step, the scope of damage identification was narrowed down to the nodes linked with the suspected

 https://doi.org/10.1051/e3sconf/20183803030

damage nodes, and the specific location of the damaged bars was determined. The test method of single-point excitation and multi-point output was used. Node 19, as shown in Figure 3, was selected as the excitation point.

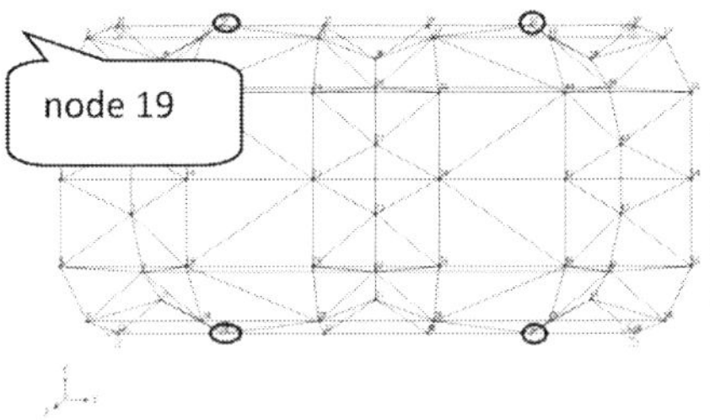

Figure 3 Excitation Node

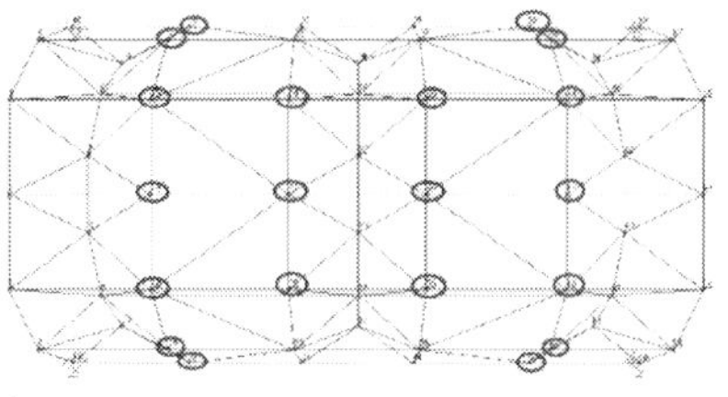

Figure 4 Sensor Arrangement

4.2 Experimental Results

The first step: The FRF matrix was input into statistical analysis software SPSS and the dimensionality was reduced using PCA. The results of dimensionality reduction are shown in Table 3 and Table 4. From Table 3 and Table 4, we can see that the cumulative variance contribution rate of the 11th matrix component under the two working conditions exceeded 85%; therefore, the first 11 principal components of the FRF were extracted, then, according to equation (3), the principal components $Z_1 、Z_2 、\Lambda 、Z_P$ were calculated.

Table 3 Result of Dimensionality Reduction Under Condition I

Total Variance Explained						
	Initial Eigenvalues			Extraction Sums of Squared and Loading		
Component	Total	% of Variance	Cumulative %	Total	% of Variance	Cumulative %
1	100.670	19.662	19.662	100.670	19.662	19.662
2	60.527	11.822	31.484	60.527	11.822	31.484
3	49.786	9.724	41.208	49.786	9.724	41.208
4	47.151	9.209	50.417	47.151	9.209	50.417
5	38.138	7.449	57.866	38.138	7.449	57.866
6	36.437	7.117	64.982	36.437	7.117	64.982
7	26.678	5.211	70.193	26.678	5.211	70.193
8	23.720	4.633	74.826	23.720	4.633	74.826
9	19.772	3.862	78.687	19.772	3.862	78.687
10	18.528	3.619	82.306	18.528	3.619	82.306
11	15.861	3.098	**85.404**	15.861	3.098	**85.404**
12	14.127	2.759	88.163	14.127	2.759	88.163

Table 4 Result of Dimensionality Reduction Under Condition II

Total Variance Explained						
	Initial Eigenvalues			Extraction Sums of Squared and Loading		
Component	Total	% of Variance	Cumulative %	Total	% of Variance	Cumulative %
1	88.848	17.353	17.353	88.848	17.353	17.353
2	59.473	11.616	28.969	59.473	11.616	28.969
3	52.610	10.275	39.244	52.610	10.275	39.244
4	50.236	9.812	49.056	50.236	9.812	49.056
5	38.182	7.457	56.514	38.182	7.457	56.514
6	32.147	6.279	62.792	32.147	6.279	62.792
7	30.520	5.961	68.753	30.520	5.961	68.753
8	27.994	5.468	74.221	27.994	5.468	74.221
9	25.622	5.004	79.225	25.622	5.004	79.225
10	18.122	3.540	82.765	18.122	3.540	82.765
11	17.141	3.348	**86.113**	17.141	3.348	**86.113**
12	13.606	2.657	88.770	13.606	2.657	88.770

The difference between the FRF vector angle and the upper confidence limit based on the principal components calculated in the first step is shown in Figure 5. It can be seen from Figure 5 that the difference reached its maximum at the node 16; therefore, it was concluded that damage had occurred in the vicinity of node 16.

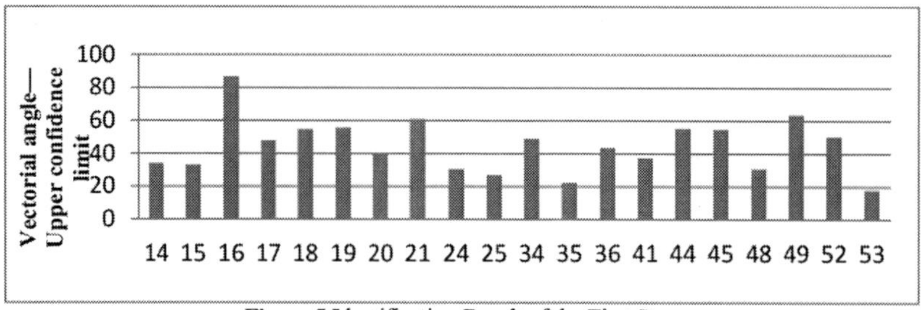

Figure 5 Identification Result of the First Step

The second step: The nodes connected to node 16 are numbered 3, 7, 11, 14, 18, 21 and 29. Time domain signals of the 7 nodes before and after the damage were acquired, the FRF matrix was input into statistical analysis software SPSS and the dimensionality was reduced using PCA. The results of dimensionality reduction are shown in Table 5 and Table 6. From Table 5 and Table 6, we can see that the cumulative variance contribution rate of the 5th matrix component has exceeded 85%; therefore, the first five principal components of the FRF were extracted, then, the principal components Z_1、Z_2、Λ 、Z_p were calculated.

Table 5 Result of Dimensionality Reduction Under Condition I

Total Variance Explained						
	Initial Eigenvalues			Extraction Sums of Squared and Loading		
Component	Total	% of Variance	Cumulative %	Total	% of Variance	Cumulative %
1	242.035	47.272	47.272	242.035	47.272	47.272
2	96.026	18.755	66.028	96.026	18.755	66.028
3	60.220	11.762	77.789	60.220	11.762	77.789
4	52.513	10.256	88.046	52.513	10.256	88.046
5	38.374	7.495	95.541	38.374	7.495	95.541

Table 6 Result of Dimensionality Reduction Under Condition II

Total Variance Explained						
	Initial Eigenvalues			Extraction Sums of Squared and Loading		
Component	Total	% of Variance	Cumulative %	Total	% of Variance	Cumulative %
1	148.089	28.924	28.924	148.089	28.924	28.924
2	103.497	20.214	49.138	103.497	20.214	49.138
3	86.298	16.855	65.993	86.298	16.855	65.993
4	65.859	12.863	78.856	65.859	12.863	78.856
5	63.479	12.398	91.254	63.479	12.398	91.254

The difference between the FRF vector angle and the confidence upper limit based on the calculated principal components is shown in Figure 6.

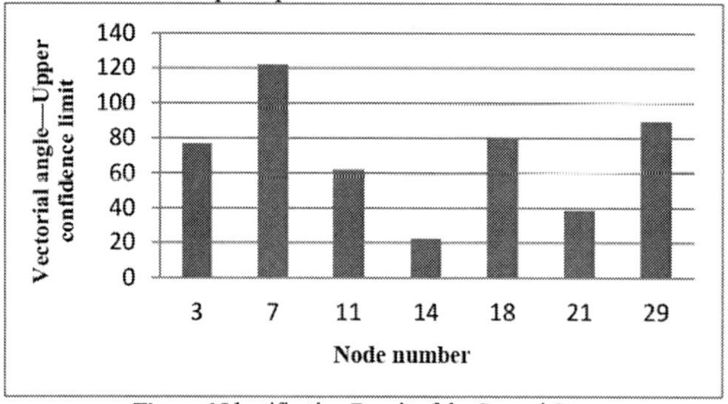

Figure 6 Identification Result of the Second Step

It can be seen from Figure 6 that the difference reached its maximum at node 7; therefore, damage was judged to have occurred in the vicinity of node 7. Combining this result with the identification result of the first step, damage occurred in the vicinity of nodes 7 and 16 and in bar 34, which was linked with nodes 7 and 16. The damage position was consistent with the experimental setup, and the result of the two step method was accurate.

5 Conclusion

This paper studied a damage detection method for a truss structure based on principal component analysis and the frequency response function. The accuracy of this method was verified by comparing the results of numerical simulation and experimental analysis. Conclusions can be drawn as follows:

(1) The structure was excited using the hammering method, and the time domain signal of the structure was extracted. The results showed that the measurement error can be reduced effectively by calculating the average value.

(2) Principal component analysis and the frequency response function were used to process all data, and numerical simulation and experimental analysis verified the effectiveness of the damage detection method. The method and its theoretical basis were provided for the study of similar problems.

(3) In the numerical simulation, time domain signals were extracted for all nodes and frequency domain signals were derived from the time domain signals based on the Fourier transform. After dimensionality reduction by SPSS, the angle between the two vectors before and after the damage was selected as the damage index. The damage location on the truss structure could be determined by the damage index in one step. The accuracy of the damage detection method has been verified.

(4) In experimental analysis, time domain signals of 20 nodes were collected using the method of optimal sensor placement based on the node strain energy participation coefficient, and then damage location detection was divided into two steps in which the scope of the damage range was narrowed gradually. The sensitivity and accuracy of the damage detection method was verified by the experimental results. Furthermore, the method has the prospect of application in practical engineering.

Acknowledgement

The authors would like to acknowledge the financial support provided by the National Natural Science Foundation of China (51578038) and Beijing National Science Foundation (8132023).

References

1. Cowley P, Adams RD. The location of defects in structures from measurements of natural frequencies. Journal of Strain Analysis 1979; 14(2): 49-57.
2. Zheng MG, Liu TX, Chen ZN. Structural damage assessment based on frequency response function. Mechanical Science and Technology 2001; 20(3): 479-481.
3. Yang HF, Wu ZY, Wu D. Structural damage detection method based on acceleration frequency response function. Journal of Vibration and Shock 2007; 26(2). 90-97.
4. Zhang CH. Study on damage identification of structures based on principal component analysis. Chongqing: Chongqing Jiaotong University;2012.
5. Li XP, Yu ZW. Structural damage identification based on frequency response function. Journal of China & Foreign Highway 2006; 26(1). 83-85.
6. Chen Y. Research on sensor optimal placement and programmatic implementation of health monitoring on truss structures. Beijing: Beijing University of Civil Engineering and Architecture; 2014.
7. Sun HM, Li HN. State-of-the-art review of the structural health monitoring in civil engineering. Journal of Disaster Prevention and Mitigation Engineering 2003; 23(3): 92-98.
8. Beijing Jintuwu Software Technology Co., Ltd. Guidance to use of SAP2000 in Chinese. Beijing, China: China Communications Press; 2012.
9. Yan D, Qi CX, Feng QH. SAP2000 structure engineering analysis and example explanation. Beijing, China: China Architecture and Building Press; 2009.

The Application of Deck Charge Technology in Hua Neng open pit Mine

Xiao Jun ZHANG [1, *], Xu Guang WANG [2], Ya Lun YU [1], De Qiang YANG [1]

[1]School of Civil and Resource Engineering, University of Science and Technology Beijing, Beijing 100083, China;
[2]Beijing General Research Institute of Ming & Metallurgy, Beijing 100160, China

Abstract: In the modern mine development and construction, how to improve the effective utilization of explosive energy, while achieving the requirements of the blasting vibration , is the key problem of blasting research.In this paper, from the actual situation of Huaneng open pit mine, for the complaints of nearby residents, the blasting program will be improved.First of all, the principle of the deck charge technology is analyzed, and then the deck charge technology is applied to the open pit mine to reduce the vibration caused by the blasting, to appease the dissatisfaction of the nearby residents. Finally It has achieved good results.

1 Introduction

To improve the effective utilization of explosive energy, With less explosives to make the ore fully broken, while meeting the requirements of blasting construction, and reducing the cost of blasting, is a basic principle of mine blasting. At present, mining intensity is increasing. How to ensure the safety of mine production and cost control is an important problem for enterprises. By optimizing the structure of blasting charge, thus improving the energy efficiency is one of the solutions to the problem of security and cost control. Huaneng open pit mine is located in Wulan Mu Lun Village,in Inner Mongolia Autonomous Region Erdos City. At present, the open-pit mine with an annual output of 600,000 tons, use continuous column charge,and by hole detonation mining technology. But the mine is near the residential area (shown in Figure 1), recently about 300m. In the existing mechanical equipment and blasting process, the impact on seismic waves generated by blasting is relatively large, and vibration is more obvious. Near the villagers reacted more intense. There are many complaints occurred. Combined with engineering practice and engineering analogy, so the blasting scheme would be improved in the open pit mine. The solution is changing the continuous column charge for the axial air separation charge and the other blasting parameters unchanged.

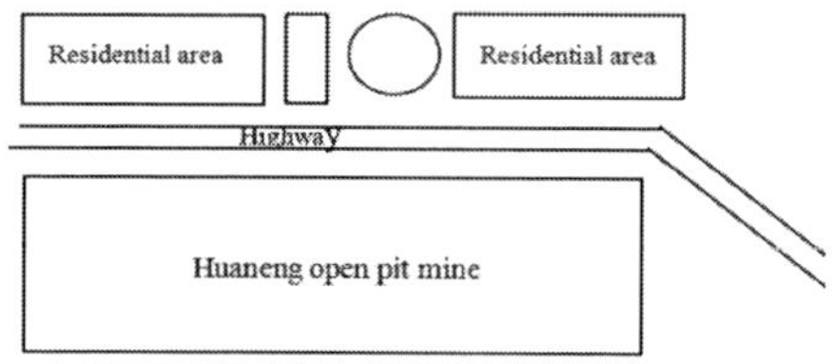

Fig.1. Schematic diagram of the mine area

Many model experiments and mine practice have proved that the axial air deck charge can be used to achieve the ideal blasting effect,compared with the continuous columnar charge. It not only can improve the effective utilization rate of the explosive energy, but also can control the blasting hazard and reduce the blasting vibration, and can reduce the cost of blasting.This paper is based on the deck charge blasting technology. The application of axial Air deck Charge in Huaneng Open pit mine is in order to reduce the vibration generated by the blasting, to appease the dissatisfaction of nearby residents, and to meet the requirements of blasting construction.

2 Principle of blasting technology for deck charge

Deck charge is divided into multiple sections of the column charge, by air, water, rock slag and other separation. According to the principle of resonance detonation of detonation energy, it is a method of rational allocation of energy.

2.1 The principle of deck charge blasting

When use continuous column charge structure in the blasting, the charge surface and the rock are in direct

* Corresponding author: 1024770807@qq.com Tel: 18210136152

https://doi.org/10.1051/e3sconf/20183803031

contact. Compressive stress waves of explosive detonation and quasi-static gas pressure are directly on the rock, usually there would be forming crushing zone in around the wall. In the case of the explosion of 5000-10000MPa produced by the ultra-high pressure, the crushing zone rock not only produces serious crushing damage, consume a lot energy, and its temperature also suddenly increased, resulting in further loss of explosive energy.By the mine blasting Standard, the generation of crushed areas marks the waste of energy from explosives.

Deck charge technology can effectively improve the energy utilization of explosives. Former Soviet Union Melnikow N V et al. have studied this technique in theory. Their research shows that due to the presence of the air layer, it causes the secondary and subsequent series of load waves during the explosive process and at the same time cause further damage to the fractured rock mass caused by the previous pressure wave. Although the use of air-spaced charge, the average pressure acting on the borehole is lower than the continuous column charge method, it can be broken by the subsequent action of the loaded wave. The use of air gap charge method, through the intermediate medium to change the contact between the charge and the walls of the hole, can reduce the initial pressure of the stress wave and detonation gases products in the hole wall. So that the surroundings do not produce particularly obvious powder-like rock particles. Meanwhile the action time of the initial pressure extended (as shown in Figure 2), and obtain a greater blasting impulse (burst pressure and burst pressure action time product). In the of case of reducing the amount of explosive, it can improve the energy efficiency of explosives and achieve the purpose of broken rock.

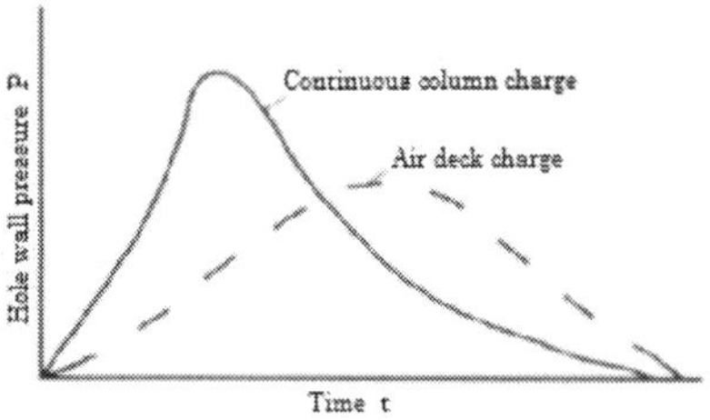

Fig.2. Hole wall pressure and action time

2.2 Deck ratio determination

During the explosion, the energy of the explosive is stored in the cavity, forming a plurality of loads, extending the time of stress acting on rocks, thereby reducing the energy of the broken hole and increasing the energy delivered to the surrounding medium. Therefore, we should make full use of the advantages of air separation blasting structure. Firstly, the reasonable air compartment ratio R_a must be determined.

$$R_a = L_a / (L_a + L_c) \qquad (1)$$

Where: L_a - charge length cm;
L_c - deck length cm;
R_a - deck ratio;

In engineering practice, the determination of the proportion of the air layer often depends on the empirical data, but the value is changing relatively large. Melnikow N V thought that taking 11% to 35% of the volume of the chamber is better. Moxon N T thought that taking 15% to 35% of the volume of the chamber is better. Some domestic scholars believe that the blasting can achieve better results when the proportion of the air layer is 30% to 40% of the volume of the chamber. Zhu Hongbing and others analysis on the air layer ratio through the shock wave theory and think that the upper limit of the appropriate proportion of the air layer is 30% to 42%.

2.3 Blasting vibration effect of deck charge

Blasting vibration is one of the effects of engineering blasting operations, which can not be eliminated. Enterprises would as much as possible avoid blasting vibration hazards in the construction process. Due to the presence of the air layer, the initial pressure on the hole wall is reduced. The initial pressure is from the stress wave and the detonation gas product generated during the blasting, Which can play the role of energy buffer, further reduce the vibration generated by the blasting, meet the requirements of blasting construction. So the in deck charge blasting technology has been widely used in the project.

3 Blasting scheme and parameter design of deck charge

Through the analysis of the principle of deck charge, combined with the actual situation of Huaneng open pit mine, the ore body rock fracture is very developed. It is easy to produce the foundation when the air gap is at the end of hole. Meanwhile it is easy to produce large and explosive decomposition when the air gap is at the upper part of hole. So here the exploration of the central air separation charge blasting is the main one. The middle air gap charge blasting used Orica high-precision non-electric detonator and hole-by-hole detonation technology. The interval time between of the holes is 25ms.The interval time between the rows is 65ms. Hole extension time is 400ms. There are some measurement points for blasting vibration monitoring in the vicinity of the mining area near the residential. The blasting parameters for the open pit are set as follows.

3.1 Borehole diameter and borehole length

The bench height is 9m. There are vertical blasting and plum blossom hole in the blasting area.The hole spacing is 5m and the row span is 4m. Borehole length is 10m.The excess drilling is 1m and borehole diameter is 110mm. There are ammonium oil explosives and down-the-hole drill. The above parameters are consistent with the blasting parameters of the original blasting scheme, which is the continuous column charge.

3.2 Charge and tamping

The axial air separation charge technology is that each hole set two kits. The first pack is placed at the bottom of the hole and the second pack is placed on the top of the hole. The explosive will be separated from the center of the hole by the barrier made by PVC pipe. And then the hole will be tamped with rock slag. Charge structure is shown in Figure 3 and the parameters are as follows:

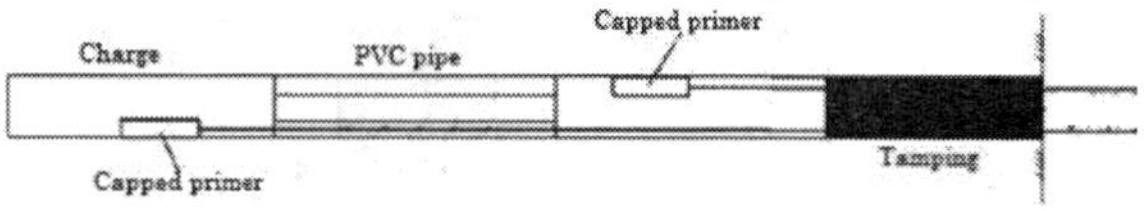

Fig.3. Charge structure diagram

The tamping length is about 2.8m. The spacer is placed in the middle of the blasting hole and the length of the air separator is 2.3m. The upper dose is 15kg and the charge length is 2.45m. The bottom dose is 15kg and the charge length is about 2.45m. The total length of the pillar and air interval was 7.2m and the total length of the pill was 4.9m. The deck ratio Ra was 32% and the maximum dose of single hole was 45 kg;

4 Analysis of blasting effect on site

The blasting effect improved significantly in accordance with the above blasting parameters using the deck charge technology in Hua Neng open pit mine, as shown in Figure 4.

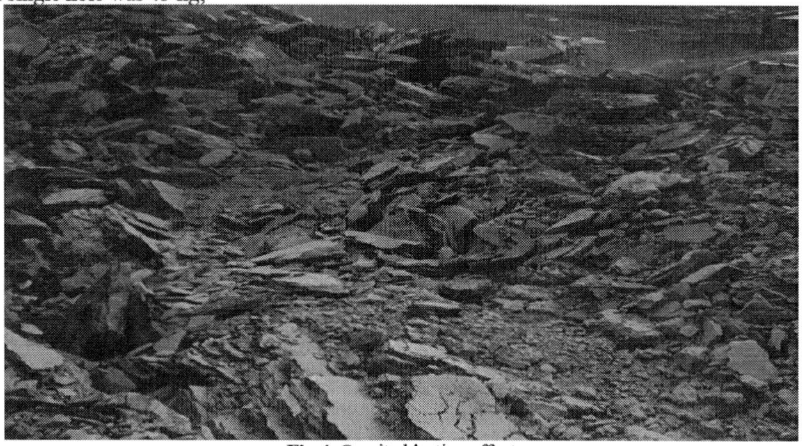

Fig.4. On-site blasting effect

By comparing the blasting effect of the continuous column charge before and the blasting effect of the axial air separation charge scheme, and the analysis and calculation, the following blasting effect comparison table is obtained.

Table1 Comparison of blasting effects

Program	large chunks rate /%	Second broken number	Bootleg rate /%	Explosive factor kg/m^3	Resident floor vibration speed cm/s	Number of resident complaints
Former	23.71	4	4%	0.34	0.40	8
Improved	14.95	1	3%	0.23	0.21	0

From Table 1 shows, the use of deck charge technology could ensure that the rock blasting crush effectively. Rock blasting loose could meet the basic requirements of excavator loading. There are few particularly large chunks. And excavator loading efficiency could maintain the normal level. At the same time, we can see that the effect of air gap charge technology for reducing the chunk rate is more obvious, but the bootleg rate is basically consistent with the previous continuous column charge program. Through the blasting effect analysis, the explosive factor of air deck charge is about 0.23Kg/m^3, and the explosive factor of continuous column charge is 0.34 Kg/m^3. It can be seen the air gap charge technology could reduce the explosive consumption in the premise of ensuring the blasting effect, and reduced by about 32.35%. It can save the cost of explosives and improve the usage efficiency of explosive energy. The air gap charge has a good vibration reduction effect, and vibration speed decreased by 47.5%. The vibration within the scope of the Residents is not very strong, and the residents can basically accept it. The number of complaints is reduced from 8 person to no one, and the program improved got a good effect.

5 Conclusion

Through the analysis of the blasting effect of the axial air separation charge scheme, the following conclusions are drawn:

(1) Air deck charge blasting method can significantly reduce the chunk rate and ensure the normal efficiency of the excavator loading compared to continuous column charge blasting method.

(2) Deck charge blasting can effectively reduce the consumption of explosives, save explosives costs, and

improve the usage efficiency of explosive energy.

(3) The axial air separation charge has a good effect of reducing the blasting vibration, which can effectively solve the problem that the blasting vibration requirement is relatively high in the actual project.

References

1. LIU Quan-xing,WANG Wei-zheng.Application and effect analysis of spaced loading blasting[J].Blasting.2013. 30(2):115-117.
2. XU Wen-wen,YE Hai-wang,ZHANG Ying-ji,et al.Experimental study on air-decking charging controlled blasting to reduce vibration in jinduicheng open-pit mine[J].Blasting,2013.30(3):70-74.
3. ZHU Hong-bing.Study on the mechanism and application of air-decking blasting[D].WU Han:Wuhan University,2006.
4. WU Liang,XU Feng,LI Feng,et al.Study on vibration law on slope produced by air-decking blasting[J].Blasting. 2016.33(3):31-35.
5. QU Shi-jie,CHEN Guang-ping,SHANG Feng-hua.Theoretical analysis and field tests on effect of decked charges in vertical holes of bench blasting[J].Engineering Blasting.1999.5(4):57-61.
6. ZHANG Zhi-yi,YANG Nian-hua,LU Wen-bo,et al.Progress of blasting vibration control technology in china[J]. Blasting.2013.30(2):25-32.
7. LIU Shi-bo.Study on deep-hole blasting interval charging technique[J].Engineering Blasting.2010.16(4):39-43.

https://doi.org/10.1051/e3sconf/20183803032

The analysis of dynamic characteristics and wind-induced displacement response of space Beam String Structure

Yong Jian CHEN [1], Zhen Fa FENG [1], Ai QI[1], Ying HUANG [2]

[1]College of Civil Engineering, Fuzhou University, Fuzhou, Fujian 350108, China;
[2]Fujian Chuanzheng Communications College, Fuzhou, Fujian 350007, China

Abstract: The Beam String Structure structural system, also called BSS, has the advantages of lighter dead weight and greater flexibility. The wind load is the main design control factor. The dynamic characteristics and wind-induced displacement response of BSS are studied by the finite element method. The roof structure of the stadium roof of the Fuzhou Olympic Sports Center is the engineering background. 1)The numerical model was built by ANSYS, by shape finding, determine the initial stress state of structural members such as external cables; 2)From the analysis of dynamic characteristics, the main mode of vibration is the vibration of cables; 3)The wind speed spectrum of MATLAB generation structure is obtained by AR method, the structural response of the structure under static wind load and fluctuating wind load is calculated. From the analysis result, considering the equivalent static wind load of BSS , the design of adverse wind is not safe, and the fluctuating wind load should be taken into account.

1 Introduction

BSS is a new type of spatial structure, it has been applied more and more widely in recent years [1-3]. The structure is mainly composed of a brace and a prestressed cable to lift the arch truss, it will cause the structure to produce the anti arch, also may reduce the vertical deflection in the use stage. The BSS shows a strong nonlinearity, and the structural design is generally controlled by stability rather than strength. The natural vibration period of the BSS is close to the predominant period of the natural wind, frequency dense, less damping, the structure more flexible, it belongs to wind sensitive structure. It is very important to consider the wind-induced vibration coefficient in the design, but the research on wind-induced vibration response of long-span BSS is less. The static wind and fluctuating wind loads on the BSS of the Fuzhou Olympic Sports Center Stadium are numerically simulated, finite element method is used to calculate and analyze, the dynamic characteristics and the structural displacement response under static wind load and fluctuating wind load are studied.

2 Numerical Investigation

2.1 Engineering Background

The reinforced concrete frame shear wall structure is adopted in the lower part of the Fuzhou Olympic Sports Center Stadium, the superstructure of the steel roof is composed of a quadrilateral ring, a string chord and a chord and a chord structure, and a plane primary and secondary truss structure, the steel wall is made of single-layer rhombus crossed grid structure. The plane primary and secondary truss structure in the competition area is distributed in all the roof structures outside the grandstand of the competition area, it can be divided into two kinds: radial truss and ring truss, the maximum span of radial truss is 35m; the maximum span of the X axis in the training area is 120m, the maximum span of Y axis truss is 73m. Fuzhou Strait Olympic Sports Center Stadium as shown in Figure 1.

* Corresponding author:cyjordan@foxmail.com

https://doi.org/10.1051/e3sconf/20183803032

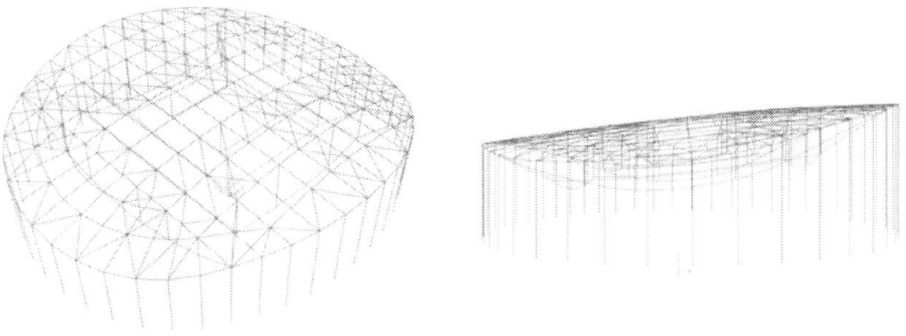

Figure 1 Fuzhou Strait Olympic Sports Center

2.2 The Finite Element Model

A three-dimensional calculation model is established by using ANSYS, upper cord, belly Bar and stay bar of the Link8 element, the Link10 element is used in the cable element, The prestressing force in the prestressed cable is applied to the structure by defining the initial strain. The total number of elements is 643, the total number of nodes is 821. The finite element model is shown in figure 2.

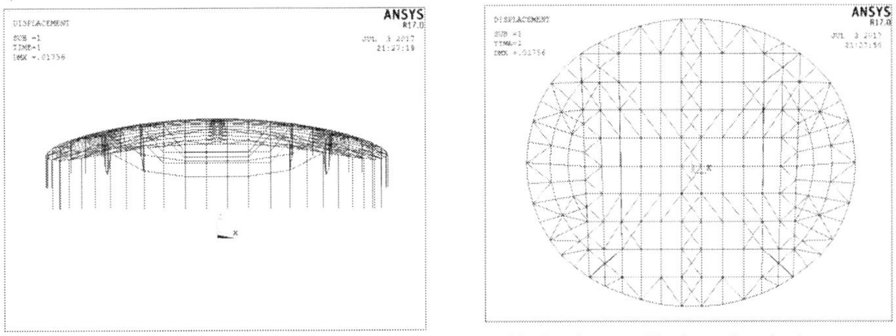

a) Sketch map of elevation b) Sketch map of horizontal projection

Figure 2 The Finite Element Model

To establish a computational model that conforms to the initial state of the actual structure, the shape of BBS must be formfinding. Because the BSS belongs to the prestressed beam cable structure system, The geometry and dimensions of the system in the initial state are generally given by the design unit: that's the size of the blueprint. In the process of exerting prestressing force, Zero State(setting out state without self weight or prestress, the equivalent of BSS placed on the temporary support, no amount of prestressed tension) of BSS gradually deforms to the initial state(the equilibrium state of gravity and prestress acting together, after the end of tension tensioning, the temporary support has exited the working state). When the deformation reaches the geometric requirement of the initial state, the prestress of the external cable reaches the tensile design requirement, this means that the tension of the external cable ends. The spatial position of each node of the structure and the overall stiffness of the structure can be determined by means of formfinding. formfinding has a great influence on the dynamic characteristics of the structure and the wind load response[8].

3 The Analysis Of Dynamic Characteristic

The structure is analyzed by modal analysis after the structure is formfinding, after removal of out of plane (lateral) vibration modes, the first 10 order natural frequencies as Table 1 and the first 6 modes are shown in Figure 3.

Table 1 Structural Natural Frequency

Formation	1	2	3	4	5	6	7	8	9	10
Frequency (Hz)	0.79	0.87	1.04	1.07	1.36	1.42	1.71	1.78	1.93	2.185

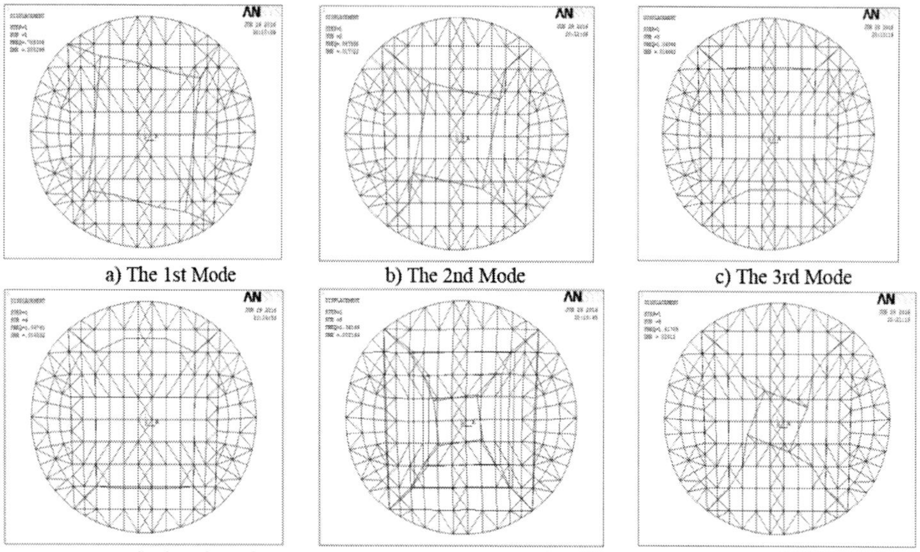

a) The 1st Mode b) The 2nd Mode c) The 3rd Mode

d) The 4th Mode e) The 5th Mode f) The 6th Mode

Figure 3 Vibration pattern of BSS in Olympic Sports Center

From Table 1, the fundamental frequency of the structure is about 0.785, the frequency distribution tends to rise uniformly. The structure is similar to some large span space structures such as reticulated shell, its frequency spectrum is close, the mode characteristic frequency is shown in Figure 4.

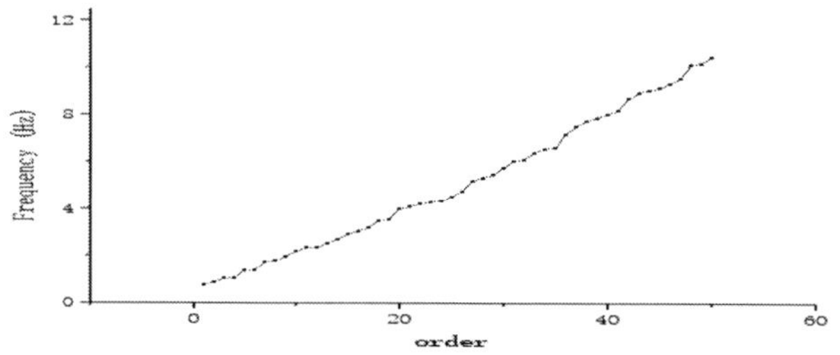

Figure 4 Spectrum Characteristic Of Structure

Figure 4 shows the ladder like distribution of frequency, at the section of the formation ladder, the vibration modes of structures also exhibit obvious characteristics of change. The frequency distribution is relatively dense, and the order reaches 40, when the structure is subjected to dynamic analysis, the number of vibration modes is best not less than 40 order. The vibration modes of structures are mainly the vibrations of cables, the upper space truss structure does not participate in vibration, Therefore, it is necessary to pay attention to the structure of the ring cable.

4 Simulation of Wind Load

4.1 Static Wind Loads

According to the regulations, the basic wind pressure in Fuzhou area is 0.7 kN/m^2, the horizontal average wind speed at the h=10m is 26.8 m/s. The site roughness class is B, the elevation of the constraint position is 32.3m, the direction of the wind is 0°(90°). In this design, the wind load on the windward (leeward) side of the structure is selected along the direction of the wind direction along 0°(90°). The calculation is carried out according to the standard.

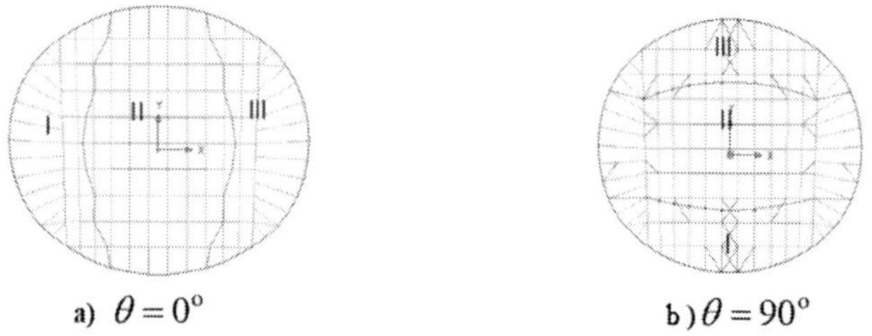

a) $\theta = 0^{\circ}$ b) $\theta = 90^{\circ}$

Figure 5 Structural Region Division Under Different Wind Directions

According to the code for load of building structural load regulation[4]: Calculate according to the following formula when standard value of wind load perpendicular to the surface of a building.

$$\omega_i = \mu_s \mu_z \omega_0$$

The building is divided into three zones according to the type and shape of the building, as shown in Figure 5.

4.2 Fluctuating Wind Load

The wind speed vector at any time can be considered as two parts: mean velocity and fluctuating velocity,

for the average wind speed $\bar{v}$, the wind speed and wind direction data recorded at the wind speed station near the engineering site can be fitted by the extreme wind speed respectively, and the extreme wind speed during the return period of the structure is predicted. The mean wind speed last night is [5]. The basic wind pressure ω_0 can be determined by $\omega = 0.5\rho v^2$, and then the average wind speed of the structure is obtained by the formula conversion.

For pulsating wind velocity vectors, it is assumed that the elements have the same spectrum and have some spatial correlation, and can be further written in the following form:

$$\vec{v}(t) = C \cdot \vec{u}(t)$$

In the formula, C is the correlation matrix, which can be obtained by the inverse transformation of the cross spectral density of the wind speed. It is a vector Fourier composed of the uncorrelated fluctuating wind speed time histories[6]. This paper uses MATLAB programming to simulate the fluctuating wind speed of structure [7], the values of the relevant parameters are as follows: According to the basic wind pressure once every 50 years in Fuzhou area; at the height of 10m wind speed is 31m/s; Cx=16, Cy=8, Cz=10; the roughness coefficient y=0.185, K=0.003; the total time step N=1024, when the distance is 204.8s, the time step is 0.2s. The autoregressive order of the AR model is 6. The calculated wind speed time history is shown in Figure 6.

a) 44 wind spectrum time history curve b) 38 wind spectrum time history curve

Figure 6 Wind Speed Time Curve

5 Structural Displacement Response

In the dynamic response analysis of structures, the static analysis gives the statistical moments of structural responses. In the analysis process, many mathematical simplifications of the structure are needed [9]. In particular, for the nonlinear system with BSS, its stiffness varies with the load, whereas in static analysis it is generally assumed to be invariant and its computational reliability is low. The time history analysis method for vibration frequency of intensive and large flexibility, obvious non-linear structure, the analysis of wind-induced response, the spectrum function for a given wind load, the wind load simulation as a function of time, and then use the finite element discrete structure and, in the role of simulation nodes the wind load, obtained the response of the structure through the method of time domain directly solving the differential equation of motion, correcting the stiffness of the structure in each time step, so the nonlinear factors have also been considered.

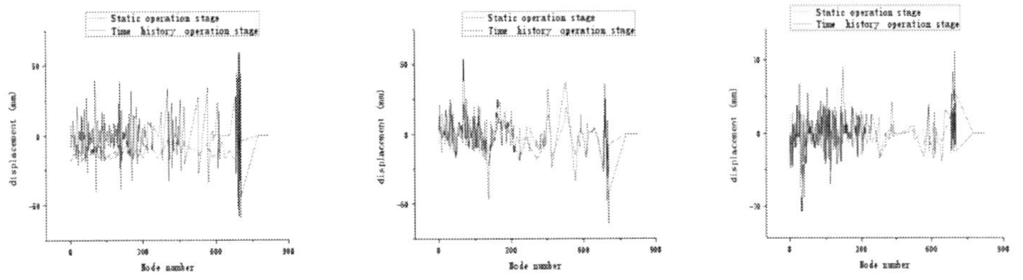

a) x displacement under X wind load b) y displacement under X wind load c) z displacement under X wind load

https://doi.org/10.1051/e3sconf/20183803032

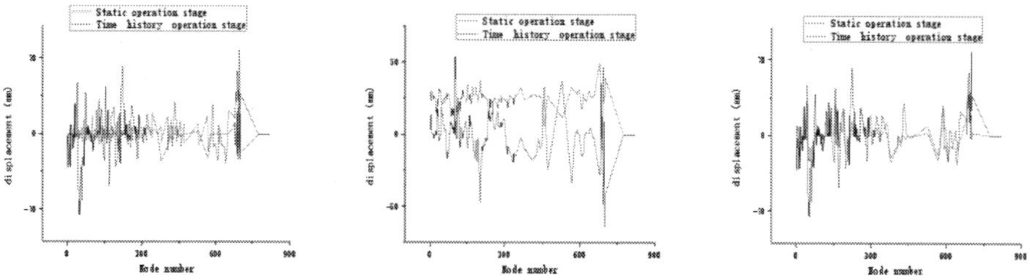

d) x displacement under Y wind load e) y displacement under Y wind load f) z displacement under Y wind load

Figure 7 Displacement Response Of Structures Under Wind Load

According to the analysis of Figure 7, with the static wind load, the response of each node under the action of fluctuating wind load is greater. And the displacement response of the node is almost envelope. When the structure is in use, the frame is dismantled, the prestressed cable plays a role, the structural rigidity becomes smaller, and the flexibility is larger. The structure moves larger under the action of the wind load, but it is still in a safe state.

6 Conclusion

In this paper, the dynamic characteristics and wind-induced vibration response of the space BSS structure of Fuzhou Olympic Sports Center are studied by the finite element method. The main conclusions are as follows:

1. The structural dynamic characteristics are studied. The fundamental frequency of the structure is 0.785, and the frequency distribution tends to rise uniformly. Like some long-span space structures, such as reticulated shell, the structure is dense and has no big jumps. The vibration modes of the structure are mainly the vibration of cables, the overall stiffness of the upper space grids is larger, and the formation of the structures is not involved in the formation, so it is necessary to pay more attention to the structure of the ring cables.

2. Comparison of structure wind vibration displacement response of key node, that node displacements the maximum value appears in the central axis, time-domain analysis frequency domain method of wind vibration displacement response is greater, trend and displacement response time Tagmemic frequency shift response similar trend. It can be seen that for the wind sensitive structure like BSS, the effect of fluctuating wind must be taken into account. The design of wind resistant structure is only unsafe when considering the equivalent static wind load only.

Acknowledgements

The authors acknowledge the support from the Project of Science and Technology Research of Department of Education, Fujian Province(Grant Nos. JK2012004, Grant Nos. JK2013001, Grant Nos. JA14039), Project of Natural Science Foundation of Fujian Province(Grant Nos. 2016J05122).

References

1. Zhou Yin. Theoretical and experimental investigations on effective static wind loads on high rise buildings [D]. Shanghai: Tongji University, 1998. (in Chinese).
2. Li Shouying,Chen zhengqing,Huang Lei,etal.Test study of wind load on roof of Yiwu Swimming Museum [J].Journal of Hunan University, 2007 , 34(5): 10-14.(in Chinese).
3. hang Xiangting. The current Chinese Code on wind loading and comparative study of wind loading codes [J]. Journal of Wind Engineering and Industrial Aerodynamics, 1988, 30(1-3): 133-142.
4. huang Bencai. Wind-resistance theory of structure sand applications [M]. Shanghai: Tongji University Press, 2001. (in Chinese).
5. Zhang Zhihong, Zhang Mingshan, Dong Shilin. Discussion on some problems of beam string structures [J]. Engineering Mechanics, 2004, 21 (6) :26-30 (in Chinese).
6. Wang Xiuli, Ding Nansheng, Chai Hong. Seismic response and parametric analysis of beam string structures [J]. Spatial Structures, 2006, 12 (2):34-38 (in Chinese).
7. Killen G P, Letchford C W. A parametric study of wind loads on grandstand roofs [J]. Engineering Structures, 2001, 23:725-735.
8. Qiao Wentao, Han Qinghua. Analysis of wind-induced vibration response of plane beam string structure [J]. Journal of Hebei University of Technology, 2006 , 35 (6):92-98 (in Chinese).
9. He Yanli,Dong Shilin, Gong Jinghai. Wind-induced response of spatial reticulated structures with mode compensation in frequency domain [J]. Engineering Mechanics, 2001, 19 (4):1-6 (in Chinese).

Modal Parameter Identification and Numerical Simulation for Self-anchored Suspension Bridges Based on Ambient Vibration

Bing LIU [1], Li Guo SUN [2]

[1]Department of Civil Engineering, Nanjing Technical Vocational College, Nanjing, Jiangsu 210019, China
[2]College of Mechanics and Materials, Hohai University, Nanjing 210098, China

Abstract: This paper chooses the Nanjing-Hangzhou high speed overbridge, a self-anchored suspension bridge, as the research target, trying to identify the dynamic characteristic parameters of the bridge by using the peak-picking method to analyze the velocity response data under ambient excitation collected by 7 vibration pickup sensors set on the bridge deck. The ABAQUS is used to set up a three-dimensional finite element model for the full bridge and amends the finite element model of the suspension bridge based on the identified modal parameter, and suspender force picked by the PDV100 laser vibrometer. The study shows that the modal parameter can well be identified by analyzing the bridge vibration velocity collected by 7 survey points. The identified modal parameter and measured suspender force can be used as the basis of the amendment of the finite element model of the suspension bridge. The amended model can truthfully reflect the structural physical features and it can also be the benchmark model for the long-term health monitoring and condition assessment of the bridge.

1 Introduction

The main cable of the self-anchored suspension bridge is anchored directly at both ends of the stiffening beam, and the horizontal force of the main cable and the vertical force of the anchor cable are directly supported by the stiffening beam. The self-balancing system is formed in the horizontal direction of the stiffening beam. In addition to its general ground anchor suspension bridge beautiful shape and good mechanical properties, it need not build the large volume of anchorage, which can be applied to poor geological conditions and urban areas without anchorage. However, since the anchorage point of the main cable is on the main beam, the strength, stiffness and stability of the main girder are critical to the structural safety of the full bridge. As the main force component, under the load of vehicle et al., the main beam is prone to undergone, crack, the effectiveness reduce of prestress and the overall stiffness reduce of the structure.Therefore, regular health monitoring of such bridges is an important method for ensuring the safe operation of the bridge during the design life. With the rapid development of structural health monitoring technology, more and more structures have established a health monitoring system [1, 2]. Due to the complex structure of the bridge and the large number of external factors, there are many uncertainties. How to use the monitoring data of a limited number of sensors to evaluate the health status of the structure is an urgent problem facing the academic and engineering circles. With the increasing ability of large-scale numerical calculation, it is possible to establish a complex and realistic numerical model, and it is possible to simulate the structural response of large-scale civil structure under various working conditions [3, 4]. Under the condition that the simulated structural response and the actual monitoring data have good compatibility, the numerical simulation results can make up the deficiency of the actual monitoring data, and realize the damage identification and state assessment of structures. In order to achieve the above objectives, the primary task is to establish with a sufficient accuracy, reflecting the bridge dynamic characteristics and physical characteristics.

Because of the errors of the model, the error of the structural parameters and the simulation difficulties of the boundary conditions, the bridge structure response and the actual response will be different, even if the finite element model is established strictly according to the

* Corresponding author: 67689704@qq.com

design drawings. The model must be corrected and verified, in order to get the actual state of the structure of the benchmark finite element model [5]. So far, many researchers have used different methods to study the different bridge structures based on the measured data [6-8]. However, the structure of the self-anchored suspension bridge is rarely studied. Based on this, the initial finite element model of the bridge structure is established by using the large-scale finite element software ABAQUS, which is based on the Nanjing-Hangzhou high-speed cross-bridge and learning eight-way self-anchored suspension bridge. The finite element analysis model of the bridge structure under the working environment was modified by the measured structural response, the internal force of the boom and the main cable, and so on. The results of the excitation experiment with the ambient environment were as close as possible and finally established In the large-scale bridge structure of the finite element model to amend the practical method to serve the bridge of health monitoring and status assessment.

2 Project background

Learning eight way self-anchored suspension bridge is located in Nanjing Chunhua Town, and the bridge is 160 meters long. The bridge across the Nanjing-Hangzhou Expressway from west to east, is the double-sided tower beam consolidation self-anchored concrete suspension bridge. The main span is 80 m, edge is 40 m, and cross span ratio is 1/5 (16 m). Full bridge length is 40 + 80 + 40 = 160 m. The main cable transverse bridge from the center is 22 m. The sling along the bridge to the spacing is 5 m. Design load standard: city-A level, sidewalk live load is 3.5 KN/m2. The main beam adopts C50 prestressed concrete box girder. The standard section is single box four rooms, the beam width is 29 m, the standard beam section bottom width is 22.65 m, and the cantilever width is 3.175 m on each side. The tower is a "column" type reinforced concrete structure, the bridge above the high is 20.6 m, and the full height is 28.2 m. There are total 50 bridge booms. The four booms near the anchor is Ø100 mm rigid booms, and the material is the galvanized 40Cr steel. The total eight booms on both sides of the tower are 187Ø5 mm high-strength galvanized steel cable, and the remaining booms are 139Ø5 mm high-strength galvanized steel wire products. The standard strength is 1670 MPa, double PE protective layer, cold anchor anchor system. The lower part of the main bridge structure: the abutment for the bridge plate plus rock-filled pile, the tower under the pillar for the rectangular pile. The bridge structure is shown in Figure 1.

(a) The view under the bridge (b) The view on the bridge

Figure 1 Learning eight way self - anchored suspension bridge

3 Establishment of Three-dimensional Finite Element Model

The finite element model of the learning eight-way self-anchored suspension bridge was established according to the construction drawing using the general software ABAQUS, which is shown in Figure 2. The numerical model (so-called fishbone model) is used to simulate the real situation of the suspension bridge. The rod unit and the solid element are applied in the calculation model at the same time, and the components with similar mechanical properties are simulated to improve the accuracy of the numerical model. The main cable and the boom are modeled by the rod unit. The bridge deck and the pagoda are modeled by hexahedral solid elements. The total number of points in the model is 67216, the total number of units is 40572, and the number of hexahedral entities is 40188. The number of boom units is 394, the connection between the main cable and the boom. The main cable and the bridge deck and between the boom and the bridge deck is connected by a JOIN connection, and the hinge connection established by such a connection restrains the same line displacement. The finite element model of this paper does not take into account the contribution of the following parts of the bridge pile. The degree of freedom of the bottom of the pier is all constrained, and the side span and the platform are connected by simply supporting the beam. In order to make the calculation result more accurate, the calculation model simulates the true box girder structure of the bridge, as shown in Figure 3 (the finite element model) and Figure 4 (boom number diagram). As the suspension tension of the suspension bridge has great influence on the linear and vibration characteristics of the bridge, the cable force of the test is applied to the corresponding boom by the cooling method in order to ensure the reliability of the calculation results. Figure 5 shows the measured cable force of the learning eight-way self-anchored suspension bridge. Since the test is to use the vibration method to measure the cable force of the boom, the four rigid boom cable forces near the abutment can not be accurately determined. The design boom power is still used in the model.

4 Modal parameter identification

Bendat et al. [9] published the "correlation and spectral analysis engineering application" in detail the peak picking method (Peak Picking, PP). The peak method based on the structure of the frequency response function at the natural frequency will have a peak principle. When the output response is used, the frequency spectrum function is used to replace the frequency response function. The modal characteristic frequency is determined by

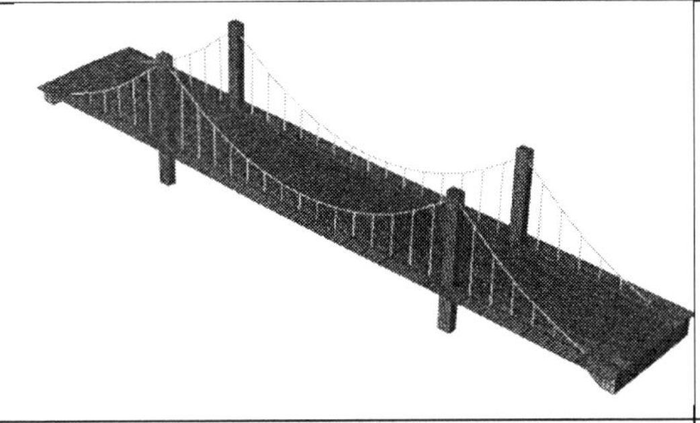

Figure 2 Finite Element Global Model of learning Eight-way Self-anchored Suspension Bridge

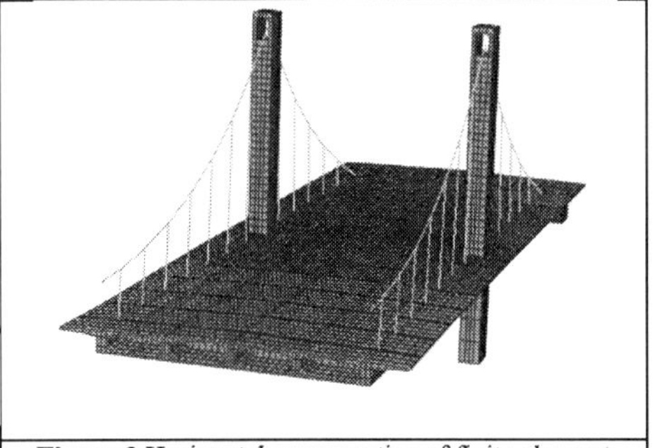

Figure 3 Horizontal cross-section of finite element model of learning Eight-way self-anchored suspension bridge

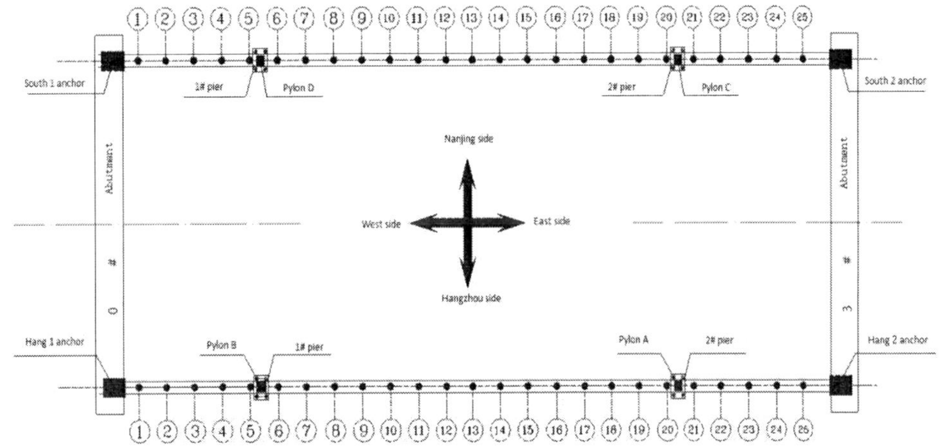

Figure 4 Boom number diagram

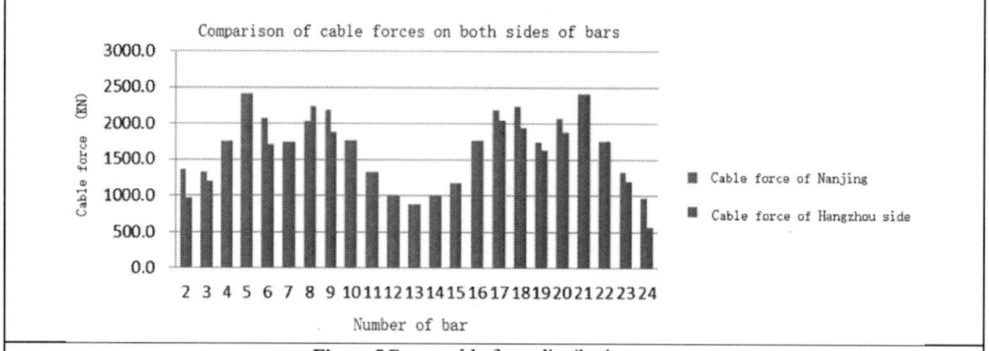

Figure 5 Boom cable force distribution map

The peak of the average regularized power spectral density curve. The modal damping ratio is determined by the half power band method.

4.1 The basic theory of peak method

For the multi-degree of freedom damping system, the dynamic equation of the system vibration is:

$$M\ddot{x}(t)+C\dot{x}(t)+Kx(t)=F(t) \tag{1}$$

Laplace transformation can be converted into

$$H(s)=\frac{X(s)}{F(s)}=[Ms^2+Cs+K]^{-1} \tag{2}$$

$H(s)$ called the system transfer function. If $s=j\omega$, the transformation process is equivalent to the Fourier transform process, the transfer function $H(s)$ is the frequency response function:

$$H(s)|_{s=j\omega}=H(j\omega)=H(\omega)=\frac{1}{K-M\omega^2+j\omega C} \tag{3}$$

For a real-mode system, the frequency response function is further decomposed by Equation (3) :

$$H_{ik}(\omega)=\frac{x_i}{f_k}=\sum_{r=1}^{N}\frac{\phi_{ir}\phi_{kr}}{k-\omega^2m_r+j\omega c_r}=\sum_{r=1}^{N}\frac{\phi_{ir}\phi_{kr}}{(j\omega-\lambda_r)(j\omega-\bar{\lambda}_r)} \tag{4}$$

Where N is the order of the modal, ϕ_{ir} and ϕ_{kr} are the modes vector at the i and k points of the r-order modal, respectively. λ_r and $\bar{\lambda}_r$ are the conjugate eigenvalues of the structure, respectively.

The dynamic displacement response $x_i(\omega)$ of any point i on the structure can be expressed by the k-point excitation force $f_k(\omega)$ and the structural frequency response function $H_{ik}(\omega)$:

$$x_i(\omega)=\sum_{k=1}^{m}H_{ik}(\omega)f_k(\omega) \tag{5}$$

For environmental excitations, it can be approximated as a white noise signal with uniformly distributed frequency energy. It means that the power spectral density function is uniform distribution over all frequency ranges, so the input at each location satisfies the condition:

$$f_k(\omega)=f(\omega)=C_1 \tag{6}$$

C_1 is a constant. Thus, equation (6) can be simplified as:

$$x_i(\omega)=f(\omega)\sum_{k=1}^{m}H_{ik}(\omega)=C_1H_i(\omega) \tag{7}$$

As can be seen from the above equation, the response of the structure $x_i(\omega)$ and the total frequency response function of the structural system are equivalent, so the natural frequency of the structure can be obtained directly from the corresponding spectrum $x_i(\omega)$.

Select a point of the structure as a reference point, the transfer rate between the response point and the reference point was established to identify the modal vibration of the structural system. The identified vibration mode is not the true mode of the structure, but the work deflection of the structure. It need to set a point which has a clear response in all the frequency as the reference point. The transfer rate between the response point and the reference point can be expressed as:

$$\alpha_i(\omega)=\frac{x_i(\omega)}{x_p(\omega)} \tag{8}$$

The sequence $\alpha_i(\omega)$ is the working deformation of the frequency ω corresponding to the structure, and it can be regarded as the modal shape of the system. Equation (4) can be taken into the above formula, obtain:

$$\alpha_i(\omega)=\frac{x_i(\omega)}{x_p(\omega)}=\frac{f(\omega)\sum_{k=1}^{m}H_{ik}(\omega)}{f(\omega)\sum_{k=1}^{m}H_{pk}(\omega)}=\frac{\sum_{k=1}^{m}H_{ik}(\omega)}{\sum_{k=1}^{m}H_{pk}(\omega)} \tag{9}$$

Assuming that the solid modes of the structures can be effectively separated from each other and there is no coupling or only weak coupling between them, the system response at the rth order natural frequency is dominated by the vibrations of the rth order mode. Equation (4) and equation (9) can be simply written as:

$$\alpha_i(\omega)=\frac{\frac{\phi_{ir}}{(j\omega-\lambda_r)(j\omega-\bar{\lambda}_r)}\sum_{k=1}^{m}\phi_{kr}}{\frac{\phi_{pr}}{(j\omega-\lambda_r)(j\omega-\bar{\lambda}_r)}\sum_{k=1}^{m}\phi_{kr}}=\frac{\phi_{ir}}{\phi_{pr}} \tag{10}$$

In the above equation, since the point p is fixed as a reference point, it is a constant value ϕ_{pr} for a certain natural frequency ω_r. So the above equation can be simplified as:

$$\alpha_i(\omega)=\frac{\phi_{ir}}{\phi_{pr}}=C_2\phi_{ir}$$

(11)

From the above equation, the modal mode of the structure at the frequency ω_r can be directly read the transfer rate curve $\alpha_i(\omega)$ to obtain. The magnitude of the mode is obtained from the amplitude of the transmission rate at the natural frequency, and the direction of the mode at the frequency or the real part of the transfer rate can be obtained by the crossfire spectrum.

4.2 Modal Parameter Identification of Suspension Bridge Deck

The above-mentioned theory shows that the modal parameters can be obtained under the condition that the excitation is unknown, by estimating the self-power spectrum and the cross-power spectrum between the measured points, combining the phase and the transfer rate. Because the environmental excitation is close to the white noise excitation in a certain frequency band, the grid structure is linear small damping vibration under the environment excitation. Based on the assumption of the peak method, the vibration response of the reference point can be used as input excitation. In this way, the frequency response function can be directly replaced by the self-power spectrum and the cross-power spectrum of the vibration response to estimate the parameters such as frequency, damping and mode. According to the identification method of the peak method, the cantilever bridge deck modal parameters can be identified only by the vibration response of the bridge deck.

(1) Construction of the suspension bridge deck model. According to the analysis, the bridge deck composed of the stiffening beams was the main force component of the self-anchored suspension bridge. The structure of the box girder composed of the deck system is directly related to the overall safety of the whole bridge. The change of the internal force of the boom and the local damage of the box girders directly lead to the change of the vibration characteristics of the deck system. Therefore, this paper only studies the vibration characteristics of the deck system. As the transverse stiffness of the deck system is much larger than the vertical stiffness, vibration is mainly expressed in vertical vibration. According to the principle of less and fine, the measurement points on the deck system are as follows.

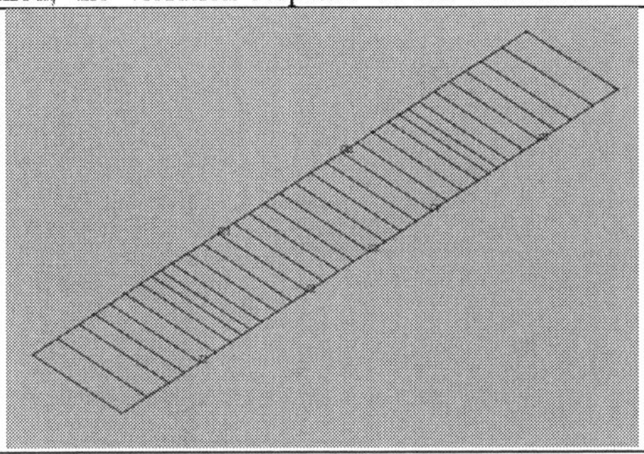

Figure 6 Design of Measuring Points for Modal Recognition of Deck System

Figure 7 Test Scene Photos

(2) Create a data file and import the acquired modal data. There were seven measurement points in the modal analysis, where C4 was taken as the base point. The collected data were imported into three batches, and each node of the model was assigned. The real value of the measurement is given to the measuring point. For non-measuring points, it is interpolated by the measuring point value, see Figure 6 and Figure 7.

(3) FFT transformation of the data. Since the data acquisition time of each measuring point in the modal test was about 1800 s, and the sampling frequency was 50 Hz, so the total data were about 90000 points. Therefore, the size of each FFT data was 8192 (213), and the signal leakage can be effectively reduced by Hanning window (hanning) function for window processing. It is possible to improve the accuracy of spectral estimation, increase the average number of times, reduce the variance and make full use of the start and end data of the segment by using the overlap processing and averaging processing in the spectrum analysis. The overlap ratio was set for 50%, and the processing was selected linear averaging processing. After FFT transformation of the collected data, we can get the modal parameters of the self-power spectrum amplitude, phase and coherence function to identify the modal parameters of the system, as shown in Figure 8.

(4) Carry out the discrimination of the real mode. The screening of frequency was subjective, and the above frequency of screening with a lot of false patterns were also identified. It need to be identified. Because the frequency and mode of the multi-degree of freedom system are the characteristic relations of eigenvalues and eigenvectors in mathematical problems. According to the vibration analysis theory, the vibration vector of the same order is completely related, and the vibration vector of different order should be unrelated. This principle can be used as a tool for modal model validation. The modal confidence factor matrix can be constructed. If it is true, the non-diagonal elements in the matrix should be close to zero. According to these criteria, the natural frequencies of vibration in the vertical direction of the bridge are 1.44 Hz and 3.09 Hz. The learning eight-way self-anchored suspension bridge deck vertical vibration mode which was identified out can be seen in Figure 9.

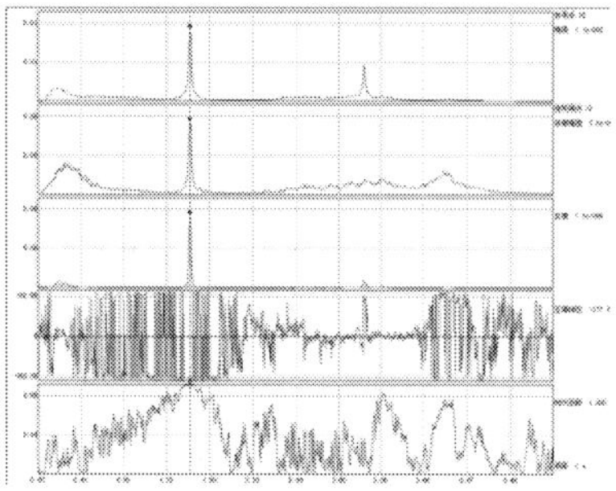

(a) C1 Point spectrum analysis

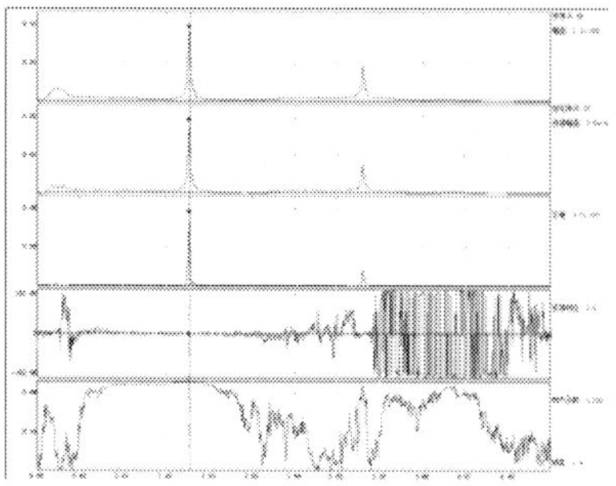

(b) C2 Point spectrum analysis

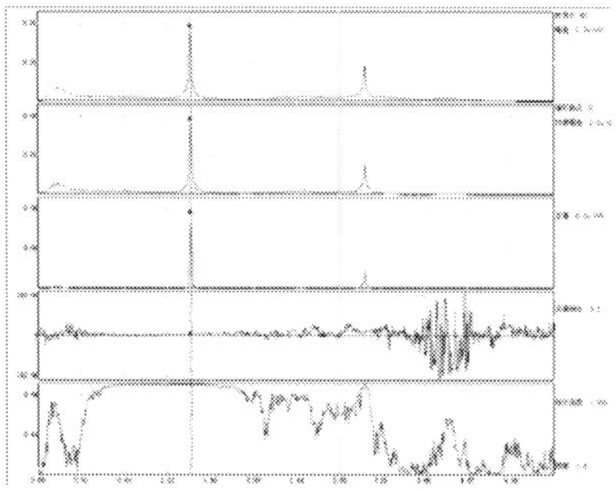

(c) C3 Point spectrum analysis

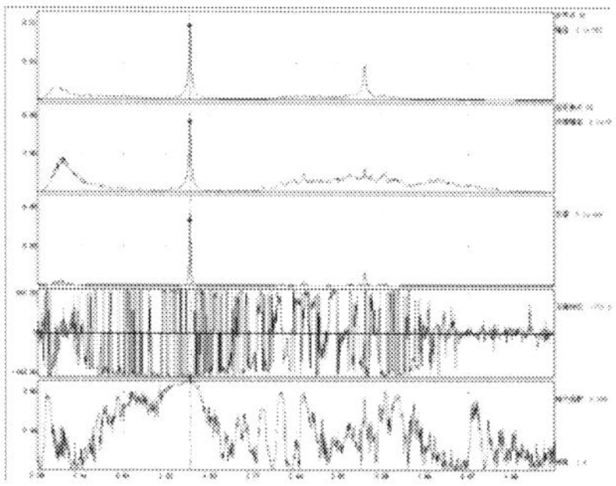

(d) C5 Point spectrum analysis

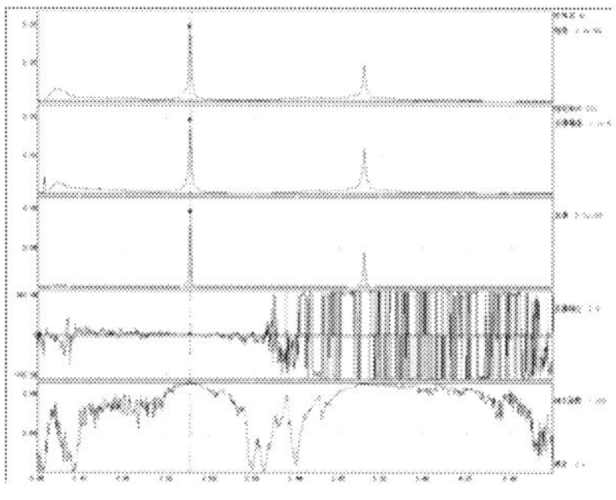

(e) C21 Point spectrum analysis

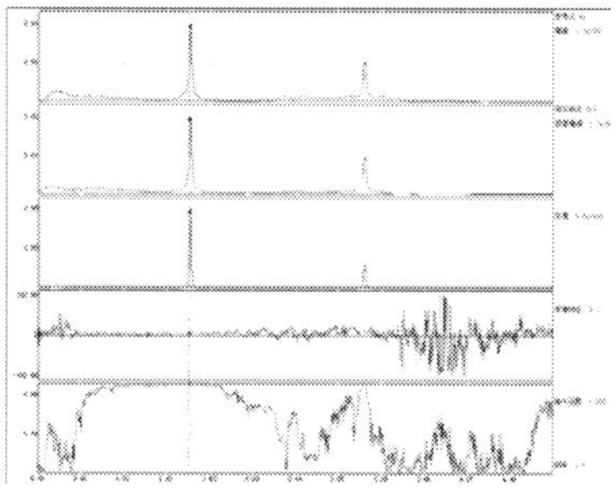

(f) C41 Point spectrum analysis

Figure 8 The self-power spectrum amplitudes, phases and coherence function curves between each measuring points and reference points

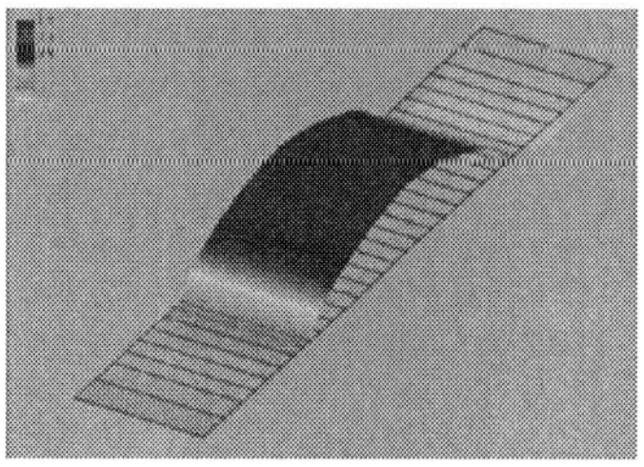

(a) Deck 1.44 Hz corresponding to the vibration

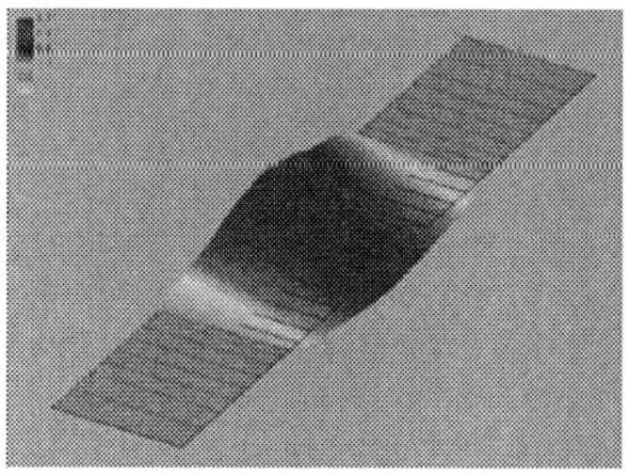

(b) Bridge diagram of 3.09 Hz corresponding to the vibration

Figure 9 Measured bridge vibration pattern

5 Finite Element Simulation Analysis of Bridge Vibration Characteristics

The analysis of the dynamic characteristics of the structure can be also called modal analysis of the structure. The modal of the structure refers to the basic vibration characteristics of the structure in the event of free vibration. It usually includes parameters such as natural frequency, natural mode, modal mass, modal stiffness and modal damping ratio. The most important parameters are the frequency, vibration mode and damping ratio. For the damping system, they are the natural frequency and vibration mode. The main reason for the analysis of the self-vibration characteristics of the structural system is as follows: 1. the natural frequency and mode of the structure are an inherent property of the structure itself. It depends only on the stiffness and mass distribution of the structure itself. Therefore, the vibration force and vibration mode of the structure can be used to analyze the stiffness of the structure and the distribution rationality of the stiffness and mass. 2. During the forced vibration, the correlation between the internal force and the displacement of the structure and the vibration frequency and vibration form of the structure is closely related. The natural frequency and vibration mode are the main parameter to calculate the forced vibration. 3. Therefore, analyzing the natural frequency and vibration mode is the basis of the dynamic response analysis of the structure. The finite element analysis of this bridge testing project will be carried out on commercial software ABAQUS. The pre-processing function of ABAQUS is relatively wonderful, and the function to solving non-linear problems is also better.

5.1 Numerical model

In this numerical modeling, the fishbone model [10] is used to simulate the real situation of the suspension bridge. The so-called fishbone model is used in the calculation model, which simulates the components with similar mechanical properties and improves the calculation accuracy of numerical model. The main cable and the boom are modeled by the rod element. The bridge deck and the pagoda are modeled by the hexahedral solid element. The total number of points is 67216, the total number of the units is 40572, the number of hexahedral entities is 40188, and the number of boom units is 394, as shown in Figure 10. Material parameters are in accordance with the relevant norms of value.

Figure 10 Finite Element Calculation of Integral Model of learning Eight-way Suspension Bridge

In order to make the calculation result more accurate, the calculation model simulates the true box girder structure of the bridge, as shown in Figure 11 and Figure 12.

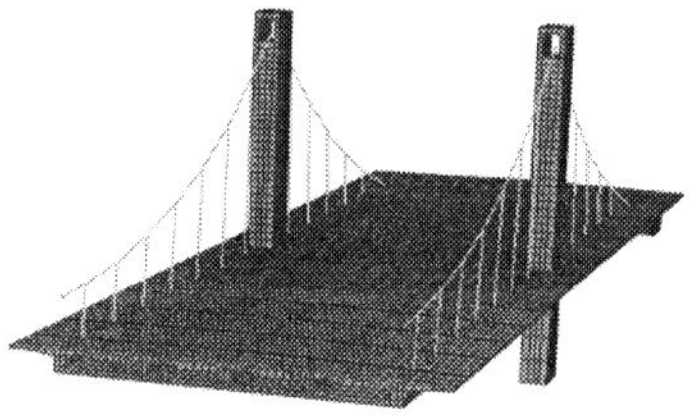

Figure 11 The calculation model transverse section of learning eight-way suspension bridge

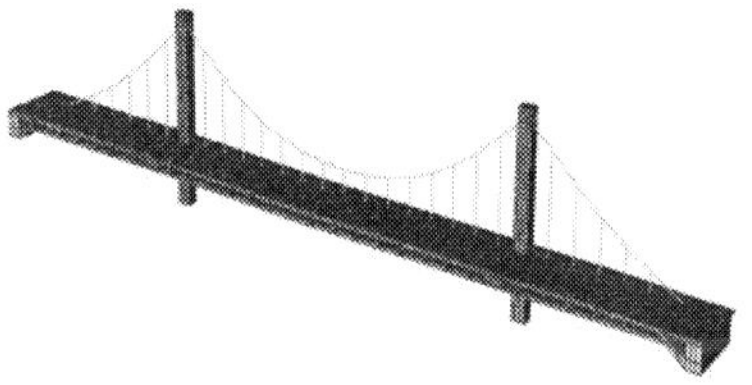

Figure 12 The calculation model longitudinal section of learning eight-way suspension bridge

5.2 Finite element analysis

The first-order natural frequency of the bridge deck is 1.17 Hz, and the analysis result of the self-interacting modal is 1.44Hz, which means that the error is 18.8 %. The first-order mode is shown in Fig. 13, and the results are consistent with modal analysis results. The second-order natural frequency is 2.45 Hz, and the error is 20.7 % comparing with the result of modal analysis 3.09 Hz. The second-order mode is shown in Figure 14, which is consistent with the result of modal analysis. The third and fourth order natural frequency are 3.14 Hz, 3.61 Hz, respectively, corresponding to the vibration mode as shown in Figure 15 and Figure 16. The two modes are not self-phantom modal analysis identified, indicating that the actual vibration of the environment did not stimulate the excitation of the bridge of these modes. As the actual material parameters of the bridge need to be obtained by the relevant experiments, the calculation of the material parameters of the model are in accordance with the relevant norms. The calculated natural frequency is not so accurate.

Figure 13 The first-order mode diagram

Figure 14 The second-order mode diagram

Figure 15 The third-order mode diagram

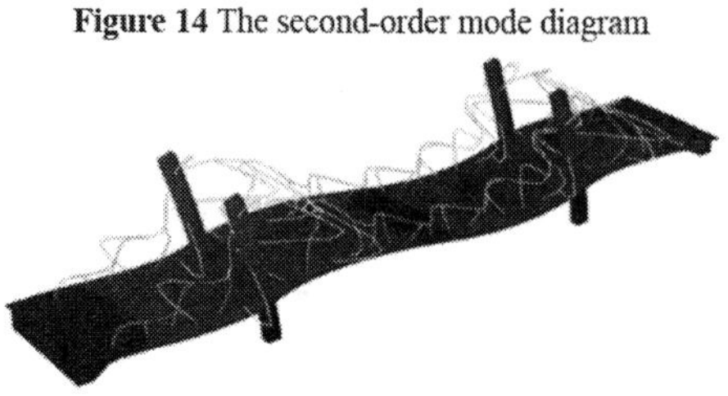

Figure 16 The fourth-order mode diagram

6 Conclusion

Using field testing and finite element simulation, the paper shows the below results: 1) The fishbone model is used to simulate the real situation of the suspension bridge. 2) Peak method is simple and rapid, which have clear physical meaning and is suitable for analysis the modal parameters of the structure under the excitation environment. It can obtain more accurate frequency and vibration results. 3) Using the optimized method to modify the frequency of the grid, the revised design parameters can be well corresponding to the actual state of the grid. The corresponding finite element model reflects the dynamic characteristics of the grid more truly. The modified model can be used as a reference model for the vibration reduction of the grid.

Due to the complex structure of the bridge and the cumbersome material used in the construction process, a good model correction process must be based on the accurate data and the actual material parameters obtained from the experiment. This is the key to the success of the model correction technique and the direction of further study.

References

1. Miao Chang-qing, Han Hui-ting, Li Ai-qun et al. Analysis on the Original SHM System of Jiangyin Bridge [J]. Journal of Highway and Transportation Research and Development, 2007, 24 (11): 81-86.
2. Zhou Wen-song, Li Hui, Ou Jin-ping et al. Design Method of Data Acquisition System of Health Monitoring Systems for Long Span Bridges [J]. Journal of Highway and Transportation Research and Development, 2006, 23 (3): 83-87.
3. Ji Wei-hong, Song Yu-pu, Liang Bing. Numerical Simulation of Damage Diagnosis of Space Truss Structures by ANSYS [J]. Journal of Wuhan University of Technology (Traffic Science and Engineering), 2006 (2): 285-288.
4. Gao Wei-cheng, Liu Wei. Numerical Simulation and Experimental Study on Damage Identification of Reticulated Shell [J]. Engineering Mechanics, 2006 (10): 111-117.
5. Wang Hao, Wang Fu-quan, Li Ai-qun et al. Multi-phase FE Model Updating on Long-span Cable-supported Bridges [J]. Engineering Mechanics, 2009, 26 (10): 111-117.
6. Jaishi B, Ren W X. Structural finite element model updating using ambient vibration test results [J]. Journal of Structural Engineering, 2005, 131(4): 617-628.
7. Li Zhao-xia, Li Ai-qun, Chen Hong-tian, et al. Finite Element Modeling for Health Monitoring and Condition Assessment of Long-Span Bridges [J]. Journal of Southeast University: Natural Science Edition, 2003, 33 (5): 562-572. (in Chinese)
8. Ren Wei-xin, Chen Hua-bin. Response-Surface Based on Finite Element Model Updating of Bridge Structures [J]. China Civil Engineering Journal, 2008, 41 (12): 73-78. (in Chinese)
9. Bendat J. S. Piersol A. G. Engineering applications of correlation and spectral analysis [M]. New York: Wiley-Inter science, 1980. (Peak Value Method)
10. Pei Xing-zhu, Liu Zheng-wei, Zhou Xiao-shong. Basic research on fishbone vibration model of double wings [J]. Journal of Jiangsu University of Science and Technology: Natural Science Edition, 2007, 06.

Research and Development Status of Grouted Splice Sleeve in Precast Concrete Structure

Ying Chang WU, Liu Peng SHENG, Wei Jian ZHAO

School of Civil Engineering, Shenyang Jianzhu University, Shenyang 110168, China

Abstract: The development of precast concrete structure is an effective way to promote industrialization of construction industry and housing industrialization, and is also a inevitable choice for green development. Grouted splice sleeve connector, regarded as one of the key technologies of precast concrete structure, has the advantages of high strength and good seismic performance. In this paper, the development of grouted splice sleeve connector is systematically summarized both home and abroad, existing problems in the domestic research of sleeve are evaluated and analyzed, and corresponding countermeasures are also further proposed.

1 Introduction

Compared with cast-in-place concrete structure, precast concrete structure has many advantages such as high efficiency,high quality, energy saving and environmental protection and so on. To achieve sustainable development in economic, environmental and social benefits, it rises gradually as a new structure system in China. Precast concrete structure is mainly made up of precast components, through the process of on-site assembly and connection. Previous investigation showed that connection failure between those components is the main reason to the collapse of the whole building[1]. The bearing capacity of building structures is directly affected by connecting quality,and the most critical part is undoubtedly the reinforced bar connection between the precast components.

Grouted splice sleeve was invented by Alfred A. Yee in late 1960 and first used for joining reinforcement bars of columns in 38 Ala Moana hotel (Honolulu, Hawaii). Then the patent of grouted sleeve connector was acquired by Japan TTK. Over the next several years, lots of improvements had been made to improve its effectiveness in Japan and North American. It was listed as a major reinforced bar connection technology in the report of the American Concrete Institute in 1983.In 1986, the work to develop the x-type sleeve was successfully completed. A large number of experimental studies and oretical analysis was conducted on it in Japan and America. Through a series of real seismic tests,It had also been proved reliable enough that can be used for high-rise construction even for ultra-high-rise concrete building. In china, many experiments had also been conducted to study this technology in the past decade. Lots of important data was obtained and some relevant specifications were also published to provide technical support for it. Now the sleeve was gradually applied to the actual projects.

2 Research and Development Status of Grouted Sleeve Connector in Foreign Countries

Grouted splice sleeve connector can be defined as a connection system using non-shrinking high strength grout to splice two discrete reinforced bars to ensure continuity during stress transmission[2]. Depending on the bond-slip resistance between the reinforced bars and the grout, it provides resistance to prevent reinforced bars pulling out from the sleeve. According to the connection type of reinforced bar and the sleeve, it can be divided into two groups: full-grouted splice sleeve, half-grouted splice sleeve.

2.1 Current Development of Full-Grouted Splice Sleeve.

Full-grouted splice sleeve can also be divided into two types from it's function: vertical full-grouted splice sleeve and horizontal full-grouted sleeve. The first kind of sleeve is applied to reinforced bar connection between precast components such as wall to wall or column to column, see Fig. 1. The second kind of sleeve is often used for reinforced bar connection between precast components such as beam to beam, see Fig. 2. Full-grouted ductile Iron sleeve connector is widely used in earthquake-prone place such as Japan and Taiwan. The length of such sleeve is slightly longer than half-grouted splice sleeve, but it has many advantages such as reliable load transferring, good toughness and energy absorption effect, which are helpful to resist earthquakes. This technology is becoming more and more perfect through

* Corresponding author: 1187562430@qq.com

https://doi.org/10.1051/e3sconf/20183803034

continuous research and improvement. NMB Japanese company and Tokyo Steel company are more skilled at this technology. Full-grouted sleeve connector produced by Tokyo Steel company was applied to The Vanke Spring River which was regarded as the representative project in china. But due to high imports prices, this type of sleeve was not scaled up to wide use.

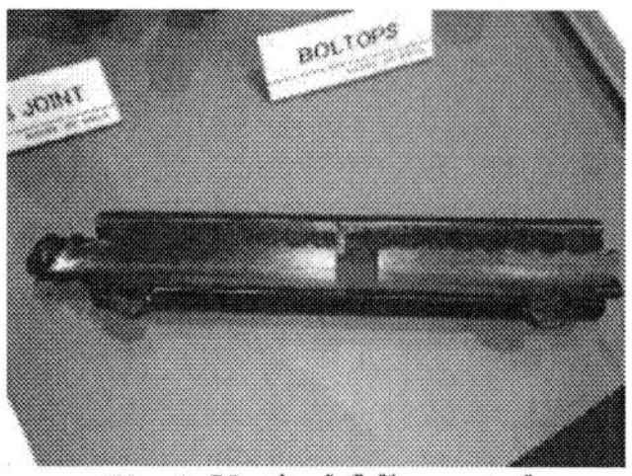

Fig. 1: Vertical full-grouted sleeve connector in Japan

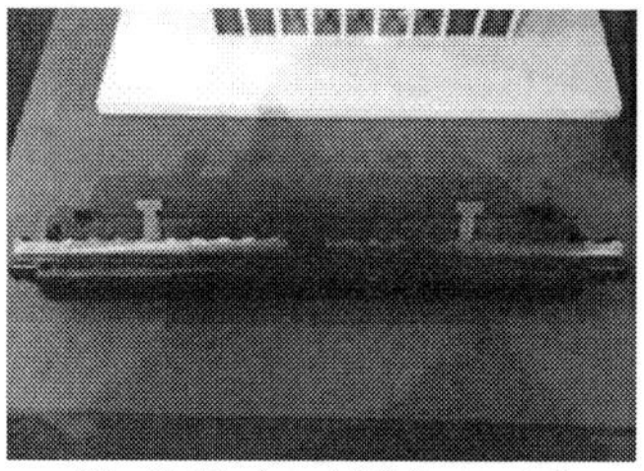

Fig. 2: Horizontal full-grouted sleeve connector in Japan

Since shear keys of the sleeve are complex and difficult to machine, grouted splice sleeve is made of ductile iron both in the United States and Japan. As for this sleeve, despite the fact that production process has been very mature, technicians still continue to improve the technology. Combined with research and development of ultra high strength grouting materials, Japan tends to develop thinner and shorter types of sleeve in the future, see Fig. 3. Right below (see Fig. 4)is a new squashed- cylinder shape sleeve invented by some company in the United States, two reinforced bars could be able to connected side by side through the sleeve.

Fig. 3: NMB sleeve in Japan

Fig. 4: Bar splicing sleeve in US

2.2 Current Development of Half-Grouted Splice Sleeve. 2.3 Parametric Studies

Connection mechanism: firstly, prepare for processing internal threaded at one end of sleeve; secondly, prepare for processing external thread at one end of the bar; and then connect the bar to the sleeve[3]; finally, at the other end of the sleeve, connect them together by pouring grout materials into the sleeve. This type of connection is a composite form of straight thread connection and grouting connection. It was mostly used for reinforced bar connection between components such as wall to wall or column to column. Cross-sectional dimensions of this sleeve produced by the company Lenton Interlok was slightly larger. It's material was Ductile Iron. In order to clamp the sleeve firmly when connecting bar and sleeve together, hexagonal structure form was utilized in the threaded end of the sleeve, see Fig. 5. Japan's Tokyo Steel companies have similar half-grouted sleeve connector, see Fig. 6. What makes it different is the special external screw bar at the corresponding threaded end of the sleeve. The reinforced bar can be directly screwed into the threaded end of the sleeve, The gap between the two will be filled by pouring grout .

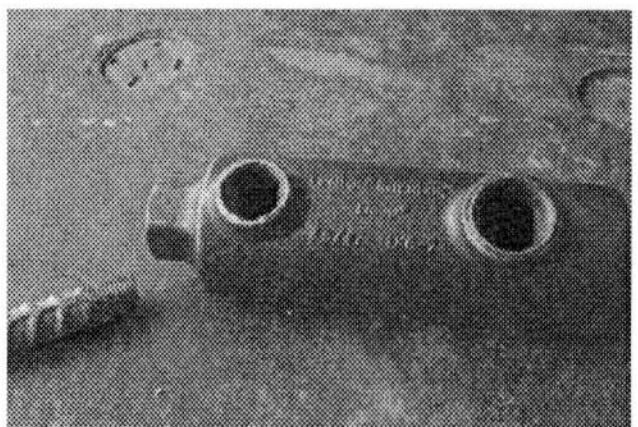

Fig. 5: Half-grouted sleeve connector in US

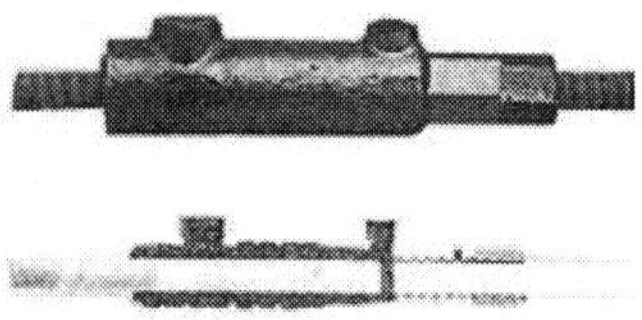

Fig. 6: Half-grouted sleeve connector in Japan

3 Research and Development Status of Grouted Splice Sleeve in China

3.1 Current Development of Full-Grouted Splice Sleeve.

Zheng Yongfeng[4]developed grouted deformed pipe

 https://doi.org/10.1051/e3sconf/20183803034

splice(GDPS sleeve) with standard seamless steel tube through cold rolling operation, see Fig. 7.And five coupler specimens were prepared and tested under incremental tensile load. The results show that if the anchorage length equals to eight times of the bar diameter, bar fracture failure mode can be achieved, see Fig. 8. and the ultimate tensile strength of the splices does meet the requirement for Class I splice. The inner cavity of sleeve structure of the grouted sleeve has a significant influence on the confining mechanism sleeve and strain distribution. The deformed section and smooth section of the sleeve show different strain distribution rule. However, due to the limitation of cold rolling operation, the distance from the first ring of the convex to the port of the sleeve is slightly larger, resulting in grouting material splitting easily at the port. In addition, the circular rib is shaped like a snakeskin tube. Deformation of the sleeve is very large under large deformation and high stress of the repeating of tension and compression, leading to joint stiffness degradation.

Fig. 7: GDPS sleeve

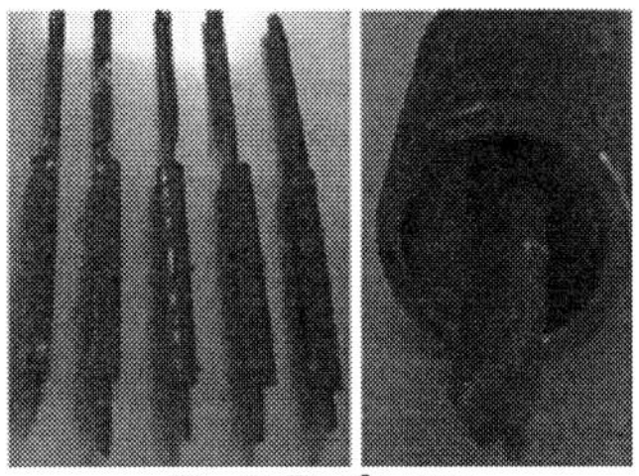

Fig. 8: Bar fracture

3.2 Current Development of Half-Grouted Splice Sleeve

When PC market had just started in China, due to the high cost of molding for nodular iron sleeve, faulty materials and expensive imported sleeve, so Vanke and Beijing Jian Mao used solid steel with the method of machining, produced China's first generation of half-grouted splice sleeve, see Fig. 9and Fig. 10. This sleeve had been applied to many engineering applications such as Half-step bridge Beijing Public Security Bureau. Although meeting the needs of the market at that time, the machining methods also had some disadvantages such as difficult machining procedures, more waste materials, insufficient production costs.

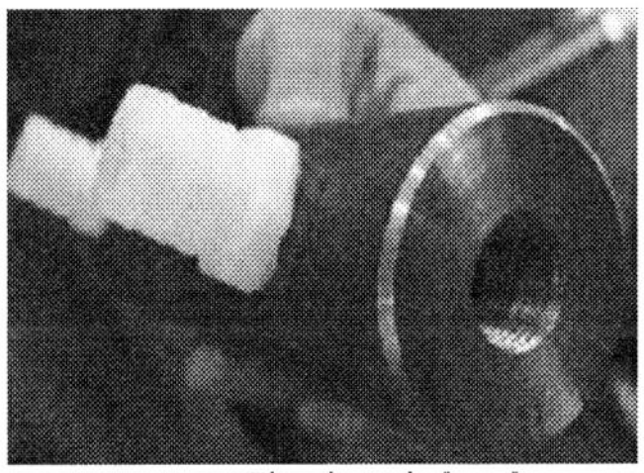

Fig. 9: The threaded end

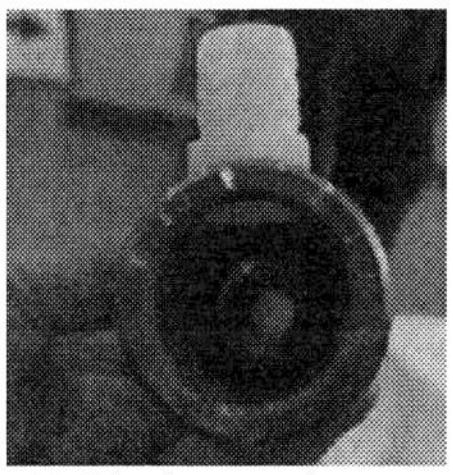

Fig. 10: The inner cavity structure

On the basis of sleeve produced by Beijing Jianmao, a new type of sleeve named split half-grouted splice sleeve was developed by Beijing Building Construction Research Institute, see Fig.11. The cylindrical tube body was made of low alloy thick-walled steel tube. some shear keys were machined in the inner wall of steel tube, and internal threads were processed at one end of the sleeve. Middleware (see Fig. 12) with internal and external threads was screwed into the cylinder thread ends, thus forming a half-grouted sleeve connector. In order to reduce material waste and the difficulty of processing, the solid steel bar is replaced by the steel pipe. However, this type of sleeve has two disadvantages: (1)a hidden peril of quality might be added because of threaded connection between middleware and tube body;(2)producing middleware was also relatively cumbersome.

Fig. 11: Split half-grouted sleeve connector

Fig. 12: Middleware

With the development of the domestic PC market in recent years, nodular iron sleeve has also appeared on the market, see Fig.13. This type of sleeve has some circular convex ribs which are substantially similar to sleeve' in Japan. In order to avoid stress mutation between concrete and port of sleeve, variable cross-section form is machined at the threaded end. But it may lead to risk of fracture of the sleeve due to the stress concentration in variable cross-section. Taking into account the high cost of nodular iron and high technology content, there are also domestic manufacturers processing sleeve with steel, see Fig.14. Compared with nodular iron, cast steel has some disadvantages such as poor toughness, easy to get rust and lower energy absorption efficiency. Moreover, it's performance is not as good as nodular cast iron when experienced fatigue test and earthquake.

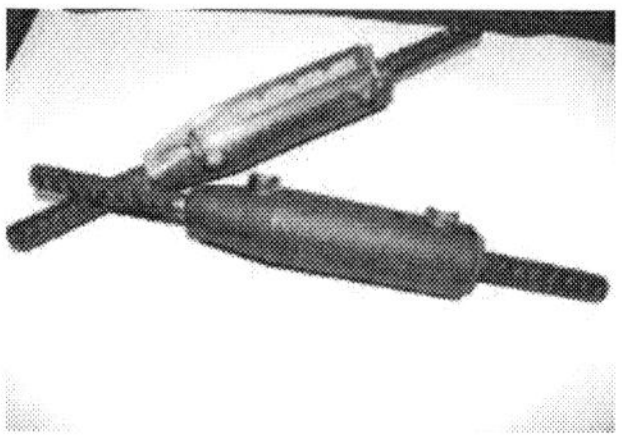

Fig.13: Nodular iron sleeve

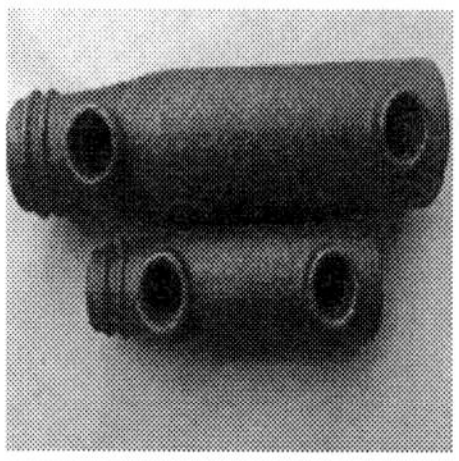

Fig. 14: Sleeve of steel casting

4 Problems & Countermeasures

As for grouted sleeve connection technology, the research on it starts relatively later in China and the relevant theoretical analysis and experimental study is insufficient. There are some problems in the theoretical study and practical application of prefabricated structures. The problems are as follows:

(1)The fundamental research on core technology of grouted sleeve is woefully inadequate both in quantity and depth. Although a number of experiments have been conducted mainly in domestic colleges and universities, these specimens mainly based on existed sleeve both home and abroad are tested through simple uniaxial tensile test to verify it's macro-performance. However, It is the insufficient fundamental research for Micro-stress transfer mechanism of the sleeve, failure mechanism, affecting relationship among dimension parameters that leads to the slow development of the optimization design of sleeve due to lack of theoretical basis.

(2)At present, domestic sleeve is developed mainly by mechanical metallurgy companies. However, it is insufficient to study it's connection performance and the effects on the whole structure. The development for grouted sleeve requires many disciplines such as materials, metallurgy, mechanical engineering and concrete structure and so on. But the ultimate goal is to serve for the concrete structure. How to meet production and construction requirements of concrete members on the basis of identity intensity of reinforced bars? What impact does increasing the local stiffness of splice have on the seismic behavior of concrete structures? From the perspective of structural engineering, These questions requires extensive research, and should be fed back into the design of grouted splice sleeve connector.

(3)Half-grouted sleeve connector is widely used in prefabricated concrete structure in china. However, it has many problems in the practical engineering. As for full-grouted sleeve connector, we lack comprehensive understanding and in-depth study. Compared with half-grouted sleeve connector, full-grouted sleeve connector is longer and it's material consumption is a little larger. So the half-grouted sleeve connector with a relatively smaller size was developed at the start in china. As for this type of sleeve, there are many problems with its threaded end in practice. Firstly, It is not easy to control the number of threads which is screwed into the threaded end of sleeve at the construction site. Secondly, cross-section area of the reinforced bar is reduced due to the process of rib-stripped rolling straight thread. At the same time, stress concentration is more likely to happed at the thread end of sleeve when it is subjected to tension. Thirdly, Thread failure may also lead to reinforced bar pulling out.

In summary, on the basis of absorbing foreign experience and technology, much more attention should be paid to the fundamental research on grouted splice sleeve connector and core technology. it is of significance to develop and improve new sleeve which is suitable for our national situation. Meanwhile, establish and improve it's design theory, which have important theoretical and practical significance to future development of prefabricated concrete structure.

References

1. J.K.Iverson,N.M.Hawkins:Performance of Precast/Prestressed Concrete Building Structures During Northridge Earthquake, edited by PCI Jouranal(1994)
2. J.H.Ling, A. B. A. Rahman and I. S. Ibrahim: Tensile performance of grouted splice connectors in precast concrete structures(Precast Concrete Elements, Malaysia 2010)
3. W.X.Zhu,L.Shao,Y.Wang and W. Fu: Study on Semi-Grouting Coupler Connecting Technique for Steel Reinforcement(Architecture Technology,China 2015)
4. Y.F.Zheng,Z.X.Guo and Z.C.Sun: Experimental Study on the Behavior of Grouted Deformed Pipe Splice(Constuction Technology,China 2014)

Experimental Research of FRP Composite Tube Confined Steel-reinforced Concrete Stub Columns Under Axial Compression

Ji Zhong WANG, Lu CHENG, Xin Pei WANG

Faculty of Instructure Engineering, Dalian University of Technology, Da'lian, Liaoning 116024, China

Abstract: A new column of FRP composite tube confined steel-reinforced concrete (FTCSRC) column was proposed. This paper elaborates on laboratorial and analytical studies on the behavior of FCTSRC columns subjected to axial compressive load. Eight circular FTCSRC stub columns and one circular steel tube confined concrete (STCC) stub column were tested to investigate the failure mode and axial compression performance of circular FTCRSC columns. Parametric analysis was implemented to inquire the influence of confinement material (CFRP-steel tube or CFRP-GFRP tube), internal steel and CFRP layers on the ultimate load capacity. CFRP-steel composite tube was composed of steel tube and CFRP layer which was wrapped outside the steel tube, while CFRP-GFRP composite tube was composite of GFRP tube and CFRP layer. The test results indicate that the confinement effect of CFRP-steel tube is greatly superior to CFRP-GFRP tube. The ductility performance of steel tube confined high-strength concrete column can be improved obviously by encasing steel in the core concrete. Furthermore, with the increase in the layers of FRP wraps, the axial load capacity increases greatly.

1 Introduction

Fiber-reinforced polymer (FRP) composites have been found wide practical applications in repair and restoration of concrete structures. In recent years, the application of FRP has increasingly in new construction as an efficient confinement material on account of its high strength-to-weight ratio and operability [1, 2]. It has been demonstrated in many studies that the load capacity and ductility of concrete structure can be improved by wrapping of FRP. However, the discrepancy between different types of FRP has a significant influence on the mechanical performance of structures [3]. GFRP has a good corrosion resistance with an inferior strength [4]. Hence, the use of combining the advantages of CFRP and GFRP with reinforced concrete has been widely explored as a new construction. FRP composite tube confined steel-reinforced concrete columns can resist the corrosion of steel efficiently [5, 6]. Confined tube can play a role of a shuttering which is beneficial to accelerate the construction process and reduce the construction cost.

The concept of FRP tube confined concrete was first proposed by Mirmiran A and Shahawy M [7, 8]. They conducted the parametric study, including concrete strength and the thickness of FRP tube, to inquire the influence of FRP. The results suggested that the load capacity of columns was enhanced remarkably. After that, there are a large number of studies in static properties of FRP confined concrete [9, 10]. The exploration about GFRP tube confined concrete has also made great process. While in the existing studies, the investigation on FRP composite confined high-strength concrete is few. The precious studies showed that the shear failure can be prevented by encasing steel in concrete.

This paper presents laboratorial and analytical studies on the compressive behavior of FTCSRC columns. The experimental program consists of 9 stub specimens to investigate the working mechanism and failure modes of composite columns. These test parameters considered include confinement material (CFRP-steel tube or CFRP-GFRP tube), internal steel and CFRP layers.

2 Experimental program

2.1. Materials and specimen preparation

In total, nine circular specimens, including eight FRP tube confined steel-reinforced concrete (FTCSRC) stub columns and one circular steel tube confined concrete (STCC) stub column, were designed and tested under axial compression. All the circular specimens had a diameter of 180mm (refer to the concrete core) and a height of 540mm. The type of inner steel is constant as

* Corresponding author: wangjz@dlut.edu.cn

No. 10a I-beam. The main tested parameters included the confinement material (FRP-steel tube or FRP-GFRP tube), the existence of internal steel and CFRP layers (0 (i.e., specimens without FRP sheet), 1, 2, 4 layers). The details of all the specimens are displayed in Table 1.

In the nomenclature of all specimens, for example, specimen "S-FTCSRC-1", S-FTCSRC represents CFRP-steel composite tube confined steel-reinforced concrete, 1 represents the CFRP layers. While G-FTCSRC represents CFRP-GFRP composite tube confined steel-reinforced concrete. STCC represents the steel tube confined concrete column.

Table 1 Specimen parameters and test results

Specimen	CFRP layers	N_e /KN	N_e / N_0	ε_u/$\mu\varepsilon$	ε_u/ε_0
STCC	0	2789	——	3.20	——
S-FTCSRC-0	0	3068	1	2.78	1
S-FTCSRC-1	1	3184	1.04	3.65	1.31
S-FTCSRC-2	2	3476	1.13	3.73	1.34
S-FTCSRC-4	4	4415	1.44	8.27	2.97
G-FTCSRC-0	0	2195	0.72	5.15	1.85
G-FTCSRC-1	1	2826	0.92	5.56	2
G-FTCSRC-2	2	3032	0.99	11.68	4.20
G-FTCSRC-4	4	3469	1.13	11.83	4.26

Standard tensile tests of flat coupons were conducted to identify the material properties of steel. The selection of steel type of inner steel was same as inner wrapping steel tube. The average values of the elastic modulus and yield stress were 212 GPa and 298 GPa. The circular steel tubes and GFRP tubes had the same inner diameter of 180mm and a thickness of 3mm. The steel tubes were manufactured by mild steel sheet and formed by modules. The GFRP tube in this paper was manufactured by Hebei Province. These GFRP tubes had a tensile strength of 240 MPa. In order to make sure that the confined tube does not take axial load directly, two small groves were carved at both edges of each specimen.

The used UT70-30 carbon fiber sheets were provided by Japan Toray Company. The CFRP sheets had a thickness of 0.167mm/ply (manufacturer data). The tensile strength and volume modulus of CFRP were provided by the manufacturer were 4216MPa and 252Gpa, respectively. JGN-T, a type of epoxy resin was adopted for gluing CFRPs together and adhering to the CFRP to the steel tube or GFRP tube by a hand-up method. To ensure the entire confinement, the finishing end of a sheet overlapped the starting end by 200mm.

The design level of concrete was C80 which belongs to high-strength concrete, and a group of cubes were cast and cured in similar conditions to the related specimens. Prior to casting, a steel plate with thickness of 20mm was welded to the steel tube at one end. Then these specimens were placed upright to air-der until testing. Fig. 1 shows the concrete configuration of specimens.

2.2. Test setup and instrumentation

The specimen tests were conducted at Dalian university of Technology using a 10000 kN capacity universal testing machine. The load was added on the specimen directly by hydraulic jack and monitored. Axial deformation of the column was measured by using four linear variable displacement transducers (LVDT) which were kept at top of the jack. Sixteen strain gauges, including twelve in the longitudinal direction and another four in the transverse direction, stuck on the external surface of each specimen at top, mid-height and bottom. The loading process can be divided into three phases. The loading was applied with load control at 1.5kN per second at the initial stage. After that, displacement control was used at approximately 0.3mm per second until specimen failure. All test data, including strains, displacements and loads, were completely recorded. The load at which the CFRP starts rupturing and the failure symptom of specimens were also marked for every specimen. Fig. 2 shows the loading arrangement.

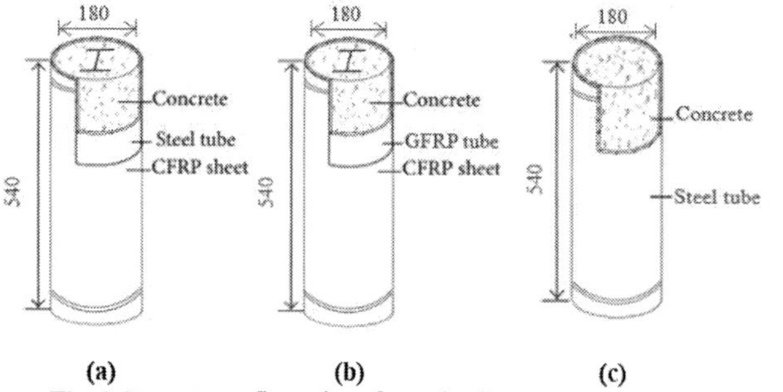

Fig. 1 Concrete configuration of tested columns
(a) S-FTCSRC; (b) G-FTCSRC; (c) STCC

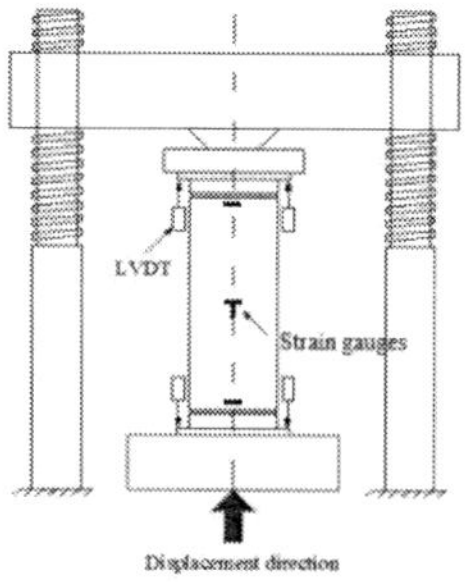

Fig.2 load system

2.3. Experimental Results and Failure modes

For the two specimens with steel tube (i.e., specimens S-FCTSRC-0 and STCC), there is not obvious phenomenon in the appearance during the initial loading process. When the load attains about 85% of the maximum load (Ne), the remarkable buckling can be observed at the mid-height cross-section. For the G-FTCSRC-0 specimen, the change in the outward appearance is also invisible until to the ultimate load. The white hairline crack can be observed and an abrupt fracture occurs with a big noise.

In the case of CFRP strengthened specimens, as shown in Fig.3, snapping sound of the oxide layers peeling can be heard as axial load increase. For CFRP-steel tube, the rupture of fibers occurred after steel tube yielded. The wrapping CFRP sheet was activated by the significant lateral deformation. However, the CFRP sheets of G-FCTSRC stub columns rupture randomly accompanied with the fracture of GFRP tube. The reason for this different failure modes about two different confinement material could be that GFRP is more brittle than steel [11].

Fig.3 also shows failure modes of the concrete after cutting off the composite specimens. It can be observed that CFRP sheets provide effectively restraint against the lateral deformation of core concrete. Similar shear failure was observed in all columns tested. It is interesting that the cracks on the surface of the concrete was tinier in S-FTCSRC-0 than STCC because of the existence of inserted steel shape. It can be explained that the inserted steel shape restricted the development of the slip in core concrete within half side of the cross-section.

Fig. 3 Failure modes of specimens

3 Parametric study

3.1. The effect of confinement material

In the following analysis, the axial load-displacement relationship is used to investigate the influence of different parameters. Fig.5 shows the influence of confinement material on the specimens under different CFRP layers. The relative values about ultimate load capacity and axial deformation are also summarized in Table.1. Axial load capacities of G-FCTSRC columns is lower than S-FCTSRC-0 except at the 4 CFRP layers, which means that ultimate load of S-FCTSRC columns is obviously higher than G-FCTSRC. In the contrary, the axial deformation of G-FCTSRC is superior to S-FCTSRC columns at peak load. The relative values about $\varepsilon_u/\varepsilon_0$ of S-FTCSRC-4, G-FTCSRC-4 were 2.97 and 4.26, respectively.

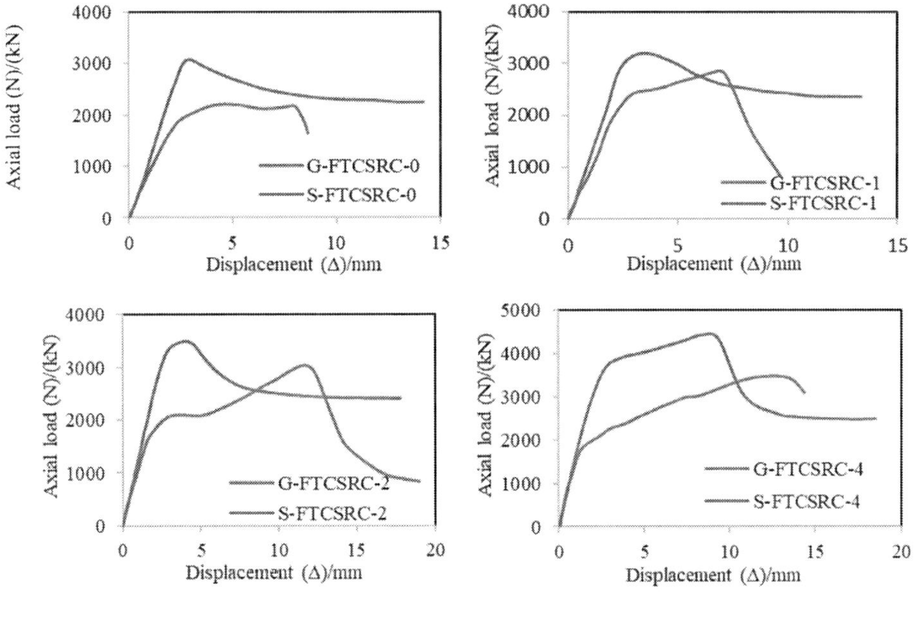

Fig.5 load-displacement curves about different confinement material

3.2. The effect of internal steel

Fig.6 shows the impact of inner steel on the specimens whose parameters are the same except an internal steel shape. It can be demonstrated that encasing internal steel has a small increment in elastic module and load capacity. After the peak load, the downward trend of S-FTCSRC-0 is gentler than STCC. For S-FTCSRC-0, the load remains a stable value approximately about 2300 kN in the later period, which is higher than STCC about 1700 kN correspondingly.

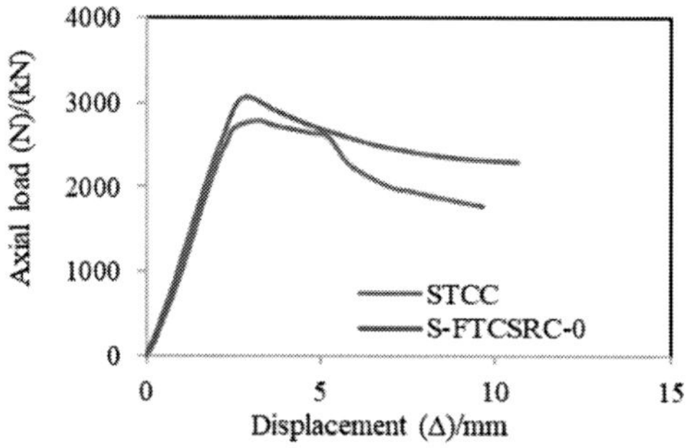

Fig.6 load-displacement curves about internal steel

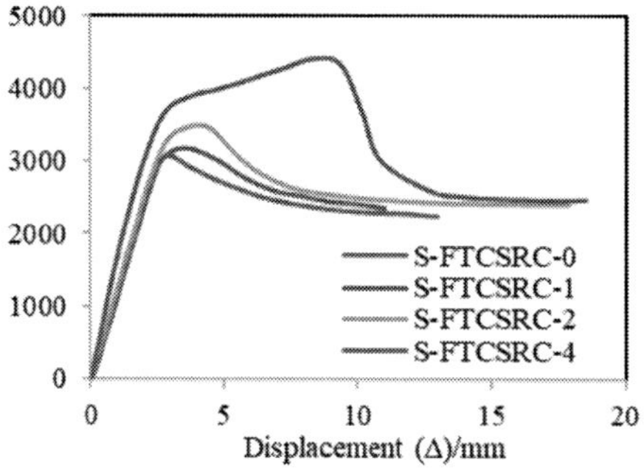

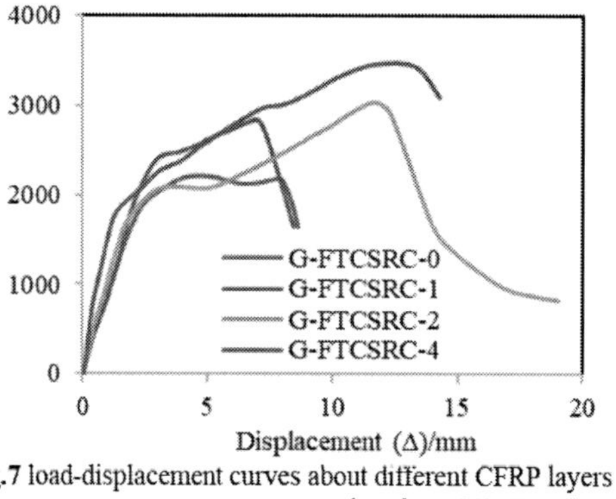

Fig.7 load-displacement curves about different CFRP layers

3.2. The effect of CFRP layers

The tested axial load versus displacement curves of the specimens are shown in Fig.7. It demonstrates that the specimens are under elasticity in the initial stage because the curves are straight. As the load increases, the CFRP sheets were gradually activated by remarkable convex deformation of concrete. The amount of the axial load and the axial displacement where maximum load capacity are increased as the number of CFRP sheet layers increased. The use of a thicker CFRP sheets leads to a second linear ascending portion. In particular, the G-FCTSRC column has a larger and more distinct slope for the second portion of the curves because CFRP sheets restrain the GFRP tube more valid with its lower strength.

4 Conclusion

(1) The buckling of core concrete was well confined in F-CTSRC columns, leading to a very strength and ductile response under axial compression.

(2) The axial load capacity of S-FTCSRCs is higher than G-FCTSRCs, but the ductility of the S-FCTSRCs is inferior to the G-FTCSRCs.

(3) The existence of internal steel has an obvious influence in the failure modes of STCCs, in which the shear failure could be deterred.

(4) The load capacity and axial deformation of FTCSRC columns was enhanced as the number of CFRP sheet layers increased, however, the induction can be concluded that G-FTCSRC columns perform a better performance both in strength and ductility when the wrapping CFRP sheets are thick enough.

References

1. Saadatmanesh H. Fiber composites for new and existing structures[J]. Aci Structural Journal, 1994, 91(3): 346-354.
2. Tanwongsval S, Maalej M, Paramasivam P. Strengthening of RC wall-like columns with FRP under sustained loading[J]. Materials & Structures, 2003, 36(5): 282-290.
3. Lau K T, Zhou L M. Mechanical performance of composite-strengthened concrete structures[J]. Composites Part B Engineering, 2001, 32(1): 21-31.
4. Sheikh S A, Jaffry S A D, Cui C. Investigation of glass-fibre-reinforced-polymer shells as formwork and. [J]. Canadian Journal of Civil Engineering, 2007, 34(3): 389-402.
5. Xiao J, Huang Y. On the seismic behavior and damage assessment of recycled aggregate concrete filled GFRP tube column[J]. China Civil Engineering Journal, 2012, 45(11): 112-120.
6. Mirmiran A, Shahawy M. A new concrete-filled hollow FRP composite column[J]. Composites Part B Engineering, 1996, 27(3–4): 263-268.
7. Mirmiran A, Shahawy M. Behavior of concrete columns confined by fiber composites[J]. Journal of Structural Engineering, 1997, 123(5): 583-590.
8. Zhu Z, Ahmad I, Mirmiran A. Effect of Column Parameters on Axial Compression Behavior of Concrete-Filled FRP Tubes[J]. Advances in Structural Engineering, 2005, 8(4): 443-450.
9. Ozbakkaloglu T, Saatcioglu M. Seismic Behavior of High-Strength Concrete Columns Confined by Fiber-Reinforced Polymer Tubes[J]. Journal of Composites for Construction, 2006, 10(6): 538-549.
10. Wang L G, Chen B L. Steel Reinforced Concrete-Filled Glass Fiber Reinforced Polymer or Steel Tubular Structures[M]. Liaoning Scientific & Technical Publishers, Shenyang, 2013.
11. Wang L G, Qin G P, Zhou L. Experimental research of gfrp tube columns filled with steel-reinforced high-strength concrete subjected to axial loading[J]. Engineering Mechanics, 2009.

Numerical Analysis on Tensile Properties of Grout-filled Splice Sleeve Rebars under ISO 834 Standard Fire

Yong Jun LIU, Chao LI, When Jun ZHOU

School of Civil Engineering, Shenyang Jianzhu University, Shenyang, Liaoing 110168, China

Abstract: This paper presents some numerical simulation results of tensile properties of reinforcing bars spliced by grout-filled coupling sleeves under fire conditions to identify the effect of load ratio on fire resistance time of spliced reinforcing bars, which provide a useful base for predicting structural behaviors of pre-cast reinforced concrete buildings in fires. A spliced rebar system investigated in this paper consists of two equal-diameter steel reinforcing bars with 25mm diameter and a straight coupling sleeve with 50mm outer and 45mm inner diameters. As a result, the thickness of grout between steel bars and sleeves are 20mm. Firstly, the temperature distributions in steel bars connected by grout- filled coupling sleeves exposed to ISO 834 standard fire were calculated utilizing finite element analysis software ANSYS. Secondly, the stress changes in heated steel bars connected by grout-filled coupling sleeves under different constant tensile loads were calculated step by step until the rebar system failed due to fire. Thus, the fire resistant time of rebar spliced by grout-filled coupling sleeves under different axial tensile loads can be determined, further, the relationship between fire resistance time and axial tensile loads ratio can could be obtained. Finally, the fire resistant times versus axial tensile load ratios curve of grout-filled splice sleeve rebars exposed to ISO 834 standard fire is presented.

1 Introduction

In the late 1960s, Dr. Alfred Alphonse YEE invented grout-filled splice sleeve (vide Figure 1) in Hawaii and got some patents successively. In 1973, grout-filled splice sleeves were first used in precast concrete column-tree connections for 38 stories high Ala Moana Hotel, Honolulu, Hawaii. From then on, the use of grout-filled splice sleeve has been significantly increasing in precast concrete components worldwide due to their reliability, quality, and durability. The connections between two precast components are key positions for ensuring the integrity of the completed structure under such hazard conditions as earthquake, fire, exploration, and flood, et al, confronted during whole life cycle. Grout-filled sleeves are currently being used widely to splice reinforcing bars of the adjacent two precast components (vide Figure 2).

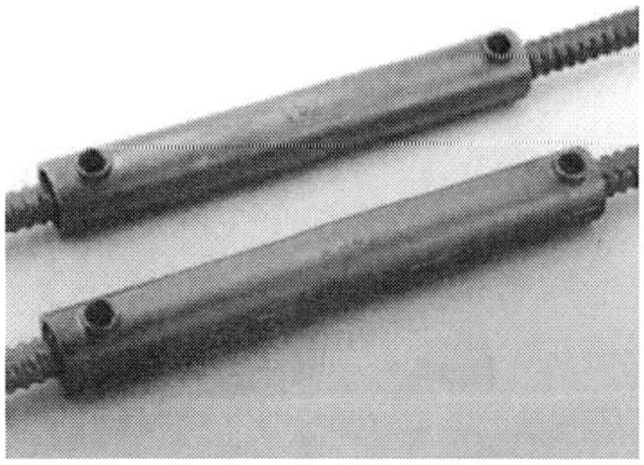

Figure 1 Splice Sleeves and bars before Grout-filling

Figure 2 Grout-filled Sleeves Connecting Bars

Up to the present, a large number of studies focus on structural behavior of precast concrete structures splicing reinforcing bars in adjacent two precast components with grout-filled sleeves [1,2]. In March 2011, all buildings using NMB Splice Sleeves in the Tohoku area withstood the 9.0 Richter scale earthquake without any structural damage, that demonstrated the high reliability of splice sleeves. It is well known that fires could damage concrete components severely (vide Figure 3) even result in total collapse of whole structures (vide Figure 4) [3-5].

* Corresponding author: ceyjliu@sjzu.edu.cn

 https://doi.org/10.1051/e3sconf/20183803036

Relatively, very few studies on fire behavior of precast reinforced concrete structures were conducted all over the world. It is urgent needed to investigate the fire behavior of precast reinforced concrete structures using splice sleeves. As a small pilot study, this paper presents some numerical simulation results of tensile properties of reinforcing bars spliced by grout-filled coupling sleeves under fire conditions.

Figure 3 Reinforced Concrete Column Seriously Damaged by Fire

Figure 4 Totally Collapsed Reinforced Concrete Building after Fire

2 Geometrical Dimensions and Material Properties of the Grout-filled Splice Sleeve

A spliced rebar system (vide Figure 5) investigated in this paper consists of two equal-diameter steel reinforcing bars with 25.0mm diameter and a straight coupling sleeve with 50.0mm outer and 45mm inner diameters. As a result, the thickness of grout between steel bars and sleeves are 20.0mm. The length of sleeve is 285.0 mm.

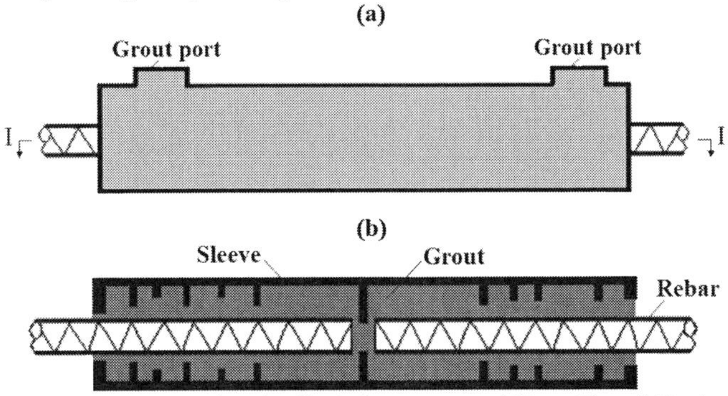

Figure 5 Investigated Spliced Sleeves System: (a)Front View; (b)Section I-I

The investigated spliced sleeves system consists of three kinds of materials, namely, nodular graphite, steel bar, and pouring mortar, the materials thermal and mechanical properties of which are displayed in Table 1 to Table 3.

Table 1 Material Properties of QTB600-3 Nodular Graphite Cast Iron Changing with Temperature

Temperature T/°C	Conductivity K/ (W/m·°C)	Density J/(kg/m^3)	Specific Heat C/(J/kg·m^3)	Modulus of Elasticity E/GPa	Poisson's Ratio μ	Yield Strength f_y /MPa	Thermal Expansion α/ °C^{-1}
20	29.91	7300	495	169.0	0.3	370	1.4×10^{-8}
100	29.79	7300	511	165.1	0.3	370	1.4×10^{-8}
300	29.17	7300	565	158.0	0.3	298	1.4×10^{-8}
600	28.76	7300	665	139.8	0.3	164	1.4×10^{-8}
≥700	27.65	7300	924	134.6	0.3	110	1.4×10^{-8}

Table 2 Material Properties of HRB400 Steel Bar Changing with Temperature

Temperature T/°C	Conductivity K/ (W/m·°C)	Density J/(kg/m^3)	Specific Heat C/(J/kg·m^3)	Modulus of Elasticity E/GPa	Poisson's Ratio μ	Yield Strength f_y /MPa	Thermal Expansion α/ °C^{-1}
20	47.56	439.00	7850	200.0	0.3	400	1.22×10^{-8}
100	45.8	487.62	7850	198.0	0.3	400	1.22×10^{-8}

300	41.4	564.74	7850	183.0	0.3	350	1.22×10^{-8}
≥600	34.8	700.00	7850	34.6	0.3	177	1.22×10^{-8}

Table 3 Material Properties of Poured Mortar Changing with Temperature

Temperature T/°C	Conductivity K/ (W/m·°C)	Density J/(kg/m³)	Specific Heat C/(J/kg·m³)	Modulus of Elasticity E/GPa	Poisson's Ratio μ	Yield Strength f_y /MPa	Thermal Expansion α/ °C⁻¹
20	1.8	900	1800	33.0	0.3	1.725	9×10^{-9}
100	1.8	900	1800	32.0	0.3	1.575	9×10^{-9}
300	1.8	900	1800	31.4	0.3	1.208	9×10^{-9}
≥600	1.8	900	1800	31.4	0.3	0.69	9×10^{-9}

3 Numerical Investigations on Temperature Distribution in Splice Sleeve System

3.1 Finite Element Mesh

Finite element models for analyzing temperature distribution in splice sleeve system were developed by utilizing the commercial finite element software ANSYS, in which all parts of the system were divided into 62834 Solid70 elements connecting at 12436 nodes (vide Figure 6). It is worth to note that the rebar were cut and the bottom surfaces of sleeve, mortar, and rebar lie within the same plane for convince.

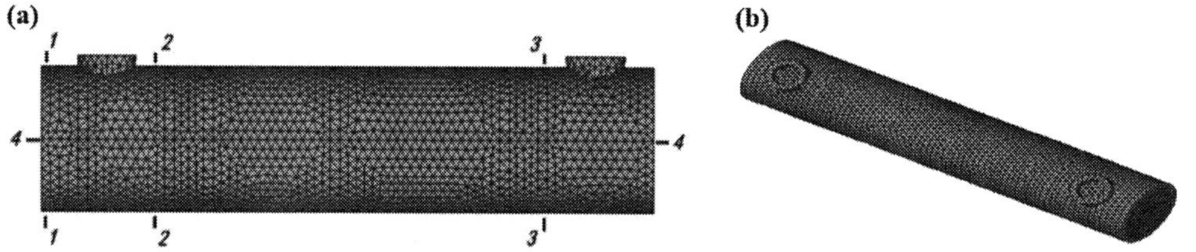

Figure 6 Finite Element Mesh of Grout-filled Splice Sleeve System: (a)Front View; (b)Axonometric Drawing

3.2 Thermal Boundary and Initial Conditions

Generally, there are at least two approaches to determine the nominal temperature-time curves of a fire and they incorporate various factors, such as fuel types and ventilation conditions[6]. For simplicity and to facilitate comparison of analysis results from different researchers, standard time-temperature relationship according to ISO 834 (vide Figure 7) is employed in numerous studies[7]. The temperature-time curve of ISO 834 fire can be described as

$$T - T_0 = 345\lg(8t+1) \tag{1}$$

In this expression, variables have meanings as: T [°C] – temperature in the compartment, t [min.] – time, and T_0 [°C] – temperature at the start of the fire.

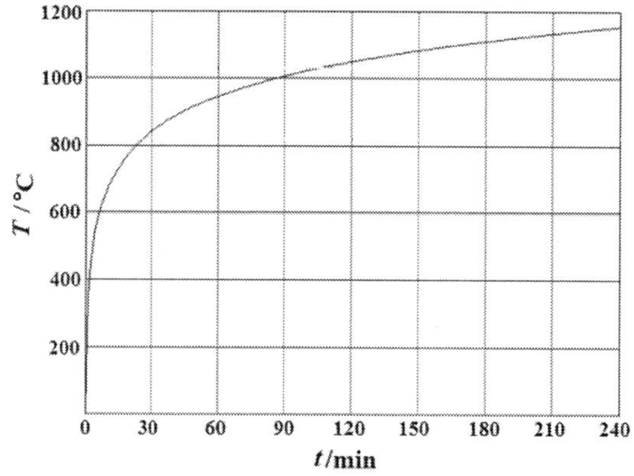

Figure 7 Temperature-time Curve of ISO 834 Fire

The heat flow across the fire-exposed surface is caused by both convection and radiation. Convection coefficient is taken as 28(W/(m²·K)), and resultant emissivity is taken as 0.9. The initial temperature of the spliced rebar system equals to 20°C.

3.3 Finite Element Analysis Results of Temperature Distribution

Temperature history of nodes in finite element mesh of splice sleeve system under ISO 834 were calculated until the time arrived 180 minutes, which would be used to determine temperature-dependent materials properties of splice sleeve system in structural behavior analysis. Contour plots of temperature distribution on four typical sections at t=7200s are shown in Figure 8.

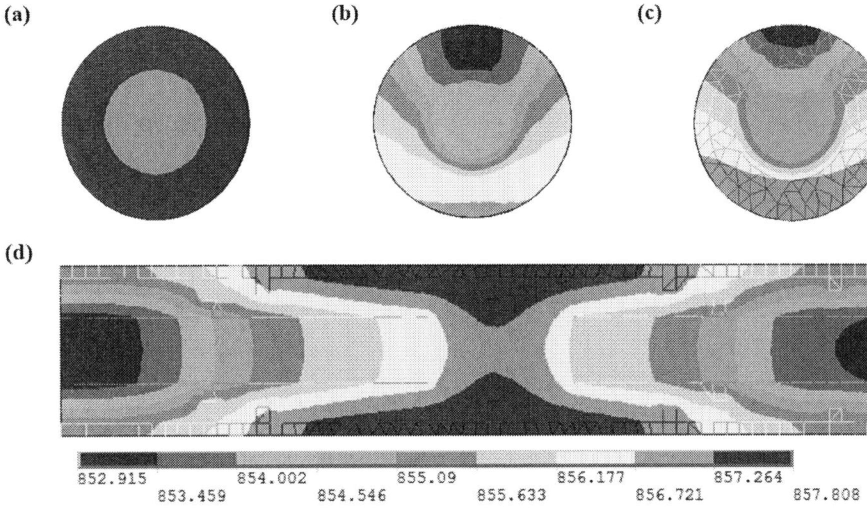

Figure 8 Contour Plots of Temperature Distribution on Four Sections at t=7200s: (a)Section 1-1; (b)Section 2-2; (c)Section 3-3; (d)Section 4-4

4 Fire Resistance Time VS Load Ratio Curve

4.1 Finite Element Model for Structural Analysis

After temperature history in splice sleeve system under ISO 834 have been obtained, the structural behavior should be analyzed to determine the fire resistant time of splice sleeve system under different load ratios. The same finite element mesh (vide Figure 6) are used, and Solid45 element were adopted instead of Solid70 element to simulate structural behavior of splice sleeve system. It is assumed that interfaces between sleeve and motor and between motor and rebar are perfectly bonded and no slips are considered.

At room temperature, two identical axial tension loads are applied on the two ends of the sleeve connected rebars, the loads are increased step by step and increment of each step is 20kN, until the splice sleeve system fracture, the total loads are referred to as ultimate loads under room temperature, expressed as F_u. For the splice sleeve system investigated in this paper, the F_u equal to 221.00 kN.

4.2 Results of Structural Analysis

If two identical loads F_P, which are less than ultimate load under room temperature, is applied on two ends of rebar in splice sleeve system, then the value of F_p over F_u is called as load ratio, expressed as L_r. Obviously, L_r will influence the fire resistance time, R_t, of splice sleeve system. Generally, fire resistance time, R_t, should decrease with the increase of L_r. The main purpose of this paper is to find the relationship between R_t and L_r.

Under several load ratios, the fire resistance time, R_t, of splice sleeve system is calculated using software ANSYS. Figure 9 and Figure 10 show the Von Mises stress contour and first principal tensile strain at t=4200s with L_r=80%. Figure 11 shows the relationship curve between R_t and L_r. Ones can obtain the fire resistant time of a splice sleeve system under ISO 834 standard fire according to the load ratio via linear interpolation.

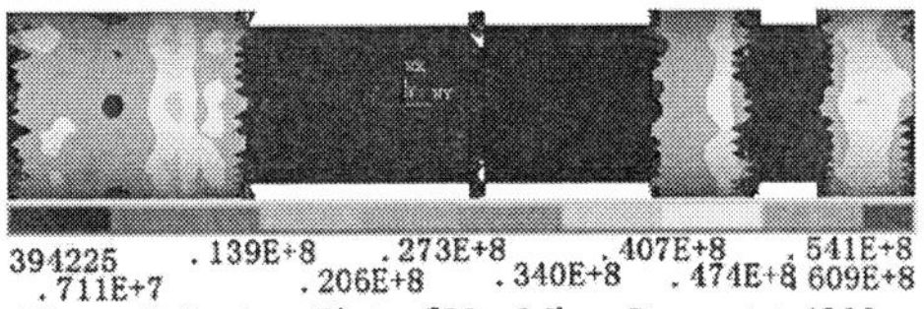

Figure 9 Contour Plots of Von Mises Stress at t=4200s, with L_r=80% (Unit: MPa)

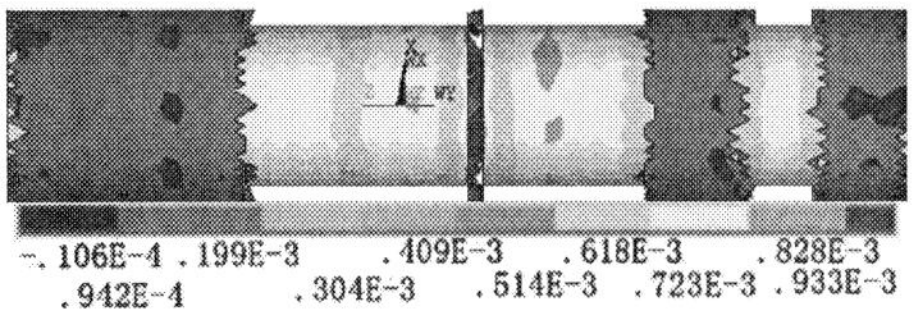

Figure 10 Contour Plots of First Principal Tensile Strain at t=4200s, with L_r=80%

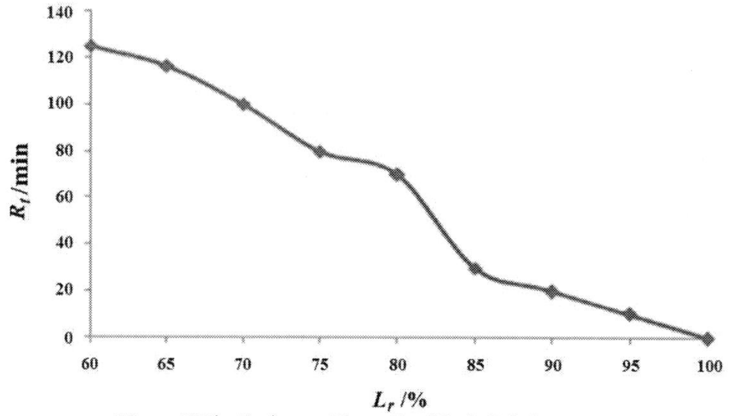

Figure 11 Fire Resistance Time – Load Ratio Relation Curve

5 Conclusions

This paper presents some numerical simulation results of tensile properties of reinforcing bars spliced by grout-filled coupling sleeves under fire conditions to identify the effect of load ratio on fire resistant time of spliced reinforcing bars, with the emphasis on relationship curve of between fire resistance times and load ratios. In addition to the spliced rebar system with straight coupling sleeve with 50mm outer and 45mm inner diameters investigated in this paper, a plenty of spliced rebar system with different sizes have been simulated using software ANSYS. Numerical simulation results indicate that the load ratio has significant effects on fire resistance time of reinforcing bars spliced by grout-filled coupling sleeves. Excessively higher load ratio should be avoided to prevent collapse of precast reinforced concrete buildings in fire conditions. Some complementary tests on fire behavior of reinforcing bars spliced by grout-filled coupling sleeves have been conducted, and results of which will be presented in subsequent papers.

Acknowledgments

This work is supported by Nature Science Foundation of Liaoning Province of China under grant No. 2015060200.

References

1. Haber Z B, Saiidi M S, Sanders D H. Seismic performance of precast columns with mechanically spliced column-footing connections[J]. ACI Structural Journal, 2014, 111(3): 639-650.
2. Belleri A, Riva P. Seismic performance and retrofit of precast concrete grouted sleeve connections[J]. PCI Journal, 2012, 57(4): 97-109.
3. Liu Yongjun, Li Hongnan. Review and prospect of the study on behavior of structures in fire conditions[J]. Journal of Disaster Prevention and Mitigation Engineering, 2004, 24(2): 219-226.
4. Huang Z, Andrew P. Nonlinear finite element analysis of planar reinforced concrete members subjected to fires[J]. ACI Structural Journal, 1997, 94(3): 272-282.
5. Mohamad J T. Numerical modeling of the behavior of concrete structures in fire[J]. ACI Structural Journal, 1998, 95(2): 183-193.
6. Tom Lennon, David Moore. The natural fire safety concept — full-scale tests at Cardington[J]. Fire Safety Journal, 2003, 38(7): 623–643.
7. International Organization for Standardization. ISO 834, fire resistance tests elements of building construction[S]. Geneve, 1999.

 https://doi.org/10.1051/e3sconf/20183803037

Flexural Behavior of Ultra-High Performance Fiber Reinforced Concrete Scale Tunnel Segment

Kun NI, Fa Sheng ZHANG, Yun Xing SHI, Yan Gang ZHANG, Jing Bin SHI

Technical Center, China State Construction Engineering Corporation, Beijing, China

Abstract: To reduce the weight of precast tunnel segment, ultra-high performance fiber reinforced concrete (UHPFRC) was studied to cast the segment. The flexural performance of UHPFRC scale tunnel segments were tested in this work. The weight of the UHPFRC thinner scale tunnel segment was only 80% of reinforced concrete (RC) segment. The segments were loaded as per CJJ/T 164-2011, and the four-point bending system was used. The results showed that the cracking load increased 50%, and 0.2 mm crack width load increased 22%, and the yield load increased 11%, and the ultimate load only decreased 1%. The stiffness of elastic stage of UHPFRC segment looked the same compared to RC segment. In a word, the UHPFRC thinner segments showed excellent flexural performance beyond the traditional RC segment.

1 Introduction

Shield tunneling method is rapidly growing in metro construction of Chinese city. Precast tunnel segment is structure unit, which was assembled through tunnel boring machine (TBM) to build the tunnel. It can be manufactured with reinforced concrete, steel, iron or composite materials. Reinforced concrete (RC) is widely used due to its good durability and low-cost among these materials. And precast concrete tunnel segment has higher manufacturing quality and building effectiveness compared to cast in situ concrete. However, RC segment is heavy; the steel rebar reinforcing cage of RC segment is complicated; the concrete is susceptible to spalling in fire. And during the installation process of segments and tunneling process of TBM, accidental thrust and impact loads are dominant and may result in segment cracking.

The fiber reinforced concrete (FRC) segment has excellent toughness, crack arrestment, durability and fire resistance. And the process of steel rebar reinforcing cage is more efficient, because the steel fibers could partially or totally substitute traditional reinforcement [1-3]. However, the FRC segment is still heavy.

Ultra-high performance fiber reinforced concrete (UHPFRC) performs a high mechanical strength[4, 5] and durability[6], improved resistance against various chemicals as well as higher penetration resistance[7]. To decrease the weight of concrete segment, UHPFRC was used to manufacture the segment and the flexural performance of the segment was tested and discussed in this study.

2 Experiment

2.1. Raw materials.

The constituent materials used to produce concrete comprised of 42.5R and 52.5R Portland cement, fly ash, manufactured sand, 5-20 mm gravel, polycarboxylate SP (solid content≈40%), silicone defoaming agent, polypropylene fiber and hooked end steel fiber. The polypropylene fiber length and diameter are 18 mm and 26 μm, and the steel fibers length and diameter are 60 mm and 0.9 mm, respectively. The main rebar is HRB400, and its diameter is 16 mm, and the yield tensile strength and ultimate tensile strength is 470 MPa and 640 MPa. All the materials are produced in China.

Table 1 The mix proportions of concrete Unit: kg/m^3

	Cement 42.5R	Cement 52.5R	Fly ash	Silica fume	Sand	Gravel	Water	SP	Steel fiber	PP fiber	defoaming agent
HSC	440	-	110	-	800	940	187	2.2	-	1	-
UHPFRC		440	110	38.5	800	940	147	6.8	55	1	0.55

* Corresponding author: nikunun@126.com

 https://doi.org/10.1051/e3sconf/20183803037

2.2. Concrete.

The mix proportions are shown in Table 1 that was used for UHPFRC and high strength concrete (HSC) as match. The HSC was used to produce the traditional RC segment, and the UHPFRC was used to make the thinner UHPFRC segment.

The strengths of concrete are shown in Table 2. The strength of both concrete were reached 60MPa and 100MPa. The flexural performance of the concrete was obtained according to the standard RILEM TC 162-TDF recommendation. The parameters are limit of proportionality (f_L), ultimate flexural strength (f_u), and equivalent tensile strength ($f_{eq,2}$, $f_{eq,3}$). It was shown that the f_L increased 50%, and the f_u increased 200%, and the post cracking strengths ($f_{eq,2}$, $f_{eq,3}$) were more than double f_u of HSC.

Table 2 The strength of concrete Unit: kg/m^3

	Compressive strength	Flexural performance			
		f_L	f_u	f_{eq2}	f_{eq3}
HSC	68.3	3.87	4.04		
UHPFRC	106	5.95	11.92	9.33	10.67

2.3. Precast segment description.

The segment ring is composed of eight segments. The tested segment geometry is shown in Figure1. The thickness of comparison RC segment was 270mm. The thickness of UHPFRC segment was 225mm. The thickness reduced by 16.7%. The rebar cages are shown in Figure2. Longitudinal rebar is 16 mm, and the other rebar is 12 mm. The longitudinal reinforcement in UHPFRC segment is 80% of that in RC segment and the hooping is 50%.

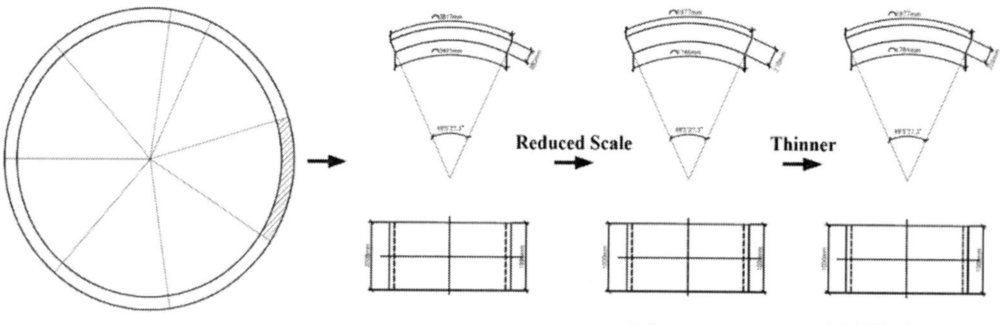

Figure 1 Tunnel segment geometry

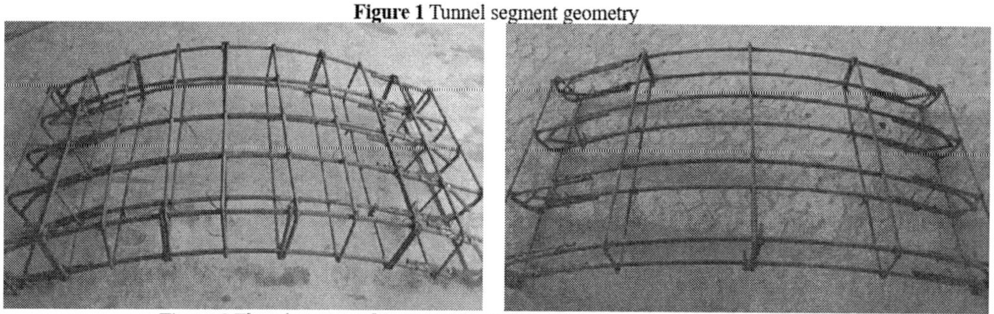

Figure 2 The rebar cage of segments (Left: RC segment; Right: UHPFRC segment)

2.4. Flexural of segment test.

The segments were loaded as per CJJ/T 164-2011. The loading system was four-point bending. It is shown in Figure3. The distance between two loading line was 635mm. The testing segment was support on sliding block. The reaction frame having a maximum bearing capacity was 2000 kN. Each type of segment has two specimens. The segment was under multi-stage loading. When the loading kept the stage, the crack openings were observed and measured using the electronic crack width tester. After the width of crack was beyond 0.2mm, continuous loading was running until the load dropped suddenly. In whole process, the deformations were tested by linear variable displacement transducers (LVDTs).

3 Results and discussion

3.1. Load-deformation relationships of segment.

The load-middle deformation curves of two types segments are shown in Figure 4. The curves of the RC segment showed obvious yield point, and the yield came suddenly. But the yield process of UHPFRC segment was mildly because of the work of steel fiber.

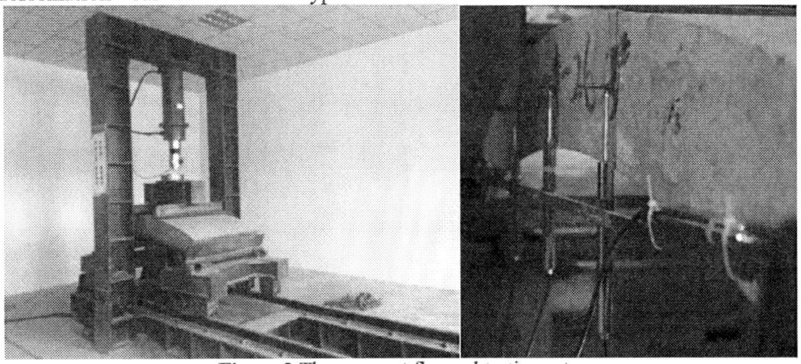

Figure 3 The segment flexural testing setup

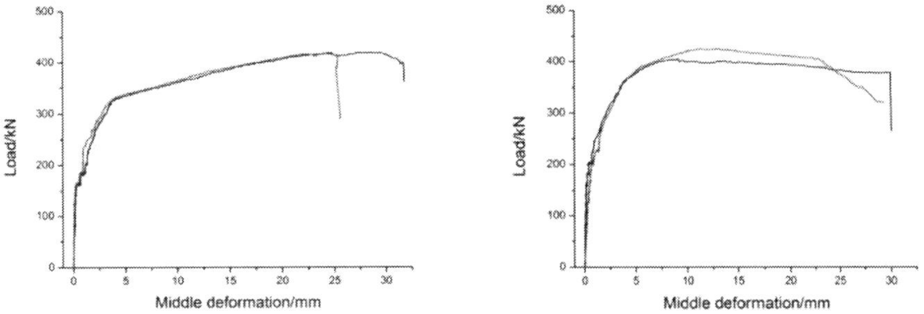

Figure 4 The load-middle deformation curves (Left: RC segment; Right: UHPFRC segment)

3.2. Load-cracking response.

The load-maximum crack width curves of RC and UHPFRC segments are shown in Figure 8. For UHPFRC segment, the first crack was observed at 120 kN. And for RC segment, the first crack was at 80 kN. For each load stage the crack of UHPFRC segments were narrower than that of RC segments. UHPFRC segments was excellent at the limiting the development of cracks.

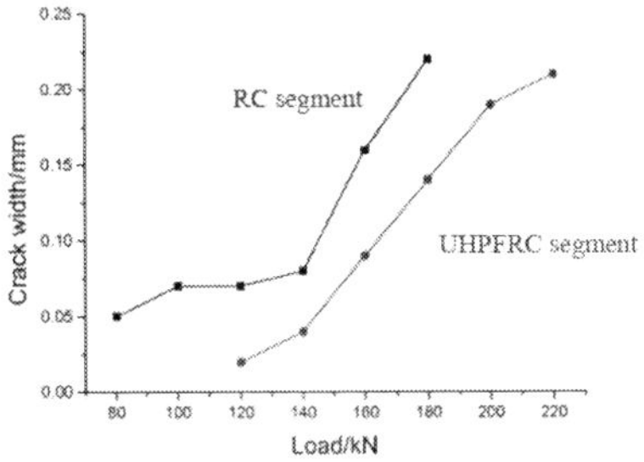

Figure 5 The load-crack width curves of RC and UHPFRC segments

The crack pattern and failure morphology are shown in Figure6. The photograph indicated that all tested scale tunnel segments exhibited a flexural failure mechanism. But there were some differences between two types segments. On the bottom of segments, there were many vertical cracks on the RC segment, but there were only one wide crack on the UHPFRC segment. On the side of segments, the cracks showed that the load of RC segment dropped because the concrete was broken in the top of segment, and the reason was too wide crack for the load drop of the UHPFRC segment.

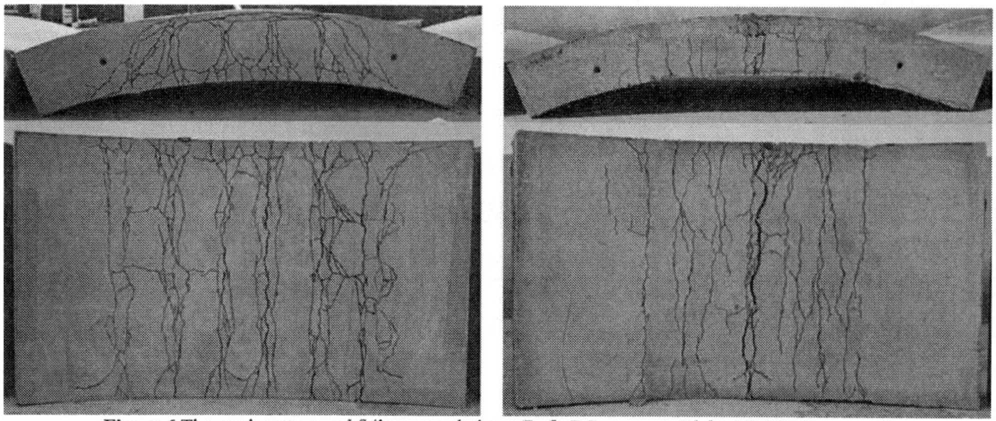

Figure 6 The crack pattern and failure morphology (Left: RC segment; Right: UHPFRC segment)

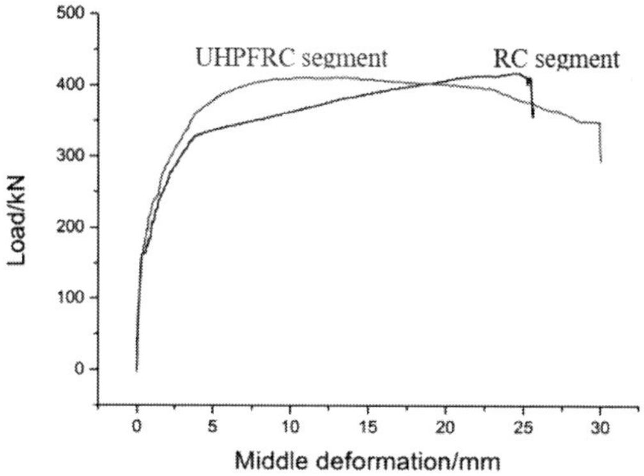

Figure 7 The compare of load-middle deformation curves of two types segments

3.2. Flexural performance of segment.

The segment is used under the ground. The environment of tunnel is complex and unpredictable. The design of segment is careful. The accepted crack width limit for serviceability conditions is 0.2 mm according to Chinese standard GB50010.

Therefore, it was defined that the 0.2 mm crack width load was the load leading to the first 0.2 mm crack. The yield loads and ultimate loads obtained from these load-middle deformation curves in Figure4. The cracking loads, 0.2 mm crack loads, yield loads and ultimate loads are listed in Table 3. The cracking load of UHPFRC segment increased 22%, and the yield load increased 11%, and the ultimate load only decreased 1%. The flexural strength of UHPFRC segment is higher than RC segment.

Table 3 The flexural performance of segment Unit:kN

	Cracking load	0.2 mm crack width load	Yield load	Ultimate load
RC segment	80	180	324-327	418-419
UHPFRC segment	120	220	360-363	404-425

To the thinner segment, the stiffness is very important. The calculation of stiffness is complicated. But it can be compared simply using the slope of the load-middle deformation curves. The curves of two types of segments were showed together in Figure 7. The slopes of two curves looked the same in the elastic stage, and the slope of UHPFRC segment was a little smaller in the elastic-plastic stage. It means that the UHPFRC segment has a similar stiffness to RC segment.

4 Conclusion

The flexural behavior of UHPFRC thinner scale tunnel segment was studied in this work. Form the result of experiments, the following conclusion was draw.

Compared to the traditional RC segment, the

longitudinal reinforcement ratio, the hooping, and the thickness and of the UHPFRC scale segment got a 20%, 50% and 17% reduction. The weight was the 80% of RC segment. However, the cracking load increased 50%, and 0.2mm crack width load increased 22%, and the yield load increased 11%, and the ultimate load only decreased 1%. The flexural strength of UHPFRC segment was higher than RC segment. And the load-middle deformation curves showed the stiffness of UHPFRC segment was similar to that of RC segments. The flexural performance of UHPFRC segment was more excellent than that of RC segment.

Acknowledgements

This work was financially supported by China State Construction Engineering Corporation TCYT-2012A-007 and CSCEC-2016-Z-20-10.

References

1. Abbas, S., Soliman, A.M., Nehdi, M.L., Experimental study on settlement and punching behavior of full-scale RC and SFRC precast tunnel lining segments[J]. Engineering Structures, 2014, 72: 1-10.
2. Caratelli, A., Meda, A., Rinaldi, Z., Romualdi, P., Structural behaviour of precast tunnel segments in fiber reinforced concrete[J]. Tunnelling and Underground Space Technology, 2011, 26: 284-291.
3. De la Fuente, A., Pujadas, P., Blanco, A., Aguado, A., Experiences in Barcelona with the use of fibres in segmental linings[J]. Tunnelling and Underground Space Technology, 2012, 27: 60-71.
4. P. Máca, R. Sovják, T. Vavřiník, Experimental Investigation of Mechanical Properties of UHPFRC, Procedia Engineering, 2013, 65: 14-19.
5. K. Wille, S. El-Tawil, A.E. Naaman, Properties of strain hardening ultra high performance fiber reinforced concrete (UHP-FRC) under direct tensile loading, Cement and Concrete Composites, 2014, 48: 53-66.
6. W. Wang, J. Liu, F. Agostini, C.A. Davy, F. Skoczylas, D. Corvez, Durability of an Ultra High Performance Fiber Reinforced Concrete (UHPFRC) under progressive aging, Cement and Concrete Research, 2014, 55: 1-13.
7. M. Alkaysi, S. El-Tawil, Z. Liu, W. Hansen, Effects of silica powder and cement type on durability of ultra high performance concrete (UHPC), Cement and Concrete Composites, 2016, 66: 47-56.

 https://doi.org/10.1051/e3sconf/20183803038

Overview of Anti-Seismic Researches of Underground Structures

Ran LIAO

Geotechnical engineering, Xipu Campus of Southwest Jiaotong University, Chengdu, Sichuan611756, China

Abstract: With the vigorous development of urban rail transit system, especially the construction of subway system, the safety of subway system draws more and more attention. The study of anti-seismic for underground structures has also become an important problem to be solved in the construction of Metro system. Based on the typical underground structure seismic damage phenomenon, this paper summarizes the seismic characteristics, research methods and design methods of underground structures to offer a guide for engineers.

1 Introduction

Underground structures have been considered to be of good seismic performance. Seismic design has not been taken into consideration in the design and construction of most of the Metro structures that have been built. Before the Great Hanshin earthquake in Japan in 1995, there were few examples of large underground structures suffered serious damage in the earthquake. People generally had the idea that the damage caused by the earthquake on the underground structure of the building was smaller than that on the over-ground structure. However, during the Great Hanshin earthquake in Kobe City, many underground structures were subject to different degrees of damage, particularly some subway stations and tunnels [1][2][3]. The fact that Kobe city subway stations and tunnelswere severely damaged had brought the huge impact to the traditional concept. This had aroused the attention of scholars and research institutions from all over the world, especially in Japan, to such issues.

In recent years, several strong earthquakes in China,such as the Wenchuan earthquake and the Yushu earthquake, have caused great damages to underground structures [4], which offered more research resources on the behavior of underground structures.With the enlargement of the scale of underground structure construction, the seismic design and safety evaluation of underground structure have been paid more and more attention.

In this paper, typical damages of underground structures are firstly presented, followed by the analysis of the difference between the seismic response of underground and that of over-ground structures. Based on which, the development in research methods and design methods are introduced in the following chapters.

2 Typical earthquake damages of underground structure

In history, countless underground structures have been damaged in different degrees during the activities of earthquake. Some typical cases in recent years are introduced in the following paragraphs [1][2][3][4][5][6].

The Tangshan 7.8 earthquake in china in 1971 caused serious damages to both over-ground and underground structures, most of which collapsed or were destroyed seriously. The earthquake also caused varying degrees of destruction to underground tunnel, coal mine tunnel and civil air defense works.

The 7.2 magnitude earthquake in Hanshin in Japan in 1995 also broughtgraveaftermath. The earthquake caused the most severe damage to underground structure in Kobe city, including underground railways, underground parking lots, multipurpose underground tunnels and underground shopping malls, etc. Among them, Daikai station and Kamisawa station were completely destroyed. More than half of the middle columns collapsed. Roof collapses and large cracks appeared on the side walks, causing the large-scale settlement of the National Highway Subgrade above the subway. Tens of meters of cracks piled up in the rest of the subway lines, causing traffic paralysis in southern Japan.

In the1999 Chi-Chi earthquake in Taiwan, only 8 out of 57 mountain tunnels of central Taiwan area were not ruined, and the remaining 49 were damaged in varying degrees. It showed cracks and spalls of the lining, damages of the entrances and exits, bulging and bending of the reinforcement, and displacement of the lining. The cracking of the floor and the collapse of the tunnel also occurred due to slope failure.

During the Wenchuan 8.0 earthquake in china in 2008, 29 of 51 tunnels detected were damaged, of which 18 tunnels were seriously destroyed. Damages of the tunnel near the epicenter were mostly severe, such as the Shaohuiping tunnel, Longxi tunnel, longdongzi tunnel and Zipingpu tunnel.

During the eastern Japan earthquake in 2011, the

* Corresponding author: 359465833@qq.com

earthquake along the Pacific coast in northeastern Japan caused ground liquefaction in vast areas from northeast to Kanto area.

3 Seismic response characteristics of underground structures

The seismic behaviors of underground structures are quite different from that of the upper structures. The characters of responses in both structures and their main differences are introduced as follows [7][8].

① The deformation of underground structure is restrained by the surrounding soil, and the dynamic responses of the structures are not largely influenced by its free vibration characters, while that of the upper structures are dependent on its free vibration characters, particularly the influence of the low modes.

② The presence of an underground structure has very little influence on the vibration of the surrounding base (when the ratio of the size of the underground structure to the seismic wavelength is small). However, the existence of the ground structure causes a great disturbance to the seismic oscillation of the free field.

③ The vibration form of underground structure is greatly influenced by the direction of seismic wave. The deformation and stress of each point of underground structure can change greatly even when the incident direction of seismic wave changes little. The vibration form of ground structure is affected relatively slight by the direction of seismic wave incidence.

④ There are obvious phase differences of the underground structure in earthquake. While the phase difference of the ground structure in the vibration is not obvious.

⑤ In general, the main strain of the underground structure in vibration is not obviously related to the magnitude of earthquake acceleration, but it is closely related to the strain or deformation of surrounding rock and soil medium under earthquake action. As for the upper structure, seismic acceleration is an important factor affecting the dynamic response of the it.

⑥ For both underground structures and over-ground structures, the interactions between them and their foundations have an important influence on their dynamic responses, yet with different extent in various ways.

In general, the seismic behavior of the over-ground structures and underground structures are influenced by different factors. The vibration characteristics of the structure and ground vibration field have important impact on their dynamic response. For the over-ground structure, its structure, shape, mass and stiffness changes, namely the self-vibration characteristics change, has great influence on the structural response, while for underground structures, the main contributing element to the reaction is the kinematic characteristics of the surrounding soil, and changes in structural shape generally have relatively little effect on the reaction.

4 Research methods of seismic resistance of underground structures

The major means to study the seismic performance of underground structures are prototype observation, model experiment and numerical calculation [7][8].

4.1 Prototype observation

The prototype observation method refers to the actual observation of the dynamic response and seismic damage of underground structures under seismic action, thus revealing the seismic response characteristics of underground structures, the seismic performance and the mechanism of earthquake damage. Prototype observation includes earthquake observation, field investigation of earthquake damage and full-scale tests. It plays a very important role in seismic research, for it can obtain not only real earthquake records, but also real dynamic responses to verify the result of the calculation method.

4.2 Model experiment

The model test method is usually used to study the response characteristics of underground structures through the shock test. It can be divided into artificial source experiment and shaking table experiment. It is the most direct method to study the structure seismic response and failure mechanism in laboratory, so it is widely used [9]. The shaking table test can be classified into ordinary shaking table model test and dynamic centrifuge model test [10], which can simulate the dynamic behavior and seismic performance of a system composed of various structures and soils.

4.3 Numerical calculation

With the development of numerical calculation techniques such as finite difference, finite element, boundary element method and discrete element, particularly dynamic numerical calculation, the dynamic numerical methods have been applied to the study of seismic analysis of underground structures. In this method, the entire site is divided into numerical grids with corresponding boundary conditions, then seismic waves are input and dynamic response analysis is performed, then the deformations, stresses and strains in the soil and underground structures will be obtained.

In summary, there is no single means to fully realize the complete and true interpretation and simulation of the dynamic response of underground structures, the seismic performance analysis is developed based on all these three methods.

5 Simplified design method for seismic design of underground structures

In order to satisfy the development of underground engineering, simplified seismic analysis method of underground structures were proposed based on the previous researches [11-21].

5.1 Seismic coefficient method

In 1899, Japanese scholar Oomori Boyoshi put forward the seismic coefficient method based on the static theory. Time dependent seismic forces were firstly equaled to static seismic load, and then static calculation models were used to analyze the internal force and deformation of the structure under seismic loading.

5.2 Free field deformation method

In 1960s, Newmark [11] argued that the seismic performance of underground structures was mainly affected by the deformation of surrounding soil, instead of the inertial force that described in the seismic coefficient method. Wang et al [12] and Hashash et al [13] proposed free field deformation method based on this feature. In this method, the free field deformation of the surrounding soil under seismic was directly applied to the structure as the structural deformation. Then the internal force of the structure was calculated, and the structure was designed.

5.3 Flexibility coefficient method

The free field deformation method assumes that the deformation of the structure is consistent with the lateral deformation of the surrounding soil, neglecting the uncoordinated deformation between the structure and the surrounding soil. However, in practical engineering, the stiffness of underground structure and the stiffness of surrounding soil are often different. The flexibility coefficient method, also known as soil-structure interaction coefficient method, was established by Penzien et al [14] based on the seismic observation results.

5.4 Response displacement method

In 1970s, Japanese scholars found that the decisive factor in the seismic response of underground structures was the deformation of surrounding rock and soil medium instead of the inertial force. And based on whichthe response displacement method was put forward. In this method, the underground structure was assumed as an elastic foundation beam, and the response of underground structure was obtained according to the method of statics.

5.5 Reaction acceleration method

The response displacement method belongs to the load-structure method, taking into account three kinds of loads: seismic earth pressure, shear force of soil-structure contact surface and inertial force. While reaction acceleration method belongs to the soil-structure method, and the integral calculation model of soil and underground structure is needed. In this method, only inertia force is considered, which depends on the mass distribution and acceleration of the system. As the mass distribution of the system is definite, it is only necessary to determine the acceleration distribution. Therefore, compared with the reaction displacement method, this method is simpler to use.

5.6 Pushover analysis method of underground structure

In early twenty-first century, the static elastoplastic analysis method for underground structures (Pushover analysis method) was proposed by Liu Jing-bo and Li Bin et al [15][16]. The method is to apply an increasing horizontal load to a structure with a certain height distribution structure until a predetermined target structural displacement to analyze the nonlinear response of the structures. This method is used to assess the deformation or capacity of structures and members.

In summary, all these methods reflect the features of behavior of underground structures under horizontal waves to some extent without considering the influence of vertical seismic action, which is an important point to be improved.

6 Conclusion

This paper presents the researches on the seismic performance of the underground structures. The main points are as follows:

1. Typical damages of underground structures under earthquakes like collapse of tunnels show the importance of anti-seismic researches.

2. There is a big difference between the seismic performance of over-ground and that of underground structures. The behaviors of underground structures show close relation with the deformation of surrounding soil.

3. Three research methods to analyze the seismic performance for underground structures are presented in this paper.

4. Simplified design methods are introduced. The future trend will be the consideration of the influence vertical seismic action.

References

1. Yoshida N, Nakamura S. Damage to Dakai subway station during the 1995 Hyogoken-Nummu earthquake and its investigation. Proceedings of Eleventh World Conference on Earthquake Engineering. 1996, 2151:283-300.
2. Iida H, Hiroto T, Yoshida N, et al. Damage to Dakai subway station. Solis and foundations, special issue on geotechnical aspects of the January 17 1995 Hyogoken-Nambu earthquake. Japanese Geotechnical Society, 1996:283-300.
3. Huo Hong-bin, Bobet A. Seismic design of cut and cover rectangular tunnels-evaluation of observed behavior of Dakai station during Kobe earthquake. Proceedings of 1st World Forum of Chinese Scholars in Geotechnical Engineering, August 20-22, 2003, Tongji University, Shanghai, 456-466.
4. Lin G., Luo S.p., Ni J., Damages of metro structures due to earthquake and corresponding treatment

measures. Modern Tunneling Technology. 2009, 46(4):36-41.

5. Chen Guo-xing, Chen Su, Du Xin-li, et al. Review of seismic damage, model test, available design and analysis methods of urban underground structures: retrospect and prospect. Journal of Disaster Prevention and Mitigation Engineering, 2016, 36(1):1-23.
6. Penzien J. Seismically induced racking of tunnel linings. Earthquake Engineering Structural Dynamics, 2000, 29(5): 683 — 691.
7. Li Bin. Theoretical Analysis of SeismicResponse of Underground Subway Structures and lts Application. PHD Thesis. Tsinghua University. 2005.
8. Sun Chao. Study on seismic capability and analysis method of underground subway structures. PHD Thesis. Institute of Engineering Mechanics, China Earthquake Administration. 2009.
9. Tao L.J., Wang P.L., BianJ.. A shaking table test on the representative subway station structure. Jorunal of Beijing Unversity of Technology. 2006, (9):798-801.
10. Kagawa T, Sato M. Centrifuge simulations of large-scale shaking table tests. Journal of Geotechnical and Geoenvironmental Engineering, 2004, 130(7):663-672.
11. Newmark N M. Problems in wave propagation in soil and rock. Proceedings of the International Symposium on Wave Propagation and Dynamic Properties of Earth Materials,1968.
12. Wang J N. Seismic design of tunnels: A simple state-of-the-art design approach. New York: Parsons Brinckerhoff Quade and Douglas Inc. 1993.
13. Hashash Y, Hook J J,SchmidtB,et al. Seismic design and analysis of underground structures. Tunneling and Underground Space Technology2001, 16(4): 247—293.
14. Pitilakis K, Tsinidis G. Performance and seismic design of underground structures in earthquake geotechnical engineering design. Springer International Publishing. 2014:279-340.
15. Fishman K L, Mander J B, Richards R, et al. Laboratory study of seismic free-field response of sand[J] Soil Dynamics and Earthquake Engineering, 1995, 14: 33-43.
16. Liu J.B., Wang W.H., Zhao D.D., et al. Integral response deformation method in seismic analysis of complex section underground structures. China Civil Engineering Journal. 2014, 47(1):134-142.
17. Liu J.B., Liu X.Q., XueY.L.. study on applicability of a pushover analysis method for seismic analysis and design of underground structures. Engineering Mechanics. 2009, 26(1):49-57.
18. Shukla D K, et al. Earthquake load analysis of tunnels and shafts. Proceeding of the 7th World Conference on Earthquake Engineering. Michigan, USA: Univ. of Michigan Press ,1980: 20-28.
19. Choi J S, Lee J S, Kim J M. Nonlinear earthquake response analysis of 2-D underground structure with soil-structure interaction including separation and sliding at interface. 15th ASCE Engineering Mechanics Conference, June 2-5, 2002, Columbia University, New York, 1-8.
20. AFPS/AFTES. Guidelines on Earthquake Design and Protection of Underground Structures. Working Group of the French Association for Seismic Engineering(AFPS) and French Tunneling Association (AF-TES) Version 1, 2001.
21. Kiyomiya O. Earthquake-resistant design feathers of immersed tunnels in Japan. Tunneling and Underground Space Technology. 1995. 10(4):463-457.

 https://doi.org/10.1051/e3sconf/20183803039

A Technique for Tunable Filters with Low Insertion Loss and Narrow Bandwidth

Chang ZHOU, Chen JI, Gen Ping WU

Wisdom Engineering, Wuhan Second Ship Design and Research Institute, Hubei 430074, China

Abstract: A technique for tunable filters with low insertion loss and narrow bandwidth is proposed in the form of comb-line structure. Both resonant capacitor with pin-diodes and resonant inductance in the tunable filter were analyzed and the main source of insertion loss was obtained. A series of filters with same pin-diodes, center frequency, absolute bandwidth and low return loss was simulated. The results showed that, by changing the values of the resonant capacitor and inductance, insertion loss of the filter can be greatly restricted. This technique will allow the design of tunable LC filters with low insertion loss and narrow bandwidth.

1 Introduction

Tunable filter is one of the suitable solutions for intelligent RF applications satisfying multi-band or multimode standards [1]. Compared to a bank of fixed filters, a tunable filter promises greater functionality, better channel selectivity, reduced size, and lower weight since the same hardware can be employed at multiple bands [2]-[5]. However, for tuning over an octave frequency by using pin-diodes switches or varactors, the insertion loss of the tunable filter is larger than normal fixed filters. Up till now, some method has been used to restrict the insertion loss of the tunable filter [6]-[10]. But few research has reported the influence of values of the resonant capacitor with pin-diodes and resonant inductance on the insertions loss of the tunable filter.

In this paper, both resonant capacitor with pin-diodes turned on and resonant inductance in the tunable filter were analyzed and a technique for restricting the insertion loss of tunable filters by changing the values of the resonant capacitor and inductance is reported.

2 Theoretical Analysis Of The Resonant Circuit With PIN-DIODES

The insertion loss and bandwidth of band-pass filter is shown in (1).

$$L(f) = -10\log\left[\frac{1+(2Q_l\frac{f-f_0}{f_0})^2}{\left[1-\frac{Q_l}{Q_u}\right]^2}\right] \quad (1)$$

$L(f)$ is insertion loss, Q_l is loaded quality factor, Q_u is unloaded quality factor, f is frequency and f_0 is center frequency.

The fractional bandwidth of band-pass filter is shown in (2).

$$BW = \frac{\Delta f}{f_0} = \frac{1}{Q_l} \quad (2)$$

According to (1) and (2), it could be concluded that, if the fractional bandwidth of the filter is fixed, the insertion loss mainly depends on Q_u of the filter and higher value of Q_u means lower insertion loss in the filter.

The original definition of quality factor is:

$$Q_u = 2\pi\frac{\text{The energy storage in the circuit}}{\text{The energy consumption in the circuit}} \quad (3)$$

In this paper, for the convenience of comparison and discussion, only the tunable filter with pin-diodes all turned on and can be equal to be a resistance was taken into considerations.

The structure of the resonant capacitors with pin-diodes is shown in Figure 1.

* Corresponding author: rabbite6@126.com

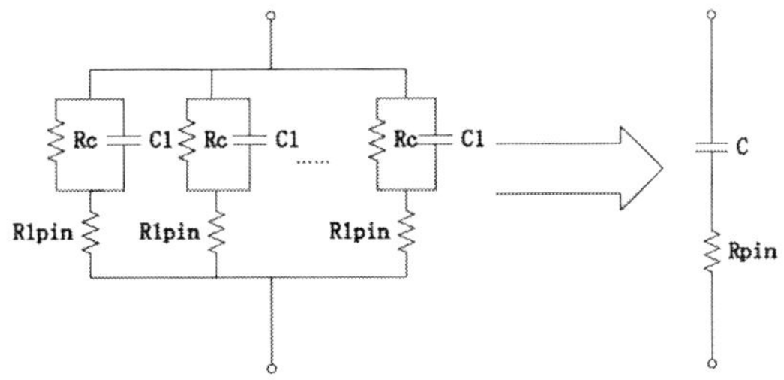

Figure 1 The structure and equivalent circuit of the resonant capacitors with pin-diodes

$C1$ is the resonant capacitor, Rc is the dissipative resistance and $R1_{pin}$ is the resistance of pin-diodes turned on. For the dissipative resistance Rc of the resonant capacitor $C1$ is much smaller than the resistance of pin-diodes, the structure can be transformed. The result is shown in (4) and (5).

$$C \approx nC1 \tag{4}$$

$$R_{pin} \approx \frac{R1_{pin}}{n} \tag{5}$$

The structure of the resonant circuit is shown in Figure 2.

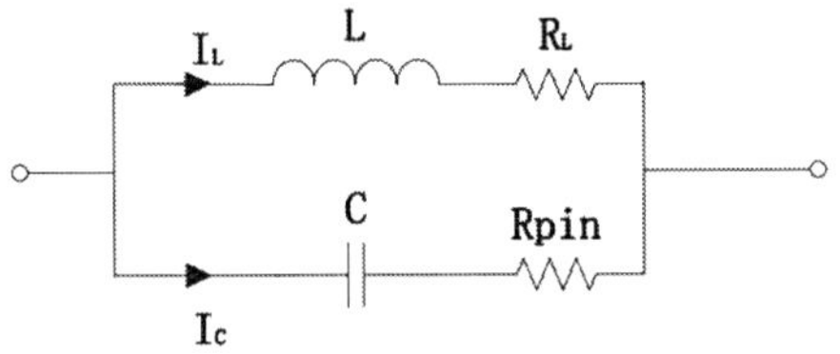

Figure 2 The structure of the resonant circuit

For the value of R_{pin} and R_L is much smaller than impedance of the resonant capacitor and inductor, I_C is almost equal to I_L and the value of unloaded quality factor could be concluded to be shown in (6).

$$Q_u = \frac{\frac{1}{2}I_C^2\frac{1}{w_0C} + \frac{1}{2}I_L^2 w_0 L}{I_C^2 R_{pin} + I_L^2 R_L} \approx \frac{w_0 L}{R_{pin} + R_L} = \frac{1}{w_0 C(R_{pin} + R_L)} \tag{6}$$

The value of R_{pin} is fixed and much bigger than R_L. So the main source of the insertion loss in the filter is the resistance of pin-diodes R_{pin}. It could be concluded that, although big resonant inductor may cause big dissipative resistance R_L, small value of resonant capacitor means higher Q_u and lower insertion loss of the tunable filter at the same center frequency.

3 Simulation Examples

According to Section 2, a series of tunable comb-line filters with different resonant capacitor and 1% bandwidth were simulated. The models of the tunable filters with the same pin-diode switches but different resonant inductors and capacitors are made and shown in Figure 3.

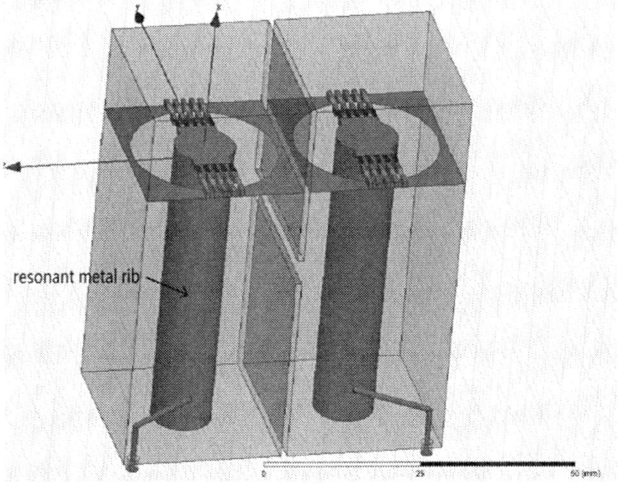

(a)

https://doi.org/10.1051/e3sconf/20183803039

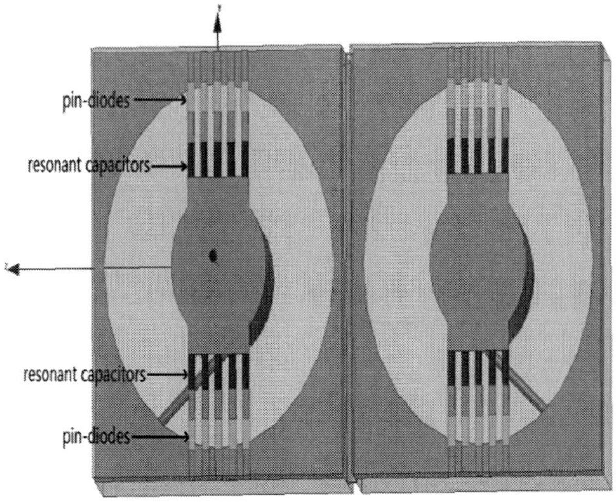

(b)

Figure 3 The models of the tunable filters with the same pin-diode switches

The pin-diodes are all turned on and can be equal to be a resistance as 0.5 Ω as shown in Fig.3. The value of the capacitors varies from 2 pf to 4 pf and the radius of the resonant metal rib varies from 1 mm to 5 mm which keeps the center frequency of the filters at about 210 MHz. The material of the metal rib is set as aluminum. By changing the coupling window and the height of the tap-line, the tunable filters with pin-diodes get the same values of bandwidth, return loss and center frequency. However, the values of the insertion loss are different. The simulation results are shown in Figure 4 and table 1.

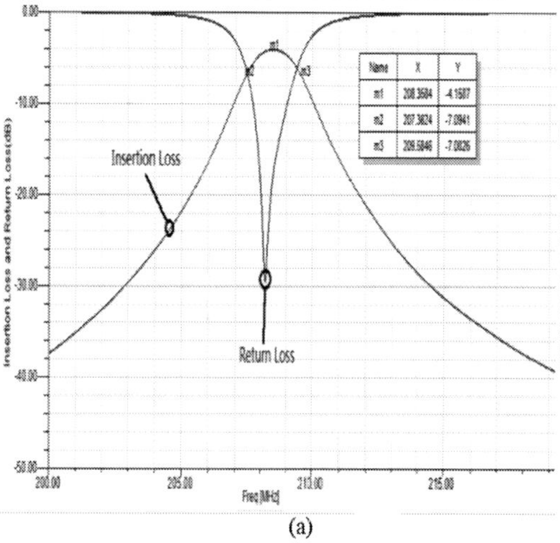

(a)

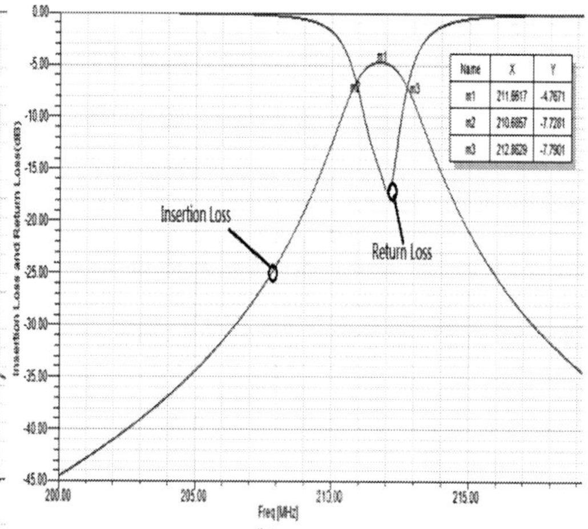

(b)

 https://doi.org/10.1051/e3sconf/20183803039

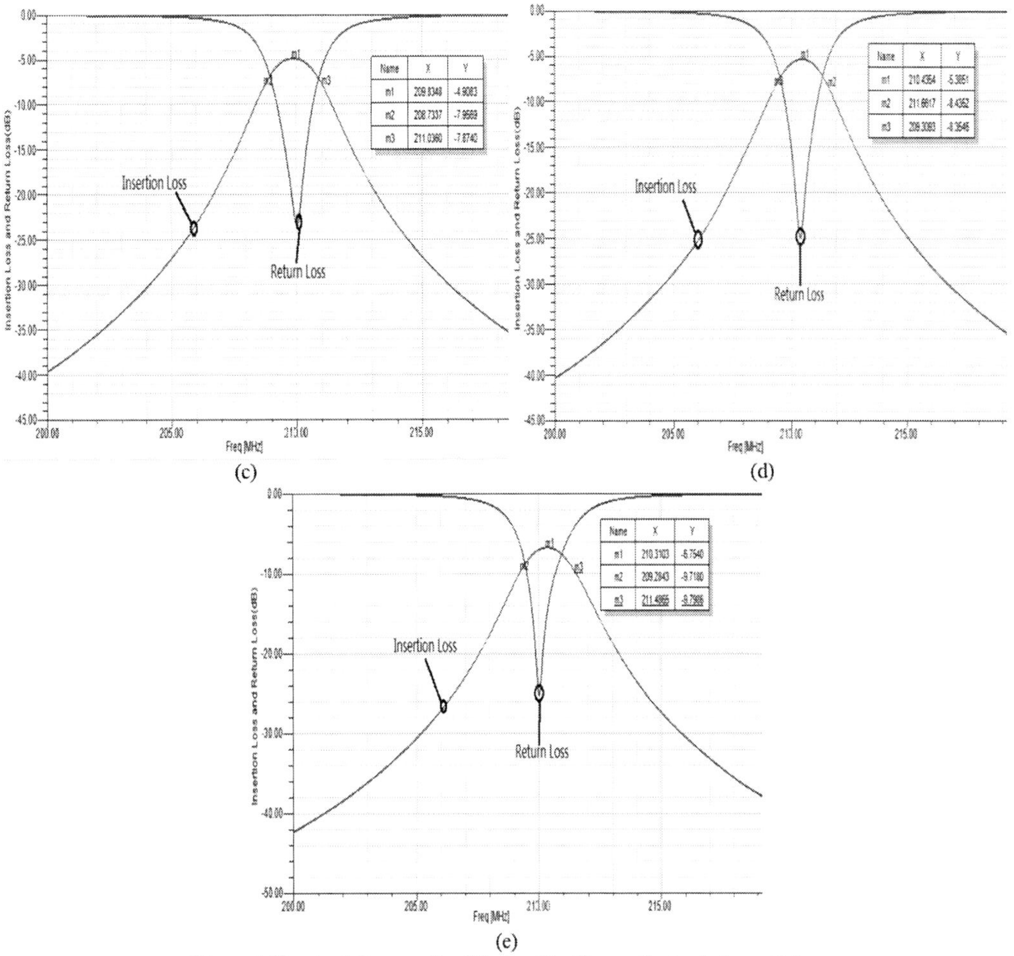

Figure 4 The simulation results of the tunable filters with pin-diodes switches

Table 1 Comparison Between Numerical and Experimental Results

Number	Single resonant capacitor	Radius of the metal rib	Center Frequency	Absolute Bandwidth	Return Loss	Insertion Loss
1	2 pf	1 mm	208.4MHz	2.2MHz	-20dB	-4.2dB
2	2.5 pf	2 mm	211.7MHz	2.2MHz	-15dB	-4.8dB
3	3 pf	3 mm	209.8MHz	2.3MHz	-25dB	-4.9dB
4	3.5 pf	4 mm	210.4MHz	2.3MHz	-15dB	-5.4dB
5	4 pf	5 mm	210.3MHz	2.2MHz	-25dB	-6.8dB

It could be concluded that, as the value of the capacitors varies from 2 pf to 4 pf and the radius of the resonant metal rib varies from 1 mm to 5 mm, the insertion loss of the filters with the same pin-diodes, center frequency, absolute bandwidth and low return loss increases from -4.2 dB to -6.8 dB. Smaller resonant capacitor means higher Q_u and lower insertion loss of the tunable filter with pin-diodes switches all turned on, which support the conclusion in section 2.

4 Conclusion

A technique for Tunable filters with low insertion loss and narrow bandwidth across the tuning range is proposed in the form of comb-line structure. The effect of values of resistant in pin-diodes, resonant capacitor and inductor on insertion loss in tunable filter has been analyzed and a series of tunable filter with the same pin-diodes, center frequency, absolute bandwidth, low return loss but different resonant capacitors and inductors

 https://doi.org/10.1051/e3sconf/20183803039

were simulated. It could be concluded that, as the value of the capacitors varies from 2 pf to 4 pf, the insertion loss of the filters with the same pin-diodes increases from -4.2 dB to -6.8 dB. Both the theory analysis and simulation result show that the small value of resonant capacitor means higher Q_u and lower insertion loss in the filter with pin-diodes turned on.

References

1. Paul S, David F, Corrado C. A Passive Tunable Matching Filter for Multiband RF Applications Demonstrated at 7 to 14 GHz[J]. IEEE Microwave and Wireless Components Letters, 2017, 27(8): 703-705.
2. J Vanhamel, S Berkenbosch, E Dekemper. Implementation of different RF-chains to drive acousto-optical tunable filters in the framework of an esa space mission[J]. URSI Radio Science Bulletin, 2016, 89(2): 37-43.
3. Tao Y, Gabriel M. Bandpass-to-Bandstop Reconfigurable Tunable Filters with Frequency and Bandwidth Controls[J]. IEEE Transactions on Microwave Theory and Techniques, 2017, 65(7): 2288-2297.
4. Tsai H Y, Huang T Y, Wu R B. Varactor-Tuned Compact Dual-Mode Tunable Filter With Constant Passband Characteristics[J]. IEEE Transactions on Components, Packaging and Manufacturing Technology, 2016, 6(9): 1399-1407.
5. Lee B, Nam S, Lee T H. Single-Filter Structure With Tunable Operating Frequency in Noncontiguous Bands[J]. IEEE Transactions on Components, Packaging and Manufacturing Technology, 2017, 7(1): 98-105.
6. Eyad A, Atif S. The Effect of Self-Heating on the Performance of a Tunable Filter With Embedded Windings in a Ferrite LTCC Package[J]. IEEE Transactions on Components, Packaging and Manufacturing Technology, 2015, 5(3): 365-371.
7. Tang C W, Tseng C T, Chang S C. Design of the Compact Tunable Filter With Modified Coupled Lines[J]. IEEE Transactions on Components, Packaging and Manufacturing Technology, 2014, 4(11): 1815-1821.
8. Oncel A, Johansen T K, Vitaliy Z. A High-Power Low-Loss Continuously Tunable Bandpass Filter With Transversely Biased Ferrite-Loaded Coaxial Resonators[J]. IEEE Transactions on Microwave Theory and Techniques, 2015, 63(10): 3425-3432.
9. Suo G N, Guo X B, Cao B S. Low Loss Tunable Superconducting Dual-Mode Filter at L-Band Using Semiconductor Varactors[J]. IEEE Microwave and Wireless Components Letters, 2014, 24(3): 170-172.
10. Chen J, Zhu X W, Ge C. Tunable bandpass filter with low-loss and enhanced selectivity based on controllable coupled negative resistance[J]. Electronics Letters, 2013, 49(24): 1544-1545.

An Antenna with Good Electrical Properties and Artistic Shape for Lampshade

Chang ZHOU, Yang YU, Chen JI

Wisdom Engineering, Wuhan Second Ship Design and Research Institute, Hubei 430074, China

Abstract: An improved transparent antenna based on transparent ceramic $MgAl_2O_4$ with artistic shape which can be used as lampshade is presented in this paper. A great peak gain and good pattern can be achieved by choosing the property shape and optimizing location of the feed of the antenna. Simulation results indicate that this antenna can achieve return loss >-15dB, peak gain >8 dBi at 2 GHz and 40% relative impedance bandwidth. Both the electrical properties and artistic shape of the antenna are realized and the lampshade which made by $MgAl_2O_4$ meets the requirement for communication, decoration and high light transmission.

1 Introduction

Recently, the modern wireless communication systems have been rapidly increased and more requirements such as special structure, small size and multi-function have been made [1]-[2]. The dielectric resonator antenna based on low loss dielectric materials has some pretty advanced features ranging from greater design flexibility, smaller size, lower loss, wide impedance bandwidth which is suitable for the design of antenna with special size and good electrical properties [3]-[7]. On the other side, new types of dielectric material have been improved and applied and the antennas based on dielectric material can have more features. Among various types of materials, the transparent ceramics with the advantages of high light transmittance, low loss and good mechanical properties has a certain development in lighting, temperature insulation and so on [8]-[10]. The antenna based on transparent ceramics with special function can be made.

In this paper, we present the method to design lampshade with multi-function by using transparent ceramic. A type of transparent ceramic with low loss and high light transmittance were chosen and the influence of the shape and feed on the electrical properties of the antenna were analyzed.

2 Models with Artistic Shape

The electrical properties of the transparent ceramics MgAl2O4 is: permmitivity (εr)=8.38, Qf=57000GHz, the temperature coefficient of dielectric constant tf=-76ppm/°C, light transmittance =70% and the structure of the antenna is chosen as shown in Figure 1 which ensures the artistic shape.

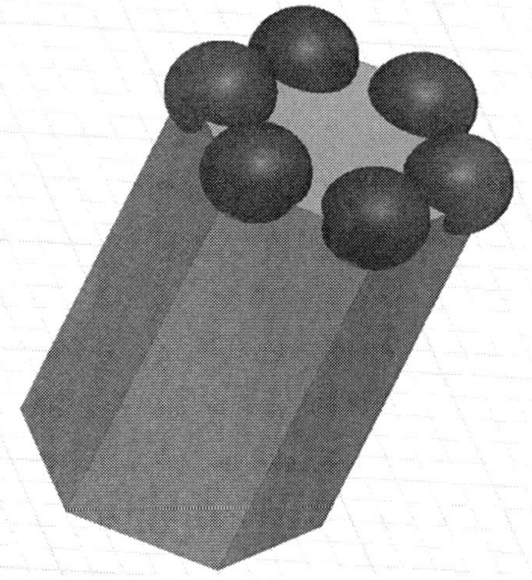

(a)

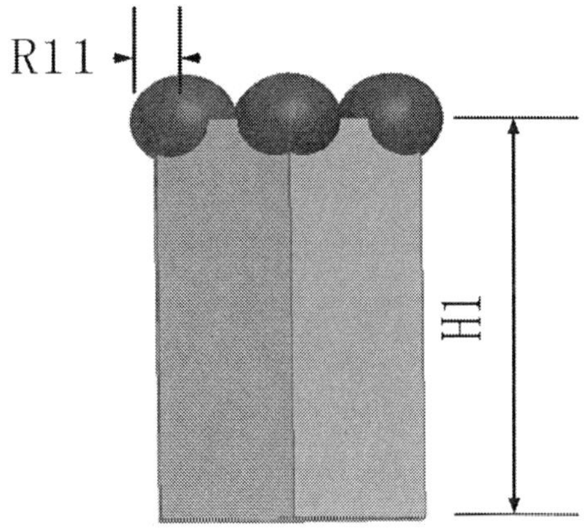

(b)

* Corresponding author: rabbite6@126.com

https://doi.org/10.1051/e3sconf/20183803040

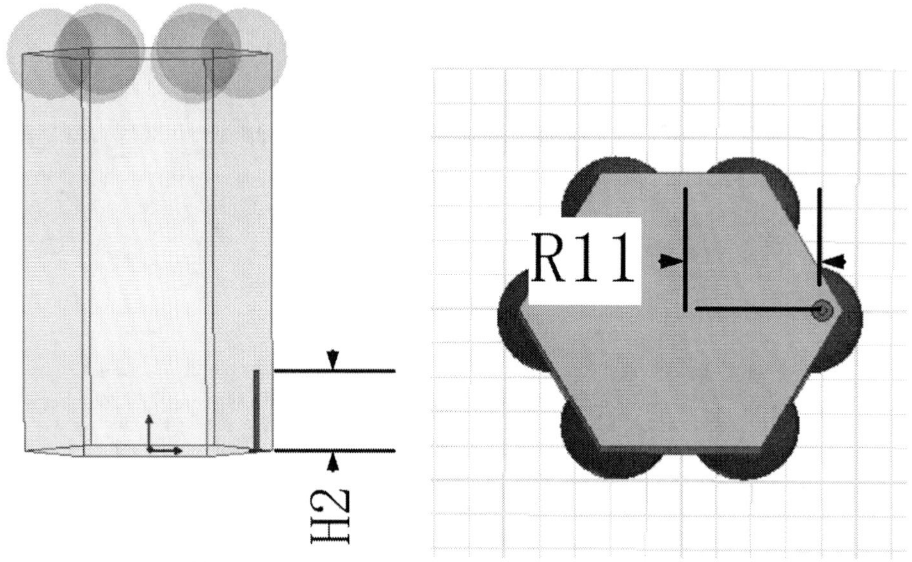

(c) (d)

Figure 1 The shape chosen from iron gate

It can be seen that the antenna is made of regular hexagonal prism and six ball which ensures the artistic shape of the lampshade. On the other side, the target electrical properties of the antenna is fo=2GHz, relative impedance bandwidth>80%, return loss>-20dB and peak gain>8dBi, respectively.

Here: the height of the ball is H1, the radius of the ball is R11, the distance of the ball from the center of the regular hexagonal prism is 23mm, the length of the side of the regular hexagonal prism is 23mm, the height of the regular hexagonal prism is H1, the height of the feed is H2 and the distance from the feed to the center of the regular hexagonal prism is R12, respectively.

3 Simulation Examples

For the artistic shape of the lampshade, some parameters of the structure is fixed such as the length of the side of the regular hexagonal prism. On the other side, the values of H1, H2, R11 and R12 are optimized to obtain the electrical properties and the simulation results are shown in Table 1.

Table 1 The simulation results of the antenna based on transparent ceramics

Number	H1 [mm]	H2 [mm]	R11 [mm]	R12 [mm]	Fo [GHz]	Return Loss [dB]	Grain Peak [dB]	Ralative Bandwidth
1	70	14	8	20	1.898	-9.7	7.0	-
2	70	14	8	15	1.898	-11.3	6.6	4%
3	70	14	8	10	1.898	-8,3	8.9	-
4	70	18	8	15	2.0	-21.3	7.8	85%
5	80	18	8	15	2.0	-22.6	8.2	90%
6	80	18	5	15	2.0	-21.6	8.8	92%

The simulated return-loss results and the radiation patterns at 2GHz are shown in Figure 2 and Figure 3.

 https://doi.org/10.1051/e3sconf/20183803040

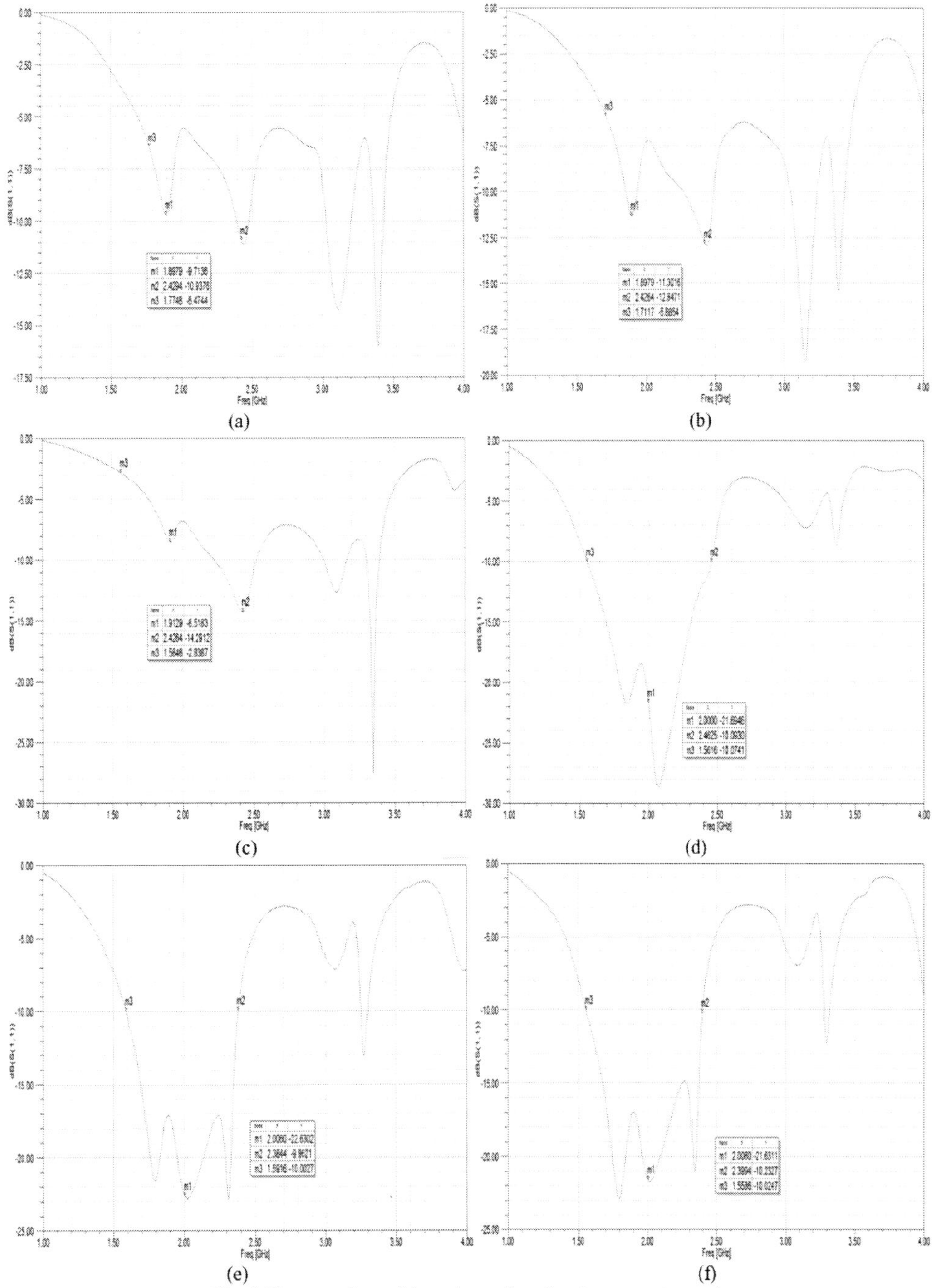

Fig. 2 The return loss of the antennas based on transparent ceramic

It could be seen that, by choosing proper material and parameter of the antenna, center frequency of the antenna could be controlled.

The radiation patterns at 2.0GHz are shown in Figure 3.

 https://doi.org/10.1051/e3sconf/20183803040

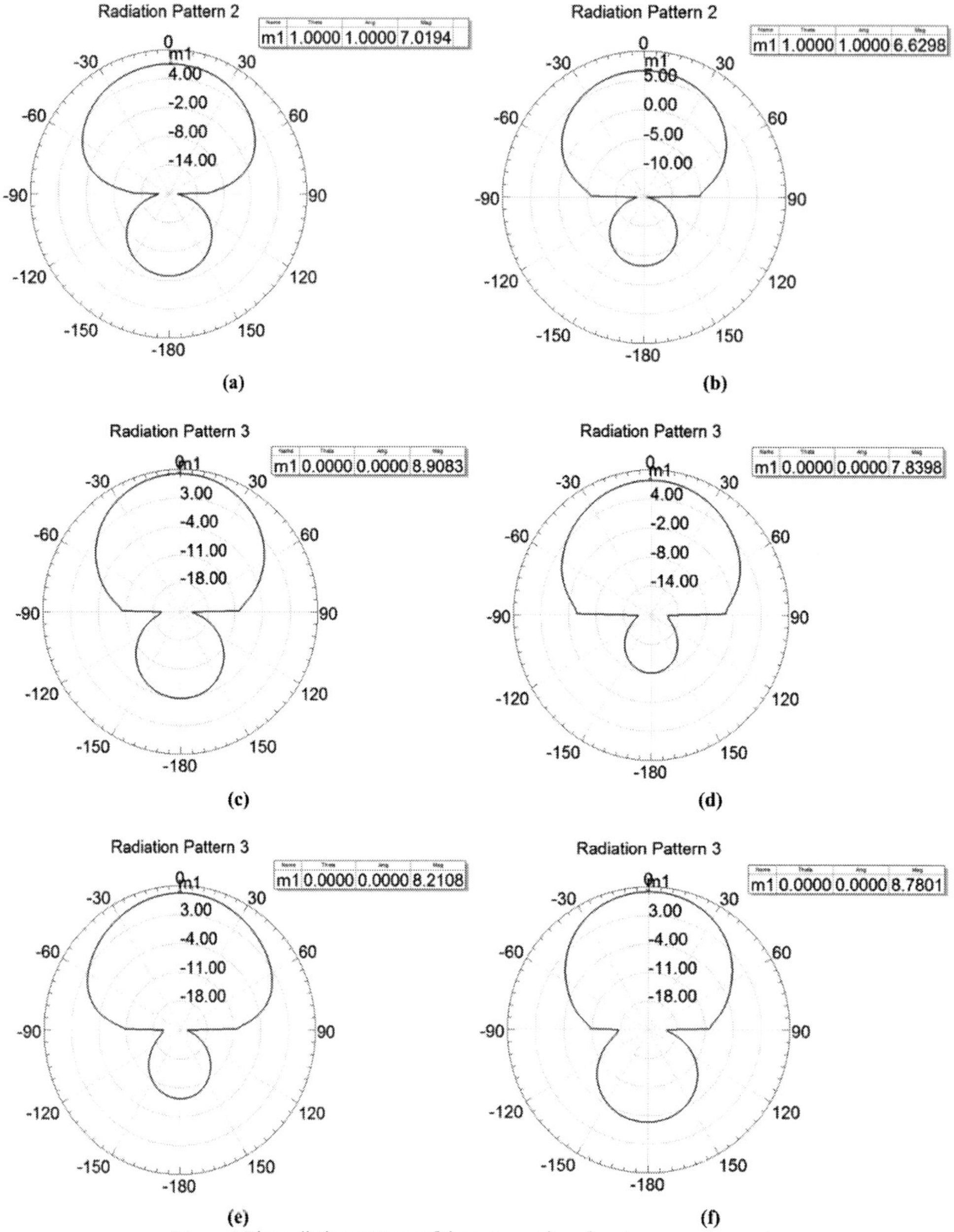

Figure 3 The radiation patterns of the antennas based on transparent ceramic

It was found that the sturcture is important for some key paramters and the satisfactory results can be obtained. As shown in Figure 2 and Figure 3, the relative impedance bandwidth of the transparent antenna are especially affected by adjusting the height of the coaxial feed and the center frequency of the antenna can be controlled by varying the height of the regular hexagonal prism. On the other side the distance of the feed from the center of the regular hexagonal prism and the radius of the ball have a less important influence on the electrical properties of the antenna.

The best combination of microwave parameters for the antenna based on transparent ceramic is got at H1=80mm, H2=18mm and R11=5mm, R12=15mm, which yields properties of: fo=2.0GHz, return loss=-21.6dB, peak gain=8.8dB and the relative bandwidth=92%.

4 Conclusion

This paper give a method to fabricate an antenna based on transparent ceramic. The antenna with good electrical properties, artistic shape and high light transmission was designed and the affection of the shape and location of tap of the antenna were summarized and presented. On the other side, it can be concluded that the lampshade and antenna can be integrated by choosing property material and structure. This method will be useful for the design of beautiful lampshade which also can be used as antenna for communication.

References

1. [1] Li Y F, Kwork W L. Dual-Fed Hollow Dielectric Antenna for Dual-Frequency Operation With Large Frequency Ratio[J]. IEEE Transactions on Antennas and Propagation, 2017, 65(6): 3308-3313.
2. Yang C, Zuping Q, Yingsong Z. A Compact Wideband SIW-Fed Dielectric Antenna With End-Fire Radiation Pattern[J]. IEEE Transactions on Antennas and Propagation, 2016, 64(4): 1502-1507.
3. Payam N, Min L, Rafael A S. A 3D Printed Dielectric Reflectarrays: Low-Cost High-Gain Antennas at Sub-Millimeter Waves[J]. IEEE Antennas and Propagation Society, 2014, 62(4): 2000-2008.
4. Nasser G, Ke W. Planar Dielectric Rod Antenna for Gigabyte Chip-to-Chip Communication[J]. IEEE Transactions on Antennas and Propagation, 2012, 60(10): 4924-4928.
5. Yonghun C, Jungyub L, Joonghee L. Quad-Band Monopole Antenna Including LTE 700 MHz With Magneto-Dielectric Material[J]. IEEE Transactions on Antennas and Propagation, 2012, 11: 137-140.
6. J B Birks. Dielectric housings for centimetre-wave antennae. Part 2: Design practice for airborne radar
[J]. Journal of the Institution of Electrical Engineers - Part IIIA: Radiolocation, 1946, 93(4): 653-657.
7. G M Whitman, C Pinthong, A A Triolo. An approximate but accurate analysis of the dielectric wedge antenna fed by a slab waveguide using the local mode theory and Schelkunoff equivalence principle[J]. IEEE Transactions on Antennas and Propagation, 2006, 54(4): 1111-1121.
8. Jian X, Lingcong F, Ying S. Scintillation and Luminescent Properties of Cerium Doped Lutetium Aluminum Garnet (Ce:LuAG) Powders and Transparent Ceramics[J]. IEEE Nuclear and Plasma Sciences Society, 2014, 61(1): 373-379.
9. Yun T W, Lai S Q, Hong S S. Preparation of Europium-Doped Gadolinium Lutetium Oxide Solid Solution Transparent Ceramics and Its Optical Properties[J]. IEEE Nuclear and Plasma Sciences Society, 2010, 57(3):1343-1347.
10. Yu S, Yong B, Nan Z. Experimental and Theoretical Investigation of Pump Laser Induced Thermal Damage for Polycrystalline Ceramic and Crystal Nd:YAG[J]. IEEE Journal of Selected Topics in Quantum Electronics, 2015, 21(1):4924-4928.

The Researches on Construction Response and Excavation Method Optimization of Extra-large Cross-section Urban Subway Tunnel

Bo WU[a, b], Wei HUANG[a, b], Yong Bo ZHAO[c]

[a]College of Civil Engineering and Architecture, Guangxi University, Nanning, Guangxi 530004, China
[b]Institute of Underground engineering, Guangxi University, Nanning,Guangxi 530004, China
[c]College of Civil Engeneering, FuJian University of Technology, Fuzhou, Fujian 350118, China

Abstract: In the extra-large cross-section urban subway underground station projects, the key problem is to choose the appropriate excavation method and grasp the mechanical behavior of the surrounding rock after excavation. The double side drift method is widely used in the urban subway underground station construction with extra-large cross-section in China. This paper presents the deformation characteristics of the extra-large cross-section tunnel of urban subway and a comparison with the numerical simulation results. In the city subway underground excavation of large section station, the mechanical behavior of surrounding rock change and its influence on the surrounding area mainly depend on the selection of construction methods. The convergent deformation of tunnel cavern, the subsidence of the surface and the force failure of the surrounding rock associated the construction approaches are demonstrated. This study provides a more in-depth demonstration of the way to optimize the excavation method of the extra-large cross-section tunnel to achieve the purpose of controlling the deformation of the surface and surrounding rock.

1 Introduction

With development and the urbanization of the city, the underground space development and the utilization of urban development have gradually become significant factors of cities' sustainable development[1]. Thus, when the excavation of extra-large cross-section tunnels or underground space is enforced, it could effectively solve the problem of urban sustainable development, but the inevitable environmental impacts and building operation bring challenges for engineers, that it inevitably disturbs the in situ stress field which causes ground movements leading to surface settlement, which may cause serious damage to adjacent structures[2]. It is essential to choose the proper excavation method for the construction of extra-large cross-section urban tunnel, because different excavation methods are of great significance for controlling the ground deformation and wall-rock convergence.

For the construction of urban subway tunnels in china, selecting a suitable design and construction method to complete the construction of the extra-large cross-section urban subway tunnel is the most critical factor to complete the whole project. It depends on the reasonable techniques and technologies adopted in the design and construction phase of the project, so as to achieve the goal of the highest construction efficiency and the lowest construction cost. These goal are strongly influenced by the excavation procedure.

In recent years, the research on the optimization of excavation methods of the underground stations is not concentrated. Tian L.F. [3] concentrates some works on the support system optimization of double-side wall excavation method for subway station tunnel under complicated condition. Yu F.C. and Zhang D.L. [4] study on excavation scheme optimization at the junction of pits and tunnels of subway station. Cheng T., Liu J.W. and Cao P. [5] have focus on optimization analysis on construction scheme of large span metro station in hard rock stratum. These above researches are not focused on the common problems. Therefore, it is significant to find out the way to optimize the tunnel excavation method of extra-large cross-section station and the construction mechanics response of large section tunnel excavation.

2 Project overview

The Central Park East station of the metro line ten is located in the east of the Yubei District of the airport town in Chongqing, on the east side of the Tongmao Avenue. The Central Park East station adopts the 14m island double decker station with a net width of 23.0m and a net height of 19.0m. The maximum excavation span of the station is 24.54m, the excavation height is 21.52m, and the section area of excavation is 468.55m^2. The vertical wall circular arch section is adopted, and the composite lining is used in the structure.

* Corresponding author: 813792833@qq.com

https://doi.org/10.1051/e3sconf/20183803041

The station surrounding rock is mainly mudstone, belongs to level IV and turns the soil thickness of 20.93 m to 30.16 m. The station adopts the double side drift method construction. It sets a construction main channel into the station hall. In the construction main channel, it sets up an auxiliary channel to enter the station platform layer, with the length of 79.702 m.

2.1 Design information of the Station

The Central Park East station adopts straight wall plus circular arch section, the excavation size of its standard section is 24.54m*21.52m, and the tunnel lining is composite lining. The design has 2 kinds of A1 and A2 type lining section (468.55 m^2). The design parameters of the main section of the station are shown in table 1 and the design section is shown in fig.1.

Table1 Support parameter of main section

Support Type	application	Supporting structure				secondary lining
		primary lining	spray concrete and steel mesh	anchor	grid arch	
IV level A1 type lining section	General location	φ42 advanced small pipe, L=4.5m,space in the circumferential direction 400 mm grouting, array pitch 3.0m	C25 concrete, thickness 30cm Φ8-@150×150mmbar-mat reinforcement, double-deck	φ25 grouted rockbolt L=4.0m, interval 1.0m×0.75m, in passing quincunx	I22b steel gantry 0.75m/interval	thickness 85cm,C40 concrete,P12 reinforced concrete
IV level A2 type lining section	Reinforced intersections	φ42 advanced small pipe, L=4.5m,space in the circumferential direction 400mm grouting, array pitch 2.0m	C25 concrete, thickness 30cm Φ8-@150×150mmbar-mat reinforcement, double-deck	φ25 grouted rockbolt L=4.0m, interval 1.0m×0.5m, in passing quincunx	I22b steel gantry 0.5m/ interval	thickness 85cm,C40 concrete,P12 reinforced concrete

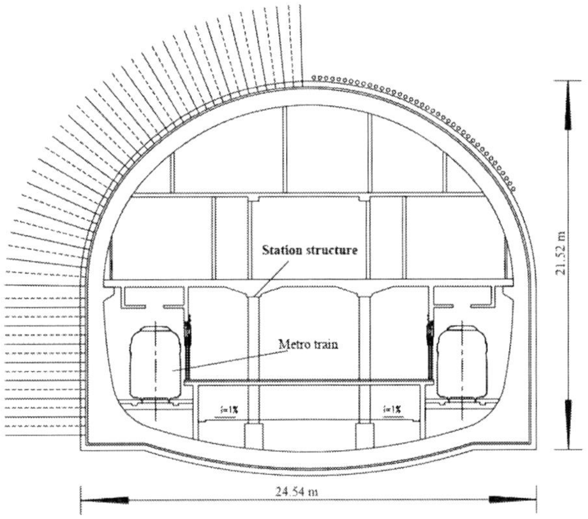

Fig.1 The section design of the Central park East station tunnel

2.2 Geological and geotechnical investigation

The original landform of Central Park east station is characterized by structural denudation and shallow mound topography. As a result of the urban area, the human engineering activities have greatly changed the topography and landscape. Along the ground, the elevation is 357 ~ 394m, and the relative elevation is about 37m. Along with the Tongmao Avenue, the whole of the vertical longitudinal slope is with the general slope angle of 2~4 degrees. Due to the transverse transformation to the original landform in the early stage of engineering construction, road or other construction scope of steep slope, is about 30~60 degrees.

3. Conventional subway station excavation methods in urban area

3.1 Finite element simulation

Selecting a construction approach that is suitable for the extra-large cross-section tunnel of urban subway is significant. For this purpose, compared with the field monitoring, the numerical simulation has better flexibility to study the mechanical response of tunnel excavation in different environments [6]. A finite difference program named FLAC3D is applied to numerical analysis to evaluate the effect of the construction condition of the double side drift method.

When FLAC3D does with numerical simulation calculation, it must firstly specify the three basic conditions including the finite difference grids, the constitutive relation and material characteristics, the boundary and initial conditions. The grids are used to define the analysis model of geometry, and the constitutive relation and corresponding material properties of the model evaluate the mechanical response characteristics under external force, besides the boundary and initial conditions are used to define the initial state of the model. After processing these conditions, the initial state of the model could be solved by running the software of FLAC3D. Then, on the basis of the initial state, the loading of other simulation conditions and the tunnel excavation simulation are carried out. In turn, the response of the model is obtained after the simulation condition changing.

According to the geological survey report, the Mohr-Coulomb constitutive model is used in the calculation process, and the mechanical indexes of the surrounding rock are calculated as shown in table2.

Table2 Material parameter list

	Elasticity modulus /*GPa*	Poisson's ratio	Cohesive force /*MPa*	Friction Angle /°
Surrounding rock	1.5	0.3	0.45	30
Double layer steel arch	36.6	0.2	/	/
Secondary lining	33.5	0.2	/	/
Rock bolt	210	0.3	0.69	30

In the calculation process, the Mohr-coulomb model is used to simulate the tunnel surrounding rock, and the elastic model is used to simulate the initial supporting concrete and the shell unit is used to simulate the secondary lining. An equivalent method is used to simulate the rigid frame of the initial support, and the elastic modulus of the steel arch is converted into shotcrete

$$E = E_0 + \frac{S_g \times E_g}{S_c}$$

In the formula: E -the elastic modulus of concrete after conversion; E_0 - the elastic modulus of the original concrete; S_g - the sectional area of the steel arch; E_g - elastic modulus for steel; S_c - Area of concrete.

Theoretical analysis[6] shows that in the circular cavern excavation of the homogeneous elastic infinite domain, on account of the load release of the hole and the change of the stress and displacement, five times in the hole diameter outside is to < 1%, and three times the hole diameter outside is about < 5%. In the calculation process of this project, considering the continuity of the calculation process and the boundary truncation error generated by the impact, the size of the model is evaluated with the horizontal width of 160 m, and the distance of vertical tunnel bottom is high hole 5 times. Besides, the length of the vertical direction is 100 meters, and the tunnel is 60 meters along the axial direction length. The calculation model for geometric dimensions is shown in fig.4.

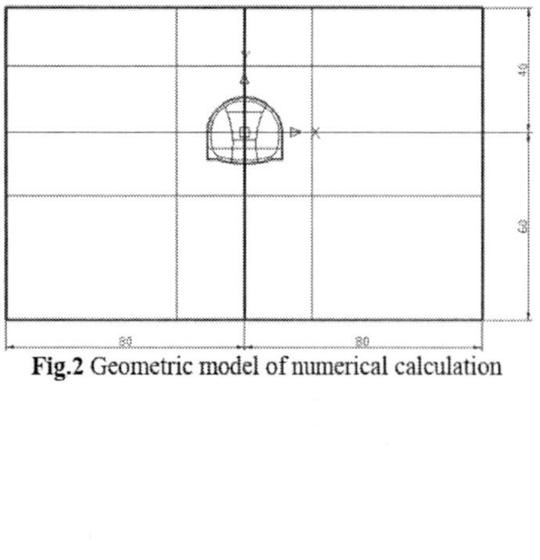

Fig.2 Geometric model of numerical calculation

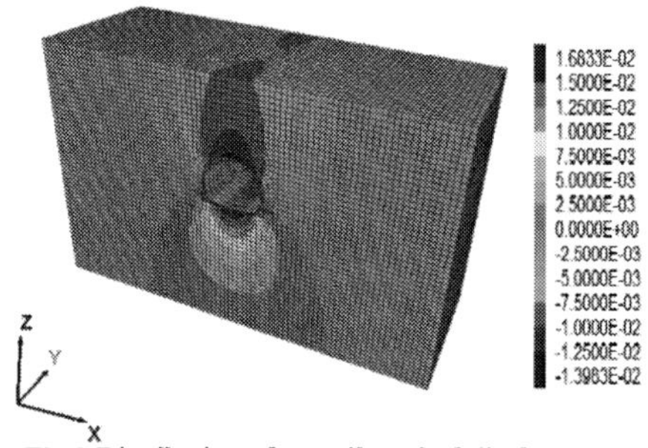

Fig.3 Distribution of overall vertical displacement

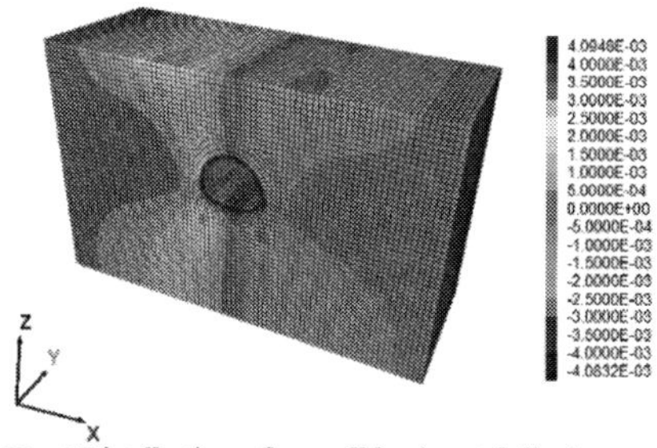

Fig.4 Distribution of overall horizontal displacement

As it could be seen from fig.4, the vertical deformation of the tunnel is concentrated in the arch and the inverted arch region, and the maximum deformation is 1.68 cm. There is a clear sedimentation tank. As for the fig.5, the horizontal deformation of the tunnel is presented with a butterfly shape. The deformation of the tunnel is as long as the upper and lower sides of the early side wall of the tunnel, and the maximum deformation is equal to 4.08 mm.

3.2 Monitoring and analysis of the stability of surrounding rock

The tunnel section of Central Park East station of the Chongqing metro line ten is super large (468.55m^2). In the excavation process, the stress distribution is so large that the tunnel surrounding rock deformation develops serious. And the tunnel is constructed by 9-steps excavation method, which is very easy to bring severe deformation of the top of tunnel in the middle stage of the excavation process and the ground subsidence which caused by tunnel excavation. Therefore, the activities and deformation of surrounding rock during the excavation of the tunnel would bring great difficulties and safety problems to the site construction. Thus, the monitoring of surrounding rock activities during tunnel excavation could predict the stability of surrounding rock in advance, and provide effective monitoring guidance for tunnel construction to reach the goals of safety, maintenance and protection. As shown in fig.2, a layout plan for surface subsidence monitoring points at the Central Park East station is shown.

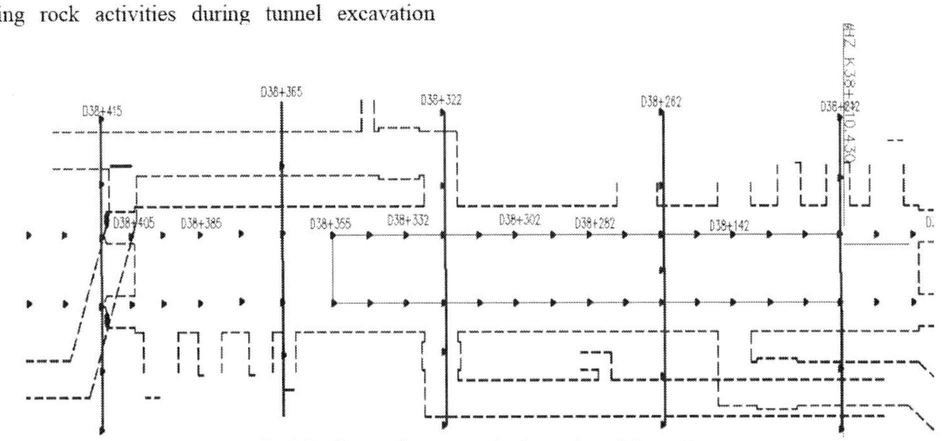

Fig.5 Surface settlement monitoring points of the station

The tunnel construction of Chongqing metro line ten is actually made from the middle to the two sides, and the actual test value of the surface of the tunnel is investigated in this phase. Thus, the results of numerical calculation could be compared and analyzed with the data of field monitoring.

(1) Deformation along the excavation direction

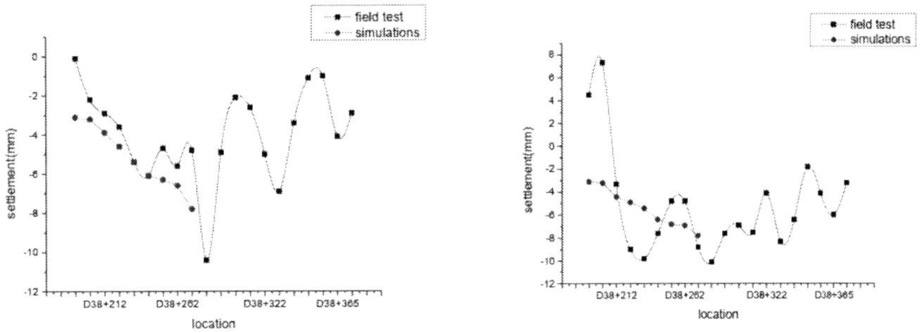

a) Subsidence along the left axis

b) Subsidence along the right axis

Fig.6 Deformation along the excavation direction

From fig.6, under the job of comparing with the measured results and the results of numerical calculation, the actual excavation is through the channel from the middle part to finish the excavation construction. Therefore, it appears symmetry along the direction of excavation deformation law, which conforms to the actual situation. Moreover, the closer the excavation is to the working face, its deformation value is relatively small, and the core earth excavation is not related to the actual construction. After completion of the whole excavation, the surface deformation is relatively large because of strengthening support.

(2) Deformation along the cross-section of the tunnel

In the cases of the tunnel excavation along the section of the subsidence curve, it could effectively reflect the actual excavation unloading of surrounding rock and the effect of the selected tunnel section of K38+212 and K38+262 through tunnel cross section of the actual monitoring results. Thus, it could compare with the results of numerical calculation.

https://doi.org/10.1051/e3sconf/20183803041

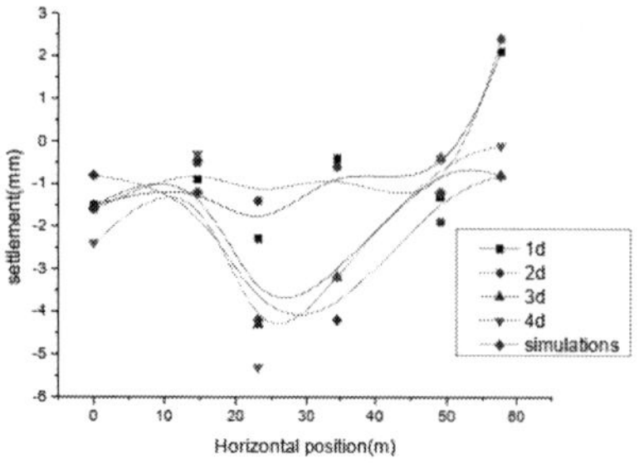

a) K38+212 sectional subsidence

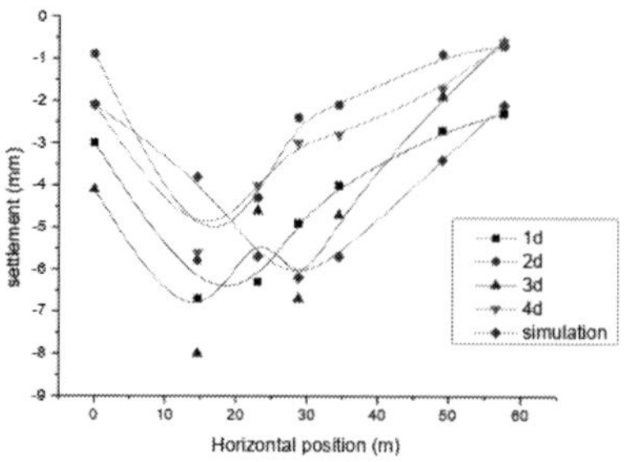

b) K38+262 sectional subsidence

Fig.7 Deformation along the cross-section of the tunnel

In order to investigate the deformation of the tunnel excavation to the D38+212 section, the numerical calculation of the circular excavation steps and the excavation response of the tunnel are compared. The settlement deformation of the K38+212 section is relatively small due to the excavation of the K38+212 section, which is controlled at the level of 5mm. Differently, the section of K38+262 is at 8mm, which is relatively large. In the sections, there are some points in the middle because of the continuous operation of the excavation support. Moreover, the local regional support force is too large, which cause some points to bulge. In other part, due to environmental factors, some point measurement values are inconsistent with actual construction mechanical reactions.

Through the comparison analysis of the above numerical simulation and monitoring measurement, it could be concluded that the method of numerical calculation would effectively simulate the excavation calculation of large section tunnel. Therefore, different excavation schemes can be designed for numerical calculation, and then comparative analysis could be made to obtain the optimal excavation method.

4 Optimization of tunnel excavation method

4.1 Large cross-section tunnel excavation method selection

The super-large section tunnel could not be full face excavation directly. With the reason that if the section size is so large that the excavation unloading of soil grows too fast. Thus, the same is the bottom of the tunnel and the stress growth of the cavern surrounding due to the cavity support adversely [7]. Therefore, the large section of the tunnel needs to be divided into smaller parts. But for the way to divide and to determine the number of partitions and the size of the division and the choice of digging order is an urgent problem which needs to be solved.

With the perspective of meet the construction cost and construction efficiency etc., it is necessarily to consider the existing construction experience and to analyze and summary the research achievements of both domestic and foreign experts and scholars to conduct the tunnel construction. The preferred method of tunnel construction is to excavate the whole section with steps. Thus, the second choose is the circular excavation core method according to actual condition. When the poor geological conditions and large cross section selection is in the next door or cross the wall, and the tunnel cross section larger needs to be divided into multiple small cavern respectively, so the double side drift method construction is the most suitable choose. Once the worse geological condition and the bigger section is, the more difficulty of the tunnel excavation is, and the construction would be more complicated. From the angle of stress deformation of construction safety, tunnel surrounding rock and supporting structure, the selection method of selecting excavation method should be matched in reverse.

It is found that on account of the construction of high efficiency, as a result, its construction section division and excavation supporting measures are relatively simple. But to guarantee the condition of rather smaller structure safety and stress, with the condition of the large cross section, it needs to divide into various complex construction parts. Therefore the demands of the way to balance the contradiction and then to choose the appropriate method are required including the problems to be settled urgently in determining the construction methods and its main performance excavation size, construction sequence, and so on.

So, for the ways to choose appropriate and optimization of large cross section tunnel construction method, the size of the excavation steps and excavation sequence, this paper would gradually deepening the study from the aspects of plane partition optimization.

4.2 Principle of dynamic planning of tunnel construction

In tunnel construction, owing to the reason that there are a few construction process of tunnel construction, and the construction process have successively, the whole process of tunnel construction could be divided into a number of interrelated construction stage namely the construction step. In every stage, it needs to make policy which is to make corresponding decision and to enter the next step. Moreover, it includes deciding what measures to take and how to carry out the excavation support, so that the whole process could achieve the best excavation supporting and economy effect.

Therefore, the decision selection of each stage construction step could not be determined arbitrarily [8]. It depends on the current construction state, and it would affect the excavation support status after the construction state. The stage when the decisions have determined after choosing the next step of construction measures for the construction is composed of a tunnel construction decision-making sequence. As a result, the process of the whole tunnel construction excavation supports implementation process route. Such as the problem of a tunnel excavation construction process associated with chain structure before and after a multi-stage process is called a multi-stage decision process in tunnel construction. This kind of problem is called the multi-stage decision optimization problem in tunnel construction. Such kinds of problems could be solved by dynamic programming theory.

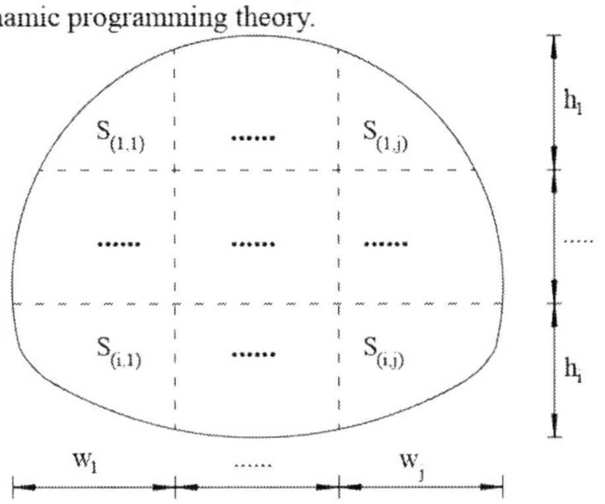

Fig.8 Plane sub-area optimization model

The theory of tunnel excavation optimization is based on dynamic programming theory. The simplified model of excavation optimization of tunnel section is shown in Fig.8. Optimization analysis process is below.

(1) According to the plane partition optimization model $f_{xy} = f[(i,h),(j,w)]$,then it needs to determine the optimal parameters i 、j 、h 、w .Among them, f_{xy} - the theoretical model of the transverse partition excavation scheme, i - horizontal layering, h - the collection of steps including $h_1....h_i$, $\sum h_i = H$ - the height of the tunnel section, j - transverse block number, w - the collection of steps including $w_1....w_j$, $\sum w_j = W$ - the width of the tunnel section.

(2) In view of the optimization parameters of the plane partition i , j , h , w , it could number the flat partition. Therefore, the dynamic model of tunnel excavation method is constructed.

(3) Through a series of calculations and analyses of comprehensive analysis of various in the tunnel excavation scheme of the optimal evaluation standards, and to determine the parameters of evaluation criteria, namely, building dynamic programming model of objective function, and to solve the optimal solution of the model with the iterative calculation method, the optimal method of excavation would be found.

(4) The optimal scheme is further analyzed and calculated, and the construction support measures and corresponding parameters are determined.

4.3 Optimization proposal analysis

Due to the construction parameters, the optimization of the construction scheme has the effect of modifying the single parameter. As shown in fig.9, the design of the design of the 9-step construction methods is found in the actual production, which is the construction method of the double side drift method construction. The construction channel through the S (1, 1) and S (2, 1) block would be divided into two parts. After the excavation of S(1,1) block, the excavation of the S(2,1) block is carried out, it is difficult to avoid the processing of the construction channel, because it is in the middle of the upper and lower layers, which is the temporary inverted arch used for layering during construction. Therefore, it is necessary to optimize the excavation method of the double side wall of the tunnel construction. For the reasons of the improvement of the construction environment, the solution of the lateral wall and the application of the temporary inverted arch, in the optimization scheme, the existing excavation support parameters are maintained, so only the size and order of excavation are optimized and analyzed.

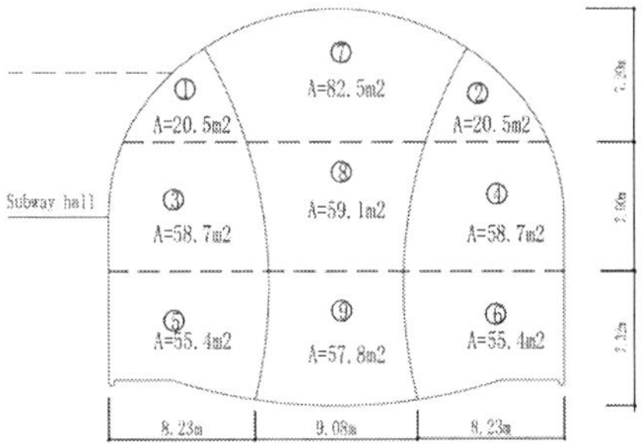

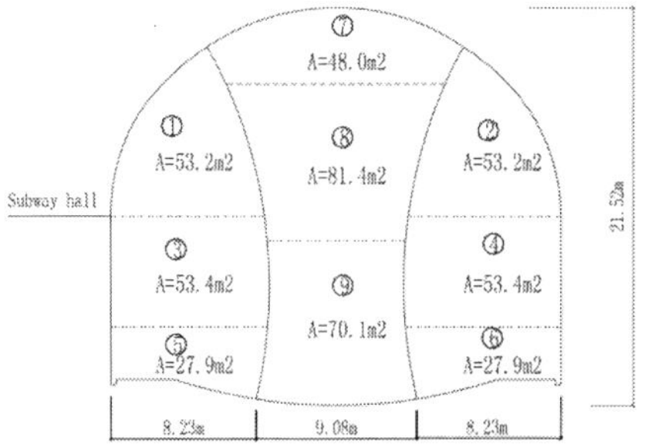

Fig.9 Optimization proposal section

The optimized scheme is shown in fig.9, and the tunnel excavation method of the double side drift method construction would continue to be used, namely $i = j = 3$. Only the parameter h collection is optimized.

Considering the actual conditions of construction, the lower part of S (1, 1) is adjusted to the station floor, so that it is easy for construction vehicles to enter through construction channels and tunnel sections through the station floor. Therefore, it could optimize the design of follow-up construction.

At the same time, the S (2, 1) region is adjusted downward, and the top block S (3, 1) of the inverted arch is reduced. Thus, the operations of reducing the S (2, 1) area which takes artificial excavation and reducing the disturbance of the large section tunnel has significant to the surrounding rock, especially for the control of the deformation of the arch. The S (2, 2) zone and the S (3, 2) area are correspondingly enlarged, and the area of the core soil is increased to facilitate the stability of the tunnel during the construction period. At the same time, the most of the core soil is reserved in the middle so as to facilitate the subsequent construction and mechanical excavation.

Table3 Parameter comparison before and after optimization

Parameter	$h_{(1,1)}$	$h_{(1,2)}$	$h_{(1,3)}$	$h_{(2,1)}$	$h_{(2,2)}$	$h_{(2,3)}$	$h_{(3,1)}$	$h_{(3,2)}$	$h_{(3,3)}$
Before (m)	5.08	7.20	5.08	7.00	7.00	7.00	6.45	7.32	6.45
After (m)	9.20	4.12	9.20	5.95	8.50	5.95	3.38	8.95	3.38

Due to changes in each partition, the actual excavation area of each partition is changed. Some of the excavation area is increased, correspondingly some of the excavation area is decreased. So the actual excavation unloading process changes, whether the design sequence can continue to be used, there needs to further optimize the demonstration.

Therefore, for the problem of excavating large section into 9 areas for multi-stage and multiple decision making, dynamic programming model and analytic hierarchy process model could be used together to analysis the optimize problem.

4.4 Optimization result analysis

Through the above optimization analysis, the main results are shown in table 2. The analysis of the first three of the comprehensive evaluation index is carried out.

Table4 Excavation optimization results

Serial number	Excavation sequence	Evaluation index	Remarks
1	S(1,1)-S(1,3)-S(2,1)-S(2,3)- S(3,1)-S(3,3)- S(1,2)-S(2,2)-S(3,2)	3.532	optimal
2	S(1,1)-S(1,3)-S(2,1)-S(2,3)- S(1,2)-S(3,1)- S(3,3)-S(2,2)-S(3,2)	3.465	Suboptimal
3	S(1,1)-S(2,1)-S(3,1)-S(1,3)- S(2,3)-S(3,3)-S(1,2)-S(2,2)-S(3,2)	3.412	Suboptimal

Table 2 lists the results of the first three of the comprehensive evaluation index, and the first row is the optimal excavation sequence. In the tunnel excavation section is divided into left and right three transverse block, upper and lower three layers, thus the construction is conducted in accordance with the cross excavation, finally the central core soil excavation would be finished.

5 Discussion and conclusion

Thesis research takes extra-large cross-section urban subway tunnel as the background, furthermore comprehensive methods are adopted including the technology research, theoretical analysis, numerical simulation, field test and other means. The optimization of super-large section tunnel excavation method, super-large section tunnel construction process simulation, super-large section tunnel surrounding rock stability monitoring measurement, etc. are carried on the thorough research systematically.

The main research work and research conclusion sum up as follows:

1. The construction of the double side drift method is used to measure the tunnel of the Central Park East station of Chongqing metro line ten. The monitoring results are compared with the numerical simulation results. The results show that the deformation law is approximately the same and the deformation quantity is at the same level by comparing the measured values and the numerical results. Therefore, the method of numerical calculation could be used to optimize the excavation method.

2. In this paper, the global optimization theory is analyzed, and the multi-stage optimization process of tunnel engineering is given priority to match the dynamic programming theory to optimize the selection of multiple schemes. In particular, the multi-objective planning theory is analyzed and the analytic hierarchy analysis model is used to evaluate and make the decision of tunnel excavation. The application range of optimization theory is studied, and the optimization of tunnel excavation method based on dynamic programming theory is finally adopted.

3. through the analysis of common construction method of the double side drift method, A unified planar partition model of $f_{xy} = f[(i,h),(j,w)]$ is constructed, which could explain the problem of the tunnel excavation plane optimization through the four parameters such as i , j h and w .

4. By means of the analysis of the Central Park East

station actual optimization problem of the Chongqing metro line ten, the construction parameters are optimized, by using the theory of dynamic programming. Finally, the optimal excavation scheme of comprehensive evaluation index is chosen.

This article mainly from the space effect of super-large section tunnel excavation method optimization and construction mechanics of material are discussed in this paper. Moreover it is necessary to do some further research concerning the effect of time and space, and to provide better guidance for similar engineering design and construction reference.

Acknowledgement

The authors acknowledge the financial support of the National Science of China (51478118, 51678164). The authors would like to express the appreciation and thanks to the managers and China Tiesiju Civil Engineering Group Co., LTD.

Reference

1. Sharifzadeha M, Kolivand F, Ghorbani M, Yasrobi S.Design of sequential excavation method for large span urban tunnels in soft ground—Niayesh tunnel[J]. Tunnel and Underground Space Technology, 2013, 35:178–188
2. P.F Li, Y Zhao. Performance of a multi-face tunnel excavated in loess ground based on field monitoring and numerical modeling [J]. Arabian Journal of Geosciences, 2016, 9 (14):640
3. TIAN Li-Feng. Support System Optimization of Double-side Wall Excavation Method for Subway Station Tunnel under Complicated Condition [J]. Railway station design, 2016, 60(7): 130-133
4. Yu F.C, Zhang D.L, Mu W.G, Wang J.C. Study on excavation scheme optimization at the junction of pits and tunnels of subway station [J].Journal of Beijing Jiao tong University, 2015, 39(6):69-73
5. Cheng T, Liu JW, Cao P. The Optimization Analysis on Construction Scheme of Large Span Metro Station in Hard Rock Stratum [J]. Chinese Journal of Underground Space and Engineering, 2014, 10(3):635-640
6. Wu ZJ, Fan LF. The numerical manifold method for elastic wave propagation in rock with time-dependent absorbing boundary conditions [J]. Engineering Analysis with Boundary Elements, 2014, 46:41–50
7. Peter J, Sellner, Schubert W. Prediction of Displacement in Tunnel [D].Graz University of Technoloy.2000.
8. Wu B, GAO B. Optimization Analysis of Excavation Sequence of Tunnel Groups between Metro Stations [J]. China railway science, 2003, 24(5):23-28

Hydrodynamic Influence Dabanhu River Bridge Holes Widening Based on Two-Dimensional Finite Element Numerical Model

Dong Feng LI, Fu Qing BAI*, Hui NIE

School of Water Resources and Environmental Engineering, Zhejiang University of Water Resources and Electric Power, Hangzhou, 310018, China

Abstract: In order to analyze the influence of bridge holes widening on hydrodynamic such as water level, a two-dimensional mathematical model was used to calculate the hydrodynamic factors, river network flow velocity vector distribution is given, water level and difference of bridge widening before and after is calculated and charted, water surface gradient in seven different river sections near the upper reaches of bridges is counted and revealed. The results of hydrodynamic calculation indicate that The Maximum and the minimum deducing numerical value of the water level after bridge widening is 0.028m, and 0.018m respective. the seven sections water surface gradient becomes smaller until it becomes negative, the influence of bridge widening on the upstream is basically over, the range of influence is about 450m from the bridge to the upstream. reach

1 Introduction

The river network flow of Shaoxing city and Keqiao district is slow and weak hydrodynamic. This is an important factor restricting the water environment of the river network, and this have a direct impact on flood control and drainage area water environment and water safety and sustainable development between local water environment and economic society. Criss cross the plain river networks is the important carrier of river water, and have the functions of flood drainage, water regulation, water supply, irrigation, shipping, transportation, landscape leisure and ecological environment. The early construction of the railway bridge holes is small, this result in water flow resistance, less flow velocity, higher water level, and this further affects the flood control, drainage and water environment of the river network.

By widening the bridge holes width and with streamlined design of revetment works, these this can reduce flow resistance, increase flow velocity, cut down the water level, increase the river network water flow hydrodynamic. This is an important measure to enhance river networks

In the past, there were many researches on the backwater of bridge and culvert, but the influence scope and intensity of backwater were not much studied from the point of decreasing the gradient of the branch reach. In this paper, based on the two-dimensional mathematical model, the river flow velocity, water level and other dynamic factors are numerical calculated and Analyzed.

2 Two Dimensional Finite Element Numerical Model

2.1 The Governing Equations and Deterministic Conditions

The equations of continuity:

$$\frac{\partial Z}{\partial t}+\frac{\partial (HU)}{\partial x}+\frac{\partial (HV)}{\partial y}=0$$

The equations of motion:

$$\frac{\partial U}{\partial t}+U\frac{\partial U}{\partial x}+V\frac{\partial U}{\partial y}+g\frac{\partial Z}{\partial x}+\frac{gn^2U\sqrt{U^2+V^2}}{H^{4/3}}-fV-\varepsilon\left(\frac{\partial^2 U}{\partial x^2}+\frac{\partial^2 U}{\partial y^2}\right)=0$$

$$\frac{\partial V}{\partial t}+U\frac{\partial V}{\partial x}+V\frac{\partial V}{\partial y}+g\frac{\partial Z}{\partial y}+\frac{gn^2V\sqrt{U^2+V^2}}{H^{4/3}}+fU-\varepsilon\left(\frac{\partial^2 V}{\partial x^2}+\frac{\partial^2 V}{\partial y^2}\right)=0$$

In which, U and V is flow velocity; H is water depth; Z is elevation of water surface level ;C is Checy's coefficient and $C=\frac{1}{n}R^{1/6}$

2.2 Boundary Conditions

The outlet and open boundary condition: water level process is given.

At the entrence, flow discharge is given, such as

* Corresponding author: Baifq@zjweu.edu.cn

$Q(x,y,t) = Q_0(x,y,t)$, $Q_0(x,y,t)$ is the discharge process.

For the closed and rigid boundary the velocity is zero, ie: $u_n|_{\Gamma_2} = v_n|_{\Gamma_2} = 0$ in which, Γ_2 expresses closed boundary; n is normal unit vector.

2.3 Initial Conditions

The initial water level and the initial velocity are given by measured tidal level or given value zero, initial conditions does not affect the precision of computed result.

2.4 The Establish of Finite Element Model

The finite element method is employed to solve the equations. this model, Triangle grids are selected to disperse calculation zones.

2.5 Verification of the Model

Verification of the model is shown in Reference.

3 Calculation Results and Analysis

3.1 Calculation Condition

3.1.1 River Network Survey

Many rivers arranged in a crisscross pattern of Keqiao District river network, Dabanhu river is an important channel connecting south and north plain river networks, the railway bridge go across on the Dabanhu river, because of the small bridge holes, water flow is obstructed. In order to speeding up the flow of the north and south river networks. widening the bridge is necessary. This is shown in Figure 1.

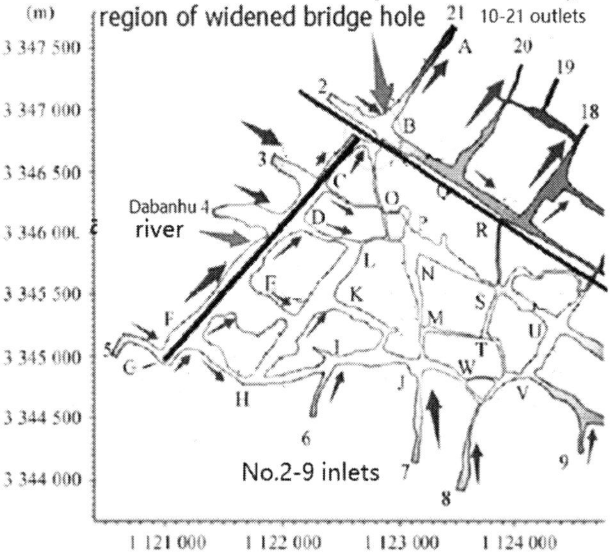

Figure 1 River Network and Boundary Conditions

3.1.2 Boundary Conditions of Model Calculation

Figure 1 is the relevant part of the whole bridge and river network model, The river network inlet is from No.2 to No. 9, the water level is controlled at 3.95 m, and the river outlet No. 18 - 21. The water level is controlled at 3.90 m, and different water level control conditions are set up respectively

3.1.3 Selection of Model Parameters

In the river networks past planning, one-dimensional unsteady flow mathematical model is established and validated, selection of model parameters is mainly on the basis of the previous planning design results, The parameters selection of the two-dimensional hydrodynamic mathematical model is reasonable。

3.2 Calculation Results and Analysis

In order to analyze the influence of bridge holes widening on hydrodynamic such as water level, The most influential reaches of the upper reaches of the bridge are selected, marking the different letters B-C-D-E-F-G at the intersection of river network. The black line segment of Figure 1, location G and B is two. From the starting point G to the position B of the bridge, the length of the river channel is 2448.56m. On the river reach of the starting point G to the bridge point B, 250 points are chosen in equidistant manner as the object of water level analysis, location (see Figure 1).

3.2.1 River Network Flow Velocity Vector Distribution

Figure 2 is river network flow velocity vector distribution nearby the bridge and along the Dabanhu river,

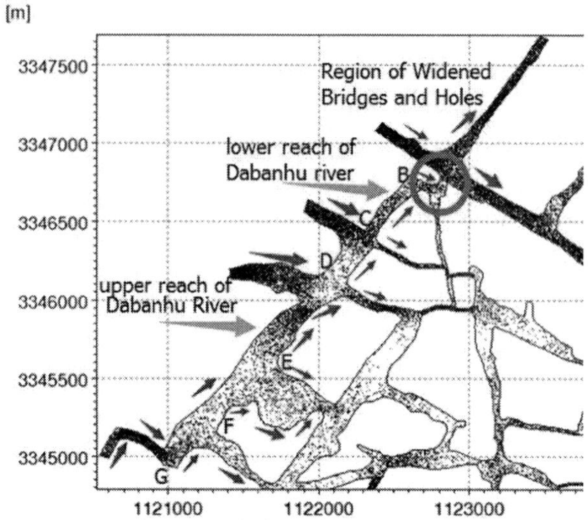

Figure 2 River Network Flow Velocity Vector Distribution

3.2.2 Effect of Water Level Analysis of Bridge and Culvert Widening

The water level of river network is directly related to river flood control and railway bridge safety. In this paper, the change of water level before and after widening of bridge is analyzed and the inflow and outflow conditions of water flow are at the same the condition (see Figure 3 and Figure 4).

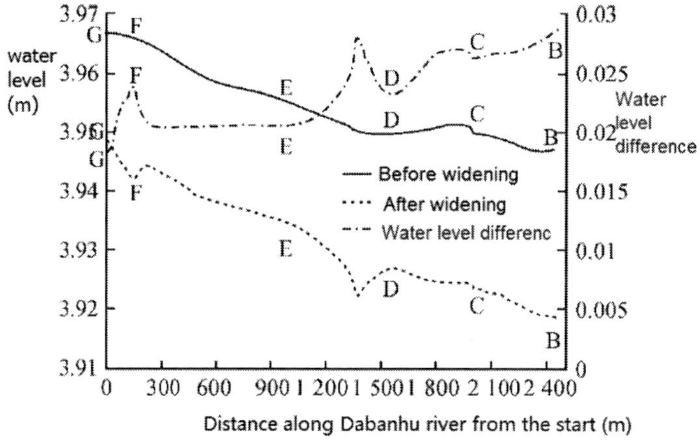

Figure 3 Water Level and Difference of Bridge Widening Before and After (Upper Reaches of the Widened Bridge Along the Dabanhu River)

There are selected and analyzed 250 points in the distance of 2 448.565m from start point G to end point B, so as to enough point to reveal to water level and water surface gradient (in Figure 5) The graphs show that the water level is different before and after the widening of bridges, although the inlet and outlet boundaries of the water flow are the same. the water level is obviously higher before widening than the after.

In Figure 3 and Figure 4, the number in the abscissa axis represents the distance, that indicates the distance from the starting point G, There are two vertical coordinates, The main coordinate is the water level before and after the widening of the bridge, sub coordinate is the difference of water level before and after widening of bridge. The Figure 3 is about upper reaches of the bridge along the Dabanhu River, Letter B-G is the location of water level analysis, Figure 4 is that nearby the widened bridge

As can be seen from the graph, as a whole, From the location B of the bridge to the upstream and the position G, the general trend of water level change is increasing, whether before the bridge widened or after; but the water level decreases at all time after bridge widening. The Maximum and the minimum deducing numerical value of the water level after bridge widening is 0.028m, and 0.018m respective.

In order to analyze the intensity and range of local backwater before and after widening of bridges, comparison chart of the water level and water level difference in the long enough upstream of the bridge are

https://doi.org/10.1051/e3sconf/20183803042

given in Figure 4. As can be seen, from the bridge to upstream, the water level difference is reduced from 0.0285 m to 0.0265m, the influence of bridge backwater on the upstream is decreasing gradually.

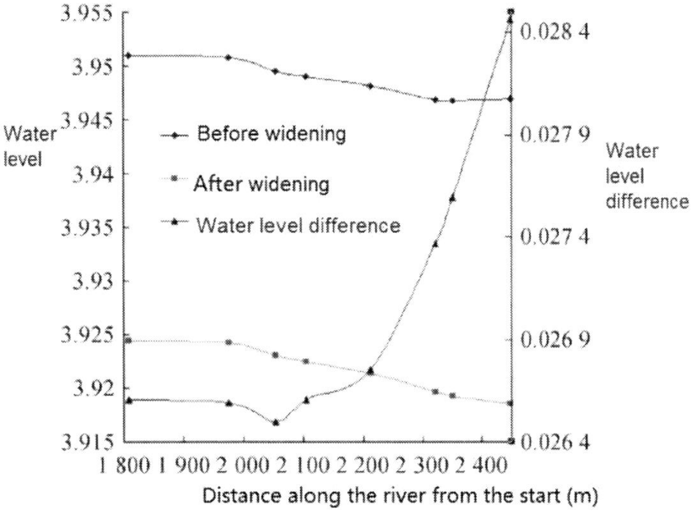

Figure 4 Water Level and Difference of Bridge Widening Before and After (Nearby the Widened Bridge)

Figure 5 is chart of water surface gradient in seven different river sections near the upper reaches of bridges. This gradient is calculated by water level and difference and distance of the points each other in 8 points. The chart shows the seven sections gradient becomes smaller until it becomes negative, It shows that the

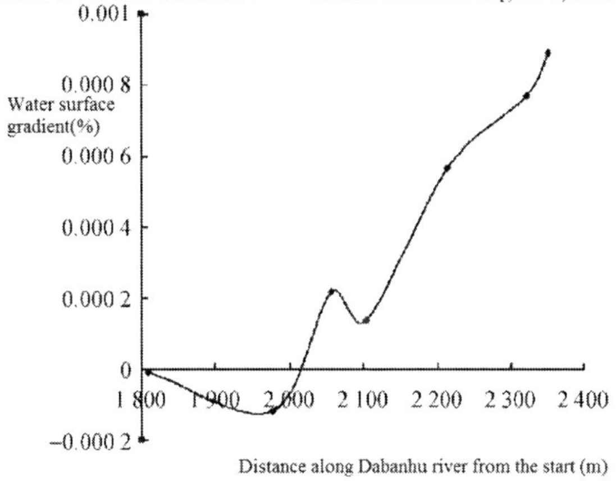

Figure 5 Water Surface Gradient

Surface water gradually Change from steep to slow, until the upper reached water level is lower than the lower. It shows that the influence of bridge widening on the upstream is basically over, the range of influence is about 450m from the the bridge to the upstream reach.

4 Conclusions

A two-dimensional hydrodynamic numerical model of river network and lake is used to calculate and analyze the influence of bridge holes widening on hydrodynamic such as water level and water surface gradient

flow velocity vector distribution is given, water level and difference of bridge widening before and after is calculated and charted, water surface gradient in seven different river sections near the upper reaches of bridges is counted and revealed. The results of hydrodynamic calculation indicate that The Maximum and the minimum deducing numerical value of the water level after bridge widening is 0.028m, and 0.018m respective. the seven sections water surface gradient becomes smaller until it becomes negative, the influence of bridge widening on the upstream is basically over, the range of influence is about 450m from the the bridge to the upstream reach.

Acknowledgements

This work was financially supported by the Project supported by the National Key Research and Development Program of China (No.2016YFC0402502); Zhejiang Natural Science Foundation(M503254); the Water Resources bureau of Zhejiang Province (RC11092011); Zhejiang provincial Education bureau key projects (No. Z200909405), the Zhejiang provincial

Education Science Plan Office Project [(2009), No. SCG220]

References

1. LI Dongfeng, Zhang Hongwu, Zhong Deyu, et al. "Numerical simulation and analysis on tidal current and sediment silting process in Yellow River estuary", Journal of Hydraulic Engineering, 2004(11) .
2. Li Dongfeng, Zou Bing, Zhang Hongwu, et al. "Two Dimensional Analysis of Tide Influence upon Estuary and Riverway" , Journal of Zhejiang Water Conservancy and Hydropower College,2006(1) .
3. Zou Bing, Li Dongfeng, Zhang Hongwu, et al. "Effect Analysis of Building Bridges on River Hydraulic Dynamics Circumstances by 2D Numerical Method", Journal of Hangzhou Dianzi University, 2007(3).
4. Lu hao, Gao Dongguang, "Bridge River Hydraulics", China Communications Press,1996
5. J.N. Bradley, Zheng Huaqian, "Bridge and River Hydraulic", China Communications Press, 1980
6. Li Dongfeng, Zhang Hongwu, Zhang Junhua, et al. "Finite Element Method of Yellow River and Sediment Movement", Journal of Sediment Research, 1999(4) (in Chinese).
7. Li Dongfeng, ZHANG Hongwu, ZHONG Deyu, et al. "2-D mathematical model for flow and sediment transport in the estuary of Yellow River", Journal of Hydraulic Engineering, 2004(6), pp 1-7.
8. Li Dongfeng, Zhang Jiyuan, Chen Bin, Flood influence evaluation of Chuanbuton bridge in the Haoxi River, Report of ZJWCHC, 2009.
9. Li Dongfeng, ZHANG Hongwu. Numerical Simulation Analysis of Bridge Construction Flood Flow Effect Based on Two Dimensional Finite Element Model [C], Proceedings of of 2010 International Conference on Modern Hydraulic Engineering,2010, xi'an, China.295-298
10. Li Dongfeng, ZHANG Hongwu. Dammed Water Level Comparison of Widening Bridge Piers Based on 2D FEM Numerical Simulation and Empirical Formula [C], Proceedings of of 2010 International Conference on Modern Hydraulic Engineering,2010, xi'an, China.255-258
11. Yang Fixiang, Pan Jie,Li Dongfeng, Zhang Hongwu. River Networks Fluidity Classification Analysis based on Two Dimensional Numerical Model and Hydraulic Characteristic. Proceedings of the 5th International Yellow River Form on Ensuring Water Right of the River's Demand and Healthy River Basin Maintenance, Volume V, pp244-252, Yellow River Conservancy Press
12. Yang Fixiang, Pan Jie,Li Dongfeng, Zhang Hongwu. Effect Research of River Networks Water Fluidity Enhancing Engineering Measures based on the 2D Numerical Model. Proceedings of the 5th International Yellow River Form on Ensuring Water Right of the River's Demand and Healthy River Basin Maintenance, Volume V, pp302-310, Yellow River Conservancy Press
13. Zhang Huaxing, Wang Lianhua, Li Dongfeng, Zhang Hongwu. Banked-up Water Analysis of Bridge Building Two Dimensional Finite Element Numerical Simulation. Proceedings of the 5th International Yellow River Form on Ensuring Water Right of the River, Demand and Healthy River Basin Maintenance, Vol.IV, pp432-437. Yellow River Conservancy Press

Discussion on runoff purification technology of highway bridge deck based on water quality safety

Sheng-guang TAN[1], *Xue-xin* LIU[2,*], *Guo-ping* ZOU[1], *Xin-zhu* XIONG[2], and *Shuang-cheng* TAO[2]

[1]Jiangxi Ganyue Expressway Co., Ltd, Nanchang, 330025, China
[2]China Academy of Transportation Sciences, Beijing, 100029, China

Abstract. Aiming at the actual problems existing, including a poor purification effect of highway bridge runoff collection and treatment system across sensitive water and necessary manual emergency operation, three kinds of technology, three pools system of bridge runoff purification, the integral pool of bridge runoff purification and ecological planting tank, are put forward by optimizing the structure of purification unit and system setting. At the same time, we come up with an emergency strategy for hazardous material leakage basing on automatic identification and remote control of traffic accidents. On the basis of combining these with the optimized pool structure, sensitive water safety can be guaranteed and water pollution, from directly discharging of bridge runoff, can be decreased. For making up for the shortages of green highway construction technology, the technique has important reference value.

1 Introduction

With the rapid development of highway construction, frequent traffic activities have raised more attention to the problem of road surface runoff pollution. Carrying large amounts of particulate matter, heavy metals, petroleum, nutrients and PAHs and other pollutants, road runoff put a negative impact on the ecological environment alongside the highway. The pollutants in road surface runoff mainly comes from the wear particles of vehicle tires and road materials, the leakage of hazardous goods in delivery, the particulate matter of snowmelt agent generated from auto parts, the pollutants in rainwater and the atmospheric dust-fall, and the erosion of road surface sediment by rainwater. [1]-[4]

The greatest potential danger to the water body lies in the serious pollution caused by the leakage from carriage of dangerous goods (CDG) due to accidents or some inner factors [5]-[6]. To prevent this kind of environmental risk, in *Notice on Strengthening Efforts on Environmental Impact Assessment of Highway Planning and construction (Doc. MEP [2007] No.184)*, requires that sedimentation tanks are required to set on both sides of the bridges over source water secondary protection area and quasi protection area and Class Ⅱ water area. Tanks are used for dealing with the bridge surface runoff after the pollution accident so as to ensure drinking water safety.

Since there is no specific design requirement for sedimentation tank, the design of bridge runoff sedimentation tank is mainly used to cope with the requirement of completion acceptance of environmental protection at present. As a result, the design is mostly too simple. Although some studies have optimized the tank structure [7], the overall pollutant precipitation grease effect is limited.

Drain valves were equipped and maintenance staff was required to regularly open the valve to empty runoff. But in fact, there was a long-term presence of sedimentation tank in the full state, resulting that tanks failed to play a role in the hazardous good leakage accident. In response to this kind of problem, researchers began to try to introduce some self-control measures into the bridge runoff emergency management system. Xu Jianping et al [8] proposed a technical system for emergency operation management based on water quality monitoring. However, it still needed to be further demonstrated whether the selected water quality indicators can cover all types of hazardous chemicals. Qin Haiwei et al. [9] proposed a selective collection system of runoff based on real-time monitoring according to different situations. But it lacks the corresponding remote automatic control mode in the disposal strategies.

In view of the above problems, three technologies are proposed to improve the water purification effect of pavement surface by adopting special measures based on smart transport means to ensure water quality and safety, aiming to provide technical support for highway water environment and water safety protection in environment sensitive areas.

2 The bridge runoff purification emergency three-tank system on different function

* Corresponding author: Xue-xin LIU, nkone@sina.com

The key point of the emergency treatment for hazardous chemical leakage runoff, in daily operations, is a better purifying function for collected runoff. In order to ensure sufficient accident response time during these two function switching processes, an intermediate buffer unit is also needed.

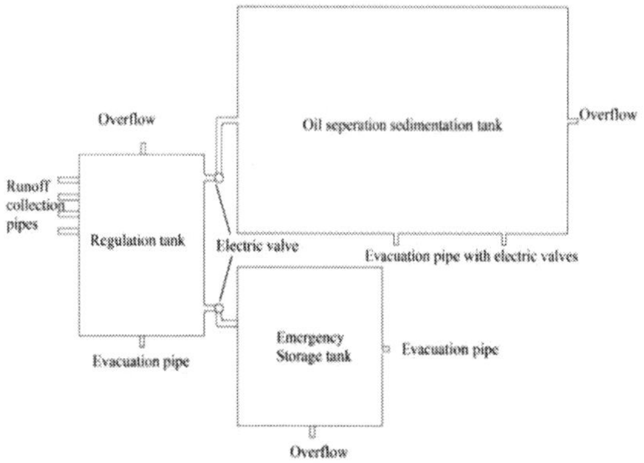

Fig. 1. Bridge runoff purification Emergency three-tank system.

Based on the considerations above, the bridge runoff purification emergency three-tank system was developed. It consists of a storage regulation tank, an oil separation sedimentation tank, an emergency storage tank and a monitoring control system. The surface runoff generated by rainfall firstly enters into the storage regulation tank, which plays a role as a buffer and provides the pretreatment for the bridge runoff.

When there is no leakage of hazardous chemicals, the runoff of bridge deck flows into the oil separation sedimentation tank through the electric butterfly valve which is normally opened. After sedimentation, filtered and purified runoff will be discharged into the surrounding water through the remote control opening the empty pipe valves of storage regulation tank and oil separation sedimentation tank.

When a traffic accident occurs on the bridge, it is first detected through the bridge head monitoring camera and the alarm telephone to determine whether there is any leakage of dangerous goods. After the leakage is confirmed, the monitoring center immediately starts the emergency control system, which orders to open the electric butterfly valve between storage regulation tank and the emergency storage tank, and close the electric butterfly valve between the storage regulation tank and the oil separation sedimentation tank at the same time. Then the leakage of dangerous goods and its dilution collected in storage regulation tank will flow directly into the emergency storage tank for post-treatment offered by the environmental protection department. This kind of technology has been applied to Jiangxi Chang Feng Expressway.

The oil separation sedimentation tank is equipped with baffle wall, pebble retaining wall and oil collection tank. Those kinds of structures are of sedimentation, oil separation and bio-degradation functions and have good removal effects on the primary pollutants in the initial rainfall runoff. Specifically, the vertical bend is set to keep folding flow and improve the sediment separation effect to stay the uniformity of water distribution.

A baffle plate is arranged at the bottom of the vertical baffle wall from top to bottom, which can improve the sedimentation effect of the system. The oil collecting tank is arranged inside the retaining wall. A set of cone mouth with a certain space is designed along the centerline of the bottom of the tank, inside which some nano-fiber oil absorption material is laid to ensure oil separation function of the oil collection tank. The pebble wall set up from the bottom is able to change the direction of water flow, and help with adsorption and biodegradation. The sand tank is set one each before and after the pebble retaining wall at the tank bottom and the blow-off pipe is emplaced one each on both sides of the pebble retaining wall. In addition, the use of electronic valve makes it possible to remote control the emissions when purification reaction time is done.

3 Coupling remote control of the bridge surface runoff purification and emergency integrated tank

Bridge runoff purification emergency three-tank system is able to perform the three functions independently, which includes regulation, purification and emergency disposal, and to protect the sensitive water quality to the fullest. However, the project quantity is relatively large and it is more difficult to implement especially when the space under short bridges is limited. So the three tanks mentioned above were combined to form a bridge runoff purification emergency integrated tank with all their functions and the remote control system was optimized.

The various control strategies for the emergency treatment of the bridge runoff are established on the basis of accurate identification of the hazardous chemicals in leakage accident. The accident identification of the three-tank system is worked through the manual confirmation by the camera on the bridge and the alarm telephone, and there is the possibility of

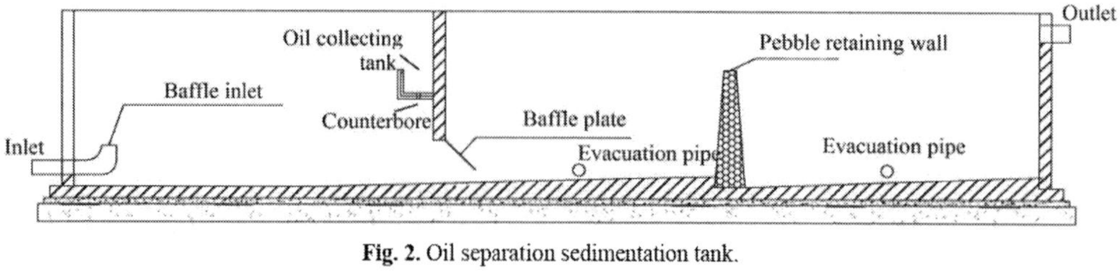

Fig. 2. Oil separation sedimentation tank.

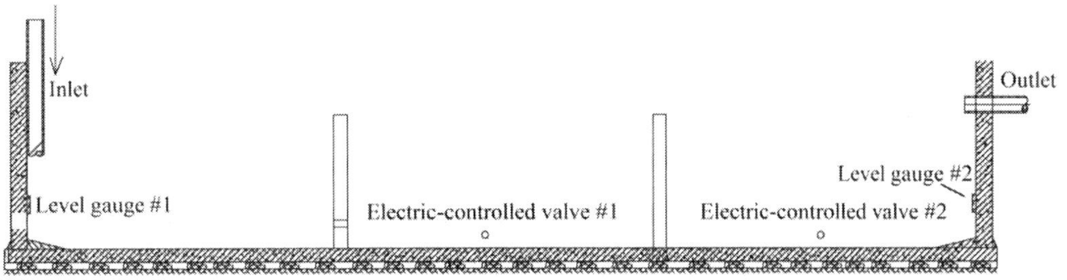

Fig. 3. Bridge runoff purification emergency integrated tank.

missing accidents. Therefore, through the bridge monitor in full coverage, the application of video-based traffic accident detection technology can achieve automatically identifying and issuing a warning to monitoring center. If manual determination ensures that the accident is involved with vehicles carrying hazardous chemicals, and then emergency operation will be confirmed to start, thus the automatic identification of accidents was optimized.

Compared with the oil separation sedimentation tank, the bridge surface runoff purification and emergency integrated tank is simpler in structure. At the same time it integrates the regulating, purifying and emergency disposal functions. There are two sets of retaining walls in the tank, dividing the tank body into three compartments. One plays a role in runoff regulation in water quality and quantity. when hazardous chemical leak accident occurs, the first two cell space with more than 100m3 area, can storage the chemicals and its flushing fluid, playing an emergency role. The water flows up and down in vertical baffle way in three cells. Normally, storm-water runoff flows into the tank from the upper part. The water level gradually increased in the first cell, and when the baffle height is reached, the runoff crosses the first baffle into the second cell. Cycling back and forth, the purpose of sedimentation purified layer by layer can be achieved. After the rainfall, the runoff purification is completed. Then the two electrically-controlled outlet drainage valves are remotely opened to realize the emptying function so as to ensure adequate storage space for the next leakage accident.

A group level gauge and electric control valve are set up one each inside the Cell 1 and Cell 3. After the rain stops, purification is also completed. According to the liquid level information, start the blow-off valve remotely to automatically discharge the treated bridge runoff. There are several overcurrent ports at the bottom of the retaining wall between Cell 1 and Cell 3. When the rainwater runoff is in a large amount, the flow passing through the overcurrent port is small. And the runoff will still pass through the first baffle and won't affect the baffle precipitation function. When the runoff in the tank settles, Cell 1 and Cell 2 will form a communication network. Therefore, only a group of level gauge and electronic controlled emptying valve are set up to save cost. This kind of technology has been applied to Heda Expressway in Jilin Province.

4 Bridge runoff purification emergency ecological planting groove Based on LID

Due to the extremely low probability of occurrence of hazardous chemicals leakage accidents, no matter the three-tank system or the purification and emergency integrated tank, they are all in masonry structures, which present a poor landscape effect even under the bridge. So they can only be a structure with temporary and limited function. At present, China is vigorously promoting the construction of "sponge city" with natural accumulation, natural infiltration and natural decontamination. Based on the concept of low impact development (LID), various ecological technologies including infiltration, retention, storage, purification, and reuse are integrated for application, thus playing a role in groundwater supplementary, and water cycle regulation [10]. Highway industry has carried out a lot of engineering practice that apply conventional treatment in the bridge runoff treatment and it has achieved initial success. But most of them fail to couple with the accident identification mechanism, so there is still the possibility of tanks full of rains when accident occurs. On the basis, guided by the LID conception, this paper puts forward the bridge runoff treatment process combining regulating tanks and ecological planting grooves, plus equipped with remote control module to achieve intelligent management.

The intelligence level of remote control technology has been further enhanced. It uses the camera with 1080P @ 30fps high-definition image full-frame detection technology. The real-time video image processing and system self-learning technology is carried to do real-time analysis of video images collected by the camera to identify road traffic accidents. After completing the initiative identification and track detection, it can initiatively remind staff through the sound and light alarming way to observe and deal with it. The alarming image will be immediately switched to the big screen, drawing attention to determine whether emergency operations should be carried out.

The treatment system is a combination of regulation ponds and ecological planting grooves. The pipeline between the regulation ponds and the ecological planting grooves are equipped with an electric-controlled connection valve, and electric-controlled emptying valves are arranged in the ecological planting grooves.

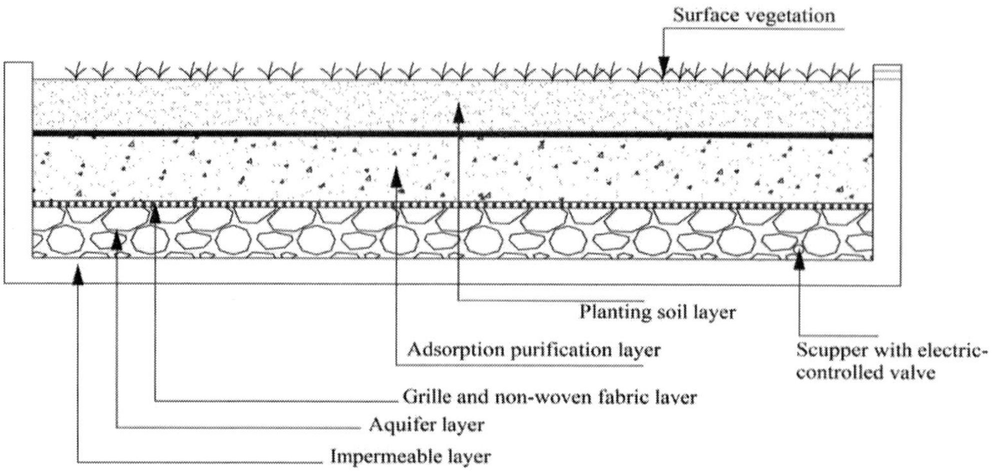

Fig. 4. Ecological planting groove unit.

When the hazardous chemical leakage occurs, staffs are required to shut down the valves remotely to fix the hazardous chemicals and their flushing liquid in the regulation tank with effective volume beyond 100m3. Under the extremely special circumstances of hazardous chemicals leakage happening in rainstorm conditions, it is possible that the regulation tank is overflowing. Then the electric-controlled connection valves are supposed to be opened while leaving the emptying valves closed. As a result, the ecological planting tanks are sacrificed in need to fully guarantee the water quality safety.

From top to bottom, each ecological planting groove unit consists of planting soil layer, adsorption purification layer, grille and non-woven fabric layer, aquifer layer and impermeable layer. Planting soil layer contains plants with good water purification effect and local soil, which needs allocation with a poor permeability. With multi-fillers of strong adsorbing and filtering ability for nitrogen and phosphorus removal, adsorption purification layer is the core unit of ecological planting groove. Grille and non-woven fabric, which plays a supporting role, separate the adsorption purification layer from aquifer layer. Filled with stones in a big size, the aquifer layer can make it possible to have reservoir of water, which can be provided for plants without rain. And anti-seepage geotextile is set in impermeable layer to prevent groundwater pollution from the leakage of hazardous contaminants.

This kind of structure has the integrated functions to deal with the bridge runoff, comprehensive utilization and accident emergency, which is an ecological technology with landscape effect. So it can meet the requirements of the green development and sponge city construction in the new era. This technology can also be applied to the road surface runoff purification emergency, which has been applied in Erhai East Lake Road.

5 Conclusions

1. The bridge surface runoff treatment three-tank system, the integrated tank and ecological planting grooves are all the new technologies of water protection especially for bridges over the sensitive water body. Compared to conventional technology, the surface runoff purification effect will be better. Due to the utilization of monitoring, accident identification, remote control and other intelligent means, emergency plan can be started in time, preventing the hazardous chemicals into the water body.

2. With the development of smart devices, accident identification and emergency response strategies adopted by the three technologies are still in the process of gradual improvement, and it can be interactively used under different situations. In the future, means of incidents identification should be further enriched to adopt video and microwave interaction technologies to further reduce the misjudgment rate, improving the accuracy of accident analysis.

3. Emergency control of the three technologies is based on accident identification. With the development of environmental monitoring technology, the three technologies mentioned in this paper could be improved with better measures, if the monitoring devices, which are of economic promotion value, could identify all kinds of hazardous chemicals and conventional bridge surface runoff.

4. From the perspective of water purification, the water purification effect increases in the order from the integrated tank, the oil separation sedimentation tank and ecological planting grooves. Besides, the discharge requirements and project cost could also be considered.

5. Under the guidance of green and innovation ideas, referring to new techniques such as sponge city and smart transport could accelerate the improvement of the technical system of water environment protection for bridge-crossing in China.

Acknowledgements

The research is supported by the science and technology project on transportation construction by Jiangxi Province (No. 2011C00005), the science and technology demonstration project of the Ministry of transport of the

People's Republic of China (No. 2016002) and the transportation science and technology project of Jilin Province (No. 2017ZDGC-2-8).

Reference

1. J.S. Wu, C.J. Allan. Characterization and Pollutant Loading Estimation for Highway Runoff. J. Environ. Eng. ASCE **124**(7): 584-592 (1998)
2. R.J. Charbeneau, M.E. Barrett. Evaluation of Methods for Estimating Stormwater Pollutant Loads. J. Water Environ. Res. **70**(7): 1295-1302 (1998)
3. J.K. Li, G.F. Du, H.E.Li. Purification Effects of Biological Filter on Urban Road Runoff. J. Soil Water Conserv. **26**(4): 1-6 (2012)
4. X. Zuo, D. Fu, H. Li. Variation Characteristics of Mercury in Speciation during Road Runoff for Different Rainfall Patterns. J. CLEAN–Soil, Air, Water **41**: 69-73 (2013)
5. F. Jia. The Protection Measures Study of Highway Operation Period on Water Environment Sensitive Points. J. Shanxi Sci. Tech. Commun. **5**: 94-96 (2015)
6. J.N. Liang, K. Liu, F. Pan. Impact and Protective Measures of Highway Construction in West Mountainous Area on Ecological. Environ. J. China Foreign Highway **30**(5): 20-24 (2010)
7. Y.H. Wang, Z.Q. Xiong, Y.T. Shan. Technical Research on Water Environmental Protection in Construction of Renhai-Chishui Expressway. J. Tech. Highway Transport **3**: 155-158 (2014)
8. J.P. Xu, S.G. Shao, W.S. Yu. Discussion of Integrated Design of Emergency Storage Facilities of Dangerous Chemicals in Expressway Transport Accidents and Treatment Facilities of Runoff on Bridge Surface. J. Highway Transport. Res. Dev. **31**(12): 139-144 (2014)
9. H.W. Qin, L.X. Zhu, X.F. Fu. Selective Collection System of Deck Runoff Based on Real-time Monitoring. J. Yellow River **31**(9): 24-25 (2009)
10. D. Zhou. The Application of the Sponge City in Modern City Construction. J. Anhui Agri. Sci. **43**(16): 174-175 (2015)

3D numerical simulation of flow field with incompletely flaring gate pier in large unit discharge and deep tail water project

Zhou Zhao, Wang Junxing*

State Key Laboratory of Water Resources and Hydraulic Engineering Sciences, Wuhan University, Wuhan 430072, China

Abstract. Limited by large unit discharge above the overflow weir and deep tail water inside the stilling basin, the incoming flow inside stilling basin is seriously short of enough energy dissipation and outgoing flow still carries much energy with large velocity, bound to result in secondary hydraulic jump outside stilling basin and scour downstream river bed. Based on the RNG k-ε turbulence model and the VOF method, this paper comparatively studies flow field between the conventional flat gate pier program and the incompletely flaring gate pier program to reveal energy dissipation mechanism of incomplete flaring gate pier. Results show that incompletely flaring gate pier can greatly promote the longitudinally stretched water jet to laterally diffuse and collide in the upstream region of stilling basin due to velocity gradients between adjacent inflow from each chamber through shrinking partial overflow flow chamber weir chamber, which would lead to large scale vertical axis vortex from the bottom to the surface and enhance mutual shear turbulence dissipation. This would significantly increase energy dissipation inside stilling basin to reduce outgoing velocity and totally solve the common hydraulic problems in large unit discharge and deep tail water projects.

1 Introduction

The seasonal rainfall requires the low head hydraulic structures to urgently response and increase discharge volume, necessarily raise up tail water depth inside stilling basin and prone to scour downstream river bed due to insufficient energy dissipation and large outgoing velocity outside stilling basin[1-2].Recently, chute sill blocks or baffle blocks are installed above the stilling basin slab to enhance turbulent dissipation and improve flow field. But compared to tail water depth, auxiliary energy dissipators are so small in sizes that almost couldn't touch the mainstream or improve flow field inside stilling basin. Worsely, large underflow velocity inside stilling basin is prone to induce cavitation or some other secondary problems to auxiliary energy dissipators[3-4].

Combined with so many low head hydraulic projects, literature 2 pointed out that flaring gate pier (FGP) would remarkably promote turbulent energy dissipation and shorten stilling basin length. Through shrinking flow section in overflow weir chamber, longitudinal high speed water jet with 3 exposure faces would take shape to continually entrain air in the process of falling down in air and violently impinge on downstream tail water inside stilling basin, which would significantly aggravate turbulent energy dissipation and collision. But referred to previous engineering practices, FGP among overflow weir should meet unique operation requirements. In order to guarantee discharge capacity for overflow weir and reserve enough room for the running down water jet, the ration of downstream water head inside stilling basin to upstream total water head should meet some requirements, which almost confine FGP in application within high water head dams. FGP is almost never seen in low head hydraulic projects for the limited running space in air for water jet.

Many researchers have proposed different optimization measures on FGP through model test, numerical simulation or prototype observation to expand FGP application domain. Yin[5] progressively studied the relationship between hydraulic jump length, height and flow energy ratio in 3D hydraulic jump through model test. Li [6-7] indicated FGP with some large convergence angle could promote water jet to longitudinally stretch and reduce impingement pressure on stilling basin slab in high arch dam projects. Li[8] changed FGP contraction ratio and pointed out the influence of water head upstream overflow weir and FGP contraction ratio on discharge capacity. Zhang [9] comparatively studied the hydraulic characteristics of X-shaped and Y-shaped FGP with RNG k-ε turbulent model, and emphasized that X-shaped FGP may not only contain the characteristics of air entrainment and water diffusion of conventional Y-shaped FGP, but also increase overflow weir discharge capacity and improve pressure distribution along overflow weir surface. Hu[10] further analyzed the influence factors of aerated cavity in stepped overflow weir combined with FGP through standard k-ε two equation turbulent model based on Suo Fengyin Hydropower.

* Corresponding author: Zhou Zhao, doctor candidate, 1021235166@qq.com

Currently, the engineering application or research about FGP are mainly focused on large unit discharge and high water head project, and the arrangement of FGP is conventional that every chamber is symmetrically installed with FGP. But none researchers have questioned the FGP arrangement and application domain, meanwhile, FGP is almost never applied in large unit discharge and low head projects. In view of previous researcher's omission, this paper modifies the FGP arrangement and proposes incompletely FGP (shown in figure 1-a and below paragraph would call such FGP arrangement as incompletely FGP). Making full use of lateral width of overflow weir and installing FGP in interval chamber, incompletely FGP could aggravate velocity gradients and water level gap between rushing inflow from adjacent chamber. Combined with RNG k-ε turbulent model and VOF method, this paper would detailedly present the flow pattern, flow field and velocity distribution inside stilling basin.

2 Engineering practice

The discharge structure of such hydropower is composed of 5 surface-outlet overflow weir, and each chamber is marked as 1#,2#,......5# from left to right bank with corresponding longitudinal chamber axis as 1-1......5-5 in this paper. The elevation of overflow weir crest reaches 364.5m with net width of inflow section reaching 5×12.5m. Downstream overflow weir is followed by 1:1 fold line underflow stilling basin. The reservoir inflow of check condition and design condition respectively reaches 13200 m^3/s、9600 m^3/s with corresponding unit discharge up to 211.2 m^2/s,153.6 m^2/s. Upstream check flood level and design flood level respectively reaches 387m, 383m, and corresponding downstream water level inside stilling basin reaches 381.64m, 375.43m, only 5.39m,7.57m lower than upstream water level. Aimed at intensifying energy dissipation and improving flow field inside stilling basin in such low water head project of typical characteristics of large unit discharge, this paper respectively arranges two different programs to comparatively study 3D flow field inside stilling basin. Program 1 is symmetrically arranged with 5 flat gate piers from 1# to 5# chamber, and program 2, call incompletely FGP, is respectively arranged with 2 flaring gate piers in 2#,4# chamber, while 1#,3#,5# chamber is flat gate pier. The contraction ration of FGP is 0.5, and the longitudinal contraction angle is 22.25° [11]. The top of FGP reaches 375.72m so as to prevent rushing outflow to inundate FGP. Incompletely FGP is shown as Figure 1 .

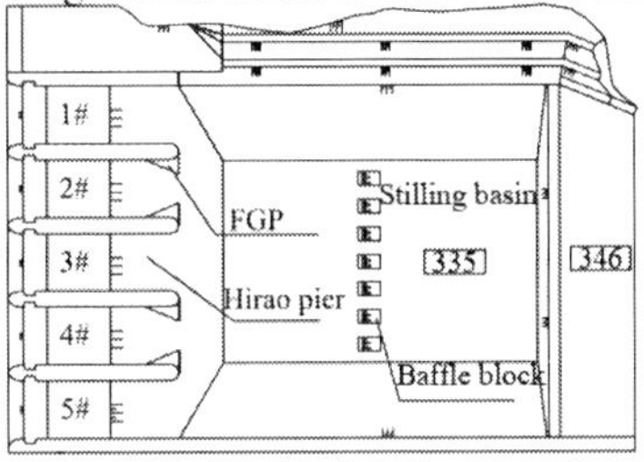

(a) Plan layout

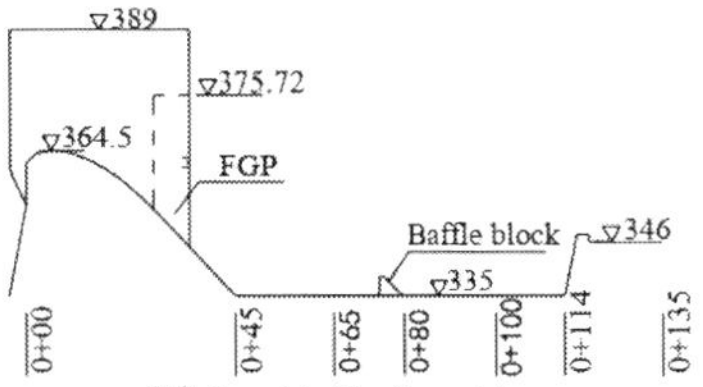

(b) Longitudinal section

Figure 1 Arrangement for incompletely FGP program (unit:m)

Notification: The unit for above elevation and stake number is m, and 0+00 is lateral axis of overflow weir.

3 Numerical model

Aimed at the full turbulent development inside stilling basin, this paper particularly chooses the most widely used RNG k-ε two-equation turbulent model in current engineering practices and adopts commercially available CFD package, Flow-3D, to divide 3D solid model with hexahedral mesh. The unique mesh processing technology, called FAVOR, is adopted in Flow-3D to evaluate the mesh quality and VOF method to be adopted to trace free surface in the process of numerical simulation

3.1 Governing equations

General governing equations for incompressible fluid are Reynolds-averaged Navier-equations. And the continuity equation, momentum equation, turbulence kinetic energy (TKE)equation, turbulence energy dissipation rate equation are respectively displayed as following.

$$\frac{\partial \rho}{\partial t}+\frac{\partial \rho u_i}{\partial x_i}=0 \quad (1)$$

$$\frac{\partial(\rho u_i)}{\partial t}+\frac{\partial(\rho u_i u_j)}{\partial x_j}=-\frac{\partial p}{\partial x_i}+\frac{\partial}{\partial x_j}[(\mu+\mu_t)(\frac{\partial u_i}{\partial x_j}+\frac{\partial u_j}{\partial x_i})] \quad (2)$$

$$\frac{\partial(\rho k)}{\partial t}+\frac{\partial(\rho k u_i)}{\partial x_i}=\frac{\partial}{\partial x_j}[\sigma_k(\mu+\mu_t)\frac{\partial k}{\partial x_j}]+G_k-\rho\varepsilon \quad (3)$$

$$\frac{\partial(\rho\varepsilon)}{\partial t}+\frac{\partial(\rho\varepsilon u_i)}{\partial x_i}=\frac{\partial}{\partial x_j}[\sigma_\varepsilon(\mu+\mu_t)\frac{\partial\varepsilon}{\partial x_j}]+C_{\varepsilon1}^*\frac{\varepsilon}{k}G_k-C_{\varepsilon2}\rho\frac{\varepsilon^2}{k} \quad (4)$$

where ρ, μ are volume fraction average density and molecular viscosity, respectively. p denotes the pressure acting on fluid element. The μt stands for turbulence viscosity and it could be calculated by the turbulence kinetic energy k and turbulence energy dissipation rate ε. Gk is generate item of turbulence kinetic energy due to mean velocity gradients.

$$\mu_t=\rho C_u\frac{k^2}{\varepsilon} \quad (5)$$

$$G_k=\mu_t(\frac{\partial u_i}{\partial x_j}+\frac{\partial u_j}{\partial x_i})\frac{\partial u_i}{\partial x_j} \quad (6)$$

$$C_{\varepsilon 1}^{*} = C_{\varepsilon 1} - \frac{\eta(1-\eta/\eta_0)}{1+\beta\eta^3} \quad (7)$$

$$\eta = (2E_{ij} \bullet E_{ij})^{0.5}\frac{k}{\varepsilon} \quad (8)$$

$$E_{ij} = \frac{1}{2}\left(\frac{\partial u_i}{\partial x_j} + \frac{\partial u_j}{\partial x_i}\right) \quad (9)$$

where Cu=0.0845 is an experimental constant. The turbulence Plandtl numbers for k and ε are all 1.39, respectively. And $C\varepsilon_1$=1.42 , $C\varepsilon_2$=1.68, η_0 =4.377, and β =0.012 are constants for ε equation.

3.2 Free surface tracking

The VOF method is a kind of means to trace free surface variation along overflow weir, which evolves from the marker-and-cell method but more computationally efficient. Recently, Hirt [12], Savage [13] et al introduced VOF method in detail.

VOF method believes both kinds of fluid wouldn't interpenetrate and its main idea is to utilize a variable F to account for how many percentages of each computational cell is occupied by the phase. The variable F is called volume fraction of the phase and is applied to equation 10. F=0 means there is no liquid in control volume, and F=1 means liquid occupies the full control volume. Furthermore, F=0~1 means control volume is free liquid surface and partly filled with liquid.

$$\frac{\partial F}{\partial t} + u\frac{\partial F}{\partial x} + v\frac{\partial F}{\partial y} + w\frac{\partial F}{\partial z} = 0 \quad (10)$$

3.3 Mesh divide

This calculation model is related with large regions and complicated structure. In order to accurately trace hydraulic characteristic in the control volume, a large scale of hexahedron grids are used to divide the computation domain. Total number of grid cells reaches 8.80E+06 in which 4.70E+06 are active. Considering the differences between upstream and downstream of overflow weir, the numerical domain is divided into three mesh blocks (shown in figure 2). Mesh block is particularly compressed with a uniform mesh size △X=△Y=△Z=0.5m among overflow weir and stilling basin regions. Upstream reservoir region and downstream natural channel region are all chosen with uniform mesh size with △X=△Y=△Z=1m. The Cartesian coordinate system is used with X, Y, Z representing streamwise, lateral direction and wall-normal, respectively. X axis directs to the right bank along lateral overflow weir axis, Y axis directs to streamwise along longitudinal overflow weir axis, and Z axis is opposite to gravity.

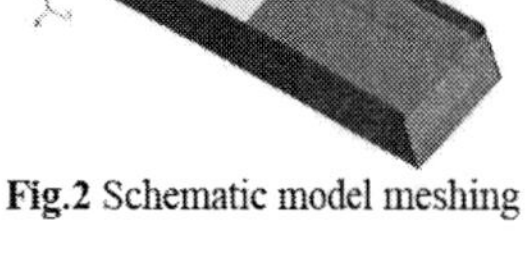

Fig.2 Schematic model meshing

3.4 Boundary condition and initial condition

To accurately simulate the given flow, the boundary condition of numerical model should be as similar as possible to what is actually occurring. There are six different boundaries in simulated domain among the Cartesian coordinate. Boundary conditions are set as following: sidewalls x-common no slip, non-porous wall; top z-pressure boundary with gauge pressure equal to zero (atmospheric),bottom z- no slip wall; left y- local stagnation pressure based on upstream total head over overflow weir crest with a hydrostatic pressure distribution, right y- local static pressure based on downstream tailwater with a hydraulic pressure distribution. 50m upstream and 400m downstream of overflow weir are used as boundaries of left and right Y, respectively.

4 Result and analysis

4.1 Model verification

In view of difference and symmetry in overflow weir chamber of each program, longitudinal water level along the axis of 2#,3# chamber from numerical simulation and measured values is particularly chosen to verify the numerical simulation accuracy. The results of water level along each chamber axis are shown in figure 3,and numerical data are quite close to practical measured values. There are only some slight discrepancy in upper region of stilling basin where rushing inflow violently fluctuates. Such results can indicate the numerical simulation to be reliable.

Further, the discharge volume of flat gate pier program in check condition and design condition respectively reaches 13820m3/s、10040 m3/s, while the discharge volume of incompletely FGP program reaches 13540m3/s、9860 m3/s. Both program all can meet the requirement of flood discharge. Comprehensively comparing above discharge volume in two crucial conditions, discharge capacity of incompletely FGP program is only 2.0%、1.7% lower than that of the flat gate pier program respectively, indicating such arrangement of incompletely FGP is of no influence to discharge capacity for overflow weir.

4.2 Flow pattern

Due to deep tail water inside stilling basin, the submerged hydraulic jump refluxes much and almost inundates overflow weir crest in flat gate pier program. There is almost no fluctuation or lateral diffusion inside the total stilling basin, but outgoing flow greatly upwells right at the end of stilling basin in view of baffle sill. Water level of outgoing flow reaches up high to 378.66m with the stable secondary hydraulic jump 30m downstream stilling basin. Above visible flow pattern indicates energy dissipation inside stilling basin is seriously insufficient and outgoing flow still carries much redundant energy. Similarly, rushing inflow from

1#,3#,5# flat gate pier chamber forms a visible submerged hydraulic jump. Yet, in view of shrunken flow section in 2#,4# FGP chamber, water jet with high speed heavily impinges on tail water and greatly diffuses towards both side due to velocity gradients and water level gap between adjacent rushing flow, forming evident V-shaped intersection with much spilt entrapped bubble and intensifying shear turbulence in upper region of stilling basin. Rushing inflow from bilateral 1#,5# chamber is also forced to offset and violently rub guide wall on both sides, inducing symmetrical large scale vertical vortexes from bottom to surface. In such program, momentum interchange and energy transfer are observably enhanced, so does the energy dissipation inside stilling basin. Outgoing flow smoothly joins downstream natural tail water and secondary hydraulic jump outside stilling basin is totally eliminated.

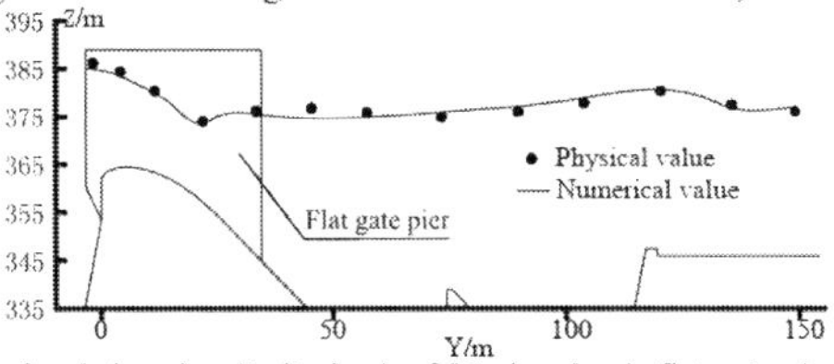

a Water level along longitudinal axis of 2# chamber in flat gate pier program

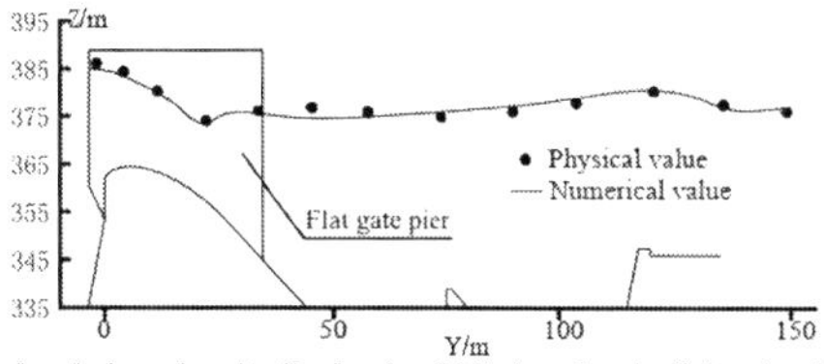

b Water level along longitudinal axis of 3# chamber in flat gate pier program

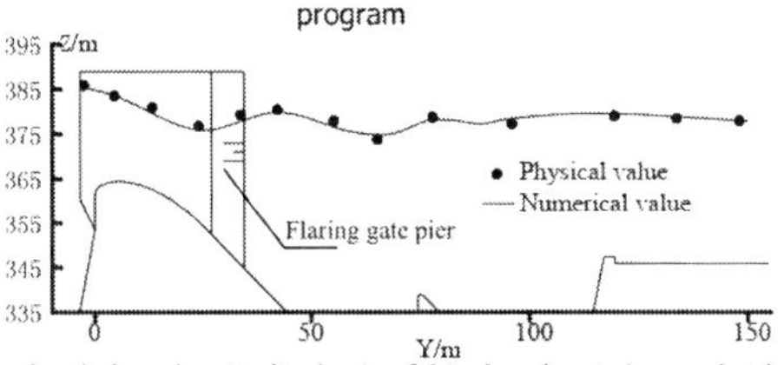

c Water level along longitudinal axis of 2# chamber in incompletely FGP program

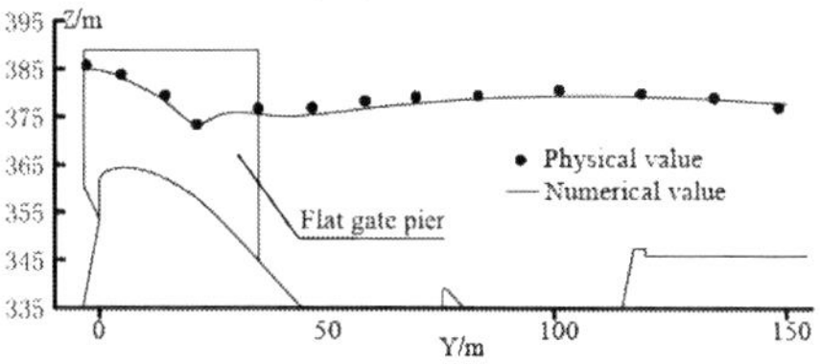

d Water level along longitudinal axis of 3# chamber in incompletely FGP program

Figure3 Water level distribution along longitudinal chamber axis（unit:m）

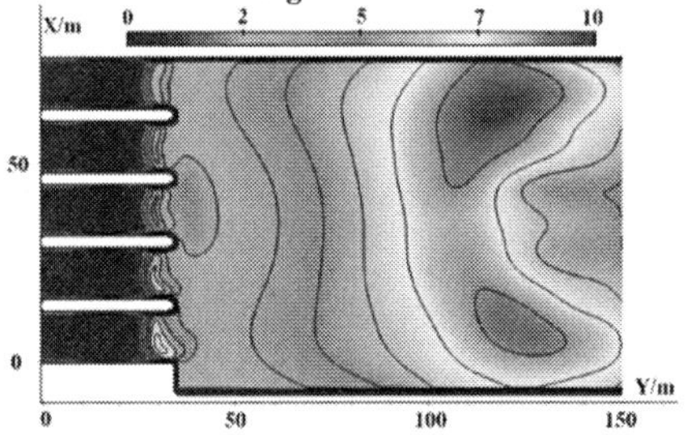

a Upper layer in flat gate pier program

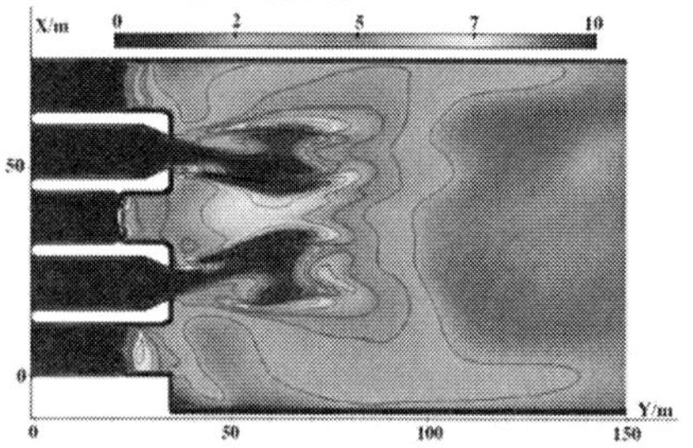

b Upper layer in incompletely FGP program

Figure 4 Turbulent kinetic energy distribution inside stilling basin (unit: m^2/s^2)

4.3 Turbulent kinetic energy distribution

Turbulence kinetic energy (T_{KE}) is an important kind of physical quantity weighting the extent of running flow fiercely moving, and T_{KE} would greatly moves up where discharge flow fiercely dissipates with turbulence. Combined with numerical calculation, T_{KE} of discharge flow inside stilling basin can been calculated with formula $T_{KE} = (\overline{\mu_x'^2} + \overline{\mu_y'^2} + \overline{\mu_z'^2})/2$, where $\overline{u_x}'$, $\overline{u_y}'$ and $\overline{u_z}'$ are mean value of pulsating flow rate in lateral, streamwise and anti-gravity direction, respectively.

Shown in figure 4, the turbulent kinetic energy of both programs differs much in upper layer of discharge flow. Rushing inflow from 5 flat gate pier chambers in program 1 evenly impinges on tail water and there is a serious lack of enough turbulence in deep layer of running flow. Maximum turbulent kinetic energy only reaches 6 m^2/s^2 in upstream the stilling basin, while outgoing flow reaches up high to $10m^2/s^2$ above the baffle sill (0+120m), consistent with the secondary hydraulic jump in above described flow pattern. But turbulence kinetic energy distribution is upper layer in incompletely FGP program is quite from that of flat gate pier program. Shown in figure 4-b, turbulent kinetic energy distribution can exactly indicate rushing inflow from FGP chamber strongly collides with that from flat gate chamber and forms 4 observable high energy turbulent areas on both side of water jet, indicating that energy dissipation and turbulent dissipation are quite significant in upstream the stilling basin due to velocity gradients and water level gap between adjacent rushing inflow. Maximum turbulent kinetic energy reaches $12.5m^2/s^2$ and turbulent regions concentrate in upstream the stilling basin. Turbulent kinetic energy rapidly drops down and values of outgoing flow is as low as 2.2 m^2/s^2, or even more lower, similarly consistent with stable flow pattern in above described in incompletely FGP program.

4.4 Aeration concentration distribution

Water jet with 3 exposed faces from FGP chamber is bound to aerate much air during the process of falling down and entrained bubble is inevitably sucked into rushing water to prevail in the whole stilling basin. Squeezed by uneven pulsating pressure from nearby running water, entrained bubble would repeat to collapse, break and shatter into smaller bubble, which would aggravate and compel rushing flow to randomly spread to intensify irregular turbulence and shear friction[14-15].

The longitudinal entrance section (0+45m) and middle section (0+70m) in stilling basin are particularly chosen to present aeration concentration distribution. Figure 5-a,5-b both indicates rushing flow aerates little and aeration concentration is as low as 0.15,0.20, respectively. Worse, aerated involved depth is seriously inadequate, only reaching 25m below free surface, and bottom flow almost finds none evident bubble. Yet, aeration concentration of each corresponding section in incompletely FGP program generally reaches 2 times or more than that of program 1 and aerated scope is much enlarged, too. Maximum aeration concentration in entrance section reaches up high to 0.6, or more, and 38m below free surface is also involved with aeration concentration high up to 0.15. There is only slight lack of aerated bubble just downstream the rushing inflow from 2#,4# chamber due to the highly shrunk and stretched water jet in such regions, further indicating lateral diffusion is strongly related to aeration. Meanwhile, with rushing flow fully offsetting and diffusing, aeration concentration in middle section (0+70m) distributes much more evenly and widely. Aeration concentration of the total middle layer running flood in such section generally reaches up high to 0.15, and even 33m below free surface could observe large scale bubble, which is far more massive than that of program 1.

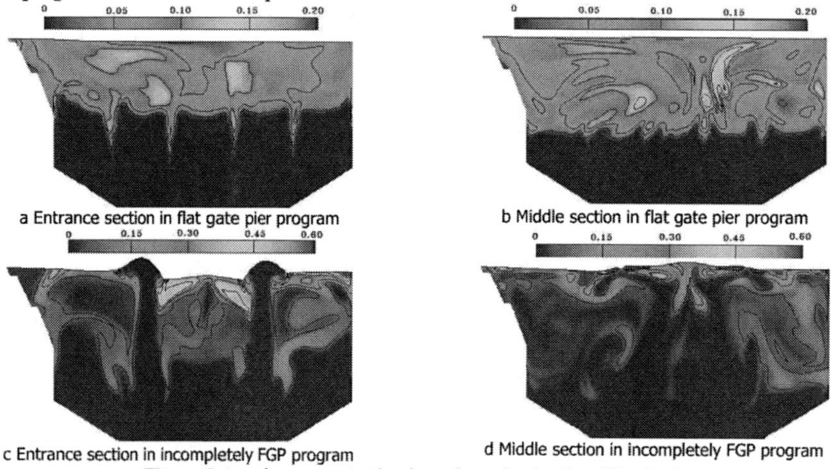

a Entrance section in flat gate pier program
b Middle section in flat gate pier program
c Entrance section in incompletely FGP program
d Middle section in incompletely FGP program

Figure 5 Aeration concentration in each section inside stilling basin

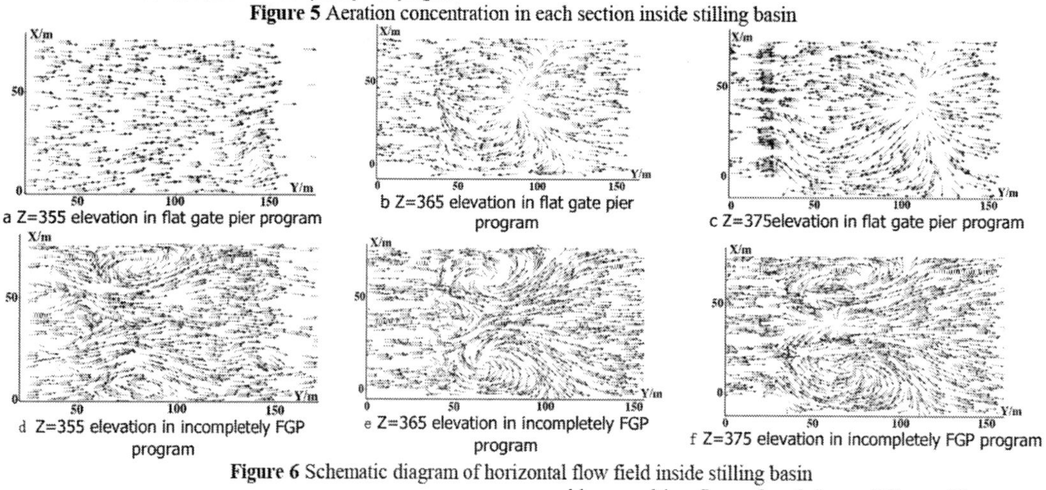

a Z=355 elevation in flat gate pier program
b Z=365 elevation in flat gate pier program
c Z=375elevation in flat gate pier program
d Z=355 elevation in incompletely FGP program
e Z=365 elevation in incompletely FGP program
f Z=375 elevation in incompletely FGP program

Figure 6 Schematic diagram of horizontal flow field inside stilling basin

4.5 Horizontal flow field inside stilling basin

Horizontal flow field is of much importance to trace trajectory of running flow inside stilling basin to reveal interaction between adjacent inflow. In view of longitudinal stream layer deviation due to submerged hydraulic jump, bottom (Z=355m), middle (Z=365m), surface (Z=375m) flow field are particularly presented to reveal how rushing flow of each layer diffuses. There are no observable differences among rushing flow from each chamber in program 1, and both bottom and middle layer inflow longitudinally impinges on tail water while upper layer inflow violently refluxes, forming submerged hydraulic jump and large scale lateral axis vortex at 0+65m section. Such flow field between adjacent rushing inflow can indicates that the flat gate pier program can only depend on conventional submerged hydraulic jump to kill superfluous energy, but actually inadequate due to deep tail water inside stilling basin. Obstructed by baffle sill at the end of stilling basin, outgoing flow observably upwells and there are evident surface reflux inside the whole stilling basin. But rushing inflow from each chamber in incompletely FGP program differs much. Inflow from 1#,3#,5# is particularly similar to that of program 1, only submerged hydraulic jump slightly moves upstream to 0+50m and lateral axis vortex dives deeper. Meanwhile, high speed stretched water jet from 2#,4# FGP chamber, shown in figure 6-d,6-e,6-f, greatly diffuses and offset towards both sides, inducing vertical vortex to aggravate shear turbulence in view of velocity gradients between adjacent rushing inflow. Little scale vertical vortex is observed near 0+80m section among bottom rushing flow field. Further, large scale vertical vortex grows and expands to involve the region from 0+50m section to 0+80m section among middle and upper rushing flow field, indicating shear turbulence to be gradually remarkable from bottom to surface in upstream regions of stilling basin.

Comprehensively comparing the difference between two programs, what can be found is that baffle blocks are so small in size relative to tail water depth inside stilling basin that almost has no influence on flow field. But installing FGP in interval chamber could observably result in large scale symmetrical vertical vortex on both side of stilling basin from bottom to surface to aggravate shear turbulence for adjacent rushing inflow in upstream the stilling basin due to enlarged lateral velocity gradients and water level gap between adjacent rushing inflow from each chamber.

4.6 Velocity distribution and energy dissipation rate

Velocity distribution of check condition in both programs is shown in table 1, in which rushing inflow of program 1 is relatively much larger with maximum velocity in entrance section reaching up high to 13.3m/s and surface reflux reaching as low as -2.7m/s. Energy dissipation inside stilling basin is of much insufficient along the longitudinal stream, underflow velocity is as much as 10-12m/s and maximum outgoing flow velocity downstream stilling basin reaches 12m/s with average velocity 5.87m/s. While due to large scale vertical vortex on both side of stilling basin in incompletely FGP program, rushing inflow velocity observably decreases down to 10.1m/s with reflux velocity as high as -4m/s. In view of diving main stream and lateral diffusion, rushing flow velocity of middle layer or upper layer swiftly decreases, so does the outgoing flow. Maximum velocity of outgoing flow decreases down to 6.8m/s with average velocity 4.44m/s, which is 24% lower than of flat gate pier program.

Energy dissipation rates of both programs are shown in table 2, in which the most traditional calculation formula for energy dissipation rate $\eta = (E_1 - E_2) / E_1 \times 100\%$ is chosen. E1、E2 means total water head for running section in upstream the overflow weir and downstream stilling basin, and contains potential energy, pressure energy and kinetic energy. Overflow weir crest section and stilling basin outlet section are particularly adopted for above calculation, and horizontal plane of Z=347.5m is chosen as reference surface for potential energy. Due to high water level inside stilling basin, potential energy accounts for a large proportion for rushing flow and kinetic energy is in relatively small proportion, leading to calculation results for energy dissipation rate to be much little. Energy dissipation rates in check condition for both programs only reach 21.09%,25.85%, respectively, and design condition to be 26.61%,30.21%. In both conditions energy dissipation rate for incompletely FGP program is 4% higher than that of flat gate pier program, and such progress is of much improvement in large unit discharge and low head project.

Table 1 Velocity distribution along longitudinal section inside stilling basin (unit:m/s)

Program	Stake / (m)	Left section bottom	Left section middle	Left section surface	Middle section bottom	Middle section middle	Middle section surface	Right section bottom	Right section middle	Right section surface
Flat gate pier	0+45	9	12.6	-2.7	6.3	13.1	-0.9	8.1	13.3	-1.8
	0+65	12.8	9	-2.7	10.8	12.6	-0.9	11.7	8.8	-1.8
	0+80	13.5	4.5	-2.7	10.8	9	-0.9	12.6	5.4	-2.7
	0+100	12.5	3.8	2.3	3.8	10.5	-1.5	11.3	5.3	-2.3
	0+114	5.8	7.5	1.5	4.5	10.5	1.5	7.5	9.8	0.8
	0+135	2.3	11.3	3.8	0.8	11.3	4.5	3	12	3.8
Incompletely FGP	0+45	9.8	12	-4.4	9.4	7	-4.4	10.1	5.5	-3.1
	0+65	12.5	2.3	-3.1	10.9	3.9	-3.1	12.5	3.1	-3.9
	0+80	12.5	-2.3	-2.3	6.2	5.5	3.9	11.7	-3.1	-3.9
	0+100	10.1	3.1	-1.6	1.6	7	3.9	9.4	2.3	1.6
	0+114	3.1	5.5	2.3	1.8	7	3.9	3.6	7.3	2.7
	0+135	1.9	6.3	4.6	1.2	6.8	5.5	2.4	6.7	4.6

Notification: Left, middle, and right section are respectively along longitudinal axis of 1#,3# and 5# chamber.

Table 2 Energy dissipation in each condition

Condition	Program	Velocity above overflow weir crest(m/s)	potential energy above overflow weir crest (m)	Outgoing velocity outside stilling basin (m/s)	potential energy outside stilling basin (m)	energy dissipation rate
Check	Flat gate pier	10.07	36.67	5.87	31.26	21.09%
	Incompletely FGP	10.43	37.15	4.44	30.65	25.85%
Design	Flat gate pier	9.22	33.39	5.51	26.14	26.61%
	Incompletely FGP	8.93	33.88	4.01	25.66	30.21%

5 Conclusion

Aimed at the common hydraulic problems of secondary hydraulic jump and river bed erosion in deep tail water projects due to insufficient energy dissipation inside stilling basin, this paper proposes a new kind of FGP arrangement—incompletely FGP to improve flow pattern and drop down outgoing velocity. The mechanism of incompletely FGP is revealed with numerical simulation and conclusions are below:

1 The new arrangement of incompletely FGP almost has no much influence on discharge capacity for overflow weir, only 2% lower than that of conventional flat gate pier program.

2 High speed water jet with 3 exposed faces would observably aerate and spread in the process of running down in air, necessarily enhancing aeration concentration and involved depth. Aeration concentration of each section is as much as 2 times than that of conventional flat gate pier program.

3 Shrinking lateral flow section in partial chamber in incompletely FGP program, high speed water jet from FGP chamber would observably offset and diffuse in upstream of stilling basin due to great velocity gradients and water level gap between adjacent inflow from flat gate pier chamber, inducing large scale vertical vortex from bottom to surface on both side and aggravating shear turbulence inside stilling basin. Energy dissipation effect is improved and secondary hydraulic jump outside stilling basin never appears again.

Comprehensively considering discharge capacity, flow pattern, aeration concentration and horizontal flow field inside stilling basin, what can be found is that the new arrangement of incompletely FGP not only guarantees discharge capacity, enhance aeration concentration and intensify shear turbulence or vertical vortex inside stilling basin with great velocity gradients, but also totally solves the secondary hydraulic jump with insufficient energy dissipation.

Reference

1. Xie Shengzong, Zhu Ronglin, Li Shiqing, et al. Hydraulics of bucket basin equipped with flaring gate piers[J], Journal of hydraulic engineering, 2:7-18,(1992)
2. Xie Shengzong, Wu Yihong, Chen Wenxue. New technology and innovation on flood discharge and energy dissipation of high dams in China[J], Journal of hydraulic engineering, 47(3):324-336,(2016)
3. An Gang, Yuan Ximin, Yang Chanzeng, et al. The experimental studyof energy dissipation structure of low water head project[J]. Journal of Tianjin University, 30(4):500-506 (1997)
4. Yang Fan, Wang Junxing, Chen Shuai, et al. Study on cavitation damages of supplementary dissipater under low-Froude-number flows[J], Journal of hydroelectric engineering, 34(2):125-129,(2015)
5. Yin Jinbu, Liang Zongxiang, Gong Honglin. Experimental study on flow characteristics of 3D hydraulic jump of flaring gate piers[J], Journal of hydraulic engineering,41(11):1333-1338,(2011)
6. Li Naiwen, Xu Weilin, Tianzong, et al. Optimal experimental study on flaring piers of surface spillways in high arch dam[J], Journal of hydroelectric engineering, 28(3):132-138,(2009)
7. Li Naiwen, Liu Chao, Deng Jun,etal. Theoretical and experimental studies of the flaring gate pier on the surface spillway in a high-arch dam[J]. Journal of Hydrodynamics. 24(4)： 496-505,(2012)
8. Li Futian, Liu Peiqing, Xu Weilin, et al. Experimental study on effect of flaring piers on weir discharge capacity in high arch dam[J], Journal of hydraulic engineering, 11:43-47(2003)
9. Zhang Ting, Wu Chao, Mo Zhengyu, et al. Comparison on hydraulic characteristics of X-shape and Y-shape flaring gate piers[J], Journal of hydraulic engineering, 38(10):1207-1213,(2007)
10. Hu Yaohua, Wu Chao, Zhang Ting, et al. Affect factors analysis of aerated cavity of X-shaped flaring gate pier[J], Journal of Sichuan University (engineering science edition), 39(3):24-28, (2007)
11. Li Zhongshu, Pan Yanhua, Han Lianchao, et al. Hydraulics of stilling basin with flaring gate piers[J], Advance in water science, 11(1):82-87,(2000)
12. Hirt C W, Nichols B D. Volume of fluid (VOF) method for the dynamics of free boundarues[J].J Compus Phys, 39(3):201-225,(1981)
13. Savage, B.M; Johnson, M.C, Flow over ogee spillway: Physical and numerical model case

study[J], Journal of hydraulic engineering, 127(8):640-649,(2001)

14. Liao Liuxia, Wang Wei, Liu Shanjun, et al. Influence of local sidewalls placed behind flaring gate pier on aeration effectiveness[J]. Journal of Sichuan University (engineering science edition), 46(2):13-16,(2014)
15. Deng Jun, Xu Weilin, Velocity distribution of self-aerated flows[J] Journal of hydrodynamics, 17(3):369-375,(2002)

Study of stability against deep sliding of gravity dam based on JC method

Zheng Si[1], *Lanfang* Xin[1], *Lingzhi* Huang[1], and *Zhengxin* Wang[1]

[1]Xi'an University of Technology, Xi'an, Shaanxi, 710048, China

Abstract. In this study, stability against deep sliding of gravity dam is analysed in order to provide guidance for ensuring stability against sliding of gravity dam during the construction and operation periods. Based on checking-point method (JC method), the study proposes an analysis model for stability against deep sliding of gravity dams. In addition, the rigid body limit equilibrium method is utilized to construct the limit state function of deep anti-sliding stability. With sliding plane anti-shear friction coefficient and anti-shear cohesion calculated as random variables, safety degree of stability against deep sliding is analysed. With JC method employed, the calculation program is compiled to calculate the stability against deep sliding of gravity dam and analyse the non-overflow section of gravity dam under normal storage level. According to calculation program, the reliability index β of stability against deep sliding of gravity dam is 4.36, and the failure risk P_f is approximately zero. Additionally, based on the traditional rigid body limit equilibrium method, the safety factor K is 3.23, which meets the requirements of the design specification. According to the above methods, stability against deep sliding of the dam section meets the requirement. In conclusion, JC method provides a calculation and analysis model for gravity dam's stability against deep sliding, serving as references in the design of gravity dams based on reliability theory.

1. Introduction

In recent years, intensive study have been conducted by domestic and foreign researchers and important progress has been made concerning the application of reliability theory in the design of gravity dams. For instance, Lan [1] delivered a preliminary discussion on the use of reliability theory to calculate the deep anti-sliding risk of gravity dams, and proposed the limit state equation as a calculation formula. Based on this structural reliability theory, Li [2] developed a reliability calculation formula for gravity dam's stability against deep sliding in relation to a deep single sliding plane within the foundations and verify its correctness. Taking advantage of equal safety coefficient method, Yang [3] established a mechanical model which considered lateral resistance in calculating the reliability index of a gravity dam's deep anti-sliding stability. Besides, Wang [4] analysed the reliability index of a gravity dam's stability against sliding along the foundation plane, and made a sensitivity analysis of several random variables affecting the reliability index. Furthermore, Duncan [5] figured out the structure reliability index with the traditional safety coefficient calculation method and variable probability values employed. Based on a real dam section, Su [6] set a detailed example to illustrate the proposed risk analysis approach while Ji [7] worked out structural reliability of gravity dam using artificial neural networks method. Zhang [8] devised a system reliability analysis method for multiple failure patterns of gravity dam on top of the Copula function. Built on the weighted regression response surface method, the reliability indexes of gravity dam foundation faults were computed [9]. Xu [10] deduced the upper limit of failure probability for failure mode and system reliability of dam on the foundation of Bayes formula and Cauchy-Schwarz inequality. Built on the previous research achievements, this study introduced the use of JC method to calculate the stability against deep sliding of the deep double sliding surface in the foundation of a gravity dam, with the aim of introducing reliability theory in the design of gravity dams.

2. Reliability analysis method

In the analysis of the structural reliability, the limit state of a structure is described by performance function when there are n random variables (x_1, x_2, Λ, x_n) that affect the reliability of the structure, the performance function of the structure can be expressed as follows:

$$Z = g(x_1, x_2, \Lambda, x_n) \quad (1)$$

When $Z > 0$, the structure is in a reliable state; when $Z = 0$, the structure is in the limit state and when $Z < 0$, the structure is in a state of failure [11].

*Correspondence should be addressed to Zheng Si; sz123hlz@163.com

If the random variables x_1, x_2, Λ, x_n obey a normal distribution, with means and standard deviations being $\mu_{x_1}, \mu_{x_2}, \Lambda, \mu_{x_n}$ and $\sigma_{x_1}, \sigma_{x_2}, \Lambda, \sigma_{x_n}$ respectively, then the performance function of the structure $Z = g(x_1, x_2, \Lambda, x_n)$ should also be a normal random variable, with mean μ_z and standard deviation σ_z being expressed as

$$\mu_z = g(\mu_{x_1}, \mu_{x_2}, \Lambda, \mu_{x_n}) \quad (2)$$

$$\sigma_z = \sqrt{\sum_{i=1}^{n} \left(\left. \frac{\partial g}{\partial x_i} \right|_{\mu_{x_i}} \right)^2 \sigma_{x_i}^2} \quad (3)$$

respectively, and probability density of Z being expressed as

$$f(z) = \frac{1}{\sqrt{2\pi}\sigma_z} \exp\left(-\frac{(z-\mu_z)^2}{2\sigma_z^2}\right) \quad (4)$$

When normal distribution of Z is converted into a standard normal distribution, the failure risk of Z shall be expressed as follows:

$$P_f = \Phi(-\mu_z/\sigma_z) = \Phi(-\beta) = 1 - \Phi(\beta) \quad (5)$$

In the formulas above, β is the reliability index of a structure, $\beta = \mu_z/\sigma_z$.

However, in most cases, many of the variables in an actual project do not obey a normal distribution. Therefore, it is necessary to transform non-normal random variables into normal random variables through equivalent normalisation, during which the advanced checking-point method (JC method) is generally recommended by JCSS. The service condition of JC method for equivalent normalisation is when the distribution function value and the probability density function value of the equivalent normal random variable $X_i^{'}$ and the non-normal variable X_i are the same, respectively, at the design checking point, x_i^*, which can be expressed with the following formulas:

$$F_{X_i}(x_i^*) = F_{X_i^{'}}(x_i^*) = \Phi\left(\frac{x_i^* - \mu_{X_i^{'}}}{\sigma_{X_i^{'}}}\right) \quad (6)$$

$$f_{X_i}(x_i^*) = f_{X_i^{'}}(x_i^*) = \frac{1}{\sqrt{2\pi}\sigma_{X_i^{'}}} \exp\left(-\frac{(x_i^* - \mu_{X_i^{'}})^2}{2\sigma_{X_i^{'}}}\right) \quad (7)$$

In the above formulas, $F_{X_i}(\bullet)$ and $F_{X_i^{'}}(\bullet)$ are the distribution function of the normal distribution variable X_i and the non-normal distribution variable X_i respectively; with $f_{X_i}(\bullet)$ and $f_{X_i^{'}}(\bullet)$ being the probability density function of the normal distribution variable X_i and non-normal distribution variable X_i, respectively.

If the mean μ_{X_i} and standard deviation σ_{X_i} of the non-normal random variable X_i are replaced by the mean $\mu_{X_i^{'}}$ and standard deviation $\sigma_{X_i^{'}}$ of the equivalent normalised random variable $X_i^{'}$, the following formulas shall be obtained:

$$\mu_{X_i^{'}} = x_i^* - \Phi^{-1}[F_{X_i}(x_i^*)]\sigma_{X_i^{'}} \quad (8)$$

$$\sigma_{X_i^{'}} = \phi\{\Phi^{-1}[F_{X_i}(x_i^*)]\}/f_{X_i}(x_i^*) \quad (9)$$

In the formulas above, $\Phi^{-1}[F_{X_i}(\bullet)]$ is the inverse function of $F_{X_i}(\bullet)$ while $\phi(\bullet)$ is the cumulative probability density function of the standard normal variable.

The reliability index β represents the length of the design checking point P^* to O (origin of the coordinates) along the normal direction of a limit state surface tangent plane in a standard normal space coordinate system. The direction cosine of the normal line, OP^*, to the coordinate vector can be expressed as follows:

$$\cos\theta_{X_i} = -\frac{\left. \frac{\partial g}{\partial X_i} \right|_{P^*} \sigma_{X_i}}{\sqrt{\sum_{i=1}^{n} \left[\left. \frac{\partial g}{\partial X_i} \right|_{P^*} \right]^2 \sigma_{X_i}^2}} \quad (10)$$

Converting the above-mentioned relationship into the original coordinate system, the coordinate of P^* in the original coordinate system can be expressed with following formulas:

$$x_i^* = \mu_{X_i} + \overline{x_i^*}\sigma_{X_i} = \mu_{X_i} + \beta\sigma_{X_i}\cos\theta_{X_i} \quad (11)$$

In the formulas above, $\overline{x_i^*}$ represents the coordinate of the design checking point x_i^* to the non-normal variable X_i in the standard normal space coordinate system.

Since the point P^* is on the limit state surface, it meets the limit state of formula (1) and can be expressed as follows:

$$g(x_1^*, x_2^*, \Lambda\ x_n^*) = 0 \quad (12)$$

The values of β and x_i^* ($i = 1,2,\Lambda, n$) can be calculated instantly by solving the formulas (10)–(12) simultaneously.

3. Calculation process of stability against deep sliding of a gravity dam (based on JC method)

(Fig. 1) shows the calculation process used to solve the stability against deep sliding of a gravity dam by using JC method, which can be briefly introduced as follows [12]:

1) Set the initial structure reliability index β_0;

2) Assume the initial design checking point $x_i^{1*} = \mu_{X_i}$;

3) Equivalent normalise the non-normal random variables through formulas (8) and (9), then obtain $\mu_{X_i'}$ and $\sigma_{X_i'}$;

4) Determine the partial derivatives of the limit state function to random variables, by calculating $\frac{\partial Z}{\partial f_1'}$, $\frac{\partial Z}{\partial f_2'}$, $\frac{\partial Z}{\partial c_1'}$, $\frac{\partial Z}{\partial c_2'}$

5) Calculate the sensitivity coefficient α_{X_i}. Sensitivity coefficient α_{X_i} reflects the effect of the random variable X_i on the standard deviation of whole the performance function, $\alpha_{X_i} = \cos\theta_{X_i}$, which can be calculated through formula (10).

6) Calculate the new design checking point x_i^{2*} through formula (11). Repeat steps 3–6 until the differences of two successive design checking point are reduced within the allowable range.

7) Substitute the new design checking point x_i^{2*} into the limit state function (12), and examine whether it can meet the limit state function. If it can't, recalculate the reliability index β through $\beta_{n+1} = \beta_n + g_n \cdot \Delta\beta / \Delta g$, and repeat steps 3–7 until $g(x_1^*, x_2^*, \Lambda\ x_n^*) \approx 0$. The final value of β represents the required structure reliability index.

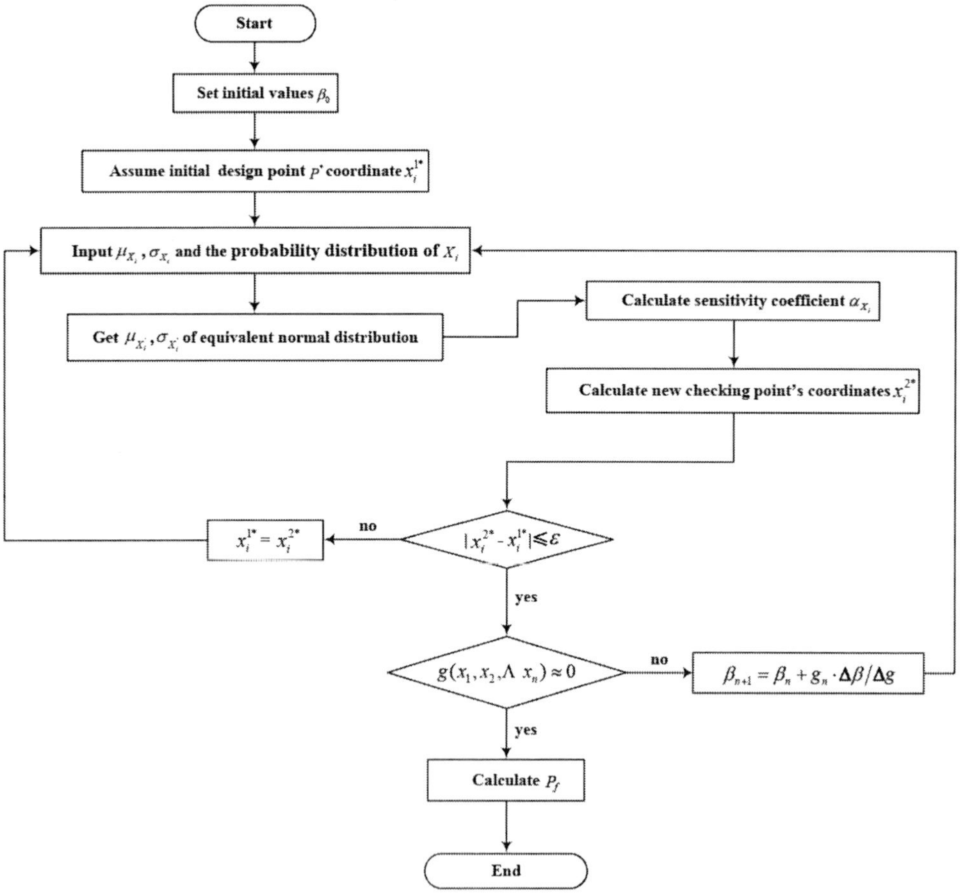

Figure. 1 Flow chart of JC method

4. Examples of Gravity Dam's Stability Against Deep Sliding Calculation

Based on the aforementioned theory and calculation procedure, this study made an analysis of stability against deep sliding in the case of a hydro project non-overflow gravity dam section (Fig. 2).

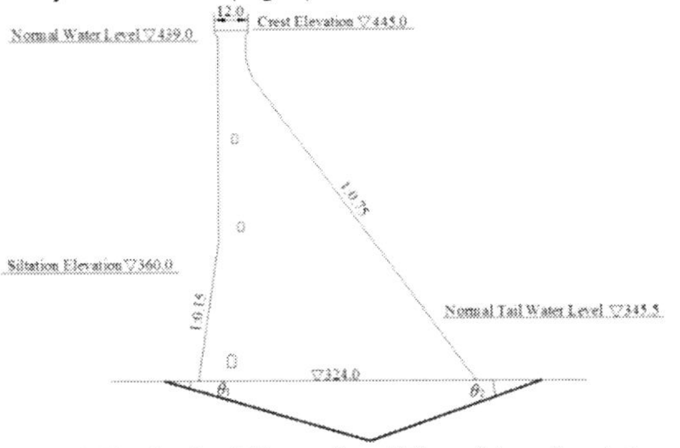

Figure. 2 Sketch of stability against sliding of deep foundations (Unit: m)

The maximum height of the non-overflow section of this dam is 121.0 m, with the foundation plane's elevation being 324.0 m, crest elevation being 445.0 m and the width of dam crest being 12.0m. In accordance with engineering geological data, the angles of the main is θ_1 = 16°while the auxiliary slide plane to the horizontal plane is θ_2 = 19°. In the calculation of stability against deep sliding of a gravity dam foundation, the shear friction coefficient $f_1^{'}$, $f_2^{'}$, as well as shear cohesion $c_1^{'}$, $c_2^{'}$ of the primary and auxiliary slide planes are taken as random variables, whose the characteristics are shown in Table 1. Moreover, bulk density of concrete is 23.8 kN/m^{3}while bulk density of bedrock is 27.0 kN/m^3. The basic combination of the normal storage level is taken as the calculation conditions when upstream water level is 439.0 m and the downstream water level is 345.5 m. The loads to be calculated include the dead weight of the gravity dam, bedrock above the slide plane and the hydrostatic and uplift pressures. Furthermore, the uplift pressure reduction coefficient is taken as 0.25. (The table 1 at the bottom of the page)

According to a self-compiled program, the reliability index β of the gravity dam's stability against sliding is 4.36 and the failure risk P_f is approximately zero. As indicated by the results calculated with reliability theory, gravity dam's stability against deep sliding can meet the requirements. Verified by the rigid body limit equilibrium method, the factor of safety against deep sliding K is found to be 3.23. For basic combination, when safety factor K is greater than 3.0, anti-sliding stability is considered to meet the requirements. Results by both methods suggest that requirements can be met by the stability against deep sliding of the gravity dam.

5. Conclusions

The existence of gently inclined soft layers in the foundations of a gravity dam may result in a deep sliding instability in the foundations. In this study, by taking anti-shear strength parameters, the most sensitive factor to the anti-sliding stability, as random variables, the author established the limit state function using rigid body limit equilibrium method and normalised the non-normal random variables with JC method in order to analyse the stability against deep sliding of the gravity dam's foundation with structural reliability theory. From the calculation results, it can be concluded that the conclusions drawn from the analysis of stability against deep sliding of a gravity dam are the same when using different methods: JC method, and the rigid body limit equilibrium method. Therefore, conclusions can be drawn that JC method can be used in analysing the deep anti-sliding stability of a gravity dam.

This study was financially supported by National Natural Science Foundation of China (51409207, 51309190), and Program2013KCT-15 for Shaanxi Provincial Key Innovative Research Team.

Table.1 Random variables statistical characteristics of gravity dam foundation deep anti-sliding stability

Variable	$f_1^{'}$	$c_1^{'}$/kPa	$f_2^{'}$	$c_2^{'}$/kPa
Mean value	0.55	450.00	0.76	500.00
Coefficient of variation	0.25	0.40	0.21	0.35
Standard deviation	0.15	120.00	0.16	180.00
Probability distribution type	Normal distribution	Lognormal distribution	Normal distribution	Lognormal distribution

References

1. Lan R.L. Reliability analysis of gravity dam stability against deep slide, *Design of Hydroelectric Power Station*, **1**, pp 26-30(2003)
2. Li S.Y and Kou X.Z. The reliability analysis for deep slide stability of gravity dam, *Journal of Hydraulic Engineering*, **1**, pp 24-27(1998)

3. Yang X.Z and Su Y. Reliability analysis on gravity dam stability against deep sliding, *Journal of Fuzhou University,* **2**, pp 80-83(2001)
4. Wang D and Chen J.K. Reliability research on sensitivity of concrete gravity dam to random variables, *Journal of Si Chuan University,* **4**, pp 1-5(2001)
5. Duncan J.M. Factors of safety and reliability in geotechnical engineering, *Journal of Geotechnical and Geoenvironmental Engineering*, **4**, pp 307-316 (2000)
6. Su H.Z and Wen Z.P. Interval risk analysis for gravity dam instability, *Engineering Failure Analysis*, **33**, pp 83-96(2013)
7. Ji Y Liu X.Q and Li T.C. Research on the Application of Radial Basis Function Neural Network to Structural Reliability Analysis of Gravity Dam, *Applied Mechanics and Materials,* **496**, pp 2505-2510(2014)
8. Zhang S.R Wang C and Sun B. Reliability analysis for multiple failure modes related sliding stability system between layers of gravity dam, *Journal of Hydraulic Engineering* , **44**, pp 426-434(2013)
9. Jiang S.H Hou J.G and He Y.M. Reliability analysis of deep sliding stability of gravity dams based on weighted regression response surface method, *Journal of Hydraulic Engineering,* **42**, pp 337-343(2011)
10. Xu Q Chen J.Y and Li J. Calculation method for system reliability of dam based on Bayes theory, Journal of Dalian University of Technology, 51, pp 84-89(2011)
11. Peyras L Carvajal C and Felix H. Probability-based assessment of dam safety using combined risk analysis and reliability methods-application to hazards studies European, Journal of Environmental and Cicil Engineering, 16, pp 795-817(2012)
12. Huang L.Z Li S.Y and Si Z. Research on Design Flood Risk Based on advanced checking-point method, Environmental Engineering and Management Journal, 13, pp 2119-2124 (2014)

Numerical simulation of pounding damage to caisson under storm surge

CHEN Yu[1,*]

[1]Key Laboratory of Harbor & Marine Structure Safety, Ministry of Communications, Tianjin Research Institute Water Transport Engineering M.O.T, Tianjin 300456, China

Abstract. In this paper, a new method for the numerical simulation of structural model is proposed, which is employed to analyze the pounding response of caissons subjected to storm surge loads. According to the new method, the simulation process is divided into two steps. Firstly, the wave propagation caused by storm surge is simulated by the wave-generating tool of Flow-3D, and recording the wave force time history on the caisson. Secondly, a refined 3D finite element model of caisson is established, and the wave force load is applied on the caisson according to the measured data in the first step for further analysis of structural pounding response using the explicit solver of LSDYNA. The whole simulation of pounding response of a caisson caused by "Sha Lijia" typhoon is carried out. The results show that the different wave direction results in the different angle caisson collisions, which will lead to different failure mode of caisson, and when the angle of 60 between wave direction and front/back wall is simulated, the numerical pounding failure mode is consistent with the situation.

1 Introduction

As the economic development in China, many large-scale offshore structures including wharf, artificial islands are growing rapidly in recent years. The Eastern and Southern Coasts in China are often attacked by typhoon, which leads storm surge attacking the offshore structures damage. In October 2016, the constructing caisson suffered pounding damage under the storm surge caused by "Sha Lijia" typhoon in southern China. Fig.1 illustrates the cracks occur in the middle and side caisson walls after "Sha Lijia" typhoon, and the pounding location of the concrete suffers serious damage.

(a)

(b)

Fig.1. Damaged caisson.

In order to explain the reason of the cracks, it needs to simulate the pounding mechanical behavior between the caissons under the storm surge. For the numerical simulation, firstly, it should get the wave force on the caisson. Morison Equation [1] or Radiation Wave Theory [2, 3] can solve wave force on the cylinder, but they are not suitable for the caisson. The literature [4] gives the static wave force distribution of the caisson, which can't be used to the pounding simulation. Because pounding simulation is based on the time depending, the wave force time history needs. Secondly, a detailed 3D finite element model of the caisson should be established to accurately simulate the pounding effects including high frequency and short duration on the caisson [5, 6]. Traditional simulation technology often needs to calculate the dynamic response of structures merging the

* Corresponding author: tjucy@tju.edu.cn

https://doi.org/10.1051/e3sconf/20183803046

wave and the structure model. The complexity and large calculating quantity made the numerical simulation impossible.

In this study, a new two-step method for the numerical simulation of pounding damage between two caissons under storm surge is proposed. The first step, fluid grids are established based on the FAVOR grid technology using high-precision fluid Flow software Flow-3D. The Continuity Equation and Navier-Stokes Equation are taken as the control equations, the coupling equations of pressure and velocity are solved by GMRES (Generalized Minimum RESidual) implicit algorithm, and then the wave force time history on the caisson structure can be calculated. The second step, a 3D finite element model of the caisson is established using LS-DYNA [7]. The time history of wave force obtained in the first step is input to the numerical simulation to calculate the pounding damage and failure process of the caissons using the LS-DYNA explicit solver. The method was applied to the pounding analysis of the collision under storm surge caused by "Sha Lijia" typhoon, and the relations between the wave direction and failure mode of the caisson is discussed.

2 Damage analysis of the caisson caused by "Sha Lijia" typhoon

2.1 Caisson details

Fig.2 shows the elevation view of the caisson. In order to clearly describe the caisson walls, each intersection point of two caisson walls is numbered. The front (C-D) and back (A-B) walls have the thickness of 400mm and 350mm respectively. The side (A-D, B-C) walls have the thickness of 300mm. All of the interior (h-f, e-g) wall thickness are 250mm. The compressive strength of the concrete used for the caisson is 40MPa. The steel reinforcement has an elasticity modulus of 206GPa and a yield strength of 400MPa, and the steel reinforcement ratio is 0.8%. The caissons are placed side by side on the foundation, and a 200 mm gap is introduced between the caissons.

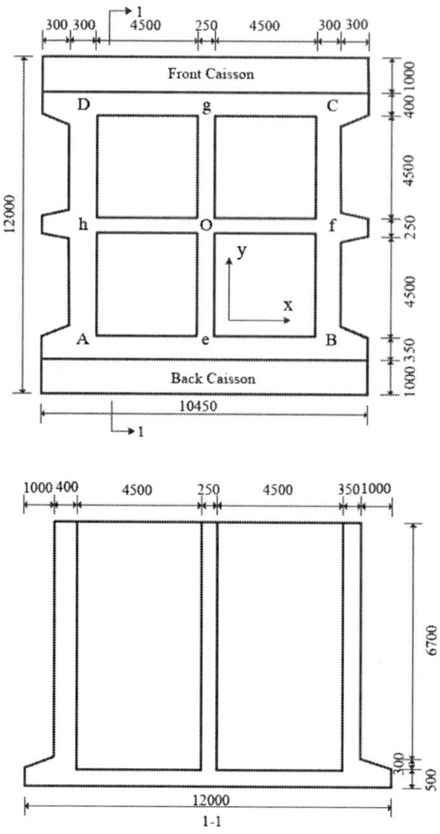

Fig.2. Caisson sizes (unit: mm).

2.2 Wave force on caisson

High-precision fluid Flow software Flow-3D [8] provides users solving the wave force on the caisson in three dimensions. In this study, three cases including the angles of 0, 30, 60 between wave direction and x direction. The x direction is defined as be parallel to the front/back wall. In the case of angle of 0 as an example, three-dimensional numerical flume model is described. As shown on Fig.3, the water channel has 12m high, 120 m long and 10m wide, and the angle between the wave direction and the front/back wall in the water channel is 0. Fig.4 shows the boundary conditions of the water channel. The front of the water channel is set to Wave boundary, the back is set to the Specified pressure boundary, the bottom is set to Wall boundary, and the rests are set to Symmetry boundary. The initial velocity of the water is set as zero, and the hydrostatic pressure along the z axis is set.

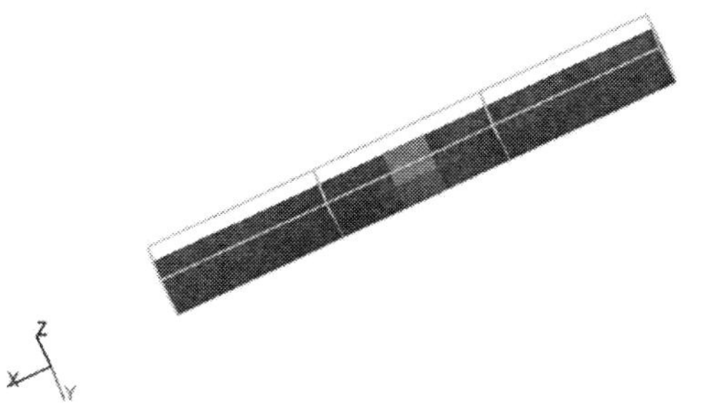

Fig.3. 3D numerical model of sink-structure.

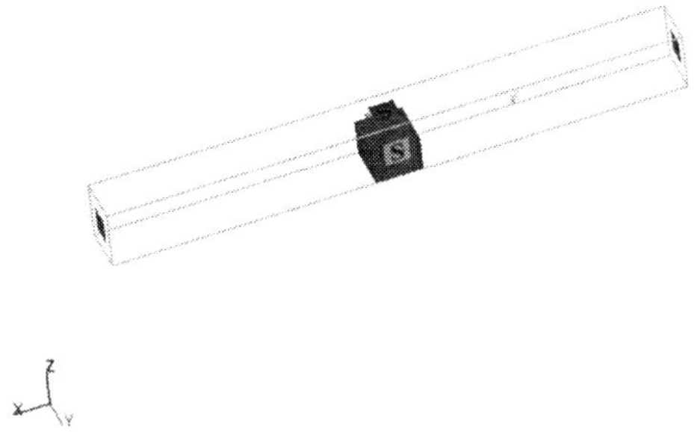

Fig.4. Sketch of boundary conditions.

According to meteorological data, the wind force of the "Sha Lijia" typhoon is 14. The significant wave height of 5.0 m and the average period of 8.8 s can be obtained corresponding to the 14 level of wind force in the literature [9]. According to the design data, the water depth is 7.35 m. The water depth, significant wave height and the average period are input to the Flow-3D wave system, the wave generation can be achieved as shown in Fig.5.

Fig.5. Wave generation.

Fig.6. Overtopping phenomenon.

Computation time is taken as 5 times of wave period (i.e., 45.5 s). Overtopping phenomenon occur after the wave acts on the caisson (see Fig.6). The wave force time history can be achieved by recording the wave pressure on the surface of the caisson. Similarly, the three-dimensional numerical flume models corresponding to the angles of 30, 60 between the wave direction and x direction are established. As shown on Fig.7, the maximum force is 2710kN, 3427kN and 3613kN corresponding to the angles of 0, 30, 60 respectively, and they occur in the first wave. As overtopping phenomenon occurs, the wave force is irregular after first wave.

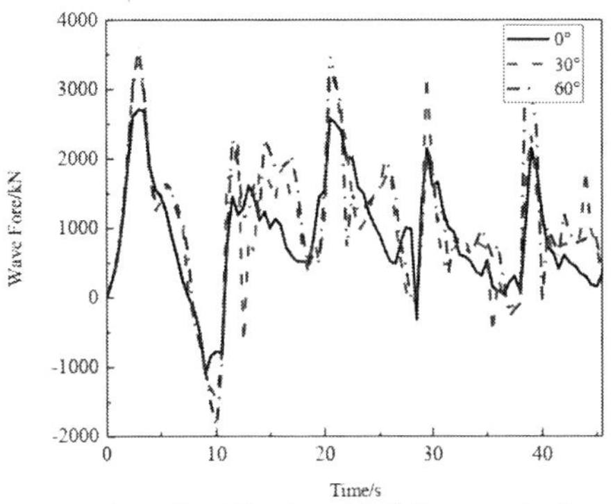

Fig.7. Wave force time histories with different angles between wave direction and x direction.

2.3. Finite Element Model

Two finite element models of 3D solid elements are established to simulate the pounding responses of the caisson using LS-DYNA, and the rigid foundation is set to simulate the rubble-bedding, as shown on Fig.8. In the finite element models, solid elements are used to simulate the behavior of concrete. The Belytschko beam elements are applied to simulate the mechanical properties of steel reinforcements. A perfect bond assumption is incorporated to model the connections between the steel reinforcements and the concrete. The 3D finite element model for the caisson is depicted in Fig.9. In order to improve calculative efficiency, the rigid model is used to simulate the concrete and steel reinforcements for the caisson B (see Fig.8), and the concrete damage model (MAT_CONCRETE_DAMAGE_REL3) and the elastic-plastic model (MAT_PLASTIC_KINEMATIC) are used to simulate the other caisson A, respectively. Table 1 tabulates the material properties adopted in this study.

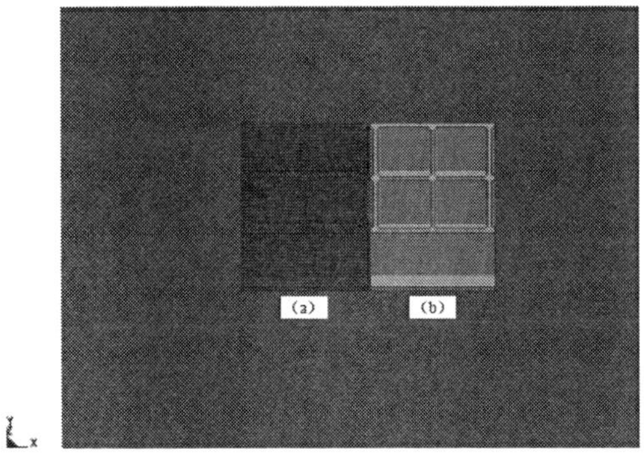

Fig.8. Model of caissons.

Fig.9. 3D finite element model of caisson.

The strengths of the structural materials are strain rate dependent as their dynamic properties can be enhanced significantly when subjected to impact loads [10, 11]. Current study employs the dynamic increase factor (*DIF*), a ratio of the dynamic to static strength against strain rate to account for the material strength enhancement with strain rate effect.The bilinear relationship developed by CEB code [12] and Malvar and Ross [13] are applied in this study for the concrete strength enhancement. In tension, the *DIF* of the tensile strength is given by the following equations [13]:

$$TDIF = \frac{f_{td}}{f_{ts}} = \left(\frac{\dot{\varepsilon}_d}{\dot{\varepsilon}_{ts}}\right)^{\delta} \quad \dot{\varepsilon}_d \le 1s^{-1} \tag{1}$$

$$TDIF = \frac{f_{td}}{f_{ts}} = \beta\left(\frac{\dot{\varepsilon}_d}{\dot{\varepsilon}_{ts}}\right)^{1/3} \quad \dot{\varepsilon}_d > 1s^{-1} \tag{2}$$

where f_{td} is the dynamic tensile strength at strain rate $\dot{\varepsilon}_d$, f_{ts} is the static tensile strength at $\dot{\varepsilon}_{ts}$ =$10^{-6}s^{-1}$, $\log\beta = 6\delta - 2$, δ =$1/(1+8f_c^{'}/f_{co}^{'})$, $f_{co}^{'}$=10MPa, $f_c^{'}$ is the static uniaxial compressive strength of concrete.

In compression, the equations are given by CEB code as follows [12]:

$$CDIF = \frac{f_{cd}}{f_{cs}} = \left(\frac{\dot{\varepsilon}_d}{\dot{\varepsilon}_{cs}}\right)^{1.026\alpha} \quad \dot{\varepsilon}_d \le 30s^{-1} \tag{3}$$

$$CDIF = \frac{f_{cd}}{f_{cs}} = \gamma(\dot{\varepsilon}_d)^{1/3} \quad \dot{\varepsilon}_d > 30s^{-1} \tag{4}$$

where f_{cd} is the dynamic compressive strength at strain rate $\dot{\varepsilon}_d$, f_{cs} is the static compressive strength at $\dot{\varepsilon}_{cs}$=30 × $10^{-6}s^{-1}$, $\log\gamma = 6.156\alpha - 0.49$, α =$(5+3f_{cu}/4)^{-1}$, and f_{cu} is the static cube strength.

The K&C model [14] is adopted to determine the *DIF* for the reinforcements which is given as

$$DIF = \left(\frac{\dot{\varepsilon}}{10^{-4}}\right)^{\alpha} \tag{5}$$

$$\alpha = 0.074 - 0.040 f_y / 414 \tag{6}$$

where f_y is the steel yield strength in MPa.

The penalty method is used to model the contact interfaces between different elements. The contact algorithm, i.e., CONTACT_AUTOMATIC_SURFACE_TO_SURFACE is used to simulate the contact between the caissons and the contact between the caisson and the foundation. By using this contact algorithm Coulomb friction needs to be specified, which are set to be 0.5 [15] and 0.6 [16] corresponding to the two contacts respectively.

Table 1. Material Parameters.

Material	Material model	Parameters	Value
Concrete	MAT_CONCRETE_DAMAGE_REL3	Mass density	2500kg/m3
		Compressive strength	40MPa
Steel reinforcements	MAT_PLASTIC_KINEMATIC	Mass density	7850kg/m3
		Young's modulus	206GPa
		Poisson's ratio	0.3
		Yield strength	400MPa
		Tangent modulus	20.6GPa

2.4. Input wave force time history

The three time histories of wave force obtained in the Section 2 corresponding to three cases of the angle of 0, 30, 60 are written to the model file in LS-DYNA. The wave forces are applied on the centre of gravity of the caisson B, and the caisson B move to collide with caisson A.

3 NUMERICAL RESULTS

The pounding damage of the caisson under storm surge is simulated in three cases using proposed method. Fig.10 shows the caisson damage modes in three cases including the angles of 0, 30, 60 between the wave direction and x direction. As shown on Fig.10, the pounding damage modes and extents of damage are not the same in the three cases. In the case of the angle of 0, Fig.10a shows that the side caisson walls (A-D, B-C) suffer serious damage, and have crack damage in the middle. The interior caisson walls (O-e, O-g) suffer serious damage, and interior walls (O-f, O-h) keep well. In the case of the angle of 30, Fig.10b shows that the side caisson walls (A-D, B-C) suffer serious damage. All of the interior caisson walls (O-e, O-g, O-f, O-h) suffer damage, and the damage extent of the walls (O-e, O-g) is relatively minor comparing to the walls (O-f, O-h). In the case of the angle of 60, Fig.10c shows that all of the front, back, side caisson walls (A-D, B-C, A-B, C-D) keep well, and the damage modes of the interior caisson walls are similar to the cases of the angle of 30. The damage extent of the interior walls (O-e, O-g) is relatively minor comparing to the case of the angle of 30, and the damage extent of the interior walls (O-f, O-h) is relatively serious comparing to the cases of the angles of 0, 30. This result occurs because the inertia force in x direction for the interior walls (O-e, O-g) is large if the angle of the collision between caisson B and caisson A is small result from the small angle between the wave and x direction. At the same time, the inertia force in y direction for the interior walls (O-f, O-h) is small. The large inertia force in x direction causes the damage in the middle and end of the wall (O-e, O-g). It is observed that the damage mode of the caisson in the case of the angle of 60 is consistent with the actual situation as shown in Section 2.1.

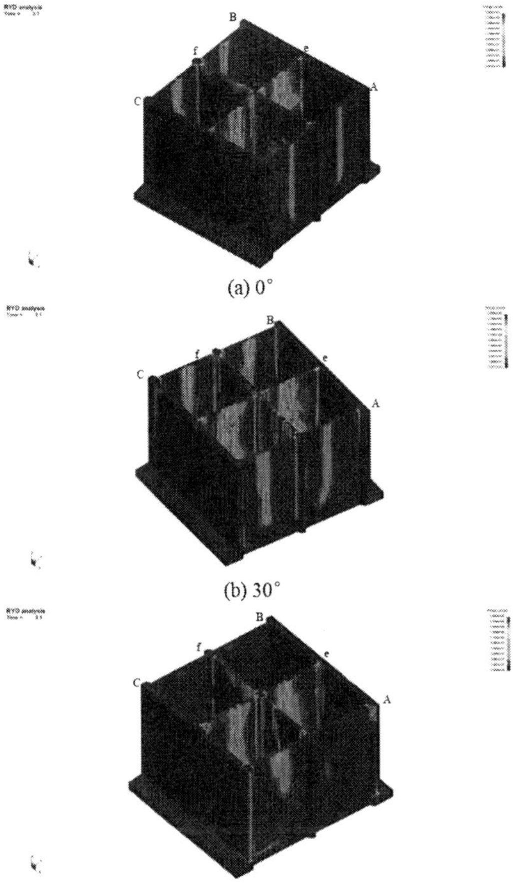

(a) 0°

(b) 30°

(c) 60°

Fig.10. Pounding failure mode of caisson under the different pounding angles between caisson B and caisson A.

4 CONCLUSION

A new method for the numerical simulation of structural model in two steps is proposed, which is employed to analyze the pounding response of caissons subjected to storm surge loads caused by "Sha Lijia" typhoon. The following conclusions are obtained.

(1)The different wave direction results in the different angle caisson collisions, and it will lead to different failure mode of caisson.

(2)When the angle of 60 between wave direction and front/back wall is simulated, the numerical pounding failure mode is consistent with the situation. The pounding failure mode is that all of the front, back and side caisson walls keep well, and the interior walls suffer damage in different extent.

ACKNOWLEDGMENT

The authors gratefully acknowledge the support for this research from the Central Public Welfare Science Foundation under Grant Nos. TKS170109 and TKS150204.

REFERENCES

1. Terro M J, Abdel-Rohman M, Terro M J. Wave induced forces in offshore structures using linear and nonlinear forms of morison's equation [J]. Journal of Vibration & Control, 2007, **13**(2): 139-157.
2. Bao W G, Fujihashi K, Kinoshita T. Interaction of a submerged elliptic plate with waves [J]. Journal of Hydrodynamics, 2010, **22**(5): 77-82.
3. Li Zhongxian, Huang Xin. Dynamic responses of bridges in deep water under combined earthquake and wave actions [J]. China Civil Engineering Journal, 2012, (11): 134-140.
4. JTS145-2015, Code of Hydrology for Harbour and Waterway [S]. Beijing: People's Communication Publishing Company, 2015.
5. Anxin Guo, Zhongjun Li, Hui Li. Point-to-Surface Pounding of Highway Bridges with Deck Rotation Subjected to Bi-Directional Earthquake Excitations[J]. Journal of Earthquake Engineering, 2011, **15**(2): 274-302.
6. Bi K, Hao H. Numerical simulation of pounding damage to bridge structures under spatially varying ground motions[J]. Engineering Structures, 2013, **46**(1): 62-76.
7. LS-DYNA. LS-DYNA user manual. Livermore Software Technology Corporation; 2007.
8. Zhang Ting. Three-dimensional numerical simulation of waves and its application [D]. Tianjin: Tianjin University, 2009.
9. Fang Zhongsheng, Jin Chengyi. Statistical analysis on wave height, wind speed and other parameters due to tropical cyclones in the northwest pacific area [J]. Journal of ship mechanics, 2003, **7**(5): 1-10. (in Chinese)
10. Shi Yanchao, Li Zhongxian, Hao Hong. Numerical analysis of progressive collapse of reinforced concrete frame under blast loading [J]. Journal of Pla University of Science and Technology, 2007, **8**(6): 652-658.
11. Bischoff P H, Perry S H. Compressive behaviour of concrete at high strain rates[J]. Materials and Structures, 1991, **24**(6): 425-450.
12. Comite Euro-International du Beton. Concrete structures under impact and impulsive loading. CEB Bulletin 187. Switzerland: Federal Institute ofTechnology Lausanne; 1990.
13. Malvar L J, Ross C A. A Review of Strain Rate Effects for Concrete in Tension[J]. Aci Materials Journal, 1998, **95**(6):735-739.

 https://doi.org/10.1051/e3sconf/20183803046

14. Malvar L J. Review of static and dynamic properties of steel reinforcing bars[J]. Aci Materials Journal, 1998, **95**(5):609-616.
15. JTS167-2-2009, Design and Construction Code for Gravity Quay [S]. Beijing: People's Communication Publishing Company, 2009.
16. Bi K, Hao H. Numerical simulation of pounding damage to bridge structures under spatially varying ground motions[J]. Engineering Structures, 2013, **46**(1): 62-76.

Urban Flood Prevention and Early Warning System in Jinan City

Shiyuan Feng, *Qingguo* Li *

School of Water Conservancy and Environment, University of Jinan, Shandong Province, 250022, China

Abstract. The system construction of urban flood control and disaster reduction in China is facing pressure and challenge from new urban water disaster. Under the circumstances that it is difficult to build high standards of flood protection engineering measures in urban areas, it is particularly important to carry out urban flood early warning. In Jinan City, a representative inland area, based on the index system of early warning of flood in Jinan urban area, the method of fuzzy comprehensive evaluation was adopted to evaluate the level of early warning. Based on the cumulative rainfall of 3 hours, the CAflood simulation results based on cellular automaton model of urban flooding were used as evaluation indexes to realize the accuracy and integration of urban flood control early warning.

1 Introduction

With the development of urbanization, the system of urban flood disaster is changing constantly. For example, the change of urban underlying surface causes the change of urban runoff [1]; the continuous expansion of urban land use has led to increased pressure on the drainage system [2]; the continuous increase of population and property in the built-up area has raised the vulnerability facing floods and other disasters. The "7.18" incident in Jinan in 2007 and the "7.21" incident in Beijing in 2012 have caused serious losses to people's lives and property [3].

Therefore, it is a commonly used and effective method to carry out flood control and early warning for preventing and controlling floods in urban areas. When more serious floods occur, emergency measures for prevention and control of floods are taken. In order to seek a positive and effective coping strategy, many scholars and management agencies have done a great deal of research and practice in the field of flood control and early warning [4]. There are a lot of experience and methods to be learnt from in flood control and early warning mode, early warning index system and the classification of early warning. Based on the analysis of the representative flood control and early warning technology at home and abroad, comparing and drawing on experiences from the early-warning model which has good effect and universality, this paper studies the construction of urban flood control early warning system in urban area of Jinan City.

2 Index System and Comprehensive Evaluation Model of Flood Warning in Jinan City

Early warning is given based on established or promulgated early warning modes. This work generally evaluates the warning level comprehensively according to the early warning indicators. The index system of early warning of flood in urban area of Jinan is the basis for flood control and early warning. Only an integrated early-warning index system which can meet the actual situation of Jinan and reflect the urban police intelligence under different conditions can be established to make accurate early warning based on real-time early warning.

First of all, we need to determine the classification of early warning mode, and then grade the various early warning indicators proposed above. According to the multistage response mechanism of national public emergency, this study divided each evaluation index into 5 levels on the basis of identifying early warning indicators, corresponding to 5 levels of response measures respectively.

2.1. Warning indicator system of Jinan urban flood

Evaluation level.

(1) Grade I: severely affected, heavy losses from flooding or serious traffic congestion or casualties.

(2) Grade II: moderately affected, serious flooding loss or serious impact but no casualties.

(3) Grade III: mildly affected, inconvenient for residents to travel, poor traffic, and slight loss in some areas.

(4) Grade IV: safety, there is water accumulation in the pavement, but the depth is shallower, so no disaster loss has occurred.

(5) Grade V: safe, no alarm or alarm clear.

Indicator system.

* Corresponding author: Qingguo Li@ffengshiyuan@126.com

https://doi.org/10.1051/e3sconf/20183803047

Determine the threshold for dividing each indicator into four levels. The indicators and indicator value standards in the city-wide unified are as follows.

(1) 3h (3 hours) cumulative rainfall (P3h)

The main driving force of urban floods is mostly heavy rainfall. Although storm surge in some coastal areas in China may also cause urban floods, it is only a case and the storm surge index is not easy to be quantified. Therefore, it is usually indicated by rainfall over a certain period of time to indicate urban floods early warning of the driving force. According to our experience in urban flood protection, generally when 3h continuous rainfall reaches a certain value, it will induce urban floods, so 3h cumulative rainfall as an indicator of urban flood warning can be used in Jinan. Of course, it is subjective to adopt the early warning index of accumulative rainfall for 3 hours. According to the actual situation in Jinan, 6h and 12h cumulative rainfall can be used in some regions for higher warning level.

The 3h accumulated rainfall, P3h, is divided into 5 levels, with reference to the current method for classifying storm rains in the domestic meteorological department. The P3h is divided according to the following criteria:

P3h <30mm, level V;

30mm≤P3h<50mm, level IV;

50mm≤P3h<100mm, level III;

100mm≤P3h<150mm, level II;

P3h≥150mm, level I.

(2) Submerged depth (Hn)

Submerged depth and range are important indicators for describing the status of floods and the submerged depth of water is the most important basis for determining whether floods occur. Under the condition that the underlying surface is invariant, the submerged depth is completely related to the submerged area. Therefore, according to the principle of appropriate number of indicators selected by evaluation index, only the depth of flood inundation is taken as the early warning indicator to describe flood disaster status.

In flood control and early warning, submerged depth, Hn, is a key indicator, which is closely related to the disaster area and disaster losses. The flowing grading standards are proposed:

Hn<20cm, level V;

20cm≤Hn<50cm, level IV;

50cm≤Hn<100cm, level III;

100cm≤Hn<200cm, level II;

Hn≥200cm, level I.

(3) Flow rate index (Vn)

Flood velocity is also a major indicator that describes the status of floods. It plays a certain role in promoting floods and is closely related to the extent of disaster and the loss of floods. Therefore, in flood control and early warning, flow rate (Vn) of the key sections (road nodes, overpasses, etc.) also is an important indicator of early warning. Referencing to the division of disaster bearing capacity of flow indicators and urban traffic, life and construction of esidents in Saint Martin Island flood disaster management research, the following grading standards are proposed:

Vn<0.5m/s, level V;

0.5m/s≤Vn< 2.0 m/s, level IV;

2.0 m/s≤Vn<3.0m/s, level III;

3.0m/s≤Vn<4.0m/s, level II;

Vn≥4.0m/s, level I.

(4) Index of Flood Control Integrated Management Efficiency (λ)

Efficiency coefficient of integrated flood control is not easy to be quantified, because there are no feasible methods to determine it with objectivity. In specific works, we can combine the practical experience of grass-roots staff to conduct a comprehensive evaluation. According to whether it has built and come into use the decision support system of three defenses and it has been tested by practice, flood control contingency plans for the operational situation, the construction of urban drainage network system, the effectiveness of urban hedge, flood management efficiency coefficient λ is divided into the following levels.

λ≥0.9, level V, Basically no effect;

0.7≤λ<0.9, level IV, Slight affected;

0.5≤λ<0.7, level III, Mild effect;

0.3≤λ<0.5, level II, Moderate effect;

λ<0.3, level I, Severe effect.

In the actual early warning work, all or part of the above indicators can be selected for early warning evaluation based on the obtained data and information, meteorological hydrology and socio-economic characteristics of different sub-regions. The standard division for flood early warning in urban area of Jinan City is detailed in Table 1.

Table 1. Warning early warning criteria division of Jinan urban area

icon	**threshold indicator of early warning**
Severe Flood Warning	P3h≥150mm; Hn≥200cm; Vn≥4.0 m/s; λ<0.3.
Flood Warning	100mm≤P3h<150mm; 100cm≤Hn<200cm; 3.0 m/s≤Vn<4.0 m/s; 0.3≤λ<0.5.
Flood Alert	50mm≤P3h<100mm; 50cm≤Hn<100cm; 2.0 m/s≤Vn<3.0 m/s; 0.5≤λ<0.7.
Flood Alert	30mm≤P3h<50mm; 20cm≤Hn<50cm 0.5 m/s≤Vn<2.0 m/s 0.7≤λ<0.9.

 No Alert	P3h<30mm; Hn<20cm Vn<0.5 m/s λ≥0.9.

2.2 Fuzzy comprehensive evaluation method for early warning level

Start-up of flood warning level in a city can not be activated only according to a certain indicator that reaches a certain warning level. We need to make a comprehensive analysis of all the indicators in order to give a warning level that truly reflects floods and floods in urban areas and provide more precise warning information for flood control departments so as to ensure the correct and efficient operation of urban flood control. Therefore, it is one of the key technologies for early warning of urban floods to establish a comprehensive evaluation model and comprehensively analyze each independent early warning index so as to give the warning level. In this paper, the method of fuzzy evaluation method is used to establish the fuzzy comprehensive evaluation model method of Jinan with the combination of subjectivity and objectivity.

The evaluation results are divided into a certain level, taking this study as an example, Grade I、 Grade II、 Grade III、 Grade IV、Grade V. The formulas are as follows. There are n precautionary zones, m precautionary indicators in the zone of the measured value x and then the evaluation indicator matrix can be obtained.

$$X = \begin{bmatrix} x_{11} & x_{12} & \cdots & x_{1n} \\ x_{21} & x_{22} & \cdots & x_{2n} \\ \vdots & \vdots & \vdots & \vdots \\ x_{m1} & x_{m2} & \cdots & x_{mn} \end{bmatrix} = (x_{ij}) \quad (1)$$

According to the index number, m, the standard value of the evaluation index is y. m indicators c-level standard value matrix is as follows.

$$Y = \begin{bmatrix} y_{11} & y_{12} & \cdots & y_{1c} \\ y_{21} & y_{22} & \cdots & y_{2c} \\ \vdots & \vdots & \vdots & \vdots \\ y_{m1} & y_{m2} & \cdots & y_{mc} \end{bmatrix} = (y_{ih}) \quad (2)$$

Here , $i = 1,2,\ldots,m; j = 1,2,\ldots,n; h = 1,2,\ldots,c$,According to the above Jinan flood warning mode, the warning level is divided into *5* levels, so $c = 5$.

Apply Formulas (3), (4):

$$r_{ij} = \frac{x_{ij} - y_{ic}}{y_{i1} - y_{ic}} \quad (3)$$

$$s_{ih} = \frac{y_{ih} - y_{ic}}{y_{i1} - y_{ic}} \quad (4)$$

Normalize the matrix X and the matrix Y to obtain the corresponding normalized precalculation index normalization matrix and precautionary classification standard normalized matrix (5), (6).

$$R = \begin{bmatrix} r_{11} & r_{12} & \cdots & r_{1n} \\ r_{21} & r_{22} & \cdots & r_{2n} \\ \vdots & \vdots & \vdots & \vdots \\ r_{m1} & r_{m2} & \cdots & r_{mn} \end{bmatrix} = (r_{ij}) \quad (5)$$

$$S = \begin{bmatrix} s_{11} & s_{12} & \cdots & s_{1c} \\ s_{21} & s_{22} & \cdots & s_{2c} \\ \vdots & \vdots & \vdots & \vdots \\ s_{m1} & s_{m2} & \cdots & s_{mc} \end{bmatrix} = (s_{ih}) \quad (6)$$

The weight of each of the early warning indicators is different, according to the previous method of determining the weight of indicators, here assume that the index weight vector is as fllows.

$$\omega = (\omega_1, \omega_2, \ldots \omega_m) \text{ and meet } \sum_{i=1}^{m} \omega_i = 1 \quad (7)$$

According to the theory of relative fuzzy comprehensive evaluation, this study gives the fuzzy comprehensive evaluation model of the warning level:

$$u_{hj} = \begin{cases} 0, & h < a_j, h > b_j \\ \dfrac{1}{\sum_{t=a_j}^{b_j} \left\{ \dfrac{\sum_{i=1}^{m} [\omega_i (r_{ij} - s_{ih})]^p}{\sum_{i=1}^{m} [\omega_i (r_{ij} - s_{it})]^p} \right\}^{\frac{2}{p}}}, & a_j \leqslant h \leqslant b_j, d_{hj} \neq 0 \\ 1, & d_{hj} = 0 \end{cases} \quad (8)$$

Here, the relative membership degree, uhj, that indicates that the alarm of the j-th pre-alarm zone of the system belongs to the relative membership degree of flood warning level h. *P* is the distance coefficient, $p = 1$ in this study, which means the generalized weighted distance.

Thus, according to uhj, the final warning level of partition j can be obtained.

$$h_j = (1, 2, 3, 4, 5) \cdot (u_{1j}, u_{2j}, \cdots, u_{5j})^T \quad (9)$$

Here, h_j indicates which level I~V partition *j* belongs to, namely its integrated warning level.

3 Simulation and Early Warning of Rain Flood in Jinan City

On the basis of reasonable set of the early warning and starting criteria beforehand, according to the simulation scheme of standard design flood at all levels, the rain-flood simulation model is used to simulate the design scheme, and the warning and response of each scheme

 https://doi.org/10.1051/e3sconf/20183803047

design flood scenario are studied based on the calculation results. The validity and rationality of Jinan urban flood early warning indicator system and comprehensive grading evaluation method are tested and need to be demonstrated further. The simulation model of urban rainstorm uses Caflood[5,6], a two-dimensional flood simulation software of Water Center of Exeter University. By using the established flood simulation model, the four-color rainstorm warning corresponding flood simulation plan, the historical typical storm and the floodplain simulation plan corresponding to the municipal design rainstorms with different recurring periods were made. Flood simulation of the flood process is dynamic and continuous, and in order to show the most important feature of the entire submergence process, the representative moment is intercepted. The relevant warning level is shown in Figure 1.

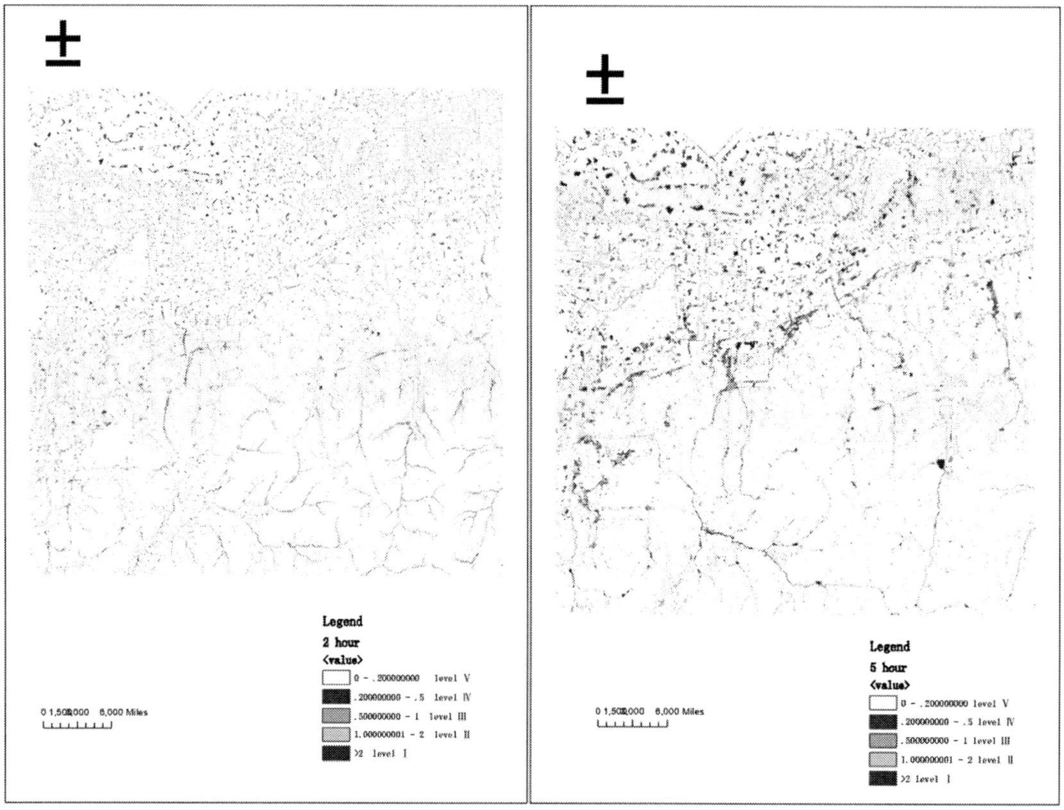

Fig. 1. 3h-50mm Rainstorm warning simulation

From the simulation results, the urban area began to accumulate water two hours after the rainfall for 2 hours. The condition of water accumulation was not very serious. Most of the areas were level Ⅴ-Ⅳ warnings and no level Ⅰ warning areas. Urban water when the most serious 5 o'clock occurred. The key flood areas is in the central and northern regions, which also confirms the argument of " South floods North Waterlogging" after rainstorm in Jinan.

4 Conclusion

The application of urban desert simulation technology for flood early warning research can provide detailed flood risk information for urban flood control and disaster relief and provide accurate information for urban flood control organizations to carry out flood control and emergency management of zoning, time-sharing, classification and grading reliable flood warning information. It can be seen from the simulation results that the fuzzy comprehensive evaluation method has the ability to refine warning waterlogging, and the results of early warning are in accordance with the actual situation.

References

1. B Shi, D Wan, et al, Water Resources and Power, 06:57-60 (2014)
2. N Liu, X Bing, et al, Water Resources and Power, 06:48-52 (2016)
3. G Hou, ShanDong University, (2012)
4. Y Chen, H Zhang, G Du, CHINA FLOOD & DROUGHT MANAGEMENT, **25**;4:25-32 (2015)
5. M Guidolin, A S Chen, B Ghimire, et al. Environ. Modell. Softw. 84:378-394 (2016)
6. R J Austin, A S Chen, D A Savić, et al. J. Hydroinform. **16**;6: 1359-1374 (2014).

Numerical Simulation Study of Booming Effect in Fast Currents of Inland River

Rongchang Chen[1,*], *Chen* Liu[1], *Xiaofeng* Luo[2] and *Wei* Shen[1]

[1]China Waterborne Transport Research Institute, Beijing, 100088, P.R. China
[2]Nanjing Hydraulic Research Institute, Nanjing, 210029, P.R. China

Abstract. In the downstream tidal section of the Yangtze River, nine kinds of combinations of hydrological environmental conditions are considered, including the annual average runoff flow, the annual average peak flow and the flood control design flow, as well as the three conditions of spring, medium and neap tides. By means of the numerical simulation method, the effective performance parameter values for conventional intercepting boom under different environmental conditions are obtained by simulating 9 kinds of maximum current speed to withstand, Max.CS, respectively. The results show that, in the downstream fast current tidal section of the Yangtze River, for the boom performance index of Max.CS, the relatively extensive applicability value should be 3.0kn under the condition of the annual average runoff flow; 4.0Kn should be selected under the condition of the annual average peak flow; and 4.5Kn should be selected under the flood control design flow. This study can provide technical support for the design, selection and use of booms in downstream waters of the Yangtze River.

1 Introduction

A marine oil spill accident from ship would severely pollute water environment and water resources due to the polluting characteristic of ship fuel oil. Compared with the coastal waters, inland waters are relatively closed and most are used for drinking water sources, so inland waters are far heavily affected by oil spill accidents in the damages to ecological environment and negative social impacts. For example, the oil spill of thousands of tons from the explosion of the tanker Jiangyou 423 in Qixia, Nanjing, on 3 June 1997, caused heavy losses to the ecological environment of the waters [1].

Booms for oil spill are major measures for emergency response in inland waters. The functions of booms in emergency response can be mainly divided into booming interception and diversion. The boom deploying schemes can be divided into wrapping, dragging, inducing, intercepting, V-shaped deploying and other methods. The interception effects of the boom depend mainly on whether the performance parameters of the boom suite for the environmental conditions of the waters. At present, the type specifications and technical parameter designs of the conventional booms are mainly based on the coastal environmental conditions. The booms for the inland waters can usually select from the types of the coastal booms directly. So, there are not yet conventional booms fully applicable to tidal section of inland rivers that can work effectively in fast currents.

There are more than 25000 operating berths in inland waterway of China, including more than 400 berths of ten thousand tons and above. Daily average standard ship flow in the main channel of the Yangtze River is more than 600 ships in a year. With the increase of ship traffic volume and the development of large-scale ships, the risk of oil spill accidents trend to increase in the Yangtze River. Therefore, it is of great significance to study the oil spill emergency response in the Yangtze River and other inland waters, especially the booming effects.

2 Analysis on performance indexes of booms

The technical performances and basic quality requirements of the oil booms are specified in *Oil Boom* (JT/T 465-2001), a transport industry standard, which is suitable for the solid float, fencing, external tension, inflatable, shoal, fireproofing and other conventional types of oil booms. The waters in the environmental conditions of the waters used can be divided into four categories of calm water, calm water-significant current, protected water and open water according to the wave height and water current speed. See Table 1.

Table 1. Division of water types.

Water	Wave height and current speed
calm water	Wave height 0-0.3m or current speed < 0.4m/s (0.78Kn)
calm water-significant current	Wave height 0-0.3m and current speed ≥ 0.4m/s (0.78Kn)

* Corresponding author: chenrongchang@wti.ac.cn

 https://doi.org/10.1051/e3sconf/20183803048

protected water	Wave height 0-1m
open water	Wave height 0-2m or wave height >2m

Commonly used booms include solid float, solid float PVC, fast deploying and other types, of which the solid float boom with wrapping material of rubber has relatively stronger current speed to withstand. The booms currently being provided for terminals handling dangerous cargo in the Yangtze River downstream are mainly composed of solid float rubber booms and solid float PVC booms. The total height of the solid float boom used in inland waters is generally 600-1100mm, and the boom with a total height of 1200mm or above is mostly used in coastal and open sea areas with heavy stormy waves. For parameter series of maximum current speed to withstand (Max.CS), maximum wave height to withstand (Max.WH), maximum wind speed to withstand (Max.WS) and overall height (OH) of the conventional rubber solid float boom, see Table 2. For the relationship between the maximum current speed to withstand and the overall height of the conventional solid float boom, see Figure 1.

Table 2. Main performance parameters of conventional rubber solid float boom

No.	Model	Max.CS (Kn)	Max.WH (m)	Max.WS (m/s)	OH (mm)
1	WGJ600	1.5	1.5	10	600
2	WGJ800	2	2	15	800
3	WGJ1000	2.5	2	15	1000
4	WGJ1100	2.5	1.5	16	1100
5	WGJ1200	3	2.5	20	1200
6	WGJ1300	3	2	18	1300
7	WGJ1400	3	3	20	1400
8	WGJ1500	3	2.5	20	1500
9	WGJ1600	3-4	3	20	1600
10	WGJ1800	3-4	3.5	25	1800
11	WGJ2000	3-4	3	22	2000

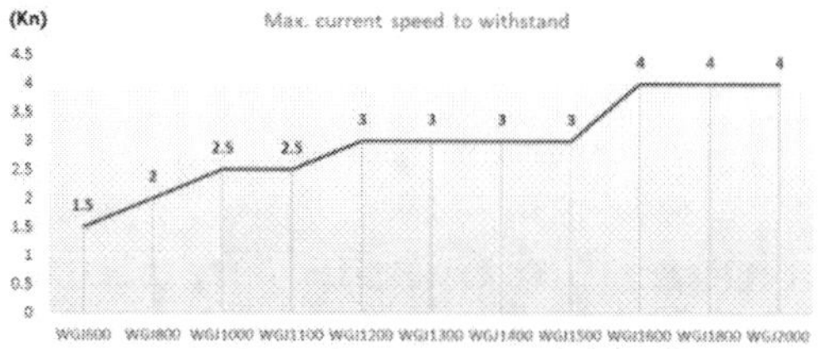

Figure 1. Relationship between Max.CS and OH of conventional boom

The current speed in inland river tidal section is affected not only by the changing runoff, but also by the tides. At the ebb tide, fast currents caused by the superposition effect would have a significant influence on the deploying and booming effect of the oil booms. Maximum effective current speed is between 1.5-6 kn, i.e. 0.77m/s -3.1m/s, according to the evaluating results of various types of oil booming technologies by USCG [2]. No dividing the current speed above 0.4m/s (0.78kn) in the industrial standard of Oil Boom will cause confusion to the type selecting of the booms in fast currents inland rivers affected by tides.

By numerical simulation method, the downstream sections of the Yangtze River affected by strong tides was selected to study the booming effect under various environmental conditions, which includes different combination of Max.CS and current states. The adopted values of Max.CS are 0.78Kn, 1.5Kn, 2.0Kn, 2.5Kn, 3Kn, 3.5Kn, 4Kn, 4.5Kn and 5Kn.

3 Oil spill simulation scene analysis

3.1 Hydrodynamic simulation of the Yangtze River downstream

The seasonal distribution of runoff in Datong of the Yangtze River was counted, and the frequency of open seas tidal range was analyzed. On this basis, a large-scale two-dimensional mathematical model of tidal current from Datong to the Yangtze River Estuary was established. Simulation results of tidal currents in various scenes combining different runoffs and tidal range were calculated. In the simulation of booming effect, 9 kinds of currents scenes are made up, including the annual average runoff flow, the annual average peak flow and the flood control design flow [3], as well as the 3 conditions of spring, medium and neap tides in coastal observing station. See table 3. The static wind condition was used for the simulation.

Table 3. Simulated flow scenes

Scene name	Runoff in Datong (m^3/s)	Tide (m) in coastal observing station
L1	28500	3.9
L2	28500	2.5
L3	28500	1.2

L4	57500	3.9
L5	57500	2.5
L6	57500	1.2
L7	85400	3.9
L8	85400	2.5
L9	85400	1.2

3.2 Scene analysis of oil spill accident

3.2.1 Oil spill accident location

The downstream tidal section of the Yangtze River was selected as the simulated oil spill accident location. The high risk area of oil spill was determined by the ship Automatic Identification System (AIS for short). Composed of shore-based (base station) facilities and shipboard equipment, AIS is a digital navigation aid system and equipment which integrates network technology, modern communication technology, computer technology and electronic information display technology. Along with GPS, AIS can provide ship static data including ship name, call sign, MMSI, ship type, ship length and width, as well as ship dynamic data including ship's longitude & latitude, heading, track and speed. The AIS activity status of the ships in the studying river section on one day in November 2017 is shown in Figure 2. It can be seen that the ships on Zhangjiagang north side waters (within the orange circle in the figure) are grouped into one route from two routes of the north and the south, and the intersection of waterways appears, which is a high risk area of ship traffic accidents and oil spill accidents. In the numerical simulation of this study, the central point of the high risk area is taken as the ship oil spill accident location to be simulated.

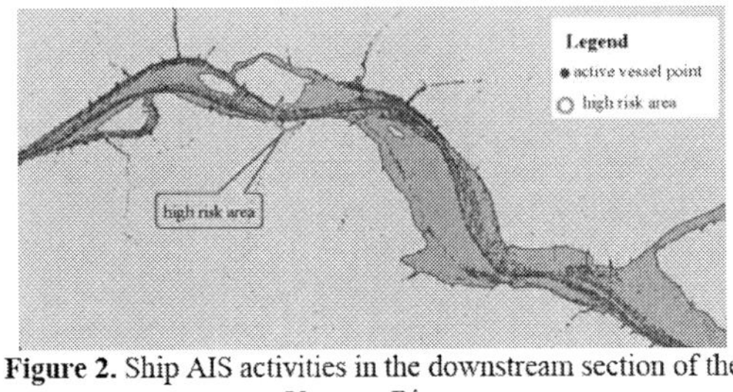

Figure 2. Ship AIS activities in the downstream section of the Yangtze River

3.2.2 Oil spillage

Extra serious and major oil spills on water are usually caused by transport ships carrying crude oil and its products. Leakage of ship fuel is oil spill accidents in ships with the highest frequency. This study assumes that a ship pollution accident occurs in the downstream water of the Yangtze River, and simulates the booming effect. A 10000 gross ton ship is generally equipped with 6 fuel tanks, with total fuel of 991.5 tons [4]. If a single tank leaks out, possible oil spillage in the accident is 165 tons.

3.3 Numerical simulation method of oil spill trajectory and fates

Oilmap is an oil spill trajectory and fates model developed by Applied Science Inc., USA. The trajectory model of Oilmap generalizes spilled oil into oil particles having mass and each oil particle represents a proper fraction of total oil spillage. Joint action of wind force, water current, wave and gravity current on floating oil is considered in the oil slick drift algorithm applied in the model, and the current-driving process is calculated with Lagrange particle tracking while the dissipation process is calculated with the Random Walk method. For the particle's drifting speed equation, see Equation (1) [5].

$$U_{oil} = U_w + U_t + U_r + \alpha U_e + \beta U_p \qquad (1)$$

Where, U_w is the speed component generated due to action of wind and wave, in m/s; U_t is the speed component due to the action of water current, in m/s; U_r is the speed component due to residual currents (e.g. gravity current), in m/s; U_e is the speed component due to action of Ekman flow, in m/s; U_p is the speed component due to action of squirt flow, in m/s; α is 0 for floating particles and 1 for underwater particles; β is 0 for non-squirt spill and 1 for squirt spill.

The fates model of Oilmap is used to calculate the results of oil spill weathering process which includes extension, evaporation, water carrying, emulsion and shoreline adsorption. Following the Law of Mass Conservation, the calculation process covers oil spills on water surface, in water bodies and substrates, in atmosphere, those absorbed on shorelines, and those manually contained and removed.

4 Numerical simulation result analysis

4.1 Simulation results without booming interception measures

The trajectory of pollutants after the oil spill accident under the conditions of L1 scene and static wind was simulated. Under the action of tide and runoff, the pollutants drift downstream rapidly at ebb tide; and the pollutants drift upstream at rising tide. In general, pollutants drift towards downstream under the action of the upstream runoff and tides and contaminate the northern shorelines.

48 hours after the accident, 1% of the leaked pollutants will volatilize into the atmospheric, 49.6% will float on the water surface and continue to move downstream, and 49.4% will be adsorbed on the shoreline and cause pollution. Pollutant drift trajectories 3 hours and 24 hours after the accident are shown in

 https://doi.org/10.1051/e3sconf/20183803048

Figures 3 and 4. The oil spill fate curve 48 hours after the accident is shown in Figure 5.

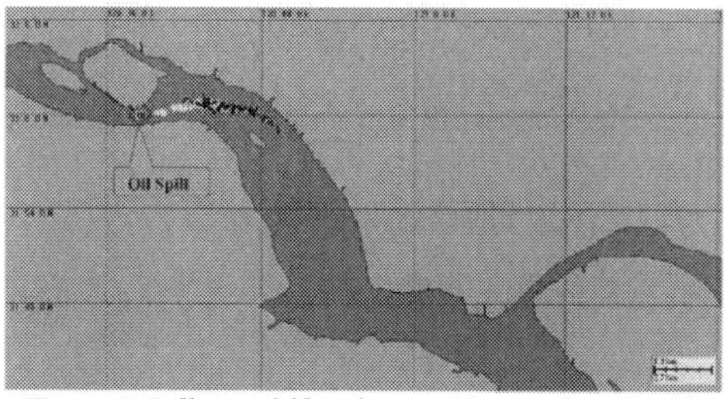

Figure 3. Pollutant drift trajectory 3 hours after oil spill accident.

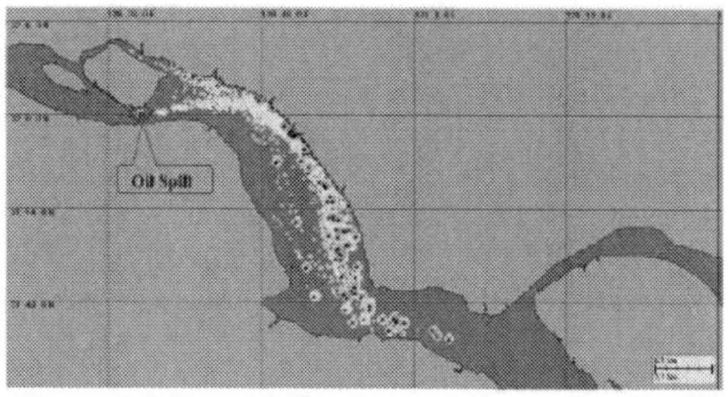

Figure 4. Pollutant drift trajectory 24 hours after oil spill accident.

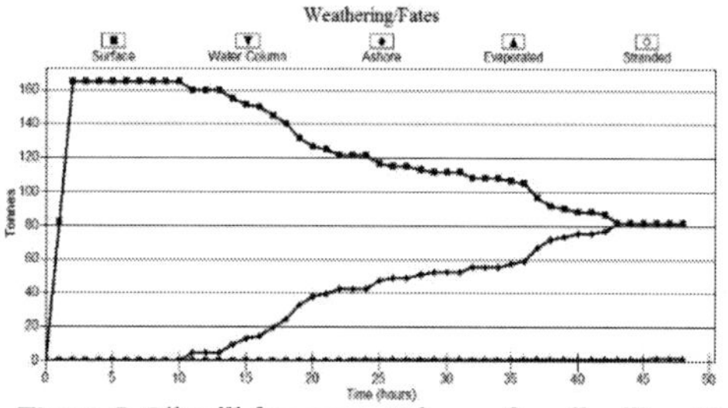

Figure 5. Oil spill fate curve 48 hours after oil spill accident.

4.2 Simulation results of booming effects of different Max.CS

The trajectory of pollutants after the oil spill accident under the conditions of L1 scene and static wind was simulated. According to the trajectory in Figure 3, deploying booms 3 hours after the accident was simulated. Because the influences of Max.CS on the booming effect are mainly investigated, the full-interception boom deploying scheme with both sides of a boom fixed on the shore was adopted in order to eliminate interference of other parameters.

The simulation results show when Max.CS of the boom is 3.0 Kn under the L1 scene, the booming failure appears 15 hours after the accident, as shown in Figure 6. When Max.CS is 3.5Kn or above, the booming is very effective, as shown in Figure 7.

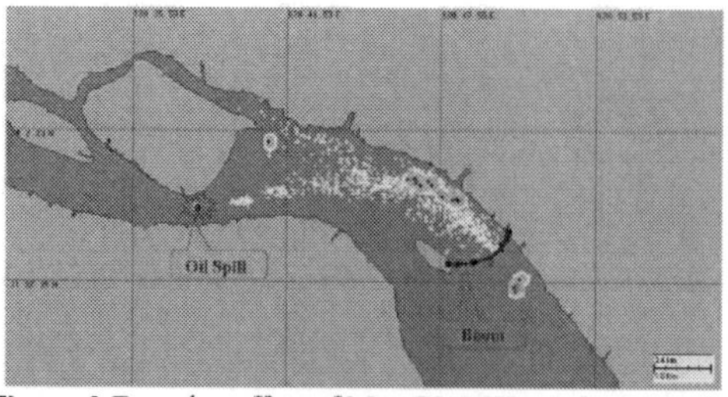

Figure 6. Booming effect of Max.CS 3.0Kn under L1 scene.

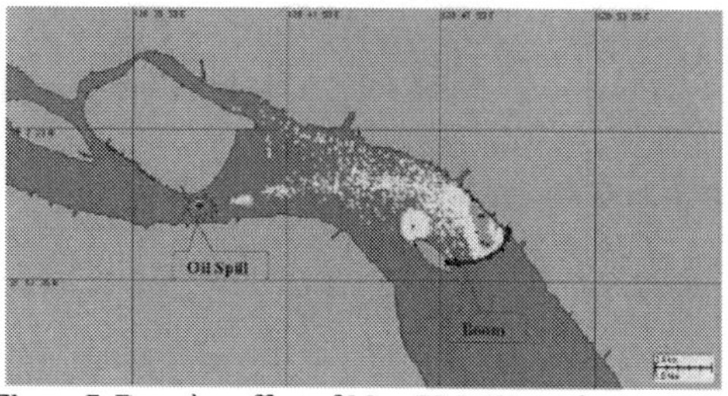

Figure 7. Booming effect of Max.CS 3.5Kn under L1 scene.

4.3 Result analysis and recommendations

In this study, the booming effects of 9 Max.CSs under 9 current scenes were simulated (see Table 4). The results are analyzed as follows:

(1) Under the conditions of annual average runoff flow, annual average peak flow and flood control design flow, the boom Max.CS should be no less than 2.5Kn, up to 5.0Kn. 3.0kn with relatively extensive applicability should be chosen under the condition of the annual average runoff flow; 4.0Kn should be chosen under the condition of the annual average peak flow; and 4.5Kn should be chosen under the flood control design flow.

(2) Normally, total height of the boom with 3.0Kn is 1200-1500mm, and total height of the boom with 4.0Kn is 1600-2000mm. In practice, the boom with total height more than 1200 mm is mostly used in coastal and open sea waters with heavy stormy waves. Total height of the boom designed for use in inland waters is generally less than 1200 mm. At present, the conventional boom types are short of fast current resistant types with suitable total height.

(3) At present, there is no conventional boom meeting the requirement of Max.CS not less than 4.5Kn. When an oil spill accident occurs under the condition of flood control design flow, most booms for oil spill interception will fail.

Table 4. Simulation results

Scene name	Max. CS (Kn)								
	0.78	1.5	2.0	2.5	3.0	3.5	4.0	4.5	5.0
L1	×	×	×	×	×	√	√	√	√
L2	×	×	×	×	√	√	√	√	√
L3	×	×	×	√	√	√	√	√	√

L4	×	×	×	×	×	×	√	√	√
L5	×	×	×	×	×	×	√	√	√
L6	×	×	×	×	√	√	√	√	√
L7	×	×	×	×	×	×	×	×	√
L8	×	×	×	×	×	×	×	√	√
L9	×	×	×	×	×	×	×	√	√

According to the above simulation results, this study puts forward the following recommendations on the design selection and use of the oil booms: (1) For the design selection of the booms applicable to fast current section of inland rivers tidal affected, the key performance parameter considered should be the maximum current speed to withstand; (2) In the standard of Oil Boom, the range of the boom maximum current speed to withstand should be expanded, and the selectable range of the parameter value should be given according to different water environments, with maximum value no less than 5Kn; (3) The development of both high current speed resistant and easy deploying with total height less than 1000 mm should be encouraged; (4) In the boom deploying, the capacity to resist fast current is increased by adjusting the angle, anchor position and balance weight of the deployed boom; (5) For different fast current sections of the inland river tidal affected, numerical simulation of the booming effect should be carried out before type selection to prevent the boom failure.

5 Conclusions

This study simulated 9 kinds of maximum current speed to withstand, Max. CS, of the conventional booms under 9 different combinations of runoffs and tidal states in the downstream waters of the Yangtze River by means of the numerical simulation method, and obtains the boom intercepting effectiveness under different environmental conditions.

The risks of oil spill accidents tend to increase with the growing traffic volume and ships in the downstream waters of the Yangtze River, so the studies on oil spill emergency responses are of importance. Due to fast current, complex waters, obvious effect by tides and other characteristics of the Yangtze River downstream waters, the correct selection of the oil boom, especially the performance parameter of maximum current speed to withstand, is playing a great role in enhancing the effects of the emergency responses. This study provides technical support for the design, selection and apply of the boom in the downstream waters of the Yangtze River.

Acknowledgements

This paper is based on the relevant achievements in Basic Scientific Research Projects of China Waterborne Transport Research Institute of *Study on Integration Technologies Associated with Emergency Response to Oil and Chemical Spill* (WTI 61706), *Study on Establishment of National Emergent Scientific Research Platform for Waterborne Oil Spill* (WTI 61708), and *Annual Report on the Development of Marine Oil Spill Emergency Response Technology* (WTI 61627).

References

1. J Hui, *Environmental protection in transportation*, **21,** 43-44 (2000)
2. U.S. Coast Guard Research & Development Center, *Oil Spill Response in Fast Currents - A Field Guide (Report No. CG-D01-02)*, Washington (2001)
3. J Ou, J R Ma, X N Zhang, X Y Zhang, *Journal of Hohai University (Natural Sciences)*, **37**, 258-262 (2009)
4. R C Chen, N Shi, K S Wang, *Collection of Papers of the 1st International Global on Renewable Energy and Development* Singapore 012172 (2017)
5. Applied Science Associates, Inc., *Technical Manual Oilmap for Windows* South Kingstown (2004)

Effect of deflector curvature on hydrodynamic performances of a double-slotted cambered otter-board

Lei WANG[1], *Lu Min WANG*[1], *Jian Gao SHI*[1], *Wen Wen YU*[1], *Guang Rui QI*[1], *Xun ZHANG*[1,*], *Rong Jun ZHANG*[2], and *Tian Shu ZHANG*[2]

[1]Key Laboratory of Oceanic and Polar Fisheries, Ministry of Agriculture; East China Sea Fisheries Research Institute, Chinese Academy of Fishery Sciences, Shanghai 200090, China
[2]China National Fisheries Corporation, Beijing 100032, China

Abstract. The effect of deflector curvature on hydrodynamic performances of a double-slotted cambered otter-board was investigated using engineering models in a wind tunnel. Four different curvature (0.06,0.09, 0.12 and 0.15) were evaluated at a wind speed of 28 m/s. Parameters measured included: drag coefficient C_x, lift coefficient C_y, pitch moment coefficient C_m, center of pressure coefficient C_p , over a range of angle of attack (0° to 70°). These coefficients were used in analyzing the differences in the performance among the four otter-board models. Results showed that the maximum lift coefficient C_y of the otter-board model with the curvature (0.06) of two deflectors was highest (2.020 at α=55°). The maximum C_y/C_x of the otter-board with the curvature (0.12) of two deflectors was highest (3.655 at α=22.5°). A comparative analysis of C_m and C_p showed that the stability of otter-board model with the curvature (0.12) of two deflectors is better in pitch, and the stability of otter-board model with the curvature (0.06) of two deflectors is better in roll. The findings of this study can offer useful reference data for the structural optimization of otter-boards for trawling.

1 Introduction

Otter-boards are an important part of fishing gear for spreading the trawl mouth. The merits of the hydrodynamic performance of otter-boards can be measured on the basis of the lift coefficient of the otter-board, the drag coefficient of the otter-board, and the pitching moment coefficient of the otter-board [1]. Optimizing the structure of otter-boards may improve the hydrodynamic performance of the otter-board [2-3]. Extensive studies on the hydrodynamic performance of otter-boards have been conducted in the United States, Japan, Norway, and other countries [4-8]. In China, researchers have studied the relevant hydrodynamic performance of otter-boards since the early 1980s, including the hydrodynamic performance and optimization of different otter-boards with various structure types [9-13]. Improvements in the hydrodynamic performance of otter-boards has become a major research interest as the development of offshore trawler fleets increasing globally in recent decades. Some studies have shown that the slit in otter-boards can reduce the resistance and improve the stability of otter-boards [14-19]. The following study investigates the importance of the curvature of the deflector within the otter-board. We describe an experiment using otter-board models (n=4 designs) in a wind tunnel in which we measured various hydrodynamic coefficients for a range of angles of attack. The results are relevant as a reference for the study of the structural parameters of the deflector curvature of otter-boards.

2 Material and Methods

2.1 Design and manufacture of otter-board model

The otter-boards evaluated in this study were double-slotted curved structures comprising two deflectors and a main-panel (figure 1). This structure design was based on the used bottom trawl otter-board currently, and was simplified in order to meet objectives and requirements of the study. Only the curvature of two deflectors was modified.

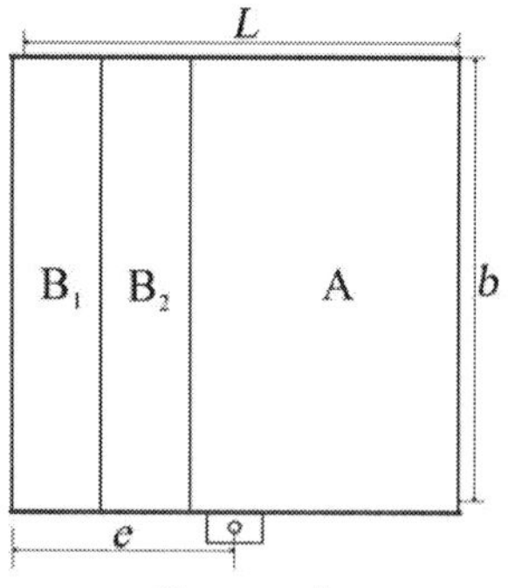

* Corresponding author: zhangxun007@hotmail.com

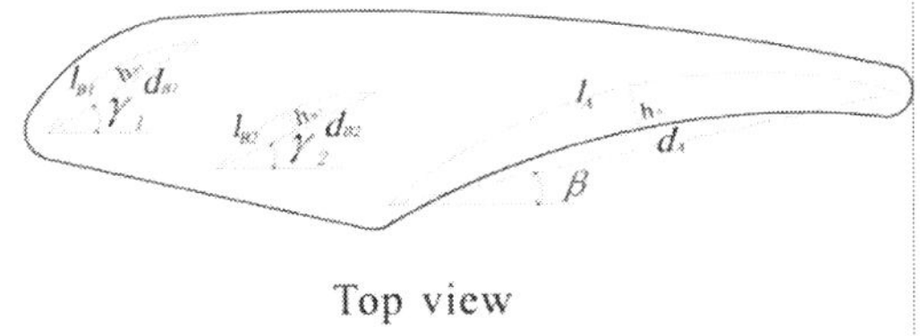

Fig. 1. Structure and parameters of otter-board

Note: L: chord; b: span; e: distance between fulcrum and the front end of model; B_1, B_2: deflector; A: main-panel; γ_1, γ_2: angle of deflector; β: angle of main-panel; l_{B1}, l_{B2}: arc length of deflector; l_A: arc length of main-panel; h_{B1}, h_{B2}, h_A: distance from the vertex of the arc to the arc chord; d_{B1}, d_{B2}, d_A: length of arc chord.

Each of models had an aspect ratio of 1.0, surface area of 0.25 m^2, and were identical in many structural parameters and dimensions (table 1). The curvature of the main-panel was 12% and was consistent in all models. The only parameter that varied between the models was the curvature of two deflectors (δ_{B1}, δ_{B2}). The models are made of steel with painted surfaces (figure 2).

Table 1. Descriptive characteristics of the four model otter boards evaluated in this study.

No.	**1**	**2**	**3**	**4**
***L*/m**	0.5	0.5	0.5	0.5
***b*/m**	0.5	0.5	0.5	0.5
λ	1	1	1	1
***S*/m²**	0.25	0.25	0.25	0.25
***e*/m**	0.25	0.25	0.25	0.25
γ_1	30°	30°	30°	30°
γ_2	25°	25°	25°	25°
β	6°	6°	6°	6°
l_{B1}	0.127	0.127	0.127	0.127
l_{B2}	0.12	0.12	0.12	0.12
l_A	0.313	0.313	0.313	0.313
δ_{B1}	0.06	0.09	0.12	0.15
δ_{B2}	0.06	0.09	0.12	0.15
δ_A	0.12	0.12	0.12	0.12

Note: *L*: chord; *b*: span; λ (*b*/*L*): aspect ratio; *S* (*L*·*b*): surface area; *e*: distance between fulcrum and the front end of model; γ_1, γ_2: angle of deflector; β: angle of main-panel; l_{B1}, l_{B2}: arc length of deflector; l_A: arc length of main-panel; δ_{B1}(h_{B1}/d_{B1}), δ_{B2}(h_{B2}/d_{B2}): curvature of deflector; δ_A(h_A/d_A) : curvature of main-panel.

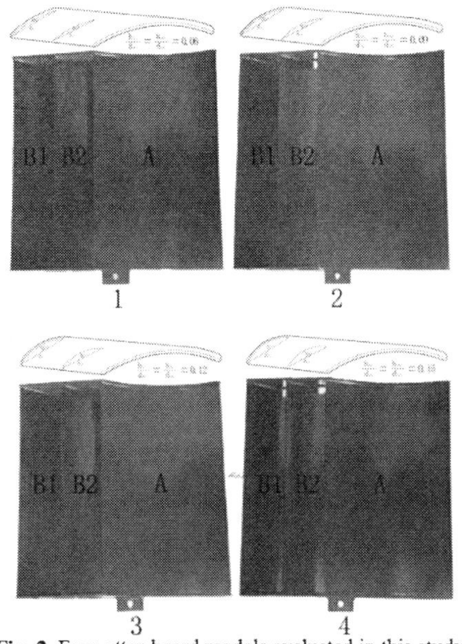

Fig. 2. Four otter-board models evaluated in this study.

Note: A: main-panel; B_1, B_2: deflector.

2.2 Test facility

The wind tunnel used for this experiment was the NH-2 wind tunnel located at Nanjing University of Aeronautics and Astronautics, China. The tunnel is a closed reflux wind tunnel with a double-string test section. The experiment was conducted in a small test section. Dimensions of the test section were 6 m (length) × 3 m (width) × 2.5 m (height). The cross-sectional area was 7.18 m^2. The minimum and maximum wind speeds of the tunnel were 5 m/s and 90 m/s, respectively. Figure 3 illustrates the experimental setup inside the wind tunnel. The otter-board models were attached to a dynamometer compromising a six-component mechanical tower-balance to measure forces in all directions. The data acquisition and processing system used is composed of a pre -amplifier and a four - networked computer system.

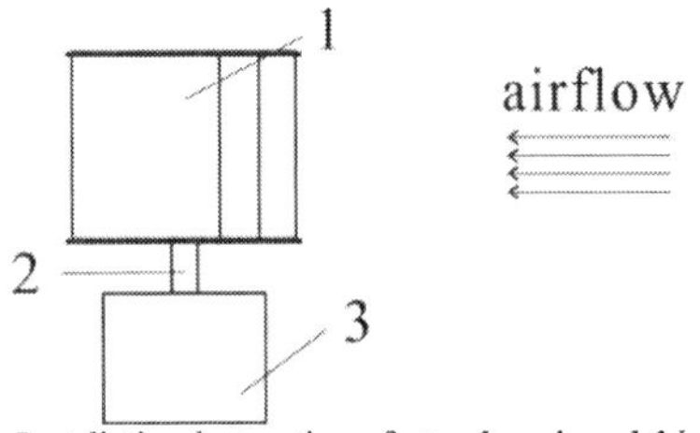

Fig. 3. Installation instruction of otter-board model in wind tunnel.

Note: 1.otter-board model 2.model connection 3.six-component force balance

2.3 Test method

2.3.1 Parameter definition of test model

 https://doi.org/10.1051/e3sconf/20183803049

Test model need to be installed on the wind tunnel in six-component balance mechanical base according to the order, angle of attack of model rotates by the 0 ° -70 ° when the wind speed reaches 28m / s (room temperature 20 ℃),wherein the angle of attack in the range 0 ° -50 °, 2.5 ° intervals to record a measurement data point, after the attack angle 50 °, each measurement interval of 5 ° to record data points, there are 25 sets of data totally, including the drag coefficient C_x, the lift coefficient C_y, the pitch moment coefficient C_m and the center of pressure coefficient C_p.

The relevant parameters of models in the wind tunnel test section are defined as shown in figure 4. In figure 4, O is torque reference point, which is the punch of the model at the bottom. During the test, the resistance of the model is provided by the force of balance along the X-axis direction, the lift is provided by the force of balance along the Z-axis direction, and the pitch moment is provided by the M_y of balance along the Z-axis direction.

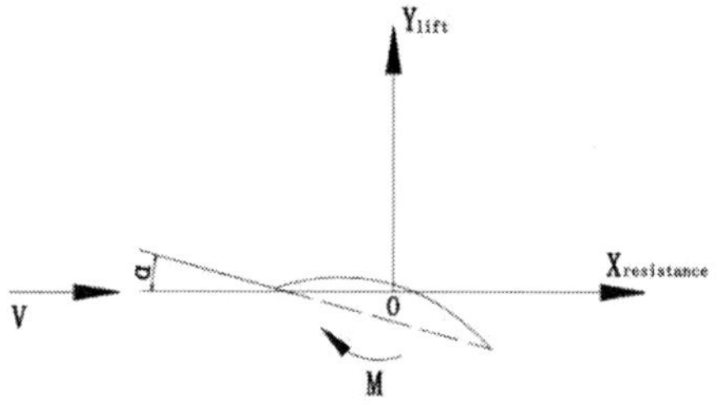

Fig. 4. Parameter definition diagram of test model in wind tunnel.

For this test, Wind speed V=28m/s, when the Reynolds number $R_e=VL/\upsilon=0.93\times10^6$ (coefficient of viscosity $\upsilon=15\times10^6\,m^2\cdot s^{-1}$) [14].

2.3.2 Parameter definition of test measurement

Three components: lift Y, drag X, pitching moment M (around the fulcrum), while the distance from the center of pressure to the front-end otter-board $d=e-(M/N)$ [16], (N is the normal force).

Lift coefficient $C_y = \frac{Y}{\rho V^2 S/2}$ [3]; drag coefficient $C_x = \frac{X}{\rho V^2 S/2}$; pitch moment coefficient $C_m = \frac{M}{\rho V^2 SL/2}$; center of pressure coefficient $C_p = \frac{d}{L}$.

Air density ρ=1.225 kg/m^3 in above formula; S is otter-board area (m^2); L is the otter-board chord length (m).

All the experimental data have been carried out the stent disturbance correction which is completed by the method of taking out light pole directly.

3 Results and Discussion

3.1 Drag coefficient and lift coefficient

Data from the experiment included the drag coefficient C_x, the lift coefficient C_y, the pitch moment coefficient C_m, and the center of pressure coefficient C_p. The lift–drag ratio was computed (C_y / C_x), which is an important factor for determining the merits of the hydrodynamic performance of otter-boards. An otter-board with excellent hydrodynamic properties can achieve higher lift–drag ratio and improved stability; such performance can be analyzed by comparing the pitching moment coefficient C_m stencil and the center of pressure coefficient C_p. The test data were divided into groups, yielding C_x-α, C_y-α and C_y / C_x-α graphs shown in figure 5. These graphs are used for analyzing the differences in the hydrodynamic properties of the four otter-board models.

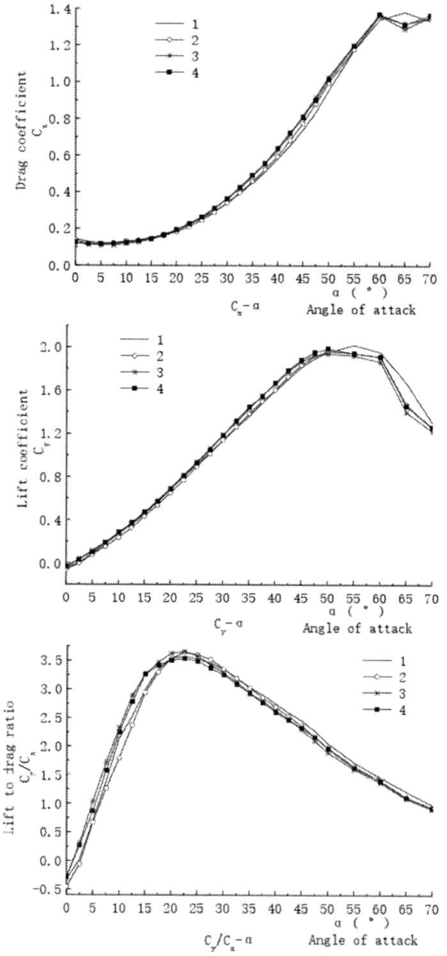

Fig. 5. Hydrodynamic properties of four otter-board models across a range of angle of attack.

In figure 5, C_x-α and C_y-α graphs show the variation curve of the drag and lift coefficient of the four models while the angle of attack α changes. The maximum lift coefficients C_y of No. 1, No. 2, No. 3, and No. 4 otter-board models were 2.020 (α = 55°), 1.965 (α = 50 °), 1.942 (α = 50 °), and 1.988 (α = 50 °), respectively. It shows that the maximum lift coefficient of No.1 model with the deflector curvature 0.06 is largest.

3.2 Lift-drag ratio

In figure 5, the maximum lift–drag ratios of No.3 model with the deflector curvature 0.12, there is 3.655 (α = 22.5 °). The maximum lift–drag ratios of No. 1, No. 2 and No. 4 models are 3.571 (α = 22.5 °), 3.644 (α = 22.5 °) and 3.534 (α = 22.5°), respectively. The higher or lower curvature of defectors may result in a lower maximum lift–drag ratio.

3.3 Stability of otter-board

Pitching moment can be divided into upper and lower pitching moments, which are usually distinguished by positive and negative values, Positive means the otter-board tilts backward, and negative means tilts forward. Its absolute value represents the level of pitching moment; and the pitching moment coefficient tending to 0 represents the more excellent stability of otter-board. In practice, comparing the absolute value C_m corresponding to the operation angle of attack of otter-board may determine the stability level of otter-board. For comparison, the angle of attack corresponding to the maximum lift–drag ratio C_y/C_x is selected [20]. The corresponding absolute value of C_m is shown in table 2. The absolute value of C_m of No. 3 model is 0.144, so hence, the stability of No. 3 otter-board model is better in pitch.

Table 2. Parameters of four otter-board models for stability analysis.

No.	Angle corresponding to the maximum lift –drag ratio α	$\lvert C_m \rvert$	Variable coefficient of C_p
1	22.5°	0.183	6.83%
2	22.5°	0.163	7.85%
3	22.5°	0.144	8.13%
4	22.5°	0.160	7.81%

The stability in roll of otter-board can be measured according to the center of pressure coefficient C_p; and the way of comparison is analyzing the coefficient of variation in C_p within the range of angle approximately 5° of the angle of attack corresponding to the maximum lift–drag ratio; a smaller coefficient results in the improved stability [21]. The calculated data are shown in table 2. The minimum variation coefficient of C_p is 6.83%; this value also means that the stability of No. 1 otter-board model is better in roll of otter-board.

4 Conclusion

Test analysis shows that the curvature of two deflectors has a point for equilibrating the hydrodynamic performances of otter-board. No.3 otter-board with the deflector curvature of 0.12 has the higher maximum lift–drag ratio at the angle of attack 22.5°, and at this angle of attack, the lift coefficient (0.813) is also higher than the other three models. In the meanwhile, the stability of No.3 otter-board model in pitch is also better comparatively, and the stability of No.1 otter-board model in roll is better. The data and conclusions of this study can provide a reference for the design of otter-board.

Acknowledgements

This work is financially supported by the National Key Technology R&D Program (No. 2013BAD13B04).

References

1. G. X. Guo, T. Y. Liu, X. H. Huang, F. L. Gu. Theory and Practice of trawl doors Kinetic. Guang. Sci. & Tech. Pub. (2008)
2. Y. Q. Zhou. Mechanics of fishing gear. Chi. Agr. Pub. (2001)
3. X. Z. Chen, X. C. Huang. Theory and method of gear model test. Sh. Sci. & Tech. Pub. (2011)
4. A. Sala, J. Prat, J. Antonijuan, A. Lucchetti. Fis. Res. **156**, 100(2009)
5. Y. Takahashi, Y. Fujimori, F. X. Hu, X. L. Shen, N. Kimura. Fis. Res. **400**, 161(2015)
6. M. K. Broadhurst, D. J. Sterling, R. B. Millar. Fis. Man. & Eco. **407**, 22(2015)
7. K. Fukuda, F. X. Hu, T. Tokai, K. Matuda. Ni. Su. Gak. **97**, 66(2000)
8. C. D. Park, K. Matuda, F. X. Hu. Ni. Su. Gak. **920**, 62(1996)
9. X. L. Shen, F. X. Hu, T. Kumazawa, D. Shiode, T. Tokai. Fis. Sci. **433**, 81(2015)
10. L. Wang, L. M. Wang, C. L. Feng, A. Z. Zhou, W. W. Yu, Y. Zhang, X. Zhang. Aqu. & Fis. **234**, 2(2017)
11. L. Wang, L. M. Wang, J. G. Shi, Y. Zhang, Y. L. Liu, W. W. Yu, X. Zhang. MAT. W. Con. **5004**, 128(2017)
12. L. Wang, L. M. Wang, A. Z. Zhou, J. G. Shi, Y. Zhang, G. D. Xu, X. Zhang. MAT. W. Con. **5003**, 128(2017)
13. L. Wang, L. M. Wang, C. L. Feng, X. Zhang, A. Z. Zhou, Y. Zhang, Y. L. Liu, G. R. Qi. Mar. Fis. **682**, 39(2017)
14. X. Zhang, J. H. Wang, M. Y. Wang, Y. F. Yu, B. S. Xu. Jou. Fis. Sci. Ch. **5**, 11(2004)
15. J. H. Wang, M. Y. Wang, X. Zhang, Y. F. Yu, B. S. Xu. Jou. Fis. Sci. Ch. **9**, S1(2004)
16. M. Y. Wang, J. H. Wang, X. Zhang, Y. F. Yu, B. S. Xu. Jou. Fis. Ch. **311**, 3(2004)
17. C. C. Li, Z. L. Liang, L. Y. Huang, W. F. Zhou, P. Sun, L. Wang. Mar. Sci. **69**, 11(2013)
18. L. Wang, L. M. Wang, C. L. Feng, X. Zhang, J. G. Shi, Y. Zhang, Y. L. Liu, W. W. Yu. Fis. Mod. **55**, 6(2015)

https://doi.org/10.1051/e3sconf/20183803049

19. L. Wang, L. M. Wang, W. W. Yu, C. L. Feng, J. G. Shi, Y. L. Liu, X. Zhang. Con. In. For. En. Sus. De. **530**, 75(2016)
20. L. Yang. Fis. Sci. Tec. **38**, 2(1996)
21. L. Yang. Fis. Sci. Tec. **42**, 5(1996)

Effect of panel shape on hydrodynamic performances of vertical v-shaped double- slotted cambered otter-board

Lei WANG[1], *Xun ZHANG*[1], *Lu Min WANG*[1,*], *Hong Liang HUANG* [1], *Yu ZHANG* [1], *Yong Li LIU* [1], *Wei Dong FENG*[2] and *Rong Jun ZHANG*[2]

[1]Key Laboratory of Oceanic and Polar Fisheries, Ministry of Agriculture; East China Sea Fisheries Research Institute, Chinese Academy of Fishery Sciences, Shanghai 200090, China
[2]China National Fisheries Corporation, Beijing 100032, China

Abstract. The effect of panel shape on hydrodynamic performances of a vertical v-shaped double-slotted cambered otter-board was investigated using engineering models in a wind tunnel. Three different shape panels (rhomboid, left trapezoid and isosceles trapezoid) were evaluated at a wind speed of 28 m/s. Parameters measured included: drag coefficient C_x, lift coefficient C_y, pitch moment coefficient C_m, center of pressure coefficient C_p , over a range of angle of attack (0° to 70°). These coefficients were used in analyzing the differences in the performance among the three otter-board models. Results showed that the maximum lift coefficient C_y of the otter-board model with the isosceles trapezoid shape panels was highest (2.103 at α=45°). The maximum C_y/C_x of the otter-board with the rhomboid shape panels was highest (3.976 at α=15°). A comparative analysis of C_m and C_p showed that the stability of otter-board model with the isosceles trapezoid shape panels is better in pitch, and the stability of otter-board model with the left trapezoid shape panels is better in roll. The findings of this study can offer useful reference data for the structural optimization of otter-boards for trawling.

1 Introduction

Otter boards are an important part of fishing gear for spreading the trawl mouth. The merits of the hydrodynamic performance of otter-boards can be measured on the basis of the lift coefficient and the drag coefficient of the trawl door [1]. Optimizing the structure of otter-boards may improve the hydrodynamic performance of the otter-board and reduce the energy consumption of fishing vessels [2-3]. Extensive studies on the hydrodynamic performance of otter-boards have been conducted in the United States, Japan, Norway, and other countries [4-8]. In China, researchers have studied the relevant hydrodynamic performance of otter-boards since the early 1980s, including the hydrodynamic performance and optimization of different otter-boards with various structure types [9-13]. The development of offshore trawler fleets has increased globally in recent decades, raising the demand for improved otter-board designs. Accordingly, improvements in the hydrodynamic performance of otter-boards has become a major research interest. Some studies have shown that the slit in otter-boards can reduce the resistance and improve the stability of otter-boards [14-19]. The following study investigates the influence of three kind of shapes of the panel within the otter-board, including the rhomboid, left trapezoid and isosceles trapezoid. We describe an experiment using model otter-boards (n=3 designs) in a wind tunnel in which we measured various hydrodynamic coefficients for a range of angles of attack. The results are relevant as a reference for the study of the structural parameters of otter-boards.

2 Material and Methods

2.1 Design and manufacture of otter-board model

The otter-boards evaluated in this study were vertical v-shaped double-slit curved structures comprising two deflectors and a main-panel (figure 1). This design was based on the structure of the otter-board used in bottom trawl currently. Only the shape of the panels was modified with three kind of shapes.

* Corresponding author: zhangxun007@hotmail.com

 https://doi.org/10.1051/e3sconf/20183803050

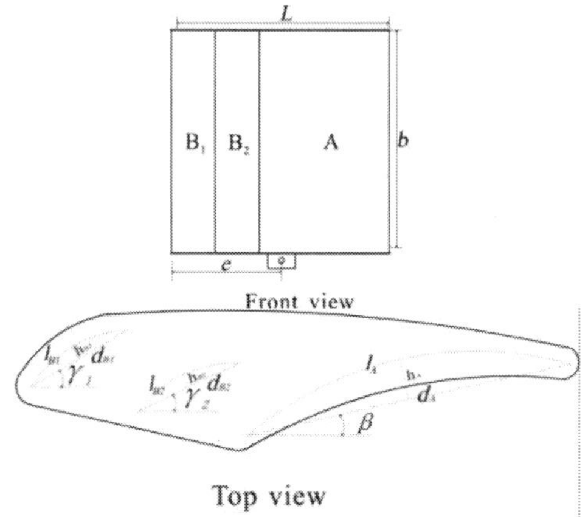

Fig. 1. Structure and parameters of otter-board

Note: L: chord; b: span; e: distance between fulcrum and the front end of model; B_1, B_2: deflector; A: main-panel; γ_1, γ_2: deflector angle of mid-plate; β: angle of main-panel of mid-plate; γ'_1, γ'_2: deflector angle of up-plate; β': angle of main-panel of up-plate; l_{B1}, l_{B2}: arc length of deflector of mid-plate; l_A: arc length of main-panel of mid-plate; l'_{B1}, l'_{B2}: arc length of deflector of up-plate; l'_A: arc length of main-panel of up-plate.

Each of models had an aspect ratio of 1.0, surface area of 0.25 m2, and were identical in many structural parameters and dimensions (table 1). The curvature of the deflector and main panels was 12% and was consistent in all models. The only parameter that varied between the models was the arc length of panels (lB1, lB2, lA; l B1, l B2, l A), and the parameters of panels of up-plate was same to down-plate. The models are made of steel with painted surfaces (figure 2).

Table 1. Descriptive characteristics of the four model otter boards evaluated in this study.

No.	1	2	3
L/m	0.5	0.5	0.5
b/m	0.5	0.5	0.5
λ	1	1	1
S/m^2	0.25	0.25	0.25
e/m	0.25	0.25	0.25
$\gamma_1 = \gamma'_1$	30°	30°	30°
$\gamma_2 = \gamma'_2$	25°	25°	25°
$\beta = \beta'$	6°	6°	6°
$l_{B1} = l_{B2}$	0.127	0.149	0.171
l_A	0.313	0.336	0.357
$l'_{B1} = l'_{B2}$	0.127	0.105	0.083
l'_A	0.313	0.292	0.27
$S_{B1} = S_{B2}$	0.032	0.032	0.032
S_A	0.078	0.078	0.078

Note: L: chord; b: span; λ: aspect ratio (b/L); S : surface area (L·b); e: distance between fulcrum and the front end of model; γ_1, γ_2: deflector angle of mid- plate; β: main-panel angle of mid-plate; l_{B1}, l_{B2}: arc length of deflector of mid- plate; l_A: arc length of main-panel of mid- plate; γ'_1, γ'_2: deflector angle of up- plate; β': main-panel angle of up- plate; l'_{B1}, l'_{B2}: arc length of deflector of up- plate; l'_A: arc length of main-panel of up-plate.

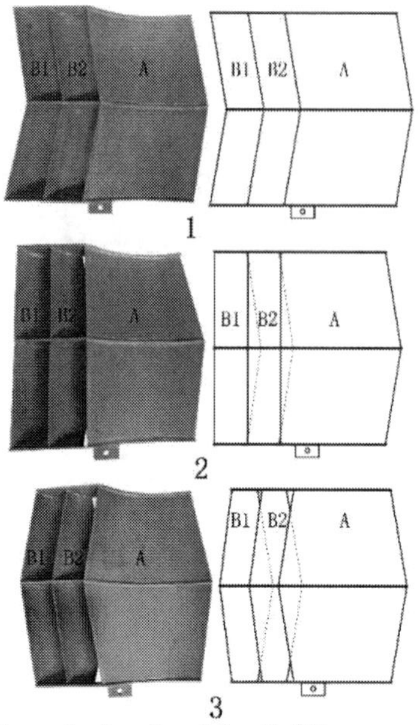

Fig. 2. Three otter-board models with different panel shapes evaluated in this study.

Note: 1: rhomboid; 2: left trapezoid; 3: isosceles trapezoid; A: main-panel; B_1, B_2: deflector.

2.2 Test facility

The wind tunnel used for this experiment was the NH-2 wind tunnel located at Nanjing University of Aeronautics and Astronautics, China. The tunnel is a closed reflux wind tunnel with a double-string test section. The experiment was conducted in a small test section. Dimensions of the test section were 6 m (length) × 3 m (width) × 2.5 m (height). The cross-sectional area was 7.18 m^2. The minimum and maximum wind speeds of the tunnel were 5 m/s and 90 m/s, respectively. Figure 3 illustrates the experimental setup inside the wind tunnel. The otter-board models were attached to a dynamometer compromising a six-component mechanical tower-balance to measure forces in all directions. The data acquisition and processing system used is composed of a pre -amplifier and a four - networked computer system.

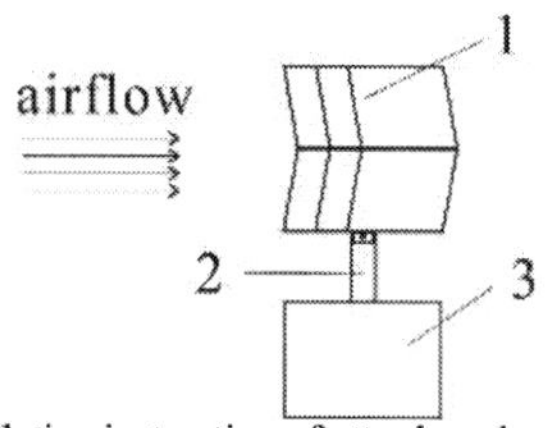

Fig. 3. Installation instruction of otter-board model in wind tunnel.

Note: 1.otter-board model 2.model connection 3.six-component force balance

2.3 Test method

2.3.1 Parameter definition of test model

Test model need to be installed on the wind tunnel in six-component balance mechanical base according to the order, angle of attack of model rotates by the 0 ° -70 ° when the wind speed reaches 28m / s (room temperature 20 ℃),wherein the angle of attack in the range 0 ° -50 °, 2.5 ° intervals to record a measurement data point, after the attack angle 50 °, each measurement interval of 5 ° to record data points, there are 25 sets of data totally, including the drag coefficient C_x, the lift coefficient C_y, the pitch moment coefficient C_m and the center of pressure coefficient C_p.

The relevant parameters of models in the wind tunnel test section are defined as shown in figure 4. In figure 4, O is torque reference point, which is the punch of the model at the bottom. During the test, the resistance of the model is provided by the force of balance along the X-axis direction, the lift is provided by the force of balance along the Z-axis direction, and the pitch moment is provided by the M_y of balance along the Z-axis direction.

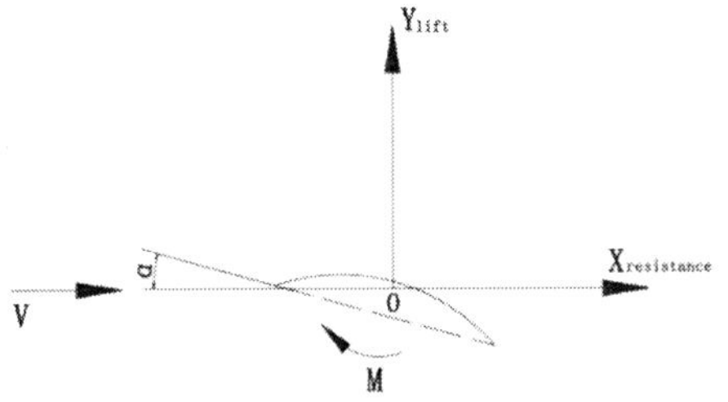

Fig. 4. Parameter definition diagram of test model in wind tunnel.

For this test, Wind speed V=28m/s, when the Reynolds number $R_e=VL/\upsilon=0.93\times10^6$ (coefficient of viscosity $\upsilon=15\times10^6$ $m^2 \cdot s^{-1}$) [14].

2.3.2 Parameter definition of test measurement

Three components: lift Y, drag X, pitching moment M (around the fulcrum), while the distance from the center of pressure to the front-end otter-board $d=e-(M/N)$ [16], (N is the normal force).

Lift coefficient $C_y = \frac{Y}{\rho V^2 S/2}$ [3]; drag coefficient $C_x = \frac{X}{\rho V^2 S/2}$; pitch moment coefficient $C_m = \frac{M}{\rho V^2 SL/2}$; center of pressure coefficient $C_p = \frac{d}{L}$.

Air density ρ=1.225 kg/m^3 in above formula; S is otter-board area (m^2); L is the otter-board chord length (m).

All the experimental data have been carried out the stent disturbance correction which is completed by the method of taking out light pole directly.

3 Results and Discussion

3.1 Drag coefficient and lift coefficient

Data from the experiment included the drag coefficient C_x, the lift coefficient C_y, the pitch moment coefficient C_m, and the center of pressure coefficient C_p. The lift–drag ratio was computed (C_y / C_x), which is an important factor for determining the merits of the hydrodynamic performance of otter-boards. An otter-board with excellent hydrodynamic properties can achieve higher lift–drag ratio and improved stability; such performance can be analyzed by comparing the pitching moment coefficient C_m stencil and the center of pressure coefficient C_p. The test data were divided into groups, yielding C_x-α, C_y-α and C_y / C_x-α graphs shown in figure 5. These graphs are used for analyzing the differences in the hydrodynamic properties of the three otter-board models.

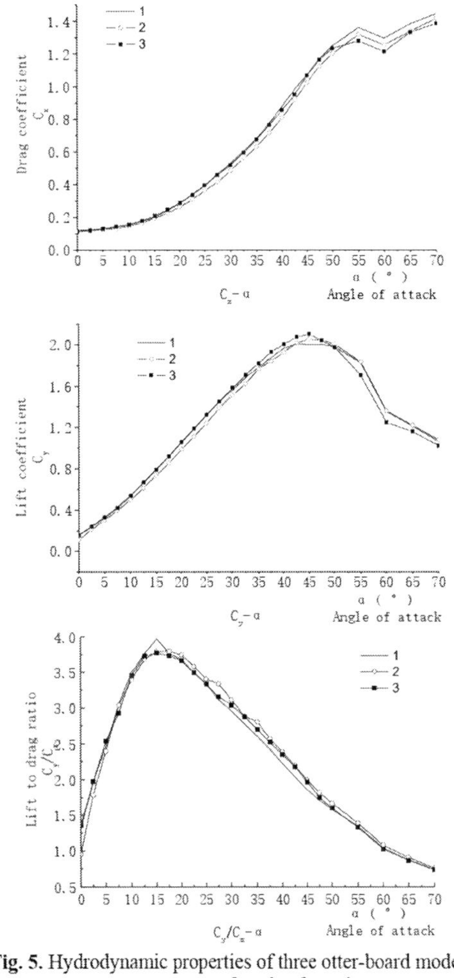

Fig. 5. Hydrodynamic properties of three otter-board models across a range of angle of attack.

In figure 5, C_x-α and C_y-α graphs show the variation curve of the drag and lift coefficient of the three models while the angle of attack α changes. The relationship between C_x and α is proportional almost but a turning point at the angle of attack 60°. The maximum lift coefficients C_y of No. 1, No. 2, and No. 3 otter-board models were 2.009 (α = 42.5°), 2.053 (α = 45 °), and

2.103 (α = 45 °), respectively. It shows that the maximum lift coefficients of No. 3 otter-board is higher.

3.2 Lift-drag ratio

In figure 5, the maximum lift–drag ratios of No.1 model with the rhomboidal panels is the highest, there is 3.976 (α = 15 °). The maximum lift–drag ratios of No. 2 and No. 3 models are 3.799 (α = 17.5 °) and 3.774 (α = 15°), respectively.

3.3 Stability of otter-board

Pitching moment can be divided into upper and lower pitching moments, which are usually distinguished by positive and negative values, Positive means the otter-board tilts backward, and negative means tilts forward. Its absolute value represents the level of pitching moment; and the pitching moment coefficient tending to 0 represents the more excellent stability of otter-board. In practice, comparing the absolute value C_m corresponding to the operation angle of attack of otter-board may determine the stability level of otter-board. For comparison, the angle of attack corresponding to the maximum lift–drag ratio C_y/C_x is selected [20]. The corresponding absolute value of C_m is shown in table 2. The absolute value of C_m of No. 3 model is 0.126, so hence, the stability of No. 3 otter-board model is better in pitch.

Table 2. Parameters of three otter-board models for stability analysis.

No.	Angle corresponding to the maximum lift –drag ratio α	$\lvert C_m \rvert$	Variable coefficient of C_p
1	15°	0.174	7.10%
2	17.5°	0.147	6.69%
3	15°	0.126	7.55%

The stability in roll of otter-board can be measured according to the center of pressure coefficient C_p; and the way of comparison is analyzing the coefficient of variation in C_p within the range of angle approximately 5° of the angle of attack corresponding to the maximum lift–drag ratio; a smaller coefficient results in the improved stability [21]. The calculated data are shown in table 2. The minimum variation coefficient of C_p is 6.69%; this value also means that the stability of No. 2 otter-board model is better in roll of otter-board.

4 Conclusion

Test analysis shows that a vertical V-shaped double-slotted cambered otter-board with isosceles trapezoid shaped panels has a higher maximum lift coefficient but a lower maximum lift-drag ratio, and the otter-board with rhomboid shaped panels has a higher maximum lift-drag ratio, but a lower maximum lift coefficient. The test results also show that the otter-board with isosceles trapezoid shaped panels is better in lateral stability and should not be tumbled easily. The otter-board with left trapezoidal shaped panels is better in the vertical stability and should not be dumped easily. The results of this study can provide some reference for the structural design of different types of otter-board.

Acknowledgements

This work is financially supported by the National Key Technology R&D Program (No. 2013BAD13B04).

References

1. G. X. Guo, T. Y. Liu, X. H. Huang, F. L. Gu. Theory and Practice of trawl doors Kinetic. Guang. Sci. & Tech. Pub. (2008)
2. Y. Q. Zhou. Mechanics of fishing gear. Chi. Agr. Pub. (2001)
3. X. Z. Chen, X. C. Huang. Theory and method of gear model test. Sh. Sci. & Tech. Pub. (2011)
4. A. Sala, J. Prat, J. Antonijuan, A. Lucchetti. Fis. Res. **156**, 100(2009)
5. Y. Takahashi, Y. Fujimori, F. X. Hu, X. L. Shen, N. Kimura. Fis. Res. **400**, 161(2015)
6. M. K. Broadhurst, D. J. Sterling, R. B. Millar. Fis. Man. & Eco. **407**, 22(2015)
7. K. Fukuda, F. X. Hu, T. Tokai, K. Matuda. Ni. Su. Gak. **97**, 66(2000)
8. C. D. Park, K. Matuda, F. X. Hu. Ni. Su. Gak. **920**, 62(1996)
9. X. L. Shen, F. X. Hu, T. Kumazawa, D. Shiode, T. Tokai. Fis. Sci. **433**, 81(2015)
10. L. Wang, L. M. Wang, C. L. Feng, A. Z. Zhou, W. W. Yu, Y. Zhang, X. Zhang. Aqu. & Fis. **234**, 2(2017)
11. L. Wang, L. M. Wang, J. G. Shi, Y. Zhang, Y. L. Liu, W. W. Yu, X. Zhang. MAT. W. Con. **5004**, 128(2017)
12. L. Wang, L. M. Wang, A. Z. Zhou, J. G. Shi, Y. Zhang, G. D. Xu, X. Zhang. MAT. W. Con. **5003**, 128(2017)
13. L. Wang, L. M. Wang, C. L. Feng, X. Zhang, A. Z. Zhou, Y. Zhang, Y. L. Liu, G. R. Qi. Mar. Fis. **682**, 39(2017)
14. X. Zhang, J. H. Wang, M. Y. Wang, Y. F. Yu, B. S. Xu. Jou. Fis. Sci. Ch. **5**, 11(2004)
15. J. H. Wang, M. Y. Wang, X. Zhang, Y. F. Yu, B. S. Xu. Jou. Fis. Sci. Ch. **9**, S1(2004)
16. M. Y. Wang, J. H. Wang, X. Zhang, Y. F. Yu, B. S. Xu. Jou. Fis. Ch. **311**, 3(2004)
17. C. C. Li, Z. L. Liang, L. Y. Huang, W. F. Zhou, P. Sun, L. Wang. Mar. Sci. **69**, 11(2013)

18. L. Wang, L. M. Wang, C. L. Feng, X. Zhang, J. G. Shi, Y. Zhang, Y. L. Liu, W. W. Yu. Fis. Mod. **55**, 6(2015)
19. L. Wang, L. M. Wang, W. W. Yu, C. L. Feng, J. G. Shi, Y. L. Liu, X. Zhang. Con. In. For. En. Sus. De. **530**, 75(2016)
20. L. Yang. Fis. Sci. Tec. **38**, 2(1996)
21. L. Yang. Fis. Sci. Tec. **42**, 5(1996)

E3S Web of Conferences **38**, 03051 (2018) https://doi.org/10.1051/e3sconf/20183803051
ICEMEE 2018

Quantitative analyzing for As's transferring process in coastal waters in Jiaozhou Bay

Dongfang Yang[1,2,3,*], *Qiang* Wang[1,2], *Wei* Zhou[1,2], *Chunhua* Su [1,2], and *Sixi* Zhu*1,2*

[1]Research Center for Karst Wetland Ecology, Guizhou Minzu University, Guizhou Guiyang, Guizhou Guiyang, China
[2]College of Chemistry and Environmental Science, Guizhou Minzu University, Shanghai, 550025, China
[3]North China Sea Environmental Monitoring Center, SOA, Qingdao 266033, China

Abstract. This paper quantified the horizontal and vertical transferring processes of As in coastal waters in the southwest and the bay mouth of Jiaozhou Bay using investigation data in July and October 1982. Results showed that the horizontal absolute loss amounts of As contents in surface waters were 0.20-2.44 μg L^{-1}, and the horizontal relative loss amounts in surface waters were 19.23%-87.14%, respectively. The horizontal absolute loss amounts of As contents in bottom waters were 0.27-2.44 μg L^{-1}, and the horizontal relative loss amounts in bottom waters were 25.00%-54.46%, respectively. The vertical absolute dilution amounts of As contents were 0.02-0.28 μg L^{-1}, and the vertical relative dilution amounts were 1.78-16.66%. The vertical absolute accumulation amounts of Pb contents were 0.10-0.86 μg L^{-1}, and the vertical relative accumulation amounts were 7.35%-35.83%. The vertical absolute dilution amount and vertical relative dilution amount of As's contents were 0.20-0.76 μg L^{-1} and 19.23%-27.14%, respectively. The vertical absolute accumulation amount and vertical relative accumulation amount of As's contents were 0.28-4.12 μg L^{-1} and 25.00%-91.96%, respectively. The changes of As contents in surface and bottom waters were determined by the source inputs of As and the transferring distance of As from the terrisgenous sources.

1 Introduction

As has been widely used industry and agriculture, yet the excessive existence of As in the environment is harmful to organism and ecosystem since As is high toxic ([1-2]). Nowadays, many marine bays have been polluted by As ([3-6]). Understanding the transferring processes of As in marine bays is essential to pollution control. Jiaozhou Bay is a semi-closed bay located in Shandong Province China, and has been polluted by various pollutants including As in the past three decades ([1-2]). This paper quantified the horizontal and vertical transferring processes of As in coastal waters in the southwest and the bay mouth of Jiaozhou Bay using investigation data in July and October 1982. The aim of this paper was to better understand the transporting processes of As in marine bay, and provide basis for scientific research and environment remediation.

2 Materials and method

2.1 Study area

Jiaozhou Bay is located in the south of Shandong Province, eastern China (35°55′-36°18′ N, 120°04′-120°23′ E). The total area, average water depth and the width of the bay mouth are 446 km2, 7 m and 3km, respectively. The bay mouth is connecting to the Yellow Sea in the south, and is surrounded by Qingdao, Jiaozhou and Jiaonan in the east, north and west, respectively (Fig. 1). There are a dozen of rivers, all of which are seasonal rivers ([7-8]).

2.2 Data collection

The investigation on As in surface and bottom waters in coastal waters in Jiaozhou Bay was carried on in July and October 1982 in four sampling sites (i.e., 083, 084, 122 and 123) (Fig. 1). As in waters was sampled and monitored follow by National Specification for Marine Monitoring ([9]).

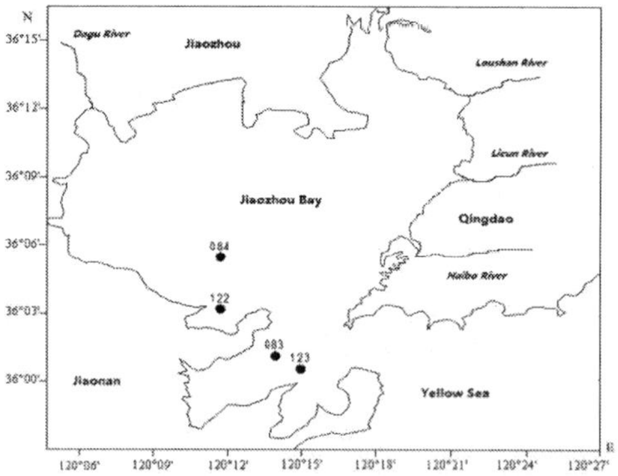

Fig. 1. Geographic location and monitoring sites in Jiaozhou Bay.

* Corresponding author: dfyang_dfyang@126.com

2.3 Modelling for horizontal transferring processes of As

Supposed that certain substance's contents in surface and bottom waters in the coastal waters in the southwest of bay were A and a, in the center waters in the southwest of the bay was B and b, respectively ([10]). In surface waters, and from the coastal waters in the southwest of bay to the center waters in the southwest of the bay, and the calculation formula for transferring process is:

$$D=A-B,\quad E=(100\times\mid A-B\mid/\max(A,B))\% \quad (1)$$

where, D is the horizontal absolute loss amount in surface waters from the inner of the bay to the open waters, E is the corresponding horizontal relative loss amount in surface waters.

In bottom waters, and from the inner of the bay to the open waters, substance's contents are changing, and the calculation formula for transferring process is:

$$d=a-b,\quad e=(100\times\mid a-b\mid/\max(a,b))\% \quad (2)$$

where, d is the horizontal absolute loss amount in surface waters, e is the corresponding horizontal relative loss amount in surface waters.

2.4 Modelling for vertical transferring processes of As

Supposed that certain substance's contents in surface and bottom waters in Site n in the coastal waters in the southwest of the bay are A and a, respectively. From surface waters to bottom waters, the calculation formula for this transferring process is:

$$V_{na}=A-a,\quad V_{nr}=(100\times\mid A-a\mid/\max(A,a))\% \quad (3)$$

where, V_{na} is the horizontal absolute dilution amount, V_{nr} is the corresponding horizontal relative dilution amount. While from bottom waters to surface waters, V_{na} refers to the horizontal absolute accumulation amount, and V_{nr} refers to the corresponding horizontal relative accumulation amount.

3 Results and discussion

3.1 Horizontal and vertical transferring processes of As

As's contents in Site 122 and 084 were used to quantify the horizontal and vertical transferring processes of As. In according to Eq. (1) and (2), the horizontal loss of As in surface and waters were calculated listed in Table 1. As's contents in surface and bottom waters in Site 122 and 084 were used to quantify the vertical dilution and accumulation processes of As. In according to Eq. (3), the vertical dilution and accumulation of As were calculated listed in Table 2.

Table 1. Horizontal loss of As in surface and bottom waters in Jiaozhou Bay 1982

Month	Surface waters		Bottom waters	
	D/μg L^{-1}	E/%	d/μg L^{-1}	e/%
July	2.44	87.14	-2.44	54.46
October	0.20	19.23	-0.28	25.00

Table 2. Vertical dilution and accumulation processes of As in Jiaozhou Bay 1982

Time	Waters	V_{na}/μg L^{-1}	V_{nr}/%
July	Coastal waters in the southwest	0.76	27.14
	Center waters in the southwest	-4.12	91.96
October	Coastal waters in the southwest	0.20	19.23
	Center waters in the southwest	-0.28	25.00

In July 1982, the horizontal absolute loss amount in surface waters was as high as 87.14%, while the horizontal absolute loss amount in bottom waters was 54.46% (Fig. 2). The vertical relative dilution amount was relative low in coastal waters (27.14), yet in the center of waters in the southwest of the bay the vertical relative accumulation was very high (91.96%) (Fig. 2).

In October 1982, the horizontal absolute loss amount in surface waters was relative low as 19.23%, and the horizontal absolute loss amount in bottom waters was also relative low as 25.00% (Fig. 3). The vertical relative dilution amount was relative low in coastal waters (19.23%), while in the center of waters in the southwest of the bay the vertical relative accumulation was also relative low (25.00%) (Fig. 3).

In general, the horizontal absolute loss amounts of As contents in surface waters were 0.20-2.44 μg L^{-1}, and the horizontal relative loss amounts in surface waters were 19.23%-87.14%, respectively. The horizontal absolute loss amounts of As contents in bottom waters were 0.27-2.44 μg L^{-1}, and the horizontal relative loss amounts in bottom waters were 25.00%-54.46%, respectively.

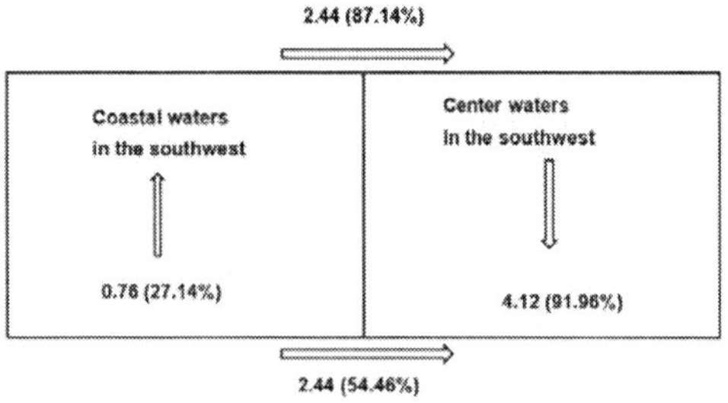

Fig. 2. Block diagram for horizontal-vertical changes of As in Jiaozhou Bay in July 1982

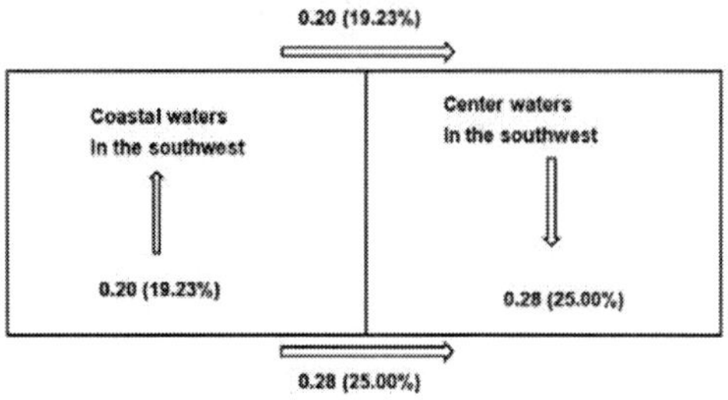

Fig. 3. Block diagram for horizontal-vertical changes of As in Jiaozhou Bay in October 1982

3.2 Horizontal loss of As in coastal waters

In July 1982, the horizontal loss of As's contents in surface waters from the coastal waters in the southwest of the bay to the center waters in the southwest was very high (87.14%), and in bottom waters was also relative high (54.46%). The vertical dilution amount was relative low in coastal waters in the southwest of the bay (27.14%), while the vertical accumulation amount was very high in center waters in the southwest of the bay (91.96%).

In October 1982, the horizontal loss of As's contents in surface waters from the coastal waters in the southwest of the bay to the center waters in the southwest was relative low (19.23%), and in bottom waters was also relative low (25.00%). The vertical dilution amount was relative low in coastal waters in the southwest of the bay (19.23%), and the vertical accumulation amount was also relative low in center waters in the southwest of the bay (25.00%).

In general, there was relative dilution process in coastal waters in coastal waters, while in the center of the southwest of the bay there were relative high or relative low accumulation processes. There were horizontal loss processes in both surface and bottom waters from coastal waters in the southwest to the center waters in the southwest of the bay. The horizontal loss of As's contents were relative high in both surface and bottom waters in July 1982, while in October 1982 the horizontal loss was relative low in bottom waters.

3.3 Spatial variations of As in coastal waters

In July 1982, the horizontal loss of As's contents in surface waters from the coastal waters in the southwest of the bay to the center waters in the southwest was very high (87.14%), and in bottom waters was also relative high (54.46%). The vertical dilution amount was relative low in coastal waters in the southwest of the bay (27.14%), while the vertical accumulation amount was very high in center waters in the southwest of the bay (91.96%). In order to reveal the spatial variations of As, for each sampling site (i.e., 083, 084, 122 and 123), As content in surface waters was subtracted by which in bottom waters.

In July 1982, the differences ranged from -4.12-1.72 μg L-1. The difference was negative in Site 084 in the center waters in the southwest of the bay, while the differences in the other three sampling sites were all positive (Table 3). In October 1982, the differences ranged from -0.54-0.46 μg L-1. The differences were negative in Site 083 and 084 in the center waters in the southwest of the bay, while the differences in the other three sampling sites were all positive (Table 3).

In general, the changes of As contents in surface and bottom waters were determined by the source inputs of As and the transferring distance of As from the terrisgenous sources. In locations closed to the terrigenous sources and the sedimentation of As was still in low level, the dilution amount was small, resulted in As contents in surface waters were higher than in bottom waters. Once As had been transferred fay away from the terrisgenous sources, the sedimentation of As was abundant, resulted in As contents in surface waters were lower than in bottom waters.

Table 3 Spatial variations of As in Jiaozhou Bay 1982

Time	Site 122	Site 084	Site 083	Site 123
July	Positive	Negative	Positive	Positive
Positive	Positive	Negative	Negativ e	Positive

4 Conclusions

The horizontal and vertical transferring processes of As in coastal waters in the southwest and the bay mouth of Jiaozhou Bay 1982 were quantitatively analyzed. The horizontal absolute loss amounts of As contents in surface waters were 0.20-2.44 μg L^{-1}, and the horizontal relative loss amounts in surface waters were 19.23%-87.14%, respectively. The horizontal absolute loss amounts of As contents in bottom waters were 0.27-2.44 μg L^{-1}, and the horizontal relative loss amounts in bottom waters were 25.00%-54.46%, respectively.

There was relative dilution process in coastal waters in coastal waters, while in the center of the southwest of the bay there were relative high or relative low accumulation processes. There were horizontal loss processes in both surface and bottom waters from coastal waters in the southwest to the center waters in the southwest of the bay. The horizontal loss of As's contents were relative high in both surface and bottom waters in July 1982, while in October 1982 the horizontal loss was relative low in bottom waters. The changes of As contents in surface and bottom waters were determined by the source inputs of As and the transferring distance of As from the terrisgenous sources.

Acknowledgments

This research was sponsored by Research Projects of Guizhou Nationalities University ([2014]02), Research Projects of Guizhou Province Ministry of Education (KY [2014] 266), Research Projects of Guizhou Province Ministry of Science and Technology (LH [2014] 7376).

References

1. D.F. Yang, W.P. Song, S.T. Chen, J.H. Guo, L. Zheng, Coastal Engineering, **31**, 4(2012)
2. D.F. Yang, Y.H. Zhao, Z.G. Piao, Ocean Development and Management, **31**, (2014)
3. D.F. Yang, S.X. Zhu, F.Y. Wang, S.T. Chen, Z.J. Xu, S.X. Zhu, Y.J. Wu, Meterological and Environmental Research, **5**, (2014)
4. D.F. Yang, F.Y. Wang, H.Z. He, Y.J. Wu, S.X. Zhu, Proceedings of the 2015 international symposium on computers and informatics, (2015)
5. D.F. Yang, F.Y. Wang, X.L. ZhaoY.J. Wu, S.X. Zhu, Sustainable Energy and Enviroment Protection, (2015)
6. D.F. Yang, F.Y. Wang, X.Q. Yang, Y.J. Wu, S.X. Zhu, Advances in Computer Science Research, **3252**, (2015)
7. D.F. Yang, Y. Chen, Z.H. Gao, J. Zhang, F. Wang, Chinese Journal of Oceanology and Limnology, **23**, 1(2005)
8. D.F. Yang, F.Y. Wang, Z.H. Gao, W.L. Cui, S.X. Huo, Marine Science, **28**, 6(2004)
9. China's State Oceanic Administration, *The specification for marine monitoring* (Ocean Press, Beijiang, 1991)
10. Yang DF, Miao ZQ, Xu HZ, Y. Chen, J.Y. Sun, Marine Environmental Science, **32**, 3(2013)

 https://doi.org/10.1051/e3sconf/20183803052

Analysis of the typical small watershed of warping dams in the sand properties

LiLi [1,2,*], *Ji Shan Yang* [2], *Wei Ying Sun* [2], and *Sha Sha Shen* [2]

[1]Xi'an University of technology, Xi'an, Shanxi 710048, China
[2]Key Laboratory of Soil and Water Loss Process and Control on the Loess Plateau of Ministry of Water Resources, Yellow River Institute of Hydraulic Research, Zhengzhou 450003, China

Abstract. Coarse sediment with a particle size greater than 0.05mm is the main deposit of riverbed in the lower Yellow River, the Loess Plateau is one of the concentrated source of coarse sediment, warping dam is one of the important engineering measures for gully control. Jiuyuangou basin is a typical small basin in the first sub region of hilly-gullied loess region, twenty warping dams in Jiuyuangou basin was selected as research object, samples of sediment along the main line of dam from upper, middle to lower reaches of dam fields and samples of undisturbed soil in slope of dam control basin were taken to carry out particle gradation analysis, in the hope of clearing reducing capacity on coarse sediment of different types of warping dam through the experimental data. The results show that the undisturbed soil in slope of dam control basin has characteristics of standard loess, the particle size are mainly distributed in 0.025~0.05mm, and the 0.05mm particle size of Jiuyuangou basinof loess is an obvious boundary; Particle size of sediment in 15 warping dam of Jiuyuangou basin are mainly distributed in 0.031~0.05mm with the dam tail is greater than dam front in general. The separation effect of horizontal pipe drainage is better than shaft drainage for which particle size greater than 0.05mm, notch dam is for particle size between 0.025~0.1 mm, and fill dam is for particle size between 0.016 ~ 0.1 mm, they all have a certain function in the sediment sorting.

1 Introduction

Nearly 50% of the silt in the river channel of the lower reaches of the Yellow River is coarse sediment with a particle size > 0.05mm, the main channel sediment and coarse sediment particle size greater than 0.05mm over 70%. Coarse sediment is the main source of sediment in the main lower reaches of the Yellow Rive[1]-[3], Jiuyuangou basin is a source of one of the coarse sand in the middle reaches of the Yellow River[4], The warping dam is one of the main maintain soil engineering measures in the basin[2]-[4], It plays an important role in reducing sediment deposition in the lower reaches [5]-[9].Sediment in 15 warping dams of Jiuyuangou dam sample through the probe hole method[10], At the same time,the original soil sampling in the dam control basin,then particle gradation analysis of samples, the difference of different types of silt Engineering Dam in Jiuyuangou dam to clear sediment sorting, in particular, the separation difference between coarse sediment with a particle size greater than 0.05mm. On the transformation of warping dam structure in the future, extend the dam of the deposition period and service life, all have a positive effect.

2 The Study Area and The Sampling Method

2.1 The Study Area

Jiuyuangou watershed's a ditch that is located middle and downstream area in the middle reaches of the Yellow River's Wuding River on the left bank. Jiuyuangou watershed is The typical small watershed about first sub region of Loess Hilly Region.Channel section is in a V shape.The main channel length is Eighteen km. Average gradient of the channel is 1.15%,gully density is 5.34 km/km^2, sea level elevation is 820-1180m, the catchment area is 70.10km^2. Jiuyuangou watershed is extremely severe erosion area.

2.2 The Sampling Method

Selection twenty dams in Jiuyuangou watershed(refer with: Table 1), along the line of dam in middle thread of channe, upstream, midstream and downstream of the layout of a sampling point, in the upstream, midstream and downstream of the layout of a sampling point, through the adjustable length of Luoyang shovel sampling, common deep 2m, the weight of each sample sample is about 0.5kg, and control of basin slope sampling sampling dam, sampling once per 0.05m, common deep 0.15m.

* Corresponding author: 47994494@qq.com

Table 1. The Main Sampling Jiuyuangou Watershed Check Dam Situation

Name of the dam	Basin unit Watershed	Type of dam	control area of watershed (km²)	Drainage works	Influence of upstream dam
Weijiayan 3# dam	Hejiagou	Small dam	0.85	Nothing	no
Hejiagou 2# dam	Hejiagou	Key dam	1.69	Aven	yes
Hejiagou 1# dam	Hejiagou	Small dam	0.11	Nothing	no
Weijiayan 1# dam	Hejiagou	Small dam	0.16	Nothing	yes
Laolimo dam	Maliangou	Medium dam	1.36	Horizontal pipe	yes
Maliangou key dam	Jiuyuangou	Key dam	2.43	Horizontal pipe	yes
Haojialiang dam	Xiyangou	Key dam	2.69	Horizontal pipe	yes
Xiyangou Mizoguchi dam	Xiyangou	Key dam	2.43	Spillway	yes
Xiyangou Village front dam	Xiyangou	Medium dam	1.81	Horizontal pipe	yes
Tanyanggou dam	Xiyangou	Gap dam	0.29	Nothing	no
Mazhangzui dan	Lijiazhai	Key dam	0.99	Horizontal pipe	no
Fanshan dam	Lijiazhai	Medium dam	2.16	Aven	no
Beitagou dam	Wangmaogou	Small dam	0.19	Nothing	no
Guandigou 3# dam	Wangmaogou	Small dam	0.12	Aven	no
Guandigou 2# dam	Wangmaogou	Small dam	0.2	Nothing	no
Guandigou 4# dam	Wangmaogou	Medium dam	0.41	Aven	no
Wangbogou 2# dam	Wangmaogou	Small dam	0.17	Aven	no
Nianyangou 4# dam	Wangmaogou	Small dam	0.21	Nothing	no
Sdizui 2# dam	Wangmaogou	Small dam	0.59	Aven	no
Guandigou 1# dam	Wangmaogou	Medium dam	1.11	Aven	yes

3 Analysis of Grain Size of Original Loess in Jiuyuangou Watershed

Table 2 shows that the Jiuyuangou watershed of loess gives first place to silt particles（0.05~0.005mm）.The change range of the soil mass percentage of fine sand（>0.05mm）percentage of the silt size is 17.3 ~ 36%, the change value is 18.7%. The change range of the soil mass （0.05~0.005mm）percentage of the silt particle size is 47 ~ 62.2%, the change value is 15.2%. The clay is 12.8 to 23.8%, the change value is 11%. The difference between the maximum and the minimum of the content of different particles is close.

The differences of average content of different sizes of Jiuyuangou watershed of loess in great. The average value of the soil mass（>0.05mm）percentage of fine sand particle size is 27.6%. The silt（0.05~0.005mm）is 55.5%. The clay（<0.005mm）17%. In Jiuyuangou watershed of loess in silty soil, followed by sand, clay is again. The average soil mass percentage of 0.05~0.01mm particle in silt is 46.4%.The average soil mass percentage of 0.01 ~ 0.005mm is 9.1%., The coarse silt is more than 5 times that of the fine silt soil, and has the main characteristics of the standard loess.

Table 2 Analysis results of particle gradation of undisturbed soil in Jiuyuangou basin

Sampling point of original loess in Dam control basin	The particle size of soil mass less than a particle size (mm)		
	>0.05mm	0.05-0.005mm	<0.005mm
Ge jia gou Dam	27.8	56.6	15.6
Wang mao gou Men hu lu Dam	36.0	47.3	16.7
Wang mao gou Shu jing Dam	25.0	62.2	12.8
Wei jia yan 3# dam	28.3	50.0	21.7
He jia gou 2# Dam	30.8	54.0	15.2
He jia gou 1# Dam	20.4	61.3	18.3
Xi yan gou dan	24.4	58.5	17.1
Hao jia liang dam	35.4	47.0	17.6
Ma lian gou dam	20.2	56.0	23.8
Lao li mao dam	28.7	58.4	12.9
Ma zhang zui dam	25.3	57.5	17.2
Fan shan dam	32.6	52.9	14.5
Xia qiao gou 3# dam	28.8	55.6	15.6
Xia qiao gou 2# dam	32.3	54.0	13.7
Xiang ta gou dan	17.3	60.6	22.2
minimum value	17.3	47.0	12.8
Maximum value	36.0	62.2	23.8
Average value	27.6	55.5	17.0

4 The Analysis of Different Types of Engineering Dam Sediment Particle

According to the particle size analysis results(refer with: Table 3), the silt dams that the average particle size's in front of the dam is less than the dam tail in Jiuyuangou Watershed are seventeen,81% of the total number; the silt dams that the average particle size's in front of the dam is close to the dam tail are three, 15% of the total number; the silt dams that the average particle size's in front of the dam staggeres distribution in the dam tail and intermediate are one,5% of the total number. Therefore,the one silt dams will be get rid of the sampling analysis.

Table 3 Particle gradation analysis of sediment from 15 sampling warping dam in Jiuyuangou basin

Dam name	position	The particle size of soil mass less than a particle size (mm)			
		<0.001	0.001~0.025	0.025~0.050	>0.050
Xi yan gou cun qian dam	The front of the dam	4.71	55.59	25.06	14.64
	Dam tail	3.14	32.46	32.77	31.63
Hao jia liang dam	The front of the dam	3.02	33.01	30.21	33.76
	Dam tail	3.12	29.77	28.28	38.83
Xi yan gou gou kou dam	The front of the dam	3.69	40.88	30.8	24.63
Tan yang gou dam	The front of the dam	3.5	36.24	31.92	28.34
Ma lian gou dam	The front of the dam	3.27	33.12	30.23	33.38
	Dam tail	2.92	26.83	30.65	39.61
Lao li mao dam	The front of the dam	3.36	42.91	30.36	23.36
	Dam tail	3.52	39.42	29.44	27.62
Ma zhang zui dam	The front of the dam	4.15	47.41	28.64	19.81
	Dam tail	3.21	35.31	30.5	30.99
Fan shan dam	The front of the dam	4.32	48.75	27.11	19.82
	Dam tail	3.98	43.23	28.61	24.18
Bei ta goui dam	The front of the dam	4.38	57.63	25.64	12.35
	Dam tail	3.17	32.59	33.36	30.88
Guan di gou 3# dam	The front of the dam	4.4	60.25	24.63	10.71
	Dam tail	3.01	30.42	33.43	33.15
Guan di gou 2# dam	The front of the dam	4.5	55.88	23.88	15.74
	Dam tail	3.69	37.03	29.48	29.8
Guan di gou 4# dam	The front of the dam	4.21	56.57	26.55	12.67
Wei jia yan 3# dam	The front of the dam	3.67	36.22	28.6	31.51
	Dam tail	3.31	31.1	28.27	37.33
He jia gou 2# Dam	The front of the dam	4.67	55.02	25.08	15.24
	Dam tail	3.8	40.63	28.3	27.27
He jia gou 1# Dam	The front of the dam	3.89	40.97	27.98	27.15

Sediment's size in front of the dam about Jiuyuangou Watershed's the 20 silt dams > 0.05mm and > 0.10mm,particle size volume arithmetic average percent respectively 22.41% and 3.07%, Sediment's size in the dam tail>0.05mm and>0.10mm,particle size volume arithmetic average percent respectively 32.07% and 4.89%. Jiuyuangou Watershed's the average particle size distribution curve about in in front of the dam and the dam tail is Figure 1. According to the particle size analysis results, the coarse sadiment size of Jiuyuangou Watershed's silt dam concentrats mainly in 0.05mm.

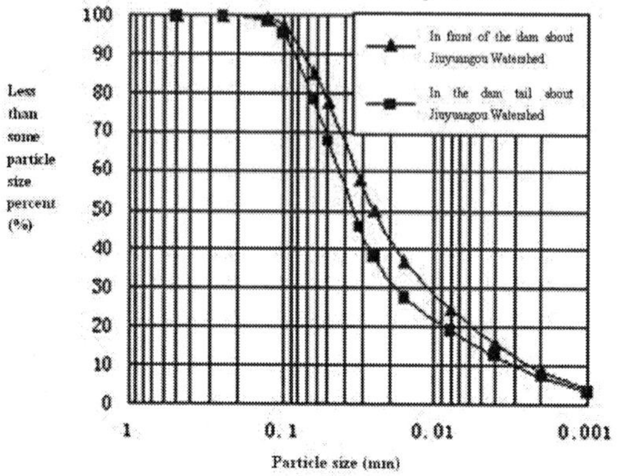

Figure 1 Average sediment grading curve in front of the dam and the dam tail about Jiuyuangou Watershed

According to the particle size analysis results(refer with: Table 3), in addition to tanyanggou que kou dam and xiyangou goukou dam, the average grain size distribution curve of the remaining 17 warping dams in the sediment of the display in front of the dam is located below the dam end curve curve shows that the average grain size of sediment in front of the dam is less than the tail dam (refer with: Figure 2). The proportion of sediment particle size at 0.031mm ~ 0.001mm, the drainage dam is greater than no drainage dam. The proportion of sediment with a sediment particle size larger than 0.05mm, the drainage dam is less than no drainage dam. The proportion of sediment particle size at 0.05mm ~ 0.031mm, no drainage dam and drainage dam is basically the same. Thus it can be seen,In front of the dam, the drainage dam is greater than no drainage dam in separation of coarse sediment with a particle size larger than 0.05mm (refer with: Figure 3).

At the front of the dam, the difference of sediment sorting is mainly because there is no drainage engineering dam for the upstream flood carrying sediment basically is retaining, the sediment concentration flood after the dam can not be discharged in time, the fine sediment carried by the flood is fully deposited, the drainage dam project is through the shaft or horizontal pipe part of the late deposition of fine sediment transported downstream. Therefore, sand in front of the drainage dam is crude than the no drainage dam.

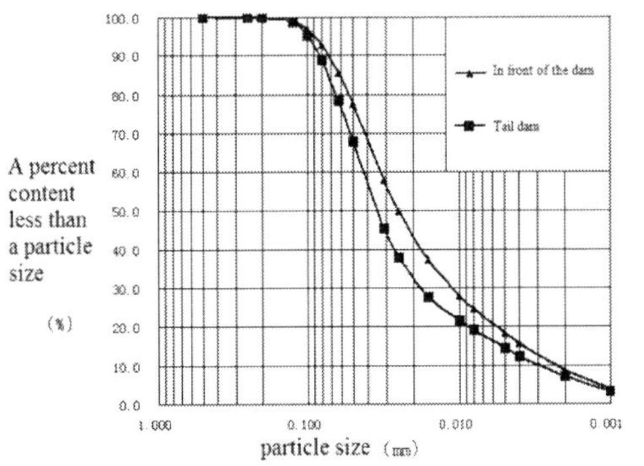

Figure 2 Average particle gradation curve of sediment from dam front and tail of 13 warping dam

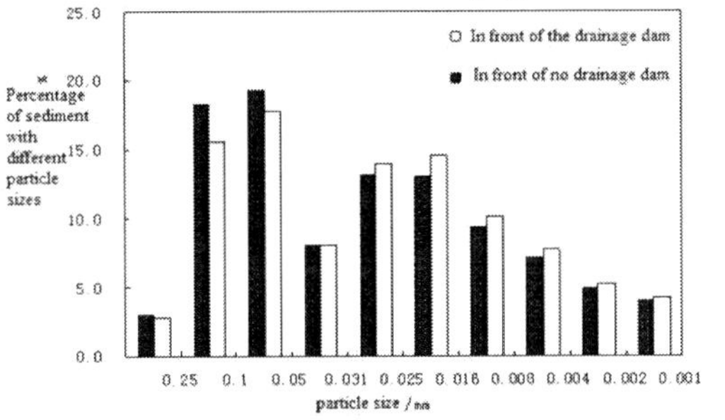

Figure 3 Variation in proportion of different particle size of sediment from dam in front of warping dam with drainage works and no drainage works

The proportion of grain size of mud and sand in front of the tanyangou dam from 0.025mm to 0.001mm, lower than the original soil in front of the dam. The proportion of particle size in 0.1mm ~ 0.025mm, the front of the dam is larger than the original soil. The proportion of sediment particle size from 0.25mm to 0.1mm, lower than the original soil in front of the dam.So, in front of the tanyanggou dam, on the particle size in 0.1mm ~ 0.025mm has a certain role in the sediment sorting. The proportion of sediment in the mud and sand grains in the front of the xiyangou goukou dam from 0.1mm to 0.016mm, the front of the dam is larger than the original soil. The proportion of sediment particle size from 0.016mm to 0.001mm and 0.25mm to 0.1mm, Lower than the original soil in front of the dam. So, the silt filled dam has a certain sorting effect on the 0.1mm ~ 0.016mm sediment in the upper reaches.

Table 4 shows that the particle size is greater than 0.05mm of sediment, horizontal pipe dam drainage optimal separation effect.

Table 4 Proportion sorting of sediment in different types of warping dam

The type of dam engineering	d_{yu}(%)		d_{yuan}(%)		d_{yu}/d_{yuan}(%)	
	d>0.05 mm	d>0.1 mm	d>0.05 mm	d>0.1 mm	d>0.05 mm	d>0.1mm
Drainage pipe drainage	26.0	4.4	21.9	3.5	118.7	125.7
Shaft drainage	20.7	3.5	21.2	3.5	97.6	100.0
No drainage	23.0	4.2	24.8	5.5	92.7	76.4
project						

Explain: d_{yu} is the arithmetic mean of the percentage of the sediment particle size greater than 0.05mm and 0.1mm in dam d_{yuan} is the arithmetic mean of the percentage of the sediment particle size greater than 0.05mm and 0.1mm in basin of Original loess

5 Conclusion

First, Jiuyuangou undisturbed soil particle size was mainly distributed in 0.025 ~ 0.05mm, the coarse silt is more than 5 times the fine silt, with standard features of loess; warping dams in sediment particle size mainly distributed in 0.031 ~ 0.05mm.

Second, At the end of the dam, there is little difference in the average particle size of silt in the dam drainage project danm and no drainage project dam. In front of the dam, dam separation effect is better than coarse sand dam drainage engineering for particles of size greater than 0.05mm without drainage engineering. The dam dam is better than vertical drainage drainage sorting horizontal coarse sediment particle size is greater than 0.05mm of the tube.

Third, the tanyanggou dam has certain sorting effect on the sediment with particle size ranging from 0.1mm to 0.025mm. The silted xiyangou goukou dam has certain sorting effect on the sediment with particle size from 0.1mm to 0.016mm.

Acknowledgments

This paper is based on work supported by the Central Nonprofit Research Institutions Basic Scientific Research Special Fund(HKY-JBYW-2018-08).

References

1. HUI Bo, LI Peng, ZHANG Wei, et al. Distribution characteristics of soil particles and organic carbon on check-dam system in Wangmaogou watershed[J]Research of Soil and Water Conservation, .2015, 22(4): 1-5.
2. ZUO Zhongguo, CHEN Hong, WANG Xiaobing, XU Jianhua, et al. Effect of warping dam in silt-laden and coarse sand region of the middle yellow river to sediment grading[J]. Yellow River, 2007, 29(2): 64-66.
3. LI Mian, YANG Jianfeng, HOU Jiancai, SHEN Zhongyuan. Sediment deposition process for a silt dam in a small watershed in Loess Hilly Region [J]. Transactions of the CSAE, 2008,24(2): 64-69.
4. XU Jianhua, LIN Yinping, WU Chengji, YU Quangang, et al. Research on boundary division of coarse sand concentrated source region of the middle Yellow River[M]. The Yellow River Water Conservancy Press, 2006.
5. QIAN Yunping, LIN Yinping, DONG Xuena, et al. Analysis of coarse sediment modulus change and sediment yield in the coarse sediment area in the

https://doi.org/10.1051/e3sconf/20183803052

Yellow River from 1855 to 1995[J]. Yellow River, 1998, 20(40): 15-18.

6. HAN Peng, NI Jinren. Grain size variation of suspended sediment due to soil conservation in the middle reach of Yellow River[J]. SHUILI XUEBAO, 2001(8): 69-74.
7. RAN Dachuan, ZUO Zhongguo, SHANGGUAN Zhouping. Effect of check dam on retaining and reducing coarse grain sediment in middle reaches of Yellow River [J]. SHUILI XUEBAO, 2006, 37(4): 443-450.
8. ZHANG Wei, YANG Mingyi, ZHANG Fengbao,ZHANG Jiaqiong,ZHAO Tianyin et al. Profile distribution of particle size of sediment at a check dam in a small watershed of the Loess Plateau[J]Research of Soil and Water Conservation, .2015, 22(2): 17-21.
9. LIU Huifang, CAO Wenhong, QIN Wei, ZHANG Xiaoming. The Contribution of Silt Dam for Soil and Water Conservation Measures [J]China Rural Water and Hydropower, 2011(1): 55-59.
10. LI Li, SUN Weiying, SHI Xuejian. Analysis of sedimentgrain size of warping dams in Jiuyuangou and Tuwei river watersheds [J]. Yellow River, 2014, 36(5): 79-81.Luigi T. De Luca, *Propulsion physics* (EDP Sciences, Les Ulis, 2009)

Study on the Evaluation Scheme of Emergency Response Capacity of The Three Gorges - Gezhouba Dam

Renbo Li[1,*]

[1]Wuhan University of Technology, WUT, Wuhan, China

Abstract. according to different navigation condition and environment of traffic emergency strategy, using fuzzy algorithm to the actual situation of ship parameters of the actual situation assessment, the result will be according to the instruction from the data graph for the actual operation, effectively avoid risk. Ships that should be properly arranged in a non-emergency session should avoid the concentration of traffic in heavy traffic and cause unnecessary security risks. The general function of ship safety is established by mathematical modeling, and the application function is used to control the ship. Ensure safe navigation.

1 Introduction

Now in the rapid development of water transport industry, which is associated with a lot of economic growth and profits, and the accident, although in recent years, since the number of shipping accidents in gradually reduce, but in the future, not to establish a good set of emergency response system will probably shipping development, especially about the future a certain impact on the development of inland waterway transport.

Experience in recent years, many of the accident emergency response mechanism in China has been the universal enhancement, but emergency corresponding need further broaden the width, as a construction, shipping, transportation, inland water transport power we want in the whole reverse improve the transport ability of our country, in the process of transportation it is also necessary to ensure the safety of the ship.

While at home and abroad, the study of Marine emergency response method is relatively more, but in view of the emergency response of the inland river ships, especially of the three gorges project - gezhouba study the key areas of safety and emergency measures in China is relatively small, according to local specific situation, set a reasonable emergency response mechanism is particularly important.

As an important part of China's Yangtze river waterway, the three gorges gezhouba is strategically important in the middle reaches of the Yangtze river. However, the general situation of the three gorges area in recent years is not optimistic. The three gorges - gezhouba area is a sensitive area, and it is imperative to maintain the security here.

2 Research on underwater search and rescue capability.

2.1 requirements of divers

Divers, hard as a high-risk industry, in recent years, the faster, diving industry construction in our country, but most diver rescue in search and rescue in our country is still limited to the range of small rivers, small DAMS in our country. In the three gorges valley complex natural environment, in terms of quality, to develop a diver

- To have the very strong psychological work under pressure and ability to cope with sudden special circumstances
- The relevant departments should strengthen the diver training to improve the efficiency of the divers explored - To strengthen the construction of the diving industry and support, the diver to have a glorious, their heart of courage. Divers in addition to complete my PADI training after the success, also need to be in when the accident happened, to make the right judgment of the accident is within a few minutes to ten minutes before the rescue of the golden age, in the process of search and rescue, divers actions must be quick and efficient, this would require the relevant rescue departments timely orders and diving rescue team timely response.

In the process of search and rescue, it is necessary to form a good system of search and rescue, in the process of search and rescue, we must be on the water in the process of search and rescue, we should draw lessons from the domestic and foreign advanced underwater detection technology, and cultivate good diver rescue technology and technique. The equipment of the rescue should be equipped with the best possible weight loss, the new equipment on the new equipment more convenient and faster and more efficient.

2.2 unmanned rescue equipment

* Corresponding author: 517909060@qq.com

https://doi.org/10.1051/e3sconf/20183803053

In recent years, the development of underwater search and rescue technology in the unceasing development, search and rescue equipment is tend to be intelligent and kind, there is no search and rescue equipment is mainly unmanned aerial vehicle (UAV) for the water to search for, and for search and rescue and localization of underwater unmanned ship, no equipment of the advantages of high efficiency, the scope of the search and rescue, can undertake some divers to work, for example, a lot of unmanned underwater search and rescue equipment such as the ROV and sonar radar, one after another, now of the underwater robot can be the deepest underwater search and rescue of 6000 meters. And it can take three to six kilometers of fast walking. However, the technical research and development required on the equipment of unmanned equipment need to be further strengthened, and the maintenance of the equipment should be timely and effective.

2.3 research on underwater plugging

In the selection of underwater plugging tools and materials, reasonable selection should be made according to the ship's or dam leakage. Main plugging method: the stopper plugging, soft patch plugging, metal box to seal leak, t patch plugging, screw frame cooperate plugging and plugging box, t activities activities patch plugging, quick dry cement plug cracks, electric welding equipment repair, waterproof seats leak and so on. According to different situations, we should choose suitable methods and materials. In underwater plugging required materials should be able to quickly fill the gaps, and can quickly under water condensation and keep underwater can have certain structural strength, rapid hardening of plaster [1]is a very good choice, then on the choice of injection plugging equipment will choose to adapt to complex underwater environment of high intensity of transmitting equipment, to ensure the work of the leak can in the shortest possible time to complete, and ensure the strength of plugging parts; On flood protection scheme, in the face of the randomness of the accident, damage to the ship or is the hole shape and position of the dam body may vary, at this moment we need to in a certain way to deal with the damaged holes. For vulnerabilities, we can use the method of combination of explosion plugging plug and perforator to deal with broken holes in different situations.[2]

3 Research on emergency rescue of ships

3.1. research on ship subtraction

In order to ensure the safe passage of the ship in the area of the three gorges gezhouba, it is necessary to reduce the load in the former port. It is necessary to ensure the speed of loading and unloading without causing a lot of damage when mixing the weight of the goods. Therefore, the method of reducing load is to be optimized and improved.

The following points should be made in the aspect of the reduction:

• Select advanced cargo cranes to improve the efficiency ratio of the selected machines,

• Prior notice of the disaster reduction precondition to be ready to prepare, calculate the tidal range and adjust the ballast water properly,

• Select different port of barge for different port of destination,

• Quick load off,

■ modified the existing equipment, or introduce using today's most advanced unloading equipment

■ strengthen training, improve the loading and unloading workers' professional skills

■ need to loading unloading equipment distribution, makes reducing load on the steps of operation optimization.

• in order to further the lightening of time, space, to ensure the efficiency, speed, necessary to establish a platform for lightening[3] but in the Yangtze river shelter-forest such a complex case, it is difficult to building a platform of lightening, should not only consider the platform's location, and higher in the three gorges region, surrounded by mountains all around is not easy to choose, and to guarantee the size of the platform, and unable to meet this section of the channel width. So to sum up, as must request for platform construction, the need is artificial on lightening platform construction, the establishment of a complete set of emergency response and under the precondition of shipping environment, to gradually to the implementation of the construction.

3.2 research on the configuration of the tug wheel

The tugboat plays an important role in the occurrence of emergency, which is mainly divided into two aspects.

3.2.1 Configuration of port tug and auxiliary tug in inbound port

In according to the characteristics of the three gorges water, especially in flood season due to the instability flow a lot of ship to port is a very difficult thing, when exam outline the need for effective tug assistance, in terms of tug assistance should do the following:

• The number of towed wheels is not as good as the size of the ship, according to the size of the ship

• Improve the working efficiency of the tug

• Strive for the ability to allocate the tug, and the tug's response should be timely and effective.

3.2.2 Emergency tug configuration

In accident happened or what will happen in an emergency, tug of high-power, high-powered features in the loss of the ship's stability or missing will guarantee stability of the vessel, or is in the ship ran aground and not under the condition of the high tide of cases gives help to the ship. Tug scheduling needs to improve the tug ship is the port scheduling, tug preparations are to be prepared ahead of time, in the period of accidents, especially in flood season to drill work from time to time about the accident, at the same time to ensure the efficiency of rescue ship, ship it is necessary to the pleiotropic, hull structure more in line with reduce the flow resistance of body and improve the speed of development trend[4].on the basis of improving the ship's condition and optimizing the training on dispatching and strengthening the crew, the configuration of the emergency tug will be more efficient.

3.3 Research on ship search and rescue capability and emergency response capability

For each vessel, in the search and rescue ability to make correct evaluation of whole search and rescue is of important significance to use data envelopment analysis (DEA) and the fuzzy algorithm, first of all, establish multiple membership functions, the synthetic evaluation of the influence of various factors

There are several factors that determine search and rescue capability.

3.3.1 Density of vessel traffic flow

The difficulty of the rescue after the accident often depends on the size of the density of the ship, the ship denser emergency dangerous waters could further aggravate the accident caused by the loss, or cause secondary the happening of the accident, so the research of vessel traffic flow density, get information about the density of the ship.

3.3.2 Emergency tug configuration

Relevant departments of the emergency response speed is reaction of a department's ability to direct factors, at the beginning of the accident is the perfect time to rescue, is the most critical factor decided to rescue ability.

3.3.3 The relevant measures and measures are in accordance with the relevant provisions.

Aid is another key factor of the rescue ability, have a certain search and rescue emergency response speed, effective means of search and rescue is also a great influence on the success of search and rescue

To sum up, the evaluation of search and rescue capability is a multi-objective problem, so the fuzzy comprehensive evaluation method based on data envelopment is used to evaluate.

4 Research on ship search and rescue capability and emergency response capability

For each vessel, in the search and rescue ability to make correct evaluation of whole search and rescue is of important significance to use data envelopment analysis (DEA) and the fuzzy algorithm, first of all, establish multiple membership functions, the synthetic evaluation of the influence of various factors.

4.1 The analysis process

Data enveloping analysis method and fuzzy comprehensive evaluation method[5-7] are relatively mature and widely used evaluation methods in comprehensive evaluation. Data envelopment analysis (DEA) can be based on objective data shows the relative efficiency evaluation unit and put forward the corresponding improvement direction, and fuzzy comprehensive evaluation is for the quantitative factors in the evaluation system of index is good for processing. Evaluation results of this paper, by using data envelopment analysis (DEA) as a secondary fuzzy comprehensive evaluation assessment indexes evaluation, using data envelopment analysis (DEA) optimization results instead of fuzzy evaluation of expert evaluation, make the fuzzy comprehensive evaluation more objective and persuasive. The evaluation process of fuzzy comprehensive evaluation method based on data envelopment analysis is as follows:

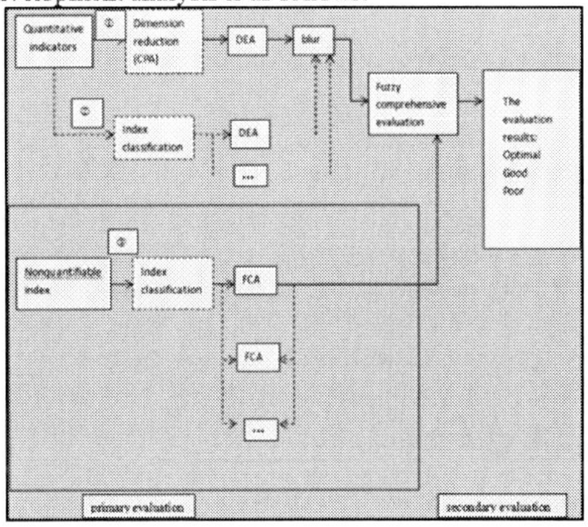

4.1.1 primary evaluation

When both exist quantitative indicators into the indicator system, there are also the quantitative indicators, the evaluation of quantitative indicators can be according to the statistics, rather than use the fuzzy comprehensive evaluation of the quantitative indicators to deal with the data standardization processing. At this point, we need to evaluate the two indicators separately. The primary evaluation of the D-FEC method is the separation of quantitative indicators and non-quantitative indicators.
The processing of quantitative indicators

When the number of indexes is too high, we use Principal component Analysis (PCA) to treat it. In statistics, principal component analysis is a simplified data set technique. It's a linear transformation. This transform the data transform into a new coordinate system, makes any data projection of the first big variance on the first coordinate is called the first principal component, the second big variance in the second coordinate, the second principal components so on. Principal component analysis can reduce the dimension of the data space studied. Using the y space of m dimension instead of the x space of p dimension, and the low dimensional y space to replace the high-dimensional x space loss of information is very little. Even if there is only one principal component, y1 (that is), the y1 is still obtained using all the x variables (p). So for example, if you want to compute the mean of Y1, you have to use the mean of all of the x's. In the first m principal component of the selection, if the coefficient of a certain coefficient approximates to zero, this can be deleted as well as a way to delete redundant variables.

4.1.2 secondary evaluation

On the evaluation of the secondary fuzzy comprehensive evaluation can effectively deal with people in itself with the subjectivity in evaluation process, as well as the objective of fuzziness, the fuzzy comprehensive evaluation is usually in the following steps:
•determine the collection of evaluation factors

$$U=\{u_1,\ u_2,\cdots,\ u_N\}$$

•establish a set of evaluation criteria

$$V=\{v_1,\ v_2,\cdots,\ v_M\}$$

•determine membership matrix
Assuming the ui of the No.i evaluation factor, a single factor evaluation is given to a fuzzy vector relative to vj

$$R_i=(r_{i1},r_{i2},\cdots,r_{ij}),i=1,2,\cdots,N;j=1,2,\cdots,M$$

•conduct multi-level comprehensive evaluation
According to the principle of maximum subordination, the evaluation grade of the evaluation object is determined and the evaluation conclusion is given.

5 Conclusion

Through the processing of quantitative indicators and the combination of fuzzy algorithm to work out the comprehensive evaluation scheme used in dealing with the accident emergency response, thus draws the evaluation level for us in the practical work is more effective to predict the type of accident to choose the right means to cope with the situation, but in the future development, the evaluation method will still new, more outstanding emergency response method is built better service in the field of ship safety, makes the ship safety better security.

References

1. Xing, Yangfeng, Research and application of rapid coagulant paste in the water plugging of foundation cofferdam [J]. Railway construction technology. 2011(09)
2. Chen Xiaoqiang, Zhang Keyu, Zhan Famin, Zhou Fangyi, Research on the technology of underwater patch repair [A] 2008(09)
3. Tang xiaogan, The safety operation of the platform for reducing loading and loading of green huashan mountains [J]. Navigation technology. 2009 (06).
4. GaoLlifeng, Key technical research on the overall design of emergency rescue and salvage tugs [J]. . Dalian university of technology. 2016 (03)
5. Liu Shun, Fuzzy comprehensive evaluation method based on data envelopment analysis and its application [D]. Zhejiang university, 2010
6. Huang Zhaofeng. Evaluation method and application of university education benefit based on fuzzy DEA. [D]. National defense science and technology university 2005
7. Gao Shubin, Liu Zi.based on fuzzy DEA's service government performance evaluation method [J]. Science and technology management. 2011(12)

Study on bridge checking evaluation based on deformation-Stress data

Jing Xian SHI [1,*], Ying Jie CHENG [2]

[1]Oxbridge College, Kunming University of Science and Technology,KunMing 650106Yunnan, China;
[2]Yunnan Aerospace Engineering Geophysical Detecting Co.Ltd, Kunming 650217, Yunnan, China

Abstract. Bridge structure plays a very important role in human traffic. The evaluation of bridge structure after a certain period of operation has always been the focus of the bridge. Based on the data collected from the health inspection system of a continuous rigid frame bridge on a highway in Yunnan, China, it is found that there is a certain linear relationship between the deformation and stress of the bridge structure. In view of a specific section of the structure, the stress value of this section can be derived according to its deformation value. The coefficient K can be calculated by comparing the estimated value to the actual measured value. According to the range of the K value, the structural state of the bridge can be evaluated to a certain extent.

1 Introduction

The research on the monitoring of bridge in the operation period abroad began in 70s. By the mid and late 80s, the monitoring system for the specific installation and operation period of some bridges has been started. Domestic concerns began in 90s, carried out monitoring during operation in jiang yin san jiang bridge, The Second Nanjing Yangtze River Bridge (2001), Runyang Yangtze River Bridge (2000-2004), Su Tong Yangtze River Highway Bridge (2007) [1] and some other large bridge has also established the structural health monitoring system. The main functional goal sof the health monitoring system are to realize data compression, index extraction, online warning, formation of maintenance advice and guidance for operation and maintenance. This paper conducts signal monitoring on a bridge located in Yunnan, China, and analyzes directly. According to the analysis results, it finds a suitable assessment method for such bridges to realize the assessment and determination of the structure.

2 Test section

There is some complicated consistency between the vertical deformation of the continuous steel bridge and the structural stress, because there are many kinds of functions involved in the measured structural response, such as live load, shrinkage and creep, relaxation of prestress, bearing Subsidence, temperature and strain and other variable effects. However, the responses caused by these loads can be separated from the measured data. Finally, the response of the structural response only to the live load can be obtained. The bearing capacity of the bridge is mostly measured by the load carrying capacity of the bridge, so this paper only studies the relationship between the stress and the deformation of the bridge under the action of live load.

The structure influence line is used to characterize the change of the internal force and deformation of a certain section with the change of the load position. At present, the research based on the impact line has been applied to model correction, damage identification and so on. According to this train of thought, combining static load test with point by point loading, the relation between displacement and stress under different section, different live load coefficient and different loading position is analyzed. The bridge structure is regarded as a "steelyard". The loading unit is not “1”, but it is scaled according to the first level standard load specified by the code, and then the comparison result is to find out the rule of live load. The study of this paper adopts the highway - I load, and the test section of the loading test section is shown in Figure1.

* Corresponding author: sara_shivip@163.com

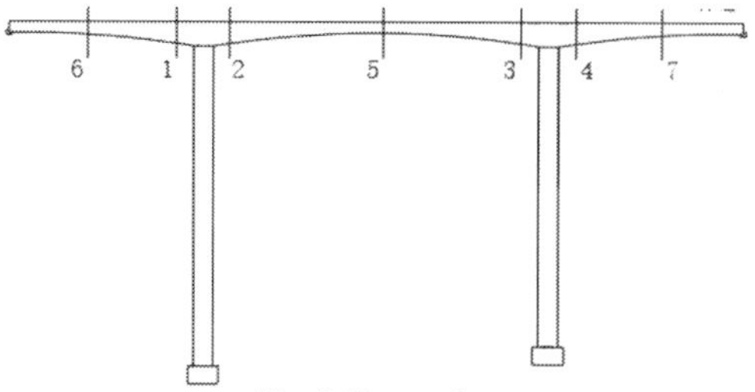

Fig. 1. Test section

3 Relationship analysis

For No. 1 section position, when the live load coefficient is 1, The variation curve of the stress and displacement of the roof as the loading position changes as Figure 2. It can be seen that the variation of structural stress and displacement under the action of live load is in good agreement (The location of the load in the diagram is the edge span of 1/4, the root, and the middle span of 1/2)

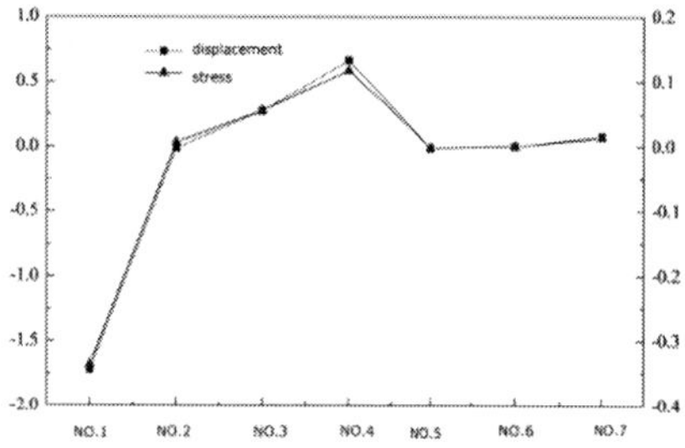

Fig. 2. Structural response under different loading position of No. 1 Section
x-loading position/m; y1-Expression of displacement/mm; y2-Expression of stress/MPa

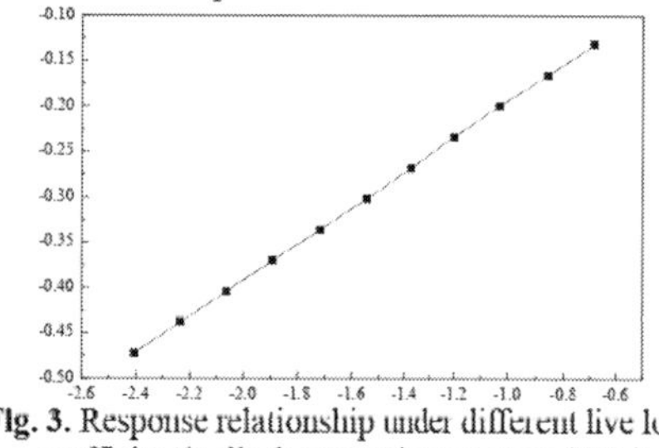

Fig. 3. Response relationship under different live load coefficient(x-displacement/mm;y-stress/MPa)

Also for No. 1 section, consider the different loading coefficients at the same loading section to see the structural response. It is found that from the normal serviceability limit state, the long-term combined live load coefficient 0.4 starts to 1.4 of the limit state of the load capacity, the strain and displacement of structure under load is linearly related to the live load coefficient, and both of them are also linearly consistent, as shown in Figure 3.

Table 1. Stress and displacement of different live load coefficient in No. 1 section No. 1 loading position

Live load coefficient	displacement(mm)	stress(MPa)
0.4	-0.687	-0.132
0.5	-0.859	-0.166
0.6	-1.031	-0.200
0.7	-1.203	-0.234
0.8	-1.375	-0.268
0.9	-1.547	-0.302
1.0	-1.719	-0.336
1.1	-1.891	-0.370
1.2	-2.063	-0.404
1.3	-2.235	-0.438
1.4	-2.407	-0.472

According to figure 3 and table 1, it is found that the response of the vehicle load at the same loading position is positively related to the multiple of the load. There is a linear relationship between the displacement and the strain, that we can ignore the effect of the load on the load. Based on this property, the influence of load multiplier can't be considered. on the other hand, the assumption of "bridge is steelyard" and the loading vehicle is "unit vehicle" is reasonable.

It is found that the displacement and strain of the same measuring point are not completely linear, by using the static load simulation to study the action of different loading positions. This nonlinearity is mainly shown in the root section, and the stress and displacement in the middle span are better. Considering the loading position is from one of the sides to middle of the middle span.

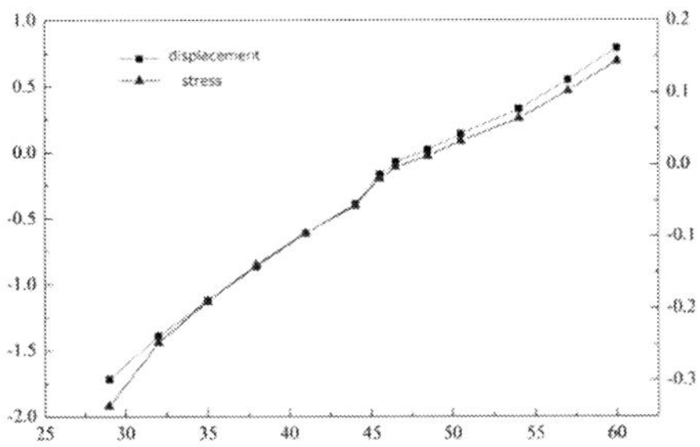

Fig.4. Stress and displacement changes from center of side to center of middle span
(x- loading position;y2-displacement/mm;y1-stress/MPa)

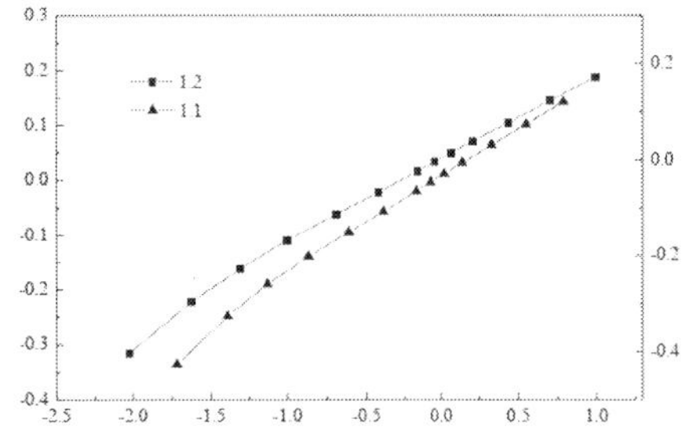

Fig.5. Response diagram under different coefficients (x-displacement/mm; y1-stress1.1/MPa;y2- 1.2/MPa)

Table 2. Stress and displacement relation of different loading positions in center of side span

Loading position /m	Loading coefficient1.0		Loading coefficient1.1		Loading coefficient1.2	
	Displacement/mm	Stress /MPa	Displacement /mm	Stress /MPa	Displacement /mm	Stress /MPa
29	-1.719	-0.336	-1.891	-0.37	-2.063	-0.404
32	-1.391	-0.248	-1.530	-0.273	-1.669	-0.297
35	-1.128	-0.190	-1.240	-0.209	-1.353	-0.228
38	-0.866	-0.140	-0.952	-0.154	-1.039	-0.168
41	-0.614	-0.095	-0.676	-0.104	-0.737	-0.114
44	-0.389	-0.057	-0.428	-0.062	-0.466	-0.068
45.5	-0.168	-0.020	-0.184	-0.022	-0.201	-0.024
46.5	-0.071	-0.004	-0.078	-0.004	-0.085	-0.004
48.5	0.020	0.012	0.022	0.013	0.025	0.014
50.5	0.138	0.032	0.151	0.035	0.165	0.038
54	0.328	0.064	0.361	0.071	0.394	0.077
57	0.549	0.102	0.604	0.112	0.658	0.122
60	0.789	0.143	0.868	0.157	0.947	0.171

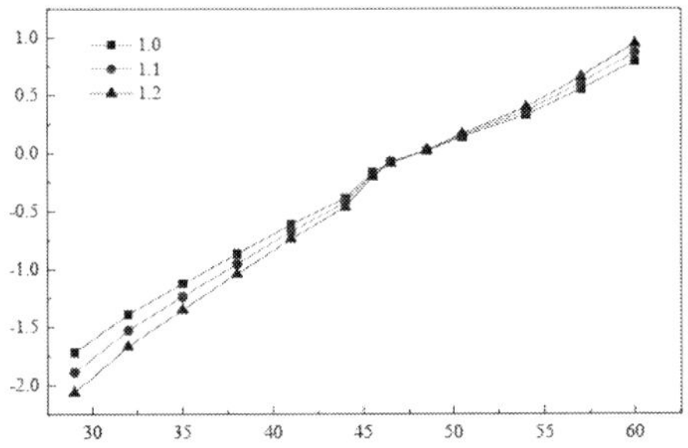

Fig.6. Displacement diagram under different coefficients(x-loading position/m;y-displacement/mm)

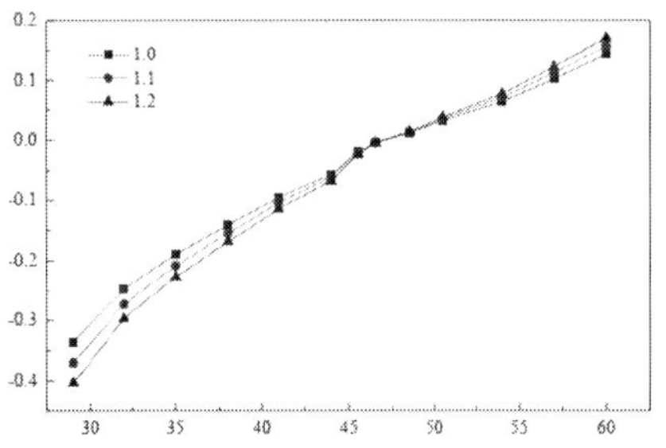

Fig.7. Stress in different coefficients(x-Loading position/m;y-stress/MPa)

Table 3. Relationship between stress and displacement of different loading position of root section

Loading position /m	Loading coefficient1.0		Loading coefficient1.1		Loading coefficient1.2	
	Displacement /mm	Stress /MPa	Displacement /mm	Stress /MPa	Displacement /mm	Stress /MPa
29	-0.238	0.156	-0.262	0.171	-0.285	0.187
32	-0.227	0.125	-0.250	0.137	-0.272	0.150
35	-0.213	0.980	-0.234	0.107	-0.255	0.117
38	-0.194	0.068	-0.214	0.075	-0.233	0.082
41	-0.172	0.035	-0.189	0.039	-0.206	-0.042

44	-0.147	-0.015	-0.162	-0.017	-0.177	-0.018
45.5	-0.111	0.009	-0.122	-0.010	-0.133	-0.011
46.5	-0.094	-0.004	-0.104	-0.004	-0.113	-0.005
48.5	-0.080	0.009	-0.087	0.010	-0.095	0.011
50.5	-0.059	0.025	-0.065	0.027	-0.071	0.030
54	-0.028	0.050	-0.031	0.055	-0.034	0.060
57	0.008	0.080	0.009	0.088	0.010	0.096
60	0.049	0.111	0.054	0.123	0.058	0.134

It is found that the linear relationship between deformation and stress is consistent with the previous one, although the nonlinearity of the root is very obvious as table 3. But because the consistency of the deformation and strain is mainly studied in the cross section, this linear relationship model can be used.

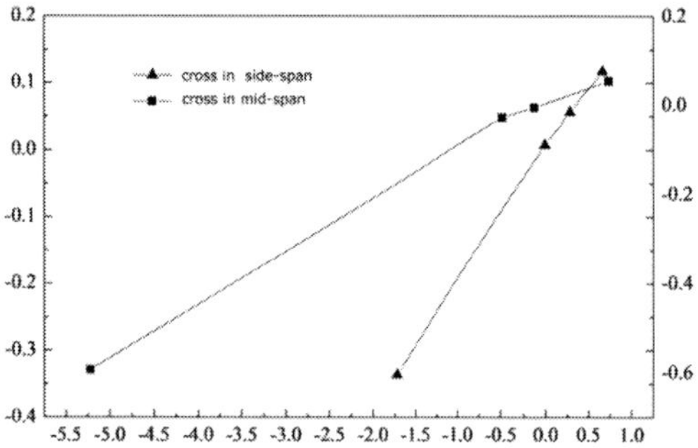

Fig.8. The relationship between stress and displacement under different loading position of No. 1 Section (x-displacement/mm; y1-center of middle span stress/MPa y2-center side span stress/MPa)

4 Evaluation method Conclusion

Based on the analysis of the above data: There is a certain linear relationship between the deformation and stress of bridge structure. Then the stress value of this section is deduced for a specific section of a structure (the predicted value of stress). The coefficient K can be calculated by replacing the predicted value with the actual measured values in the following formula.

$$k = \left| \frac{\sigma_{\text{actual value}} - \sigma_{\text{predicted value}}}{\sigma_{\text{actual value}}} \right|$$

Table 4 Technical grade evaluation of deformation and stress data checking and analysis

Bridge Category	Ratings range	Structure status	Management tools
1	5%>k≧0	Satisfy the design state	Safety or enhanced monitoring
2	10%≧k≧5%	Meet the standard state	Adjustment limit load
3	k≧10%	Internal forces need to be adjusted	Traffic control or bridge closure

According to the above K value in Table 4, the structural state of the bridge can be determined. Managers can take corresponding management measures for this state. To some extent, this evaluation method can reflect the structural state of bridge, but it can't completely reflect the state of structure, so it can only be used as a part of bridge comprehensive evaluation. The study of comprehensive evaluation will be studied in detail in the other papers of the author.

Acknowledgments

Project Source: Yunnan Provincial Department of Education(2017ZZX319);

References

1. FENG Liang-Ping, LI Na. Status Quo and Trend of Research and Development of Long - span Bridge Structural Safety Monitoring System in China[J]. Highway, 2009, 05:176-181
2. Zong Zhou-hong , Gao min-lin , Xia zhang-hua . Finite element model validation of the continuous rigid frame bridge based on structural health monitoring Part I: FE model updating based on the response surface method[J] , China Civil Engineering Journal, 2011. 44(2): 90-98.
3. AASHTO(2003), Guide Manual for condition Evalua-tion and Load and Resistance Factor Rating (IRFR) of Highway Bridges[s].
4. Ming Liu,Dan M. Frangopol, and Sunyong Kim, Bridge Safety Evaluation Based on MonitoredLiveLoad Effects[J]Journal of Bridge Engineering,2009
5. REN Yuan, HUANG Qiao , LIN Yang-zi.Study and Developmentof Comprehensive Assessment System of Long Span CablestayedBridges [J] .Journal of Nanjing University of Aeronautics&Astronautic , 2007 , 39 (4):535-539.
6. ZHAO Ji-guang , LU Jian , XIANG Qiao-jun, et al.Researchon Highway Maintenance Quality Evaluation System [J] .Journalof Highway and Transportation Research and Development,2006, 13 (4):1-6.

Analysis and Research on the Optimal Allocation of Regional Water Resources

Xi rui-chao,Gu yu-jie

Zhejiang Cuangchuan Engineering Consulting Co.,Ltd.,Hangzhou, Zhejiang 310020, China

Abstract: Starting from the basic concept of optimal allocation of water resources, taking the allocation of water resources in Tianjin as an example, the present situation of water resources in Tianjin is analyzed, and the multi-objective optimal allocation model of water resources is used to optimize the allocation of water resources. We use LINGO to solve the model, get the optimal allocation plan that meets the economic and social benefits, and put forward relevant policies and regulations, so as to provide theoretical which is basis for alleviating and solving the problem of water shortage.

1 Introduction

According to the WMO and UNESCO in the definitions of water resources, water resources is available source of water, the water should have sufficient quantity and the appropriate quality, and meet a specific use in a period of time. Taking the allocation of water resources in Tianjin as an example, by multi-objective optimal allocation and LINGO to solve the model, so as to provide theoretical which is basis for alleviating and solving the problem of water shortage.

2 Water Resources Current Status in Tianjin

First, the average annual value of surface water resources in Tianjin is 1055 million m^3, the groundwater resource content is 832 million m^3, the repeated calculation is 70.8 million m^3, and the total water resources mean is 1816 million m^3 in Tianjin[1]. Second, because of urban sewage discharge and pesticide pollution, and the annual river water volume decreasing year by year, the water self purification ability is reduced, and the water pollution of each river is serious, and the water quality is deteriorating. Third, the groundwater is overtaken seriously.There is salt water area in the south of Tianjin because that surface water resource is lack and it is explored seriously that it is difficult to restore the deep groundwater[2].

Through the understanding of the current situation of water resources in Tianjin, the optimal allocation model of water resources in Tianjin is established, which lays the foundation for the rational allocation of regional water resources.

3 Establishing the optimization model of water resources

The optimal allocation of water resources is the guarantee of the sustainable utilization of water resources, and it is the basic to realize the rational exploitation and utilization of water resources [3]. It refers to the basin area, by using the theory of system engineering and optimization method, a variety of available in regional and the water sector between the optimal allocation and allocation of water resources utilization, and strive to comprehensive benefits of social and ecological economy and realize the limited water resources maximum.

3.1 Objective Functions

The general goal or the highest level goal of water resources system optimization is to achieve rational allocation of water resources and support the comprehensive and coordinated development of the whole society and economy. [4]. Therefore, water resource optimal allocation is a complex multi-objective decision problem.

Its objective function can be described as follows:

$$M=\max\{f_1(X),f_2(X)\} \quad (1)$$

where X is decision vector made up of different number and quality water resources; $f_1(X)$ and $f_2(X)$ are the social and economic benefit of water resources development and utilization.

3.1.1 Objective 1: Economic Benefit

It is described by the maximum region water supply net benefit:

$$\max f_1(X)=\max\sum(B_j-C_j)X_j \quad (2)$$

where X_j is water supply volume (10^4 m^3); B_j is benefit coefficient (yuan/ m^3); C_j is cost benefit coefficient (yuan/ m^3).

* Corresponding author email: happyxrc@126.com

3.1.2 Objective 2：Social Benefit

It is described by the minimum total volume of region water shortage:

$$\max f_2(X) = -\min\sum\sum(D_i^k - \sum x_{ij}^k) \qquad (3)$$

where D_j^k is water demand volume of *j* user in *k* subarea (10^4 m^3); x_{ij}^k is water supply volume (10^4 m^3).

3.2 Constraint Condition

Water supply capacity constraint of regional water supply system, water demand capacity constraint of regional water system and regional harmonious development constraint are mainly considered in the water resources optimal allocation model of Tianjin city.

3.2.1 Regional Water Supply Capacity Constraint

Water supply capacity constraint of regional water supply system, that means water volume that obtained by *j* user in *k* subarea from water source should be meet the user's water supply:

$$\sum x_{ij}^k \le W_i^k \qquad (4)$$

where W_i^k is the water supply volume of *i* water resources in k subarea respectively.

3.2.2 Regional Water Demand Capacity Constraint

Water volume that obtained by *j* user in *k* subarea from water source should be between the upper and lower limit of user's water demands:

$$\sum x_{ij}^k \le H_i^k \qquad (5)$$

$$\sum x_{ij}^k \le L_i^k \qquad (6)$$

where H_j^k and L_j^k are the upper and lower limit of water demand volume of *j* user in *k* subarea respectively.

3.2.3 Variable Non-negative Constraint

The variable non-negative constraint, which means that the supply water are non-negative, and as following:

$$x_{ij}^k \ge 0 \qquad (7)$$

3.3 Model Establishment

The whole water resources optimal allocation model of Tianjin city can be obtained to combine the objective functions with the constraint conditions.

$$obj: M(X) = opt\{f_1(x), f_2(x)\}$$

$$\max f_1(X) = \max\sum_j (B_j - C_j)X_j$$

$$\max f_2(X) = -\min\sum_{k=1}^{K}\sum_j (D_j^k - \sum_i x_{ij}^k)$$

$$s.t. \qquad \sum_{j=1}^{J(k)} x_{ij}^k \le W_i^k$$

$$L_j^k \le \sum_{i=1}^{I(k)} x_{ij}^k \le H_j^k$$

$$x_{ij}^k \ge 0$$

where X is decision vector; f_1(X) and f_2(X) are the social and economic benefit of water resources development and utilization; D_j^k is water demand volume of *j* user in *k* sub area (10^4 m^3); W_i^k is the water supply volume of *i* water resources in k sub area respectively.

4 Solving Model

Water resources optimization model of Tianjin city is multi-objective optimal model, in order to solve model, we use the optimal solver of LINGO.

4.1 The upper and lower limits of Living Water Demand

Considering the characteristics of domestic water, the upper and lower limits are taken as the water requirement.

$$H_1^k = L_1^k = D_1^k \qquad (k=1,2,3) \qquad (8)$$

where H_1^k and L_1^k are the upper and lower limit of water demand volume respectively; D_1^k is the life water volume of *k* subarea in planning year. Thus, the upper and lower limit of life water of Beisan river mountain area is H_1^1= L_1^1= H_1^1=64 million m^3, and the upper and lower limit of life water of Beisi river plain is H_1^2=L_1^2=D_1^2=534 million m^3, and meanwhile, the upper and lower limit of life water of Diandongqingnan plain is H_1^3= L_1^3= H_1^3=453 million m^3.

4.2 Upper and Lower Limits of Industrial Water Demand

Considering the characteristic of industrial water demand, the upper and lower limits of industrial water demand volume is calculated by the following formula:

$$\begin{cases} H_2^k = D_2^k \\ L_2^k = 0.8\,D_2^k \end{cases} \qquad (9)$$

where $H_2^k$and $L_2^k$are the upper and lower limit of life water consumption respectively; $D_2^k$is the industrial water demand volume of *k* subarea in planning year.

Therefore, the upper limit of industrial water consumption in Beisan river mountain area is H_2^1=0.0720 billion m^3, and the lower limit is L_2^1=57.6 million m^3; the upper limit of industrial water consumption in Beisi river plain is H_2^2=600 million m^3, and the lower limit is L_2^2=480 million m^3; the upper limit of industrial water consumption in Diandongqingnan plain is H_2^3=508 million m^3, and the lower limit is L_2^3=406 million m^3.

4.3 Upper and Lower Limits of Agricultural Irrigation Water Demand

The upper and lower limits of agricultural irrigation water demand should be decided by effective irrigation area, guaranteed irrigation area and comprehensive irrigation quota:

$$\begin{cases} H_3^k = S_{yx}^k \times G^k = D_3^k \\ L_3^k = S_{bx}^k \times G^k \quad (k=1,2,3) \end{cases} \qquad (10)$$

where H_3^k and are the upper and lower limit of life water consumption respectively; S_{yx}^k and S_{bz}^k are effective irrigation area and guaranteed irrigation area of k subarea; D_3^k is the industrial water demand volume of k subarea in planning year.

Therefore, the upper limit of agricultural irrigation water demand volume in Beisan river mountain area is H_3^1=156 million m^3, and the lower limit is L_3^1=140 million m^3; the upper limit of agricultural irrigation water demand volume in Beisi river plain is H_3^2=1300 million m^3, and the lower limit is L_3^2=1157 million m^3; the upper limit of agricultural irrigation water demand volume in Diandongqingnan plain is H_3^3=1102 million m^3, and the lower limit is L_3^3=992 million m^3.

4.4 Solving model Based on LINGO

New model can be obtained by combining goal programming in optimization solver of LINGO with the above conditions and inducing into original model. And we may get the data from table 1 to table 3, the result from table 4.The LINGO model is as follows:

$$\min = P_1(3d_1^- + 3d_2^+ + 3d_3^- + 3d_4^+ + 3d_5^- + 3d_6^+) + P_2(2d_7^- + 2d_8^+ + 2d_9^- + 2d_{10}^+ + 2d_{11}^- + 2d_{12}^+) + P_3(d_{13}^- + d_{14}^+ + d_{15}^- + d_{16}^+ + d_{17}^- + d_{18}^+) + P_4 d_{19}^+$$

$$s.t.\begin{cases} \sum_{j=1}^{J(K)} x_{i,j}^k \ge W_{,s}^k \\ \sum x_{i,j}^k \le W_{,D}^k \\ \sum_{j=1}^{J(K)} x_{i,j}^k + d_s^- - d_s^+ = L_j^k \quad s = 1,3,5,\dots,17 \\ \sum_{j=1}^{J(K)} x_{i,j}^k + d_t^- - d_t^+ = H_j^k \quad t = 2,4,6,\dots 18 \\ \sum_j (B_j - C_j) X_j \ge 0 \\ x_{i,j}^k \ge 0 \quad i = 1,2,3;\ j = 1,2,3;\ k = 1,2,3 \end{cases}$$

It can be shown that water distribution principles of mountains Zone of Beisan River, Plain of Beisi River and Diandongqingnan are satisfied basically by the above calculation and analysis. Life water requirement should be met firstly, and surface water should be used to supply water at the beginning, and shadow groundwater follows, and at last it is the other water sources(include wastewater reuse and cross border water resources etc.); according to water deficiencies the shortage of water is serious in Diandongqingnan plain that local water source is 224 million m^3 which cannot meet local water demand of 1851 million m^3 and it needs to divert a large scale of other sources to satisfy the demand water of normal life, industrial and agricultural water.

Table 1. The annual water supply scale of 2030 level in Tianjin under the condition of 75% guarantee rate（$10^8 m^3$）

Subarea	Surface water resource	Groundwater resource	Other water volume	Total
Beisan river mountain area	1.00	0.34	2.13	3.47
Beisi river plain	2.30	3.79	7.43	13.52
Diandongqingnan plain	1.87	0.37	0.00	2.24
Total	5.17	4.50	9.56	19.23

Table 2. Water demand table of Tianjin in 2030 under 75% guarantee rate （$10^8 m^3$）

Subarea	Life water	Industrial water	Agriculture water	Total
Beisan river mountain area	0.64	0.72	1.56	2.92
Beisi river plain	5.34	6.00	13.00	24.34
Diandongqingnan plain	4.53	5.08	11.02	20.63
Total	10.51	11.80	25.58	47.89

Table 3. Water resources benefit and cost table in Tianjin

Subarea	Benefit (Yuan/m^3)	Cost (Yuan/m^3)
Life water	500	2.38
Industrial water	280	3.78
Agriculture water	1	0.30

Table 4. The results of optimal allocation of water resources in Tianjin under the 75% guarantee rate in 2030（$10^8 m^3$）

Subarea	Life water			Industrial water			Agricultural water		
Beisan river mountain area	X_{111}	X_{112}	X_{113}	X_{121}	X_{122}	X_{123}	X_{131}	X_{132}	X_{133}
	0.30	0.34	0.00	0.15	0.00	0.57	0.00	0.00	1.56
Total	0.64			0.72			1.56		
Beisi river plain	X_{211}	X_{212}	X_{213}	X_{221}	X_{222}	X_{223}	X_{231}	X_{232}	X_{233}
	5.34	0.00	0.00	4.80	0.00	0.00	0.35	3.79	7.43
Total	5.34			4.80			11.57		
Diandongqingnan plain	X_{311}	X_{312}	X_{313}	X_{321}	X_{322}	X_{323}	X_{331}	X_{332}	X_{333}
	1.37	0.00	3.16	0.50	0.00	3.56	0.00	0.37	9.55
Total	4.53			4.06			9.92		

Note :$X_{ijk}$is the water volume used for *j* user of *i* water source in *k* sub area, where i represents the subarea; j represents the user, k represents the water source。

5 Measures for optimal allocation of water resources

5.1 Rational allocation of water resources

The available water resources distribution principle is tapping and saving [5]. The following measures should be taken to rationally optimize the allocation of water resources: ① Control of surface water by means of government macro-control. ② The exploitation of deep groundwater can be reduced so that it can be used as a backup water source or emergency water source as far as possible. ③ Diversion water with a comprehensive

https://doi.org/10.1051/e3sconf/20183803055

means.

5.2 Reinforcement of Water Resources Protection

To ensure the safety of water resources in China, water pollution should be prevented from the source, and the ecological protection of the water body of the reservoir, river and canal system. The specific measures are as follows: ① To combine civil air defense with technological prevention. ② Increasing publicity and enforcement of water pollution prevention and treatment.

5.3 Improving the efficiency and efficiency of water resources utilization

China's water resources are seriously short. Strengthening the rational allocation of water resources is the objective requirement for the sustainable development of the economic society. Water saving is also the top priority in the allocation of water resources. Therefore, it is imperative to improve the efficiency and benefit of water resources utilization.

6 Conclusions

China's water resources are seriously short, and it is resources water shortage and its exploitation and utilization ratio is large, and the water pollution is grave, meanwhile groundwater is explored excessively and water quality standard is exceeded. Based on this situation, the water resources of Tianjin city as an example, the model of multi-objective optimization is established economic and social benefits as the goal, through the LINGO optimization method of goal programming solver to solve the model, the first to meet the living water, then the second to content the industrial water, finally meet the principle of optimization scheme of water supply of agricultural water. Combined with the shortage of water resources in our country, we must not rely on water saving alone, we should take the policy measures that both open source and throttle should be equally emphasized. At the same time, we should pay attention to preventing and controlling water pollution and strengthening the protection of water resources.

References

1. Zhang Kai. About Tianjin water resources value and water value research. College of management of Tianjin university.2006,1.
2. Yao Yutian. The optimal allocation of water resources and water save research of Tianjin. The construction technology of Tianjin. 1996,4.
3. Han Haiyan, Yu Sujun. Study on the optimal allocation of water resources in Rong County based on Multi-objective optimization model. The agricultural science of Anhui. 2010,38 (19) :10200～10202.
4. Zhang Chengfeng. Study on optimal allocation of Yulin water resources based on genetic algorithms. Thesis for Master Degree Northwest A &F University .2008:21～29.
5. Zhou Shumei, Jia Liye, Zhou Jinda. Some kinds of suggests of Tianjin City about optimal allocation of water resources and water saving. Hydrology. 2009, 29 (1) :36～37

Construction Simulation Analysis of 60m-span Concrete Filled Steel Tube arch bridge

Jing Xian SHI*, Qing Hua DING

Oxbridge College, Kunming University of Science and Technology, KunMing 650106Yunnan, China

Abstract. The construction process of the CFST arch bridge is complicated. The construction process not only affects the structural stress in the installation, but also determines the form a bridge and internal force of the bridge. In this paper, a 60m span concrete filled steel tube tied arch bridge is taken as the background, and a three-dimensional finite element simulation model is established by using the MIDAS/Civil bridge structure analysis software. The elevation of the main arch ring, the beam stress, the forces in hanger rods and the modal frequency of the main arch during the construction stage are calculated, and the construction process is simulated and analyzed. Effectively and reasonably guide the construction and ensure that the line and force conditions of the completed bridge meet the design requirements and provides a reliable technical guarantee for the safe construction of the bridge.

1 Introduction

The engineering background of this paper is a simply supported concrete-filled steel tube tied arch bridge, which is located on the new urban road of Suijiang County in Yunnan, China. It is part of the relocation project of Xiangjiaba Hydropower Station. The layout of the main bridge is shown in Figure 1. The theoretical calculation of the bridge has a 60m span, a calculated calculated rise of arch is 12M, and ratio of rise to span is 1/5. The bridge deck structure adopts the vertical and horizontal beam system and integral bridge deck. There are 2 concrete-filled concrete arches in the entire bridge, and the cross section of the arch rib is dumbbell type. The design speed of 20km/h, bridge deck width: 0.25m (railing)+1.75m (pavement)+8.0m (roadway)+1.75m (pavement)+0.25m (railing), The basic intensity is 7 degrees, the peak acceleration of earthquake motion is 0.112g, and the category of anti-seismic is the C class.

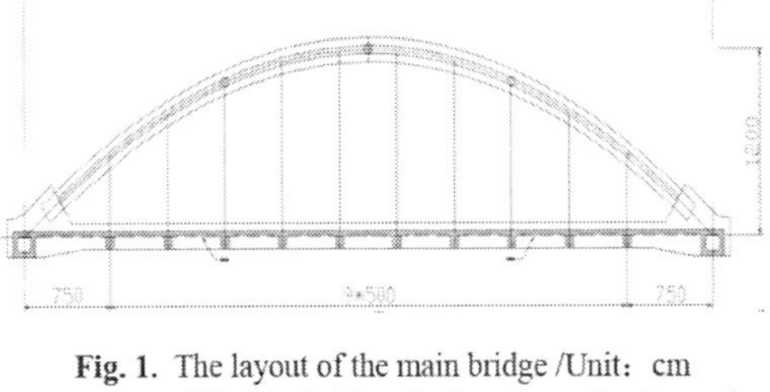

Fig. 1. The layout of the main bridge /Unit: cm

Concrete filled in bridge deck and steel tube arch rib is C50 micro-expansive concrete, the crossbeam of tie beam with C50 concrete; all ordinary steel bars are HRB400. The hanger rod adopts the OVM.GJ15-12 Steel Strand Full-bundle-extruding cable system, the standard strength is fpk=1860MPa, breaking force is 3125kN, the outer color HDPE sheath of the hanger rod is recommended to use orange.

2 The brief description of the analysis model

The simulation analysis of construction uses the MIDAS/Civil bridge structure analysis program. The structural analysis model is shown in Figure 2. The bridge has 516 units and 1324 nodes, in which the main arch, the wind bracing, the tie beam, the end floor beam and the inner beam unit unit are the space beam units. The hanger rod adopts the space truss element, and the bridge panel is analyzed by the space plate element. The boundary conditions in the analysis are as follows: the hinges of the full framing bottom(only compression, stiffness is 100kN/mm), the arch springing is hinged at both ends(longitudinal is movable), and the cast-in-place bridge deck unit is consolidated with the tie beam and cross beam.

There are 10 main kinds of calculation section types of the unit (variable cross-section is automatically calculated by the program), in which the concrete filled steel tube is used in the joint section of the construction stage, and there are 3 types of material types. The specific parameters are shown in table 1 and table 2. In the material parameters, the self weight and the constant load of the main arch are determined by the relation between the actual bulk density and the sectional area.

* Corresponding author: sara_shivip@163.com

 https://doi.org/10.1051/e3sconf/20183803056

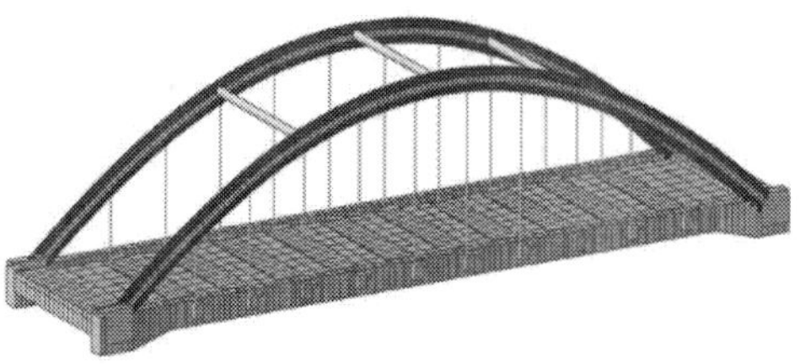

Fig.2. Simplified calculation model of the whole structure

Table 1. Cross section parameter summary of calculation model

No of Cross section	Location	Cross section area m^2	Izz	Czm
1	Boom	0.0059	0	0.0435
2	Tie beam1	3.5	0.5717	1.25
3	Tie beam2	1.36	0.0725	0.85
4	Variable cross-section of tie beam	3.5	0.5717	1.25
5	End floor beam	6.38	4.4713	1.1
6	Variable cross-section of end floor beam	2.3871	1.3599	0.7565
7	Inner cross beam	0.7	0.0286	0.5
8	Arch--steel tube	0.069	0.0035	0.8
9	Arch—concrete	0.7569	0.0172	0.7848
10	Wind bracing	0.3318	0.0088	0.325

Notes： **Izz**-The moment of inertia to the Z axis of the unit local coordinate system(Moment of Inertia) (m^4):**Czm**-In the direction of the -z axis along the local coordinate system of the unit, the distance between the unit section and the shaft to the edge fiber(m).

Table 2. Material parameter summary of calculation model

No.	Unit type	Material type	Elasticity modulus (MPa)	unit weight (kN/m^3)	thermal coefficient of expansion1/℃
1	Concrete in plate and steel tube arch	Concrete	3.45e4	25	1.0e-5
2	arch rib and wind bracing	rolled steel	2.10e5	78.5	1.2e5
3	prestressing force boom	rolled steel	1.95e5	78.5	1.2e-5

3 Construction simulation analysis

3.1 Division of construction stage

The concrete filled steel tubular arch section is constructed by bracket method. The arch ribs, tie beam, beams and deck plates are cast-in-place on the support formwork, and the steel structures such as arch ribs, wind bracing and so on are all fabricated in factories and assembled on site. Each arch rib is divided into three sections and assembled on the spot. According to the design drawing and construction management plan, the whole bridge construction is divided into 30 calculation stages to simulate, the total time period is 236 days, and table 3 is detailed.

3.2 Load calculation

The unit weight of empty steel pipe and hanger; each section of the concrete; tie beam and crossbeam are $78.5kN/m^3$; $25kN/m^3$; $26kN/m^3$, respectively. The self weight is assigned to each node in the form of the equivalent concentrated load. The shrinkage and creep are calculated at 3000 days after the bridge is formed. The temperature load is considered at heating-up 20°C and cooling 20°C uniformly. The hangers and arch ribs are considered in accordance with the individual temperature difference of ±10°C. This structure is an external static structure so that the settlement of the support is not considered. Vehicle live load is bi-directional 2 lanes, secondary road, the impact coefficient is automatically calculated by program. Crowd load is3.5KN/M, width of the sidewalk is 1.75m. The second constant load is applied to the bridge deck with uniform load, considering the constant load of bridge deck pavement, edge stone, railing and sidewalk.

$q=(1051.848+0.3767+48.3633+2119.53)kN/404.46\ m^2= 7.69\ kN/m^2$

Table 3. Division of bridge construction stage

construction stage	Duration	start time	terminal time	Name of construction component	Working condition description
1	30	1	31	Tie beam	Cast-in-place beam
2	30	31	61	End crossbeam, Inner crossbeam	Cast-in-place end crossbeam and inner beam
3-7	1	61	64	End crossbeam	Tension prestressed steel bar N1-N3 - N2 - N4

 https://doi.org/10.1051/e3sconf/20183803056

9	30	64	94	arch rib	Installing steel pipe with arch ribs
10	5	94	99	wind bracing	Installing wind bracing
11	10	99	109	boom	Jib attachment
12	0	109	109	arch rib	Pumpcrete under the lower pipe of arch rib
13	15	109	124	arch rib	Pumpcrete under the above the arch rib
14-15	15	124	154	arch rib	Pumpcrete in arch rib
16-20	1	154	159	Boom1-5	Tensioning boom 3-3'(325KN) ,4-4'(325KN) , 2-2'(325KN), 5-5'(280KN), 1-1'(280KN)
21	1	159	160	tie beam	Tensioned prestressed steel barN2、N3
22	1	160	161	end crossbeam	Tensioned prestressed steel barN2
23	30	161	191	bridge deck	Pouring of the bridge deck unit
24	30	191	221		Secondary dead load
25-29	1	221	226	Boom1-5	Tensioning boom3-3'(650KN) ,4-4'(650KN), 2-2'(650KN), 5-5'(650KN), 1-1'(560KN)
30	10	226	236		Supports disassembly
31-32	3000	236	10236	Operating for 10 years and 30 years	Considering the long term shrinkage and creep

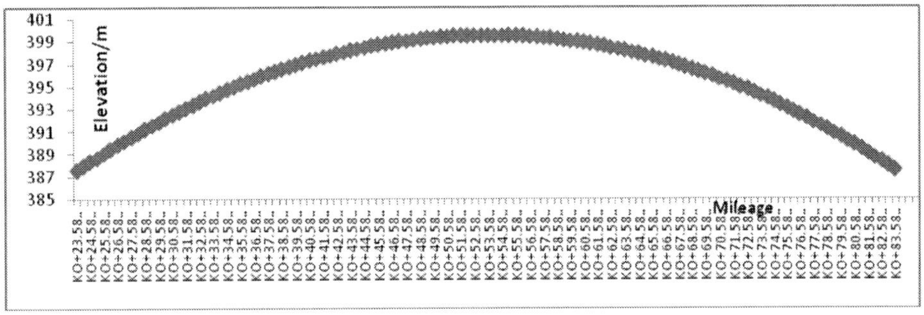

Fig.3. Arch ring linear of bridge

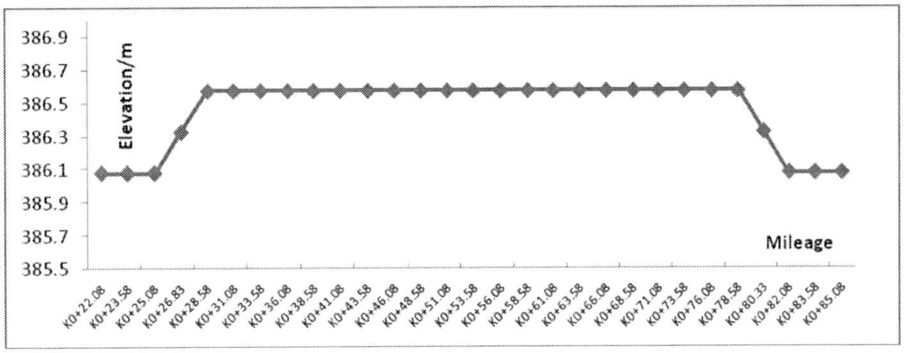

Fig. 4. Tie beam linear of bridge

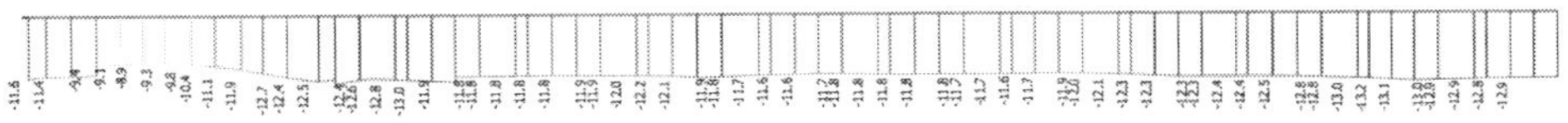

A Compression stress envelope of the upper edge of a tie beam(half bridge)

B Compression stress envelope of the lower edge of a tie beam(half bridge)

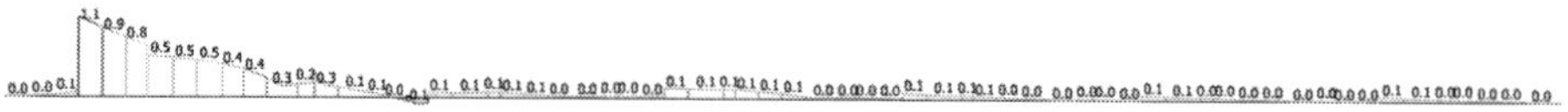

C The tensile stress envelope of the upper edge of a tie beam(half bridge)

D The tensile stress envelope of the lower edge of a tie beam(half bridge)

Fig.5. Calculation result of tie beam stress

4 Main results of construction phase calculation

4.1 The elevation of arch ring and tie beam

Calculation of the design elevation of the tie beam and arch based on the design drawings. According to the calculation results of the monitored calculation model and the formula of the specification, the pre arch degree and the bridge line form are calculated, the long term deflection is positive after calculation. The anti pre arch is not required according to the specification, so the design elevation of the girders and arches is shown in figures 3 and 4. The accumulative displacement of each node in the finished bridge is downward 3mm, the tie beam is upward 5mm and the crossbeam is upward 12mm. Therefore, it is not necessary to set up the pre - arch of the construction, and the elevation of the model is the sum of the design elevation and the deformation of the support.

4.2 Checking calculation of tie beam stress

According to the monitoring and analysis results, the stress envelope diagram of tie beams in all stages of the construction process is shown in Figure 5, the maximum compressive stress is 13.2MPa, and the maximum tensile stress is 1.8MPa (In the upper edge drawing, the 2.7MPa-6.2MPa near the support is due to the failure of the whole modeling to consider the local force, not the real stress of the structure.), which meets the specification requirements.

4.3 Checking calculation of cable force of boom

According to the calculation results, the maximum cable force at each stage of the boom in the whole construction process is shown in Table 4. It can be seen from the table that the safety factor is more than 4.0, which meets the requirements of the safety factor of the cable in the construction state of the specification <Guidelines for Design of Highway Cable-stayed Bridge>(JTG/T D65-1-2007)3.4.2.

Table 4. Calculation result of cable force of boom

Boom No.	Boom quantity	sectional area(mm^2)	cable force of the end beam(KN)	Cable force at the arch end(KN)	safety factor
101 102	4	1680	572.2	572.4	5.46
201 202	4	1680	661	661.3	4.73
301 302	4	1680	649.8	650.2	4.81
401 402	4	1680	650.8	651.2	4.80
501 502	4	1680	666.3	666.7	4.69

4.4 Frequency and mode of vibration

The results of the frequencies and modes of each stage are shown in Figure 6.

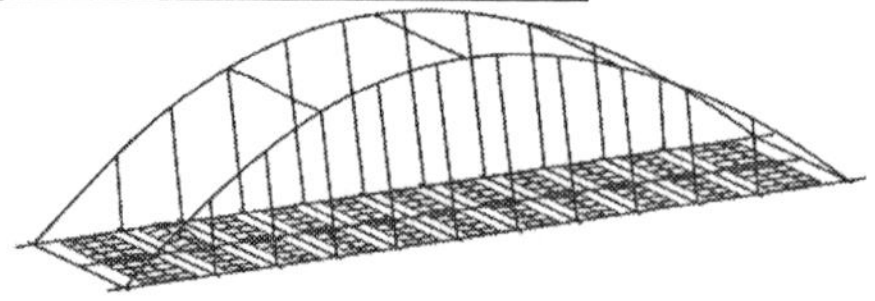

A First stage of mode of vibration(f=0.689 HZ)

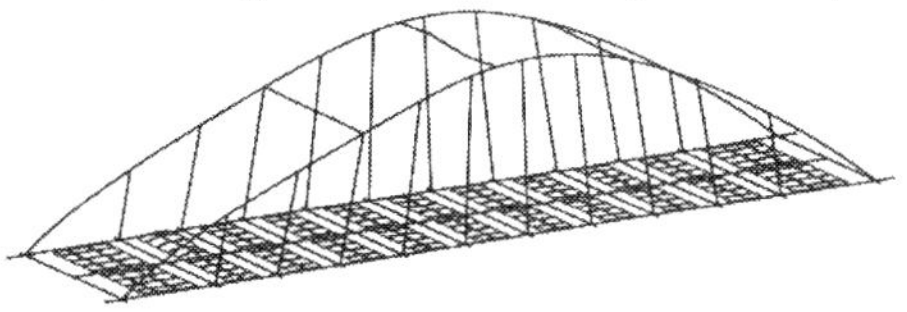

B Second stage of mode of vibration(f=1.774 HZ)

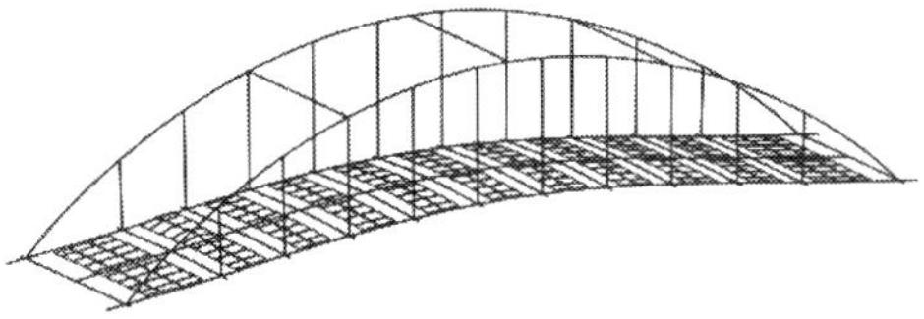

C Third stage of mode of vibration(f=2.156 HZ)

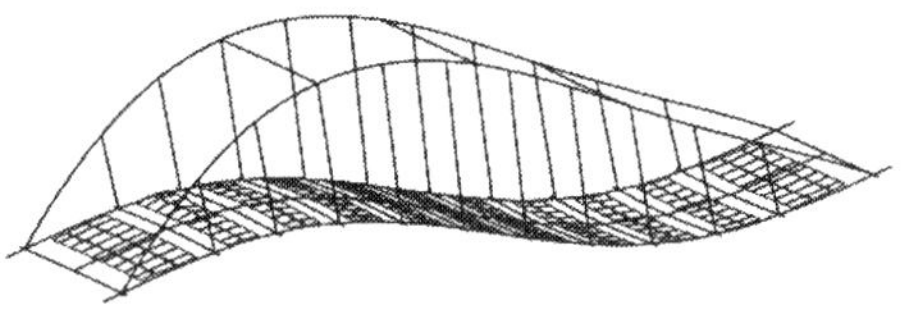

D Forth stage of mode of vibration(f=2.712 HZ)

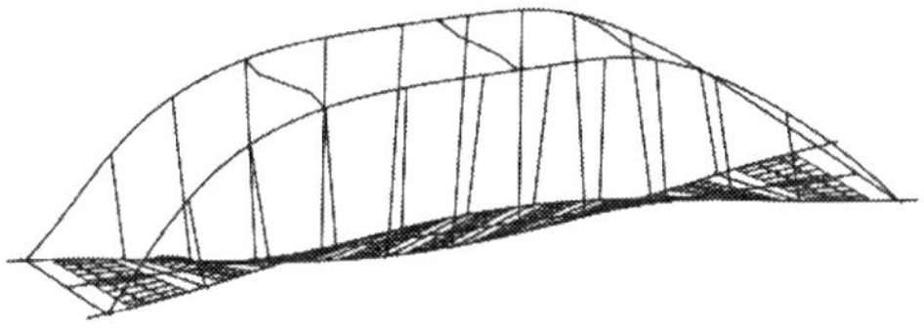

E Fifth stage of mode of vibration(f=3.424 HZ)
Fig.6. The frequencies and modes of each stage

4 Main results of construction phase calculation

This paper combines the example of the construction monitoring of the steel tube concrete tied arch bridge with 60m. The selection of load calculation parameters, the division of construction stages, the calculation of the vertical formwork elevation of the arch ring and the girder are mainly elaborated, and a reasonable bridge line is determined. The stress of the tie beam and the cable force of the boom, as well as the frequency and mode of the bridge are calculated.

According to the simulation calculation of the construction, the data of the height and stress of the structure are obtained, so that the construction of the bridge is carried out in order. In order to ensure the reasonable theoretical basis of the bridge alignment and stress, it provides a reliable technical guarantee for the safe construction of the bridge.

Acknowledgements

Project Source: Yunnan Provincial Department of Education(2017ZZX319);

References

1. People's Republic of China Ministry of Transportation. JTGD60-2015General Code for Design of Highway Bridges and Culverts Beijing: People's Communications Press.2015
2. People's Republic of China Ministry of Transportation. JTGD62-2004 Code for Design of Highway reinforced concrete and prestressed concrete and Culverts. Beijing: People's Communications Press.2004.
3. People's Republic of China Ministry of Transportation. JTG/T D65-1-2007Design Specifications of Highway Cable-stayed Bridge. Beijing: People's Communications Press.2004.
4. Midas IT (Beijing)Corporation.MIDAS/CIVIL Analys for Civil strueture2006.
5. ZHAO Hui,HAO Chao, Analysis of Mechanical Performance of Tied Arch Bridge with Multiple Arch Ribs [J]. World Bridge2013(41)52-56.
6. Wang Zhijun,Li Nianwei. Construction Monitoring for Concrete-filled Steel Tube Rigid Tied Bar Arch Bridge [J] . CONSTRUCTION TECHNOLOGY 2010(10) 63-66.

Damage Model of Reinforced Concrete Members under Cyclic Loading

Bo Chen Wei[1,2], *Jing Shu* Zhang[1,2,*], *Yin Hua* Zhang[3], and *Jia Lai* Zhou[1,2]

[1]Key Laboratory of Mechanics on Disaster and Environment in Western China of the Ministry of Education, Lanzhou University, Lanzhou 730000, China
[2]School of Civil Engineering and Mechanics, Lanzhou University, Lanzhou 730000, China
[3]Gansu Institute of Architectural Design and Research, Lanzhou 730000, China

Abstract. Based on the Kumar damage model, a new damage model for reinforced concrete members is established in this paper. According to the damage characteristics of reinforced concrete members subjected to cyclic loading, four judgment conditions for determining the rationality of damage models are put forward. An ideal damage index (D) is supposed to vary within a scale of zero (no damage) to one (collapse). D should be a monotone increasing function which tends to increase in the case of the same displacement amplitude. As for members under large displacement amplitude loading, the growth rate of D should be greater than that of D under small amplitude displacement loading. Subsequently, the Park-Ang damage model, the Niu-Ren damage model, the Lu-Wang damage model and the proposed damage model are analyzed for 30 experimental reinforced concrete members, including slabs, walls, beams and columns. The results show that current damage models do not fully matches the reasonable judgment conditions, but the proposed damage model does. Therefore, a conclusion can be drawn that the proposed damage model can be used for evaluating and predicting damage performance of RC members under cyclic loading.

1 Introduction

The failure of structure subjected to earthquake action is essentially caused by the accumulation of irreversible damage. Seismic performance of structure is the damage performance during earthquakes in essence [1]. Among various evaluation methods for seismic performance, the damage index D involved in damage models is more rational for evaluating damage state of structures and members. In addition, a damage model is established on the basis of performance-based seismic design and seismic damage assessment.

There mainly exist current failure criteria such as strength-based failure criteria, deformation-based failure criteria, energy-based failure criteria, deformation and energy-based failure criteria, and performance-based failure criteria [2]. Based on different failure criteria and experimental researches as well as seismic damage investigations, a variety of damage models are proposed. However, the damage index D, which ranges from 0 to 1 as per the definition, has to be determined for whatever damage model. D is a value of 0 in the absence of damage and a value of 1 in the case of complete failure. The larger the value D is, the more serious damage extent will become.

The damage state of RC members has not been described accurately by current damage models due to the complexity of damage development. There exist common problems that damage index D is larger than 0 when members are in elastic range, and the D is not converged to 1 when they approach to collapse.

In order to investigate above several key aspects, taking into consideration characteristics of current damage models at home and abroad, a damage model that is capable of a real reflection of RC members under cyclic loading is proposed on the basis of slabs, walls, beams and columns, 30 RC members in total.

2 Current main damage models

Damage models can be divided into two types: non-cumulative damage models and cumulative damage models. Taking into account the effect of first excursion, the former is simpler generally and mostly based on displacement [3], stiffness [4,5] and so forth. Loading histories and loading paths can not be characterized by the models, which suggest that they are not suitable for RC members subjected to cyclic loading.

Main cumulative damage models are listed in Table 1, most of which belong to double variables damage models based on dual deformation-energy failure criteria, and take into consideration the interactive effects on failure boundaries of maximum responses (e.g. strength, displacement or ductility) and cumulative damages (e.g. cracks, energy dissipation). These damage models are more suitable for RC members under cyclic loading.

* Corresponding author: jshzhang@lzu.edu.cn

Table 1. Cumulative damage models.

Scholars	Damage models	Characteristics
Park, Ang [6]	$D=\frac{X_m}{X_{u,mon}}+\beta\frac{E_h}{F_y X_{u,mon}}$	A Simple form, and having wide experimental foundation
Ou J P, Niu D T [7]	$D=\left(\frac{X_m}{X_u}\right)^\beta+\left(\frac{E_h}{E_u}\right)^\beta$	Same weights of energy dissipation component and displacement component
Fajfar [8]	$D=\frac{E_h}{F_y\left(X_{u,mon}-X_y\right)}$	One variable form, enabling to take into account the effects of low-cycle fatigue
Kumar, et al [9]	$D=(1-\beta)\left(\frac{X_m-X_y}{X_{u,mon}-X_y}\right)^c+\beta\left(\frac{E_i}{F_y\left(X_{u,mon}-X_y\right)}\right)^c$	Regarding the damage caused by displacement and energy dissipation as nonlinearity
Niu D T, Ren L J [10]	$D=\frac{X_m}{X_u}+\alpha\left(\frac{E_h}{E_u}\right)^\beta$	Various combinations with displacement and energy dissipation components, good upper convergence ability
Lu D G, Wang G Y [11]	$D=(1-\beta)\frac{X_m-X_y}{X_u-X_y}+\beta\frac{E_h}{F_y\left(X_u-X_y\right)}$	Convenient calculation and performance design for seismic damage

Where X_m, X_y denote the maximum displacement and yielding displacement through whole loading process, respectively. $X_{u,mon}$ is the ultimate displacement of the structure subjected to monotonic loading. X_u is the ultimate deformation. E_h is the cumulative dissipated hysteresis energy. E_u is the ultimate dissipated hysteresis energy. E_i denotes the dissipated energy of ith half circle. c, α and β are constants involved in those models.

3 The proposed damage model

Kumar and Usami [9] studied the mechanical properties of steel members under cyclic loading and put forward a suitable damage model, where displacement component and energy dissipation component were regarded as nonlinearity, and these two components were a combination of the model parameter β. It covered a wide range of applications and was capable of taking into account damage exacerbation due to large displacement. Aimed at making it suited for RC members, the damage model is modified appropriately.

3.1 Parameters involved in Kumar damage model

The Kumar damage model seemed to be derived from equation 1 and 2. By employing experimental data of Usami[12], Kumar et al assumed damage index was equal to 1 when members were in the ultimate state to calculate the exponential item c by means of trail-and-error method, and the results are given in Table 2. With the purpose that large displacement circle occupied larger weight, Kumar suggested the value c should be equal to 2. Then, the parameter β was obtained by trail-and-error method.

$$D=\sum_{i=1}^{N}\left(\frac{X_m-X_y}{X_{u,mon}-X_y}\right)^c \tag{1}$$

$$D=\sum_{i=1}^{N}\left(\frac{E_i}{E_{mon}}\right)^c \tag{2}$$

Table 2. The parameters involved in Kumar damage model.

Loading methods	Usami's [12] Experiment of steel columns		
	c (Eqn. 1)	c (Eqn. 2)	β
Loading three times with same amplitude of displacement per cycle	2.9	3.7	0.11
Loading with constant amplitude	3.0	3.1	0.12
Mean values	2.9	3.5	0.11

The Kunnath's experimental data of reinforced concrete bridge piers was used for figuring out the model parameters. The results are given in Table 3. Calculation results indicated that the Kumar damage model maintained better parametric stability when applied to RC members. Compared with steel members in Table 2, the value of the parameter c of RC members was smaller, while the parameter β was approximately equal to 0.3.

Table 3. The parameters involved in Kumar damage model for RC members.

Loading methods	Kunnath's [13] Experiment of RC members		
	c (Eqn. 1)	c (Eqn. 2)	β
Loading three times with same amplitude of displacement per cycle	2.8	3.0	0.29
Loading with constant amplitude	2.9	3.0	0.27
Mean values	2.9	3.0	0.28

On the basis of parameters involved in the model calibrated by Kunnath's [13] experimental data, the calculative results of the experimental specimen A2 under changeable amplitude loading conditions are given in Figure 1, from which it can be seen that the growth of damage index has no significant changes, nevertheless, maximum loads lead to the obvious growth of damage index. It is also observed that damage aggravation causes the decrease of bearing capacity. Kumar [9] has once observed the phenomenon in Usami's experiment.

A key aspect from above calculative results is, differing from steel members, that concrete is a kind of heterogeneous material within which there exist inevitable flaws such as pores, cracks between porosity and interface. RC structures are not ideal elastic-plastic body. A small amount of cumulative plastic damage of RC members is caused prior to yielding. However, the results calculated by the Kumar damage model do not reflect that phenomenon.

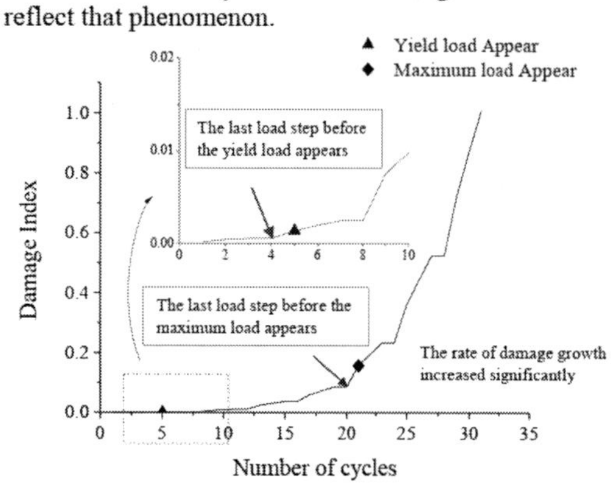

Fig. 1. Damage index of specimen A2.

3.2 The proposed damage model

For the members having similar mechanical properties when subjected to positive and reverse loading, dissipated energy of *i*th half circle E_i is transformed into that of a circle to simplify Kumar damage model. At the same time, dissipated energy E_{mon} multiplies by two when members are under monotonic loading. The parameters involved in the model are figured out by trail-and-error method based on Kunnath's empirical data. The results are given in Table 4.

Table 4. The Parameters involved in simplified Kumar damage model for RC structures.

Loading methods	c (Eqn. 1)	c (Eqn. 2)	β
Loading three times with same amplitude of displacement per cycle	2.1	2.3	0.61
Loading with constant amplitude	2.4	2.4	0.54
Mean values	2.3	2.4	0.58

In order to enhance convergence ability of Kumar damage model, dissipated energy component is rewritten and displacement component X_y is omitted at the same time. Then, the damage index of the proposed model can be written by the form:

$$D=(1-\beta)\left(\frac{X_m}{X_u}\right)^{c_1}+\beta\left(\frac{E_h}{E_u}\right)^{c_2} \tag{3}$$

Where, β a combination parameter evaluating the importance of displacement component and energy component, can be taken as 0.6 from Table 4 to keep the proportion of two components constant when members are at failure. Both of c_1 and c_2 are the exponential items' parameters. c_2 can be taken as 2.0, so as to give more importance to the larger plastic deformation, as Kumar suggested [9].

When at yielding, members are considered to approach to the upper limit of good condition. As the code [14] suggests, D=0.1 is an appropriate value for members at yielding. Hence, the exponential items' parameter c_2, the last unknown parameter, can be figured out by means of trail-and-error method. By employing experimental data of Zhang [15], the value of c_2 can be calculated as 1.2.

In this model, cumulative dissipated hysteretic energy E_h is normalized by ultimate dissipated hysteretic energy E_u to improve the upper limit convergence ability of the model. By employing different parameters c_1 and c_2, the relationship between total damage and displacement component damage, as well as the relationship between total damage and dissipated energy component damage can be evaluated to improve applicability of the model for different kinds of members.

4 Damage model rationality

A variety of current damage models differ significantly, and it is of much necessity to judge whether the damage model are rational according to some conditions.

4.1 Judging conditions for damage models rationality

For RC members subjected to cyclic loading, damage model rationality can be judged based on the following conditions:

Condition 1: D should have upper and lower limit convergences. D is a value of 0 when no damage occurs and a value of 1 in the case of complete failure.

Condition 2: The damage of members is irreversible. D is supposed to be a monotone increasing function.

Condition 3: Damage models should exhibit the effects of low-cycle fatigue when members are under same displacement cyclic loading, which means that D should tend to increase in this condition.

Condition 4: Damage models should exhibit that large displacement amplitude loading has larger influence on members' damage, compared with small displacement amplitude loading [16]. The growth rate of D under large displacement amplitude loading, should be greater than that of D under loading conditions with small amplitude of displacement.

Both of conditions 1 and 2 belong to the qualifications for basic framework of the whole damage models, which can be stated that all of damage models should match conditions 1 and 2. Conditions 3 and 4 refer to requirements of damage models' performance, which aim at guaranteeing the rationality of damage models for assessing the damage state of members.

4.2 Rationality judgments for damage models

The analysis of classical Park-Ang damage model [6], Niu-Ren damage model [10], Lu-Wang damage model

[11] and the proposed damage model, is based on the following parameter values:

(1) For the Park-Ang damage model, $X_{u,mon}$ is taken as $X_u/0.62$ which is obtained by analyzing quantities of experiments in that paper [17]. The parameter β is a value of 0.05 for well-confined RC members [18].

(2) For the Niu-Ren damage model, α=0.1387 and β=0.0814 [10].

(3) For the Lu-Wang damage model, parameter β is equal to 0.15 [19], and a value of 0 when the damage index is calculated to be negative.

(4) For the proposed damage model, parameters are involved in Section 3.

By employing experimental data of slabs [15], walls [20], beams [21,22] and columns [23], 30 RC members in total, damage state of members applied to various damage models are analyzed, as plotted in Figures 2 to 6. The results of D calculated by different damage models when members are at yielding or in the ultimate state, are given in Table 5.

Table 5. Damage indices involved in various RC members.

Damage models	At yielding		In ultimate state	
	Damage index	Mean values	Damage index	Mean values
Park-Ang	0.102-0.417	0.196	0.690-1.066	0.822
Lu-Wang	0.005-0.244	0.094	1.366-3.390	2.152
Niu-Ren	0.253-0.747	0.404	1.138-1.139	1.139
Prop	0.041-	0.	1.000-	1.

On the basis of calculation results, considering the proposed rationality judgments for damage models, the rationalities of various damage models are analyzed as follows:

Condition 1: From Figures 2 to 6 and Table 5, combination forms and normalized forms of components in damage models have a significant impact on their convergence abilities. When members are in the ultimate state, the average value of damage index of the Park-Ang damage model is equal to 0.822, slightly below a theoretical value of 1. And the average value of damage index of the Lu-Wang damage model is equal to 2.152, beyond a theoretical value of 1. The damage index of the Niu-Ren damage model ranges from 1.138 to 1.139, which is close to theoretical value of 1. It is worth noting that the proposed damage model whose average value of D is equal to 1.005, has better upper limit convergence ability owing to the normalized form of dissipated energy.

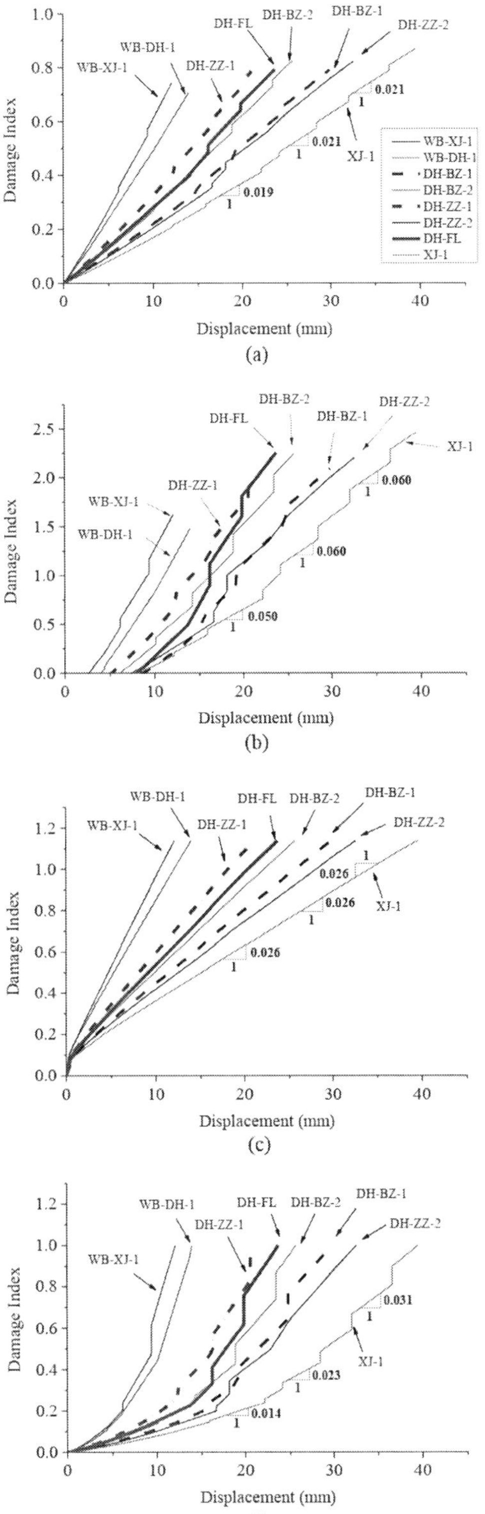

Fig. 2. Damage state of RC composite slabs with precast ribbed bottom panel. (a) Park-Ang damage model. (b) Lu-Wang damage model. (c) Niu-Ren damage model. (d) The proposed damage model.

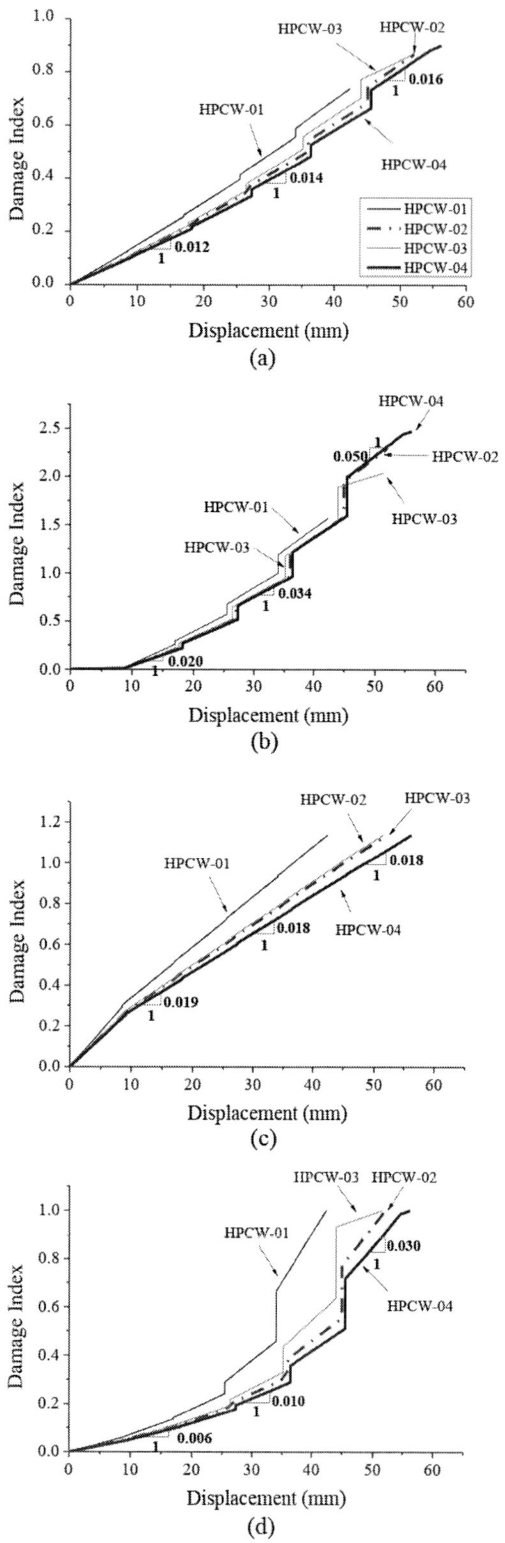

Fig. 3. Damage state of high-performance concrete shear walls. (a) Park-Ang damage model. (b) Lu-Wang damage model. (c) Niu-Ren damage model. (d) The proposed damage model.

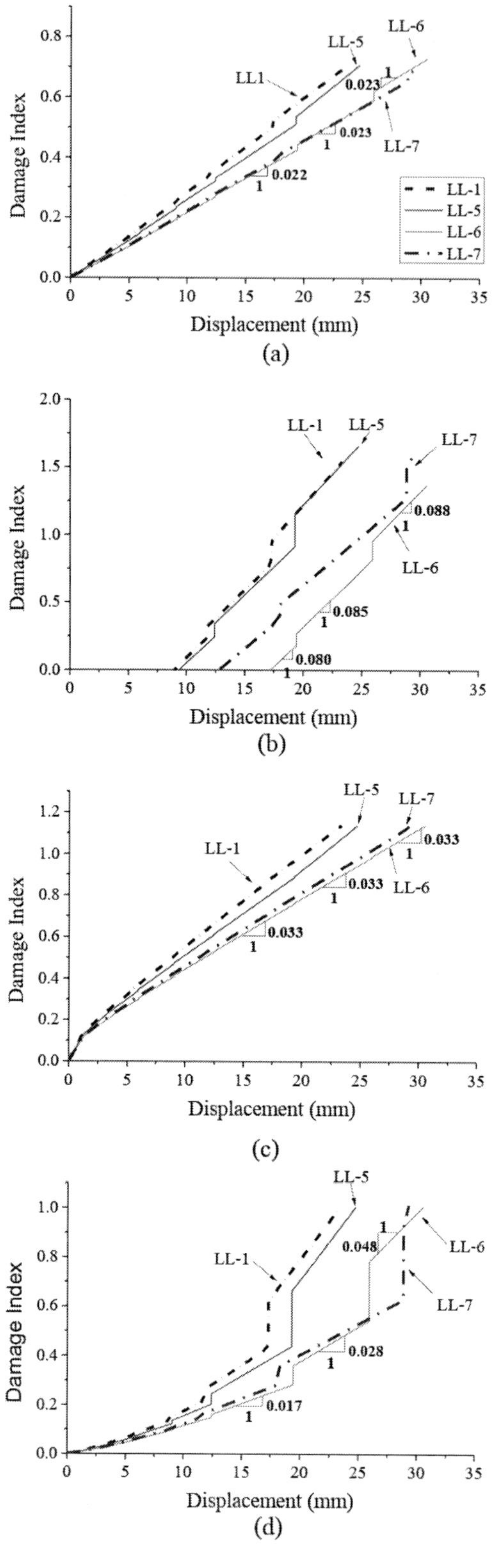

Fig. 4. Damage state of deep coupling beams with new reinforcement layouts. (a) Park-Ang damage model. (b) Lu-Wang damage model. (c) Niu-Ren damage model. (d) The proposed damage model.

https://doi.org/10.1051/e3sconf/20183803057

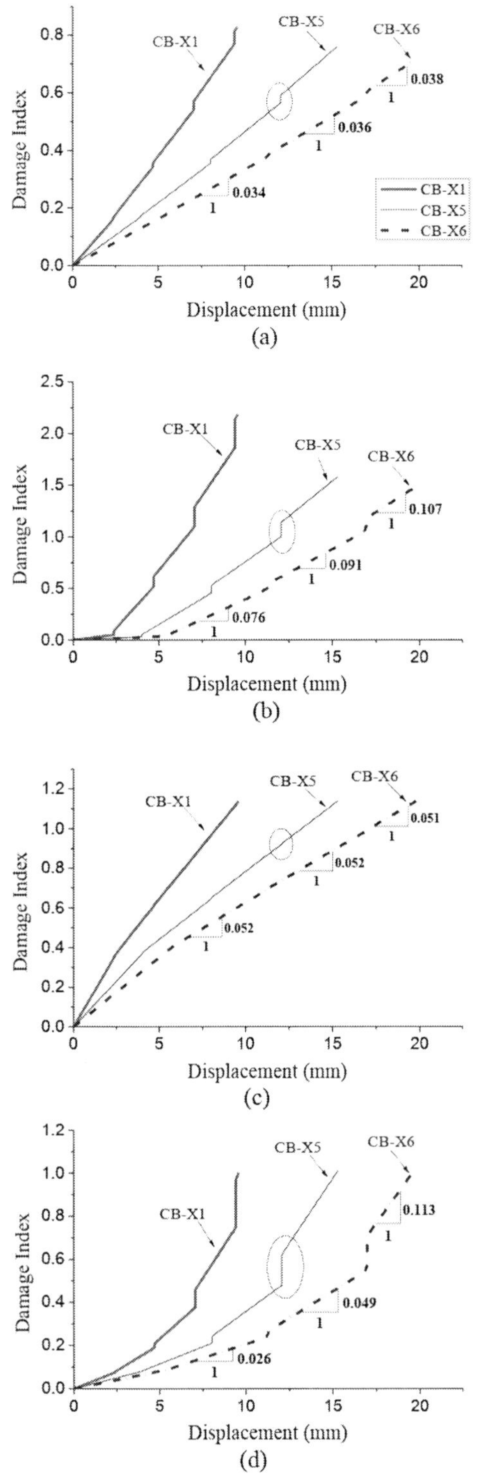

Fig. 5. Damage state of short-span coupling beams with bidiagonally reinforced bar. (a) Park-Ang damage model. (b) Lu-Wang damage model. (c) Niu-Ren damage model. (d) The proposed damage model.

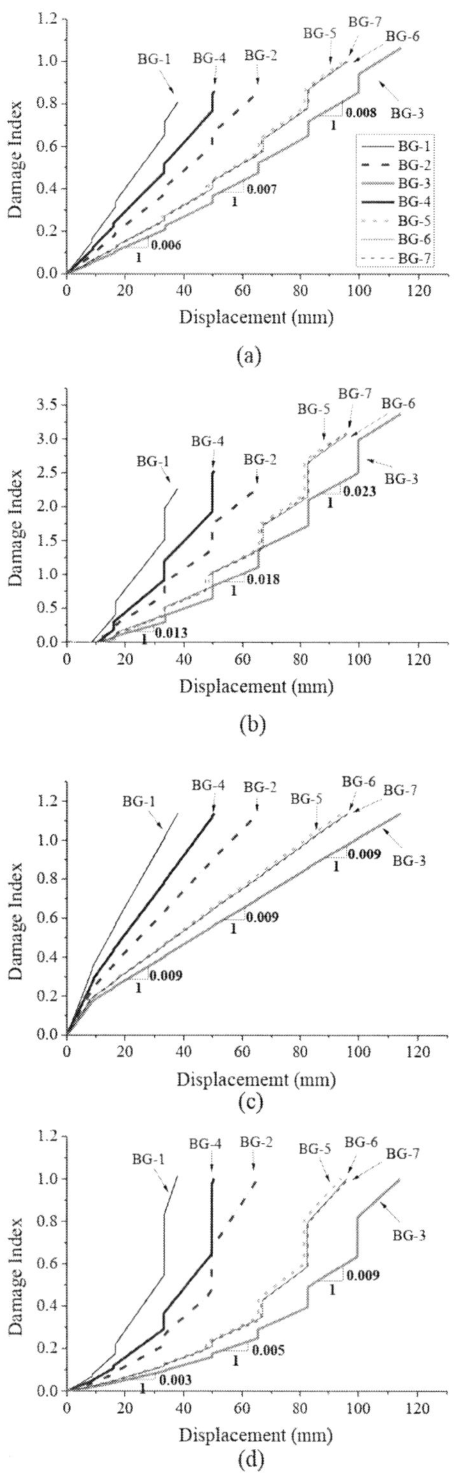

Fig. 6. Damage state of reinforced concrete columns with welded reinforcement grids. (a) Park-Ang damage model. (b) Lu-Wang damage model. (c) Niu-Ren damage model. (d) The proposed damage model.

Condition 2: As shown in Figures 2 to 6, the damage index tends to increase. Namely, the whole damage models exhibit a monotone increasing.

Condition 3: From Figures 2 to 6, it is not observed for Niu-Ren damage model that there emerges obvious low-cycle fatigue phenomenon called "damage index jump". The reason is that dissipated energy component makes little effect on the whole, which means the exponential item β is an absolutely small value. However, the phenomenon can be observed significantly for the others. Except for Niu-Ren damage model, the "damage index jumps" of other models under the same displacement is illustrated by the circles in Figure 5.

Condition 4: The slopes of displacement-damage index curves are partially labeled in Figures 2 to 6, in which the Park-Ang and Lu-Wang damage models exhibit fainter growth of the slopes, and the Niu-Ren damage model don't exhibit any growth of slope. The three above models estimate the damage state of members as a linear trend in the mass, which does not indicate an increasing rate of members' damage development under large displacement loading. Nonetheless, the proposed damage model exhibits significant nonlinear characters.

The adaptabilities of judging conditions for various models are collected in Table 6. On account of the complexity of damage development in RC members, most of current damage models incompletely satisfy above judging conditions. However, the proposed damage model is capable of meeting the conditions all above. Thus, it can be used for description the damage development of RC members.

Table 6. The definitions of rationality for various damage models based on judging conditions.

Damage models	Judging condition			
	1	2	3	4
Park-Ang	○	●	●	○
Lu-Wang	¤	●	●	○
Niu-Ren	●	●	¤	¤
Proposed	●	●	●	●

Where fully satisfying the judging conditions is denoted by •. Incompletely satisfying the conditions is denote by ○. Not satisfying the conditions is denoted by ¤.

5 Conclusions

In this paper, the analysis of the Park-Ang damage model, the Niu-Ren damage model, the Lu-Wang damage model and the proposed damage model by employing the experimental data of slabs, walls, beams and columns, 30 RC members in total, allows the following conclusions to be drawn:

(1) Four judging conditions for rationality of damage models are put forward: Damage index D is a value of 0 in the absence of damage and a value of 1 in the case of complete failure; D is supposed to be a monotone increasing function; D should tend to increase under the same displacement cycles; and the growth rate of D under large amplitude of displacement loading should be greater than that of D under small amplitude of displacement loading.

(2) Most of current damage models incompletely satisfy rationality judging conditions. However, the proposed damage model based on Kumar damage model, which matches the rationality judging conditions, can be used for evaluating and predicting damage performance of RC members under cyclic loading.

Acknowledgements

This work was supported by the National Natural Science Foundation of China [grant number 51078176, 51678283]. The Kunnath and Saatcioglu empirical data used in this paper are from the PEER Structural Performance Database. The authors are grateful for that.

References

1. G. Y. Wang, *Practical Methods of Optimum Aseismic Design for Engineering Structures and Systems* (China Achitecture&Building Press, Beijing, 1999)
2. G. H. Li, X. Zhu, Earthq Eng. Eng. Vibrat. **26** 67-69 (2006)
3. E. Cosenza, G. Manfredi, R. Ramasco, Earthquake Eng. Struct. Dyn. **22** 855-868 (1993)
4. A. Ghobarah, H. Abouelfath, A. Biddah, Earthquake Eng. Struct. Dyn. **28** 79-104 (1999)
5. T. H. Kim, K. M. Lee, Y. S. Chung, H. M. Shin, Eng. Struct. **27** 576-592 (2005)
6. Y. J. Park and H. S. Ang, J. Struct. Eng. **111** 722-739 (1985)
7. J. P. Ou, D. T. Niu, G. Y. Wang, Earthq Eng. Eng. Vibrat. **10** 9-20 (1990)
8. P. Fajfar, Earthquake Eng. Struct. Dyn. **21** 837–848 (1992)
9. S. Kumar, T.Usami, J. Struct. Engng. **40** 177-188 (1994)
10. D. T. Niu, L. J. Ren, Earthq Eng. Eng. Vibrat. **16** 44-54 (1996)
11. D. G. Lu, G. Y. Wang, Chin. Civi.l Eng. J. **34** 44-49 (2001)
12. T. Usami, S. Banno, H. Zetsu, T. Aoki, Jpn. J. Struct. Eng. **39A** 235-247 (1993)
13. S. K. Kunnath, A. El-Bahy, A. Taylor, W. C. Stone, *Cumulative seismic damage of reinforced concrete bridge piers* (NIST Interagency/Internal Report, Buffalo, 1997)
14. J. J. Sun, Y. F. Hong, Z. P. Wen, *The chinese seismic intensity scale* (Standards Press of China, Beijing, 2008)

15. J. S. Zhang, *Research on seismic performance of concrete composite slabs with precast ribbed bottom panels* (Lanzhou University, Lanzhou, 2013)
16. J. P. Fu, M. Wang, S. L. Bai, Earthq Eng. Eng. Vibrat. **25** 44-54 (1996)
17. CEB-FIP, *Displacement based seismic design of reinforced concrete buildings* (CEB-FIP, Lausanne, 2003)
18. CEB, *Seismic design of reinforced concrete structures for controlled inelastic response* (CEB, Lausanne, 1998)
19. Y. H. Zhang, *Damage Index Evaluation Method for Seismic Performance of Structural Members* (Lanzhou University, Lanzhou, 2013)
20. X. W. Liang, M. K. Deng, X. H. Zhang, S. F. Tian, J. Build. Struc. **28** 80-88 (2007)
21. C. Z. Xia, *Study on seismic behavior & shear capacity of deep coupling beams with new reinforcement layouts* (China University of Mining and Technology, Xuzhou, 2015)
22. L. Zhan, *Experimental study on short-span coupling beams with bidiagonally reinforced bar and design advise* (Chongqing University, Chongqing, 2007)
23. M. Saatcioglu, M. Grira, ACI Struct. J. **96** 29-39 (1999)

Calculation of crack stress density of cement base materials

Sui Chun-e[1,*]

[1]Transportation Institute, Inner Mongolia University, Inner Mongolia 010000, China

Abstract. In this paper, the fracture load of cement paste with different water cement ratio, different mineral admixtures, including fly ash, silica fume and slag, is obtained through experiments. the three-dimensional fracture surface is reconstructed and the three-dimensional effective area of the fracture surface is calculated. the effective fracture stress density of different cement paste is obtained. The results show that the polynomial function can accurately describe the relationship between the three-dimensional total area and the tensile strength

1 Introduction

In recent years, the quantitative description of the geometric characteristics of fracture surface has become the research object of many people through various experimental methods and calculation methods. According to the role of various structures and external factors in the fracture process, it is a good method to establish the relationship between the geometric characteristics of fracture surface and material properties, and to calculate the fracture stress. So far, however, most of the analysis has been directed to planar analysis. According to the analysis of the occurrence and extension of cracks and other related factors, there are some problems in the study of three-dimensional effective fracture stress according to the plane characteristics [4,5].

According to the different water cement ratio and tensile strength of cement paste with different mineral admixtures, the three-dimensional effective area of crack surface is reconstructed and the effective crack stress of different cement paste is calculated.

2 Test

2.1 Materials

The cement used in the study was type I/II Portland cement with fineness 381 m^2/ kg. 28d flexural strength of the samples was 7.5MPa. Fly ash (FA) with apparent density of 2.2 g/cm^3 was used, and it is a low-calcium ash with fineness 454 m^2/ kg. The fineness of silica fume (SF) was about 20000 m^2/ kg. Ground blast furnace slag (GBFS) with fineness of 740 m^2/kg and apparent density of 2.88 g/cm^3 was used

2.2 Test mixture ratio

Controlling the water cement ratio of the cement slurry to be 0.3 and 0.35 respectively; The mixed materials are fly ash, ground slag and silica fume respectively, and the content of the mixed materials is 15 %, 25 % and 5 % of the total amount of the cementing materials respectively. The mix design is shown in table 1.

Table 1. Mix designs of cement mixtures prepared.

Series	Mix	W/B	C(g)	W (g)	FA (g)	GBFS(g)	SF (g)
A1	Control	0.3	2600	780	0	0	0
A2	Control	0.35	2600	910	0	0	0
B1	15% FA	0.3	2210	780	390	0	0
B2	15% FA	0.35	2210	910	390	0	0
C1	25% GBFS	0.3	1950	780	0	650	0
C2	25% GBFS	0.35	1950	910	0	650	0
D1	5%SF	0.3	2470	780	0	0	130
D2	5%SF	0.35	2470	910	0	0	130

2.3 Test methods

The specimen is 40x40x160 mm, after being demoulded the specimens are put into water for curing to test age, the curing temperature is (20±2)℃,. Then the following tests are carried out: firstly, the tensile strength of each group of test pieces is measured, and the average value is taken; Protect section well, using KEYENCE super depth of field three-dimensional micro system to shape the fracture surface of each specimen three-dimensional morphology, and through the mathematical method to calculate the effective cross-sectional area of fracture

* Corresponding author: sce2008_tl@163.com

 https://doi.org/10.1051/e3sconf/20183803058

surface of tensile fracture specimen, finally combined with tensile strength to calculate the effective cracking stress of different cement paste was broken, the three-dimensional total area and tensile strength according to polynomial function, linear function, logarithmic function, power function and exponential function fitting regression analysis, comparative analysis. The KEYENCE ultra-depth-of-field three-dimensional microscopic system is as shown in figure 1.

Fig.1. KEYENCE ultra-depth three-dimensional microscopy system

3 Test results and analysis

3.1. The tensile strength of each group of specimens is shown in table 2.

Table 2. Tensile strength

Specimen number	A1	A2	B1	B2	C1	C2	D1	D2
Tensile strength(Mpa)	3.6	3.2	2.53	2.48	2.84	2.65	2.3	2.2

3.2 Three - dimensional morphology of fracture surface of remolded specimen

The actual morphology of the fracture surface of the C1 sample is shown in figure 2. The three-dimensional morphology of the fracture surface of the specimen was modeled by KEYENCE, as shown in figure 3. Because the use of this instrument has the requirement of depth of field range, the overall section area of the test is relatively large, and the surface of the concave and convex condition is relatively small, so the observation will be a specimen section is divided into four parts, each part of the measured area is the effective section area of the cross section. Figure 4 shows a processed three-dimensional sectional perspective view

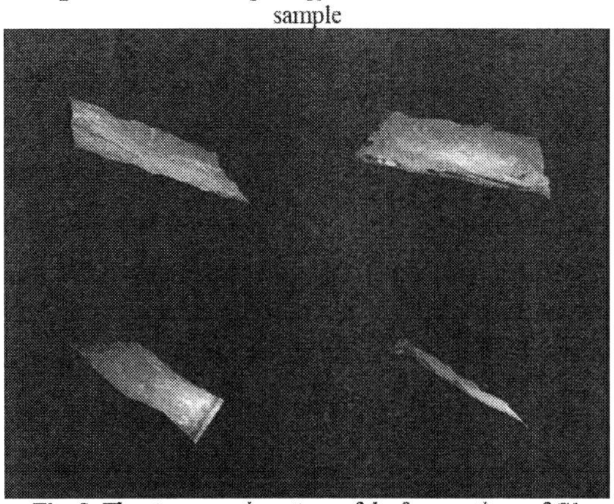

Fig. 2. The actual morphology of fracture surfaces of C1 sample

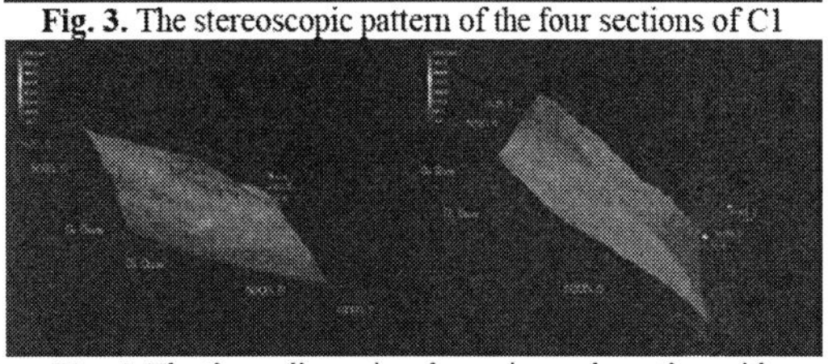

Fig. 3. The stereoscopic pattern of the four sections of C1

Fig. 4. The three-dimensional map in crack section with elevation

3.3 Calculation of three-dimensional effective cross-sectional area of tensile test piece

Due to the requirements of the depth of field of the test instrument, and in order to improve the test accuracy, each 40 mm x 40 mm interface is divided into eight partitions as shown in Fig. 5, the plane area of each partition is 10 mm x 10 mm = 100 mm.. Ten elevation difference curves are measured from top to bottom in each partition, and the length of any curve is calculated by using Fotran language and principle of calculus. Assuming that the width of each curve is 1 mm, the three-dimensional effective fracture area of this part is A1 = s × 1 mm × 10, and the three-dimensional effective area of section a1 is A = A1 +A2 + … + A8.

The three-dimensional effective cross-sectional area of A1 is calculated using the above method and is shown in table 3. Table 4 lists the three-dimensional effective cross-sectional area of all specimens.

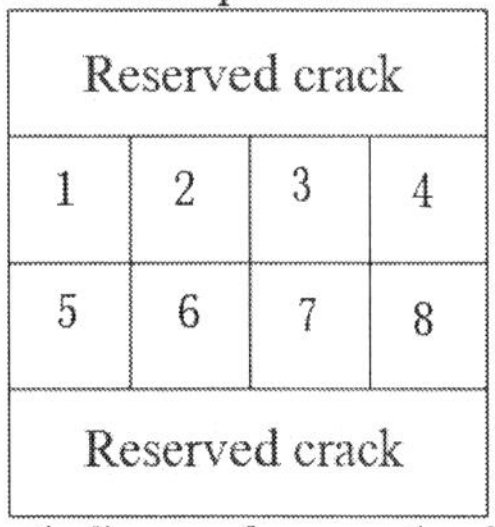

Fig. 5. Schematic diagram of cross-section division of A1

Table 3. Effective cross-sectional area of A1

Name of each part	Area per partition (mm^2)	Three - dimensional total area (mm^2)	Total area of plane (mm^2)

1	368.381	2541.644	800
2	347.838		
3	314.061		
4	225.863		
5	218.479		
6	493.138		
7	246.195		
8	327.688		

Table 4. Effective cross-sectional area of A1

Specimen number	Tensile strength (Mpa)	Three - dimensional total area (mm^2)	Cracking stress density(Mpa/ mm^2)
A1	3.6	2541.644	1.42E-03
A2	3.2	2202.400	1.45E-03
B1	2.53	2123.390	1.19E-03
B2	2.48	2230.400	1.11E-03
C1	2.84	2073.750	1.37E-03
C2	2.65	2000.000	1.33E-03
D1	2.3	2858.610	8.05E-04
D2	2.2	3004.800	7.32E-04

3.4 Analysis of the relationship between the three-dimensions total area and tensile strength

According to polynomial function, linear function, logarithmic function, power function and exponential function, the tensile strength and the total area of fracture surface were fitted and regressed respectively, and the results are compared and analyzed.

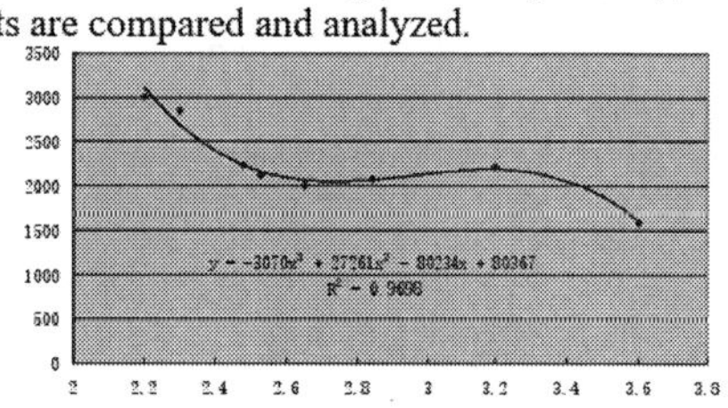

Fig. 6. The polynomial function relationship between three-dimensional areas and tensile strength

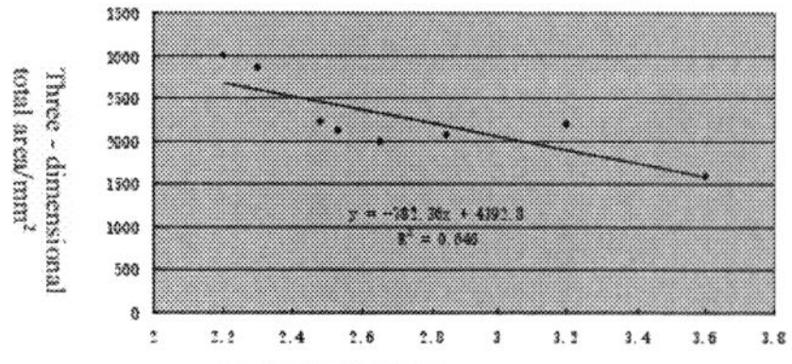

Fig. 7. The linear function relationship between three-dimensional areas and tensile strength

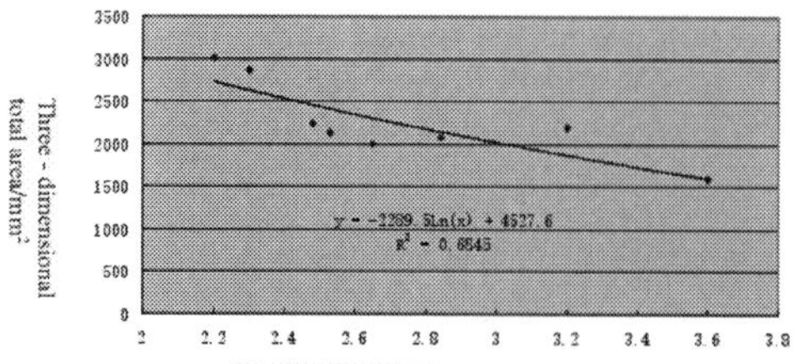

Fig. 8. The logarithmic function relationship between three-dimensional areas and tensile strength

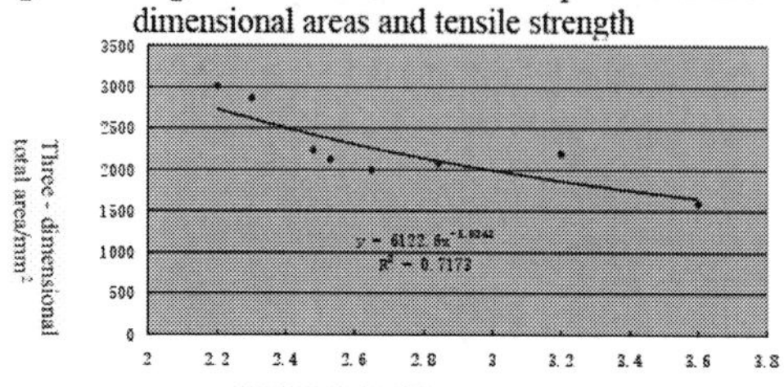

Fig. 9. The power function relationship between three-dimensional areas and tensile strength

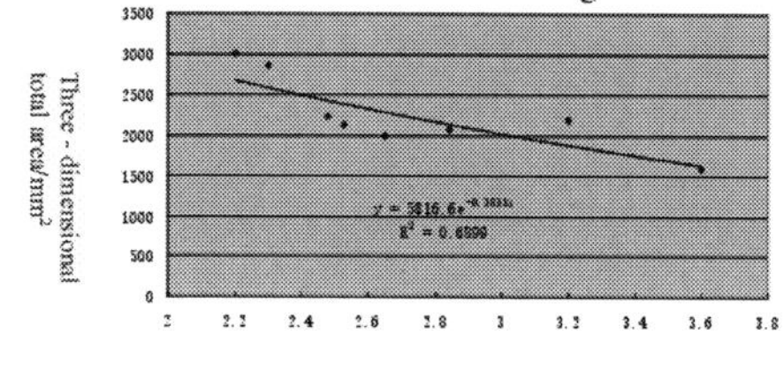

Fig. 10. The exponential function relationship between three-dimensional areas and tensile strength

4 Conclusions

1) The three-dimensional morphology of fracture surfaces of different specimens can be formed by using KEYENCE ultra-depth-of-field three-dimensional micro-system; The three-dimensional effective area of tensile fracture specimen section can be calculated by mathematical method, and combined with tensile strength, the effective cracking stress of different cement pastes can be calculated.
2)Analysis and fitting regression the relationship between the three-dimensional total area and tensile strength by polynomial function, linear function, logarithmic function, power function and exponential function, it can be found that the polynomial function fitting effect is the best, its correlation coefficient is the largest.

Acknowledgements

Authors wishing to acknowledge the financial support provided by Inner Mongolia autonomous region natural science foundation project(2015BS0505).

References

1. Coster M and Chermant JL. Recent developments inquantitative fractog raphy. I nt Metals Rev. 28 228-250 (1983)
2. Zhu P, Chen B and Tang W. Three dimensional topography reconstruction and quanti tative analysis of metallic fracture surfaces.Chi. J. of Stereology and image analysis. 2 1-5(1997)
3. Pentland A. Shape information from shading:A theory about human perception. Proc. Intl. Conf.CV.TAMPA.404-413 ()1988
4. Zen C and Der-Baau Perng. Automatic reconstruction of 3-D solid objects from 2-D orthographic. Patten Recogition. 21 155-162(1988)
5. Kimmel R and Bruckatein A M. Goble shape from shading. CVIU. 62 360-369(1995)

Electricity market liberalization under the power of customer value evaluation and service model

Hong Kun BAI[1], *Jiang Bo* WANG[1], and *Da Wei* SONG[1]

[1]National Institute of Economic and Technology Research, He Nan Electric Power Company, Zhengzhou, Henan450000, China

Abstract: After the power reform No. 9 was released in March 2015, the state officially released the Opinions on the Implementation of the Reform on the Power Sales Side. From this document, we can see that the openness of sales of social capital to the electricity business, the sales side of the market competition through multiple ways to train the main competitors, the result is more users have the right to choose, sales service quality and user energy levels will significantly improve. With the gradual promotion of the electricity sales market, the national electricity sales companies have been established one after another. In addition to power grid outside the power generation companies, energy-saving service companies and distributed power companies may become the main selling power, while industrial parks, commercial complex, large residential area, industrial and commercial users, large industrial users in the new electricity demand appearing The new changes, some power customers have also self-built distributed power supply, installation of energy storage devices or equipment to participate in the transformation of the electricity market. The main body of the electricity sales market has gradually evolved from the traditional electricity generation main body to the multi-unit main body and emerged new value points. Therefore, the electricity sales companies need to establish a power customer value evaluation method and service mode to adapt to the new electricity reform, Provide supportive decision support.

1 Introduction

Nowadays, the theories and methods of power customer value both at home and abroad are mainly focused on the definition of customer value, the index system of customer value evaluation, the model and method of customer value evaluation, and the classification of customer value. When defining customer value, customer value is generally evaluated in two ways: one is the value that the business provides to the customer and the other is the value of the customer to the business. In the power customer value evaluation model and method, a comparatively complete matrix model is formed both at home and abroad. The matrix model is compiled on the basis of a more complete data. This method is applied to segment the customer so that a better service In the high-quality customers. The classification of customer value is generally based on the evaluation results after the completion of the customer value rating based on the level of value, and finally divided into important customers, general customers and low value customers, different types of customers to adopt different marketing strategies.

When the electricity sales market is liberalized, the trading activities of both the main power suppliers and the main power suppliers will change. The services required by the customers are not limited to the power supply services with small differences in the past. The service demands of different customers on the power sales agents Not the same, at the same time liberalization of customer power options, you can choose to serve more high-quality electricity sales of the main. Therefore, the main power sales should pay more attention to customer needs, and strive to improve customer satisfaction in order to achieve a win-win situation. The customer value evaluation index system constructed in this paper integrates the value-added value of customers after the liberalization of the electricity sales market, designs two indexes of customer demand side value and energy management consulting service value, and the weight design of the index system adopts fuzzy analytic hierarchy process Entropy method of subjective and objective combination of methods, the use of TOPSIS evaluation method for comprehensive evaluation of customer value. In the service mode, the design of the customer stickiness index is considered considering the customer power option. The customer characteristics are analyzed according to the customer value rating result to better provide the decision support for the differentiated service of the power company.

2 Electricity customer value evaluation index systems

This article builds the customer value evaluation system

* Corresponding author: zhaoningning_538@126.com

as shown in Figure 1 below. The demand-side peak clipping potential depends mainly on the customer's load characteristics, which mainly reflects the customer's peak clipping potential during peak load periods, so as to provide reference for the selling company to implement the interruptible load service. The calculation method is as follows:

If the customer's load value at peak time $t_1,t_2,t_3,\mathrm{L}\ t_Z$ is $l^{t1},l^{t2},l^{t3},\mathrm{L}\ l^{tZ}$, the potential demand response potential of the customer is:

$$f=(l^{t1}+l^{t2}+l^{t3}+\mathrm{L}\ +l^{tZ})\lambda/Z \tag{1}$$

Where λ is the average load drop rate under the customer's demand response measure and the other is the ratio of the average load reduction to the maximum load when the demand response is implemented.

The value of energy management consulting services mainly includes customer's electricity management costs, distributed power supply commissioning and maintenance wishes, energy-saving solutions, commissioning of electricity engineering projects and purchase of electricity. Customer's energy management consulting commissioned demand reflects the added value of customers, the higher the commission will, the higher the value. Due to the different types of customers of the sales company, different types of customers are not the same in terms of characteristics, and evaluation of various types of customer value can't be made using a unified value evaluation index.

Table.1. The index system of power customer value evaluation

Electricity customer value	A1 Sales economic contribution	A11 Annual sales of electricity
		A12 Average annual selling price
		A13 Annual electricity growth
	A2 Customer credit conditions	A21 The number of default electricity
		A22 Electricity pay timely rate
		A23 Contract compliance rate
		A24 Annual contract deviation of electricity
	A3 Customer management	A31 Assets and liabilities
		A32 Liquidity turnover
		A33 Total asset turnover
		A34 ol
		A35 Industry market share
	A4 Electricity management level	A41 Equipment reliability and operating conditions
		A42 Electricity safety management level
		A43 Work with the situation
	A5 Customer demand side value	A51 Demand side clipping potential
		A52 Energy storage capacity
		A53 Charging pile number
		A54 Controllable micro-source installed capacity
		A55 Electricity demand elasticity
	A5 Energy Management Consulting Services	A61 Customer electricity management costs
		A62 Distributed power consignment operation and maintenance wishes
		A63 Energy-saving solution needs
		A64 Electrical Engineering Commission situation
		A65 Purchase electricity willingness to consult

3 Customer stickiness index

Considering that customers have the right to choose in the electricity sales market, this paper designs the customer stickiness index, which reflects the satisfaction of different customers to the sales company and on the other hand, can reflect the persistence of customer value. Customer satisfaction is determined by two factors, namely, the customer's expectation and the customer's perception. The lower the customer's expectation, the

 https://doi.org/10.1051/e3sconf/20183804001

easier it is to satisfy. The worse the actual perceived customer is, the more difficult it is to satisfy. It can be deduced that customer satisfaction is inversely proportional to expectation, and customer satisfaction is proportional to perceived. The evaluation criteria of customer satisfaction can be described by a simple function as follows:

$$c = b / a \tag{2}$$

Where, A represents customer satisfaction; B represents the customer's perceived value; C represents the customer's expectations. In evaluating the customer stickiness index, the sales company should take the initiative to understand the core needs of customers, whether their own products and service water products meet customer expectations, and the evaluation process is based on the judgment of the sales company itself and the customer questionnaire. The core requirements are different for different types of customers. This article lists only some of the indicators used to reflect customer stickiness (Figure 2). In practice, sales companies should strengthen customer interaction and understand the real core needs of customers.

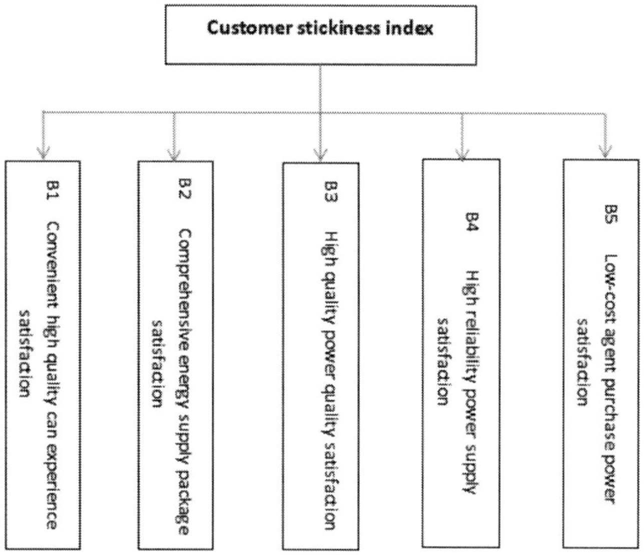

Figure.1. Customer stickiness index system

4 Electricity customer value evaluation model

In this paper, the design of customer value index weight is based on the combination of fuzzy analytic hierarchy process and entropy method, subjective fuzzy analytic hierarchy process for the second-level indicators, objective entropy method for the third-level indicators, subjective and objective A comprehensive weight, with some scientific rationality.

Design index weight steps are as follows:

4.1. Design two indicators of weight

Establish a fuzzy consistency judgment matrix, U is a set of elements A_i (secondary indicators), that is $U = \{A_1, A_2, \cdots, A_n\}$, $i = 1 \sim n$,its fuzzy relation matrix expressed as

$$R = \begin{bmatrix} r_{11} & r_{12} & \cdots & r_{1n} \\ r_{21} & r_{22} & \cdots & r_{2n} \\ \vdots & \vdots & \ddots & \vdots \\ r_{n1} & r_{n2} & \cdots & r_{nn} \end{bmatrix} \tag{3}$$

r_{ij} indicates the importance of A_i relative to A_j. If $r_{ij} > 0.5$, then A_i is more important than A_j ; if $r_{ij} < 0.5$, then A_j is more important than A_i . If there is consistency in the process of determining the degree of membership of an element that is more important than the other element, then when $r_{ij} > 0.5$, $\forall k (\text{k} = 1 \sim n)$, $r_{ik} > r_{jk}$, then $r_{ij} = r_{ik} - r_{jk} + 0.5$. From the literature [3] that the two indicators weight:

$$w_i = \frac{1}{n^2 - n}(2\sum_{k=1}^{n} r_{jk} - 1), i = 1 \sim n \tag{4}$$

4.2. Design three indicators of weight

Select n power customers, m three under the two indicators, x_{ij} is the i th customer j index value($i=1\sim n, j=1\sim m$). Which can be quantified using the actual data of electricity customers, non-quantifiable values scored by experts, this article uses the percentage system.

1) The normalization of indicators

Because they can't unify the units of measurement of each indicator, they should be standardized before they can be used to calculate the comprehensive indicator. Processing calculation is as follows:

For the benefit index:

$$x_{ij}^{'}=\frac{x_{ij}-\min\{x_{1j},x_{2j},...,x_{nj}\}}{\max\{x_{1j},x_{2j},...,x_{nj}\}-\min\{x_{1j},x_{2j},...,x_{nj}\}}\times 100 \quad (5)$$

For cost indicators:

$$x_{ij}^{'}=\frac{\max\{x_{1j},x_{2j},...,x_{nj}\}-x_{ij}}{\max\{x_{1j},x_{2j},...,x_{nj}\}-\min\{x_{1j},x_{2j},...,x_{nj}\}}\times 100 \quad (6)$$

2) Calculate the i-th customer's j-th indicator of the proportion of the indicator

$$p_{ij}=\frac{x_{ij}^{'}}{\sum_{i=1}^{n}x_{ij}^{'}},(i=1,2...,n,\ j=1,2...,m) \quad (7)$$

3) Calculate the entropy of the j-th indicator:

$$e_j=-k\sum_{i=1}^{n}p_{ij}\ln(p_{ij}) \quad (8)$$

Where, $k=1/\ln(\mathrm{n})>0$, satisfy $e_j\geq 0$;

4) Calculate the information entropy redundancy of the j-th indicator:

$$d_j=1-e_j \quad (9)$$

$0\leq d_j\leq 1$ for the j-th index, the greater the redundancy of information entropy, the greater the effect on the evaluation program, the greater the entropy.

5) Calculate the weight of each indicator:

$$w_j=\frac{d_j}{\sum_{j=1}^{m}d_j},\qquad j=1,2,...,m \quad (10)$$

4.3. Comprehensive evaluation method based on TOPSIS method

1) Construct the matrix of weighted norms

By formula (4) and formula (5) respectively, the weight of the second-level index and the third-level index are obtained. The final weight of the third-level index can be obtained by multiplying

$$w_t=w_iw_j \quad (11)$$

We construct a weighting rule matrix $Y=(y_{ij})_{n\times m}$:

$$y_{ij}=w_tx_{ij}^{'},(i=1,2...,n,\ j=1,2...,m) \quad (12)$$

2) Determine the positive and negative ideal solution

The positive and negative ideal solutions are respectively the combination of the optimal value and the worst value of each solution.

Positive ideal solution:

$$x_j^{+}=\max_i x_{ij}^{'} \quad (13)$$

Negative ideal solution:

$$x_j^{-}=\min_i x_{ij}^{'} \quad (14)$$

3) Calculate the Euclidean distance between the solutions and the positive and negative ideal solutions

Distance to the ideal solution:

$$d_i^{+}=\sqrt{\sum_{j=1}^{n}w_j(x_{ij}^{'}-x_j^{+})^2} \quad (15)$$

Distance to negative ideal solution:

$$d_i^{-}=\sqrt{\sum_{j=1}^{n}w_j(x_{ij}^{'}-x_j^{-})^2} \quad (16)$$

4) Calculate the relative closeness between the index value and the ideal solution of each evaluation object

$$c_i=d_i^{-}/(d_i^{-}+d_i^{+}) \quad (17)$$

According to the relative close evaluation of the various programs (that is, each customer) the pros and cons, the closer the greater the greater the customer value, and vice versa lower.

5 Case Analyses

In this paper, we choose five large industrial customers for empirical analysis, the statistics of the customer value data can be quantified after normalization of indicators, non-quantifiable data based on objective and reasonable experts scoring obtained by the formula (5) and (6) After processing the data shown in Table 2:

Table.2. Each indexes scoring of customer

	Evaluation index	A	B	C	D	E
Sales economic contributio n	Annual sales of electricity	92	68	40	68	98
	Average annual selling price	80	93	50	70	50
	Electricity growth rate	45	61	30	20	57
Credit conditions	The number of default electricity	95	47	30	90	70
	Electricity pay timely rate	96	60	34	98	80
	Contract compliance rate	92	70	36	99	78
	Contract deviation power	90	80	57	98	80

Operating conditions	Assets and liabilities	64	40	49	69	50
	Liquidity turnover	59	37	67	89	66
	Total assets growth rate	70	45	46	80	76
	ol	65	50	50	68	48
	Industry market share	34	21	14	40	69
Electricity management level	Equipment reliability and operating environment	67	80	59	70	40
	Safety management level	45	80	48	78	80
	Work with the situation	89	78	30	78	94
Customer demand side value	Customer demand side clipping potential	92	90	20	40	95
	Energy storage capacity	20	30	0	60	0
	Charging pile number	12	0	0	0	30
	Controllable micro-source installed capacity	0	40	0	0	45
	Electricity demand elasticity	44	67	40	70	30
Energy Management Consulting Services	Customer electricity management costs	78	30	20	30	80
	Distributed power entrusted operation and maintenance of the situation	66	45	0	80	70
	Energy-saving solution needs	88	20	30	20	90
	Electrical Engineering Commission situation	58	20	0	0	50
	Purchase electricity willingness to consult	80	10	0	20	40

For the second-level index element $\{A_1, A_2, A_3, A_4, A_5, A_6\}$, by contrasting the importance of each two, the experts form a fuzzy judgment matrix after empowerment, as follows:

$$R = \begin{bmatrix} 0.5 & 0.7 & 0.8 & 0.9 & 0.6 & 0.7 \\ 0.3 & 0.5 & 0.6 & 0.7 & 0.4 & 0.5 \\ 0.2 & 0.4 & 0.5 & 0.6 & 0.2 & 0.3 \\ 0.1 & 0.3 & 0.4 & 0.5 & 0.1 & 0.3 \\ 0.4 & 0.6 & 0.8 & 0.9 & 0.5 & 0.6 \\ 0.3 & 0.5 & 0.7 & 0.7 & 0.4 & 0.5 \end{bmatrix}$$

Using the consistency condition $r_{ij} = r_{ik} - r_{jk} + 0.5$, adjust the fuzzy matrix to form R_{new}

$$R_{new} = \begin{bmatrix} 0.5 & 0.7 & 0.8 & 0.9 & 0.6 & 0.7 \\ 0.3 & 0.5 & 0.6 & 0.7 & 0.4 & 0.5 \\ 0.2 & 0.4 & 0.5 & 0.6 & 0.3 & 0.4 \\ 0.1 & 0.3 & 0.4 & 0.5 & 0.2 & 0.3 \\ 0.4 & 0.6 & 0.7 & 0.8 & 0.5 & 0.6 \\ 0.3 & 0.5 & 0.6 & 0.7 & 0.4 & 0.5 \end{bmatrix}$$

Using formula (4) calculate the weight of two indicators are 0.167, 0.127, 0.087, 0.207, 0.167. Use formula (7-10) to calculate the weight of the third-level indicator. Enter the formula (11) and obtain the comprehensive weight of the indicator. As shown in table 3:

Table.3. The weight of customer value indexes

Secondary indicators	Secondary indicator weight	Three indicators	Three-level indicator weight	Comprehensive weight
Sales economic contribution	0.247	Annual sales of electricity	0.431	0.106
		Average annual selling price	0.416	0.103
		Electricity growth rate	0.154	0.038
Credit conditions	0.167	The number of default electricity	0.146	0.024
		Electricity pay timely rate	0.272	0.045
		Contract compliance rate	0.253	0.042
		Contract deviation power	0.330	0.055
Operating conditions	0.127	Assets and liabilities	0.181	0.023
		Liquidity turnover	0.240	0.030
		Total assets growth rate	0.264	0.033
		ol	0.190	0.024
		Industry market share	0.125	0.016
Electricity	0.087	Equipment reliability and operating	0.303	0.026

manageme nt level		environment		
		Safety management level	0.357	0.031
		Work with the situation	0.340	0.029
Customer demand side value	0.207	Customer demand side clipping potential	0.297	0.061
		Energy storage capacity	0.163	0.034
		Charging pile number	0.111	0.023
		Controllable micro-source installed capacity	0.125	0.026
		Electricity demand elasticity	0.304	0.063
Energy Managem ent Consultin g Services	0.167	Customer electricity management costs	0.255	0.043
		Distributed power entrusted operation and maintenance of the situation	0.217	0.036
		Energy-saving solution needs	0.289	0.048
		Electrical Engineering Commission situation	0.110	0.018
		Purchase electricity willingness to consult	0.129	0.021

Use formula (12-17) to determine the positive and negative ideal solutions, and calculate the Euclidean distance of each customer, and finally get the relative closeness of each customer, as shown in Table 4 below.

Table.4. The weight of customer value indexes

Electricity customers	Proximity		The relative post progress
	Positive	Negative	
client1	2.611	9.150	0.778
client2	5.071	6.691	0.569
client3	11.124	0.638	0.054
client4	4.949	6.813	0.579
client5	3.794	7.968	0.677

Through the relative closeness of the size of the customer value can be drawn as A> E> D> B> C. Close to the value of more than 0.6 for the high value customers, 0.4 to 0.6 for the value of the customer, 0.4 following the low value customers, according to grading standards A ~ E customers were high value, medium value, low value, medium value, high value. As all five electric power customers are large industrial users, the core customer needs are as follows: (1) in the research, we use the satisfaction of low-cost agent purchase as the index to evaluate the customer's viscosity. According to the sensitivity of 5 electric power customers to the price, The customer's viscosity is high viscosity, medium viscosity, medium viscosity, high viscosity, low viscosity. The resulting two-dimensional customer evaluation chart 5, as shown in Figure 2

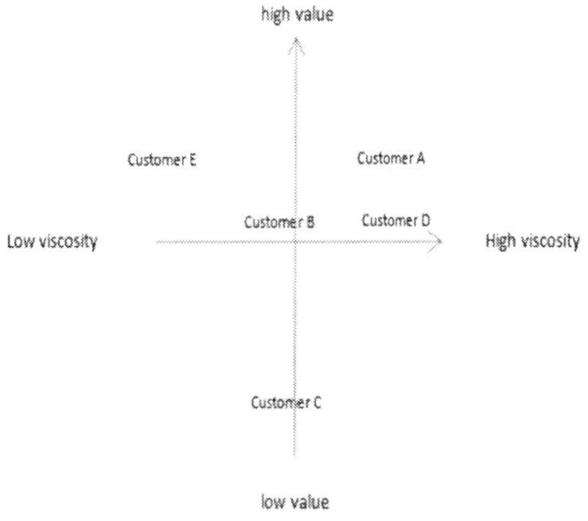

Figure.2. Customer two-dimensional evaluation chart

For customers with high value and high viscosity, such customers are the best customers of the sales company and achieve a win-win result. The service company of the sales company should maintain a good relationship with the customers and provide high quality VIP service. For high-value low-viscosity power customers, sales companies should continue to improve their product and service quality, as much as possible to improve customer satisfaction and achieve win-win results. For low-value but highly-viscous customers, the main strategy of a sales company is to increase customer value by tapping customer value added. For the low-value and low-viscosity customers, the sales company needs to analyze the reasons that cause its low value while providing the basic services, and in what ways can help the enterprise improve its own value, meanwhile, it should pay attention to prevent the operation risk.

6 Conclusion

With the liberalization and gradual maturation of the electricity sales market, sales companies are increasingly important to customer relationship management. This paper mainly constructs the power customer value evaluation method. This method mainly adapts to the current electricity market environment, and puts forward the customer evaluation method of customer value and customer stickiness. It not only evaluates the customer value from the perspective of the enterprise, but also from the perspective of the customer Analysis of customer satisfaction, more specific reflection of the characteristics of customers, for the sale of electricity companies to provide more comprehensive management information. After the market is further matured, the sales and after-sales service company can realize the accurate positioning and marketing of customers by observing the customers in more dimensions from customer value evaluation, customer demand identification, customer stickiness identification, customer transaction monitoring, etc., and realize customer-centric , The ultimate power company and customers achieve a win-win situation.

Acknowledgements

State Grid Electric Power Company funding projects(5217L017000M)

References

1. National Energy Administration. Opinions on the Implementation of the Reform of the Electricity Market [EB / OL] http://www.nea.gov.cn/2015-11/30/c_134867851.htm, 2015-11-26
2. Poplar, Xie Le, Xia Qing, et al. Design and suggestion on the promotion of marketization of sales side in China [J] .Automation of Electric Power Systems, 2015 (14): 1-7.
3. Wang Songtao. Evaluation of electricity customer value under market conditions [J] Power System Technology, 2010,02: 155-158.
4. Angus J. Valuing your customers: from quality information to quality relationship through database marketing [M]. London: McGraw Hill Book Co., 1995: 326, 336-338.
5. Berger P D, Nasr N I, Customer lifetime value: marketing models and applications [J]. Journal of Interactive Marketing, 1998, 12 (1): 17-30.
6. Bailey F. Getting the most out of CRM [M]. Hoosier Banker, 2004:
7. Chuang Y F, Chia S H, Wong J Y. Customer value assessment of pharmaceutical marketing in Taiwan [J]. Industrial Management & Data Systems, 2013, 113 (113): 1315-1333.
8. Qu Bin, Li Cunbin, Tian Huiying. Construction of industrial electricity customer credit comprehensive evaluation index system [J]. Power System Technology, 2007,31 (1): 75-78,83.28-29.
9. Zeng Ming, Yang Suping, Yang Pengju, Tang Jian, Tian Weiao.Evaluation of power customer value under social energy-saving environment [J] .East East China Power, 2008,06: 15-19.
10. Song Cairua, Lan Yuanjuan, Wang Yongcai, Zhai Hongrong, Li Bintao. Application of Synthetic Value Evaluation Method in Customer Segmentation of Power Enterprise [J], Electronic Design Engineering, 2014,12: 111-113 + 116.
11. Li Junfei. Electricity customer value of integrated rough set and support vector machine [D]. North China Electric Power University (Hebei), 2007.
12. Li Hongze, Guo Sen, Wang Bao. Evaluation of power customer value based on genetic improvement ant colony clustering algorithm [J]. Power System Technology, 2012, 12: 256-261.
13. Wang Lei. Distribution model of power industry based on data mining [D]. Shanghai Jiao Tong University, 2007.
14. Ren Bingli, Zhang Zhengao, Wang Xuejun, et al.An evaluation method of demand response peak clipping potential based on electricity collection data [J] .Electric Power Construction, 2016,37 (11): 64-70.
15. BAI Yin-hao, FENG Yi-zhan. Application of Customer Satisfaction in Direct Power Purchase by Large Users [J] .H Henan Power, 2015 (4): 22-25.Journal of North China Electric Power University (Natural Science), 2010, 37 (6): 61-64.
16. Wang Jingmin, Sun Yanfu, Kang Junjie. Evaluation of competitiveness of power enterprises based on entropy method and improved TOPSIS method [J] Journal of North China Electric Power University (Natural Science Edition), 2010, 6:61-64.
17. WANG Zhi-jie, LIU Shanshan, XUE Song, et al. Evaluation model of demand side response resource value based on entropy weight and TOPSIS [J] .East East China Power, 2014, 42 (1): 143-149.

E3S Web of Conferences **38**, 04002 (2018) https://doi.org/10.1051/e3sconf/20183804002
ICEMEE 2018

Solution of Thermoelectricity Problems Energy Method

Muheyat Niyazbek [1], Nogaybaeva M.O [3], Kuenssaule Talp[2,*], Kudaikulov A.A[3]

[1]College of Information Science and Engineering, Xinjiang University, Urumqi, China ,830046
[2]College of Chinese Medicine of Xinjiang Medical University, Urumqi, Xinjiang, China 830011
[3]Al–Farabi Kazakh National University, Kazakhstan， Almaty

Abstract: On the basis of the fundamental laws of conservation of energy in conjunction with local quadratic spline functions was developed a universal computing algorithm, a method and associated software, which allows to investigate the Thermophysical insulated rod, with limited length, influenced by local heat flow, heat transfer and temperature

1 Introduction

Supporting elements of modern gas-power plants [1], nuclear and thermal power plants, hydrogen and jet engines, internal combustion engines, deep processing of mineral raw materials and oil rigs are operating in the complex field of power and heat. The reliability of the above mentioned plants depend on the characteristics of the hot-bearing elements. The elements considered as bearing rods of limited length and of constant cross section [1] [2]. In these problems, based on the fundamental laws of Thermal Physics [3] was determined the temperature distribution of the field along the length of the rod of limited length in view of the existing types of heat sources. Other similar problems were considered in [4-6]. This study was to determine the law of temperature distribution along the length of the test bar depending on the types of specified heat sources. In this paper, a horizontal rod of limited length and constant cross section is considered as a carrier element of construction. And the lateral surface of the test rod is completely insulated. The cross-sectional area of the left end of the rod is under heat flux with constant intensity. The cross-sectional area of the right end of the rod has a heat exchange with the ambient environment area. But, in the simultaneous presence of different types of heat sources in the supporting elements of the strategic structures there are axial compressive forces, displacement distribution field, elastic, thermal and thermo-elastic deformations of components and related stress, in addition to the distribution of temperature field. Therefore, the development of fundamental methods for deep study of complex thermo-physical states of bearing structural elements is an actual problem of Applied Thermophysics. In this work following statements were defined with the help of the fundamental laws of conservation of energy [7]

1) The law of the temperature distribution along the length of the rod;
2) The value of its thermal expansion;
3) The value of the arising axial compressive force;
4) The law of distribution of elastic, thermal and thermo-elastic components of the strain and stress;
5) The displacement field.

We consider the horizontal rod, with limited length L [см] and constant cross-sectional area F [см2]. The cross-section of the rod can be circular, quadrilateral, triangula

2 Conclusion of resolving equations

Thermal properties of the rod material characterized by a coefficient of thermal expansion of the material of the rod α $[\frac{1}{°C}]$, thermal conductivity кх $[\frac{Вт}{см*°C}]$, as well as the modulus of elasticity кх $[\frac{Вт}{см*°C}]$.

The cross-sectional area of the left end of the rod is under heat flux q $[\frac{Вт}{см^2}]$. Under the influence of the local heat flow the surface value of heat flux is negative, that is q<0. The cross-sectional area of the right end of the rod is under a convective heat transfer with the surrounding environment. Here, heat exchange coefficient is h $[\frac{Вт}{см^2 °C}]$, and ambient temperature is $T_{°C}$ [°C]. First must be determined the temperature distribution along the length of the test rod depending on the type of existing heat sources, thermal and geometrical characteristics of the rod. To do this, first we build a local approximational quadratic spline function.

Design scheme of the problem is shown in Figure 1.

Figure 1 - Diagram of the problem

Suppose that

Corresponding author: Kuenssaule Talp@163.com ; Muheyatn@xju.edu.cn, Kudaykulov2006@mail.ru

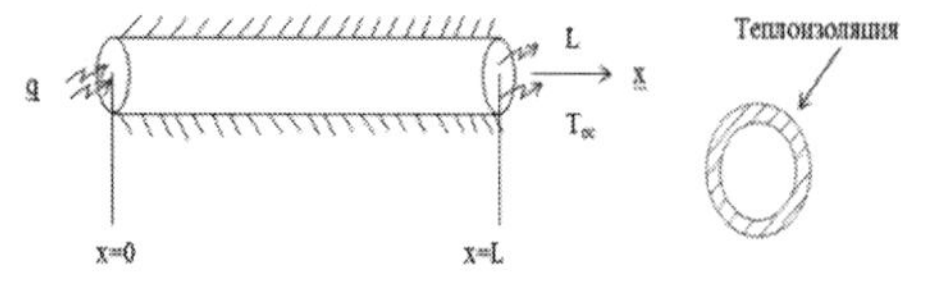

$$T(x = 0) = T_i; T\left(x = \frac{L}{2}\right) = T_j; T(x = L) = T_k \quad (1)$$

Approximate the full second-order polynomial spline functions of the distribution of temperature along the length of the test rod [8]

$$T(x) = ax^2 + bx + c = \varphi_i(x) * T_i + \varphi_j(x) * T_j + \varphi_k(x) * T_k; \ 0 \le x \le L \quad (2)$$

Where:

$$\varphi_i(x) = \frac{2x^2-3Lx+L^2}{L^2}; \ \varphi_j(x) = \frac{4Lx-4x^2}{L^2}; \ \varphi_k(x) = \frac{2x^2-Lx}{L^2}; \ 0 \le x \le L \quad (3)$$

Now we write functional for the problem that characterizes the law of conservation of energy [7]

$$J = \int_{S(x=0)} qTdS \int \ + \int_V \frac{kxx}{2}\left(\frac{\partial T}{\partial x}\right)^2 dV + \int_{S(x=L)} \frac{h}{2}(T - T_{OC})^2 dS, \quad (4)$$

It should be noted that the dimension of each member is [Вт°C]. This is the work carried out by temperature, work force [кГ*см]. Due to the physical nature of the phenomenon, we have:

$$J_1 = \int_{S(x=0)} q * TdS = FqT_i \quad (5)$$

$$J_3 = \int_{S(x=L)} \frac{h}{2}(T - T_{OC})^2 dS = \frac{Fh}{2}(T_k - T_{OC})^2 \quad (6)$$

For the calculation of the integral over the volume of equation (4), it is necessary to determine the temperature gradient

$$\frac{\partial T}{\partial x} = \frac{\partial \varphi_i}{\partial x}T_i + \frac{\partial \varphi_j}{\partial x}T_j + \frac{\partial \varphi_k}{\partial x}T_k = \frac{4x-3L}{L^2}T_i + \frac{4L-8x}{L^2}T_j + \frac{4x-L}{L^2}T_k, \ 0 \le x \le L \quad (7)$$

Then substituting (7) in terms of J, and using the known formula

$\int_V f(x)dV = F\int_0^L f(x)dx$ we have

$$J_2 = \int_V \frac{k_x}{2}\left(\frac{\partial T}{\partial x}\right)^2 dV = \frac{Fk_x}{\partial x}(7T_i^2 - 16T_iT_j + 2T_iT_k - 16T_jT_k + 16T_j^2 + 7T_k^2) \quad (8)$$

Then, an integrated form of the complete thermal energy functional is:

$$J = J_1 + J_2 + J_3 = = FqT_i + \frac{Fk_i}{\partial x}(7T_i^2 - 16T_iT_j + 2T_iT_k - 16T_jT_k + 16T_j^2 + +7T_k^2) + \frac{Fh}{2}(T_k - T_{OC})^2 \quad (9)$$

It should be noted that for determining the values of T_i, T_j и T_k We need to get the corresponding system of linear algebraic equations, which take into account all existing natural boundary conditions, varying from J to T_i, T_j и T_k.

$$\frac{\partial J}{\partial T_i} = 0; \rightarrow Fq + \frac{Fk_x}{6L}(14T_i - 16T_j + 2T_k) = 0$$
$$\frac{\partial J}{\partial T_j} = 0; \rightarrow \frac{Fk_x}{6L}(-16T_i + 32T_j - 16T_k) = 0$$
$$(10)$$
$$\frac{\partial J}{\partial T_k} = 0; \rightarrow \frac{Fk_x}{6L}(2T_i - 16T_j + 14T_k) + FhT_k - FhT_{OC} = 0$$

After simplification, we have

$$\left.\begin{aligned} &7T_i - 8T_j + T_k = -\frac{3Lq}{k_x} \\ &T_i - 2T_j + T_k = 0 \\ &7T_i - 8T_j + 7T_k + \frac{3Lh}{k_x}T_k = \frac{3LhT_{OC}}{k_x} \end{aligned}\right\} \quad (11)$$

After solving this system we define that

$$T_i = T_{OC} - \frac{q}{h} - \frac{Lq}{k_x}; \ T_j = T_{OC} - \frac{q}{h} - \frac{Lq}{2k_x}; \ T_k = T_{OC} - \frac{q}{h} \quad (12)$$

Then substituting (12) into (2-3) and after simplifying it we define the law of temperature distribution along the length of the test rod in view of simultaneous presence of the thermal insulation, heat flux and heat transfer. It will have the following form:

$$T = T(x, T_{OC}, q, h, L, k_x) = \left(T_{OC} - \frac{q}{h} - \frac{qL}{k_x}\right) + \frac{q}{k_x}x; \ 0 \le x \le L \quad (13)$$

This shows that in this case the law of the temperature distribution along the length of the test bar is linear

3 Determination of thermophysical state of the rod

Now we solve next problem. Because of the temperature field, the rod will extend. It is required to determine the elongation of the rod while it is under different sources of heat. To do this, we assume that the left end of the rod is rigidly fixed, and the right is free. From the general laws of thermodynamics [7-8] we known that the value of elongation of the rod from the temperature field is defined as:

$$\Delta l_T = \int_0^L \alpha T(x)dx \quad (14)$$

If assume that α=const, then

$$\Delta l_T = \int_0^L \alpha T(x)dx = \alpha L(T_{OC} - \frac{q}{h} - \frac{qL}{2k_x}) \quad (15)$$

Next step is to solve the third problem. If both ends of the rod is rigidly-clamped, the rod can neither lengthen nor shorten. In this case appears an axial compression force R[кг]. It is defined as the solution of statically indeterminate problems, while applying strain compatibility conditions:

$$\frac{RL}{EF} + \Delta L_T = 0 \rightarrow R = -\frac{\Delta L_T EF}{L} = -\alpha EF(T_{OC} - \frac{q}{h} - \frac{qL}{2k_x}) \quad (16)$$

After this the solution for the fourth problem is easily determined, defining the emerging field of thermo-elastic stress$\sigma \left[\frac{кг}{см^2}\right]$. It is determined in accordance with Hooke's law: [9]

$$\sigma = \frac{R}{F} = -\alpha E(T_{OC} - \frac{q}{h} - \frac{qL}{2k_x}) \quad (17)$$

This shows that the thermoelastic component of the stress field distribution σ is straight and parallel to the shaft axis and the x-axis. Using again the generalized Hooke's law we find a solution for the fifth problem of determining the thermo-elastic deformation of the component

$$\varepsilon = \frac{\sigma}{E} = -\alpha(T_{OC} - \frac{q}{h} - \frac{qL}{2k_x}) \quad (18)$$

From solutions it is clear that it linear and parallel to the axis OX. If we consider that $q<0$, then (16-18) shows that the R, σ, ε will compress. Next, using the fundamental laws of thermodynamics we can solve the sixth problem of determining the field of thermal strain

$$\varepsilon_T(x) = -\alpha T(x) = -\alpha\left[\left(T_{OC} - \frac{q}{h} - \frac{qL}{k_x}\right) + \frac{q}{k_x}x\right], \quad 0 \leq x \leq L \quad (19)$$

This shows that ε_T- will compress and distribution of the field will be linear.

The seventh problem can be determined by using the generalized Hooke's law. Then the field distribution of the temperature stress component will be:

$$\sigma_T(x) = E\varepsilon_T(x) = -\alpha E\left[\left(T_{OC} - \frac{q}{h} - \frac{qL}{k_x}\right) + \frac{q}{k_x}x\right], \quad 0 \leq x \leq L \quad (20)$$

From the decision it is clear that it has linear and squeezing character. The eighth problem of determining the field of elastic deformation of the components is determined from the fundamental law

$$\varepsilon_x(x) = \varepsilon - \varepsilon_T(x) = \frac{\alpha}{k_x}\left(-\frac{qL}{2} + qx\right) = \frac{q\alpha}{k_x}(-\frac{L}{2} + x), \quad (21)$$

This shows that $\varepsilon_x(x)$ is linear. At the length $0 \leq x \leq \frac{L}{2}$, it has expansive character. In section $x=\frac{L}{2}$, $\varepsilon_x\left(\frac{L}{2}\right) = 0$. Further it compresses.

The relevant law of Hooke determines the decision of the tenth problem

$$\sigma_x(x) = E\varepsilon_x(x) = \frac{q\alpha E}{k_x}(-\frac{L}{2} + x) \quad (22)$$

It is similar to $\varepsilon_x(x)$.

Now, finally, we decide the tenth problem of determining displacement field U(x). It is determined from the Cauchy relation

$$\varepsilon_x = \frac{\partial U}{\partial x} = \rightarrow U(x) = \int \varepsilon_x(x)dx = \frac{q\alpha}{k_x}\left(-\frac{L}{2}x + \frac{x^2}{2}\right) + C, \text{ где } C=\text{const}.$$

The value of C is determined from the condition of pinching the two ends, ie U(x=0)=U(x=L)=0. Then we have that C=0. Then the field of movement is

$$U(x) = \frac{q\alpha x}{k_x}(\frac{x}{2} - \frac{L}{2}), \quad 0 \leq x \leq L \quad (23)$$

This shows that U (x) has a quadratic form. The cross section located on the interval $0 < x \leq L$ moves in the direction of OX.

Naturally, clamped ends do not move, ie, U(x=0)=U(x=L)=0.

4 Conclusion

Based on the fundamental laws of energy conservation was developed computational algorithm and method of steady thermo-physical condition of the insulated rod of limited length, which is under the heat flow and heat transfer. It was found that the temperature distribution of elastic and temperature components are linear. While the value of the thermoelastic component of the strain and stress will be constant. The distribution law of movement will have a quadratic character, and all the section of the rod will move from left to right if q<=0

References:

1. Tikhonova Zh.M , Nurlybaeva E.N , Kudaykulov, A, Zhumadillaeva A.K. Numerical study of stablished Thermo-mechanical state of rods of limited length, with the presence of local heat flows, temperatures, heat Insulation and heat transfer [J]. Advanced Science Letters, 2013, 19 (8) : 2395-2397.
2. Tashenova Zh.M., Nurlybaeva E.N, Kudaykulov A. Developing a Computational Modeling Algorithm for Thermostressed Condition of Rod made of Heat-resistant Material ANB-300 type [J]. Advanced Materials Research, 2013, 19 (1): 4562-4566.
3. Nicolas X, Benzaoui A, Xin S. Numerical simulation of thermoconvective flows and more uniform depositions in a cold wall rectangular APCVD reactor [J]. Crystal Growth, 2008, 310 (1): 174-186.
4. Chen W.R. A numerical study of laminar free convection heat transfer between inner sphere and outer vertical Cylinder [J]. International Journal of Heat and Mass Transfer, 2007, 50 (13-14) : 2656-2666.
5. Gamrat G, Favre-Marinet M, Le P.S. Numerical study of heat transfer over banks of rods in small Reynolds number cross-flow [J]. International Journal of Heat and Mass Transfer, 2008, 51 (3-4): 853-864.
6. Nouri-Borujerdi A, Lavasani A.M. Experimental study of forced convection heat transfer from a cam shaped tube in cross flows [J]. International Journal of Heat and Mass Transfer, 2007. 50 (13-14): 2605-2611.
7. Yalcin H.G, Baskaya S. Sivrioglu M. Numerical analysis of natural convection heat transfer from rectangular shrouded fin arrays on a horizontal surface[J]. International Communications in Heat and Mass Transfer, 2008, 35 (3): 299-311.
8. Das A, Das B. Thermo-elastic stress distribution in three layered system [J]. Proc. Nat. acad Sci, 2001, 71(1): 21-31.

Typical Construction Technology and Method of UHV Main Transformer

CHEN Kerui[1], HAN Yang[2], WANG Lei[3], PEI Ying[4], HE Dengsen[5], ZHANG Yan[6], GAO Nannan[7], LI Hongbo[8], LI Pei[9], GUO Liangfeng[10]

[1-10]State Grid of China Technology College 250002, Jinan, China

Abstract. According to the latest experience of UHV transformer on-site installation work, according to the rules and regulations requirements, the construction process and process of UHV transformer are summarized, and a typical UHV transformer construction method commonly used at present is obtained. Operation comes down to the arrival of acceptance, annex installation, vacuum oiling and hot oil circulation, sealing inspection, standing five processes. In accordance with the above process steps, this paper is divided into five parts to introduce the process and method of UHV transformer construction in order to guide the on-site construction and training teaching.

1 Introduction

UHV is in the promotion of the use of power grids in China greatly enhanced power transmission capacity and efficiency. The data show that a 1000kV line is equivalent to 500kV line transmission capacity 4-5 times the transmission distance is 500kV lines 2-3 times. At the same time, in accordance with the national grid company data, transmission of the same energy situation, through the 1000kV line transmission land corridor occupies only 500kV line area of 60% [1]

The core equipment of UHV transmission substation is 1000kV main transformer, its installation technology and construction quality directly affect the substation and even the whole UHV project construction quality and level. In order to improve the installation quality of UHV main transformer and ensure the safe running and stable operation of the main transformer, the author has compiled this article according to the relevant requirements and can provide the reference for 1000kV main transformer installation and construction. [2-4]

2 Main Transformer Construction Workflow

UHV main transformer construction work is more complex, with clear steps and strict process. The process includes the arrival of the acceptance, annex installation, vacuum oiling and hot oil circulation, sealing inspection, standing and other steps [3-7] , As shown below.

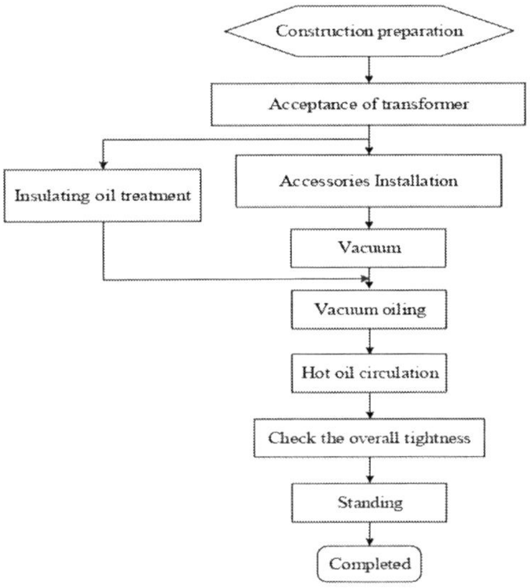

Fig.1. Main transformer construction work flow chart

3 Arrival acceptance

3.1. Acceptance of the transformer body arrival

According to equipment manufacturers to verify the size of the transformer base, and draw the basis of the design drawings based on the central line for the use of equipment transport unit in place.

After installing the transformer in the collision recorder, check the record of every collision recorder.

The level acceleration of the transformer during transportation is not more than 3g and the vertical acceleration is less than 3g, and the collision recorder and collision records should be kept properly.

After the body transported to the scene, immediately check whether the moisture after shipping the product. Whether the product is moisture based on the initial:
a) The nitrogen pressure of the nitrogen filling body is greater than 0.01MPa at normal temperature;
b) Take the oil residue in the main test, pressure, micro-water, dielectric loss should be consistent with the relevant provisions.

3.2. Accessories out of the box acceptance

1) In accordance with the demolition table and packing list out of the box inventory components, accessories, the provisions of the contract spares and special tools, the number of physical objects are consistent with the packing list, is damaged.

2) Check all valve operation is flexible, no jamming; opening and closing position is indicated on the valve position corresponds to the valve and the flange connection is sealed good tank should be no deformation damage.

3) Accessories should be complete, no corrosion and mechanical damage and seal well. Cooling device, CT elevated seat, casing and other visual inspection intact. The surface of the oil conservator should be free from collision deformation, corrosion, paint layer intact. Oil-filled casing of the oil level should be normal, no leakage of oil, porcelain body without injury.

4) Spare parts and complete information.

3.3. Parametric Studies

With the finite element models validated, further numerical analyses were done to generate results over a wider range of geometries and local slenderness to evaluate whether DSM is suitable for hot-rolled RHS/SHS stub columns. A total of 864 different RHS/SHS were modeled, which cover three kinds of yield strengths (S275, S355, S460) and four kinds of cross-sections defined in EN 1993-1-1 [1]. The value of aspect ratio (h/b) varies from 1mm to 3mm, the value of B varies from 40mm to 80mm, the value of H varies from 40mm to 200mm and the value of inner radius which equals to the value of thickness varies from 1.5mm to 5mm. The material properties used in the parametric studies are presented in table 2 and other properties that are not mentioned in this paper can refer to EN 1993-1-1 [1] and ECCS [7].

4 Accessories Installation

4.1 Installation of oil conservator

1) The oil conservator should be cleaned and kept dry before installation. The capsule inside the oil conservator should be complete without damage, the capsule mouth is well sealed and the breathing is smooth.

2) Lifting of oil conservator: a. The angle between strap and oil conservator should be greater than 45 degrees.

3) Assemble the tube before the heat with the transformer oil flush tube inside.

4) According to the drawings, manufacturers mark, the tube assembly; tighten the flange connection bolt force moderate, uniform, according to manufacturers torque requirements fastening, and pay attention to whether the rubber band inside the pad is positive.

5) Oil level indicator should be correct, no false oil level. In order to shorten the exposure of the body to the air, the assembly of the external accessory is carried out at the same time as the internal inspection - that is, the total assembling work. The total assembly, the inflatable hose changed to the main transformer oil injection valve to the tank and the continuous filling into the dry air.

4.2 Cooler installation

1) The cooler has been installed in accordance with the provisions of the manufacturer before the pressure value seal test qualified

2) The external oil pipeline should be thoroughly cleaned before installation and clean.

3) The cooler bracket, the upper and lower part of the first tube installed to the transformer body. Before installation, remove the corresponding cover plate from the joint cover and cooler tube, and wipe the oil on the port with a clean cloth.

4) Cooler hoisting: the choice of two cranes, the main transformer before and after the installation of a Taiwan. 1. Sling with a hanging hook hanging hook on the top of the cooler ring, the cooler stand up slowly, cooler end (with the end of the oil plug) to cushion rubber to prevent the cooler stand up and the ground bumps and damage. 2. The angle between the strap and the cooler should be greater than 45 degrees.

5) Open the cooler lower part of the oil plug, let cooler internal residual oil tighten the oil drain plug.

6) Open the cooler oil channel cover, install the cooler on the bracket and dock with the body oil channel, fasten the bolt and then remove the harness.

Fig.2. Cooler and the body docking

7) Install the oil pump on the cooler oil line.

8) After the installation of all the fans check whether the normal operation, the oil pump steering is correct, no abnormal noise.

 https://doi.org/10.1051/e3sconf/20183804003

4.3 Installation of high - pressure outlet

1) Lifting the lower support frame of the high-pressure outlet device to the installation base.

2) Remove the connection between the high-voltage lead and the transport cover, and then remove the transport cover on the mounting flange on the fuel tank.

3) High-pressure outlet device filled with dry air transport, it should first release its positive pressure inside the dry gas, and then remove the transport side of the connection with the body cover.

4) Prepare 4 5t manual hoists Press (Fig. 3) to connect the sling and manual hoist. The lifting hook makes the main sling slightly force.

5) Tighten the manual hoist, so that auxiliary sling force.

6) Lifting hook, and adjust the manual hoist, so that the outlet device is adjusted to the installed state.

7) The high-pressure outlet device hoisting to the tank, and connect the lead.

8) Install the support fastener between the outlet device and the bracket, and tighten the outlet device.

9) Remove the upper transport cover from the outlet unit.

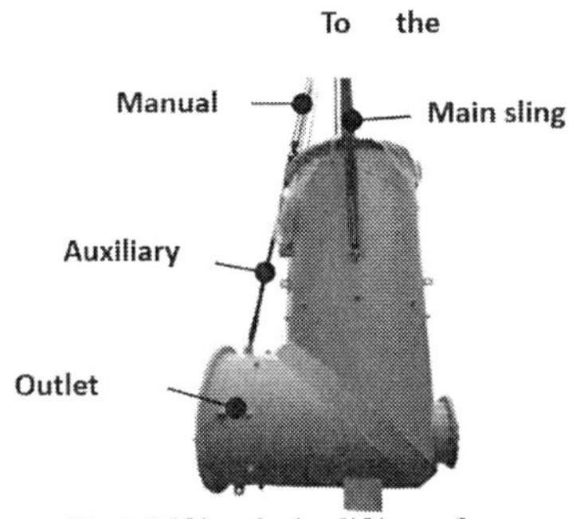

Fig.3. Lifting device lifting reference

4.4 Raise the seat mount

1) Before the installation of elevated seat, should complete the current transformer test.

2) Elevated seat lifting: the angle between the sling and the elevated seat should be greater than 45 degrees.

Fig.4. Higher installation

3) When installing the elevated seat, the venting position should be at the highest point of the raised seat.

4) The current transformer and the center of the elevated seat should be consistent, the insulation tube should be installed firmly to ensure that the transformer lead wire will not collide with it.

5) Before installing the elevated seat, store the dry test strip in the elevated transport cylinder and check whether the color of the test paper is changed to determine whether the outlet insulation cylinder is damp during transportation.

4.5 High - pressure casing hoisting

1000kV high-voltage casing hoisting and installation work, pay attention to the weight of high-pressure casing, length, transformer height, and 1000kV outlet device.

1) The casing installation direction for the vertical installation, lifting with two cranes.

2) The casing dedicated spreader fixed to the casing, so that both ends of the soft strap, respectively, through the spreader fixed to the casing flange rings, the other end hanging on the main crane hook. Also take a harness, one end hanging in the flange rings, the other end hanging in the auxiliary crane.

3) The beginning of a certain level of lifting height of the casing, the auxiliary crane hook brake, the main crane hook slowly rise, so that the casing slowly erected until the auxiliary crane hook is not force, take off and auxiliary crane connection Sling, remove the casing tail protection tube, continue to rise until the casing test rack.

4) Casing hoisting flow chart.

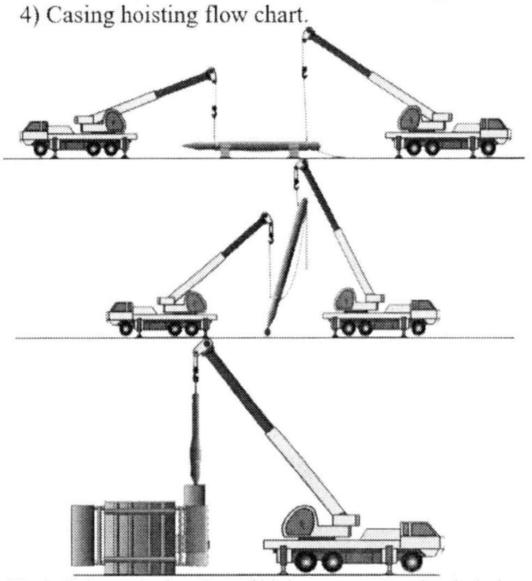

Fig.5. Schematic drawing of high - pressure casing hoisting

4.6 Internal lead connection

(1) High-voltage casing of the lead installation, the use of guide-type installation, must be strictly in accordance with casing instructions for. Advance the lower part of the guide rod and lead wire connection, from the elevated seat into the hole with a wrench to tighten the guide rod connection, with a locking fluid locking. Fasten the casing head bolts with a torque wrench as per the casing instructions. The casing lead connections are as follows:

 https://doi.org/10.1051/e3sconf/20183804003

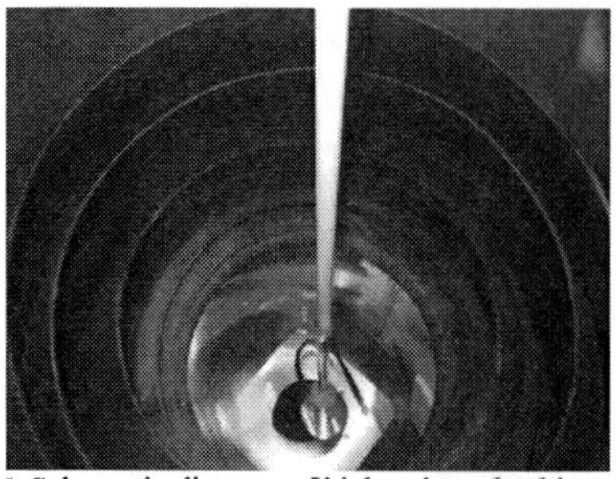

Fig.6. Schematic diagram of high voltage bushing

2) When installing and dropping the casing, observe the cooperation between the casing and the lead wire and the outlet device, avoid damage, and make the lower part of the casing and the lead wire fully enter the insulated ball of the outlet insulation. The specific size complies with the drawing requirements. Casing tail and pressure ball position shown below:

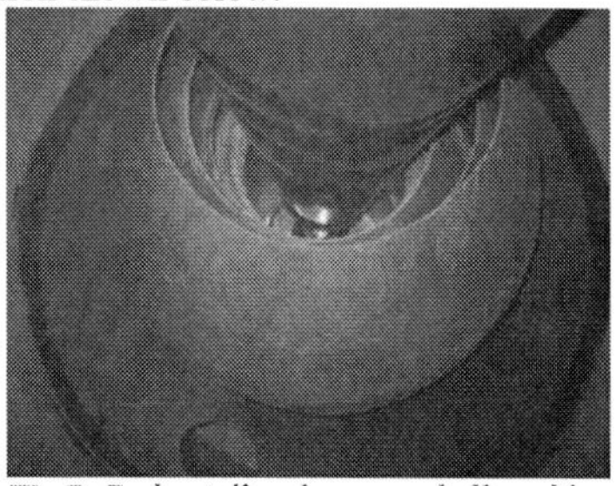

Fig.7. Casing tail and pressure ball position

3) The casing flange and the top of the current transformer flange evenly bolted symmetrical tightening, revealing the deduction for the 2 ~ 3 buckle and pay attention to the casing oil table outward.

4) The contact surface of the terminal to be wiped, no dirt, oxide film and other impurities impeding the existence of electrical contacts. The connecting piece of the connecting terminal should be straight, no burr, flash. Fastening bolts on both sides with a disc spring washer. The tightening bolts are tightened according to the tightening torque of Table 1. To ensure good electrical contact.

5) After installing the casing head are installed pressure ring.

4.7 Other accessories installation

4.7.1 Seal ring installation

1) General requirements Manufacturers will be transported to the scene when the installation has been installed in the ring with a new ring replacement of all.

2) All flange connections are sealed with oil-resistant seals. The seals must be free from distortion, distortion, cracks and burrs, and the gasket should be dimensioned to match the flange face.

3) Keep the sealing ring and the sealing ring surface clean.

4.7.2 Tank sealing

1) Clean the inside of the fuel tank.

2) According to the record check, inventory has been removed all the tools and materials.

3) Close the dry air generator outlet valve.

4) with the cover closed into the hole and other flange

4.7.3 Installation of gas relays

1) Gas relays should be inspected before installation.

2) Gas relay should be installed horizontally, the arrow on the top of the logo should point to the conservator, the connection with the tube should be sealed well.

3) Gas relay should have moisture and anti-water function.

4) Cable leads in access to the gas relay should be dripping bend, into the line hole should be sealed tightly

4.7.4 Installation of pressure relief valve

Pressure relief valve should be installed in the correct direction, the valve cover and the internal rise should be clean, sealed well, electrical contacts should be accurate, insulation should be good.

4.7.5 Temperature measurement device installation

1) Thermometer installed before the check, the signal contact action is correct, conduction is good; winding thermometer should be set according to the provisions of the manufacturer.

2) The top of the thermometer seat should be noted with transformer oil, the seal should be good, no oil leakage phenomenon.

3) Expansion of thermometer fine metal hose shall not be crushed or sharp twist, bending radius of not less than 50mm.

5 Vacuum oiling and hot oil circulation

5.1 The transformer is evacuated

1) the transformer tank sealed with a qualified transformer oil to clean the bottom of the tank, no debris left over. Then seal all open covers. Check the vacuum pump, vacuum line transformer around the valve status and sealing conditions, open the tank and the radiator between the valves.

2) Connect the vacuum pump tubing to the special butterfly valve located at the top of the tank. Connect the vacuum gauge.

3) Before evacuating, the valve between the accessories (eg oil pillow) and the tank which can not bear the vacuum must be closed. For parts that allow vacuuming, the radiator should be evacuated at the same time.

4) Start the vacuum pump and slowly open the vacuum suction valve. Pumped to about 2 hours after the suspension (first close the exhaust valve, then stop the vacuum pump), check the seal of the tank connection. If

suction sound, should be promptly handled. When the elimination can not afford to break vacuum treatment, the air must be filtered through the dry silica gel before being placed in the transformer. In the vacuum process, the vacuum rose slowly or a large pressure leak, that may leak, should be promptly dealt with.

5) Vacuum, it should be observed at any time record the deformation of the tank, the maximum deformation of not more than 2 times the wall thickness, while attention to the deformation of the radiator, if the problem promptly contact with the manufacturer representatives.

6) Vacuum to 13Pa, the transformer leakage rate test. The results of the test results do not meet the requirements, check all the joints and tighten, do leakage rate until qualified.

7) After completion of the leak test, continue to vacuum to 13Pa or less, to maintain ≥48 hours (pressure to 24 hours).

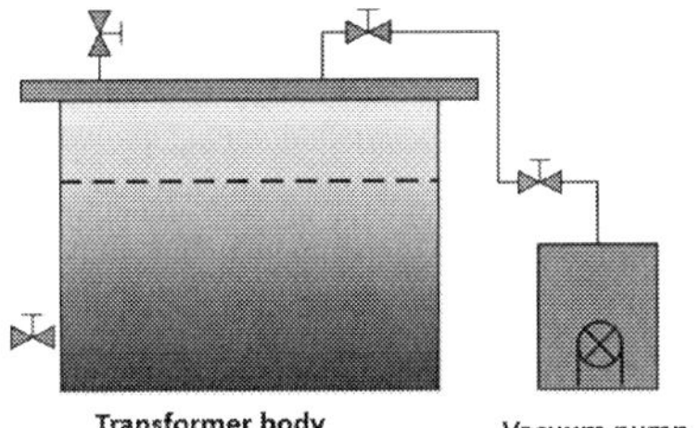

Figure 3-12 Vacuum tooling diagram

5.2 Vacuum oiling

1) Start the vacuum oil filter The transformer oil is injected into the injection valve, continuous vacuum during oil filling, and adjust the filling speed to maintain the vacuum within ≤20Pa (modulation is not greater than 133Pa), but the maximum flow rate shall not exceed 5t / h, oil temperature 60 ± 5 C, until the oil surface under the vault to 200mm.

2) Continue to vacuum 2h, close the butterfly valve and the unit, the vacuum gauge and vacuum line switch to the conservator breathing tube Department, open the conservator breathing mouth valve, exhaust port and breathing port Unicom valve, close the exhaust valve, Install the gas relay, open the 80B butterfly valve at both ends, and draw the vacuum from the breather port of the conservator. Continue to fill the oil tank to half the oil level, stop the vacuum, stop oiling, close the exhaust port and breathing port Unicom valve, close the vent valve.

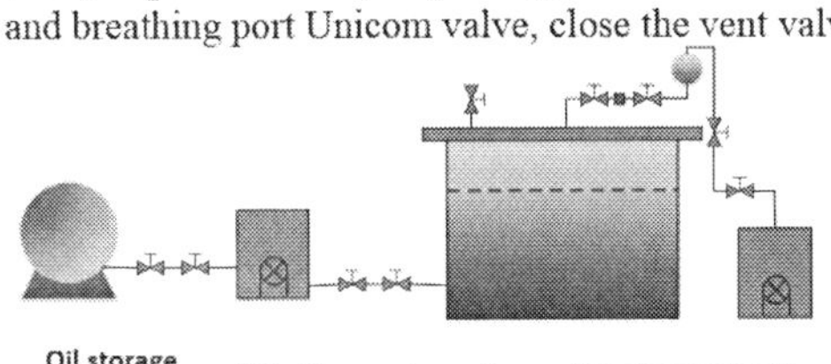

Fig.8. Vacuum oiling equipment diagram

5.3 Hot oil circulation

1) In order to eliminate the installation process of insulation of the surface of the body moisture, the use of the way up and down the hot oil circulation.

2) Hot oil circulation before the tubing should be vacuum, the tubing in the air pumping clean.

4) Cycle time should not be less than 96h, cycle flow 8 ~ 10m3 / h.

5) During the hot oil circulation, the outlet temperature of the oil filter should be controlled within 60 ± 5 °C, and the outlet temperature of the transformer should be controlled within 60 ± 5 °C.

6) The hot oil circulation should meet the following conditions before the end:

Hot oil cycle time of not less than 96h, the cycle time of not less than 3 × total transformer oil / oil through the oil hourly.

7) The transformer outlet temperature is reached every 2 hours after opening a group of cooler valve for hot oil circulation, to prevent the cooler affect the hot oil circulation effect.

8) After the end of the hot oil circulation, should close the oiling valve, open the transformer all components, accessories and piping bleed valve exhaust.

9) The use of machinery and equipment and insulating oil processing part of the same.

5 The whole sealing test and standing

The overall sealing test: the transformer as a whole (including the cooling system) and the oil conservator can withstand the highest oil level of 30kpa static pressure on the oil seal test, the test time for 24h, no leakage and damage.

After the hot oil circulation, must be in the transformer before the high-voltage test must be placed more than 168h (adjustable voltage compensation more than 72h) during the static, the product static interval of 24 hours, the use of all components of the transformer, accessories and piping All deflate venting pieces, oil spill immediately tighten the venting plug, wipe the spilled oil. Gloves of the gasket is small, so to moderate force, it is necessary to seal good and do not twist bad pads.

6 Conclusion

In this paper, the construction of UHV main transformer in China is fully considered, and the construction process and process of the UHV main transformer are summarized and the construction process of a typical UHV transformer is put forward with reference to relevant rules and regulations. The construction process is attributed to the acceptance of the arrival of the annex to install the vacuum oil and hot oil circulation, sealing inspection, standing five parts, the steps of convergence reasonable and tight, in line with the general UHV transformer installation process customary, High-voltage main transformer on-site construction operations, and provide reference for related fields of training and teaching.

References

1. Liu Zhenya "Innovation of UHV AC transmission technology in China" Power System Technology vol. 37 No. 3 pp. T1 – T8 2013.
2. Dept. of Science & Technology: New Technology Catalog of SGCC (2014) Beijing: China electric power press 2014.
3. Liu Zhenya Equipment Research of UHV AC Pilot Demonstration Project in China Beijing: China Electric Power Press 2010.
4. H. Weng X. Lin "Studies on the unusual maloperation of transformer differential protection during the nonlinear load switch-in" IEEE Trans. Power Del.</em> vol. 24 no. 4 pp. 1824-1831 Oct. 2009.
5. R. Hunt J. Schadfer B. Benter "Practical experience in setting transformer differential inrush restraint", Proc. 61st Annu. Georgia Tech Protect. Relay. Conf.</em> pp. 118-141 2007-May.
6. D.Shao X. Yin Z. Zhang W. Chen D. Chen "Experiment research on differential protection for UHV transformer in China", Proc. IEEE Power Energy Soc. Gen. Meeting, pp. 1-5 2008-Jul.-20–24.
7. H. Weng X. Lin P. Liu "Studies on the operation behavior of differential protection during a loaded transformer energization", IEEE Trans. Power Del. vol. 22 no. 3 pp. 1386-1391 Jul. 2007.

Cause Analysis and Countermeasures of a Breakdown Failure of Flexible 110 kV Cable Terminal

Feng Huang

Electric Power Research Institute of Guangxi Power Grid Co., Ltd., Nanning, Guangxi 500023, China

Abstract. disintegration examination and analysis are employed in flexible terminal breakdown of 110 kV XLPE insulated cables. It is considered that the main reason of breakdown is the separation of the stress cone of the terminal and the fracture of the semi- conductive layer of the cable insulation. Therefore, the finite element method is used to electric field model and simulate the dislocation fault of internal stress cone and outer semiconductor layer of cable insulation. The distribution of the electric field intensity is calculated and compared. The simulation and calculation results verify the validity of the breakdown mechanism analysis, and put forward some practical suggestions.

1 Introduction

It is predicted that the clean energy installation in China will reach 570 million kW in 2020, which exceed 1/3 of the total installed capacity. The emission of CO_2 can be reduced 13.8 million tons per year [1]. The rapid increase of power generation and the requirements of friendly environment has a profound impact on power transmission, and brings development opportunities for the application of cable. The demand for power cable is increasing rapidly, and the application scope of cable is also gradually expanding. Because of power cable using closed compact structure, the solid extrusion material insulation or oil paper composite insulation is wrapped in closed enclosure. Once fault occurs, it will lead to power failure directly, and it is difficult to locate and maintain. Therefore, although the failure rate of the cable is much lower than the overhead transmission line in the actual operation, the accidental cable accident still brings great trouble. Compared to cable, cable accessories (terminal and joint) insulation have complex structure and are compose by multilayer material, so the electrical stress concentrates on the insulation shield fracture which is prone to failure. In addition, the cable accessories are generally required to complete the installation on site, it will cause safety problems due to the poor sealing, bad construction technology and inadequate dust prevention measures. So the probability of cable accessory failure is much higher than cable.

The flexible cable terminal is one kind of prefabricated cable terminal, it solves many shortcomings of the traditional high voltage XLPE cable terminal such as complicated structure, inconvenient installation and oil spill. Besides, it has many advantages, such as high consistency of the manufacturing process, the simple installation process, reliable operation and no need to maintain [2-3]. It has become the ideal high voltage XLPE cable accessory and been widely used. In recent years, more and more urban overhead lines have been converted to cable, and the number of flexible cable terminals has increased greatly. With the gradual extension of the running time, the breakdown of prefabricated flexible terminal has appeared all over the country, which brings serious hidden danger to the security and stability of the power grid. As the most concentrated place of the electrical stress, the insulation shield fracture of the high voltage cable is the main part of insulation fault. If the electrical stress isn't evacuated fully, then the part will be the breakdown location on power cable insulation, and the stress cone can't play its due role [4]. The finite element method is a numerical method based on the variational principle [5]. Using this method can effectively calculate the electric field intensity. It is an effective tool for electric field simulation [6].

In this paper, a flexible terminal of 110 kV XLPE cable is disassembled, checked and analysed. The possible cause of the failure is pointed out. Finally, the finite element electric field simulation method is applied to verify the cause of the failure.

2 Failure Information and Cause Analysis

2.1. Failure Profile

In September 2016, a breakdown of outdoor flexible terminal of 110 kV XLPE cable occurred. The basic information of the failure cable is as follows: the cable line was put into operation in October 2006, the cable terminal type is YJZWG4-64/110-1×500mm2, and the cable type is ZR-YJLW02-64/110-1×500mm2. The circulation test of the cable lines has been carried out before the failure, and the three phase circulation values

* Corresponding author: huang_f.sy@gx.csg.cn

https://doi.org/10.1051/e3sconf/20183804004

are all in accordance with the requirements of the regulations. When the failure occurs, there are no external forces such as construction and excavation on the cable channel. The weather is clear, and there is no overvoltage phenomenon such as lightning.

2.2 Disintegration Examination of Failure Terminal

The cable terminal surface has obvious flexible burn holes after disassembling, and most of the region outside the surface have been blackened, careful inspection shows that the main insulation has complete breakdown, as shown in figure 1. The fault point is located in the insulation shield fracture, and the stress cone has a certain distance of abnormal displacement with the insulation shield fracture.

Fig. 1. Failure Cable Terminal.

Inspection and measurement showed that there was a 2~3 mm gap between the cable aluminium bellows sheath and the longitudinal water blocking buffer layer (the water blocking zone), they were not tightly integrated, as shown in Figure 2.

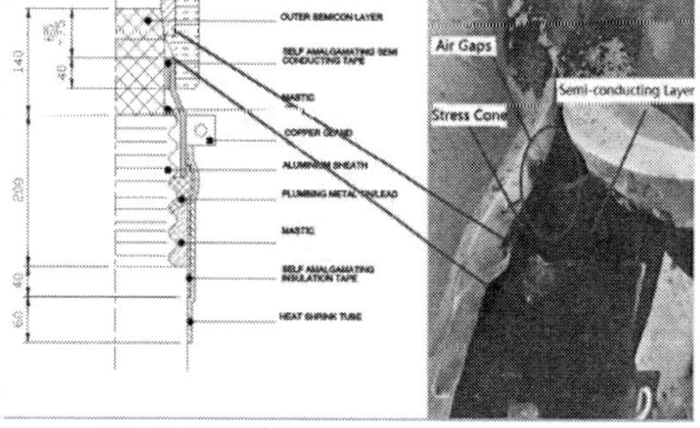

Fig. 2. Internal Gap in the Failure Terminal.

2.3 Analysis of Failure Causes

Because the installation position of the cable terminal is about 15~20 meters from the ground, the cable terminal is manufactured on the ground and then hoisted to the installation position. When the hoisting is fixed, only the aluminium sheath of the cable is fixed. For the cable whose type is ZR-YJLW02-64/110-1×500mm2, its weight is about 10kg per meter. Then the carrying weight of cable terminal can reach more than 150kg. If the cable aluminium corrugated sheath is not closely combined with the longitudinal water blocking buffer layer (the water blocking zone) of the cable, it may cause the phase displacement of the cable body and the aluminium bellows sheath due to the weight of the cable.

As we all know, f is integrated with corrugated aluminium, so it will generate relative displacement between cable body and flexible terminal when the cable body has relative displacement with the aluminium ripple sheath. And it will lead to that the stress cone of flexible terminal can't get in touch with the insulation shield fracture normally because of the abnormal displacement. Then the suspended discharge can be generated because the electric stress can't be evacuated. Finally, the flexible terminal breakdown happened.

3 Simulation Analysis with Finite Element Method

3.1. Simulation Model Establishment

Referring to the structural parameters of the cable terminal, a complete cable terminal model is established according to the actual size (1:1). The finite element method is applied to simulate and calculate the electric field numerical value of the terminal when the applied voltage is 64 kV. The structure of the terminal model is shown in Figure 3, and the dielectric constant of the material is shown in Table 1.

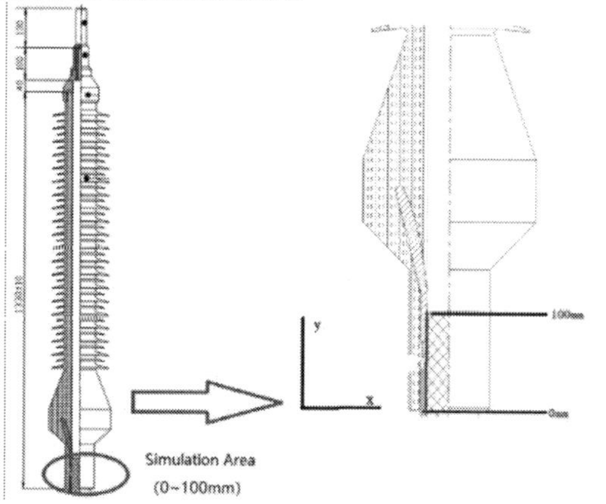

Fig. 3. Model of Flexible Terminal.

Table 1. Dielectric constant of terminal material

Material	Relative Dielectric Constant
XLPE	2.25
Semi Conductive Layer	100
Copper Core	10000
Stress Cone	2.78
Air	1

3.2 Simulation Results

According to the disintegration result and cause analysis of the failure, the simulation of two contact situation of stress cable and semi conductive layer were carried out respectively. The first case is that stress cone gets in touch with semi conductive layer normally, and the second case is that stress cone has 20mm displacement with semi conductive layer. The distribution of electric field was observed, and the electric field value was calculated. The simulation results are shown in Figure 4 to figure 6.

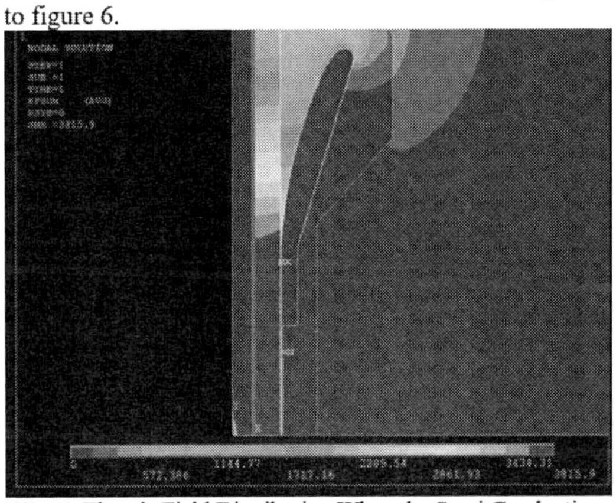

Fig. 4. Electric Field Distribution When the Semi Conducting Layer Overlap the Stress Cone Completely.

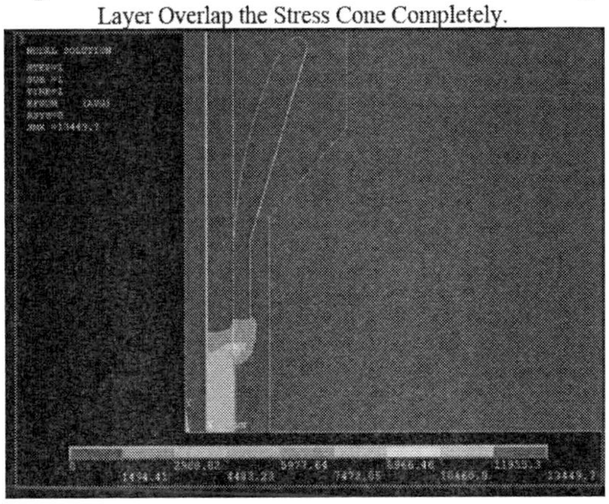

Fig. 5. Electric Field Distribution When Semi Conducting Layer Dislocate the Stress Cone 20mm.

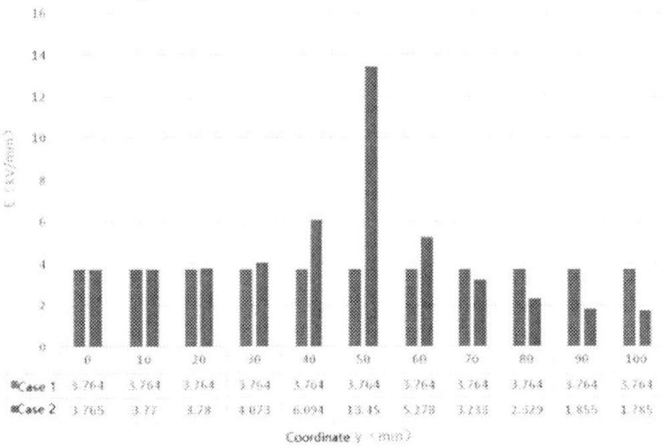

Fig. 6. Diagram of Electric Field Variation Trend.

3.3 Simulation Results Analysis

According to Fig. 6, the electric field is seriously distorted when the semi conductive layer is misaligned with the stress cone. When y=50mm, the maximum electric field value reaches 13.45kV/mm, which is 3.57 times of the normal case. It is far more than the breakdown electric field intensity of air (about 3kV/mm), and it also exceeds the electric field intensity of design interface between the cable insulating layer and the stress cone (about 11kV/mm). It can be seen that the main cause of the cable terminal breakdown is that the inner semi conductive layer of the stress cone and the fracture of the semi conductive layer of the cable insulation can't lap normally.

4 Conclusions and Countermeasures

4.1. Conclusion

According to the analysis of the failure terminal and the simulation results of the electric field, the following conclusions can be drawn:

1) when the cable terminal is installed, only the cable aluminium sheath layer is fixed. The weight of the cable causes the phase displacement of the body and the aluminium bellows sheath, and the stress cone of the flexible terminal can't connect with the insulation shield fracture properly. It resulted in the breakdown of the terminal.

2) The electric field simulation results verify the correctness of the accident mechanism analysis. The electric field distortion is more intense under the condition of the misalignment of the outer semiconducting layer and the stress cone, then partial discharge is generated. When the electric field intensity reaches a certain level, it will cause weakness breakdown.

4.2 Countermeasures

According to the above reasons analysis and practical experience, the following countermeasures are put forward:

1) Cable hoisting is carried out before the installation of the flexible terminal. Before installing cable accessories, cable should be pre hung to install position. Then cable will be put down, so repeated 3 times. Thus the displacement of cable body and the aluminium corrugated sheath will be generated in advance so as to prevent the relative displacement between the two after installation.

2) Improve the space layout of the flexible cable terminal. The flexible cable terminal on the tower should avoid bending as far as possible. Under ideal condition, the cable terminal should be perpendicular to the ground. In cases where the site conditions are not allowed, the proper fixed method should be used to ensure that the bending angle of the terminal is not more than 30 degrees in the long run.

3) The insulation state detection of the flexural flexible terminal should be strengthened. For the flexural flexible terminal, the operation and maintenance personnel can evaluate the running state through infrared thermal imaging, partial discharge test and ultraviolet imaging in order to prevent terminal breakdown.

References

1. Zhou Y, Zhao J, Liu R, et al. Key Technical Analysis and Prospect of High Voltage and Extra-high Voltage Power Cable[J]. High Voltage Engineering, 2014, 40(9): 2593-2612.
2. Yin Q. Technical progress of the accessories for high and extra high voltage XLPE cables[J]. Electric Wire & Cable, 2000(1): 3-11.
3. Wang W. Technical foundation of XLPE insulated power cable[M]. Northwestern Polytechnical University press, 2005.
4. Si C. Power Cable[M]. China Electric Power Publishing House, 2006.
5. Wu X, Zhou L. Application and development of time-domain finite element method on EM analysis[J]. Chinese journal of radio science, 2008, 23(6): 1209-1216.
6. Li H, Zhang L, Zhou Z. Optimizing design of stress controlled corn by FEM[J]. High Voltage Engineering, 2005, 31(11): 55-57.

Temperature Drift Compensation of Eddy Current Sensor under High Temperature Environment

Ping He[1], *YunKai* Ma[2], and *Hui* Chen[3]

[1]HIT.hepinghit@126.com, Harbin, Heilongjiang, China
[2]HIT, 1127345182@qq.com,Harbin,Heilongjiang, China
[3]SACTI, 201109, Shanghai, Shanghai, China

Abstract. Eddy current sensor is an sensor based on eddy current effect. In practical engineering applications, the ambient temperature of eddy current sensor may be up to 135 ℃. The temperature drift of eddy current sensor magnifies the error of displacement detection. In this paper, the main factors that cause temperature drift are analyzed in detail, and the results show that the compensation based on single parameter can not meet the demand of high-precision measurement. For this reason, this paper proposes an external compensation method which applies mathematical fitting to realize compensation for temperature drift. The experimental results show that the measurement accuracy of the external compensation method reaches 0.25% in the working temperature range, which greatly improves the measurement accuracy of eddy current sensor under high temperature.

1 Introduction

The eddy current sensor is a special inductance sensor that is based on the eddy current effect [1]. Eddy current sensor has many advantages, such as high sensitivity, fast dynamic response and non-contact detection, which makes the sensor widely be used in displacement measurement, speed measurement, temperature measurement, vibration measurement, thickness measurement, eddy current testing(ET) and so on. However, eddy current sensors are very susceptible to temperature changes,especially in high temperature environment. The measurement error, which can't be ignored in the high temperature area, makes the use of the eddy current sensor be limited in the high temperature environment. Analysis shows that temperature affects almost all the materials of the sensor system, so it is difficult to achieve a good effect by traditional single parameter compensation. In this paper, the influence of temperature is estimated by the method of mathematical fitting. Experiment proves that the external compensation method may achieve high precision.

2 Working Principle

When the metal panel is in a changing magnetic field, the induced electromotive force inside the metal panel will generate an electric current.The current is closed inside the metal panel and become an eddy current [2]. As shown in figure 1 , X is the distance between the coil and the metal conductor,the eddy current density changes as the X changes, and the intensity of the eddy current varies with the variation of the distance X.

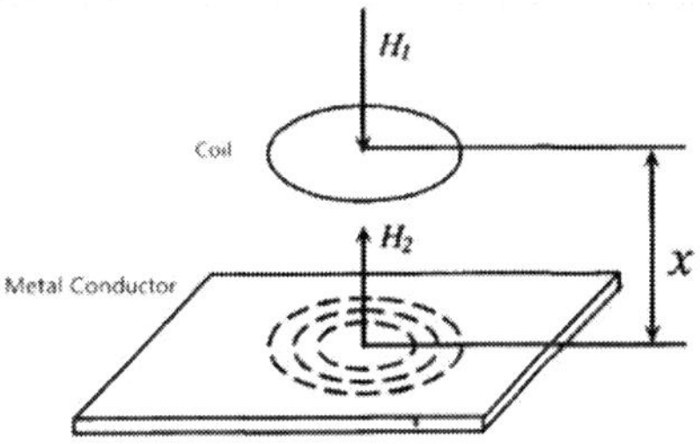

Fig. 1. Principle of Eddy Current Effect

The change of eddy current will cause the change of coil impedance. The factors related to the coil impedance change are: resistivity ρ of the tested material, the distance X between the coil and the metal conductor, the permeability μ of the metal conductor, and the frequency f of the coil excitation current f. To control factors ρ, μ, ω do not change, the impedance of the coil is only related to the distance X. We need to measure the distance, that is to say, if we kept ρ, μ, ω constant, and the impedance Z is only a function of distance X as

$$Z = f(X) \tag{1}$$

The eddy current displacement sensor can be made as long as the suitable probe coil and circuit parameters are selected [3]. The structure of the eddy current sensor is simple, mainly composed of the probe, the extended cable and the proximitor. The above three are combined with the tested mechanism to form a basic working system. Working principle diagram of eddy current displacement sensor as shown in Figure 2 .

Corresponding author: 1127345182@qq.com

https://doi.org/10.1051/e3sconf/20183804005

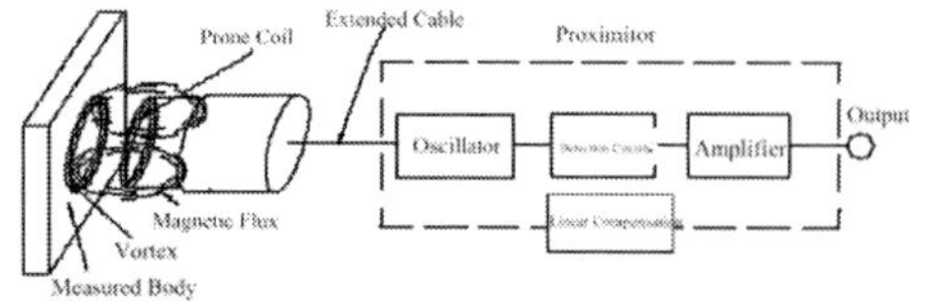

Fig. 2. Basic Principle of Eddy Current DisplacementSensor

The high frequency oscillating current in the preamplifier flows into the coil in the probe through an extended cable and produces alternating electromagnetic fields in the coil of the probe. If there is no metal material around the magnetic field, the magnetic field energy is actual total loss; when the metal material near the magnetic field,the induced current is produced on the metal surface. At the same time, the current field also produces an alternating magnetic field that is opposite to the direction of the head coil.Due to the reaction of magnetic field, the amplitude and phase of the high frequency current of the probe coil has changed .This change is related to the permeability and the conductivity of a metal conductor, the shape and the size of the coil, the frequency of the current and the distance from the probe coil to the metal material and so on.We maintain a constant in all conditions except the distance from the probe metal material, the impedance Z of the coil and the distance X have a function relation. Through the proximitor, the change of the distance between the probe and the metal becomes a change in the output of the voltage.The eddy current sensor is based on this principle to measure the displacement of metal materials.

3 Compensation for Tempereature Drift

The output voltage of the eddy current sensor is affected with temperature drift.So some scholars apply temperature compensation for the eddy current sensor. A common method is to use multi-strand lines to reduce the influence of temperature on the coil impedance by reducing the AC resistance of the coil, other methods include optimizing the parameters of the sensor, so as to improve the sensor's ability of restraining temperature drift. These methods have achieved effective results, but the factors affecting the temperature drift is varied, the single parameter compensation can only meet the general requirements. If the project requires high precision, external compensation method that uses mathematical fitting will achieve comprehensive compensation for temperature drift.

In order to determine the effect of high temperature on the sensor, the displacement sensor is installed on the displacement calibration table which is in the incubator. Adjust the distance of the target with the center point of the sensor. The temperature of the incubator is set at room temperature of 20°C, and change the temperature according to 20°Cinterval as a measurement point. Then change the displacement between the sensor and the target, 21 displacement points are measured at each temperature, the interval between every two displacement points is 0.1mm, read the sensor output voltage of each measurement point, output voltage data of eddy current sensor as shown in table 1and table 2.

Table 1. Output Voltage of Eddy Current Sensor 1

Displacement (1/10mm)	**20℃ (v)**	**40℃ (v)**	**60℃ (v)**
1	-9.989	-10.054	-10.096
2	-8.997	-9.051	-9.21
3	-8.004	-8.049	-8.213
4	-7.015	-7.044	-7.212
5	-6.024	-6.049	-6.193
6	-5.031	-5.041	-5.169
7	-4.068	-4.016	-4.134
8	-3.071	-2.998	-3.093
9	-2.086	-1.962	-2.057
1	-1.091	-0.928	-0.983
11	-0.094	0.131	0.074
12	0.91	1.176	1.152
13	1.915	2.226	2.225
14	2.93	3.294	3.33
15	3.94	4.36	4.415
16	4.94	5.423	5.553
17	5.96	6.493	6.613
18	6.961	7.575	7.755
19	7.968	8.665	8.9
20	8.968	9.77	10.067
21	9.989	10.137	10.137

Table 2. Output Voltage of Eddy Current Sensor 2

Displacement (1/10mm)	**80℃ (v)**	**100℃ (v)**	**120℃ (v)**	**135℃ (v)**
1	-10.096	-10.096	-10.096	-10.096
2	-9.512	-9.761	-10.086	-10.096
3	-8.509	-8.76	-9.123	-9.437
4	-7.498	-7.805	-8.152	-8.493
5	-6.483	-6.804	-7.177	-7.532
6	-5.46	-5.801	-6.194	-6.583
7	-4.45	-4.788	-5.215	-5.616
8	-3.408	-3.782	-4.229	-4.651
9	-2.36	-2.747	-3.226	-3.664
1	-1.296	-1.706	-2.221	-2.691
11	-0.217	-0.663	-1.199	-1.698
12	0.848	0.37	-0.199	-0.721
13	1.911	1.417	0.786	0.231
14	2.987	2.456	1.799	1.208
15	4.075	3.515	2.827	2.195
16	5.178	4.594	3.882	3.189
17	6.299	5.685	4.926	4.215
18	7.45	6.792	5.97	5.207
19	8.603	7.918	7.038	6.219
20	9.77	9.02	8.094	7.225
21	10. 137	10.137	9.167	8.251

From table 1 and table 2, we can draw the curve of the eddy current sensor output voltage varying with the detection displacement at different temperatures, and draw all the temperature curves on a map for comparison, as shown in figure 3 .

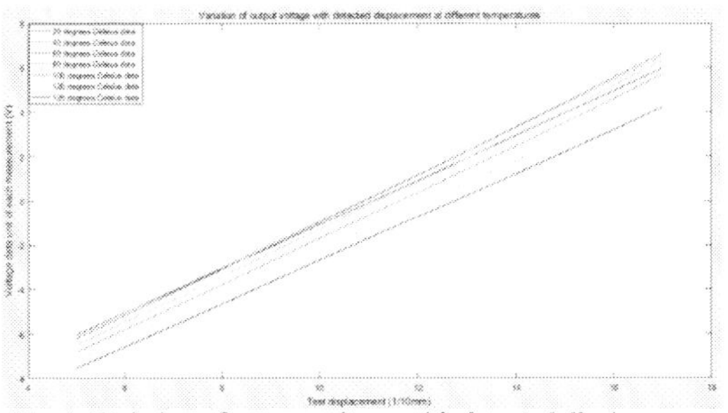

Fig. 3. Variation of output voltage with detected displacement at different temperatures

For 21 data at each temperature, the following seven sets of voltage-displacement equations are obtained by using the MATLAB fitting curve.

$$U=\begin{cases}0.99867*X-11.047\ ;\ T=20℃\\ 1.04560*X-11.325\ ;\ T=40℃\\ 1.06920*X-11.610\ ;\ T=60℃\\ 1.06600*X-11.888\ ;\ T=80℃\\ 1.04080*X-12.066\ ;\ T=100℃\\ 1.00770*X-12.259\ ;\ T=120℃\\ 0.97794*X-12.451\ ;\ T=135℃\end{cases} \quad (2)$$

In the formula (2),
U-sensor output voltage (V);
X-10 times of the measured displacement (mm);
T - temperature (℃);

From the above seven equations, it can be found that the relationship between the output voltage of the sensor and the measured displacement is highly linear at the same temperature. But the temperature drift of the sensor is quite obvious at different temperatures. If the displacement at high temperature is calculated by the equation at normal temperature,the error is relatively large. So we should seek a better compensation method for temperature drift error under the high temperature environment.

Through the analysis, compared with 20℃, the output characteristics of the eddy current sensor have changed considerably at 135 ℃ high temperature. Take the displacement data of 135℃ into temperature of 20℃ voltage-displacement equation, the error voltage is above 1V, the accuracy of error is above 5%, the traditional method is to compensate for a single parameter or several parameters, and often can only apply to a lower temperature range (less than 100℃). The traditional method for high temperature zone (above 100 ℃) compensation can not achieve good results. The influence of high temperature on the temperature drift of sensors is critical. So, In this paper, a mathematical fitting method is proposed to get the relationship between all kinds of temperature and the slope and intercept, and the external compensation is done through the formula calculation.

First, fitting the function relation of voltage-displacement straight slope at different temperatures. In order to make the fitting error smaller, higher order fitting is applied,the final fitting result is shown in figure 4 below.

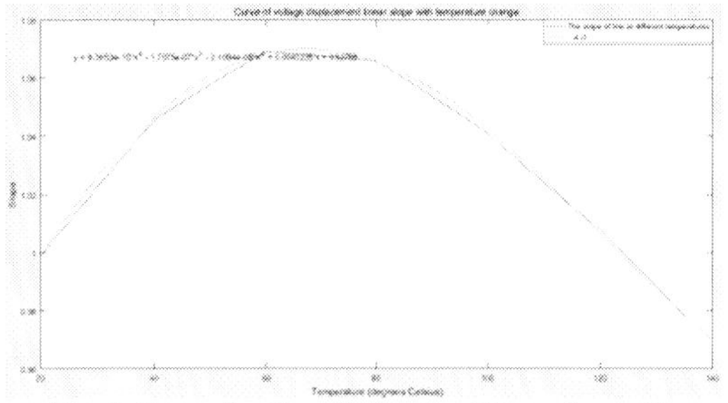

Fig. 4. Curve of voltage displacement linear slope with temperature change

The equation of the relation between the line slope and the temperature is

K=9.3953*10^(-10)*T^4 -1.7075*10^(-7)*T^3-2.1094*10^(-5)*T^2+0.0040235*T+0.92766 (3)

In the formula(3), K - line slope;
T - temperature (℃);

It can be concluded that the influence of temperature on the slope is nonlinear, and gradually increases with the increase of temperature, and then gradually decreases with the increase of temperature.

Then fitting the relationship between the linear intercept at different temperatures and fitting results as shown in figure 5 below

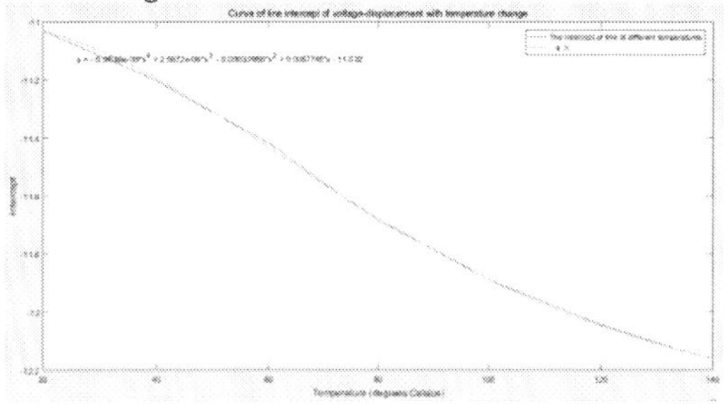

Fig. 5. Curve of line intercept of voltage-displacement with temperature change

The relation equation between the linear intercept and the temperature is

B= -1.5984*10^ (-10)*T^4 + 4.7603*10^(-10)*T^3 -0.00044815*T^2 + 0.0018916*T-10.942 (4)

In the formula (4), B - straight line intercept;
T - temperature (℃);

It can be seen that the influence of temperature on intercept is almost linear, and the intercept gradually decreases with the increase of temperature. This is because the resistance increases with the increase of temperature, and the output of sensor decreases with the increase of resistance. By fitting the relationship between the function coefficient of the output voltage of the sensor and the temperature, the influence of temperature on the temperature drift of the sensor can be compensated.

 https://doi.org/10.1051/e3sconf/20183804005

Finally, get the relationship between the output voltage and the displacement of the sensor

$$U = K*X+B \quad (5)$$

In the formula (5),

K - line slope; B - linear intercept;

U - sensor output voltage (V);

X - measurement displacement *10 (mm).

4 Conclusion

First, the error analysis of the external compensation of mathematical fitting is carried out in this paper.

The project required precision is 0.25%, and the sensor output voltage range is -10~10V, so the output voltage error is no more than 0.05V, after compensation. When the measurement range is less than 1.2mm, the sensor can meet the precision of 0.25%, reaching the goal of the project. The error under the different temperature of each measurement point within the 1.2mm is shown in figure 6.

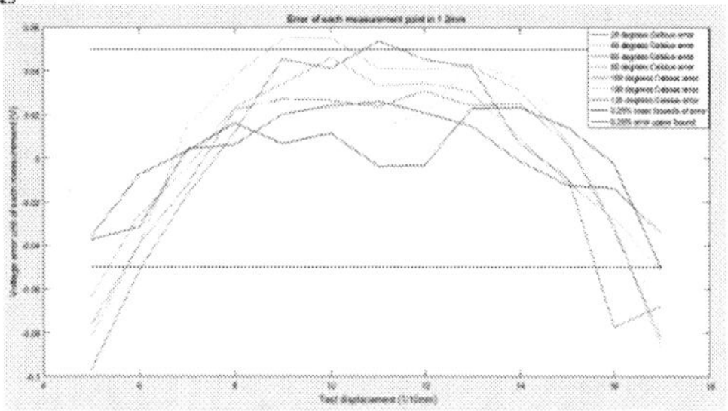

Fig. 5. Error of each measurement point within 1.2mm

According the above analysis, the main factors affecting the output voltage of electric eddy current sensor temperature drift is the resistance of the test coil, and there are many factors affecting the resistance parameters. Due to its dispersion,the use of conventional compensation that only optimizes the parameters as much as possible to reduce temperature drift can not achieve full compensation. In this paper, we propose to compensate the temperature drift of the sensor by fitting the experimental data. Experiments show that the external compensation can achieve up to 0.25% accuracy, which is obviously better than the general compensation method for the similar sensor (2.34%) [6].

References

1. Xianjun Wang, Design of differential temperature drift compensation of eddy current sensor and transducer (Zhengzhou University,2006)
2. Baoqiang Du, Comprehensive compensation for temperature drift of eddy current sensor, Semiconductor Technology,50(2004)
3. Qianqian Si,Stability evaluation and experimental study of eddy current sensor,,Beijing University of Chemical Technology,7(2011)
4. Qianqian Si ,Coil material of the eddy current sensor temperature drift , Sensor World,26(2011)
5. Qiuhua Fang.Temperature drift compensation of eddy current sensor. Journal of Southeast University, 5(1995)
6. Baoming Ling, Xiangbin Zhuge, Yun Ling, Study on the temperature stability of the eddy current sensor, Eddy Current Sensor 345(1994)

Structural Statics Analysis and Optimization Design of Regulating Device for Air Conveyer Outlet in Coal Mine

Xiaoyan GONG Ying LI Yongqiang ZHANG

School of Mechanical Engineering, Xi'an University of Science and Technology, Xi'an, Shanxi 710054, China

Abstract: In view of the enlargement of fully mechanized face excavation and long distance driving, gas emission and dust production increase greatly. However, the current ventilation device direction angle, caliber and front-back distance cannot change dynamically at any time, resulting in the serious accumulation in the dead zone. In this paper, a new device were proposed that can solve above problems. Finite element ANSYS software were used to simulate and optimize the structural safety of the control device' key components. The optimization results showed that the equivalent stress decreases by 49%; after the optimization of deformation and mass are 0.829mm and 0.548kg, which were 21% and 10% lower than before.The quality、safety、reliability and cost of the control device reach the expected standards perfectly, which can meet the requirements of safe ventilation and down-dusting of fully mechanized face.

1 Introduction

With the expansion of fully mechanized excavation face and long distance tunneling, the gas emission and dust production increased significantly. At present, the installation of the ventilation device was located on the sidewall of the roadway. The outlet angle, diameter and front-back distance could not be changed at any time. Resulted in the serious accumulation of gas dust in the dead zone, safety and environmental pollution increased risk.[1-3]. For this reason designed digging duct outlet control device, through the scientific adjustment angle, diameter and front and rear distance to change the direction speed of the wind flow, thus diluting the gas and dust concentration. In this paper, ANSYS software were used to analyze the blades and frame of the control device. The simulation was carried out to key parts during the actual application of the force situation. Small deformation of the leaves for optimal analysis. It provides a certain basis for the optimization design of regulating device and has important significance for improving ventilation environment of fully mechanized heading face.

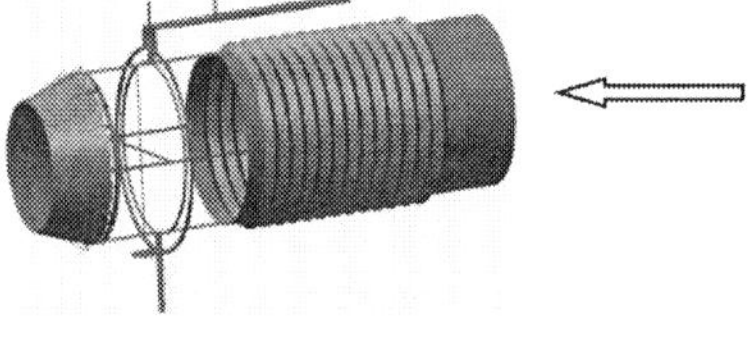

Figure1 outlet control device three-dimensional model

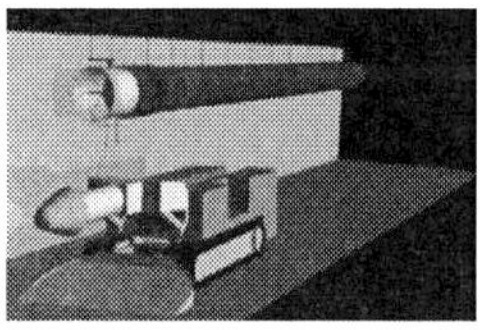

Figure 2 manual duct outlet airway model

2 Device Structure Safety Analysis

2.1 Structural Design

Control device of the three-dimensional model shown in Fig1. The inner diameter adjustment blade has a thickness of 0.5mm, a length of 400mm and an arc length of 400mm; the outer diameter adjustment blade has a thickness of 0.5mm, a length of 400mm and an arc length of 400mm; the main section of the duct frame upper beam thickness 10mm, width 15mm, length 530mm; dial fixed frame width 20mm, inner diameter 1171mm, thickness 40mm;

2.2 Leaf Structure Safety Analysis

(1) Finite Element Model

* Corresponding author: 541657551@qq.com

https://doi.org/10.1051/e3sconf/20183804006

In the Ansys Workbench to establish a contact relationship, in the outer part of the blade and the inner blade contact with the displacement contact, Fig 3shows a Finite element mesh model. According to the structural design size, structural characteristics and force situation. Lastly, the mesh used chosen tetrahedral configuration and fine enough to contour the curved surfaced of the blade model properly. There were 316531 total element in place. The material of the blade is Stainless steel, and density: 7.90g / cm^3 ; elastic Modulus:195 G_{pa} ; poisson's ratio: 0.247; tensile strength: σ_b (MPa) ≥ 520; yield Strength: $\sigma_{0.2}$ (MPa)≥205, linear elastic material model was used in numerical simulation.

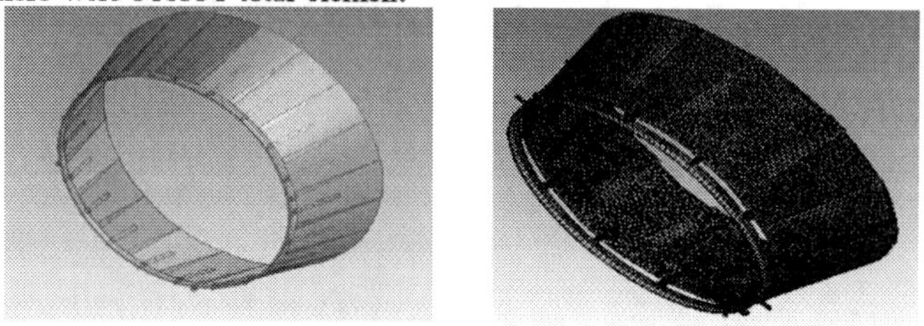

(a) geometry model (b)Finite Element Mesh Generation

Figure 3 Blade geometry and meshing

(2) Constraints and Loads

The total weight of the air duct outlet device blade is 22.7 kg (based on creo simulation software, quality calculation) about 227 N. According to data of Ningtiaota coal mine shows the wind speed of air duct are 10m/s [4-5]. The flaring diameter is 1.2m, and the diameter at the maximum limit is 1m. The cross-sectional area is $F = 0.785m^2$ Air volume $Q = 7.85m^3 / s$. When caliber of the control device is reduced to 0.7m, the cross-sectional area of the duct is $F = 0.38465\ m^2$; the head wind speed $V = Q / F = 20.4\ m / s$. In this limit state, the wind pressure generated by the wind speed is $P = P V^3 / 2 = 269Pa$.The blades are fixed on the frame of the control device connect by pins. The inner and outer blades are in displacement contact, the outer blade and the traction rope are connected at a fixed constraint. The maximum wind pressure value under the limit state of 269Pa.The blade is loaded with a force parallel to the ring with a size of 67.25 Pa according to the actual situation under flaring, as shown in Fig 4.

(a)Blade necking state (b)Flared flaps

Figure 4 blade necking, flaring state loading load

(3) Safety Analysis of Blade 's State Structure

Base on CAE analysis software Ansys Workbench calculate solution, we can obtain the stress and strain diagram of control device, as shown in Fig.5.The results showed that maximum deformation of the leaf strain region occurs at the point where the blade has no fixed restraint. The maximum equivalent strain is 0.1345mm/mm. The maximum stress of the blade is at the fixed constraint of the outer blade link spring, the maximum stress of the blade is 21.09 pa.The stress of fixed restraint at the bottom of the blade is smaller, about 12.33 pa.The stress at the fixed constraint of the blade's bottom link is about 0.76Pa. According to the calculation results of blade stress, with the maximum stress of blade as the actual force check(at the most dangerous locations of structural stress value), the safety factor of blade ultimate stress structure strength reaches 9.7 times. According to the performance index of strength and stiffness of the hair dryer, the existing design size is too large, the stress concentration is obviously existed in the blade, and the structure needs to be further optimized.

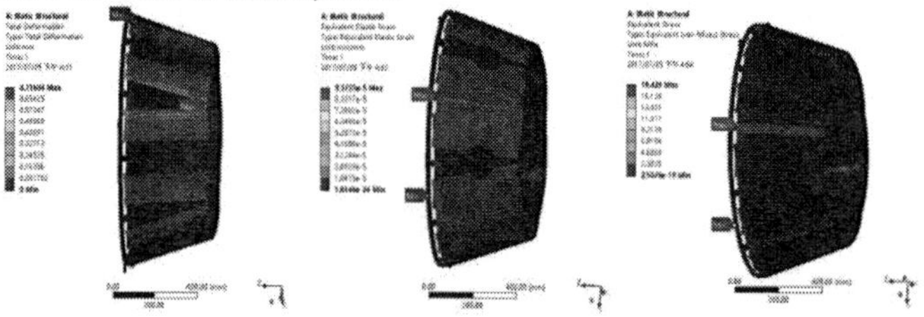

a contraction of the total b blade shrinkage c blade shrinkage

Figure 5 blade necking

2.3 Framework Security Analysis

(1) Finite Element Model

According to the design dimensions, structural characteristics and stress conditions, the outer frame adopts a shell element for the discrete grid, the total number of divided elements is 87719. The material of frame and the control rod is Q235 steel, the density is 7.85g / cm^3, Elastic modulus 200 G_{pa} , Poisson's ratio 0.3, tensile strength σ_b (MPa) $\geq$ 460, yield strength $\sigma_{0.2}$ (MPa) $\geq$ 235, linear elastic material model was used in numerical simulation.

(2) Constraints and Loads

The angle steering mechanism to withstand total weight, load 227N. Frame base to withstand the lifting of the recoil when the air cylinder, when the lifting angle of 6 degrees, the maximum recoil, frame subjected to force F1 = mgsin6 ° = 22.7N. Because of the outer frame horizontal rotating device is fixed on the slide, a fixed constraint is added to the connection of the horizontal frame of the outer frame; According to the actual situation, the weight of the intelligent is loaded, an overall gravitational acceleration is applied to the whole outer frame. With working conditions, the push rod in the frame contact position is to add a horizontal force, the size of 22.7N; the overall acceleration of gravity; as shown in Figure 6;

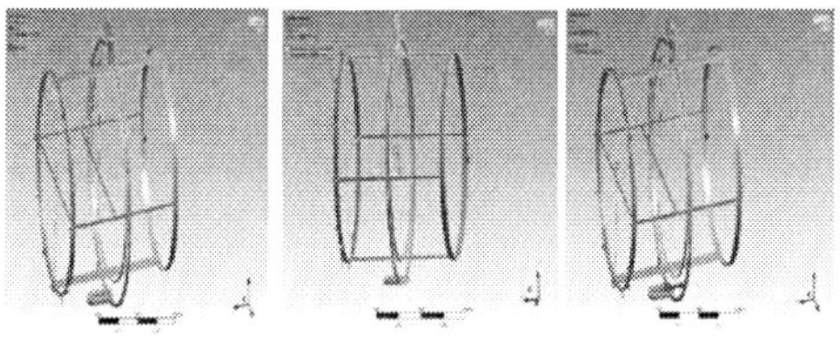

a fixed b gravity c

Figure 6 trame boundary constraints

(3) Frame Structure Safety Analysis

Ansys Workbench calculation solution can get the control device frame stress and strain diagram, shown in Fig 7. The results showed that the main deformation area of the outer frame is at the bottom of the connecting ring between the inner frame and the blade, and the maximum value is about 4.93 mm. The upper middle beam of the outer frame has the largest equivalent strain in the left suspension. The maximum value is about 0.000446mm / mm. According to the calculation results of the outer frame stress, the maximum stress value at the connection between the frame and the angle steering mechanism is 17.3Pa and the maximum stress value at the fixed constraint of the upper beam is 33.8Pa. The maximum stress of the frame (the stress value at the most dangerous position of the structure) for the actual strength of the check, the outer frame of the two locations of the ultimate stress strength of the safety factor of 6.9 times.

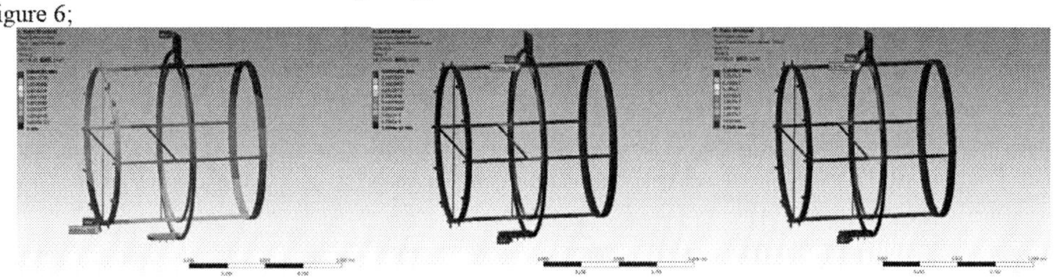

a frame total deformation b frame equivalent strain diagram c frame equivalentt stress

Figure 7 framework cloud

3 structure optimization design

3.1 Optimization Program Analysis

The blade is the key part of the entire device. There is a stress concentration phenomenon in the blade from the finite element analysis. The material cost of the blade structure accounts for about 50% -65% of the whole machine, so as to reduce the weight of the blade as much as possible under the premise of meeting the stiffness and strength. In this paper, the size parameters of the blade, the maximum equivalent stress, the total deformation and the mass of the blade are set as the target of the optimization parameters. When the strength requirement is met and the deformation amount is the minimum, the geometrical dimension is the smallest, that is, the thickness is the thinnest.

3.2 Optimization Analysis

(1) Parameter Sensitivity Analysis

The results of global sensitivity analysis of this model are shown in Fig.9.The results show that thickness of blade has a great influence on the model stress. The deformation of the blade decreases as the thickness increases. The blade quality and deformation are affected by the thickness, with the increase thickness, the blade mass obviously increases and the blade deformation decreases. Sensitivity analysis results show that in the optimization process the input parameters selected have a significant impact on the output parameters of the model, and optimization was feasible.

 https://doi.org/10.1051/e3sconf/20183804006

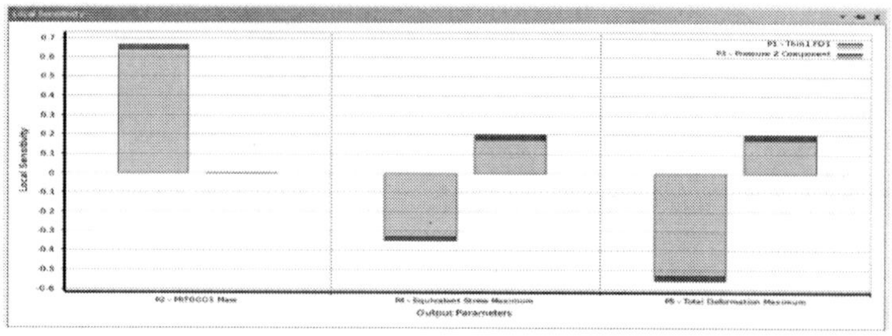

Figure 8 Sensitivity Analysis

(2) Response Surface Analysis

In this paper, used the response surface analysis method, ANSYS Workbench was to find the equilibrium relationship between input and output parameters. The fitting of wired points by adding response points [7-8].For the blade model, the thickness is taken as the parameter, the output parameters of the blade mass, total deformation and equivalent stress corresponding to the response surface is obtained through the response surface analysis, as shown in Fig 9.

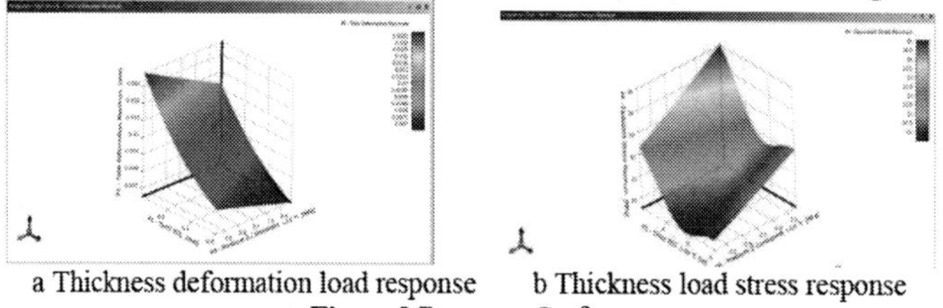

a Thickness deformation load response b Thickness load stress response

Figure 9 Response Surface

The responses of the input parameters are obtained intuitively from the response surface and the blade thickness has a linear relationship with the deformation. As the blade thickness increases, the amount of deformation gradually decreases. The relationship between the blade thickness and the maximum wind pressure is linear. With the increase of the blade thickness, the maximum wind pressure gradually increases.

(3) Leaf Optimization Analysis

Through the sensitivity and response surface analysis, we can get the relationship between input and output parameters. Analysis shows that: in ensuring the blade stiffness and strength of the conditions, there is still room for further optimization of blade thickness. The optimization results are shown in Table 1 below. It can be seen from Table 1 that the optimum design points of A, B and C are obtained through optimization calculation under the conditions of satisfying the maximum wind pressure not exceeding 269 Pa, the minimum total deformation and the lightest mass.

Table 1 Optimization Design Calculation Results

Parameter	A	B	C	original plan
P1-thin1.FD1 /mm	0.5	0.45	0.3	0.6
P3-pressure /MPa	2.421*-4	2.69*-4	2.69*-4	2.959*-4
Max total deformation /mm	7.9421	6.892	13.417	4.075
Max equivalent stress /MPa	6.9	11.104	16.849	52.655
Max equivalent strain /mm/mm	0.352*-5	5.6288*-4	9.4829*-4	0.2898*-4
Mass /kg	0.567856*-4	0.54866*-4	0.36583*-4	0.73144*-4

According to the optimization results in Table 1, the rationality of the scheme is evaluated comprehensively and the best scheme was selected. Above optimization results are in line with the requirements. Compared with the original scheme, the quality of scheme A is reduced by 10%, the overall deformation is reduced by 21%, the stress concentration is reduced by 49%, the intensity is strengthened, and the initial optimization goal is achieved.

4 Conclusion

In this paper, the structural safety numerical simulation and optimization analysis were carried out on the key blade of newly developed control device. The conclusions are as follows:

(1) The outlet vane in the deflated state is 0.1345mm / mm, the maximum stress is 21.09 Pa and the structural strength safety factor is 9.7 times. The maximum equivalent stress in the flared state of the blade is 0.48mm / mm, the maximum stress is 6.60 Pa, and the structural strength safety factor is 31 times;

(2) It can be obtained by sensitivity analysis that the overall deformation of the blade decreases as the thickness of the blade increases. The blade quality and deformation are affected by the thickness of the leaves greatest impact. The response surface analysis results are consistent with the sensitivity analysis. According to the analysis, the A group scheme is the best, the maximum

equivalent stress is reduced by 49%, the total deformation is reduced by 21% and the quality is reduced by 13.2%, to achieve the purpose of optimization.

Acknowledgements

Project Support: The paper was supported by the Key Research and Development Plan of Shaanxi Province (2017GY-170) and the Industrialization Program of Shaanxi Provincial Education Department.

References

1. Jiao Chenlei, Ma Shitai, Yang Xinde, et al.Analysis and optimization of NGW51 planetary gear set based on Ansys and Matlab [J]. Hoisting and transport machinery 2017 (3): 21-23.
2. Gong Xiaoyan, Li Gen, Jiao Wanying et al. Fully mechanized face air duct outlet all-directional control device [P]. Shaanxi: 201610574176.7, 2016-10-13.
3. Zhou L, Pritchard C, Zheng Y. CFD modeling of methane distribution at a continuous miner face with various curtain setback distances [J]. International Journal of Mining Science and Technology, 2015, 25 (4): 635-640.
4. JIA Xiao-quan. Ventilation Distribution and Prevention of Gas Accumulation in Drivage Face [J] .Journal of Shanxi Coal Mining Administrators College, 2012, 25 (2): 38-39.
5. Li Rongrong, Ding Jianwen, Cao Pingxiang.Optimization Design of Large Cross-section Glulam Press Frame with ANSYS Workbench [J]Journal of Northeast Forestry University 2017,14 (11): 62-66.
6. Yao Daozhuang, Yang Jianguo, Lv Zhijun.Optimization design of shelf column section based on ANSYS Workbench [J]Journal of Donghua University (Natural Science Edition), 2011,37 (4): 438-441.
7. Liu Chengjie, Luo Peng, Zhao Lei and other optimization design based on ANSYS Workbench crankpin [J]. Chinese Journal of Applied Mechanics 2017,34 (12): 1141-1144.

https://doi.org/10.1051/e3sconf/20183804007

Complementary Characteristics Analysis on Cumulative Load of Asia - North America Energy Interconnection

Jun-shu Feng[1,*]

[1]State Grid Energy Research Institute, Energy Strategy and Planning Research Department, 102209 Changping Beijing, China

Abstract. The idea of Global Energy Interconnection will facilitate a large-scale development and exploitation as well as a wide-scope coordination and balance of clean energy, and Asia-North America energy interconnection is an important link of it. This paper discusses the Asia-North America energy interconnection conception with power grid as the backbone at first. Then, comprehensive analysis and forecast of power load of both Asia and North America has been given respectively. After the comparison between cumulative load characteristics of Asia and North America and their respective characteristics, benefits from interconnection of Asia and North America can be identified preliminary.

1 Introduction

Global Energy Interconnection (GEI), which proposed in [1], is a globally interconnected strong and smart grid, which will serve as a platform for extensive development, deployment and utilization of clean energy globally. The construction of GEI will facilitate a large-scale development and exploitation as well as a wide-scope coordination and balance of clean energy, promote energy revolution and sustainable development, and thoroughly solve the problems that restrict the development of human society involving energy security, environmental pollution and greenhouse gas emission.

An important link of GEI is Asia and North America energy interconnection, which is focused on in this paper. Firstly, this paper discusses the Asia and North America energy interconnection conception with power grid as the backbone. Then, comprehensive analysis and forecast of power load of both Asia and North America has been given respectively. Lastly, comparison between cumulative load characteristics of Asia and North America and their respective characteristics has been done to identify the benefits of energy interconnection.

2 Conception of Asia and North America Energy Interconnection

Both Asia and North America are the most important load centres in the world, also are abundant with solar, wind and hydropower resources. The idea of Asia and North America energy interconnection, which shown in figure 1, mainly has two influences, including transporting the power from Arctic large-scaled wind power bases to Asian and North American load centres, and smoothing power load curve and renewable power output curve benefitted from its hemisphere complementation characteristics, which can improve ability and efficiency of global energy allocation.

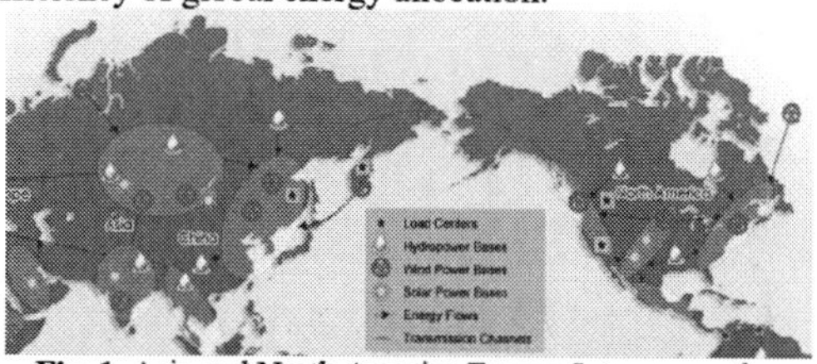

Fig. 1. Asia and North America Energy Interconnection

Asia and North America can both take advantages of time zone differences, the energy interconnection will start from northeast of China and Siberia, across Bering Strait, connecting to Alaska, and then into Canada and the United States, which also called west coast load center. Transmission distances of north-eastern china to Siberia, Bering Strait to Alaska and Alaska to west coast of the United States are less than 3500, 2000 and 4500 km respectively, and underwater electric wire will be only used in Bering Strait, which about 90 km. East Asia and west coast are main load center of Asia and North America separately. Time zone difference between East Asia and west coast is about 9 hours, and about 12 hours between East Asia and eastern part of the United States, which lead to a strong complementary characteristic of their load curves.

In view of power load of Asia is large, considering complementary needs of peak and valley load, overall size of Asia-North America energy interconnection is expected to be over 500 million kilowatts. As the wind farm of Bering Strait is located in the channel between Asia and North America, exploitation and transmission of wind power in Bering Strait can be taken into account and taken good use of. An initial interconnection idea is shown in Figure 1. In consideration of many difficulties,

* Corresponding author: author@e-mail.org

such as long distance, alpine region, undersea wire, interconnection is expected to achieve near 2050.

3 Analysis and Forecast of Power Load of Asia and North America

East Asia and west coast of North America are two main load centres, their power load will be analysed and forecasted below.

3.1 the United States & Canada Load Center

The United States and Canada load center mainly includes Canada and the United States East and West Coast load center, which the United States East Coast Coastal load center refers to the US East Coast New York, Pennsylvania and West Coast California. Due to latitude differences, Canada and the United States there are significant seasonal differences in electricity load, Canada's winter heating load is large, the monthly electricity curve showed significant "winter summer small" features, the US summer air-conditioning load and winter heating load , The monthly electricity consumption curve showed obvious "summer and winter big autumn" features.

Taking into account the annual electricity consumption characteristics of Canada and the United States, get the United States and Canada load center in 2015 the monthly electricity consumption. As the US electricity consumption is about 8 times that of Canada, the annual load characteristics of the United States and Canada load center and the US annual load characteristics are basically the same, summer and winter high electricity consumption, spring and autumn low electricity consumption.

According to the historical data from International Energy Agency (IEA), the polynomial autoregressive model is used to forecast the annual electricity consumption in 2020 and 2050, and the seasonal load of the medium and long-term US and Canada regions is assumed to be the same. The maximum annual maximum load and the minimum monthly maximum load Forecast, the forecast results in figure 2.

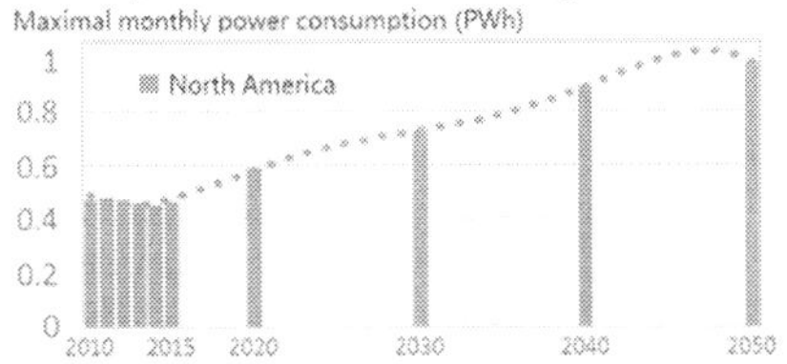

Fig. 2. Forecast of Maximal Monthly Power Consumption of North America

Figure 3 shows typical daily load curve for the summer of 2015 in the United States and Canada. The load of the United States and Canada load center showed obvious "single-peak and single-valley", the daily maximum load appeared around 6 o'clock in the day, daily minimum load appeared at 17:00, peak-valley difference could reach 338.85 GW and peak-valley ratio reach 39.7%.

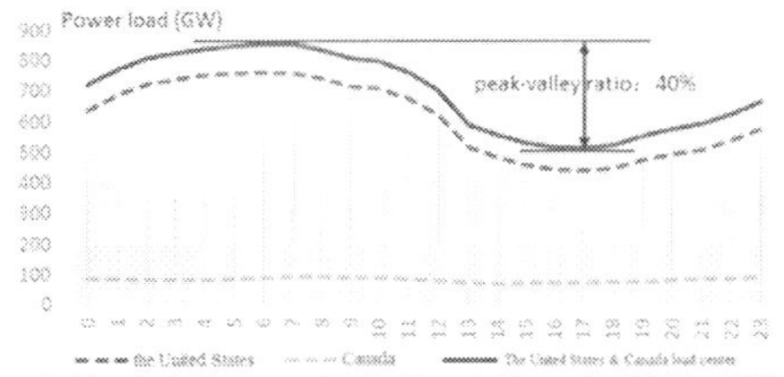

Fig. 3. Typical Summer Daily Load Curve in 2015 of North America Load Center

It is assumed that the medium and long-term the U.S. and Canada load curves keep the characteristics of "single peak single valley", and the peak and valley time is basically the same. According to the multi-factor regression model, consider the future increase with the degree of electricity market and the demand side management level, and consider Energy storage technology in the grid and the user side of the popularity and application of the United States and Canada load center peak valley rate will continue to decline, to 2050 peak rate difference from 40% in 2015 to 30%. Furthermore, the peak and valley differential rates of typical daily load are predicted. The results are shown in table 1.

Table 1. Forecast of Daily Typical Load Characteristics of North America Load Center in 2020-2050

	Peak Load (GW)	Peak valley rate
2015	853	40%
2020	948	39%
2030	1181	37%
2040	1427	34%
2050	1529	30%

3.2 Northeast Asia Load Center

Northeast Asia load center mainly includes China, Japan and South Korea. China's electricity consumption in addition to the Spring Festival in February, the monthly electricity consumption is more stable, slightly higher in summer and winter; Japan and South Korea's annual electricity show high winter, low spring and low characteristics.

Taking into account the annual electricity consumption characteristics of China, Japan and South Korea, the electricity consumption of the Northeast Asia Load Center is formed. As China's electricity consumption accounts for about 80% of the total electricity consumption in Northeast Asia, the annual load characteristics of the Northeast China load center are basically consistent with China's annual load characteristics.

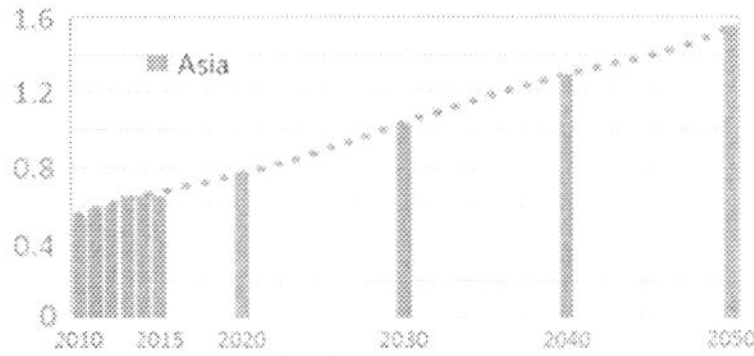

Fig. 4. Forecast of Maximal Monthly Power Consumption of Asia

According to the historical data, the polynomial autoregressive model is used to forecast annual electricity consumption in 2020 and 2050, and seasonal load of the medium and long term Northeast Asia is assumed to be the same. The forecast results are shown in figure 4.

Figure 5 shows typical daily load curve for the summer of 2015 in Northeast Asia. Load curve of load center in Northeast Asia showed characteristics of "multi-peak and single-valley". Maximum daily load was about 11 o'clock in the daytime. Daily minimum load appeared at about 4:00 am, and peak-valley difference could reach 201.6 GW, and its ratio reach 21.6%.

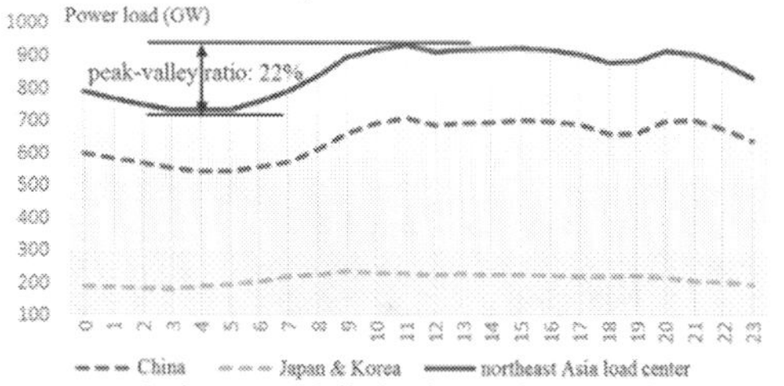

Fig. 5. Typical summer daily load curve in 2015 of northeast Asia load center

It is assumed that the characteristics of the mid-long-term Northeast Asia load curve "multi-peak and single-valley" are unchanged, and the peak-valley time is basically the same. According to the multi-impact factor regression model, considering the future increase with the degree of electricity market and the demand side management level, and consider the energy storage technology in the grid and the user side of the popularity and application of peak and valley rate will continue to decrease; Economic and social development, the proportion of residents in Northeast Asia and the tertiary industry increased, the peak rate will increase. On the whole, it is expected that the peak-valley rate of the load center in Northeast Asia will remain unchanged by 2030. After 2030, the peak-valley rate will decrease slowly, and the peak-valley difference will decrease to 20% by 2050. Furthermore, the peak-valley difference of typical daily load is predicted. The results are shown in table 2.

Table 2. Forecast of Daily Typical Load Characteristics of North America Load Center in 2020-2050

	Peak Load (GW)	Peak valley rate
2015	932	22%
2020	1142	22%
2030	1522	22%
2040	1898	21%
2050	2267	20%

4 Cumulative Load Complementary Characteristics Analysis

Cumulative load is the summing daily load curve of interconnected areas. Figure 6 shows cumulative load curve of typical summer day after Asia-North America energy interconnection in 2050. It can be seen that, because of time zone differences and its complementary characteristics, interconnection of North America and Asia can achieve the goals that reducing both peak load and peak-valley ratio. In 2050, assuming that scale of North America and Asian power grid interconnection is big enough, the peak-valley difference can be reduced from 493.9 GW (North America) and 641.7 GW (Asia) separately, down to 615.6 GW after interconnection, and peak-valley ratio drops to only 16.4%.

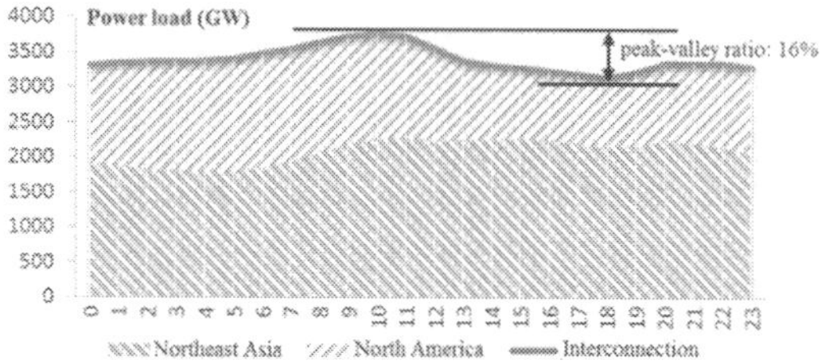

Fig. 6. Typical Summer Daily Cumulative Load in 2050 of Asia-North America Interconnection

The change before and after the interconnection of daily load characteristics in 2050 has been summarized as shown in table 3. It can be seen that after Asia-North America energy interconnection, load characteristics of interconnected power grid have been improved obviously. All of peak daily load, daily peak-valley difference and ratio are significantly reduced.

Table 3. Load Characteristics Comparison of before and after Interconnection in 2050

	Before			After
	Asia	North America	Sum	
Daily peak Load (GW)	1,529	2,267	3,796	↘ 3,699
Daily peak-valley difference (GW)	494	642	1,136	↘ 616
Daily peak-valley ratio	30%	20%	/	↘ 16%

5 Conclusion

As we all know, with the expansion of interconnection, there can be a lot of benefits, including resource optimization configuration, promotion of secure and reliable energy use, and so on. This paper has proved the possibility of these benefits by analysing the decrease of peak cumulative load and its peak-valley ratio due to power load complementary characteristics after Asia-North America energy interconnection. Furthermore, some detailed calculations of the interconnection benefits will be discussed in further study.

References

1. Zhenya Liu 2015 Global Energy Interconnection (New York: Wiley–Interscience).
2. International Energy Agency 2016 WORLD ENERGY STATISTICS (2016 edition) (Paris).

Experiment Comparison between Engineering Acid Dew Point and Thermodynamic Acid Dew Point

Jinghui Song[1], Hui Yuan[2]，jianhua Deng[1]

[1]Guangdong Power Grid Co., Ltd. Electric Power Research Institute, Guangzhou 510000,China
[2]North China Electric Power University (Baoding), Baoding 071000,China；

Abstract： in order to realize the accurate prediction of acid dew point, a set of measurement system of acid dew point for the flue gas flue gas in the tail of the boiler was designed and built, And measured at the outlet of an air preheater of a power plant of 1 000 MW, The results show that: Under the same conditions, with the test temperature decreases, Nu of heat transfer tubes, fouling and corrosion of pipe wall and corrosion pieces gradually deepened. Then, the measured acid dew point is compared with the acid dew point obtained by using the existing empirical formula under the same coal type. The dew point of engineering acid is usually about 40 ℃ lower than the dew point of thermodynamic acid because of the coupling effect of fouling on the acid liquid, which can better reflect the actual operation of flue gas in engineering and has certain theoretical guidance for the design and operation of deep waste heat utilization system significance.

1 Introduction

With the continuous improvement of the national energy-saving and emission reduction targets, it is urgent to reduce the temperature of the boiler exhaust gas and improve the utilization of the tail flue gas waste heat [1]. Excessive flue gas temperature is not only not conducive to the improvement of boiler efficiency, but also increases the risk of boiler failure. Studies have shown that the boiler efficiency can be increased by 0.6% -1.0% for every 10°C reduction in boiler tail flue gas temperature [2] However, the boiler exhaust gas temperature can not be reduced, when the exhaust gas temperature is too low, the heat exchanger is operated under the acid dew point of the flue gas, the SO_3 in the flue gas will form a H_2SO_4 droplet with the condensed water vapor, Low temperature corrosion and ash deposition on the surface will inevitably increase the flow resistance of the flue gas and decrease the heat transfer efficiency [3]. Therefore, the acid dew point is the key index restricting whether coal-fired power plants can utilize the waste heat of the tail flue gas in depth.

At present, domestic and foreign institutions for the study of low temperature corrosion and acid dew point can be divided into two categories: a class of empirical formula focusing on the acid dew point [4], such as China's power industry commonly used in the former Soviet Union Computational Standard Method "empirical estimation formula, which is based on experimental data of All-in-one Institute of Thermal Engineering in the 1950s. It is mainly applied to solid and liquid fuels. In 1959, Muller [5] based on the mixture of SO_3 and H_20 In the gas phase system, the acid dew point curve of flue gas containing low concentration of H_2SO_4 vapor is obtained by using the thermodynamic theory. The other type focuses on the acid dew point corrosion test bench for building simulated flue gas The simulation results show that the law of acid corrosion and the acid dew point of flue gas come into being. For example, two different kinds of flue gas acid dew point measuring devices have been designed and designed by Bai Xiang [6]. The acid dew point under different flue gas conditions were measured, The results show that with the increase of partial pressure of sulfuric acid vapor or steam in the flue gas, the change trend of acid dew point of flue gas gradually slows down. Although the acid dew point obtained by these two methods can prevent boiler tail replacement Surface is subject to the risk of low temperature corrosion, but its design reliability parameters are too high, can not truly reflect the dew point characteristics of coal-fired flue gas, that is, within a certain range of thermodynamic acid dew point containing acid ash particles did not show the film should Sediment characteristics, and therefore has some limitations. Shi Yuetao, Shandong University [7] put forward the concept of engineering acid dew point for the first time. Compared with the traditional thermodynamic acid dew point, express the influence law of acid condensation on the heat exchanger in ash-containing gas flow and enlarge the low temperature waste heat temperature range.

Based on the experimental study of acid dew point, a set of new experimental system of acid dew point measurement for flue gas was established in this paper. The casing with corrosion sheet was used as the experimental section to measure the acid dew point of the air preheater exit of a power plant. In addition, the article also compares the measured acid dew point with the acid dew point calculated from each empirical formula under the same coal conditions.

2 engineering acid dew point test system and calculation method

2.1 Test System

The entire project acid dew point measurement system consists of three parts: water temperature system, test section, measurement and acquisition system. Experimental system diagram shown in Figure 1.

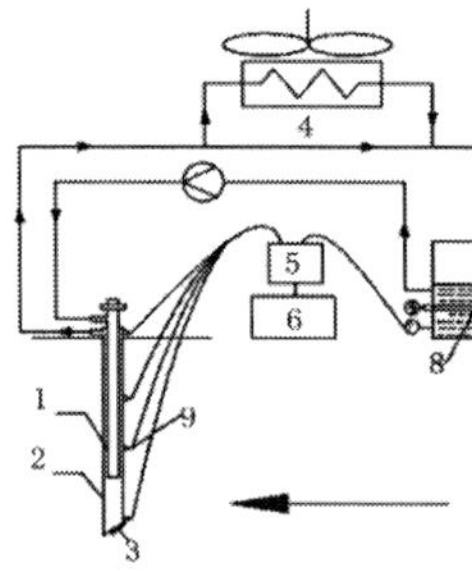

1-inner sleeve; 2-outer sleeve; 3-corrosive sheet; 4- air-cooled coil;
5- temperature recorder; 6- control cabinet; 7- constant temperature water tank; 8- electric heating pipe;
Figure 1 Engineering acid dew point measurement test system diagram

The test section is a double-sleeve structure, as shown in Figure 2, which can be installed in the tail flue measuring hole. The device has a simple structure, including an inner sleeve, an outer sleeve, inlet and outlet water pipes and corrosion patches. The measuring device has three sets of cannulas (length of 69cm, 79cm and 89cm, respectively), which can conduct research on the performance of low-temperature corrosion at different depths of the same position of the flue respectively; the top of each cannula is connected with the ND steel corrosion- The inner casing of the gun casing is connected to the outlet section of the circulating pump through the plastic hose and the radiator casing is connected to the outlet of the casing. The bent pipe is used to cool the working medium in the pipeline with air to ensure the constant inlet water temperature.

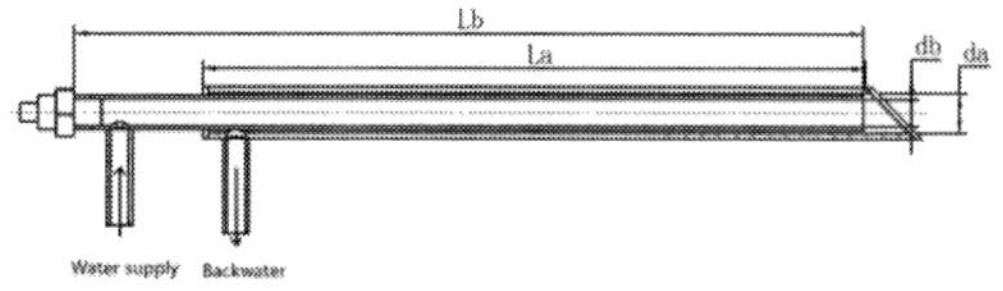

Figure 2 Test section device diagram

The main targets of the data acquisition system are the temperature points, including the temperature measurement points of the corrosion wall on the test section and the temperature measurement points on the water return pipeline and the water tank, the temperature data collected by the temperature recorder is directly reflected on the control cabinet test interface, Through the control system can set the tank temperature, return water temperature and electric heating pipe.

2.2 test process

The test point selection of a power plant 1 000 MW ultra-supercritical coal-fired turbine generator air preheater outlet to the entrance of the low-pressure economizer horizontal flue. During the test, the load of the unit is stable at 1 000 MW. At the beginning of the test, the electric water heater and the fan are started and the circulating pump is turned on to preheat the test piece to 80°C. Then, the test piece heat exchanger tube and the corrosion piece are inserted Test hole and sealed with cotton cloth, to prevent leakage of air.

In the control cabinet were set heat exchanger wall temperature, observe and record the test temperature, when the test section in the flue gas flue gas after 2h, the test section is pulled out, the first observation and recording of corrosion and heat exchangers wall area gray and corrosion The weight of the fly ash deposited on the wall of the heat exchange tube at the temperature was weighed and the heat transfer tube Nu was calculated at the temperature. The corrosive sheet was taken out of the top of the smoking gun and placed on the numbered A4 paper, respectively, Test paper on the surface of the corrosion test piece of acid to observe the discoloration of the pH test paper and compared with the original pH paper, if the pH test paper no discoloration, the temperature point is not in the flue gas acid dew point range, polished corrosion sheet to the clean state, the test Section back into the flue, the other conditions remain unchanged, lower the wall temperature of the heat transfer tube, repeat the above steps until the end of the test, draw the heat transfer wall sediment weight and Nu with the outer wall temperature curve, and finally according to Nu number Interval and pH test paper color change, get the project acid dew point roughly change range.

2.3 Theoretical Calculation

Test tube section and high temperature flue gas convection heat transfer, and the heat transfer to the water, test tube enhanced heat transfer characteristics can be Nu number to represent, as follows:

$$Nu=\frac{h\times d_a}{\lambda} \quad (1)$$

Where da is the outer diameter of the outer tube of the heat exchange tube, h is the convective heat transfer coefficient of the heat exchange tube, and λ is the thermal conductivity of the flue gas.

Heat exchange tube flue gas side convection heat transfer:

$$Q_y = hA\Delta t = hA(t-t_a) \quad (2)$$

Where Qy is the heat absorption on the flue gas side, A is the wall area of the heat transfer outer pipe, t is the flue gas temperature, and ta is the temperature of the wall of the heat transfer outer pipe.

 https://doi.org/10.1051/e3sconf/20183804008

Exchange heat pipe water side convection heat transfer:

$$Q_S = C_p m\Delta t = C_p m(t_{out} - t_{in}) \quad (3)$$

Cp is the specific heat of circulating water, m is the circulating water quality flowing through the heat exchange tube in unit time, tout is the backwater temperature, and tin is the feed water temperature.

Heat transfer tube wall area:

$$A = \pi \times d_a \times l_0 \quad (4)$$

Since the entire heat exchange process is in steady state, the heat exchange on the flue gas side is equal to the heat exchange on the water side. which is:

$$Q_y = Q_s \quad (5)$$

By the formula (1) ~ (5), finishing can be obtained Nu number:

$$Nu = \frac{C_p m d_a (t_{out} - t_{in})}{A\lambda(\text{t-}t_a)} \quad (6)$$

3 test results and discussion

3.1 wall sediment weight and Nu number changes with the outer wall temperature

After each experiment, different wall temperature heat pipe fouling was weighed, sediment weight and wall temperature under different theoretical calculations obtained Nu tube circulating water number changes as shown in Figure 4. As can be seen from the figure, as the outer wall temperature decreases, the Nu number of the heat transfer tube decreases, the weight of the outer wall sediment increases, and the Nu number falls sharply with the outer wall temperature in the 67~72°C and 55~60°C, The temperature of the sediment rising steeply with the outer wall temperature is in the range of 67-72°C, indicating that a large amount of fly ash is deposited on the wall of the heat exchange tube during the temperature interval, and the thickness of the fly ash layer increases sharply, resulting in increased heat transfer resistance of the heat exchange tube, Greatly affecting the heat transfer performance, resulting in deterioration of the heat transfer tube heat transfer characteristics.

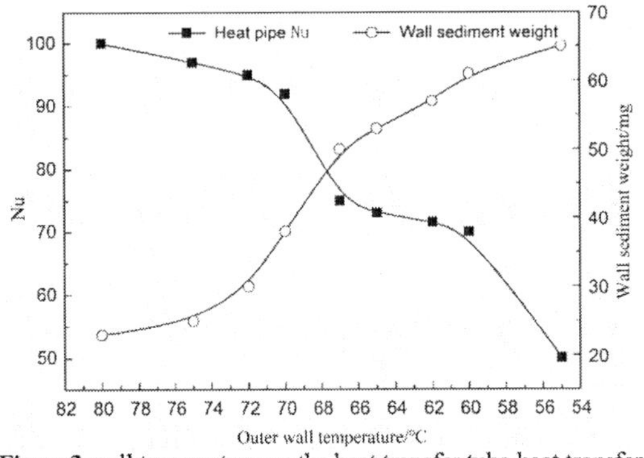

Figure3 wall temperature on the heat transfer tube heat transfer characteristics

In general, when the coal species and combustion conditions remain unchanged, the low-temperature economizer inlet flue ash concentration does not change much, when the wind speed is constant, the ash to achieve a dynamic balance, does not Infinite increase, that is, 80°C~72°C temperature range, the thickness of fouling stabilized at a certain value. When the outer wall temperature dropped to 72~67°C, the number of Nu sharp decline, and the fly ash contains a lot of water vapor and acidic substances, which can be judged in the temperature range, due to lower smoke temperature, fly ash and flue gas The acid-ash coupling phenomenon occurs in the precipitated sulfuric acid vapor, resulting in the further increase of the thickness of the ash layer. As shown in Fig. 4, the ash of the pipe wall area varies at three different temperatures. First, at 80°C, the windward and leeward surfaces of the pipe wall are clean , And good heat transfer capacity; and then reduce the wall temperature to near 70°C and found that the surface of the heat exchange tube surface appeared dry ash, but there is no evidence of corrosion, can be removed by a simple soot blowing, indicating that at this temperature a small amount of sulfuric acid vapor condensation , Wall heat transfer capacity began to decline; continue to reduce the wall temperature to 60°C, can be found in the windward and leeward surface of the pipe are a large number of cohesive fouling, and windward ash more difficult to remove, indicating that the temperature, smoke A large amount of sulfuric acid vapor is precipitated in the gas, and the sour ash is coupled with the ash, which greatly affects the heat transfer capacity of the heat exchange tube, thus indicating that the temperature near the temperature has changed within the range of the acid dew point. , The heat exchange tubes should avoid the wall temperature condition, to prevent acid corrosion.

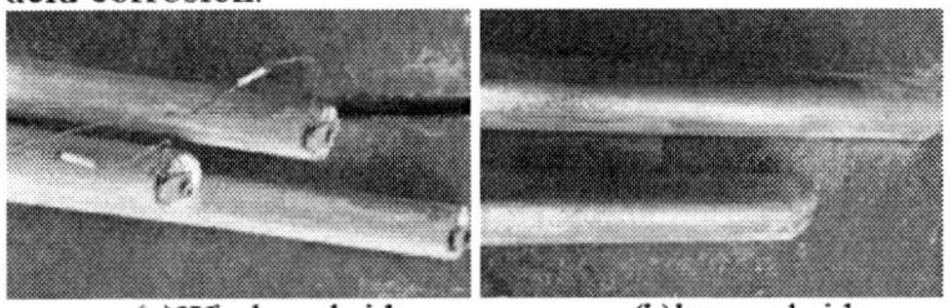

(a)Windward side (b)leeward side

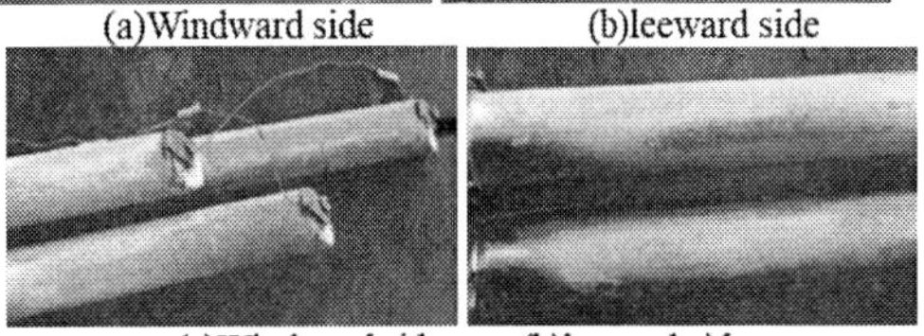

(a)Windward side (b)leeward side

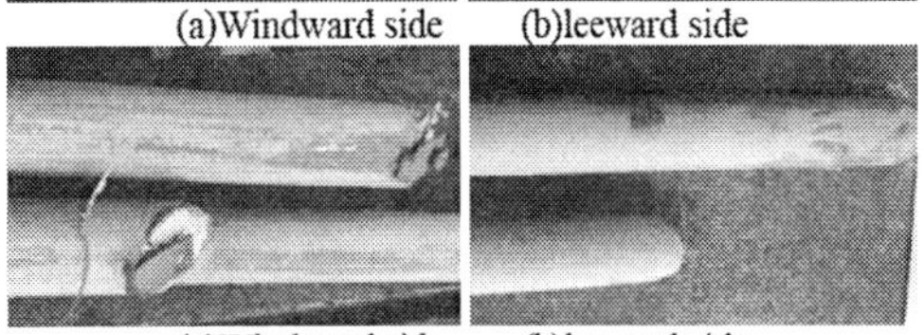

(a)Windward side (b)leeward side

Figure4 Pipe wall gray area conditions in the temperature of 80°C/ 70°C/ 60°C

3.2 Corrosive pieces of fouling and pH test paper test phenomenon

Figure 5 (a) shows the initial corrosion sheet before the experiment, the surface is smooth, and Figure 5 (b) (c) (d) shows the corrosion after two hours of flue gas scrubbing at three sets of temperature The pictures of the pH test strips and the pictures of the pH test strips after the test show that the degree of acid ash adhesion on the surface of the corrosion strips gradually increases and the color of the pH test strips changes from shallow to deep.

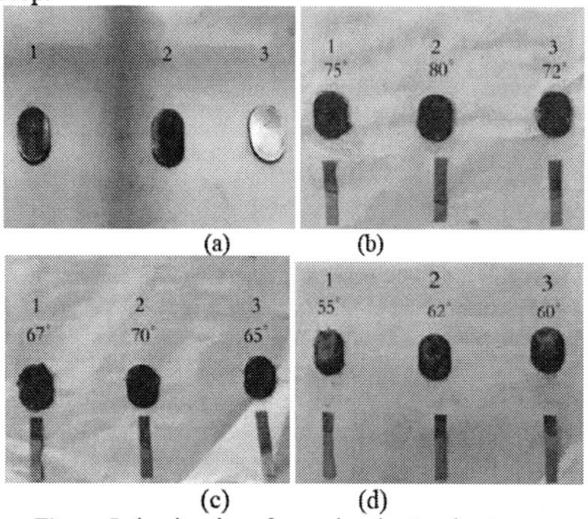

Figure 5 the situation of corrosion sheet and test paper discoloration in different groups of temperature

By observing the corrosion condition of the corrosion sheet at the temperature of 72°C/75°C/80°C in Figure 5(b), it is not difficult to find that the corrosion condition of the corrosion sheet at this temperature point is not much different from that in Figure 5 (a) Gray, scratched ash layer is difficult to observe the corrosion with the naked eye, compared to pH discoloration test conditions, we can find 75°C and 80°C pH test paper almost no discoloration, and 72°C pH test paper began to occur slight discoloration, indicating at 72°C A small amount of sulfuric acid vapor was precipitated in the flue gas, but most of them were absorbed by the ash particles in the flue gas. As a result, the surface of the corrosion sheet did not show any signs of corrosion. However, when the temperature was further lowered to the second set of temperature points (65°C/67°C/70°C) and the third set of temperature (55/60°C /62°C), it can be clearly observed that the corrosion degree of the corrosion sheet surface is gradually aggravated, the thickness of the ash layer is increased, and the color of the pH paper is deepened. Further reducing the amount of precipitated sulfuric acid increased significantly, while the ability of ash particles to absorb acid tends to saturation, acid and ash at this time the acid-ash coupling phenomenon, making the surface ash viscosity changes, the balance of the ash is broken, and thus corrosion Sheet surface Knot of constantly fouling and thickening occurs more severe acid corrosion.

Based on the above analysis of the Nu number of different tube wall temperature, the gray area of the wall of the heat exchange tube, the surface corrosion phenomenon of the corrosion sheet and the acidity of the pH test paper, it shows that when the test tube section is at 72°C. and above, However, the heat transfer characteristics of the heat transfer tube did not deteriorate, and no corrosion occurred on the heat transfer tube and the surface of the corrosion film, indicating that most of the sulfuric acid vapor in the flue gas is absorbed by the fly ash at this time However, when the wall temperature of test pipe section is lower than 72°C, Nu number of heat exchange tube will drop sharply and acid corrosion of pipe wall and corrosion plate will occur. The heat transfer characteristics will be deteriorated, which shows that the power plant Air preheater exit tail flue engineering acidity point in the vicinity of 72°C.

4 different acid dew point comparison results and bias analysis

The power plant boiler coal quality analysis and design conditions tail flue raw data in Table1,Table 2.

Table1 Analysis of coal quality and ash analysis of the power plant

heading	M_t	Mad	Aad	Vad	FCad	St,ad	Qnet,ar
data	20.2%	10.94%	11.08%	29.56%	48.42%	0.58%	20.60 MJ/Kg

Table 2 Original data for the tail flue under the design of the plant

heading	V_{H_2O}	P_{H_2O}	α	V_{SO_2}	P_{SO_2}	P_{SO3}
data	9.91	10042.1	194.00	0.0155	15.76	0.322

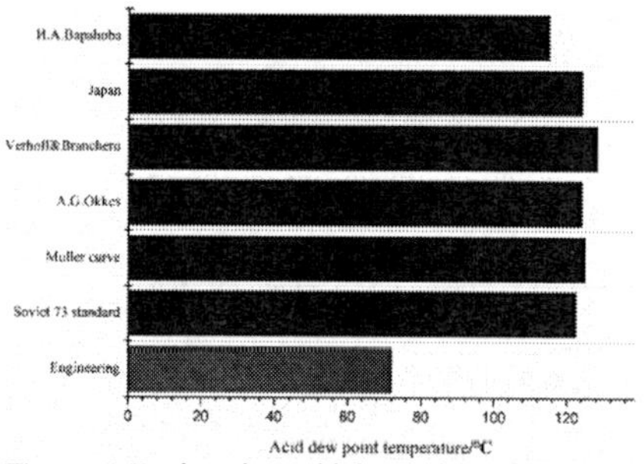

Figure 6 Engineering acid dew point and thermodynamic acid dew point comparison

Engineering acid dew point and thermodynamic acid dew point calculation formula obtained by the comparison of the calculated results shown in Figure 6, can be seen from the figure, the selected thermodynamic acid dew point formula is not much difference, were maintained at 120°C up and down, but are far Engineering acid dew point under the same coal combustion conditions, the formula listed in the calculation of acid dew point and project acid dew point minimum error of 39 °C, the maximum error of 53°C, resulting in engineering acid dew point far lower than the thermodynamic acid dew point is the actual operation The composition of ash in boiler flue gas is complex.

When the flue gas temperature decreases, a large amount of gas phase components such as $SO_3/H_2O/H_2SO_4$ contained in the fly ash are precipitated. These components can easily diffuse to the surface or pore of fly ash particles, Physical adsorption, and SO_3/H_2SO_4 and fly ash Ca and Mg and other alkaline oxides between the chemical adsorption, resulting in acid-ash coupling phenomenon, resulting in the SO_3 vapor content in the flue gas greatly reduced compared to Under the traditional thermodynamic acid dew point, the complicated flue gas is often simplified as the ideal mixture of SO_3 and H_2O, ignoring the effect of ash in the flue gas on acid dew point, thus making the engineering Acid dew point is generally lower than the thermodynamic acid dew point.

5 Conclusion

In this paper, a new set of acid dew point measurement system for flue gas project was established. Field experiments were carried out on the tail flue of a 1 000 MW power plant. Using the thermodynamic method, the variation law of Nu number and wall temperature of circulating water in the pipe was studied. Different pipe wall temperature, analysis of fouling and the condensation of sulfuric acid vapor on the heat transfer performance of the heat exchange tube, the acid dew point of the project fluctuates around 72°C, that is the temperature at which Nu number plunges for the first time, and the resulting project acid dew point The results show that the acid dew point of the project is at least 39°C lower than the thermodynamic acid dew point. Finally, the reasons for the deviations between the measured and calculated results are analyzed. The results show that Engineering acid dew point in practical engineering application value, the design of the power plant low temperature economizer heat exchanger cold water inlet temperature should be controlled as much as possible above 72°C, to guide the boiler tail flue gas deep waste heat utilization system design and operation of a very Significance.

References

1. LI Jun, Yan Weiping, Gao Baotong, *et al.* The Calculation of the Gas Acid dew-point in Utility Boiler. Boiler Technology, 2009; 40 (5): 14-17
2. Xiang B, Zhang M, Wu Y, et al. Experimental and modeling studies on sulfur trioxide of flue gas in a coal-fired boiler. Energy & Fuels, 2017; 31(6): 6284-6297
3. Li Pengfei, Tong Huiling. Comparison and Analysis on the Calculation Methods of Acid Dew Point of Flue Gas. Boiler Technology, 2009; 40 (6): 5-8
4. Tang Zhiyong, Jin Baosheng, Sun Keqin, et al. The contrast and evaluation of the calculation formulas of acid dew point temperature of the flue gas after the WFGD. Journal of Power Engineering,2005: 18-21
5. Muller P. Contribution to the problem of the effect of sulphuric acid upon the dew-point temperature of flue gas. Chem.Eng.Tech., 1959; 31: 345-390
6. Xiang Bai Xiang, Zhao Congzhen, Ding Yanjun, et al. Measurement and prediction model for theacid dew point in flue gases. Journal of Tsinghua University (Science and Technology), 2015; (10):1117-1124
7. Shi Yuetao, Zhang Xiangyu, Li Fei, et al. acid dew pointand other projects-a new criterion for the use of low temperature waste heat boiler limit. Chinese Science Bulletin, 2014; (28): 2810-2816

 https://doi.org/10.1051/e3sconf/20183804009

Numerical Simulations of SCR DeNOx System for a 660MW coal-fired power station

Deng Yongqiang[1], Mei Zhongming[1], Mao Yijun[2], Liu Nianping[2], Yin Guoming[1]

[1] Electric Power Research Institute of State Grid Jiangxi Electric Power Co., Ltd., Nanchang 330096, Jiangxi Province, China
[2] Huaneng anyuan power generation co. LTD. Pingxiang 337200, Jiangxi Province, China

Abstract. Aimed at the selective catalytic reduction (SCR) DeNOx system of a 660 MW coal-fired power station, which is limited by low denitrification efficiency, large ammonia consumption and over-high ammonia escape rate, numerical simulations were conducted by employing STAR-CCM+ (CFD tool). The simulations results revealed the problems existed in the SCR DeNOx system. Aimed at limitations of the target SCR DeNOx system, factors affecting the denitrification performance of SCR, including the structural parameters and ammonia injected by the ammonia nozzles, were optimized. Under the optimized operational conditions, the denitrification efficiency of the SCR system was enhanced, while the ammonia escape rate was reduced below 3ppm. This study serves as references for optimization and modification of SCR systems.

1 Introduction

Owing to the "poor in oil and gas, rich in coal" characteristic of China, the coal-fired power generation dominates the domestic power industry [1]. In 2012, the domestic power generation was 4.97 trillion kW·h and the coal-fired power generation was 3.91 trillion kW·h (78.6%) [2]. The dominant role of coal-fired power generation makes significant reductions of NOx by coal-fired furnaces extremely challenging. Nevertheless, optimization methods of coal-fired furnaces reported previously do not meet requirements by current standards [3]. Among various NOx treatment methods, the SCR denitrification technology is one of the most widely applied and most effective methods [4]. The mechanism of SCR denitrification is that reducers (e.g., NH_3 and urea) selectively react with NOx to generate N_2 and H_2O under prescribed catalysts.

In denitrification processes by the selective catalytic red uction (SCR), the concentration distribution of reducers in S CR reactor and airflow distribution has a direct effect on its denitrification efficiency. Currently, the flow fields of the sy stem were analyzed using computational fluid dynamic (CF D) methods [5-6].

2 Equipment

The denitrification of the target coal-fired station was achieved by SCR with double flue, double reactor, and 2+1 catalyst layers (one spare layer). Liquid ammonia was used as reducer and the denitrification rate was no less than 80% (NOx at standard status, 6% oxygen content, dry basis). Each reactor in the denitrification system has a size of 13950mm (L) ×11200 mm (W) ×19080mm (H) (requirements by flue gas under 50%BMCR working conditions). Table 1. summarizes key design parameters of the proposed SCR denitrification system.

Table 1. Key design parameters of the SCR DeNOx system.

NOx concentration at denitrification system inlet	$350mg/Nm^3$
Denitrification efficiency	80%
Reducer	liquid ammonia
NH_3 escape rate	<3 ppm
SO_2/SO_3 conversion rate	<1%
Systematic resistance of SCR	≤1100 Pa(with spare catalyst)
NH_3/NOx mole ratio	0.81
Flow velocity in reactor	~5m/s
Number of catalyst layers	2+1 (one spare layer)

3 Numerical simulations of SCR DeNOx system

3.1 Establishment of numerical simulations model

A 1:1 model from the economizer outlet to the air preheater inlet was established according to the structure of the SCR reactor. The SCR reactor has a length of 13950mm, width of 11200mm, and height of 19080mm, and it contains two catalyst layers. Deflectors were designed according to the actual arrangement, ammonia injection mixer and AIG were designed at the flue inlet, and rectifier grills were designed on top of the catalyst layer.

Owing to the complicated structure of SCR, which contains various deflectors and rectifier grills, the model was divided into seven areas and the 3D model was divided by polyhedral grid [7]. Meshes at certain locations (e.g., small AIG nozzles) were encrypted to observe flow fields and ammonia injections around nozzles. Boundaries at deflectors and side walls are divided into two layers to enhance the accuracy of numerical simulations.

For numerical simulations of SCR, several assumptions should be proposed. First, the hot flue gas is assumed to be Newtonian fluid and uniformly distributed at the model inlet. Second, the catalyst layer of the reactor is assumed to be porous medium. Third, chemical reactions (ammonia and NOx) are assumed to be negligible as the temperature is below the critical value. The standard k-Epsilon model[8], which has been widely used for calculations of Reynolds number of high turbulence and fluid flows far from side walls[9], was selected. The boundary conditions at the flue gas inlet were aligned with those at the velocity inlet and the compositions of flue gas at inlet were designed accordingly. The boundary conditions at AIG were aligned with those at the velocity inlet and the compositions of flue gas at inlet were designed according to actual ammonia/air ratio (see Table 2.).

Table 2. SCR boundary conditions.

Temperature of flue gas at inlet (°C)		38.4
Flue gas velocity of flue gas at inlet (m/s)		17.3
Compositions of flue gas at inlet	O_2(Vol%)	3.2
	N_2(Vol%)	74.1
	H_2O(Vol%)	8.4
	CO_2(Vol%)	14.2
Temperature at ammonia nozzle (°C)		320
Velocity at ammonia nozzle (m/s)		15.0
Compositions of flue gas at ammonia nozzle	NH_3(Vol%)	5.0
	O_2(Vol%)	74.8
	N_2(Vol%)	20.2

3.2Numerical simulation results and discussion

As the distributions of ammonia concentration and flue gas velocity at the first catalyst layer inlet have significant effec ts on denitrification by SCR, the first catalyst layer is the fo cused area for study of flow field. Moreover, the set-up cont ains two catalyst layers and distributions of flue gas velocity and ammonia concentration at the second catalyst layer outl et should also be paid attention. Herein, the standard deviati on coefficient (C_V) was used to evaluate performance of the SCR system. C_V is defined as the percentage of standard de viation over average for velocity/concentration at different c ross sections in the SCR reactor. It can be calculated by Eq (1) and (2):

错误!未找到引用源。错误!未找到引用源。(1)

where σ_V错误!未找到引用源。 refers to the standard deviation,

错误!未找到引用源。(2)

Fig. 1 shows distributions of flue gas velocity and concentration at the first catalyst layer inlet under BMCR working conditions obtained by numerical simulations. As observed, C_V of flue gas velocity at the first catalyst layer inlet was 8.30%, which is lower than the critical value(15%). As shown in Fig. 1 (b), the ammonia concentration field was high at the region close to X-axis. Indeed, C_V of ammonia concentration at this location was 14.9%, while the critical C_V of ammonia concentration is lower than 10.0%.

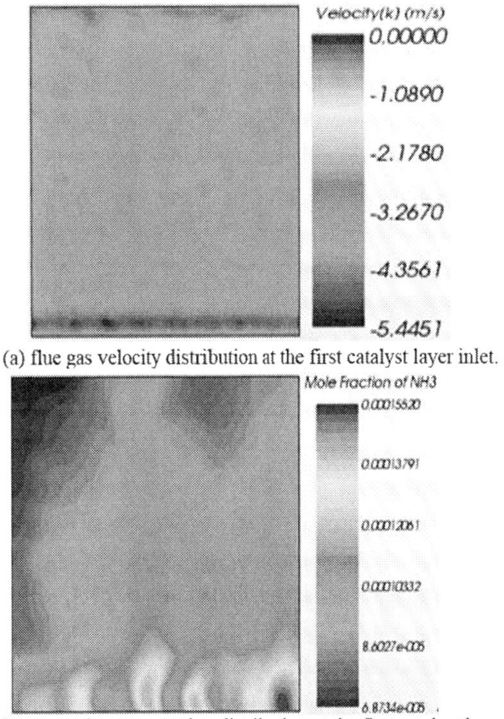

(a) flue gas velocity distribution at the first catalyst layer inlet.

(b) ammonia concentration distribution at the first catalyst layer inlet (gas volumetric ratio).

Fig 1. Flow field at the first catalyst layer inlet in initial design.

Fig 1. (a) shows the distribution of flue gas velocity at the first catalyst layer inlet. As observed, the flue gas velocity at the nozzle was up to 4.5m/s, which can be attributed to the angle of the static mixer.

Fig 1. (b) shows the distribution of ammonia concentration field at the first catalyst layer inlet. As observed, the ammonia concentration was high at bottom

https://doi.org/10.1051/e3sconf/20183804009

nozzle which can be attributed to the angle of the static mixer. More specifically, the flue gas flow in this area was relatively high, resulting in high ammonia injections and ammonia concentrations in this area as the gas involved was a mixture of ammonia and flue gas.

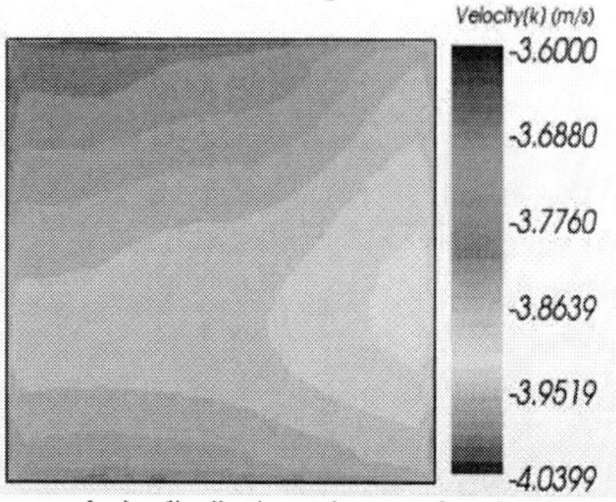

(a)flue gas velocity distribution at the second catalyst layer outlet.

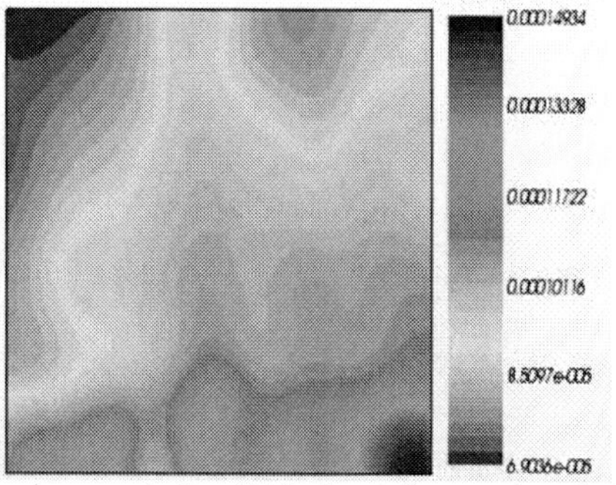

(b)ammonia concentration distribution at the second catalyst layer outlet.

Fig. 2 Flow field at exit of the second catalyst layer in initial design.

Fig 2. shows numerical simulations of flue gas velocity distribution and ammonia concentration distribution at the second catalyst layer outlet. As observed, flue gas velocity was uniformly distributed, while slightly high at the top. Meanwhile, ammonia concentration uniformly distributed, although it was relatively high at top left corner and relatively low at bottom left corner.

The distributions of velocity and concentration obtained by the actual operational data of the SCR DeNOx system were consistent with those obtained by numerical simulations, despite slight errors induced by equipment. Owing to this phenomenon, potential problems in the design of SCR system can be predicted by numerical simulations. In other words, non-uniform distributions of flue gas in the catalyst layer can be relieved in design, thus optimizing the performance of the SCR DeNOx system. Additionally, vector diagrams of numerical simulations velocity illustrate flow field distributions, thus avoiding non-uniform flue gas distribution and unexpected wear of the catalyst layer.

It is, therefore, can be concluded that the distributions of concentration and velocity in SCR are affected by factors such as structure and angle of the mixer, location and angles of deflectors, and ammonia injections by each nozzle.Hence, these factors are optimized by numerical simulations to obtain optimized design parameters. In this study, ammonia injections by each nozzle were optimized with the overall ammonia injection staying constant.

4 Ammonia injection optimization

Before entering AIG, the flue gas distribution was non-unif orm and the flow of ammonia was determined by the flue ga s velocity and the flue gas/ammonia ratio. Therefore, ammo nia injection by each nozzle in AIG was adjusted to achieve uniform mixing of flue gas and ammonia in AIG. With cons tant overall ammonia injection, define the upper and lower n ozzles before AIG as nozzles in Row A and B, respectively; adjust ammonia injections by nozzles in Row A and B succ essively. In order to enhance denitrification efficiency and r educe ammonia consumption and ammonia escape rate, C_V of ammonia concentration field at the first catalyst layer inle t of SCR should be minimized and C_V of velocity field shou ld be no larger than15%.

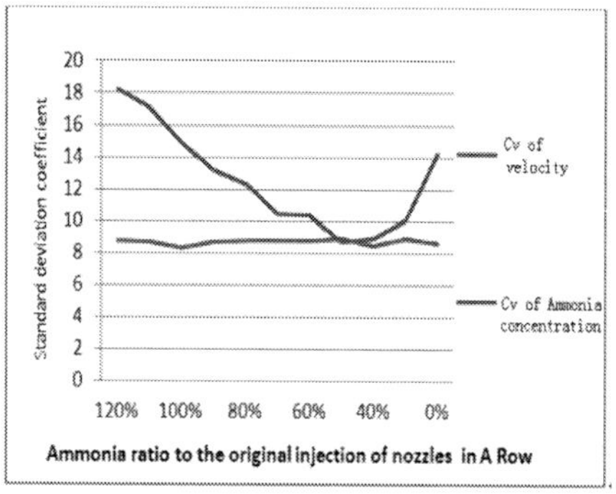

Fig 3. Cv of ammonia injections by nozzles in Row A and B.

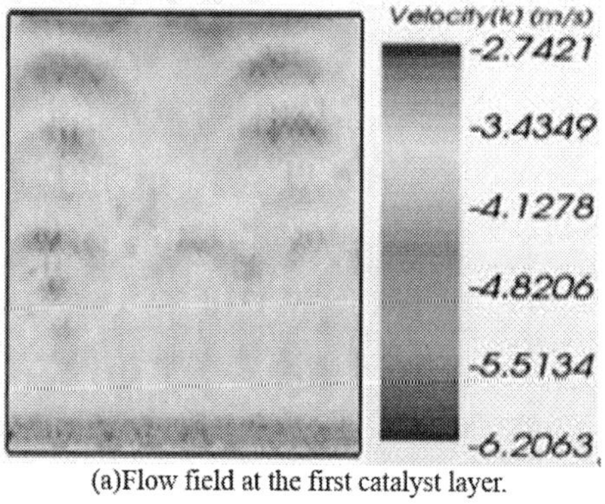

(a)Flow field at the first catalyst layer.

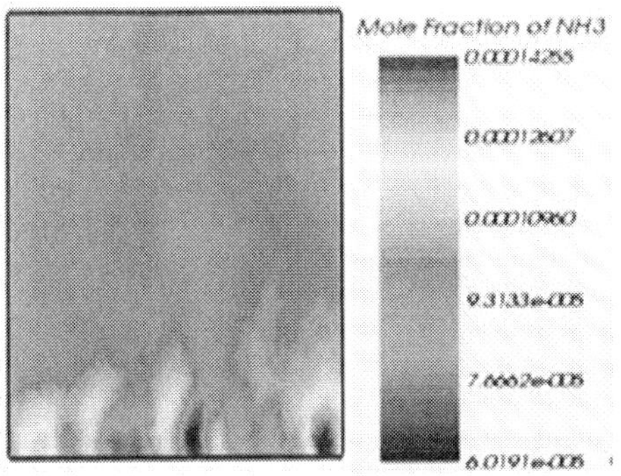

(b)concentration field at the first catalyst layer.

Fig 4. Flow field and concentration field at the first catalyst layer inlet after optimization.

Under BMCR working conditions, the ammonia injections by nozzles in Row A and B were adjusted. Fig 3. summarizes the results of numerical simulations. As observed, denitrification by SCR is optimized by adjusting ammonia injections by nozzles in Row A and Row B to 55% and 145% of the initial values, respectively.

Under BMCR working conditions of ammonia injection optimization, the ammonia concentration distribution at the first catalyst layer inlet was uniform (see Fig 4.). Indeed, CV of ammonia concentration distribution was reduced from 14.9% to 8.7% and CV of flue gas velocity was 8.2%, which was slightly enhanced.

5 Hot tests

With constant overall ammonia injection, optimizations of a mmonia injections by nozzles in Row A and B were adjuste d successively. Table 3. summarizes results of hot tests in th is thermal power set under conventional working conditions.

Table 3. Performances of Furnace 1.

Item	Unit	Working Conditions 1	Working Conditions 2
Electrical load	MW	652	666
NOx A at SCR inlet	mg/m^3	449	456
NOx B at SCR inlet	mg/m^3	445	450
NOx A at SCR outlet	mg/m^3	88	84
NOx B at SCR outlet	mg/m^3	85	82
SCR denitrification efficiency A	%	80.40	81.58
SCR denitrification efficiency B	%	80.90	81.78
NH_3 at SCR outlet A	ppm	3.92	2.93
NH_3 at SCR outlet B	ppm	3.33	2.85
Ammonia consumption A	kg/h	137.5	135.9
Ammonia consumption B	kg/h	136.2	135.2

To verify the ammonia injection optimization, under typical working conditions, NOx concentration distributions at Outlet A before (see Table 4) and after ammonia injection optimization (see Table 5) were obtained. Under working conditions 1 (before ammonia injection optimization), the average NOX at Outlet A of the SCR system was 87.7mg/Nm3 (deviation = 16.86%), resulting in large ammonia consumption, high ammonia escape rate, and low denitrification efficiency. Under working conditions 2 (after ammonia injection optimization), the average NOx at Outlet A of the SCR system was 83.95mg/Nm3 (deviation = 9.89%) and concentration distributions (<10%) meet requirements. Indeed, the ammonia consumption was reduced from 137.5kg/h to 135.9kg/h, the ammonia escape rate was reduced from 3.92ppm to 2.93ppm, and the denitrification efficiency was enhanced from 80.4% to 81.58%.

Table 4. NOx concentration distribution at SCR system Outlet A under working conditions 1.

Location	Hole 1	Hole 2	Hole 3	Hole 4
1	92.4	61.2	87.1	101.7
2	94.2	62.8	85.9	102.1
3	91.4	62.4	86.3	105.2
4	94.9	65.2	88.3	105.4
5	98.2	66.3	89.4	102.1
6	99.5	67.7	75.8	104.3
7	96.6	65.6	70.5	102.1
8	97.7	63.4	88.4	108.6
9	98.4	66.8	89.3	105.3
10	97.5	68.5	91.1	108.2
Average (mg/Nm^3)	87.7	Mean square deviation (%)	16.86	

Table 5. NOx concentration distribution at SCR system Outlet A under working conditions 2.

Location	Hole 1	Hole 2	Hole 3	Hole 4
1	82.5	72.2	85.5	96.5
2	83.6	73	83.2	96.4
3	83.5	74.5	84.7	95.2
4	85.2	75.8	86.9	97.2
5	85	70.2	88.1	95.3
6	87.5	72.9	87.3	96.5
7	86.6	71.6	83.1	93.1
8	82.8	70.9	84.4	97.6
9	81.5	68.9	81.1	92.9
10	83.5	69.5	78.5	93.6
Average (mg/Nm^3)	83.95	Mean square deviation (%)	9.89	

6 Conclusions

https://doi.org/10.1051/e3sconf/20183804009

In this study, flue gas velocity field and concentration field distributions of SCR DeNOx system for a 660MW coal-fire d power station were simulated ,the results of the proposed SCR system before and after optimization were compared w ith each other. The following conclusions can be obtained:

(1) Field distributions obtained by numerical simulations are consistent with the actual operational data, indicating th at numerical simulations can predict distributions of flue gas in the catalyst layer for design of SCR systems, thus facilita ting optimization of denitrification by SCR systems by elim inating non-uniform flue gas distribution and unexpected w ear.

(2) With a constant overall ammonia injection, denitrific ation by SCR is optimized by adjusting ammonia injections by nozzles in Row A and Row B to 55% and 145% of the in itial values, respectively. As a result, Cv of concentration di stribution was reduced from 14.9% to 8.7% and that of flue gas velocity was 8.2%.

(3) Hot tests indicated that the mixing of ammonia and flue gas is affected by the ammonia injection ratio. With a constant overall ammonia injection, optimization can be further enhanced by adjusting ammonia injection by each nozzle and the denitrification efficiency of the SCR system is improved. Indeed, the ammonia escape rate was reduced below 3ppm and the denitrification ammonia consumption was reduced. This study provides references for optimization of SCR systems.

Refences

1. ZHU Jun The development of IGCC power generation technology in China [J] Shaanxi power 2006-8 36 （1）46-49.
2. China electricity council In 2012 the national electric power industry run situation [N] China electric power news
3. Yu Tingfang Zhu Hongzhen, chun-huapeng. The NSGA - II algorithm in the application of multi-objective optimization of boiler combustion. Journal of nanchang university. Engineering [J], 2013, 35 (1): 58-61.
4. WANG Shuxiao YU Chao HAO Jiming Power plant NOx control policy and technology: the experience and enlightenment to our country [N] Journal of environmental engineering, 2011-6 5 (6).
5. YANG Guoshi Thermal power coal-fired boiler NOX formation mechanism and combustion optimization technology research [N] Huainan normal university journal 2011 13(3) 19-21.
6. CENG Kefa YAOQiang Combustion theory and pollution control [M] Mechanical industry press, Beijing (in Chinese) 2004.
7. STAR-CCM+ User's Guide，Chapter 13, Volume5, 2013.
8. LIN Jianzhong RUAN Xiaodong Fluid mechanics [M], Higher education press,2013.7.
9. SCR denitration reaction tower flow field numerical simulation research JIANZhongzhao; Li Bao energy application technology [J], 2014(4).

E3S Web of Conferences **38**, 04010 (2018) https://doi.org/10.1051/e3sconf/20183804010
ICEMEE 2018

Analysis and Assessment of Operation Risk for Hybrid AC/DC Power System based on the Monte Carlo Method

Xiaojing Hu[1,*], *Qiang* Li[1], *Hao* Zhang[2], *Ziming* Guo[2], *Kun* Zhao[1] and *Xinpeng* Li[2]

[1]China Electric Power Research Institute Co., Ltd, Beijing 100192, China
[2]State Grid Jibei Electric Power Co., Ltd, Beijing 100032, China

Abstract. Based on the Monte Carlo method, an improved risk assessment method for hybrid AC/DC power system with VSC station considering the operation status of generators, converter stations, AC lines and DC lines is proposed. According to the sequential AC/DC power flow algorithm, node voltage and line active power are solved, and then the operation risk indices of node voltage over-limit and line active power over-limit are calculated. Finally, an improved two-area IEEE RTS-96 system is taken as a case to analyze and assessment its operation risk. The results show that the proposed model and method can intuitively and directly reflect the weak nodes and weak lines of the system, which can provide some reference for the dispatching department.

1 Introduction

With the increasing flexible DC system connected to AC system, the scale of AC/DC power system with VSC stations (Voltage Source Converter) is constantly expanding, the power grid structure is becoming complex and the power grid is facing more and more uncertainties. The potential operation risk of power grid cannot be underestimated and ignored. Once the fault occurs, it will bring serious economic and social losses to the society [1]. Therefore, it is extremely urgent to carry out the research on the operation risk for hybrid AC/DC power system with VSC stations.

In recent years, the reliability analysis and risk assessment for hybrid AC/DC power system gradually gained the attention of domestic and foreign scholars. However, the research is mostly in the early stage at present, and there is no reliability research on hybrid AC/DC power grid with VSC stations [2][3]. At the same time, the existing research evaluation period is long and mostly based on the planning perspective, which is not suitable for the dispatching operation risk assessment [4][5]. Therefore, based on the Monte Carlo method and the sequential AC/DC power flow method, this paper presents a risk assessment model and risk indices from the perspective of dispatching operation, and then carries out the research on operation risk assessment for the hybrid AC/DC power system with VSC stations.

2 Operation risk assessment method

2.1 Operation risk assessment model

The hybrid AC/DC power grid with VSC stations can divided into generators, AC lines, VSC stations, DC lines, which all adopt the classic model of power system stability analysis [6]. According to the risk model proposed in [7] and [8], a more elaborate operation risk assessment model for hybrid AC/DC power system is proposed, as shown in Eq.1.

$$R_{isk} = \sum_{k=1}^{N}[p_{gen}(E_k)\,p_{bran}(E_k)\,p_{convs}(E_k)\,p_{brandc}(E_k)\times S_{ev}(E_k)] \tag{1}$$

Where: R_{isk} represents a risk value, related to the severity function $S_{ev}(E_k)$; $p_{gen}(E_k)$, $p_{bran}(E_k)$, $p_{convs}(E_k)$ and $p_{brandc}(E_k)$ respectively represents the status probability of generator, AC line, VSC station and DC line under the Monte Carlo sampling time k ; N represents the total sampling time.

Generators, AC lines, VSC stations and DC lines can be modeled as a two-state model, that is, the normal operation status and the outage status [9]. Taking a VSC station as an example, the forced outage rate of a VSC station h is $U_{convs\_h}^{FOR}$, which can be obtained by the state-space diagram method or the frequency-duration method [9], and the random number under the Monto Carlo sampling time k is $R_{convs\_h}^{k}$. Then comparing the relationship between the size of $U_{convs\_h}^{FOR}$ and $R_{convs\_h}^{k}$ (shown as Eq.2) can be obtained the s_h^k , that is the operation status of the VSC station. According to the s_h^k , the operation status of that VSC station can be changed into operation or outage.

* Corresponding author: huxiaojing0394@163.com

$$s_h^k = \begin{cases} 0 & R_{convs\_h}^k > U_{convs\_h}^{FOR} \quad (Normal\ operation\ status) \\ 1 & 0 \le R_{convs\_h}^k \le U_{convs\_h}^{FOR} \quad (Outage\ status) \end{cases} \tag{2}$$

Due to the short operation risk assessment period, the operation status between the various components can be considered as independent [9]. Then the status probability of the VSC station under the sampling time k can be calculated, as is shown in Eq.3.

$$p_{convs}(E_k)=\prod_h U_{convs}^{FOR}(h)\prod_{h'}\left(1-U_{convs}^{FOR}(h')\right) \tag{3}$$

Where: h represents the VSC station in outage; h' represents the VSC station in operation.

Similarly, $p_{gen}(E_k)$, $p_{bran}(E_k)$ and $p_{brandc}(E_k)$ can be obtained in this way.

2.2 Operation risk assessment index

In order to fully evaluate the operation risk for hybrid AC/DC power system with VSC stations, the risk indices of node voltage over-limit and line active power over-limit are proposed in this paper. The smaller the risk value, the more reliable of the power system. The purpose of operation risk assessment is to guide the power system dispatching. If the results after dispatching are used to guide the power system, it is unreasonable. Therefore, the node voltage and line power in this paper are the results of the ground-state power flow without corrective action.

2.2.1 Line active power over-limit risk

The risk index of the node voltage over-limit is to characterize the risk that the power system will exceed the voltage upper limit or voltage lower limit under certain operation condition. The larger the risk value is, the weaker the node is. The formula is as shown in Eq.4.

$$R_{vol} = \sum_{k=1}^{N} p_{gen}(E_k)\, p_{bran}(E_k)\, p_{convs}(E_k)\, p_{brandc}(E_k) \times \left[\sum_{i}^{n} \frac{e^{\max(U_i - U_i^{\max},\, U_i^{\min} - U_i,\, 0)} - 1}{e-1}\right] \tag{4}$$

Where: R_{vol} represents the risk value of the node voltage over-limit; e represents the natural logarithm; U_i represents the node voltage in per unit value of node i ; $U_i^{\max}$, $U_i^{\min}$ represents the maximum permissible voltage and the minimum permissible voltage in the per unit value; n represents the node number.

2.2.1 Node voltage over-limit risk

The risk index of the line active power over-limit is to characterize the risk that the power system will exceed the active power upper limit under certain operation condition. The larger the risk value is, the weaker the line is. The formula is as shown in Eq.5.

$$R_{pow} = \sum_{k=1}^{N} p_{gen}(E_k)\, p_{bran}(E_k)\, p_{conv}(E_k)\, p_{brandc}(E_k) \times \sum_{m}^{line} \frac{e^{\max(P_m - P_m^{\max},\, 0)} - 1}{e-1} \tag{5}$$

Where: R_{pow} represents the risk value of the line active power; P_m represents the line active power in per unit value of line m ; $P_m^{\max}$ represents the maximum permissible active power in the per unit value; *line* represents the line number.

2.3 Operation Risk Assessment Process

The operation risk assessment of power system mainly includes system state selection, system state analysis and system risk value calculation. The assessment process is shown in Figure 1.

As the Figure 1 shows the main step:

1) Input the original data

The original data includes the VSC control mode, network topology, network parameters, component forced outage rate, the maximum number of sampling and so on.

2) System state selection

The operation status of generators, VSC stations, AC lines and DC lines are obtained based on the Monte Carlo method. Then to calculate the status probability of each part.

3) System state analysis

Change the operation status of each part and the network topology of system based on the step (2), then calculate the node voltage and line active power by the sequential AC/DC power flow algorithm.

4) System risk value calculation

When the variance coefficient is less than the given accuracy or the simulation number reaches the maximum allowed, calculate the system risk index according to the Eq.4 and Eq.5.

5) Sort risk value and output

After calculating the operation risk value, sort the risk value and output.

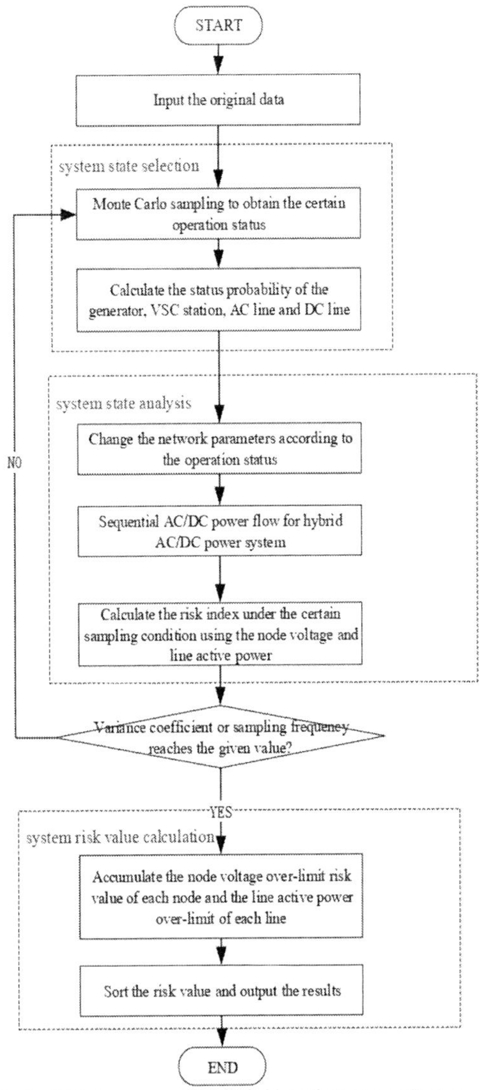

Fig.1. Risk Assessment for Hybrid AC/DC Power System with VSC Station

3 Case study

3.1 Case description

Based on the simulation platform of Matlab, a modified two-area IEEE RTS-96 reliability test system is used to evaluate the hybrid AC/DC power system with VSC station. The two-zone has achieved asynchronous interconnection with three-terminal VSC-MTDC and four-terminal VSC-MTDC. The single line diagram is shown in Figure 2.

In the case, all node voltage and line power flow are used per unit value. Some data is chosen from [10]. And set the sampling time is 5000. The case has a total of 50 AC nodes and 7 DC nodes, 77 AC lines and 7 DC lines.

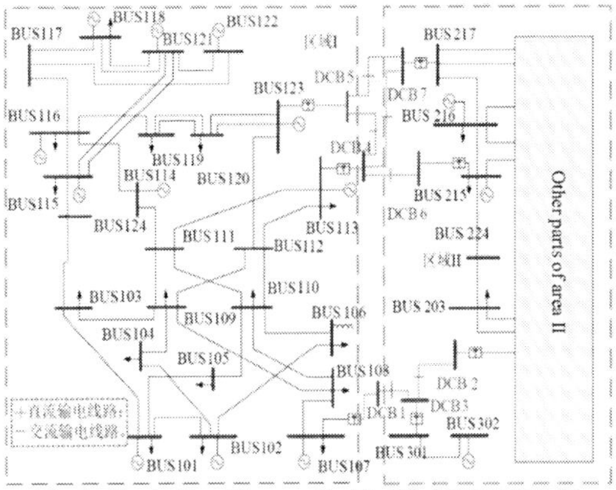

Fig.2. Single Line Diagram for Modified IEEE RTS-96

3.2 Case results

After simulation and sorting the risk value, the results (the top 5) are as shown in Table 1 and Table 2.

Table 1. Risk Value of Node Voltage Over-limit.

Node	Risk Value (10^{-6})
106	34.003
206	8.913
124	6.835
103	4.744
203	4.312

Table 2. Risk Value of Line Active Power Over-limit.

Line	Risk Value (10^{-6})
1-3	54.566
107-108	16.711
211-213	5.552
111-113	5.010
216-217	2.302

Through the risk value sorting, the weak nodes and the weak line of the power system are visually shown, which can provide reference for the dispatching department. In addition, it is not difficult to dine the DC lines1-3, AC lines 107-108, 111-113, 216-217 around the VSC station with higher risk value. The reason is that the access of the VSC station to the AC/DC power system. In order to verify this conjecture, the modified IEEE RTS-96 system is named System 1, and the system without VSC station is named System 2. The comparison results of the above weak nodes and weak lines after simulation are shown in Table 3 and Table 4.

Table 3. Risk Value of Node Voltage Over-limit.

Node	System 1 (10^{-6})	System 2 (10^{-6})
106	34.003	10.969
206	8.913	5.628
124	6.835	4.442
103	4.744	3.321
203	4.312	2.687

Table 4. Risk Value of Line Active Power Over-limit.

Line	System 1 (10^{-6})	System 2 (10^{-6})
1-3	54.566	-
107-108	16.711	≈0
211-213	5.552	0.007
111-113	5.010	4.313
216-217	2.302	0.008

From the above table, the corresponding risk value of System 2 is always lower than that of System 1, which proves that the access of VSC station increases the risk value of system. Therefore, dispatchers should pay special attention to the risk changes of nodes and lines around the VSC station.

4 Conclusion

Based on the Monte Carlo method, this paper establishes an operation risk model for hybrid AC/DC power system with VSC station. Then the risk indices are proposed to characterize the weak nodes and weak lines of the system. The case reflects that the model and risk indices proposed in this paper can clearly show the weak nodes and weak lines, which is achieved the risk assessment for hybrid AC/DC power system with VSC station. The result can provide reference for the dispatching department, and it is also of great significance to ensure the safe and reliable operation of the power grid.

Acknowledgements

This work was financially supported by the Study on the Key Technology for Major Science and Technology Demonstration Project of Zhangjiakou Renewable Energy Demonstration Area in Operation and Maintenance way(52130A17005H).

References

1. Baojie Li, Jinbo Li, Hongjie Li, et al.Analysis of Turkish Blackout on March 31, 2015 and lessons on China Power Grid[J]. Proceedings of the CSEE, **36**(21): 5788-5795(2016).
2. Zhen Ren, Jianjun He, Jun Chen. Monte Carlo-FD comprehensive method for reliability evaluation of AC/DC hybrid network systems[J]. Power System Technology, **24**(5): 13-19, 51(2000).
3. Shaoyu Xie, Xiuli Wang, Xifan Wang. Reliability estimation of AC/DC hybrid power system[J]. Electric Power Automation Equipment, **31**(7): 10-16(2011).
4. Lin Cheng, Ying Wang, Fulong Song, et al. Reliability evaluation of flexible AC/DC hybrid power grid[J]. Electric Power Construction, **38**(3): 55-62(2017).
5. Haipeng Xie, Jian He, Zhaohong Bie, et al. Reliability evaluation of large-scale hybrid AC/DC grids based on SOCP load shedding model[J]. Power System Technology, **40**(12): 3761-3767(2016).
6. Xifan Wang. Modern power system analysis[M]. Beijing: Science Press(2003).
7. J.D.McCalley, V.Vittal, N.Abi-Samra. An Overview of Risk Based Security Assessment[C]. Power Engineering Society Summer Meeting, **1**: 173-178(1999).
8. Jinghua Li, Junjun Zuo, Sai Wang. Analysis and assessment of operation risk for power system with large-scale wind power integration[J]. Power System Technology, **40**(11): 3503-3510(2016).
9. Wenyuan Li. Power system risk assessment model, method and application[M]. Beijing: Science Press(2006).
10. The IEEE Reliability Test System-1996: a report prepared by the reliability test system task for force of the application of probability methods subcommittee[J]. IEEE transactions on Power Systems, **14**(3): 1010-1020(1999).

Connection Analysis of Different Modes in Multimodal Transport

Zhi Zhao[1,*], Ya Ya Lu[1], Xing Hua Liu[1], Ying Jiang[1], and Yan Zhou Zhang[1]

[1]Shandong Jiaotong University, Ji'nan, Shandong 250300, China

Abstract. As the most advanced way of transport organization, container multimodal transport provides high quality and efficient systematic logistics transportation services in a wide range of freight transport activities. So it has been widely promoted worldwide. China is in a period of sustained and rapid economic development, which needs greater support from logistics, while the rationalization of multimodal transport enables the best transport area of each transportation mode to be reflected. This paper makes an analysis of the connection between united transportation of railway and highway, rail and water transport and untied transportation of highway and water.

1 Introduction

Container multimodal transport is a high stage in the development of transportation. It is formed and developed on the basis of the post-industrialization background of the economic development of developed countries, the complete infrastructure and the continuous improvement of the integrated transportation system. Worldwide, as an advanced mode of freight transport, container transportation represents the development of waterway transport, railway transport and road transport. To meet the needs of modern logistics industry and develop container multimodal transport is the inevitable requirement of building a socialist harmonious society and building an economical comprehensive transportation system.

2 Development of Multimodal Transport in China

Multimodal transport refers to the process of systematic transportation, which is carried out by two kinds of means of transportation, which are connected and transported by each other. And each mode of transportation has its own most rationalized economic transport interval, in which the reasonable collocation of various modes of transport will be carried out, that is, the rationalization of multimodal transportation. At present, China is in the period of structural overcapacity, the cooperation between industries is particularly important. January 4, 2017, China's Ministry of Transport, the National Development and Reform Commission, and other 18 departments of the joint study on the development of a package of multimodal transport policy measures on further encouraging the implementation of the work of multimodal Transport (2016) Number No. 232). China has made clear the strategic positioning of multimodal transport from the national level for the first time. Therefore, it is an important task for multimodal transport in China to exert the comparative advantages of various modes of transportation, form a reasonable division of labor and effective cohesion, and allocate resources scientifically according to market rules.[1]

In recent years, with the rapid development of China's economy, the traffic road facilities have also achieved rapid growth, in which the development of highways is very rapid. The main trunk lines of the "five vertical and seven horizontal" national highways have basically been built, and there have also been large-scale construction of railways, waterways, aviation, and port stations. The improvement of transportation infrastructure has greatly promoted the development of container multimodal transport the rapid development of multimodal transport in China is also facing great challenges..

3 The Connection of Several Modes of Transport in Container Multimodal Transport

3.1 Analysis of the Rail-water Combined Transports

The combined transportation of waterway and railway is the main body of container multimodal transport, because the distribution of China's railways and ports has its own characteristics, so that China's waterway and railway joint transport development model is concentrated on "one-port-line multi-city" and "one-port multi-area". "The" Hong Kong line multi city "mode," port "means the port" line "for the connection between the port and inland city railway transport, A development model of combined water-iron transportation by using the extension of the main railway

* Corresponding author: 1808708221@qq.com

https://doi.org/10.1051/e3sconf/20183804011

line to cover many cities. In the "one-port multi-area" model, "Hong Kong" means the main coastal ports, and "area" means the service area of the inland water free port. The development mode of "one port and multi area" is to take the coastal port as the centre, take the inland water port as the dot, and through the connection of the railway Transportation channel, form the Inland collection and transport network system with the port as the node. Among them, "waterless port" refers to a logistics centre with the functions of customs declaration, inspection, bill of lading and other port services established in inland areas, extensible range of port hinterland. Through the continuous exploration of two modes of water-iron transport development, china's water-iron-transport mode has been continuously improved.

First, Insufficient railway capacity restricts the connection between water and iron transport. The proportion of container traffic in port container throughput of Chinese railway is far lower than that of waterway in Europe and America, The lack of the capacity of railway trunk line leads to the lack of connection between port and railway, which becomes the direct reason of restricting the water and rail transport. According to incomplete statistics, the proportion of ports without railway connection in China is higher. Second, The container standardization process is slow, which restricts the connection between waterway and railway joint transportation. China is currently in the domestic container shipping transport ratio is very low, the US standard container size 53 feet, 45 feet in Europe, but China has yet to establish a standard system of inland container. First of all, the Chinese models are extremely complex, resulting in a very low degree of wishful thinking and poor compatibility between railway boxes and water containers, which makes it difficult to achieve large-scale general-purpose, thus creating a series of problems. The resulting cost of water-rail transport is much higher than expected and the figure below will quote the Chinese national standard for container external dimensions and rated quality standards.

Table 1. National Standard Container External Dimensions and Ratings

Container type	Height /mm	Width /mm	Length /mm	Rated quality /Kg
1EEE	2896	2438	13716	30480
1EE	2591	2438	13716	30480
1AAA	2896	2438	12192	30480
1AA	2591	2438	12192	30480
1A	2438	2438	12192	30480
1AX	<2438	2438	12192	30480
1BBB	2896	2438	9125	30480
1BB	2591	2438	9125	30480
1B	2438	2438	9125	30480
1BX	<2438	2438	9125	30480
1CC	2591	2438	6058	30480
1C	2438	2438	6058	30480
1CX	<2438	2438	6058	30480
1D	2438	2438	2991	10160
1DX	<2438	2438	2991	10160

Third, The port infrastructure construction is insufficient, which restricts the connection of water and iron transport. First, China's large part of the port transport equipment, the lack of technical backwardness, resulting in a large number of human consumption. Secondly, the unreasonable construction of the station restricts the connection efficiency between the port and the railway.[2]

Fig.1. Combined Waterway and Rail Transport

3.2 Analysis of Highway and Railway Combined Transport Convergence Problem

As one of the transport modes of multimodal transport, it is an important content of China's development of integrated transportation strategy and modern logistics service model. It plays an import role in the global transportation market competition in the 21st century. It also combines the advantages of two modes of transportation, highway and railway. Though the integration of resources, the traditional and single two modes of transportation are organically linked together to provide economical, reasonable, rapid, safe and convenient transportation services for customers. This mode of transportation service not only reduces the cost of transportation, but also improves the efficiency of transportation. It can also save energy, reduce traffic congestion and environmental pollution. In recent years, it has attracted much shortage and urban traffic congestion. At the same time, along with the rapid development of economy, the improvement of technology, information technology and advanced automation transportation equipment is widely used in the transportation industry, also has laid the foundation for the sustainable development of rail transport.

At present, the two modes transportation of China' s roads and railways seem to be seamlessly connected. but since China's road transport and railway transport are managed by the Ministry of Communications and Transport and the Ministry of Railways. There are differences in the formulation of transport policies and the entry of industry, resulting in poor coordination in planning and management, and the separation of road transport and rail transport. Multitudinous transport has a variety of modes of transportation, complicated transportation routes and the full tracking of containers is also of vital importance. In order to ensure the security of the multimodal transport service, the whole transportation process can be monitored visually. However, the current Chinese road and rail transport does not realize the full tracking of the container.

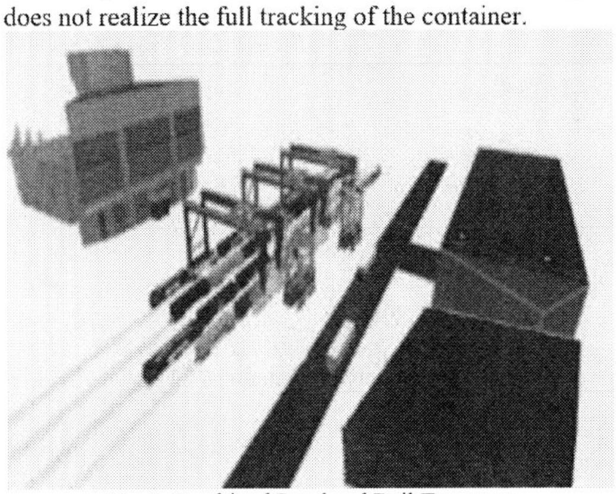

Fig.2. Combined Road and Rail Transport

3.3 Road and Waterway Combined Transport Convergence Problem Analysis

Public water transport is mainly by waterway transport, and road transport is a form of multimodal transport. It has the characteristics of low carbon, environment protection and sustainable development. It is a promising form of transportation organization in modern logistics industry. The adoption of combined mode of transport by road and waterway will make transport a whole, make the highway transportation and water transportation give full play to the advantages, make up the disadvantages can reduce the transportation cost while improve the efficiency of transportation, better meet the needs of customers realize the door to door service to solve the last mile distribution problem. Analysis of the connection between Public and Water Transport

Fig.3. Combined Road and Waterway Transport

As a form of multimodal transport, there are mainly two modes of connection between public and water intermodal transport. One is to use port infrastructure in transit ports to complete transport mode changes. The other is to use the automatic force of the truck to cycle and roll so as to complete the mode of transportation change. As most of the road and waterway transport are operated independently, customs clearance and other inspection procedures are often cumbersome in the bridging process, and the lack of links between the various departments leads to a lack of coordination in planning and operation and a lack of full services. It is easy to cause delay and loss of goods, etc. which is not conducive to the low cost and efficient transportation of goods.

Because the adaptability of road transportation to freight volume, the truck's load tonnage is different, so that the container specifications used in road transportation are different, which make it more difficult to connect with waterway transportation.

4 Recommendations and Measures

4.1 Advancing the Process of Container Standardization

China's container multimodal transport is in the development stage. The unified standard has not yet been formed in the process of road, railway and water transportation. So it is impossible to realize the generalization of transport vehicles, which seriously restricts the depth of the development of multimodal transport. There is a need to further raise awareness of the standardization of container multimodal transport, to coordinate the views of all parties, and to develop and improve as soon as possible the standards for carrying units, including dimensions and tolerances, performance requirements and experimental selection standards, Marking standards and quality acceptance standards etc. That provides support for the improvement of multimodal transport to realize perfect interlinking between the links.

4.2 Promoting Institution-building for Multimodal Transport

Create a consistent policy environment and industry rules, concluding standardize market access, service standards and other management systems. For less open modes of transport, such as railway transport and air transport, the characteristics of public transport should be noted. In order to safeguard fairness and public interest, shippers contain freight forwarders, should be treated equally in terms of shipping qualifications, freight tariff policies and contract terms. Special inspection and demonstration should be conducted. According to the industry characteristics of multimodal transport, the scope of deductible input tax should be set up, and the chain of value-added tax deduction should be perfected. The appropriate tax rate should be set, and the system of value-added tax in transportation industry should be perfected so as to achieve structural reduction of burden. Under the guarantee of the system, the multimodal transport in China can develop towards the goal.[3]

4.3 Urgent Needs for the Construction of Information Platforms

Information system is the blood of the development of multimodal transport. At present, China has a large number of enterprise groups, but most of the enterprise information systems have not been docked. And the enterprise information system and the national information platform have not realized the docking. In order to standardize the information of multimodal transport, we should first establish the enterprise multimodal transport information system, and implement the standards such as data interface, data exchange and so on. It is of great significance to realize the docking of information system between different transportation modes and different transportation enterprises, to promote the docking between enterprise information system and national multimodal transport information platform, and to realize the timely sharing of information.

4.4 Establishment of Policies and Regulations to Facilitate Customs Clearance

The Government should provide legal support for the development of multimodal transport and the convergence of various modes of transport. Aim at shortening the time for customs clearance, improving the efficiency of customs clearance, reducing transaction costs and promoting international cooperation. To promote the "one-stop operation" of international container transport, to carry out a new customs clearance mode such as "joint inspection, one release", and to carry out joint inspection and inspection of goods for the means of transport which are required to be inspected or inspected by the relevant management departments at ports. Internal Coordinate the working relationship between the railway department of the international port station and the department of "one customs; inspection and quarantine" (customs, border inspection, inspection and quarantine) departments, do parallel operations and coordinate their actions; external, coordinate well overseas To facilitate efficient transit, in particular to achieve substantial progress in the harmonization of rail waybills.[4]

5 Conclusion

At present, multimodal transport is more important and urgent than ever for the development of China's logistics industry. From the point of view of China's domestic factors, the transport infrastructure should be increasingly improved, and the comprehensive transportation network should be continuously enriched and perfected. The problem of insufficient capacity in a single mode of transport has ceased to exist. The initial convergence of container transport between roads, waterways and railways has been realized, and the ability to dock trailers between roads and railways has been gradually enhanced. In the administrative system, it is hoped that in the near future, the standardization of container construction will be promoted and the information level will be levelled. The construction of Taiwan, the construction of government laws and regulations, and the perfection of multimodal transport system witnessed the rise of multimodal transport in China.

References

1. Zhao Ying, Zhou Chai Yun. Research on the current situation and Development of Multimodal Transport Standardization in China [J]. Research on quality and Technology Supervision, 2017(5).
2. Yan Xiang Bin. Deng Jun. Study on the future development strategy of water-rail combined transport [J]. Studies on the Theory of Urban Construction (Electronic version), 2011(21).
3. Luoqiao. Study on the development countermeasures of public-water transport logistics [J]. Construction technology and design, 2017(23).
4. Yang Lei. Research on the development of container multimodal transport in China [J]. Railway transportation and Economy, 2016 (7).

The Analysis for Energy Consumption of Marine Air Conditioning System Based on VAV and VWV

Sai Feng XU [1,*], *Xing Lin YANG* [1], and *Zou Ying LE* [1]

[1]School of Energy and Power, Jiangsu University of Science and Technology, Zhenjiang 212003, China

Abstract. For ocean-going vessels sailing in different areas on the sea, the change of external environment factors will cause frequent changes in load, traditional ship air-conditioning system is usually designed with a fixed cooling capacity, this design method causes serious waste of resources. A new type of sea-based air conditioning system is proposed in this paper, which uses the sea-based source heat pump system, combined with variable air volume, variable water technology. The multifunctional cabins' dynamic loads for a ship navigating in a typical Eurasian route were calculated based on Simulink. The model can predict changes in full voyage load. Based on the simulation model, the effects of variable air volume and variable water volume on the energy consumption of the air-conditioning system are analyzed. The results show that: When the VAV is coupled with the VWV, the energy saving rate is 23.2%. Therefore, the application of variable air volume and variable water technology to marine air conditioning systems can achieve economical and energy saving advantages.

1 Introduction

More than 85% of ship air conditioners operate at partial load. The load is usually 60% to 80% of rated load[1]. However, the selection of a ship's equipment is usually based on the maximum load of the equipment. During actual operation, this situation results in waste energy and causes inefficient operation of the equipment, leading to the equipment not achieving its basic requirements for cabin comfort [2,3].

Zhou Hongyu [4] applied variable air volume and variable water volume systems to a large commercial air-conditioning system. The results show that the average energy-saving rate of chilled water pump can reach 47% when the variable air volume and variable water volume are coupled. This shows that VAV, VWV regulation not only allows the vessel to maintain the thermal comfort of the ship's cabin while sailing, but also saves energy [5]. Land and ship air conditioning systems have something similar. However, in the whole sea area, the load is more complicated to Research due to the change of the external environment and seawater temperature, and the research on the energy consumption of air conditioners based on VAV and VWV is lacking. Therefore, this paper conducts in-depth research and analysis on this issue.

2 Design of ship air conditioning and cabin

2.1 Design of the seawater source heat pump air conditioning system based on VAV and VWV

Conventional ship air conditioning system uses air as the cold [6]. The seawater source heat pump air conditioning system uses sea water as a cold. When the ship is in a different area, the temperature of the air varies greatly, but the temperature of the seawater is relatively stable.

Conventional ship air conditioner uses constant air volume (CAV) and constant water volume (CWV) systems. The variable air volume (VAV) and the variable water volume (VWV) air conditioning system can adapt the change of the cabin load by adjusting the volume of air supply and the flow of the chilled water pump. The reduction of air volume means the reduction of energy consumption and the improvement of energy efficiency.

Route from Asia to Europe through the Mediterranean Sea spans large area which can better be used as research path, therefore, this paper chose three typical shipping areas along this route for load calculation, the ship navigation is divided into Malacca Strait of Kuala Lumpur, Jaffna of Indian Ocean and the Gulf of Aden Obok.

The distribution of the cabin and the main parameters of the superstructure of a ship are as shown in Table1.

Table.1 Main parameters of multifunctional cabins

Cabin name	Size (length * width * height) ($m \cdot m \cdot m$)	crew
101Restaurant	6×6×2.1	10
102Kitchen	4.5×6×2.1	1
103Restaurant	6×6×2.1	10
104meeting room	6×6×2.1	10
105Office	4.5×6×2.1	5
106meeting room	6×6×2.1	10

201 skipper	4.5×4×2.1	1
202 crew	4.5×4×2.1	2
203 crew	4.5×4×2.1	2
204 Captain's Office	4.5×4×2.1	4
205 crew	4.5×4×2.1	2
206 crew	4.5×4×2.1	2
301 bridge	12×9×2.1	5

According to the relevant ship design standards, when the air conditioning system is cooling, the designed dry-bulb temperature of indoor air in each cabin is 27℃, and the relative humidity is 50%.

3 Mathematical model

3.1 The mathematical model of cabin load

According to the formula of heat and humidity load in air-conditioning room[7]:

$$\frac{dt_r}{dt}=\frac{Q_m+Q_{ex}}{v_r\rho_a C_p}+\frac{\rho_a V\left(h_o-h_i\right)}{v_r\rho_a C_p} \quad (1)$$

Including: Q_λ is the energy for entering the room per unit time, kJ ; Q_0 is the energy that flows out of the room in unit time, kJ ; Q_m is the cooling load caused by the internal heat source, Q_{ex} is the amount of heat that is brought into the room; the volume of the room; v_r is the amount of fresh outdoor air; m^3 ; V is the amount of fresh outdoor air; m^3/h ; h_o is the value of the outside air, kJ/kg ; h_i the value of the air inside the cabin; kJ/kg .

3.2 Compressor model

Compressor indicated power:

$$P_{th}=P_s\cdot V\cdot\frac{k}{k-1}\cdot\left[\left(P_d/P_s\right)^{\frac{k-1}{k}}-1\right] \quad (2)$$

Including: P_d , P_s represents the compressor exhaust, the suction pressure, Pa ; k is the thermal insulation.

3.3 Fan and pump model

The fan and the water pump are very similar in power and operation characteristics, but used in different places., the model is only used in the case of a fan. This model is also suitable for water pumps [8] According to the law of similarity, there is a relationship between the flow rate, pressure head and speed of a fixed fan.

$$\left(\frac{n_d}{n}\right)^2 SQ^2=f\left(\frac{n_d}{n}Q\right) \quad (3)$$

Fan shaft power:

$$W_m=\frac{P\cdot q_v}{\eta_m\eta_n} \quad (4)$$

Including: W_m is the Shaft power of the fan, kW ; η_m is the total pressure efficiency of the fan, η_n is the fan efficiency, P is the total wind pressure of the fan, q_v is the air volume of the fan, m^3/s .

4 Analysis of Ship Air Conditioning System

4.1 Analysis of the energy consumption of the Marine VAV and VWV Air Conditioning System

Marine air conditioning system mainly includes chiller, water system and wind system. The water system includes chilled water pump and cooling water pump [9,10].Analysis of table 2 shows that the VAV+VWV air conditioning system, the compressors energy consumption accounted for the proportion of total energy consumption of the air conditioning system remaining above 50%, the average proportion was 53.8%, the average proportion of fans accounted for 29.2%, the cooling water pump and chilled water pump accounted for 8.8% ,8.16%, respectively. From the above, energy consumption of compressor is the most important factor leading to the change of total energy consumption of ship cabins, followed by fans, chilled water pumps are also a factor affects the total energy consumption, while the proportion of cooling water pumps is small, which has little effect on total energy consumption.

Table.2 The percentage of energy consumed in total energy consumption under the VAV+VWV

	Kuala Lumpur		
	max	min	average
compressor	54.47	40.00	53.40
Fan	34.72	28.23	29.13
Cooling water pump	11.95	7.85	9.29
Chilled water pumps	13.32	7.87	8.19
	Jaffna		
	max	min	average
compressor	55.13	52.10	53.77
Fan	29.83	28.19	29.09
Cooling water pump	11.85	6.72	9.02
Chilled water pumps	8.31	7.86	8.11
	Obock		
	max	min	average
compressor	55.55	52.19	54.28
Fan	30.06	28.24	29.37
Cooling water pump	11.71	6.02	8.17
Chilled water pumps	8.38	7.87	8.18

4.2 Changes in total Energy Consumption of VAV and VWV Air Conditioning Systems at Different Time of the Whole Day

Figures 1 and 2 show that the changes in the total energy consumption of the VAV system and the VAV+VWV system at different times of the whole day. During the period of 10:00~19:00, the energy consumption of the marine air conditioning system increased first and then decreased, and the total energy consumption reached the peak at 17:00. In Obock, which is typical the area waters, the total energy consumption of the maximum variation, diurnal energy consumption maximum difference was 11.3℃ and the average difference was 8.2℃. As the variation curves of ship air conditioning system show that energy consumption is different with time within the same area, routes within the area also has different energy consumption. The estimation of traditional ship air conditioning system load is obviously larger, so the total energy consumption is also higher than the energy consumption when using VAV+VWV technology. Therefore, by adjusting the air volume and water volume to adapt to the change of cabin load, economic and energy saving can be achieved.

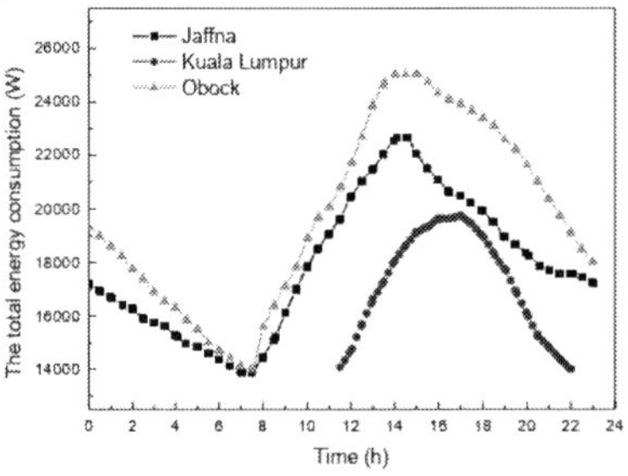

Fig. 1 The energy consumption of each equipment under the VAV at different time of the whole day Lumpur

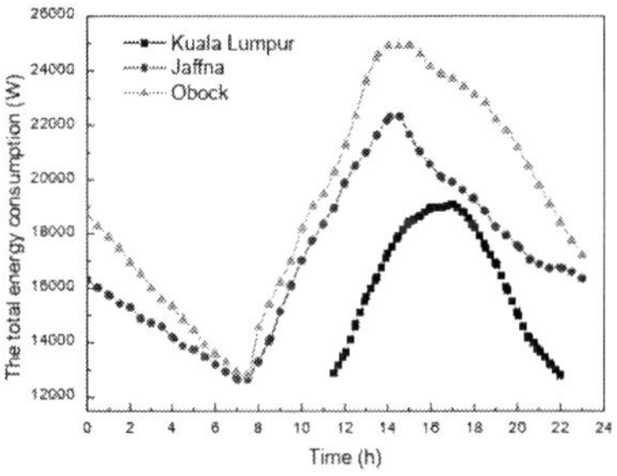

Fig. 2 The energy consumption of each equipment under the VAV+VWV at different time of the whole day

4.3 Calculation of energy consumption of traditional ship air-conditioning system

The heat load Q of a ship's air conditioner is mainly composed of five parts: incoming heat Q_1, human body Q_2 heat, lighting heat Q_3, fresh air heat Q_4, and heat penetration of the air duct. Q_5

$$Q_1 = hA_1(t_o - t_n) + hA_1\Delta t_r + h_g A_g(t_o - t_n) + G_s A_s \quad (5)$$

Including: A_1 is the area after deduction of the glass window, m^2; t_o, t_n is the design temperature outside or inside of the cabin, ℃; Δt_r is the solar radiation temperature rise on the sun surface; h_g is the coefficient of heat transfer on the solar window, $W/(m^2 \cdot K)$.

$$Q_4 = \rho_a V(h_o - h_i) \quad (6)$$

The vessel is closed tightly therefore penetration load is not considered, the heat of the body and the heat dissipation are only rough estimates. The formula above is used to calculate the available items and the total load as shown in Table 3.

Table.3 cabin load steady state calculation results

Q_1/W	Q_2/W	Q_3/W	Q_4/W	Q/W
9542	7767	5781	12712	35802

According to the calculation example of this paper, the total energy consumption of the traditional ship air conditioning system under the refrigeration condition of a certain navigation area is shown in table 4.

Table.4 the total energy consumption of the traditional ship air conditioning system under the cooling condition

$W_t(W)$	$W_m(W)$	$W_s(W)$	$W_q(W)$	$W(W)$	$Q(W)$
12268	6563	2200	1500	22531	32646.3

4.4 Comparison between VAV, VWV ship air conditioning system and traditional ship air conditioning system

The energy saving rate calculation formula is:

$$\varphi = \frac{\Delta E}{E} \quad (7)$$

Including: ΔE is the energy that can be saved, kJ, E is the energy consumed before the conversion of the air-conditioning system, kJ.

Including: In view of energy consumption analysis, the following three comparative analyses are proposed: 1. CAV+CWV; 2. VAV+CWV; 3. VAV+VWV;

From Table 5, it can be seen that the energy saving rates when using VAV technology and VAV + VWV in the Kuala Lumpur are 23.24%, 27.04%, respectively. Therefore, it is better to use conversion technology for ship air conditioning than traditional ship air conditioning system.

Table.5 The economical analyses of ship air conditioning system

	VAV	VAV+VWV
Kuala Lumpur	23.24%	27.04%
Jaffna	20.58%	24.18%
Obock	12.43%	15.24%

5 Conclusion

In this paper, the model of ship air conditioning system was established. The cabin load of certain areas was calculated by cabin model and then the energy consumption of water chiller, fan, chilled water pump and cooling water pump were calculated; by analyzing the relationship between energy consumption of VAV, VWV and CAV, CWV, the following conclusions can be drawn:

1) In the ship air-conditioning system, the largest energy consumption area is the refrigeration unit, which accounts for 52.4% of the total energy consumption, whiles the fan accounts for 28.9%. The energy consumption of the cooling water pump and the frozen water pump account for 7.4 %, 11.3%, respectively.

2) The energy saving rate is about 23.1% when the combination of the air-changing and the water-changing were used. According to the energy saving calculations, the ship air-conditioning system which uses frequency conversion technology is obviously superior to the conventional ship air conditioning system. Therefore, the application of the air-changing and the water-changing technology in ship air conditioning system can achieve economic and energy saving advantages, and this also has important significance in engineering application.

References

1. Cheng Hua, Zhao Zhongchao, Shi Ce. The analysis for Dynamic load of ocean-going ship air conditioning system [J]. Journal of Jiangsu University of Science and Technology(Natural Science), 2015, 29(1):64-69.
2. Li Weiguang, Li Anbang, Xu Xinhua, et al. Discussion on the Value of Heat Transfer Coefficient of Complex Ship Surrounding Wall[J]. Chinese Ship Research, 2014(2):78-83.
3. Mu Qinghao. Research on Energy Saving of Variable Frequency Regulation of Ship Air Conditioning System[D]. Dalian Maritime University, 2016.
4. Zhou Hongyu, Chen Xiaojian, Chen Yuhu. Central air-conditioning energy-saving control strategy for changing water volume and variable air volume [J]. Control Engineering. 2011, 18(3):
5. Zeng Qinyang. Research on Energy Consumption Distribution and Energy Efficiency Evaluation of "Yanyu" Ship[D]. Dalian Maritime University, 2012.
6. Li Jun. A new type of marine air conditioning system and its characteristics analysis [D]. Jiangsu University of Science and Technology, 2016.
7. Han Baoqi. Principles and Applications of Refrigeration and Air Conditioning [M]. Mechanical Industry Press, 2002.
8. Yu Longqing. Variable flow analysis of air conditioning cooling water system[J]. Energy Conservation, 2014(11):53-55.
9. Zhao Zongzhou. Energy consumption modeling and optimization simulation of VAV air conditioning system [D]. Qingdao Technological University, 2016.
10. Wang Shenglong, Li Guoping. Analysis of variable flow energy saving for centrifugal chillers and air conditioning systems [J]. Refrigeration and Air Conditioning, 2016, 16(4): 61-64.

The Fractal Characterization of Mechanical Surface Profile Based on Power Spectral Density and Monte-Carlo Method

Jingfang SHEN[1,2,3], Yuqi GONG[2], Hang MENG[3], Jiajun YANG[1,a]

[1]School of Mechanical Science and Engineering, Huazhong University of Science and Technology, 430074 Wuhan, China
[2]College of Science, Huazhong Agricultural University, 430070 Wuhan, China
[3]Faculty of Engineering, University of Waterloo, N2L 3G1 Waterloo, Canada

Abstract. The analysis of rough surface morphology plays an important role in the functional characteristics of the contact surface of mechanical parts. Fractal geometry method is more accurate and sensitive than classical statistics model. For fractal representation of rough surface, it is necessary to determine the proper fractal dimension calculation method. In this research, the effect of power spectral density method is studied based on Monte-Carlo method. The fractal dimensions are calculated, the theoretical and the calculated values are compared with paired samples. And the results are compared by non-parametric test. The result shows that power spectral density method has good characterization effect on fractal simulation contour curve. In addition, the precision of fractal dimension of power spectral density is related to fractal dimension of contour theory. The estimation methods of classical power spectral density have different application range.

1 Introduction

Contact mechanics of rough surfaces is important in studying and modeling physical phenomena such as thermal and electrical conductivity, friction, adhesion, wear, etc. Obviously, the ability to characterize surface profile by adequate parameters is crucial in these cases. The research of rough surface characteristics is one of the crucial subjects in engineering. The traditional surface morphology has many parameters. It can be divided into three types, geometric parameters, shape parameters and random process parameters [1,2]. The contour arithmetic mean deviation is a geometric parameter. It is the arithmetic mean value of absolute value of contour deviation in the sampling length. The disadvantage is that it can only describe the roughness of the same kind of surface obtained by the same processing method. The skewness and kurtosis coefficient are the shape parameter, which can reflect the shape of the convex peak, as shown in table 1 [3]. The structure function and power spectral density is parameter of random process. However, geometric parameters and shape parameters change as the sample size and measurement scale change. In other words, geometric parameters and shape parameters are parameters of scale. The fractal dimension of surface profile can effectively overcome scale correlation of traditional roughness parameters.

Table 1. Skewness And Kurtosis.

Coefficient	Value	Shape
Skewness	A negative number	The contour convex peak is blunt peak.
	Zero	The convex peak is a composite shape with a peak on the blunt peak.
	A positive number	The profile has a convex peak.
Kurtosis	Less than 3	Temple kurtosis
	Is equal to 3	Normal distribution
	More than 3	Leptokurtic

However, it is found that only fractal dimension cannot determine the unique surface, and the surface of different topography may have similar or even same contour fractal dimension. By measuring rough contour curve of grinding and turning surface, GE Shirong analyses fractal characteristics of rough surface. The results show that rough surface has fractal characteristics and fractal dimension can be calculated. The characteristic roughness is combined with similar

* Corresponding author: [a]j97shen@uwaterloo.ca

measurement and absolute measurement. The numerical value can not only represent the scaling law characteristics of surface multi-scale measurement, but also can hold the sensitivity and accuracy [4]. In the study, ZHU Hua calculates the fractal dimension and scale coefficient of wear surface profile by structural function method. He gives the characteristic roughness calculation expression and does the wear test on the test machine. The results show that fractal dimension of feature roughness is more regular [5].

In order to apply the characteristic roughness parameter to the tribological research, the first step is to choose the appropriate fractal dimension calculation method. There are many different methods for calculating fractal dimension, such as box counting, root-mean-square method, structural function method and power spectrum method. In the book, GE Shirong points out that the method of root-mean-square method and structural function method have the highest calculation accuracy. In the literature, there are many applications of structural function method and fewer researches on power spectrum method relatively. The Monte-Carlo method is also called a stochastic simulation method, which can also be called a random sampling technique or a statistical experimental method. The basic idea of Monte-Carlo method is to establish a probabilistic model or stochastic process, to calculate the statistical characteristics of the desired parameters in the model sampling test, and to give the approximate value of the solution. Monte-Carlo method can solve the problems of physics, mathematics, engineering and so on. The accuracy of the solution can be expressed by the standard error of the estimate. The Monte-Carlo method solves the problem of randomness. The subjects studied were influenced more by randomness. In general, the direct simulation method is used which is a sampling experiment with a digital machine.

In this paper, the simulated surface contour curve of different fractal dimension is based on W-M function, and the Monte-Carlo method. Meanwhile, statistical hypothesis tests are used to study the effect of power spectral density fractal characterization. The results can provide a theoretical reference for the accurate characterization of the characteristic roughness.

2 PSD Fractal Dimension Calculation Method

2.1 Power Spectral Density Method

The W-M function is introduced by Majumdar and Tien in 1989 to describe the height distribution function of surface topography[6,7]. The relational expression is given as

$$z(x)=G^{D-1}\sum_{n=n_1}^{\infty}\frac{\cos\pi\gamma^n x}{\gamma^{(2-D)n}};1<D<2,\gamma>1 \quad (1)$$

Where D denotes the similarity of functions $z(x)$ on different scales. G is the magnitude of the amplitude. V is the spectrum, generally $V=1.5$. Therefore, D is one of the parameters of the decision. The power spectrum density method is derived to solve the fractal dimension of the isotropic rough surface.

The rough surface with fractal characteristics can be characterized by power spectrum density method. Autocorrelation function and power spectral density describes the statistical characteristics of stochastic process from two aspects, time domain and frequency domain. According to Wiener-Khinchin theorem, autocorrelation function and power spectral density function is a pair of Fourier transform. So the power spectral density of stationary random process is the Fourier transform of its autocorrelation function[8]. The relationship of autocorrelation function $R(\tau)$ and power spectral density function $S(\omega)$ is shown as

$$R(\tau)=\int_{-\infty}^{\infty}S(f)e^{j2\pi f\tau}df \quad (2)$$

$$S(f)=\int_{-\infty}^{\infty}R(\tau)e^{-j2\pi f\tau}d\tau \quad (3)$$

The two-dimensional autocorrelation function of isotropic rough surface is given by

$$R(\tau)=\lim_{L\to\infty}\frac{1}{L}\int_0^L z(x)z(x+\tau)dx \quad (4)$$

The Fourier transform of the autocorrelation function $R(\tau)$ is the discrete power spectral density. The approximate continuous power spectral density $\hat{S}(\omega)$ can be written simply as

$$\hat{S}(\omega)\propto\omega^{-(5-2D)} \quad (5)$$

Fitting the line $\lg S(\omega)-\lg\omega$, the slope of the linear region is α, then $D=(5+\alpha)/2$.

2.2 Classical Spectral Density Estimation

There are many techniques to estimate the power spectrum density. In addition to the autocorrelation method used here, there are also classical estimation methods such as periodogram method, Bartlett method, and Welch method.

The simulated rough surface profile height is a finite long random signal sequence. The direct method is to take the sampled data of the simulated rough surface profile to the Fourier transform to solve the power spectral density. The relationship between the Fourier transform and the power spectral density estimation $\hat{S}(\omega)$ is as follows

$$\hat{S}(x)=\frac{1}{N}\left|FFT[z(x)]\right|^2 \quad (6)$$

Where $FFT[z(x)]$ is Fourier transform of the sequence $z(x)$. This method is also known as the periodogram method. Because it is a fast Fourier transform to solve the power spectral density function with a finite long sample sequence, the result obtained is only an estimation way, and there will be a certain amount of error.

The Bartlett method is an improvement to the periodogram method. The sample data $z(x)$ of the rough surface profile height is divided into S segments that do not overlap each other (can be overlapped), and each segment has K sample value and $N = S \times K$. The power spectrum of each segment is calculated by the periodogram method. Then, the average of each segment is calculated. Finally, the average value is estimated as the power spectrum of the entire contour height sampling data $z(x)$. The computed result of the overlapping section is better than that of the non-overlapping section.

The Welch method develops Bartlett method. Firstly, the sample data of rough surface profile height is segmented. Then, the window function is used to preprocess each segment. Finally, the power spectral density function of the sampled data of the whole contour height is estimated by using the Bartlett method. Using the proper window function can improve the spectral resolution and reduce the error.

The periodogram method, also called the direct method. This estimation method is simpler than the calculation of autocorrelation method. Bartlett method and Welch method are the improvement of the autocorrelation method and the periodogram method, and the estimation results are more accurate. Welch method makes two modifications to the Bartlett method. First, select the appropriate window function to make the spectrum estimation non-negative. The second is that there is overlap between the segments, so that the variance decreases. The power spectrum density method is suitable for self - affine fractal curve.

3 Characterization of Power Spectral Density Method

3.1 The Fractal Representation of Simulated Contour Curve

Based on the principle and application characteristics of Monte-Carlo method, the Monte-Carlo method is used to randomly simulate 300 numbers as D [9]. The value of contour height $z(x)$ is calculated with the given parameter, and the W-M function is given to simulate the rough contour curves of multiple groups, as shown in the Fig1-4. Given the fractal roughness $G = 0.01$, $\gamma = 1.5$, $n_1 = 1$, the simulation length is $L = 25mm$, and the rough contour curves are simulated.

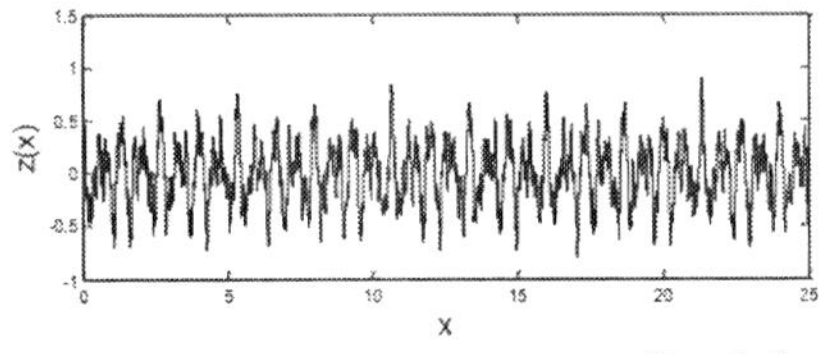

Fig. 1. The simulated contour curve($D = 1.2$).

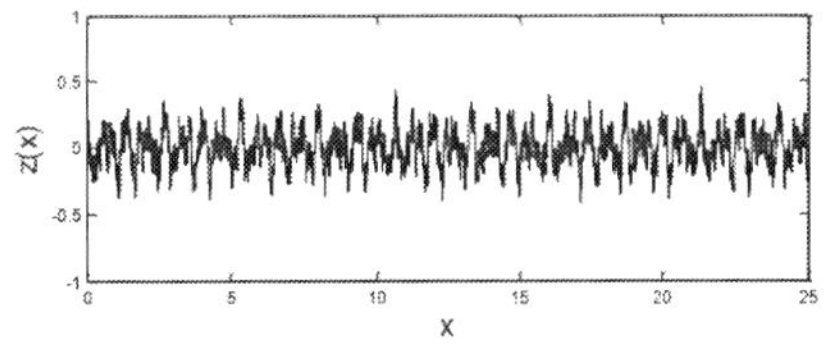

Fig. 2. The simulated contour curve($D = 1.4$).

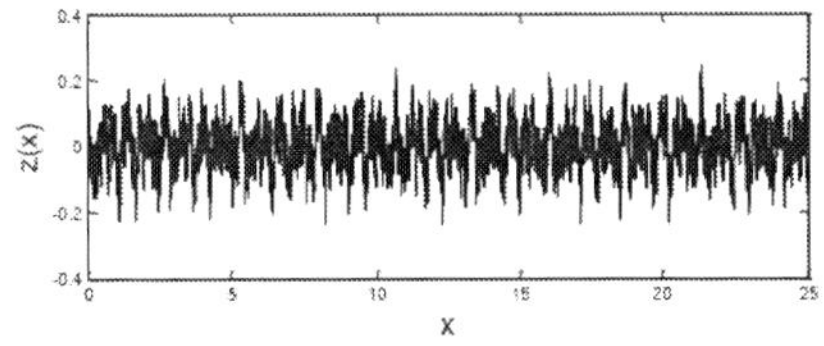

Fig. 3. The simulated contour curve($D = 1.6$).

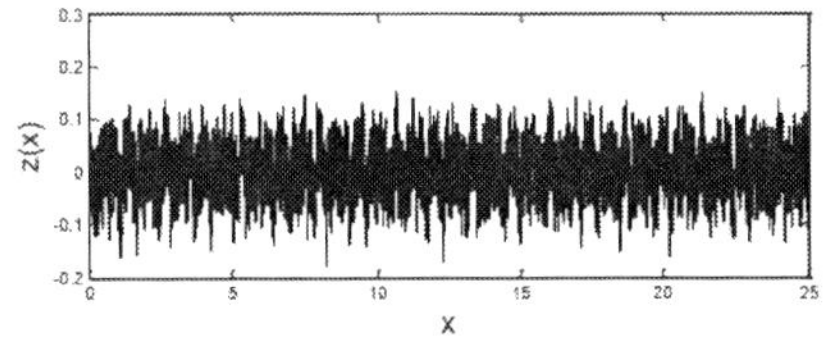

Fig. 4. The simulated contour curve($D = 1.8$).

3.2 Characterization Effect of Power Spectral Density Method

3.2.1 Power Spectral Density Method to Solve Fractal Dimension

The fractal dimension of simulation curves is calculated by power spectral density method. The sampling frequency is $fs = 1/0.00025$, the amounts of sampling points are $N = 2^{13}$. Sampling the above contour curves. The fractal dimension is calculated by using the power spectral density method, and the range of fractal features is determined. The theoretical and calculated values of fractal dimension are calculated and the outliers are filtered. The calculation results are obtained by MATLAB R2014a. We choose the classic power spectrum density estimation method of autocorrelation method, periodogram method, Bartlett method, and Welch method to estimate power spectral density. The estimation methods are shown in Fig. 5.

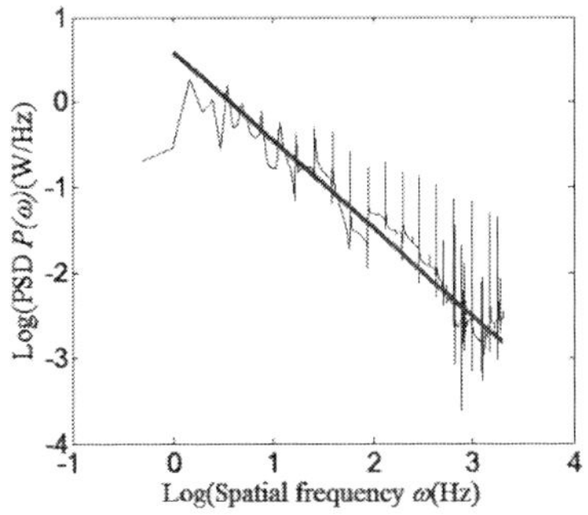

(a) Autocorrelation method

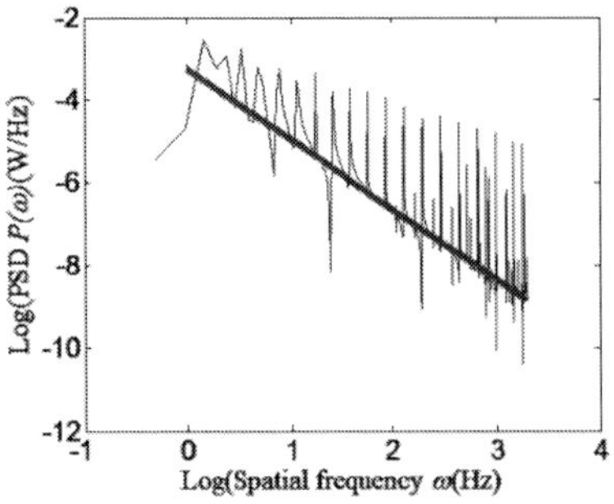

(b) Periodogram method

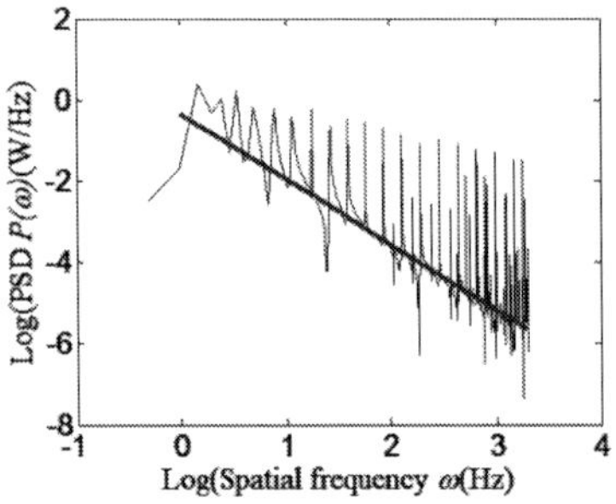

(c) Bartlett method

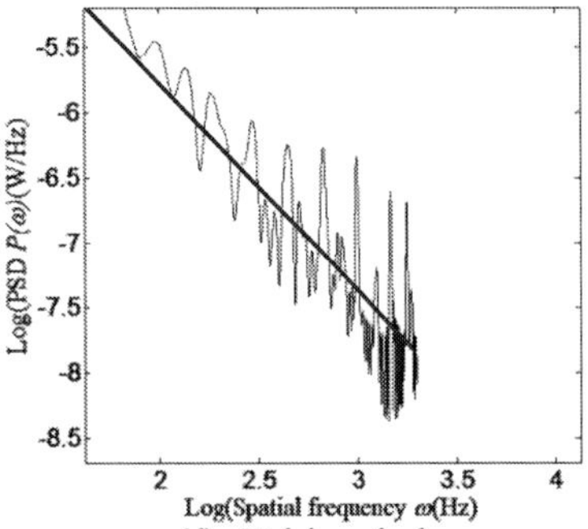

(d) Welch method

Fig. 5. The fractal dimension is calculated by PSD method.

The calculated value of fractal dimension is the slope α of the straight line, $D=(5+\alpha)/2$. The simulation data is shown in fig. 6. The results illustrate that the estimation value of autocorrelation method is meticulous in the range of the theoretical fractal dimension between 1.6 and 1.9, and the calculated value is larger than theoretical value. The periodogram method is relatively accurate in the range of theoretical fractal dimension from 1.7 to 1.85, and the calculated value is smaller than theoretical value, and the calculated value of both ends is larger. The Bartlett method is correct in the theoretical fractal dimension and is estimated to be exact in the range of 1.7 to 1.9, and the calculated value of both ends is large. The Welch method is precise in the theoretical fractal dimension of 1.7 to 1.8, and the calculated value is small, and the calculated value of both ends is large. The estimation error of autocorrelation method is the largest in those methods. Although the calculation precision is high, the periodogram method has great fluctuation. The range that Bartlett method and Welch method can accurately estimate fractal dimension is larger than autocorrelation method and periodogram method.

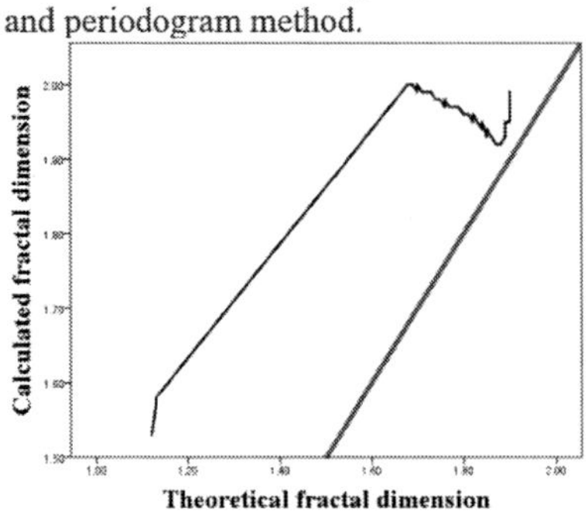

(a) Autocorrelation method

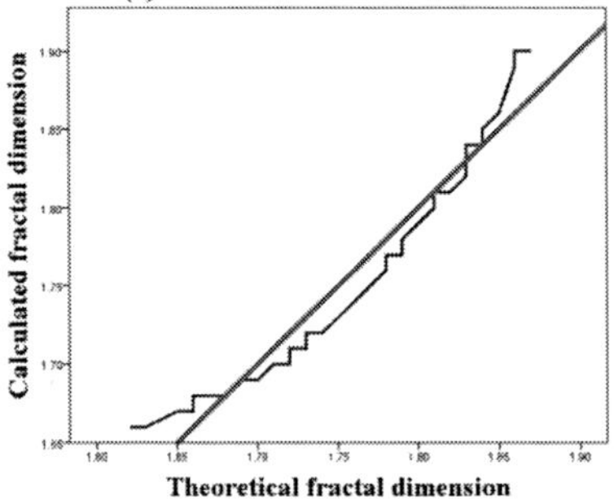

(b) Periodogram method

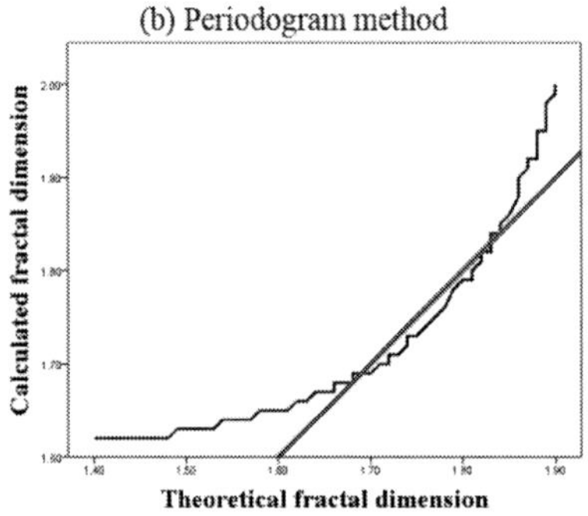

(c) Bartlett method

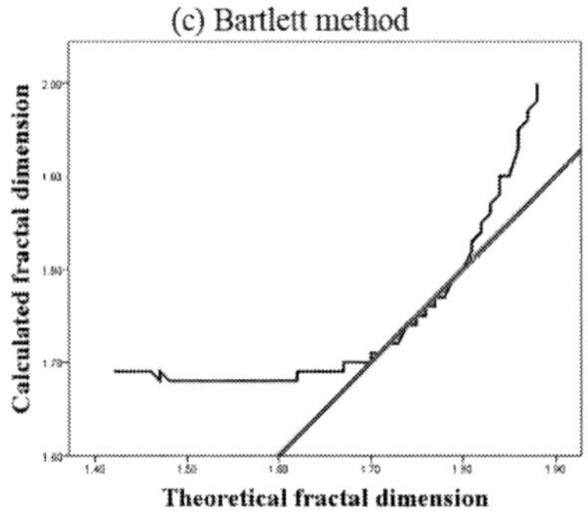

(d) Welch method

Fig. 6. Comparison between theoretical value and fractal dimension.

3.2.2 Statistical Testing of Characterization Results

A batch of theoretical values and calculated values are listed below. The outliers is cleaned. Because of the large amount of data, part of the data is shown here, as shown in table 2. D is theoretical value, $D^{'}$ is calculated value.

Table 2. Partial Data of Fractal Dimension.

	Autocorrelation method		Periodogram method		Bartlett method		Welch method	
	D	$D^{'}$	D	$D^{'}$	D	$D^{'}$	D	$D^{'}$
1	1.80	1.96	1.51	1.63	1.77	1.75	1.67	1.69
2	1.84	1.94	1.49	1.63	1.48	1.62	1.67	1.70
3	1.69	2.00	1.88	1.93	1.68	1.68	1.53	1.68
4	1.87	1.92	1.35	1.61	1.42	1.62	1.46	1.69
5	1.90	1.99	1.45	1.62	1.47	1.62	1.58	1.68
6	1.13	1.58	1.96	2.32	1.56	1.64	1.58	1.68
7	1.83	1.95	1.04	1.60	1.85	1.86	1.72	1.72
8	1.80	1.96	1.97	2.36	1.41	1.62	1.86	1.93
9	1.69	2.00	1.19	1.60	1.46	1.62	1.77	1.76
10	1.80	1.96	1.67	1.68	1.83	1.82	1.62	1.68
11	1.74	1.98	1.59	1.65	1.52	1.63	1.60	1.68
12	1.80	1.96	1.68	1.68	1.74	1.72	1.70	1.71
13	1.82	1.96	1.36	1.61	1.57	1.64	1.74	1.74
14	1.12	1.53	1.62	1.66	1.82	1.82	1.47	1.68
15	1.82	1.95	1.81	1.80	1.65	1.67	1.65	1.69
16	1.90	1.95	1.02	1.60	1.48	1.62	1.52	1.68
17	1.76	1.98	1.08	1.60	1.47	1.62	1.65	1.69
18	1.86	1.93	1.97	2.37	1.54	1.64	1.86	1.94
19	1.73	1.99	1.65	1.67	1.66	1.67	1.62	1.68
20	1.84	1.94	1.23	1.60	1.89	1.96	1.83	1.86
21	1.83	1.95	1.40	1.62	1.44	1.62	1.66	1.69
22	1.13	1.57	1.12	1.60	1.61	1.65	1.55	1.68
23	1.88	1.92	1.27	1.61	1.55	1.64	1.63	1.69
24	1.69	2.00	1.26	1.61	1.83	1.83	1.76	1.75
25	1.73	1.99	1.33	1.61	1.84	1.84	1.47	1.68
26	1.89	1.95	1.15	1.60	1.83	1.83	1.81	1.83
27	1.71	1.99	1.35	1.61	1.54	1.64	1.48	1.68
28	1.87	1.92	1.12	1.60	1.87	1.92	1.88	1.98
29	1.75	1.98	1.88	1.95	1.86	1.88	1.64	1.69
30	1.70	2.00	1.09	1.60	1.90	2.00	1.69	1.70
31	1.68	2.00	1.93	2.12	1.47	1.62	1.50	1.68
32	1.85	1.93	1.40	1.62	1.84	1.84	1.76	1.76
33	1.90	1.97	1.05	1.60	1.47	1.62	1.73	1.72
34	1.70	1.99	1.34	1.61	1.41	1.62	1.70	1.71
35	1.86	1.93	1.74	1.72	1.50	1.63	1.46	1.69
36	1.74	1.98	1.79	1.78	1.87	1.91	1.58	1.68
37	1.85	1.94	1.54	1.64	1.60	1.65	1.51	1.68
38	1.68	2.00	1.69	1.69	1.86	1.90	1.83	1.87

The simulation results are estimated and the results are demonstrated in table 3. The calculation results and confidence range of fractal dimension are different with different estimation methods. The calculated value of the Periodogram method is the closest to the theoretical value.

Table 3. Fractal Dimension And Its Estimation.

Test		Autocorrelation method		Periodogram method		Bartlett method		Welch method	
		D	$D^{'}$	D	$D^{'}$	D	$D^{'}$	D	$D^{'}$
Mean		1.80	1.96	1.51	1.63	1.77	1.75	1.67	1.69
confidence interval	Upper limit	1.84	1.94	1.49	1.63	1.48	1.62	1.67	1.70
	Lower limit	1.69	2.00	1.88	1.93	1.68	1.68	1.53	1.68

In order to verify the validity of the calculation method, we test whether the theoretical value and the calculated value are consistent. And we can use the Paired Sample T test [10]. Null hypothesis is that there is no significant difference between the calculated value and the theoretical value. Alternative hypothesis is that there is a significant difference between the calculated value and the theoretical value. Inspection statistics is given as

$$y_i = x_{1i} - x_{2i} (i = 1,2,\cdots n) \quad (7)$$

$$t = \frac{\bar{y}}{s_y / \sqrt{n-1}} \sim t(n-1) \quad (8)$$

Table 4. Paired Sample T Test.

Test	Mean	Standard deviation	standard error of the mean	95% confidence interval for the difference		t	P
				Lower limit	Upper limit		
Autocorrelation method	-0.165	0.105	0.011	-0.187	-0.142	-14.593	0.000
Periodogram method	-0.002	0.017	0.002	-0.006	0.001	-1.190	0.238
Bartlett method	-0.065	0.068	0.005	-0.075	-0.054	-12.163	0.000
Welch method	-0.085	0.079	0.007	-0.099	-0.072	-12.807	0.000

The results of the paired sample test showed that $P = 0.238 > 0.05$, when the periodogram method is adopted. At the significance level of 0.05, we could not reject the null hypothesis. There is no significant difference between the calculated value and the theoretical value. The theoretical data is mainly concentrated in $[1.6, 1.8]$. This indicates that the fractal dimension calculation of rough contour curves in this range is practically no difference with the theoretical value.

K-S testing method can use sample data to infer whether sample from the overall to obey a certain distribution theory, is a kind of goodness-of-fit test method, applicable to explore the distribution of continuous random variables. The K-S test can not only test whether a single population is subject to a theoretical distribution, but also can test whether there are significant differences between the two general distributions.The data does not obey the normal distribution. So the theoretical values and calculated values are respectively tested by Wilcoxon Signed Rank Test in non-parametric tests.

Table 5. Wilcoxon Signed Rank Test

Test	null hypothesis	Test	p	Decision
Periodogram method	The median of the difference between the theoretical and the calculated values is equal to 0	Wilcoxon Signed Rank Test	0.482	Accept the null hypothesis
Autocorrelation method			0.000	
Bartlett method			0.000	Reject the null hypothesis
Welch method			0.000	

There are no basic assumptions about non-parametric methods. The test results show that the null hypothesis cannot be rejected when the periodogram method is adopted and the significance level is 0.05. There is no significant difference between the calculated value and the theoretical value. The accuracy of the fractal dimension calculation of rough contour curves is also verified by the PSD method.

Acknowledgements

This work was financially supported by Natural Science Foundation of Hubei Province (2017CFB759), Fundamental Research Funds for the Central Universities Project (2662017JC024), National university student innovation project (20170504094, 2017BC027, 2015310200731). Thanks to Prof. Jiajun Yang, he is the corresponding author.

References

1. GE Shirong. *Research on fractal features and fractal expressions of rough surfaces*. Journal of tribology, (1997).
2. ZHU Hua, GE Shirong. *Study on the characterization of the surface topography of friction pairs during wear process with fractal theory*. Science In China, **44**(2001).
3. GE Shirong, ZHU Hua. *Fractal of tribology*. Mechanical industry press, (2005).
4. ZHU Hua, GE Shirong. *Experimental study on the characterization of worn surface topography with characteristic roughness parameter*. Wear, **255**(2003).
5. Robert L. Jackson,Jeffrey L. Streator. *A multi-scale model for contact between rough surfaces*. Wear, **261**(2006).
6. A.Majumdar, B.Bhushan. *Role of Fractal Geometry in Roughness Characterization and Contact Mechanics of Surfaces*. Journal of Tribology, **112** (1990).
7. A.Majumdar, B.Bhushan. *Fractal Model of Elastic-Plastic Contact Between Rough Surfaces*. Journal of Tribology, **113**(1991).
8. Houjin Chen. *Digital signal processing (version 2)*. Higher education publishing house, (2008).
9. Ma Runbo, Xu Shimeng, Du Jianhua, Qiu Jianbin. *Statistical analysis and research on the fractal characteristics of the structure of copper matrix composites*. Practice and cognition of mathematics, **42**(2012).
10. Sheng, Xie Qian, Pan Chengyi. *Probability theory and mathematical statistics (fourth edition)*. Higher education publishing house, (2008).
11. Jiunn-Jong Wu. *The Properties of Asperities of Real Surfaces*. Journal of Tribology, **4**(2001).
12. Wenwei Liu. *Study on the contact characteristics of mechanical interface based on fractal theory*. Huazhong university of science and technology, (2016).
13. Jeng Luen Liou,Jen Fin Lin. *A modified fractal microcontact model developed for asperity heights with variable morphology parameters*.Wear, (2010).
14. Hongliang Tian. *Dynamic modeling of virtual materials in fixed joint part of mechanical structure* [D]. Huazhong university of science and technology, (2011).
15. Melih Eriten, Chul-Hee Lee, Andreas A. Polycarpou. *Measurements of tangential stiffness and damping of mechanical joints: Direct versus indirect contact resonance methods*. Tribology International, (2012).
16. Xi Shi,Andreas A. Polycarpou. *Measurement and Modeling of Normal Contact Stiffness and Contact Damping at the Meso Scale*.Journal of vibration and acoustics: Transactions of the ASME, **1**(2005).
17. Yahav Morag, Izhak Etsion. *Resolving the contradiction of asperities plastic to elastic mode transition in current contact models of fractal rough surface*. Wear ,(2007).

Fixed-speed and Variable-speed Pumped Storage Dispatch Model in Power Systems with High Renewable Penetration

Bo Yuan[1,*], *Jin* Zong[2], and *Zhicheng* Xu[1]

[1]State Grid Energy Research Institute, Beijing, 102209, China
[2]State Grid Jibei Electric Power Company, Beijing, 10020, China

Abstract. According to different operating characteristics of pumped storage fixed speed unit and variable speed unit, a joint dispatching model of pumped storage unit and other types of units based on mixed integer linear optimization is constructed. The model takes into account the operating conditions, reservoir capacity, cycle type and other pumped storage unit constraints, but also consider the frequent start and stop and the stability of the operation of the unit caused by the loss. Using the Cplex solver to solve the model, the empirical example of the provincial power grid shows that the model can effectively arrange the pumping storage speed and the dispatching operation of the variable speed unit under the precondition of economic life of the unit, and give full play to the function of peak shaving and accommodating new energy. Because of its more flexible regulation characteristics of power generation and pumping conditions, the variable speed unit can better improve the operating conditions of other units in the system and promote the new energy dissipation.

1 Introduction

Pumped storage power station is the most mature energy storage system in power system and a flexible power supply with excellent performance. It is an effective means to solve the problem of power system peak regulation, reduce new energy discard and ensure the safe and stable operation of the system.

The traditional fixed-speed pumped storage unit can only operate at rated power at fixed speed, and it may appear "overshoot peak" phenomenon in the period of low load, and even need to start or increase output of thermal power unit to balance the difference between pumping output and breathing energy. , and the variable speed unit can be more flexible to cope with the load and intermittent power fluctuations because of the stable operating range under the pumping conditions, and the related scheduling model and scheduling strategy are also different. In [1], a low carbon scheduling model of pumped-storage variable speed unit is established by using a mixed integer quadratic programming model and a minimum cost flow algorithm. The scheduling strategy of pumped-storage variable speed unit is discussed beneficially; however, the quadratic programming model needs further improvement due to its long solving complexity and time.

In this paper, based on mixed integer linear programming model, the scheduling model of fixed speed and variable speed unit of pumped storage power station is established, and the loss caused by frequent start and stop of pumped storage unit and the operation of stability zone is fully considered. By using the Cplex and analyzing the example of the provincial power grid, the difference of the scheduling result between the variable speed and the fixed speed unit is compared, and the validity of the proposed model is validated. The results show that the model can arrange the speed of pumped storage and the operation mode of variable speed unit reasonably, and give full play to its peak regulation, promote the new energy consumption and reduce the operating cost of the system under the premise of considering the economic life of the unit.

2 Model of Pumped Storage Unit in Optimal Dispatching

2.1. Output Range Constraint

Under the condition of generating mode, the output range of fixed speed and variable speed pumped storage unit is

$$U_{pump}P_{g,\min} \le P_g \le U_{pump}P_{g,\max} \tag{1}$$

Where U_{pump} is the working condition of pumped storage, 1 represents generating mode and 0 represents pumping mode. $P_{g,\min}$ and $P_{g,\max}$ are the minimum and maximum output range of generating mode.

Under the condition of pumping mode, for variable-speed pumped storage, we give

$$(1-U_{pump})P_{p,\min} \le P_p \le (1-U_{pump})P_{p,\max} \tag{2}$$

Where $P_{p,\min}$ and $P_{p,\max}$ are the minimum and maximum output range of pumping mode.

* Corresponding author: yuanbo@sgeri.sgcc.com.cn

For fixed-speed pumped storage, we give

$$P_p = (1 - U_{pump}) P_{p,\max} \tag{3}$$

Which means the output under pumping mode can only be 0 or rated power.

2.2. Reservoir Capacity Constraint

When a pumped storage unit is in operation, the water level needs to be maintained between the maximum water level and the dead water level.

$$W_{u,\min} \le W_u \le W_{u,\max} \tag{4}$$

$$W_{d,\min} \le W_d \le W_{d,\max} \tag{5}$$

Where W_u and W_d are the water level of the upper and the lower reservoir; $W_{u,\min}$ and $W_{u,\max}$ are the minimum and maximum water level of the upper reservoir. $W_{d,\min}$ and $W_{d,\max}$ are the minimum and maximum water level of the lower reservoir.

2.3. Water Level Dynamic Equilibrium

$$W_u^t = W_u^{t-1} - \lambda_g P_g^t + \lambda_p P_p^t \tag{6}$$

$$W_d^t = W_d^{t-1} + \lambda_g P_g^t - \lambda_p P_p^t \tag{7}$$

Where λ_g is the generating conversion coefficient and λ_p is the pumping conversion coefficient.

2.4. Daily/weekly Reservoir Cycle

For the day/week cycle of reservoir scheduling type units, assuming that one day/week is divided into s slices, the initial and final period of the upper reservoir water level should be consistent.

$$W_u^s = W_u^0 \tag{8}$$

Where W_u^0 is the water level of initial time period and W_u^s is the water level of final time period.

2.5. Overcall Losses of Generating Units

The output range of pumped storage unit can be divided into stable operation zone($P_{p,\text{stable}}$ and $P_{g,\text{stable}}$) and minimum technical output zone($P_{p,\min}$ and $P_{g,\min}$), which can reflect the unit loss caused by the carrying load outside the stable operation zone. An auxiliary binary decision variable is introduced to represent the operating zone. $k = 1$ represent the pumping mode stable operation zone, $k = 2$ represent the pumping mode minimum technical operation zone, $k = 3$ represent shut down, $k = 4$ represent the generating mode minimum technical operation zone, $k = 5$ represent the generating mode stable operation zone.

A continuous decision variable is also introduced to represent the output in each operation zone. The following constraints should be met.

$$\sum_k w_{i,k}^t = 1 \tag{9}$$

$$\sum_k P_{\text{pump},k}^t = P_p + P_g \tag{10}$$

$$w_{i,3}^t = 1 - U_{pump}^t \tag{11}$$

$$P_{\text{pump},k,\min} w_{i,k}^t \le P_{\text{pump},k}^t \le P_{\text{pump},k\max} w_{i,k}^t \tag{12}$$

(9)guarantees that each output state is mutually exclusive; (10) (11) makes the auxiliary decision variable and the output range of the original decision variable coupled; (12) is the upper and lower bound of output in each operation zone.

After the auxiliary decision variable is introduced, two methods can be used to deal with the loss modeling of operating outside the stable operation zone. One is to specify in the constraint condition that the running time outside the stable operation zone of the unit per day cannot be greater than the given value $t_{\min}$, which is

$$\sum_t (w_{i,2}^t + w_{i,4}^t) \le t_{\min} \tag{13}$$

The other method is the operation loss cost function for a given unstable operating region, which is included in the objective function of the scheduling model as follow

$$C_{pump} = h \cdot \sum_t (P_{pump,4}^t - P_{pump,2}^t) \tag{14}$$

Where h is the cost coefficient of operating outside the stable zone.

3 Optimal Dispatching Model of the System

3.1. Decision Variables

The decision variables include the output of thermal power, hydropower, wind power, solar energy, pumped storage unit, and the on/off status of the thermal units, and the operating conditions of pumped storage units.

Table 1. Decision Variables of the Dispatching Model

	Type	Illustraion
$P^t_{[illegible],i}$	continuous	the output of unit *i* in period *t*
$U^t_{\text{thermal},i}$	binary	on/off state of unit *i* in period *t*
$P^t_{g,i}$	continuous	generating output of pumped storage unit
$P^t_{p,i}$	continuous	pumping output of pumped storage unit
$U^t_{pump,i}$	binary	operating mode pumped storage unit, 1 represent generating and 0 represent pumping

3.2. Objective Function

The objective function is the minimum operating cost of the system:

$$C = \sum_t \sum_i C_i^{SU} (1 - U_{\text{thermal},i}^{t-1}) U_{\text{thermal},i}^t + \sum_t \sum_i F(P_{\text{thermal},i}^t) + C_{pump} \tag{15}$$

Where C_i^{SU} is the start-up cost of thermal unit i; $F()$ is the fuel cost function of thermal unit; C_{pump} is the overcall operation loss cost function of pumped storage.

3.3. Constraints

(1) Power balance constraint

$$\forall t,\ \sum_{s \in unittype}\sum_{i} P_{s,i}^t + \sum_{i} P_{g,i}^t - \sum_{i} P_{p,i}^t = L_t \quad (16)$$

Where L_t is the load demand in period t.

(2) Maximum and minimum output range

$$\forall t, \forall i, \forall s \in \text{unittype}, \quad P_{s,i,\min} \le P_{s,i}^t \le P_{s,i,\max} \quad (17)$$

Where $P_{s,i,\min}$ and $P_{s,i,\max}$ are the maximum and minimum output of unit i. For wind and solar power, the maximum output at each moment is the predicted available output at that moment.

(3) Minimum on/off time constraint

$$\sum_{l=t}^{t+T_{i,on}-1} U_{s,i}^l \ge T_{i,on}(U_{s,i}^l - U_{s,i}^{l-1}) \quad (18)$$

$$\sum_{l=t}^{t+T_{i,off}-1} (1-U_{s,i}^l) \ge T_{i,on}(U_{s,i}^{l-1} - U_{s,i}^l) \quad (19)$$

Where $T_{i,on}$ and $T_{i,off}$ are the minimum on and off time of unit i.

(4) Ramp rate constraint

$$P_{s,i}^t \ge P_{s,i}^{t-1} - RD_{s,i} - (1-U_{s,i}^t)P_{s,i,\max} \quad (20)$$

$$P_{s,i}^t \le P_{s,i}^{t-1} + RU_{s,i} + (1-U_{s,i}^{t-1})P_{s,i,\max} \quad (21)$$

Where $RD_{s,i}$ and $RU_{s,i}$ are the up and down ramp rate of unit i.

(5) Spinning reserve constraint

$$\sum_{s}\sum_{i} R_{s,i}^t \ge L_t(1+R_L) \quad (22)$$

$$\sum_{s}\sum_{i} R_{s,i}^t \le L_t(1+R_{VG}) \quad (23)$$

Where R_L and R_{VG} are the reserve requirement of load and variable generations; $R_{s,i}^t$ are the available spinning reserve provided by unit i.

(6) Pumped storage constraint

The pumped storage constraints are (1)-(14).

4 Case Study

4.1. Test System

The objective function of the proposed model involves that the starting and stopping costs of thermal power units and pumped storage units are quadratic functions, which can be converted into linear functions according to the method in [15]; In this case, the whole model is a mixed integer linear optimization model, which can be programmed by using OPL language developed by IBM and solved by CPLEX.

Based on the 2030-year planning system of a real power grid in the eastern province, this paper studies a typical day 24-hour power system dispatch situation. The typical daily system load curve is shown in Fig. 2.

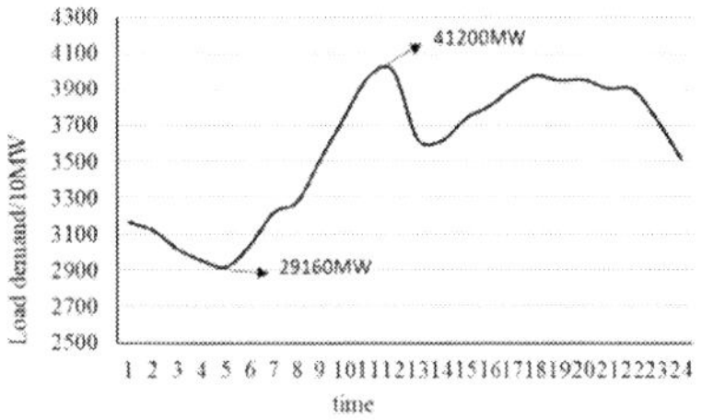

Fig. 1. Load Curve

4.2. Results

If there is no pumped-storage power station in the system, the unit output arrangement is shown in Figure 3, and the wind/solar curtailment is shown in Table 3. Wind/solar curtailment is 66 million Kwh, the curtailment rate is 24.2%.

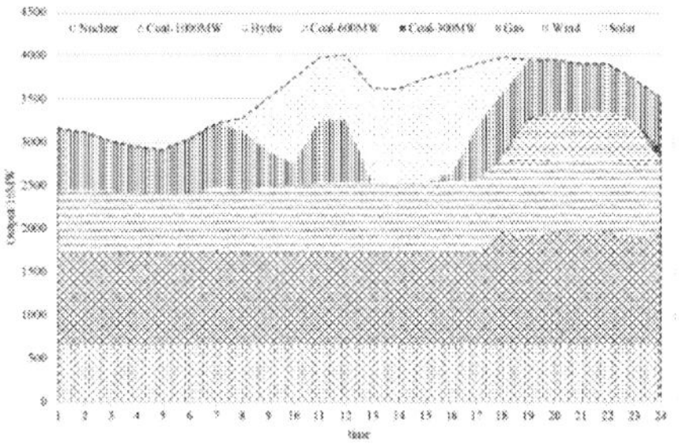

Fig. 2. Dispatch result without pumped storage

Eight fixed speed pumped storage units of 300,000 kW were added to the system, unit loss is not take into consideration, and the dispatch result is shown in Figure 3. Results shows that the pumped storage unit can pump water during the period of low net load and generate electricity during the peak period, so that the tasks of peak regulation and new energy consumption can be well realized. After adding a fixed-speed pumped storage unit, the total amount of the system curtailment is about 54.43 million kw h, and the curtailment rate is 19.9%, which is 4.3% lower than the system without pumped storage.

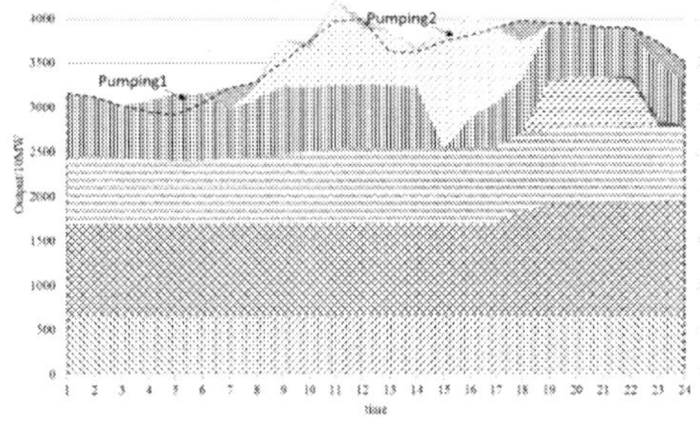

Fig. 3. Dispatch result with fixed-speed pumped storage

The fixed speed unit is replaced by variable speed unit of the same capacity. The model also arranges the output of the variable speed unit well, effectively

https://doi.org/10.1051/e3sconf/20183804014

realizes the peak shaving and new energy consumption function; at the same time, due to the stepless adjustment of pumping conditions, the system operation is more flexible, the amount of curtailment is reduced to 46 million kW h, and the curtailment rate is 16.9%.

When taking into account the loss of pumped storage, the total cost of the system will rise, but the operating cost of the system is still lower than that of no pumped storage, see table 2;

Table 2. Curtailment and utilization hour of pumped storage in different scenarios

		Curtailment/%	Daily utilization hour/hour
no pumped storage		24.2	-
\ no loss	fixed-speed	19.9	5.2
	variable-speed	16.9	6.5
loss	fixed-speed	20.6	4.3
	variable-speed	17.3	6.0

From table 2, it can be found that the operating cost of the system and the curtailment rate of are slightly increased, but the operating range is controlled in a better interval. This shows that the unit can operate in a more secure and stable range after considering the excessive call of pumped storage.

5 Conclusion

In this paper, the loss model of pumping storage with operating outside stable zone is constructed, the joint dispatch model of pumped storage and other types of units based on mixed integer linear programming is built, and the model is solved using Cplex, considering the various operating characteristics of fixed-speed and variable-speed unit. Through to the provincial power grid demonstration analysis obtains the following conclusion:

(1) The proposed model can effectively arrange the output of fixed-speed and variable-speed pumped storage unit under the premise of considering the economic life of the unit, give full play to the function of peak shaving and new energy consumption;

(2) In terms of promoting new energy consumption, the variable speed unit has stepless regulation characteristics due to its large working range of power generation and pumping condition, which makes the system run more flexibly and can effectively solve the phenomenon of "over peak shaving" of the fixed speed pumping unit, better improve the operating conditions of other units in the system and reduce curtailment.

(3) After taking into account the overcall loss, the pumped storage unit can operate in a safer and more stable zone. Although the system curtailment has increased slightly in the short term, in the long run, the failure frequency and maintenance time of the pumped storage unit have been reduced, the life of the unit has been extended, and the unit can be better guaranteed to give full play of the new energy consumption and peak-shaving function.

Acknowledgements

This work is supported by Science and Technology Project of State Grid Corporation of China (B3670116A005) : Study on the evaluation model and application of peak shaving technology, economy and policy for power system wiht large-scale renewable energy integration.

References

1. Bueno C., Carta J. A.. Technical-economic Analysis of Wind powered Pumped Hydro storage Systems. Part 2: Model Application to the Island of EI Hierro [J]. *Solar Energy*, 2005, 78(3): 396-405.
2. Kenneth B , Yury D , Erik D , et al . Coupling pumped hydro energy storage with unit commitment[J] . *IEEE Tansactions on Sustainable Energy*, 2016, 7(2) : 786-796 .
3. Garcia-Gonzalez J., de la Muela R. M. R., Santos L. M.. Stochastic Joint Optimization of Wind Generation and Pumped-storage Units in an Electricity Market[J]. *IEEE Transactions on Power Systems*, 2008, 23(2): 460-468
4. Lin-Her Jeng, Yuan-Yih Hsu, Chang B. S.. A Linear Programming Method for the Scheduling of Pumped-storage Units with Oscillatory Stability Constraints[J]. *IEEE Transactions on Power Systems*, 1996, 11(4): 1705-1710.
5. Huang Shu, Lin Shunjiang, Liu Mingbo, Multi-objective Security Constrained Dynamic Optimal Dispatch with Wind Farms and Pumped Storage Stations[J]. *Proceedings of the CSEE*, 2016, 36(1): 112-120.

https://doi.org/10.1051/e3sconf/20183804015

Reforming the Exhaust Passage of Low-pressure Cylinder for 330MW Steam Turbine

YAN Tao[1], CAI Wen[1], CHEN Wen[1], LU Jin[1], Yang Hong-yan[2]

[1]Jiang Xi Electric Power Science & Research Institute of National Grid, Nanchang 330096, Jiangxi Province,China
[2]Shanghai Pengken Energy Technology Co.LTD, Shanghai 200090, China

Abstract: In concern of the velocity distribution of the exhaust passage of 330MW turbine is not uniform, which results in higher the upper temperature difference of the condenser and higher exhaust pressure. It is introduced in this article that based on mathematical simulation, steam-equalizing equipment is augmented at the exhaust area of the condenser which makes the decrease in the steam resistance, much more uniform velocity distribution, and the increase of the heat transfer coefficient. By comparison of the condenser performance test before the amending and after, the result shows that after the amending, the upper temperature difference of the condenser and the exhaust pressure decreases dramatically.

1 Introduction

A steam turbine manufactured by Beijing BEIZHONG Steam Turbine Generator Co.,Ltd, whose model is N330-17.75/540/540, is equipped at a power plant, auxiliary with N-18000 condenser, whose cooling area is 18000m^3, and the length of each cooling tube is 12420mm. Since the turbine was put into operation in 2008, there have been problems, such as big terminal temperature difference of condenser and high exhaust pressure of the low-pressure cylinder(low vacuum), which leads to an increase in both heat rate and coal consumption rate of the turbine unit.

2 Analysis

The possible reasons for the phenomena presented above are insufficient cooling water flow, dirty cooling tubes of the condenser, and uneven flow field at the throat outlet of the condenser. The exhaust passage model of the turbine is established by a professional software Gambit in this article, and the flow field model of The exhaust passage model is simulated by a software Fluent. The simulation results show that the flow field distribution of the exhaust steam in the condenser is not reasonable (shown in Figure 1), which restricts the cooling effect of the condenser to a certain extent and has the potential for further optimization. Fig. 1 is the velocity contour map of steam turbine cross-section at the outlet of the lower half of the condenser (condenser cooling pipe inlet) before the exhaust passage is optimized. The straight line on the axis of the turbine is the semi-side symmetrical exhaust passage. As can be seen from the figure, there is a large variety of low-velocity area (dark blue area in the figure, whose speed is below 10m/s) at the inlet of the tube bundle under the original channel, and the heat transfer coefficient in this area must be small.

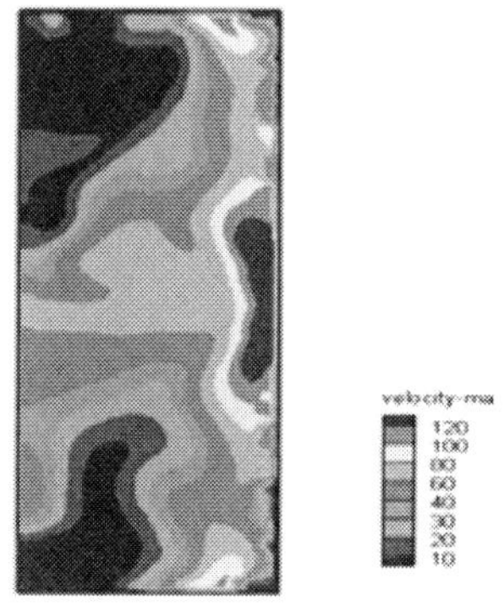

Figure 1. Original exhaust steam velocity chart of LP cylinder

3 Optimal reform for the exhaust passage

If measures are taken to distribute part of the steam in the high-speed steam (velocity above 50m/s) to the above low-speed zone (corresponding to "peak load shifting"), the heat transfer coefficient of the pipe bundle in the area will increase and the overall heat transfer of the condenser will increase. The coefficient and heat transfer effect must be strengthened.

According to the simulated calculation, the distribution of the steam flow field at the inlet of the condenser tube under the original structure is targeted at the appropriate position in the neck of the condenser, and the streamlined oblique, flat vertical, and straight inclined are installed. The flow guiding and equalizing devices were set up to properly guide the original steam flow and optimize the steam flow field, so that the turbine exhaust steam flow into the flow field of the cooling pipe bundle tends to be reasonable. Exhaust passage optimization measures should include the following:

(1) All baffles are arranged in the upper half of the condenser in the connecting neck, in which a row of inclined plates is perpendicular to the axis of the turbine, and four rows of vertical plates are parallel to the axis of the turbine (Figure 2). Both governing end and electrical end of the LP cylinder are symmetrically arranged.

(2) Inclined plates are set as follows. The inlet side is vertical, and the angle between the outlet side and the vertical direction is 30 degrees. The exit side of the swash plate is located on the cross-section of the square neck outlet, and the horizontal distance of the entrance side from the end wall (tune end or electric end) is 650mm; the total length of the swash plate (length direction of the cooling pipe extension) is 6600mm, and it is symmetrically arranged according to the center.

(3) Vertical plates are set as follows. The vertical plate is divided into four pieces, whose height is 940mm, and length is 1550mm, upright arranged in the square connecting neck between the upper and lower two rows of struts. There are steam extraction on the side of the tube, the distance between the two vertical plates from the center line is 1500mm and 2900mm differently, while the distance is 1600mm and 2900mm from the center line for non-extraction side. The distance between the four vertical plates from the electrical end (or governing end) is 1610mm. The above is just an arrangement on the outlet side, and the other outlet arrangement is symmetrical with this. Each of the four vertical plates on both sides is actually a two-by-two combination of 940×3100 mm.

(4) The material of the diversion device is made of stainless steel and is resistant to erosion. The rooting position of the deflector is the frame support tube in the throat of the condenser, supported by the φ63.5*5 tubes.

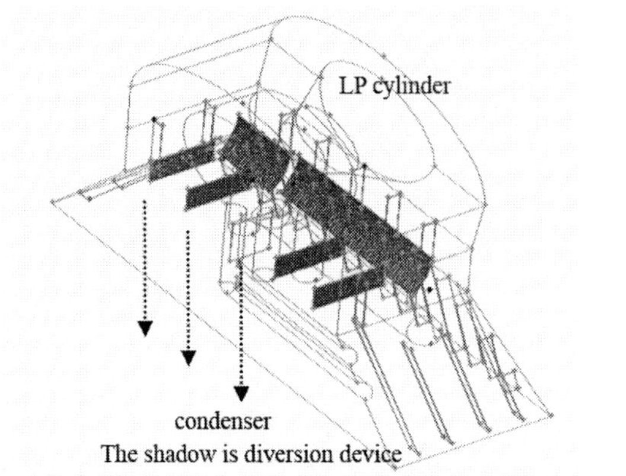

Figure 2. reform drawing for the exhaust passage

After optimization measures are taken, steam flow of the exhaust passage is more uniform, while the thermal load distribution of condenser cooling tube bundles in the longitudinal and transverse directions is reasonable, and the area of the low-speed area is significantly reduced, as shown in figure 3.

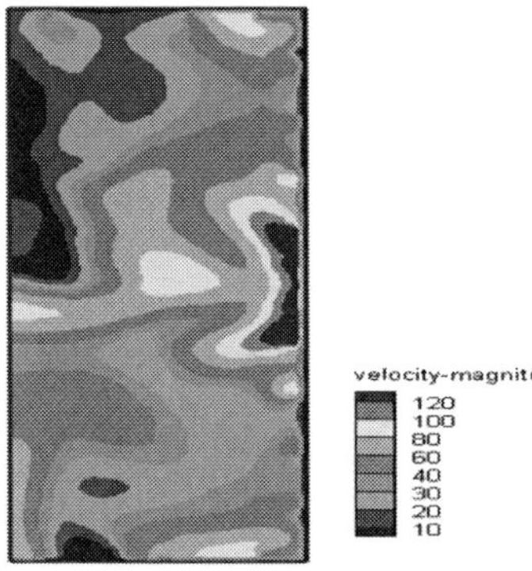

Figure 3. Exhaust steam velocity chart of LP cylinder after reform

4 Results

The results in Table 1 show that after the reform measures are fulfilled on the turbine, both the terminal temperature difference of the condenser and the exhaust pressure of LP cylinder dropped significantly under various operating conditions, most of which happens at 320 MW condition, t the terminal temperature difference of the condenser decreased by 1.22°C while exhaust pressure by 0.51 kPa.

Table 1. comparison of the condenser performance before and after reform

Power	MW	320	300	270	240
Revised Temperature difference before	°C	6.73	4.83	4.55	4.23
Revised Temperature difference after	°C	5.51	4.03	3.69	3.28
Difference between before and after	°C	1.22	0.80	0.86	0.95
Revised exhaust pressure before	kPa	7.25	6.38	5.97	5.54

Revised exhaust pressure after	kPa	6.74	5.97	5.62	5.24
Difference between before and after	kPa	0.51	0.41	0.35	0.30

Table 2. calculated back pressure under various conditions before and after the reform(designed water flow of 39200m³/h)

Heat load	item	unit	Circulating water inlet temperature						
			10℃	15℃	20℃	23.5℃	25℃	30℃	33℃
100%	Back pressure after	kPa	3.48	4.35	5.53	6.57	7.08	9.08	10.55
	Back pressure before	kPa	3.78	4.68	5.90	6.98	7.50	9.59	11.11
	difference	kPa	0.30	0.32	0.36	0.40	0.42	0.50	0.56
90%	Back pressure after	kPa	3.15	3.98	5.10	6.08	6.56	8.45	9.83
	Back pressure before	kPa	3.40	4.25	5.40	6.42	6.91	8.87	10.30
	difference	kPa	0.25	0.27	0.30	0.34	0.35	0.42	0.48
80%	Back pressure after	kPa	2.85	3.64	4.69	5.62	6.07	7.85	9.15
	Back pressure before	kPa	3.06	3.86	4.94	5.90	6.36	8.20	9.55
	difference	kPa	0.20	0.22	0.25	0.28	0.29	0.35	0.40
70%	Back pressure after	kPa	2.58	3.32	4.32	5.19	5.61	7.29	8.51
	Back pressure before	kPa	2.74	3.50	4.52	5.41	5.85	7.58	8.84
	difference	kPa	0.16	0.18	0.20	0.23	0.24	0.29	0.33
60%	Back pressure after	kPa	2.33	3.03	3.96	4.78	5.18	6.76	7.91
	Back pressure before	kPa	2.46	3.17	4.13	4.97	5.38	6.99	8.18
	difference	kPa	0.13	0.14	0.16	0.18	0.19	0.23	0.26
50%	Back pressure after	kPa	2.10	2.76	3.64	4.41	4.79	6.27	7.35
	Back pressure before	kPa	2.20	2.87	3.76	4.55	4.93	6.45	7.56
	difference	kPa	0.10	0.11	0.12	0.14	0.15	0.18	0.20

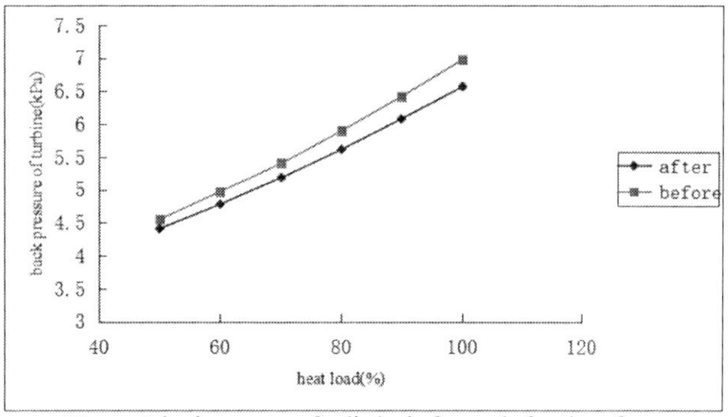

Figure 4. back pressure of cylinder before and after the reform
(designed water flow of 39200m³/h and inlet water temperature of 23.5°C)

Under the designed circulating water flow volume at various conditions, the back pressure of the turbine listed in table 2 decreases dramatically by comparing before and after the reform that the performance of the condenser increases corresponding. For example, When the circulating water inlet water temperature is 23.5℃, the back pressure drops by 0.40kPa, while it drops by 0.56kPa at 33℃. Figure 4 shows the back pressure curve by data fitting before and after the modification of the #9 steam turbine exhaust passage under the designed circulating water inlet flow, inlet water temperature of 23.5℃, and at different thermal loads of the condenser.

5 Conclusion

By optimizing and reforming the steam turbine exhaust passage, a flow equalizing device is installed at the condenser throat. Simulation results show that the low velocity area of the exhaust passage can be significantly reduced and the overall heat transfer coefficient can be improved. The test results show that after the reforming, the terminal temperature difference of the condenser is reduced, the exhaust pressure is also reduced, and the economic efficiency of the unit is improved, which has important significance for promotion.

References

1. Fu wenfeng, etc. Three-dimensional Numerical Flow Simulation and Structure Optimization of Low-pressure Last Stage Exhaust Passage in a 600MW Steam Turbine, Proceedings of the CSEE,2014,13(14):2267-2272.
2. 2.Song Zhen, etc. Numerical simulation of coupled flow in exhaust passage of steam turbine, Journal of Southeast University (Natural Science Edition), 2015,45(4):701-706.
3. CAo Lihua, etc. Numerical Study on Effectiveness of Installing Guide Device into Exhaust Passage of a 600MW Steam Turbine, Journal of Chinese Society of Power Engineering, 2015,35(5):366-371.
4. ZHOU lan-xin, etc. Numerical Study on the Performance of Exhaust Steam Passage with Different Guide Baffles, Turbine Technology,2015,57(1):33-36.
5. White A J. A Comparison of modelling methods for polydispersed wet-steam flow, International Journal for Numerical Methods in Engineering, 2003,57(6):819-834.

Study on pricing mechanism of pumped hydro energy storage (PHES) under China's electricity tariff reform

Fuqiang Zhang*, *Zhicheng* Xu, *Bingqi* Jiao and *Junshu* Feng

State Grid Energy Reasearch Institute CO., LTD., Beijing, 102209, China

Abstract. This paper presents a pricing mechanism for pumped hydro energy storage (PHES) to promote its healthy development. The proposed pricing mechnism includes PHES pricing mechanism and cost sharing mechanism. Regarding the PHES pricing mechanism, the existed two-part tariff is still recommended to implement at the current and future stages. Regarding the cost sharing mechanism, it is suggested that the electricity capacity fee and pumping-loss fee should be all included in the allowable transmission and distribution costs of the regional power grids, which can be further transmitted to the provincial power grids. At the same time, we also suggest to exploring other funding sources to relieve the pressure of PHES cost sharing totally through the electricity sales price.

1 Introduction

In China, the capacity fees and pumping losses of the PHES were included in the operating costs of the local provincial power grid (or regional power grid) according to the policy document *Notice on Improving Price Formation Mechanism of Pumped Hydro Energy Storage* (No.1763), which was released in 2014 by the Development and Reform Commission of China (NDRC) [1]. The settlement of electricity charge between the PHES and power grid enterprises will be conducted in accordance with the price standards approved by the government price authorities. In December 2016, the National Development and Reform Commission of China promulgated the *Measures for Electricity Tariff Determination for Transmission and Distribution of Provincial Power Grid (Trial)* with the promotion of transmission and distribution tariff reform [2]. It clearly states that the relevant costs of a PHES were not included in the allowable income of the power grid company, but how to share the cost was not stipulated in detail. Consequently, the new-built PHESs are now facing the challenge that the annual operating costs (AOC) cannot be properly settled, which results in affecting the healthy development of the PHES industry. For example, at present, 25% of the capacity fees of *Hei Mifeng* PHES cannot be settled normally, which results in a loss of 445 million Yuan accumulately. For other PHESs which are built or put into operation recently in Zhejiang and Jiangxi provinces, their capacity fees still have not been normally settled.

In order to develop a suitable pricing mechanism and electricity tariff revenue mechanism under the electricity market environment, this paper tries to put forward a reasonable scheme for the PHES. We first analyze and conclude the present price formation and tariff revenue mechanism for PHES. Secondly, an in-depth analysis of digging China's existing problems for the current electricity system reform has been conducted. Next, the feasibility of participation of various ancillary services of PHES into electricity market system has been analyzed. Then, an electricity tariff scheme for different stages of the electricity market is proposed for PHES under the electricity market. The proposed pricing mechanism will provide policy reference and detailed implementation measures for the healthy development of PHES in the electricity market environment.

2 International PHES revenue mechanisms in different market phases

2.1 Existing international markets and their revenue mechanisms

According to literature review and research, about 85% of the 150 GW PHES in the world adopt the grid unified management mode or leasing mode, such as France, Japan and several states in the United States. In this way, the PHES is either built and operated by the power grid or built by a third party and leased out by the power grid and the related costs are incorporated into the unified accounting of the power grid and are shared through the sales price [8,9].

About only 15% of the world's PHESs earn revenue by participating in the electricity market competition, typically such as RTO/ISO in the United Kingdom and the United States. PHESs are mostly funded by independent power plants (IPPs) or distribution (sales) utility companies that generate revenue by participating in the energy markets and ancillary service markets. However, the income received in the market can only

* Corresponding author: zhangfuqiang@sgeri.sgcc.com.cn

cover 20% to 30% of its costs. It also needs to negotiate with the trading agencies to sign a medium and long-term contract for sale ancillary services to obtain a stable income, so as to guarantee the investment return. For the PHESs of Britain's First Hydroelectric Company, their annually fixed income parts of 70% to 80% are compensated by the British National Grid, and the surcharge (uplift) approach to user guidance; through participation in the energy market arbitrage variable obtained only 20% to 30%, this part also covers the pumping station hair loss.

The main reasons why the global PHES is not involved in the competition in the electricity market are as follows: 1) Difficult to determine the benefits: The ancillary services such as emergency backup and black start provided by the PHES have a great effect on the safe and stable operation of the power system. However, the beneficiaries are extensive and difficult to quantify and characterize, so it is not easy to determine the compensation standard. 2) Market risk: In electricity market competition, PHES will face the risk brought by the volatility of electricity price, while investors generally prefer to invest in projects with a lower risk of investment. In the recent 10 years, with the rapid development of renewable energy such as wind and solar, the demand for storage power plant such as PHES has greatly increased. However, a majority of countries in the world face the problem of low enthusiasm for investment in PHES due to market risks. 3) Historical inertia: The vast majority of owners of PHESs built in the last century are vertically monopolized electricity companies, and still use the existing revenue mechanism that integrates PHES costs into the electricity sales price.

2.2 PHES revenue mechanism in China and existing barriers to PHES development

At present, the price of the existing PHESs in the operating area of the State Grid Corporation of China (SGCC) mainly consists of the two-part tariff and the single-capacity tariff. The relevant charges of the PHES have basically been covered by the sales price. As of the end of 2016, two of the 20 PHESs that belong to SGCC have put into operation mainly implemented two power price modes: one is the two-part tariff scheme, which consists of capacity-tariff and energy-tariff. The PHESs implement this kind of tariff including Zhejiang Tianhuangping, Liaoning Pushihe, Fujian Xianyou, Beijing Ming Tombs, Henan Huilong, Zhejiang Xianju, and Jiangxi Hongping PHESs. The capacity fee in the two-part tariff is approved by the NDRC of China and mainly reflects the ancillary service value of the PHES that provides power system reserve, regulation, and black start. The tariff will be settled between the provincial power grid company and the PHES, and the capacity fee is included in the local provincial level (or regional grid) operating expenses. The energy fee is mainly designed to make up for the variable costs such as PHES pumping and generation losses. In China, the pumping electricity for PHES is the 75% of local coal-fired units benchmark electricity price (considering desulfurization, denitrification, dust and other environmental protection electricity prices, the same below). Other PHES plants adopt the single capacity-tariff scheme at present n China, including Shandong Taishan, Zhejiang Tongbai, Jiangsu Yixing, Anhui Langya Mountain, Hebei Zhang Hewan, Henan Baoquan, Shanxi Xilong Chi, Hunan Hemifeng, Hebei Panjiakou, Jilin Baishan, Hubei Bailian River, Anhui Xiangshui, and Anhui Xiang Hongdian [10].

With the progress of power market reform, there are three main challenges and barriers to the healthy development of PHES, which are listed as follows:

(1) There is no reasonable pricing mechanism to ensure the healthy development of PHESs

The measures pointed out in [2] for "provincial power transmission and distribution pricing determination method " indicated that the PHES shall not be included in the effective asset of power systems. However, there is no clear regulation of what channel to pass through the costs of PHESs, so the newly put into operation and the unregulated PHESs are faced with the difficulty of the normal settlement of the electricity price. Therefore, it is necessary to study the price dredging mechanism of PHES which is suitable for the present stage.

(2) Few ancillary services can participate in market competition

Taking China's currently running ancillary services market as an example, such as Northeast Power Market, only two ancillary services (peak shaving and black-start) can participate in the market. There are too few ancillary services that can participate in the market competition, and the general ancillary services such as AGC and reserve service do not involve in present power market in China. Secondly, there is no discrepancy between the requirements of the adjustment rate of peak regulation service in the ancillary service market in Northeast China, and the advantages of the fast adjustment of the power plant can not be reflected. Thirdly, for the excess pumping power and black-start service provided by PHES, the administrative pricing method is used to give a small amount of compensation, which does not reflect the true market value.

(3) The sharing mechanism of "who benefits, who pay" has poor operability

According to the requirements of *No. 9* issued by the Central Government of China, the principle of "who benefits, who pay" requires a sharing mechanism for each participated user should be established to offer and share the services. However, the implementation of the above mechanism has not yet been implemented in practice timely during the period of the reform of transmission and distribution pricing mechanism. There is a vacancy period of policy convergence, resulting in the situation that the newly-started PHESs cannot be settled normally. At the same time, the operational characteristics of PHES also determine that this mechanism is difficult to implement in the future ancillary service market. On the one hand, the benefits provided by PHES such as frequency regulation, phase modulation, accidental backup and black start are difficult to be accurately measured and calculated. On

https://doi.org/10.1051/e3sconf/20183804016

the other hand, the ancillary services for PHES have a variety of beneficiaries, including various types of thermal power plants, nuclear power, and renewable energy, as well as power grids and load consumers. In fact, the ancillary services offered by PHESs serve the entire power system and are difficult to accurately target specific beneficiaries. The wide range of beneficiaries led to the lack of maneuverability.

3 The revenue scale calculation in different market phases

In order to estimate whether the PHES is suitable for participating in the electricity market competition, we conduct a revenue scale calculation for a specific PHES if it is in a free electricity market in this part. In this way, we can prove that whether the main part of the investment and operation cost can be covered or not when it participates in the market competition.

3.1 Market-oriented transitional phase

In the transition period of the electricity market, the ancillary service market has not yet been fully established, so it is suitable for using fixed compensation mode to obtain ancillary services. According to the successful establishment of electric energy spot market or not, the electricity price and revenue mechanism in the transitional period can be divided into two situations: 1) only the medium and long-term electricity energy market; 2) the medium- and long-term electricity energy markets exist at the same time.

(1) Only the medium and long-term electric energy market

Due to the limited reservoir capacity of PHES, the plant cannot provide a continuous and stable power output. Therefore, it can only be profitable in the spot market of electric energy by short-term (intra-day and inter-period) price fluctuations, and cannot participate in medium- and long-term electric energy market.

(2) Both medium- and long-term electricity energy markets exist

If the spot market for electricity energy is established, the PHES can participate in the competition in the electricity energy market and obtain some benefits through providing peak shaving service. The ratio of the revenue scale from electricity energy market to the capacity fee is estimated below to provide a reference for the government price makers to make a reasonable pricing mechanism. In order to calculate the portion of benefits revenue from the electricity energy market, first of all, we need to determine the peak-valley electricity price ratio. Reference to the domestic and international electricity market operation experience shows that the market peak spot electricity price ratio is only about 1.5 to 2 times [11-13]. See Table 1 for details. It is noted that, when PHES participates in the competition of electricity energy market, it will reduce the peak-valley ration of origin power system. The main reason for this phenomenon is that PHES will ease the shortage of supply and demand at the peak of load and the oversupply situation at the time of low load as the PHES has the function of peak shaving. In general, the peak-valley electricity price difference can be reduced, and the reduction range can reach 10% to 30%. Based on the above considerations, in the subsequent calculation of the profitability of the electricity energy market, it is assumed that the ratio of the peak load price to the valley price is controlled within 1.5 times. Take Pushihe PHES in Liaoning Province as an example. It is assumed that the PHES can benefit from not only the electricity energy market, but also the peak load shaving. It means that the revenue of this PHES is composed of two parts if it participates in the electricity market. In addition to the revenue of electricity energy brought by peak-valley price difference; it is also possible to obtain income from providing peak load shaving.

For the revenue obtained from the electricity energy market, the measured results (Table 2 and Table 3 show the parameters and corresponding values) show that in the ideal situation, when the PHES totally undertakes the peak peaking task of the Northeast Power Grid, the total revenue of the PHES still has a large gap with the current capacity fee (about 600 million Yuan). Under the assumption that the generating price is 1.5 times of the pumping price, the proportion of the revenue obtained from the energy market accounts for the capacity fee within 20%.

For the revenue obtained from the peak load shaving, according to the latest rules of Northeast Ancillary Power Service Market listed in table 4, the revenue scale depends on the depth of peak regulation and the utilization hours. According to the comparison between the utilization hours of pumping water in the previous year and the national average, the pumping power amount higher than the national average shall be compensated after the fact in the current year. The specific compensation principles are as follows: Pumping hours is higher than the national average utilization hours 0 to 100% range, this part of the electricity will be compensated by 0.01 ￥/kWh; 100 to 200% range, 0.02 ￥/kWh to compensate; 200% of the above range, 0.03 ￥/kWh to compensate.

In 2016, the average pumping utilization hours of PHESs belong to the State Grid Xinyuan Company Ltd. was 2000h, assuming that the annual utilization hours of the Pushihe PHES in Northeast China Power Grid are 2 times of the national average, namely 4,000 hours (for very extreme cases). According to the price compensation principle shown in table 4, the extra 2000 hours will be compensated by 0.01 ￥/kWh, so the proceeds will be 1200MW×2000h×0.01 ￥ /kWh=24 million Yuan. To summarize, if Pushihe PHES participate in the electricity market, the revenue scale is only 0.736×100+24=97.6 million Yuan. However, the capacity fee will reach to 600 million Yuan annually, the revenue scale obtained in the competition market only accounts for 16.2%. The results indicate that only participating in the market to obtain revenue cannot meet the demand for capacity fee recovery.

Table 1. Comparison of average peak-valley pool purchase price in major electricity markets.

Country	Average peak price	Averaege valley price	Peak-valley price ratio
Australia	41 AUD /MWh	21 AUD /MWh	1.95
USA PJM	40.97 $ /MWh	28.11$/MWh	1.46
Zhejiang province	0.5 ¥/kWh	0.25 ¥ /kWh	2

Table 2. Important arguments values for case study.

Parameters	Capacity (MW)	Design power generation hours (hr)	PHES efficiency (%)	Generating electricity price (¥/kWh)	Pumping electricity price (¥/kWh)
Value	1200	1500	75	0.3685[a]	0.2457[b]

Table 3. PHES annual revenue profile (unit: MW, 100 million yuan).

Attributes	Value
capacity (MW)	1200
Annual power generation revenue	6.633
Annual pumping cost	5.897[a]
Electricity energy market revenue	0.736

Table 4. The price compensation principe for PHES.

Range	Subsidy Price
0～100%	0.01¥/kWh
100～200%	0.02¥/kWh
Above 200%	0.03¥/kWh

3.2 Mature electricity market stage

With reference to foreign experience, at the mature stage, the PHES can participate in the electric energy market and the ancillary energy markets such as AGC and reserve and other ancillary services markets. In the electric energy market, PHES draws water when the spot price is low and generates electricity when the spot price is at its peak. In this way, it gains profit from the "low buy and high sell" of electric energy considering the pumping loss. In the ancillary service market, the PHES can participate in AGC and compete in the reserve market to provide responsive service and gain market benefits. In addition, services such as black-start, emergency backup that are unsuitable for the short-term market can be procured by the system's security management authority and reimbursed for expenses through cost-based rates.

Taking the Pushihe PHES in northeastern China as an example, we calculate the profitability of the storage power station through the electric energy market and various ancillary service markets. According to international experience, the price of ancillary services is very low, which is generally only 1/50 to 1/10 of the price of electricity. Therefore, the size of the entire ancillary service market is limited. In particular, at present, the National Energy Administration of China is promoting the reform of thermal power flexibility across the country and the flexibility of thermal power generation will be greatly increased in the future. Compared with conventional thermal power, PHES plants have no significant advantages in providing peak shaving, AGC, and routine back-up, and the benefits obtained are relatively low even if they participate in ancillary service market,. With reference to foreign experience, assuming that all types of ancillary services undertaken by the load bear the cost of 0.01/kWh Yuan and the scale of Liaoning's power generation is about 200 billion kWh in 2015, the total ancillary service cost will be 2 billion Yuan. The installed capacity of Pushihe PHES is 1200MW, and it accounts for nearly 3% of the total province installation capacity of 43,000 MW. Assuming the Pushihe PHES can receive ancillary service compensation proportional to its capacity share, it can only get 60 million Yuan. Considering the revenue from the electric energy market, the ratio between the revenue and the required electricity tariff is between 22% and 70%, and there is still a large funding gap with the current capacity electricity tariff. The results are illustrated in table 5, and two other scenarios are also listed in the table for different peak-valley price ratios in order to model various electricity market environmental situations. It can be found that the peak-valley price ration has an obvious impact on the revenue of PHES.

It is noted that Case 2 and Case 3 here are difficult to achieve in practice. On the one hand, it is assumed that the calculation result is that PHES pumping and generating time 100% are used for peak shaving and valley filling. In actual operation, this is impossible to achieve the ideal situation; on the other hand, in order to reduce market risk and ensure the enthusiasm for investment in a power station, the mainly power plant investors hope that most of its proceeds in the early stage are guaranteed.

a. Liaoning Province unified coal-fired unit benchmark electricity price (tariff included).
b. According to respectively 2/3 of the on-grid electricity price, that is, the on-grid electricity price is 1.5 times of the pumping electricity price
c. The calculation of the electricity price of the Internet is 1.5 times as much as the price of the pumped electricity.

* Corresponding author: zhangfuqiang@sgeri.sgcc.com.cn

Table 5. Total revenue estimation when phes participated in the energy and ancillary services market (Unit: 100 Million Yuan).

Cases	Peak-valley price ratio	Energy market revenue	Ancillary services market revenue	Total revenue	Ratio of total revenue/capacity fees
Case 1	1.5	0.736	0.6	1.336	22%
Case 2	2	2.211	0.6	2.811	47%
Case 3	3	3.685	0.6	4.285	70%

4 Recommendations for future pricing mechanism

According to the full-featured, unique advantages and universal service features of PHES, the following suggestions are made for the electricity pricing mechanism of PHES in combination with the current construction of power market in China.

(1) The pricing mechanism: According to the No. 1763 Document of the National Development and Reform Commission in 2014, the two-part tariff is still recommended to implement in current stage. Among them, the capacity-tariff mainly reflects the value of ancillary services such as capacity reserve, frequency regulation, phase modulation and black start provided by the PHES, which is set to make up the fixed costs and allowable profits of the PHES. The energy-tariff basically reflects the energy value provided by the PHES such as peak shaving, and its price is determined by the pumped storage power generation losses and other variable costs.

(2) Cost sharing mechanism: It is suggested the cost of PHES should be included in the allowable cost of power transmission and distribution, and be recovered together with the transmission and distribution tariff. Considering that the beneficiaries of PHES are not rather specific power plants, but the whole power systems, so it is not easy to distinguish accurately the beneficiaries. In order to ease the pressure of PHES cost recovery, the capacity-tariff and energy-tariff should be shared by all users in order to reflect the principle of equitable sharing. The cost of PHES can be taken along with the transmission and distribution tariff to be recovered. The allowable cost of the power grid is specified in the transmission and distribution tariff determination method. It includes the "ancillary service" fee (that is, the electricity capacity of the capacity) purchased by the grid enterprises from PHES. To ensure the normal operation, the PHES in operation and forthcoming operation should be accurately included in the transmission and distribution tariff approval cycle. In order to reflect the principle of "who benefits, who pay", we should clarify the scope and extent of the benefits of PHES so that we can extend the PHES tariffs to all beneficiary provinces in a region.

(3) Explore diverse funding sources. Three possible ways can be used to revenue the cost of PHES, and they are illustrated as follows. First, the PHES can stabilize the power output of renewable energy such as solar and wind power and consequently reduce the curtailment scale of renewable energy. The benefits produced by reducing the curtailment of solar and wind energy can cover part of the cost of PHES. For example, in the first half of 2017, the curtailment rate of wind power in China decreased by 7% and 4.5% respectively over the same period of previous year, and the corresponding economic benefits were about 15.4 billion Yuan. At present, ultra-high voltage transmission technique is the most effective way to solve the high-level energy curtailment. Besides, PHES plays an important role in the power supply side, grid side and load side to improve the efficiency of renewable energy utilization. Therefore, the benefits from wind power and photovoltaic power generation can pay part of the service fee to PHES. Second, it is suggested that economic benefits from increasing utilization hours of nuclear power plants or not participating in peak regulation should be deducted, and an appropriate proportion should be deducted from the nuclear power price to compensate the PHES operate costs. Third, explore the establishment of a nationwide dispatch service fund to solve the problem of compensating for operating expenses before the newly-started PHES are not diverted.

(4) Multi-channel electricity pricing mechanism adopting "fixed purchasing+market bidding" in the market environment should be recommended. On the one hand, considering that the PHES affects the emergency reserve function when participating in the market competition, the main function such as black start means that it cannot compete freely with other power sources in the market. On the other hand, some functions provided by PHES such as black start-up and reactive power support have no corresponding market, which means the market cannot fully reflect the benefit of the PHES in power systems.

In view of the special properties and attributes of PHES, it is suggested to adopt multiple channels of "fixed purchasing+market bidding" to solve the cost-revenue of PHES in electricity market environment. First, about 80% of the PHES costs are secured through long-term fixed contracts with the electricity trading agencies, and the electricity trading agencies will charge the fixed service ancillary services with the power to all users. It is a compensation for ancillary services that are not suitable for market-oriented such as emergency reserve, black-start and reactive power support for PHES. For this part, it is recommended these functions should adopt the mode of "fixed purchasing" to ensure a reasonable and reliable revenue.

Second, for the remaining about 20% of the cost of PHESs, they can participate in the energy market and ancillary service market in accordance with market rules, obtain revenue through market competition, and compensate for the shortage of fixed contracts.

5 Conclusion

This paper first stated the challenges the PHES faced in the progress of transmission and distribution tariff

 https://doi.org/10.1051/e3sconf/20183804016

reform. It is necessary to propose a reasonable pricing mechanism for the healthy development of PHES, we review and summarize international PHES development experience and conduct a comparative analysis of revenue mechanism in different markets. The main challenges and barriers to PHES development in China are also analyzed. In order to measure the impact of electricity tariff reform on the PHES, the revenue scale is estimated respectively in China's different market-oriented reform phase, namely the transitional phase and the mature power market stage. It indicates that the revenue gained by participating in the electricity market competition is not enough for PHES, and the majority of the cost shoul be covered by capacity-tariff. Based on the research, four recommendations are proposed for the policy makers. The two-part tariff scheme is proposed in China considering the special characteristics of PHES, of which the capacity-tariff is used to cover the about 80% of the cost and the energy-tariff is designed to compensate the variable cost of PHES.

Acknowledgements

This work is supported by Science and Technology Project of State Grid Corporation of China (B3670116A005): Study on the evaluation model and application of peak shaving technology, economy and policy for power system wiht large-scale renewable energy integration.

References

1. Development and reform commission of China (NDRC), Notice on improving price formation mechanism of pumped hydro energy storage (No.1763) http://jgs.ndrc.gov.cn/zcfg/201408/t20140812_622012.html (2014)
2. Development and reform commission of China (NDRC), measures for electricity tariff determination for transmission and distribution of provincial power grid (Trial) http://www.ndrc.gov.cn/gzdt/201701/t20170104_834330.html (2014)
3. C. Chusheng, Hydropower Pumped Storage, **3**, 3 (2017)
4. L. Zhiwei, L. Qinghua, H. Shuangcheng, Water Power, **43**, 5 (2017)
5. C. Shangfa, Electr. Power Syst. Equip., **4** (2008)
6. S. Zhongming, J. Wuhan Polytech. Coll., **4** (2014)
7. Z. Quanren, Hunan Electr. Power, **1** (1998)
8. L. Shasha, L. Yun, L. Guozhong, Sino-Global Energy, **18** (2013)
9. E. Barbour, I. A.G. Wilson, J. Radcliffe, Renew. Sustain. Energy Rev., **61** (2016)
10. China Smart Grid : Pumping stations: the reasonable price mechanism has not been how to make the best use, http://www.chinasmartgrid.com.cn/news/20150807/607957-4.shtml. (2014)
11. M. Hamada, V. D. H. John, *Social Science Electronic Publishing* (2011)
12. M. Analytics, *State of the Market Report for PJM* (2016)
13. Z. Fuqiang, Z. Hao, Autom. Electr. Power Syst. **30**, 4 (2006)

Modeling and Simulation of Hydropower Station Diversion System's characteristic line method by introducing water head to flow feedback

Xu Guangwen, *Li* Xi*, and *Yao* Ze

Electric Power Research Institute of Guangdong Power Grid Co., Ltd., Guangzhou 510080, Guangdong, China

Abstract. To solve the damping problem of water hammer wave in the modeling method of water diversion system of hydropower station, this paper introduces the feedback regulation technology from head to flow, that is: A fixed water head is taken out for flow feedback, and the following conclusions are obtained through modeling and simulation. Adjusting the feedback coefficient F of the water head to the flow rate can change the damping characteristic of the system, which can simulate the attenuation process of the water shock wave in the true water diversion pipeline. Even if a small feedback coefficient is introduced, the damping effect of the system is very obvious, but it has little effect on the amplitude of the first water shock wave after the transition process. Therefore, it is feasible and reasonable to introduce water head to flow rate feedback coefficient F in hydraulic turbine diversion system.

1 Introduction

In order to make rational use of water energy and take account of the feasibility and economy of hydropower development, the layout and structure of diversion system of hydropower stations are usually varied. Generally speaking, it is mainly composed of upstream and downstream reservoirs, pressure conduit, turbine runner and draft tube [1]. A simple single pipe and single machine diversion system is simply shown in Figure 1.

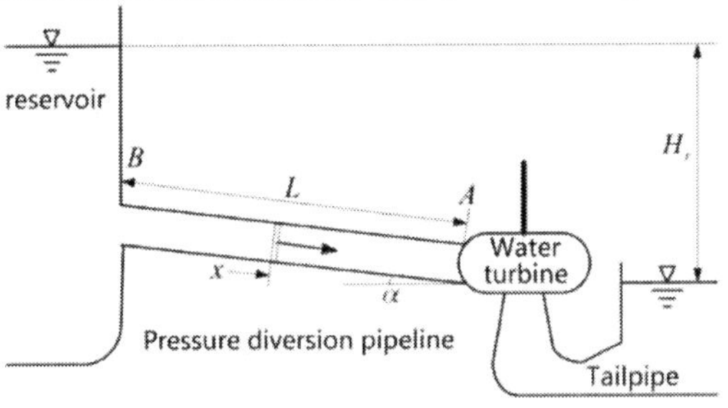

Fig. 1. Water diversion system of a typical hydropower station.

For the modeling or calculation method of the water diversion system, the linear transfer function description method [2] can be used when the length of the pipe is less than 600~800 meters. If the diversion pipe is longer, the effect of the elastic water hammer should be considered [3-5], and the characteristic line method is usually used. It is pointed out that when the flow of water in the diversion pipeline is rapidly cut off, the pressure wave in the pipeline is very slow, which is not consistent with the actual situation. In order to overcome these defects, in the nonlinear simulation calculation of the hydraulic turbine governing system, this paper improves the characteristic line method, and introduces the water head to the flow feedback to the modeling of the water diversion system of the hydropower station, and the simulation is carried out. The results are reasonable and credible.

2 Basic characteristic line method and calculation example

2.1 An example of the basic characteristic line method

The basic characteristics of line method for modeling as shown in Figure 2, figure symbol "| - |" absolute value calculation

* Corresponding author: lilixi5217@163.com

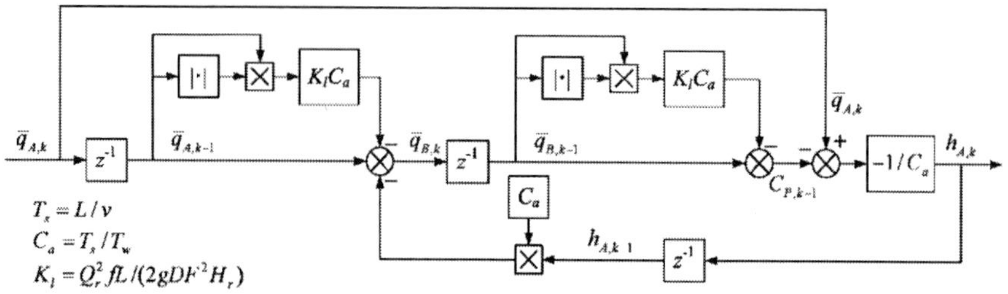

Fig. 2. The basic characteristic line block diagram for single section calculation of simple water diversion pipeline.

2.2 Calculation example

The parameters of a hydraulic turbine guide system are as follows: the length of the equivalent water diversion pipe: L=200 meters; water hammer wave speed: V=1000 M / sec; calculation step length: T_s=L/v=0.2; the speed constant of water flow: T_w=2 seconds; the water head loss coefficient is K_l=0.02 and K_l=0.1, respectively, assuming that the flow of the turbine is proportional to the opening of the relay, and is linear from full open to full or full off in 5 seconds. The initial conditions of all the variables were 0. The simulation results are shown in Figure 3 and Figure 4.

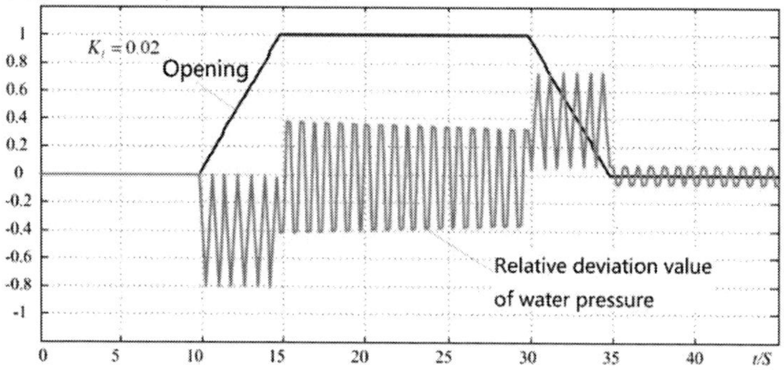

Fig. 3. Water pressure deviation response process of basic characteristic line method at K_l=0.02.

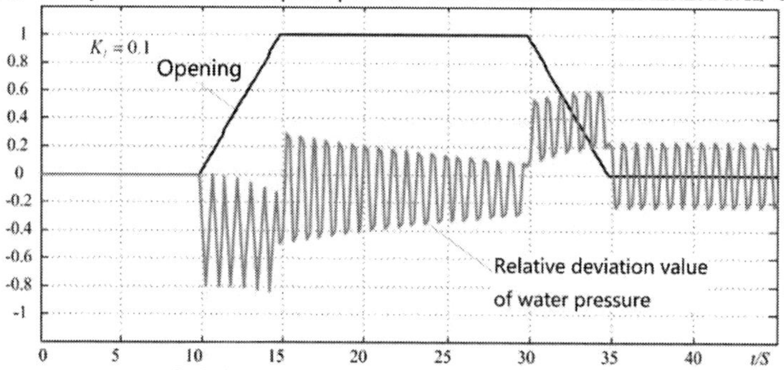

Fig. 4. Water pressure deviation response process of basic characteristic line method at K_l =0.1.

As can be seen from Figure 3 and Figure 4, the basic characteristic line method discussed can show the properties of elastic water hammer, but the elastic water strike wave attenuates slowly and the damping is smaller, especially after the water flow is cut off. This situation is usually not in conformity with the reality. It is generally considered that the attenuation of elastic water hammer wave is related to the head loss coefficient Kl, and the larger the head loss coefficient is, the faster the attenuation is. However, the two diagram shows that the underdamped condition has a certain relationship with the water head loss coefficient, but the correlation is not significant. Even if the K_l is increased from 0.02 to 0.1, the damping condition does not improve when the flow is cut off. This further verifies the damping problem of the head wave in the basic characteristic line method proposed at the beginning of this section, especially after the sudden flow cut-off.

3 Improved characteristic line method and its calculation example

3.1 A linear discrete model with water head to flow feedback

Since the damping of the head wave has little relation with the loss coefficient of the head, the water head loss item in the characteristic line algorithm can be removed and the linear model based on the characteristic line method is established. Moreover, based on the linear model, it is easy to control damping and transform the sampling period. On the basic characteristic line block diagram of single section calculation of water diversion pipeline, after removing the head loss item and introducing the feedback from water head to flow, the algorithm block diagram is changed to figure 5.

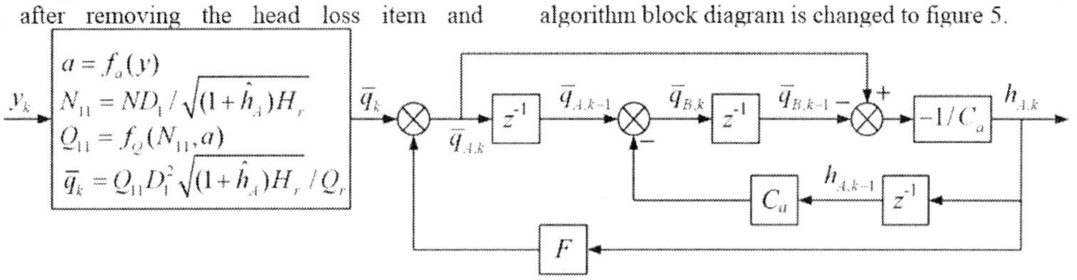

Fig. 5. Water pressure deviation response process of basic characteristic line method at K_I=0.1.

In Figure 5, F is the feedback coefficient of additional water head to flow, $\bar{q}_k$ is the relative flow volume of turbine and $\hat{h}_A$ is the estimated value of water head increment during iteration. Here we only discuss the linear model from the $\bar{q}_k$ to $h_{A,k}$ .First, we write the discrete form of the system (Z transfer function) equation according to the 5 column of block diagram.

$$Q_B(z) = Q_A(z)z^{-1} - C_a H_A(z)z^{-1} \tag{1}$$

$$H_A(z) = -\frac{1}{C_a}(Q_A(z) - Q_B(z)z^{-1}) \tag{2}$$

$$Q_A(z) = FH_A(z) + Q(z) \tag{3}$$

For the sake of derivation, the Z transformation of all variables is expressed by capitalized letters of variables. Take (3) into (2) and (1).

$$Q_B = (FH_A + Q)z^{-1} - C_a H_A z^{-1} \tag{4}$$

$$H_A = -\frac{1}{C_a}(FH_A + Q - Q_B z^{-1}) \tag{5}$$

Then bring (5) into (4)

$$H_A = \frac{1}{C_a}(-FH_A - Q + FH_A z^{-2} + Qz^{-2} - C_a H_A z^{-2})$$

After arranging

$$(F + C_a - (F - C_a)Z^{-2})H_A = Q(-1 + Z^{-2})$$

The corresponding Z transfer function is

$$G(z) = \frac{H_A(z)}{Q(z)} = \frac{-1 + z^{-2}}{(F + C_a - (F - C_a)z^{-2})} \tag{6}$$

If F = 0, it is equivalent to the case of no additional water head to the feedback coefficient of flow, which is equivalent to the basic characteristic algorithm model without considering head loss.

$$G(z) = \frac{H_A(z)}{Q(z)} = \frac{H_A(z)}{Q_A(z)} = \frac{1}{C_a}\frac{-1 + z^{-2}}{1 + z^{-2}} \tag{7}$$

3.2 Calculation example

In order to verify the feasibility of the proposed method, the improved characteristic line method is simulated by combining 2.2 sections of turbine intake system parameters.

The feedback coefficients of the head to flow are taken as F = 0, F = 0.001, F = 0.01 and F = 0.1, and assume that the flow of the turbine is proportional to the opening of the relay and linearly from full open to full or full off in 5 seconds. The initial conditions of all the variables were 0. The simulation results are shown in figure 6~9.

When the feedback coefficient is F = 0, the system is unstable and its time domain response is an obvious equal amplitude oscillation. With the increase of the feedback coefficient, the damping of the response output increases and the attenuation accelerates. When the feedback coefficient is F = 0.1, the system has the characteristic of over damping and has no oscillation characteristics; the adjustment of the feedback coefficient F can change the damping characteristic of the system and thus can change the damping characteristic of the system. The attenuation process of water hammer wave in real water diversion pipeline is simulated. Even if a small feedback coefficient is introduced, it is very obvious to increase the damping of the system, but it has little effect on the amplitude of the first water shock wave after the transition process. Therefore, it is feasible and reasonable to introduce water head to flow rate feedback coefficient F in hydraulic turbine diversion system.

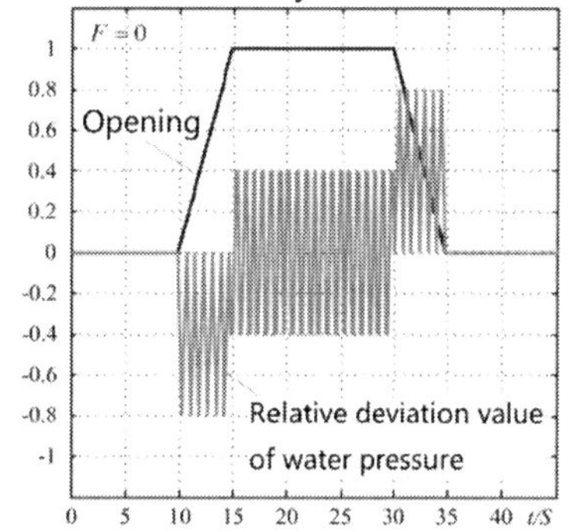

Fig. 6. Response of the system to feedback, F=0.

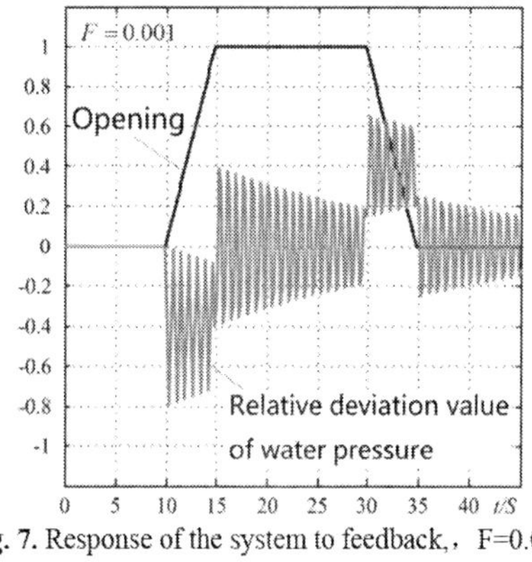

Fig. 7. Response of the system to feedback, F=0.001.

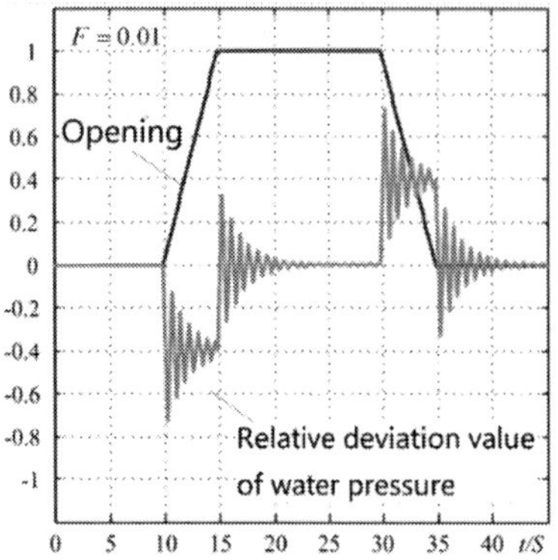

Fig. 8. Response of the system to feedback，F=0.01.

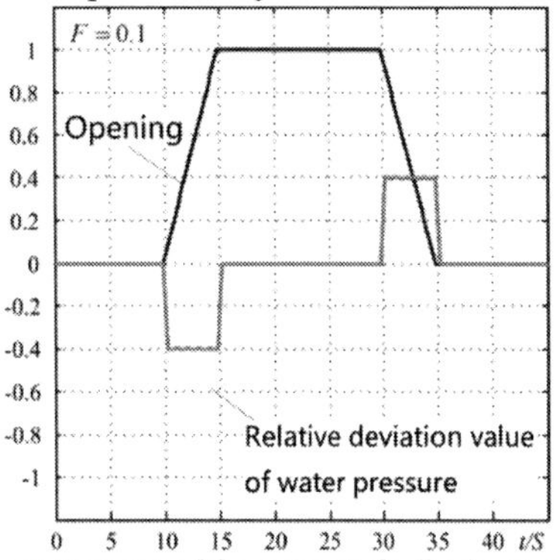

Fig. 9. Response of the system to feedback，F=0.1.

4 Conclusion

The modeling method of water diversion system of hydropower station, that is, the characteristic line method, even to the solution of a pair of nonlinear hyperbolic partial differential equations, is to solve the damping problem of water shock wave. In this paper, the feedback regulation technology of water head to flow is introduced in this paper, and a fixed head of water head is taken out to feed back the flow. The results show that the reasonable feedback coefficient is helpful to the damping of water shock wave. The modified model is reasonable. The adjustment of the feedback coefficient F can change the damping characteristic of the system, which can simulate the attenuation process of water shock wave in the true water diversion pipeline. Even if a small feedback coefficient is introduced, the damping effect of the system is very obvious, but the amplitude of the first water shock wave after the transition process has little effect. Therefore, it is feasible and reasonable to introduce water head to flow rate feedback coefficient F in hydraulic turbine diversion system.

Acknowledgments

The research work was supported by Science and technology project of Guangdong Power Grid Co., Ltd. (GDKJQQ20152035).

References

1. Liu H, Lai C, Yuan Q, et al. Study on Optimization Plan of Water Diversion System with High-head and Small Flow Hydropower Station[J]. South-to-North Water Transfers and Water Science & Technology. 2010, 8(1672-1683(2010)8:5<149:GSTXLL>2.0.TX;2-S5): 149-152.
2. Cai D, Zhao J. ANALYSIS OF AXISYMMETRIC FORWARD/BACKWARD EXTRUSION BY THE CHARACTERISTIC LINE METHOD.[J]. Ku Ti Li Hsueh Hsueh Pao/Acta Mechanica Solida Sinica. 1987(2): 116-126.
3. Chang Y, Xi Y, Fan Y. Analysis on transient flow of unpressurized water diversion system for Manweilai Hydropower Station in Cameroun[J]. Water Resources and Hydropower Engineering. 2017, 48(1000-0860(2017)48:10<31:KMLMWL>2.0.TX;2-O10): 31-36.
4. Guo W, Yang J, Shan X. Pipeline system's hydraulic design of multi-water-source diversion-type hydropower station[J]. Engineering Journal of Wuhan University. Engineering Edition. 2013, 46(1671-8844(2013)46:1<73:DSYYSS>2.0.TX;2-#1): 73-77, 83.
5. Abi E, Zheng Y, Feng X, et al. Modification and development of characteristic line method of tunnel[J]. Yanshilixue Yu Gongcheng Xuebao/Chinese Journal of Rock Mechanics and Engineering. 2015, 34: 3067-3073.

Research on the Direction of China's Energy Development and Coping Strategies Based on the Trend of World Energy Development

Lixin Zhang, *Ye* Tao, *Yan* Jiang, and *Ju* Ma

School of Management, Qingdao University Of Technology, Qingdao, Shandong266520, China

Abstract. To realize the modernization of the national economy, it is necessary to develop energy science and technology for China,which is third largest countries in the world.The rapid development of science and technology has promoted the continuous transformation of the global energy industry. By analyzing the trend of energy development in the world today, this paper discusses the challenges that the global energy development facing and the situation and tasks faced by China's energy sustainable development, and looks forward to China's strategies to cope with the development of the world's energy.

1 Introduction

Energy science and technology is an important development strategy that is closely watched by all countries in the world today. It covers a wide range of areas and has a large impact. The world is now undergoing a new round of information revolution. Whoever seizes the opportunity will be able to become a superpower. The global production organization model and lifestyle are undergoing fundamental changes.There is an urgent need to give a clear answer that in the face of the challenge of the information revolution, what kind of strategies should be adopted by energy science and technology to adapt and support the development of the energy industry.

At present, the development of the world's energy science and technology is becoming a trend, and the old energy sources are drying up. There is an urgent need for science and technology to inject new energy into new energy sources. However, the world's energy development trends are facing many problems and challenges. In response to this situation, how should China adapt itself to the new situation in combination with its own level of development? The development of new energy has been an important development strategy at present.

2 Major issues and challenges facing the current global energy development strategy

On March 31, 2017, BP China released the 2017 World Energy Outlook, which assesses the long-term development trends of energy sources in the world and regions, and forecasts the development of the world energy market in the next 20 years. The BP report pointed out that by 2035, oil, natural gas and coal will still be the dominant energy of the world economy, accounting for more than 75% of the total energy supply in 2035, and the growth rate of natural gas will exceed that of petroleum and coal by 2035. The share of primary energy will exceed coal and become the second largest source of fuel. [1]

The following characteristics can be drawn based on the analysis of BP energy outlook:

2.1Global total energy demand continues to grow slowly

BP's "World Energy Outlook (2017 Edition)" predicts that the global energy demand from 2015 to 2035 is expected to increase by 30%, with an average annual increase of 1.3%. This increase in energy demand is significantly lower than the expected growth rate of 3.4% per year in global GDP, reflecting the improvement in energy efficiency due to technological progress and environmental concerns. The growth of the world economy requires more energy, although the degree of its growth is tempered by the reduction of energy intensity (energy per unit of GDP): the global GDP has doubled and the energy demand has only increased by 30%.

Due to the rebalancing of the Chinese economy and the global emphasis on improving energy efficiency, the global energy intensity decline rate is expected to accelerate. China will be the largest energy growth market, although it may be overtaken by India at the end of the horizon.

2.2The development of energy supply towards multi-polarization

* Corresponding author email: 2498566339@qq.com

https://doi.org/10.1051/e3sconf/20183804018

According to BP statistics, the growth rate of primary energy consumption in the world in 2012, 2013, 2014 and 2015 was 1.4%, 2.0%, 1.1% and 1.0% respectively, which was significantly lower than the average annual growth rate of 2.7% in the previous 10 years. [2] The shale oil and gas revolution has led to major changes in the structure of the world's energy region. For example, the phenomenon of "production shifting westward and consumption moving eastward" appears in the oil field. Non-OPEC countries such as Russia and the United States have gradually exceeded their OPEC oil production. [3] With the rise of unconventional oil and gas such as shale oil and shale gas, the energy supply is increasingly polarized and diversified. Driven by the growth of LNG in the United States, shale gas production accounts for two-thirds of the increase in natural gas supply. Due to increased supply in Australia and the United States, LNG growth is expected to lead to the formation of a globally integrated natural gas market based on US gas prices.

As a cleaner source of energy, renewable energy plays an increasingly important role in the diversified development of energy supply. [4]Driven by the increasing competitiveness of solar and wind power, the growth of renewable energy will quadruple. In the next 20 years, China will be the largest source of renewable energy growth, and its incremental renewable energy will exceed the sum of the EU and the United States. In summary, the development of energy supply in the direction of diversification is a general trend.

2.3Certain measures still need to be taken to reduce carbon emissions

The goal set by the Paris Climate Change Conference is to keep the global average temperature within 2°C above the pre-industrial level and work hard to keep the temperature within 1.5°C. To achieve the temperature rise control target of 1.5°C, in 2040, zero emissions will be achieved in 2040. However, the effect of greenhouse gas emission reduction comes from cooperation and common emission reduction. [5]

BP estimates that carbon emissions are expected to increase at a rate of less than one-third of the past two decades, with an average annual growth rate of 0.6%, and an average annual growth rate of 2.1% over the past two decades, reflecting energy efficiency and fuel structure are constantly changing. If this goal is achieved, it will be the lowest emission growth rate in any one of the two decades since it was recorded in 1965. However, it is expected that during this period, carbon emissions from energy use will still increase by about 13% under the basic scenario. This far exceeds the forecast of the International Energy Agency (IEA) 450 scenario - that by 2035 carbon emissions need to drop by about 30% before they have the opportunity to achieve the goals of the Paris Climate Agreement. BP's forecast of global carbon emissions growth is shown in Figure 1.

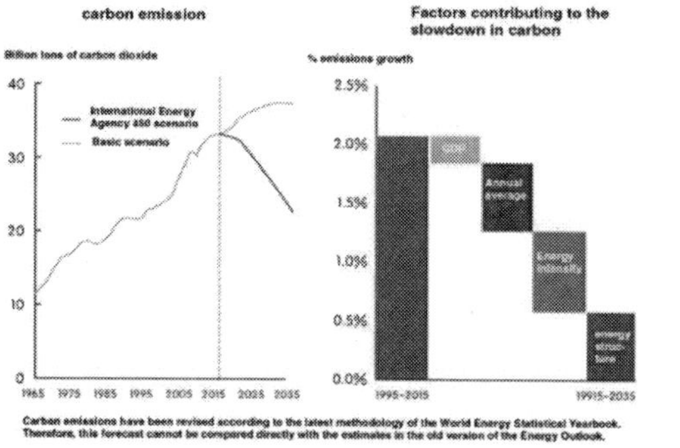

Fig. 1 Forecast of global carbon emissions growth over the next 20 years Source: BP p.l.c.2017

Therefore, global energy-related carbon emissions will peak around 2030.

2.4The future energy consumption structure is still dominated by fossil energy

BP predicts that non fossil fuels will account for half of the growth of energy supply in the next twenty years, but oil, gas and coal will remain the main source of energy for the world economy, accounting for more than 75% of the total energy supply in 2035, compared with 85% in 2015.BP's forecast of global primary energy consumption and share is shown in Figure 2.

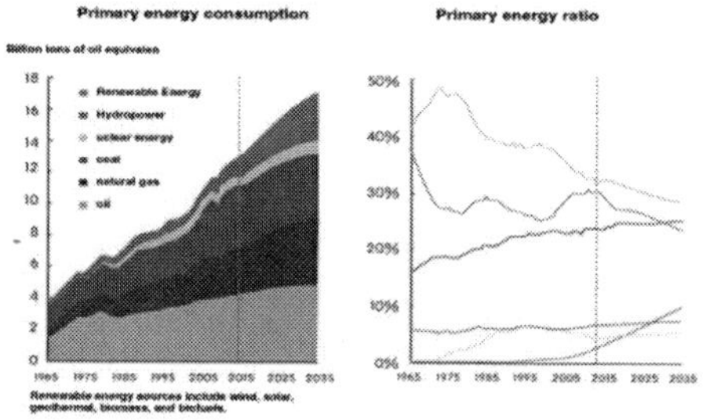

Fig. 2 Forecast of world primary energy consumption and proportion in 2035 Source: BP p.l.c.2017

Although oil demand is expected to slow gradually during the horizon, it will still grow at an average annual rate of 0.7%. The transportation sector continues to consume most of the world's oil, and its share of global demand is still close to 60% in 2035. However, by the early 1930s, the non-fuel use of petroleum, especially for the production of petrochemicals, would be the main source of oil demand growth. As China turns to cleaner, lower-carbon fuels, coal consumption is expected to peak by the mid-2020s. India is the largest growth market for coal demand. Its share of world coal demand has doubled from about 10% in 2015 to 20% in 2035. [2]

Although there is a certain difference in forecast, as long as countries take certain measures to solve the problem of climate change, the trend of energy structure changes in the future is about the same, and fossil fuels will still dominate the energy composition.

2.5 Natural gas is increasingly emerging

Thanks to energy policies that encourage industry and

power generation to transform, coal's share continues to shrink and shift to natural gas. The main growth comes from China, the Middle East and the United States.

According to the BP report, the net increase in global LNG exports (2.1 billion cubic meters) is from Australia (19 billion cubic meters). The LNG exports of the United States increased from 700 million cubic meters in 2015 to 4.4 billion cubic meters in 2016. [6]

Finally, BP predicts that by 2035 China will account for 26% of the world's total energy consumption and 35% of global net growth. With the continuous evolution of China's energy structure, China's energy intensity will decline by 45% in the horizon, with an annual decline of 3%. BP's forecast of global gas supply growth and some of its consumption are shown in Figure 3.

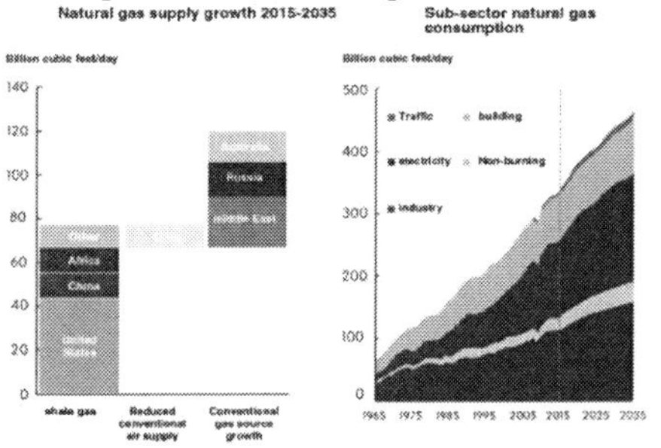

Fig. 3 Forecast of global gas supply growth and partial consumption 2015-2035 Source: BP p.l.c.2017

2.6Transfer of energy trade centers to non-OECD countries

In terms of energy trade, BP issued its own prediction that the global energy demand of non-OECD countries will increase significantly by 2040.

According to BP's forecast, by 2035 Asia will account for nearly 80% of the net imports between regions, and more than 40% of primary energy demand will depend on imports, which will basically contribute to all new energy trade. According to the prediction of China Energy Research Association, China's foreign oil dependence in 2030 will reach around 68%. [7]

3 China's main countermeasures in the face of global energy development strategy

(1) Energetically develop energy science and technology, and follow the energy development path in line with China's national conditions. Currently, world-wide energy is dominated by high-quality oil and gas, and China is still dominated by coal. From the perspective of the energy structure of all countries in the world, all industrialized countries use oil-vapor fuels as their mainstay. This is a choice to increase energy efficiency, reduce energy system costs, reduce environmental pollution, and provide quality services. It is also a basic trend of energy development in the world today. However,resources are rich in coal and poor in oil in China, and economically underdeveloped,so China is continue to be one of the countries in the world with a coal-based energy structure for quite some time. Therefore, there is an urgent need for energy science and technology innovation.

(2) Build a global energy internet. The global energy internet is a global energy allocation platform that uses UHV grids as backbone grids and smart grids to deliver clean energy. [8] The core idea of the global energy Internet construction is clean development, the use of clean energy such as wind, light, and water to replace fossil energy for power generation, and UHV transmission technology is used to connect these clean electricity through various regions, countries, and continents. The power grid is delivered to millions of households in a safer, more reliable and more economical way. The two alternative is to transform energy structure from fossil energy to clean energy and to implement electric energy substitution on energy consumption.

(3) Encourage environmental protection. Globally, fossil energy sources emit large amounts of pollutants every year, which not only cause serious pollution and damage to the atmosphere, water quality, soil, etc., but also cause great changes to the climate. The replacement of fossil energy with clean energy will effectively reduce the discharge of various pollutants. Promote ecological restoration and environmental protection.

(4) Promote economic recovery. Globalization is not only conducive to the upgrading of energy structure, the promotion of industrial restructuring, the development of upstream and downstream industries, and the promotion of the development of green industries; it can drive infrastructure construction and greatly increase employment; it can also promote the optimal allocation of energy elements for development. Provide new power to provide a new engine for economic growth; it can also promote the development and utilization of renewable energy, turn resource advantages into economic advantages, and promote the country's accelerated development.

(5) Making full use of the leading role of strategic planning. We should actively respond to the call of the times, further strengthen ideological awareness,and integrate the global energy concept of green, low-carbon, interconnected, and jointly built and shared energy into China's energy development strategy. Strengthen the top-level design, scientifically plan global energy development, guide and promote the development of various types of clean energy, and promote the close integration of power planning in various countries with global energy planning.

(6) Giving full play to the supporting role of innovation driven. We will focus on promoting breakthroughs in key technologies, equipment development, and standards construction, and comprehensively improve the safety, economics, and reliability of global energy development. At present, China has made breakthroughs in UHV, smart grid, clean energy, and operation control of large power grids.

4 Conclusion

China is a country of economy and population. To realize the modernization of the national economy, energy is the foundation. Only ensuring the sustainable development of the energy industry can promote the sustainable development of the Chinese economy. Strengthening and developing energy science and technology is the key to ensure China is transiting to a high-quality, high-efficiency, low-consumption country with environmentally-friendly energy system. China's energy science and technology is in its infancy and it has a long way to go. We should strongly encourage the development of new theories and technologies and make a contribution to the revitalization of China's energy industry.

References

1. BP. World Energy Outlook (2017 Edition) [R/0 L]
2. BP. World Energy Outlook (2016 Edition) [R/0 L]
3. Wang Lu. Analysis of Energy Development Trends at Home and Abroad [J]. Modern Economic Information, 2017(14): 297-300.
4. Kaveri K. Iychettira,Rudi A. Hakvoort,Pedro Linares,Rob de Jeu. Towards a comprehensive policy for electricity from renewable energy: Designing for social welfare[J]. Applied Energy,2017,187.
5. Shi Dan. China's lead in energy transformation, risk evolution and response ideas [J]. China Energy, 2017, 39(11):19-23+13.
6. Jiang Tianhai. Global Energy Market: Slowing Consumption Accelerating Transformation [J]. Science News, 2017(06):20-21.
7. Qiu Lijing. Prospects for global energy development trends from 2030 to 2040[J]. New Energy Economics and Trade Review, 2017(05):36-41.
8. Cheng Zhiqiang. China's Wisdom to Promote World Energy Transformation and Development [J]. Wisdom China, 2017(07):10-12.

Construction of Multimodal Transport Information Platform

Ya Wang[1], *Yu Cheng*[1], *and Zhi Zhao*[1,*]

[1]Shandong Jiaotong University, Ji'nan, Shandong 250300, China

Abstract. With the rapid development of economy, the volume of transportation in China is increasing, the opening process of the market is accelerating, the scale of enterprises is expanding, the service quality is being improved, and the container multimodal transport is developing continuously.The hardware infrastructure of container multimodal transport is improved obviously, but the network platform construction of multimodal transport is still insufficient.Taking Shandong region of China as an example, the present situation of container multimodal transport in Shandong area can no longer meet the requirement of rapid development of container, and the construction of network platform needs to be solved urgently. Therefore, this paper will briefly describe the conception of construction of multimodal transport network platform in Shandong area.In order to achieve the rapid development of multimodal transport.

1 Introduction

The construction of multimodal transport information platform can effectively avoid the traditional mode of strip management.The innovation of the operation mode of container multimodal transport management can completely change the split state of container multimodal transport.Based on the complete public information service system, a safe and reliable multimodal transport information service network platform is constructed to improve the barriers of information transmission among various departments.In this way, all kinds of transportation modes and transportation resources can be effectively coordinated.

2 Analysis on the Necessity of the Development of Multimodal Transport Information Platform in Shandong Province

2.1 Status quo and Policy Analysis of Multi-modal Transport Information platform Construction in Shandong Province

In developed countries such as Europe and America, multimodal transport, especially container seaborne transport, has reached a high level. Developed countries such as the Netherlands, Germany and the United States attach particular importance to the application of information technology in container transportation, especially EDI technology.These developed countries have already established the EDI information network system, realized the high efficiency transmission and the sharing multimodal transport information, has ensured each kind of transportation mode close connection, has brought the advanced container transportation.

In China, various departments attach great importance to the informatization construction in their respective fields in multimodal transport, and also have a relatively high degree of informatization. However, because the information coordination platforms among the systems are not unified, each department system operates independently.There is a big gap between China and the developed countries.In the long run, ports, railway departments, shippers and agents have great constraints on information sharing and communication, which not only seriously affect the advantages of multimodal transport, but also do not meet the development momentum and needs of modern logistics.

In 2016, Shandong Provincial Transportation Department held a forum in Qingdao to promote the development of multimodal transport. The competent transportation departments and key freight, logistics, water and port enterprises in 14 cities of Shandong Province participated in the meeting.At the meeting, Inspector Gao Hongtao stressed that the development of multimodal transport and the construction of a new system of transportation and logistics integration and development are the "breakthrough points" for improving the efficiency and efficiency of the logistics industry.Aiming at the imperfect construction of multimodal transport information platform at present, the paper puts forward some important tasks such as strengthening information sharing of multimodal transport, innovating application of technology and equipment, unifying standard and so on.In addition, in recent years, Shandong Province under the leadership of the Department of Communications, Shandong Province port and shipping management to promote water-water-water-water-water-water-water-water-water-water-water transport.At the provincial level, the related policies to

* Corresponding author: 1808708221@qq.com

https://doi.org/10.1051/e3sconf/20183804019

support multimodal transport should be put forward, and the information sharing platform of hot metal intermodal transport should be set up in a unified way to further improve the construction of comprehensive transportation system.It can be seen that the construction of multimodal transport information platform in Shandong Province receives great attention from the Transportation Department.

2.2 Meet the needs of China's national policy development

Informatization is the trend of the development of the contemporary world and the main force to promote the economic and social change.China has formulated the 《National Informatization Development Strategy from 2006 to 2020》, which focuses on transforming and improving the development of traditional industries by using information technology, and making full use of information technology to promote the transformation of industries with high energy consumption, high material consumption and high pollution.Promote supply chain management and customer relationship management.

2.3 Meet the needs of the current development of sea-rail intermodal transport

The advantages of sea-rail transport such as different customs declaration, high punctuality, less intermediate links, large transportation capacity, etc., it can effectively reduce the transfer time and cost in freight transportation.As a green and low-carbon transport mode, sea-rail transport is more and more popular.Since 2003, China's container throughput has ranked first in the world, accounting for more than 40% of the world's total.The port container transportation volume is huge, the highway carries about 84%, the water and water transport accounts for 14%, the sea rail transport is less than 1.5%, and in the major ports of Europe and the United States, the proportion of sea rail combined transport accounts for 20% or 30%.

2.4 Meet the need to reduce the cost of multimodal logistics

At present, more than 80% of freight transportation in China still depends on road transportation, but the network transport capacity between road, railway, water transport and aviation is very poor, which also leads to higher logistics costs.The proportion of logistics cost to China's GDP is about twice that of developed countries. The high logistics cost weakens the competitiveness of Chinese manufacturing industry and hinders the development of Chinese economy.The public information platform of container multimodal transport can link logistics enterprises, industrial manufacturing enterprises and other enterprises to form a seamless alliance to transform the situation of logistics enterprises with many, small and dispersed.By enhancing the matching degree of capacity resources and sources of goods, increasing the number of Intensification, scale and professional level of intermodal transport to improve the efficiency and efficiency of multimodal transport and reduce the total cost of social logistics.

3 Multimodal transport information platform architecture in Shandong

For the construction of multimodal transport information platform in Shandong area and the optimization of the whole transportation process, its core is the maximum integrated utilization of information.The following are mainly from the perspective of the transport chain, in order to obtain accurate information for the purpose of integrating the information of all participants included in the transport chain.

3.1 Center Platform Objectives and Functions

Taking the core node of the container multimodal transport center in Weifang Port as an example, a high efficiency and low cost operation platform is established on its transportation chain.

Objective: to build an information management system that connects all participants in the transport chain from Weifang to all major hinterlands. The information platform is composed of an independent information management system, and accompanied by a certain degree of comprehensiveness and complexity. In the process of practical operation, these information management not only includes the advantages of modularization and networking of common information systems, but also has the advantages of integration, dynamic and real-time. In the process of constructing multimodal transport information platform, using the current advanced computer technology and cloud computing, a cross-system, multi-source heterogeneous, real-time integrated multimodal transport information system is established. Information platform. In this way, the inter-departmental information transmission can be realized, the coordination of various transport departments and the automation level of document flow can be improved, and then the interactive operation can be realized, and the efficiency and quality of multimodal transport can be greatly improved. Therefore, the multimodal transport information platform should meet the following conditions in practical use: the information management system for multimodal transport in the port transport area should contain the basic information of all transport participants, the basic information of the port area, Receive information, etc. Install docking ports in all information management systems, so that each system can share and exchange information data, so that data information can play the most important role. Great use value. Cloud computing technology can be used to ensure the real-time sharing of information, to strengthen the basic functions of multimodal transport information management system, to realize the shipper or consignee of the entire transportation of goods query. In a word, the purpose of multimodal transport information platform is to enhance the quality of information service and efficiency of intermodal transport system by information integration and sharing.

(2) Functions: basic information sharing includes highway network information, inland river and channel information, port comprehensive information, railway network information, railway container transportation information, shipping company information, transportation chain related geographic information, traffic management information,Information on all participants involved in the transport chain.Transport information sharing includes information about means of transport (state of transport, location, etc.), information of goods (state of goods, location of goods in transit, etc.), transportation schedule (difference from actual to outgoing time).The information sharing of supply and demand includes the demand and supply information of various transportation vehicles, warehouse demand information and cost calculation.

3.2 Business logic structure construction

The logical structure is based on user service, which aims to provide information transmission between different functions to meet the needs of users.The logical structure of the central information platform is shown in figure 1.

3.3 Information platform physical structure

Transportation chain Management system (TCMS) is the main external information processing system of the information platform of the central station, which plays a key role in the fast and stable operation of the whole transportation chain.As the central database and central data processing system of the Ministry of Railways, the Transportation Information Management system is responsible for grasping the dynamic information of railway transportation in real time, reflecting the dynamic data report of the transportation in real time, and providing service for the relevant business departments and customers.Therefore, the establishment of multimodal transport information system in Shandong can try TCMS system. Taking Weifang Port Intermodal Transport Center as an example, the physical structure of the information platform can be set up as shown in the figure 2.

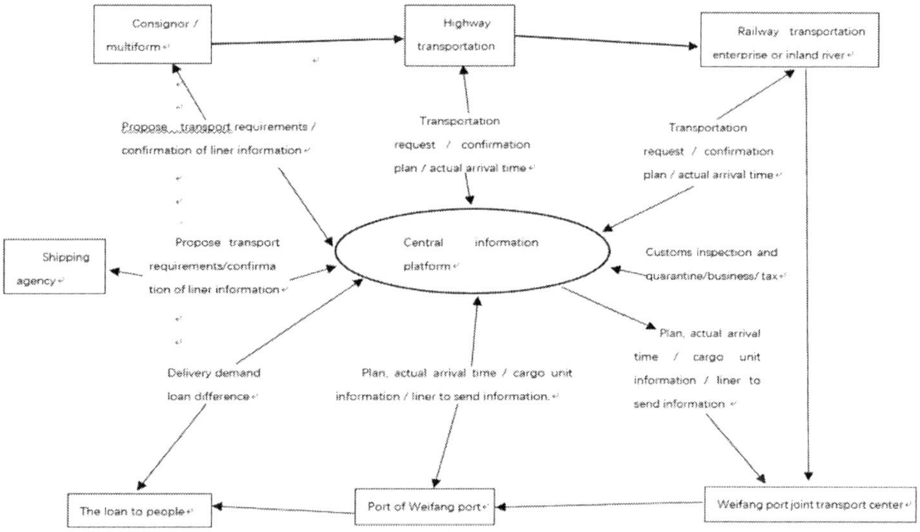

Fig. 1. Logical structure diagram of central information platform

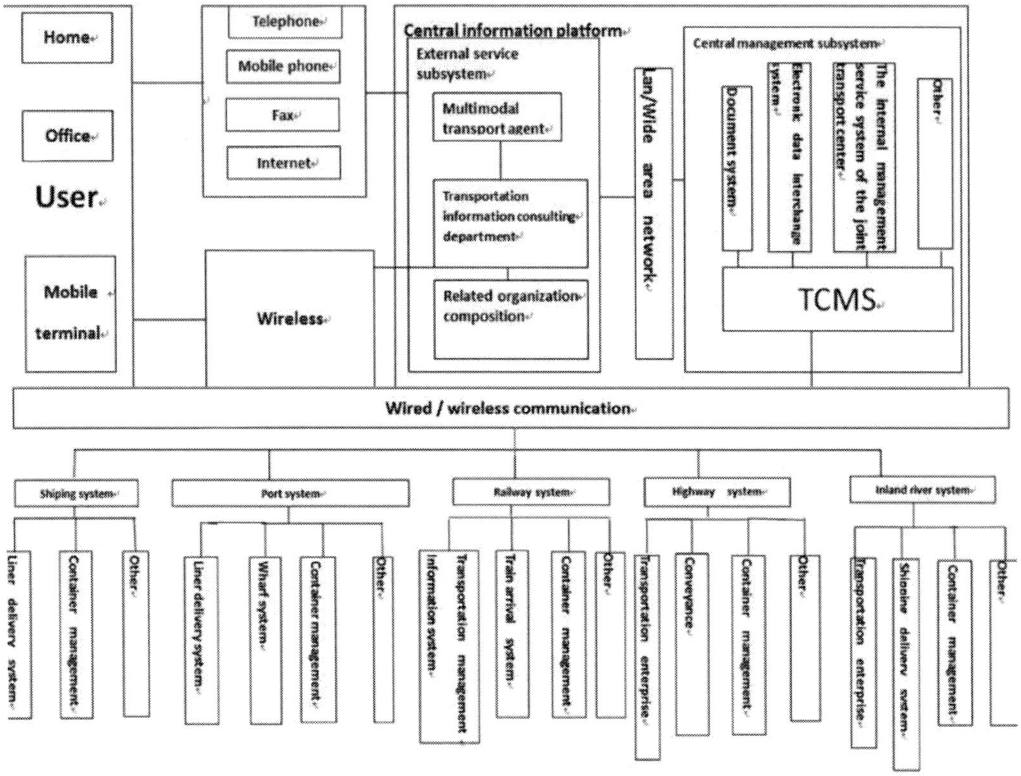

Fig. 2. Physical structure of information platform

4 Conclusion

China multimodal transport has entered a rapid development stage, but the level of information is not high, the departments are relatively independent.This paper takes the core node of the container multimodal transport center in Weifang Port, Shandong Province as an example to illustrate how to construct the business structure and overall framework of the information platform.Through the establishment of multimodal transport information platform, various modes of transportation are effectively connected based on information technology, so as to realize the continuous, safe and efficient multimodal transport of goods.

References

1. Jia Jingjing. On the construction of multimodal transport information platform in port[J]. Journal of Chongqing Jiaotong University: Social Sciences, 2010.10.
2. Fang Tiange, Zhang Xiuyang. On the construction of multimodal transport information platform[J]. Xi'an: School of Economics and Management, Chang'an University, 2015.
3. Zhang Heng, Zhang Jiguo. Strategy of Information Platform for Port Logistics Center [J]. Logistics Technology, 2003(8).
4. Jia Limin, Li Ping, Qin Yong. Research on the Framework of China's Railway Intelligent Transportation System[J]. Chinese Railways, 2003(11).
5. Zhang Hao, Liu Zhiqi. Research on the information platform framework of container multimodal transport center [J]. Shanghai: School of Transportation Engineering, Tongji University, 2016.2.

 https://doi.org/10.1051/e3sconf/20183804020

Study on the construction of Intelligent Courier Station Model

Ce zhao[1,*],Jia xin lu[1], Zhuang zhuang li[1] ,Zi rong shao[1],Kun yi pi[1]

[1]Shan dong Jiao tong University, Ji nan, Shandong 250300, China

Abstract. Campus Express is an important window to observe the city consumption logistics service "last kilometer".The research on Campus Express service is not only conducive to campus environment improvement and service quality promotion, but also provides all types of community, agglomeration areas such as urban terminal "last kilometer" logistics with reference.This article first proposed the main problems of campus express service,analyzed the mode of smart express station and finally built a smart express station.

1 Introduction

In recent years, electronic commerce in China is developing rapidly,and the number of express business is increasing by more than 40% each year[1], which also causes great environmental pressure. In 2016, electronic commerce logistics used approximately 20,700 million express waybills, consumed about 9.922 billion crates, consumed about 3.1 billion woven bags, about 8.268 billion plastic bags, about 3.105 billion envelopes, and about 16.988 billion meters of tape.Electronic commerce packaging garbage in large and medium-sized cities accounts for 85% of the total amount of clean-up domestic waste[2], which has become the focus of attention of the whole society.In such environment, campus express delivery is rapidly developing with its unique market demand and living environment.However, the"last kilometer" distribution of campus express delivery has become a common problem for many colleges[3].

2 The main problems of campus express service

2.1Taking long time to take a piece

At present, the rookie station uses self-raising express mode, customers need to find their own goods in the post, and then wait in line to scan code.Due to the unfamiliarity with the location of the goods and the relatively concentrated collection time of the students, the space was congested and chaotic, resulting in long time for picking up the parts and easy loss of goods.

2.2 Frequent of truck in and out

At present, the campus express is distributed from each express delivery company to the school's agents such as the Nojiri Station and the Neighboring Po.The large number of delivery vehicles and the inconsistent vehicle models have caused chaos in schools and increased the security risks and distribution costs on campus.

2.3 A backlog of goods

Due to the unreasonable layout of outlets, It is inconvenient for customers to pick up the parts, and it is easy to cause the backlog of goods.

2.4 Low express packaging recycling

Used courier boxes have not been properly recycled and are not easily reusable.

3 Analysis on the mode of smart express station

3.1 Project plan

* Corresponding author: 954190932@qq.com

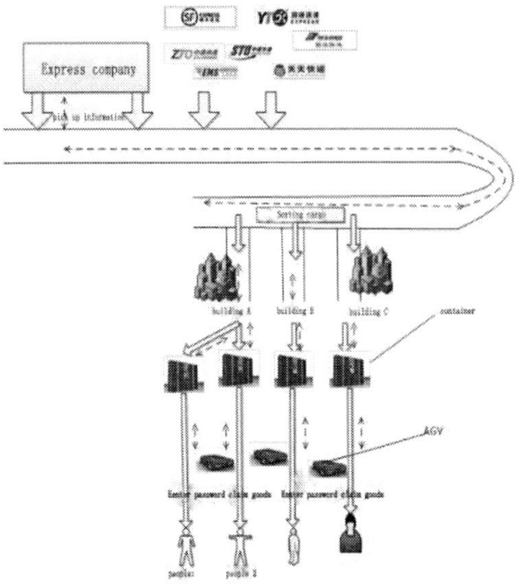

Fig. 1. Wisdom "e" Operation Mode

Integrate all the outlets in the school and establish a total distribution point in the school.In the smart express station, automatic sorting equipment is installed to automatically sort the goods.The goods will be sorted according to the building number information on the bar code to different shipping ports, and the loading personnel will carry out the loading according to the electronic tag-assisted picking system.The conveyor belts under the unit containers will dock the unit containers with AGV trolleys, and the AGV trolley will transport the unit containers to the designated location.Cell containers will be connected to the base.The AGV trolley transports empty cabinets or cabinets that have reached the set time to the recycling destination.The staff will inspect the lockers and classify and check the sent goods and the recovered cartons.The empty cabinet is circulated through the conveyor below to the sorting area.

Automating the distribution process and inspection process can improve efficiency and reduce manual labor. Staff concentrates on a single point to increase labor efficiency.Through the "shipping cabinet" mailing function, it is convenient for customers to send goods, through the recycling function, to increase the secondary use of recyclables.

3.2 The Location of the Post Station and the Location of the Pick-up Outlets

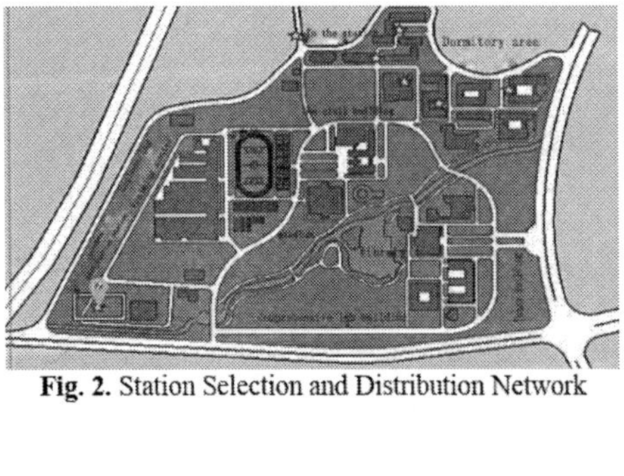

Fig. 2. Station Selection and Distribution Network

Note (the yellow box in the lower left corner is the new station set up, and the yellow part in the upper right corner is the location of the new self-introduction point)

Facilities and equipment layout have a close relationship with the optimization of express delivery services.The layout of facilities and equipment affects the optimization of express services.A good spatial layout is more conducive to the implementation of express delivery service optimization.The rational layout we have designed aims to achieve the following goals:The most efficient use of space; the most effective reduction in pick-up distances; the most efficient use of existing personnel.To achieve low-cost, high efficiency, high-quality express delivery.Through the investigation of the actual express transport volume on campus and the distribution of the actual number of people, we have selected the appropriate station locations and self-evaluation sites in accordance with the relevant regulations of the school.The smart express station is located at the south gate of the campus, away from the densely populated area of the campus to reduce potential safety hazards.Large area available in the area and can be placed into automatic sorting equipment.Close to the school gate, it is convenient for transportation of goods

3.3 Route Selection

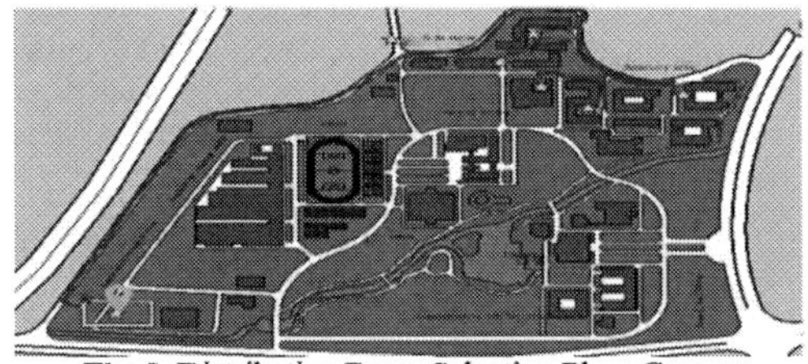
Fig. 3. Distribution Route Selection Place Campus

Distribution route selection as shown in figure 3, figure in the red line is the AGV conductor. Route selection is edge route for the school, the number of students is less, it can not only ensure the safety of the goods,but also can reduce the traffic impact.

4 The Construction of Intelligent Post Station

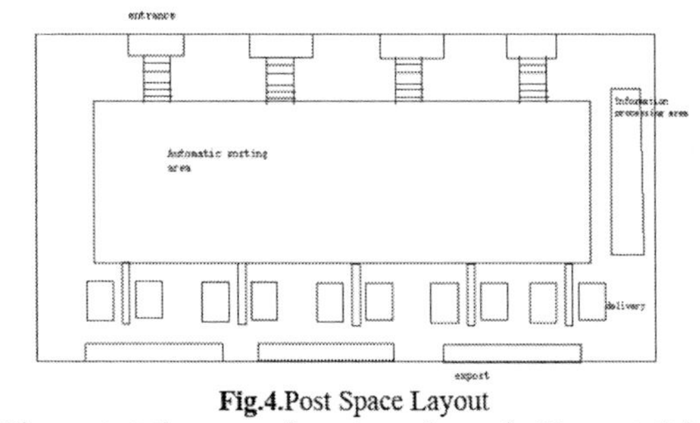

Fig.4. Post Space Layout

The post station space layout, as shown in Figure 4, it is based on the "smart campus green logistics platform" to build the campus smart station. The smart station is divided into four modules, it is include the collecting and sorting module, the delivery and recovery module, the distribution module and the information module. Using the automatic sorting system and its management platform, electronic label technology and RFID technology, it can improve the efficiency of the campus distribution service, promote the green recycling mode of innovative express packaging, develop the intelligent green logistics and create a new green new campus logistics model.

4.1 Sorting Module

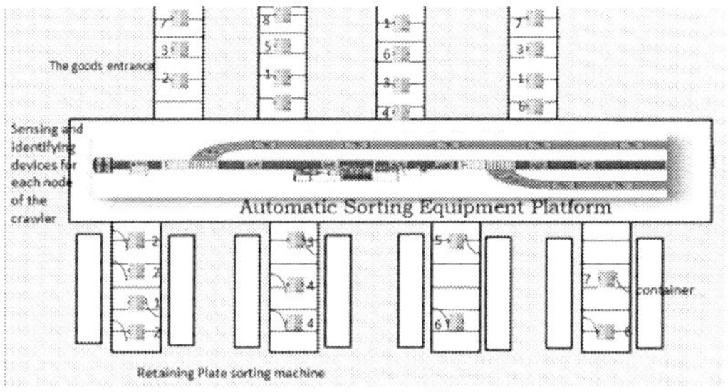

Fig. 5. the Sorting Facility Layout

The layout of the sorting facilities is shown in Figure 5. The goods are transported from the logistics company to the school, and the goods enter the sorting system through the track. After sorting, the goods enter the mobile unit express cabinet, and automatically connect with the AGV， it can achieve the aim that the power of unit, standardization, efficient and automated for service process.[5]. Using bar code identification technology to record and integrate information data and useing professional automatic sorting equipment to carry out automatic sorting processing for the aim we must to achieve according to user needs.Using automation facilities and resources advantages, the existing artificial processing mode is automated, efficient, and reduced manpower cost. The sorting process is shown in figure 6:

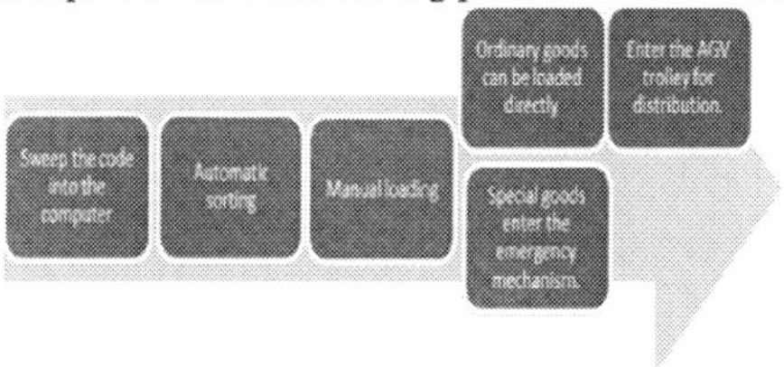

Fig. 6. Sorting Process

Goods will enter into the sorting system after ensure the information of the goods, and determine the size of goods according to the set of system data, then the belt sorting machine will classify the goods according to the number of floors, the goods will enter into the corresponding area according to corresponding number of address.For ordinary goods, they will be processed directly, and for special goods, such as fragile goods and so on will be handled through the emergency mechanism.

4.2 Information Module

Information management platform is the important guarantee of wisdom campus express operation mode, it will combine the relevant express information such as customer information, cargo information, then realize the uninterrupted seamless sharing and real-time monitoring and management of express information in all parts of the distribution line.

4.3 Shipping and Recycling Module

Delivery and recycling is an important part of the intelligent Courier Station, through the "ark" of the delivery and recovery function, convenient for customers to send goods, and more efficient recycling of resources. The staff check the cargo in the container, check and print and post the waybill, and return the deposit. For recycled cartons, the information management system will automatically set up the "environmental honor" file according to the information that the customer fills in.Then we will give customers reduce the cost of transportation .

4.4Distribution Module

The entire distribution process is through the AGV and Information system operation completed, the entire distribution process to achieve a full automation and greatly reducing the loss of human factors.distribution flow as shown in Figure 7, the Unit container packing transport as shown in Figure8.

Fig.7. distribution process

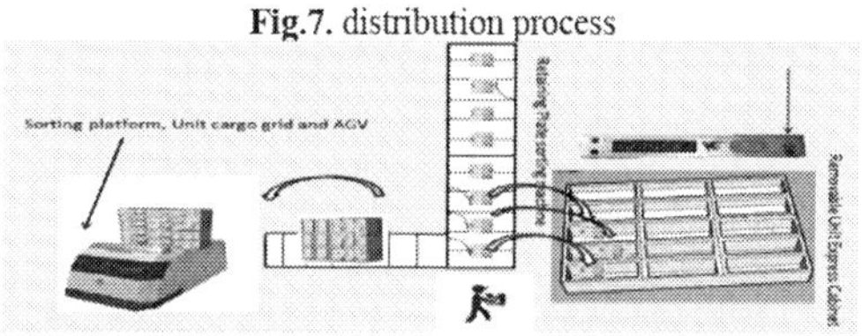

Fig. 8. transportation unit container packing

When the goods arrive at the sorting end, they are loaded by the staff according to the electronic tag-assisted sorting system.The unit locker moves along the conveyor and completes the docking with the AGV car through the operation of the back-end information system.The AGV car travels along the preset line and arrives at the designated location, and then connects the interactive terminal with the express counter via RFID.The information management platform transmits the cargo information to the interactive terminal and sends a message to inform the customer of the arrival of the goods.For special goods, the staff will take the initiative to contact the customer or the customer can contact the staff through a fixed telephone, through the agreement agreed time and place, the staff will offer unified delivery or collection to special goods.

5 Conclusion

The project always adheres to the "green, safe, low-carbon and high-efficiency" development concept.Based

https://doi.org/10.1051/e3sconf/20183804020

on the existing research results, we will continue to pay attention to the green logistics issues and "last kilometer" distribution issues that are urgently needed by the campus and the society, in-depth study of key technical equipment and application issues, and strive to build this project into teaching, research, and social service integration. The interactive demonstration project will guide the development of campus distribution to standardization, safety, and standardization.Through the demonstration and promotion of the project, the school will be built into a domestic campus logistics brand with influential characteristics.Afterwards, the project will be promoted and designed according to the specific conditions of other campuses. Eventually, all campus models will be integrated and the operation mode of large loops and small loops will be established.

References

1. Feng Xueyuan, Zhang Rujia, Gu Yuefang.Investigation and improvement suggestion on campus express logistics of comprehensive university [J].Hong Kong and Macao economy.2014 11: 63-64.
2. Lei Yun, Tu Qinghua, Song Junfei, Zhong Yuan.Construction and Research of efficient and Intelligent Campus Service platform in big data era [J].Communications World. 2017: 275-276.
3. Ding Yuanqing, Guo Xing, Wang Xiaotian, et al.Current situation and Development of Logistics recovery of waste goods in Colleges and Universities [J].Modern Commerce. 2017: 2: 188-189.
4. Xi Peng. Analysis and improvement plan of university campus express mode [J].2017, (2): 104-105
5. Yang Shiquan, Zhou Yanrong, Gu Shuhong.Research on the Distribution Mode of College Campus Express delivery in the era of Internet [J].Logistics Technology 2: 89-92

The Researches on Food Traceability System of University takeout

Jia xin lu [1,*], Ce zhao [1], Zhuang zhuang li[1] ,Zi rong shao[1],Kun yi pi[1]

[1]Shan dong Jiao tong University, Ji nan, Shandong 250300, China

Abstract. In recent years, campus takeaway has developed rapidly, and all kinds of online ordering platforms are running. The problem of distribution in the campus can not only save the time cost of the businessmen, but also guarantee the effective management of the school, which is beneficial to the construction of the standard health system for the takeout. But distribution according to the existing mode will cause certain safety and health risks. The establishment of the University takeaway food traceability system can solve this problem. This paper first analyzes the sharing mode and distribution process of campus takeaway, and then designs the intelligent tracing system for the campus takeaway; the construction of the food distribution information platform and the problem of the recycling of the green environment of the dining box. Finally, the intelligent tracing system of the school takeout is analyzed with the braised chicken as an example.

1 Introduction

In recent years, the O2O online ordering platform represented by the US group takes the first place to grab the campus market, and is welcomed by the students. The reasons for the rapid expansion of this new consumption pattern in the campus include three points: first, the rapid development of the information technology, the popularization of electronic commerce, the student groups in Colleges and universities are also easy to accept new things, and more convenient for the promotion of new consumption patterns; second, the phenomenon of regional gathering in universities and colleges is remarkable, and the catering industry has also gathered around it in order to meet the needs of students. Therefore, it is easy to concentrate the distribution of logistics management; third, the students' lifestyle is diverse, busy learning or weather reasons, such as unwilling to go to the dining room, the takeout provides convenience [1]. The delivery of logistics in the campus can not only save the time cost of the businessmen, but also guarantee the effective management of the school, help the construction of the standard health system, and reduce the cost of the students' living and facilitate the life of the students.

2 Analysis of Campus Takeout Sharing Mode and Distribution Process

2.1 Analysis of Campus Takeout Sharing Mode and Distribution Process

The takeaway E station mode we use is a smart common distribution mode, as shown in Figure 1.

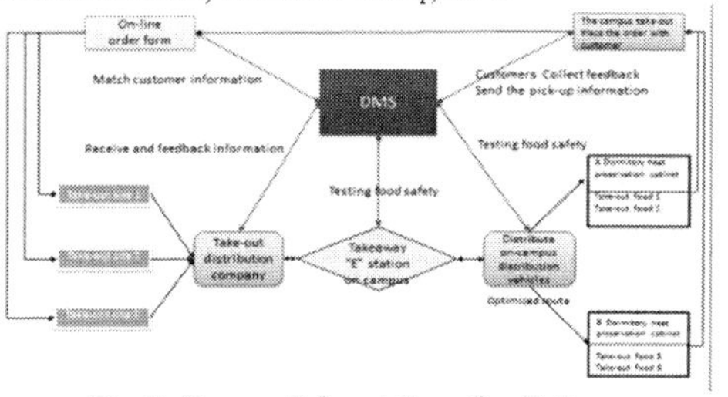

Fig. 1. Campus Takeout Operation Pattern

After the order information is completed and the order information is generated. This order information is sent to the store for takeout and distribution, and then the goods are taken away from the logistics company to the takeout merchant and sent to the E station for the campus takeaway. Through the campus cloud platform management system, the order information is matched with the campus customer information (including address information, telephone, etc.), and then assigned to the school distribution personnel who is to distribute the takeaway to the insulation cabinet of the various school floors. This kind of access cabinet is usually set in the dormitory building, which separates the student's living area from the learning area. In addition, it also separates the problems arising from the arrival of the foreign distribution personnel after the dormitory building.When the takeaway food is distributed to the insulation cabinet, the system automatically identifies the order information and sends SMS prompts to the customers. When the customers take out, they can pick up the goods according to the two-dimensional code or password provided by the SMS, which not only ensures

* Corresponding author: 971362610@e-mail.org

https://doi.org/10.1051/e3sconf/20183804021

the punctuality, safety, optimization of the road, but also creates a good takeout environment for the students in the school. To automate the process of distribution and sub inspection, so as to improve the efficiency and reduce the artificial link, and the staff will concentrate on a work point, which can improve the efficiency of work.

2.2 Analysis of the operation mode of E station in campus takeout

Analysis on the operation mode of the campus takeaway e station as shown in Figure 2.

Try to ensure that lines are no thinner than 0.25 point.

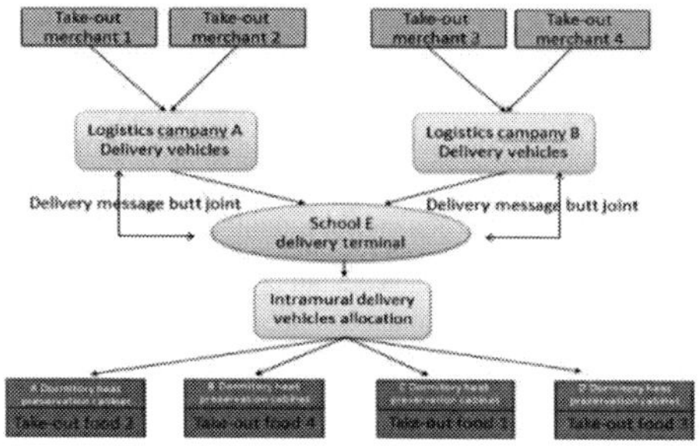

Fig. 2. "One-to-many" recycling distribution pattern

The distribution information of the day is displayed on the GIS electronic map by GIS, and then the position of student dormitories, the number of orders and the location of the order are analyzed with the GIS spatial analysis function. Finally, the location of the distribution center, the traffic situation of the road, the number of vehicles and so on are analyzed, and the genetic algorithm is used to advance the distribution route.

After the delivery route is determined, the number of customers on each route, the number of orders, the total volume of distribution goods and so on will also be determined. According to these parameters and the status of vehicles and personnel in the distribution center, the loading plan can be decided.

3 Design of intelling traceability system for campus takeout

3.1 System concept design

As the whole takeout begins, the visualization and traceability of information must be embodied, so that we can reasonably control the problems of each link and trace the whole process. When a problem occurs in a process, stop distribution in time, so as to minimize the crisis and improve service quality. At present, many takeaway systems are complex, and food safety is not guaranteed. After the food crisis is found, the users do not know where the source is, and the responsible person of each link has unloaded the responsibility, which leads to the weakness of the awareness of rights, the lack of means of safeguarding rights and the higher cost of safeguarding rights. Most customers take a tolerant attitude, and just few people really pursue it, so a complete traceability system is needed for the current takeaway system. Figure 3 is the system concept design, which is divided into one system and seven processes.

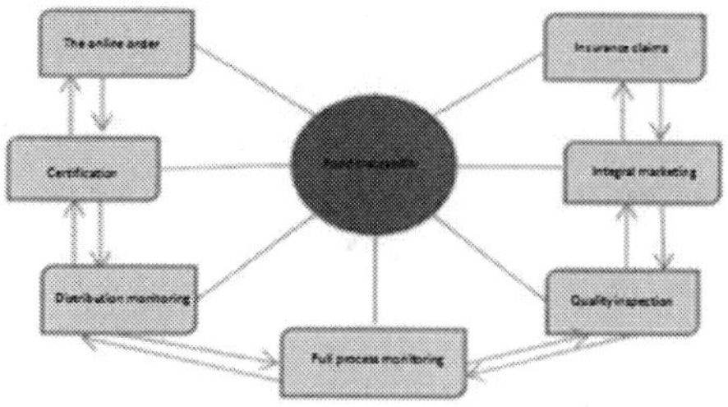

Fig. 3. Campus Takeout Operation Pattern

First, a traceability system is built based on cloud platform data, which is divided into seven parts based on online monitoring. The emphasis is on the traceability of the intramural takeout, encoding from the takeaway merchants to the completion of the distribution. If the potential crisis is found, the distribution is terminated immediately, and the safety monitoring of the material, the inspection of the distribution link and the inspection of the service process are carried out. Besides, it can also score marketing. If the takeaway business successfully completes an order, its score will rise. The higher the score is, the higher the credit value of the takeaway business will be, and then the merchant will be promoted to the app home page, so that more people can buy at ease. If the food safety problem appears in the link, it can be traced back to every link of distribution, and can reduce the loss according to the insurance claim, realize the minimization of the impact, and control the problem of distribution. The whole distribution and the traceability process are shown in Figure 4.

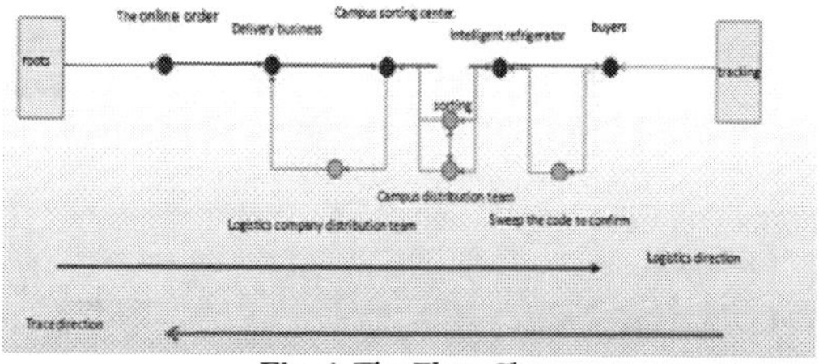

Fig. 4. The Flow Chart

3.2 Two dimensional code tracing process

Product traceability is to record its history, and the location should be querying. The recording system is carried out well and executed well, and the traceability of products is also realized. First, load information, combine the feature encryption logo with the product, and give the unique barcode and encrypted digital information of each takeaway by software. Next, we use bar code as a carrier to input and upload every link information to the traceability system of product quality information. Before the transport, sorting, and distribution links, ensure that each link has generated data and uploaded, and then supervised by the specialized personnel. In the final arrival of the intelligent takeaway cabinet, the takeaway generates a bar code containing all the links. Finally, the whole process can be traced by scanning code, and consumers can scan the whole process of takeaway food distribution by scanning WeChat.

Through two dimensional code, the information of each circulation node can be interconnected to form a chain of quality safety traceability, and "point supervision" is transformed into "chain supervision" "face supervision", which will be more conducive to the quality supervision of product quality and improve the quality of products.

3.3 Safety traceability guarantee the rights and interests of consumers

The quality safety traceability system of takeaway food is an information management system that can connect production, inspection, supervision and consumption, so that consumers can understand the process of regular production and circulation, and improve the degree of consumers' reassurance. The system provides the traceability model of "from distribution to table", extracts the elements of supply chain links such as production, processing, circulation and consumption, and establishes a food safety information database. Once problems are identified, effective control and recall can be traced back to source, so as to protect consumers' legitimate rights and interests from the source.

(1)The advantage of two-dimensional code is used to achieve the management of food safety and traceability, so that it is more efficient, real-time and convenient.

(2)Using the two dimensional is to provide completely transparent management ability in food supply chain, to ensure the whole process of food safety visual control, and to call back the problem food.

(3)It can provide basis for effective assessment and scientific early warning of potential potential food safety hazards.

(4)The data can be reported in real time and accurately through the network, so it is convenient to do deeper analysis and research quickly and efficiently.

(5)Through the Internet, consumers can query the complete tracking information of the food purchased.

4 Taking the braised chicken as an example to analyse the intelligent tracing of the campus takeout

The data show that the "braised chicken set meal" of the takeout food of Shandong Jiaotong University is the most, and because of various factors that lead to the safety problems such as decay and deterioration, we use the "braised chicken set meal" as an example to reflect the intelligent tracing of the scheme. The basic process is shown in Figure 5.

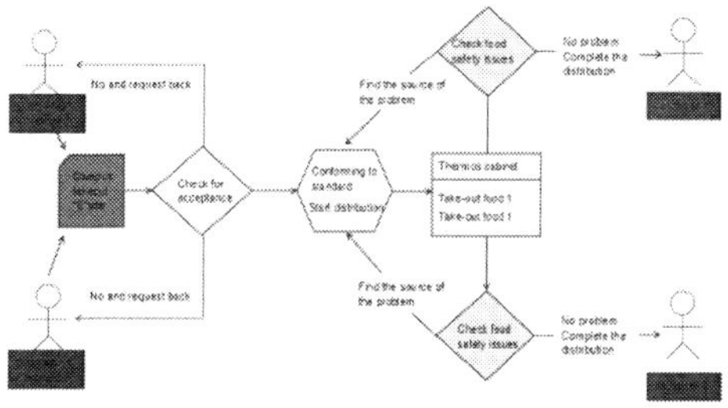

Fig. 5. Retrospective flow chart

The safety problems of "yellow braised chicken set meal" can be generally divided into food safety, safety of production process, packaging safety and distribution freshness. Food safety can only be controlled by takeaway businesses, which is only suitable for food distribution in campus. If this link detects the security problem, then go back to the last step. So this plan is tracking the freshness of distribution, packaging safety and food safety.

The scheme uses two-dimensional code to complete the collection of tracking information. The takeaway from outside school enters the sorting center of the school. We scan two-dimensional code to obtain food information and detect food safety issues, such as food that meets the provisions of the national food safety law and the business license of the business. The order that will affect the safety of takeaway will be returned in time, and the "yellow braised chicken set" that meets the food safety regulations will enter the school delivery section. When sorting points confirm that "yellow braised chicken set meal" is safe, we start distributing, updating APP's information and two-dimensional code information. If there are problems in the delivery process, the reason for the problem should be in the school delivery process, and the loss is borne by the company responsible for the school distribution. Another important safety problem is that's major safety problems such as discomfort or food poisoning after eating "yellow braised chicken set meal". Tracking system should be traced throughout the whole process to find the source of food safety problems. We will use relevant laws to safeguard students' rights, and solve the problem of takeaway food safety through this plan.

The common distribution mode of campus disrupts the principle of separate distribution by the takeaway company, and adopts the unified sorting according to the criteria of regional distribution, and generates two-dimensional code of user related information and is attached to the package of the takeaway. Sorting is done through the sorting device according to the takeaway place. Arrange vehicles to distribute the cabinets reasonably. The cabinets can identify the user's personal information by two-dimensional code and send messages automatically to inform the students to take out, or send a meal to the door, and to identify whether the type of takeaway needs to be cooled or heated and treated accordingly. Students can get their takeaway food only by certification and recognition. It can also recycle the lunch box, make use of environmental protection, and solve the pollution problem of a large number of lunch boxes.

5 Conclusion

The project always adheres to the development concept of "green, safe, low carbon and efficient". On the basis of the existing research results, the project will continue to pay attention to the need of the campus and the society for the delivery mode and the "last mile" distribution problem. The project studies the key technical equipment and application problems, and endeavours to build the project into a demonstration project of the integrated and interactive operation of the teaching, scientific research and social services, and to guide the development of the distribution of the campus takeout to the standard and safe direction. Through the demonstration and popularization and application of the project, we will build our domestic influential campus takeaway brand. After that, the project is popularized and designed according to the specific circumstances of other campus. Finally, all the campus modes will be integrated and the operation mode of the large ring is set up.

References

1. Liu Junhua, Wang Yanfeng, du Juan, Tian Zhihong.Research on Food traceability system based on Information sharing.2006.12.16.
2. Feng Xueyuan, Zhang Rujia, Gu Yuefang.Investigation and improvement suggestion on campus express logistics of comprehensive university [J].Hong Kong and Macao economy.2014 11: 63-64.
3. Li Xiaojuan.Study on the implementation of Green Logistics [J].Shandong textile economy.2017,(1):5-6.
4. Lei Yun, Tu Qinghua, Song Junfei, Zhong Yuan.Construction and Research of Intelligent Campus Service platform in big data era [J].Communications World. 2017: 275-276.
5. Huo Wen-jin.Research on Logistics Distribution Management system based on RFID.2010.

 https://doi.org/10.1051/e3sconf/20183804022

A New Test Method of Circuit Breaker Spring Telescopic Characteristics Based Image Processing

Huimin Huang, *Feifeng* Wang, *Yufeng* Lu, *Xiaofei* Xia, and *Yi* Su

Electric Power Research Institute of Guangxi Power Grid Co., Ltd, Nanning, Guangxi 530023, China

Abstract. This paper applied computer vision technology to the fatigue condition monitoring of springs, and a new telescopic characteristics test method is proposed for circuit breaker operating mechanism spring based on image processing technology. High-speed camera is utilized to capture spring movement image sequences when high voltage circuit breaker operated. Then the image-matching method is used to obtain the deformation-time curve and speed-time curve, and the spring expansion and deformation parameters are extracted from it, which will lay a foundation for subsequent spring force analysis and matching state evaluation. After performing simulation tests at the experimental site, this image analyzing method could solve the complex problems of traditional mechanical sensor installation and monitoring online, status assessment of the circuit breaker spring.

1 Introduction

The reliability of high-voltage circuit breaker operation is vital to the protection and control of the power grid. According to the statistics, most of the failures of high-voltage circuit breakers are faults of the operating mechanism, and the spring of the operating mechanism is an important component of the circuit breaker. In normal operation, the spring of high voltage circuit breaker mainly bears changing load or is in long-term tight state, so most of the failure modes are reduction in rigidity or fatigue failure by Dai Wen-fang [1]. Accidents of circuit breaker failure due to sudden breakage caused by long-term fatigue work of the spring often occur. Therefore, it is of great significance to study the fatigue condition monitoring of spring in operating mechanism.

At present, the monitoring of the high voltage circuit breaker operating mechanism focuses more on the overall mechanical performance of the operating mechanism. Through direct or indirect measurement of moving contact opening and closing stroke curve, switching coil current, etc. There are few studies on spring monitoring and status evaluation methods. Some new circuit breakers have pressure sensors installed at the bottom of the spring in the manufacturing process, but most circuit breakers can only rely on manual regular inspection, which is not only time-consuming but also labor - consuming as follows [2] and [3]. The function of the general spring fatigue tester is to calculate the fatigue curve of the spring, which needs to be tested after disassembling the circuit breaker spring, so the on-line analysis of the spring force and deformation state of the circuit breaker operating mechanism cannot be achieved.

Zhao Si-yang [4] proposes that the decrease of the rigidity of the switching energy-storing spring of the circuit breaker will cause the eigenfrequency of the spring to decrease. Using this point, the acceleration sensor is used to detect the state of the spring energy storage, but it does not leave the range of contact measurement, and different circuit breakers need to be designed with different installation equipment, which is not conducive to popularization.

This study uses a high-speed camera to capture the deformation of the operating mechanism spring when the circuit breaker is opened and closed. Based on the deformation characteristics of the spring during the operation of the circuit breaker, the first quarter of the movement is used as the motion recognition target. A normalized cross correlation (NCC) algorithm with adaptive ROI region is proposed by Zhao shu-tao [2], which can continuously adjust the ROI region. This method not only guarantees the matching accuracy, but also can obtain the parameters of the spring expansion characteristics of the circuit breaker more quickly and efficiently.

2 Circuit Breaker Spring Telescopic Deformation Detection

In this paper, the high-speed camera is used to replace traditional contact sensors, which obtains the image sequences of the deformation of the circuit breaker operating mechanism springs during the opening and closing process. Image matching technology are used to detect the spring deformation. This "non-contact" circuit breaker spring test method effectively solves the problem of complex installation of the traditional sensor.

* Corresponding author: 397163797@qq.com

When the circuit breaker coil current reaches the set threshold, the high-speed camera is triggered to capture the action or energy storage process of the circuit breaker operating mechanism spring; Image preprocessing are used to improve the quality of the image to be analyzed, which reduces the interference of field light change and camera electronic noise on the captured images, and influence on the recognition accuracy of the target on the spring; then a suitable region of interest (ROI) is set, and the NCC algorithm is used to capture the motion of the coil closest to the motion end and the position of the fixed-end coil. The positional relationship between the two above is used to determine the real-time length and deformation of the spring, and various characteristic parameters such as the free length of the spring, the effective number of turns, the deformation, the maximum stroke, and the wire diameter are calculated [5] and [6], which lays the foundation for spring mechanical analysis, fatigue condition monitoring and defect fault diagnosis of the circuit breaker operating mechanism. The specific detection process is shown in Fig. 1.

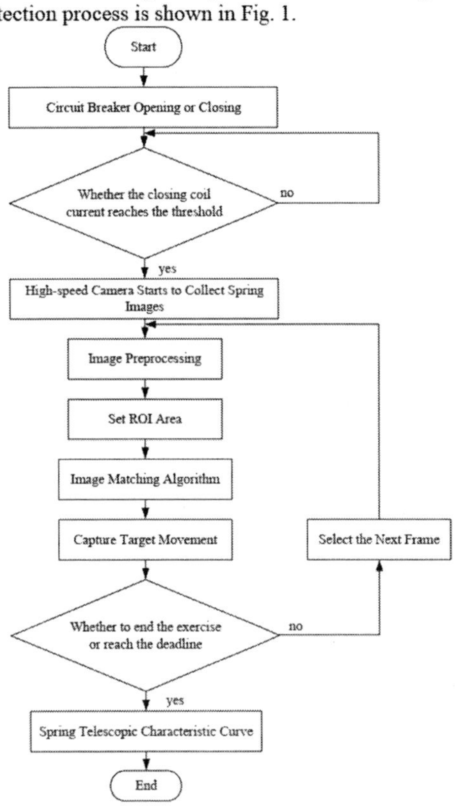

Fig. 1. Flow Chart of Circuit Breaker Spring Telescopic Deformation Detection

3 Spring Image Matching Algorithm

3.1 Normalized Cross-Correlation Algorithm

Normalized cross-correlation matching algorithm is a classical matching algorithm based on image grayscale, often written as NCC algorithm [8]. This method matches the template image and the original image one by one and calculates the cross-correlation values of the template image and the image to be matched to determine the degree of matching. The position of the search point with the largest cross-correlation value determines the position of the template image in the image to be matched. The algorithm is as follows.

$$NCC(m,n)=\frac{\sum_x\sum_y\left(G(x,y)-\bar{G}(x,y)\right)\left(F(x+m,y+n)-\bar{F}(x+m,y+n)\right)}{\sqrt{\sum_x\sum_y\left(G(x,y)-\bar{G}(x,y)\right)^2\sum_x\sum_y\left(F(x+m,y+n)-\bar{F}(x+m,y+n)\right)^2}} \quad (1)$$

$$\bar{G}(x,y)=\sum_{x=1}^{r}\sum_{y=1}^{s}G(x,y)\Big/(r\times s) \quad (2)$$

$$\bar{F}(x+m,y+n)=\sum_{x=1}^{r}\sum_{y=1}^{s}F(x+m,y+n)\Big/(r\times s) \quad (3)$$

Here we use F to represent the original image, whose pixel value is $R\times S$; the template image is G, whose pixel value is $r\times s$. F(x+m,y+n) is the gray value of (x+m,y+n) point in the original image F, while G(x,y) is the gray vaule of (x,y) point of the template image G. (m,n) is the position of the top left vertex of the template image G in the original image F when searching. When the coordinates of the top left corner vertex of the template image G is (m,n), the point G(x,y) in the template image G is corresponding to the point F(x+m,y+n) in the original image F. Through the above formula (1), the value of NCC between the template image G and the sub-image of the original image F under its coverage are calculated. When the correlation values of all positions throughout the original image F and the template G is obtained, the sub-image which is correspond to the maximum value is the destination image, and its position is the coordinate to be determined.

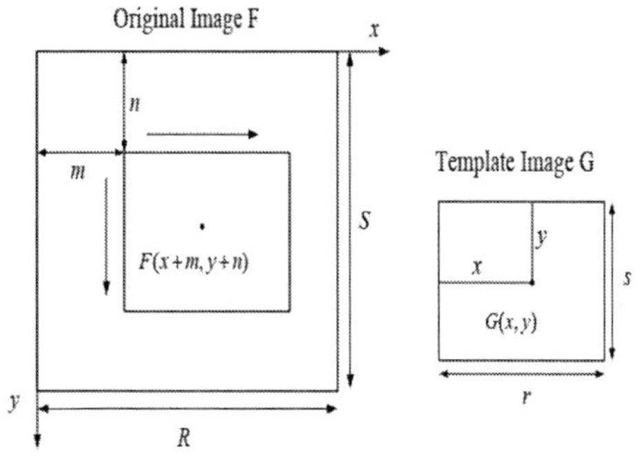

Fig. 2. Principle of NCC algorithm

Although the NCC algorithm has high matching accuracy, the disadvantages such as large amount of calculation and slow operation speed affect the application in moving target recognition of the circuit breaker. In order to solve these problems, the following improvement is adopted to reduce the amount of calculation and increase the efficiency to meet the real-time requirements of on-line monitoring.

3.2 Adaptive ROI Area

Before the image matching starts, a suitable area is set, ie, the target motion range, is selected as the region of interest (ROI), which is generally a rectangle along the pixel direction. In the process of image processing, only the ROI area needs to be analyzed. The approach can not only greatly decrease the computational complexity of the image analysis process, but also eliminate the interference of the structure outside the region. In order to further reduce that area of the ROI region and reduce the amount of calculation, the position of the ROI region is adjusted from the target position of the previous frame when the image to be matched of the next frame is processed, starting with the process of the second frame image. The specific adjustment method is as follows.

The center of the new ROI area is decided by target position of the previous frame. The length and width of the new frame satisfy the following relationship.

$$L_i = kl + c + \delta_{i-1} \tag{4}$$

The value of k is determined according to the spring deformation direction. It is 1 when to non-main direction of deformation, and become 1.2 when to the main direction of deformation (telescopic direction). l is the length and width of the template image. c is a fixed value of 10 to 20 pixels, (here take 10 pixels). δ_i is the distance moved from the previous frame compared to the target. This method is equivalent to prediction, which greatly limits the size of ROI region and improves the matching efficiency

4 Circuit Breaker Spring Flexibility Measurement Example

A test platform is set up at the experimental site. The closing process of a ZN65-12 high-voltage circuit breaker is selected as the subjection. The coil current was set to 0.5A in order to trigger the high-speed camera to capture the motion of the circuit breaker operating mechanism spring at a velocity of 3,500 fps. The initial frame of the target motion is searched and then the fixed-end pixel coordinates is recorded as the initial point A, which is used to conveniently calculate the actual working length and movement speed of the spring in the circuit breaker deformation process. The coil closest to the moving end of the spring, B, is selected as the moving target which needs to be identified in the image. Finally, the target is identified by the adaptive ROI region image matching algorithm to obtain the pixel position in each frame of the image, which represent the actual working position of the operating mechanism spring.

a) Spring Fixed End

b) Spring moving End

Fig. 3. Circuit Breaker Spring Image Matching Result

The distance between the target position B and the initial point A in each frame is calculated to plot the displacement-time curve and velocity-time curve of the moving target during the deformation of operating mechanism spring. The speed of the operating mechanism spring is the point-by-point slope value of the displacement-time curve. The result is as follows.

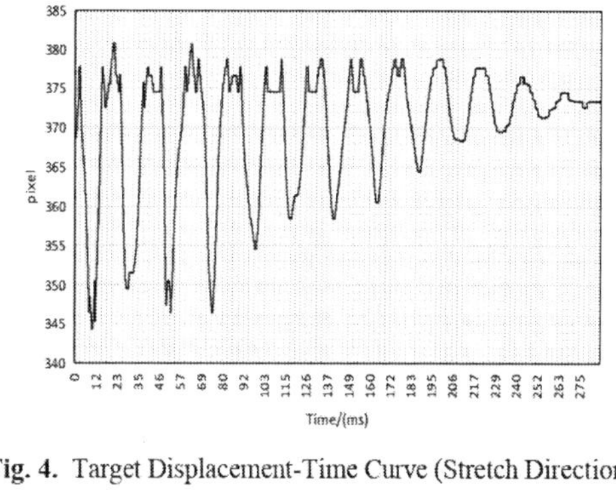

Fig. 4. Target Displacement-Time Curve (Stretch Direction)

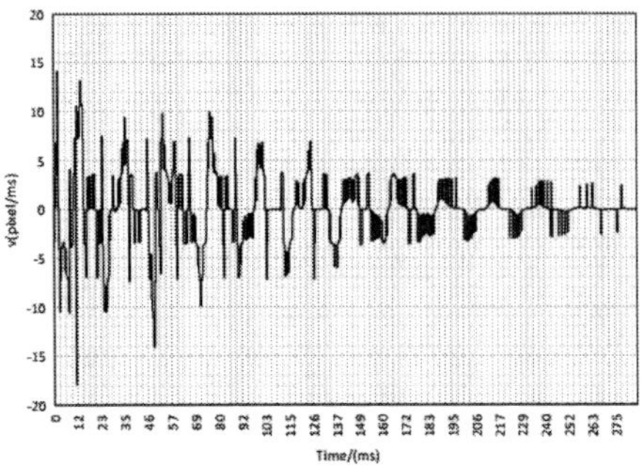

Fig. 5. Target Speed-Time Curve (Stretch Direction)

It can be seen from the figure that the spring is continuously reciprocating during the circuit breaker closing processed by Li Pengfei [7]. According to the motion curve of the circuit breaker spring, the initial frame target pixel position s_0 was determined to be 370.62, and the fixed-end pixel position s was 927.5. The initial spring height H_{t0} was calculated to be 556.9. The final working height H_t was calculated as 554.1 which is determined by the difference between the ending frame target pixel postion s_t at 373.36 and the fixed-end pixel position s (the numerical units above are all pixels).

The maximum x-coordinate value of the intersection point between the straight line ($y = s_t \pm \varepsilon$)and the spring displacement time curve is taken as the spring stretching time t. ε is used to judge the maximum amplitude of the spring still in expansion or compression motion, (default can take 0~2 pixels, one pixel value is taken in this experiment). Such setup is to eliminating the action caused by the identification error due to the too small spring stretching range. Finally, the spring stretching time t was calculated to 274ms.

The difference between the upper and lower bounds of the spring displacement-time curve is calculated as the maximum movement amplitude which was obtained to be 36.4 pixels. Through the displacement-time curve, the times that the coordinates of the first coil of the spring crossed its final coordinate during spring retraction is counted as the number of radial reciprocations. It was obtained to be 26 times. Finally, the maximum operating speed of the circuit breaker spring was 16 pixels/ms by analyzing the target speed-time curve.

Table 1. Circuit Breaker Operation Mechanism Spring Extension Characteristics.

Parameter Name	Unit	Value
Initial Working height	Pixels	556.9
Final Working height	Pixels	554.1
Telescopic time	ms	274
Maximum stroke	Pixels	36.4
No. of reciprocations	times	26
Maximum speed	Pixels/ms	16
Telescopic Frequency	Hz	47.4

Processing 1000 frames of image by utilizing the adaptive ROI area's NCC algorithm takes about 22.8 s. Compared to using the NCC algorithm alone, the same process required 69.4 s to complete. Obviously, the former algorithm is much faster on the speed of calculation. On the premise of not to decline the calculation accuracy, it greatly accelerates the calculation speed and saves the analysis time, and provides a new method for the online detection of the spring telescopic characteristics on the circuit breaker.

5 Conclusion

1) Image processing technology is utilized to analyze the spring telescopic characteristics of the circuit breaker operating mechanism. It can reflect spring expansion and contraction status in the energy storage and release process and extract the spring deformation parameters. This work is a new method of circuit breaker spring detection.

2) The NCC algorithm has a high matching accuracy and is suitable for use in the spring detection of the circuit breaker operating mechanism. However, a large amount of calculation requires stronger hardware support. The improved method improves the speed by 3 times without reducing the calculation accuracy, which is conducive to the promotion of the algorithm.

3) Identifying the circuit breaker spring fatigue based on computer vision technology is a kind of "non-contact" test method. It can effectively solve problems in sensor installation, such as time-consuming and labor-consuming. It has a broad application prospect in the circuit breaker spring online monitoring and electrification test.

References

1. W.F.Dai, S.Y.Zhao, G.Zeng, etal. *Prediction of state of closing spring energy storage based on characteristic parameters of circuit breakers*[J]. Instrument Technology and Sensors,**08** (2015) .
2. S.T.Zhao, C.J.Wu,M. Li, etal. *Study on testing method of circuit breaker mechanical characteristics based on NCC-P-S optimization algorithm*[J]. Proceedings of the CSEE,**37** (2017) .
3. G. Guo,P.F. Li, J.Li, etal. *Status Monitoring of Energy Storage Springs for Circuit Breaker Closing*[J]. Instrument Technology and Sensors, **04** (2015).
4. S.Y.Zhao, A.B.Wang, etal. *Evaluation of Spring Energy Storage State of Circuit Breaker Operating Mechanism Based on Fuzzy Synthetic Evaluation*[J].High Voltage Apparatus, **52** (2016).

5. J.H.Sohn, S.K.Lee, S.O Kim, etal.*Comparison of Spring Models for Dynamic Analysis of a High Voltage Circuit Breaker with a Spring Operating Mechanism*[J]. Mechanics Based Design of Structures and Machines,**36**(2008).
6. W.S.Yoo,S.O.Kim,J.H.Sohn. *Dynamic analysis and design of a high voltage circuit breaker with spring operating mechanism*[J]. Journal of Mechanical Science and Technology, **21**(2007).
7. P.F.Li,W.J.Zhou,G.Zeng,etal.*Dynamic characteristics of the closing spring of high voltage circuit breaker and detection method of energy storage state*[J].Transactions of China Electrotechnical Society, **31**(2016).
8. A.Pourmohammad, S.Poursajadi, Karimifar S. *Scene matching NCC value improvement based on contrast matching*[C]//Proceedings of the 2013 8th Iranian Conference on Machine Vision and Image Processing.Zanjan:IEEE,(2013).
9. Y. Fouda, K.Ragab. *An efficient implementation of normalized cross-correlation image matching based on pyramid*[C]//Proceedings of 2013 International Joint Conference on Awareness Science and Technology and Ubi-Media Computing. Aizuwakamatsu:IEEE,(2013).
10. L.J.Sun, X.G.Hu,Y.Ji.*Mechanical fault classification of high voltage circuit breakers based on support vector machine*[J]. Transactions of China Electrometrical Society, **08**(2006).
11. J.A.Martinez-Velasco,J.Magnusson.*Parametric analysis of the hybrid HVDC circuit breaker*[J]. International Journal of Electrical Power & Energy Systems, 84(2016).

Study on the optimization allocation of wind-solar in power system based on multi-region production simulation

Zhicheng Xu*, *Bo* Yuan, and *Fuqiang* Zhang

State Grid Energy Reasearch Institute CO.,Ltd. , Beijing, 102209, China

Abstract. In this paper, a power supply optimization model is proposed. The model takes the minimum fossil energy consumption as the target, considering the output characteristics of the conventional power supply and the renewable power supply. The optimal capacity ratio of wind-solar in the power supply under various constraints is calculated, and the interrelation between conventional power supply and renewable energy is analyzed in the system of high proportion renewable energy integration. Using the model, we can provide scientific guidance for the coordinated and orderly development of renewable energy and conventional power sources.

1 Introduction

Power planning plays an important role in the development of power system. The traditional mixing configuration refers to the simple combination, in which the system lacks the detailed calculation model so that the guarantee rate is low, and the service life is short. The real meaning of the renewable energy system configuration is to maximize the use of resources such as wind and solar energy. When designing the system, we should choose the appropriate capacity configuration according to the local meteorological conditions. The target of hybrid system configuration is to determine the rated power, solar battery power and battery capacity of wind turbines in hybrid power system according to local meteorological data and load demand design algorithm, so that the configuration results match each other, and the system cost is lowest while satisfying the requirements of power balance and minimum boot capacity. The paper [1] uses the method of production simulation to calculate the optimal configuration of scenery, and the document [2,3] introduces the power programming modeling and solving method considering the minimum target cost. The influence of uncertain factors on power supply planning is considered in the literature [4], and the influence of uncertainty and electricity price on the load forecasting is taken into account in the objective function. The paper [5,6] considers the environmental protection benefit in the planning, analyzes the harm of various kinds of units emissions to the environment and establishes the multi-objective power supply optimization model with the minimum current value of total cost and the least CO_2 emission. The paper [7] proposes a power supply planning model which pursues the profit maximization of each generation company under the market environment.

The power supply planning problem is a high dimensional non-convex nonlinear optimization problem, so that the solution is more complex. Heuristic algorithms, mathematical optimization methods and artificial intelligence methods are used to solve the algorithm. However, the existing methods are based on the total cost value and profit, take the benefit cost as the center, and seldom take into account the influence of renewable energy on the power grid, and the interrelated mechanism of various power sources in the system of high proportion renewable energy integration is not clear enough. Therefore, it is necessary to study the new programming algorithm for the multi area power supply planning.

2 Multi-region renewable energy accommodation assessment model

2.1 Model framework

In this paper, a novel multi-region renewable energy accomodation capacity assessment model is proposed. In this model, the installed capacity of renewable energy is not given in advance, but is regarded as a decision variable in the optimization process. It means that the model is a joint optimization one, and it also takes into account the characteristics of renewable energy output, the characteristics of different types of generating units, the transmission channels, and the maximum capacity of the installed renewable energy.

2.2 Decision variables

The decision variables mainly include the following parts, and the key variables are shown in table 1:

* Corresponding author: xu-zhicheng2008@163.com

https://doi.org/10.1051/e3sconf/20183804023

1) the accomodation capacity of wind and solar power;

$C_{Wind,k}$--maximum wind-installed capacity located in k-region;

$C_{Solar,k}$-- maximum solar-installed capacity located in k-region.

2) operation simulation decision variables, which are shown in table 1.

Table 1. Key variable definition and its description.

Variable	Type	Comments
$P^t_{\text{unittype},i}$	Continuous type variable	The output of i-unit at time t
$U^t_{\text{thermal},i}$	0-1Variable	The state of i-thermal power unit at time t, 0 for shutdown, 1 for startup
$U^t_{pump,i}$	0-1Variable	The state of i-pump power unit at time t, 0 for pumping, 1 for generating

2.3 Objective function

The model is aimed at the optimal economy, and the objective function includes the following parts: 1) fuel cost and variable operating cost, 2) unit starting and shutdown cost for the thermal power, 3) fixed operating cost, 4) electricity outage cost, 5) pollutant emission cost, namely

$$C=\sum_t\sum_i C_i^{SU}(1-U^{t-1}_{\text{thermal},i})U^t_{\text{thermal},i}+\sum_t\sum_i F(P^t_{\text{thermal},i})+C_{fix}+C_{eens}+C_{emission} \tag{1}$$

The first right part of the equal sign is the start/shutdown cost of the thermal power unit, C_i^{SU} for startup charges, in the following notation:

$$C_i^{SU}=\begin{cases}C_i^h, T_{i,off}\le t_{i,off}\le t_{i,h}\\ C_i^c, t_{i,off}\ge t_{i,h}\end{cases} \tag{2}$$

which, C_i^h for hot start-up costs, C_i^c for cold start-up costs, $T_{i,off}$ for minimum continuous shutdown time, $t_{i,\,off}$ for the actual continuous shutdown time, t_{iH} is the hot start decision time.

The second part is the fuel cost of thermal power unit, $F(\cdot)$ for the thermal power unit operating cost function, can be expressed by linear or piecewise linear function, if divided into k segments, namely

$$F=\begin{cases}a_{i,1}P^t_{\text{thermal},i}, P^t_{\text{thermal},i}\in[0,P_{\text{thermal},i,1}]\\ a_{i,2}P^t_{\text{thermal},i}, P^t_{\text{thermal},i}\in(P_{\text{thermal},i,1},P_{\text{thermal},i,2}]\\ \cdots\\ a_{i,k}P^t_{\text{thermal},i}, P^t_{\text{thermal},i}\in(P_{\text{thermal},i,k-1},P_{\text{thermal},i,k}]\end{cases} \tag{3}$$

The third, fourth and fifth parts are fixed operating cost, electricity outage cost and pollutant discharge cost respectively. It is noted that the unit power operation cost of wind power plant / solar power station is set to zero. The amount of curtailement wind and solar power is minimized, and priority scheduling of renewable energy power generation is ensured.

2.4 Constraint conditions

(1) Power balance constraint:

All types of units and line output and equal to the load.

$$\forall t,\ \sum_{s\in unittype}\sum_i P^t_{s,i}+\sum_i P^t_{\text{line},i}=L_t \tag{4}$$

Which L_t denotes the load demand at moment t.

(2) Upper and lower limit constraint of unit output:

$$\forall t,\forall i,\forall s\in \text{unittype},\quad P_{s,i,\min}\le P^t_{s,i}\le P_{s,i,\max} \tag{5}$$

Among them, $P_{S,i,\text{Min}}$ and $P_{S,i,\text{Max}}$ represent the minimum and maximum output of the first S class i units. For wind power and solar power generation, the maximum output of each moment is the product of the predicted output value and the decision variable of the installation capacity.

(3) Constraint of starting and shutdown of thermal power unit:

For thermal power units, especially coal and nuclear units, the minimum continuous opening and minimum continuous shutdown time constraints need to be met.

$$\sum_{l=t}^{t+T_{i,on}-1} U^l_{s,i}\ge T_{i,on}(U^t_{s,i}-U^{t-1}_{s,i}) \tag{6}$$

$$\sum_{l=t}^{t+T_{i,off}-1}(1-U^l_{s,i})\ge T_{i,on}(U^{t-1}_{s,i}-U^t_{s,i})) \tag{7}$$

which, $T_{I,on}$, $T_{I,off}$ represents minimum continuous opening and minimum continuous shutdown time for the unit.

(4) Unit ramp rate constraint:

It is necessary to meet the ramp up constraints for the thermal power unit, that is, the normal operation state can not exceed the speed of the ramp rate.

$$P^t_{s,i}\ge P^{t-1}_{s,i}-RD_{s,i}-(1-U^t_{s,i})P_{s,i,\max} \tag{8}$$

$$P^t_{s,i}\le P^{t-1}_{s,i}+RU_{s,i}+(1-U^{t-1}_{s,i})P_{s,i,\max} \tag{9}$$

which, $RD_{S,I}$ and $RU_{S,I}$ denote the unit's lower ramp and upper rate.

(5) Constraint of pumped storage power station:

When the pumped storage unit is running, it is necessary to maintain the water level between the top and bottom of the reservoir between the highest and the dead water level:

$$W_{u,\min}\le W_u\le W_{u,\max} \tag{10}$$

$$W_{d,\min}\le W_d\le W_{d,\max} \tag{11}$$

Which W_u and W_d represent the water content of the upper reservoir and the lower reservoir respectively; $W_{u,\text{Max}}$ and $W_{u,\min}$ are the maximum and minimum allowable water content of the upper reservoir respectively; $W_{d,\text{Max}}$ and $W_{d,\min}$ represent the maximum and minimum allowable water quantity of the lower reservoir respectively. There is a dynamic coupling relationship between the upper and lower reservoir water at different time, namely:

$$W^t_u=W^{t-1}_u-\lambda_g P^t_g+\lambda_p P^t_p \tag{12}$$

$$W^t_d=W^{t-1}_d+\lambda_g P^t_g-\lambda_p P^t_p \tag{13}$$

For the day/week cycle type unit, if one day/week is divided into S time, then the initial and final period of the upper reservoir water level should be consistent, that is:

$$W_u^s = W_u^0 \tag{14}$$

which W_u^0 and W_u^s represent the water level at the start period and the final period.

(6) Transmission lines constraints:

$$C_{line,\min} \leq P_{line,i}^t \leq C_{line,\max} \tag{15}$$

(7) Spinning reserve constraints:

Spinning reserve generally includes the reserve required by the load and the reserve to cope with the uncertainty of renewable energy sources.

$$\sum_s \sum_i R_{s,i}^t \geq L_t(1 + R_L) \tag{16}$$

$$\sum_s \sum_i R_{s,i}^t \leq L_t(1 + R_{VG}) \tag{17}$$

(8) Curtailment ratio constraints:

Curtailment ratio must be less than the given value:

$$\frac{\sum_i \sum_t Cur_{wind,i}^t + \sum_i \sum_t Cur_{solar,i}^t}{\sum_i \sum_t P_{wind,i}^t + \sum_i \sum_t P_{solar,i}^t} \leq \alpha \tag{18}$$

After the construction of the evaluation model of the renewable energy accomodation capacity of the multi-area, the CPLEX commercial solver is invoked to optimize the calculation of the model, and the renewable energy accomodation capacity and the simulation results of the system can be obtained.

3 Scenario Design

The following are the important constraints, such as load demand, power supply capacity, renewable energy output characteristics and the scale of transmission channels.

3.1. Load demand

According to the medium and long term energy demand forecast model in the literature [8], it is estimated that in 2050, the whole society electricity consumption, the maximum load is 11.7×10^6 GWh, 18.7×10^6 GW respectively, other characteristic parameters as shown in table 2.

Table 2. National electricity consumption and maximum load forecast in 2050.

Regional	Category	2050 (GWh, GW)
National	Power consumption	11.7×10^6
	Maximum Load	18.7×10^6
	Average load factor	64.37%
	Average peak and valley ratio	25.84%

3.2. Power supply development (except for renewable energy)

Combining the assessment of resource potential and the technical progress of various types of power sources (except renewable energy sources) and the foreseeable national planning and prediction, the general situation of the installation of various types of power sources is obtained. The results are illusrtrated in table 3 and figures 1, 2, and the specific prediction model can be referred to [9]. The overall situation of the development of power sources is described from the total amount, structure and layout of power supply units in three aspects.

Table 3. Installation capacity of various power supply in China in 2050.

Power type	Capacity /(GW)
Coal power	671
Gas and electricity	201
Nuclear power	340
Hydropower	519
Pumped storage	138
Biomass	160
Total	**1358**

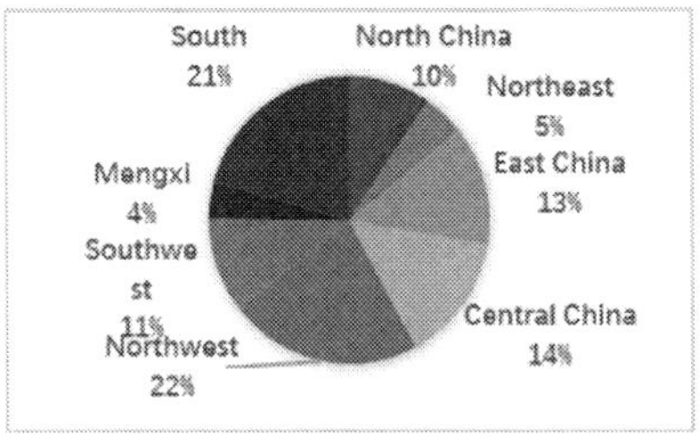

Fig. 1. Regional distribution of power capacity in China in 2050 (except RE power generation).

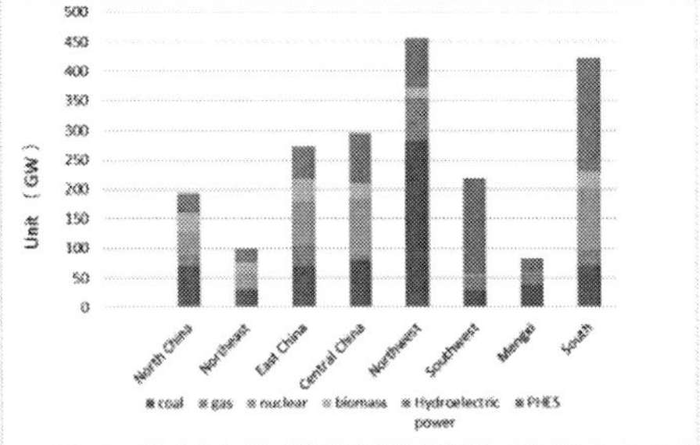

Fig. 2. National power supply structures in China in 2050 (except RE power generation).

3.3. Inter-regional channel power flow direction and scale

With the accelerated development of renewable energy in the northern part of the western region, the scale of power flow is gradually expanding. Considering the capacity of the transmission channels, and the coordination of inter-district and inter-provincial transmission corridors and routes, the power flow from the "sending end" in the northern part of the western region to the receiving provinces (cities) is expected to reach 553 GW in 2050, as shown in the graph below.

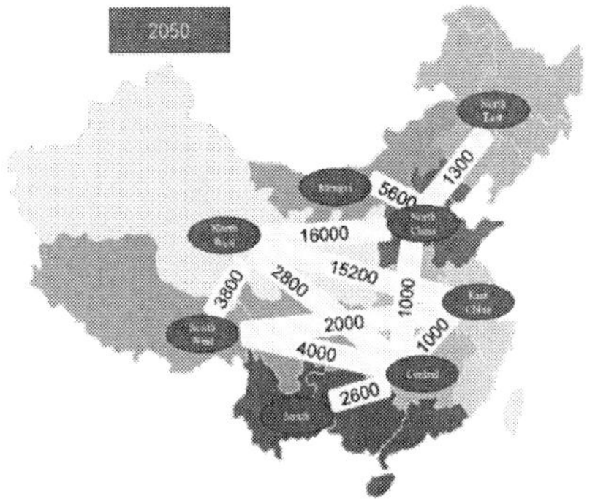

Fig. 3. Inter-regional channel power flow direction and scale in 2050.

4 Case Study

This section studies the renewable energy accomodation capacity in all regions of China in 2050, that is, the optimal installation capacity of wind and solar power generation that can be friendly integrated by all provinces and regions when the overall renewable energy curtaiment rate in each area is controlled below 5%. The process is conducted based on the multi-region renewable energy accomodation assessment model proposed in Section 2, and the various constraints proposed in Section 3. The renewable energy construction and its distribution are listed in table 4 and illustrated in figure 4.

Table 4. Renewable energy allocation in different regions in China in 2050 (Unit: GW).

	Renewable energy	Wind	Solar
Total	2367.23	1432.02	935.21
North	149.841	89	60.841
Northeast	165.542	77.157	88.385
East	132	48	84
Central	259.55	35	224.55
Northwest	1286.87	906.44	380.43
Southwest	37.423	19.423	18
Inner mongolia	227	201	26
South	109	56	53

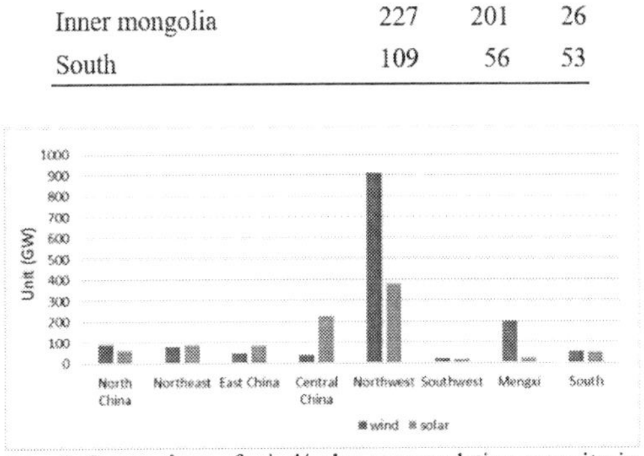

Fig. 4. Comparison of wind/solar accomodation capacity in different regions of China in 2050.

In order to evaluate the operation condition, the proposed production simulation is conducted to calculate the renewable energy curtailment rate and some important performance indicators such as the utilization hour of wind and solar power generation under the given renewable energy planning.

Table 5. Analysis of RE accomodation capacity in different regions of China in 2050.

	Curtailment rate			Utilization hours		
	Average curtailment rate	Wind curtailement	Solar curtailement	Average utilization hour	Wind utilization hour	Solar utilization hour
Total	**11.36%**	**11.39%**	**11.27%**	1796	3759	2184
North	5.53%	3.79%	9.72%	1551	3623	1877
Northeast	10.29%	8.23%	13.20%	1556	3252	2003
East	8.25%	8.60%	7.90%	1548	3914	2099
Central	5.78%	3.54%	6.40%	1154	3391	1890
Northwest	12.85%	12.49%	14.27%	1964	3532	2232
Southwest	11.93%	11.70%	12.30%	1537	4037	1850
Inner mongolia	12.14%	12.53%	5.80%	2166	3548	2295
South	8.32%	7.80%	9.30%	1659	3759	2130

5 Conclusions and Recommendations

The renewable energy accommodation is the key factor that affects the healthy development of renewable energy. This paper puts forward an innovative assessment model to evaluate the optimal wind-solar hosting capacity. What is more, we also quantifies and analyzes the scale of renewable energy installed in different regions of China in 2050. The cross-provincial transmission capacity for renewable energy bases and the corresponding optimization power flow schemes are of important significance for the healthy development of renewable energy in China. An optimal allocation of wind-solar in power system is meaningful for the decision-maker for system's planning and manegement.

As to China's medium and long term power supply planning, the reverse distribution of energy production and load center will be more obvious, and the scale of energy trans-regional transportation, especially the amount of renewable energy transmission, will be further enlarged. In order to meet the large-scale renewable energy integration, the following suggestions are put forward to ensure the healthy and sustainable development of power system. The specific suggestions are as follows: (1) Carry out special research on renewable energy integration, and the renewable energy planning plan should be optimized and updated by rolling. (2) Speed up the construction of HVDC transmission projects to increase the transmission capability of renewable energy. (3) Motivate the enthusiasm of power generation enterprises to participate in peak shaving and power plant flexibility transformation, and to accelerate the transformation of thermal energy flexibility, and to implement the transformation target proposed by the national "Thirteen-Five" electricity planning. (4) Establish a mature market mechanism to break the renewable energy accomodation barriers between provinces so that the renewable energy can be accomodated in a larger region.

Acknowledgements

This work is supported by Science and Technology Project of State Grid Corporation of China (B3670116A005): Study on the evaluation model and application of peak shaving technology, economy and policy for power system wiht large-scale renewable energy integration.

References

1. A.S.O. Ogunjuyigbe, T.R. Ayodele, O.A. Akinola, Appl. Energy **171** (2016)
2. H. Hassanzadehfard, A. Jalilian, Int. J. Green Energy **15**, 2 (2018)
3. Y. Yuan, Y. Cao, X. Zhang, C. Liu , Y. Huang, L. Peng, Q Kejun, J. Mod Power Syst. Clean Energy **3**, 1 (2015)
4. Y. Zhu, P. Jiang, S. Yang, *2014 IEEE Conference on Energy Conversion (CENCON)* (2014)
5. Y. C. Kim, B. H. Ahu, IEEE Trans. Eng. Manage **40**, 2 (1993)
6. C. T. Su , G. R. Lii , J. J.Chen, *Proceedings of IEEE International Conference on Industrial Technology* (2000)
7. X. Zhao, X. Wang, H. Chen, Proc. CESS **24**, 8 (2004)
8. State Grid Energy Institute Co., Ltd. Research on China's medium and long term power demand forecast and forecasting model (2017)
9. State Grid Energy Institute Co., Ltd. "Thirteen-Five"Analysis of renewable energy dissipation and prediction of trans-regional power flow (2017)

The researches on energy sustainability in Northern China

Ping Wang[1], *Zhiqiang* Zhu[1,*], and *Shuang* Zhang[2]

[1]Postdoctoral Workstation, Harbin Sport University, 150001 Harbin Heilongjiang, China
[2]Scientific Research Centre, Harbin Sport University, 150001 Harbin Heilongjiang, China

Abstract. Energy, which accounts for two-thirds of today's greenhouse gas emissions, is the key to reducing greenhouse gas emissions and slowing global warming. In this paper, the IPCC-recommended reference approach and scenario analysis were applied to evaluate dynamic change of the energy supply and energy-related carbon dioxide emissions within the period of 2000-2025 in Northern China (NC). The results show that energy importing reliance reached 85% in 2015 and the energy structure has become more diversified in NC. In addition, the per-capita CO_2 emission is significantly higher while carbon intensity is lower than those of the national average. Under the LC scenario, CO_2 emissions begin to fall for the first time in 2022. Hence, if Energy-Saving and Emission-Reduction strategy and regional planning for NC are implemented fully, NC will achieve the national emission reduction targets in 2025 and will have a large CO_2 mitigation potential in the future.

1 Introduction

For a century, both carbon intensity and global temperature have been on the rise. And at the same time human society has gradually come to know the consequences of climate change due to the "greenhouse effect", especially carbon dioxide emissions, making it a top international political of the day. Energy, which accounts for two-thirds of today's greenhouse-gas emissions, is at the heart of problem and so must form the core of the solution[1]. In recent years, the famous international institution such as IPCC (Intergovernment Panel on Climate Change), IEA (International Energy Agency), WB (World Bank), and the Energy Department of German, England, America issues energy report periodically to mitigate greenhouse-gas emissions and climate change. The detailed inventory of energy and future projections in light of recent economic and policy developments come from them and other institutions and countries have more and more influence on the other countries. But the regional-scale studies of energy and CO_2 emissions are comparatively scarce[2]. The Northern China (NC), the China's much developed region (GDP was 22.1% of the nation's total), is now in the middle periods of industrialization with the feature of energy-intensive and capital-intensive. The rapid urbanization and industrialization are still sending NC's rigid demand of energies upward. At the moment, enhancing energy security, maintaining economic growth and reducing greenhouse gas emission are becoming the major issues challenging the future of the region. Therefore, we have estimated the past and the present energy supply and CO_2 emissions in NC by applying the IEA and IPCC recommended approaches and projected the future trends of energy demand and CO_2 emissions based on China's Medium and Long Term Energy Strategy and 2016-2025 regional planning for NC. According to the results, we have put forward a sustainable development projects for NC's future energy planning and policy making, which may provide some general insights on the effectiveness of regional-level energy conservation and GHG reduction for other regions as well.

2 Numerical investigation

2.1 Estimate of energy supply

Total primary energy supply (TPES) is made up of production, imports, exports, international marine bunkers, international aviation bunkers and stock changes. Starting with the 2009 IEA edition, international aviation bunkers is subtracted out of supply in the same way as international marine bunkers. According to energy balance table data of NC, energy supply can be calculated by the following express

$$E = \sum_i \left(P_i + R_i + I_i \pm S_i\right) \tag{1}$$

Where E is the total primary energy supply (Mtce, million tonnes of coal equivalent), the subscript i represents the various fuels, P_i, R_i, I_i, E_i and S_i denote primary production, recoveries, imports, exports and stock changes of fuel i respectively.

For this analysis of the final energy consumption in NC, the total and sectoral energy data for 2000–2015 were obtained from the Energy Balance Tables of the Yearbook database (http://www.soshoo.com.cn). The final energy consumption sector is divided into four parts:

* Corresponding author: 31777669@qq.com

primary industry (such as agriculture, forestry, animal husbandry and fishery), secondary industry (such as industry and construction), tertiary industry (such as transport, storage, post and communications, wholesale, retail sales, catering trade and others) and household consumption. This study considers mainly three types of energy, including primary energy (such as coal, oil, natural gas), secondary energy (Washed coal, other coal washing, mould coal, coke, coke oven gas, other gas, gasoline, kerosene, diesel oil, fuel oil, liquefied petroleum gas, refinery gas, other petroleum products, other coking products, heat, and electricity), and other energy in the energy balance tables of NC.

2.2 Estimation of CO_2 emissions

Based on the energy data, the following expression provides the detailed calculation for estimating carbon dioxide (CO_2) emissions from fossil fuel combustion from NC using the IPCC-recommended Reference Approach (1996).

$$E_{CO_2} = \sum_i (A_i e_i c_i \times 10^{-3} - S_i) o_i M_i \quad (2)$$

Where Eco_2 is the total carbon dioxide emitted from f ossil fuel combustion in kilograms (kt CO_2); A_i is the app arent consumption of fuel i (kt or 100 million m^3 for natu ral gas); e_i is the net calorific value of fuel i (TJ/kt); c_i is t he carbon emission factor of fuel i (t C/TJ), $S_i = A_{si} e_i c_i s_i$, is the carbon in products from non-en ergy uses of fossil fuels (kt C), o_i is the carbon oxidation rate of fuel i, and M is the molecular-to-atomic weight ra tio of CO_2 to carbon (44/12). Calorific values and emissi on factors of fuels were taken from "the Revised 1996 IP CC Guidelines for National Greenhouse Gas Inventories (The IEA is still using the 1996 IPCC Guidelines)".

2.3 Scenario analysis

Scenarios are self-consistent story lines of how a future energy system might evolve over time under a particular set of conditions[3]. In order to analyze possible effects of series policies for energy savings and emissions mitigation, two scenarios were set up: the business as usual (BAU) scenario and the low carbon (LC) scenario. The BAU scenario assumes that the indicators of energy-saving and emission reduction will be the same as before. Considering the development of new and renewable energy, the rates of energy-saving and emission reduction during 2016-2025 under the LC scenario will be increased by 0.5-1 percentages compared with BAU. The calculation of energy consumption per unit of output value[4] and the major assumptions and for the BAU and LC scenarios are as follows (table 1):

$$\frac{E_o}{G_o}(1-m)^Y = \frac{E_t}{G_t} = \frac{E_t}{G_o(1+n)^Y} \quad (3)$$

$$E_t = E_o\left[(1-m)^Y \times (1+n)^Y\right] \quad (4)$$

$$\frac{C_O}{G_O}(1-m')^Y = \frac{C_X}{G_X} = \frac{C_X}{G_O(1+n)^Y} \quad (5)$$

$$C_X = C_O\left[(1-m')^Y \times (1+n)^Y\right] \quad (6)$$

Table 1. The basic assumptions for key variables in NC

Key variables		2015	2020	2025
Population (millions)		149.91	157.71	165.92
Population growth rate (%)		1.02	1.02	1.02
GDP per capita (yuan)		53434	82000	110000
GDP (billions of yuan)		8010	12932	18251
GDP growth rate (%)		14	10	7
Saving rate of energy intensity (%)	BAU	4.0	4.0	4.0
	LC	4.5	5.0	5.0
Reduction rate of carbon intensity (%)	BAU	6.5	6.5	6.5
	LC	7.0	7.5	7.5

3 The energy status in NC

3.1 Energy supply in NC

Based on energy balance table, the NC's energy demand rose rapidly with the continuing growth in population and economic. During 2000-2015 period, the energy supply increased from 170.51 to 436.55 Mtce, representing an annual rate of increase of 8.11%. However, primary energy production was only 40.45 Mtce in 2010, an increase of 60.32% over 2000 levels but only 9.26% of NC's total supply, compared to 14.79% in 2000. An average of nearly 85 percent of energy supply for the 2000-2015 period must be met by imports from overseas and domestic other regions. Oil imports would be 79.73 Mtce in 2015, which was ten times as large as in 2000, as shown in figure 1.

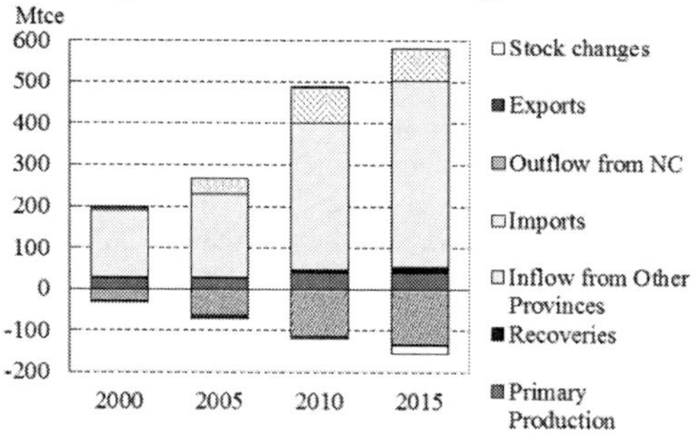

Fig.1. Energy supply by source in NC

3.2 The final energy consumption in NC

The analysis showed that the final energy consumption in NC rose from 169.32 Mtce in 2000, to 399.73 Mtce in 2015, accounting for 16.7 percent of total final energy consumption of China (2538.61 Mtce in 2015). In recent years, the energy structure has become more diversified in NC: the coal share has decreased from 74% in 2000 to

63% in 2015, and the share of crude oil, natural gas and outdoor electricity has increased 1%, 2% and 6% in 2015, respectively (Figure 2). From 2000 to 2015, the energy consumption from industry contributed the largest shares, followed by household consumption and transport, as shown in Figure 3. In 2015, the total final energy consumption of the industry sector in NC reached 274.87 Mtce and accounted for 69 percent of the total. This was a substantial decline of 8 percentage points from 77 percent in 2000. Still, construction sector was the smallest energy consumer in the energy consumption. The construction sector consumed 6.82 Mtce in 2015, accounting for just 2 percent of total final energy consumption. In contrast to its small share, however, the construction sector's energy consumption grew at a leading rate of 16 percent a year on average during 2000-2015 period.

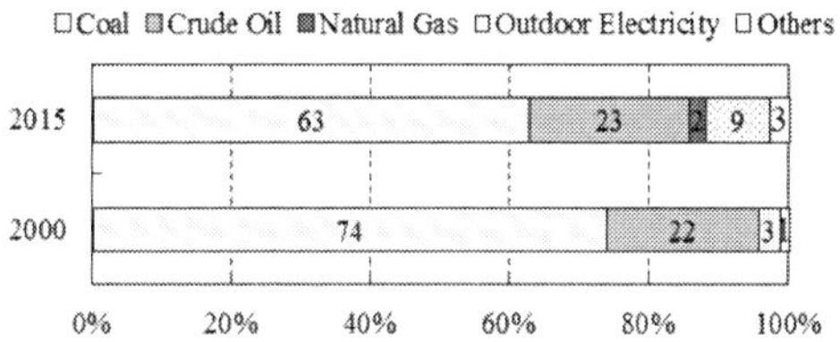

Fig.2. Shares of fuel types in final energy consumption

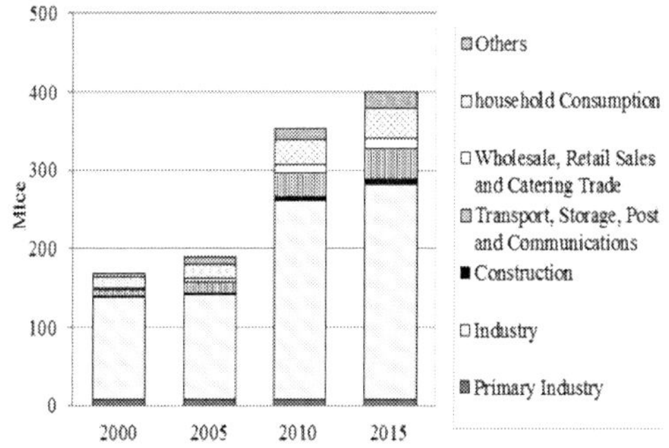

Fig.3. Final energy consumption by sector in NC

3.3 CO_2 emissions in NC

CO_2 emissions from fuel combustion in NC increased fro m 361.29 million tons in 2000 to 901.86 million tons in 2015, with an average annual growth rate of 7.88%. The emission of CO_2 per capita in the year of 2015 was 6.3 to ns (based on resident population), which is higher than th e world average and the national average in China. In co ntrast to per-capita emissions, carbon intensity, measured as emissions per unit of GDP, was 37% lower than that of China average and 25% higher than that of the world average level at the same period (table 2). On the other h and, due to the rapid urbanization and economic develop ment of urban areas in NC, CO_2 emissions from urban ar eas were 623.36 million tons in 2015, which was 69% of the total.

Table 2. Comparisons of CO_2 emissions indicators, 2015

	World	China	India	USA	NC
Population (million)	6614	1327	1123	306	144
Share of world population (%)	100	20	17	5	11
GDP($ 2016 trillion, PPP)	67.2	7.6	3.1	14.1	1.8
Share of world GDP (%)	100	11	5	21	22
Share of world CO_2 emissions (%)	100	21	5	20	15
CO_2 emissions per capita (t)	4.4	4.6	1.2	18.7	6.3
CO_2 intensity (kg/ $ 2016, ppp)	0.4	0.8	0.4	0.4	0.5

Sources: IEA analysis (2017)

4 The energy future in NC

4.1 Energy demand

Based on the assumptions of socio-economic development in NC and the various parameters for scenarios in the calculation, the total energy demand under the BAU scenario will increase to 814.88 Mtce in 2025, which is twice as large as in 2015 (figure 4) . Under the LC scenario, it will reach 722.46 Mtce in 2025, suppressing the growth of total energy demand by 0.98% annually compared to BAU. The energy intensity was 0.76 tce/10,000 yuan in 2015, and it would be 0.44 and 0.39 tce/10,000 yuan in 2025 for the BAU and LC scenarios, respectively. While the energy intensity has improved in NC, the improvements have only been able to slow demand growth to some extent, but not to halt it, owing to rapid economic and population growth.

4.2 CO_2 emissions

Two important turning points occurred in 2022 (figure 5): under the BAU scenario, CO_2 emissions will maintain a 0.03% lower average annual growth rate during the period 2022-2025 compared to 3.54% during 2015-2022. Due to the series of energy-saving and emission-reduction policies and measures, for the first time in 2022, CO_2 emissions under the LC scenario will begin to fall. CO_2 emissions will reach 1055.70 million tons in 2025, suppressing the growth of total CO_2 emissions by 0.95% annually compared to BAU. In per capita terms, this means 7.20 and 6.36 tons in 2025 under the BAU and LC scenarios, respectively, compared to 6.30 tons in 2015. The CO_2 intensities of the economy for the BAU and LC scenarios will be reduced by 58% and 63% in 2025 compared with that in 2015, respectively. The results show that NC will achieve the national emission reduction targets (A 40-45% reduction target)[5] only when it continues its past emission reduction targets and regional plans related to energy saving and environmental protection are implemented.

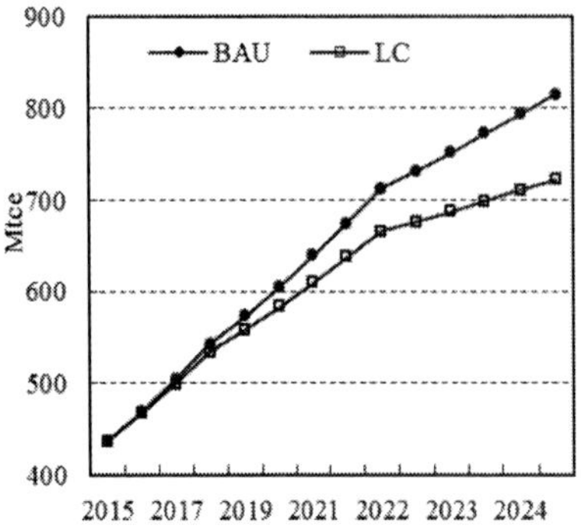

Fig.4. Trends of energy demand for two scenarios

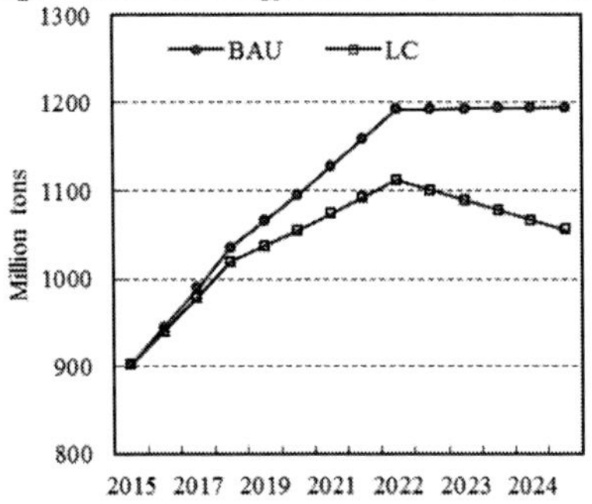

Fig.5. Trends of CO_2 emissions for two Scenarios

5 Conclusion

In this study, the IPCC-recommended reference approach and scenario analysis were applied to NC to analysis dynamic change of the energy supply and energy-related CO_2 emissions for 2000-2015 and project the energy demand and CO_2 emissions up to 2025 under the BAU and BP scenarios. Based on the above research results, the following strategies should be undertaken to increase geographic and fuel-supply diversity, to curb the growth in energy demand and to mitigate greenhouse-gas emissions: 1) In NC, energy security has emerged as a central policy issue and is increasingly affecting regional economic and development policy. The government should ensure regional supply by diversifying the geographic sources of oil and physical supply route, strengthening energy infrastructure construction and increasing domestic production of convertional fuels. 2) We must continue recent energy-saving and ejection-decreasing policies to promote the adjustment and upgrading of industrial structure, as prescribed in China's Medium and Long Term Energy Conservation Plan. Development and deployment of clean energies in a more cost-effective manner to phase in the fuel switch from coal-dominance to more shares of clean energy types such as natural gas and electricity[6]. 3) Despite a rapid declining in carbon intensity in NC, there is still much potential for carbon intensity to decline further, especially in some urban areas. Therefore, pursuing closer cooperation between cities and building integrated regional management platform are becoming the major tasks, which will further accelerate the spread of the information technique and ensure that energy policy challenges facing the region are addressed in a consistent manner.

Acknowledgments

Foundation item: Heilongjiang province philosophy and social science research planning project (No. 17TYH41), Heilongjiang province higher education scientific research project (No. 16G317), Project Funded by China Postdoctoral Science Foundation.

References

1. International Energy Agency. France: IEA Publications. *World Energy Outlook 2017,* **25** (2007)
2. P Moriarty, D Honnery. Energy Policy. *What energy levels can the Earth sustain* **37**, 2469 (2009)
3. R Kadian, R P Dahiya, H P Garg. Energy Policy. *Energy-related emissions and mitigation opportunities from the household sector in Delhi,* **35** 6195 (2007)
4. National Bureau of Statistics of China. Beijing: China Statistics Press, *China Statistical Yearbook 2017,* **15** (2018)
5. M Uwasu, Y Jiang, H P Garg. Sustain.Sci. *On the Chinese carbon reduction target* ,**23** ,1553 (2015)
6. J W Hannah, M O Hari. Ecol.Law.Quar. *Regional Energy Governance and U.S. Carbon Emissions.* **43**, 22 (2016)

The Study on the Optimization of Container Multimodal Transport Business Process in Shandong

Fengmei Wang[1,*], Xiaoyi Gong[1], Yingying Ni[1],Jun Zhan[1],and Huiping Che[1]

[1]Shan dong Jiao tong University, Ji nan, Shan dong, China

Abstract. Shandong is a coastal city with good location advantages. As a hub port for international trade goods and a port of transhipment, shandong's demand for multimodal transport is more urgent. By selecting the suitable non-water port and the multimodal transport carrier to improve the efficiency of multimodal transport, the purpose of saving the time of logistics is achieved, thus reducing the logistics cost.It branch out through Shandongt, and it can reach the central region of China, can reach the Western remote area ,too. This paper puts forward the optimization scheme of the business process of container multimodal transport. The optimization of freight forwarding business process is analyzed. The multimodal transport model in Shandong was designed. Finally, the optimal approach of multimodal transport in Shandong is put forward.

1 Introduction

The container multimodal transport is an advanced form of transport organization in comprehensive cargo transport. It is organized in the form of a container as a transport unit, combining the different modes of transport with units to form a continuous and comprehensive integrated transport of goods. The so-called "organic combination of different modes of transport" is the focus of the integrated freight transport system, which effectively connects the different modes of transport through containers, so that the whole process can be integrated through containers, realize the efficiency and integrity of the transport products, and form a truly comprehensive cargo. Container multimodal transport is the key to solve the bottleneck of China's comprehensive transport services. Now, the state has put forward the "Belt and Road" and other policy support, Container multimodal transport has received unprecedented attention and has become the trend of logistics industry development in our country.

2 Container multimodal transport business process in Shandong

At present, the Shandong region is slower than the sea rail transport, but the development potential is greater. The port development of the sea rail transport will help to promote the expansion of the port throughput and the development of the port industry, and also help to transform the port to the diversified ports, which is beneficial to the long-term development of the port.[1] In addition, the Fifth China multimodal transport cooperation and development conference clearly shows that the link between the railway and the port is an important trend in the future. Therefore, it is a necessary direction and an important opportunity for the Shandong region to focus on the development of the sea rail joint transport.

Take the example of the exportation of the sea rail link. The shipper sends a consignment application, the multimodal transport operator (the freight forwarder) accepts and signed a multimodal transport contract with it. After that, it makes a rational use of the box plan, reservations to the railway department, and arranges for the application, extraction and delivery of empty boxes by the highway transportation enterprises;and the cargo is made to the shipping company, then the cargo owner is ready to stock up and pay more. The owner after delivery (or by forwarding instead) sent to the office of departure from the customs surveillance zone, sealing clearance. They are transported to the railway container freight station by container trucks, marshalling, loading and Issuing the waybill, and pay the first carrier (railway carrier). After the railway line for long-distance transportation to the railway container marshalling station, and then sent to the railway container freight station, and delivery of goods. The cargo is packed into port and transported to the rear yard of the port. The second carrier (Sea Carrier) is responsible for maritime transport, and the bill of lading is issued after shipment and the multimodal carrier is paid.

During the course of the sea rail transport business, the multimodal transport operator is responsible for the entire transportation process, including participating in the shipping company, handling the secondary school procedures, issuing and delivering the bill of lading and so on. The specific sea rail transport business process is shown in Figure 1.

* Corresponding author: 1423185475@qq.com

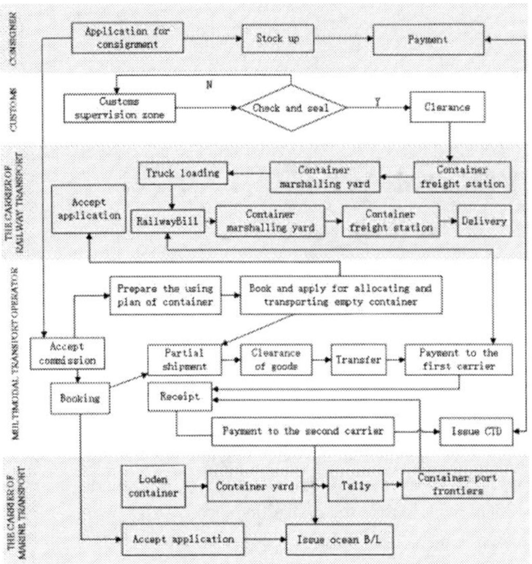

Fig.1. Export flow chart of sea rail multimodal transportation.

Besides, business process optimization needs policy guidance. Taking Qingdao port as an example, compared with ports such as Lianyungang, Tianjin Xingang, the transport distance is longer and the freight rate is higher. In the fierce competition of logistics market, price competitiveness is one of the important factors of port development. In view of Qingdao port, Rizhao Port and other ports with certain supply and location advantages, we can appropriately increase policy and financial support.

(1) Subsidize the costs of the two ends of the rail transport (railway station and port loading and unloading, storage, and port short-term berth) to reduce the cost of comprehensive transport of customers.

(2) Give tax relief to port enterprises.

(3) The export source of China's coastal ports is about 1.6 times as much as the import source, and the internal and external trade structure is very different. In addition, the unequal flow of the goods is caused by the unequal factors of regional economic development. Accordingly, we can directly subsidize the shipping companies and railway enterprises to make up for the uneven of ports. [2]

(4) In terms of short distance transportation, sea rail intermodal transport and intermodal transport still have no advantages. In particular, there are some unreasonable subsidies to highway transportation, which leads to the lack of advantages of the sea rail transport in the comprehensive freight rate. Through the policy adjustment, it can attract the medium and short distance goods and promote the development of the sea rail transport.

3 Business process optimization of freight forwarders

Freight forwarding enterprises undertake the importance of connecting all transportation links in the process of multimodal transport.While most of the companies are smaller and less competitive,they can achieve cooperation and win-win cooperation on through the establishment of alliance groups through business division and information sharing.The consortium business process is shown in Figure 2.

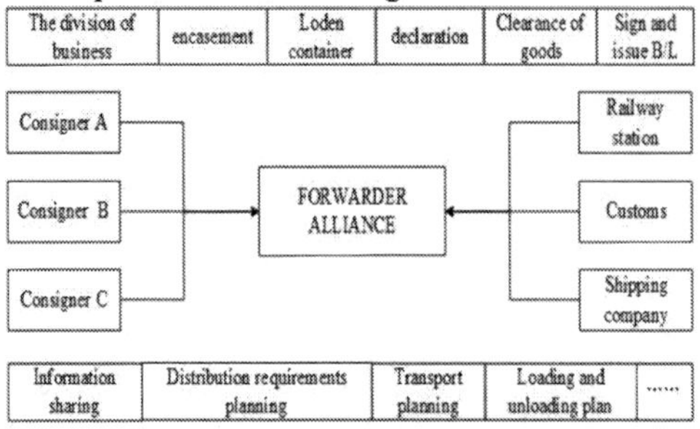

Fig.2. The operation flow chart of forwarder alliance.

In the process of multimodal transport, receive and collect the consignment, due to the train or ship shipping space, or yard, loading and unloading the goods at the factory, to the station or the ship generation to handle the declaration, issued a bill of lading, submitted to the customs and other business carriers on the reasonable division of labor. And sharing the information of distribution plan, transportation plan, loading and unloading plan, customer demand information, loading and unloading plan, market demand forecast and so on, can efficiency of the consortium. By concentrating the volume of the same route of each company to get the tariff preference and priority of loading and unloading,it can reduce the customer's comprehensive freight rate and effectively reduce the transportation time. In addition, on the basis of the establishment of the alliance, the freight forwarders can promote the coordinated development of the logistics supply chain more quickly.

4 Design of multimodal transport mode in Shandong area

The container transport dominated by railways and highways and foreign trade container transport dominated by sea are linked together by multimodal transport. [3] The multimodal transport process involves many principals, including suppliers, multimodal transport operators,customers, agents,segment carriers and customs inspection and other individuals.

Take a Shandong intermodal company as an example,The establishment of business outlets in the economic hinterland, the inland port is responsible for the transfer, the headquarters is responsible for the coordination and freight settlement mode.The division of business of general company, branch company and inland port company is shown in Table 1.

Table 1. The division of individual and business of a company in Shandong.

Name of individual	Business divisions
Shandong local multimodal transport corporation	The organizer of the through carriage.
	Coordinate with transportation department and settlement of freight with all transportation departments.
Branch office	To organize and handle multimodal transport of goods.
	Hand over and take over goods at local road station.
	Issue of multimodal transport documents.
	Collect full road, rail and sea freight.
Inland port company	Coordination "one customs and three inspections(custons: commodity inspection, animal and plant inspection, sanitary inspection)"

In this mode, relying on the perfect information system of the logistics center, the effective link between the inland port and the short road transportation of the railway container can be realized. And the container from the road trailer to the railway vehicle can be transferred directly. The storage time of the container is compressed in the port and the whole logistics of the container door to the door will be saved. And the profits of the enterprises and the society are added up. [4]

The specific operational processes can be represented in Figure 3 and Table 2.

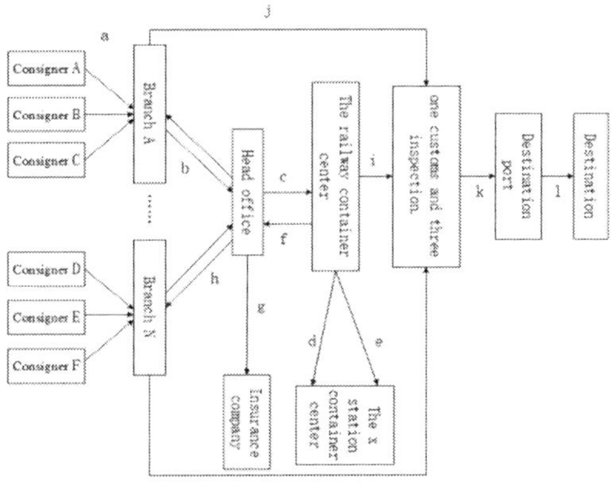

Fig.3. The Flow chart of multimodal transportation.

Table 2. The meaning of each order number.

order number	Meaning
a	Each branch office collects the source information in the economic hinterland and summarizes the information, conclude a multimodal transport contract, collects freight.
b	Upload the summarized information to the multimodal transport company.
c	The multimodal transport company informs the railway container center of the container optimum volume information.
d	The railway container center calculates the required number of empty boxes and requests the x station container center to arrange the empty boxes.
e	The x station container center inform the time of arrival in railway container center and quantity of the empty container to the railway container center .
f	The railway container center informs the multimodal transport company of this information.
g	Handle insurance and contact the shipping space of the destination port.
h	The multimodal transport company notifies the branches to load the goods at the right time and issue a multimodal transport bill of lading to the shipper.
i	The goods arrive at customs by road and carried out one customs and three inspection.
j	The railway container center informs the time of arrival and quantity of the cargo and empty container to the Jinan station.
k	After packing the goods, the goods reach the port of destination (Qingdao or Yantai) by railway.
l	From the destination port to the destination by sea.

5 Optimization ways of multimodal transport in Shandong area

5.1 Strengthening the joint transport of sea rail to promote regional economic development

In Shandong Province, the coastal ports can establish a waterless port to reach the direct connection of the inland ports through the way of strengthening the sea rail joint transport, and establish a smooth and modern logistics transportation system to enhance the radiation capacity of the port. The circulation of goods will make the hinterland close to the outside world. At the same time, strengthening the connection between the Shandong port and the hinterland area outside the province will also bring new opportunities for the economic development of the hinterland in the province.

5.2 Establish multimodal transport information platform

Shandong should set up a multimodal transport information platform at the core of the port. According to the provisions of relevant laws and regulations of port management, the port administrative department should fulfill its rights and obligations in port planning, construction, maintenance, management, and safety supervision. The construction of port information management platform can improve work efficiency and provide more comprehensive and efficient services for port operators, ship owners and shippers. The improvement of the working level of the port administrative department, the construction of the port management information platform, and the exchange of data for other port management departments to provide interface to improve the efficiency of customs clearance.

5.3 Vigorously develop the freight forwarding agent

The freight agent understands all kinds of business links, understands all kinds of transportation business and is familiar with relevant laws and regulations. Through the loading and unloading of goods, storage and packaging, the way of transportation, transportation and transportation, the insurance of the goods, and so on, the insurance of the goods is put forward to the client with more correct advice. To assist clients in designing and selecting appropriate solutions so as to avoid unnecessary losses.

5.4 Improve the level of multimodal transport of container

To improve the organization level of container multimodal transport, we must: A unified transport management mechanism shall be established, and the renewal of container type should be accelerated, so that the container will be closer to the national standard as soon as possible. Promote the development of information system. To achieve the coordination between technical equipment and container multimodal transport system. As a result, there is still a lot to do in improving the level of multimodal transport of container. Improving the capacity of container organization will play a great role in improving the economic benefit of multimodal transport.[5]

6 Conclusion

Shandong province is a coastal province, which is very advantageous to carry out multimodal transport and has achieved relatively satisfactory results in multimodal transport, but it is still far from developed countries. Therefore, Shandong province should be more actively respond to the policies of the state, vigorously develop multimodal transport, improve the original traditional logistics enterprises operating model, and promote long-term and flexible development of the economy.

References

1. Zhou J J. The necessity of the development of the combined transport of sea iron by the port of sheep port[J]. Water transport management, 2014(4):20-21.
2. Qin L. A review on the development strategy and research evaluation of Marine railway transportation[J]. Logistics science and technology,2017(5):84-88.
3. Wei J G, Rong Z H. Analysis of macroeconomic factors in the development of container multimodal transport in China[J]. Chinese soft science.2000.8
4. Ma Y P. Multiple modes of transport connection operation management and organizational technology research[D]. Chang 'an university,2012
5. Zhang Y L. Shandong will build a multimodal transport development enterprise alliance, and container traffic volume increased by 10%[N]. China Network, 2017.

Thermal distribution analysis of inner defects in TSV

Chuan Kai Jiang[1], *Lei* Nie[1], *Wen* Jia, and *Yu Ning* Zhong

[1]School of Mechanical Engineering, Hubei University of Technology ,Wuhan, China

Abstract. In order to uncover the external manifestations of TSV internal defects, the finite element models of typical internal defects, which were filling missing, axial cavity and end cavity, were established. The thermal analysis was carried out using thermoelectric coupling method. The temperature distribution of TSV with and without defects were obtained. And the temperature variation profiles on the defined paths of TSV layer were also analyzed. The analysis indicated that all the defective TSV showed distinct temperature distribution with the defect-free TSV. Among three typical defects, TSV with filling missing showed the most obvious difference on the temperature distribution and path variation. TSV with end cavity has relatively weak affect and the slightest defect was TSV with axial cavity. Therefore, it could be seen that the external temperature difference caused by the internal defects of TSV could provide effective information for the identification and detection in TSV with internal defects.

1 Introduction

In the process of photoelectric manufacturing, microelectronic packaging was one of the most critical and difficult steps, which determined the electrical and mechanical properties of the device. In order to meet the requirements of modern IC product packaging, 3D packaging technology was presented. Among current 3D packaging technologies, TSV (Through Silicon Via) had attracted extensive attention due to its unique technological characteristics. Because the interconnect channel passed directly through the wafer (or chip), it not only improved the device integration, but also reduced the interconnection delay and raised the speed of the device[1]. However, because of the development trend of small aperture, high density and aspect ratio, the defects, such as filling missing, end cavity and axial cavity, were easily found which would lead to serious reliability problems. While most of these defects were located within the inner part of device package, conventional methods were hardly able to detect them directly. Thus, the inner defect inspection was becoming a research focus. In 2016, Shen J extracted seven features representative of TSVs from the images, and then inputted them into a self-organizing map (SOM) network for classification and testing. The results demonstrated that the normal TSVs and defective TSVs could be distinguished obviously by SOM network[2]. In 2016, Li F established a numerical model of Cu-filled TSV and analyzed the effect of defects for TSV thermal mechanical stress. The results showed that thermal stress distribution in TSV was distinct with the different shape and location of defect[3]. In 2017, in order to optimize TSV design and the quality of via filling, Pan Y established a numerical model of Cu-filled TSV to simulate and analyze the effect of diameter, aspect ratio (AR) and defects on TSV thermal stress and deformation. Simulation results showed that the equivalent stress and total deformation of TSV increased as the increase of the diameter of TSV. The effect of aspect ratio on equivalent stress was very little[4]. These researchers investigated the influence of some characteristic parameters of TSV on the temperature distribution, but did not solve the problem of some information about TSV with inner defects.

For TSV 3D-packaging, the TSV with internal defects showed abnormal temperature distribution under the electric-thermal composite load at work. If the corresponding temperature signal could be captured, the internal defects of TSV could be identified to realize the active online detection of internal defects. This paper was based on thermoelectric coupling method. And the temperature distribution inside the TSV work had been carried on the simulation analysis, which fully revealed the trend of the internal temperature distribution with different defects and provided a theoretical basis for discovery and orientation of internal defects in TSV.

2 Principle of thermoelectric coupling analysis

The temperature distribution of the chip module was an external result of the direct coupling of two physical fields, heat and electricity, which had a high degree of nonlinearity.

The general expression of the finite element of the heat conduction equation was as follows:

* Corresponding author: [1]jiangchuankai123@163.com

$$C^T \dot{T} + K^T = Q \quad (1)$$

$$Q = Q^{nd} + Q^c + Q^g + Q^j$$

The general expression of the finite element of the electric conduction equation was as follows:

$$K^V \bullet V = I^{nd} \quad (2)$$

Among, C^T was specific heat matrix, K^T was heat conduction matrix, K^V was conductivity matrix, T was temperature coefficient of node, $\dot{T}$ was temperature vector that changed with time on a node, V was node voltage vector, Q^{nd} was thermal flow vector on the node, Ind was current vector, which was applied on a node, Q^c was surface thermal convection vector, Q^g was external heat flow vector, Q^j was heat generation rate vector of Joule heat of internal heat source.

Backward difference was used in time, there was the following:

$$\left(\frac{1}{\Delta t}C^T + K^T\right)T_n = Q + \frac{1}{\Delta t}CTT_{n-1} \quad (3)$$

This formula was used to calculate the temperature of the model node at each time Δt.

When chip module was working, the temperature of the device increased due to the working current. Along with the temperature rising, the material properties, such as the resistance, thermal conductivity and specific heat, were changed correspondingly, resulting in the thermal-electric coupling effect. The internal heat source vector Q^j was calculated by the following form:

$$Q^j = \int_{vol} \{N\} V_g \sigma d(vol) \quad (4)$$

Among, N was element shape function, V_g was voltage gradient vector, σ was conductivity matrix. The formula for calculating the voltage gradient vector was as follows:

$$V_g = BV_{n-1} \quad (5)$$

Among, B was the matrix of the shape function x, y, z to the direction differential. V_{g-1} was node last iteration voltage. Because of the direct coupling between the heat conduction equation and the electric conduction equation, the solution of the whole equation group was accomplished through iteration.

3 Modeling

3.1. TSV chip interconnection structure model

The cross section of 3D-TSV stack chip packaging structure was shown in Fig. 1. In Fig. 2, the top layer was the chip, the second layer and the fourth layer were the welding ball and resin filling material between the chips (solder layer), the third layer was the pillars and the Si filling layer, and the substrate layer was the bimaleimide resin substrate. The schematic diagram of finite element mesh was shown in Fig. 3. The feature sizes of the model were respectively chip, TSV pillars, solder, substrate, and TSV layer. And the model dimension parameters were shown in Table 1. The simulation parameters (density, resistivity and thermal conductivity) used in the finite element analysis were shown in Table 2.

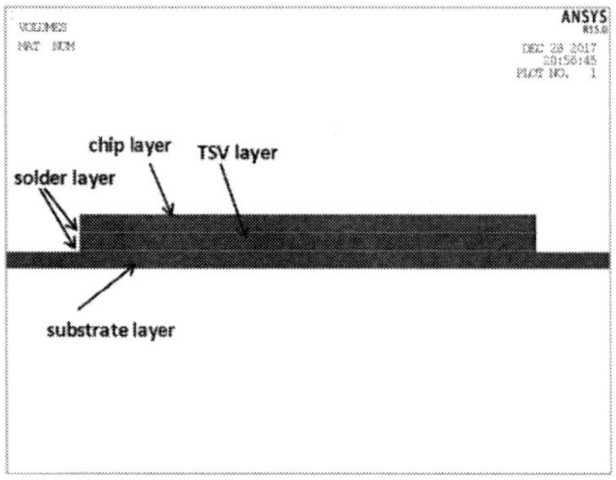

Fig. 1. The schematic diagram of the cross-section structure

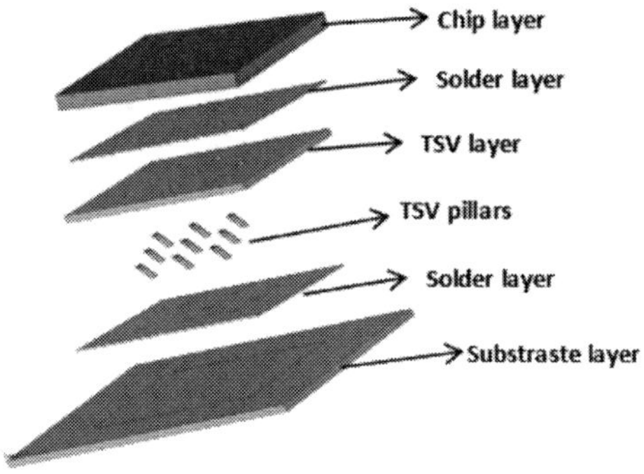

Fig. 2. Structure of TSV structure

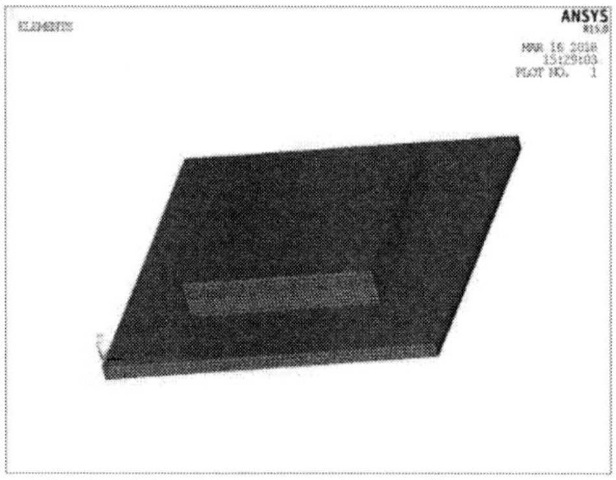

Fig. 3. Finite element mesh its overall structure

Table 1. Feature size of each module

	Length (mm)	Width (mm)	High (μm)	Radius (μm)
1.Copper cylinder	---	---	300	100
2.Silicon chip	4	4	300	---
3.Substrate	7	7	300	---
4.Solder	4	4	100	---
5.TSV layer	4	4	300	---

Table 2. Simulation parameters

	Thermal conductivity [W/(m·k)]	Electrical resistivity (Ω·m)	Specific heat capacity [J /(kg·℃)]	Density (kg/m³)
1	390	1.75e-8	385	8950
2	124	1	700	2330
3	0.3	0.1	---	1550
4	57	0.1	---	8740
5	124	0.1	700	2330

3.2. Common defect types

In the process of TSV manufacturing, three types of inner defects in TSV usually arose, which were shown in Fig. 4, as (a) filling missing (radius 100μm), (b) axial cavity (radius 80μm) and (c) end cavity (radius 80μm, height 100μm). All the defects located in the same pillar as indicated in Fig. 5.

In order to investigate the influence of different defects and pillars self-heating on the temperature distribution, thermoelectric coupling method was used in TSV structure thermal analysis. Thus, the solid 226 thermal unit was selected. And the unit was high order form with including 20 nodes, which had 5 degrees of freedom. It could improve the analysis precision and was often used in thermoelectric coupling analysis. The grid division adopted the sweep method directly. The grid contained 178948 grid cells in Fig. 3. The same load was respectively applied to the TSV pillars with different defects on the chip module. The zero potential was applied to the bottom surface of the 9 TSV pillars. And the voltage of the 1.5V was applied to the upper surface of the chip layer. The boundary conditions of all models were following: the ambient temperature was 20℃. And the surface of each part was convective heat transfer with the air. The convective heat transfer coefficient was 15W/(m²·℃). And the corresponding material properties were applied to different materials.

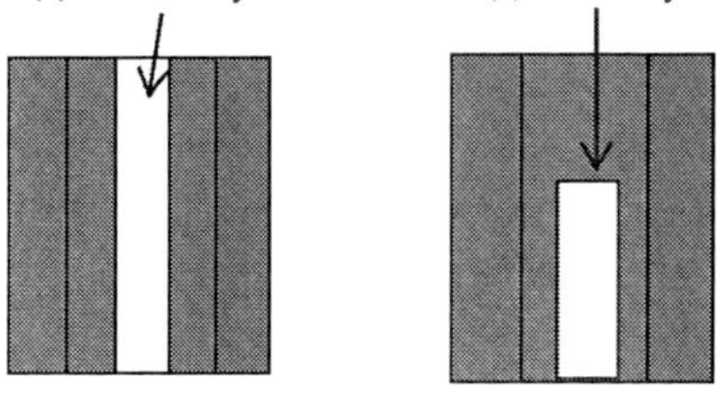

Fig. 4. Common defect profiles of TSV interconnect structure

4 Simulation results and analysis

The finite element analysis models of the TSV with defects were established, according to the above three kinds of inner defect types in TSV. The corresponding load and boundary conditions were applied. The nephogram results of the TSV with inner defects were obtained. The temperature distribution of the key layers was also given and compared with that of TSV without defects.

4.1. Temperature distribution of TSV layer under various defects

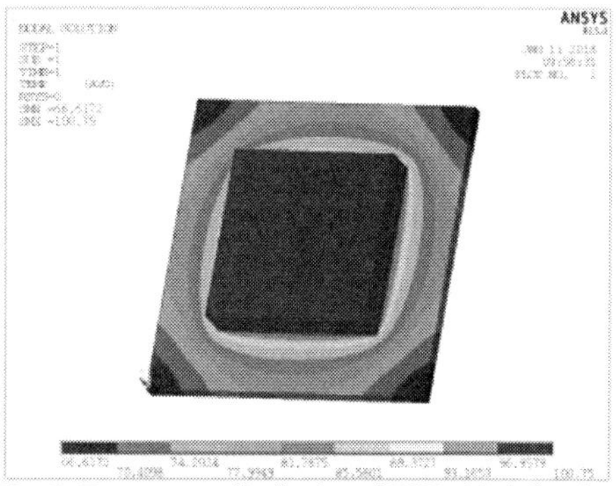

Fig. 5. Temperature distribution of TSV with defect-free

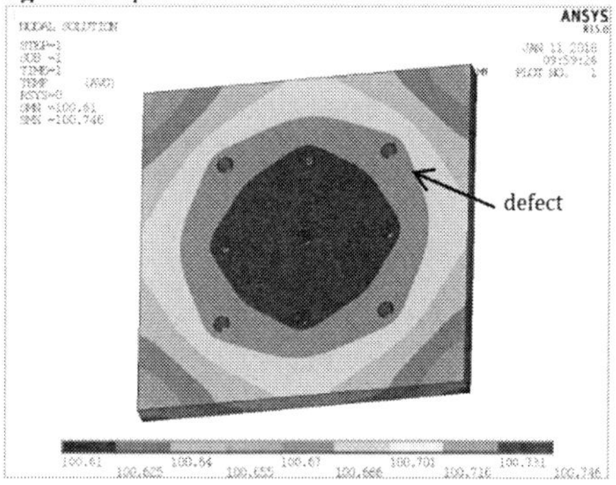

(a) Temperature distribution of the chip layer without defects

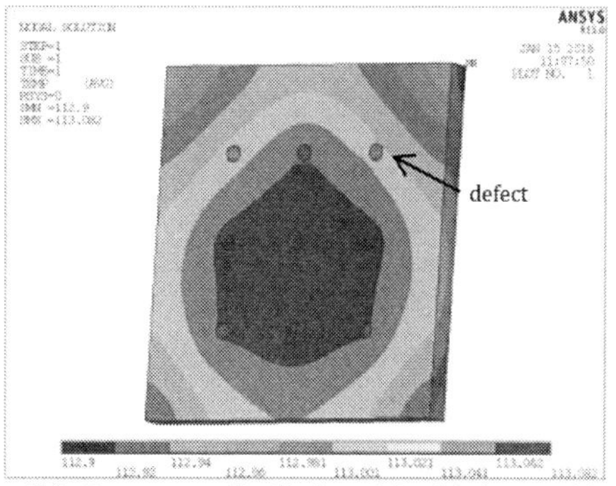

(b) Temperature distribution of the chip layers with filling missing

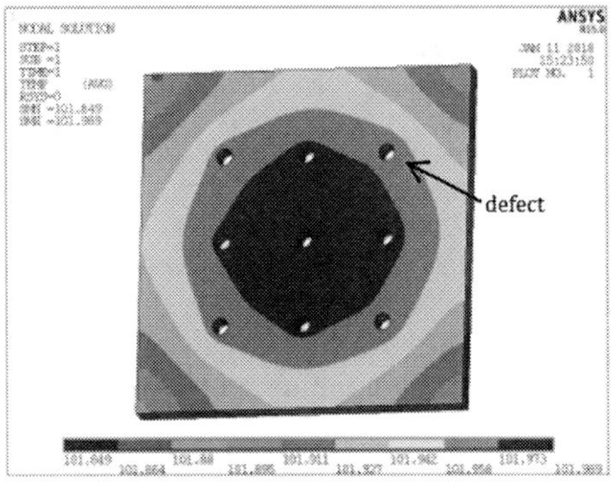

(c) The temperature distribution of the chip layer under the axial cavity

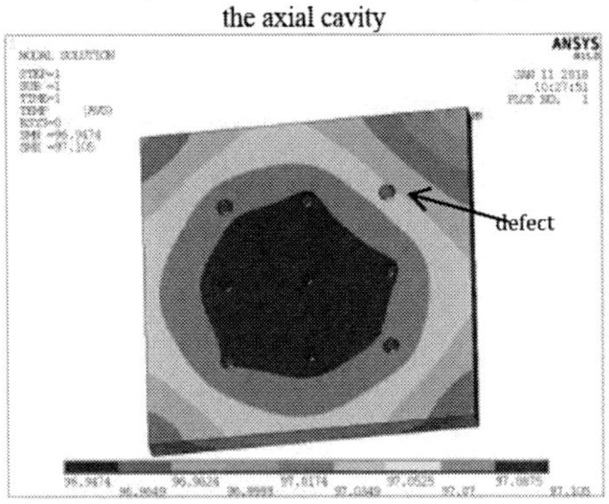

(d) Temperature distribution of chip layer under the end cavity

Fig..6. Temperature distribution of TSV layer under defects

In Fig. 5, the overall temperature distribution of defect-free TSV was given. The surface of the distribution nephogram was radial. It was easily found that the heat gradually transmitted from the center to the edges. The highest and lowest temperature were 100.75 ℃ and 66.62 ℃ respectively. The temperature distribution of TSV layer under various defects was shown in Fig. 6. The TSV layers of all three kinds of defects had obvious differences compared with that of defect-free TSV. Furthermore, as seen from Fig. 5 and Fig. 6, the peak temperature of TSV with filling missing, end cavity and axial cavity were 113.082℃, 101.99℃ and 97.105 ℃ respectively. That was to say, when compared to defect-free one (100.732 ℃), TSV with filling missing defect led to the highest temperature difference (12.332℃) as shown in Fig. 6(b). The second highest difference was 3.645℃ caused by the end cavity defect as shown in Fig. 6(d). And the lowest difference due to the axial cavity was 1.239℃ in Fig. 6(c).

4.2. Temperature profile in observing path

As shown in Fig. 7, an observation path of was defined from point A to B on the substrate to investigate the node temperature variation. In comparison with the node temperature profile of the defect-free TSV in Fig. 8, the temperature variation profile of TSV with filling missing had the most obvious different, and that of axial cavity was the most inconspicuous. Another observation path was defined from point C to point D on the TSV layer in Fig. 9. The path temperature variation profiles showed in Fig. 10. T0 was the node temperature at C point and had different values for defect-free, filling missing, axial cavity and end cavity defect in order to facilitate the comparison, which were 100.61℃, 112.94℃, 101.85℃ and 96.982 ℃ respectively. It was obvious that the temperature profiles of various defects showed different patterns.

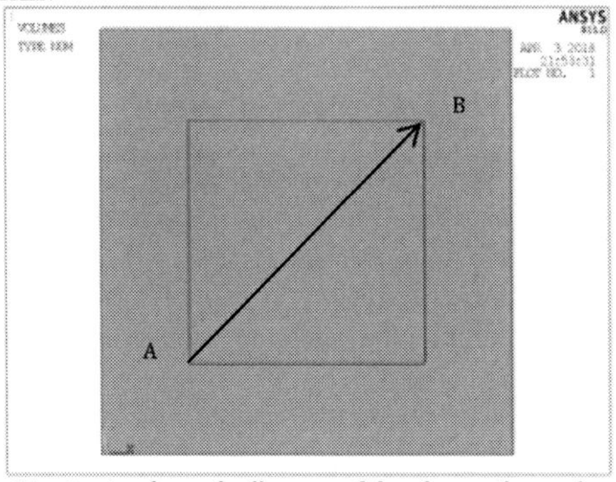

Fig. 7. A schematic diagram of the observation path on the substrate layer

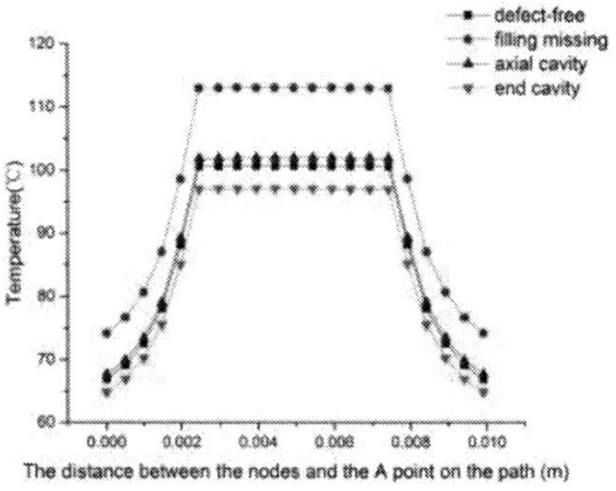

Fig. 8. Temperature profiles with various defects on the substrate layer

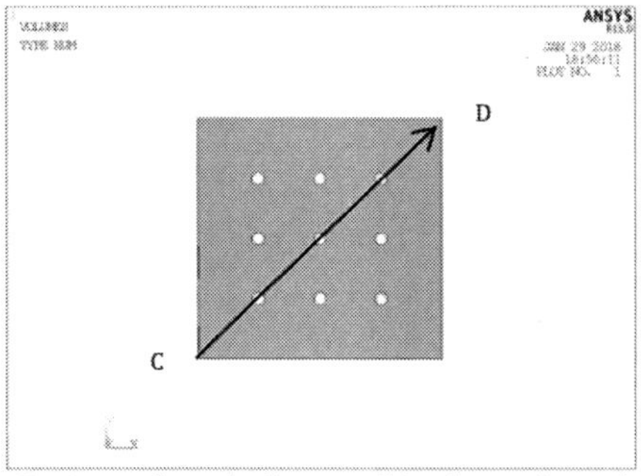

Fig. 9. A schematic diagram of the observation path on the TSV layer

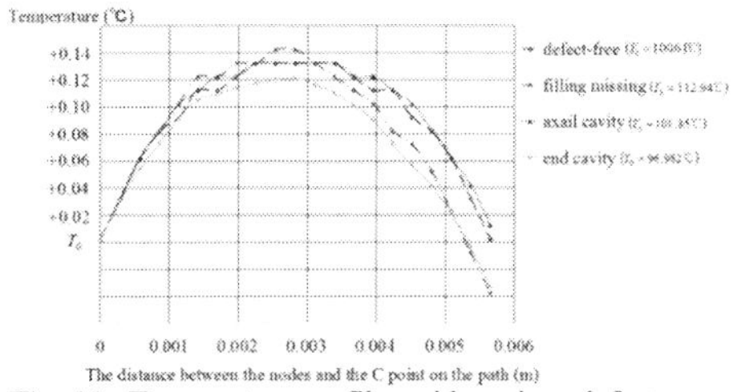

Fig. 10. Temperature profiles with various defects on the TSV layer

5 Conclusion

The finite element models of TSV packaging with three typical defects were established. These models were analyzed using thermoelectric coupling method and compared with the TSV without defect under the same condition. The corresponding temperature distribution and variation profiles were obtained. All kinds of defects showed unique temperature patterns which were obviously different from the defect-free TSV. Among these defects, filling missing had the most evident effect on temperature pattern. The end cavity defect took the second place and axial cavity was the weakest one. According to the analysis result, different defects obeyed different temperature distribution rules, which could be used to monitor the temperature of TSV packaging and realize the detection and positioning of TSV internal defects.

Acknowledgments

This work was funded by the open Foundation of National Key Laboratory of Precision Testing Techniques and Instrument (pilab1708), Key Project of Educational Commission of Hubei Province of China (D20131407), and Special Fund in the Public interest of General Administration of Quality Supervision Inspection and Quarantine of P.R.China (201310004).

References

1. Zhao Z, Tong Z. 3D-TSVthe continuation of Moore's law effective way of [J]. Electronic industry special equipment,2011,40(03):10-16.
2. Shen J, Chen P, Su L, et al. X-ray inspection of TSV defects with self-organizing map network and Otsu algorithm[J]. Microelectronics Reliability, 2016, 67:págs. 129-134.
3. Li F, Xiao C, He H, et al. Investigation on the defect induced thermal mechanical stress for TSV[C]// International Conference on Electronic Packaging Technology. IEEE, 2016:713-715.
4. Pan Y, Li F, He H, et al. Effects of dimension parameters and defect on TSV thermal behavior for 3D IC packaging[J]. Microelectronics Reliability, 2017, 70:97-102.
5. Yuan Y. Thermodynamic analysis of the three dimensional package of silicon through hole [D]. Shanghai Jiao Tong University, 2013.
6. Li C, Xie X, Shang Y, Huang H, Zou M. Transmission performance analysis of TSV interconnect structure based on electrothermal coupling effect [J]. modern electronic technology, 2017,40 (08): 4-7.
7. Huang Q, Shang Y, Zhang M, Li C, Yue K. TSV thermal structure coupling analysis [J]. Electronic components and materials, 2015,34 (04): 74-78.
8. Wang H. TSV modeling, heat transfer and loading analysis for three-dimensional laminated package [D]. Xi'an Electronic and Science University, 2012.

 https://doi.org/10.1051/e3sconf/20183804027

Research on the Mode of University-Enterprise Cooperation to Promote Engineering Students' Employment

Yan Hui[1*], Sun Lihua[2]

1.Shandong Women's University, Teaching Affairs Office, Shandong, China
2.Shandong Women's University, School of Data and Computer Science, Shandong, China

Abstract: The employment of university students has become a hot issue of concern to the whole society. Promoting the employment of university students is a top priority for higher education institutions. University-enterprise cooperation is an important trend in the development of modern higher education. It is also an important channel for promoting the employment of university students, especially for engineering students. Through an in-depth analysis of the status quo of employment of university graduates, this paper proposes four modes of university-enterprise cooperation to promote university graduates' employment: The post-employment cooperation model, the professional internship cooperation model, the second classroom expansion cooperation model and the enterprise-oriented recruitment model, and further proposed the countermeasures to strengthen the cooperation between university and enterprise in order to promote the employment of university students.

1. Introduction

After decades of development, higher education in China has achieved remarkable results. With the continuous development and growth of higher education, it has also brought about some problems. Although the number of university graduates in engineering universities has increased significantly in recent years, it is very scarce to be qualified for professional and technical personnel in the field of science and technology.

One outstanding problem is that professional qualities and technical skills of engineering university students cannot meet the needs of the job. For this reason, universities and universities are actively exploring and reforming the existing personnel training model, hoping to find ways and means to solve the problem, in which the university-enterprise cooperation model is increasingly recognized by universities and enterprises. University enterprise cooperation mode can achieve the goal of personnel training, enhance the vitality and strength of running universitys, and improve the level of running universitys and the quality of education. At the same time, it will help to meet the needs of social and economic development, provide high-quality resources for enterprises, and solve the problem of employment of engineering students in universities.

2. Analysis of University Students' Employment Status

The employment of university graduates is an important part of the employment of the whole society. The employment of university graduates is an essential requirement for improving the quality of higher education and maintaining the sustained and healthy development of higher education.

Fig. 1 Number of university graduates in China (2010-2017)

As showed above, the number of university graduates has increased year by year. With the continuous increase in the size of enrollment, the number of graduates entering the job market every year has increased year by year. It can thus be seen that the employment situation of university graduates has become increasingly severe. This is an inevitable result of China's higher education entering the popularization stage from the elite stage. At the same time, it also reflects the importance that Chinese families attach to higher education from another perspective.

* Corresponding author: yanhuibangong@126.com

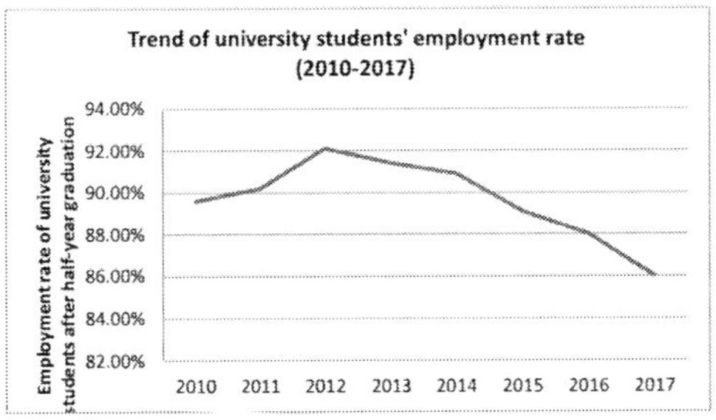

Fig. 2 Employment rate of university graduates in China (2010-2017)

It can thus be seen from Fig 2 that the employment situation of university graduates has become increasingly severe. This is an inevitable result of China's higher education entering the popularization stage from the elite stage. At the same time, it also reflects the importance that Chinese families attach to higher education from another perspective.

3. Advantage of University Enterprise Cooperation in Promoting University Engineering Students' Employment

Firstly, university-enterprise cooperation is conducive to the promotion of university curriculum development and promotes the integration of engineering students' learning and enterprise production. Secondly, university-enterprise cooperation is conducive to promoting the construction of university training bases. Students can participate in the actual work of the company during the study period, make up for the lack of professional knowledge, and strengthen practical operation skills, so as to enhance innovation and practical ability.

From the enterprise perspective, university-enterprise cooperation can enhance the recognition of the quality of personnel training. Under the university-enterprise cooperation model, companies combine their actual situation and school needs, to sign the talent development agreement. University develops or adjusts its own teaching plans and curriculum settings in conjunction with the agreement so that students can contact the company during their studies. Under the university-enterprise cooperation model, higher education is no longer limited to the theoretical teaching of the campus. Engineering students can integrate into the production team within a short period of time and continue to enhance their comprehensive professionalism. University-enterprise cooperation places great emphasis on the matching of school education with the current industry and enterprise skill requirements. Both the teaching arrangement and teaching evaluation incorporate the opinions of industry and enterprise experts, and timely make relevant adjustments in light of changes in the social and industrial human resources markets. This not only communicates the theoretical knowledge of the university and the company's production practices, but also promotes the transformation of the traditional talents from the traditional knowledge transfer to the improvement of the overall professional quality, which not only improves the quality of education, but also innovates the mechanism for employing.

4. Influence of University Enterprise Cooperation on Promoting Engineering Students' Employment

4.1 Improving the pertinence of engineering talent training

The cooperative education model between universities and enterprises can further improve the function of higher education, optimize the educational model, and fully integrate the functions of education and training with the functions of the service enterprise, so as to give full play to its advantages in resources and meet the needs of enterprise development, school teaching and student personal growth. Therefore, colleges and universities should combine the actual needs of the human resources market with their own reality, optimize their own education plans, adjust curriculum settings, and strengthen links with current social needs so as to achieve the best results in running schools.

4.2 Promoting all-round cooperation and establish stable employment channels

There are often significant differences between the operating mechanism of enterprises and university teaching models, which results in limited participation of enterprises and inefficient cooperation. The main reason is that universities do not pay attention to the linkages and cooperation with enterprises, and there is no perfect long-term mechanism for university-enterprise cooperation. To achieve the institutionalization and standardization of university-enterprise cooperation, it is necessary to construct a series of management systems. First of all, universities must work together to formulate and implement the talent training model through university-industry enterprises. Secondly, it should build a practical training base through university-enterprise collaboration to truly realize the integration of universities and enterprises. Finally, it can build a teaching team and set up business scholarships to reward university engineering students who are highly qualified. Through all-round cooperation between universities and enterprises, a stable employment channel will eventually be established.

5. University-Enterprise Cooperation Model

5.1 Zero distance training model

Universities shall vigorously advance education and teaching reform according to market demand, adhere to the concept of zero distance training in teaching, internship, and graduation, and realize the integration and cooperation of university-enterprise education and teaching, practice training, and employment, and establish the zero distance training mode between university and enterprise. By integrating high-quality resources such as teaching, research, and base, universities and enterprises will establish a perfect practice teaching system to achieve seamless connection between university and enterprise cooperation.

5.2 Building a dual-qualified teacher team

Under the conditions of university-enterprise cooperation, colleges and universities need to establish a team of dual-qualified teachers who understand both theory and practice. Colleges and universities should send teachers with rich professional teaching experience and solid foundation teaching to go deep into enterprises to participate in production practices or joint scientific and technological research to improve teachers' practical ability and technological innovation ability to meet the needs of practical teaching. At the same time, companies should select practical and technically experienced business and technical backbones to participate in the practice of teaching activities in colleges and universities, and specifically guide students to practice training. Through the establishment of a two-way exchange mechanism for university-enterprise talents, a "dual-qualified" faculty team is created.

5.3 Pursuing the integration of "theory and practice" teaching

Colleges and universities should effectively change the teaching concept, abandon the traditional theoretical teaching model, and gradually establish the theory plus practice integrated teaching model. In the professional training program, it is necessary to make adjustments and revisions in accordance with market demand; in terms of curriculum setting, and to pay attention to the ratio of professional basic courses, practical courses, and public basic courses, to highlight practicality and application. In terms of teaching methods and means, it is necessary to constantly carry out innovative reforms in accordance with the characteristics of the curriculum, and to improve the effectiveness of practical teaching.

5.4 Cooperating to establish an innovation base for production, education and research

The innovation base of production, education and research established by cooperation between universities and enterprises is an important place for implementing quality education to cultivate students' innovative ability and practical ability. It is also a direct platform for cultivating students' innovative ability, and provides a transition and preparation for students' employment.

6. Conclusion

Under the background of difficult employment for university students, the university-enterprise cooperation model enables students to understand the company's needs for talents in the school. This is a university-running road worthy of continuing research and exploration. Through the two different learning environments, different knowledge and skills are acquired, and comprehensive applied talents with innovative spirit and abilities are developed to satisfy the needs of society and enterprises for professional and technical personnel, and the employment competitiveness and career development potential of undergraduates are improved.

References

1. Ministry of Education of the People's Republic of China Higher Education Department National Higher Vocational School Principals Association. Click Core-Higher Vocational Education Professional Setup and Curriculum Development Guide [M]. Beijing: Higher Education Press, 2017 : 130.
2. Wu Hongyu, Yang Qunxiang. Study on the factors affecting enterprises' cooperation between schools and enterprises-Based on the analysis of 910 questionnaire surveys[J]. Vocational and Technical Education, 2012 (16): 15-20.
3. Chen Shenhan, Li Ke. Integrated teaching is a new educational revolution[J]. Abroad and Employment, 2009, (10): 89.
4. Wang Yani. Misunderstandings and roles of school-enterprise cooperation [J]. Vocational and Technical Education, 2007, (7): 40.

AUTHOR INDEX

AUTHOR INDEX

AUTHOR INDEX

AUTHOR INDEX

AUTHOR INDEX